T0379647

Steel, Concrete, and Composite Design of Tall and Supertall Buildings

About the International Code Council®

The International Code Council is the leading global source of model codes and standards and building safety solutions that include product evaluation, accreditation, technology, codification, consulting, training and certification. The International Code Council's codes, standards and solutions are used to ensure safe, affordable and sustainable communities and buildings worldwide.

The International Code Council family of solutions includes the ICC Evaluation Service (ICC ES), S. K. Ghosh Associates, the International Accreditation Service (IAS), General Code, ICC NTA, ICC Community Development Solutions, Alliance for National & Community Resilience (ANCR) and American Legal Publishing.

Office Locations:
Headquarters:
200 Massachusetts Avenue, NW, Suite 250
Washington, DC 20001
888-ICC-SAFE (888-422-7233)
www.iccsafe.org

Eastern Regional Office
900 Montclair Road
Birmingham, AL 35213

Central Regional Office
4051 Flossmoor Road
Country Club Hills, IL 60478

Western Regional Office
3060 Saturn Street, Suite 100
Brea, CA 92821

MENA Regional Office
Dubai Association Centre Office, One Central
Building 2, Office 8, Dubai World Trade Centre Complex
PO Box 9292, Dubai, UAE

OCEANIA Regional Office
Level 9, Nishi Building
2 Phillip Law Street
Canberra ACT 2601 Australia

Steel, Concrete, and Composite Design of Tall and Supertall Buildings

MUSTAFA MAHAMID

BUNGALE S. TARANATH

Third Edition

Library of Congress Control Number: 2025932470

Steel, Concrete, and Composite Design of Tall and Supertall Buildings, Third Edition

1 2 3 4 5 LBC 29 28 27 26 25

ISBN 978-1-260-45315-7
MHID 1-260-45315-4

Sponsoring Editor
Lara Zoble

Production Supervisor
Richard Ruzycka

Project Manager
Nitesh Kumar Singh,
KnowledgeWorks Global Ltd.

Copy Editor
KnowledgeWorks Global Ltd.

Proofreader
KnowledgeWorks Global Ltd.

Illustration
KnowledgeWorks Global Ltd.

Composition
KnowledgeWorks Global Ltd.

Contents

Preface

This book has been developed to serve as a comprehensive reference for designers of tall building structures. Structural design aspects of concrete, steel and composite tall buildings, and all structural loads are discussed with particular reference to wind and seismic loads. Methods for providing gravity and lateral resistance, including the state-of-the-art structural systems, are discussed as well as many facets of structural elements design.

This publication is intended to serve as a practical book that is useful for engineering students, consulting engineers, architects, engineers employed by federal, state, and local governments, and educators. The material is presented in an easy-to-understand form to make it useful for young engineers with their first high-rise design, and to offer new approaches to experienced tall building engineers. Many case studies for tall, super-tall, and mega-tall buildings from around the world with numerous examples illustrating design procedures are worked out in detail.

The book begins with a description of case studies of structural systems of tall, super-tall, and mega-tall buildings from around the world within the past five decades. The purpose of these case studies is to familiarize the reader with the information that currently resides in these designs, for the engineering mind constantly needs past solutions and tried formats as anchors before it can break new ground or differ markedly from conventional wisdom.

Chapter 2 presents different approaches for evaluating wind loads appropriate for building design. Building code step-by-step procedures and wind tunnel procedures are discussed, including analytical methods for determining building response related to occupant comfort.

Chapter 3 outlines seismic design, highlighting the dynamic behaviour of tall buildings. Static, dynamic, and time-history analysis are described. Seismic vulnerability study and retrofit design of buildings not meeting current building code detailing requirements are also discussed.

The design of framing systems for lateral forces is the subject of Chapters 4, 5, and 6. Traditional, and newer-type bracing systems in steel, concrete, and combinations of the two, known as composite construction, are analyzed.

Chapters 7, 8, and 9 are dedicated to gravity design of vertical and horizontal systems. In addition to common gravity systems, novel techniques such as composite stub girders are also discussed.

Chapter 10 focuses on the analysis of structural systems and components. Approximate methods are discussed first, followed by computer modeling techniques for two- and three-dimensional analyses. Torsional analysis on warping behavior of open section shear walls is covered in detail. This information is particularly useful in making preliminary designs and verifying three-dimensional computer models.

Chapter 11 introduces the performance-based seismic design (PBSD), also known as PBD, which represents a shift in engineering design methods, building upon insights gained from studying how structures perform in earthquakes. PBSD intends to support code-based methods, not to replace them.

The last chapter, Chapter 12, is devoted to the discussion on various topics unique to the design of tall buildings. Differential shortening of columns, design of curtain walls, mechanical damping systems for reducing wind-induced sway accelerations, drilled pier and mat foundations, and earthquake mitigation technologies are some of the subjects covered in this chapter. Brittle fracture of welded moment connections subjected to large inelastic demands is discussed. Unit structural quantities for estimating preliminary steel tonnage of high-rise steel and composite buildings are described.

The book attempts to achieve several objectives; it is intended to bridge the gap between a novice and an experienced engineer while serving simultaneously as a comprehensive resource document. The first and the foremost audience is the practicing structural engineers ranging from structural engineering students, young students who just entered the profession, and those with considerable experience in the field.

Mustafa Mahamid, PhD, SE, PE, P.Eng., F.SEI, F.ASCE, F.ACI, LEED AP
University of Illinois at Chicago
Technion—Israel Institute of Technology

Acknowledgment

I would like to express my deepest gratitude to my colleagues and friends in the field of structural engineering who have provided invaluable insights and support throughout the creation of this book. Your expertise and encouragement have been instrumental in shaping this comprehensive reference on tall building design.

I am profoundly grateful to my late father and mother for their unwavering belief in me and for instilling in me the values of perseverance and dedication. To my wife, Rawan, thank you for your boundless patience, love, and encouragement, which have been my source of strength during this journey. To my children, Ahmad, Mariam, and Maram, your joy and enthusiasm have been a constant reminder of what truly matters, and I am forever thankful for your support. To my late father, whose wisdom and guidance continue to inspire me, I dedicate this work to his cherished memory.

This book is a testament to the collective contributions of everyone who has stood by me, and I dedicate it to all of you with heartfelt appreciation.

Mustafa Mahamid, PhD, SE, PE, P.Eng., F.SEI, F.ASCE, F.ACI, LEED AP
University of Illinois at Chicago
Technion—Israel Institute of Technology

Chapter **1**

General Considerations and Case Studies

Table 1 Lateral Structural Systems Used in Each Case Study

Case study	Lateral structural system
1.3 Empire State Building	Braced steel frame with semi-rigid connections, Shear Walls, Belt Trusses.
1.3.1 The Museum Tower, Los Angeles	Tubular ductile concrete frame with perimeter columns spaced interconnected with upturned spandrel beams
1.3.2 Bank One Center, Indianapolis	Two large vertical flange trusses in the north–south direction and two smaller core braces in the east–west
1.3.3 Two Union Square, Seattle	Braced frame in the core, perimeter columns, 10 feet steel pipes filled with high-strength concrete and dampers
1.3.4 Bank of China Tower, Hong Kong	Cross-braced space truss and composite columns at the corners of the building
1.3.5 Dallas Main Center	Three-dimensional moment-resisting frame and composite columns
1.3.6 The Miglin-Beitler Tower, Chicago, Illinois	Core walls, outrigger walls and cruciform tube system
1.3.7 The NCNB Tower, North Carolina	Reinforced concrete perimeter tube—closely spaced columns connected by deep spandrel beams
1.3.8 The South Walker Tower, Chicago	Core shear walls with shear wall-frame interaction
1.3.9 AT&T Building, New York City	Rigid-frame steel tube at the building perimeter, vertical steel trusses, steel plates,
1.3.10 Trump Tower, New York City	Slender reinforced concrete core with large cap girder at the roof level
1.3.11 Metro-Dade Administration Building	Core truss and braced frame
1.3.12 Jin Mao Tower, Shanghai, China	Central reinforced concrete core linked to exterior composite mega-columns by outrigger trusses
1.3.13 Petronas Towers, Malaysia	Reinforced concrete system consisting of a central core and perimeter columns and ring beams.
1.3.14 Tokyo City Hall	Super steel box columns connected to K braces.
1.3.15 Leaning tower; a building in Madrid, Spain	A triangulated structural system consisting of super diagonals for both stiffness and strength with post-tensioning in the columns on the exterior and interior concrete core.

Table 1 Lateral Structural Systems Used in Each Case Study (*Continued*)

Case study	Lateral structural system
1.3.16 Hong Kong Central Plaza	Triangular reinforced concrete core and closely spaced columns with deep spandrel beams
1.3.17 Fox Plaza, Los Angeles	Special steel moment-resisting frames located at the building perimeter
1.3.18 Bell Atlantic Tower, Philadelphia	Braced core linked to four super columns via a shear-resisting system consisting of 2-story Vierendeel girders.
1.3.19 Norwest Center, Minneapolis	Composite super-columns, alternating Vierendeel, and core X bracing.
1.3.20 First Bank Place, Minneapolis	Composite columns, braced core, and multistory Vierendeel.
1.3.21 Figueroa at Wilshire, Los Angeles	Steel super-columns at the perimeter interconnected in a crisscross manner to an interior braced core with moment-connected beams acting as outriggers at each floor.
1.3.22 One Detroit Center	Composite concrete columns, Vierendeel welded frames.
1.3.23 One Ninety-One Peachtree, Atlanta	Composite partial tube consists of concrete columns encasing steel erection columns with cast-in-place concrete spandrels.
1.3.24 Nations Bank Plaza, Atlanta	Braced Core, composite super-columns at the interior core and at the perimeter. Steel girders are moment-connected between the composite columns. Diagonal truss between levels 56 and 59 used to tie the core columns to the perimeter super-columns.
1.3.25 Allied Bank Tower, Dallas, Texas	Perimeter trussed frame that performs a dual function by providing the required lateral resistance, and an architecturally desired free-span at the base (Mega Truss and sub trusses), Vierendeel truss, Hat truss, and welded moment frame.
1.3.26 First Interstate World Center, Los Angeles	Dual system consisting of square braced core interacting with a perimeter ductile moment resisting frame.
1.3.27 Singapore Treasury Building, Singapore	Circular reinforced concrete core wall.
1.3.28 City Spire, New York	Shear walls in the spine connected to exterior jumbo columns with staggered rectangular concrete panels.
1.3.29 City Corp Tower, Los Angeles	Steel perimeter tube with closely spaced columns and deep spandrels.
1.3.30 Cal Plaza, Los Angeles	Ductile steel moment-resisting frame at the perimeter.
1.3.31 MTA Headquarters, Los Angeles	Perimeter tube with widely spaced columns tied together with spandrel beams.
1.3.32 The 21 Century Tower	Exterior braced frame with super column & super braced, and moment frame. Interior braced core.
1.3.33 Burj Khalifa, Dubai, UAE	Top steel spire: diagonally braced lateral system. Main Tower: buttressed-core system with outrigger walls to link the perimeter columns to the interior wall.
1.3.34 Shum Yip Upperhills Tower 1, Shenzhen, China	Ladder-core system consists of a center reinforced concrete core and eight perimeter steel reinforced composite mega columns aligned with the outer core walls; and Belt truss at mechanical floors.
1.3.35 Cayan Tower (Formerly Infinity Tower), Dubai, UAE	A combination of a reinforced concrete moment-resisting perimeter tube frame and a circular central core wall. A twisting tower where each column has a unique position on every floor.
1.3.36 Lotte Super Tower, Seoul, South Korea	Perimeter steel diagrid with a ductile concrete core.
1.3.37 South Dearborn, Chicago, Illinois, USA	"Stayed-mast" structural system. Central square reinforced concrete core, multi-story outrigger trusses, and exterior steel columns (the "stays") engaged by multi-story outrigger trusses to stiffen the concrete core.

Table 1   Lateral Structural Systems Used in Each Case Study (*Continued*)

Case study	Lateral structural system
1.3.38 Nanjing Greenland Financial Center	Triangular interior reinforced concrete core coupled with exterior composite columns via steel outrigger and belt trusses.
1.3.39 Shenzhen Rural Commercial Bank HQ, Shenzhen, China	Perimeter diagrid frame and a reinforced concrete core.
1.3.40 CITIC Financial Center Tower 1, Shenzhen, China	Composite steel/concrete perimeter frame and reinforced concrete central core. The perimeter frame is an innovative system comprising a series of tie-braced frames.
1.3.41 Tianjin CTF Financial Center, Tianjin, China	A stepped reinforced concrete core and eight groups of sloped perimeter mega-columns, and partial belt trusses.
1.3.42 Pearl River Tower, Guangzhou, China	Dual system consists of a central reinforced concrete core wall system which is linked to the exterior columns by a series of structural steel outrigger and belt trusses, and composite mega-columns linked by diagonal mega-bracing at the two narrow faces of the building.
1.3.43 One Manhattan West, New York City, USA	Reinforced concrete core with steel perimeter columns and moment frame.
1.3.44 A25 Xinyi Tower	Highly ductile Buckling Restrained Braces (BRBs) with Elastic Primary Braces (EPB) and secondary brace (SB) combined with a nominal moment frame.
1.3.45 181 Fremont, San Francisco, California	Perimeter mega-brace system with viscous damping within the mega-brace system.
1.3.46 Atrio	Perimeter Special Concentric Braced Frame (SCBF), composite columns and a central concrete core with Special shear walls.
1.3.47 CITIC Tower, Beijing	Perimeter mega frame and central core, composite walls, and "multi-cell" mega column.
1.3.48 The F5 Tower	Perimeter brace system and a central concrete core.
1.3.49 Torre BBVA	External, fully exposed, eccentrically braced megaframe around the perimeter of the tower, mega steel box section columns, mega steel box section filled with concrete.
1.3.50 Beijing CCTV Headquarters, China	Unconventional continuous tube structure, a "mesh" of steel or reinforced concrete columns, diagonal bracing and beams which wrap around every surface.
1.3.51 MahaNakhon, Bangkok	Central concrete core with outriggers at transfer levels to engage perimeter columns. A unique system with setback columns into the floor plate.
1.3.52 Shanghai Center Tower	Core-Outriggers-Mega Frame system including a composite core, super columns, outrigger trusses, and belt trusses.
1.3.53 The New York Times Building, New York, USA	Central braced core and outriggers at mid-height and at the roof to engage all perimeter columns. Pretensioned rods to provide tension-only X-bracing to supplement the lateral load-resisting system of the building.
1.3.54 Federation Tower East, Moscow, Russian Federation	Concrete core, Outrigger and Belt Trusses engaging the perimeter columns.
1.3.55 Jeddah Tower, Kingdom of Saudi Arabia	Bearing wall system with deep coupling beams.
1.3.56 Wilshire Grand Center Tower, Los Angeles, California USA	Concrete core, outriggers to engage perimeter columns, and multi-story steel belt trusses. Buckling-Restrained Braces (BRBs) were used as outrigger diagonals. Performance Based Seismic Design (PBSD) was used.

Table 1   Lateral Structural Systems Used in Each Case Study (*Continued*)

Case study	Lateral structural system
1.3.57 ThyssenKrupp Innovation & Qualification Center Case Study	Concrete walls forming a box around the entire perimeter.
1.3.58 Market Square Tower	Reinforced concrete core walls, four concrete perimeter super columns connected to the core through concrete outrigger walls.
1.3.59 KAUST Solar Chimney, Saudi Arabia	Diagrid tube with precast, prestressed composite concrete strut.
1.3.60 The Shard, London	Concrete Core.
1.3.61 22 Bishopsgate, London	Concrete core with outrigger trusses and belt to engage perimeter columns.
1.3.62 Torre Reforma, Mexico City	Three architecturally exposed concrete shear walls, and the perimeter double-V steel diagrid.

1.1   INTRODUCTION

Ancient tall structures such as the pyramids of Giza in Egypt, Mayan temples in Tikal, Guatemala, and the Kutab Minar in India are just a few examples testifying to the human aspiration to build increasingly tall structures. These buildings are primarily solid structures serving as monuments rather than space enclosures. By contrast, contemporary tall structures are human habitats, conceived in response to rapid urbanization and population growth although the sheer audacity in their vertical scale may often give them the dubious title of monuments. The difference in the usage of buildings, from solid monumental structures to space enclosures, in itself has not changed the basic stability and strength requirements; the structural issues are still the same, but the materials and methods are different.

In the design of early monuments, consideration of spatial interaction between structural subsystems was relatively unimportant, because their massiveness provided for strength and stability. In comparison, the size and density of structural elements of a contemporary tall building are strikingly less, and continue to diminish motivated by the real estate market, aesthetic principles, and innovative structural solutions. Thus the trend in high-rise technology can be thought of as a progressive reduction in the quantity of structural material used to create the exterior architectural enclosure and the spaces within.

To be successful, a tall building must economically satisfy the often conflicting demands imposed by various trades such as mechanical, electrical, structural and architectural. In doing so, from a structural point of view, a building can be defined as tall when its height creates different conditions in the design, construction, and use than the conditions that exist for its lower brethren. These conditions are manifest when the effects of lateral loads begin to influence its design. For example, in the design of tall buildings, in addition to the requirements of strength, stiffness, and stability, the lateral deflections due to wind or seismic loads should be controlled to prevent both structural and nonstructural damage. Also, the wind response of top floors in terms of their accelerations during frequent wind storms, should be kept within acceptable limits to minimize motion perception and discomfort to building occupants.

The trend in high-rise architecture is to create an overall spatial form with an intricate detailing of the cladding system (Fig. 1.1). The reason is uniquely to define a tower within an urban environment, and at the same time, provide interior spaces that are highly desirable to the building tenants. More often, the resulting structural solution is complex. However, the engineer, who until the early 1970s exercised considerable influence on the building's architectural shape, no longer deems it necessary to do so. Instead, with the immense analytical backup provided by computers, the structural engineer has freed the architect of structural restraints, especially in seismically benign regions. Needless to say, free-form architecture has demanded closer scrutiny of proven systems, challenging the engineer to either modify the proven systems or to come up with new structural solutions altogether. Although it is possible to arrive at a number of structural solutions that are equally applicable to a particular high-rise building, the final scheme more often depends on how best it meets other nonstructural requirements. Optimization of structural systems is thus a task that is studied in concert with other building disciplines.

Although the form of the building exterior plays a large role in how the building behaves under wind and seismic forces, few engineers are given the opportunity (and rightly so, otherwise all our buildings would be prismatic, and either be square or round) to influence the shape of the building. Instead, their role is confined to the

Figure 1.1 Spatial form of modern high-rise architecture. Fox Plaza, Los Angeles Johnson, Fain and Pereira; Architects John A. Martin & Asso. Inc., Structural Engineers.

optimization of the structure for the particular shape that the architect and the owner provide.

1.2 STRUCTURAL CONCEPTS

The key idea in conceptualizing the structural system for a narrow tall building is to think of it as a beam cantilevering from the earth (Fig. 1.2). The laterally directed force generated, either due to wind blowing against the building or due to the inertia forces induced by ground shaking, tends both to snap it (shear), and push it over (bending). Therefore, the building must have a system to resist shear as well as bending. In resisting shear forces, the building must not break by shearing off (Fig. 1.3a), and must not strain beyond the limit of elastic recovery (Fig. 1.3b). Similarly, the system resisting the bending must satisfy three needs (Fig. 1.4). The building must not overturn from the combined forces of gravity and lateral loads due to wind or seismic effects; it must not break by premature failure of columns either by crushing or by excessive tensile forces; its bending deflection should not exceed the limit of elastic recovery. In addition, a

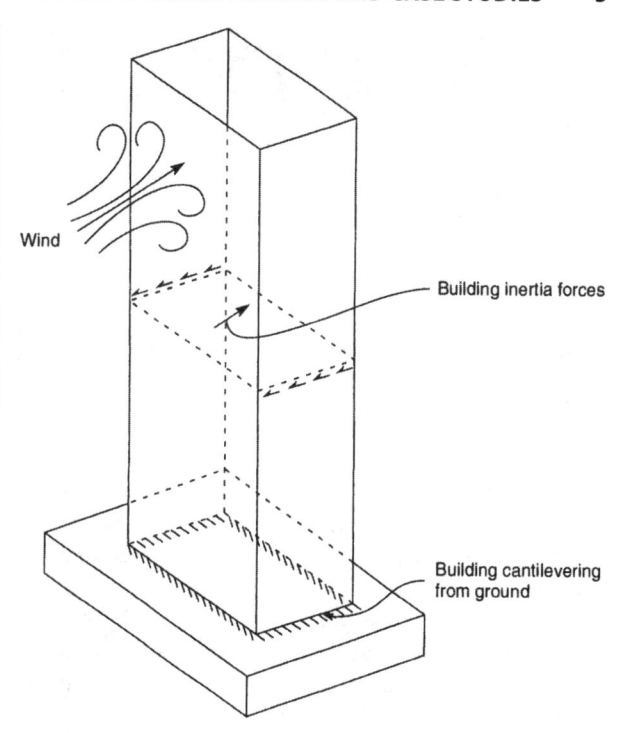

Figure 1.2 Structural concept of tall building.

building in seismically active regions must be able to resist realistic earthquake forces without losing its vertical load-carrying capacity.

In the structure's resistance to bending and shear, a tug-of-war ensues that sets the building in motion, thus creating a third engineering problem; motion perception or vibration. If the building sways too much, human comfort is sacrificed, or more importantly, non-structural elements may break resulting in expensive damage to the building contents and causing danger to the pedestrians.

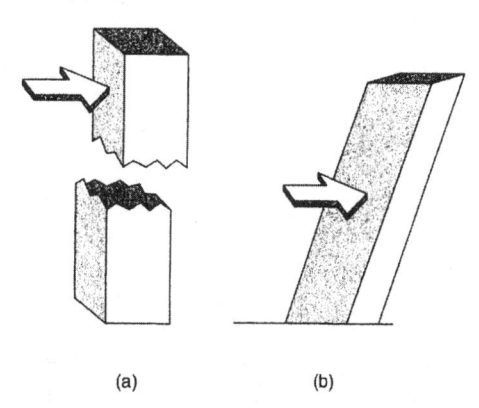

(a)   (b)

Figure 1.3 Building shear resistance: (a) building must not break; (b) building must not deflect excessively in shear.

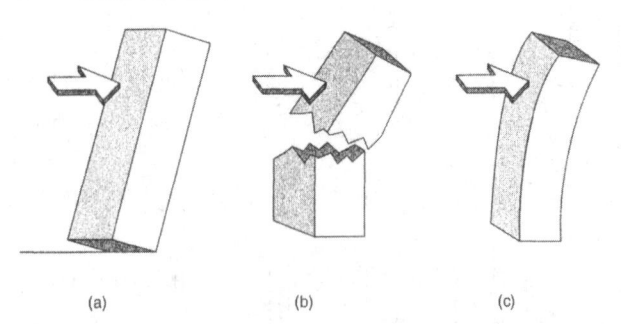

(a) (b) (c)

Figure 1.4 Bending resistance of building: (a) building must not overturn; (b) columns must not fail in tension or compression; (c) bending deflection must not be excessive.

A perfect structural form to resist the effects of bending, shear and excessive vibration is a system possessing vertical continuity ideally located at the farthest extremity from the geometric center of the building. A concrete chimney is perhaps an ideal, if not an inspiring engineering model for a rational super-tall structural form. The quest for the best solution lies in translating the ideal form of the chimney into a more practical skeletal structure.

With the proviso that a tall building is a beam cantilevering from earth, it is evident that all columns should be at the edges of the plan. Thus the plan shown in Fig. 1.5(b) would be preferred over the plan in Fig. 1.5a. Since this arrangement is not always possible, it is of interest to study how the resistance to bending is affected by the arrangement of columns in the plan. We will use two parameters, Bending Rigidity Index BRI and Shear Rigidity Index SRI, first published in Progressive Architecture, to explain the efficiency of structural systems.

The ultimate possible bending efficiency would be manifest in a square building which concentrates all the building columns into four corner columns as shown in Fig. 1.6a. Since this plan has maximum efficiency it is assigned the ideal Bending Rigidity Index (BRI) of 100. The BRI is the total moment of inertia of all the building columns about the centroidal axes participating as an integrated system.

The traditional tall building of the past, such as the Empire State Building, used all columns as part of the lateral resisting system. For columns arranged with regular bays, the BRI is 33 (Fig. 1.6b).

A modern tall building of the 1980s and 90s has closely spaced exterior columns and long clear spans to the elevator core in an arrangement called a "tube." If only the perimeter columns are used to resist the lateral loads, the BRI is 33. An example of this plan type is the World Trade Center in New York City (Fig. 1.6c).

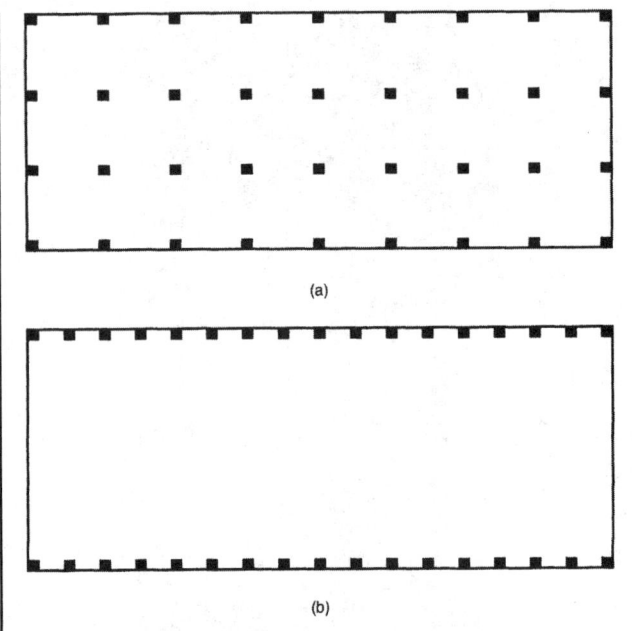

(a)

(b)

Figure 1.5 Building plan forms: (a) uniform distribution of columns; (b) columns concentrated at the edges.

The Sear Towers in Chicago uses all its columns as part of the lateral system in a configuration called a "bundled tube." It also has a BRI of 33 (Fig. 1.6d).

The Citicorp Tower (Fig. 1.6e), uses all of its columns as part of its lateral system, but because columns could not be placed in the corners, its BRI is reduced to 31. If the columns were moved to the corners, the BRI would be increased to 56 (Fig. 1.6f). Because there are eight columns in the core supporting the loads, the BRI falls short of 100.

The plan of Bank of Southwest Tower, a proposed tall building in Houston, Texas, approaches the realistic ideal for bending rigidity with a BRI of 63 (Fig. 1.6g). The corner columns are split and displaced from the corners to allow generous views from the office interiors.

In order for the columns to work as elements of an integrated system, it is necessary to interconnect them with an effective shear-resisting system. Let us look at some of the possible solutions and their relative Shear Rigidity Index (SRI).

The ideal shear system is a plate or wall without openings which has an ultimate Shear Rigidity Index (SRI) of 100 (Fig. 1.7a). The second-best shear system is a diagonal web system at 45° angles which has an SRI of 62.5 (Fig. 1.7b). A more typical bracing system that combines diagonals and horizontals but uses more material is shown in Fig. 1.7c. Its SRI depends on the slope of the diagonals and has a value of 31.3 for the most usual brace angle of 45°.

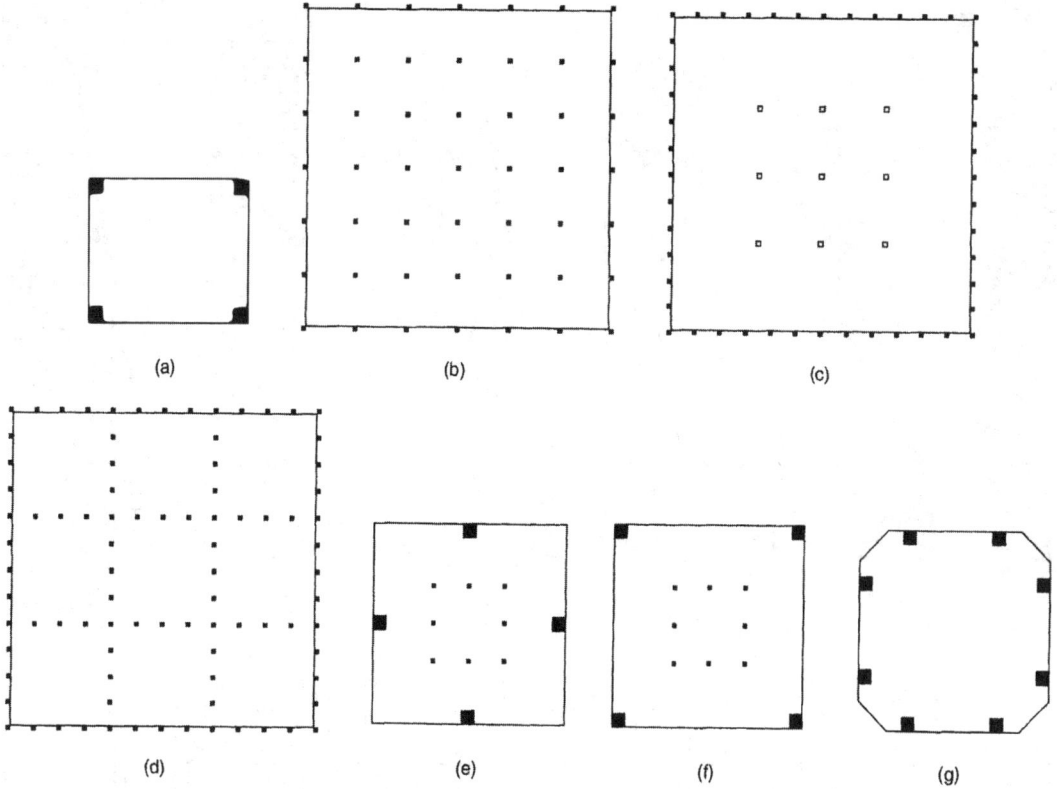

Figure 1.6 Column layout and Bending Rigidity Index (BRI): (a) square building with corner columns: BRI = 100; (b) traditional building of the 1930s, BRI = 33; (c) modern tube building, BRI = 33; (d) Sears Towers, BRI = 33; (e) City Corp Tower: BRI = 33; (f) building with corner and core columns, BRI = 56; (g) Bank of Southwest Tower, BRI = 63.

The most common shear systems are rigidly joined frames as shown in Figs. 1.7d-g. The efficiency of a frame as measured by its SRI depends on the proportions of members' lengths and depths. A frame, with closely spaced columns, like those shown in Fig. 1.7e-g, used in all four faces of a square building has a high shear rigidity and doubles up as an efficient bending configuration. The resulting configuration is called a "tube" and is the basis of innumerable tall buildings including the world's two most famous buildings, the Sears Tower and the World Trade Center.

In designing the lateral bracing system for buildings it is important to distinguish between a "wind design" and a "seismic design." The building must be designed for horizontal forces generated by wind or seismic loads, whichever is greater, as prescribed by the building code or site-specific study accepted by the Building Official. However, since the actual seismic forces, when they occur, are likely to be significantly larger than code-prescribed forces, seismic design requires material limitations and detailing requirements in addition to strength requirements. Therefore, for buildings in high-seismic zones, even when wind forces govern the design, the detailing

and proportioning requirements of seismic resistance must also be satisfied. The requirements get progressively more stringent as the zone factor for seismic risk gets progressively higher.

1.3 STRUCTURAL VOCABULARY; CASE STUDIES

Having noted that a building must have systems to resist both bending and shear, let us visit some of the world's tall buildings to explore how prominent engineers have exploited the concept of SRI and BRI in their designs. In describing the designs, an attempt is made to present the structural scheme descriptions in a doctored form. This serves the educational purposes of this book more effectively than a prosaic recounting of the design data. Although some examples include run-of-the-mill designs that a large number of engineers have to solve on a day-to-day basis, other studies recounted are somewhat poetic, high-profile projects, even daring in their engineering solutions. Many are examples of buildings constructed or proposed in seismically benign regions requiring careful examination of their ductile behavior

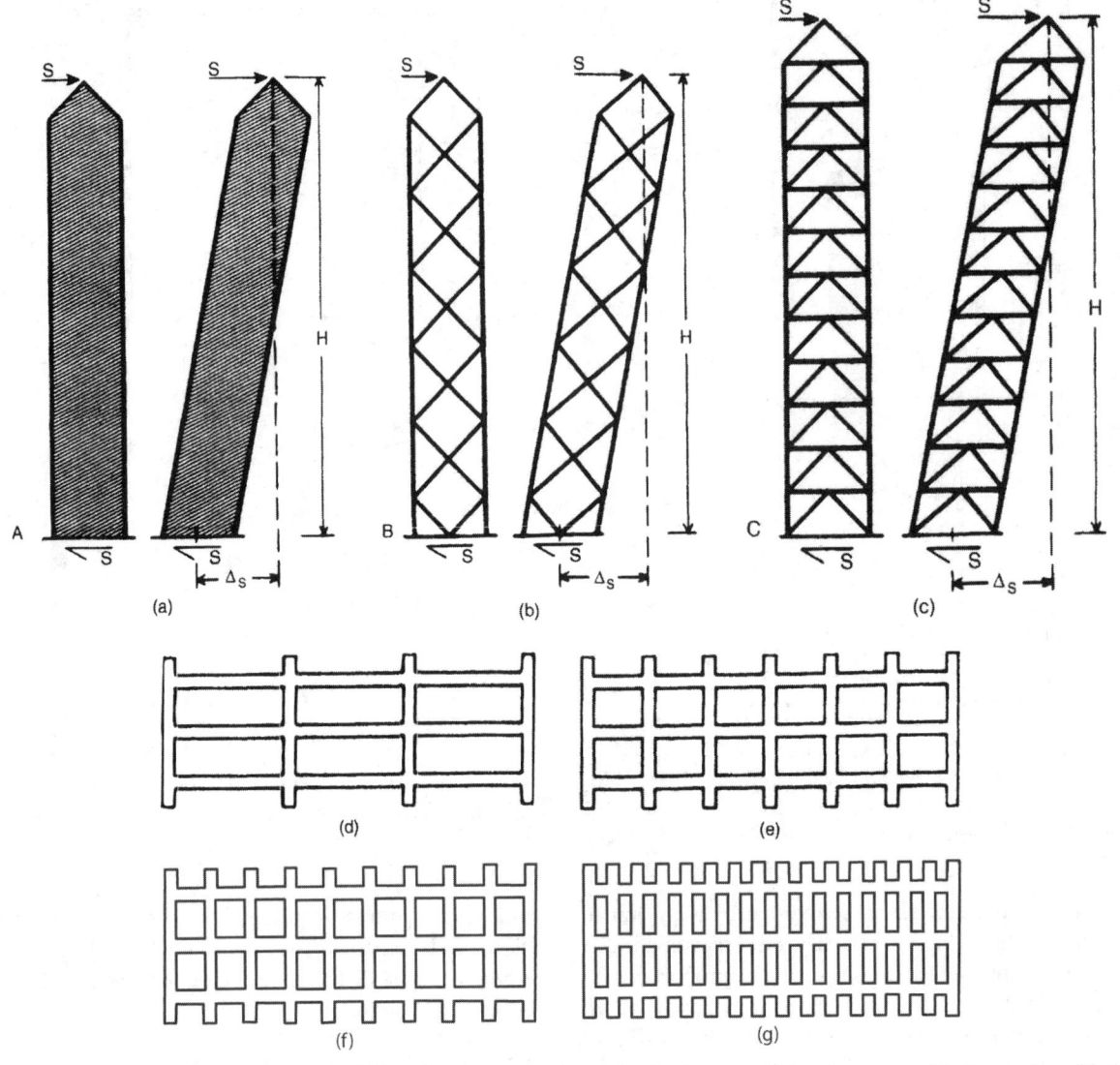

Figure 1.7   Tall building shear systems: (a) shear wall system; (b) diagonal web system; (c) web system with diagonals and horizontals. (d–g) Rigid frames. (h) Empire State Building bracing system; riveted structural steel frame encased in cinder concrete.

and reserve strength capacity before they are applied to seismically active regions.

The main purpose of this section is to introduce the reader to the existing and new vocabulary of structural systems normally considered in the design of tall buildings. Structural design is in a period of mixing and perfecting structural systems such as megaframes, interior and exterior super diagonally braced frames, spine structures, etc., to name a few. The case studies included in this section illuminate those aspects of conceptualization and judgment that are timeless constants of the design process and can be as important and valuable for understanding structural design as are the latest computer software. The case histories are based on information contained in various technical publications and periodicals. Frequent use is made of personal information obtained from the structural engineers-of-record. Table 1 outlines the case studies presented in this chapter with the lateral structural system used for each.

We start our world tour in New York City to pay homage to the Empire State Building which was the tallest building in the world for more than 40 years, from the day of its completion in 1931 until 1972 when the Twin Towers of the New York's World Trade Center exceeded its 1280 ft (381 m) height by almost 120 ft (37 m) (Fig. 1.7h). The structural steel frame with riveted joints, while encased in cinder concrete, was designed to carry 100% of gravity and 100% wind load imposed on the building. The

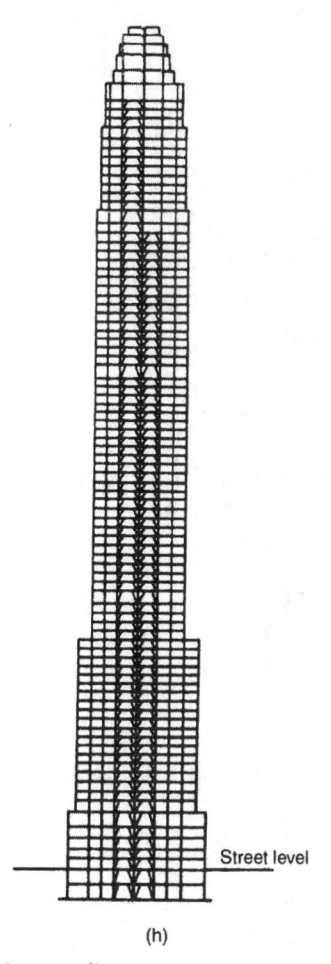

(h)

Figure 1.7 (*Continued*)

encasement, although neglected in strength analysis, stiffened the frame, particularly against wind load. Measured frequencies on the completed frame have estimated the actual stiffness at 4.8 times the stiffness of the bare frame.

1.3.1 The Museum Tower, Los Angeles
The Museum Tower, a 22-story residential complex, shown in Fig. 1.8a, is part of the California Plaza complex which is one of the largest urban revitalization projects in a zone of high seismic activity in North America. The structural system for the building, located in downtown Los Angeles, consists of a tubular ductile concrete frame with perimeter columns spaced at 13 ft (3.96 m) centers interconnected with upturned spandrel beams (Fig. 1.8b). The exterior frame is of exposed painted concrete.

The gravity system for the typical floor consists of an 8 in (203 mm) thick post-tensioned flat plate with banded and uniform tendons running in the short and long directions of the building respectively as shown in Fig. 1.8a.

Although the building is regular both in plan and elevation, and is less than 240 ft (73 m) in height, because of transfers at the base (Fig. 1.8b), a dynamic analysis using site-specific spectrum was used in the seismic design. The dynamic base shear was scaled down to a value corresponding to the 1992 UBC static base shear. To preserve the dynamic characteristics of the building, the spectral accelerations were scaled down without altering the story masses. The structural design is by John A. Martin & Associates, Inc., Los Angeles. The architecture is by Fujikawa Johnson Asso. Inc., and Barton Myers Asso. Inc.

1.3.2 Bank One Center, Indianapolis
Bank One Center in Indianapolis is a 52-story steel-framed office building that rises to a height of 623 ft (190 m) above the street level. In plan, the tower is typically 190 × 120 ft (58 × 37 m) with setbacks at the 10th, 13th, 23rd, 45th, and 47th floors (Fig. 1.9a).

The structural system resisting the lateral forces consists of two large vertical flange trusses in the north–south direction and two smaller core braces in the east–west acting as web trusses connecting flange trusses. The flange trusses which provide maximum lever arm for resisting the overturning moments, also serve to transfer gravity loads of the core to the exterior columns. The resulting equalization of axial stresses in the truss and the nontruss perimeter columns keeps the differential shortening between them to a minimum. To ensure a direct load path for the transfer of gravity load from the core to the truss columns, the core column is removed below the level of braces at every 12th level, as shown in Fig. 1.9b. In addition, the step-back corners are cantilevered to maximize the tributary area of gravity load, to compensate for the tensile force due to overturning moments. The structural design is by LeMessurier Consultants, Inc., Cambridge, Massachusetts.

1.3.3 Two Union Square, Seattle
This 50-story office tower (Fig. 1.10c) has a curved façade with widely spaced perimeter columns. Lateral resisting elements are placed in the interior core walls enabling the perimeter columns to be spaced approximately 44 ft (13.42 m) rather than a more typical 10 to 15 ft (3.05 to 4.58 m).

Four 10 ft diameter (3.05 m) steel pipes filled with high-strength, 19,000 psi (131 mPa) concrete are the primary lateral load-resisting elements (Fig. 10.a,b). To reduce the perception of lateral movement in the upper levels of the building, the building's structural system incorporates 16 dampers. Structural design is by Skilling Ward Magnusson Barkshire Inc.

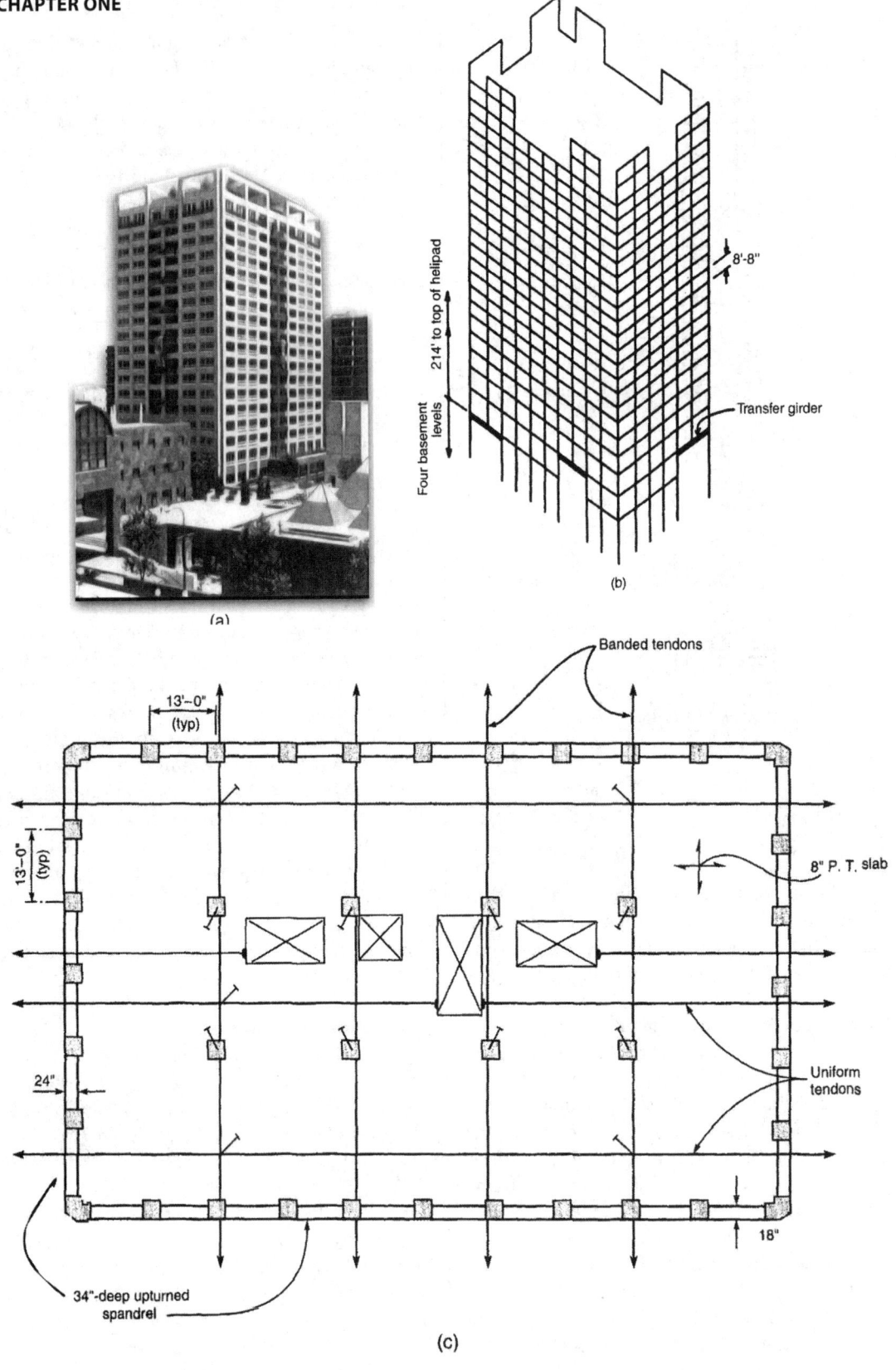

Figure 1.8 The Museum Tower, Los Angeles: Architects: Fujikawa Johnson Asso. Inc. and Barton Myers Asso. Inc. Structural engineers: John A. Martin & Asso. Inc., Los Angeles. (a) building elevation; (b) lateral system; (c) typical posttensioned floor framing plan.

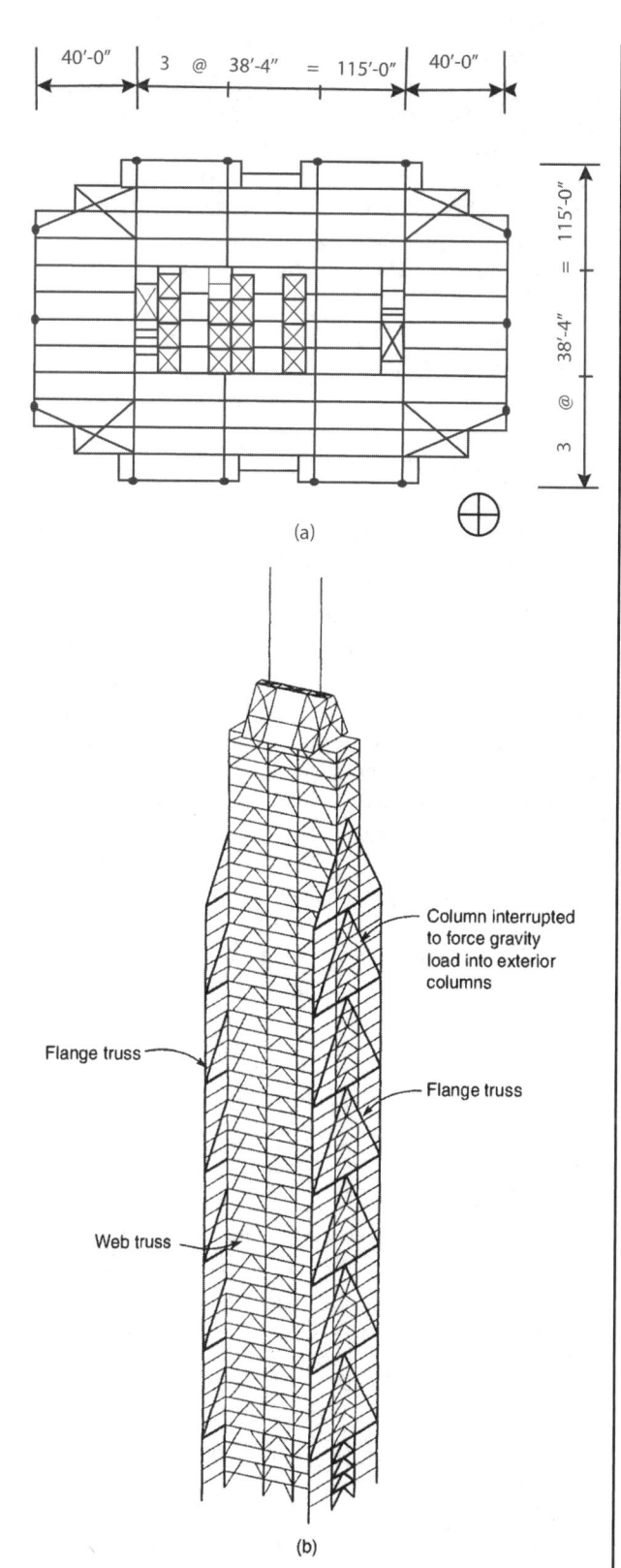

(a)

(b)

Figure 1.9 Bank One Center, Indianapolis: (a) plan; (b) lateral system.

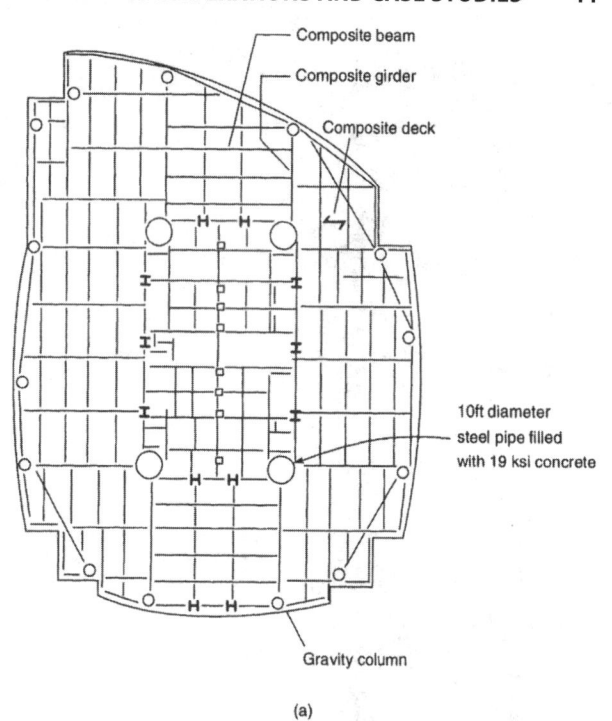

(a)

(b)

Figure 1.10 Two Union Square, Seattle: (a) plan; (b) construction photograph; (c) building elevation.

(c)

Figure 1.10   (*Continued*)

1.3.4   Bank of China Tower, Hong Kong

The structural system for the 70-story, 1209 ft (368.5 m) Bank of China Tower in Hong Kong consists primarily of a cross-braced space truss. The space truss supports almost the entire weight of the building while simultaneously resisting the lateral loading of typhoon winds. Both the lateral and gravity loads are carried to four composite columns at the corners of the building, allowing a 170 ft (51.82 m) clear span at the base of the building.

A fifth composite column in the center of the building begins at the 25th floor and extends to the top. The loads on this column are transferred to the corner columns at the 25th level. At the foundation level, the corner columns are 14×26 ft (4.3×7.93 m). The size of the steel section of the composite columns varies, and the concrete portion gets progressively smaller as it rises, varying by more than 10 ft (3.05 m). Compositing of frame elements by enclosing the steel members with reinforced concrete eliminated the need for expensive three-dimensional steel connections at the building corners. The structural design is by Leslie Robertson & Associates. The structural system is shown schematically in Fig. 1.11.

1.3.5   Dallas Main Center

The 921 ft (280.7 m) building is of composite construction consisting of 73 stories of office space. A three-dimensional moment-resisting frame made of highly repetitive 36 in. (0.30 m) rolled shapes spans the entire building to sixteen composite columns consisting of light steel columns encased in 10,000 psi (68.95 mPa) concrete in sizes up to 7 ft (2.14 m) square (Fig. 1.12).

The steel verticals in the core are web members of a vierendeel system and are not carried to the foundation. All gravity loads and the overturning forces from the wind are resisted by exterior concrete columns. Wind shear is resisted by the steel frame, which connects the columns across the building. The distance across the building is 127 ft (38.71 m) between columns, giving a height-to-width ratio of the frame of more than 7:1. The structural design is by LeMessurier Consultants, Inc., Cambridge, Massachusetts.

1.3.6   The Miglin-Beitler Tower, Chicago, Illinois

The Miglin-Beitler Tower designed by the New York Office of Thornton-Tomasetti Engineers, if built; will establish a new record as the world's tallest building as well as the world's tallest non-guyed structure surpassing the Sears Tower and the CN Tower which rise to heights of 1454 and 1822 ft (443.2 and 555.4 m), respectively. Rising to 1486.5 ft (453 m) at the upper sky room level, 1584.5 ft (483 m) at the top of the mechanical areas, and finally to 1999.9 ft (609.7 m) at the tip of the spire, the project will provide a regal landmark to the Chicago skyline. An elevation and schematic plans of the building are shown in Fig. 1.13a, b.

A cruciform tube structure has been developed to achieve structural efficiency, superior dynamic behavior, simplicity of construction, and unobtrusive integration of structure into leased office floor areas (Fig. 1.13f).

The tube consists of five major components as shown in Fig. 1.13f. These components are, in order of construction sequence:

1. A 62 ft 6 in. × 62 ft 6 in. (19×19 m) concrete core with walls varying from a maximum thickness of 3 ft 0 in. (0.91 m) to a minimum thickness of 1 ft 6 in. (0.46 m).

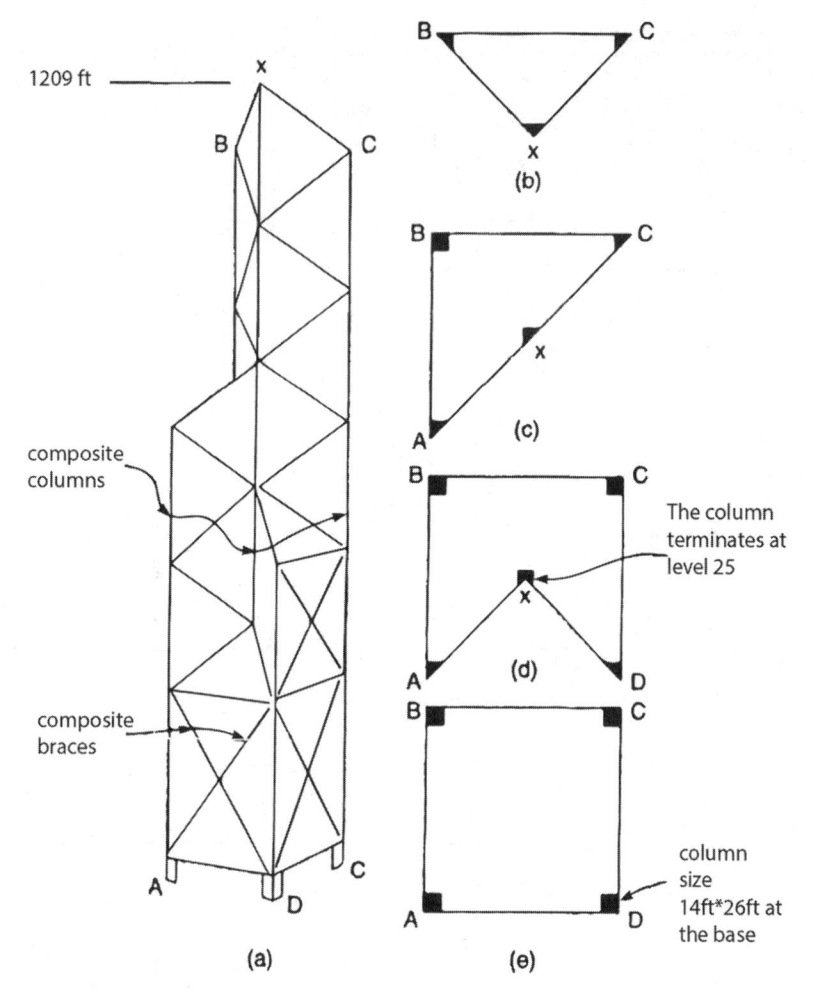

Figure 1.11   The Bank of China Tower, Hong Kong: (a) schematic elevation; (b-e) floor plans.

2. A conventional structural steel composite floor system utilizing 18 in. (0.46 m) deep rolled steel sections spaced 10 ft (3.05 m) on the center with 3 in. (74 mm) deep corrugated metal deck spanning the 10 ft (3.05 m) between the beams and 3½ in. thick (89 mm) normal weight concrete topping. The steel floor system is supported on light steel erection columns which allow the steel construction to proceed 8 to 10 floors ahead of the next concrete operation.

3. The concrete fin columns, each of which encases a pair of steel erection columns, are located at the face of the building. These fin columns, which extend 20 ft (6.10 m) beyond the 140 × 140 ft (42.7 × 42.7 m) footprint at the base of the building, vary in dimension from 6½ × 33 ft. (2.0 × 10 m) at the base, 5½ × 15 ft (1.68 × 4.6 m) at the middle, to 4½ × 13 ft (1.38 × 4 m) near the top.

4. The next components of the cruciform tube system are the link beams which interconnect the four corners of the core to the eight fin columns on every floor. These link beams are comprised of reinforced concrete placed simultaneously with floor concrete. They become the concrete link between the fin columns and the core to make the full structural width of the building resist lateral forces. In addition to the link beams on each floor there are three two-story deep outrigger walls located at the 16th story, the 56th story, and the 91st story. These outrigger walls further enhance the structural rigidity by linking the exterior fin columns to the concrete core.

5. The last structural components of the cruciform tube are the exterior vierendeel trusses which are comprised of horizontal spandrel and two vertical columns at each of the 60 ft (18.3 m) faces on the four sides of the building.

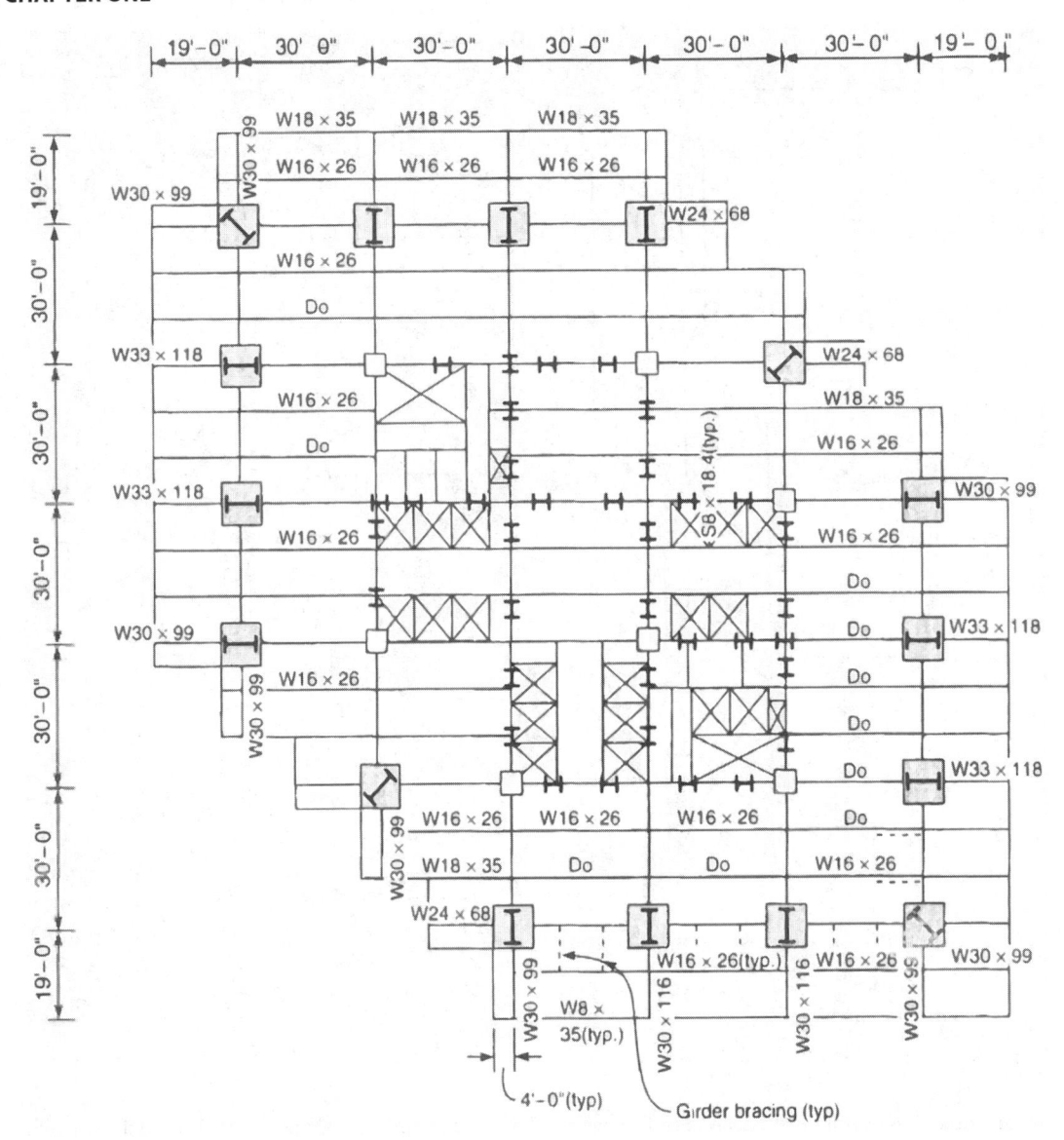

Figure 1.12 Dallas Main Center: (Inter-first plaza), 26th–43rd floor framing plan.

These vierendeels supplement the lateral force resistance and also improve the torsional resistance of the structural system. In addition, gravity loads are transferred out to the fin columns eliminating uplift forces.

The design wind pressures, shear, and overturning moments are shown in Fig. 1.13c, d, and e. The axial force distribution in the fin columns and the horizontal shear in the fins are shown in Fig. 1.13g.

The proposed foundation system for the project is rock caissons varying in diameter from 8 to 10 ft (2.44 to 3.0 m). The caissons will have a straight shaft, steel casing and will be embedded in rock a minimum of 6 ft (1.83 m).

The length of these caissons is 95 ft (29 m). A 4-ft thick (1.22 m) concrete mat ties the caissons and provides a means for resisting the shear forces at the base of the building. The bottom of the mat will be cast in a two-directional groove pattern to engage the soil in shear. Passive pressure on the edge of the mat and on the projected side surface of the caisson provides additional resistance to shear at the base.

1.3.7 The NCNB Tower, North Carolina

The NCNB tower in Charlotte, North Carolina, is an 870 ft (265.12 m) high, concrete office building with a 100 ft

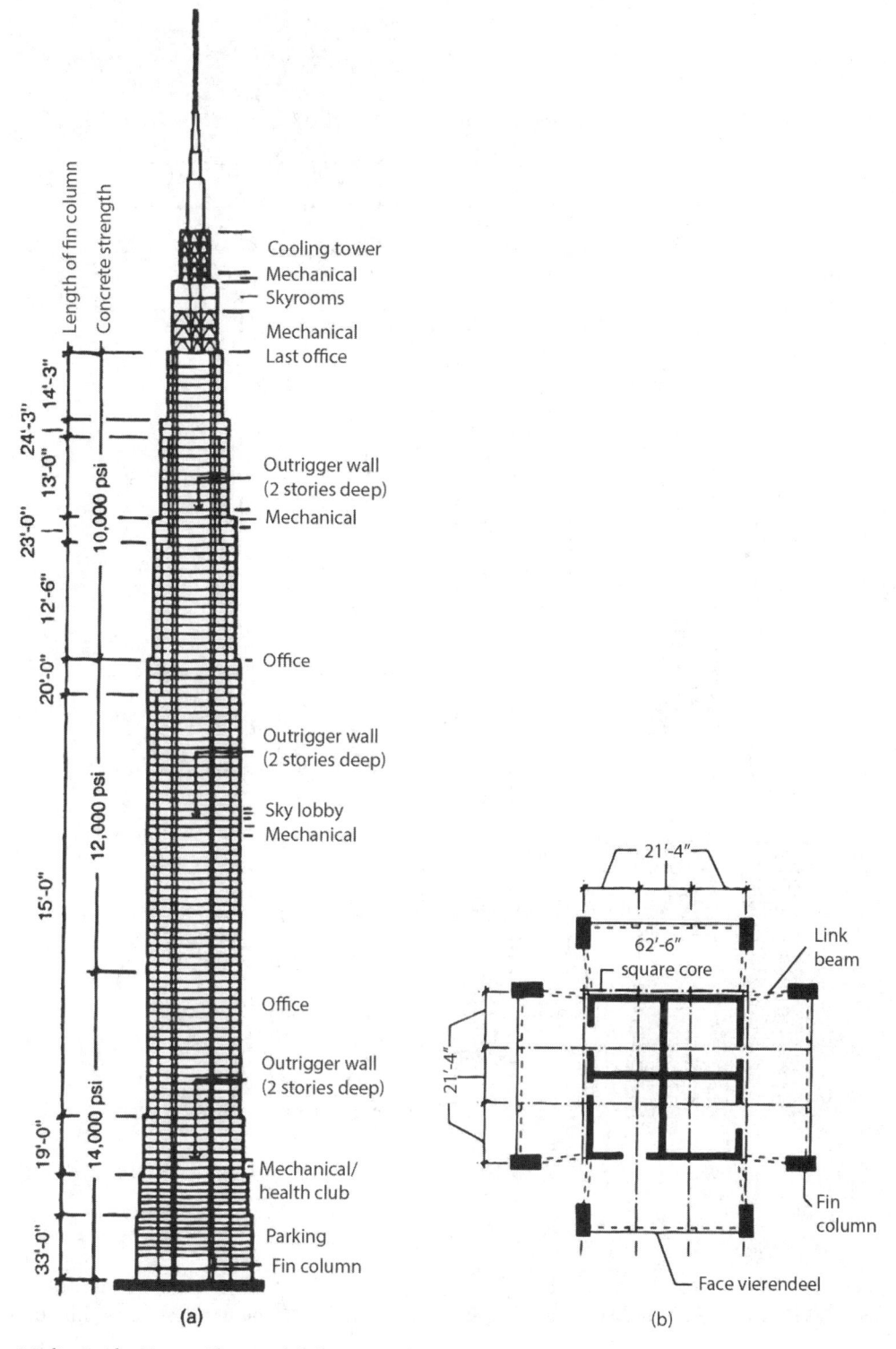

Figure 1.13 Miglin-Beitler Tower, Chicago: (a) elevation; (b) plan; (c) wind pressures; (d) wind shear; (e) wind moment; (f) typical floor framing plan; (g) wind force distribution in fin columns.

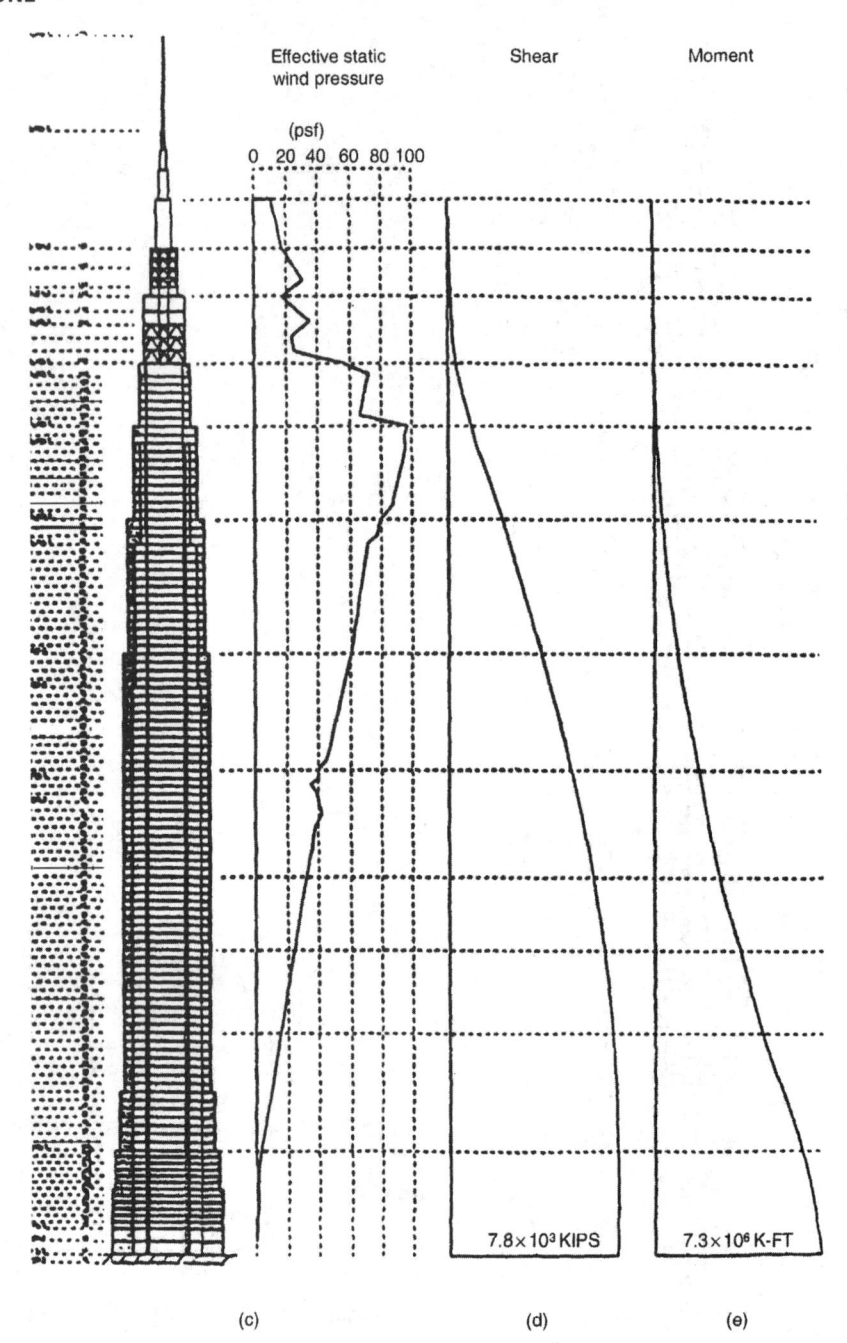

Figure 1.13 (*Continued*)

(30.5 m) crown of aluminum spires (Fig. 1.14a). The building has 12 ft 8 in. (3.87 m) floor-to-floor heights and 48 ft (14.63 m) column-free spans from the perimeter to the core.

The structural system for resisting lateral loads consists of a reinforced concrete perimeter tube with normal-weight concrete ranging in strength from 8000 psi (55.16 mPa) near the building's base to 6000 psi (41.37 mPa) at the top. Typical column sizes range from 24 × 38 in. (0.61 × 0.97 m) at the base to 24 × 24 in. (0.61 × 0.61 m) at the top. The floor system (Fig. 1.14b) consists of a 4⅝ in. (118 mm) thick lightweight concrete slab supported on 18 in. (458 mm) deep post-tensioned beams spaced at 10 ft (3.05 m) on centers. Lightweight concrete was used to reduce the building's weight and give the floors

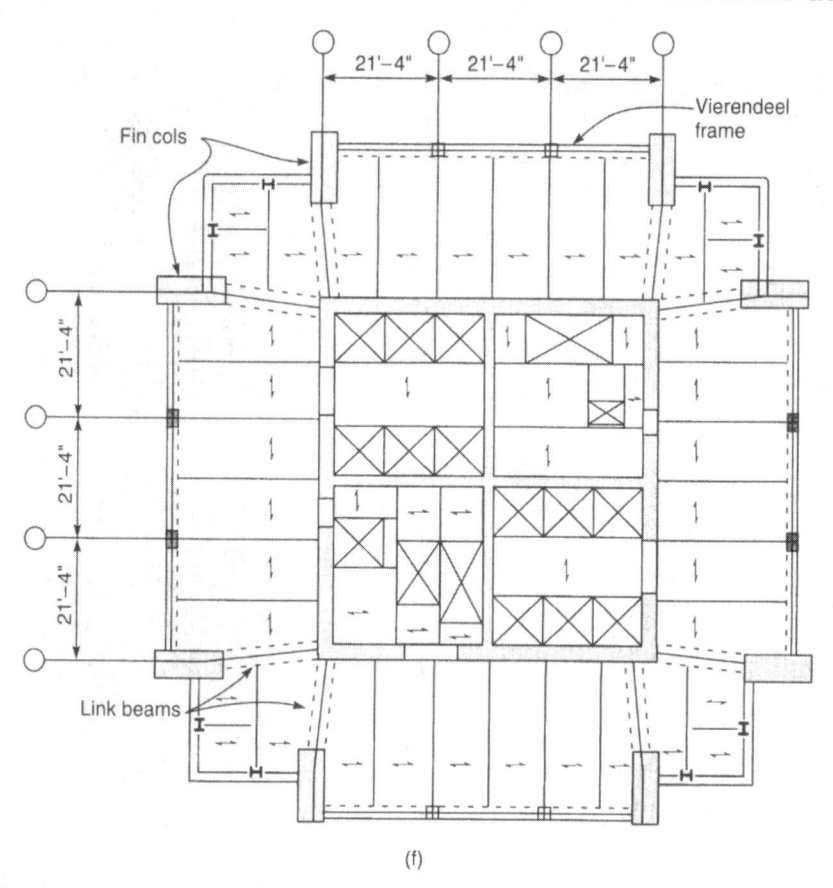

(f)

Figure 1.13 (*Continued*)

the required fire rating. The beams span as much as 48 ft (14.63 m) providing column-free lease space from the core to the perimeter (Fig. 1.14b).

The tower's columns are spaced 10 ft (3.05 m) on center and are connected by 40 in. deep (1.01 m) spandrel beams. The building has a roughly square plan at the base, but above the 13th floor it resembles a square set over a slightly larger cross, with the building's four corners recessed and its four major faces bowed slightly outward.

To maintain tube action between the 13th and 43rd floors, engineers used L-shaped vierendeel trusses to continue the tube around the corners. Instead of using transfer girders at the building step-backs, the building's column-and-spandrel structure is used to create multi-level vierendeel trusses on the building's main façades. These vierendeels transfer loads using another set of vierendeel trusses perpendicular to the façade at the edges of recessed corners (Fig. 1.14c).

Differential shortening between the core and perimeter columns was a concern during design because the core columns will be under significantly higher stresses than the closely spaced perimeter columns. To compensate for this, the core columns were constructed slightly longer than the perimeter columns.

Aware that Charlotte is located in an area of moderate earthquake risk, engineers considered seismic provisions applicable to moderate seismic zones. While wind loads controlled the design of the lower two-thirds of the building, seismic forces controlled the design of the upper third. To meet the seismic requirements additional ties and stirrups were added to columns and beams. The shear reinforcing for critical beams, particularly at setback levels was designed with a diagonal configuration to provide greater ductility (Fig. 1.14d,e).

Both standard and lightweight concrete were used simultaneously. The normal-weight concrete was used for the perimeter columns, which ranged in size from 24×38 in. (610×965 mm) at the bottom to 24×24 in. (610×610 mm) at the top, as well as the core columns, ranging from 2×18 ft (0.61×3.5 m) at the base to 2×3 ft (0.61×0.92 m) at the top.

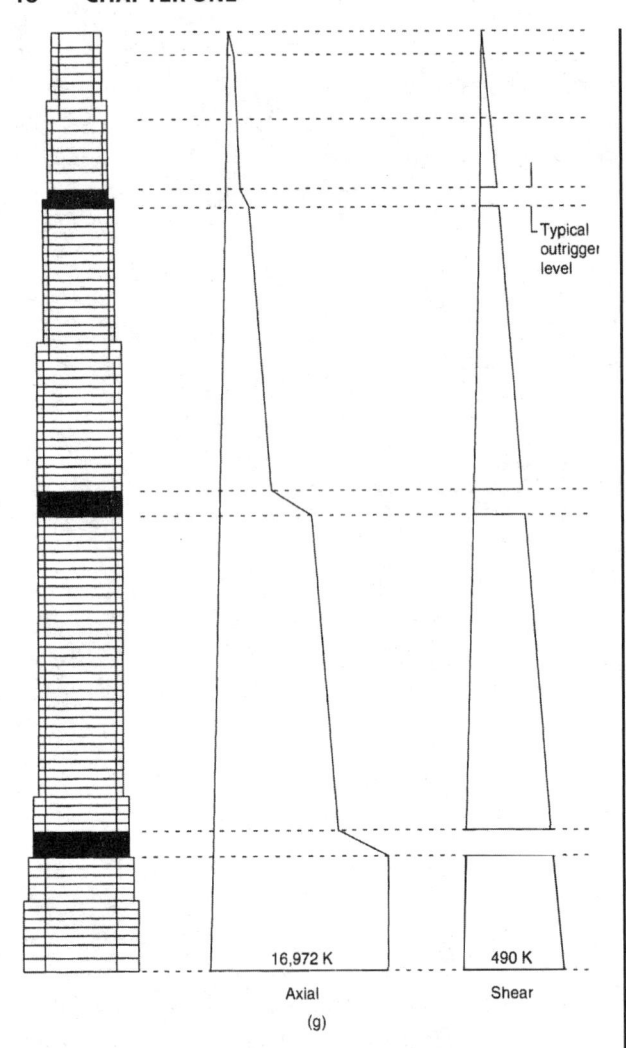

Axial   Shear

16,972 K   490 K

Typical outrigger level

(g)

Figure 1.13   (*Continued*)

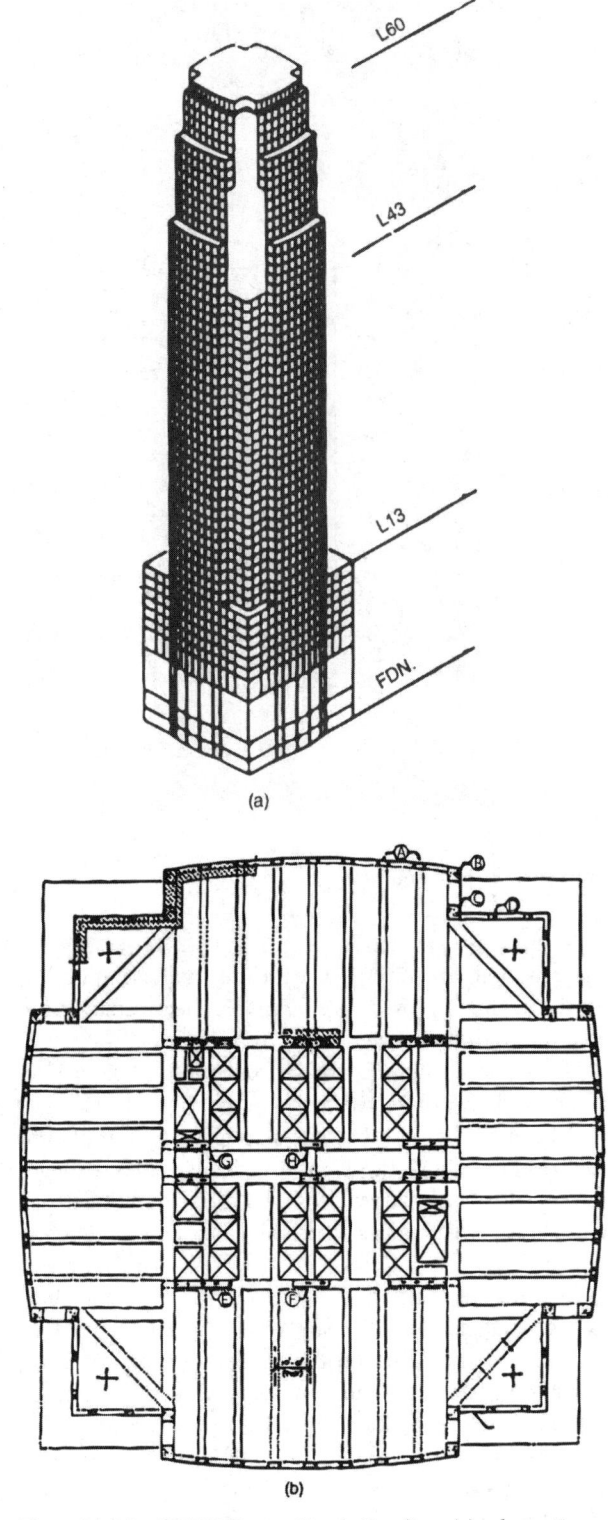

(a)

(b)

Figure 1.14   NCNB Tower, North Carolina: (a) schematic elevation; (b) typical floor framing plan; (c) vierendeel action at perimeter frame; (d) location of ductily-reinforced beams; (e) ductile diagonal reinforcement detail.

Normal-weight concrete was also used for post-tensioned spandrels at the perimeter of each floor, but 5000 psi (34.5 mPa) lightweight concrete was used to form the 4⅝ in. thick (118 m) floor slabs and the 18 in. deep (0.46 m) post-tensioned beams, spaced 10 ft (3.05 m) on center. The two types of concrete were poured in quick succession and puddled to avoid a cold joint.

The foundation system for the Tower consists of high-capacity caissons under the perimeter columns and a reinforced concrete mat for the core columns.

The high-capacity caissons were designed for a total end-bearing pressure of 150 ksf (7182 kN/m²) and skin friction of 5 ksf (240 kN/m²). The high bearing pressure required that the caisson be advanced through the fractured and layered rock zones into high-quality bedrock. Full-length casing was provided to prevent intrusion of soil and groundwater into the drilled hole and for

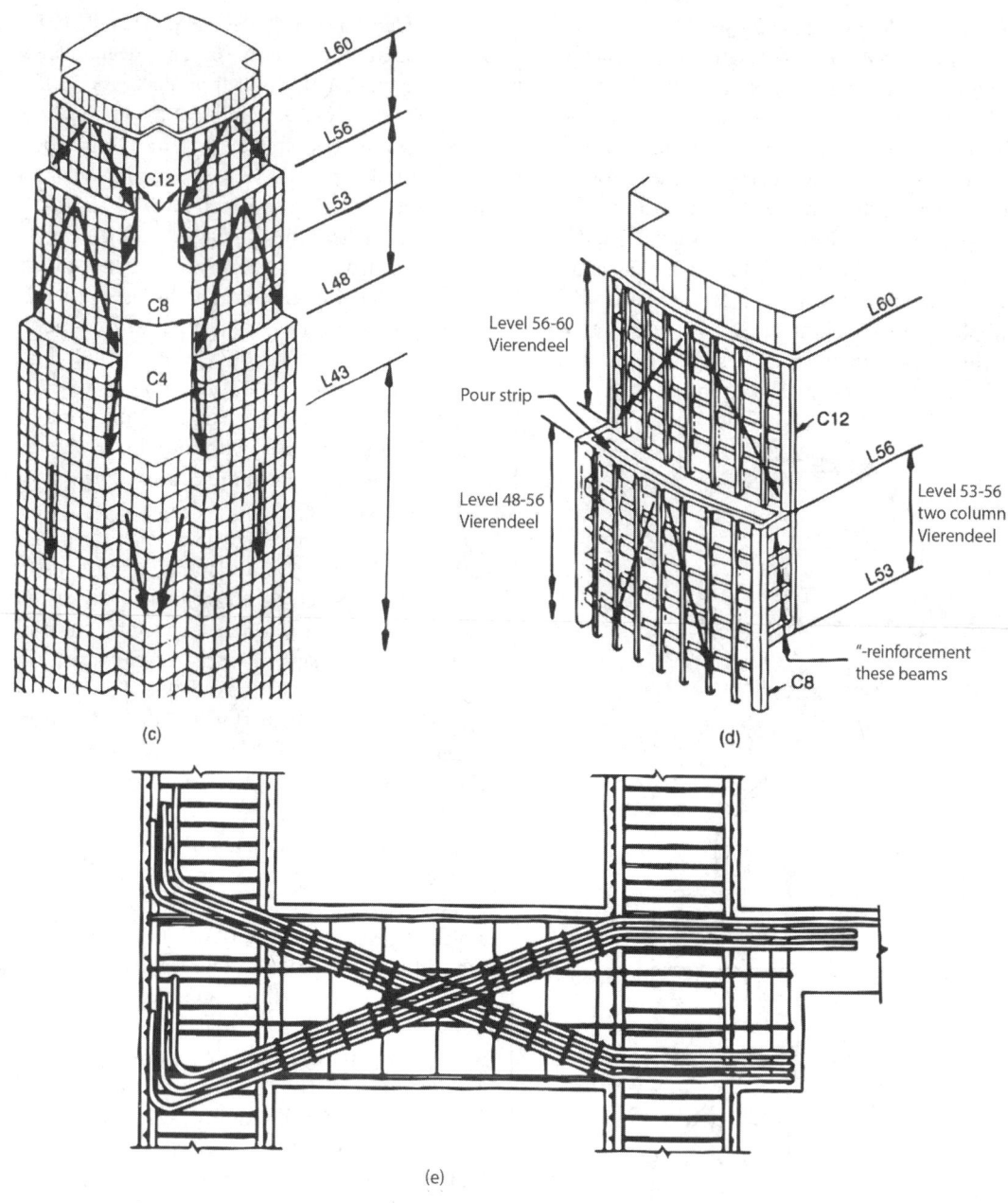

Figure 1.14   (*Continued*)

the safety of inspectors. Each caisson excavation was inspected in the field by the project geotechnical engineer. A 1½ in. (38 mm) diameter pilot hole was drilled to verify rock quality immediately below the caisson. Caisson diameters ranged from 54 to 72 in. (1.37 to 1.83 m), and the length ranged from 30 to 100 ft (9.15 to 30.5 m) (1.37 to 1.83 m). The concrete strength was 6000 psi. (41.37 kN/m²).

The core columns were supported on a foundation mat bearing on partially weathered rock. The mat dimensions were $83 \times 93 \times 8$ ft ($25.3 \times 28.35 \times 2.44$ m). The average total sustained bearing pressure under the mat is equal to 20 ksf (958 kN/m²). The mat was predicted to settle ½ in. (12.7 mm), mostly during construction. The structural design is by Walter P. Moore and Associates, Inc., Houston, Texas.

1.3.8 The South Walker Tower, Chicago

This tower, 946 ft (288.4 m) in height, has a changing geometry with the east face rising in a single plane from street level to the 65th floor while the other three faces change shape. To the 14th level, the structure is basically a trapezoid in plan 135 × 225 ft (41.15 × 68.6 m) overall. The building steps back at the 15th floor on three faces to provide ten corner offices on each floor. There are additional setbacks at the 47th floor. At the 51st floor, the sawtooth shape is dropped and the tower becomes an octagon in plan with 70 ft long (21.4 m) sides. The slenderness ratio of the structures is 7.25:1. The schematic floor plans at various levels are shown in Fig. 1.15.

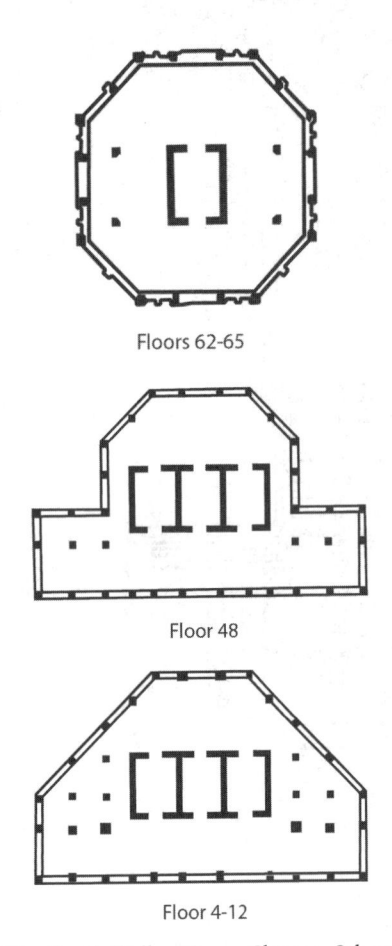

Floors 62-65

Floor 48

Floor 4-12

Figure 1.15 South Walker Tower, Chicago. Schematic plans.

The core shear walls in the tower's lower floors carry much of the lateral loading with shear wall-frame interaction. There are four main shear walls—two I shapes and two C shapes—on a typical floor. These interact with the perimeter columns and perimeter spandrel beams through girders that span from the core to the perimeter.

The girders have 39 in. deep (1.0 m) haunches at the columns. Spandrels are 36 in. (0.92 m) deep. Core wall concrete design strength varies from 8000 psi (55.121 mPa) at the base to 4000 psi (27.6 mPa) at the upper levels.

There is a 40 to 48 ft (12.2 × 14.63 m) span between the core and the perimeter. The spacing between the perimeter columns is fairly short, about 14 ft (4.3 in.) except at two corners where the spacing is 32 ft (9.76 m). Column loads range from 12,000 kips (53,376 kN) to 30,000 kips (133,440 kN). Concrete strengths used are 12,000 and 10,000 psi (82.74 and 68.95 mPa) at low- and mid-rise areas, and columns at upper levels include both 10,000 and 4000 psi (68.95 and 27.58 mPa) concrete. The largest columns, which are 5 ft square (1.53 m), contain 52 # 18, grade 75 bars, compared with the 66 # 18 bars that would have been needed if the more conventional, grade 60 steel had been used.

The original floor design had 16 in deep (406.4 mm) pans with 4 in. (101.6 mm) slabs. By using a post-tensioned system, the required depth of floor joists was reduced to 10 in. deep (254 mm) with a 4.5 in. (114.3 mm) slab.

The foundation system for the tower consists of a combination of caisson and mat foundation. The caissons range in size from 36 to 108 in. (0.91 to 2.75 m) in diameter. The mat under the core walls is 8 ft (2.44 m) thick. The structural design is by Brockette, Davis, Drake Inc., Dallas, Texas.

1.3.9 AT&T Building, New York City

The basic structural system for the building shown in Fig. 1.16a consists of a rigid-frame steel tube at the building perimeter. Additional stiffness is added along the width of the building by means of four vertical steel trusses. At every eighth floor, two I-shaped steel plate walls, with holes cut for circulation, extend from the sides of the trusses to the exterior columns on the same column line. The steel walls act as outrigger trusses mobilizing the full width of the building in resisting lateral forces. The horizontal shear at the base of the building is transferred to two giant steel plate boxes (Fig. 1.16b). Structural design is by Lesley Robertson and Associates.

1.3.10 Trump Tower, New York City

The 61-story Trump Tower, a reinforced concrete building shown in Fig. 1.17a, is an example of a multi-use complex that required extensive column transfers. The first seven floors are dedicated to commercial space, including a six-story atrium, followed by eleven floors of office space and a mechanical floor, and capped at the top with apartment dwellings.

The building consists of a slender reinforced concrete core which serves as the primary lateral load-resisting

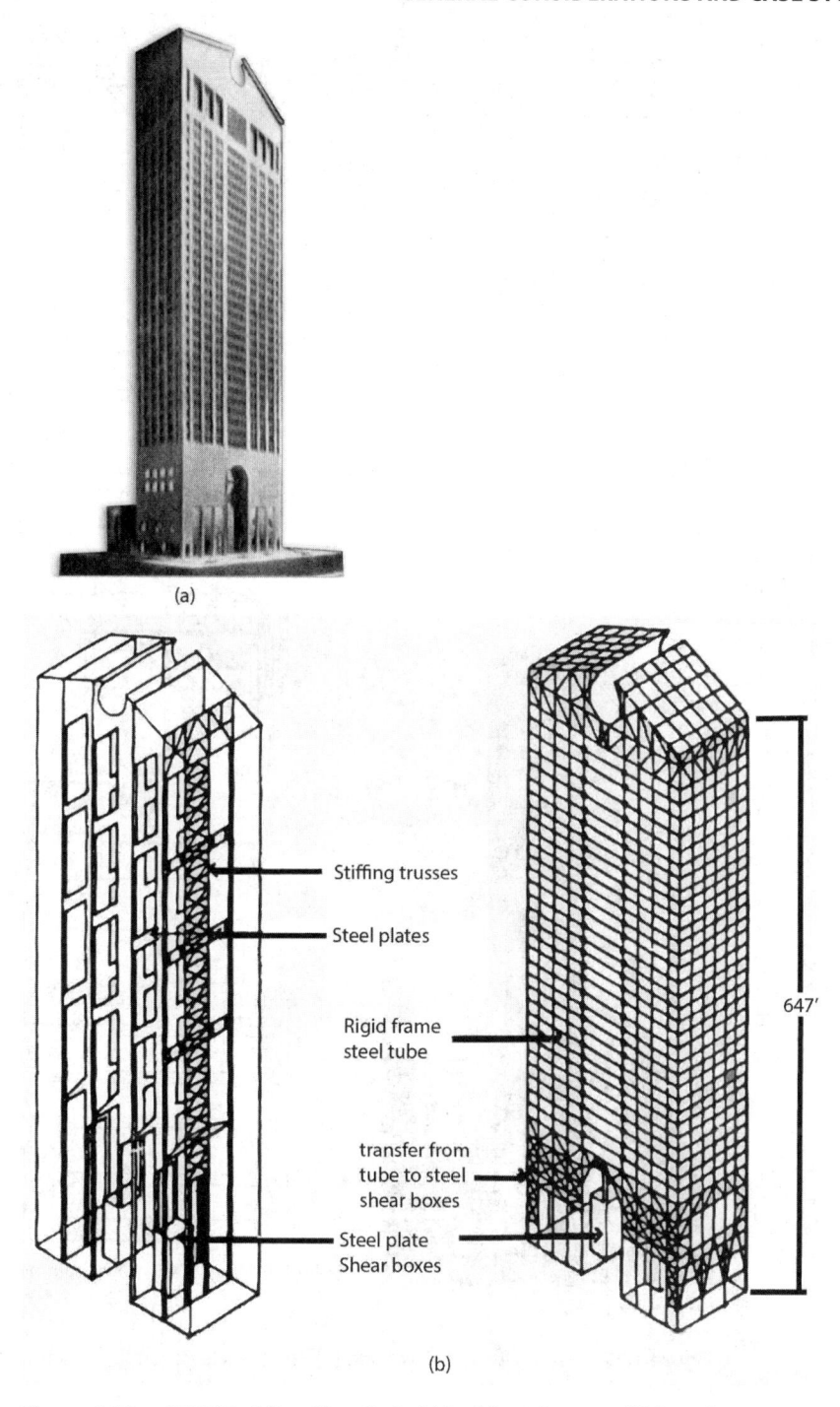

Figure 1.16   AT&T Building, New York: (a) building elevation; (b) lateral system.

element. The core is tied to the exterior columns through a large cap girder at the roof level to reduce wind overturning moments in the core and lateral deflections of the building.

The vertical loads are shared by the columns and core, but many of the columns acceptable in the apartment floors would be obstructive in the office floors. Therefore, deep concrete girders are used on the

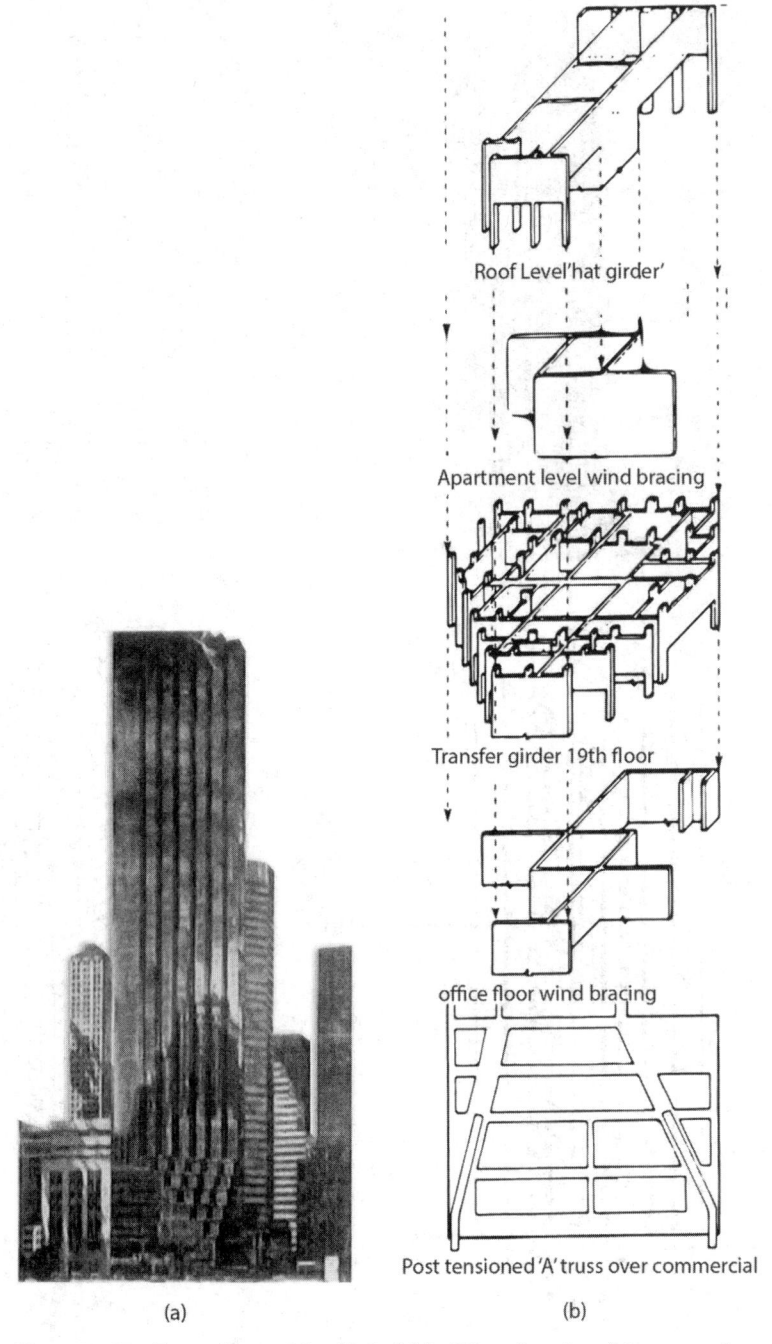

Roof Level 'hat girder'

Apartment level wind bracing

Transfer girder 19th floor

office floor wind bracing

Post tensioned 'A' truss over commercial

(a)   (b)

Figure 1.17   Trump Tower, New York: (a) building elevation; (b) structural system.

19th floor to transfer these columns. Similarly, many of the columns permitted in the offices must not drop into the commercial space. They are, therefore, transferred onto a huge post-tensioned "A" frame structure (Fig. 1.17b). The structural design is by the Office of Irwin G. Cantor, New York.

1.3.11   Metro-Dade Administration Building

Hurricane wind loads which may be as high as four times the U.S. Code values for inland regions played a large role in the selection of a structural system for this 500 ft (152.4 m) tall building in Miami, Florida (Fig. 1.18a).

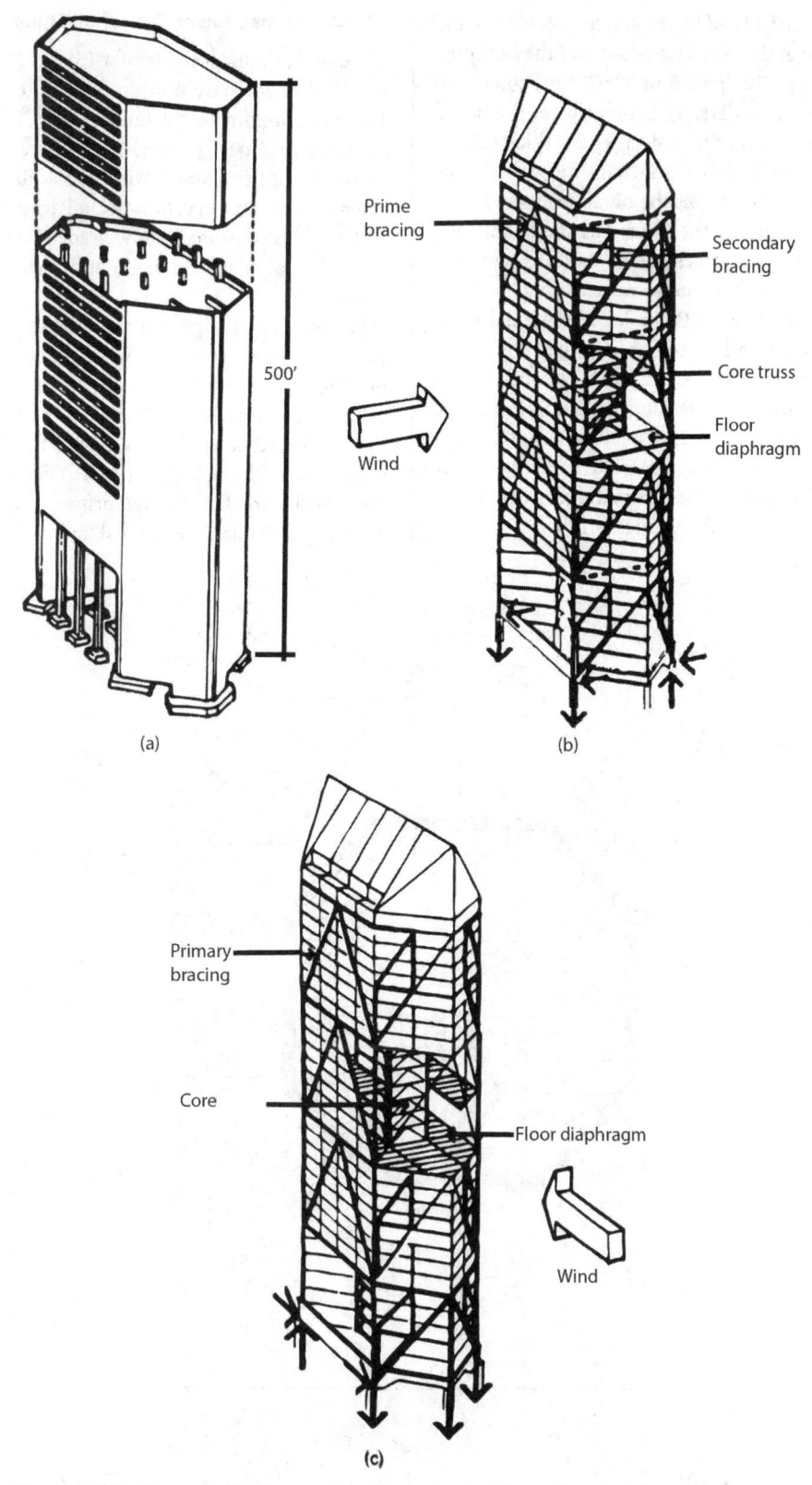

Figure 1.18 Metro Dade County Administrative Building, Florida: (a) building elevation; (b) lateral system: transverse direction; (c) lateral system, longitudinal direction.

The gravity loads are resisted by the concrete columns and shear walls located at the two narrow sides of the building. The floor system consists of a 6 in. (152 mm) thick slab supported on haunch girders. The columns on the broad face of the building rest on a 60 ft deep by 2 ft thick (18.3 × 0.61 m) concrete transfer girder that spans the width of the building. When the wind strikes the broad face, the lateral load is transferred through the floor diaphragm into the end shear walls. The axial forces due to overturning in the columns on the broad face due to overturning moment are transferred to the end walls similar to the axial loads due to gravity loads. Wind on the short face of the building is resisted above the transfer girders by a combination of the end shear walls acting as cantilevers and the frame action of the exterior columns and spandrels on the broad face. The structural design is by LeMessurier Consultants, Inc., Cambridge, Massachusetts. Schematic representations of the structural system are shown Figs. 1.18b, c.

1.3.12 Jin Mao Tower, Shanghai, China

Jin Mao Building consists of a 1381 ft (421 m) tower and an attached low-rise podium with a total gross building area of approximately 3 million sq ft (278 682 m²). The building includes 50 stories of office space topped by 36 stories of hotel space with two additional floors for a restaurant and observation deck. Parking for automobiles and bicycles is located below grade. The podium consists of retail spaces as well as an auditorium and exposition spaces.

The superstructure is a mixed-use of structural steel and reinforced concrete with many major structural members composed of both steel and concrete. The primary components of the lateral system include a central reinforced concrete core linked to exterior composite megacolumns by outrigger trusses (Fig. 1.19a,b). A central shear-wall core houses the primary building functions, including elevators, mechanical fan rooms and washrooms.

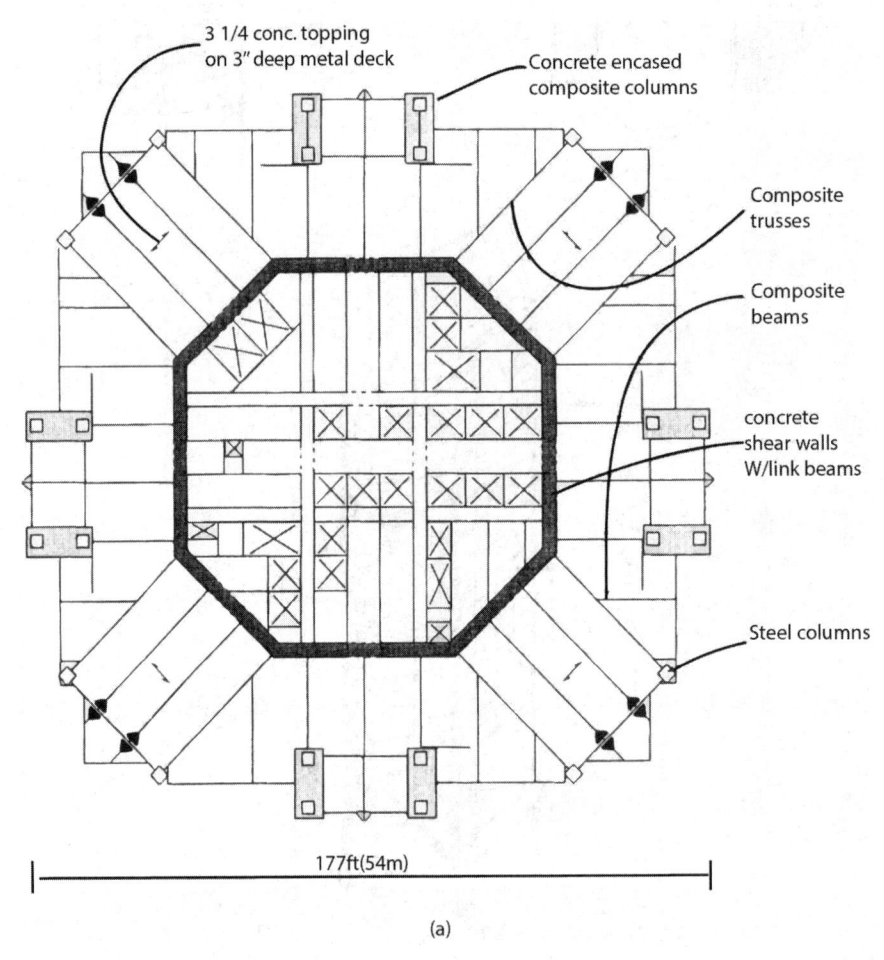

(a)

Figure 1.19 Jin Mao Tower, Shanghai, China: (a) typical office floor framing plan; (b) structural system elevation.

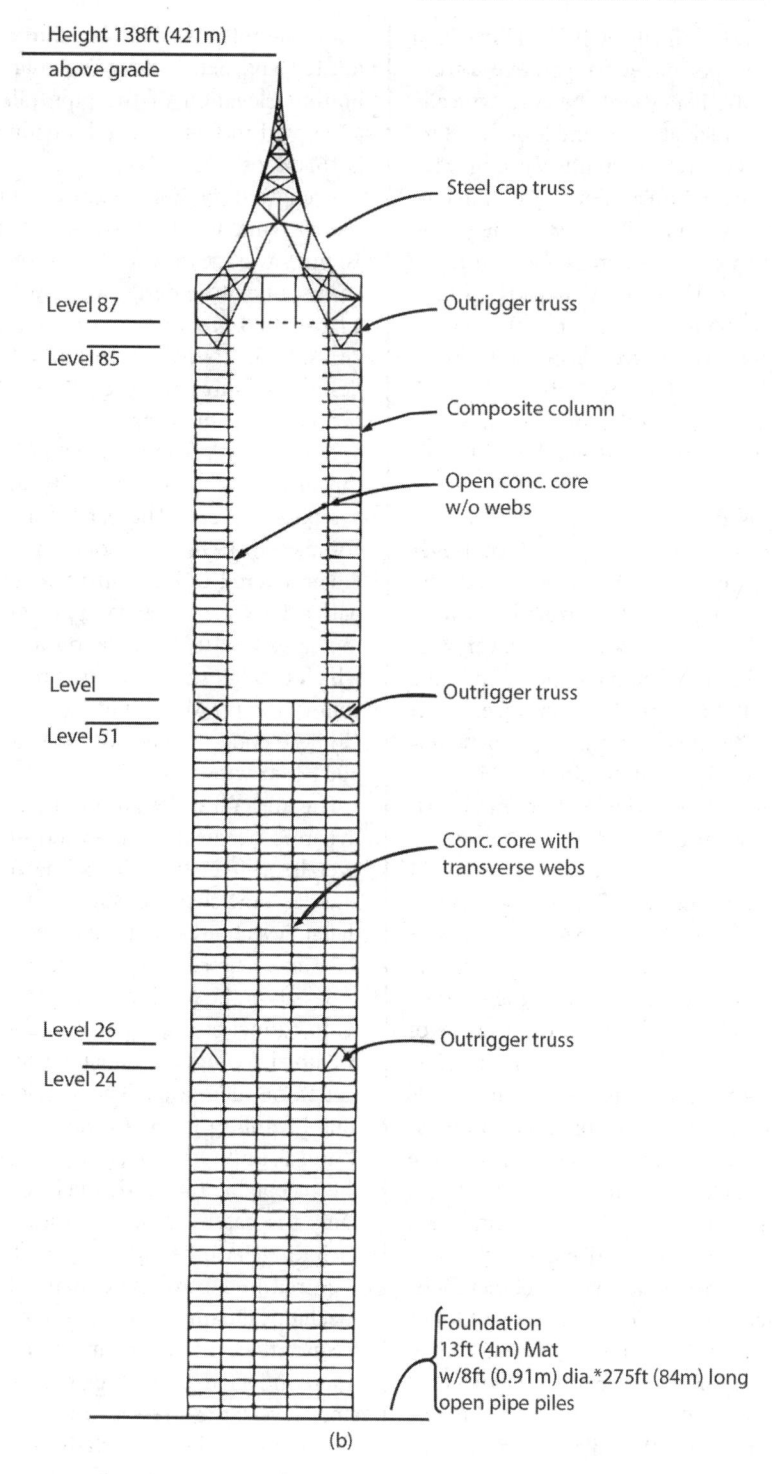

Height 138ft (421m) above grade

Steel cap truss

Level 87

Outrigger truss

Level 85

Composite column

Open conc. core w/o webs

Level

Outrigger truss

Level 51

Conc. core with transverse webs

Level 26

Outrigger truss

Level 24

Foundation 13ft (4m) Mat w/8ft (0.91m) dia.*275ft (84m) long open pipe piles

(b)

Figure 1.19 (*Continued*)

The octagon-shaped core, nominally 90 ft (27.43 m) from centerline to centerline of perimeter flanges, exists from the foundation to level 87. Flanges of the core typically vary from 33 in. (84 cm) thick at the foundation to 18 in. (46 cm) at level 87 with concrete strengths varying from 7500 to 5000 psi (51.71 to 34.5 mPa). Four 18 in. (46 cm) thick interconnecting core wall webs exist through the office floors. The central area of the core is open throughout the hotel floor, creating an atrium that leads into the spire with a total height of approximately 675 ft (206 m). The composite megacolumns vary from a concrete cross-section of 5 × 16 ft (1.5 × 4.88 m) with a concrete strength of 7500 psi (51.71 mPa) at the foundation, to 3 × 11 ft (0.91 × 3.53 m) with a concrete strength of 5000 psi (34.5 mPa) at level 87.

The shear-wall core is directly linked to the exterior composite megacolumns by structural steel outrigger trusses. The outrigger trusses resist lateral loads by maximizing the effective depth of the structure. Under bending, the building acts as a vertical cantilever with tension in the windward columns and compression in the leeward columns. Gravity load framing minimizes uplift in the exterior composite megacolumns. The octagon-shaped core provides exceptional torsional resistance, eliminating the need for any exterior belt or frame systems to interconnect exterior columns.

The outrigger trusses are located between levels 24 and 26, 51 and 53, and 85 and 87. The outrigger truss system between levels 85 and 87 is capped with a three-dimensional steel space frame which provides for the transfer of lateral loads between the core and the exterior composite columns. It also supports gravity loads of heavy mechanical spaces located in the penthouse floors.

The structural elements for resisting gravity loads include eight structural steel built-up columns. Composite wide-flange beams and trusses are used to frame the floors. The floor-framing elements are typically 14 ft 6 in. (4.4 m) on-center with a composite 3 in. (7.6 cm) deep metal deck and a 3¼ in. (8.25 cm) normal weight concrete topping floor slab spanning between steel members.

The foundation system for the Tower consists of high-capacity piles capped with a reinforced concrete mat. High-water conditions required the use of a 3 ft 3 in. (1 m) thick, 100 ft (30 m) deep, continuous reinforced concrete slurry wall diaphragm along the 0.5 mile (805 m) perimeter of the site.

The high-capacity pile system consists of a 3 ft (0.91 m) diameter structural steel open-pipe piles with a ⅞ in. (2.22 cm) thick wall typically spaced 9 ft (2.75 m) on-center capped by a 13 ft (4 m) deep reinforced concrete mat. Since the soil conditions at the upper strata are so poor, the piles were driven into a deep, stiff sand layer located approximately 275 ft (84 m) below grade. The bottom elevation of the pipe piles is the deepest ever attempted in China. The individual design-pile capacity is 1650 kips (7340 kN).

Strength design of the structure is based on a 100-year return wind with a basic wind speed of 75 mph for a 10 min. average time for the Tower. The basic wind speed corresponds to a design wind pressure of approximately 14 psf (0.67 kN/m²) at the bottom of the building and 74 psf (3.55 kN/m²) at the top of the spire. Exterior wall-design pressures are in excess of 100 psf (4.8 kN/m²) at the top of the building.

Wind speeds can average 125 mph (56 m/s) at the top of the building over a 10-min. time period during a typhoon event. The earthquake ground accelerations compare to 1994 UBC Zone 2A.

The overall building drift for a 50-year return wind with a 2.5% structural damping is H/1142. The drift value increases to H/857 for the future developed condition in which two tall structures are proposed to be located adjacent to the Jin Mao Building. The drift based on specific Chinese code-defined winds which were equivalent to a 3000-year wind is H/575.

The structural design for the tower is governed by its dynamic behavior under wind and not by its strength, overall or interstory drift. The inherent mass, stiffness, and damping characteristics of the Tower lead to achieving dynamic stability with fundamental translational periods of 5.7 s for each principal axis and a torsional period of 2.5 s.

Based on the results from force-balance and aeroelastic wind-tunnel study, the accelerations at the top floor of the hotel zone were evaluated using a value of 1.5 percent structural damping. The accelerations are between 9 and 13 milli-g for a 10-year return period and between 3 and 5 milli-g for a 1-year return period—well within the acceptable ranges defined by international standards. Only the passive characteristics of the structural system including its inherent mass, stiffness, and damping are required to control the dynamic behavior. Therefore, no mechanical damping systems are used.

Since the central core and composite megacolumns are interconnected by outrigger trusses at only three two-story levels, the stresses in the trusses due to differential shortening of the core relative to the composite columns were of concern. Therefore, concrete stress levels in the core and megacolumns were controlled in an attempt to reduce relative movements. To further reduce the adverse effect of differential shortening, slotted connections are used in the trusses during the construction period of the building. Final bolting with hard connections is made after completion of construction to relieve the effect of differential shortening occurring during construction.

The architecture and structural engineering of the building is by the Chicago office of Skidmore, Owings, and Merrill.

1.3.13 Petronas Towers, Malaysia

Two 1483 ft (452 m) towers, 33 ft (10 m) taller than Chicago's Sears Tower, and a sky bridge connecting the twin towers define the new record-breaking tall buildings in Kuala Lumpur, Malaysia.

The towers have 88 numbered levels but are in fact equal to 95 stories when mezzanines and extra-tall floors are considered. In addition to 6,027,800 ft² (560,000 m²) of office space, the project includes 1,507,000 ft² (140,000 m²) of retail and entertainment space in a six-story structure linking the base of the towers, plus parking for 7000 vehicles in five below-ground levels.

The lateral system for the towers is of reinforced concrete consisting of a central core and perimeter columns and ring beams using concrete strengths up to 11,600 psi (80 mPa). The foundation system consists of pile and friction barrette foundations with a foundation mat.

Typical floor system consists of wide flange beams spanning from the core to the ring beams. A two-inch-deep composite metal deck system with a 4¼ in. (110 mm) concrete topping completes the floor system.

Architecturally, the towers are cylinders 152 ft (46.2 m) in diameter formed by 16 columns. The facade between columns alternates pointed projections with arcs giving unobstructed views through glass and metal curtain wall on all sides. The floor plate geometry is composed of two rotated and superimposed squares overlaid with a ring of small circles. The towers have setbacks at levels 60, 73, 82, 84, and 88 and circular appendages at level 44. Concrete perimeter framing is used up to level 84. Above this level, steel columns and ring beams support the last few floors and a pointed pinnacle.

The towers are slender with an aspect ratio of 8.64 (calculated to level 88). The design wind speed in the Kuala Lumpur area is based on 78 mph (35.1 m/s) peak 3 s gust at 33 ft (10 m) above grade for a 50-year return. In terms of U.S. Standard of fastest mile wind, the corresponding wind speed is about 63 mph (28.1 m/s).

The mass and stiffness of concrete is taken advantage of in resisting lateral loads while the advantages of speed of erection and long span capability of structural steel are used in the floor framing system. The building density is about 18 lb/cu ft (290 kg/m³).

As is common for buildings of high aspect ratios, the towers were wind tunnel tested to determine the dynamic characteristics of the building in terms of occupant perception of wind movements and acceleration on the upper floors. The acceleration is in the range of 20 mg, well below the normally accepted criteria of 25 mg. Figure 1.20a shows a photograph of an early rendering.

The periods for the primary lateral modes are about 9 s, while the torsional mode has a period of about 5 s. The drift index for lateral displacement is of the order of 1/560.

Because the limestone bedrock lies 200 ft (60 m) to more than 590 ft (180 m) below dense silty sand formation, it was not feasible to extend the foundations to bedrock. A system of drilled friction piers was designed for the foundation but barrettes (slurry-wall concrete segments) proposed as an alternative system by the contractor were installed. A 14.8 ft (4.5 m) thick mat supports the 16 tower columns and 12 bustle columns and building core.

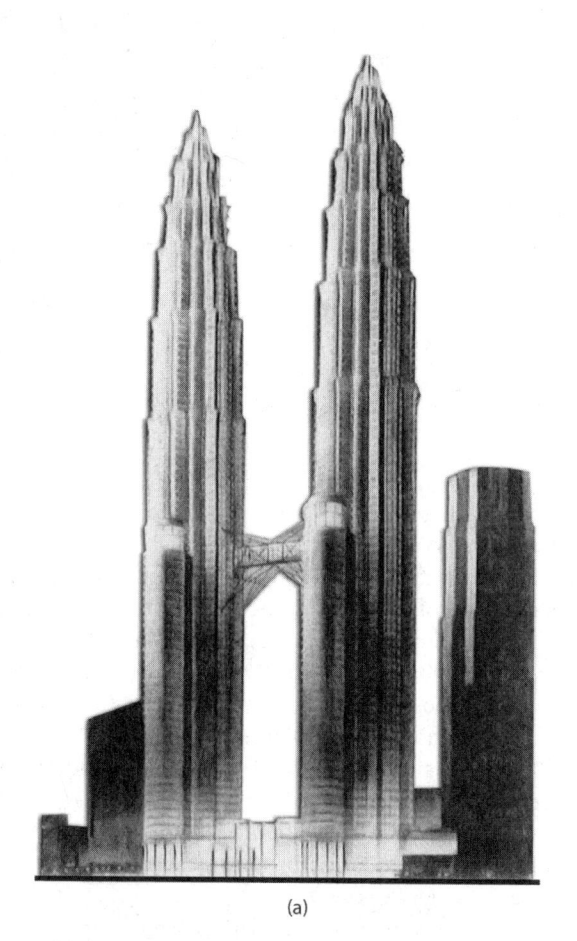

(a)

Figure 1.20 Petronas Twin Towers, Kuala Lumpur, Malaysia; (a) an early rendering; (b) structural system; (c) height comparison, (1) Petronas Tower, (2) Sears Tower.

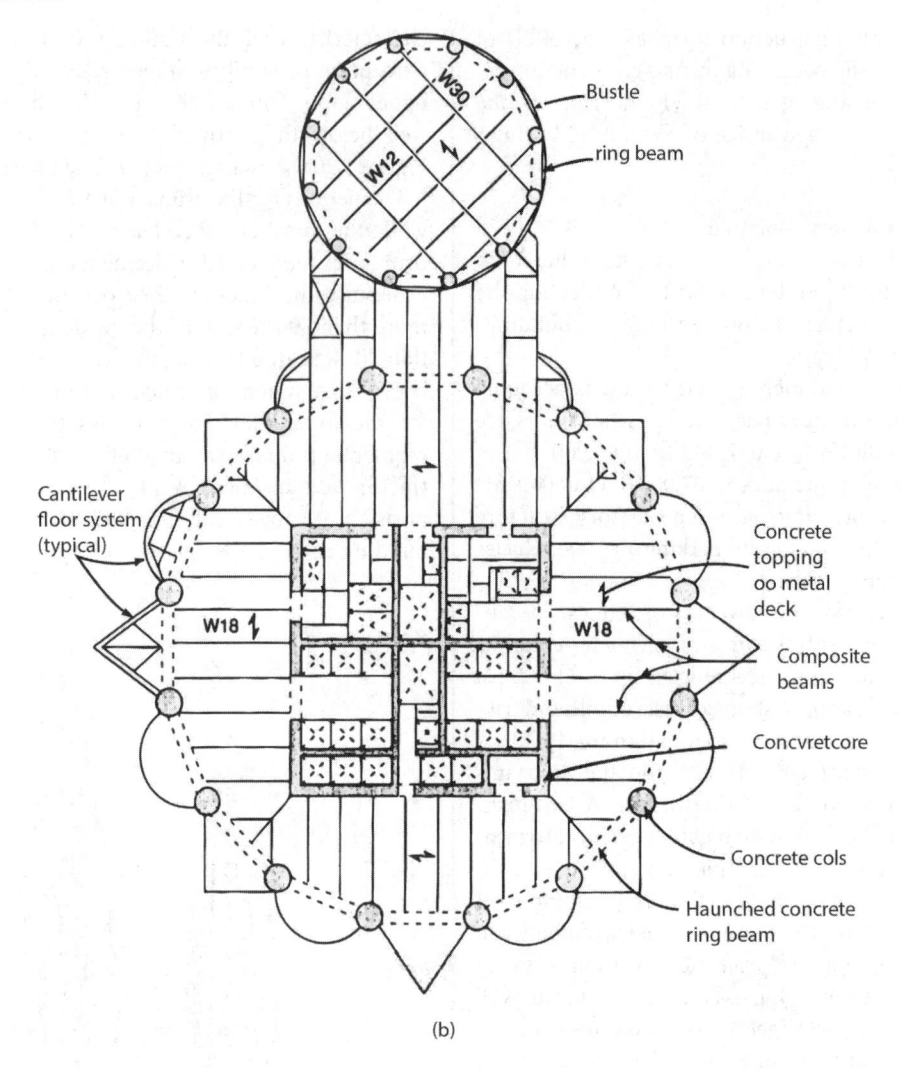

Figure 1.20   (*Continued*)

The tower columns vary in size from 7.8 ft (2.4 m) in diameter at the base to 4 ft (1.2 m) at the top. In the bustles, eight of the 12 columns vary from 4.6 to 3.28 ft (1.4 to 1.0 m) in diameter; the four facing the tower, being more heavily loaded, are slightly larger. Because all columns are exposed to view, they are cast with reusable steel forms, smoothed and painted.

The setbacks at floors 60, 73, and 82 are made with sloped columns over three-story heights. The method of transfer eliminates the need for deep transfer girders that would interrupt the constant floor height necessary for double-deck elevators.

The floor corners of alternating right angles and arcs are cantilevered from the perimeter ring beams. Haunched ring beams varying from 46 in. (1168 mm) deep at columns to 31 in. (775 mm) at midspan are used

to allow for ductwork in office space outside of the ring beams. A similar approach with a midspan depth of 29 in. (737 mm) is used in the bustles. The haunches are used primarily to increase the stiffness of ring beams.

The central core for each tower houses all elevators, exit stairs and mechanical services. The core walls carry about half the overturning moment at the foundation level.

Each core is 75 ft (23 m) square at the base, rising in four steps to 62×72 ft (18.8×22 m). Inner walls are a constant 14 in. (350 mm) thick while outer walls vary from 30 to 14 in. (750 to 350 mm). The concrete strength varies from 11,600 to 5800 psi (80 to 40 mPa).

To increase the efficiency of the lateral system, the interior core and exterior frame are tied together by a two-story deep outrigger truss at the mechanical equipment room (level 38). A vierended type of truss with

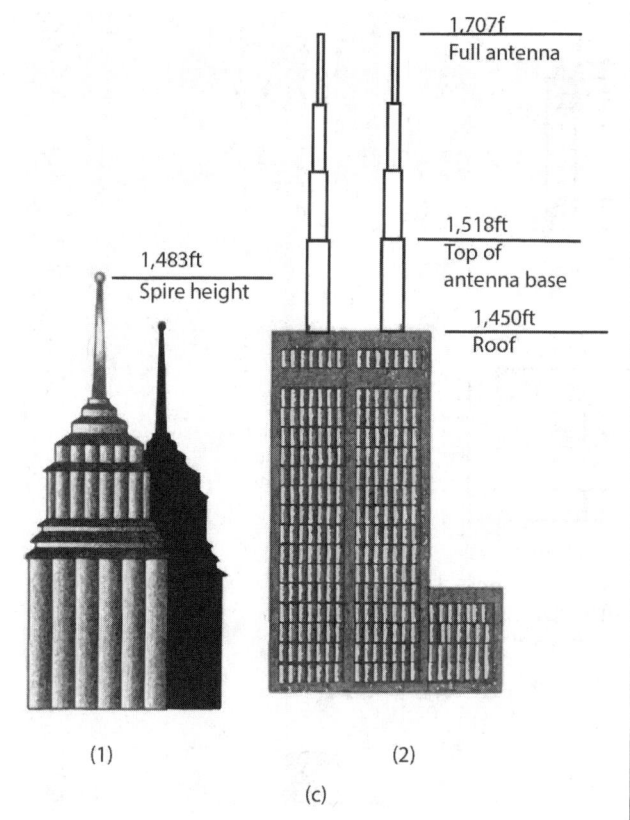

Figure 1.20 (*Continued*)

three levels of relatively shallow beams connected by a midpoint column is used to give flexibility in planning of building occupancy.

The tower floors Fig. 1.20b are typically composite metal deck with concrete topping varying from 4½ in. (110 mm) in offices to 8 in. (200 mm) on mechanical floors, including a 2 in. (53 mm) deep composite metal deck. Wide-flange beams frame the floors at spans up to 42 ft (12.8 m), and on most floors are W18 or shallower to provide room for ductwork, sprinklers, and lights.

Cantilevers for the points beyond the ring beams are 3.28 ft (1 m) deep prefabricated steel trusses. For the arcs, the cantilevers are beams propped with kickers back to the columns. Trusses and beams are connected to tower columns by embedded high-strength bolts. The structural engineering is by Thornton-Tomasetti Engineers with Ranhill Bersekutu Sdn. Bhd., engineer of record.

Although the Sears Tower's 110-stories dwarf the Malaysian twin skyscrapers' 88 floors (Fig. 1.20c), an engineering panel from the Council on Tall Buildings and Urban Habitat says that the Sears Tower is no longer the world's tallest building. This panel which sets international building height standards, contends that the Petronas Towers'

242-ft high ornamental spires are part of their height while the radio antennas of the Sears Tower are not. This is because traditionally the measurement from the ground floor entrance to the highest original structural point has been the criterion for assessing the height of skyscrapers for over a quarter of a century. Executives of the Chicago skyscrapers disagree, and say their building is actually 35 ft taller if the radio bases are considered as part of the height. The debate once again confirms that the vanity and desire to build taller skyscrapers is alive and well, as evidenced by even taller skyscrapers planned in Shanghai, China, and in Melbourne, Australia.

1.3.14   Tokyo City Hall

Tower no. 1 is a high-rise building with a height of 800 ft (243.4 m) consisting of 48-stories above ground and three-stories underground. The basic structural element in the vertical direction consists of a 21 ft (6.4 m) square super column. The super column is made up of four 40×40 in. (1020×1020 mm) steel box columns linked by K-braces. Eight such columns run through the building from the foundation to the top. These columns are connected by an orthogonal system of beams at each floor level. The super columns are interconnected by a system of one-story deep belt trusses at the 9th, 33rd, and 44th floors.

A large and flexible column-free space of 63×357 ft (19.2×108.8 m) is established on every floor with the use of deep beams. The typical floor-to-floor height is 13.13 ft (4.0 m). Figure 1.21 shows schematic plans and sections.

1.3.15   Leaning Tower; a Building in Madrid, Spain

Beyond the needs of a typical building, rising straight up from the ground, a leaning building requires an enhanced lateral-force system in the direction of the cantilever and an enhanced torsional system on account of eccentricity of transverse lateral forces. Note that although the overhang creates a substantial gravity-induced overturning moment, the gravity induced shear force is zero. In addition to resisting normal lateral forces from wind and earthquakes, the structural system for a leaning building must resist the gravity-induced overturning moment—and must do so at very low levels of lateral deflection. This additional requirement is particularly severe since the lateral deflection may not be recoverable, i.e., such deflection may remain permanently in the structure. The problem is even more severe in concrete structures because the deflections may increase with time, on account of the long-term creep properties of concrete.

A triangulated structural system consisting of super diagonals offers unique advantages in terms of both

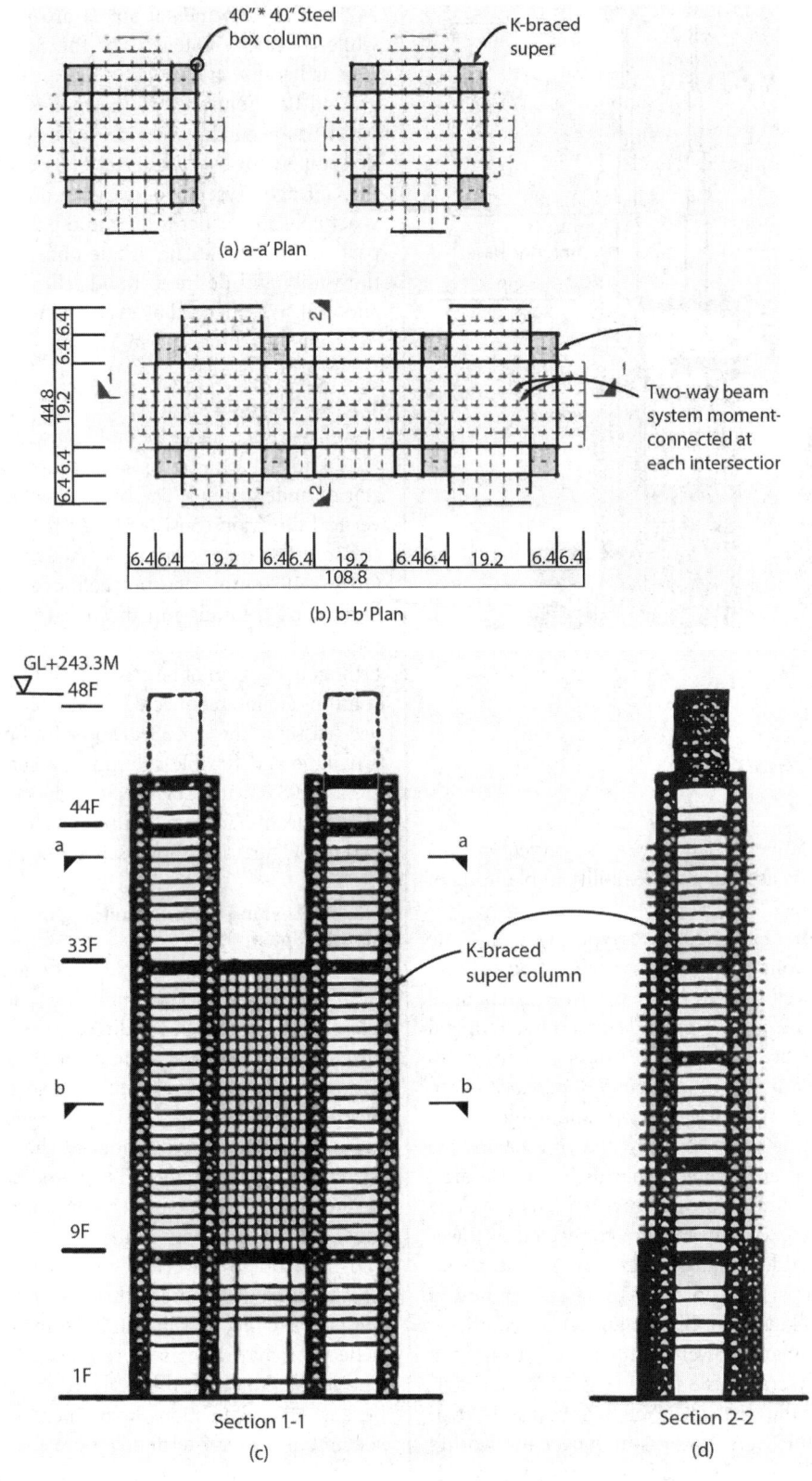

Figure 1.21   Tokyo City Hall: (a,b) schematic plans; (c,d) schematic sections.

stiffness and strength. Additionally, it may offer opportunities to incorporate the structural system into visual expression of the building.

Observe that in the basic triangular unit shown in Fig. 1.22a, the slopping exterior columns carry symmetrical vertical loads without bending. To carry unsymmetrical loading as is required for live loads and lateral loads, a secondary system acting within the basic triangular unit is necessary.

A parallelogram shape consisting of two or more triangular blocks as shown in Fig. 1.22b more or less balances the dead load on one lower corner. However, under lateral forces and for unbalanced live loads the unit has to

be tied down in the other lower corner. The tie-down can be composed of rock or earth anchors, tension piling or a ballast consisting of concrete mass.

To compensate for gravity-induced lateral deflection, many approaches are possible. These include cambering the structure, post-tensioning and increasing the stiffness of the structure. Any of the approaches can be combined with the other.

Post-tensioning is a proven technique for the control of gravity-induced deflections. In concept, post-tensioning is introduced into those elements of the structural system that are under gravity-induced tension so as to place these members in compression. By post-tensioning the

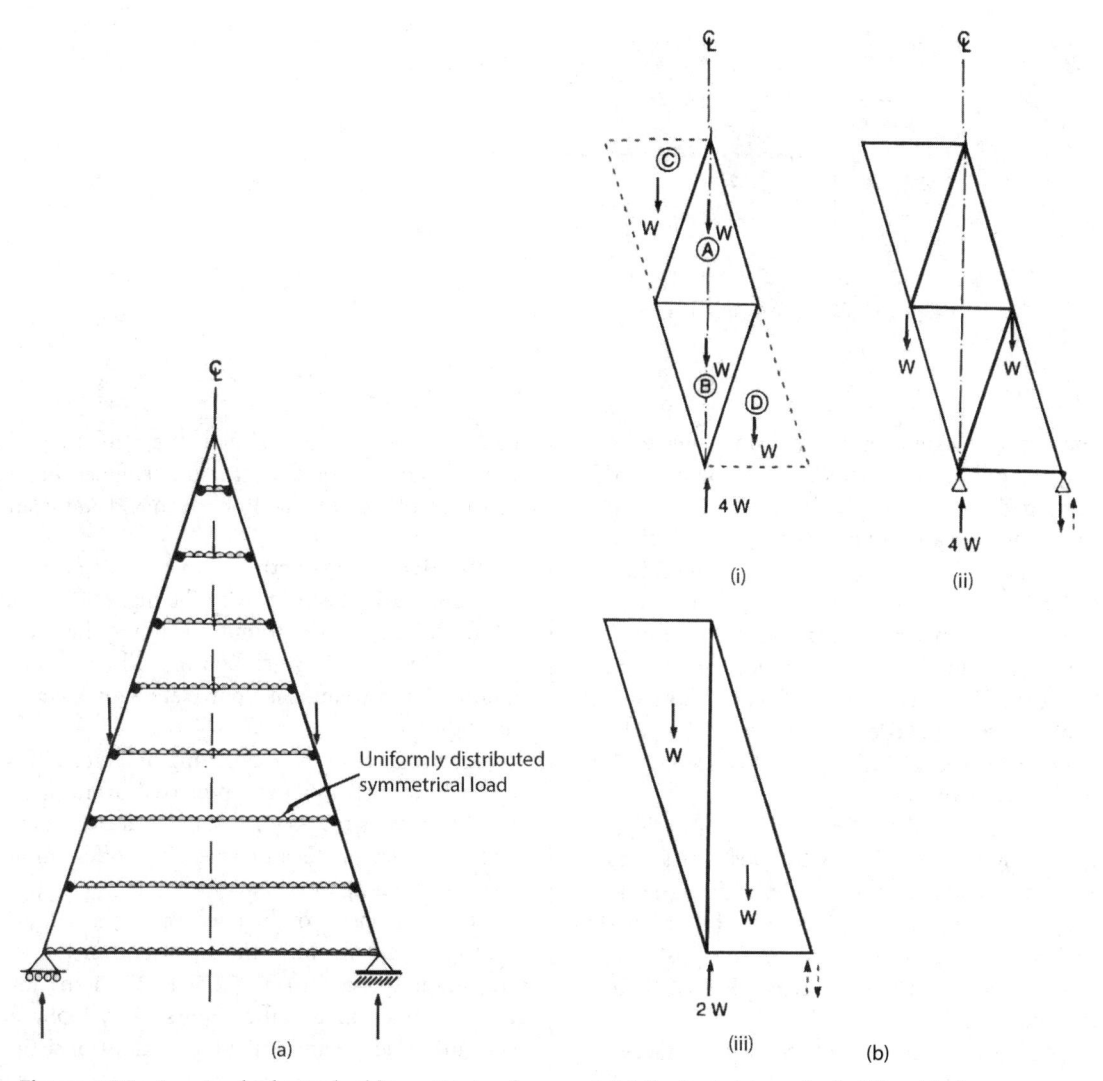

Figure 1.22 Leaning high-rise building, structural concept: (a) the basic triangular building block; (b) interconnected blocks: for symmetrical gravity loads no uplift exists at the base; (c) structural schematics, Twin Towers of Puerta de Europe.

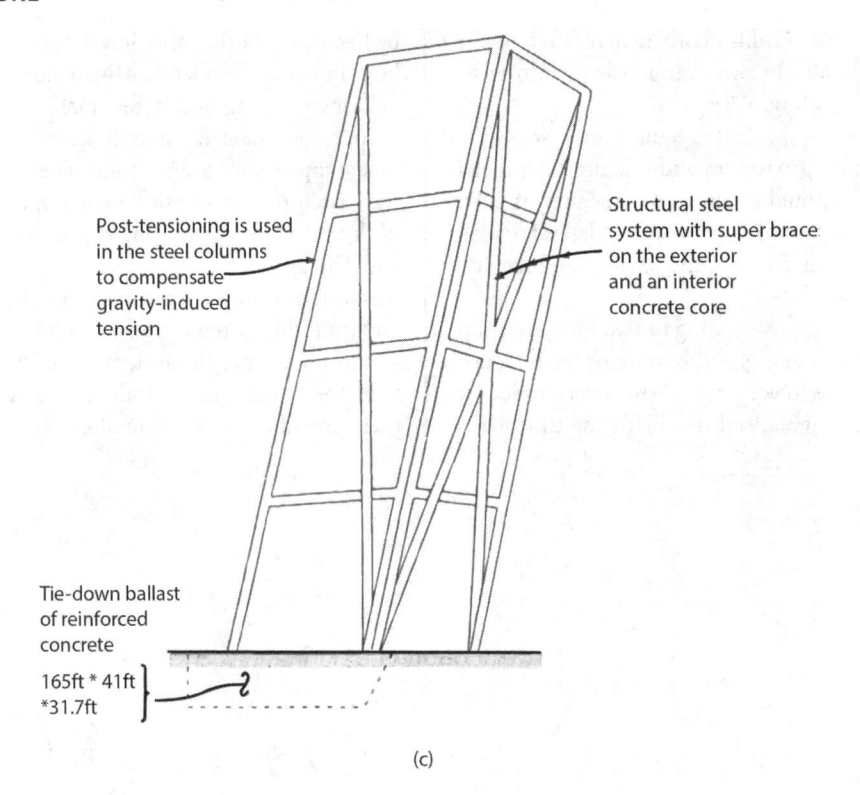

Post-tensioning is used in the steel columns to compensate gravity-induced tension

Structural steel system with super brace on the exterior and an interior concrete core

Tie-down ballast of reinforced concrete

165ft * 41ft *31.7ft

(c)

Figure 1.22 (*Continued*)

outer columns of a leaning building, it is possible to compensate for all or part of the lateral deflection induced by gravity loading.

As an example of a leaning tower, Fig. 1.22c shows a schematic framing system for a high-rise building in Madrid, Spain.

The building perimeter framing and the interior floor beams are of structural steel. The service core is of reinforced concrete. Tie-down ballast of reinforced concrete is used to counteract the overturning effect. The ballast is 165 ft long, 45 ft wide, and 31.7 ft deep ($50 \times 13.7 \times 9.67$ m) and weighs 15,400 tons.

The post-tension used on the steel for this project is a conventional system popular in the post-tensioning of concrete construction. To protect the tendons, the post-tensioning is carried in steel pipes, not post-tensioning ducts used in concrete construction. The post-tension system is anchored into the concrete ballast at the base of the building.

Transverse to the direction of the sloping faces, the lateral system is fairly straightforward. It consists of super diagonal braces running for the full height of the building. A 25 ft deep (7.62 m) truss is used in the roof-top

mechanical space to mobilize the stiffening effect of triangulated facades. The structural engineering is by the New York office of Leslie Robertson and Associates.

1.3.16 Hong Kong Central Plaza

The building is 78 stories with the highest office floor at 879 ft (268 m) above ground. Including the tower mast, the building is 1207.50 ft (368 m) tall (Fig. 1.23a). The building has a triangular floor plate with a sky lobby on the 46th floor.

The triangular design consisting of a typical floor area of 23,830 ft^2 (2214 m^2) was preferred over a more traditional square or rectangular plan, because the triangular shape has very few dead corners and offers more views from the building interiors.

The tower consists of three sections: (i) a 100 ft (30.5 m) tall tower base forming the main entrance and public circulation spaces; (ii) a 772.3 ft (235.4 m) tall tower section containing 57 office floors, a sky lobby, and five mechanical floors; and (iii) a top section consisting of six mechanical floors and a 334 ft (102 m) tall tower mast.

The triangular building shape is not truly triangular because its three corners are cut off to provide a better

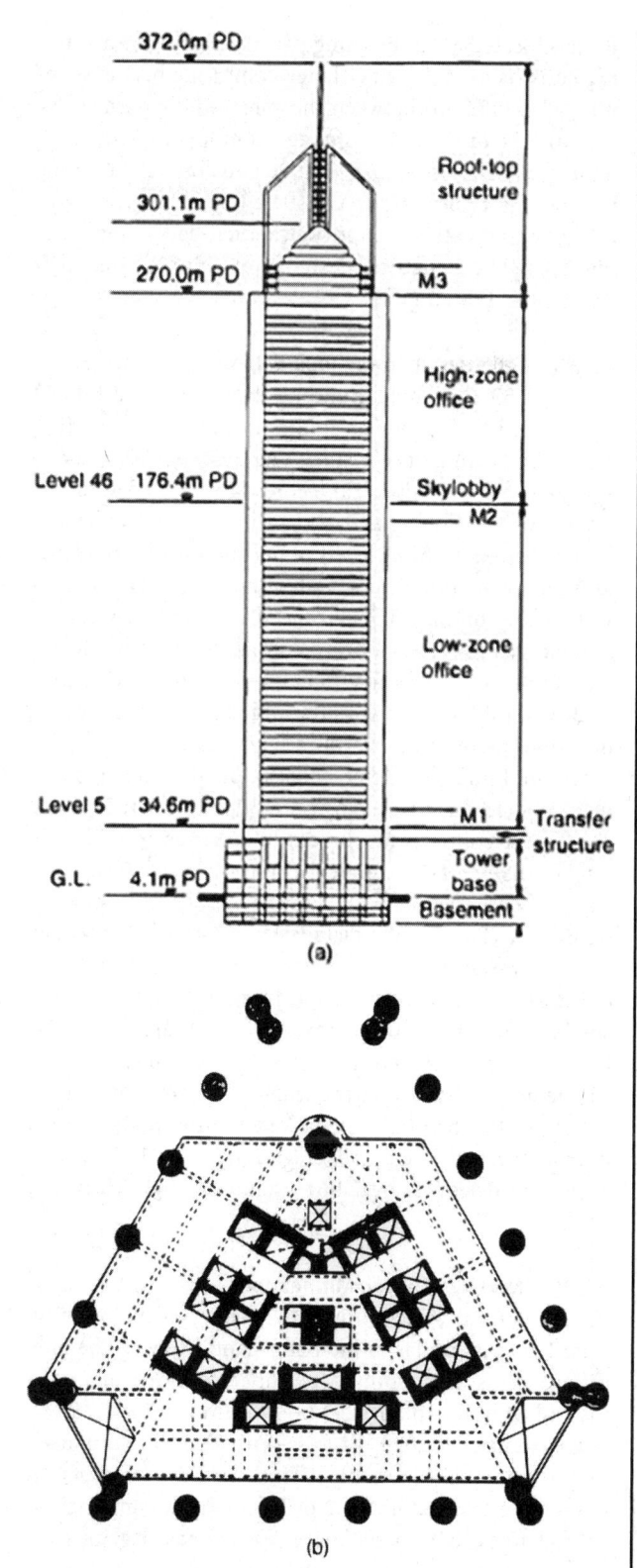

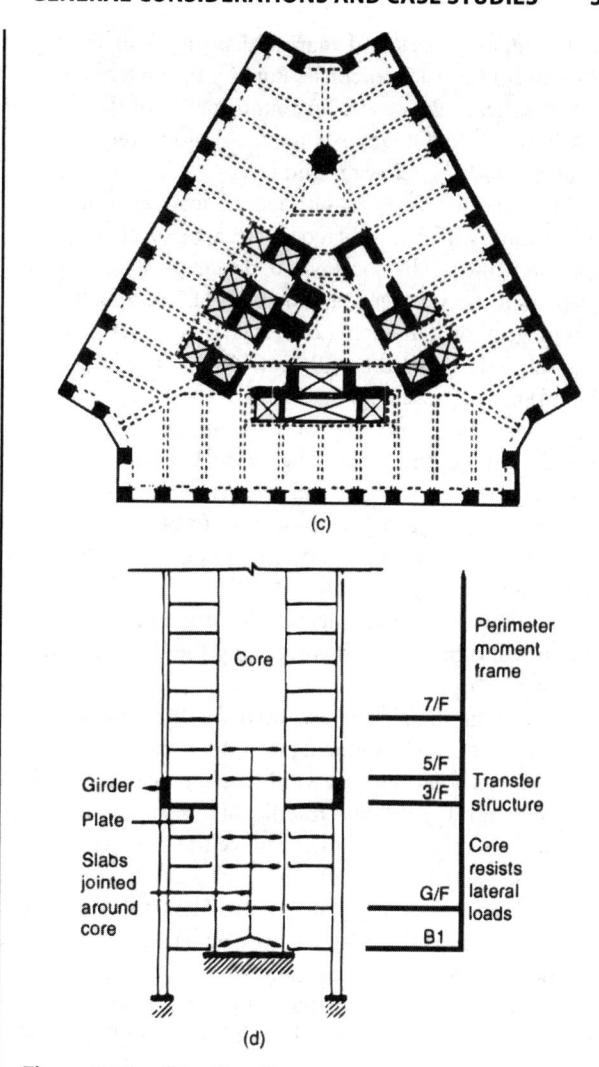

Figure 1.23 (*Continued*)

Figure 1.23 Central Plaza, Hong Kong: (a) elevation; (b–c) floor plans; (d) lateral load transfer system.

internal office layout. The building façade is clad in insulated glass. The mast is constructed of structural steel tubes with a diameter up to 6.1 ft (2 m).

The triangular core design (Fig. 23b,c) provides a consistent structural and building services configuration. A column-free office space, with 30.84 to 44.3 ft (9.4 to 13.5 m) depth is provided between the core and the building perimeter.

To enhance the spatial quality of the tower at the base, the 15 ft (4.6 m) column grid of the tower is transformed into a 30 ft (9.2 m) column grid by eliminating every other column. An 18 ft (5.5 m) deep transfer girder facilitates column termination.

The building site is typical of a recently reclaimed area in Hong Kong with sound bedrock lying between 82 and 132 ft (25 and 40 m) below ground level. This is overlaid by decomposed rock and marine deposits with the top 33 to 50 ft (10 to 15 m) consisting of a fill material. The allowable bearing pressure on sound rock is of the order of 480 ton/ft^2 (5.0 kN/m^2). The maximum water table is about 6.1 ft (2 m) below ground level.

Wind loading is the major lateral load criterion in Hong Kong which is situated in an area susceptible to typhoon winds. The local wind design is based on a mean-hourly wind speed of 100 mph (44.7 m/s), a three-second gust of 158 mph (70.5 m/s) and gives rise to a lateral design pressure of 86 psf (4.1 kN/m^2) at 656 ft (200 m) above ground level.

The basement consisting of a diaphragm slurry wall extends around the whole site perimeter and is constructed down to and grouted to rock. The diaphragm wall design allowed for the basement to be constructed by the "top down" method. This method has three fundamental advantages:

1. It allows for simultaneous construction of superstructure and basement thus reducing the time required for construction.

2. Basement floor slabs are used for bracing of diaphragm walls thereby reducing lateral tie-backs.

3. Creates a watertight box within the site enabling installation of hand-dug caissons, traditional in Hong Kong.

The lateral system for the tower above the transfer girder consists of external façade frames acting as a tube. These consist of closely spaced 4.93 ft (1.5 m) wide columns at 15 ft (4.6 m) centers and 3.6 ft (1.1 m) deep spandrel beams. The floor-to-floor height is 11.82 ft (3.6 m). The core shear walls carry approximately 10 percent of the lateral load above the transfer level. The transfer girder located at the perimeter is 18 ft (5.5 m) deep by 9.2 ft (2.8 m) wide, allowing alternate columns to be dropped from the façade, thereby opening up public area at ground level. The increased column spacing together with the elimination of spandrel beams in the tower base, results in the external frame no longer being able to carry the lateral loads acting on the building. Therefore, the wind shears are transferred to the core through the diaphragm action of 3.28 ft (1 m) thick slab located at the transfer level. The wind shear is taken out from the core at the lowest basement level, where it is transferred to the perimeter diaphragm walls. In order to reduce large shear reversals in the core walls, the floor slabs and beams are separated horizontally from the core walls (Fig. 1.23d) at certain levels. Structural engineering for the project is by Ove Arup and Partners.

1.3.17   Fox Plaza, Los Angeles

The structural system for resisting lateral loads for the 35-story building consists of special moment-resisting frames located at the building perimeter. The floor framing consists of W21 wide flange composite beams spanning 40 ft (12.2 m) between the core and the perimeters. A 2 in. (51 mm) deep 18-gauge composite metal deck with a 3¼ in. (83 mm) lightweight concrete topping is used for typical floor construction. A typical floor framing plan with sizes for typical members is shown in Fig. 1.24. The structural design is by John A. Martin & Associates, Los Angeles.

1.3.18   Bell Atlantic Tower, Philadelphia

This is a 53-story steel building consisting of a braced core linked to four super columns via a shear-resisting system consisting of two-story Vierendeel girders placed alternately at each side of the core. The varying configurations of the floor plates shown in Fig. 1.25a,b would have required multiple levels of perimeter column transfer in a conventional tube system making the structural system uneconomical. The braced core system working in concert with the exterior columns, on the other hand, maintains the structural efficiency close to that of a conventional tube system, with the added benefits of providing column-free corner offices (Fig. 1.25e).

The building perimeter consists of four major built-up steel columns measuring 54×30 in. (1.37×0.76 m). These box columns are linked by a series of vertically stacked five-story vierendeel frames (Fig. 1.25d). Each stack of vierendeels is linked to the one below by a series of hinges (Fig. 1.25d, Detail A) designed to transfer horizontal shears only. This detail prevents the build-up of gravity loads in the vierendeel frame by systematically shedding the loads to the box columns, thereby increasing their efficiency in resisting the overturning moments. The lateral resistance in the transverse direction is provided by the braced core linked to the box columns by the two-story vierendeel girders (Fig. 1.25e). The structural design is by CBM Engineers, Inc., Houston, Texas.

1.3.19   Norwest Center, Minneapolis

The structural system for the 56-story bank building (Fig. 1.26a) is similar to the Bell Atlantic Tower. The only difference is that composite super-columns are used instead of built-up all-steel box columns (Fig. 1.26b,c). High strength, 10,000 psi (68.94 mPa) concrete is used in the composite columns. This type of construction has been estimated in the North American construction market to be five to six times less expensive than steel columns of equivalent strength and stiffness. A representative detail of the connection between the steel beam and composite column, and the structural actions associated

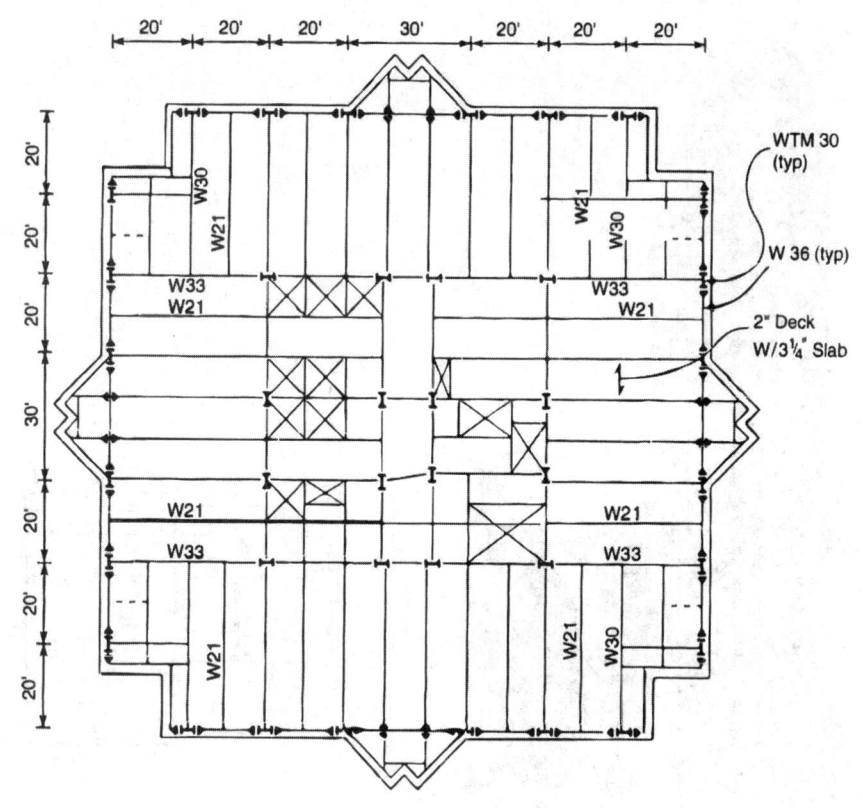

Figure 1.24 Fox Plaza, Los Angeles. Architects: Johnson, Fain & Pereira Inc. Structural engineers: John A. Martin & Asso. Inc., Los Angeles.

with moment transfer between the two are shown in Fig. 1.26d. Details of the hinge used in the vierendeel frame are shown in Fig. 1.26e. The structural design is by CBM Engineers, Inc., Houston, Texas.

1.3.20 First Bank Place, Minneapolis

This is a 56-story, granite-clad building (Fig. 1.27a) comprising of an array of changing floor plans (Fig. 1.27b) with an added architectural stipulation that the north-east corner of the building, line A^1B^1 in Fig. 1.27b, should have a minimum of structural columns to provide unobstructed views of the city. The response was to come up with an uninterrupted cruciform-shaped structural spine as shown in Figs. 1.27c,d,e,f. A combination of steel bracing and moment frames consistent with interior space planning is provided along the two spines, AA^1 and BB^1 (Fig. 1.27c) to act as shear membranes. Composite columns with concrete strengths up to 10,000 psi (68.44 MPa), and varying in size from 75 sq. ft (7 m²) at the base, to 50 sq. ft (4.65 m²) at the top, are used at the spine extremities as shown in Fig. 1.27c.

Since the cruciform shape is torsionally unstable, the spine is stabilized by providing braced frames along the perimeter BC and B^1C^1, and by a series of three-story tall vierendeel trusses along lines CC^1 and B^1D^1. A 12-story tall vierendeel girder along A^1D^1 links the composite concrete super-column at A^1 to the perimeter steel column D^1 (Fig. 1.27d). It also supports the perimeter circular vierendeel along BAB^1E above the 45th floor. This circular vierendeel provides both torsional and lateral resistance to the entire frame. The entire system is an outstanding example of a strategic structural response to complex building geometry and leasing demands without having to pay an undue premium for the structural system. The structural design is by CBM Engineers, Inc., Houston, Texas.

1.3.21 Figueroa at Wilshire, Los Angeles

Floor framing plans at various step-backs and notches for the 53-story tower (Fig. 1.28a) are shown in Fig. 1.28b. The structural system, designed by CBM Engineers, Inc., Houston, Texas, consists of eight steel super-columns at the perimeter interconnected in a crisscross manner to an interior braced core with moment-connected beams acting as outriggers at each floor (Figs. 1.28c,d). The floor framing is structured such that the main columns

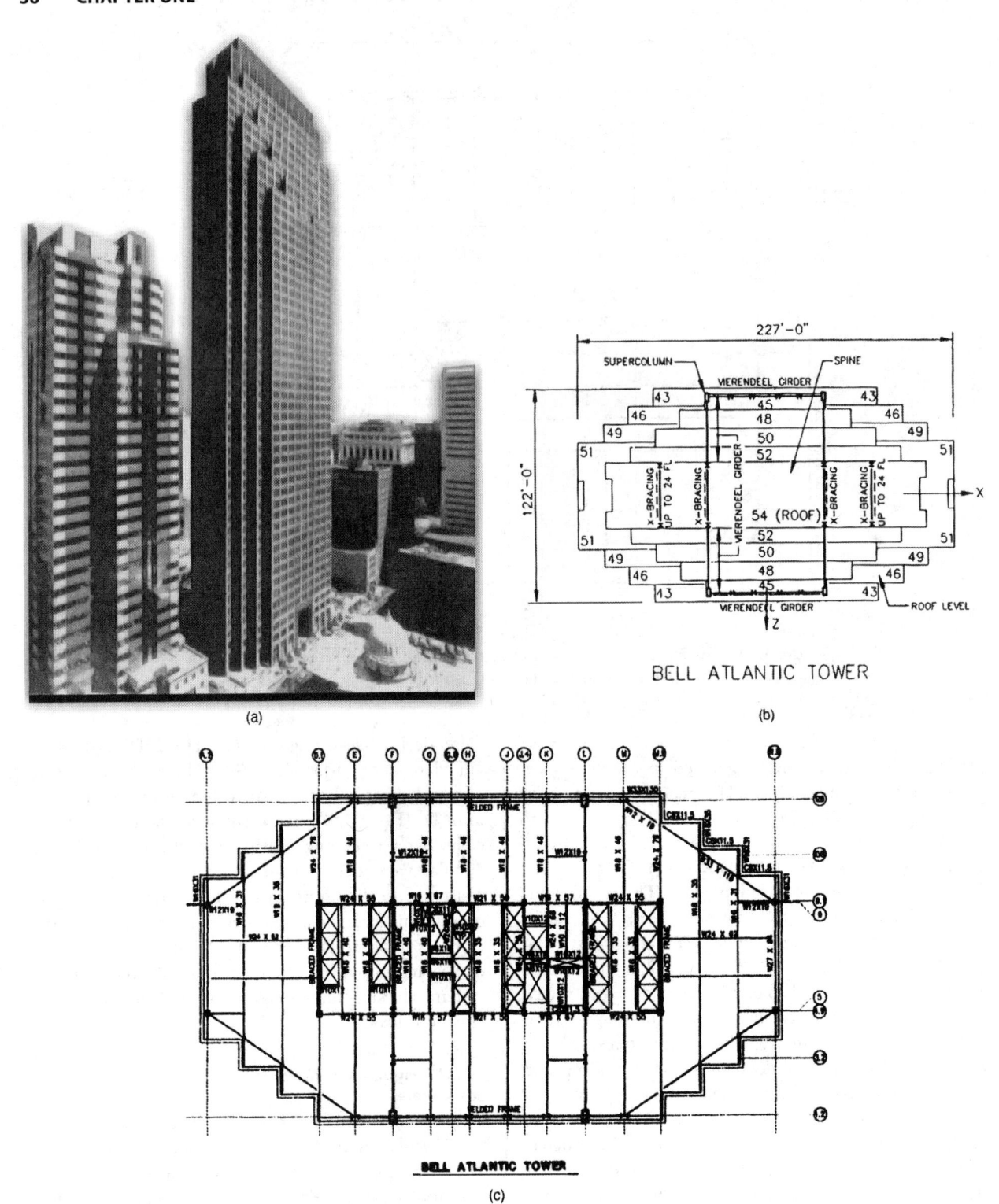

Figure 1.25   Bell Atlantic Tower: (a) building elevation; (b) composite floor plan; (c) floor framing plan; (d) lateral system; (e) section.

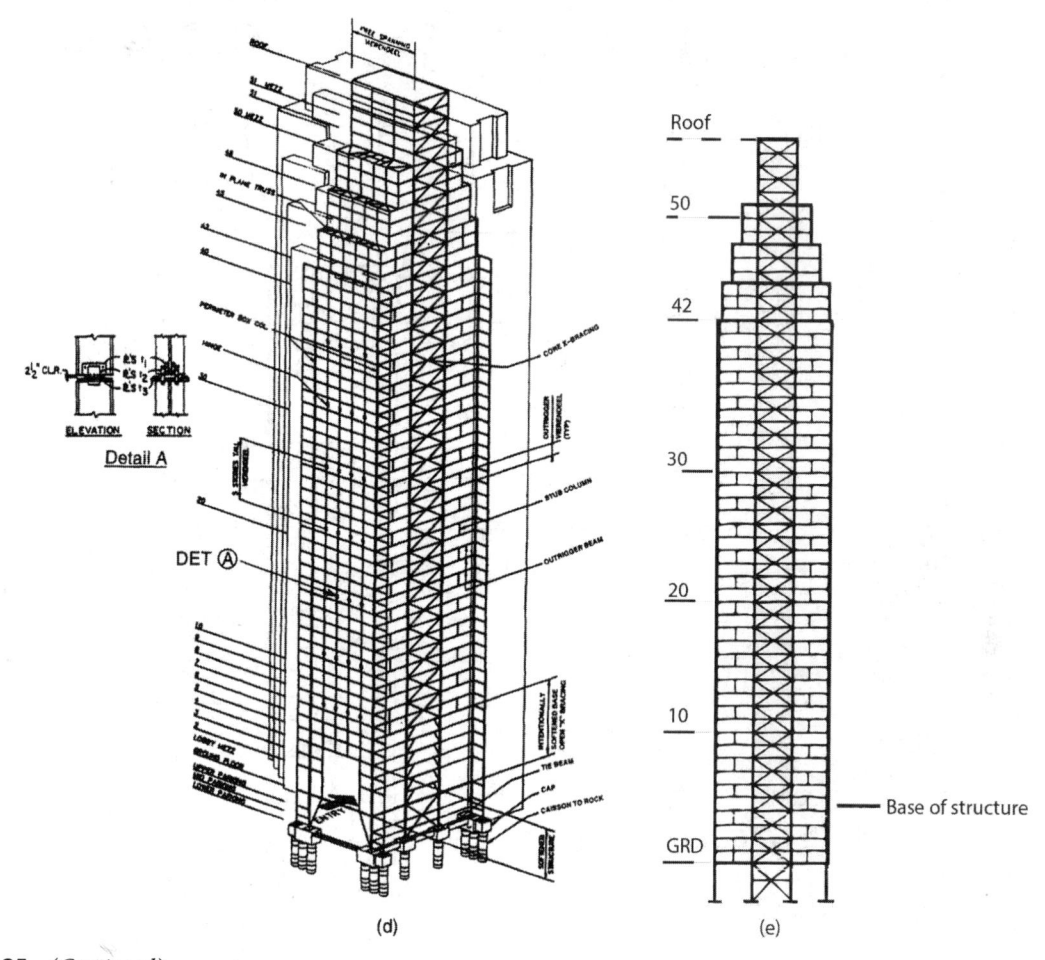

Figure 1.25 (*Continued*)

participating in the lateral loading system are heavily loaded by gravity loads to compensate for the uplift forces due to overturning. The structural system consists of three major components.

1. Interior concentrically braced core.

2. Outrigger beams spanning approximately 40 ft from the core to the building perimeter. The beams perform three distinct functions. First, they support gravity loads. Second, they act as ductile moment-resisting beams between the core and exterior frame columns. Third, they enhance the overturning resistance of the building by engaging the perimeter columns to the core columns. To reduce the additional floor-to-floor height that might otherwise be required, these beams are notched at the center, and offset into the floor framing as shown in Fig. 1.28e, to allow for mechanical ductwork.

3. Exterior super columns loaded heavily by gravity loads to counteract the uplift effect of overturning moments.

1.3.22 One Detroit Center

This is a 45-story office tower with a clear 45 ft 6 in. (13.87 m) span between the core and the exterior (Fig. 1.29a). The structural system consists of eight composite concrete columns measuring 7 ft 6 in. × 4 ft 9 in. (2.28 × 1.45 m) at the base, placed 20 ft (6.1 m) away from the corners to provide column-free corner offices, and also to optimize the free-span of the vierendeel frames. The composite columns are connected at each face by a system of perimeter columns and spandrels acting as vierendeel frames. The vierendeels are stacked four-story high and span between composite super-columns to provide column-free entrances at the base of the tower. At each fourth level, the vierendeels are linked by hinges which transfer only horizontal shear between adjoining vierendeels and not gravity loads. The reason for this type of connection is to reduce: (i) the effect of creep and shrinkage of super-columns on the members and connections of the vierendeel, and (ii) gravity load

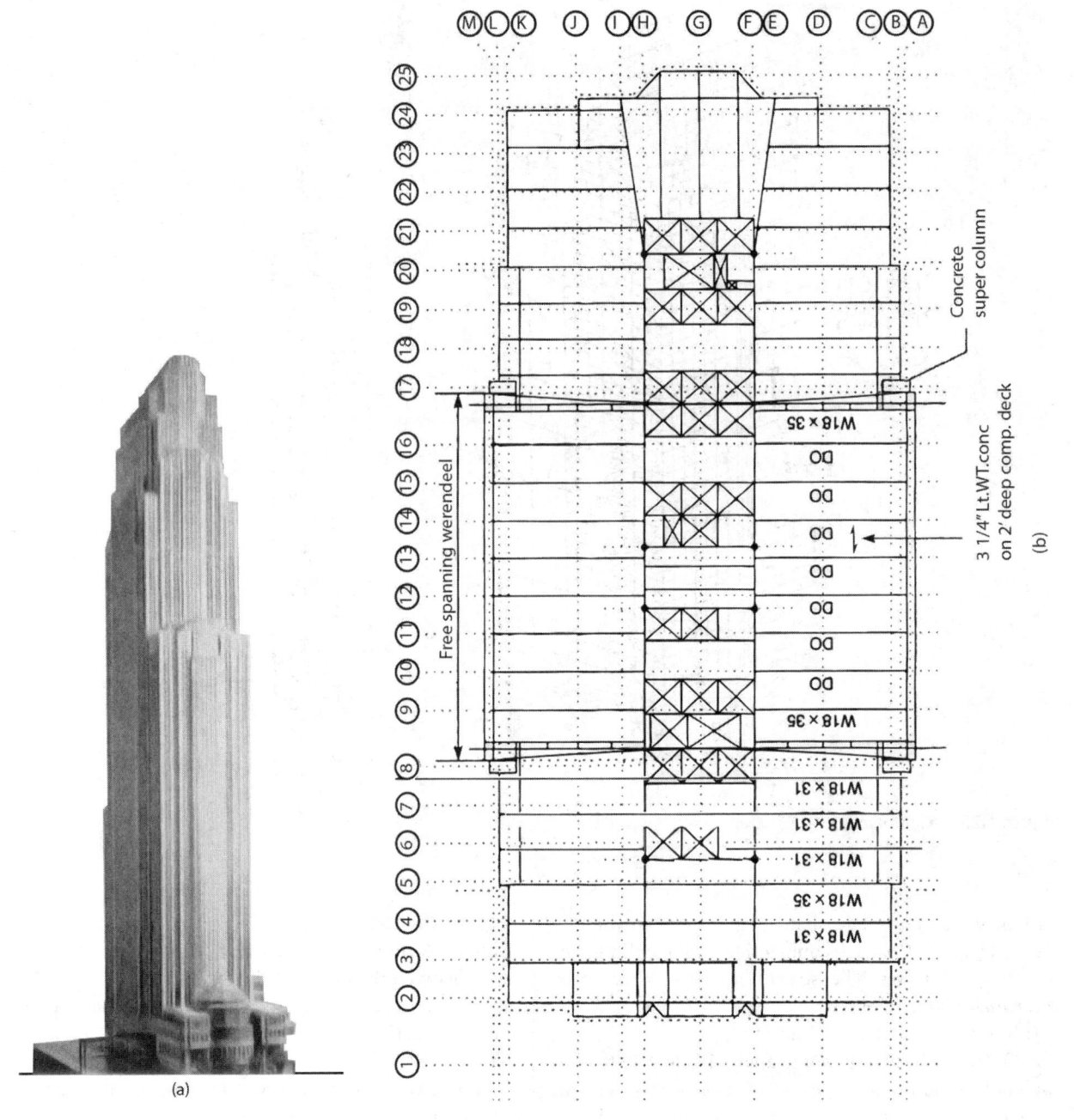

Figure 1.26   Norwest Center: (a) building elevation; (b) typical floor framing plan; (c) structural systems, isometric; (d) structural details, steel beam to concrete column connection; (e) hinge details.

transfer due to arch action of the vierendeel with associated horizontal thrusts. The four-story vierendeel achieves uniformity in the transfer of moment and shear between horizontal steel beams and composite super-columns throughout the height of the tower.

A schematic representation of the structural system is shown in Figs. 1.29b through 1.29c. Figure 1.29d shows the connection details for the vierendeel frame. The structural design is by CBM Engineers, Inc., Houston, Texas.

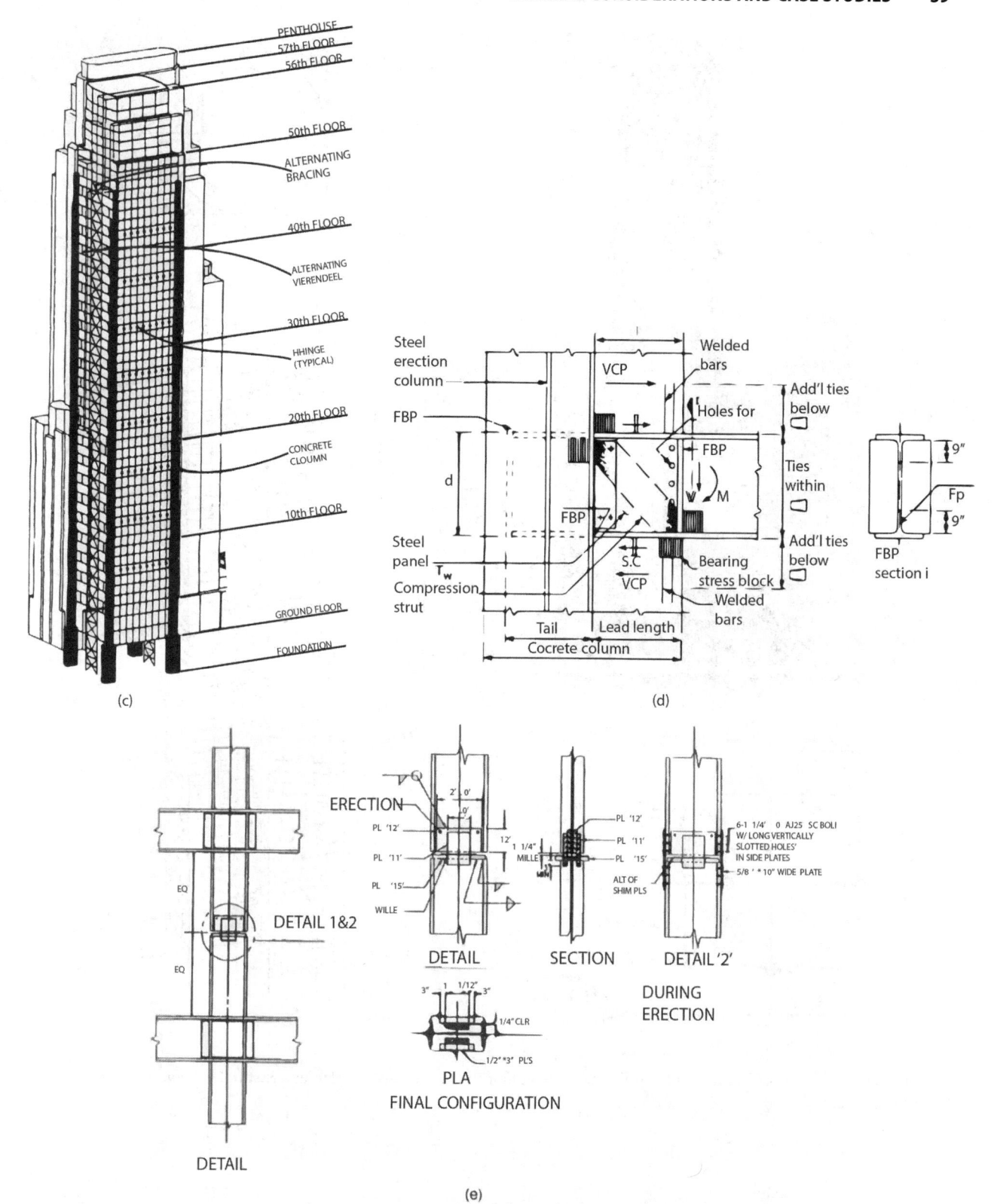

Figure 1.26   (*Continued*)

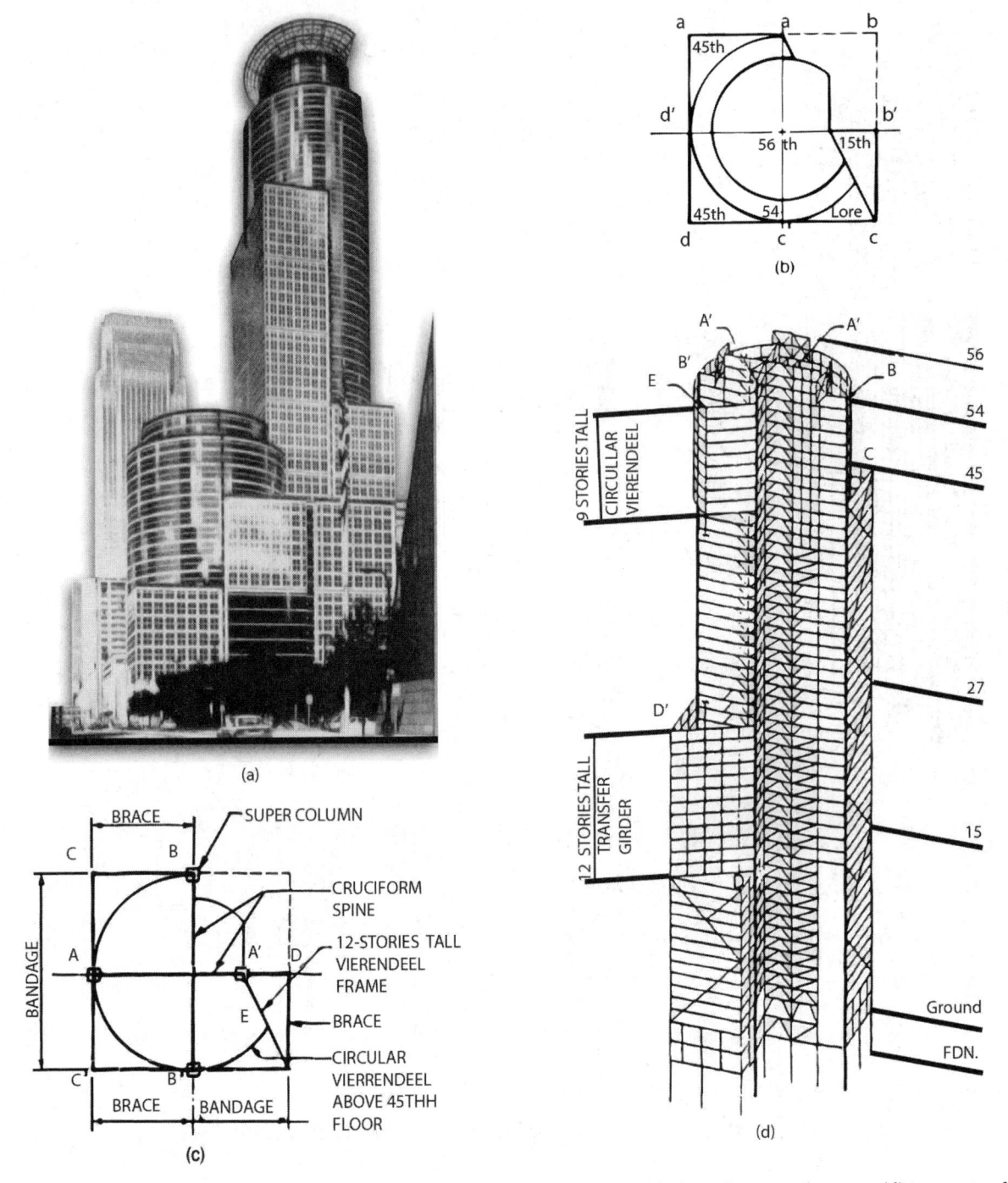

(a)

(b)

(c)

BRACE

SUPER COLUMN

CRUCIFORM SPINE

12-STORIES TALL VIERENDEEL FRAME

BRACE

CIRCULAR VIERRENDEEL ABOVE 45THH FLOOR

BANDAGE

BRACE   BANDAGE

(d)

9 STORIES TALL CIRCULLAR VIERENDEEL

12 STORIES TALL TRANSFER GIRDER

Figure 1.27   First Bank Place: (a) building elevation; (b) composite floor plans; (c) plan of structural system; (d) isometric of structural system.

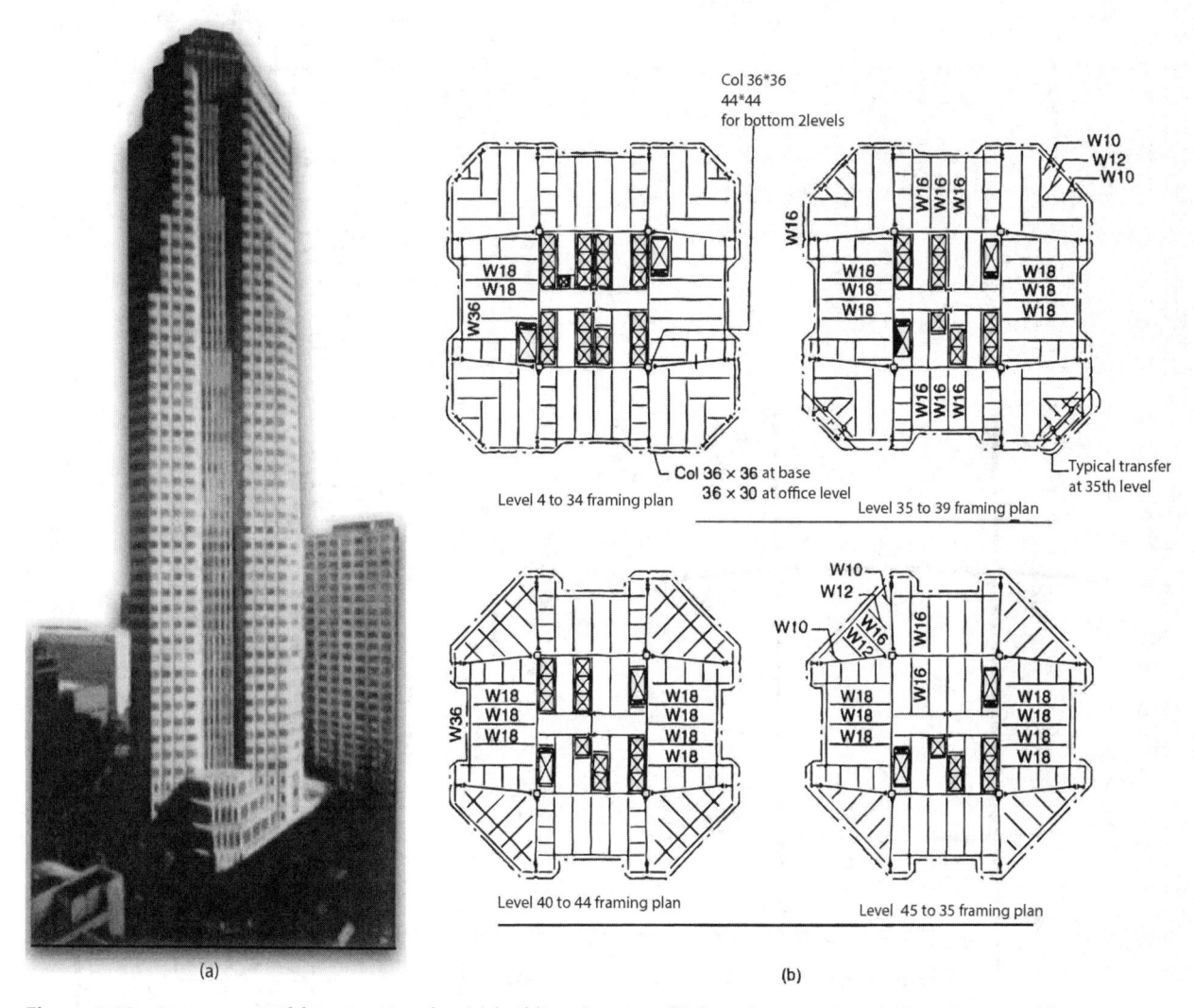

Figure 1.28 Figueroa at Wilshire, Los Angeles: (a) building elevation; (b) floor framing plans; (c) lateral system; (d) section; (e) reinforcement at beam notches, design concept. (Architects: Arthur Erickson Inc., Structural engineers: John A. Martin & Asso. Inc., Los Angeles.)

1.3.23   One Ninety One Peachtree, Atlanta

This 50-story building uses the concept of composite partial tube as shown in Fig. 1.30a. The partial tubes which extend uninterrupted from foundation to the 50th floor consist of concrete columns encasing steel erection columns with cast-in-place concrete spandrels. The building interior is an all-steel structure with composite steel beams supported on steel columns (Fig. 1.30b).

Since the building does not achieve the lateral resistance until after the concrete in composite construction has reached substantial strength, a system of temporary bracing was provided in the core. The erected steel was allowed to proceed 12 floors above the completed composite frame with six floors of metal deck and six floors of concrete floors. The structural design is by CBM Engineers, Inc., Houston, Texas.

1.3.24   Nations Bank Plaza, Atlanta

The 57-story office building has a square plan with the corners serrated to create the desired architectural appearance and to provide for more corner offices (Fig. 1.31a). The typical floor plan (Fig. 1.31b) is 162 × 162 ft (49.37 × 49.37 m) with an interior core measuring 58 ft 8 in. × 66 ft 8 in. (17.89 × 20.32 m). A five-level basement provided below the tower is of reinforced concrete construction. The foundation consists of shallow drilled piers bearing on rock.

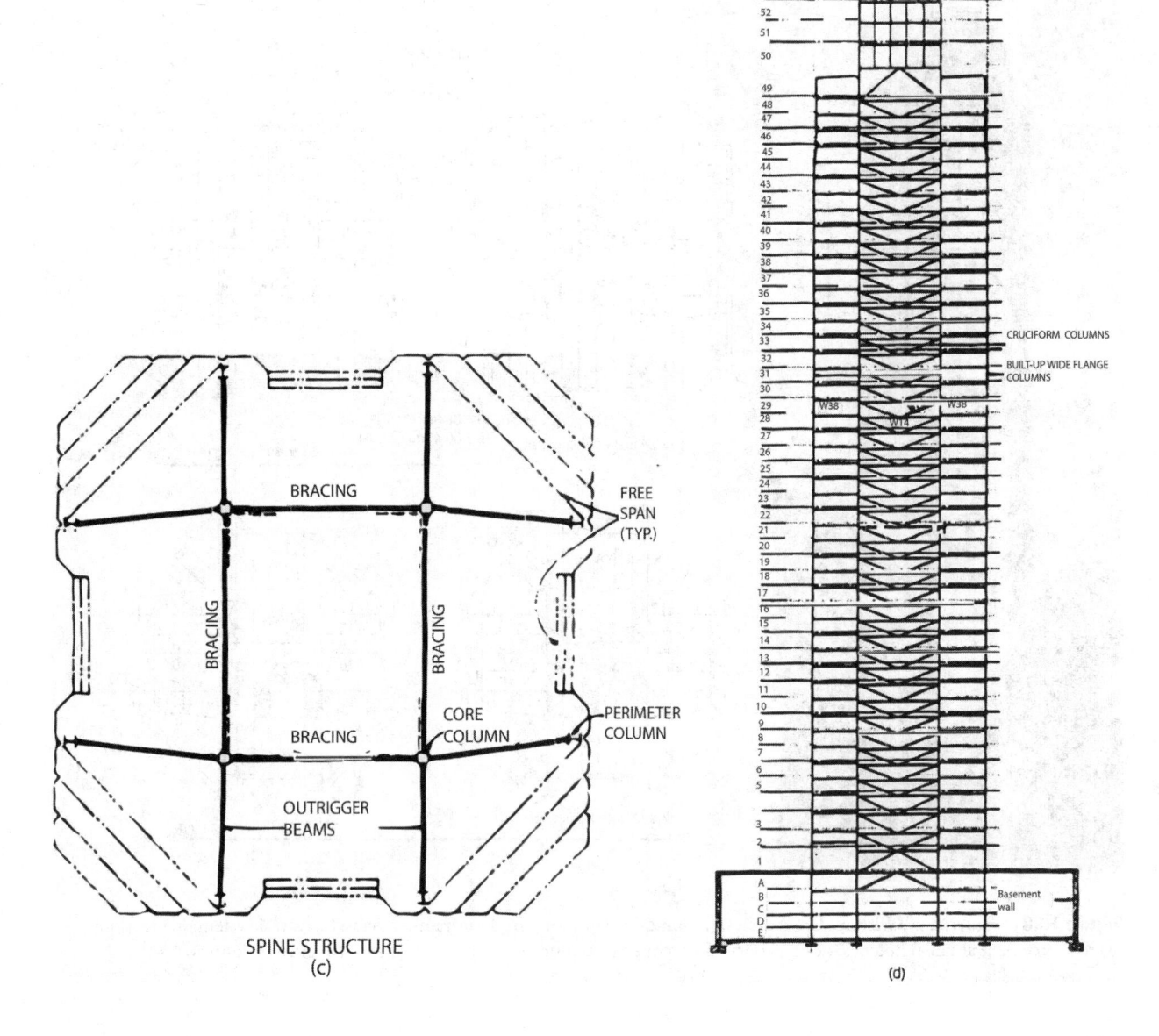

SPINE STRUCTURE
(c)

(d)

The gravity load is primarily supported by 12 composite super-columns. Four of these are located at the corner of the core, and eight at the perimeter, as shown in Fig. 1.31b. The core columns are braced on all four sides with diagonal bracing as shown schematically in Fig. 1.31c. Since the braces are arranged to clear mechanical and door openings in the core, their configuration is different on all four sides. Steel girders 36 in. deep (0.91 m) are moment-connected between the composite columns, to transfer part of the overturning moment to the exterior columns. Because the girders are deeper than other gravity beams, openings have been provided in the girders to provide for the passage of mechanical ducts and pipes. A diagonal truss is used between levels 56 and 59 to tie the core columns to the perimeter super-columns. These trusses transfer part of the overturning moment to the perimeter columns and also add considerable stiffness to the building. Above the 57th floor, the building tapers to form a 140 ft (42.68 m) tall conehead which is used to house mechanical and telecommunication equipment. The structural design is by CBM Engineers, Inc., Houston, Texas.

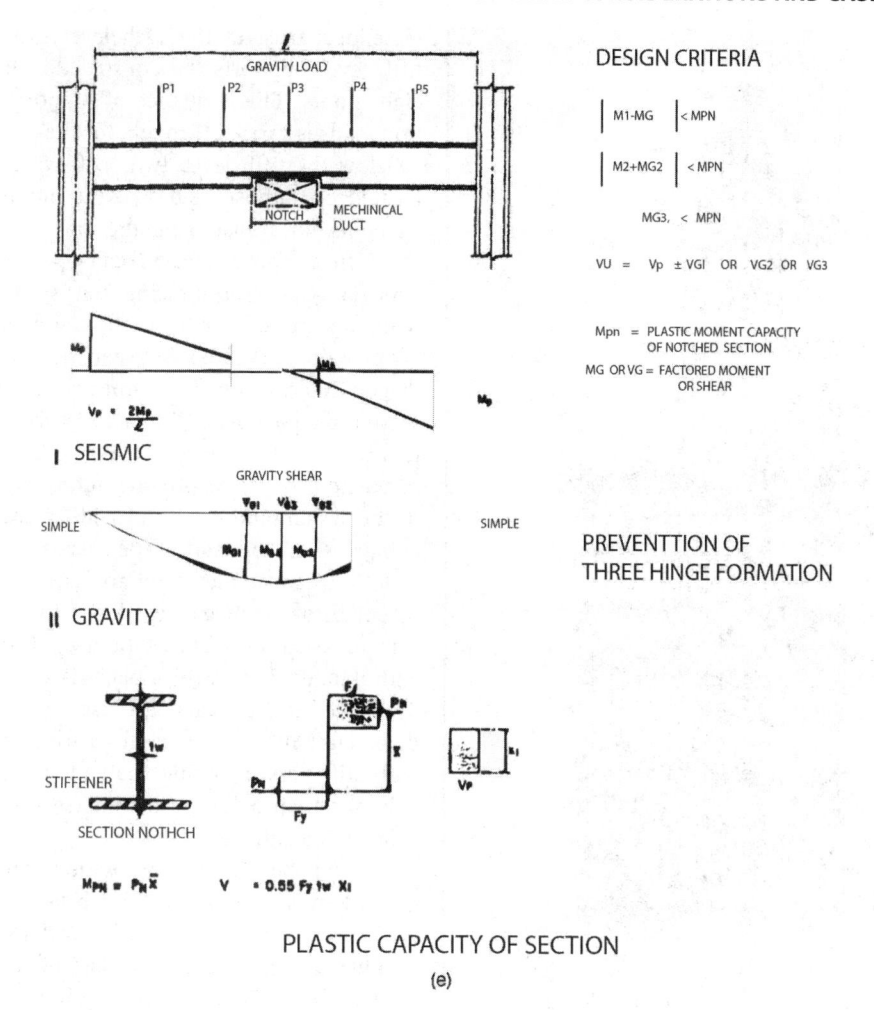

DESIGN CRITERIA

$$|M1\text{-}MG| < MPN$$

$$|M2\text{+}MG2| < MPN$$

$$MG3, < MPN$$

$$VU = Vp \pm VGI \ \ OR \ \ VG2 \ \ OR \ \ VG3$$

Mpn = PLASTIC MOMENT CAPACITY
OF NOTCHED SECTION
MG OR VG = FACTORED MOMENT
OR SHEAR

I SEISMIC

II GRAVITY

PREVENTTION OF
THREE HINGE FORMATION

STIFFENER

SECTION NOTHCH

$$M_{PH} = P_H \bar{X} \qquad V = 0.55 \, F_y \, t_w \, X_I$$

PLASTIC CAPACITY OF SECTION

(e)

Figure 1.28   (*Continued*)

1.3.25 Allied Bank Tower, Dallas, Texas

This is a 60-story building with an unconventional geometrical shape, raising to 726 ft (221.3 m) above a tree-studded plaza in downtown Dallas (Fig. 1.36a). The geometrical composition of the tower shown in Fig. 1.36b consists of: (i) a rectangular block 192 ft (58.53 m) square in plan raising to 54 ft (16.46 m) above the plaza; (ii) a 480 ft (146.3 m) tall geometric shape that wedges gradually from a square at the bottom to a 96 × 192 ft (29.3 × 58.5 m) rhombus; (iii) a 16-story high skewed triangular prism capping the building top. The combined 672 ft (204.83 m) high building is raised on a four-story high pedestal that matches the rhombus shape at the top as shown schematically in Fig. 1.36c. The resulting unconventional shape serves as an eye-catcher with a distinct exterior façade and a unique sloping top.

The structural system shown in Fig. 1.36b,c consists of a perimeter trussed frame that performs a dual function by providing the required lateral resistance, and an architecturally desired free-span at the base. This system has a 40-story deep megatruss on each of the 156 ft (47.55 m) side, and an eight-story deep truss on each of the 96 ft side as shown in Fig. 1.36c. The trussed exterior frames are setback 3 ft (0.92 m) from the building skin to clear the curtain wall. Also, the geometry of the diagonals dictated by the architectural and leasing requirements resulted in the intersection of the diagonal at mid-height of columns (Fig. 1.32d). Therefore, a conventional truss action in which the truss diagonal, the column, and the floor beam, all meet at a common node, could not be provided to transfer the unbalanced horizontal components of axial forces in the diagonals, directly to the floor members. As an alternative solution, a story-deep

(a)

Figure 1.29   One Detroit Center: (a) building elevation;
(b) typical floor framing plan; (c) free-spanning vierendeel
elevations; (d) structural details for vierendeel frame, (1) partial
elevation, (2) detail 1, (3) detail 2.

vierendeel truss was designed to resist the moments
caused by the unbalanced horizontal components of the
forces in the diagonals.

Above the 45th level, moment frames are used to
resist the lateral loads in both directions. To facilitate a
smooth flow of forces between the trussed and moment-
connected regions, an overlap zone was created by
extending the moment frames nine floors below the apex
of the trussed floor. A schematic of the structural system
is shown in Fig. 1.36c.

To eliminate the temporary shoring, otherwise required
for erecting steel four levels above the open plaza, a one-
story deep truss is used all around the perimeter between
the fifth and sixth levels. The truss is designed to support
the erection load of eight floors above the fifth level. The
eight-story deep subtruss (Fig. 1.36c) together with the

vierendeel truss at the 12th level and the truss at the
fifth level, supports the construction loads of the next
eight floors. This sequence of supporting the construction loads is carried through, for the entire structure.

Below the fifth level, two 30 ft wide (9.15 m) pylons
located on opposite sides of the building are the only
elements for transferring the entire north-south wind
shear from above. Heavy built-up columns with W14
wide flanges as cross-bracing resist shear and overturning
without significant uplift in the columns.

The floor spandrels between the intersection of the
diagonals are moment connected to the diagonals to
serve three purposes: (i) to act as a secondary lateral system to transmit wind loads applied at each level to truss
panel points; (ii) to provide additional lateral restraint
to the system and (iii) to provide lateral bracing for the
compression diagonals of the truss.

Both 36 and 50 ksi steel are used for the project. All
diagonals of the mega truss are W14 wide flange shapes.
The maximum weight of built-up trapezoidal column
with 8 in. (203 mm) thick plates is 2450 lb/ft (35 kN/m).
Two-sided gusset plates are used at the intersection of
diagonals with the box columns, while single-gusset plate
with stiffeners matching the wide flange diagonals and
columns are provided at the intersection of the diagonals
(Figs. 1.36e and 1.36f).

The typical floor construction consists of 50 ksi
(344.74 mPa) composite floor beams with a 3 in. deep
(76 mm) metal deck and 2½ in. thick (63.5 mm) normal-
weight concrete topping. Instead of welded wire fabric,
fiber-reinforced concrete is used in the slab as reinforcement. Additional mild steel reinforcement is used at vierendeel floors and at levels 13 and 45, where the shape of
the exterior façade changes.

The building consists of three basements below the
plaza level. Shallow drilled piers bearing on 50-ton capacity rock support the building. The unit weight of the
structural steel is 22 psf (1053.4 Pa). The floor-to-floor
height is 12 ft (3.66 m) as compared to 13 ft (3.96 m)
normally used for this type of office buildings. The exterior column spacing is 48 ft (14.63 m) offering generous
views from the building interior even with the presence
of diagonals. The structural design is by CBM Engineers,
Inc., Houston, Texas.

1.3.26 First Interstate World Center, Los Angeles

This 75-story granite-clad building (Fig. 1.33a) sports multiple step-backs as shown in Figs. 1.37b,c. The structural
system is a dual system consisting of an uninterrupted
73 ft 10 in. (22.5 m) square braced spline interacting with

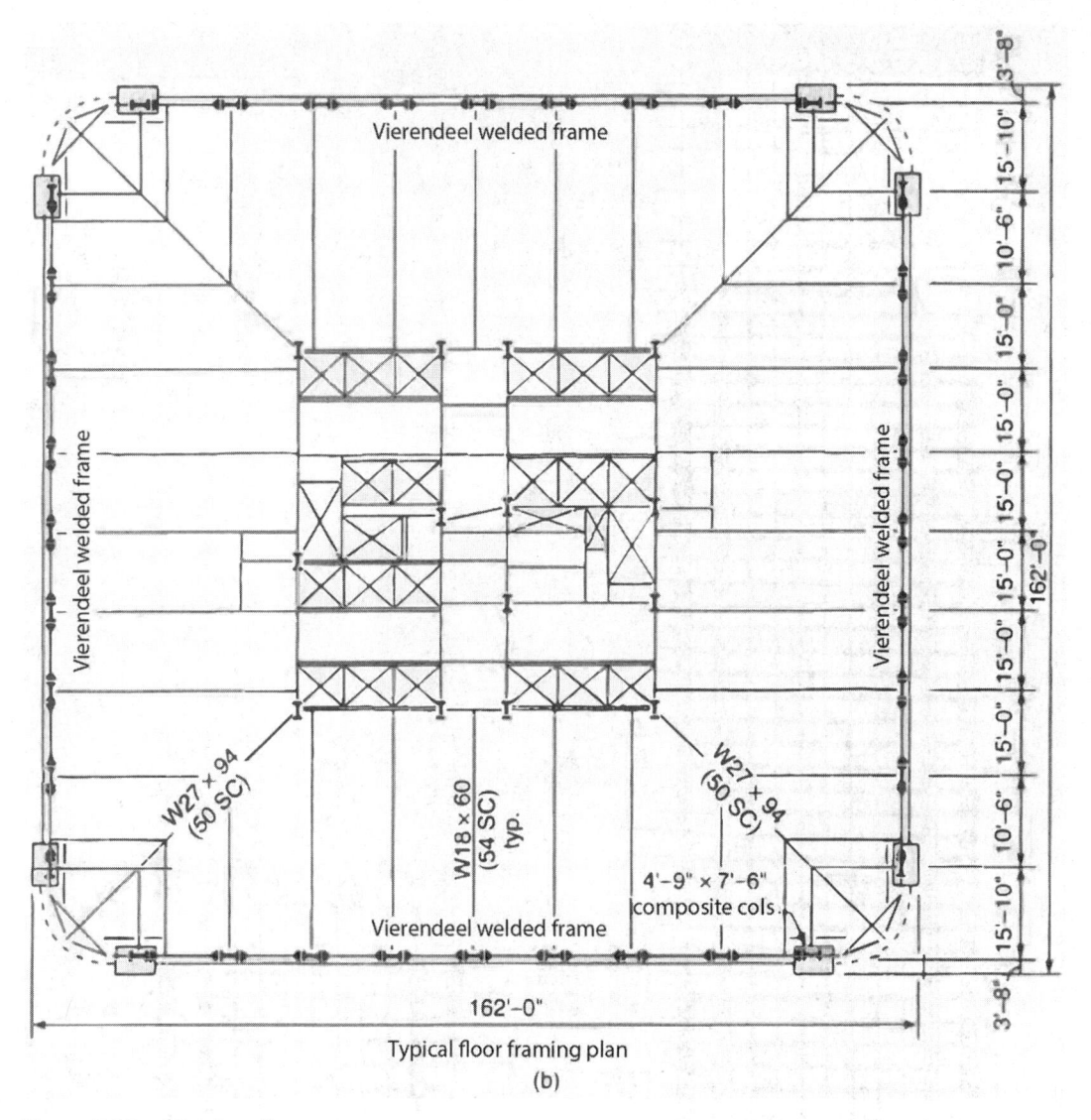

Figure 1.29 (*Continued*)

a perimeter ductile moment resisting frame. The spine is a two-story tall chevron-braced core as shown in Fig. 1.33d.

The 55 ft (16.76 m) span for the floor beams coupled with the two-story-tall free-spanning core, loads the corner core columns in such a way that the design is primarily governed by gravity design. To study the effect of buckling of chevron-braced diagonals, two types of failure modes were investigated. In the first mode, the buckled diagonal was assumed to have lost the axial load capacity, and in the second failure mode, the lower end of the diagonal was assumed to be absent. The structural members and connections were designed for the resulting overload due to the assumed modes of failure. Also, an attempt was made

to proportion the stiffness of the perimeter frame and core bracing such that the ductile yielding of the frame precedes the buckling of the diagonals.

To achieve overall economy and take advantage of the increase in stresses allowed under transient lateral loads, the columns are intentionally widely spaced to collect gravity loads from large tributary areas (Fig. 1.33e). The column design is very close to an optimum solution; the design is primarily for gravity loads with additional loads due to seismic and wind resistance by the one-third increase in allowable stresses.

The structure is designed to remain essentially elastic for an anticipated credible earthquake of magnitude 8.3 on the

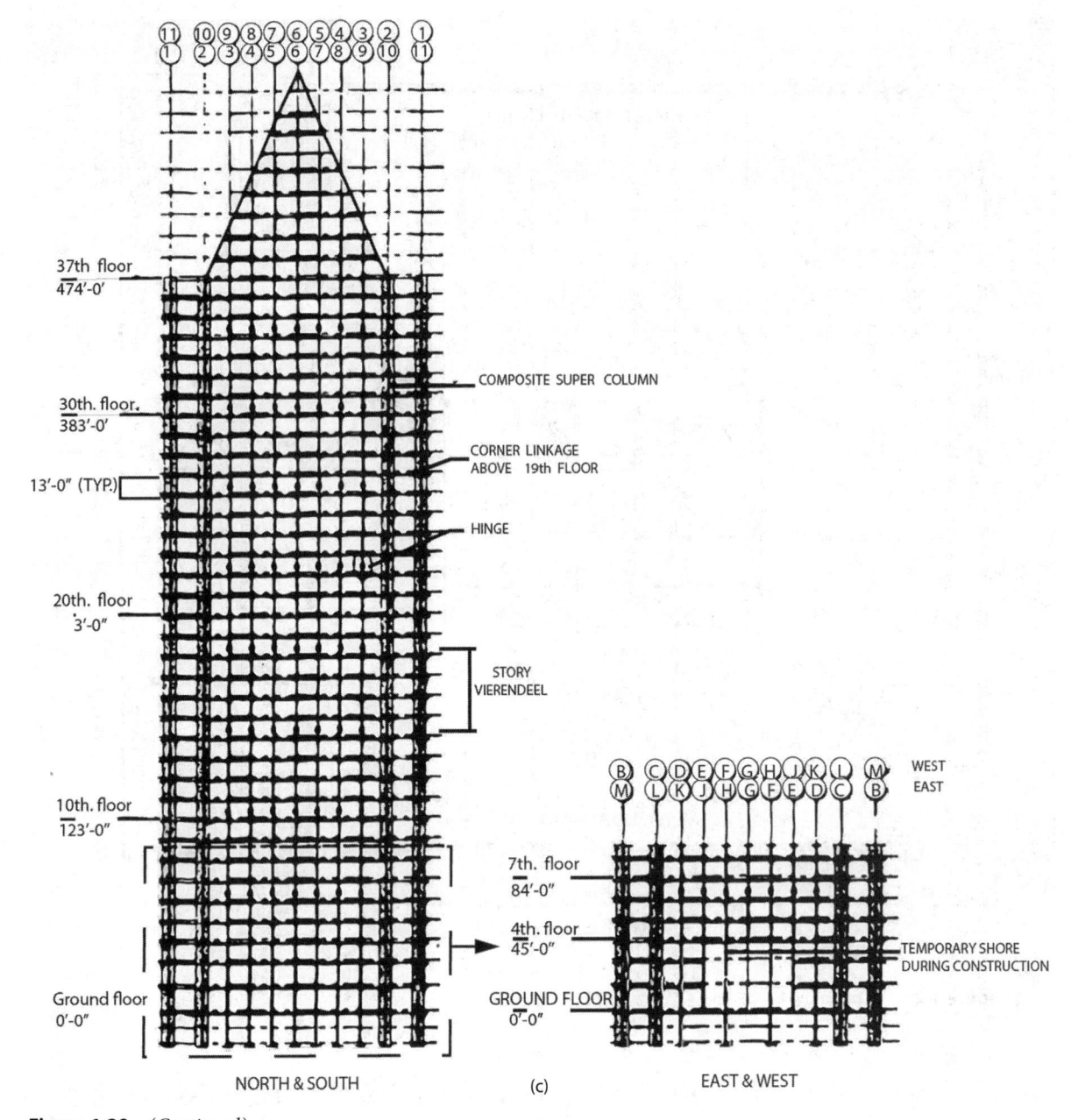

Figure 1.29 (*Continued*)

Richter scale at the nearby San Andreas fault. The strong column weak beam concept is maintained in the design of beam-column assemblies of the perimeter tube.

The sustained dead weight of the structure is 204,000 kips (927,272 kN) with the fundamental period of vibrations $T_x = 7.46$ s, $T_y = 6.95$ s, and $T_z = 3.57$ s. The interaction between the interior braced core and the perimeter ductile frame is typical of dual systems; the shear resistance of core increases progressively from the top to the base of the building. Nearly 50 percent of the overturning is resisted by the core. The maximum calculated lateral deflection at the top, under a 100-year wind is 23 in. (584 mm).

Sixteen critical joints (Fig. 1.37e,f) in the braced frame were mechanically stress relieved by using the Leonard Thompson vibration method of stress relief.

The structure is founded on shale rock with an allowable bearing capacity of 7.5 tons/ft² (720 kPa). The building core is supported on an 11.5 ft (3.5 m) thick concrete mat while a perimeter ring supports the ductile frame. Typical floor framing consists of W24 wide

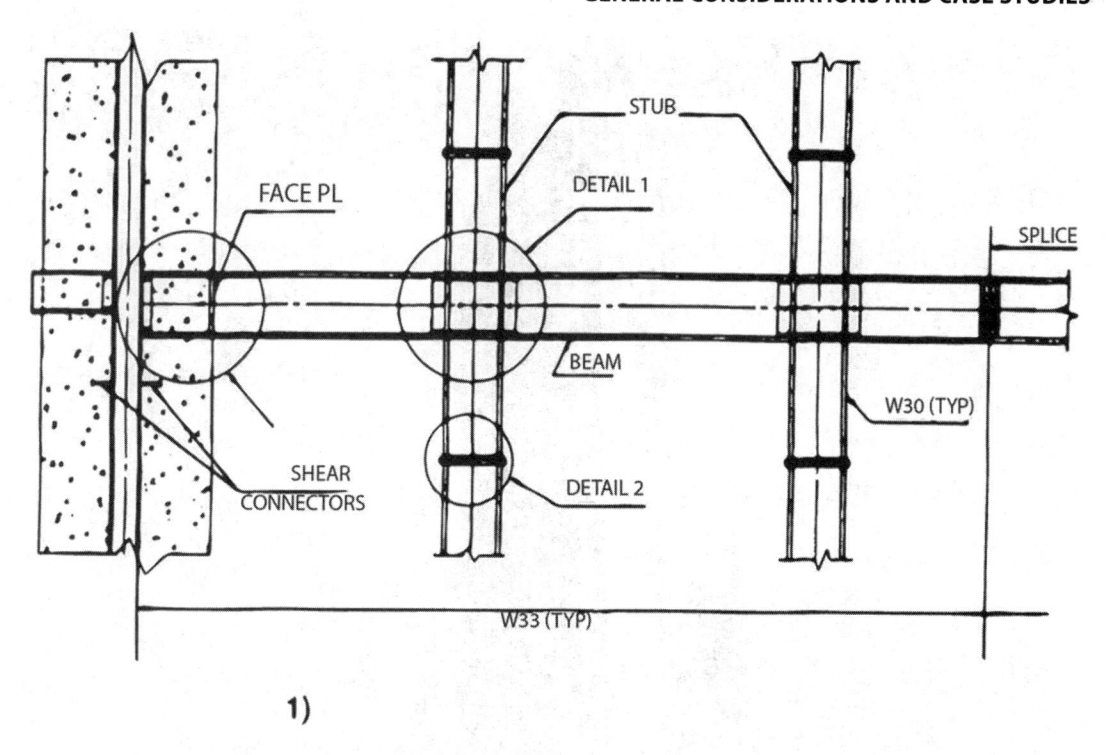

1)

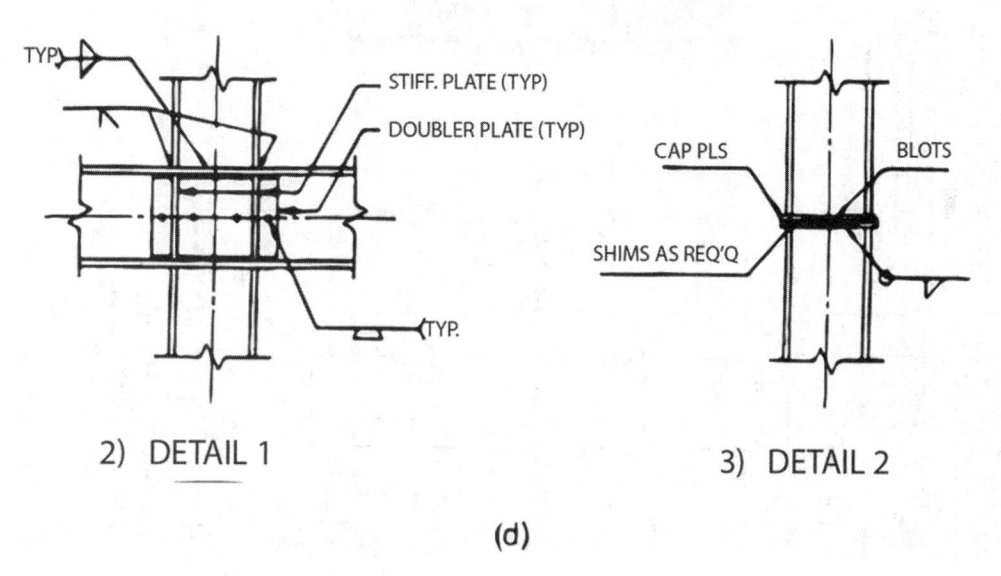

2) DETAIL 1

3) DETAIL 2

(d)

Figure 1.29 (*Continued*)

flange composite beams spaced at 13 ft centers, spanning a maximum of 55 ft (16.76 m) from the core to the perimeter. The structural design is by CBM, Engineers, Inc., Houston, Texas. Construction photographs of the chevron brace, corner columns at the base, and link beam at the lobby are shown in Figs. 1.33g-i.

1.3.27 Singapore Treasury Building, Singapore

This 52-story office tower is unique in that every floor in the building is cantilevered from an inner cylindrical, 159 ft (48.4 m) diameter core enclosing the elevator and service areas (Fig. 1.34a). Radial beams cantilever 36 ft (11.6 m)

(a)

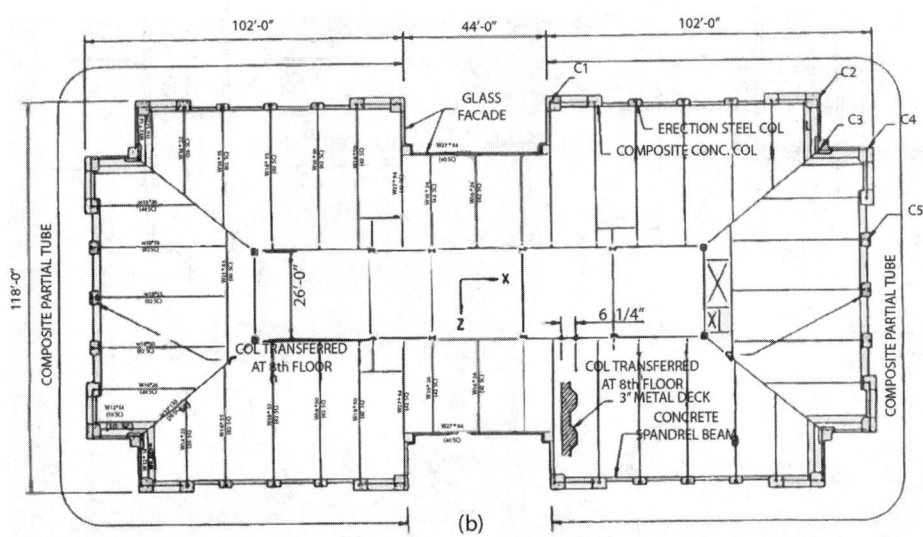

(b)

Figure 1.30   One Ninety One Peachtree-Detroit: (a) building elevation; (b) typical floor framing plan.

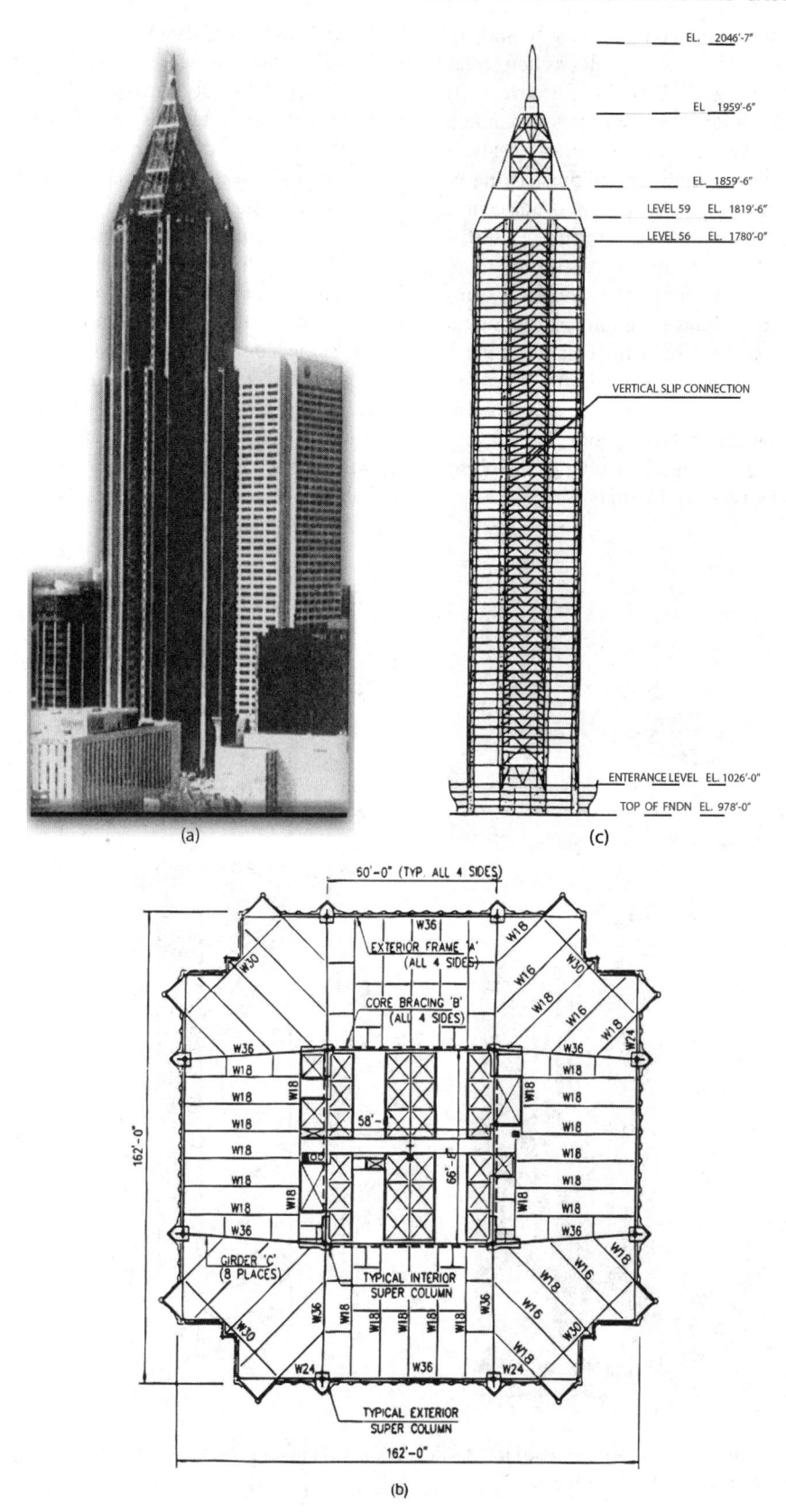

Figure 1.31 Nations Bank Plaza, Atlanta, Georgia: (a) building elevation; (b) typical framing plan; (c) section.

from the reinforced concrete core wall (Fig. 1.34b). Each cantilever girder is welded to a steel erection column embedded in the core wall. To reduce relative vertical deflections of adjacent floors, the steel beams are connected at their free ends by a 1×4 in. (25×100 mm) steel tie hidden in the curtain wall. A continuous perimeter ring-truss at each floor minimizes relative deflections of adjacent cantilevers on the same floor produced by the uneven distribution of live load. Additionally, the vertical ties and the ring beam provide a back-up system for the cantilever beams.

Since there are no perimeter columns, all gravity and lateral loads are resisted solely by the concrete core. The thickness of core walls varies from a 3.3 ft (1.0 m) at the top to 4 ft (1.2 m) at the sixteenth floor, and remains at 5.4 ft (1.65 m) below the sixteenth floor. The structural engineering is by LeMessurier Consultants, Cambridge, Massachusetts, and Ove Arup Partners, Singapore.

1.3.28 City Spire, New York

This 75-story, office and residential tower, with a height-to-width ratio of 10:1 is one of the most slender buildings, concrete or steel, in the world today. The critical wind direction for this building is from the west, which produces maximum across-wind response. Wind studies indicated possible problems of vortex shedding as well as occupant perception of acceleration. This possibility was eliminated by adding mass and stiffness to the building.

The main structural system consists of shear walls in the spine connected to exterior jumbo columns with staggered rectangular concrete panels. The structure is subdivided into nine major structural subsystems with many setbacks and column transfers as evident from the plans shown in Fig. 1.35a. The structural design is by Robert Rosenwasser Associates, New York.

(a)

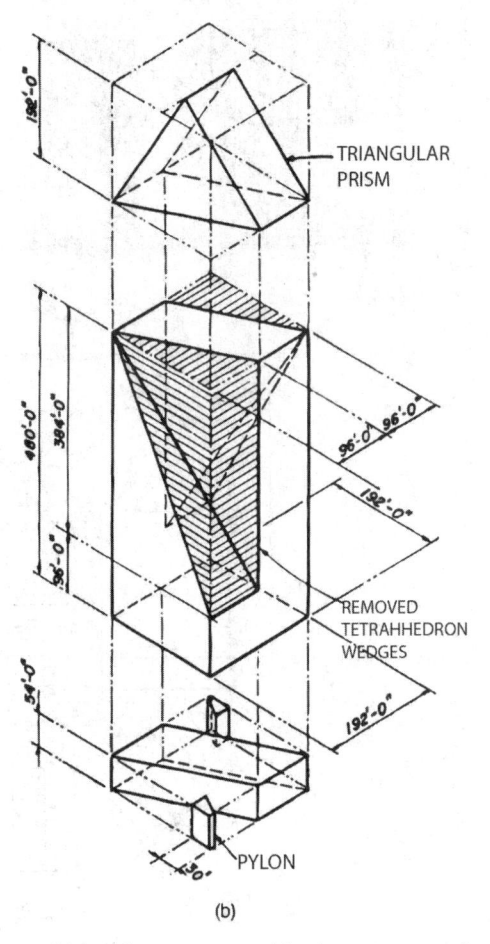

(b)

Figure 1.32 Allied Bank Tower, Dallas, Texas: (a) building elevation; (b) building geometry; (c) schematic view of structural system; (d) intersection of diagonals, detail 1, detail 2.

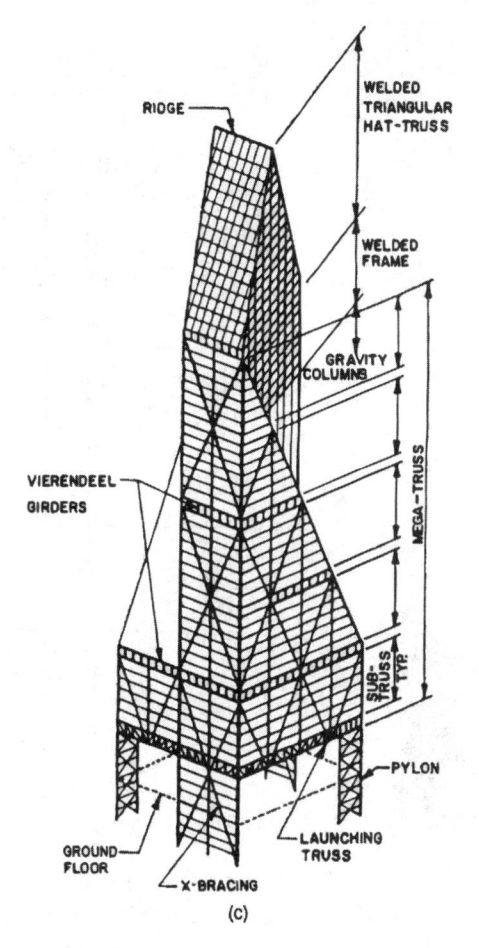

(c)

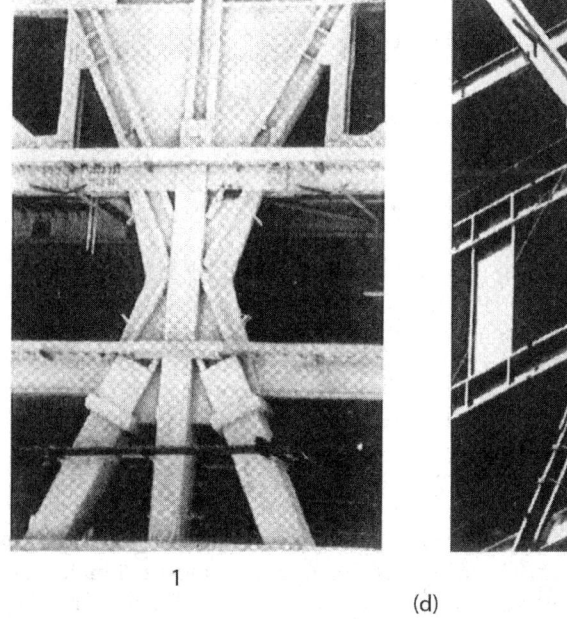

1

2

(d)

Figure 1.32   (*Continued*)

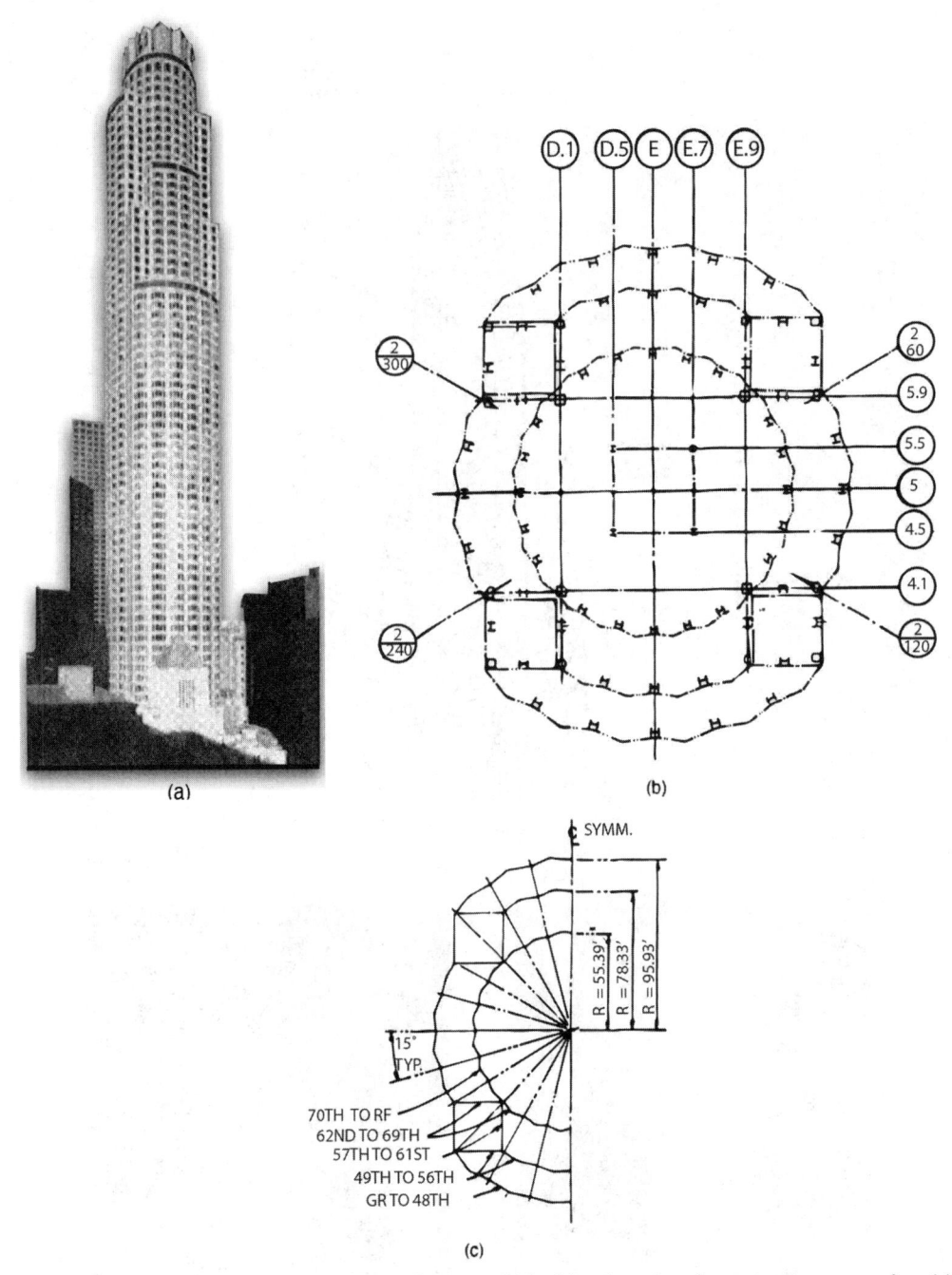

(a)

(b)

(c)

Figure 1.33   First Interstate World Center: (a) building elevation; (b) building key plan showing column transfers; (c) composite plan; (d) structural system; (e) floor framing plan; (f) location of stress-relieved joints; (g) chevron brace connection; (h) connection of corner column and braces at base; (i) link beam connection at lobby.

1.3.29   City Corp Tower, Los Angeles

This 54-story tower raises to a height of 720 ft (219.50 m) above the ground level and has a height-to-width ratio of 5.88:1 (Fig. 1.36a). The building has two vertical setbacks of approximately 10 ft (3.05 m) at the 36th and 46th floors as shown in the composite floor plan (Fig. 1.36b). As is common to most tall buildings in seismic Zone 4, this building was designed for site-specific maximum probable and maximum credible response spectrums, which represented peak accelerations of 0.28 g and 0.35 g

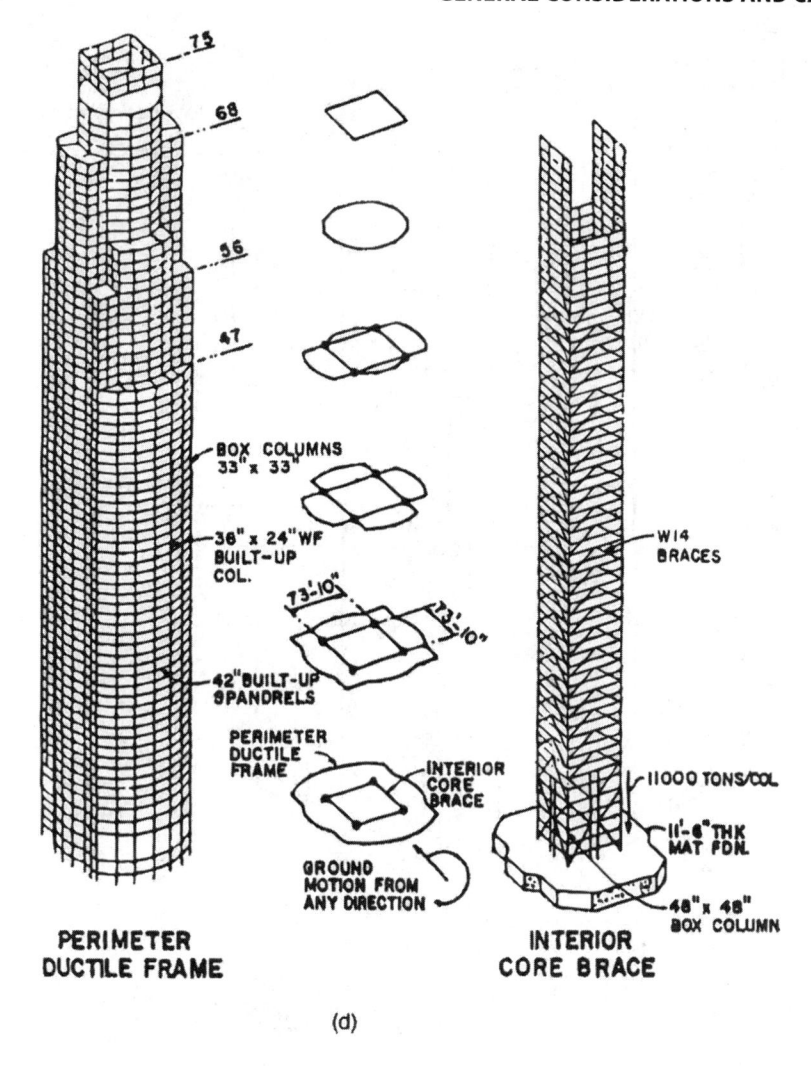

(d)

Figure 1.33   (*Continued*)

respectively. The corresponding critical damping ratios were 5 and 7.5 percent. The structural system consists of a steel perimeter tube with WTM24 columns spaced at 10 ft (3.05 m) centers, and 36 in. (0.91 m) deep spandrels. The columns at the setback levels are carried on 48 in. (1.22 m) deep transfer girders and by the vierendeel action of the perimeter frame. Typical floor plans at the setback levels are shown in Figs. 1.38c,d.

The foundation for the tower consists of a 7 ft (2.14 m) deep mat with a four-story basement for parking. The structural design is by John A. Martin & Associate, Inc., Los Angeles.

1.3.30   Cal Plaza, Los Angeles

The project consists of a 52-story office tower rising above a base consisting of lobby and retail levels, and

six levels of subterranean parking (Fig. 1.37a). A structural steel system consisting of a ductile moment-resisting frame at the perimeter resists the lateral loads. The parking areas outside of the tower consist of a cast-in-place concrete system with waffle slab and concrete columns. Figure 1.37b shows a typical midrise floor plan for the tower with sizes for typical framing elements. The structural design is by John A. Martin & Associates, Inc., Los Angeles.

1.3.31   MTA Headquarters, Los Angeles

MTA Head Quarters (Fig. 1.38a) is a 28-story office building located east of the Union Station, Los Angeles, California. The building has a gross area of 622,000 ft^2 (57,785 m^2). It has a four-level subterranean structure, which will serve as a common base for the MTA Tower,

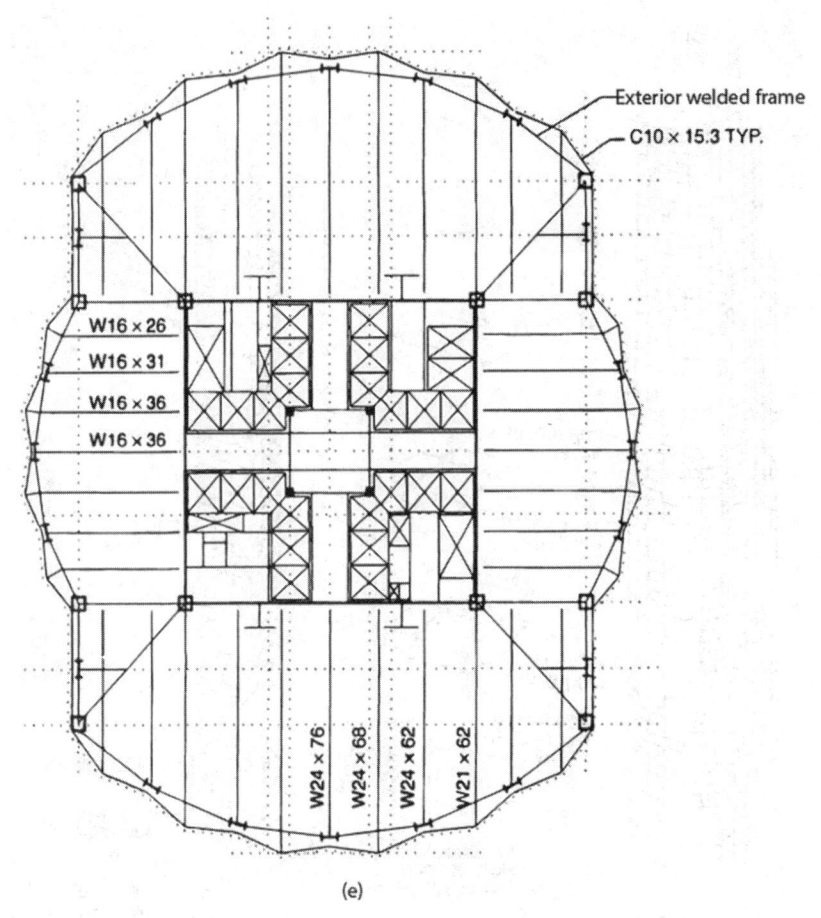

(e)

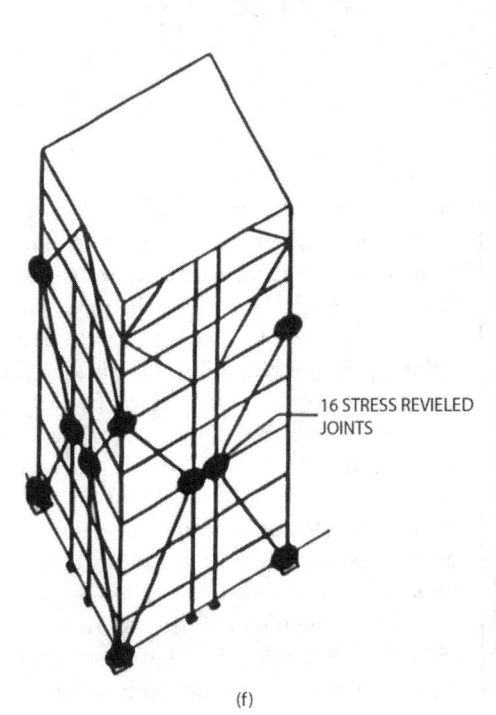

(f)

(g)

Figure 1.33   *(Continued)*

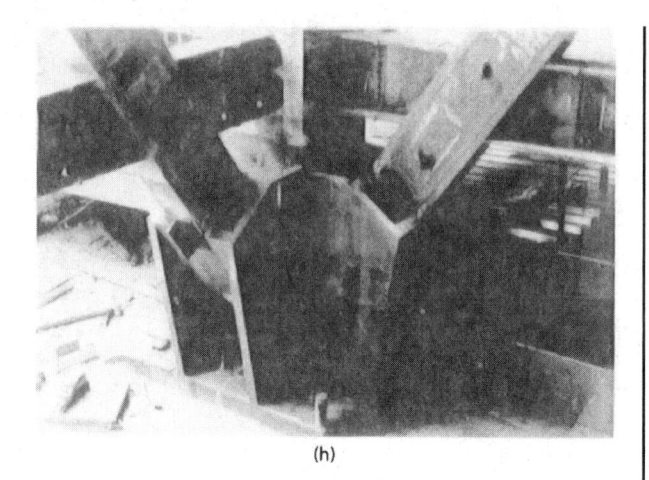

(h)

(i)

Figure 1.33   (*Continued*)

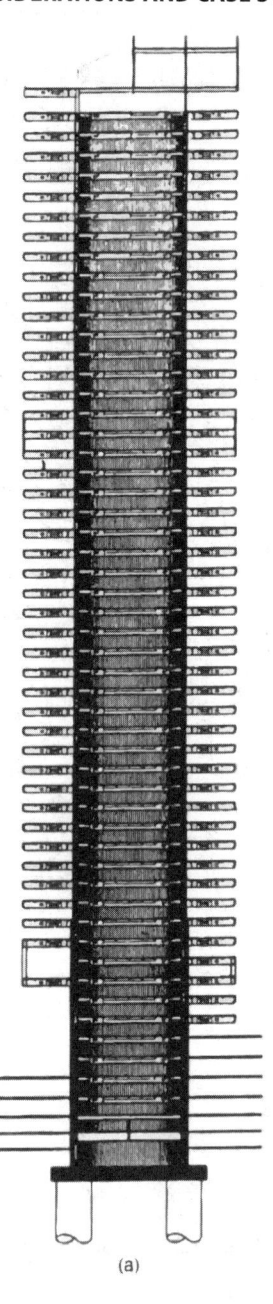

(a)

Figure 1.34   Singapore Treasury Building: (a) schematic section; (b) typical floor framing plan.

and two future office buildings. The basement levels, which serve as a parking garage, extend beyond the footprint of the tower. The construction for the parking structure consists of precast reinforced concrete columns and girders with cast-in-place concrete slab. The plaza level underneath the tower consists of a composite floor system with a 4½ in. (114 mm) normal-weight concrete topping on a 3 in. (76 mm) deep, 18-gauge composite metal deck. The metal deck spans between composite steel beams spaced typically at 7 ft 6 in. (2.29 m) outside of the tower which has heavy landscape, and 10 ft (3.04 m) on centers within the tower footprint.

The building is essentially rectangular in plan, 118 ft × 165 ft (36 m × 50.3 m) with a slight radius on the short faces. The building height is 400 ft (122 m) with a fairly low height-to-width ratio of 3.39. Typical floor framing shown in Fig. 1.38b consists of 21 in. (0.54 m) deep composite beams spanning 41 ft (12.5 m) from the core to the exterior. A 3 in. (76 mm) deep metal deck with a 3¼ in.

(83 mm) thick lightweight concrete topping completes the floor system.

The lateral system consists of a perimeter tube with widely spaced columns tied together with spandrel beams. The exterior columns on the broad faces vary from W30 × 526 at the plaza to W30 × 261 at the top. The spandrels vary from a WTM36 × 286 at the plaza level to W36 × 170 at the top floors. The columns on the curvilinear faces are

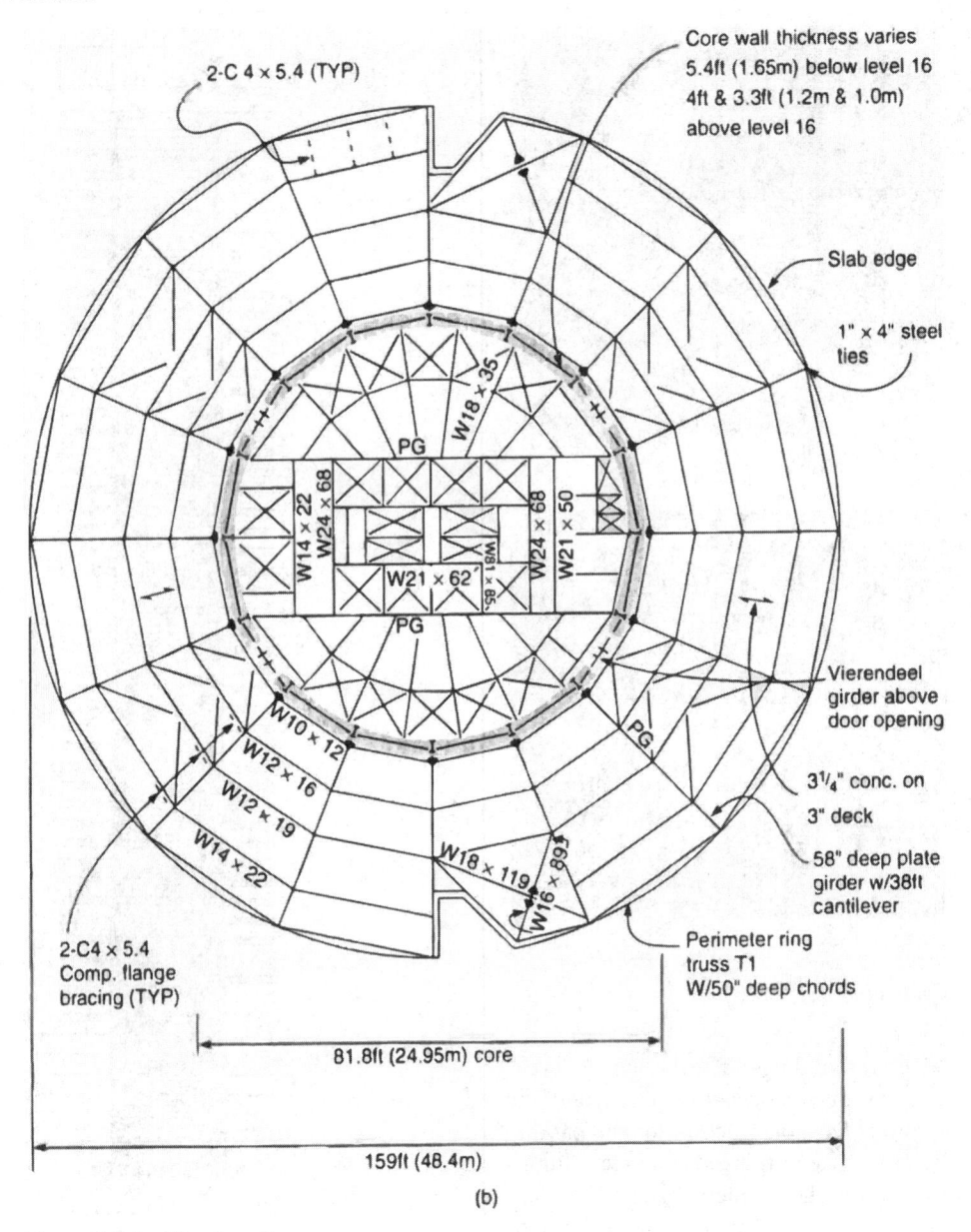

2-C 4 × 5.4 (TYP)

Core wall thickness varies
5.4ft (1.65m) below level 16
4ft & 3.3ft (1.2m & 1.0m)
above level 16

Slab edge

1" × 4" steel
ties

W18 × 35

PG

W14 × 22
W24 × 68
W24 × 68
W21 × 50
W8 × 85

W21 × 62

PG

Vierendeel
girder above
door opening

3¼" conc. on
3" deck

58" deep plate
girder w/38ft
cantilever

W10 × 12
W12 × 16
W12 × 19
W14 × 22

PG

W18 × 119
W16 × 89

2-C4 × 5.4
Comp. flange
bracing (TYP)

Perimeter ring
truss T1
W/50" deep chords

81.8ft (24.95m) core

159ft (48.4m)

(b)

Figure 1.34   (*Continued*)

built up, 34 × 16 in. (0.87 × 0.40 m), box columns while 24 × 24 in. (0.61 × 0.61 m) box columns are used at the corners. Plates varying in thickness from 4 in. (102 mm) at the bottom to 1 in. (25 mm) at the top are used for the built-up columns. As is common for most steel buildings in seismic zones 3 and 4, 50 ksi steel for columns, and 36 ksi steel for spandrels are used to satisfy the strong-column-weak-beam requirement. The salient seismic design parameters are as follows.

Building period x-direction = 4.57 s
Building period y-direction = 4.40 s

Wind shear x-direction = 1580 kips (7028 kN)
Wind shear y-direction = 2078 kips (9243 kN)
1991 UBC base shear for strength check = 2894 kips (12,873 kN)
1991 UBC base shear for deflection check = 1497 kips (6659 kN)
Unscaled UBC base shear = 5163 kips (22,966 kN) x-direction
Unscaled base shear Maximum Credible Earthquake with 7 percent damping 6460 kips (28,735 kN) x-direction

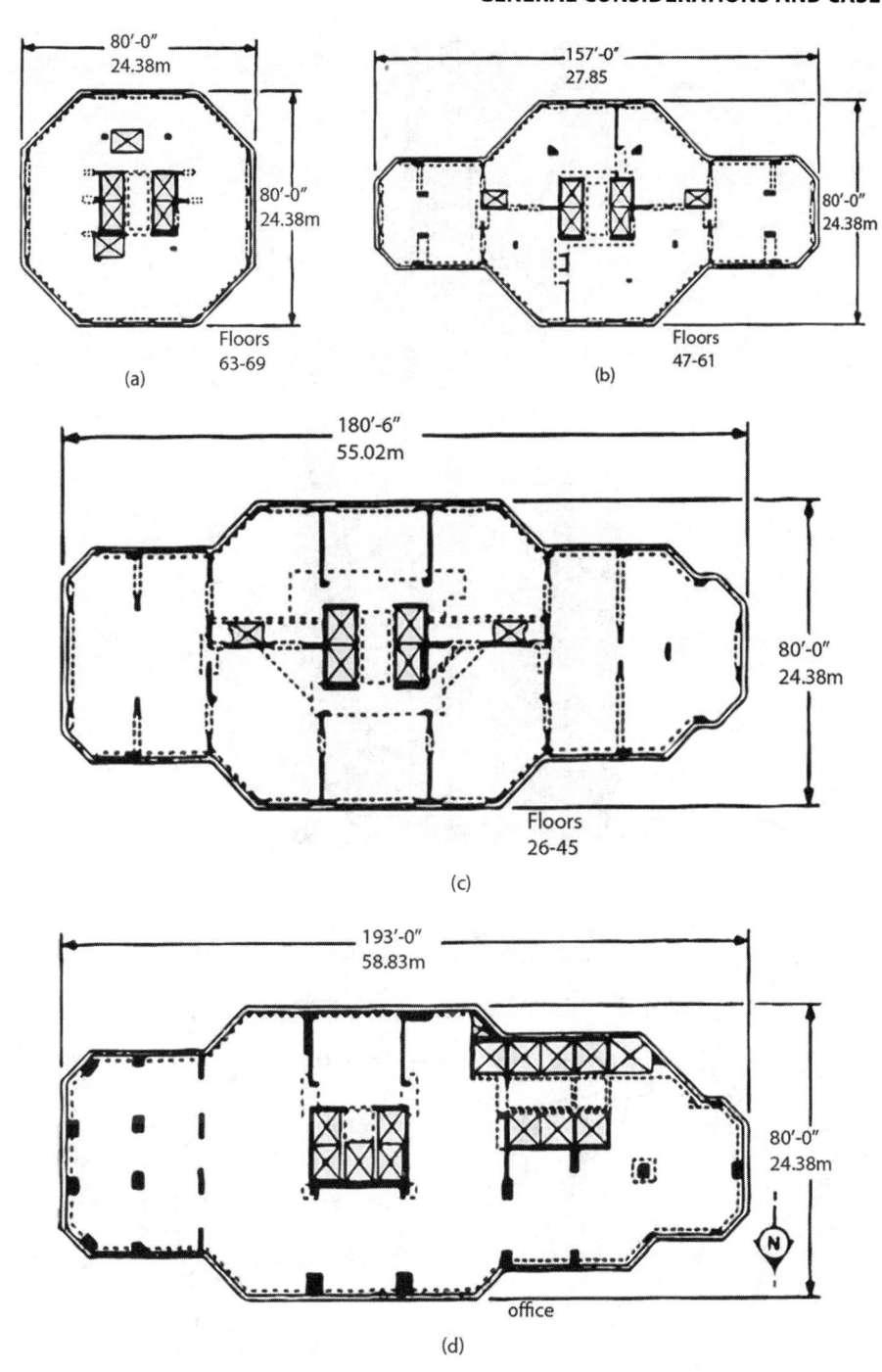

Figure 1.35 City Spite New York.

Building weight above shear base (plaza level) = 77,212 kips (343,456 kN)
Building floor-to-floor height—13 ft 4 in. (4 m)
Unit density of building = 9 lb per ft^3 (144 kg/m^3)
Unit weight of building =118 psf (5650 Pa)

Steel Tonnage = Approximately 24 psf (1149 Pa)
The building architecture is by McLarand Vasquez & Partners Inc., while the structural engineering is by John A. Martin & Asso. Inc., both of Los Angeles, California.

(a)

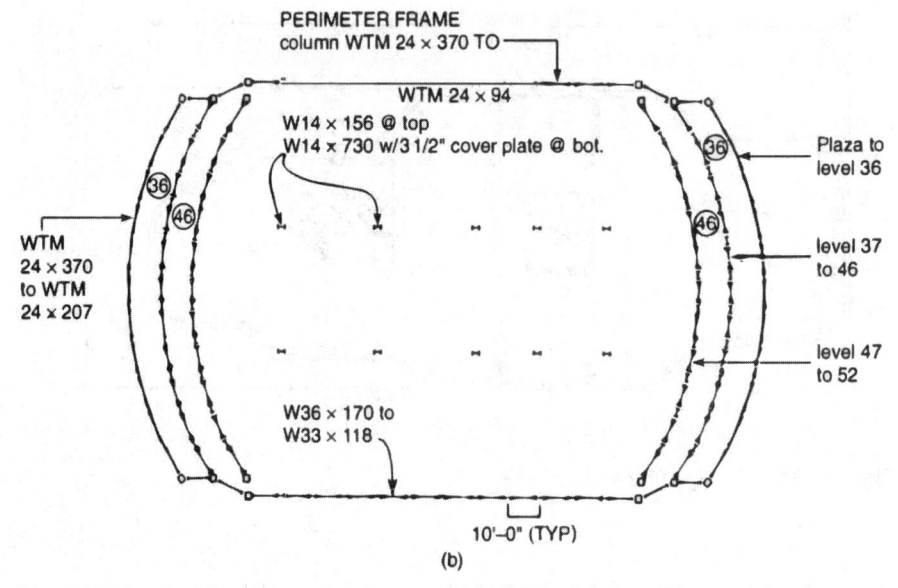

(b)

Figure 1.36   City Corp Tower, Los Angeles: (a) building elevation; (b) composite plan; (c) 36th floor framing plan; (d) 47th–52nd floor framing plan. (Architects: Cesar Pelli Asso. Inc. Structural engineers: John A. Martin & Asso. Inc., Los Angeles.)

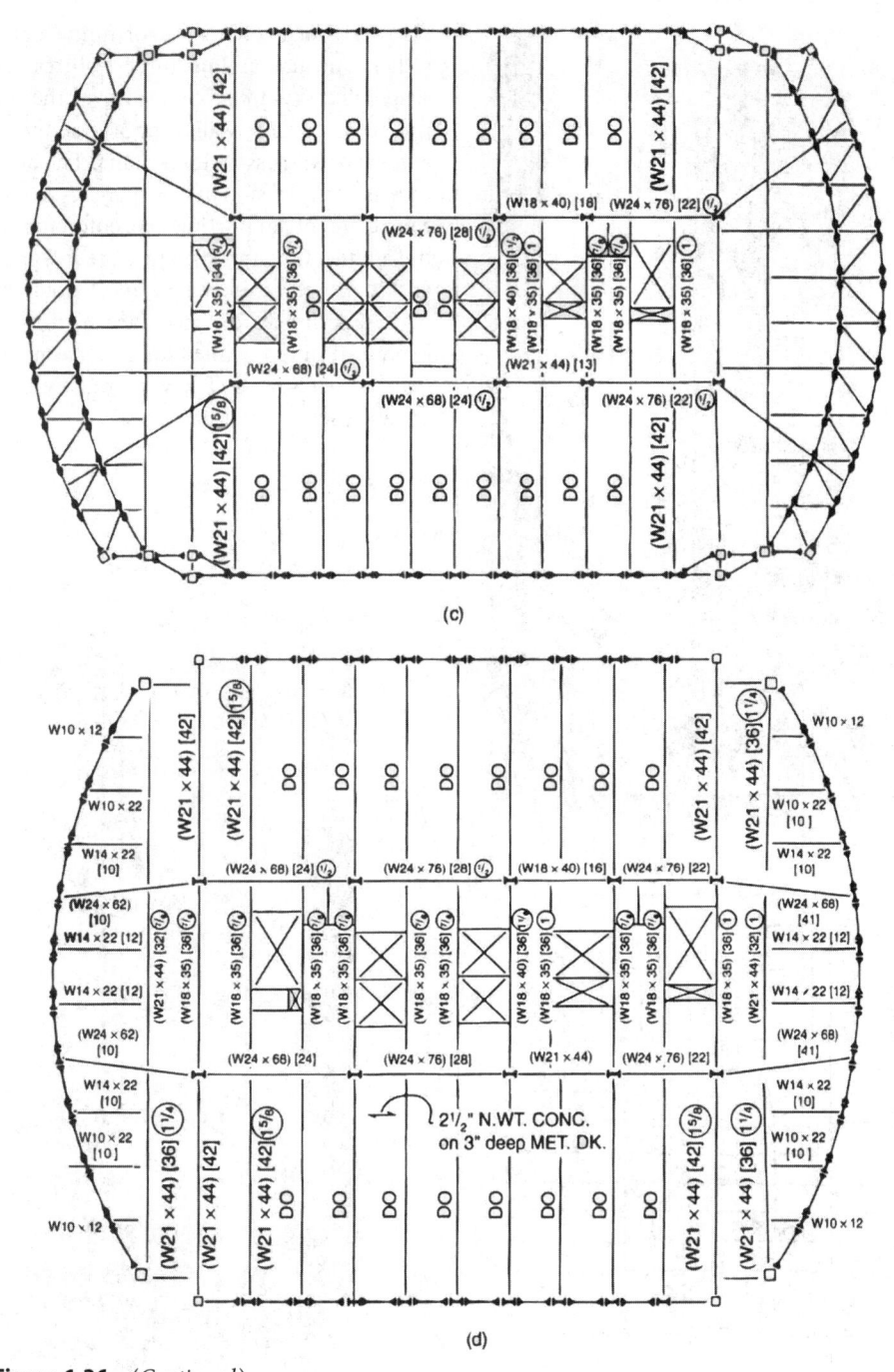

Figure 1.36 (*Continued*)

1.3.32 The 21 Century Tower

The 21 Century Tower (Figs. 1.39a, 1,2,3) when built will be a landmark 50-story, 1,000,000 sq ft (100,000 sq m) building in Shanghai. Designed by the architectural firm of Murphy/Jahn Inc., Chicago, the exterior form of the building is that of a rectangular tower; however, the structure is made unique by a setback base and a series of nine-story high wedge-shaped atria or "winter gardens," which run the full height of the tower. These features effectively remove one corner column over the full height of the building, and remove the opposite column at both the top and bottom of the tower. The structure therefore

(a)

takes the form of a stack of nine-story high chevrons. Other unusual architectural features include a cable-suspended skylit canopy roof over the podium, exposed rod-truss curtain wall supports at the winter gardens, and exposed truss-stringer stairs. The building façade is a futuristic expression of transparency, color, and structure, each element supporting and enforcing the other so that the architecture and structure are integrated into a single entity. Structural elements, most notably the nine-story high superbraces on each face of the tower, are boldly expressed in red, while blue and green solar glazing covers all office spaces. The winter gardens and the podium are enclosed in clear glass.

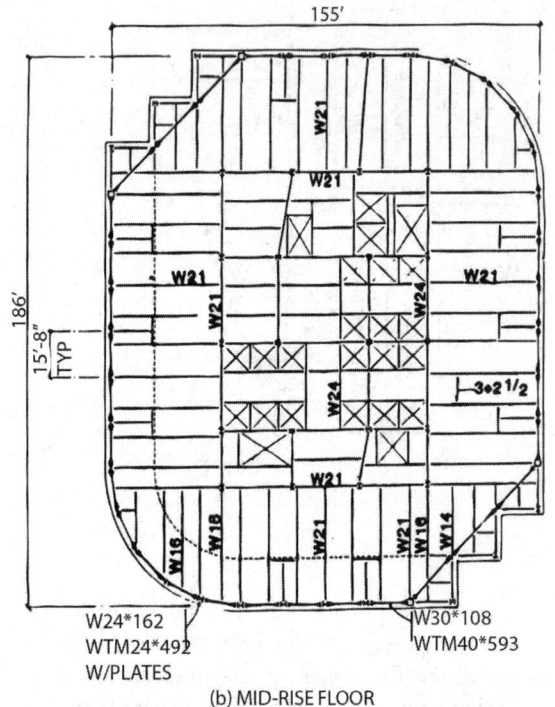

(b) MID-RISE FLOOR

Figure 1.37 Cal Plaza, Los Angeles: (a) building elevation; (b) mid-rise floor framing plan. (Architects: Arthur Erickson Inc., Structural engineers: John A. Martin & Asso. Inc., Los Angeles.)

(a)

Figure 1.38 MTA headquarters, Los Angeles: Architects: McLarand Vasquez & Partners, Inc., Structural engineers: John A. Martin & Asso. Inc., Los Angeles. (a) building elevation; (b) typical floor framing plan.

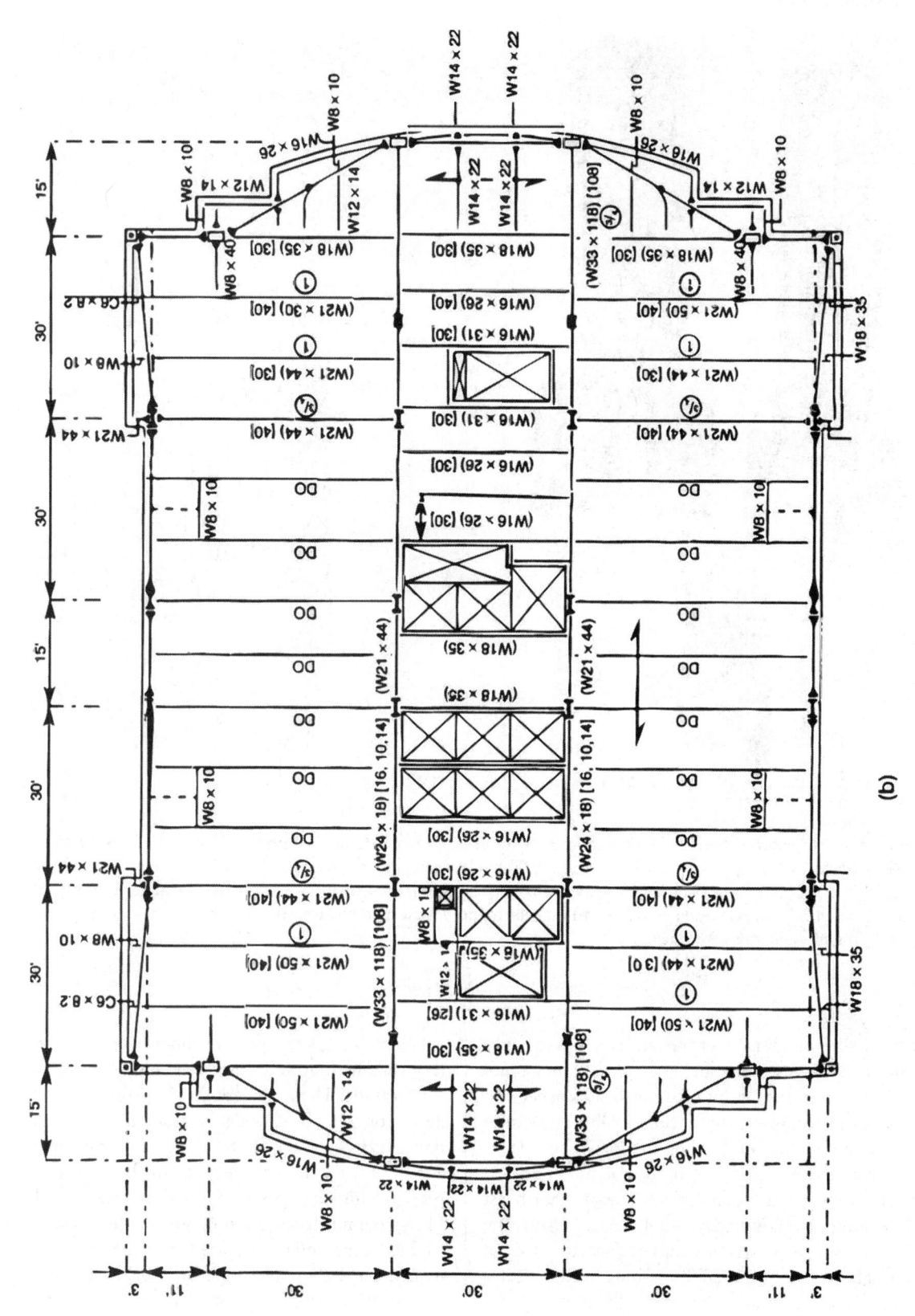

Figure 1.38 (*Continued*)

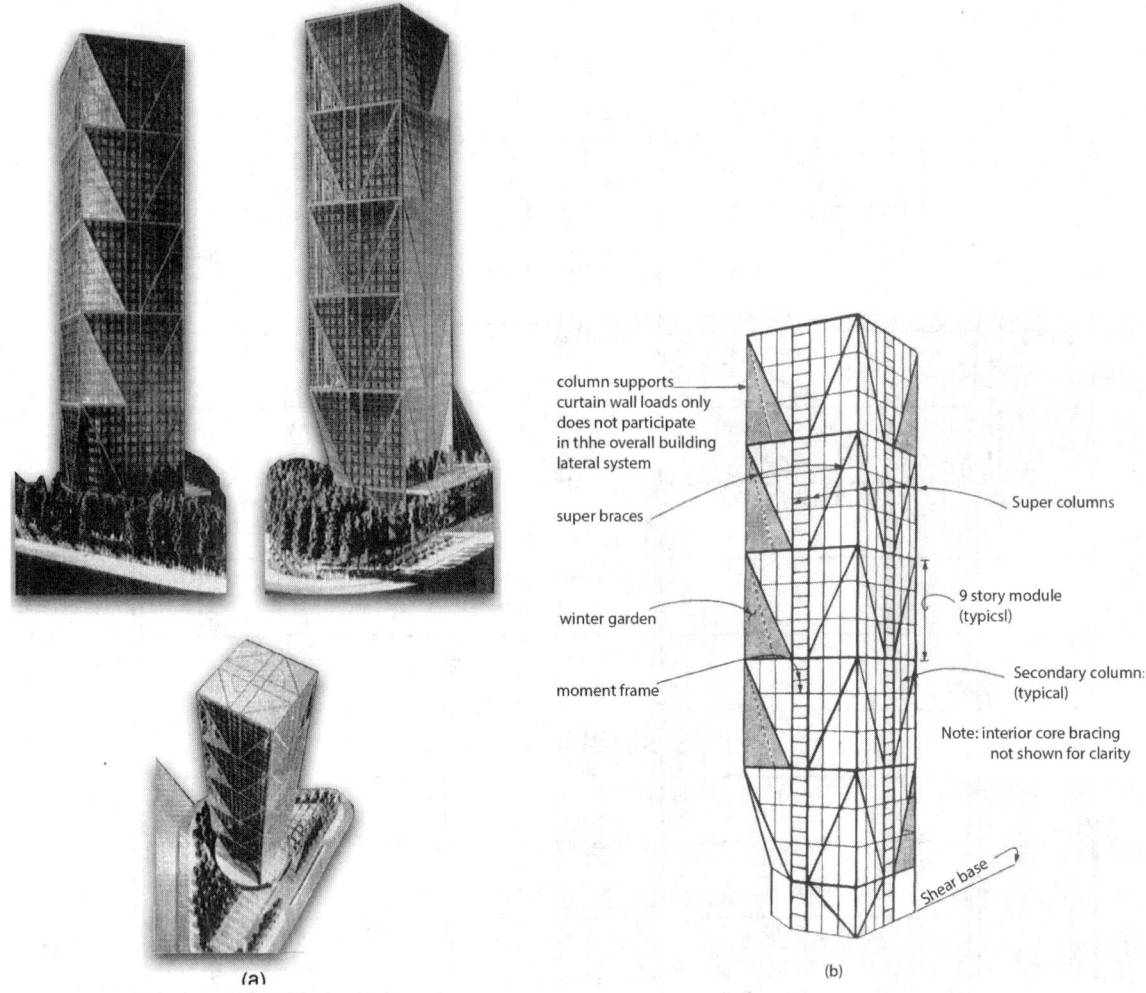

(a)   (b)

Figure 1.39   21st Century Tower: Architects: Murphy/Jahn Inc., Chicago, Structural engineers: John A. Martin & Asso. Inc., Los Angeles, Martin & Huang, International, Los Angeles: (a) model photographs 1, 2, and 3; (b) bracing system; (c) framing plan; levels 19, 28, and 37; (d) framing plan, levels 20, 29, and 37; (e) exterior frame elevation, north face; (f) exterior frame elevation, west face; (g) exterior frame elevation, east face; (h) exterior frame elevation, south face; (i) structural action in primary columns and braces; (j) typical interior core bracing.

Although the building is expressed as a square, it is punctuated by a series of four nine-story high wedge-shaped winter gardens cut into the northeast corner and two more at the southwest corner (Fig. 1.39b). The winter gardens have the effect of dividing the tower, both visually and structurally, into a stack of five modules outlined by the superbraces. At the lowest module, the northeast corner column is eliminated entirely. As a result, the tower has only a single axis of symmetry, which passes at 45° through the corners; in addition, nine different floor plans are required within each module. Plans at two representative floors are shown in Figs. 1.39c,d.

The building site consists primarily of sandy sedimentary soils with groundwater level only 1 ft 8 in. (0.5 m) below grade. The basement substructure is a three-story deep concrete box, roughly triangular in plan, with typical exterior walls 1 ft 8 in. (0.5 m) in thickness. A 6 ft 3 in. (1.9 m) wide zone between the basement walls and property line is provided for a slurry wall and backfill. The maximum excavation depth for the substructure is 44 ft (13.4 m), and is supported by a combination of 24 in. (600 mm) driven steel pipe piles below the tower and cast-in-place concrete piles of the same diameter below the plaza and podium. The piles have a maximum length

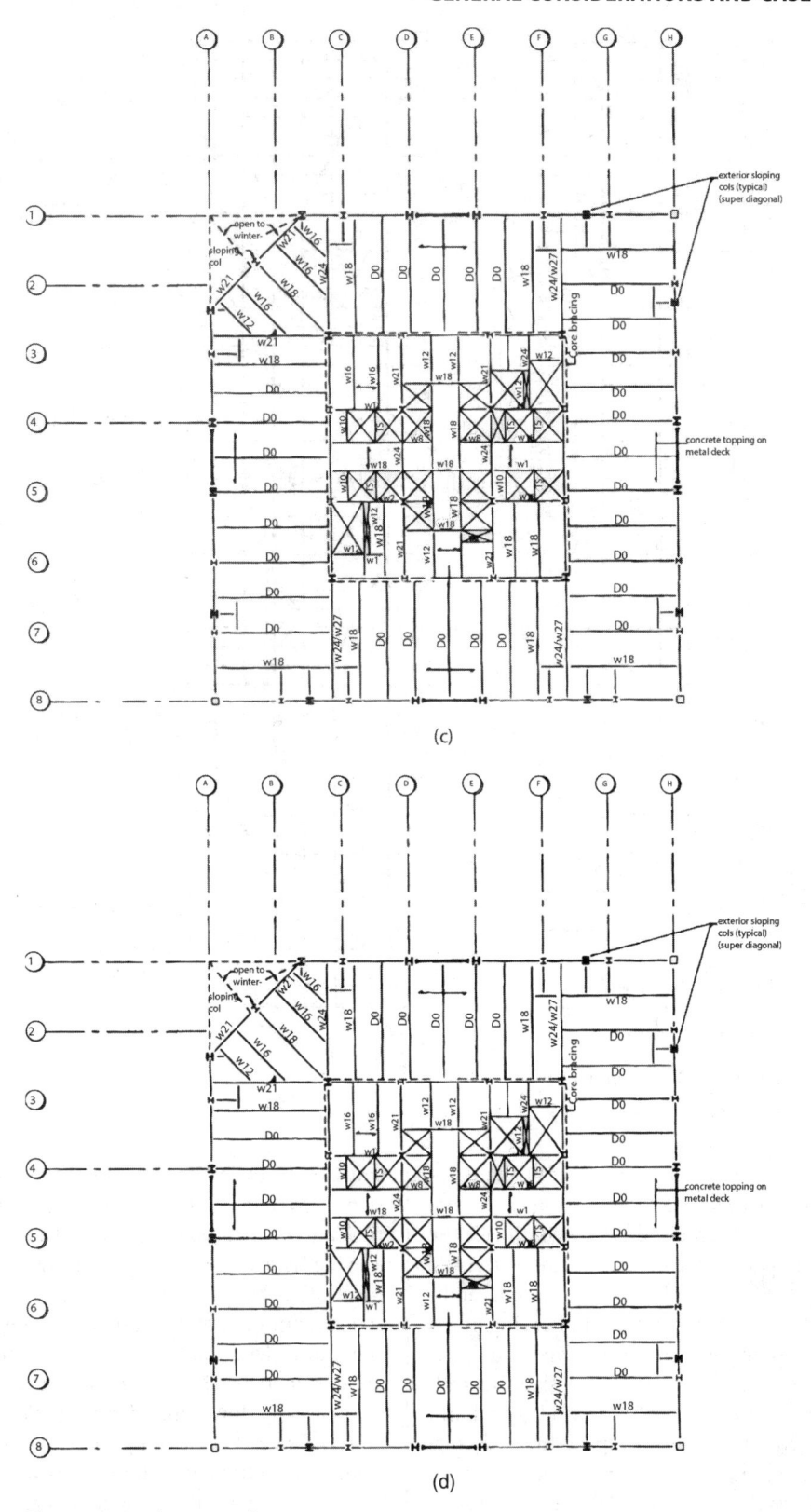

(c)

(d)

Figure 1.39   (*Continued*)

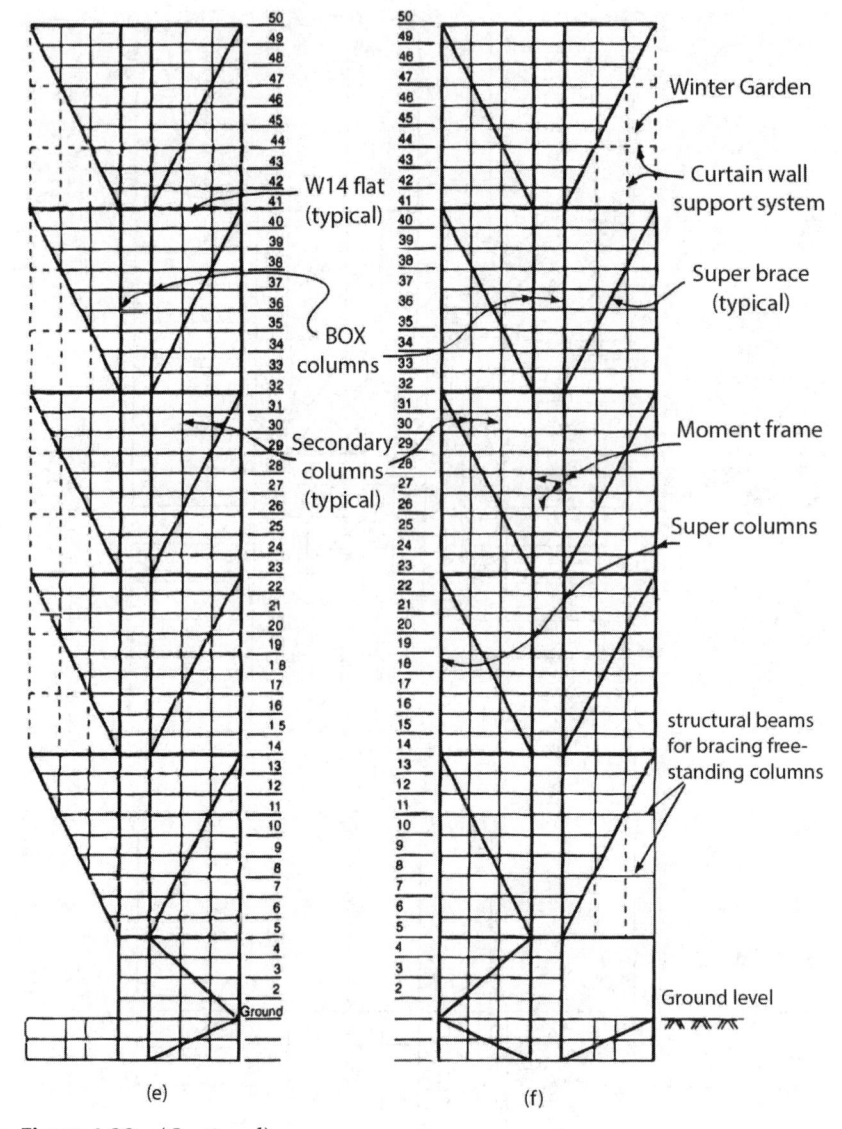

Figure 1.39 *(Continued)*

of 40 m, and are installed prior to the start of the base-ment excavation using a reusable steel driving extension. Driving from the ground level in this way eliminates the need to move and operate pile drivers on the extremely wet soils found at the bottom of the excavation; in addi-tion, the piles will pin the soil and prevent heaving of the subgrade while overburden is being removed. The piles support a 5 ft (1.5 m) thick concrete mat slab. Beneath the tower the mat is strengthened by a grid of column pedestals and inverted grade beams topped by a concrete slab: the whole system forming a complex of story high cells which also serve as water tanks. This cellular system extends beneath the plaza some distance to the north and

south of the tower. Beneath the remainder of the plaza the slab is covered by a 14 in. (350 mm) thick layer of crushed rock and a 6 in. (150 mm) slab at the lowest level, the rock serving as a capillary break for any water which might infiltrate through the mat. Under some portions of the substructure the hydrostatic uplift pressure exceeds the structural dead load; therefore the piles in these areas are designed as friction piles to prevent uplift.

Below the tower the framing consists of concrete-encased steel beams supporting cast-in-place concrete floor slabs. Other structural slabs are cast-in-place flat slabs with drop-panels. The design of the tower is driven by the unique architectural treatment to the building

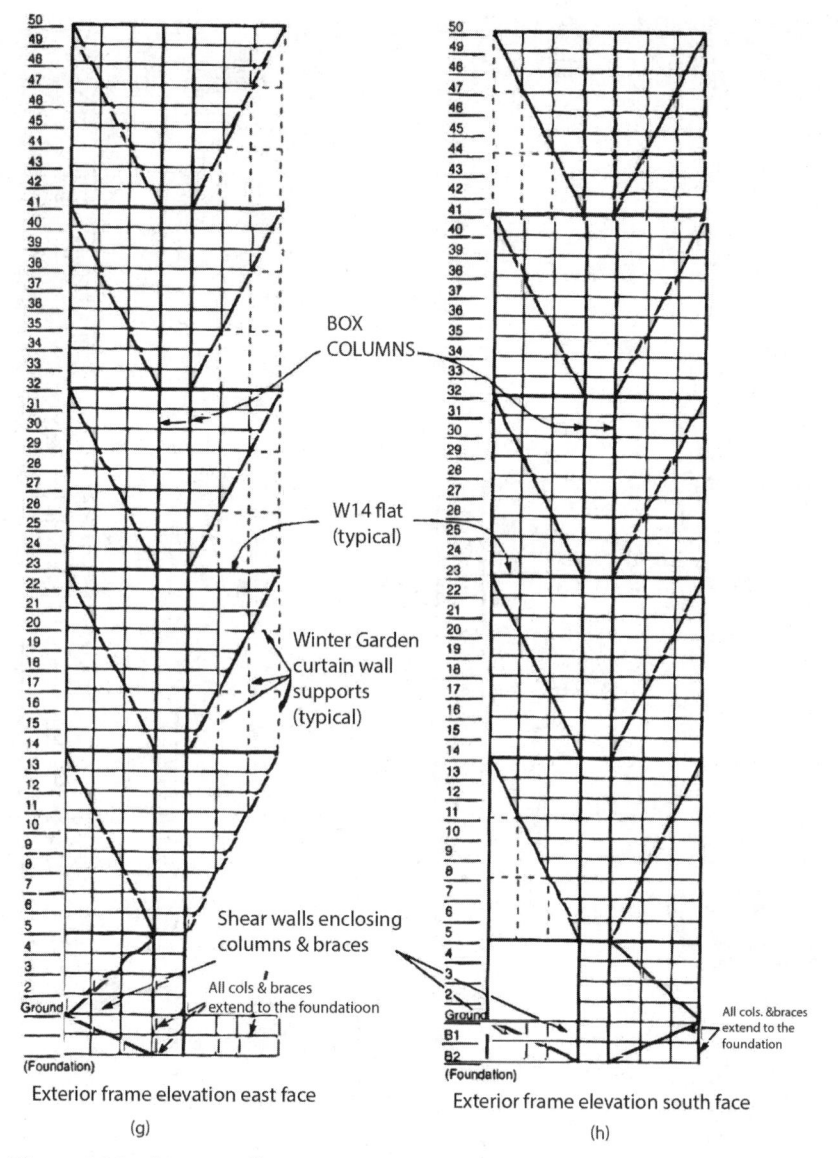

BOX COLUMNS

W14 flat (typical)

Winter Garden curtain wall supports (typical)

Shear walls enclosing columns & braces

All cols & braces extend to the foundatioon

All cols. &braces extend to the foundation

(Foundation)

Exterior frame elevation east face

(g)

(Foundation)

Exterior frame elevation south face

(h)

Figure 1.39 (*Continued*)

envelope, by the arrangement of the winter gardens, and by the high wind speeds of up to 115 mph (185 km/h) resulting in design pressures as high as 136 psf (6.5 kPa).

The structural solution for the tower consists of a superbrace system on the exterior skin supplemented by an eccentrically-braced interior service core. The superbrace system makes maximum use of the total tower width, giving optimum resistance to wind loading, maximum economy, and minimum structural intrusion into interior spaces. Schematic exterior brace elevations are shown in Figs. 1.39e–h. Structural action in the primary columns and braces due to lateral loads are shown in Fig. 1.39i for the lower three modules.

The braces generally consist of heavy W14 sections, field-spliced every three floors and connected at their ends to square steel box columns. The braces also act as inclined columns, and carry the vertical loading from all secondary columns above them. This arrangement maximizes the vertical loading carried by the box columns and minimizes corner uplift. Columns vary in size from 20 to 24 in. (520 to 610 mm) with plate thicknesses up to 5 in. (130 mm). The braces are arranged in five 9-story-high vees with a one-bay gap in the middle of each building face; stiffness in the gap is provided by one-bay wide rigid frame. Panel points for the superbraces occur at the ground, 5th, 15th, 24th 33rd 42nd, and roof levels. Horizontal members and

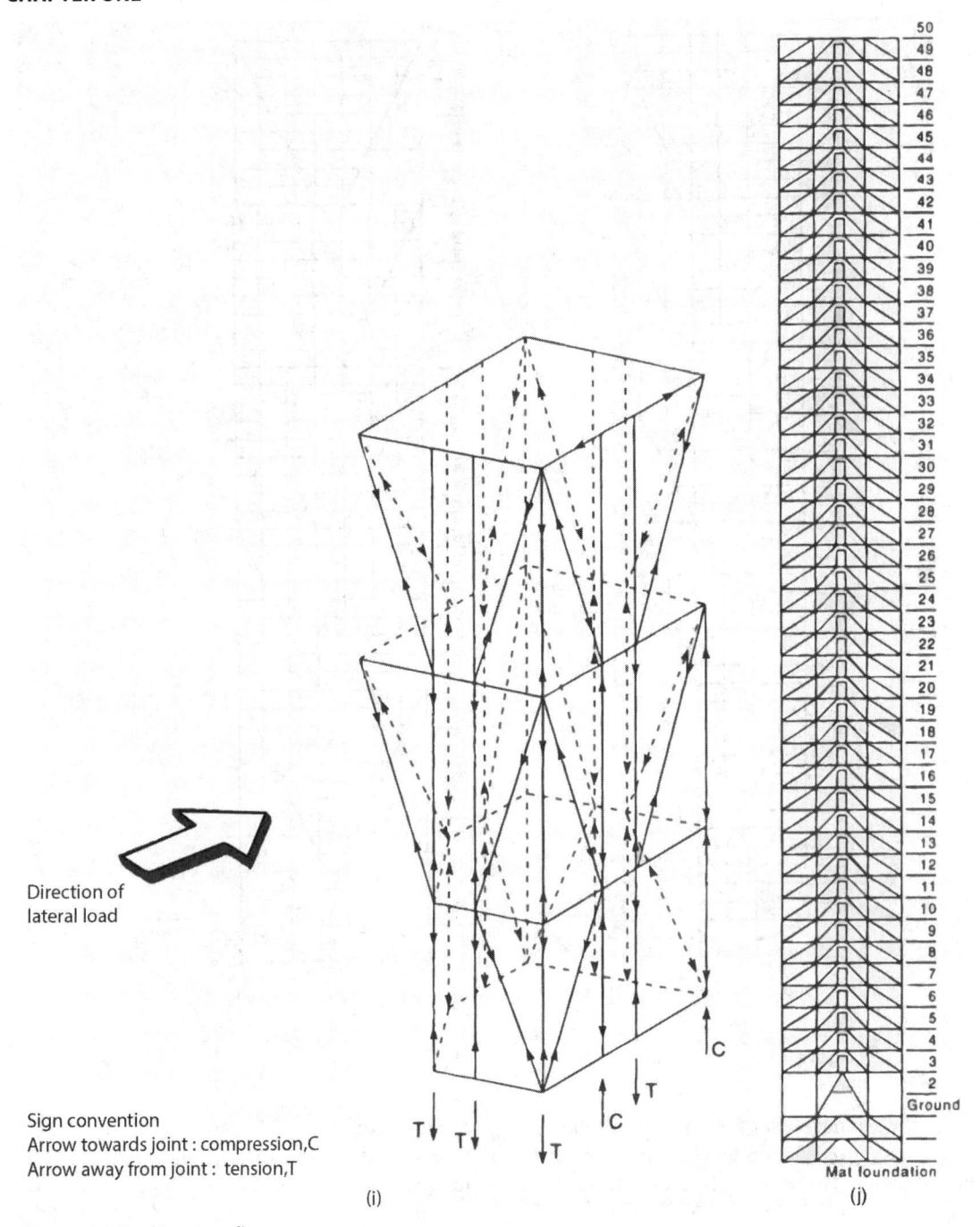

Direction of
lateral load

Sign convention
Arrow towards joint : compression,C
Arrow away from joint : tension,T

(i)

Mat foundation

(j)

Figure 1.39   (*Continued*)

diaphragms at these levels are stiffened to transfer horizontal brace forces. Service core bracing provides additional overall stiffness and gives lateral support to floors between the superbrace panel points. For architectural reasons, braces at the center bay of each core-face are eccentric (Fig. 1.39j). Although most lateral loading is transferred at the ground level to the shear walls, core bracing is extended to the foundation of the substructure. The numerous corner cutouts of the tower structure effectively rotate the principal axes of the structure by 45%, to pass through the corner columns. The lowest two modes of vibration of the tower are single-curvature bending through these axes; the third mode is torsional. Period of the first three modes are 4.93, 4.62, and 2.12 s, respectively.

The 21 Century Tower is an example of how the super-brace structural frame can be integrated with the architectural theme of the building to create an impressive architectural statement. By concentrating the primary lateral resistance of the building in a relatively small number of perimeter members, both high economy and maximum lateral strength are achieved. This is borne out by the buildings average steel weight of 29 psf (142 kg/m^2). The conceptual and preliminary design is by structural engineers John A. Martin & Associates; working drawings are by Martin & Huang, International, both of Los Angeles, California.

1.3.33 Burj Khalifa, Dubai, UAE
(Skidmore Owings & Merrill (SOM))

This 828 meter (2717 ft) reinforced concrete mixed-use tower is currently the tallest building in the world, with 163 stories above grade, and 2 levels of basement (Fig. 1.40). The top of the Tower consists of a structural steel spire utilizing a diagonally braced lateral system.

The floor plan of the tower consists of a tri-axial, "Y" shaped plan, formed by having three separate wings connected to a central core (Fig. 1.41). As the tower rises, one wing at each tier sets back in a spiraling pattern, further emphasizing its height. The Y-shape plan is ideal for residential and hotel use in that it allows the maximum views outward without overlooking a neighboring unit. The floor plan is also ideal for providing a high-performance, efficient structure.

The structural system is an innovative buttressed-core system, consisting of high-performance concrete wall construction. Each of the wings buttresses the others via a six-sided central core. This central core provides the torsional resistance of the structure. Corridor walls extend from the central core to near the end of each wing, terminating in thickened hammerhead walls. Perimeter columns and flat plate floor construction complete the structural system. At mechanical floors, outrigger walls are provided to link the perimeter columns to the interior wall systems, allowing the perimeter columns to participate in the lateral load resistance of the structure. Hence, all of the vertical concrete is utilized to support both gravity and lateral loads, resulting in a stiff and efficient structure.

As the building spirals in height, the wings setback to provide many different floor plates. The setbacks are organized with the tower's grid, such that the building stepping is accomplished by aligning columns above with walls below to avoid structural transfers. These setbacks also have the advantage of providing a different width to the tower for each differing floor plate. This stepping and

Figure 1.40   Burj Khalifa Tower.

shaping of the tower has the effect of "confusing the wind: wind vortices never get organized over the height of the building because at each new tier the wind encounters a different building shape (Fig. 1.42).

High-performance concrete with strengths ranging from C80 to C60 cube strength was used for the walls and columns. Wall thickness and column sizes were also fine-tuned to reduce the effects of creep and shrinkage on the structure. To reduce the effects of differential column shortening due to creep between the perimeter columns and interior walls, the perimeter columns were sized such that the self-weight gravity stress on the perimeter columns was equal to the stress on the interior corridor walls. The outriggers at the five mechanical floors tie all the vertical load-carrying elements together, further ensuring uniform gravity stress by essentially allowing the structure to redistribute gravity loads at five locations along the building's height, thereby reducing differential creep movements. With respect to concrete shrinkage, the perimeter columns and corridor walls were given

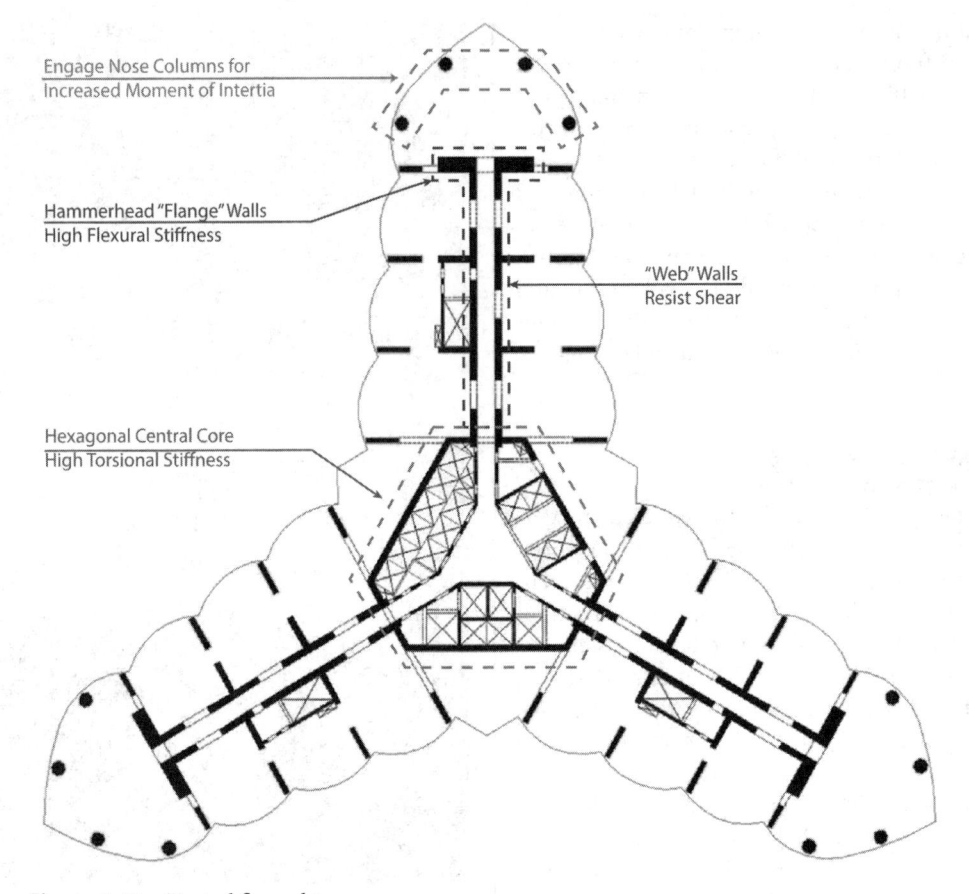

Engage Nose Columns for
Increased Moment of Intertia

Hammerhead "Flange" Walls
High Flexural Stiffness

"Web" Walls
Resist Shear

Hexagonal Central Core
High Torsional Stiffness

Figure 1.41   Typical floor plan.

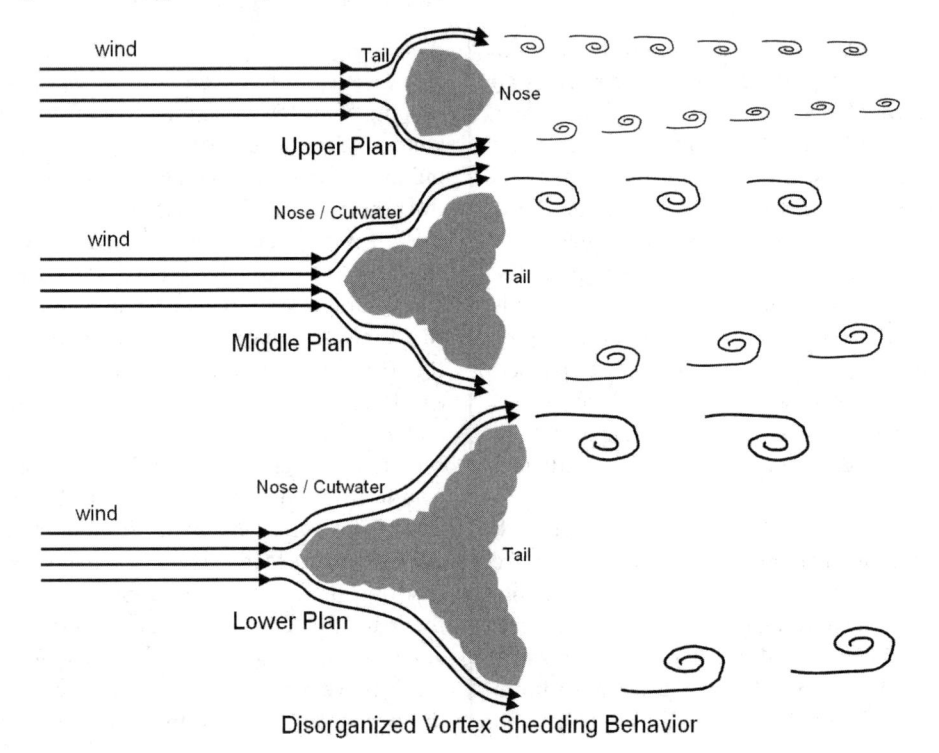

wind

Tail

Nose

Upper Plan

Nose / Cutwater

wind

Tail

Middle Plan

Nose / Cutwater

wind

Tail

Lower Plan

Disorganized Vortex Shedding Behavior

Figure 1.42   Tower wind behavior.

Figure 1.43 Tower construction photo.

matching thickness of 600 mm, which provided them with similar volume-to-surface ratios. This measure allows the columns and walls to generally shorten at the same rate due to concrete shrinkage.

The tower foundations consist of a pile-supported raft. The reinforced concrete raft is 3.7 m thick, and is supported by 194 bored cast-in-place friction piles, each 1.5 m in diameter and approximately 43 m long, with a capacity of 3000 tons. Figure 1.43 shows the tower during construction.

The structural engineering is by Skidmore, Owings, and Merrill, Chicago.

CITATIONS

Baker, W. F., Pawlikowski, J. J., and Young, B. S. (2010), *The Burj Khalifa Triumphs: Reaching toward the heavens*, Civil Engineering Magazine Archive, 80(3), 48-55.

Baker, W. F., Pawlikowski, J. J., and Young, B. S. (2009), *The Challenges in Designing the World's Tallest Structure: The Burj Dubai Tower*, Proceedings of Structures Congress 2009, Austin, TX, April 30-May 2, 2009.

1.3.34 Shum Yip Upperhills Tower 1, Shenzhen, China (Skidmore Owings & Merrill (SOM))

This 403 m (1321 ft) tall mixed-use tower, shown in Fig. 1.44, consists of class A office space in the lower 66 floors, a luxury hotel in the upper portion, and an observation deck and helipad on the top floor (Fig. 1.45).

The tower features an innovative ladder-core system (Fig. 1.46), which comprises a center-reinforced concrete core and eight perimeter steel-reinforced composite mega columns aligned with the outer core walls. Each mega column is engaged with the core by a composite coupling beam at every level, forming the ladder system. The ladder-core system allows the tower to meet its wind and seismic performance requirements without outriggers or perimeter bracing, thereby improving ductility, redundancy, and uniformity of structural stiffness.

At the office floors, deep built-up girders span between the mega columns on each face and extend as balanced cantilevers out to the corners. This directs gravity loads to the mega columns to eliminate tension produced by overturning moments and creates a column-free exterior facade (Fig. 1.47). Shallower floor framing members

Figure 1.44 Photo of tower.

span from the perimeter girders to the core, creating uninterrupted interior spaces and allowing a continuous mechanical loop around the floor. The core design is coordinated with the architectural layout to allow elevator lobbies within the core to serve as corridors, eliminating the conventional perimeter corridor and increasing floor efficiency (Fig. 1.48).

At mechanical levels, the floor framing is optimized for heavy loads by making use of the perimeter belt truss. The belt truss provides the required stiffness for mechanical floor operation, including vibration and acoustic performance. The large column-free interior spans also provide more flexibility for MEP planning. Absence of outrigger trusses within the floor plate enables more direct mechanical transfers.

The mechanical-level trusses above and below the hotel floors double as transfer mechanisms for the perimeter HSS posts at the hotel floors. The vertical members of the perimeter trusses above and below the guest rooms are aligned with the hotel guest room module, which supports the HSS posts in line with the demising partitions between the hotel rooms. The conventional framing at the hotel levels results in shallower perimeter beams and eliminates the need for long-span perimeter girders, reducing the height of the stories and the quantity of steel without compromising the architecture of the space. Figure 1.49 shows the construction sequence, and Fig. 1.50 shows a photo during construction.

The site has 2 floors of basements. The tower foundation utilizes mat bearing on slightly weathered rock.

The structural engineering is by Skidmore, Owings, and Merrill, New York City, and Capol International, Shenzhen, China.

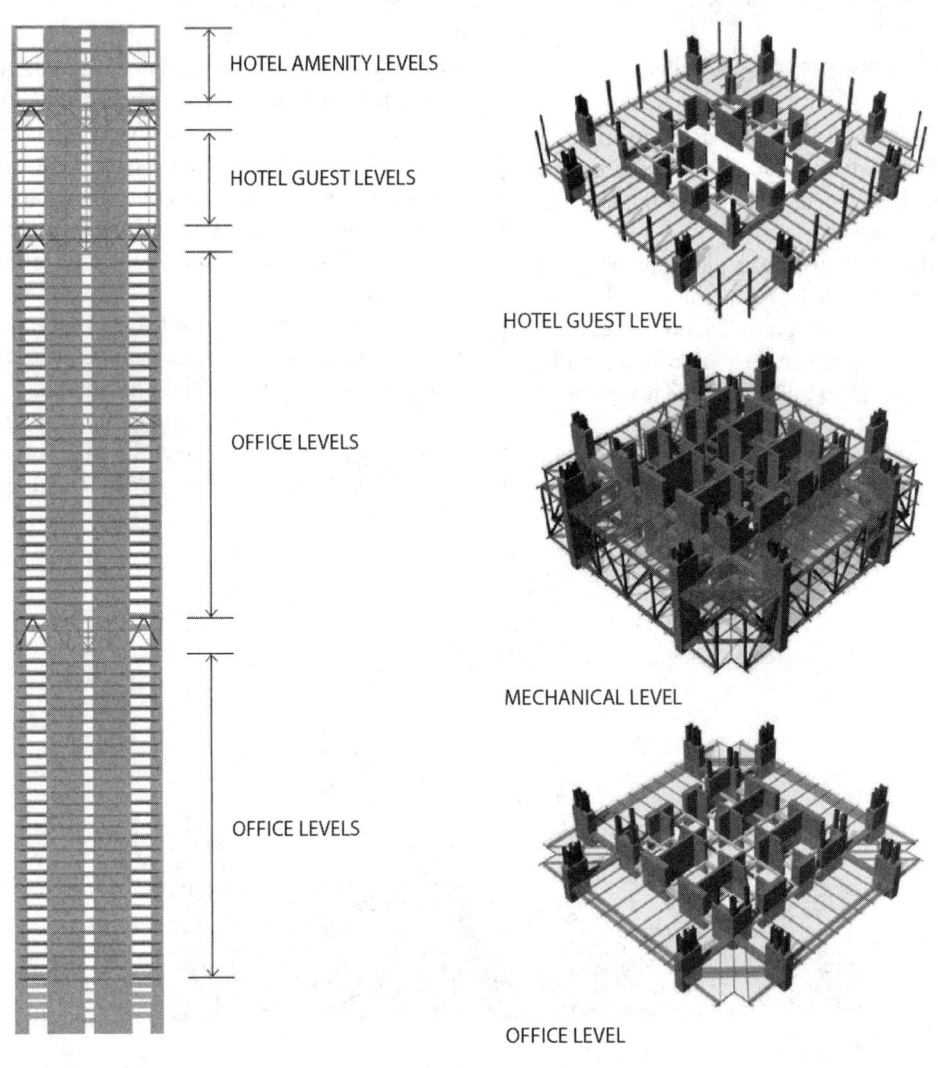

HOTEL AMENITY LEVELS

HOTEL GUEST LEVELS

OFFICE LEVELS

OFFICE LEVELS

HOTEL GUEST LEVEL

MECHANICAL LEVEL

OFFICE LEVEL

Figure 1.45   Typical framing.

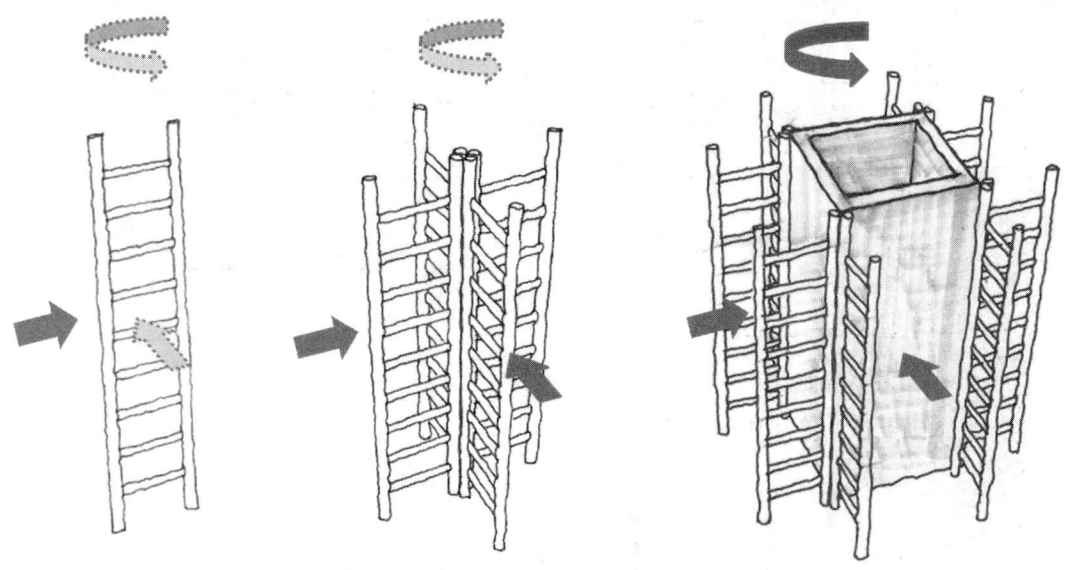

Figure 1.46   Lateral and torsional stiffness of: (a) single ladder, (b) bundled ladders, (c) ladder-core system.

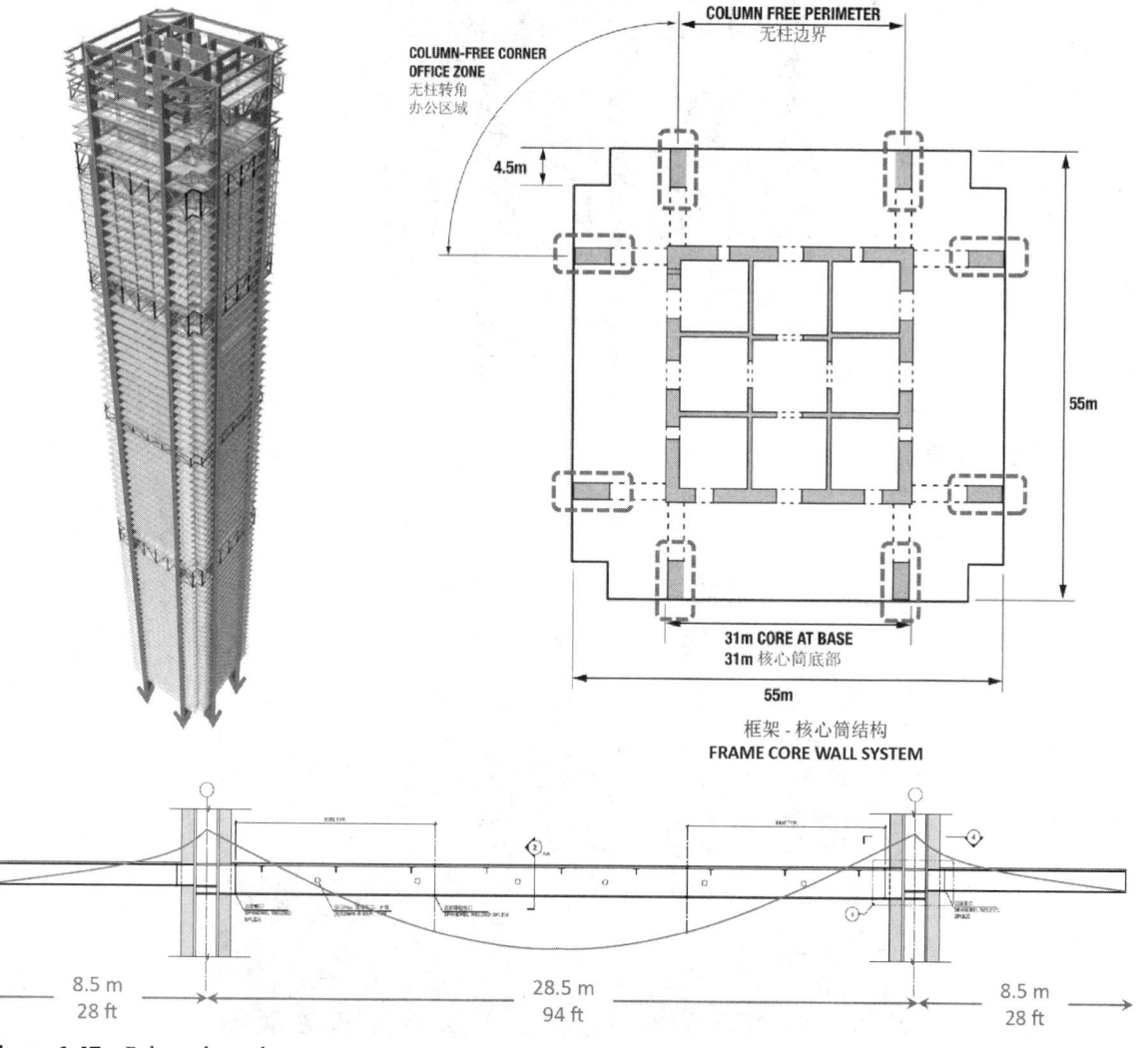

Figure 1.47   Balanced cantilever gravity system.

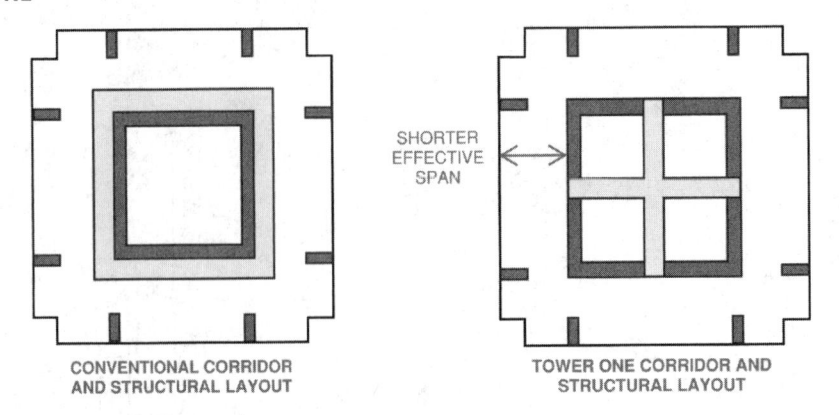

CONVENTIONAL CORRIDOR
AND STRUCTURAL LAYOUT

SHORTER
EFFECTIVE
SPAN

TOWER ONE CORRIDOR AND
STRUCTURAL LAYOUT

Figure 1.48   Tenant layouts.

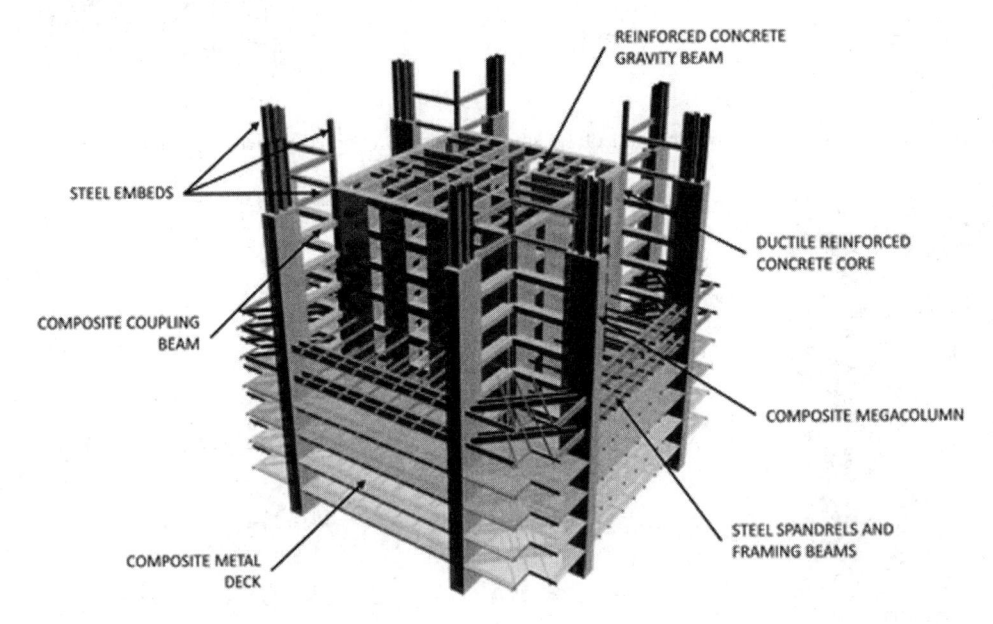

REINFORCED CONCRETE
GRAVITY BEAM

STEEL EMBEDS

DUCTILE REINFORCED
CONCRETE CORE

COMPOSITE COUPLING
BEAM

COMPOSITE MEGACOLUMN

STEEL SPANDRELS AND
FRAMING BEAMS

COMPOSITE METAL
DECK

Figure 1.49   Construction sequence.

Figure 1.50   Construction photo.

CITATIONS

Besjak, C., Haney, G., Biswas, P., Zhuang, J., and Petrov, G. I. (2019), *Shenzhen Shum-Yip: New Super Tall Systems Through A-E Collaboration.* Proceedings of AEI Conference 2019, Tysons, VA, April 3-6, 2019.

Besjak, C., Haney, G., Biswas, P., and Petrov, G. I. (2019), *A Ladder at Its Core.* Civil Engineering Magazine, 89(9), 68-76.

1.3.35 Cayan Tower (Formerly Infinity Tower), Dubai, UAE (Skidmore Owings & Merrill (SOM))

This 305-m (1000 ft) cast-in-place reinforced concrete tower consists of 73 residential stories above grade. The tower twists a full 90° from its base to its crown; this distinctive architectural and structural form is achieved through a series of incremental plan rotations at each level (Fig. 1.51).

The lateral load resisting system for the Tower consists of a combination of a reinforced concrete moment-resisting perimeter tube frame and a circular central core wall, connected at each level by the two-way spanning reinforced concrete flat plate slabs acting as diaphragms

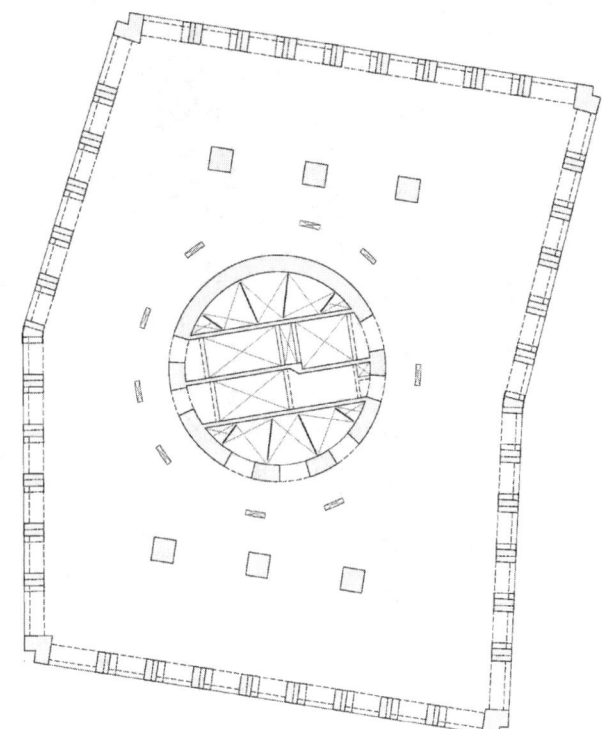

Figure 1.52 Typical floor plan.

(Fig. 1.52). This system maximizes the effective structural "footprint" of the Tower by utilizing a significant amount of the vertical reinforced concrete for lateral load resistance.

The engineers studied a series of options for the perimeter frame in order to create the unique twisting geometry of the Tower. Ultimately it was determined that there were distinct advantages from the standpoint of architectural efficiency, structural performance, and ease of construction, to stacking the columns in a step-wise manner at each level in order to generate the twisting building form.

As the perimeter columns ascend from story to story, they lean in or out, in a direction perpendicular to the slab edge. At every level, the columns make a small step to the side, shifting in position along the spandrel beams such that as the building twists, each column maintains a consistent position at each floor relative to the tower envelope. Careful consideration was given to the detailing of reinforcement at the column step (Fig. 1.53). The corner columns and the six (6) interior columns follow a different rule, twisting as they ascend.

This system offers significant construction simplification by permitting a high level of repetition in the formwork, reducing the construction cycle time. In addition,

Figure 1.51 Twisting Cayan tower.

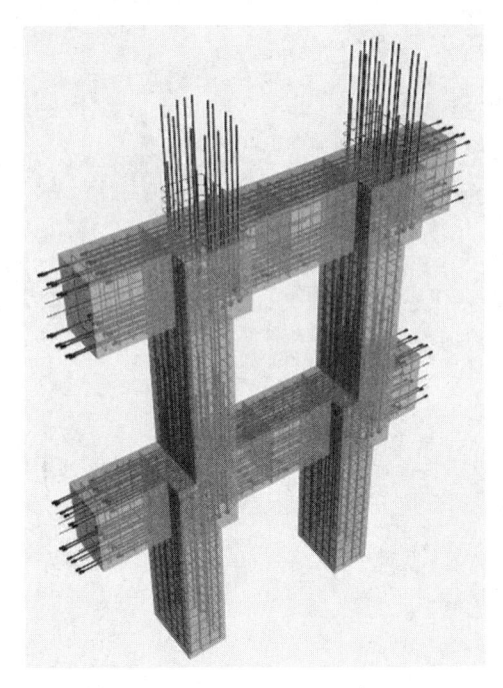

Figure 1.53   Stepped perimeter columns.

Figure 1.54   Construction photo.

the residential floor layouts are repetitive at each level despite the twisting nature of the building form.

There are six levels of below-grade parking and mechanical spaces. The tower is founded upon a 3-m thick reinforced concrete mat foundation which supported ninety-nine 1.2-m diameter bored, cast-in-place reinforced concrete piles extending approximately 30 meters below the mat foundation. The piles transfer the tower loads to the subgrade primarily through side friction. The subgrade consists of loose sands and sandstone bands overlaying cemented marine deposits and calcareous silt limestone/siltstone. Figure 1.54 shows the tower during construction.

The structural engineering is by Skidmore, Owings, and Merrill, Chicago.

CITATION

Baker, W. F., Brown, C. D., Young, B. S., and Zachrison, E. (2010). *Infinity Tower, Dubai, UAE*. Proceedings of Structures Congress 2010, Orlando, FL, May 12-15, 2010.

1.3.36   Lotte Super Tower, Seoul, South Korea (Skidmore Owings & Merrill (SOM))

Lotte Super Tower is a 555-m (1820 ft) unbuilt mixed-use tower, designed to be the tallest building in South Korea.

The tower is conceived as an architecturally expressed perimeter diagrid that transforms from a 70-m square base to a 39-m circle at the top (Fig. 1.55). Tapering of the tower in elevation decreases the wind-induced motions by varying the aeroelastic response along the height of the tower, "confusing" the wind. The tapering also decreases the wind sail area at the top of the building.

The diagrid is organized over multiple floor modules and is coupled with a ductile interior concrete core tube system which provides additional stiffness and damping to the overall system (Fig. 1.56).

Detailed studies were performed to optimize the angle of the diagrid. In its most simplistic form, a tall building is essentially a cantilever. When the trajectories of the principle stresses in a solid cantilever under wind loads are broken down, one can see that the forces at the base are primarily axial loads and thus want to be more vertical while the stressed at the top of the building are controlled by shear and thus want to be more horizontal. Applying this simple concept to the Lotte Super Tower, the optimum angles for the primary structural columns

Figure 1.55   Rendering of Lotte tower.

to strategically place the appropriate steel material at the proper location and angle were calculated. Since the diagrid runs through several floors before changing the angle at a node, the theoretical angles had to be rationalized and grouped as shown in Fig. 1.57. This led to a diagrid solution in which the angles vary over the height of the structure. The stiffness of the system eliminated the need for outriggers and belt trusses. In addition, finite elements analysis was performed to optimize the diagrid nodes, and a full-scale mockup was produced (Fig. 1.58).

At the typical office levels, wide flange structural steel floor framing beams span between the perimeter diagrid and the concrete core, supporting a reinforced slab on truss deck system (Fig. 1.59). This long-span floor framing increases the gravity loading in the perimeter, limiting tension in the perimeter columns under combined wind and gravity loads. A perimeter structural steel spandrel beam is connected to the diagrid diagonals and transfers gravity loads from the floor to the diagrid. A one-way concrete slab is utilized within the core.

The tower foundation consists of a reinforced concrete mat with a foundation wall directly under the diagrid

in order to uniformly distribute the diagrid forces while stiffening the mat against differential settlement. Structural steel columns are embedded within the foundation wall to provide tension capacity through the wall as well as to facilitate the construction of the diagrid.

The structural engineering is by Skidmore, Owings, and Merrill, Chicago.

CITATION

Besjak, C., Kim, B., and Biswas, P. (2009), *555M Tall Lotte Super Tower, Seoul, Korea*. Proceedings of Structures Congress 2009, Austin, TX, April 30-May 2.

1.3.37   7 South Dearborn, Chicago, Illinois, USA (Skidmore Owings & Merrill (SOM))

7 South Dearborn tower is an unbuilt mixed-use tower located in Chicago, Illinois. The proposed tower includes a parking structure for the first 12 floors, surmounted by 100 stories of office, residential, and telecommunications spaces above. Three structural steel interlaced antenna masts top the tower for a total height of 2000 ft (Fig. 1.60).

A unique feature of this supertall tower, which would have been the tallest in the already impressive Chicago skyline, is what was termed the "stayed-mast" structural system (Fig. 1.61). The "mast" of the tower is the central square reinforced concrete core, which consists of 48" thick walls that extend for the full height of the tower without ever changing shape. From the foundations to the top of the parking levels, a perimeter reinforced concrete wall, the "silo," stiffens the main tower core wall. From the top of the parking levels to the top of the office floors, exterior steel columns (the "stays") engaged by multistory outrigger trusses stiffen the concrete core. These four-story outrigger trusses occur only at the mechanical levels and link the core and perimeter. From the bottom of the residential (level 53) and telecommunications floors to the roof, the core alone resists all imposed vertical and lateral loads providing a column-free space. At the top of the tower, where the core wall overlaps the antennae structure, tuned liquid column damper tanks are provided to improve the service level performance of the structure.

The lower retail, ground level, and parking floors utilize a conventional reinforced concrete beam and slab system. At the office floors and in the core throughout the entire height of the tower, composite structural steel floor framing is utilized. For the residential floors, cantilevered post-tensioned framing is utilized. Small diameter steel pipe "spacer" columns connect the ends of

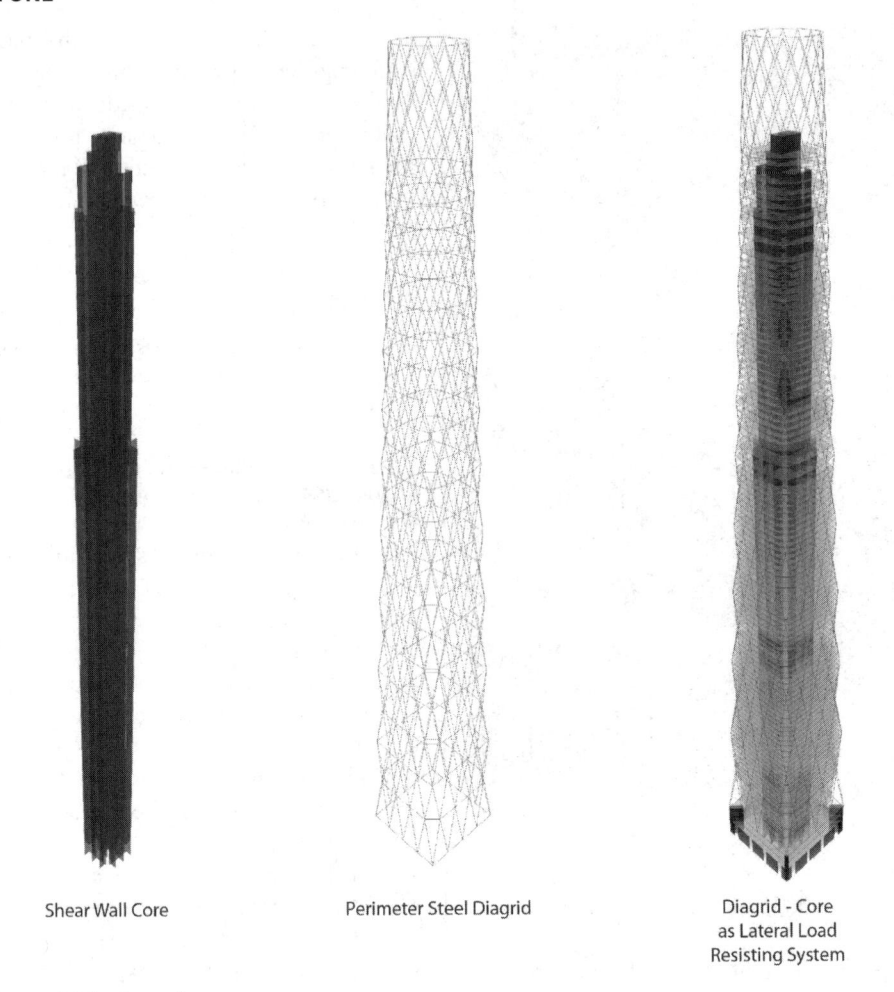

Shear Wall Core   Perimeter Steel Diagrid   Diagrid - Core
                                                                        as Lateral Load
                                                                        Resisting System

Figure 1.56   Lateral system.

the cantilevered floor beams between floors in order to reduce the differential live load deflection of the beams at the facade. The telecommunication floors consist of cantilevered reinforced concrete slabs. Figure 1.62 shows the typical framing plans.

The foundation system consists of a mat slab that sits on straight-shaft caissons socketed into the bedrock.

The structural engineering is by Skidmore, Owings, and Merrill in Chicago.

1.3.38   Nanjing Greenland Financial Center (Skidmore Owings & Merrill (SOM))

Nanjing Greenland Financial Center is a 450-m (1476 ft) mixed-use tower consisting of 70 stories of office and hotel space (Fig. 1.63). The hotel space includes a sky lobby and restaurants on the upper levels, and the tower also includes a public observation deck on level 59.

The lateral system consists primarily of a triangular interior reinforced concrete core coupled with exterior composite columns via steel outrigger and belt trusses. The closed form of the perimeter core walls provides a large amount of the overall torsional stiffness of the building. To provide structural redundancy and additional torsional stiffness, a secondary lateral system of a moment-resisting frame is utilized at the perimeter of the building (Fig. 1.64). Steel floor beams span between the core and perimeter, supporting composite metal deck floors.

Careful consideration was given to the alignment of the core and perimeter elements. The interior core walls are aligned with the columns to simplify and facilitate the connection of the outrigger trusses (Fig. 1.65). On mechanical floors, the floor framing is oriented parallel to the outrigger trusses to simplify the layout. On typical floors, the floor framing is oriented perpendicular to the

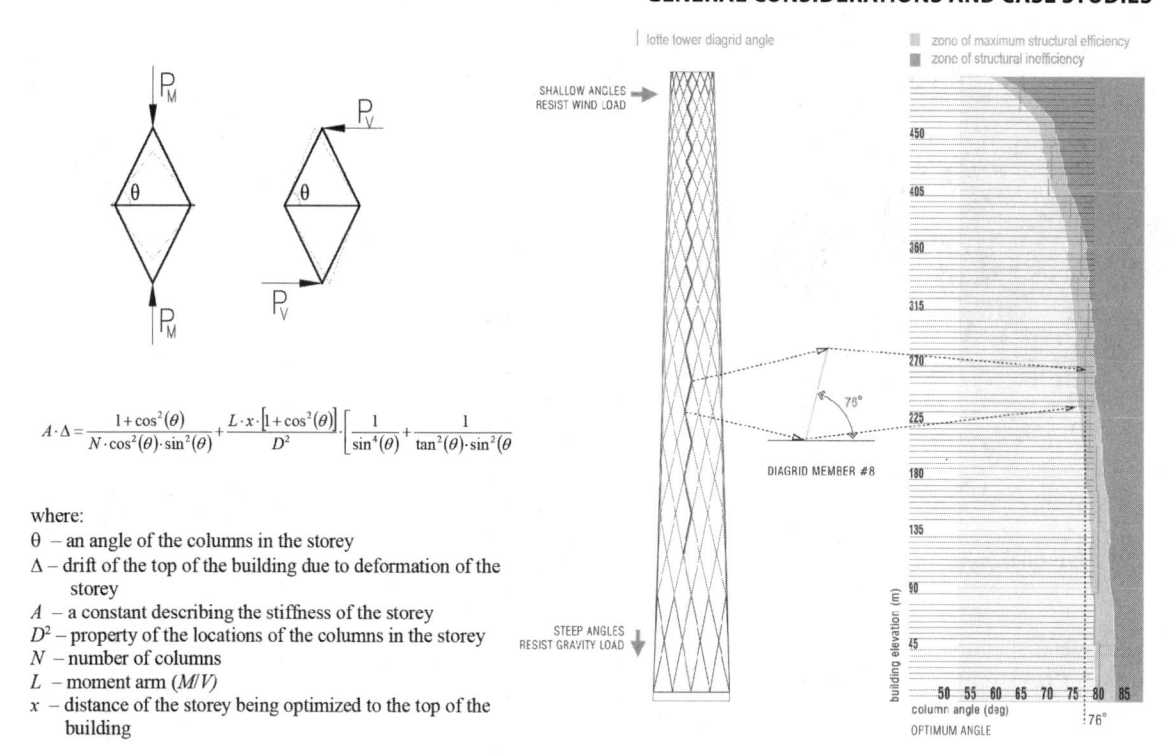

$$A \cdot \Delta = \frac{1+\cos^2(\theta)}{N \cdot \cos^2(\theta) \cdot \sin^2(\theta)} + \frac{L \cdot x \cdot \left[1+\cos^2(\theta)\right]}{D^2} \cdot \left[\frac{1}{\sin^4(\theta)} + \frac{1}{\tan^2(\theta) \cdot \sin^2(\theta)}\right]$$

where:

θ – an angle of the columns in the storey

Δ – drift of the top of the building due to deformation of the storey

A – a constant describing the stiffness of the storey

D^2 – property of the locations of the columns in the storey

N – number of columns

L – moment arm (M/V)

x – distance of the storey being optimized to the top of the building

Figure 1.57 Theoretical optimum angle solution and rationalized.

Figure 1.58 Diagrid node detailing.

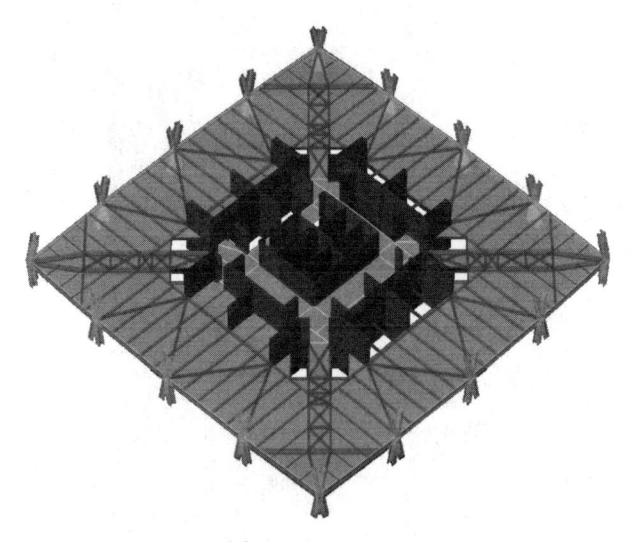

Figure 1.59   Typical framing.

core and perimeter spandrel to minimize material. As the floor plate steps back on the upper levels, the perimeter columns are aligned with interior web walls below to eliminate transfers (Fig. 1.66).

Figure 1.60   Tower rendering.

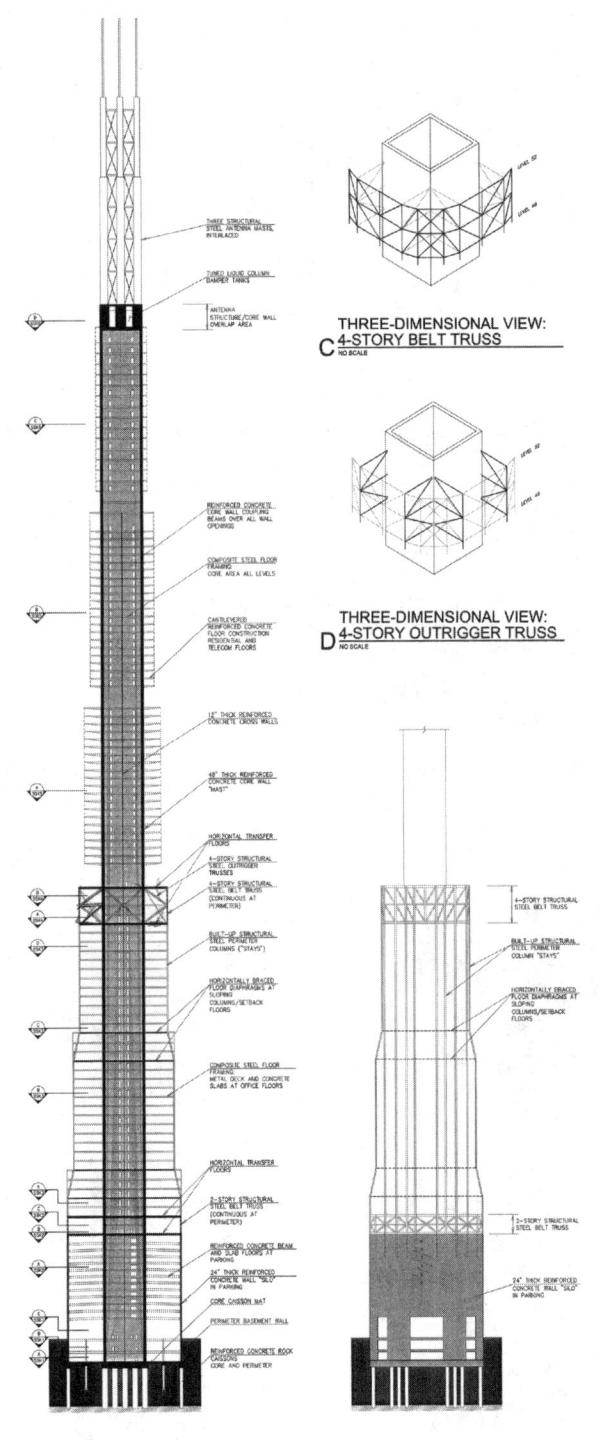

THREE-DIMENSIONAL VIEW:
C **4-STORY BELT TRUSS**
NO SCALE

THREE-DIMENSIONAL VIEW:
D **4-STORY OUTRIGGER TRUSS**
NO SCALE

Figure 1.61   Structural system.

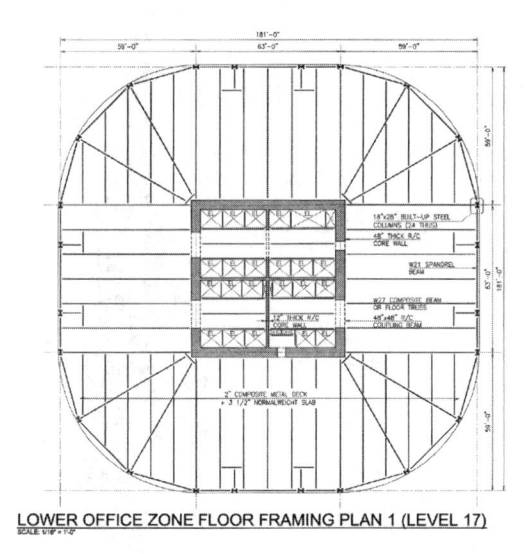

LOWER OFFICE ZONE FLOOR FRAMING PLAN 1 (LEVEL 17)
SCALE: 1/16" = 1'-0"

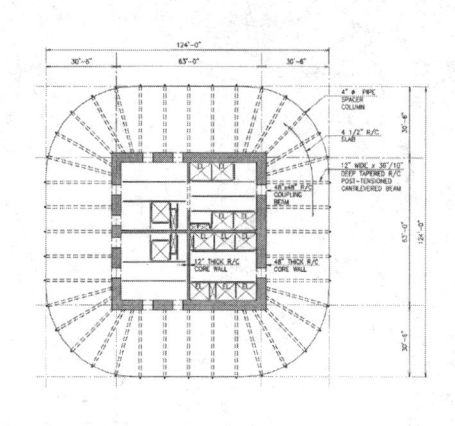

RESIDENTIAL FRAMING PLAN 1 (LEVELS 53 TO 72)
SCALE: 1/16" = 1'-0"

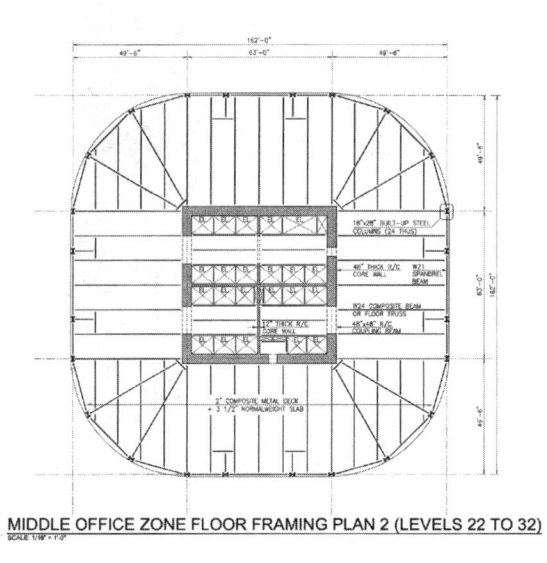

MIDDLE OFFICE ZONE FLOOR FRAMING PLAN 2 (LEVELS 22 TO 32)
SCALE: 1/16" = 1'-0"

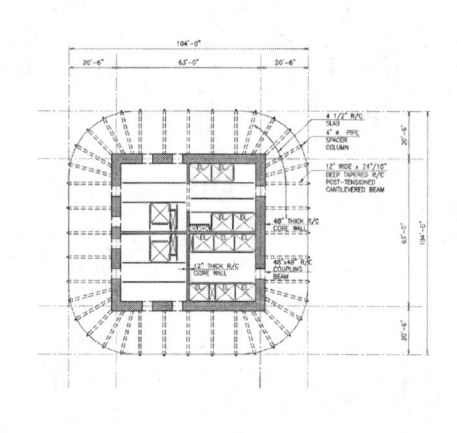

RESIDENTIAL FRAMING PLAN 2 (LEVELS 74 TO 94)
SCALE: 1/16" = 1'-0"

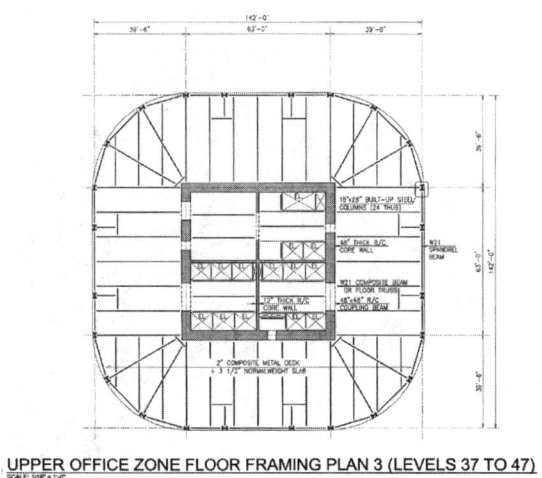

UPPER OFFICE ZONE FLOOR FRAMING PLAN 3 (LEVELS 37 TO 47)
SCALE: 1/16" = 1'-0"

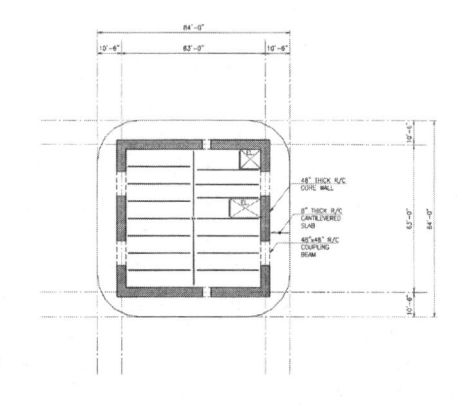

TELECOM FLOOR FRAMING PLAN (LEVELS 98 TO 111)
SCALE: 1/16" = 1'-0"

Figure 1.62 Tower plans.

Figure 1.63   Photo of tower

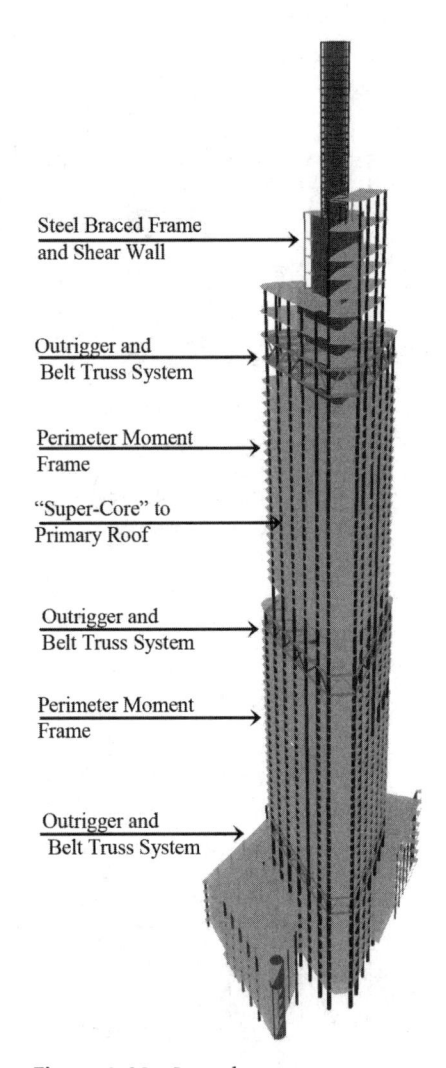

Steel Braced Frame
and Shear Wall

Outrigger and
Belt Truss System

Perimeter Moment
Frame

"Super-Core" to
Primary Roof

Outrigger and
Belt Truss System

Perimeter Moment
Frame

Outrigger and
Belt Truss System

Figure 1.64   Lateral system.

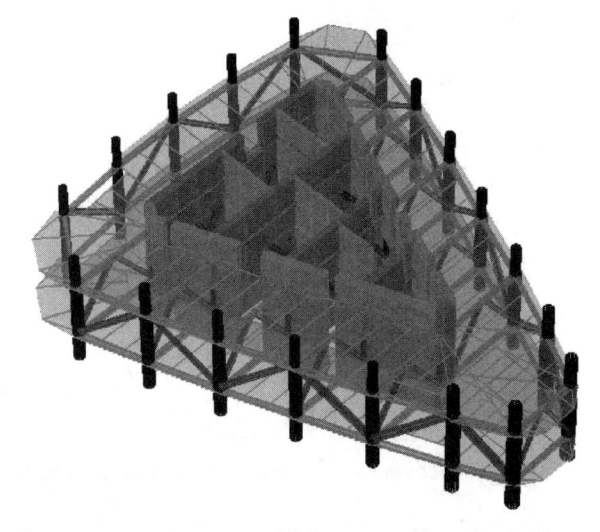

Figure 1.65   Outrigger and belt truss configuration.

The four levels below grade levels were constructed of reinforced concrete using a temporary, internally-braced slurry wall retention system. A permanent reinforced concrete foundation wall was then constructed inside of the slurry wall system. The tower foundation consists of a reinforced concrete mat supported on cast-in-place reinforced concrete belled caissons embedded in the underlying rock. Figure 1.67 shows a construction photo of the Outrigger Truss.

The structural engineering is by Skidmore, Owings, and Merrill, Chicago and East China Architectural Design & Research Institute (ECADI).

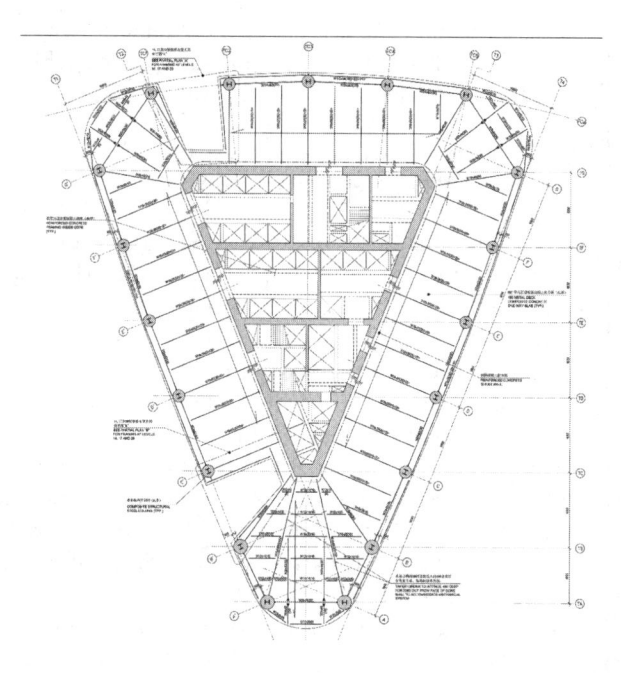

Typical Office Floor

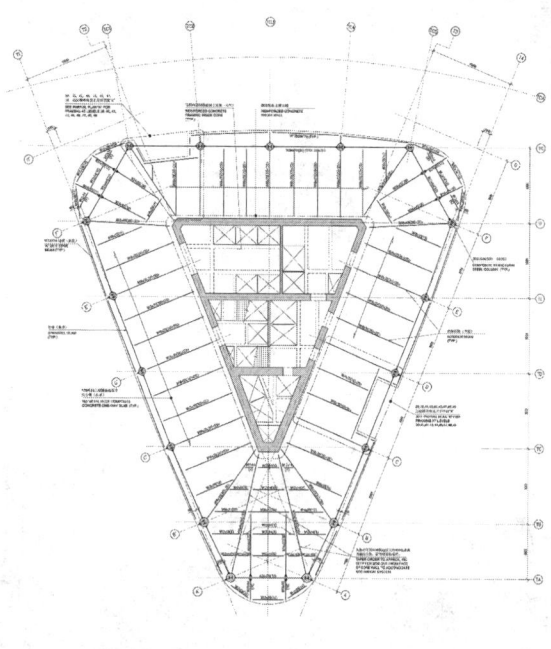

Typical Hotel Floor

Figure 1.66   Typical floor framing plans.

Figure 1.67   Photo of outrigger truss construction.

CITATION

Besjak, C. M., McElhatten, B. J., and Biswas, P. (2010), *Performance-Based evaluation for the 450m Nanjing Greenland financial center main tower.* Proceedings of Structures Congress 2010, Orlando, FL, May 12-15.

1.3.39   Shenzhen Rural Commercial Bank HQ, Shenzhen, China (Skidmore Owings & Merrill (SOM))

SHENZHEN RURAL COMMERCIAL BANK HQ, SHENZHEN, CHINA

The Shenzhen Rural Commercial Bank Headquarters is a 150 m (492 ft) tall, 34-story office tower in Shenzhen, China.

The key feature of this building is the architecturally exposed Exo-Diagrid frame (Fig. 1.68). The frame is set outside the building enclosure to allow for a column-free floor plate, and provides shading to the interior, improving the building energy performance. The lateral system of the tower consists of the perimeter diagrid frame and a reinforced concrete core (Fig. 1.69). Bracing of the diagrid system is provided by diagonal beams from the corner columns to the core and by box beams that extend from diaphragm trusses at every other floor (Figs. 1.70 and 1.71), allowing for full interaction between the core and

Figure 1.68   Photo of the tower.

the diagrid. The perimeter spandrel beams, diaphragm trusses, and corner diagonal beam resolve the hoop tension forces at the corner diagrid nodes. The diagrid is a redundant system with significant capability to redistribute loads, allowing for stability even with the removal of any one diagrid member. The diagrid is concrete-encased to provide both corrosion and fire protection.

The gravity floor framing consists of steel floor framing spanning between the building core and perimeter diagrid, supporting composite metal deck floors. Within the core, the gravity floor framing consists of reinforced concrete beams and slabs.

The foundation system for the tower is a reinforced concrete mat supported by reinforced concrete hand-dug enlarged base cast-in-situ piles. Slurry foundation walls surround the basement levels.

The structural engineering is by Skidmore, Owings, and Merrill, New York, and Beijing Institute of Architectural Design, China.

1.3.40   CITIC Financial Center Tower 1, Shenzhen, China (Skidmore Owings & Merrill (SOM))

This 312-m (1024 ft) mixed-use tower has 65 stories of office and luxury condo above grade (Fig. 1.72). The shape of the tower tapers from a square floor plate at the base to a more circular floor plate with large corner chamfers at the top to optimize wind and seismic performance (Fig. 1.73).

The lateral system consists of a composite steel/concrete perimeter frame and reinforced concrete central core. The perimeter frame is an innovative system comprising a series of tie-braced frames that simultaneously meet the stiffness and ductility requirements for the site, which has high wind and seismic loads. The perimeter frame is architecturally celebrated on the facade.

The geometry of the braced frame follows the optimal layout for a cantilever beam loaded at the top, to attain maximum stiffness for the minimum weight. Figures 1.74 and 1.75 show a series of optimal layouts with a variety of members' densities n for a cantilever with aspect ratio 1:10. The tower's braced frame configuration is based on a density of $n = 3$, and minor geometric manipulations are made for architectural requirements. The frame is composed of concrete-filled rectangular tubes (CFT), using high-strength structural steel to minimize member sizes. The configuration of the brace layout also creates opportunity for a grand entrance to the tower at the base.

At the four corners of the tower, pairs of adjacent columns at the adjoining facades are connected by ductile steel moment-connected links at each floor. These links are sized to remain elastic for wind and frequent seismic loading and to yield at the rare seismic event. The yielding of the links limits the maximum force transferred to the braced frame and dissipates seismic energy under cyclic loading.

The gravity floor framing consists of steel floor framing spanning between the building core and perimeter supporting composite metal deck floors. Within the core, the gravity floor framing consists of reinforced concrete beams and slabs.

The perimeter braced frame and core walls above-grade transform to a system of columns and shear walls within the four levels of the basement. The foundation is a reinforced concrete mat supported on piles extending down to a suitable bearing layer.

The structural engineering is by Skidmore, Owings, and Merrill, San Francisco.

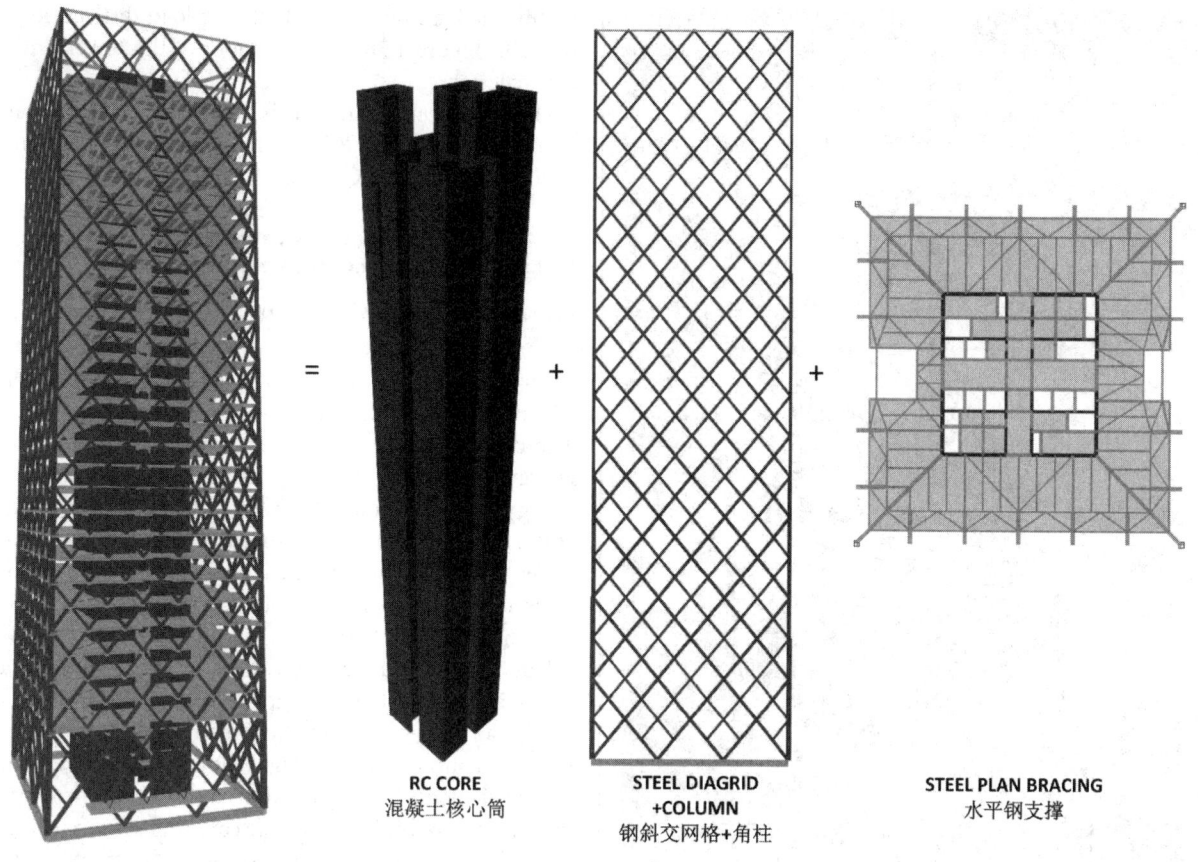

RC CORE
混凝土核心筒

STEEL DIAGRID
+COLUMN
钢斜交网格+角柱

STEEL PLAN BRACING
水平钢支撑

Figure 1.69   Structural system.

CITATION

Beghini, A., Sarkisian, M. P., Mathias, N., and Baker, W. F. (2015), *Structural Optimization for Stiffness and Ductility of High-rise Buildings*. Proceedings of Structures Congress 2015. Portland, OR, April 23-25.

1.3.41   Tianjin CTF Financial Center, Tianjin, China (Skidmore Owings & Merrill (SOM))

Tianjin CTF Financial Center is a 530 m (1738 ft) skyscraper housing a five-star hotel, service apartments, and office space (Fig. 1.76).

The lateral system of the tower consists of a stepped reinforced concrete core and eight groups of sloped perimeter megacolumns (Fig. 1.77). This hybrid system is able to withstand wind and seismic forces without the need of outriggers. The megacolumns at the perimeter slope to follow the changing floor plates, helping to define the curved geometry of the tower. They are designed as steel-reinforced concrete (SRC) elements at the upper levels and concrete-filled tube (CFT) elements at the lower levels. The steps for the core occur at the mechanical levels which also have partial belt trusses to increase the stiffness of the perimeter frame. The core stiffness transition is designed to occur at the step location while maintaining a smooth shear and axial load the rest of the way. Steel shapes are embedded at the ends and intersections of the core walls to improve the ductility of the core. Additionally, there is a belt truss located at the mid-height of the tower which acts as a transfer truss to reduce net uplift in the perimeter mega columns as well as provide increased stiffness of the perimeter.

The gravity system for the tower is a metal deck slab on structural steel framing with a reinforced concrete slab inside the core.

The tower foundation consists of a reinforced concrete mat on drilled piles, with buttress walls added in the basement to provide additional stiffness to the core. There are also reinforced concrete walls tied to the mega columns

Figure 1.70 Diagrid construction.

Figure 1.71 Photo of diagrid detail.

within the basement to facilitate the load distribution to the mat. Figure 1.78 shows a photo of the tower during construction.

The structural engineering is by Skidmore, Owings, and Merrill, Chicago and East China Architectural Design & Research Institute (ECADI).

1.3.42 Pearl River Tower, Guangzhou, China (Skidmore Owings & Merrill (SOM))

PEARL RIVER TOWER, GUANGZHOU, CHINA

Pearl River Tower shown, in Fig. 1.79, is a 71-story, 309 m (1013 ft) office building in Guangzhou, China. The overall goal for the project was to create the most energy-efficient building possible, resulting in a reduced energy usage by about 40% from a typical office building at the time. For this reason, the geometry of the building is very much defined by the two sets of wind turbines housed at roughly one-third and two-thirds of the building height.

Due to the wind loads, seismic loads and geometry of the building, resistance to lateral loads perpendicular to the broad face of the building required several systems in combination, shown in Fig. 1.80. The lateral resistance perpendicular to the broad face is a dual system comprising a central reinforced concrete core wall system which is linked to the exterior columns by a series of structural steel outrigger and belt trusses, and composite megacolumns linked by diagonal mega-bracing at the two narrow faces of the building. The cross walls of the core are aligned with the exterior columns to provide the best possible link and transfer of load between lateral system vertical elements. Lateral loads are split almost equally between the two lateral systems. The lateral resistance perpendicular to the narrow face is primarily the reinforced concrete core wall system. A secondary lateral system in this direction consists of structural steel moment frames at the office floors and the structural steel belt trusses at the mechanical levels. Additional torsional stiffness is generated by the perimeter moment frames and mega-bracing. Structural steel outriggers and belt trusses occur above the wind turbines. The belt trusses aid in providing uniform distribution between exterior columns, and coupled with the perimeter moment frames increase the redundancy and robustness of the overall structural system. The megacolumns on each end of the building consist of concrete-encased structural steel section up to Level 60 at which point the structural steel continues alone. Mega-bracing consists of structural steel sections joining the mega columns together.

Figure 1.72 CITIC Financial Center Towers 1 & 2.

The Tower gravity-load resisting system consists of steel floor framing supporting concrete slabs on metal deck. One-way reinforced concrete floor framing is utilized within the core. The central reinforced concrete core wall system along with the interior and exterior columns support the floor framing system and transmit the loads to the foundations. The steel framing on each column line allows for erection and bracing of the leaning exterior steel columns. Typical floor framing plan is shown in Fig. 1.81.

The foundation system for the Tower consists of a reinforced concrete mat bearing on slightly weathered rock under the core wall system, and reinforced concrete footings under all the columns and mega-columns.

The structural engineering is by Skidmore, Owings, and Merrill, Chicago, and Guangzhou Design Institute, Guangzhou, China. Figure 1.82 shows a photo of the tower during construction.

CITATION

Baker, W., Besjak, C., McElhatten, B., and Li, X. (2014), *Pearl River Tower: Design Integration towards Sustainability*. Proceedings of Structures Congress 2014, Boston, MA, April 3-5.

1.3.43 One Manhattan West, New York City, USA (Skidmore Owings & Merrill (SOM))

This 304-m (997 ft) office tower consists of 67 office floors above ground level. The tower is rectangular in plan with the north, south, and west faces rising vertically up from the ground. The east face bows out until the 16th floor and then tapers in uniformly to the roof (Fig. 1.83). The tower's lateral system is a reinforced concrete core with steel perimeter columns and moment frame.

The site of the tower and the larger Manhattan West development is over the active railroads to the west of New York's Penn Station. Concrete box girders cover the rails to support a public plaza and provide a suitable working platform for the construction of the tower. The tower's reinforced concrete core, measuring approximately 18 m by 45 m (60 ft by 150 ft) at the base, reaches terra-firma and is socketed into solid rock. The perimeter columns on the south side of the tower footprint cannot reach ground level and thus are supported back to the core at the tower level 6 (approximately 28 m, or 90 ft, above the plaza level). As the transfer of the gravity loads

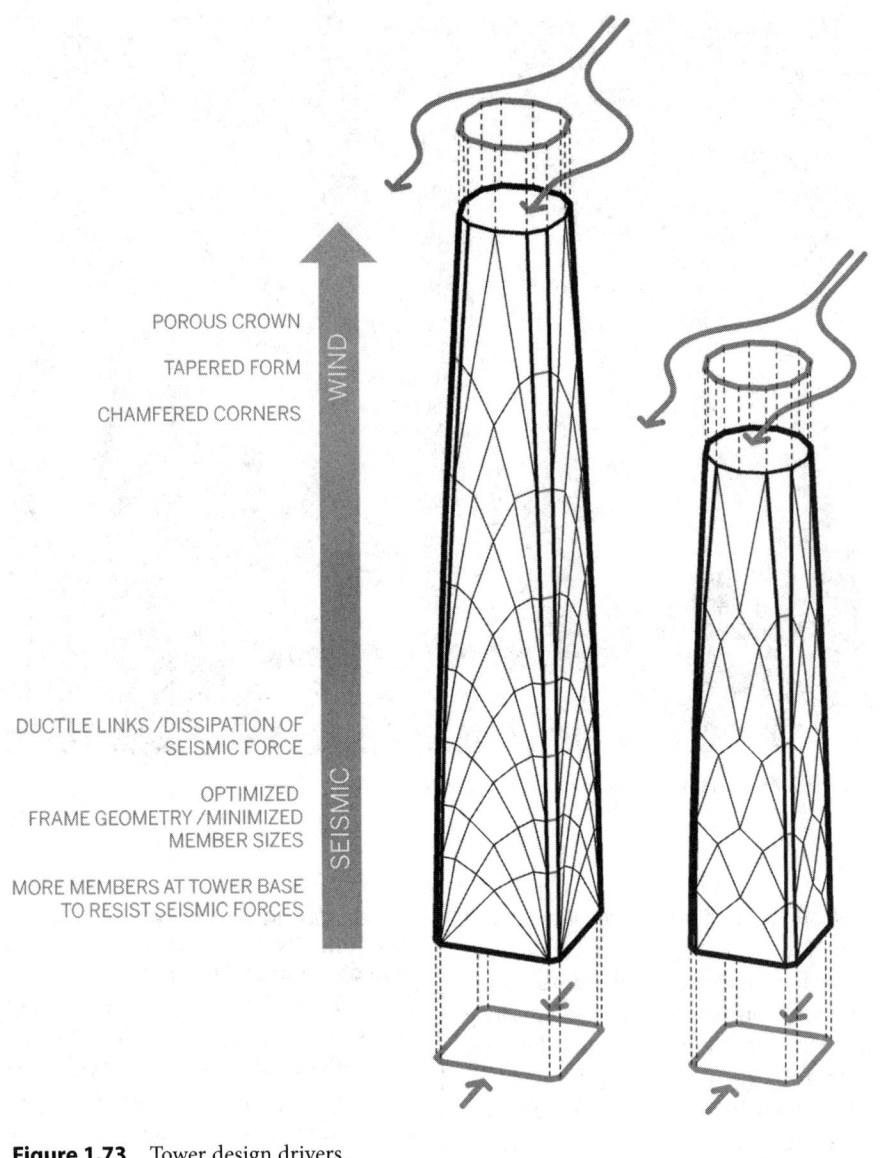

Figure 1.73 Tower design drivers.

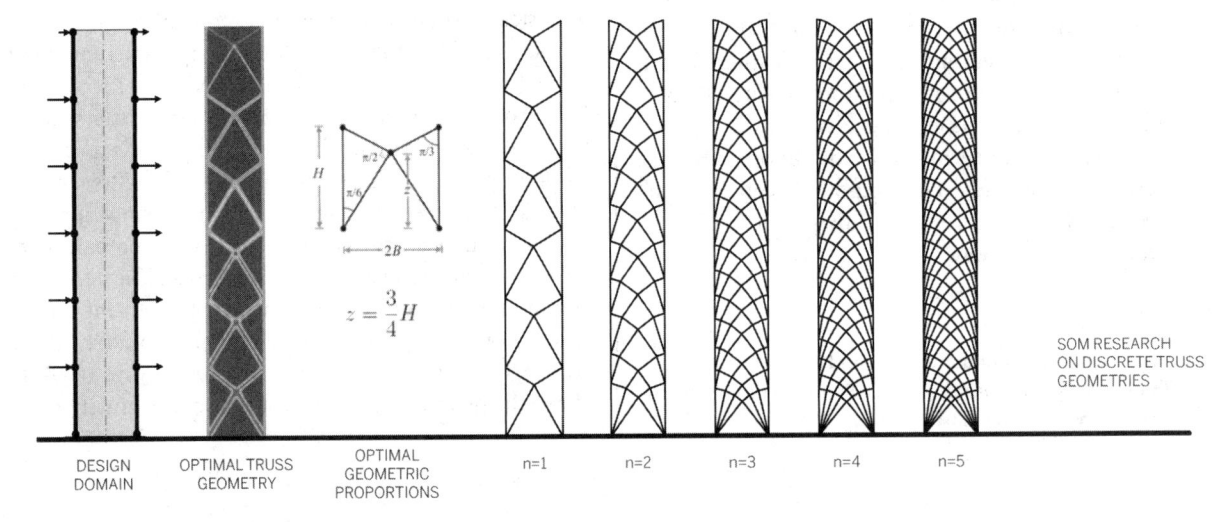

Figure 1.74 Optimal layouts for a cantilever beam for a variety of members' densities n.

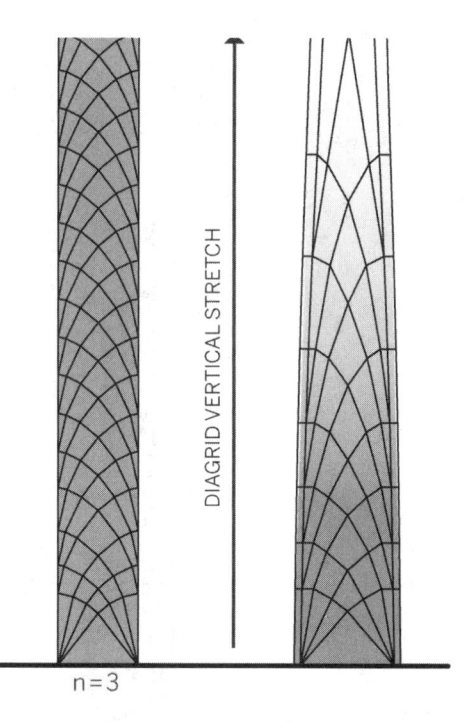

Figure 1.75 Modification of optimal brace layout for tower.

Figure 1.76 Photo of tower.

at the perimeter introduces a significant lateral load into the building core, the perimeter structure on the north side is also "kicked" back into the core above the lobby level, resulting in a balanced structure (Fig. 1.84). Thus, the lateral system at the base of this supertall tower is only the reinforced concrete core, resulting in one of the most slender structures in New York City, with an aspect ratio of 1:16. The site limitation was converted into an opportunity to create a column-free lobby at the base of the tower.

In the regular zone above level 6, 1MW is a relatively conventional tower with a reinforced concrete core and perimeter steel moment frame. A perimeter moment frame at the office floors provides additional building stiffness. To enhance structural resilience, and reduce lateral drift, a perimeter belt truss is located at the top of the tower.

The gravity and lateral loads of the core are resisted by continuous reinforced concrete strip footings embedded into bedrock, with high-strength steel rock anchors to resist net uplift.

The structural engineering is by Skidmore, Owings, and Merrill, New York City. Figure 1.85 shows a photo of the tower during construction.

CITATION

Petrov, G. I., Biswas, P., Johnson, R. B., Seblani, A., and Besjak, C. (2018). *Supertall Over the Train Tracks – One Manhattan West Tower*. Structural Engineering International, 29(1), 116-122.
Biswas, P., Petrov, G., Shen, Y., Wilson, S., and Besjak, C. *Manhattan West: Converting Site Challenges into Design Opportunities*, Proceedings of 2019 IABSE Congress New York.

1.3.44 A25 Xinyi Tower (ARUP)

A25 Xinyi Tower is a 56-story commercial office in Taipei, Taiwan (Fig. 1.86). The Design Architects are Renzo Piano Building Workshop working with Kris Yao—Artech in Taipei. Structural Engineers are Ove Arup and Partners International Ltd, London, and Evergreen Consulting Engineering in Taiwan.

Taipei is highly seismic but is also subject to regular typhoons giving rise to some of the highest wind loading on earth. Wind sculpting and careful proportioning were therefore important steps in limiting shear and overturning demand in this project. However, seismic demands and ductility requirements are also severe.

The building is 265 m tall and occupies a footprint of around 42×42 m (Fig. 1.87), giving a slenderness ratio of around 6.3. The massing includes substantially cut-back corners to reduce the cross-wind response. Nonetheless,

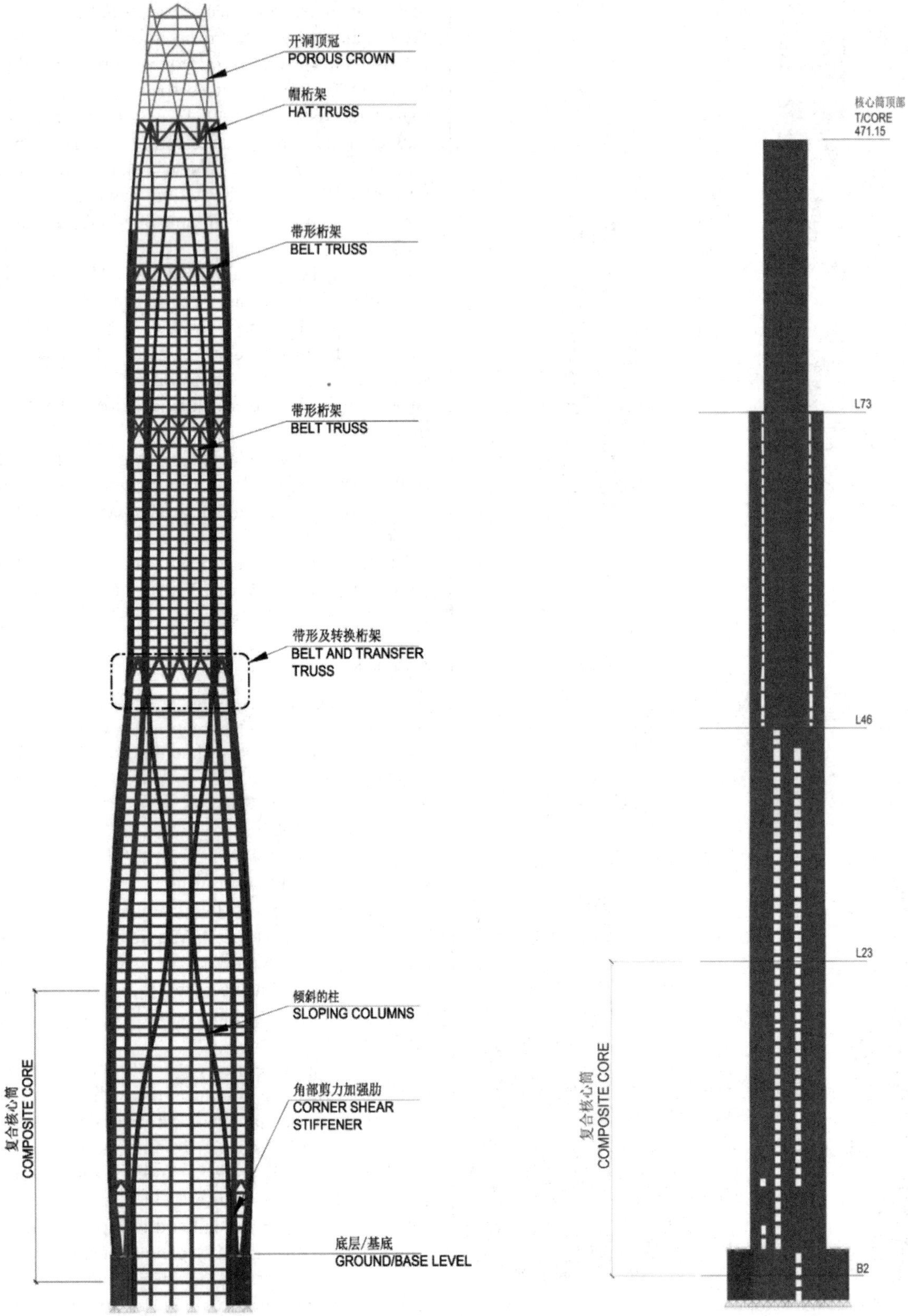

开洞顶冠
POROUS CROWN

帽桁架
HAT TRUSS

带形桁架
BELT TRUSS

带形桁架
BELT TRUSS

带形及转换桁架
BELT AND TRANSFER
TRUSS

倾斜的柱
SLOPING COLUMNS

角部剪力加强肋
CORNER SHEAR
STIFFENER

底层/基底
GROUND/BASE LEVEL

复合核心筒
COMPOSITE CORE

核心筒顶部
T/CORE
471.15

L73

L46

L23

B2

复合核心筒
COMPOSITE CORE

Figure 1.77   Lateral system.

Figure 1.78   Tower construction photo.

Figure 1.79   Photo of Pearl River tower.

net (factored) design wind loads are 8.4 kN/m² when averaged for shear and 10.8 kN/m² when back calculating from the overturning. The factored wind shear represents a substantial 11%*g*.

The applicable PGA for the location in Taipei is 0.21g, with a minimum shear strength requirement of 7.5%*g*. Wind loading therefore defines the minimum lateral strength requirement. However, elastic seismic demands at design basis earthquake, DBE, and maximum considered earthquake, MCE, are significantly higher, and a highly ductile system is required. The practical design implication of the wind and seismic loading for a tower of these proportions in that location is that the design of the ductile (yielding) elements is governed by wind. However, the ductile elements yield and strain harden during DBE and MCE-level earthquakes.

The maximum loading in the columns and foundations is therefore governed by seismic response. The effect of the high wind loading is to substantially increase the overturning demand due to the seismic response. In

effect, both wind and seismic loading govern, but only reduction in wind loading allows a reduction of required design strength and material use.

Global demands on the lateral system are summarized below. This follows the beneficial effect of sculpting the corners of the project. The building is configured to be strength governed, whereby drift limits are met with a structure optimized for strength. Damping is provided for wind comfort with a tuned mass damper (TMD).

Wind shear unfactored (factored)	60 MN (96 MN)
Wind overturning unfactored (factored)	10 GNm (16 GNm)
Seismic DBE elastic shear	139 MN
Seismic DBE elastic overturning	22 GNm
Seismic minimum strength shear	64 MN
Seismic minimum strength overturning	11 GNm
Seismic overstrength shear	150 MN
Seismic overstrength overturning	25 GNm
Weight	870 MN

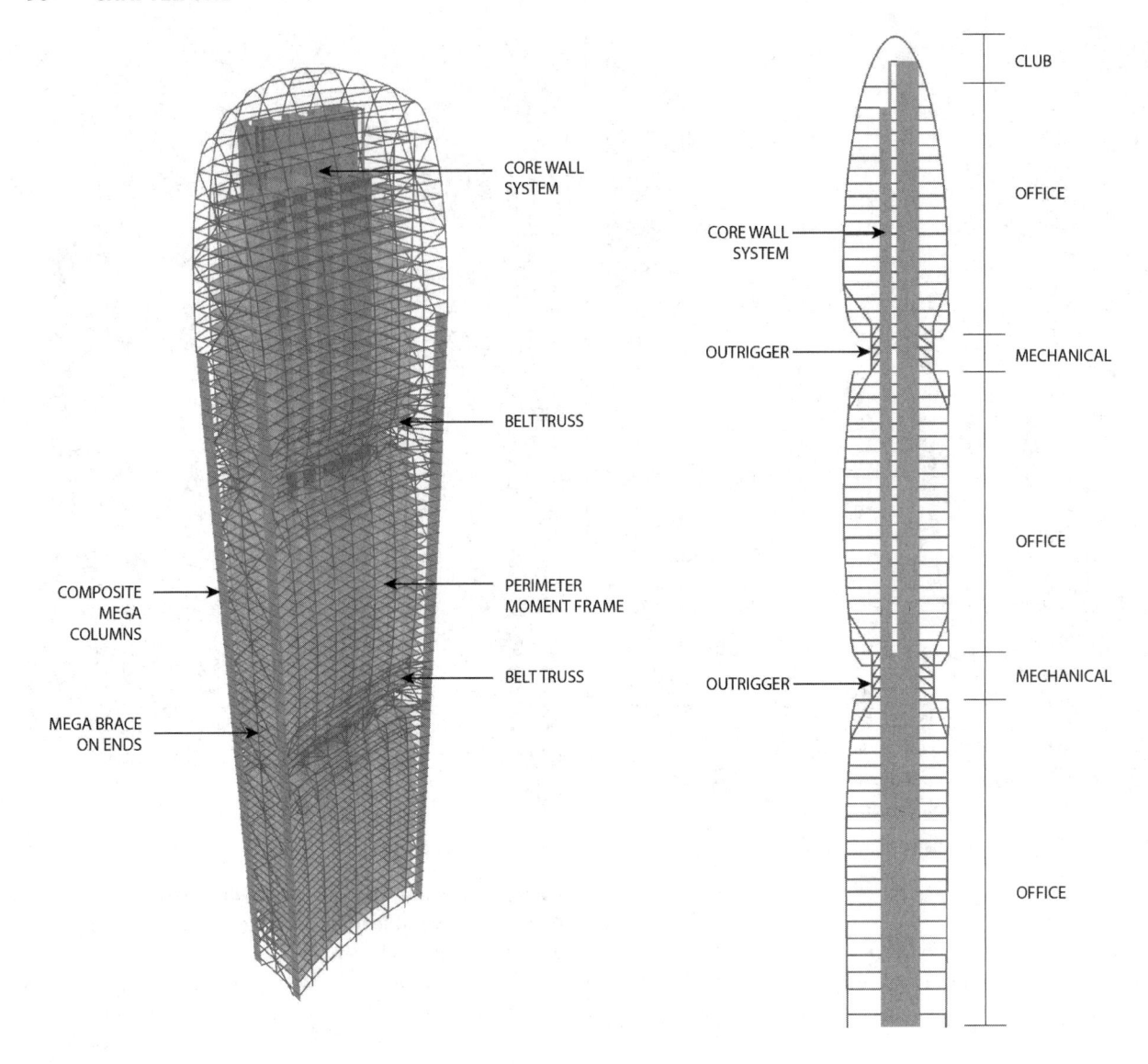

Figure 1.80 Tower structural system.

The design approach adopted to most efficiently resist the substantial shear and overturning demands is as summarized:

• Shear resistance provided by highly ductile Buckling Restrained Braces (BRBs), combined with a nominal moment frame to enable the highest code allowable ductility factor.

• Parasitic gravity loads attracted to the BRBs are discounted from the design load to minimize the overstrength that develops under-design seismic response.

• All shear carried on the full footprint of the tower to resist overturning on the largest possible lever arm.

• Large, double columns are located in the corners.

The perimeter lateral system employs a combination of BRBs and elastic braces coupling around the notched corners (Figs. 1.88, 1.89, and 1.90). To balance forces, limit demands in each element and provide robustness, braces form an X pattern on the elevation. The primary bracing is configured over 10 stories, with each brace spanning two stories and elastic braces at the cross-over at the center of the X. The braces in the notches are also over two stories with a four-story repeating module.

Despite being provided in pairs, and four sets acting in parallel in each direction, the design force for the BRB's with peak demand is 17MN. The BRBs are provided by

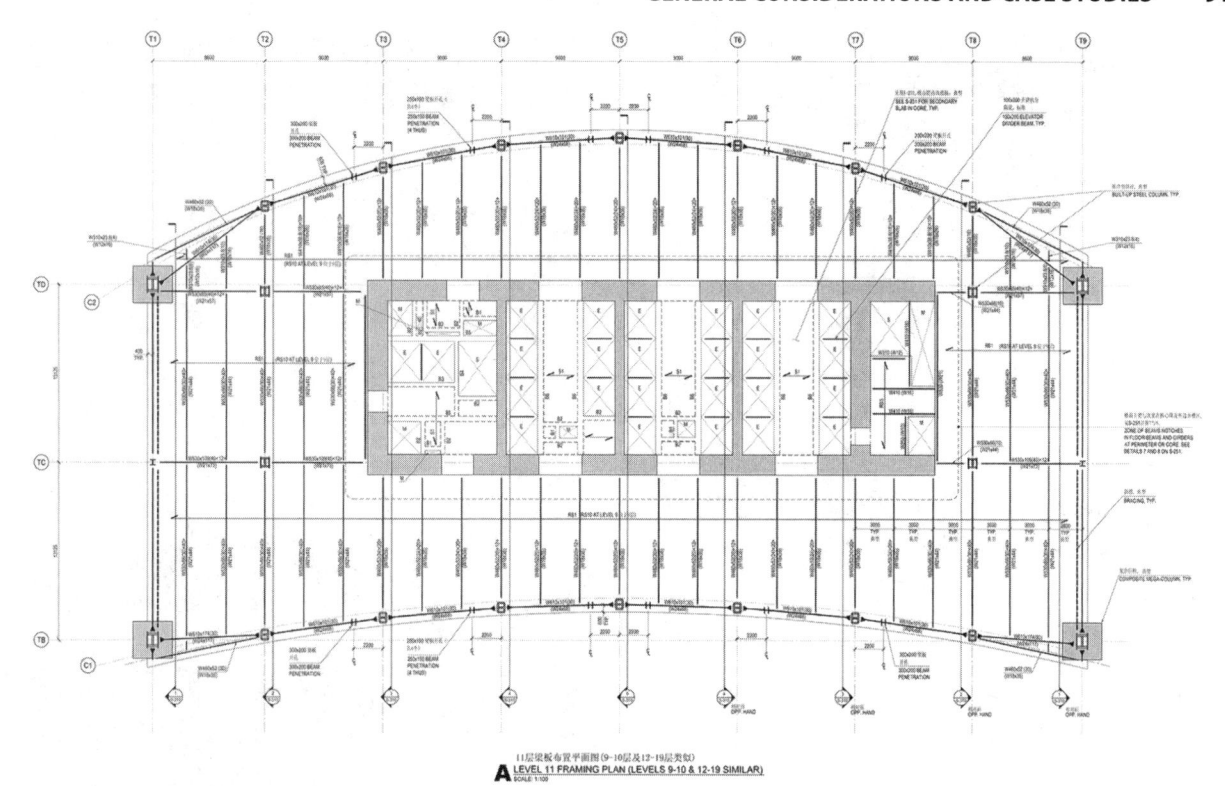

Figure 1.81   Typical floor plan.

Figure 1.82   Tower construction photo.

Figure 1.83   Manhattan west site rendering.
1 Manhattan West at right.

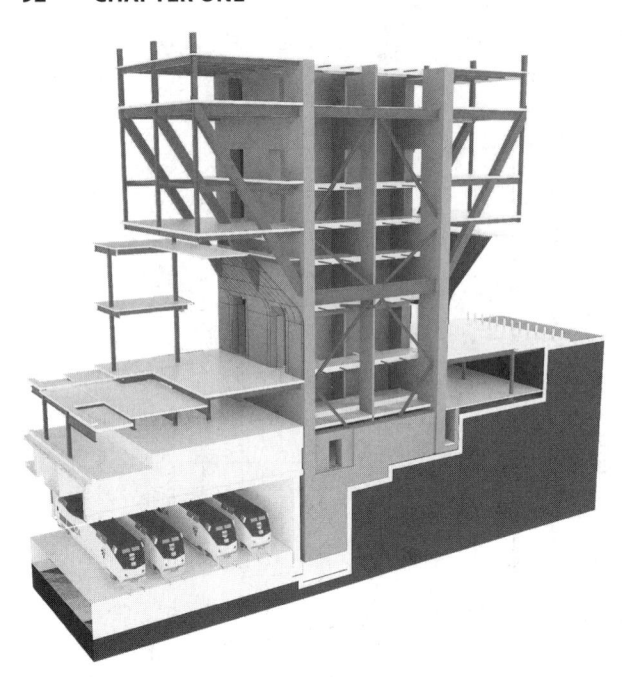

Figure 1.84   1 Manhattan west perimeter column transfer structure.

Nippon steel and are thought to be of the largest capacity employed on any project globally.

The building is doubly symmetrical with six columns on each face, at 5.8 m spacing. Columns are rectangular

Figure 1.86   A25 Xinyi Tower in Taipei. (© RPBW.)

Figure 1.85   1 Manhattan west construction photo.

concrete filled tubes (CFTs) to efficiently resist the compression and tension demands. The columns taper linearly on section and elevation, responding to the reduction in demand with height. The corner columns and elastic braces in the notches are located outside the building envelope.

The core is 17.4×17.4 m on plan and is designed to only carry a nominal proportion of the shear given the small lever arm that is available for overturning. The core is generally a flexible moment frame, except over the lower four stories where being detached from the perimeter structure some bracing is added to retain full gravity capacity.

The floors employ 600 mm deep rolled standard wide-flange sections as primary beams, spanning around 12 m orthogonally between the core and the perimeter columns. Moment connections are provided at both ends to meet minimum moment frame requirements and to provide robustness. The beams are typically notched towards the connections at the core to provide space for primary HVAC distribution. Secondary beams span 5.8 m between primary beams and are typically 250 mm deep rolled standard wide-flange sections.

The steel frame employs approximately 17,000 tons structural steel in the superstructure, equivalent to around 180 kg/m^2. This is highly competitive given the extreme combination of typhoon and seismic loading. Construction photo is shown in Fig. 1.91.

1.3.45 181 Fremont, San Francisco, California (ARUP)

The design of this tower followed a pioneering resilience-based approach to achieve "beyond code" seismic performance. The building was designed to exceed the life safety performance objective required by the building code, instead targeting immediate re-occupancy and limited disruption to functionality after a design-level earthquake. This enhanced seismic design minimizes both structural and non-structural seismic damage and, along with pre-disaster contingency planning, helped earn 181 Fremont the world's first REDi Gold rating for seismic resilience.

The 56-story, 802-ft-tall steel structure houses commercial office space on the lower two-thirds of the building with luxury condominiums on the upper floors.

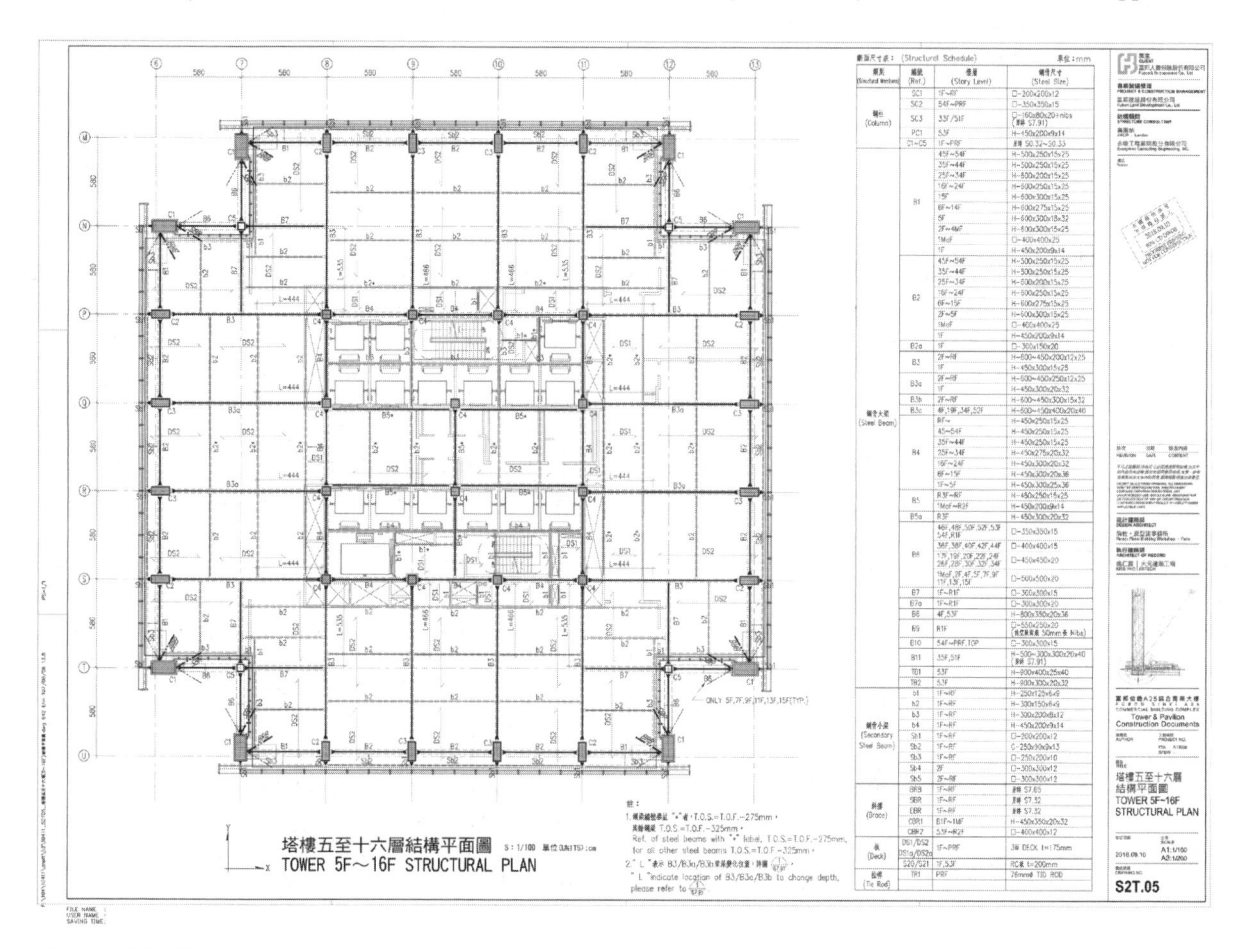

Figure 1.87 Typical floorplate.

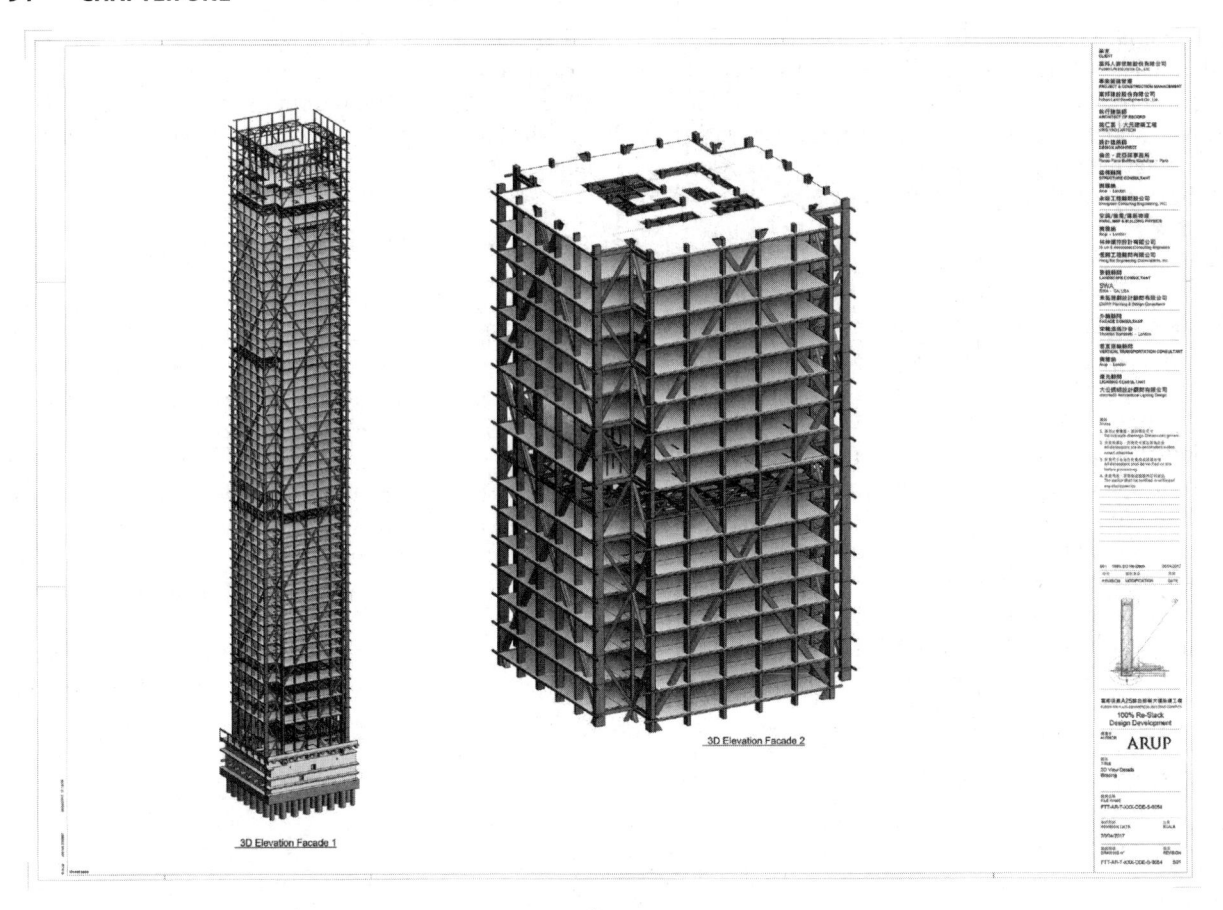

Figure 1.88   3D images indicating perimeter structure configuration.

The tower is very slender, with a 100-ft by 100-ft footprint at the base that gradually tapers along the height. The architectural design includes a faceted façade that folds along visually expressed diagonal lines. Transfer trusses at level 3 carry load to corner megacolumns to create a column-free ground floor lobby. The structural design was conceived to meet the enhanced resilience objectives while still satisfying stringent wind comfort criteria. A traditional concrete core system was not practical given the tower's small footprint. Instead, a perimeter mega-brace system is used to resist lateral forces. Viscous damping is incorporated within the mega-brace system to reduce seismic and wind forces and control vibrations in the tower. Figure 1.92 shows the architectural and the structural rendering of the tower.

Each mega-brace consists of three parallel braces—a middle built-up hollow steel box primary brace with solid steel secondary braces on either side—spanning up to 250 ft between mega-nodes. The primary brace is a built-up box section 14-in. by 16-in. with variable thickness

(roughly 2 in.) and the secondary braces are 9-in. by 9-in. solid boxes built up from steel plates. Viscous dampers are introduced at one end of each of the secondary braces. The mega-braces are restrained laterally at each floor to prevent buckling but slide freely along their length against polytetrafluoroethylene (PTFE) bearing pads. As the tower sways, strains in the primary brace are generally low but total axial deformation accumulated over the brace length can be significant. In the stiff secondary braces this deformation is purposely concentrated in the damper which generates on the order of 5–8% of critical damping. To protect the columns and secondary braces from large axial forces in very rare earthquake shaking, buckling-restrained braces (BRBs) are placed in the load path to act as fuses. In conjunction with the mega-brace system, perimeter special steel moment frames resist the inertial seismic loads at each floor and distribute the loads up and down to mega-node locations. Figure 1.93a shows mega-brace connection with BRBs, 1.93b shows the rendering of mega-node with viscous dampers and BRBs.

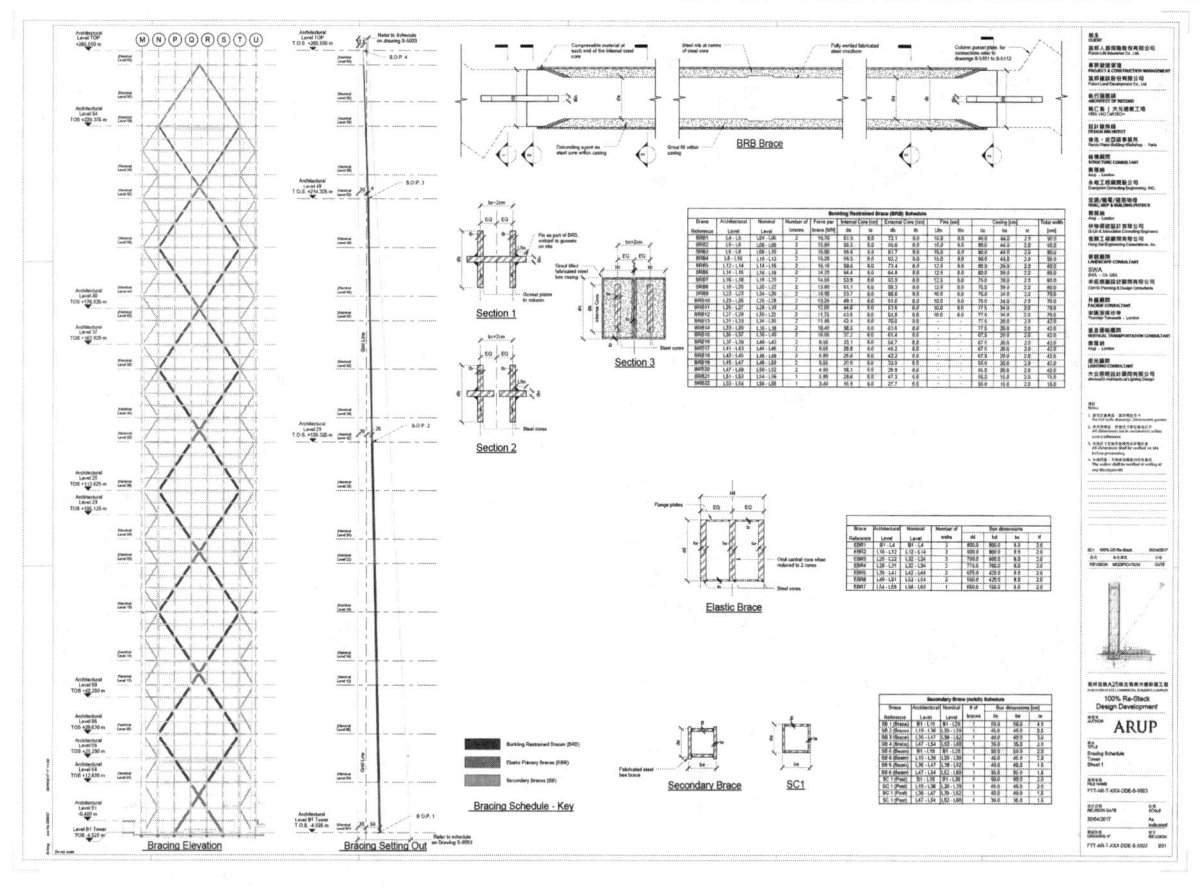

Figure 1.89   Bracing elevation.

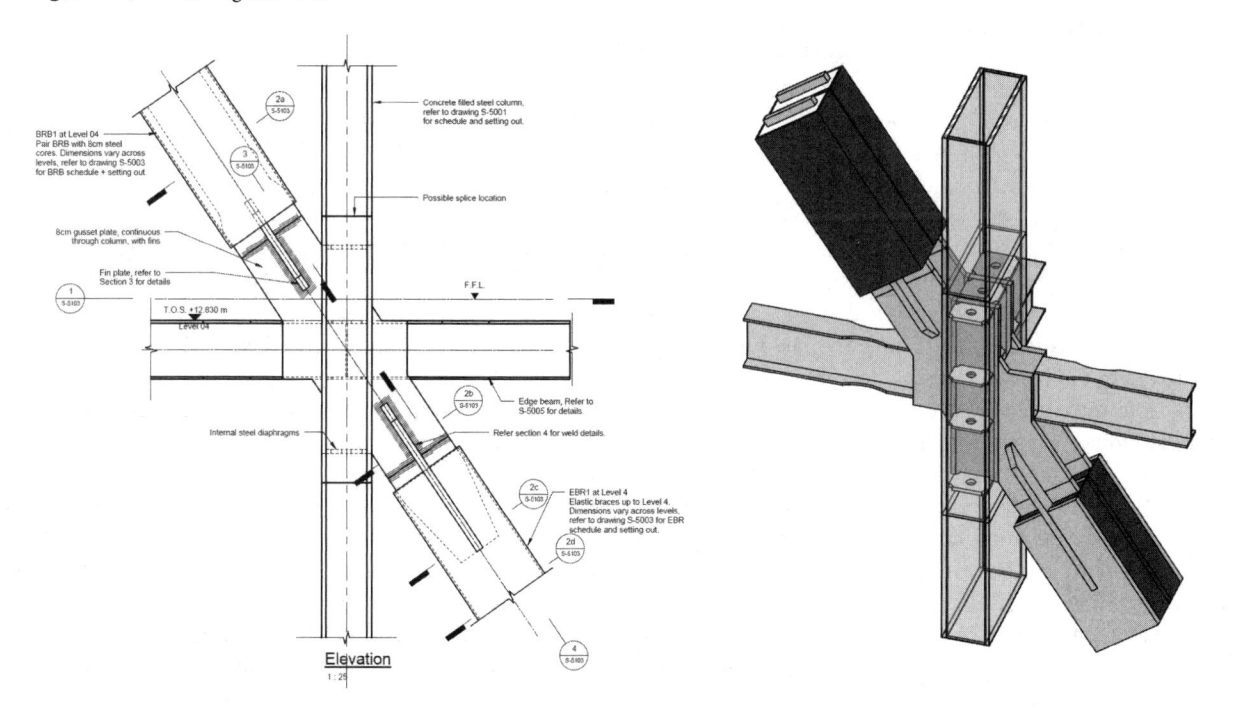

Figure 1.90   BRB detail at interface with elastic cross-over brace.

Figure 1.91 Construction status—August 2020.

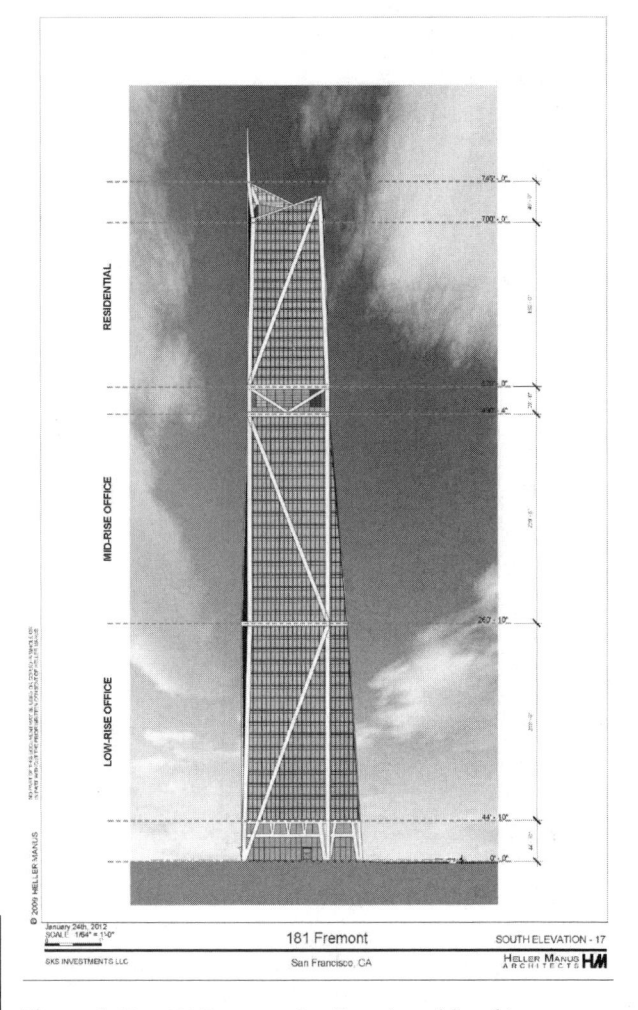

Figure 1.92 181 Fremont, San Francisco: (a) architectural rendering by Heller Manus Architects, (b) structural rendering.

The perimeter megacolumns are designed to remain elastic under maximum credible earthquake (MCE) shaking. The large seismic demands required built-up plate box columns as large as 36-in. by 36-in. by 5 in. thick at the base and prompted the use of 65 ksi material in select locations to save steel tonnage. Due to the small footprint, the axial loads become extremely high during earthquake shaking. The megacolumns are allowed to uplift slightly at their base to limit tension demands in the columns and foundation. This reduces weld requirements in the columns and also eliminates the need for a large steel truss in the foundations intended to distribute axial loads from the corners to the perimeter foundation. The megacolumns are anchored by 3-in. rods extending to the bottom of the foundation that are prestressed such that no uplift occurs under design-level earthquake or wind loads. In the event of uplift, shear is transferred across the plane by a solid steel shear key that floats inside circular holes in the mega-column base. Given that this was a crucial component, the shear key was designed for more than 6 in. of uplift, even though the nonlinear response history analysis indicated an uplift of less than 1 in. on average. The megacolumns are supported by a five-story concrete basement with 5 and 6 ft piles socketed into bedrock over 250 ft below grade. Figure 1.94 shows the megacolumn base during construction.

An earlier iteration of the design (by another engineering firm) utilized very heavy steel sections to stiffen the tower to resist wind forces. This had the unfavorable effect of increasing forces in the tower during earthquakes. Meanwhile, a tuned mass damper was required at the top floor to control wind vibrations to address occupant comfort concerns. The introduction of damping in the mega-brace system led to a virtuous cycle of material reduction. Since the damping reduced seismic forces, steel tonnage could be decreased, which reduced the building stiffness and increased its flexibility. The increased flexibility decreased the seismic demands further and this process was iterated until the

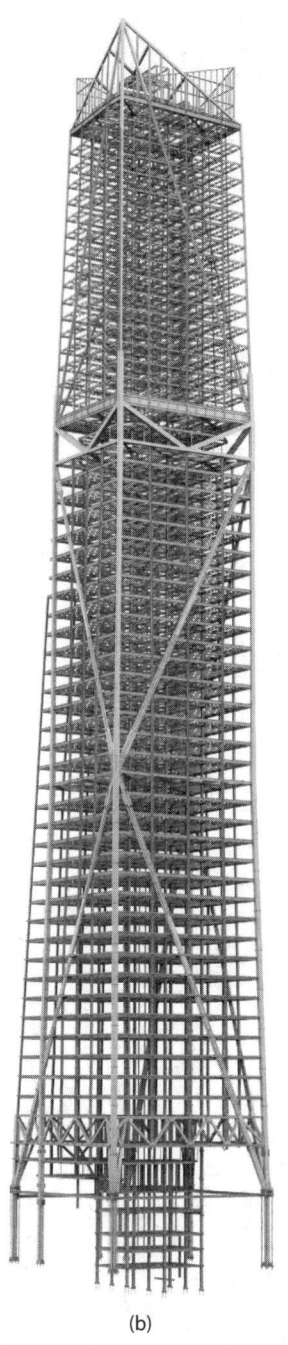

(a)

Figure 1.93   181 Fremont, San Francisco: (a) photograph of mega-brace connection showing BRBs, (b) rendering of mega-node showing viscous dampers and BRBs.

Overall, the innovative structural design saved approximately 2700 tons of steel compared to the original design—roughly 25% of the building weight—while achieving the greater resilience objectives. The structural engineering was by Arup in San Francisco.

1.3.46   Atrio (ARUP)

Atrio is a major new development in downtown Bogota, Colombia, comprising two high-rise buildings, and at ground level, the city's largest privately funded public realm area. Tower North (Fig. 1.95) was opened in 2020 and at 201 m (46 stories) is the tallest office building in Colombia. The second tower, South Tower, is proposed to reach 268 m (59 stories). The overall development area is 250,000 m² providing commercial office space, retail, public realm, basement parking and a hotel within Tower South. The architectural design is by RSHP, London and co-architect El Equipo Mazzanti, Bogota. Structural Engineers are Ove Arup and Partners International Ltd, London, and PyD, Bogota.

(b)

Figure 1.92   (*Continued*)

design was tuned to meet the seismic and wind criteria. Integrating viscous damping within the megaframe also eliminated the need for a tuned mass damper, resulting in a significant material savings and freeing the penthouse level for increased net floor area. After construction, vibrations measured at the top story during a major wind storm verified the damping system design and even indicated a peak acceleration slightly less than predicted.

(b)

Figure 1.93 (*Continued*)

Figure 1.94 181 Fremont, San Francisco: photograph of mega-column base during construction.

Figure 1.95   Atrio, North Tower. (© Rudolf/Arpro.)

The striking geometry of the towers features a dramatic cut-away over the lower 8 stories of the towers unifying it with public space at ground level (Fig. 1.96). The top of the towers also cut-back creating terrace areas on the upper floors with views to the cityscape and Andes mountain range. The typical floor plate plan geometry is formed by two overlapping squares aligned on their diagonal axis around a central lift core. Additional shuttle lifts project from the central overlapping area (Fig. 1.97).

The towers are located in an area of significant seismicity, with a peak ground acceleration of 0.26g. The Seismic Force Resisting System (SFRS) is formed by a perimeter Special Concentric Braced Frame (SCBF) and a central concrete core with Special shear walls. The SCBF is configured on a two-story module and connecting at column nodes with gusset plate connections. Braces are formed from square fabricated box sections, which in the North Tower vary between 200 mm to 475 mm wide with plate thickness up to 45 mm. Columns are typically Steel Reinforced Concrete (SRC) varying between 1 m and 1.8 m diameter with concrete strength varying between 41.3 MPa (6 ksi) and 55 MPa (8 ksi). An elevation of the braced frame is shown in Fig. 1.98.

The towers were designed to meet requirements of ACI 318, AISC 360 and Colombian NSR-10 together with a performance based seismic design approach using the PEER methodology. Performance was assessed at the service level (43y return period), design basis level (475 y) and

Figure 1.96   Atrio, North Tower ground plane cut-away and interface with public realm canopy structure. (© Rudolf/Arpro.)

the Maximum Credible Earthquake (2475 y. Non-linear time history analysis was undertaken using LS-DYNA to validate the assumed ductility demands. The floor diaphragms at SCBF node floors were designed for the out-of-balance bracing forces following brace buckling.

The braced frame forms a key feature of the building architecture with the node connections, bracing and columns being fully expressed, and in a number of locations positioned outside of the building envelope. The node detailing was subject to extensive testing and modelling during the design phase due to their importance as both a structural element and a key aesthetic feature. A parametric design tool was scripted by the engineers to enable quick 3D visualization of compliant geometry for every node in the tower and to drive the BIM model (Fig. 1.99). The construction methodology for the node was developed at early stage with contractor input, in particular the interface between SRC hoop reinforcement and embedded plates. Scaled models, mock-ups and modelling were used to test the installation.

The floors are formed from wide-flange steel beams acting compositely with a concrete slab on metal decking, with a total thickness of 165 mm (6.5") 2" deck. Perimeter columns are spaced at 10 m centers and the floorplate configured such that maximum spans do not exceed 14.14 m. A notable feature of this arrangement is that only two internal columns are required either side of the central core and a large "column-free" corner bay measuring 20×20 m is created.

The five-story basement was constructed using a local top-down technique using sacrificial concrete piles together with a perimeter slurry wall. The North Tower is founded on a 2 m thick raft slab and an array of 2 m diameter piles. Knowledge sharing between the international and local practices involved in the design and construction was a key feature of the project approach. The existing local market was heavily focused on concrete construction and as such the use of steel was reserved for areas of Atrio where it delivered optimum benefits. The resulting solution is a highly efficient and innovative composite high-rise structure with an average steel weight of 70 kg/m² and 0.3 m³/m² of concrete. The high specification fabrication elements were shipped from Canada, with typical elements fabricated in Colombia (Fig. 1.100). Field welding of the frame was undertaken by skilled workers from the Colombian oil industry. In this sense the project has made a significant impact on the local steelwork industry expertise and capability to the extent that South Tower is planned to be built using local

skills. Figure 1.101 shows an architecturally expressed SCBF node on Atrio North Tower; Fig. 1.102 shows a construction image showing steelwork elements of SRC columns prior to pouring of concrete; and Fig. 1.103 shows the SRC column node scale model showing embedded steel I-section, floor beam connection plates and hoop reinforcement. The model was used to demonstrate the installation of hoop reinforcement through holes in the vertical floor beam connection plates

1.3.47 CITIC Tower, Beijing (ARUP)

GENERAL

Located at Beijing's new CBD core area with a height of 528 m, the 108-story CITIC tower (or "China Zun") provides 350,000 m² gross floor area above ground mainly used for office space (Fig. 1.104a). The tower stands on a seven-story basement of 40 m deep which accommodates car parking and mechanical facilities. The tower's gently rising and curving profile gives the building a contemporary and elegant expression that accommodates more valuable prime-floor spaces at the top levels and provides structural stability at the base. Arguably the world's tallest building constructed in a high seismic zone, a highly efficient and cost-effective multiple lateral resisting system with extensive composite construction ensures the building's safety and provides grand space at entrance lobby and the top observatory deck.

SUMMARY

As the leading structural engineer, Arup introduced a highly efficient lateral resisting system composed of a fully braced megaframe and a concrete core (Fig. 1.104). Composite steel-concrete member like composite steel plate wall and composite mega columns are also extensively utilized to provide a cost-and-seismic-resisting-effective solution to the structural challenges.

EFFICIENT AND SAFE STRUCTURAL
SOLUTIONS

Multiple Lateral Resisting System Beijing is located in one of the highest seismic zone around the world, where there is no precedent of tall buildings that exceeds 500 m. In addition, engineers were challenged by an enlarged top of the tower. With comparison of various systems, a highly efficient multiple lateral resisting system was proposed for the 528 m tall building composed of perimeter frame with mega columns, mega braces, transfer trusses and a reinforced concrete central core (Fig. 1.105). The typical floor plan with 8 mega columns is shown in Figure 1.106.

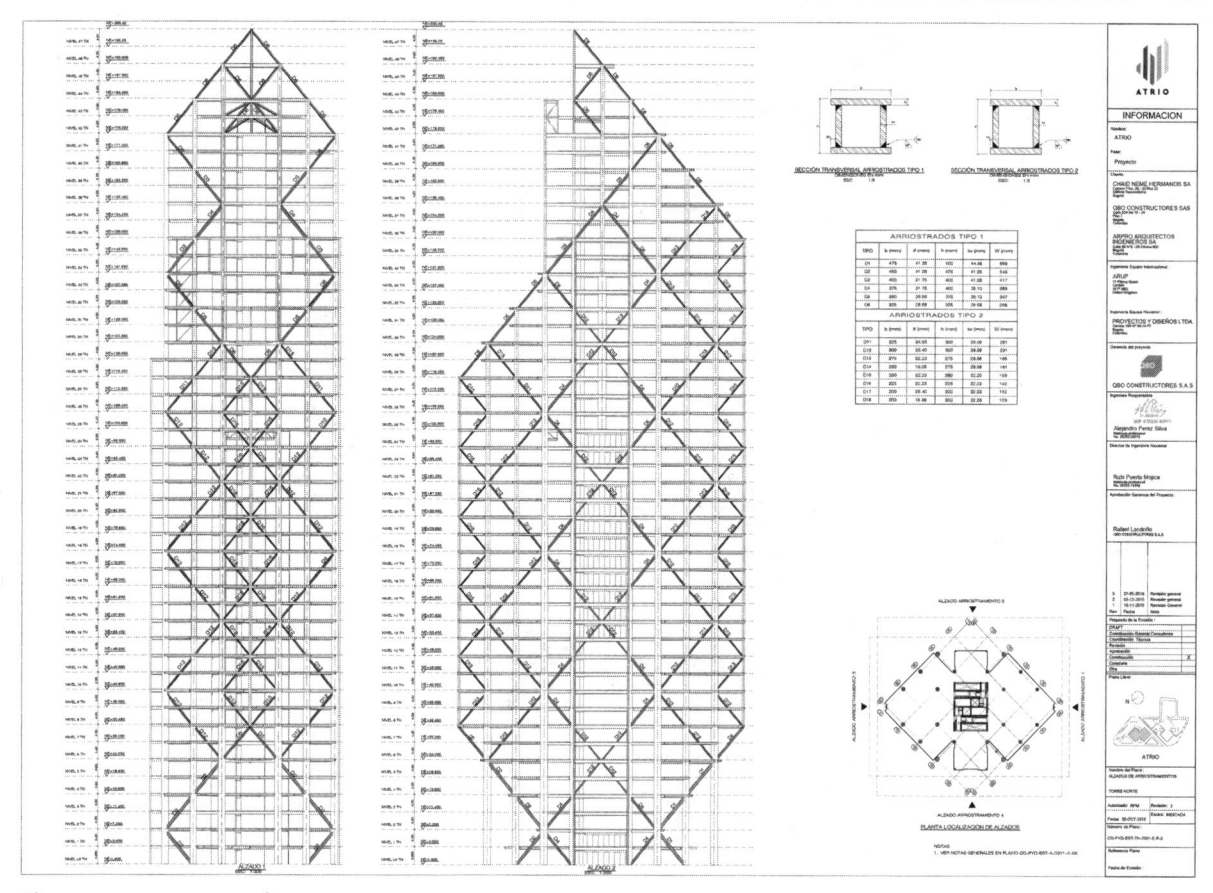

Figure 1.97 Atrio, North Tower, Structural elevation showing braced frame, composite columns and cut-away at the ground plane and at the top of the tower.

Composite Steel-Concrete Members Various composite steel-concrete members are extensively used strategically with steel and concrete members to achieve the cost-effectiveness of seismic resistance and also increase the usable floor area. The concrete core walls at bottom levels are embedded with steel plates which significantly improve seismic performance and reduce the maximum wall thickness to only 1.2 m (Figs. 1.107 and 1.108).

Innovation of the "Multi-Cell" Mega Column The perimeter megaframe makes use of the full building dimension to maximize the lateral stiffness. In typical floors eight concrete-filled-box mega columns are distributed at perimeter, sized from 4 m × 4.5 m to 1.6 m × 1.6 m along the height. At the bottom zone the two columns at one corner are merged into one super column, with 60.8 m^2 cross-section area, due to the curvature of the building. With elaborate engineering and lab test of 1:12 physical model, the "multi-cell" concrete filled steel tube section is proposed for the ease of construction. Box "cells" are shipped to site and vertically spliced together. Inside each "cell" tie bars are welded onto the vertical stiffeners to control the local buckling of the steel plates, achieving a reasonable steel ratio. The steel profile serves as the construction platform and also eliminates the rebar-fixing and formwork installation. With large volume of concrete in-filled, the thickness of the fire proofing layer can be reduced largely. The max. dimension of the columns (4 m) reducing the welding length and workload of the section. The "multi-cell" column and detailing have been incorporated into Chinese composite code and national standard. Figures 1.109–1.111 show photos of the mega column construction.

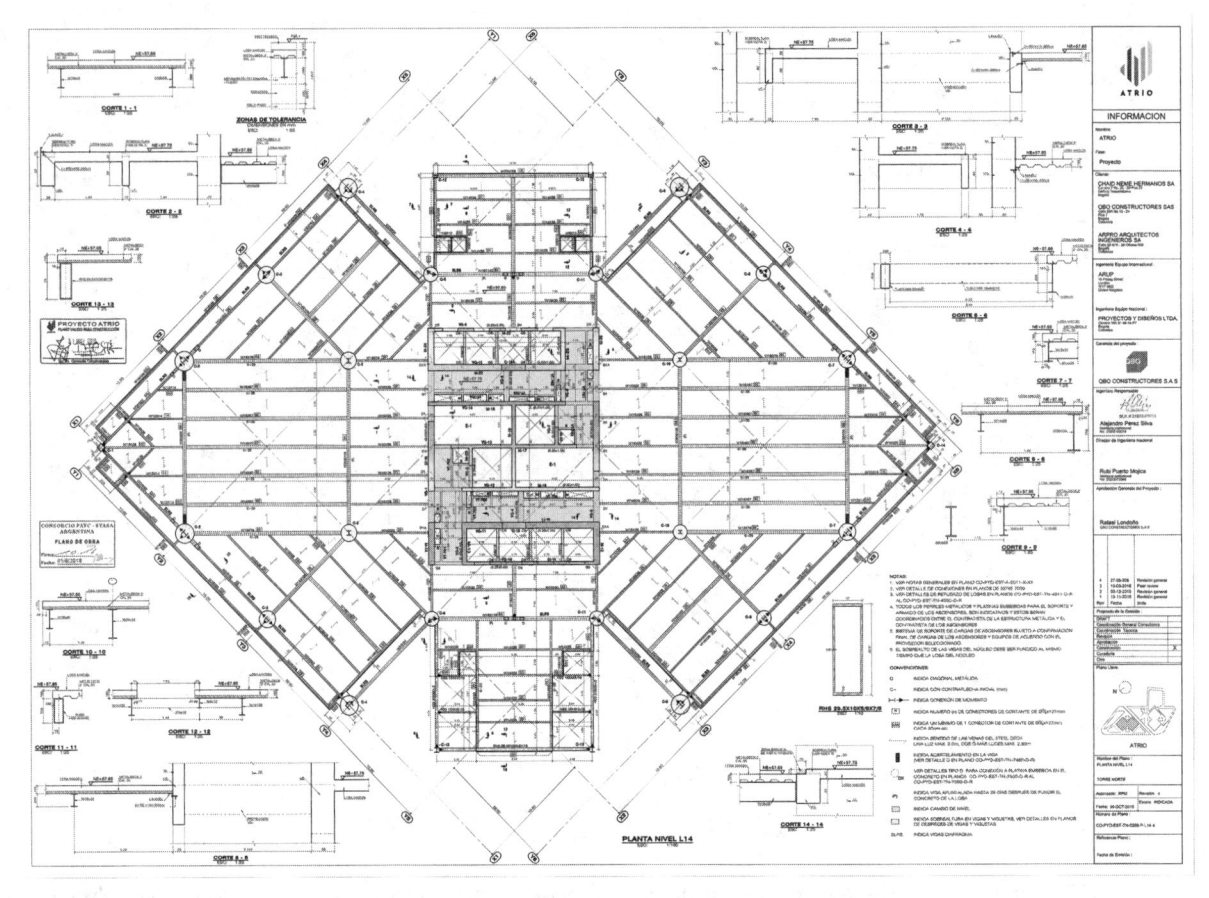

Figure 1.98   Atrio, North Tower, Structural floor plan. The typical floor plate plan geometry is formed by two overlapping squares aligned on their diagonal axis around a central lift core. Additional shuttle lifts project from the central overlapping area.

NODES:

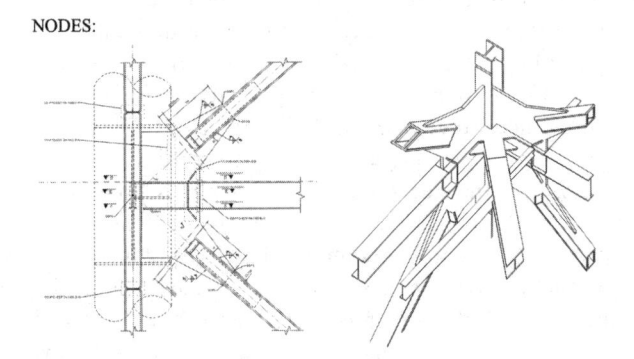

Figure 1.99   Atrio SCBF node drawings.

Figure 1.100   Shipping of SCBF node steelwork from Canada to Bogota. (© Supermetal.)

Figure 1.101   Architecturally expressed SCBF node on Atrio North Tower. (© Jason Garcia/RSHP.)

Figure 1.102   Atrio Tower North, construction image showing steelwork elements of SRC columns prior to pouring of concrete. (© Jason Garcia/RSHP.)

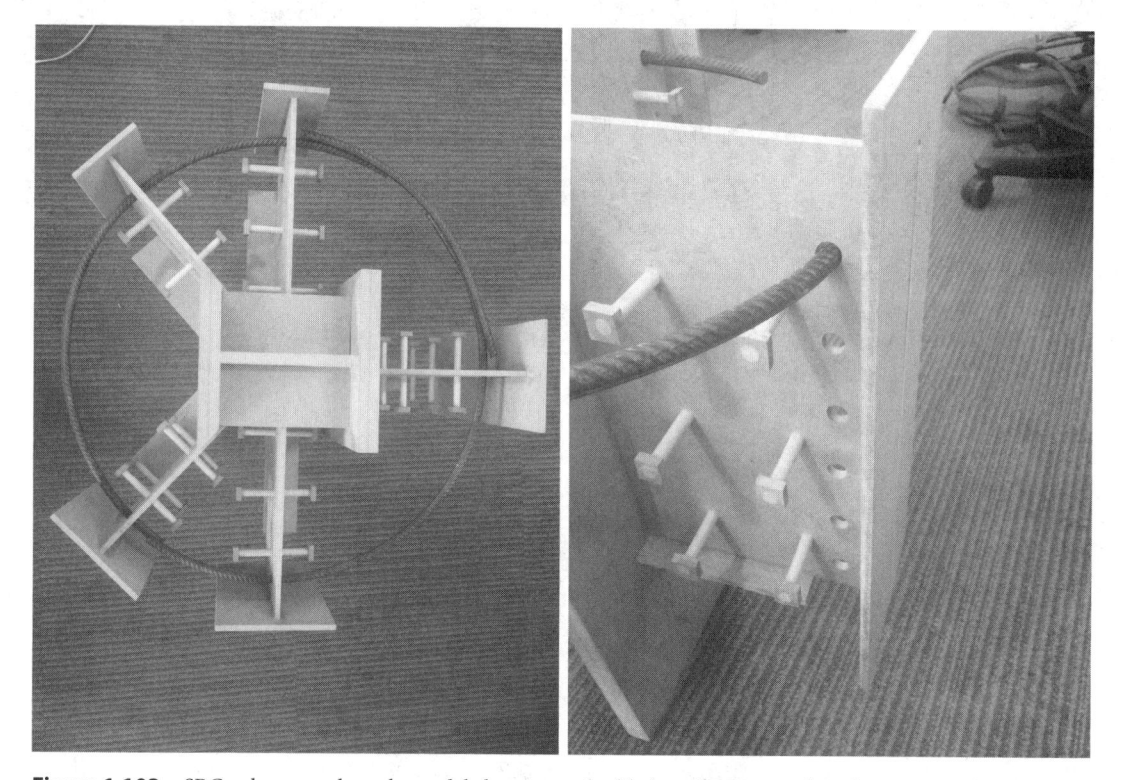

Figure 1.103   SRC column node scale model showing embedded steel I-section, floor beam connection plates and hoop reinforcement. The model was used to demonstrate the installation of hoop reinforcement through holes in the vertical floor beam connection plates. (© Arup.)

Figure 1.104   The perimeter megaframe and the central core.

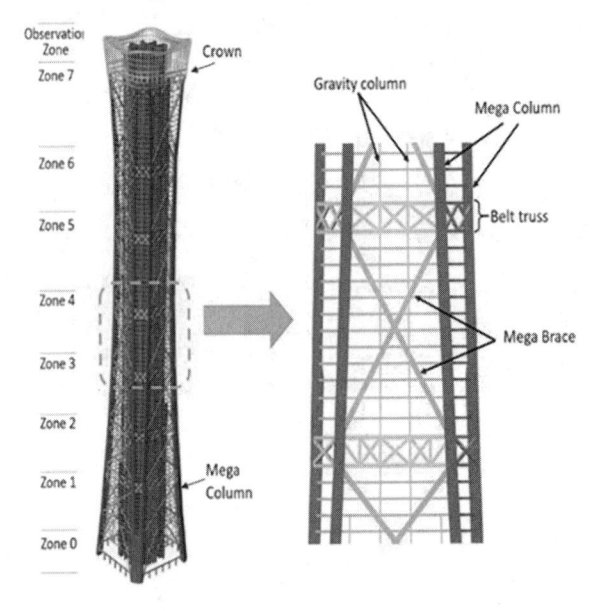

Figure 1.105   The megaframe and the gravity-resisting sub-frame at building's perimeter.

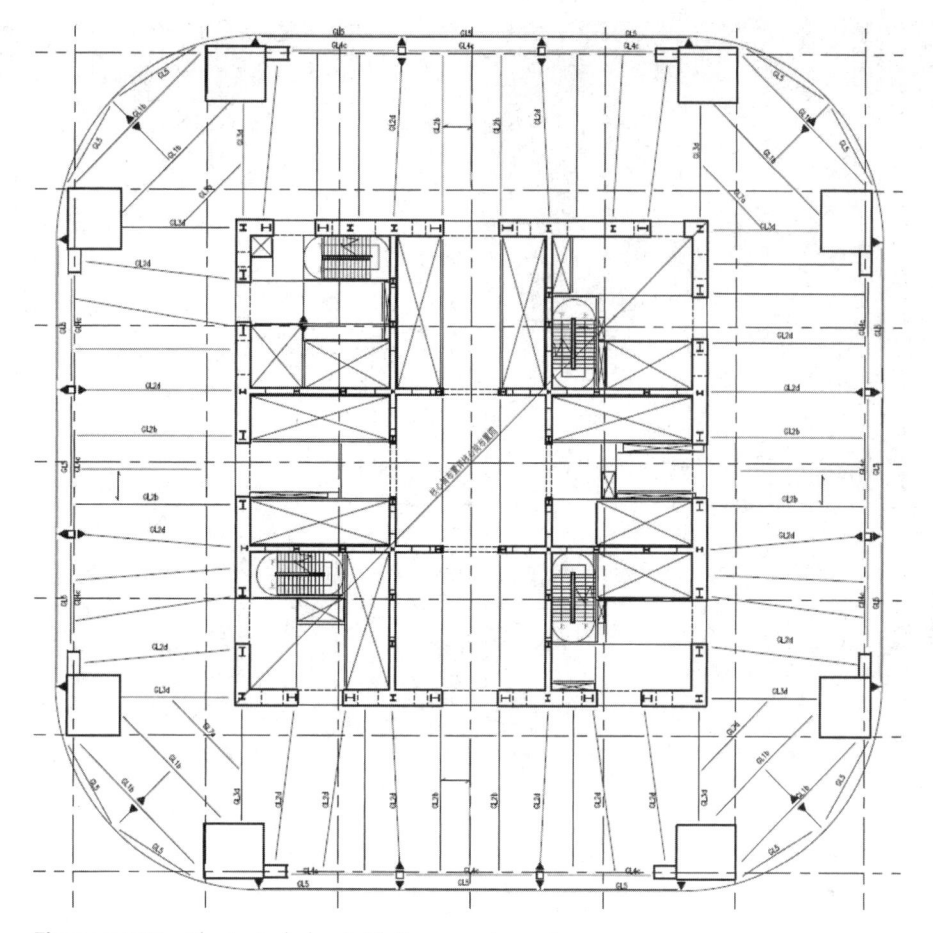

Figure 1.106 The typical plan (with 8 mega columns).

Figure 1.107 Steel plates installed before concreting the core walls.

Figure 1.108 Finished composite walls—concrete provides fire protection to the steel plates.

Figure 1.109   Construction of the mega column at basement.

Figure 1.110   Two columns merged into one super column at base.

Figure 1.111   The self-support construction of the mega column.

Perimeter Sub-Frame System   Engineers also created a high-robustness perimeter sub-frame system to provide multiple load paths for gravity load transfer, the gravity columns are located at the same plane with the megaframe, conveying the gravity load to mega braces and the belt truss located at the mechanical/refuge floors of each zone. At the top of each zone the gravity columns are also connected with the belt truss above through a "slot" connection. Once the gravity column at a specific floor fails in extreme cases, the column above becomes a hanger column and transfers the loading upwards to avoid a total collapse.

Optimization of Foundation Design   Due to the small site area (only 11,000 m²), CITIC Tower also owns the deepest basement among super high-rise buildings globally. The 40 m deep basement is supported by ~40 m long bore piles via a 6.5 m thick raft. There is no settlement late cast strip between tower footprint and other basement parts which significantly shortened the site dewatering period by two years and minimized the underground water loss in Beijing CBD area. The 56,000 m³ concrete of raft was casted continuously within 93 h, tackling unprecedented technical and logistics challenges due to the deep excavation. Figure 1.112 shows the construction photo of the raft plate.

Strengthen Measure at Weak-Story   The floor plan reduces from 78 m in width at the bottom to 54 m × 54 m at 385 m which then enlarges to 69 m × 69 m at the top. The core at the "waist" levels are thus subjected to larger forces and specially strengthened per the analysis results. To assist the energy dissipation during earthquakes, additional steel braces were installed in the concrete core at weak areas, and metallic steel lintel damper were adopted at specific floors. A 1:40 shaking table test verified the validity of the analysis and ensured the safety of the building (Figs. 1.113 and 1.114).

High-Strength Material Adopted for Savings in Material Consumption   To increase structural capacity and minimize section size of members, high-strength material was adopted. Highest concrete grade is C70 (fck = 44.5 MPa), highest rebar grade is HRB500 (fyk = 500 MPa), and the steel grade of key elements is Q390 (fy = 390 MPa). For majority of steel members, Q345 (fy = 345 MPa) grade with thickness not higher than 60 mm is adopted to decrease the difficulty of material purchasing, manufacturing and on-site welding.

Digital Parametric Design to Optimize the Form   During the scheme and preliminary design stage the function of the building has undergone several significant changes. Most of them changed the geometric setting due to the shape of the tower and demanded a whole rebuilt of the structural model. With our in-house developed parametric structural modelling and optimization environment built upon the Rhino + Grasshopper platform, the geometric

Figure 1.112   The "trough + pipe" concrete pouring of the raft plate.

Figure 1.113a   Shaking table test.

Figure 1.113b   Megaframe joint test.

and configuration relationships constituting the logic of building a structural model was coded explicitly in a visual way so that engineers were able to re-setup the whole model by just adjusting the relevant key parameters. The generated model was then assigned with member sizes automatically through structural optimization procedures. With this design automation tools engineers were able to change a new scheme from existing in a few hours, rather than a few days before. More than 800 structural schemes were created and studied for comparison purpose, all with quantitative results for structural performance. The quick valuable feedback from engineers enabled the client and the team to make the decision in a short period. Interoperability tools were further also developed to connect the parametric model with BIM and other analysis software, automating largely the

production of plan/section drawings, connection FEA models and nonlinear analysis models. This smart design framework is now commonly adopted in structural design of different projects within the firm (Fig. 1.115).

Intelligent High-Rise Building Construction Integrated Platform   CITIC Tower adopted the intelligent high-rise building construction integrated platform provided by the contractor. The platform was a steel frame structure installed on top of the concrete core walls through several embedments. Two tower cranes, construction elevators, concrete distributing machines, concrete forms, storage yard and other construction equipment were integrated into this platform, covering over 1800 square meters, four and half vertical work floors, with 4800-ton maximum jacking force. It was able to be raised up by one button-click with the intelligent monitoring system.

Figure 1.114   Steel braced concrete shear wall test.

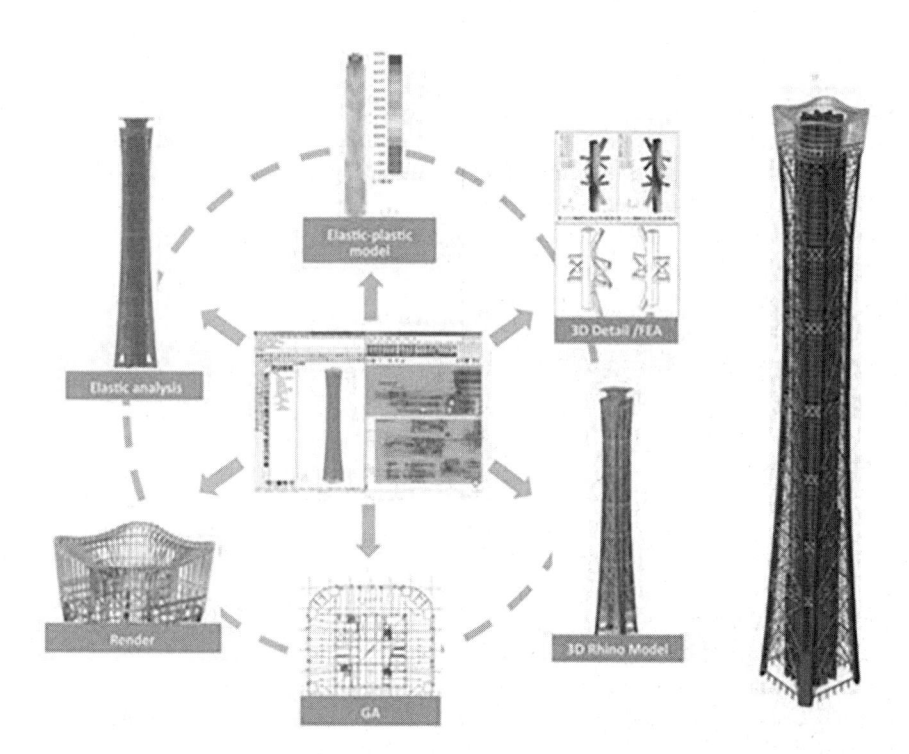

Figure 1.115   Parametric design work flow.

Figure 1.116   Intelligent high-rise building construction integrated platform.

Compared to conventional method, it could reduce 28 times of crane climbing, 56 days of related construction, 400 tons of tower crane embedded parts. In the peak time, hundreds of people worked simultaneously on the same platform, making it a skyscraper builder on the sky. The intelligent platform optimized resource distribution, decreased problems caused by equipment, and increased the construction speed by 30% (Fig. 1.116).

New Solution of Vertical Construction Transport   Focusing on the common problem of vertical worker and material transportation for high-rise building, CITIC Tower adopted a brand-new solution "JumpLift." Its operation speed is 4 m/s while the conventional hoist runs at 1 m/s, and a single unit (3600 kg model) can move nearly three times more people and material compare to conventional one. The temporary machine room of "JumpLift" could move up inside the elevator shaft under its own power and could follow the rise of concrete core by 3 to 5 floors. After main structure's topping out and construction of machine room, it could convert to permanent elevator easily by changing a few components, thus it could save 120 days of construction compare to conventional elevator. The adoption of "JumpLift" highly improved the

capacity of vertical construction transport, and reduced construction cost by saving lease expense of construction hoists (Fig. 1.117).

Major Involved Companies
Owner
   CITIC Heye Investment Co. Ltd.
Architect
   Concept: TFP Farrells
   Design: Kohn Pedersen Fox Associates
   Architect of Record: Beijing Institute of Architectural Design
Structural Engineer
   Design: Arup
   Engineer of Record: Beijing Institute of Architectural Design
Main Contractor
   China Construction Third Engineering Bureau Co. Ltd.

1.3.48   The F5 Tower

The F5 Tower is a 650-ft tall steel and concrete structure located in downtown Seattle, Washington (Fig. 1.118), which was designed by architect Zimmer Gunsul Frasca and structural engineer Arup. The design of

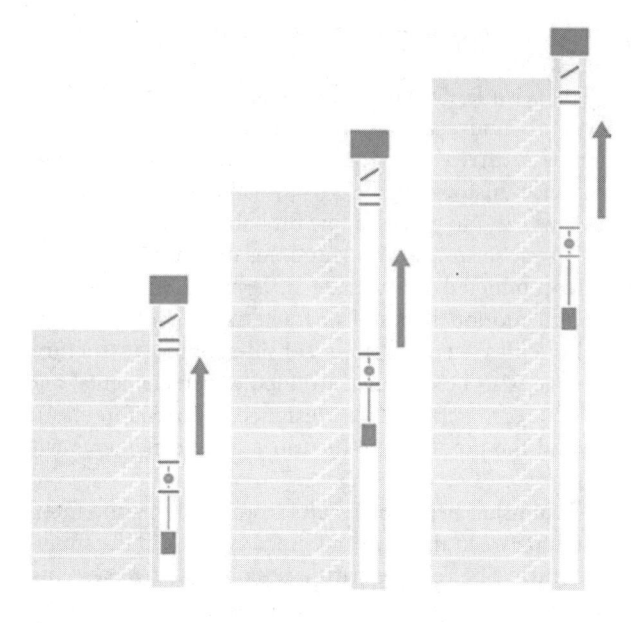

Figure 1.117 KONE JumpLift.

Figure 1.118 F5 Tower, view from southwest. (Credit: Connie Zhou.)

this tower is shaped primarily by the challenge of maximizing leasable floor area on a small site. Two defining features arose in response to that challenge. First, the owner purchased air rights above two adjacent historic buildings on the same block, so that the floor plate could expand outside its base footprint. Second, the tower is tall relative to its base, requiring a structural solution that engages the full building envelope to resist lateral forces in this seismically active area.

Close collaboration between the architect and the structural engineer resulted in the building's unique faceted façade. The subtly angled facets create variations in the floor plan up the height of the building, such that the 22nd floor has the greatest gross floor area, approximately 20% greater than the base footprint. The edges of these facets are in turn defined by structural braces which provide lateral stiffness. The brace geometry is defined both to create facets that are readable on the building scale, and to ensure that the braces intersect columns at certain floors to restrain the structure's lateral movement.

The apparent slenderness of the perimeter braces is achieved by coupling this system to a central concrete core. These two systems are tied together through the floor structure, which consists of steel beams acting compositely with concrete-filled metal deck. Force transfers between these systems can only occur at floors containing "nodes"—where a brace, beam and column intersect at a point. This occurs at approximately every six floors

Figure 1.119 Brace "nodes" on the west facade—locations where braces, beams and columns intersect at a point. (Credit: ZGF/Arup.)

(see Fig. 1.119). The concrete core restrains most lateral movement between these key floors. Because braces span from node to node rather than from floor to floor, they are termed "megabraces." Special nodes occur at four corner locations where megabraces from adjacent building faces meet at a point without a column. Corner columns are eliminated to maximize the office real estate value.

At a typical floor, steel beams act to transfer lateral shear forces into and out of concrete diaphragms. Around the perimeter of each floor is a "belt" of steel beams designed to carry axial forces from brace nodes into concrete through standard welded studs, over a length sufficient to maintain low stresses at individual studs. Other steel beams fixed into the concrete slabs by such studs are welded to embed plates at nine points around the central concrete core, thus completing the load path between core and perimeter braces, see Fig. 1.120.

The seismic design of this structure required a high level of analysis. Dynamic ground motion simulation was carried out (otherwise known as "nonlinear response history analysis") to validate the building response and to ensure a low probability of collapse in a 2475-year seismic event. Hundreds of analysis iterations were conducted to evaluate different permutations of ground motion and soil characteristics. Both strength and ductility were carefully considered; most structural damage is to be confined to ductile fuse elements, and critical connections are designed to remain elastic beyond the point where the more ductile elements will yield.

An unusual behavior results from the dynamic interaction between the core and brace systems in the superstructure. The relative stiffnesses of these two systems, and hence the proportion of the lateral force carried by each, varies with height. Overall, the core carries more than 50% of the story shear over most of the superstructure height, while the perimeter frame contribution increases at lower levels. In addition, there are large force transfers back and forth between the core and perimeter at intermediate stories (see Fig. 1.121). The core might be visualized as a vertical cantilever restrained by spring supports at the floors which have a brace node.

Other structural challenges were addressed in addition to the core/brace interaction. The perimeter structure is not connected to floors over the first five stories, so these floors are suspended from transfer girders at the fifth floor, and perimeter braces and columns have 60-ft clear spans. These elements are constructed of welded steel box sections instead of rolled wide flange shapes. There are eight stories of parking garage below the ground level, resulting in a considerable soil-structure interaction effect and requiring a bounded analysis approach to account for uncertainties in the soil response. And the faceted façade creates kinks in the perimeter columns, such that gravity loads generate additional bending moments in columns and horizontal restraining forces at floor levels.

Construction of the tower was complete by the fall of 2017.

1.3.49 Torre BBVA (ARUP)

Torre BBVA is a 235 m tall commercial office in Mexico City and is the headquarters for BBVA Bancomer, Mexico's largest bank (Fig. 1.122). The architectural design is a collaboration between RSHP, London and

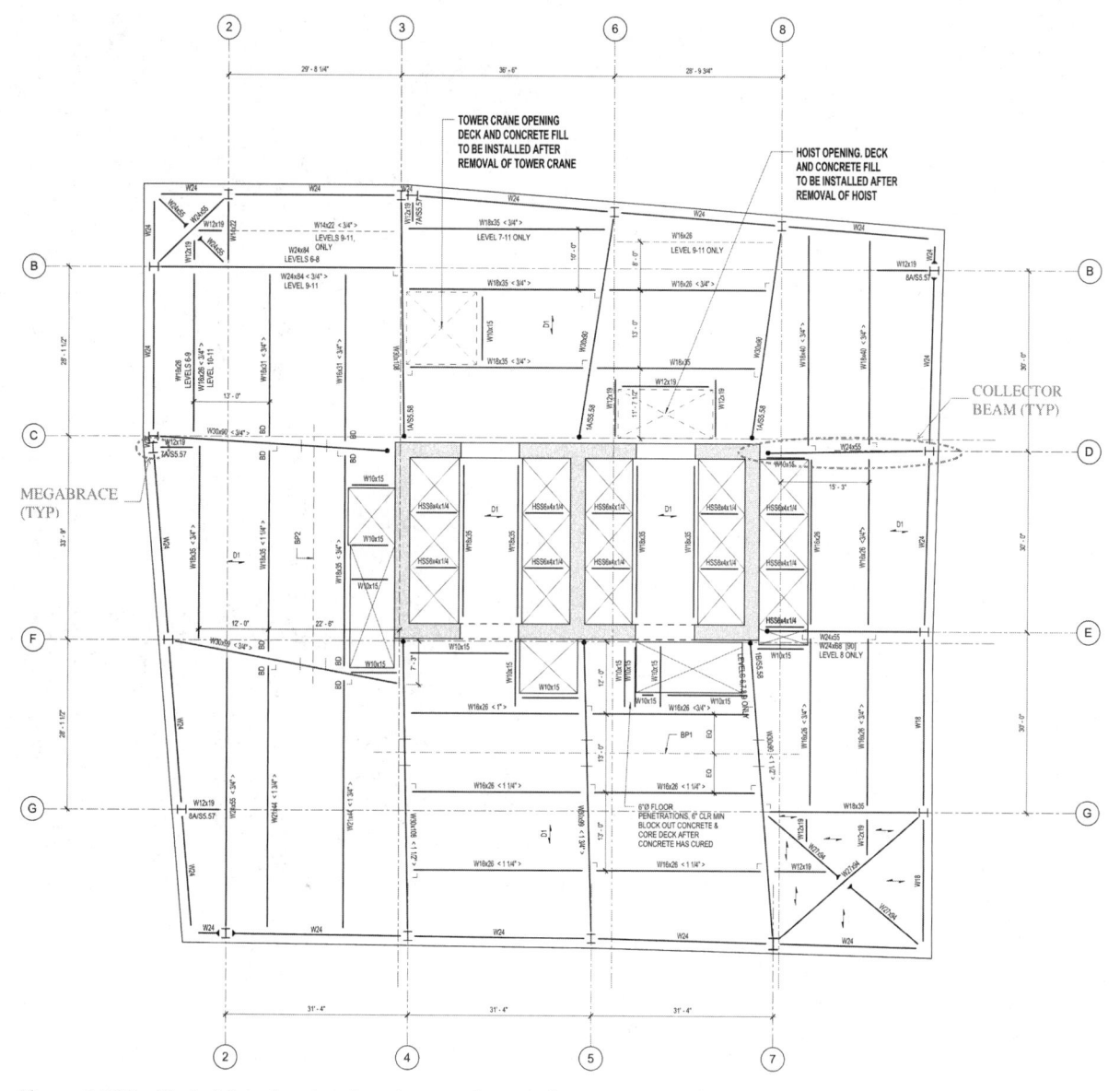

Figure 1.120 Typical floor framing plan, showing elements that connect to perimeter braces and interior core. (Credit: Arup.)

Legoretta+Legoretta, Mexico. Structural Engineers are Ove Arup and Partners International Ltd, London, and Colinas de Buen, Mexico.

Mexico City is highly seismic. The soft soils on this site give rise to the classic Mexico City seismic hazard where distant Pacific subduction-zone earthquakes are modulated and amplified to create long-duration, long-period ground motions. In a location where there were 10,000 earthquake fatalities in 1985 an explicit and rigorous approach is of critical importance.

The tower is 235 m tall and occupies a footprint of around 46 m × 46 m, giving a slenderness ratio of around

5.1. The tower is designed to house 4500 BBVA Bancomer head office staff and includes a seven-story basement, a 12-story annex building and an independent ramp structure incorporating a large auditorium atop.

The applicable PGA for the location in Mexico City is 0.20g, with a minimum shear strength requirement of 5.0%g under code-based design basis earthquake. Wind loading is moderately benign and therefore maximum loading in the bracing, columns and foundations are governed by seismic response.

The design brief generated exceptional stacking requirements resulting from city planning requirements for

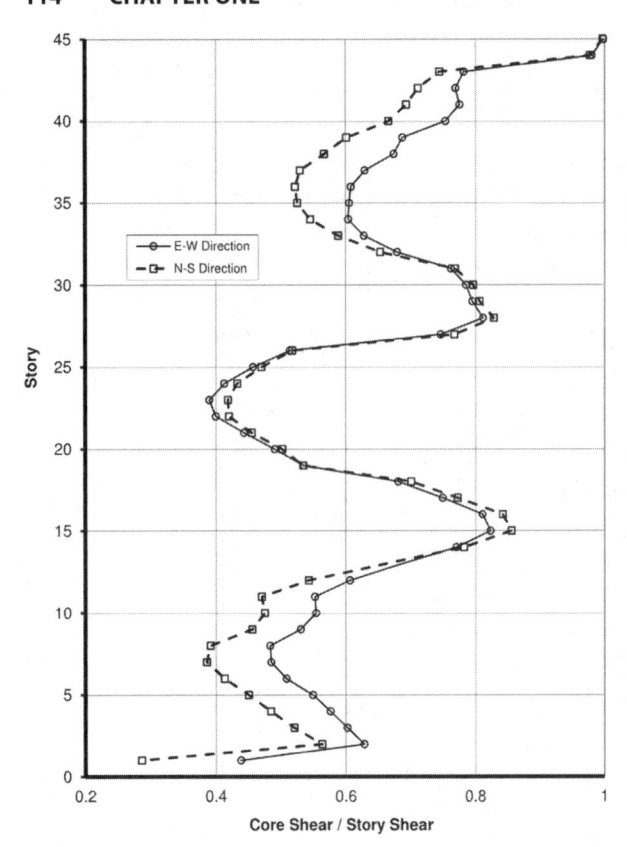

Figure 1.121 Approximate percentage of seismic story shear carried by concrete core with building height. (Credit: Arup.)

Figure 1.122 BBVA Tower in Mexico City. (© LEGORETTA+LEGORETTA.)

2800 parking spaces. The resulting design solution exhibits extraordinary flexibility for stacking functions, whereby there is no central core and the primary vertical circulation starts at Level 12 above ground. This allows efficient vehicle circulation and parking in the tower superstructure and tower footprint in the basement plus a completely open Ground Floor lobby. This is achieved by an external, fully exposed, eccentrically braced megaframe around the perimeter of the tower that carries all of the lateral seismic and wind loads and half of the gravity.

The frame is arranged on a three-story module, supported on six steel box section megacolumns 1.55 × 1.55 m, concrete filled below level 30. The eccentric braced frame extends around the full perimeter of the tower, Fig. 1.123. The principle frame members are 90 cm deep and between 60 and 90 cm wide. The eccentric seismic link fuse element is located centrally on each frame, and is independent of the floor plate and facade. The separation allows concentrated deformation of the link during seismic response to occur without impacting the floors and façade, remedying the classic drawback of eccentrically braced frames.

Gravity load from the floor plates is supported off short posts to the megaframe beams, which in turn direct the gravity load to the corners. This approach leaves the facades column-free and directs the gravity load to the corners at the global extreme fibers, minimizes column forces and foundation tensions and maximizes the column area most efficient in resisting overturning.

The floor plates are architecturally independent of the external megaframe (Fig. 1.124). Lateral stability within each module is ensured by the six perimeter megacolumns spanning across the three or four stories, transferring lateral forces from intermediate floors to the megaframe levels (Fig. 1.125). The 1.55 m square perimeter columns act as small cores effectively stabilizing the floors and interior columns within each module. The floor plates within the megaframe module attach to

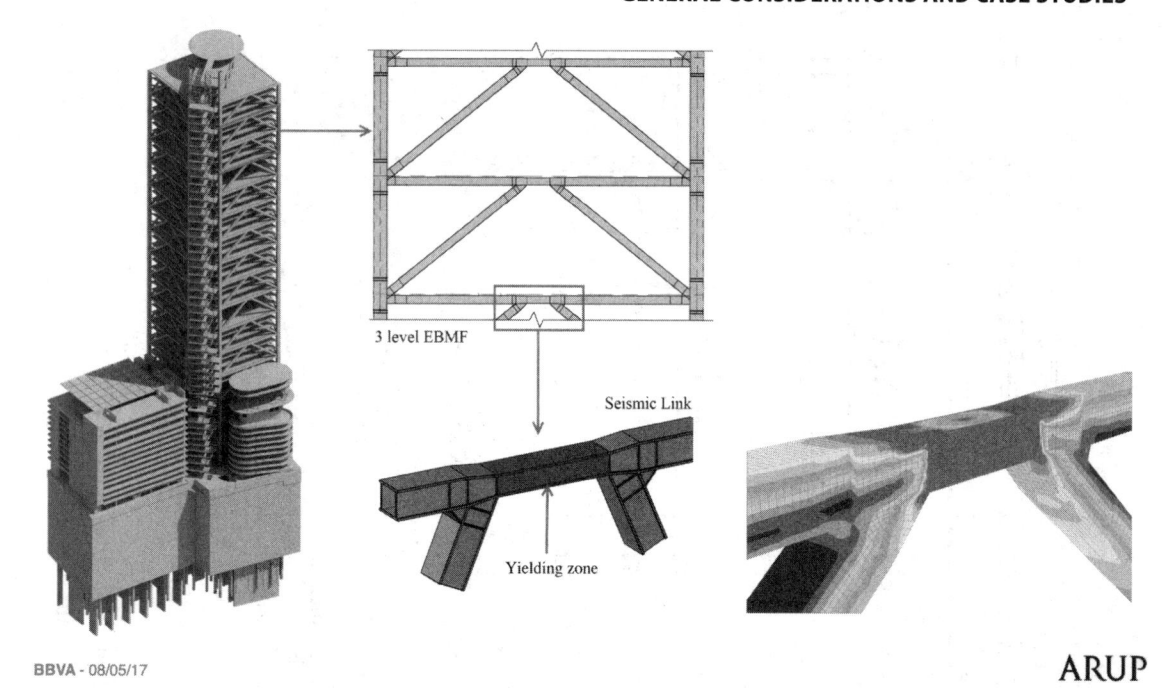

Figure 1.123 External Braced Megaframe, nonlinear plastic fatigue analysis of seismic link. (© Arup.)

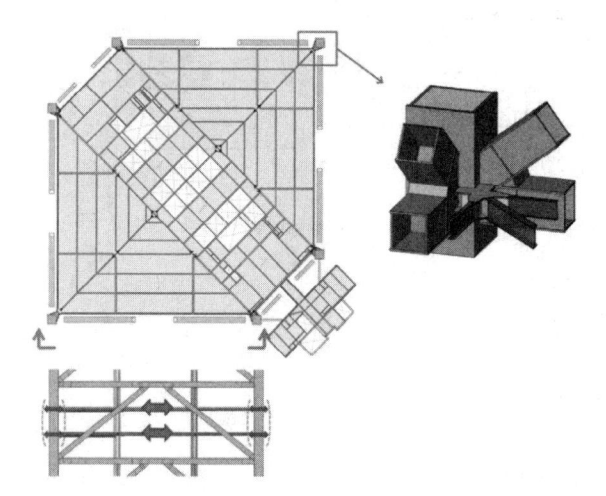

Figure 1.124 Eccentrically Braced Megaframe, stability of intermediate floors by six megacolumns in flexure. (© Arup.)

the perimeter columns through rigid connections of the primary floor beams.

The floors employs W18 section primary beams, spanning around 9 m and W21 section secondary beams spanning around 13.5 m (Fig. 1.125). Moment connections are provided on beams framing between the corner columns to provide additional robustness. The beams include notches and holes to provide space for primary HVAC distribution creating a generous 2.85 m floor-to-ceiling height.

The six megacolumns are supported directly on the basement retaining walls, 1.2 m diaphragm wall panels with a two-story capping beam. A steel shoe or "zapatta" buried in the capping beam spreads and anchors the 210 MN column force into the diaphragm panels, Fig. 1.126. Lateral loads are distributed by strut action through the diaphragm wall panels with the capping beam providing the tying action, avoiding the need for a structural lining wall.

The steel frame employs approximately 16,000 tons structural steel in the superstructure, equivalent to around 145 kg/m^2. This is highly competitive given the extreme seismic loading and lack of a conventional concrete core.

1.3.50 Beijing CCTV Headquarters, China (ARUP)

The China Central Television (CCTV) Headquarters building in Beijing is one of the most unconventional structures in the world (Fig. 1.127). The 234 m tall building, a competition-winning design by OMA, contains approximately 473,000 m^2 of floor area, comprising filming, production and administration facilities for the national broadcaster—the entire television-making process. Its unique form is made up of two leaning towers plus a cranking podium and "Overhang" which together create a single continuous "loop" (Fig. 1.128).

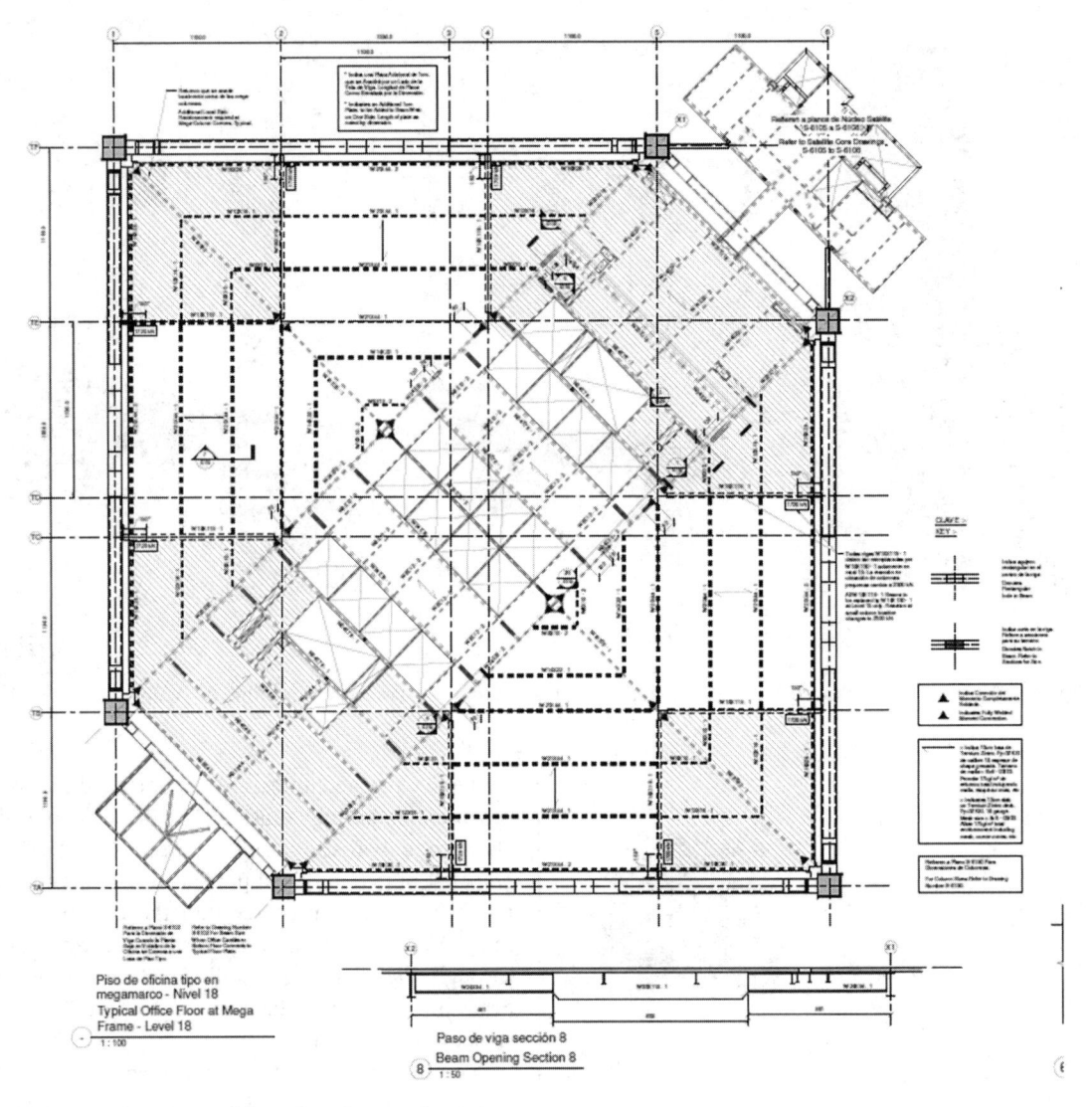

Figure 1.125 Typical floorplate. (© Arup.)

As well as the usual tall building challenges of robustness and high seismic and wind loads, CCTV's unusual shape led to unprecedented stress concentrations, multidisciplinary coordination challenges and, fundamentally, the question of whether such a building could be built safely and at an acceptable cost.

CONTINUOUS TUBE STRUCTURE

To resist the lateral forces, Arup engaged the entire skin of the building by adopting a continuous tube structure, creating a "mesh" of steel or steel-reinforced concrete (SRC) columns, diagonal bracing and beams which wrap around every surface. This robust braced tube, with multiple load paths, was then optimized to suit areas of

high and low stress, with the mesh made more dense in areas of high stress, and sparser in areas of low stress (Fig. 1.129).

INFLUENCE OF CONSTRUCTION METHOD ON DESIGN

Both the method and sequence of construction would influence the distribution of permanent loads in the building and the stresses that develop during construction and after completion. To provide flexibility for the contractor, Arup considered a number of possible construction methods from the outset, and developed a Particular Technical Specification (PTS) to communicate the assumptions that had influenced the final design and

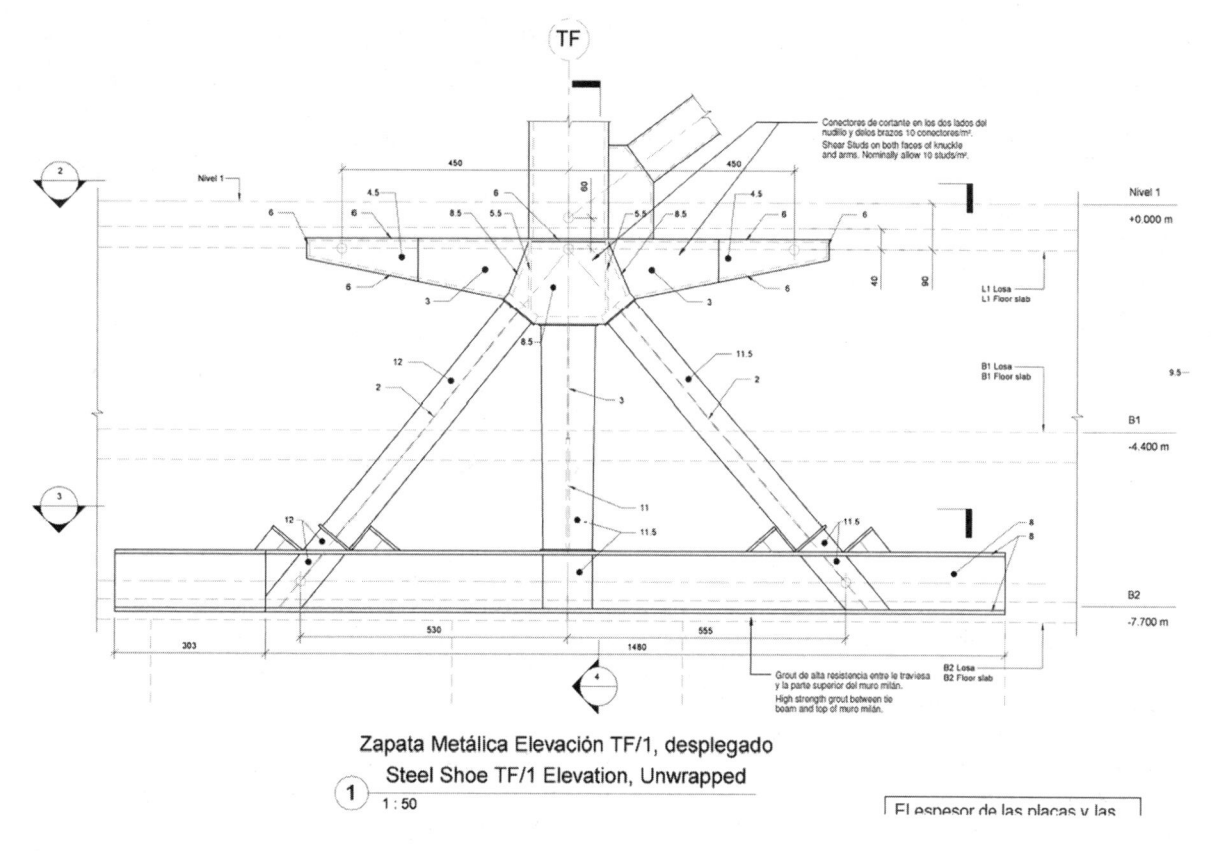

Zapata Metálica Elevación TF/1, desplegado
Steel Shoe TF/1 Elevation, Unwrapped
1 : 50

Figure 1.126   Zapata shoe detail. (© Arup.)

Figure 1.127   Beijing CCTV Headquarters, China. (© Zhou Ruogu Architecture Photography.)

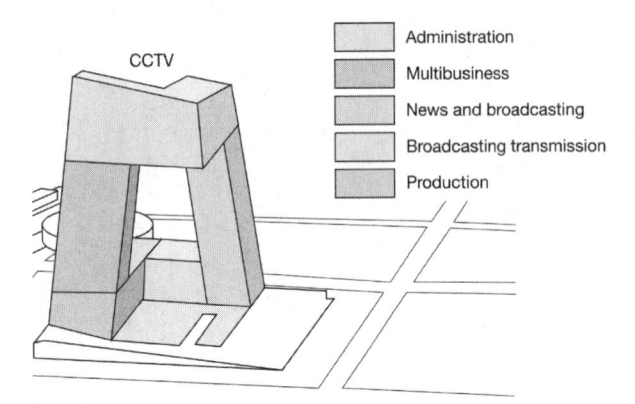

Figure 1.128 Diagram showing the four components of CCTV. (© Nigel Whale/Arup.)

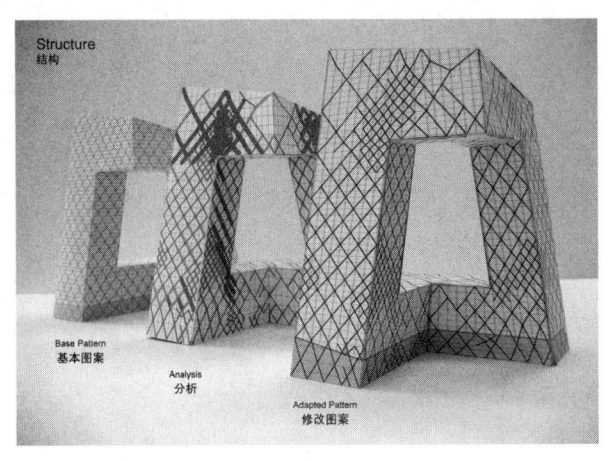

Figure 1.129 The CCTV building's optimized "tube" structure. (© OMA.)

their implications for construction. This would enable contractors and specialist designers to understand the stresses and movements during each stage of construction, particularly prior, during and after the first connection of the Overhang. The following four issues were described in detail in the PTS:

(1) Presetting to offset predicted movements

Once steelwork had been constructed to roof level, the Overhang steelwork was progressively cantilevered from each tower (Fig. 1.130).

Arup's analysis predicted that the Overhang would deflect 300 mm vertically under its own weight immediately prior to connection. The PTS stipulated that the two towers should be preset so they would meet at the correct position before being connected together as one (Fig 1.131). This was achieved by constructing each floor slightly out of position, with continuous movement monitoring verifying the contractor's detailed "construction time-history" analysis and enabling presetting adjustments as required.

(2) Postponing installation of critical columns

The analysis showed that the tower corner columns beneath the Overhang would carry the brunt of the cantilever dead load, thus limiting their capacity to resist seismic and wind loads. Since increasing the capacity of these critical elements would attract additional load, Arup instead specified that they should be installed after completion of the structure; thus the dead load would be transferred into adjacent elements, and their full capacity would become available for lateral forces (Fig 1.132).

(3) Loading scenario analyses

Arup carried an upper and lower bound analysis to explore the impact of varying the amount of dead load

added prior to connection of the Overhang. More load prior to connection (e.g. additional façade installation) would increase tower and foundation stresses in the temporary condition, whilst less load would increase the propping forces in the Overhang after connection. The worst case was used for element design as well as movement criteria, to increase construction sequence flexibility.

(4) Connecting the overhang

Linking the two 75-m cantilevers at a height of 162 m was the greatest construction challenge since the daily relative movements of each tower due to wind loads and thermal expansion risked overstressing the initial connection between the towers.

The PTS outlined the exact conditions under which connection should take place: both towers at a uniform temperature, and movements at a minimum. In practice, this meant just before dawn on a windless day. To provide sufficient immediate capacity, the contractor was required to install a series of strong permanent links which could initially allowed movement, but quickly be "locked off" to create a stiff connection (Fig. 1.133). The PTS also specified a week of movement monitoring prior to connection: this recorded daily relative tower movements of up to 10 mm.

The contractor installed seven connection elements, using pins to provide the necessary strength for the temporary condition (Fig. 1.134). The joints were later welded whilst the remaining Overhang steelwork was erected.

Despite the complexities of CCTV's design and construction, there were no major delays on site, nor any unexpected building movements. Its gravity-defying architecture challenged the future direction of tall

Figure 1.130 Cantilevered steelwork installation for the Overhang. (Image a-c © Frank P Palmer Image d © Arup.)

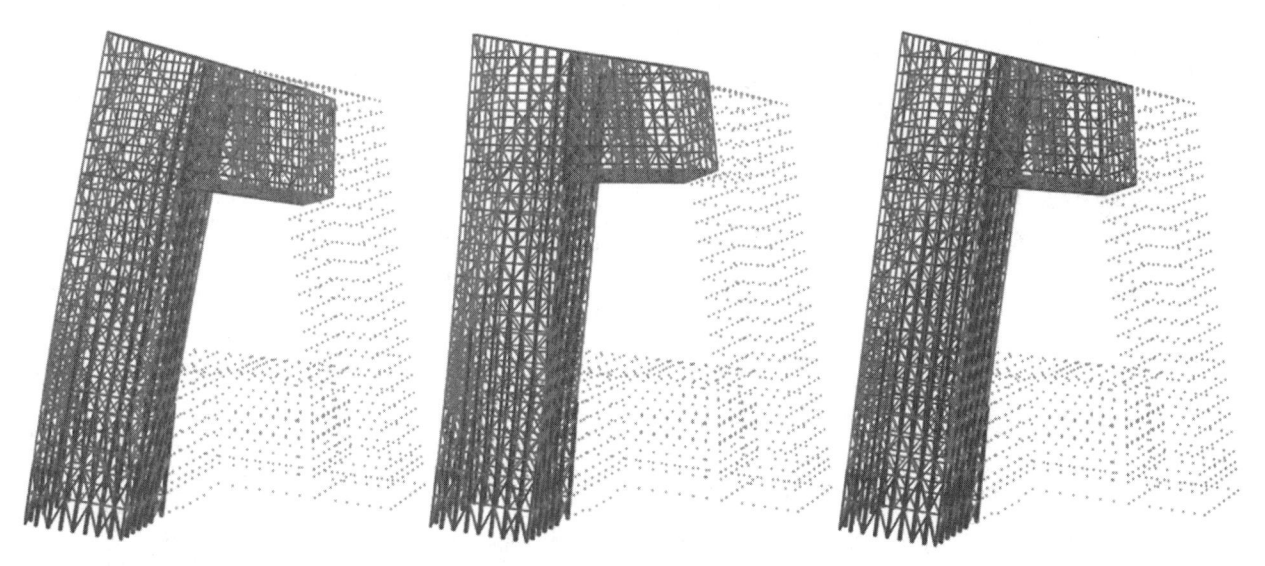

Figure 1.131 Pre-setting concept. (a) Theoretical tower deflection under dead load. (b) Presetting the tower upwards and backwards during construction. (c) No resulting deflection under self-weight. (© Arup.)

Figure 1.132   Post-installed corner columns. (© Arup.)

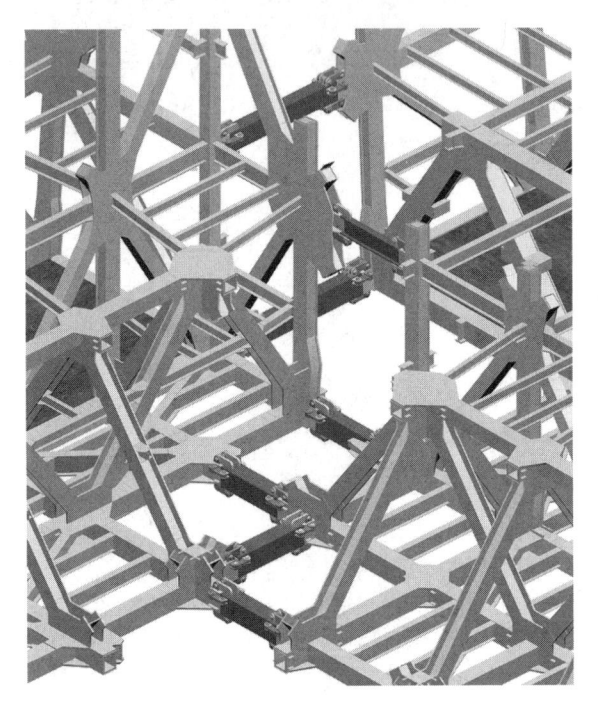

Figure 1.133   Structural model showing initial connection members (in red). (©CSCEC.)

buildings and was made possible by an enlightened design and construction team working closely together. Arup's PTS was key in communicating the assumptions on which the design was based and the outcomes required.

Client: China Central Television
Design architect: OMA Stedebouw BV
Associate architect: East China Architecture & Design Institute
Year of completion: 2012
Height: 234 m
Number of stories: 54
Gross floor area: 473,000 m^2
Use: office
Main contractor: China State Construction Engineering Corporation
Arup's role: Structural, geotechnical, security, fire, mechanical, electrical and public health engineering.

1.3.51   MahaNakhon, Bangkok (ARUP)

MahaNakhon is a 77-story tower in Bangkok and was the tallest tower in Thailand upon completion. Its most

Figure 1.134 Connection, (a) Installing connection elements (b) Elements in place prior to "locking off" the joints. (© Arup.)

striking architectural feature is a 5-m deep pixelation that occurs throughout its full 314 m height (Fig. 1.135). As well as being a dramatic visual feature, it generates balconies and floating rooms, as well as public terraces on the retail floors. The 150,000 m² mixed-use development includes a hotel, apartments and retail, as well as an observation deck and outdoor rooftop bar frequented by Engineers.

Owner PACE Development Corporation Plc. commissioned OMA / Büro Ole Scheeren, with the aim of achieving a new standard of architectural design quality for Bangkok. Arup was brought in to provide multidisciplinary engineering services, delivering a construction-led design which could be delivered by the Thailand industry whilst maintaining the unique architectural vision.

Figure 1.135 MahaNakhon. (© Arup.)

The tower needed a structural system that could achieve sufficient strength and stiffness to resist the lateral loads acting on the building, whilst accommodating the pixelation with a minimum of transfer structure.

Whilst a tower of this height would normally utilize a central core with a perimeter column frame to resist the vertical and lateral forces, MahaNakhon's unique form required the columns to be setback into the floor plate, creating a perimeter zone that could be "eroded" to form the pixelations without impacting on the overall tower performance (Fig. 1.136). Arup worked with the architect to finalize the scheme through optimizing the following parameters:

- Number of columns: minimize the impact on floor planning whilst maintaining structural performance
- Distance from columns to façade: enabling each floor to be cantilevered independently within the available floor depth without compromising the appearance of the pixelation
- Introduction of bespoke local solutions for a limited number of pixels with more complex framing requirements

Bringing the columns inwards would increase the effective slenderness of the tower, and lead to significant stress imbalances between the columns and core walls. The optimum column location—balancing structural performance and architectural/planning requirements—was found to be approx. 5 m from the perimeter, resulting in an effective slenderness ratio (height/structural width) of 9.4.

Increased slenderness makes a tower less efficient at resisting lateral forces, since the smaller lever arm reduces the lateral stiffness yet increases the push-pull forces in the columns under lateral loading. It is a particular concern for residential towers, since occupants are more likely to perceive lateral movements than in noisier transient environments such as office towers. Arup worked with the wind tunnel laboratory to verify that accelerations would remain within established international recommendations.

Outriggers were introduced at transfer levels to improve the lateral stiffness by sharing loads more effectively between core and columns, whilst the latter were kinked outwards at the lowest transfer level to increase the lever arm on the stories nearest the base of the tower.

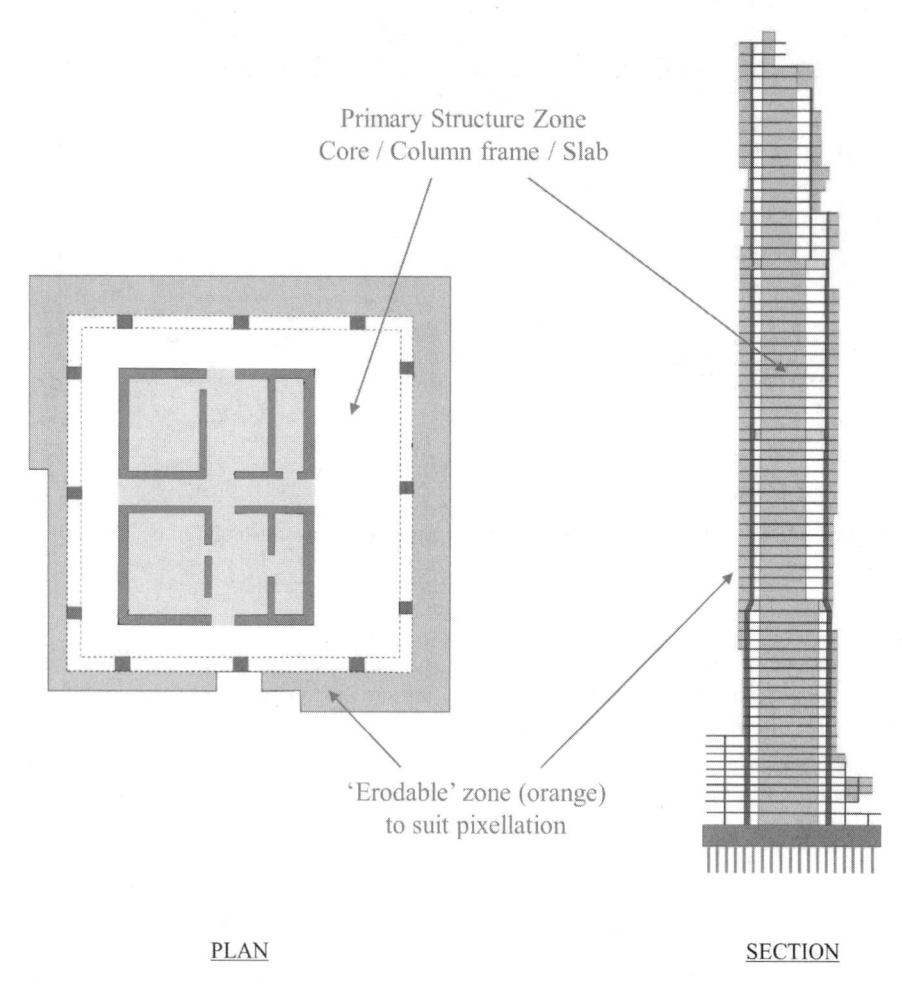

Primary Structure Zone
Core / Column frame / Slab

'Erodable' zone (orange)
to suit pixellation

PLAN

SECTION

Figure 1.136   Structural Concept. (© Arup.)

The inboard columns attract a larger tributary area and hence gravity loading than in a typical tower, whilst the core (which requires thick walls to provide lateral stiffness) carries relatively low vertical forces (Fig. 1.137). The resulting stress imbalance meant that differential axial shortening between columns and core would become a key consideration during construction, as described below.

The final scheme comprised twelve columns working with a central core. A primary beam runs between the columns on all floors to provide a stiff support for the secondary structure supporting the long cantilevers. Reinforced concrete was used for the majority of the structural elements, making MahaNakhon one of the tallest concrete buildings in the world. The floor system was further optimized by contractor Bougyues-Thai's under a Design & Build contract, whose specialist consultant VSL specified post-tensioned slabs spanning on to 600 mm band beams.

Although seismic forces in Bangkok are low, the effects of long-distance earthquakes are felt due to the soft ground in the city, especially in tall buildings (MahaNakhon's natural period is 7.1s). The tower was designed to Thai seismic codes, which follow IBC2006, but the performance was checked against *CTBUH Seismic Design Guide 2008: Appendix B - Guidance for Design of Tall Buildings in Low Seismic Areas*. This permits a simplified design approach provided all elements have a demand-capacity ratio of less than 2 when analyzed elastically (R=1) with 2% damping under a Maximum Considered Earthquake. All elements were found to remain elastic, hence non-linear analysis and seismic detailing were not required.

The stratigraphy consists of made fill over layers of clays and ultimately dense sands. The tower is founded on 129 no. reinforced concrete barrettes, 1.2 m × 3.0 m in section and with a toe depth at −65 m, within a sand layer. Barrettes have a greater perimeter than conventional

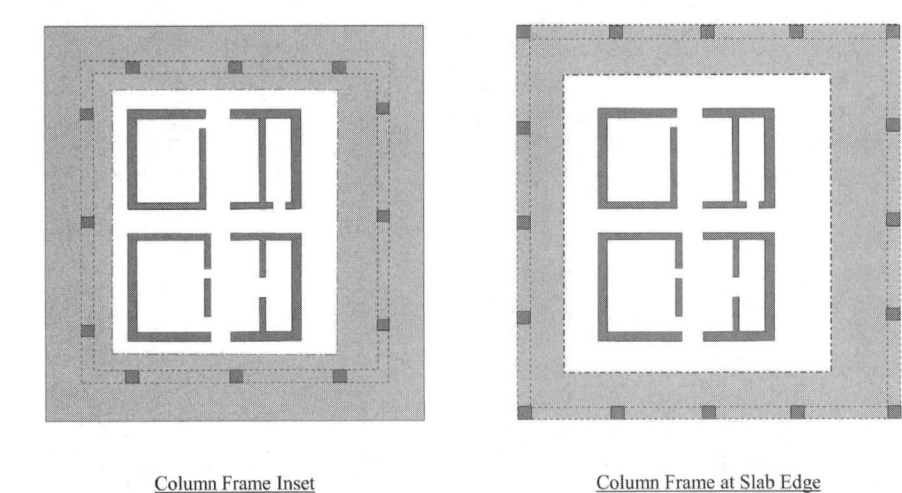

Column Frame Inset   Column Frame at Slab Edge

Figure 1.137 Tributary Area: impact of moving columns inwards (column tributary area increases by 45%). (© Arup.)

bored piles with equivalent cross-sectional area, hence more load carrying capacity (in skin friction) per unit volume of concrete. The superstructure loads are transferred into the barrettes via a pile cap up to 8.75 m thick.

For planning reasons, both the upper levels of the tower and the core step back on the east side. As a result, the tower's center of loading is approx. 450 mm west of its geometric center. This induces theoretical lateral deformations under dead loading, which are greater than those predicted under wind load. This was mitigated by presetting of the tower during construction with an incline in the opposite direction of approximately 2 mm per floor. Regular monitoring of movements, as well as creep considerations, were fed into the presetting process to keep the tower on target (Fig. 1.138).

Arup worked with Bougyues-Thai to address differential axial shortening through analysis and optimization

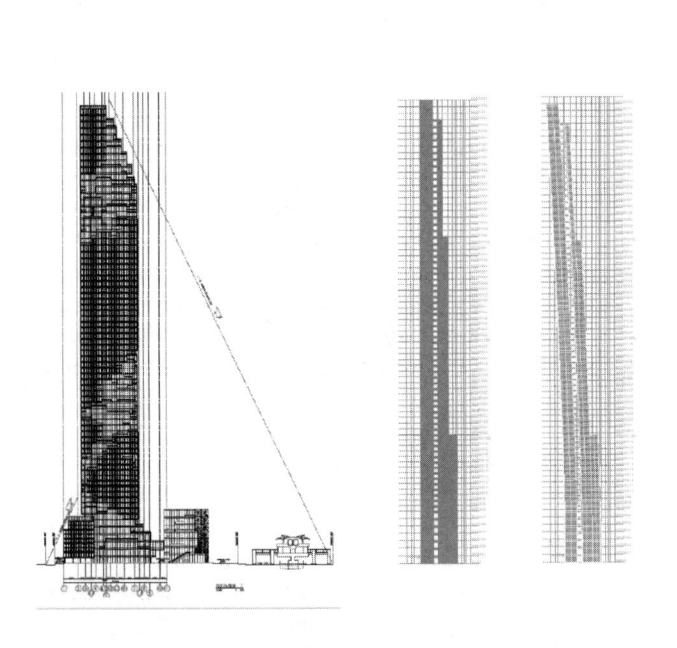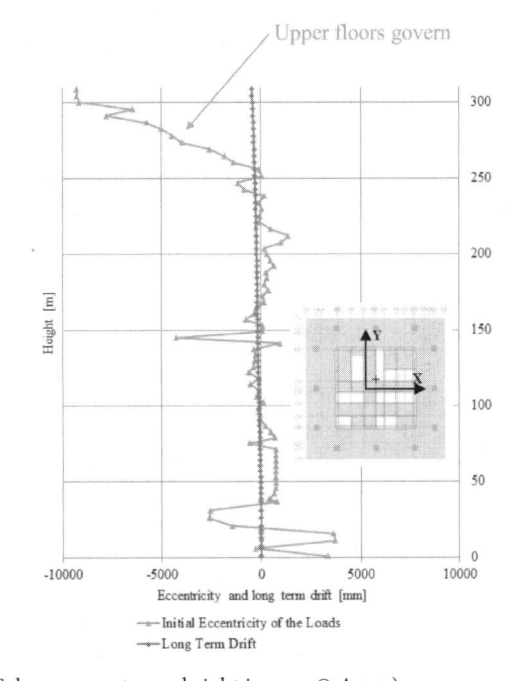

Figure 1.138 Dead Load Eccentricity. (Left image © OMA / Büro Ole Scheeren; center and right images © Arup.)

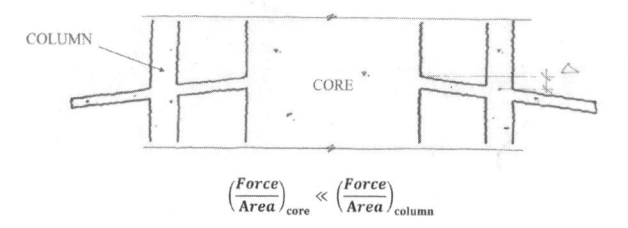

Figure 1.139 Differential Axial Shortening. (© Arup.)

$$\left(\frac{Force}{Area}\right)_{core} \ll \left(\frac{Force}{Area}\right)_{column}$$

of the construction sequence. Normally, outriggers are connected at a late stage of construction such that they do not attract gravity loads (hence maintain full capacity for resisting lateral forces). On MahaNakhon, this conventional approach was predicted to lead to relative shortening of 100 mm between core and columns (Fig. 1.139). Through construction sequence analysis, we recommended to connect each outrigger as construction progressed. This reduced shortening to acceptable levels, yet maintained sufficient lateral load capacity in the outriggers. These studies also showed that column forces would be some 10% larger than those in a simple "wished-in-place" model, demonstrating the importance of carrying out such time-based staged analysis.

MahaNakhon has become a city icon since completion in 2018, and its extensive public components make it a destination for locals and visitors alike. Arup worked successfully with the architect to turn the original concept into a practical scheme, and later with the contractor to realize the project successfully on site.

Key Facts

Year of Completion:	2018
Height:	314 m
Gross Floor Area:	150,000 m²
Use:	Hotel, Apartment, Retail
Arup Scope	Concept & Schematic Design:
	- Structural Engineering
	- Building Services Engineering
	- Vertical Transportation
	A Construction Stage Review:
	Structure
Owner	Pace Development Corporation Ltd
Architect	OMA / Büro Ole Scheeren
Main Contractor	Bougyues-Thai Limited
Other consultants	Warnes Associates; Aurecon Group; Palmer & Turner (Thailand) Ltd.

1.3.52 Shanghai Center Tower (Thornton Tomasetti)

At 632-m tall, the 118-story Shanghai Center Tower is one of the tallest buildings in the world (Fig. 1.140). The tower tapers and twists along its height, creating a unique elegant shape while helping to reduce wind and seismic load at the same time (Fig. 1.141). The twisting shape of the building is created by an innovative double-skin architectural design. An eight-stacked-zone approach, resembling tiers of a wedding cake, was adopted for interior spaces. Each zone has uniform cylindrical floor plates. Floor diameters vary from 82.2 m at Zone 1 to 46.5 m at Zone 8. At the top

Figure 1.140 Shanghai Center tower. (Courtesy Shanghai Center Development Group.)

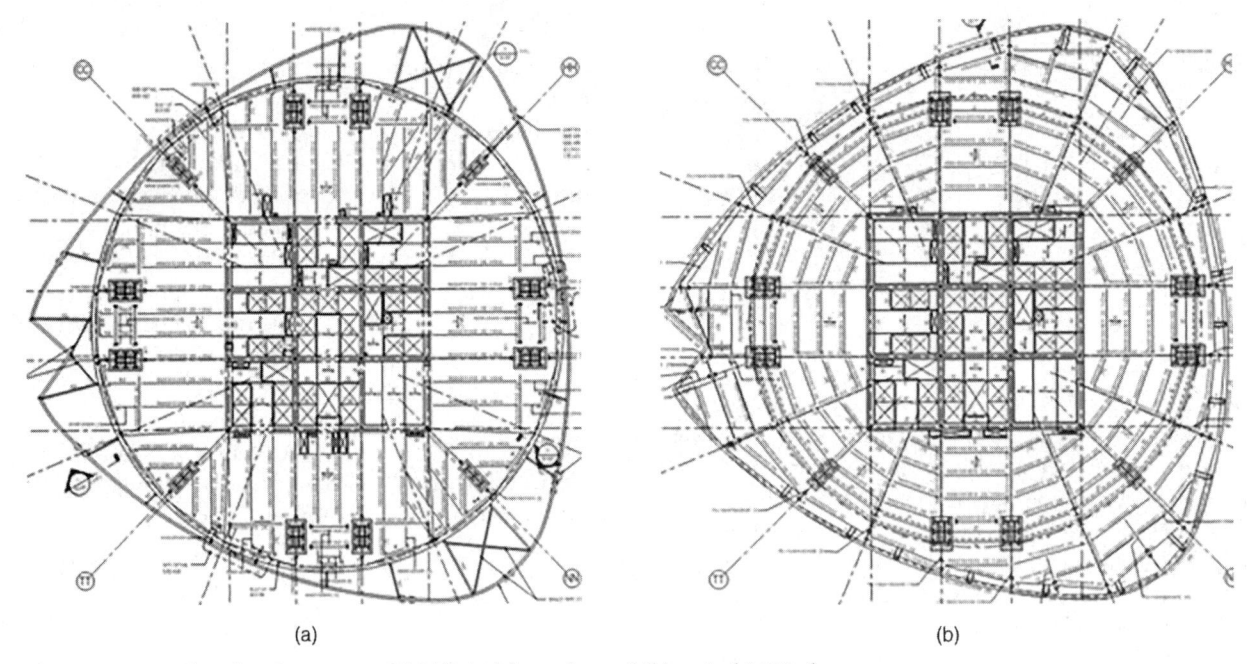

| (a) | (b) |

Figure 1.141   Shanghai Center tower: (a) Typical floor plan and (b) typical MEP plan.

of each zone, two levels of MEP and/or refuge slabs extend to the outer skin (Fig. 1.142).

The lateral force-resisting system for the tower is the "Core-Outriggers-Mega Frame" system including a composite core, super columns, outrigger trusses, and belt trusses (Fig. 1.143).

The 9-cell composite core has a square shape of 30 m × 30 m from Zones 1 through 4. At Zones 5 and 6, where fewer elevators are needed, the four corners of the core are cut off to increase the leasing span. The core further reduces to a cruciform at Zone 7 and 8. The composite core has wide flange steel columns embedded at the

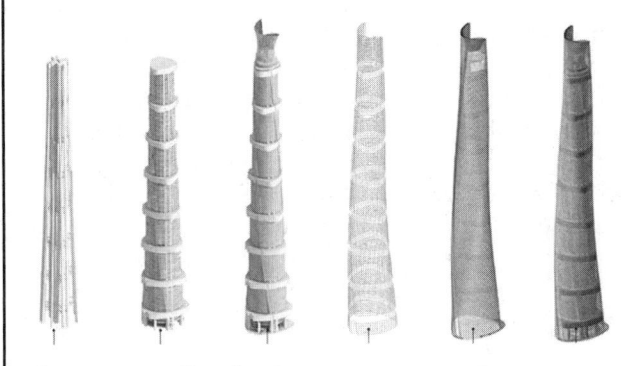

Figure 1.142   Shanghai Center tower: structural components.

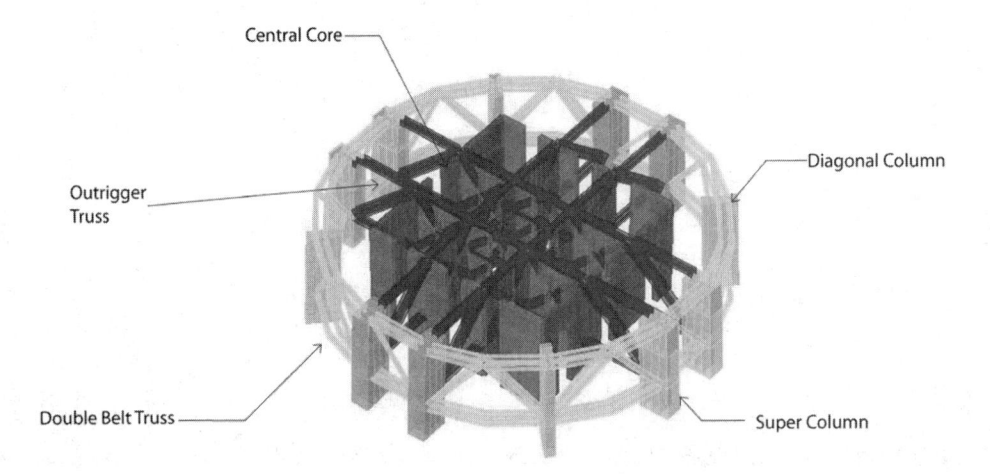

Figure 1.143   Shanghai Center tower: lateral force resisting system.

boundary zones, or most highly stressed corners and ends of core walls. They serve to both strengthen the core and provide a clear load path as well as easier connections from outrigger trusses to the core. The bottom two zones also have embedded steel plates to enhance wall ductility and reduce wall thickness.

Eight rectangular mega columns are located on the perimeter of the floor plates, two on each side to align with core web walls. They have a slight uniform inward tilt to work with the tower taper. The mega columns also have embedded steel to boost ductility and optimize member sizes. Steel outrigger trusses on MEP floors connect mega columns to the core to improve overturning resistance and lateral stiffness. In addition two-story-tall steel belt trusses at every MEP level link super columns to form a perimeter megaframe that provides additional lateral stiffness and strength to the overall tower lateral system.

Shanghai Center Tower is enclosed by a unique "double-skin" system—a cam-shaped-plan exterior skin that twists and tapers in elevation, and a circular-plan interior skin. The space between the double skins creates signature atrium spaces extending vertically more than ten stories from each Amenity/sky lobby level to the soffit of the MEP level above. The large and varying distance from exterior curtain wall to main structural system, together with the twisting and tapering geometry of the tower, requires a creative curtain wall supporting system (Fig. 1.144). The design adopted is a "bicycle wheel" system: the exterior curtain wall is supported by a hoop ring on each floor to back up the outer skin cam-shaped plan, with hoop rings hung from the MEP level above by sag rods, and braced against the inner cylindrical tower by radial struts (i.e. spokes). The unique shape of the tower's outer façade is created by a set degree of rotation between hoop rings on adjacent floors, while also decreasing in dimension incrementally as the tower rises up to fit the tapered shape of this super-tall tower.

At the top of the tower is a 58 m tall crown. The exterior façade is supported from behind by vertical trusses that deliver gravity load directly to the floor below, and transfer lateral loads from wind pressure back to the core through radial struts. Kicker trusses support the crown inner face while laterally bracing the outer trusses above the tower roof level.

The foundation system of Shanghai Tower is a reinforced concrete mat supported by nine hundred forty-seven (947) 1-m diameter reinforced concrete bored piles. To address soft soil condition, the piles have 52 to 56 m effective lengths and end grouting provided at pile tips enhances working load capacity to 1000 tons

Figure 1.144 Shanghai Center tower: curtain wall supporting system. (Courtesy Connie Zhou Studio.)

while reducing settlement. Predicted tower settlement is approximately 100 to 120 mm after 5 years. Construction of the 6-m thick mat set a record, pouring 61,000 cubic meters of concrete continuously in 60 h.

The design architect is Gensler and the structural consultant is Thornton Tomasetti, with Tongji Architectural Design (Group) as AOR and EOR.

1.3.53   The New York Times Building, New York, USA (Thornton Tomasetti)

After spending 100 years in their previous headquarters in Times Square, it was time for the New York Times to construct their new headquarters building on Eighth Avenue between 40th and 41st Streets. The New York Times partnered with Forest City Ratner Companies to develop a 52-story office building above a three-story podium. The podium houses the newsroom as well as a garden atrium and an auditorium. The lower half of the office tower is occupied by the New York Times and the upper half was developed as speculative office space. **Refer to Fig. 1.145.**

Figure 1.145   Finished building.

Renzo Piano Building Workshop won the design competition and partnered with Fox & Fowle (now FX Collaborative).

In a world where building shapes morph and never contain a right angle, the New York Times Building tower has a beautifully symmetrical skeleton with a rectangular floor plan of 150′ × 145′. The central 65′ on the north and south facades pulled away from the rectangle by 20′ to create a cruciform shape. **Refer to Fig. 1.146**. The tower has no setback and no column transfers. The tower is topped with a slender 300-ft tall mast that tapers from 8 ft at the base to 8 inches at the top.

The building is constructed of structural steel with a central braced core and outriggers at mid-height and at the roof. The outrigger trusses extend from the core to engage all perimeter columns providing a very efficient lateral load resisting system. The structure is founded on spread footings bearing on 20-ton Manhattan schist.

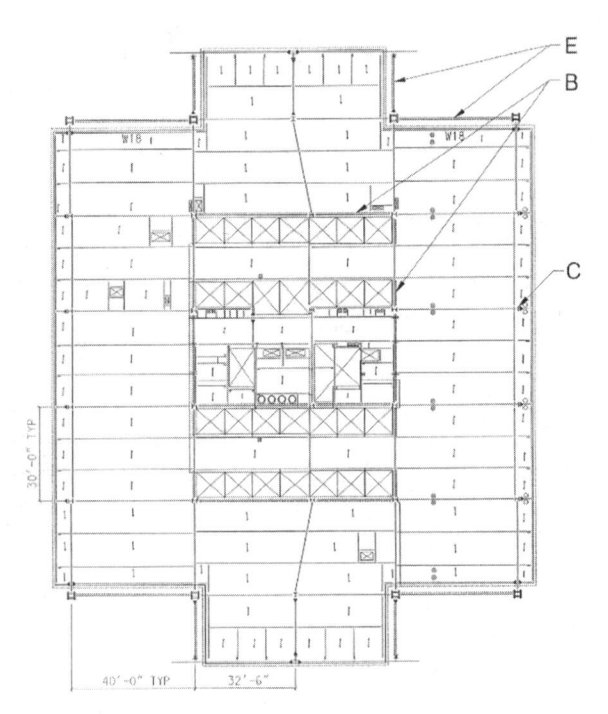

Figure 1.146   Floor plan.

By far the most unique aspect of the tower is the use of exposed structural steel. **Refer to Fig. 1.147**. Eight columns on the north and south sides of the building are pulled 3 ft outside of the façade and are connected by pretensioned rods to provide tension only X-bracing to supplement the lateral load resisting system of the building. Additional exposed steel beams and rods support the 65′ wide × 20′ long cantilevered floor areas which create the cruciform shape.

The exposed structural steel becomes part of the architecture and required significant study and coordination with the architect. Sketches led to wooden models and building information models and finally a full-size steel mockup to make sure all of the details were perfect.

Figure 1.147   Exposed structural steel.

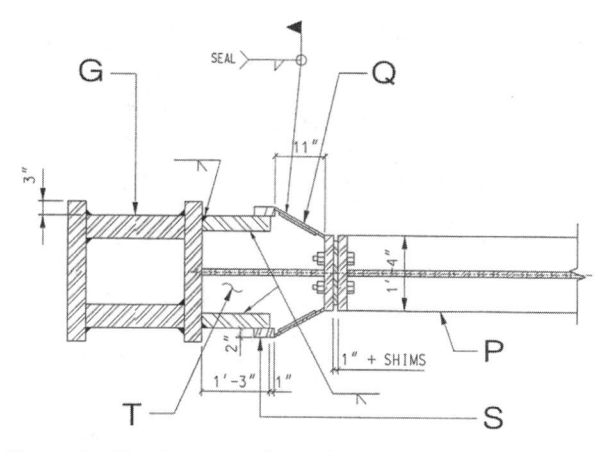

Figure 1.148 Exposed column detail.

Figure 1.149 X-bracing connection detail.

The exposed columns are built-up steel plate boxes with 30 inch × 30 inch outside dimension with the web plates slightly recessed to allow the flange tips to be exposed. **Refer to Fig. 1.148.** The outside shape of the steel columns remains consistent for the full height of the building. To maintain an efficient use of material, the column area is reduced by reducing the plate thickness of the webs, which are not visible, and stepping the flange dimension from 4 to 3.5, 3, 2.5, and 2 inches over the height of the building. As the column flange thickness changes, the X-bracing rods also reduce in diameter.

The exposed X-bracing provides significant improvement in building drift and building accelerations, but is not relied on for building structural safety and stability and as such, the rods are not required to be fireproofed. The rods improve the building drift under wind loads from height/350 to height/450.

The X-bracing rods were prestressed using the manufacturer's proprietary hydraulic system that provides an accuracy of 2%. During construction, the rods were over tensioned to account for axial shortening of the columns and to retain a final tension such that none of the rods go slack or into compression under wind loads.

The rods are oriented in a precise pattern with two rods oriented side-by-side which straddle two crossing rods oriented over-under. The rods connect with clevises to paddle-shaped gusset plates welded to the exposed columns. This highly detailed connection includes a stub to receive a field bolted horizontal beam to resolve the tension in the rods. **Refer to Fig. 1.149.**

The combination of enclosed and exposed structure creates additional challenges resulting from thermal changes that can affect member and connection forces and floor levelness. We established a design temperature differential of +70°F to −80°F based on local climate data and recommendations from the National Building Code of Canada (NBC)

that reflect radiant heating and cooling effects. For a temperature change of 70°F, the exposed column of 750 ft will grow approximately 3.5 in. This change in length will not occur for the enclosed column 30 ft within the building creating a floor slope of length/100. To mitigate this issue, the roof outriggers and additional belt trusses were used to restrain the resulting change in length of the exterior column to 1.2 in which reduced the floor slope to length/300.

Additional challenges that exposed structural steel presents are thermal bridge, condensation and weather tightness issues. Beams penetrate the façade in 16 places on every floor. At the façade line, the beam is upturned into the raised floor so that the penetration can occur centered on the façade's spandrel panel. The exterior portion of these beams are fireproofed, insulated and clad. A thermal gradient is established along the stub and condensation on the beam within the building is avoided. Special detailing was coordinated with the façade manufacturer to ensure that the façade is sealed for weather tightness. **Refer to Fig. 1.150.**

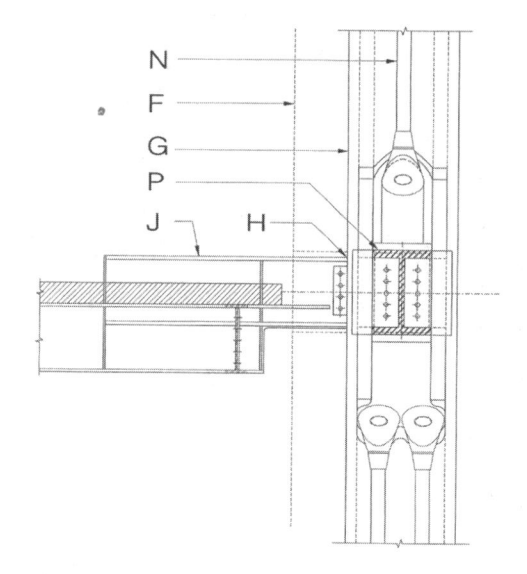

Figure 1.150 Beam detail at façade penetration.

1.3.54   Federation Tower East, Moscow, Russian Federation (Thornton Tomasetti)

Two towers rise over a sculpted podium creating the visual impression of a 374-meter-tall sailboat drifting through the Moscow International Business Center, a.k.a. "Moscow City," in the western part of Russia's capital. **Refer to Figs. 1.151, 1.152, and 1.153.** The 93 story East Tower and 66 story West Tower, designed by a collaboration between Architects ASP Schweger Assoziierte and Tchoban Voss Architekten, are triangular in plan and address each other across a narrow void space, originally intended to feature the mast of the "ship." The other two faces of each tower curve gently in plan and vertically giving the impression of billowing sails. Federation Tower East is the second tallest building in Europe and was the first in Russia to utilize high performance concrete. The "B90" material produced for the tower is roughly equivalent to 11,000 psi concrete in the United States.

Figure 1.151   Federation tower completed towers. (Courtesy Igor3188 - Own work, CC BY-SA 4.0, https://commons.wikimedia.orgwindex. phpcurid=50238600.)

Figure 1.152   Federation Tower Completed Towers. (Courtesy CTBUH.)

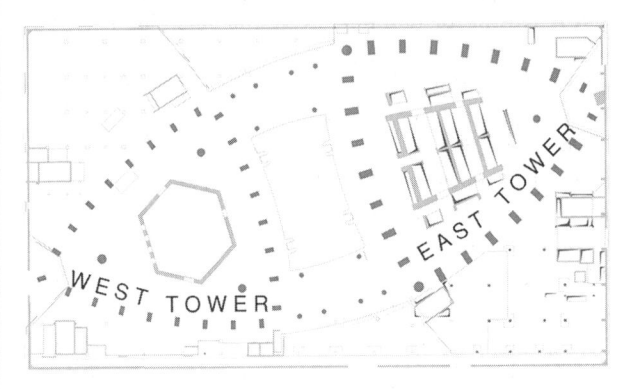

Figure 1.153   Level 60 East Tower—Typical Hotel Level Framing Plan.

The East Tower structure is centered around a 20 m × 22 m trapezoidal concrete elevator core with walls as thick as 1.4 m at the base and narrowing to 400 mm at the top. Perimeter columns are reinforced concrete and step slowly inward along with the tower façade with increasing height. These are aligned about 2 m to 3 m inboard of the exterior façade, which produces two positive outcomes:

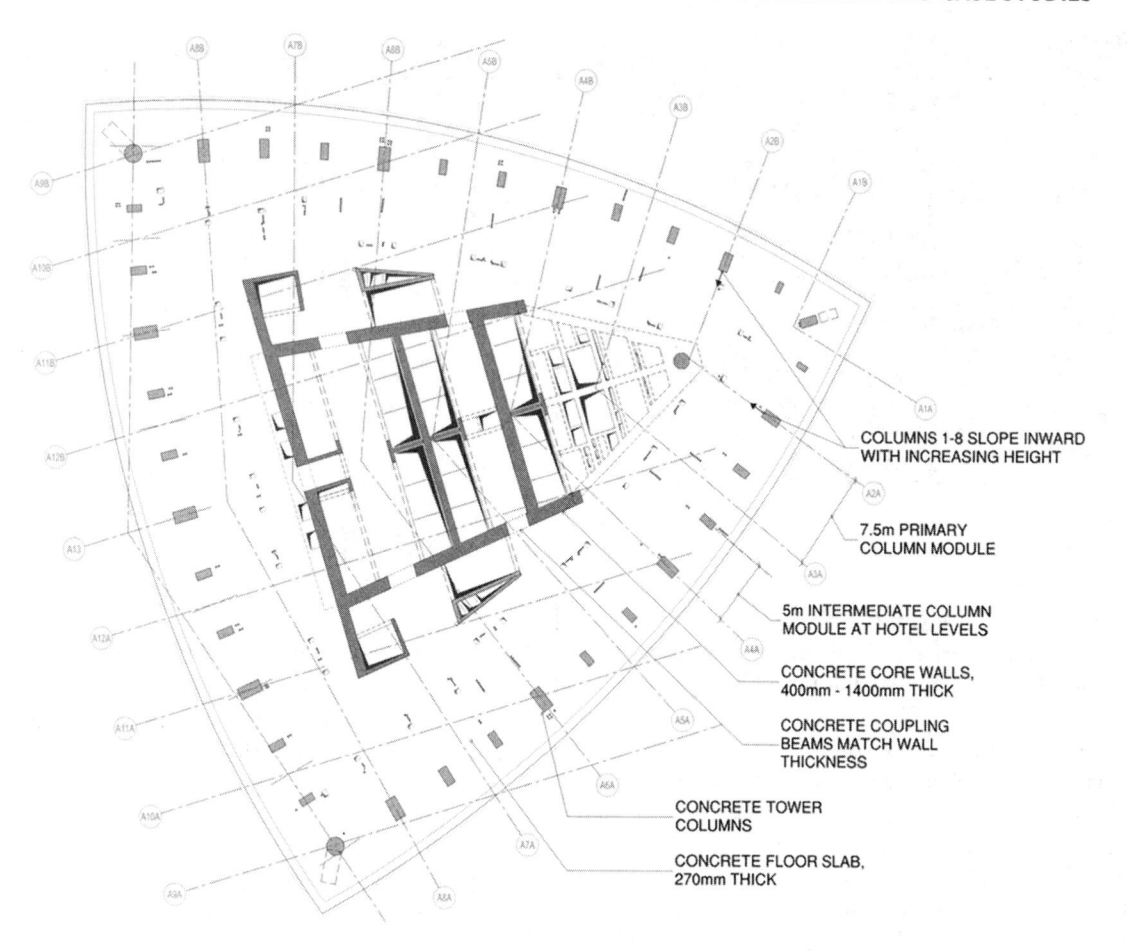

COLUMNS 1-8 SLOPE INWARD
WITH INCREASING HEIGHT

7.5m PRIMARY
COLUMN MODULE

5m INTERMEDIATE COLUMN
MODULE AT HOTEL LEVELS

CONCRETE CORE WALLS,
400mm - 1400mm THICK

CONCRETE COUPLING
BEAMS MATCH WALL
THICKNESS

CONCRETE TOWER
COLUMNS

CONCRETE FLOOR SLAB,
270mm THICK

Figure 1.154   Level 60 East Tower—Typical Hotel Level Framing plan.

(1) core to column floor spans are short enough that a 270 mm conventionally reinforced slab is sufficient to span from the core to the column line for any of office, residential or hotel occupancy, and (2) the wide space between glazing and columns allow for unimpeded views from the office and hotel levels that are laid out in rhythm with the column module. **Refer to Fig. 1.154.**

Concrete encased steel outriggers and belt trusses stabilize the tower against the wind and balance out differential shortening that could otherwise occur between the columns and walls. **Refer to Figs. 1.155 and 1.156.** Perimeter columns are interconnected by belt trusses at three levels, each of which doubles as a hub for the building's MEP systems. Using structural steel here allowed the MEP system to function in the shared space and for the building to breathe through the large open spaces between truss members in a way that would not have been viable in an all-concrete belt system. At the 1/3 and 2/3 tower heights, outrigger trusses cross the tower footprint in the

Figure 1.155   Axonometric model view of East Tower level 61 Outrigger and Belt Trusses.

Figure 1.156 East Tower Level 33 Outrigger and Belt Trusses Prior to Encasement.

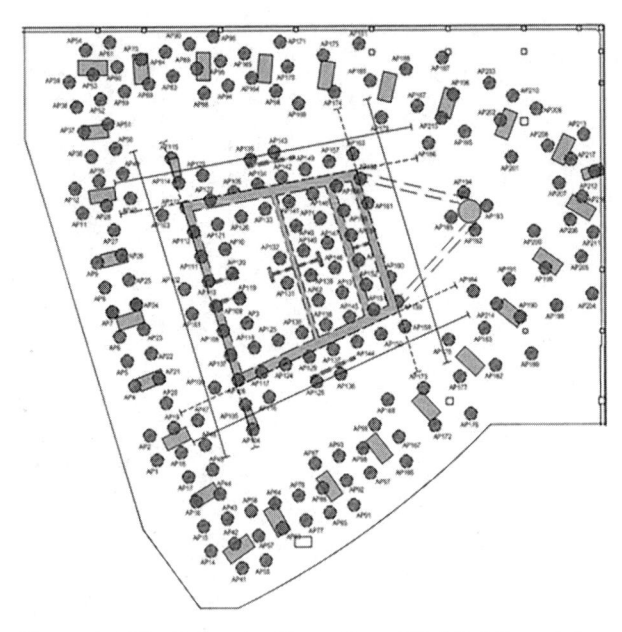

Figure 1.157 Federation Tower East—Piled Raft Foundation.

form of a tic-tac-toe board to link the core to the perimeter belt. The mid-height belt system is there to accomplish a wholesale column shift at the 47th floor intended to accommodate the change in occupancy from Office to Hotel. That belt does not connect back to the core but does act as a "virtual outrigger," participating in the overall stabilization of the tower. Thus, the gravity system and the lateral load resisting system merge in a single, fully coherent structure. Sequential construction cases were built into the 3D analysis of the full building in order to capture the evolution of the gravity load paths within the overall structure as belt and outrigger systems were added to the partially built tower with increasing height.

The original Owner of the development was not shy in touting the tower's design as an early post-9/11 example of extreme resiliency, and the structural design respects this goal of extreme resiliency. Extra steps were taken during design to ensure that a local loss of column capacity would not lead to progressive collapse. All of the perimeter column reinforcement is proportioned and mechanically coupled such that any column could hang from the belt truss above in the event of an unexpected and sudden loss of structure below. The practical benefits to the everyday occupants are a very well-behaved tower; from the standpoint of all of wind drift, differential shortening and wind-induced vibrations, Federation Tower East far exceeds the industry standard performance targets.

At the time of design, Tower A was far taller than any other occupied building in Russia, with residential units at the upper floors. Thus, serviceability criteria was held quite tight with wind drifts of about H/750 under 50-year wind gusts.

Both the East and West towers feature a fully glazed multistory open space at the top of each tower, including a downward curving glazed roof. Exposed structural steel frames support the glass creating a pair of light filled volumes

The East Tower is supported on a piled mat foundation that consists of a four-meter-thick mat that delivers about 10% of the tower weight directly into the soil at foundation level and (196) 1.5 m diameter drilled shafts which deliver 90% of the tower weight down to the underlying bedrock. **Refer to Fig. 1.157.** The pile layout reflects the sloping tower's weight distribution, with dense pile spacing along the west edge of the tower and under the core, and looser pile spacing on the eastern portion of the tower footprint. This layout resulted in a calculated differential settlement in the range of only 10 to 15 mm across the tower footprint. Piles generally follow the curved tower structure layout but were laid out on a grid to avoid congested interfaces between the piles, columns and dense reinforcement in the mat.

1.3.55 Jeddah Tower, Kingdom of Saudi Arabia (Thornton Tomasetti)

The 1-km tall Jeddah Tower is designed as a powerful three-legged, linearly tapering form combining architectural simplicity with structural engineering logic to produce a highly constructible ultra-tall tower, **see Fig. 1.158.** The program for the tower includes primarily hotel and residential occupancy with a small amount of office space near the base. The combination of

Figure 1.158 Jeddah Tower Architectural Rendering. (Courtesy Adrian Smith + Gordon Gill Architecture.)

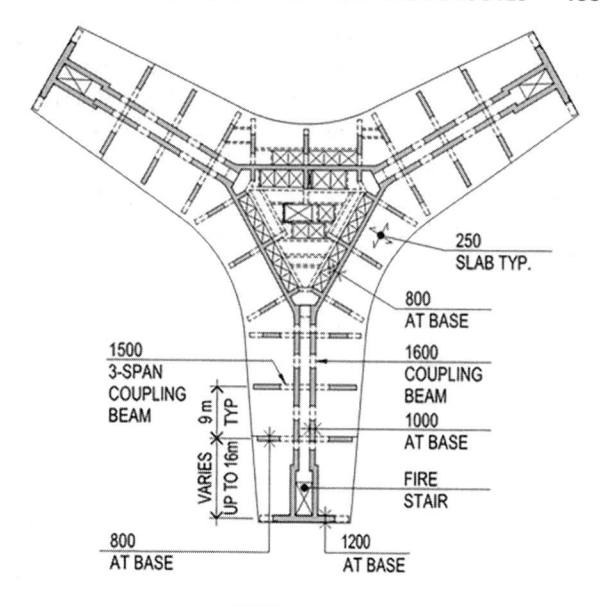

Figure 1.159 Level 7 Framing Plan.

the desired space utilization, the preponderance of cast-in-place concrete construction in the Kingdom and the benefits of high building mass in response to the crucial issue of wind effects on the tower led to the selection by the design team of an all-concrete tower.

The structural system is composed of a series of concrete wall planes organized around the central triangular core area and vertical transportation, the public corridors traversing the length of each of the three wings, and around the fire escape stairs located at the ends of the wings. The system is fundamentally a bearing wall system wherein the lateral and gravity load resistance is located coincident and in the walls alone; **see Fig. 1.159.** Stabilizing "Fin" walls are located perpendicular to the core and corridor wall lines at approximately 9 m centers which allow for simple (250 mm thick), conventionally reinforced flat plate floor construction to be employed. These fin walls reduce the slab spans and support gravity loads along the perimeter of the floor plate. The fin walls are themselves connected continuously across the

corridor space through three-span coupling beams. It is recognized that the system is uniquely suited to residential construction where the units may be effectively laid out based on the compartmentalization created by the wall lines.

The most significant features that serve to organize the wall system are as follows (**refer to Fig. 1.160**):

1. All walls are vertical with the exception of the fire stair walls which incline together as a unit along a single angle to the vertical.

2. All wall segments are fully connected by deep coupling beams over the door openings. Therefore, all walls participate in resisting both gravity and wind loads.

3. There are no outrigger or belt walls or trusses in the entire system.

4. There are no column or wall transfers. As the wall system rises in the tower, the Fin walls simply drop off when they become too near to the sloping fire stair wall units.

Figure 1.160 Structural System Isometric.

5. There are no spandrel beams along the perimeter. All floors are reinforced concrete flat plate construction.

6. The system is entirely of cast-in-place reinforced concrete, other than the uppermost 50 m of steelwork for the spire pinnacle.

Typical story heights are uniformly 4 m which allows 1.6 m deep coupling beams over the door openings while still providing very generous floor-to-ceiling heights within the residential units. All door openings are carefully organized vertically such that gravity loads are managed without abrupt discontinuities in the vertical load-path. While these structural demands impose a certain limitation on the residential unit layouts, the architects were able to incorporate these features without significant issue.

The last occupied floor is at elevation 660 m. Above this elevation, the walls at the ends of the three wings have inclined toward the center of the tower to the point that the floor plate is quite small which does not allow for suitable rentable space. Most importantly, the walls at the ends of the three wings still continue upward in an uninterrupted fashion and eventually rise to over 950 m in elevation. The vertical angle of each of the three wings is constant from the top of the foundation plane to the tower pinnacle. Each of the three wings, however, are defined by a slightly different vertical angle. Therefore, the three wings reach the top of the tower at different elevations. This has been exploited by the architectural team to render the distinct geometry at the top of the tower.

The unique issue of vertical shortening effects due to elastic, creep and shrinkage of the elements was a major driver in configuring the structural system. All bearing wall systems will tend toward equal gravity stresses near the base of the tower as long as the wall segments are interconnected and not free to shorten vertically in an independent manner. As all wall segments for the Jeddah Tower are connected by deep, stiff coupling beams, this is an ideal arrangement to combat the vertical shortening issue. Whilst there is significant redistribution of gravity loads, there is very little differential shortening between the wall segments throughout the height of the tower. Full three-dimensional sequential construction-based analytical modelling was performed on the tower to estimate the short- and long-term effects due to wall vertical shortening.

The tower is supported on 270 large diameter (1.5 and 1.8 m diameter) reinforced concrete bored (augured) piles ranging in depth from 45 m in the wings to 105 m in the center of the tower, **see Fig. 1.161.** All piling was excavated and concreted under polymer slurry with the majority of the load transferred to the surrounding rock

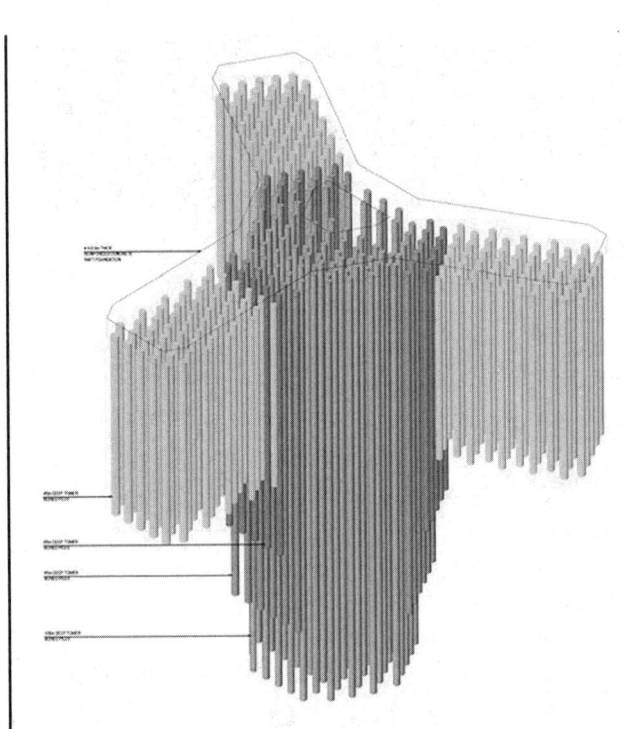

Figure 1.161 Piled Raft Foundation.

mass through friction along the sides of the piles. Working load capacity of the 1.8 m diameter piles was 4500 tons. The piling matrix extends throughout the entire footprint of the tower and is capped by a continuous raft foundation 4.5 to 5.0 m in depth. As the raft foundation itself rests directly on a rock stratum, the foundation elements together form a piled raft with a significant amount of the tower weight being transferred directly into the ground at the raft/rock interface. Long term foundation settlements are predicted to be on the order of 110 mm with no more than 25 mm differential between the center of the tower and the ends of the three wings. Construction progress in early 2020 had reached an elevation of 260 m, **see Fig. 1.162.** The architects are Adrian Smith + Gordon Gill Architecture, Chicago and the structural engineering is by Thornton Tomasetti, Chicago.

1.3.56 Wilshire Grand Center Tower, Los Angeles, California USA (Thornton Tomasetti)

Rising 1100 ft (335 m) above grade, the 73-story Wilshire Grand Center tower is the tallest building in the Western US. A downtown Los Angeles site with high seismicity added to the design challenge. The tower with gradually tapering ends has a high-end 889 room hotel stacked atop 400,000 ft^2 of office space, **see Fig. 1.163.** An adjacent podium, seismically separated by an atrium with a signature glazed roof, contains ballrooms and other amenities.

Figure 1.162 Status of Construction (January 2020).

Figure 1.163 Wilshire Grand Center tower showing tapering ends behind its podium. (Thornton Tomasetti.)

The tower and podium rise from a five-level-deep basement filling a sloping full-city-block site with a 17.6 ft thick tower mat and podium footings bearing on sandstone, **see Fig. 1.164.** Hotel functionality required a long, narrow floor plan gradually reducing in length at upper stories, with a long, narrow central core. These proportions drove key structural decisions.

The high strength (8 ksi, 55 MPa) concrete core provides stiffness and inherent damping for occupant wind comfort, constructed using self-climbing forms and rebar prefabricated into large panels. Long wall thickness varies with height from 48 inches to 24 inches. Transverse walls are 36 inches thick. Perimeter columns are steel plate boxes narrow in plan to minimize room intrusions, filled with high strength concrete contributing ¼ to ⅜ of compressive axial stiffness. Floors are concrete-filled metal deck on composite steel framing for long clear spans, fast erection

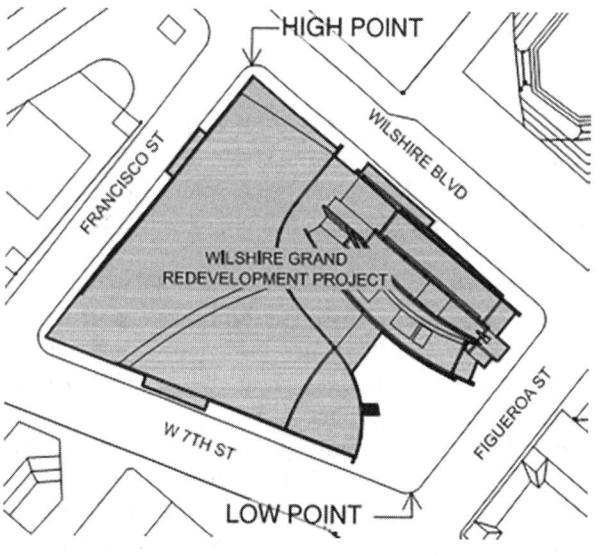

Figure 1.164 Site plan showing podium at left and tower at right separated by glazed atrium. (Brandow & Johnston.)

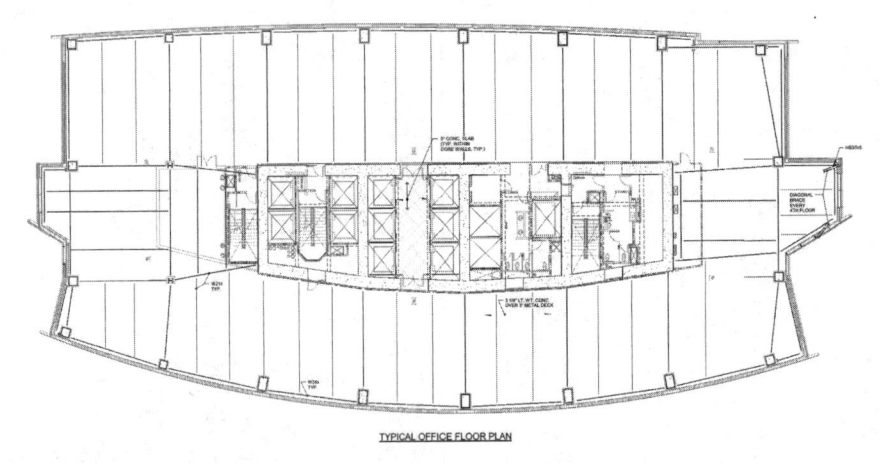

TYPICAL OFFICE FLOOR PLAN

Figure 1.165   Typical lower-level plan showing core and office floor framing. (Brandow & Johnston.)

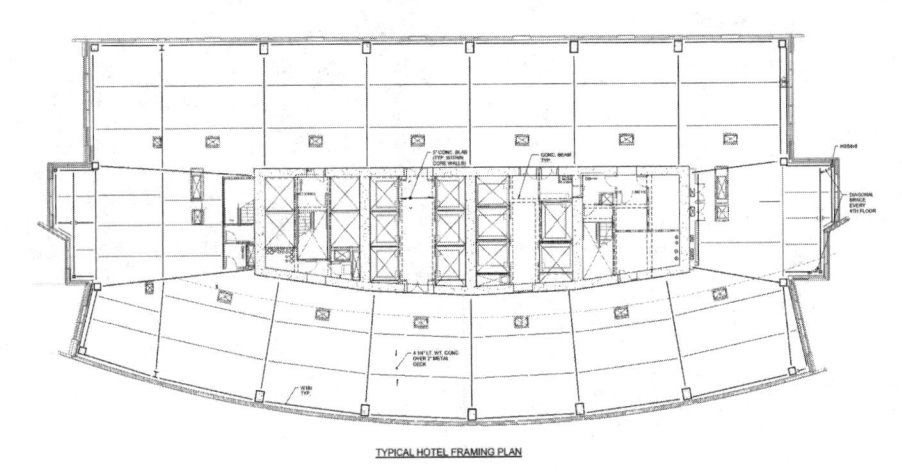

TYPICAL HOTEL FRAMING PLAN

Figure 1.166   Typical upper-level plan showing core and hotel floor framing. (Brandow & Johnston.)

and minimal seismic mass. Office floor beams span core to perimeter, see Fig. 1.165, while hotel floor girders hidden above room walls span core to perimeter, with shallower beams infilling between them, see Fig. 1.166.

Performance based seismic design (PBSD) was used for the tower to demonstrate acceptability of structural innovations rather than use prescriptive dual system moment frames. Nonlinear response history analyses (NRHA) modeled nonlinear element behaviors and subjected them to excitation by realistic earthquake time history records. Key structural design features include three different types of outriggers, belt trusses, buckling restrained braces, jacking for differential shortening, "phantom openings," and special fatigue and fracture control provisions for critical steel details.

A core 895 ft (273 m) tall but roughly 38 ft (11.6 m) wide is too slender to resist transverse wind and seismic loads by itself. Outriggers help resist overturning by connecting the core to perimeter columns at three zones along the building height, joining the core at the five transverse "web" walls, see Fig. 1.167. Multistory steel belt trusses at lower and upper outrigger zones maximize column stiffness and maintain compatible behavior by engaging all 20 perimeter columns rather than just the 10 columns with direct outrigger connections. Because conventional steel outriggers sized for wind comfort and stiffness would develop huge forces in a major earthquake, potentially damaging the walls and columns at connections, buckling-restrained braces (BRBs) are used as outrigger diagonals. BRBs are carefully-proportioned steel bars encased in mortar-filled steel jackets to accept

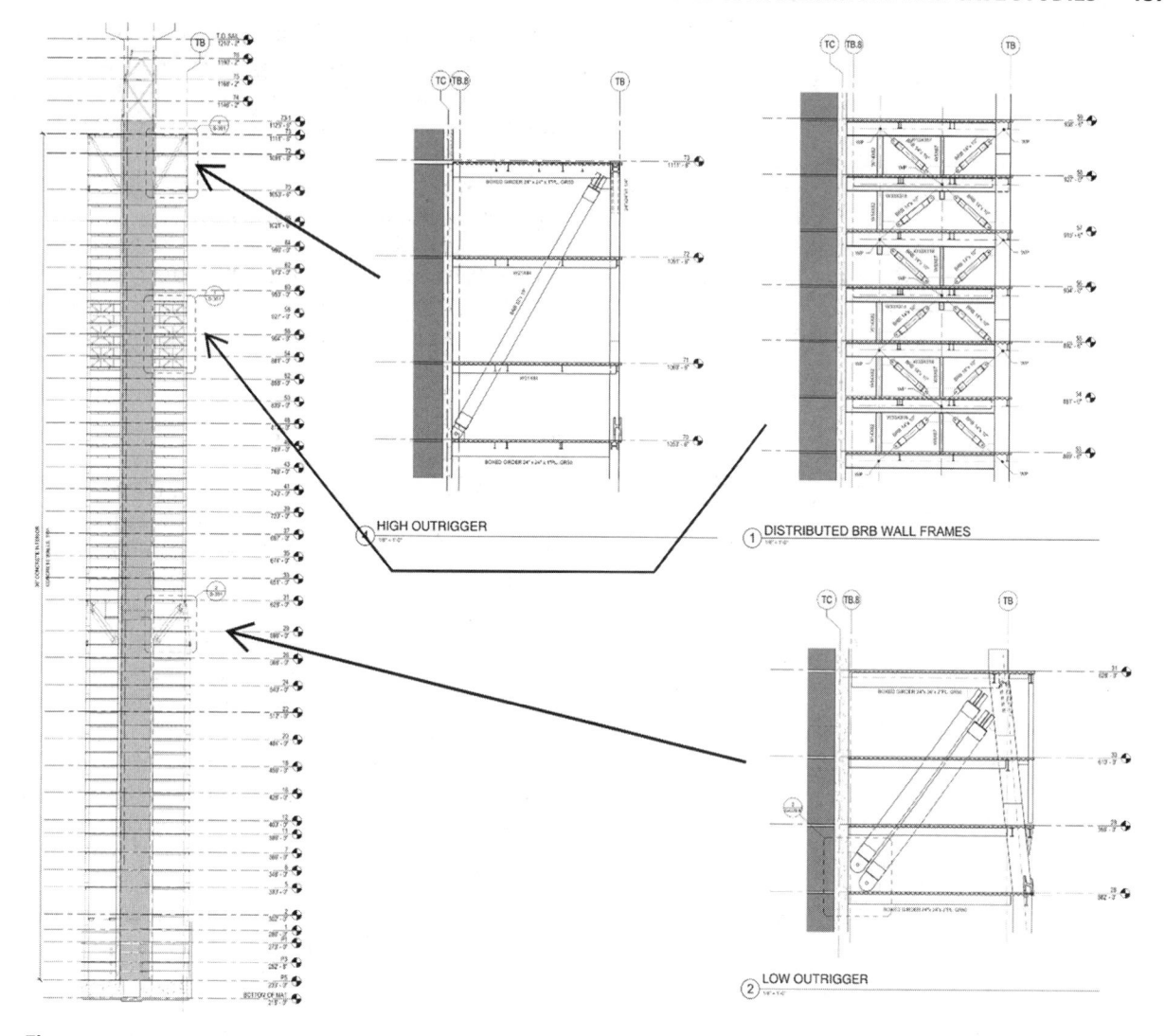

Figure 1.167   Transverse section showing locations and details of three outrigger zones. (Brandow & Johnston.)

multiple cycles of yielding in tension and compression with well-controlled behavior. The BRBs were sized to remain elastic under strength-level wind demands, and under a Service Level Earthquake (43-year return period), but to yield during a Maximum Considered Earthquake (2475-year return period).

Top outriggers, displayed in the Level 70 hotel lobby, use single BRB diagonals yielding at 2200 kips. Lower outriggers, visible on several office floors, use groups of four BRBs totaling 8800 kips that are so stiff they establish a secondary yield zone in the core, **Fig. 1.168**. Middle outriggers are unique, to avoid blocking hotel floor corridors: 7 levels of deep, stiff W33 girders reach across the corridors, backed up by a series of stacked X's using 800-kip yield BRBs. BRB diagonals with large

strains induce end rotations at outrigger beams. Upper and lower outriggers have seismically compact shallow beams for plastic rotations. Deep middle outrigger beams require a true pin at each end.

Differential shortening is always a concern when mixing steel columns and concrete core walls tall buildings. Shortening predictions from the B3 creep and shrinkage model, calibrated to values from lab testing of planned concrete mixtures, were incorporated in staged construction models by artificially shortening selected core wall segments. Although outriggers will restrain the core and columns to shorten similarly (within about 1.5 inches or 38 mm), they will experience large forces over time. The top BRBs were actively jacked to 50% of compression yield to avoid yielding in tension

Steel fracture risk was studied for several critical locations, from column bases to the tower top spire, considering fatigue from cyclic wind loads followed by a major seismic event. This risk was controlled through steel specifications with special Charpy V-Notch values and testing requirements, as well as minor adjustments to details at welded joints and splices.

The project was completed June 2017. Architect: AC Martin Partners, Los Angeles. Structural engineering consultant: Thornton Tomasetti, Los Angeles. Engineer of Record: Brandow & Johnston, Los Angeles.

1.3.57 ThyssenKrupp Innovation & Qualification Center Case Study (Walter P. Moore)

ThyssenKrupp's Innovation and Qualification Center (IQC) in Atlanta, Georgia (Fig. 1.169) includes a 420-ft elevator test tower to further develop elevator technologies. The 70-ft curtain wall glazed crown at the top of the IQC includes an Observation Level and two Conference Levels with a spectacular view of The Battery and the Atlanta Braves stadium. The low-rise facility surrounding the tower includes a product showroom, event, and laboratory spaces.

The test tower was constructed with 18" thick, 6000 psi concrete walls forming a box around the entire perimeter (Fig. 1.170). A full-size pit was blasted 32' into bedrock due to the site grading and to withstand the buffer loads of the test elevators and passenger elevators. The 398' tall concrete tower walls were formed and placed in less than 56 days using slipform wall construction, which was less than half the time estimated for the use of self-climbing or traditional jump forms. Slipform wall construction involves the placement of concrete into a continuously moving form. The wall forms are jacked upward slowly at a controlled rate until the final elevation is reached. For the ThyssenKrupp tower, the concrete was continually placed in 10" lifts around the tower footprint. Jumpform wall construction is an alternative option to slipforming, but it would have taken longer as the process involves releasing, raising, and resetting the forms between lifts of wall pours. Other advantages of the slipform construction method include tight construction tolerances of the walls (continually checked by means of control lasers at the corners) and the added safety of using entirely encapsulated working and hanging decks which eliminate the need for workers to tie-off to safety lines.

Steel floor framing with composite decks are located at a vertical spacing of 42' around the tower's nine test elevators and one freight elevator (Figs. 1.170 and 1.171), with partial catwalk levels provided at a vertical spacing of 14'

Figure 1.168   Red showing high core strains above ground floor and at lower outrigger. (Thornton Tomasetti.)

or compression under combined shortening and wind conditions, both now and in the future. NLRHA showed similar peak deformations for no-shortening, jacked, and full-shortening conditions, though the timing of those peaks differed.

At core long walls with coupling beams, fewer wall openings were needed at office floors than at hotel floors. That stiffened office levels and increased demands on hotel level coupling beams. "Phantom" openings were added at office levels, continuing the line of doorways from above, to result in more uniform demands. Unneeded openings were closed off with fire rated drywall.

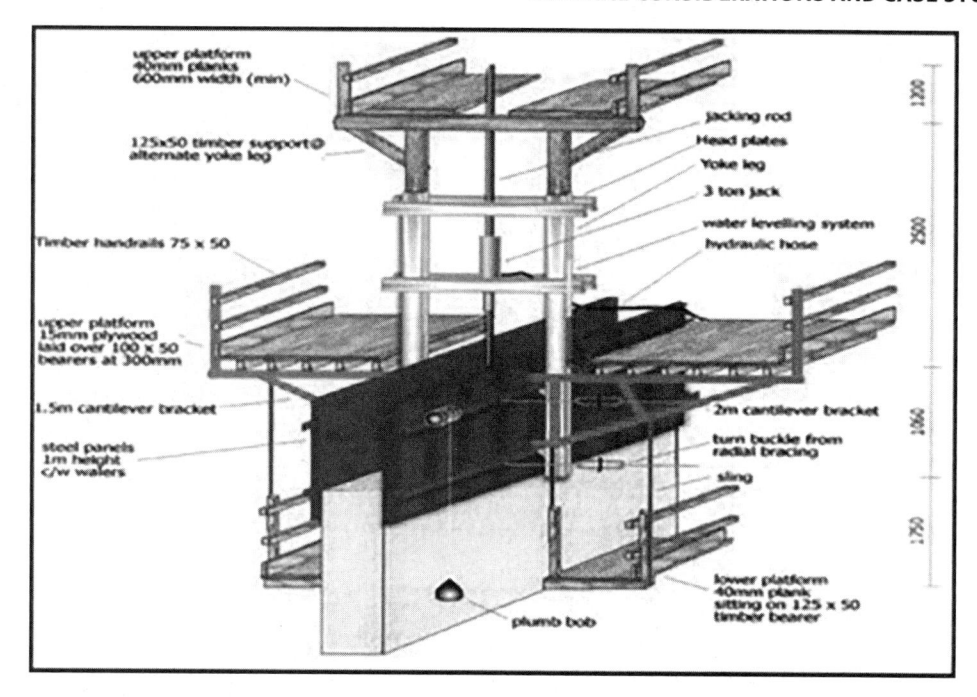

Figure 1.169 Image from State-of-the-Art Review on Application of Value Engineering on Construction Projects: High Rise Building. (**May 2015** Program Management Development; **Authors:** Ibrahim Mahdi.)

Figure 1.170 Concrete Test Tower.

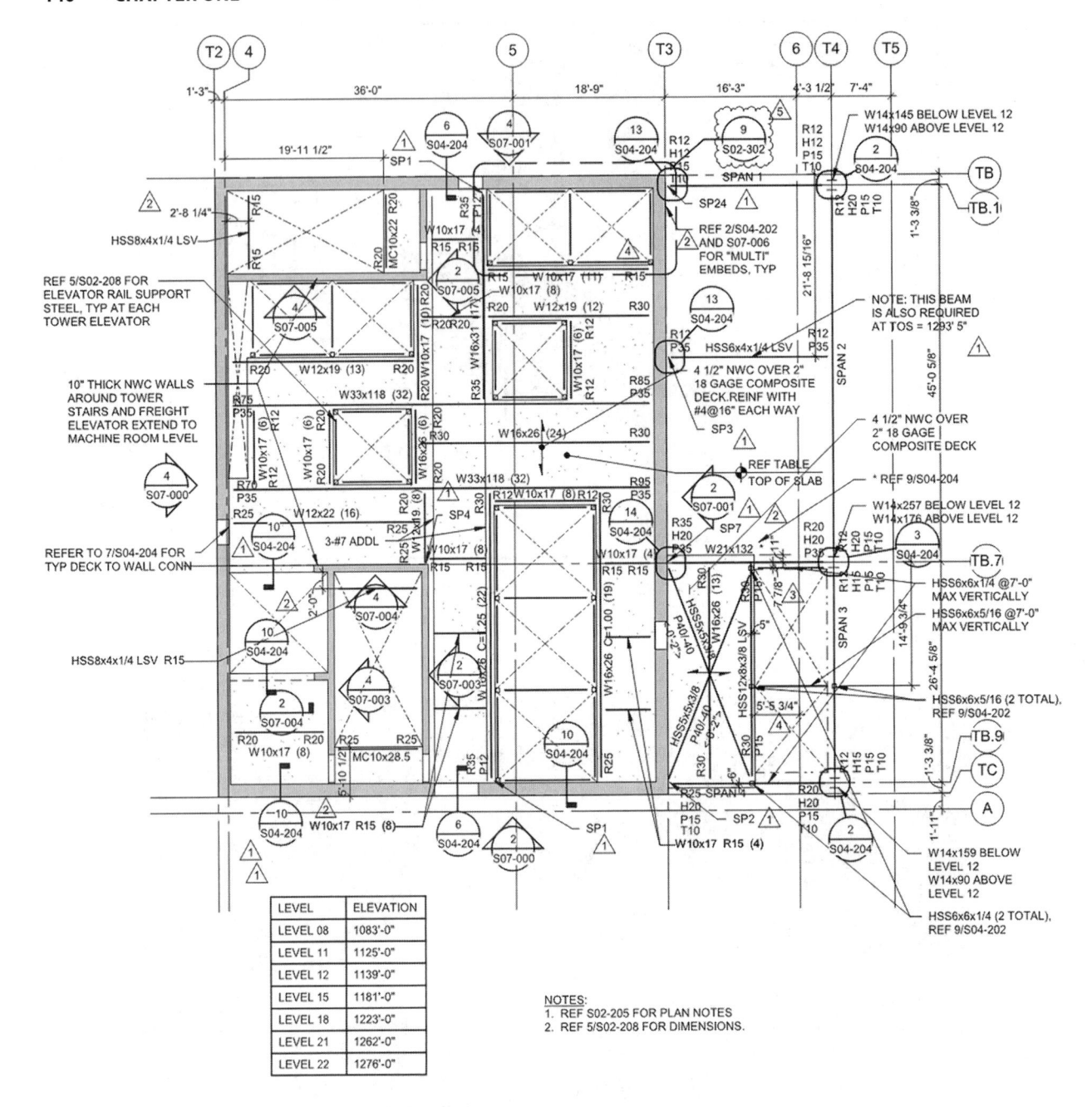

LEVEL	ELEVATION
LEVEL 08	1083'-0"
LEVEL 11	1125'-0"
LEVEL 12	1139'-0"
LEVEL 15	1181'-0"
LEVEL 18	1223'-0"
LEVEL 21	1262'-0"
LEVEL 22	1276'-0"

NOTES:
1. REF S02-205 FOR PLAN NOTES
2. REF 5/S02-208 FOR DIMENSIONS.

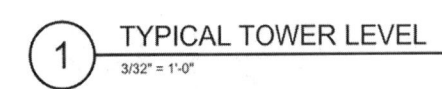

1 TYPICAL TOWER LEVEL
3/32" = 1'-0"

Figure 1.171   Typical Framing Plan.

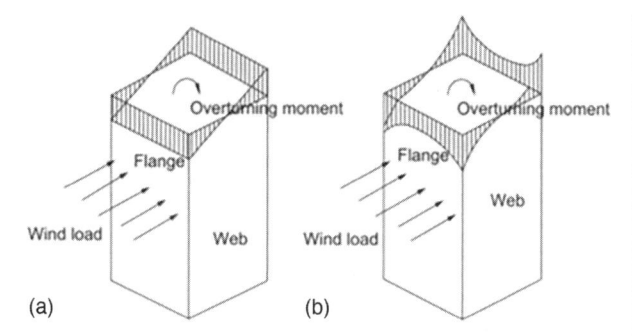

Figure 1.172 Shear Lag Effect.

between the full floors. This large spacing of the full floor diaphragms is unusual for a tower, as floor diaphragms (which brace the tower walls) are typically present at a vertical spacing of 10 to 15 ft. For the ThyssenKrupp tower, the concrete tower walls were designed to span in two directions—vertically by 42' between the floor decks and horizontally by 56' to 74' between the walls in the transverse direction. Concentrated loads applied to the tower walls from the tower crane and hoists during construction, and later from the cantilevered metal panels projecting from the tower walls, had to be checked carefully.

When considering vertical stresses in the walls due to overturning resulting from wind loads, shear lag was a key design consideration. Wall stresses near the corners of the box core are approximately 20% higher than wall stresses near the middle due to this effect (Fig. 1.172).

1.3.58 Market Square Tower (Walter P. Moore)

Market Square Tower is a 42-story cast-in-place concrete apartment building located in the historic section of downtown Houston, Texas (Fig. 1.173). This 500-ft tower features a wide array of amenities including two swimming pools—one located on the second level deck and another located on the 40th floor that features a cantilevered acrylic bottom extending out 8 ft from the face of the building. The lateral system for the tower consists of reinforced cast-in-place concrete core walls, four concrete perimeter super columns connected to the core through concrete outrigger walls at level 40 (Fig. 1.174). The floor system is an 8" lightweight cast-in-place, post-tensioned concrete flat plate system, typical floor plan is shown in Fig. 1.175. The foundation system is a soil-supported cast-in-place concrete mat ranging from 6ft thick at the perimeter to 12ft thick at the center core. The site is prone to hurricane activity and the basic wind speed is 110 mph (3 s gust, 50-year MRI) per IBC 2006.

Figure 1.173 Market Square Tower. (Photo © Hugh Hargrave.)

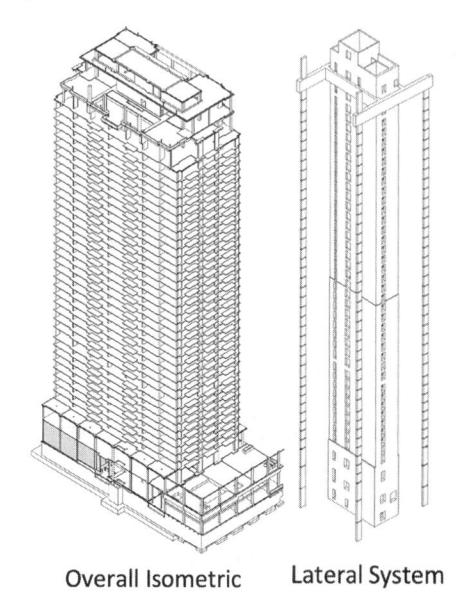

Overall Isometric   Lateral System

Figure 1.174 Overall isometric and lateral system.

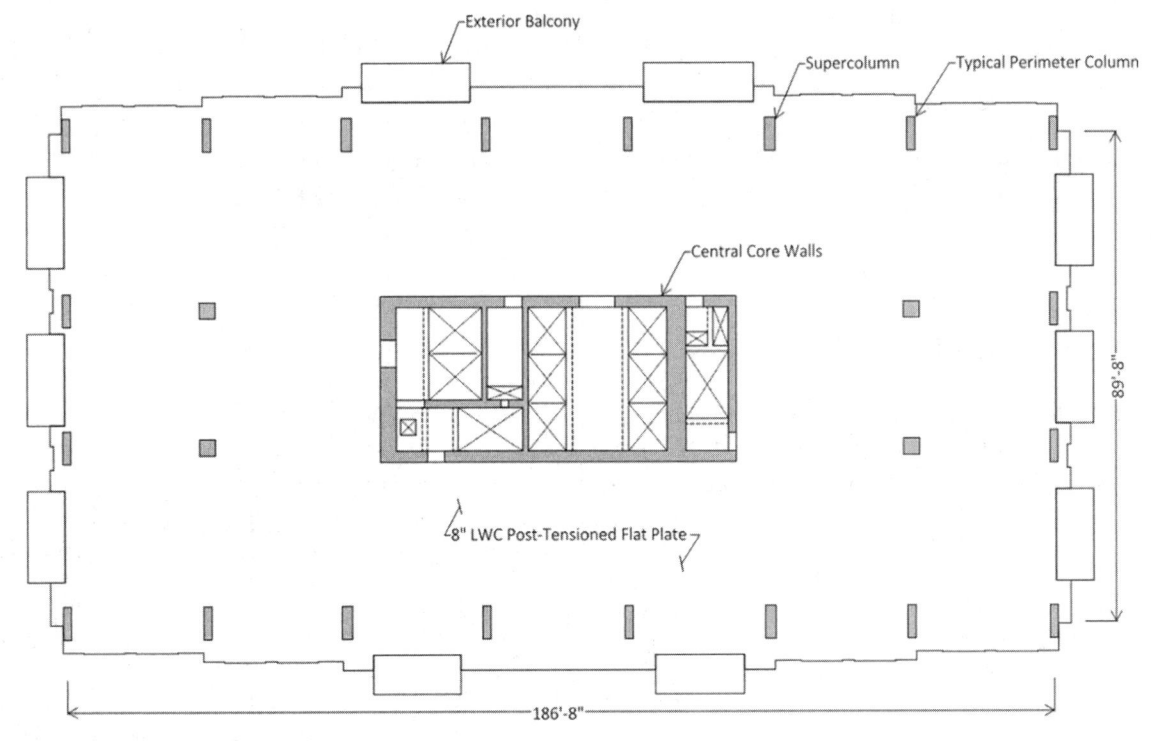

Figure 1.175 Typical Floor Plan.

The tower was constructed adjacent to an existing parking garage and extended all the way to the property lines, a section view is shown Fig. 1.176. Therefore, an excavation retention system was needed to construct the mat foundation and 20-ft tall basement. The H-pile and timber lagging excavation system with temporary tiebacks had to be carefully coordinated around the perimeter of the site to avoid any interference with existing foundations and utilities.

Houston has medium to strong clay soils, and thick mat or raft foundations are common for medium to high rise buildings in the downtown area. In order to minimize both settlement and concrete usage, the mat foundation tapered down from 6-ft thick at the perimeter to 12-ft thick at the central core. Multiple guidelines were provided for the mat foundation construction to reduce the early thermal cracks due to the heat of cement hydration. The concrete temperature was specified not to exceed 95°F at placement or 160°F during any point in the curing process. The temperature differential for any two points in the mat was also specified not to exceed 60°F at any time during hydration of concrete. The concrete mix included a 50% replacement of the cement with Fly Ash which both limited the heat of hydration in the mat and also reduced the overall embodied carbon in the structure. In order to monitor the in-place

temperature, thermocouples were installed in select locations throughout the mat. Figure 1.177 shows the mat foundation pour.

In the design development phase, the owner expressed the interest in adding four more floors to maximize the number of units on the site. To achieve this goal without exceeding the maximum bearing pressure under the mat on this very tight site, light-weight concrete was proposed as a solution for the apartment-level floors to limit the weight of the tower. All tower floors are 8" lightweight cast-in-place, post-tensioned concrete flat plate floors with stud-rails at several locations. The lobby and amenity floors, however, are normal-weight concrete with pan-formed beam and slab systems.

Due to the height of the tower and hurricane wind speeds, an outrigger-braced core wall was selected as the most appropriate and effective system to resist the significant hurricane wind loads. The central core was constructed with high-strength 9000 psi concrete at the base of the tower and 7000 psi concrete at the upper levels. Core wall thickness ranged from a maximum of 42 inches at the base to 12 and 18 inches near the top of the tower. Due to low floor to floor heights and multiple openings into the core, concrete link beams were heavily reinforced and embedded steel link beams were used in a few locations. Ultimately, the core could not provide the stiffness

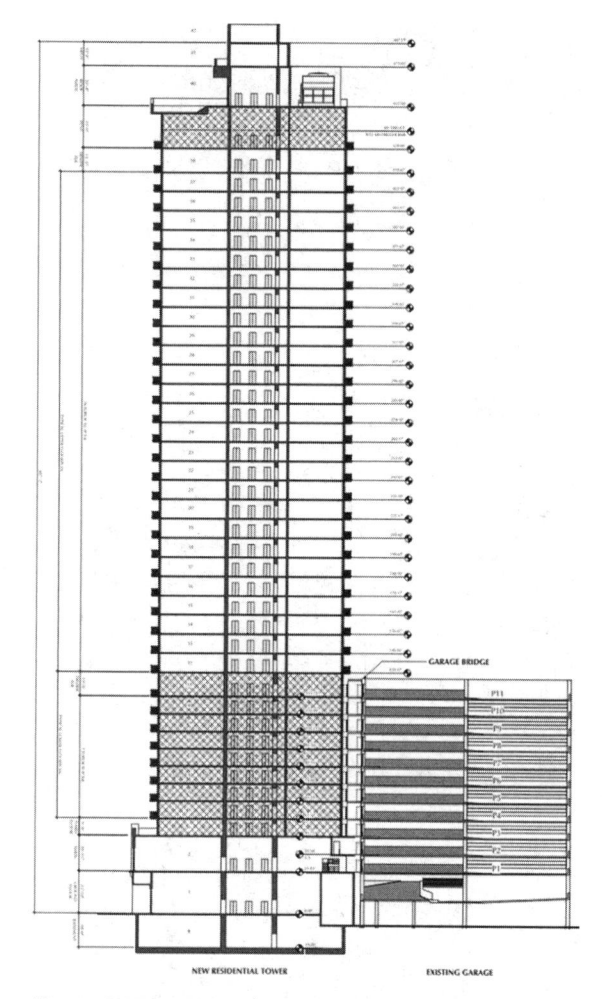

Figure 1.176   Section view.

needed for the tower and a hat outrigger was needed at Level 40. This hat outrigger extends from the central core and connects to four perimeter super columns (Fig. 1.178). However, due to the residential unit layout, the super columns could not be aligned with the corner of the elevator core. Instead of skewing the outriggers, orthogonal "outrigger links" were designed to connect the core walls and the outriggers together. Figures below show the isometric view of the outrigger and the outrigger link detail (Fig. 1.179). The outrigger and outrigger links are 14-ft deep and were designed using a strut and tie approach. Reinforcing bar-end terminators were provided to reduce the rebar congestion and to facilitate the construction, the first-pour bottom 4-ft of the outrigger was designed to support the self-weight of the upper portion without additional shoring needed. Figure 1.180 shows a construction photo of the outrigger. The central core was constructed using a self-climbing jumpform system to minimize the crane use and reduce

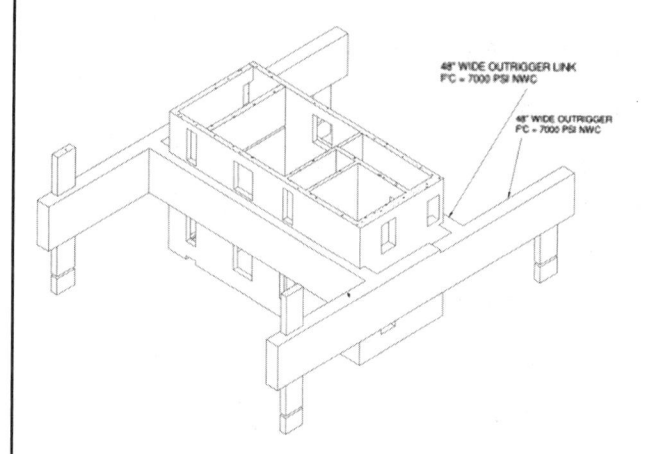

Figure 1.178   Outrigger and super column isometric.

Figure 1.177   Mat Foundation Pour.

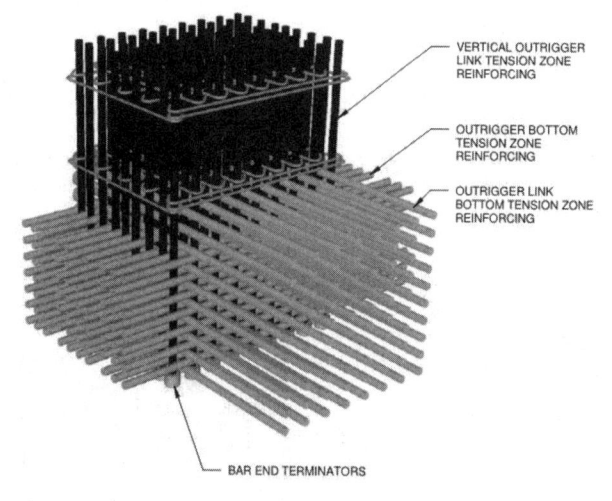

Figure 1.179   Outrigger link node isometric.

Figure 1.180   Outrigger construction.

construction time. Since this construction method requires that the core is constructed in advance of the floors, the staybox system was utilized at the flat-plate floors to provide a keyway and lap splice at the construction joints between the wall and the floor slab. This system provided a cost- effective solution for the trailing floor construction and eliminated the problematic rebar misalignment issues that result from using form savers.

One of the unique programmatic features of this building is the clear, acrylic-bottom pool on level 40 that extends out from the face of the building and cantilevers over the street below. The '10ft- cantilevered pool has an 8-inch thick acrylic floor that allows swimmers to look directly down 40 stories. The entire weight of the pool is supported by deep curved cast-in-place concrete beams on either side of the pool (Fig. 1.181).

1.3.59   KAUST Solar Chimney
(Walter P. Moore)

Located in Saudi Arabia; the King Abdullah University of Science and Technology (KAUST) makes extensive use of solar and wind collection technology (Fig. 1.182). The

Figure 1.181   Photo © Hugh Hargrave.

Figure 1.182   Solar Chimney and its integration within the KAUST Site.

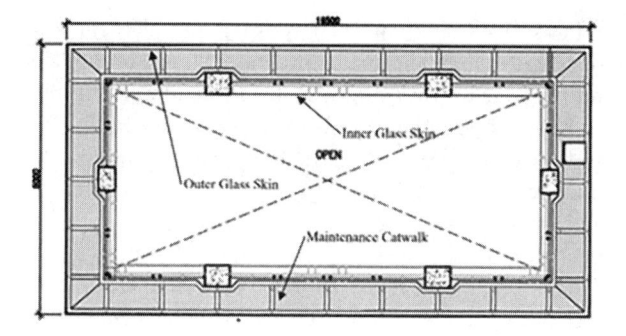

Figure 1.183 Solar chimney cross section.

campus houses two iconic solar chimneys to assist natural ventilation of courtyards and alleyways surrounding two large building clusters. The chimneys create a stack effect by absorbing solar radiation and releasing it back to the passing air stream. The chimney intends to maintain a uniform breeze through the pedestrian alleyways thereby improving thermal comfort at times during the year when conditions would otherwise be uncomfortable.

The structural design supported the holistic realization of this "wind machine" by addressing a wide variety of performance criteria. Design issues included provision for appropriate thermal mass, accommodation of elevated temperatures within the tower, 100-year durability for a highly aggressive environment 100 m from the Red Sea, and integration of a finely detailed cladding and services system.

The solar chimney is 60 m tall and is composed of double skin glass construction having an outer dimension of 16 m by 8 m (Fig. 1.183). The outer double skin primarily serves to generate heat from solar energy while the interior shaft serves as the main air conduit. There exists a 1 m separation between the outer and inner skin which houses the main structure as well as the cat walks for maintenance. The upper 4.6 m of the chimney houses

a maintenance platform and supplemental mechanical equipment. The top portion of the equipment contains dampers to stop solar tower flow in excessive winds and this is followed by a set of mechanical fans as a backup airflow device. The chimney is attached to anchor blocks cast within the podium structure that rests on mat foundations. Walter P Moore & Associates was the lead structural engineer involved in the design of this solar chimney.

The selection of the structural system is primarily driven by the requirement to have a high thermal—evenly distributed mass. This is achieved by adopting a diagrid tube that typically makes use of diagonal elements along the face of the tower which serve as both columns as well as bracing elements. Bands of horizontal members that serve as ties are required at every corner node. Another key feature of the solar chimney is its modularity. The vertical module is 9.6 m high and the horizontal module is 7 m deep along the long face and 6 m wide along the short face. It was decided to make use of concrete in order to leverage the locally sourced labor skills in Saudi Arabia. Moreover concrete provides a high thermal mass and desirable thermal lag compared to steel.

The chimney operation relies on air being pulled through the chimney, drawing air from the outside inward at the spine and courtyard entrances. Consequently to determine how the chimney would perform under winds from different directions; a set of wind tunnel tests (Fig. 1.184) and CFD simulations were performed. Cladding pressure coefficients were measured around the building entrances as well as around the upper chimney exhaust. These pressure coefficients obtained for different directions were then used as pressure boundary conditions to evaluate the wind flow distribution through the laboratory.

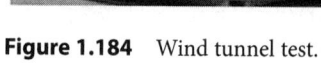
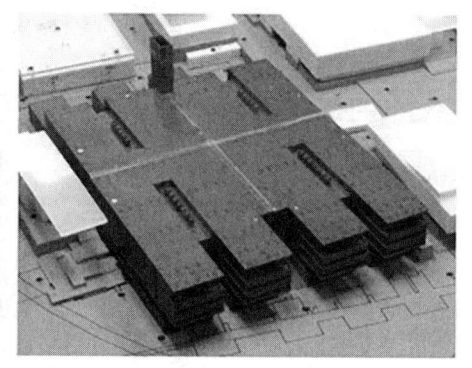

Figure 1.184 Wind tunnel test.

The aspect ratio of the chimney is close to 10:1 and it was imperative to keep drifts under control that may have potential to damage the glass façade. In-situ reinforced concrete is susceptible to cracking leading to a loss in stiffness. Furthermore cracking increases the potential for reinforcement corrosion. Aesthetically, there were also concerns over the ability to achieve a high-quality finish with in-situ casting due to complex joint geometry. To address these issues, a unique precast, prestressed composite concrete strut system was developed that took advantage of the natural stiffness and damping of concrete while also providing a joinery system that allowed the erection to be similar to that of structural steel (Fig. 1.185).

The primary struts of the diagrid consist of rectangular precast concrete sections with embedded ducts for post tensioning via a high-strength threadbar system. High-strength bars were anchored to steel plates integrated into a steel weldment that allowed the nodal connections to be made via high-strength slip-critical bolts. Each strut end was then bolted into a composite nodal element. The conceived system allows the unique attributes of the steel and concrete to work compositely for maximum stiffness of the structure. The prestressing force introduced by the threadbars provides a level of precompression in the concrete struts that gets overcome at the highest tensile forces. Thus, under an applied tension the primary effort goes into unloading the precompressed concrete rather than loading the steel bar tendon, which is much less stiff than the concrete section. The underlying principle is similar to that of pretensioned bolts, where applied tensile forces simply serve to reduce the clamping force and do not add additional load to the bolt. The complex flow of forces in the nodal zone is resolved in relatively compact steel plates. All steel plates are encased in a concrete casting. The bolted region of the joint is covered with a stainless steel sleeve and the space within the sleeve is grouted once the connection is made. The grout not only provides corrosion protection but also contributes to the compressive strength of the joint.

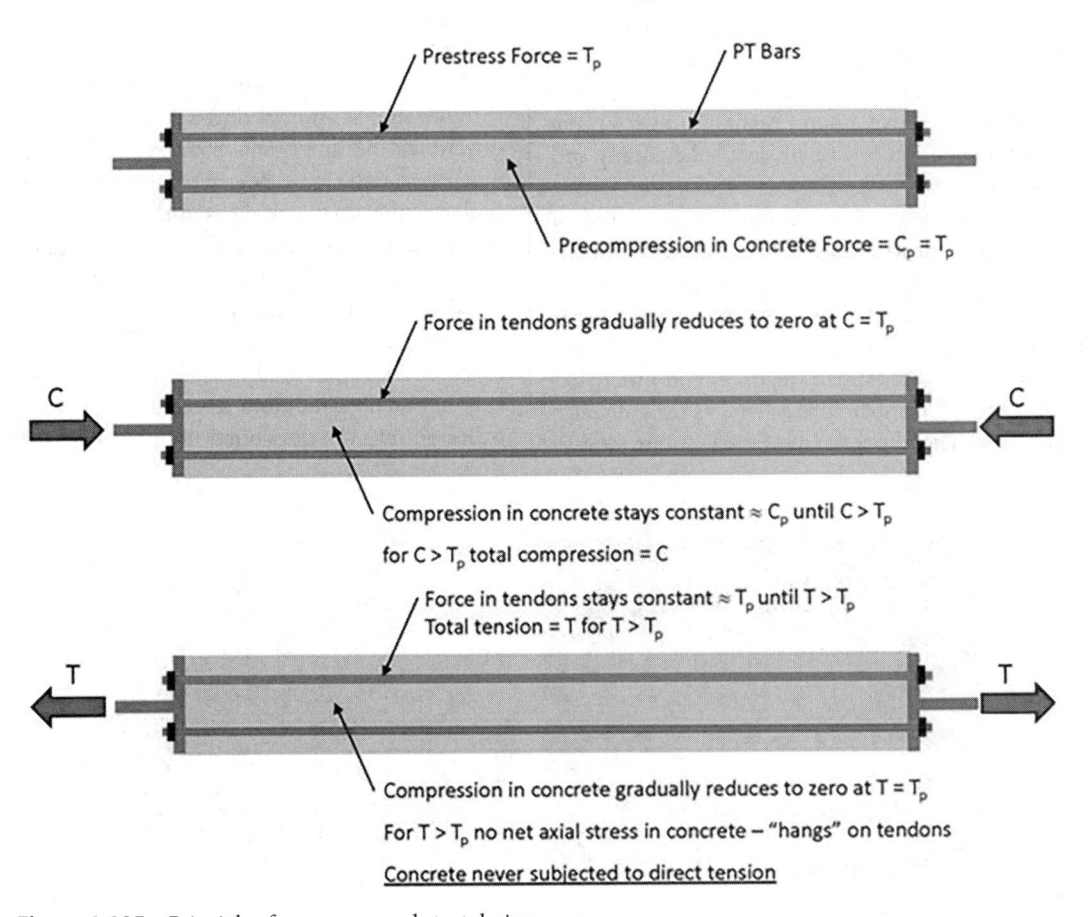

Prestress Force = T_p

PT Bars

Precompression in Concrete Force = $C_p = T_p$

Force in tendons gradually reduces to zero at $C = T_p$

C   C

Compression in concrete stays constant $\approx C_p$ until $C > T_p$

for $C > T_p$ total compression = C

Force in tendons stays constant $\approx T_p$ until $T > T_p$
Total tension = T for $T > T_p$

T   T

Compression in concrete gradually reduces to zero at $T = T_p$

For $T > T_p$ no net axial stress in concrete – "hangs" on tendons

Concrete never subjected to direct tension

Figure 1.185 Principle of precompressed strut design.

In order to provide a practical level of post-tensioning in the structure, the struts were allowed to fully decompress under ultimate loads. This approach contributed to the constructability but required an advanced level of analysis due to the non-linearity of the actual load distribution between the tendons and the concrete section. The structure was analyzed in SAP2000 using explicit separate modeling of the tendon and concrete elements. The tendons and struts were modeled in parallel as tension-only and compression-only elements respectively connecting to coincident nodes. A fully nonlinear analysis including explicit geometric second order nonlinearity was conducted to establish design forces for each element.

As in any form of diagrid structure, the joints/nodes of the connecting members needed precise coordination to achieve the required form and geometry on the site. Each plane of bar tendons was precisely coordinated with the tendons in the strut that crossed the X-node. The solar chimney has three types of nodes:

Type A: X-Crossing node at mid-point of major horizontals along the long face. They consist primarily of a concrete casting integrated into the horizontal element of the tower (Fig. 1.186).

Type B: X-Crossing node at crossing point of struts between major node locations, occurring at either the long or short side of the tower. These are accomplished by assembling a particular pair of struts in and X-configuration prior to stressing of the bar tendons.

Type C: Corner node at major nodal points—These are essentially a steel plate weldment to receive and transfer the forces from the diagonal elements coming into the corner.

The KAUST solar chimney provided an opportunity to implement a unique structural system in response to the project location, the material and labor availability as well as the requirement to have a wind generating system. The overall transparency of the skin integrated with the diagrid structural elements creates an iconic landmark on the KAUST campus that has stood the test of time.

1.3.60 The Shard, London UK (WSP)

Introduction

Completed in 2012, the Shard is the tallest building in western Europe. It is the centerpiece of London Bridge Quarter, a new development on the south side of the Thames, next to the newly-refurbished London Bridge railway station—see Fig. 1.187. In concept it is a "vertical city" containing offices, restaurants, a hotel, apartments and a viewing gallery—see Fig. 1.188. Principal data are provided in Table 2.

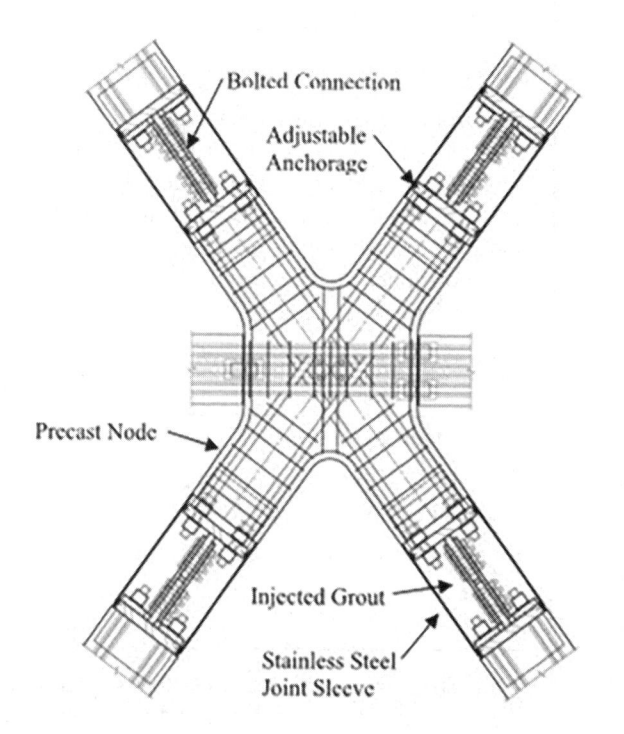

Figure 1.186 Diagrid nodes.

Figure 1.187 The Shard.

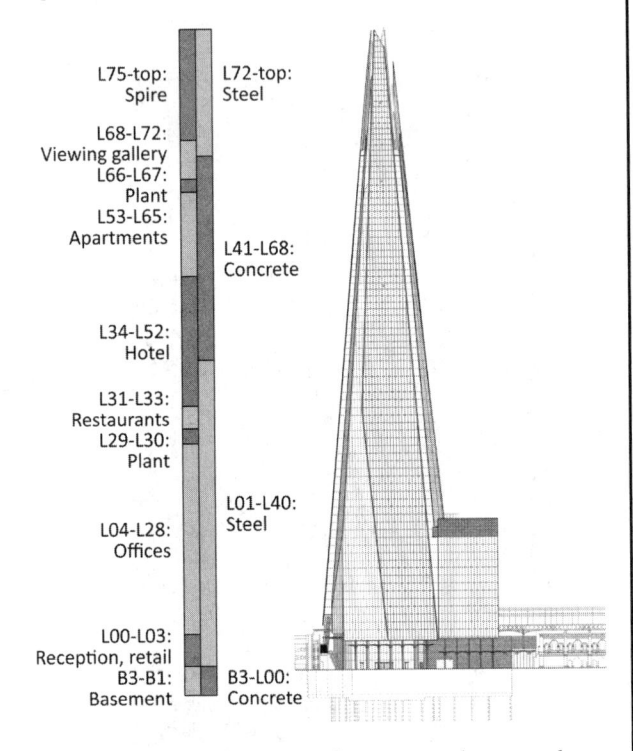

Figure 1.188 Shard elevation showing mixed usage and materials of hybrid frame.

Table 2 Building Data

Client	Sellar Property Group on behalf of London Bridge Quarter Ltd
Architect	Renzo Piano Building Workshop
Structural and geotechnical engineer	WSP
Main contractor	Mace
Height	306 m from ground level
Number of stories	72 (viewing gallery) 87 (building maintenance unit) 95 (top of glass)
Gross internal area	127 000 m²
Net internal area	85 000 m²
Residential	5 800 m²
Hotel	17 800 m²
Offices	55 200 m²
Retail	5 600 m²
Volume of concrete	54 000 m³
Weight of steel	11 000 tons
Area of facade	56 000 m²
Number of lifts	44
Number of car parking spaces	48
Building population	8 000

FOUNDATIONS

The ground conditions were typical for the area—see Table 3. A small fault crossed the site: the base of the London Clay and all the strata below were 6 m lower on the east side. The Shard piles—up to 53 m deep—were therefore longer on the east side than the west, and were positioned so as to avoid the piles from a previous 25-story building—see Fig. 1.189.

Table 3 Ground Conditions (West Side of Site)

Stratum	Depth below ground level
Made Ground / Alluvium	0–5 m
River Terrace Gravel	5–9 m
London Clay	9–30 m
Lambeth Group clays	30–45 m
Thanet Sands	45–59 m
Chalk	Below 59 m

The Shard basement was three-story (13.5 m) deep and was surrounded by secant piles, which excluded ground water from the highly-permeable gravels. These piles also

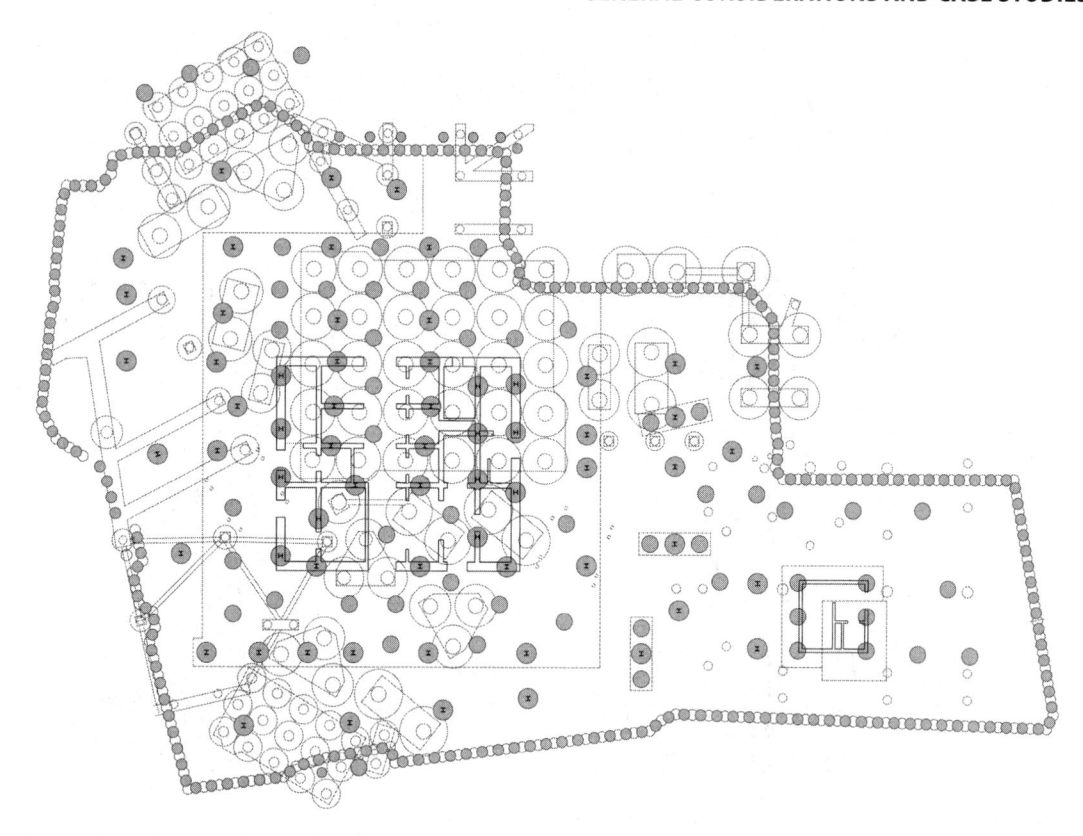

Figure 1.189   Foundation plan. Shard piles shaded grey; under-reamed piles from previous building dotted.

carried vertical loads from the perimeter columns. Internal columns and the core were supported by a 3 m-thick raft slab, which shared load between bearing piles and the ground.

The raft was very heavily reinforced—up to seven layers of bars in some areas. WSP decided not to increase the depth of the raft and reduce the reinforcement because there would have been increases in the volume of concrete, the volume of excavation and the vertical span—and therefore the reinforcement content—of the perimeter walls. The raft was cast in a single continuous pour: 5500 m³ of concrete placed over 32 h—at the time, the United Kingdom's largest ever concrete pour. The raft concrete used a cement blend containing 70% ground granulated blastfurnace slag (GGBS) to control the heat of hydration, thereby limiting maximum and differential temperatures.

In order to reduce the construction program, the basement was constructed top-down: that is, the ground floor was constructed first; then the soil was excavated beneath, first to level B2 where the slab was cast, and finally to raft level. This method—relatively common in London—avoided the need for temporary propping of the perimeter wall.

The Shard, however, introduced an innovation: the core was constructed top-down as well. Large steel plunge columns, cast into the piles, supported the core as construction progressed both below and above, until the plunge columns carried the weight of 23 stories of core. Then the level B3 raft slab was cast, the core walls were completed (using self-compacting concrete pumped up to the underside of the walls built earlier) and construction resumed upwards—see Fig. 1.190.

SUPERSTRUCTURE

The Shard was designed with a hybrid structural frame. The floors were steel-framed in the office levels, using fabricated plate girders—see Fig. 1.191 for a typical floor arrangement. The plate girders were a uniform 500 mm deep, with flange and web plate sizes adjusted according to the span of the beam and the load carried. The uniform depth allowed standard 300 mm diameter web openings at a constant level, which simplified the installation of services.

The steel frame allowed long spans where the building was widest, and used the same ceiling space for both structure and services. In the upper levels spans were shorter, services were located above the

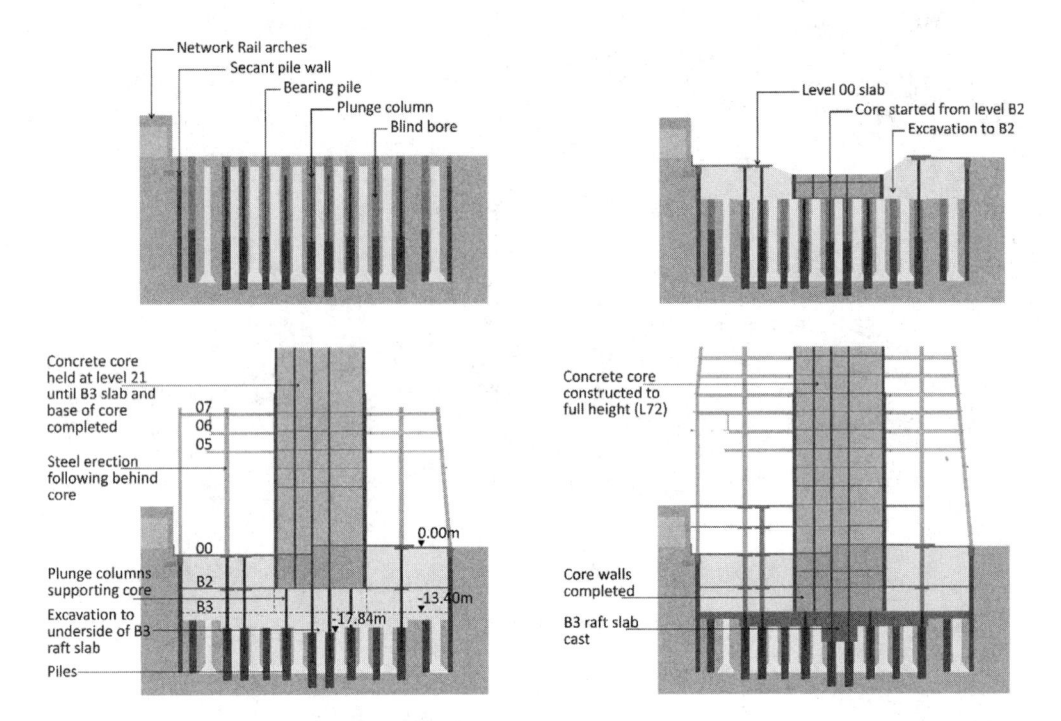

Figure 1.190   Basement and core top-down construction sequence.

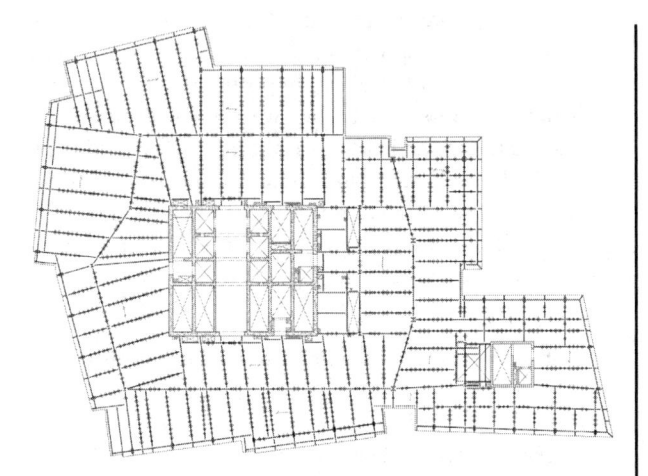

Figure 1.191   Level L09 floor plan.

corridor surrounding the core, and 200 mm deep post-tensioned concrete flat slabs were used. This reduced the floor-to-floor height by 550 mm compared to the steel levels. In addition the hybrid frame delivered more structural damping than an all-steel frame, eliminating the need for a tuned mass damper. These innovations enabled four additional floors to be provided in the building without increasing the overall height.

Columns were steel box sections up to L40—in the steel levels—and concrete above that point. Net lettable floor area was maximized by reducing the column sizes: steel columns employed plate up to 125 mm thick, and C65/80 concrete (cylinder/cube strength in MPa) was specified for the concrete columns carrying the highest loads.

The core contained numerous shafts for stairs and lifts—see Fig. 1.192. Many were re-purposed as they rose up the building—for example, low-level office lift shafts were merged above L14 and the space was used for bathrooms, increasing lettable space on the floors. Above the offices, this same space was used for lift shafts once again; these served the hotel and the viewing gallery. The plan area of the core and the thicknesses of the walls were reduced with height—see Table 4. Internal walls were 450 mm thick at the base; some were reduced to 250 mm higher up the building. The concrete specified for the core also reduced in strength with height: C50/60 below L31 and C40/50 above that level.

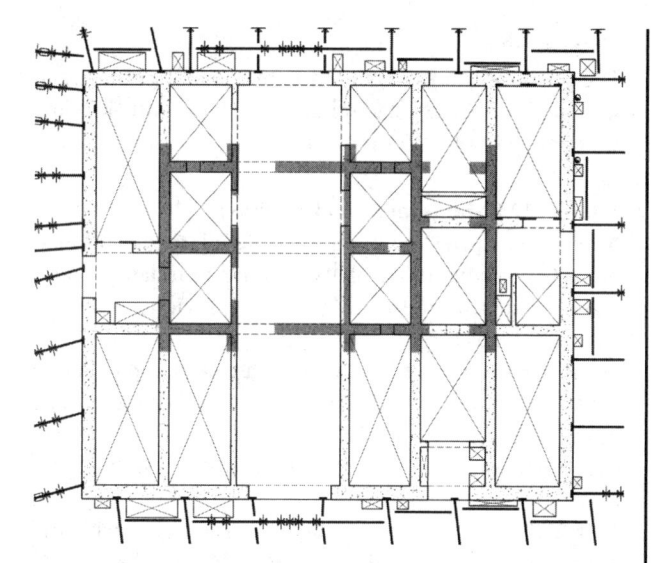

Figure 1.192   Level 09 core plan (level 59 shaded gray).

Table 4   **Principal Core Dimensions**

Level	Core dimensions	Perimeter wall thickness
L00	22.3 × 18.9 m	800 mm
L08	21.9 × 18.5 m	600 mm
L38	15.0 × 16.0 m	450 mm
L54	15.0 × 9.0 m	450 mm
L67	11.7 × 9.0 m	450 mm
L72	6.3 × 8.4 m	Steel mast

Steel beams were connected to the core using "embedment plates"—steel plates fitted with headed shear studs and reinforcement that were cast into the concrete as the core was formed. A trailing platform was provided on the slip; from this level fin plate connections were welded to the embedment plates. Slabs were connected to the core using proprietary pull-out bar systems.

The stability of the building was provided by the core, acting as a vertical cantilever. Lateral forces were taken out at ground floor (L00) where they reacted against the perimeter wall and the ground. Push-pull forces in the core walls continued down to the piled raft at basement B3 level.

Accelerations at the top of the building were limited to 15 milli-g (0.015 m/s²) using the natural damping of the concrete structure and a "hat truss." This increased the stiffness of the core by connecting it to the perimeter columns in the top plant room (L66-L67)—see Fig. 1.193. The response of the building was determined

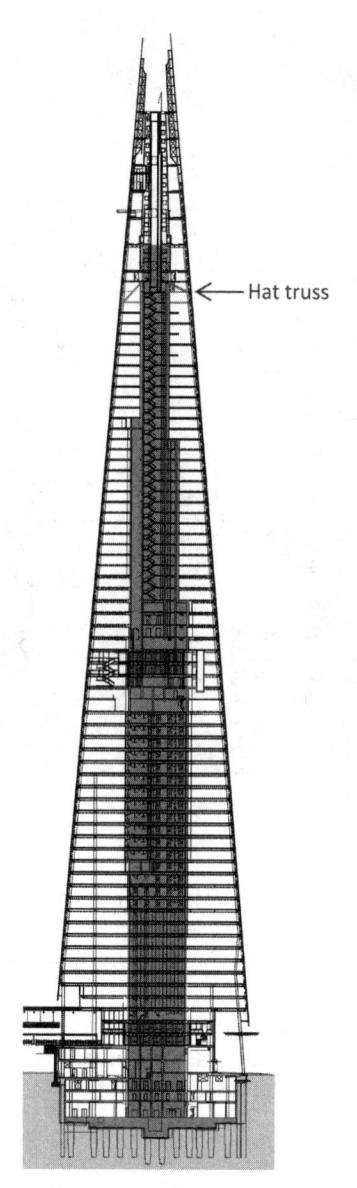

Figure 1.193   Cross section through the Shard showing core and hat truss.

through WSP's ETABS model and by wind tunnel tests, carried out by RWDI in their facilities in Toronto. The tests used the high frequency force balance technique and a 1:400 scale model; other wind tunnel tests used pressure taps to determine the wind pressure on the building façade.

The core was slip-formed. Uniquely, the tower crane was supported by the slip and rose continuously, eliminating the need to halt production while raising the crane and re-attaching it to the core higher up. The slip achieved a production rate of 3 m per day.

Figure 1.194   22 Bishopsgate, London.

SPIRE

The Shard was surmounted by a 60 m tall steel spire which did not end in a single point but reflected the architect's vision for the building to "fade into the sky." A trial assembly of the spire was carried out at the fabrication yard to train steel erectors and refine the method statements. The frame was delivered to site in modular components, which were erected quickly and safely.

CONCLUSION

The Shard was an efficient structure and a commercial success. It was completed on time and has swiftly become an iconic part of London's skyline.

1.3.61   22 Bishopsgate, London (WSP)

A 62-story tower built on the site of an abandoned project, re-using 100% of its existing foundations, and incorporating more than 40% of the previous building's basement slabs (Fig. 1.194).

The superstructure is formed by a traditional concrete core, steel columns and beams (Fig. 1.195). The core steps in at levels 27, 41 and 58. This is driven by the vertical transportation (VT) strategy, with some lifts serving different groups of floors.

At 278 m tall and being this the orientation with the largest wind force, stability was provided by a set of outrigger trusses and belts. They are located at levels 25 and 41, which are double height floors hosting plant room and other shared amenities (gym, spa, etc.). The outrigger trusses connect the core walls to the perimeter, where additional belt-trusses connect them to the main building columns, distributing the forces of each outrigger over two columns (Figs. 1.196 and 1.197).

The existing foundations (Fig. 1.198), as shown below, included a large 5 m deep raft under the concrete core, an 800 mm lower basement slab, and individual mega piles connected with deep pile caps. The capacity of the existing mega piles varied from 18 MN to 50 MN (SLS), with diameters varying from 1.8 m to 2.5 m, and depths of up to 50 m.

85 new piles were built to provide additional capacity for the larger building. Each with a capacity of 5 MN, which required a very careful geometrical assessment, trying to locate these piles between the forest of existing

Figure 1.195   Typical framing.

OUTRIGGER FRAMES

Figure 1.196 Stability system. Concrete core and steel outrigger trusses and belt.

Figure 1.197 Outrigger trusses at level 25.

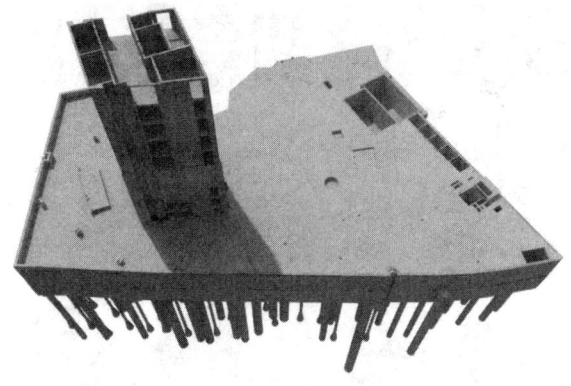

Existing basement and core

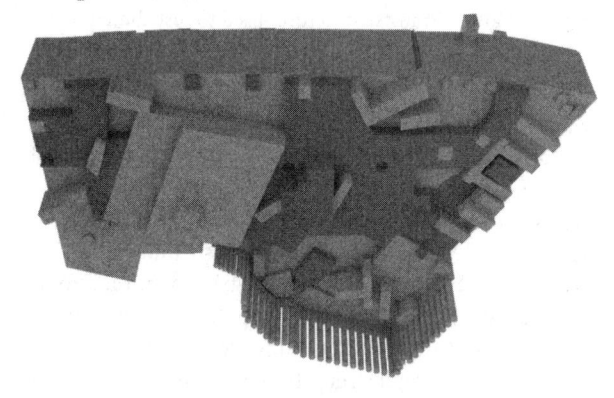

View from below. In grey, existing concrete elements (raft and pile caps). In blue, new 3m deep raft and localized additional pile caps.

Figure 1.198 Existing and new foundation.

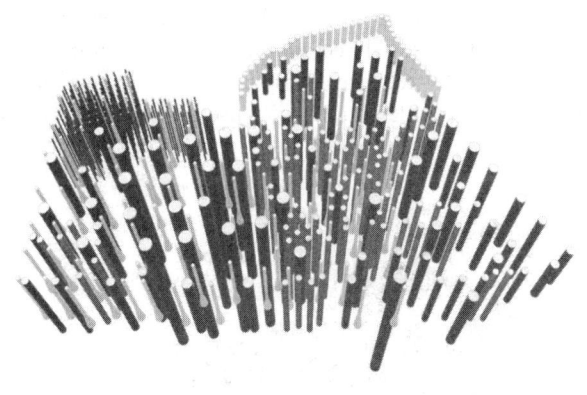

buildings, Pinnacle piles and mega piles, and 85 900mm Martello piles for the new scheme.

Figure 1.199 Pinnacle piles and mega piles, and 85,900 mm Martello piles for the new scheme.

elements, and maintaining minimum distances between piles. These new piles were built from within the lower basement level (B3) using a small telescopic rig, to allow a faster construction sequence and to minimize the amount of temporary works (Fig. 1.199).

A series of transfer structures were used to distribute the forces from the new column location towards those of the existing foundations.

Deep concrete walls span between the existing pile caps and other mega piles. Isolated columns that landed close enough to existing piles were resolved by creating small walking columns, using the existing basement slabs to provide support against the resulting horizontal push-pull forces (Fig. 1.200).

Figure 1.200   Transfer structures used to distribute the forces from the new column location towards the existing foundations.

In the south of the building, the façade line moves away from the perimeter walls, and two of the new main columns needed to find their way down into the existing foundations. Column C24 was inclined from ground floor over the three basement levels (Fig. 1.201). The inclination creates large horizontal push-pull forces. The push is directly transferred into the basement raft trough a special baseplate; the pull force is transferred into the concrete core through the ground floor slab. To create a direct load path, high strength steel cables were installed running along ducts embedded in the slab. These cables were pre-tensioned to introduce pre-compression into the slab and to ensure displacement compatibility of the superstructure.

The existing basement created constraints that directly affected both the architectural and the structural solution. At the lowest level (B3), the waste management strategy required large lorries to use a turning table to rotate before leaving the building. The only location

available for this element was directly under one of the main building columns. As shown below (Fig. 1.202), a solid steel girder was designed to support 62 stories of load, hidden between basement B1 and ground floor, transferring the loads to where the existing foundation's capacity was present. This beam is 15 m long and 3.8 m deep. At 97 tons, it required a special permit to be delivered to site, needing a police escort and some of London's most important streets to be closed.

Figure 1.201   Column C24 inclined from ground floor over the three basement levels.

Figure 1.202   Solid steel girder designed to support 62 stories of load.

Figure 1.203   A mega truss designed bringing together the loads from three of the main perimeter columns and spanning over the existing car lift.

The existing basement included a car lift connecting all the levels up to the ground floor. During construction, this was the main way of accessing the areas below ground. For this reason, any elements of the superstructure needed to bridge over the car lift; it was not possible to relocate it. A mega truss was designed, bringing together the loads from three of the main perimeter columns and spanning over the existing car lift (Fig. 1.203).

Re-using the existing foundations and basement of a different tower created a logistical challenge. As shown below, the north of the site (left) was excavated down to the lowest basement level to reach the existing 5 m deep raft and its mega piles. This was done by propping the existing slabs in the temporary condition to create the opening. The new basement slabs would be connected to the existing once the core was completed. On the south side (right), 85 new piles and a large 3 m deep raft was to be built to provide the extra capacity needed to support the larger new concrete core (Fig. 1.204).

To improve the construction program, an alternative method was proposed; create a transfer system (a large grillage of concrete walls) built within the existing upper basement level, spanning between five existing columns and engulfing all the future core walls; then build the core (up) from this grillage, whilst simultaneously working on the new foundations below the grillage (from basement B3). This system would transfer the loads from the new core down into the existing piles until the new foundations were completed underneath it. The image below (Fig. 1.205) shows the large excavation for the new raft, and the existing steel columns that supported the transfer structure and concrete core above. It also shows the existing pile caps exposed and ready to be integrated into the new raft.

Figure 1.204   Aerial view of the site. Blue-sky construction on the left. Top-down on the right.

This top-down methodology allowed for 20 levels of the concrete core to be built before the foundations were completed, improving the construction program by at least three months. The image below was taken from below the core, showing it "floating" above the basement (Fig. 1.206).

Project Team
Developer, **Lipton Rogers Developments**
Architect, **PLP**
Main Contractor, **Multiplex**
Concrete Frame, **Careys**
Steel Frame, **Severfield**
Holistic Engineering, **WSP**

Acknowledgments and References
All photos belong to the author. All images belong to WSP UK Ltd.

Figure 1.205   Lowest basement level, B3, showing the excavation for the new raft after exposing the existing piles.

Figure 1.206   The "floating" core. 20 stories supported on 5 existing columns.

1.3.62   Torre Reforma, Mexico City (ARUP)

The unique massing of Torre Reforma is a function of the site located on Mexico City's major thoroughfare, Paseo de la Reforma, and its adjacency to the cities central Chapultepec Park (Fig. 1.207). Responding to setback requirements, the 246 m tall tower takes the form of an isosceles right triangle, with a minor kink in the hypotenuse that migrates westward in plan along at each level up the height of the tower.

The floor plan is column free (Fig. 1.208), and the gravity system of the tower is supported by cast-in-place concrete walls at the open book ends on the north and east

Figure 1.207   Torre Reforma Building, Mexico City.

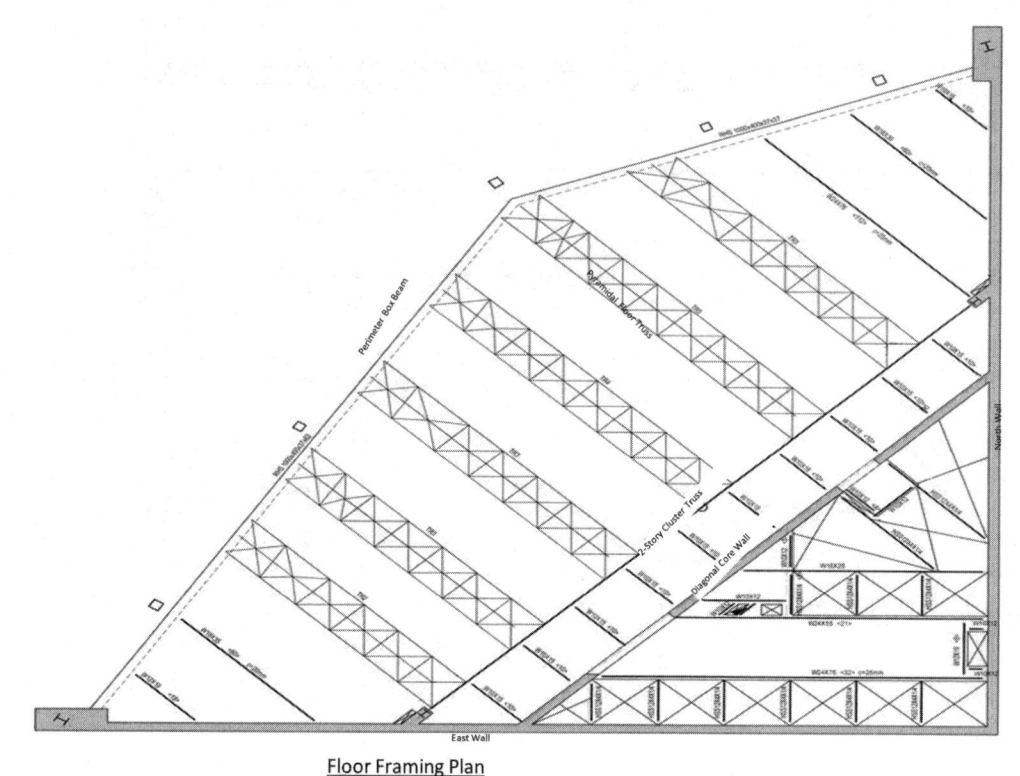

Floor Framing Plan

Figure 1.208   Typical Floor Framing.

faces of the tower, a diagonal cast in place wall separating the triangular core and the floor plate, and a double-V configured hanging steel diagrid along the hypotenuse hung from the outer corners of the shear walls.

The building is arranged into four-story clusters, a repeating unit up the height of the building with a three-story atrium adjacent to the core capped by a complete floor plate (Fig. 1.209). The floor framing is comprised of pyramidal floor trusses with a single WT bottom chord, two WT top chords and square tube diagonals acting composite with the lightweight concrete slab on metal deck (Fig. 1.210). The 2 m wide trusses are spaced 5.5 m on center, with spans as large as 25 m near the top of the tower. The highly efficient framing system maximizes structural depth and serves as a raceway for services, while minimizing the number of crane lifts during construction. A continuous welded-plate box girder supports the floor trusses along the diagrid, and the other end of the floor trusses are supported by a two-story cluster truss where the floor plate is interrupted by the atrium opening.

The lateral system is comprised of the 3 architecturally exposed concrete shear walls, and the perimeter double-V steel diagrid (Fig. 1.211). Due to the arrangement of the lateral system, the primary dynamic modes

of the building are aligned diagonally, approximately 45° off the axis of the orthogonal concrete walls. The perimeter lateral system is effective in resisting the seismic loads along these directions, and also mitigates the inherent torsional response from the building's shape.

The shear walls along the north and east faces of the building are coupled once every four floors by a full story coupling beam, leaving a three-story glass opening on the north and east faces of each atrium. During a frequent seismic event, the coupling beams behave elastically and preserve the stiffness of the building to control drift. During the Maximum Considered Earthquake event (2475-year return period), the coupling beams undergo plastic deformation and absorb the seismic energy induced by the ground motions.

The diagrid along the hypotenuse resists lateral loads similar to conventional braced frames with the shallow and steep brace on one side resisting through tension, and the braces on the opposite side resisting through compression (Fig. 1.212). The double-V orientation offers several benefits against a conventional braced system. The preloading of the hanging braces preloads the elements in tension. This preload shifts the static condition

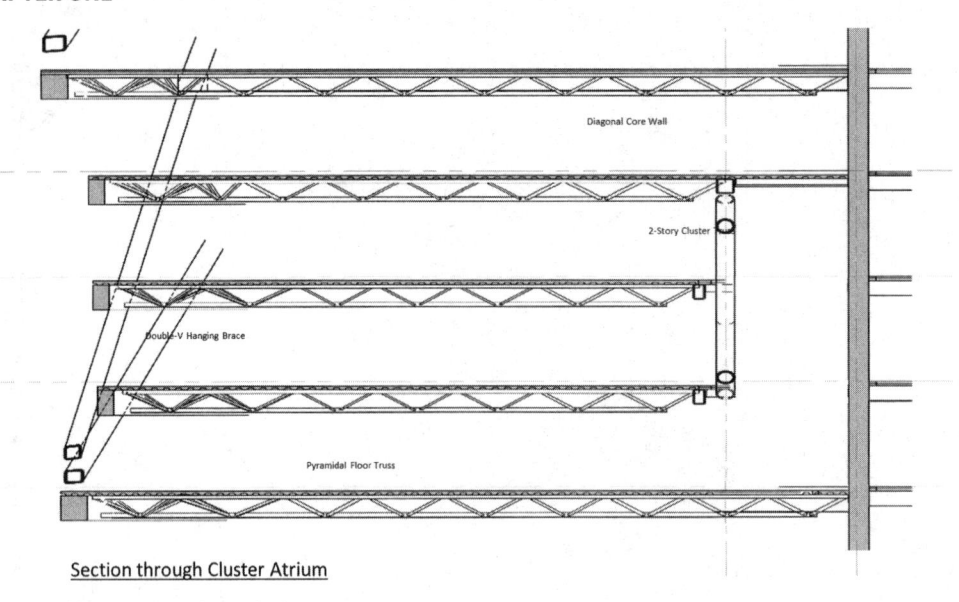

Section through Cluster Atrium

Figure 1.209   Section through cluster atrium.

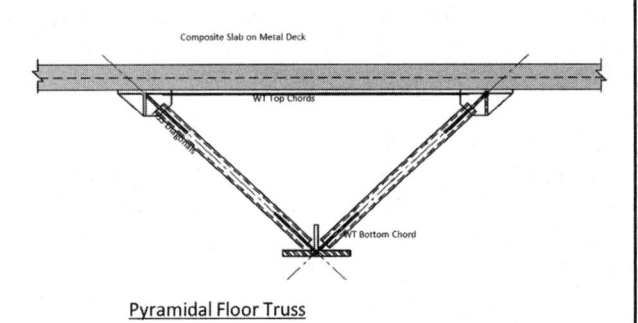

Pyramidal Floor Truss

Figure 1.210   Pyramidal floor truss.

Figure 1.211   Double-V Hung Diagrid.

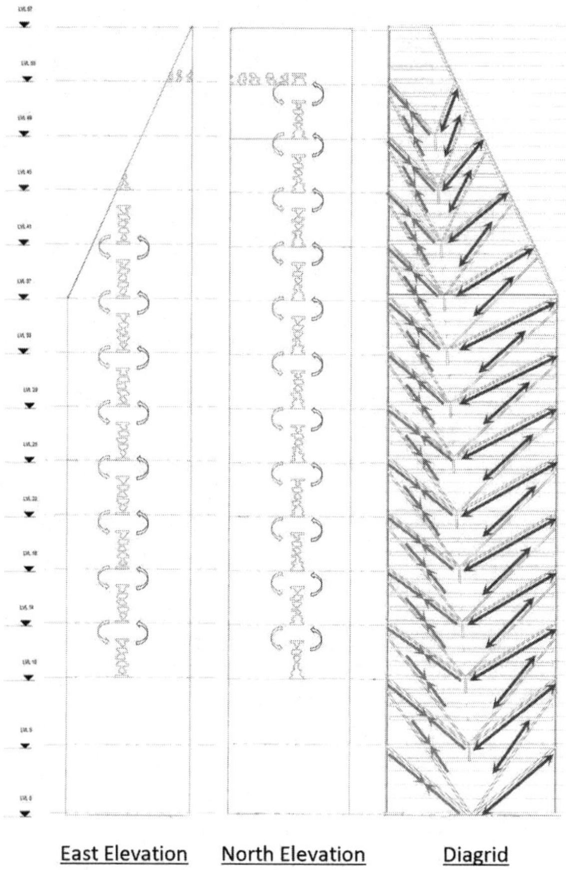

Figure 1.212   Elevations and Diagrid.

closer to the average between the larger tension capacity and lower compression capacity of the brace, permitting the use of more slender sections. The double-V also offers redundancy to the gravity system allowing for limited yielding without compromising vertical stability under the MCE event.

Chapter 2
Wind Effects

2.1 DESIGN CONSIDERATIONS

Windy weather poses a variety of problems in new skyscrapers, causing concern for building owners and engineers alike. The forces exerted by winds on buildings increase dramatically with the increase in building heights. Static wind effects increase as the square of a structure's height, and the high-rise buildings of the 1980s and 1990s, which are at times 1000 ft (305 m) tall, must be 25 times stronger than the typical 200 ft (61 m) building of the 1940s. Moreover, the velocity of wind increases with height, and the wind pressures increase as the square of the velocity of wind. Thus, wind effects on a tall building are compounded as its height increases. With the increased heights of tall buildings in the 21st century, supertall and megatall buildings, building shape optimization is almost necessary to reduce the wind forces on the buildings to achieve a reasonable design in terms of member sizes, a building cost, and a building functionality.

In designing for wind, a building cannot be considered independent of its surroundings. The influence of nearby buildings and land configuration can be substantial. The swaying at the top of a high-rise building caused by wind may not be seen by a passer-by, but its effect may be of concern to those occupying the top floors. There is scant evidence that winds, except in the case of a tornado or hurricane, have caused major structural damage to new high rises. However, a modern skyscraper, which uses lightweight curtain walls, dry partitions, and high-strength materials, is more prone to wind motion problems than the early skyscrapers, which had the weight-advantage of masonry partitions, heavy stone façades, and massive structural members.

To be sure, all buildings sway during windstorms, but the motion in earlier tall buildings with locked-in gravity loads from their enormous weight is usually imperceptible and certainly has not been a cause for concern. Structural innovations and lightweight construction technology have reduced the stiffness of modern high-rise buildings. Wind action has become a major concern for the designer of today's high-rises. In buildings prone to wind motion problems, objects in a room may vibrate, catching the occupant's eye. Doors and chandeliers swing, pictures lean, and books fall sideways off shelves. If the building has a twisting action, the occupants may get an illusory sense that the world outside is moving, creating symptoms of vertigo and disorientation. In more violent storms, windows can blow out, creating safety problems to pedestrians below. Sometimes, strange and often frightening noises are heard by the occupants as the wind shakes elevators, strains floors and walls, and whistles around the sides.

Keeping the movements in the upper levels of the building to acceptable human tolerances is the goal. Exactly what this tolerance is has been very difficult to assess. Engineers today try to design structures that have inherent stiffness achieved through engineering techniques rather than depending upon dead weight to stabilize the structure. In spite of all the mathematical and engineering sophistication possible with computers, wind has still managed to dodge complete quantitative analysis, mainly because of two major problems. First, unlike dead loads which are permanent and unchanging and live loads which are tacitly assumed to change slowly, wind loads change rapidly and even abruptly, creating effects much larger than if the same loads were applied gradually. The other problem is limiting of building accelerations below human perception. The true complexity of wind and acceptable human tolerance have been understood better during the last two decades. Still, there is a need for understanding the nature of wind and its interaction with a tall building, with particular reference to allowable deflections and comfort of occupants. In designing tall buildings to withstand wind forces, the following are important factors that must be considered:

1. Strength and stability requirements of structural system.

2. Fatigue in structural members and connections caused by fluctuating wind loads.

3. Excessive lateral deflection that may cause cracking of partitions and external cladding, misalignment of mechanical systems and doors, and possible permanent deformations.

4. Frequency and amplitude of sway that can cause discomfort to occupants.

5. Possible buffeting that may increase the magnitudes of wind velocities on neighboring buildings.

6. Effects on pedestrians.

7. Annoying acoustical disturbances.

8. Resonance of building oscillations with vibrations of elevator hoist ropes.

2.2 NATURE OF WIND

2.2.1 Introduction

Wind is the term used for air in motion and is usually applied to the natural horizontal motion of the atmosphere. Motion in a vertical or near vertical direction is called a *current*. Winds are produced by differences in atmospheric pressure, which are primarily attributable to differences in temperature. These temperature differences are caused largely by unequal distribution of heat from the sun, and the difference in the thermal properties of land and ocean surfaces. When temperatures of adjacent regions become unequal, the warmer and thus lighter air tends to rise and flow over the colder, heavier air. Winds initiated in this way are modified by rotation of the earth.

Movement of air near the surface of the earth is three-dimensional, with a horizontal motion much greater than the vertical motion. Motion of air is created by solar radiation, which generates pressure differences in air masses. Vertical air motion is of importance in meteorology but is of less importance near the ground surface. The surface boundary layer involving horizontal motion of wind extends upward to a certain height above which the horizontal airflow is no longer influenced by the ground effect. The wind speed at this height is called the *gradient wind speed* and generally occurs at an altitude greater than 1500 ft (458 m). In this boundary layer is precisely where most of the human activity is conducted, and therefore how the wind effects are felt within this zone is of great concern in engineering.

Although one cannot see the wind, it is a common observation that its flow is quite complex and turbulent in nature. Imagine taking a walk outside on a reasonably windy day. You no doubt experience the constant flow of wind, but intermittently you will experience sudden gusts of rushing air. This sudden variation in wind speed is called gustiness or turbulence. The up-and-down fluctuations of speed about the mean velocity that occur over long periods of time due to solar energy cycles are of little importance in engineering, but the shorter-period peaks resulting from surface-generated turbulence are of great importance for the human activity in the earth's boundary layer.

In describing global circulation and specific recurrences of certain types of wind, modern meteorology relies on wind terminology used by early long-distance sailors. For example, terms like trade winds and westerlies were used by sailors who recognized the occurrence of steady winds blowing for long periods of time in the same direction. A broad indication of the flow of wind in the lower levels resulting from the general circulation of the atmosphere can be obtained by considering the interface between the cold winds from the polar regions and the westerlies.

Near the equator, the lower atmosphere is warmed, by the sun's heat. The warm air rises, depositing much precipitation and creating a uniform low-pressure area. Into this low-pressure area, air is drawn from the relatively cold high-pressure regions from northern and southern hemispheres, giving rise to trade winds between the latitudes of 30° from the equator. The air going aloft flows counter to the trade winds to descend into these latitudes, creating a region of high pressure. Flowing northward and southward from these latitudes in the northern and southern hemispheres, respectively, are the prevailing westerlies, which meet the cold dense air flowing away from the poles in a low-pressure region characterized by stormy variable winds. It is this interface between cold, dense air and warm, moist air which is of main interest to the television meteorologists of northern Europe and North America.

Air above hot earth expands and rises. Air from cooler areas such as the oceans then floats-in to take its place. The process is called *circulation*. Two kinds of circulation produce wind: (i) general circulation extending around the earth; and (ii) smaller secondary circulations producing local wind conditions. Figure 2.1 shows a simplified theoretical model of the circulation of prevailing winds which result from the general movement of air around the earth. There are no prevailing winds within the equatorial belt, which lies roughly between latitudes 10° S and 10° N. Therefore, near the equator and up to about 700 mi (1127 km) on either side of it there lies a low-pressure belt in which the air is hot and calm. The air in this region rises above the earth instead of moving across it, creating a region of relative calm called the *doldrums*. In both hemispheres, some of the air that has risen at the equator returns to the earth's surface at about 30° latitude, producing no wind. These high-pressure areas are called *horse latitudes*, possibly because many horses died on the sailing ships that got stalled because of lack of wind. The winds that blow between the horse latitudes and the doldrums are called *trade winds* because sailors relied on them in sailing trading ships. The direction of trade winds is greatly modified by the rotation of the earth as they blow from east to west. Two other kinds of winds that result from the general circulation of the atmosphere are called the *prevailing winds* and the *polar easterlies*. The prevailing winds blow in two belts bounded by the horse latitudes and 60° north and south of the equator. The polar easterlies blow in the two belts between the poles and about 60° north and south of the equator. Thus the moving surface air produces six belts of winds around the earth as shown in Fig. 2.1.

2.2.2 Types of Wind

Of the several types of wind that encompass the earth's surface, winds that are of interest in the design of buildings can be classified into three major types: the prevailing winds, seasonal winds, and local winds.

1. *The prevailing winds.* Surface air moving from the horse latitudes toward the low-pressure equatorial belt constitutes the prevailing winds or trade winds. In the northern hemisphere, the northerly wind blowing toward the equator is deflected by the rotation of the earth to become northeasterly and is known as the northeast trade wind. The corresponding wind in the southern hemisphere is called the southeast trade wind.

On the polar regions of the horse latitudes, the atmospheric pressure diminishes and the winds moving toward the poles are deflected by the earth's rotation toward the east. Because winds are known by the direction from which they blow, these winds are known as the prevailing westerlies. The winds from the poles, particularly in the southern hemisphere, are deflected to become the polar easterlies. In comparison to the westerlies, the trade winds and the polar easterlies are shallow, and above a few thousand feet they are generally replaced by westerlies.

2. *The seasonal winds.* The air over the land is warmer in summer and colder in winter than the air adjacent to oceans during the same seasons. During summer, the continents thus become seats of low pressure, with winds blowing in from the colder oceans. In the winter, the continents are seats of high pressure with winds directed toward the warmer oceans. These seasonal winds are typified by the monsoons of the China Sea and Indian Ocean.

3. *The local winds.* Corresponding with the seasonal variation in temperature and pressure over land and water, daily changes occur which have a similar but local effect, penetrating to a distance of about 30 mi (48 km) on and off the shores. Similar daily changes in temperature occur over irregular terrain and cause mountain and valley breezes. Other winds associated with local phenomena include whirlwinds and winds associated with thunderstorms.

All three types of wind mentioned here are of equal importance in design. However, for purposes of evaluating wind loads, the characteristics of the prevailing and seasonal winds are analytically studied together, while those of local winds are studied separately. This grouping is for analytical convenience and to distinguish between the widely differing scale of fluctuations of the winds; prevailing and seasonal wind speeds fluctuate over a period of several months, whereas the local winds vary

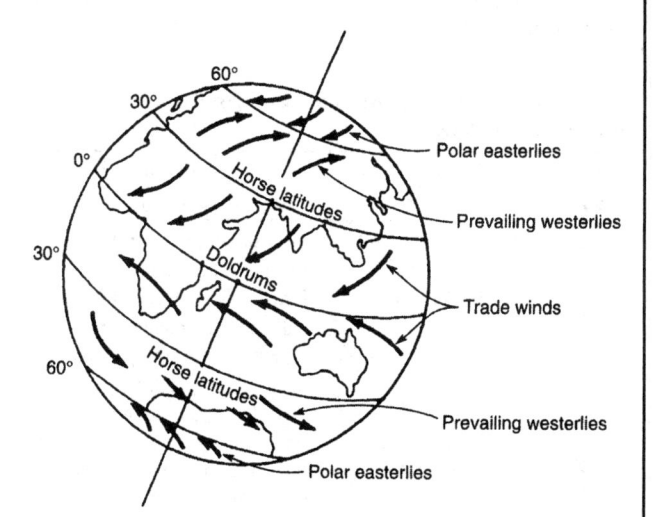

Figure 2.1   Circulation of world's winds.

almost every minute. The variations in the speed of prevailing and seasonal winds are referred to as *fluctuations* in mean velocity. The variations in the local winds, which are of a smaller character, are referred to as *gusts*.

Flow of wind, unlike that of other fluids, is not steady and fluctuates in a random fashion. Because of this, the properties of wind are studied statistically. The roughness of the earth's surface creates frictional drag in the flow of air, causing a gradual decrease in its velocity near the earth's surface. Also, turbulence is caused by surface roughness. Before these characteristics are studied in greater detail, it is of interest to have an overview of other types of extreme wind conditions and their effect on structures.

2.3   EXTREME WIND CONDITIONS

2.3.1   Introduction

Human beings and their works are subjected to hazards arising from forces and disturbances occurring in the natural environment. One of these natural hazards is the dynamic action of wind in the shape of hurricanes and tornadoes. These winds cause direct economic damages exceeding many millions of dollars in any given year. The large risk to both investment and human life which occurs from wind makes necessary an understanding of the physical phenomenon involved in the wind hazard and the development of improved planning and design methods. Wind engineering is, therefore, important, both in terms of potential economic damage and life loss and in human comfort.

Extreme winds, such as thunderstorms, hurricanes, tornadoes, and typhoons, impose loads on structures that are many times more than those normally assumed in their design. Some standards, such as the ASCE 7-16, provide for hurricane wind speeds for a specified probability of occurrence but do not consider directly the effect of other types of extreme wind conditions.

2.3.2   Thunderstorms

Thunderstorms are one of the most familiar features of temperate summer weather, characterized by long hot spells punctuated by release of torrential rain. The essential conditions for the occurrence of thunderstorms are warm, moist air in the lower atmosphere and cold, dense air at higher altitudes. Under these conditions, warm air at ground level rises, building storm clouds in the upper atmosphere. Thunder and lightning accompany downpours, creating gusty winds. Wind speeds of 50 to 70 mph (22 to 31 m/s) are typically reached in a thunderstorm and are often accompanied with swirling wind action exerting high suction forces on roof and cladding elements. Approximately, 10,000 severe thunderstorms occur in the United States each year, which is about 10 percent of the total thunderstorms more common in the southwest and midwest and occur in the spring and summer. There is approximately 16 million thunderstorms each year worldwide.

2.3.3   Hurricanes

Hurricanes are severe atmospheric disturbances that originate in the tropical regions of the Atlantic Ocean or Caribbean Sea. The diameter of the storm varies between 50 and 600 mi. The forward movement of a hurricane (translational speed) can vary between approximately 10 and 25 mph (4.5 and 11.2 m/s). The Saffir-Simpson Hurricane Scale rates the intensity of hurricanes, see Table 2.1. The five-step scale ranges from the weakest (Category 1) to the strongest (Category 5).

They travel north, northwest, or northeast from their point of origin and usually cause heavy rains. They originate in the doldrums and consist of high-velocity winds blowing circularly around a low-pressure center known as the eye of the storm. The low-pressure center develops when the warm saturated air prevalent in the doldrums interacts with the cooler air. From the edge of the storm toward its center the atmospheric pressure drops sharply raising the wind velocity. In a fully developed hurricane, winds reach speeds up to 70 to 80 mph (31 to 36 m/s), and in severe hurricanes can attain velocities as high as 200 mph (90 m/s). Within the eye of the storm, the winds cease abruptly, the storm clouds lift, and the seas become exceptionally violent. One of the most destructive hurricane in the U.S. history hit the Atlantic seaboard in June, 1972 causing at least 122 deaths and damage amounting to over $3 billion.

Table 2.1   Saffir-Simpson Hurricane Wind Scale (Source ASCE 7-16)

Hurricane category	Sustained wind speed[a] mph (*m/s*)	Types of damage due to hurricane winds
1	74-95 (33-42)	Very dangerous winds will produce some damage
2	96-110 (43-49)	Extremely dangerous winds will cause extensive damage
3	111-129 (50-57)	Devastating damage will occur
4	130-156 (58-69)	Catastrophic damage will occur
5	>157 (70)	Highly catastrophic damage will occur

[a]1-minute average wind speed at 33 ft (10 m) above open water.

The maximum basic wind speed velocity (3 s gust) for any area of the United States specified in ASCE 7-16 for buildings in Risk Categories I, II and III, and IV are 170, 180, and 200 mph, respectively. Note that these values are much less than the highest wind speeds in hurricanes.

Except in rare instances, such as defense installations, a structure is not normally designed for full hurricane wind speeds.

Hurricanes are one of the most spectacular forms of terrestrial disturbances that produce heaviest rains known on earth. They have two basic requirements, warmth and moisture, and consequently develop only in the tropics. Almost invariably they move in a westerly direction at first, and then swing away from the equator, either striking land with devastating results or moving out over the oceans until they encounter cool surface water to die out naturally. The region of greatest storm frequency is the northwestern Pacific, where the storms are called *typhoons*—a name of Chinese origin meaning "wind which strikes." The storms which occur in the Bay of Bengal and the seas of north Australia are called *cyclones*. Although there are some general characteristics common to all hurricanes, no two are exactly alike. However, a typical hurricane can be considered to have a 375 mi (600 km) diameter, with its circulating winds spiraling in toward the center at speeds up to 112 mph (50 m/s). The size of the eye can vary in diameter from as little as 3.7 to 25 mi (6 to 40 km). However, the typhoon which roared past the island of Guam in 1979 had a very large diameter of 1400 mi (2252 km) with the highest wind speed reaching 190 mph (85 m/s). Storms of such violence have been known to drive a plank of wood right through the trunk of a tree and blow straws end-on through a metal deck. Fortunately, storms of such magnitude are not common. Table 2.1 shows the Saffir-Simpson Hurricane Wind Scale adopted from ASCE 7-16 commentary.

2.3.4 Tornadoes

Tornadoes develop within severe thunderstorms consisting of a rotating column of air usually accompanied by a funnel-shaped downward extension of a dense cloud having a vortex of several hundred feet, typically 200 to 800 ft (61 to 244 m) in diameter whirling destructively at speeds up to 300 mph (134 m/s). They contain the most destructive of all wind forces, destroying everything along their path. Tornadoes form when a cold storm-front runs over warm, moist surface air. The warm air rises through the overlaying cold storm clouds intercepted by the high-altitude winds that are even colder rapidly moving above the clouds. Warm air collides with the cooler air and begins to whirl. The pressure at the center of the spinning

column of air is reduced because of the centrifugal force. This reduction in pressure causes more warm air to be sucked into it, creating a violent outlet for the warm air trapped under the storm. As the velocity increases, more warm air is drawn up to the low-pressure area created in the center of the vortex. As the vortex gains strength, the funnel begins to extend toward the ground, eventually touching it. Funnels usually form close to the leading edge of the storm. Larger tornadoes may have several vortices within a single funnel. If the bottom of the funnel can be seen, it usually means that the tornado has touched down and begun to pick up visible debris from the ground.

A typical tornado travels 20 to 30 mph (9 to 14 m/s), touches ground for 5 to 6 mi (8 to 10 km), and has a funnel 300 to 500 ft (92 to 152 m) wide. Distance from the ground to the cloud averages about 2000 ft (610 m). Tornadoes contain the most powerful of all winds, causing damage well in excess of $100 million a year in the United States. Although it is impractical to design buildings to sustain a direct hit from a tornado, it behoves the engineer to pay extra attention to anchorage of roof decks and curtain walls of buildings in areas of high tornado frequency.

Rolling plains and flat country make a natural home for tornadoes. Statistically, flat plains get more tornadoes than other parts of the country. In North America, communities in Kansas, Nebraska, and Texas have many tornadoes and are classified as "tornado belt" areas. No accurate measurement of the inner speed of a tornado has been made because tornadoes destroy standard measuring instruments. However, photographs of tornadoes suggest that wind speeds are of the order of 167 to 224 mph (75 to 100 m/s). Although there are definite tornado seasons, tornadoes can occur at any time.

Similar to a hurricane, a tornado consists of a mass of unstable air rotating furiously and rising rapidly around the center of an area with low atmospheric pressure. The similarity ends here, because whereas a hurricane is generally of the order of 300 to 400 mi (483 to 644 km) in diameter, a large tornado is unlikely to be more than 1500 ft (458 m) across. However, in terms of destructive violence no other atmospheric disturbance compares with a tornado.

Wind alone is not the only damaging element at work in a tornado. The pressure at the center of a tornado is extremely low. As the storm passes over a building, the pressure inside the structure is far greater than the outside, causing the building to literally explode. Typically, buildings are not designed to withstand a direct hit from a tornado. However, for those which are deemed essential, such as defense installations and nuclear facilities, sufficient information is available in engineering literature to implement tornado-resistant design. This information is in the form of tornado risk probabilities, wind speeds, and forces.

2.4 CHARACTERISTICS OF WIND

2.4.1 Introduction

Wind is a phenomenon of great complexity because of the many flow situations arising from the interaction of wind with structures. However, in wind engineering simplifications are made to arrive at meaningful predictions of wind behavior by distinguishing the following features:

- Variation of wind velocity with height (2.4.2).
- Turbulent nature of wind (2.4.3).
- Probabilistic approach (2.4.4).
- Dynamic action of wind (2.4.5).
- Vortex shedding phenomenon (2.4.6).
- Dynamic nature of wind-structure interaction (2.4.7).

2.4.2 Variation of Wind Velocity with Height

At the interface between a moving fluid and a solid surface, viscosity manifests itself in the creation of shear forces aligned opposite to the direction of fluid motion. A similar effect occurs between the surface of the earth and the atmosphere. Viscosity reduces the air velocity adjacent to the earth's surface to almost zero. A retarding effect occurs in the layers near the ground, and these inner layers in turn successively slow down the outer layers. The slowing down is less at each layer and eventually becomes negligibly small. The velocity increase is curvilinear varying from zero at the ground surface to a maximum at some distance above the ground. The height at which the velocity ceases to increase is called the *gradient height*, and the corresponding velocity, the *gradient velocity*. The shape and size of the curve depends less on the viscosity of the air than on the type and predominance of the turbulent and random eddying motions in the wind, which in turn are affected by the type of terrain over which the wind is blowing (see Fig. 2.2). This important characteristic of variation of wind velocity with height is a well understood phenomenon as evidenced by higher design pressures specified at higher elevations in most building codes.

The variation of velocity with height can be considered as a gradual retardation of the wind closer the ground as a result of surface friction. At heights of approximately 1200 ft (366 m) from the ground, the wind speed is virtually unaffected by surface friction and its movement

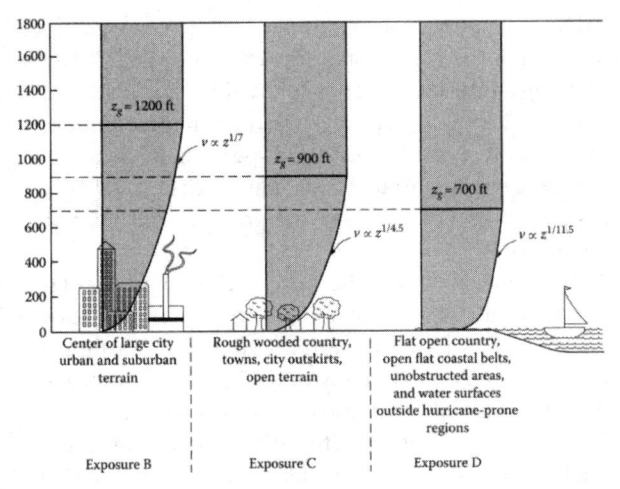

Figure 2.2   Variation of wind velocity with height.

is solely dependent on the prevailing seasonal and local wind effects. The height through which the wind speed is affected by the topography is called the *atmospheric boundary layer*. The wind speed profile within this layer is in the domain of turbulent flow and can be mathematically predicted by a logarithmic equation. However, in practice, wind speed variation is given by a simpler power-law expression of the form:

$$V_z = V_g (Z/Z_g)^{1/\alpha} \tag{2.1}$$

where V_z = the mean wind speed at height Z above the ground

V_g = gradient wind speed assumed constant above the boundary layer

Z = height above the ground

Z_g = depth of boundary layer

α = power law coefficient

By knowing the mean wind speed at gradient height and the value of exponent α, the wind speeds at height Z are easily calculated by using Eq. (2.1). The exponent α and the depth of boundary layer Z_g vary with terrain roughness. The value of α ranges from a low of 0.087 for open country to about 0.2 for built-up urban areas, signifying that wind speed reaches its maximum value over a longer height in an urban terrain than in an open country. The pressure and suction generated by wind are a function of the wind speed, and in general increase with the building height.

2.4.3 Turbulent Nature of Wind

The motion of wind is turbulent. A concise mathematical definition of turbulence is difficult to give, except to state that it occurs in wind flow because air has a very low viscosity of about one-sixteenth that of water. Any movement of air at speeds greater than 2 to 3 mph (0.9 to 1.3 m/s) is turbulent, causing particles of air to move randomly in all directions. This is in contrast to the laminar flow of particles of heavy fluids, which move predominantly parallel to the direction of flow.

The variation of wind velocity with height describes only one aspect of wind in the boundary layer. Superimposed on the mean wind speed is the turbulence or gustiness of wind, which produces deviations in the wind speed above and below the mean, depending upon whether there is a gust or lull in the wind action.

Flow of air near the earth's surface changes in speed and direction because of the obstacles which introduce random vertical and horizontal movements at right-angles to the main direction of flow. These gusts vary over a wide range of frequencies and amplitudes, both in time and space. Shown in Fig. 2.3 are the anemometer readings of wind speeds in which V_z *is mean wind speed, also denoted as v; and* $V_z^l = $ *gust speed.*

The scale and intensity of turbulence can be likened to the size and rotating speed of the eddies or vortices that make up the turbulence. It is generally found that the size of the flow affects the size of the turbulence within it. Thus, the flow of a large mass of air has a larger overall turbulence than a corresponding flow of a small mass of air. Because of its random nature, the properties of wind are studied statistically. A statistical property is the mean or the average. Because wind speed changes constantly, different averages are obtained by using different averaging times. For example, while a 1-h average of wind speed may be 30 mph (13.4 m/s), the same wind averaged for 1 min may be as high as 80 mph (35.8 m/s). Therefore, it is necessary to specify averaging time whenever a mean velocity is referenced. However, one of the strange phenomena that occurs when measuring wind speed is that averages taken over periods between 10 and 60 min adequately avoid the violent peaks and valleys due to gustiness. The averages taken over 10-, 20-, and 60-min

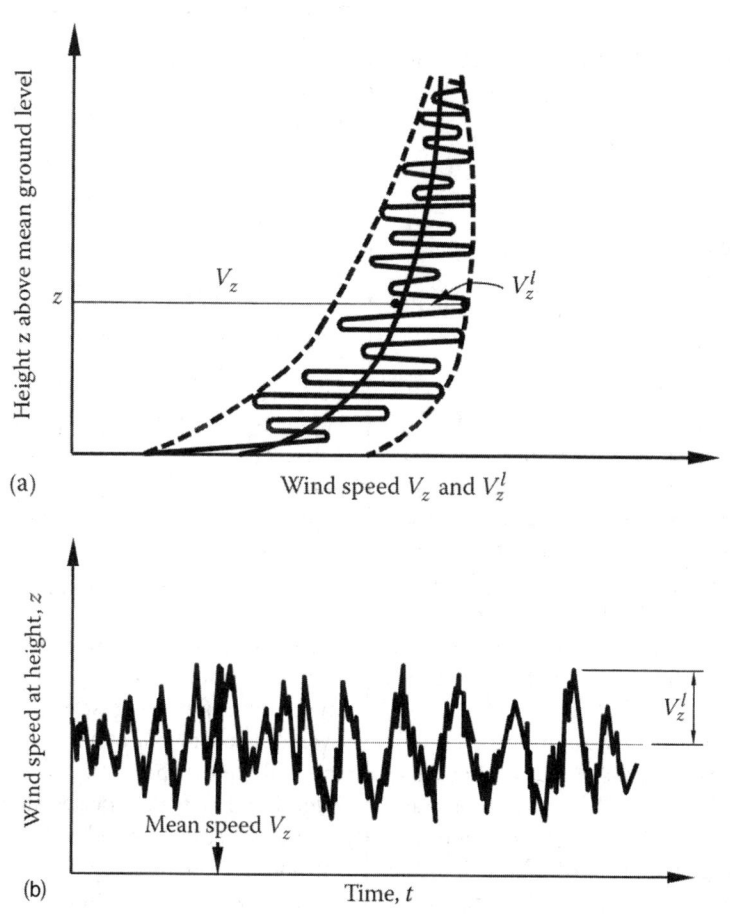

(a)

(b)

Figure 2.3 Variation of wind velocity with time.

intervals are nearly the same. This characteristic of the wind allows the comparison of mean data taken over different time periods.

For structural engineering purposes, the velocity of wind can be considered as having two components; a mean velocity component whose value increases with height and a turbulent velocity fluctuation. This can be visualized by considering the mean velocity as increasing with height as before, but subject to vortices or eddies (small currents of air spinning in space) oriented randomly to the mean direction. By assigning different size and rates of spin to these vortices, one can equate these to the wavelength and frequency of variable components.

Spectral analysis techniques provide a convenient method for dealing with the random turbulence of wind. A complete treatment of the method is beyond the scope of the present work except to state that the method has some similarities with structural problems, wherein a required solution, for example, the deflection of a simply supported beam, can be obtained by superposition of a sufficiently large number of deflection patterns with different amplitudes and shapes.

The velocity at any instant v_t can be represented as the summation of the average velocity and the instantaneous value of velocity fluctuation about the mean value as shown in Fig. 2.3. Thus

$$V_t = V_z + V_z^l \qquad (2.2)$$

where V_t = velocity at instant t
V_z = average or mean velocity
V_z^l = instantaneous velocity fluctuation about the mean velocity V_z

Figure 2.4 schematically represents the fluctuation of mean and gust pressure along the height of a building.

The longest averaging time used in structural engineering practice is 1 h. As the averaging time decreases, the maximum speed of wind increases. The average or mean wind speed used in many building codes of the United States is the *fastest-mile wind*, which can be thought of as the maximum velocity measured over 1 mi of wind passing through an anemometer. Normally, the wind speed in structural design is in the range of 60 to 120 mph (97 to 154 m/s), giving an averaging period of 30 to 60 s.

Rapid bursts in the velocity of wind are called gusts. Tall buildings are sensitive to gusts that last about 1 s. Therefore, the fastest mile wind (which has the averaging period of 30 to 60 s) is inadequate for design of tall buildings. One must use the gust speed rather than the mean wind speed in the determination of wind loads.

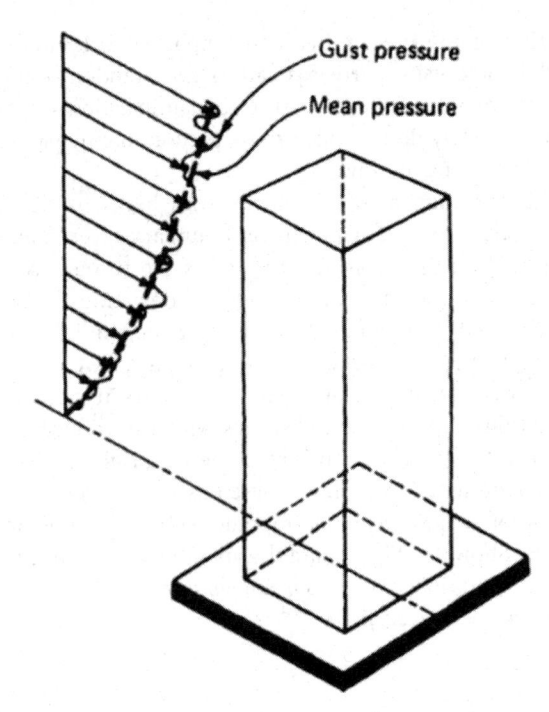

Figure 2.4   Schematic representation of mean and gust pressure.

The gust speed can be obtained by multiplying the mean wind speed by a gust factor G_v. Thus

$$V_g = G_v V \qquad (2.3)$$

where G_v = the gust factor
V_g = the gust speed
V = the mean wind speed

Not all buildings are equally sensitive to gusts. In general, the more flexible a structure is, the more sensitive it is to gusts. The only accurate way to determine the gust factor (also called gust response factor) is to conduct a wind-tunnel test. However, attempts are made in some contemporary wind load standards, such as the National Building Code of Canada (NBC 1990) and the ASCE 7, to give analytical procedures for determining gust response factors. These are considered later in this chapter.

2.4.4   Probabilistic Approach to Wind Load Determination

In many engineering sciences the intensity of certain events is considered as a function of duration recurrence interval (return period). For example, in hydrology the intensity of rainfall expected in a region is considered in terms of return period because the rainfall expected once in 10 years is likely to be less than the one expected once in every 50 years. Similarly, in wind engineering the speed of wind is considered to vary with return periods. For example, the nominal Design 3-s gust speed for most of nonhurricane areas of the United States at 33 ft (10 m) aboveground and corresponding to a mean recurrence interval (MRI) of 700 years is 107 mph (48 m/s); this is compared to 114 mph (51 m/s) for an MRI of 1700 years. A 700-year return-period wind of 107 mph (48 m/s) means that on the average, these areas will experience a wind faster than 107 mph (48 m/s) within a period of 700 years. A return period of 700 years corresponds to a probability of occurrence of 1/700 = 0.00143 ~ 0.14 percent. Thus, the chance that a wind exceeding 107 mph (48 m/s) will occur in these regions within a given year is 0.14 percent. See Figures 26.5-1A, 26.5-1B, 26.5-1C, 26.5-1D, 26.5-2A, 26.5-2B, 26.5-2C, and 26.5-2D of ASCE 7-16 for basic wind speed maps.

Suppose a building is designed for a 100-year lifetime using a design wind speed of 120 mph. What is the probability that wind will exceed the design speed within the lifetime of the structure? The probability that this wind speed will not be exceeded in any year is 49/50. The probability that this speed will not be exceeded in 100 years in a row is $(49/50)^{100}$. Therefore, the probability that this wind speed will be exceeded at least once in 100 years is

$$1-\left(\frac{49}{50}\right)^{100}=0.87=87 \text{ percent}$$

This signifies that although a wind with low annual probability of occurrence is used to design structures, there exists still a high probability of the wind being exceeded within the lifetime of the structure. However, in structural engineering practice it is believed that the actual probability of overstressing a structure is much less because of the factors of safety and the generally conservative values used in design.

It is important for design engineers to understand the notion of probability of occurrence of design wind speeds during the service life of buildings. The general expression for probability, P, that the design wind speed will be exceeded at least once during the exposed period of n years is given by

$$P=1-(1-P_a)^n \qquad (2.4)$$

where P_a = annual probability of being exceeded (reciprocal of the MRI)

n = exposure period in years

Consider a building in Dallas being designed for a 50-year service life instead of 100 years. The probability of exceeding the design wind speed at least once during the 50-year lifetime of the building is

$$P = 1 - (1 - 0.02)^{50} = 1 - 0.36 = 0.64 = 64 \text{ percent}$$

The probability that wind speeds of a given magnitude will be exceeded increases or decreases with exposure period of the building and the MRI used in the design. Values of P for a given MRI and a given exposure period are shown in Table 2.2.

2.4.5   Dynamic Action of Wind

The factors that affect the dynamic action of the wind on tall buildings are:

1. Buffeting by gusts.
2. Buffeting by turbulence and vortices shed by the structure itself.
3. Buffeting by the wake from another structure.
4. Aerodynamic damping.

2.4.6   Vortex-Shedding Phenomenon

In general, wind blowing past a body can be considered to be diverted in three mutually perpendicular directions, giving rise to forces and moments about the three directions. In aeronautical engineering, special terminology is used to describe these forces and moments as shown in Fig. 2.5. Although all six components are significant in

Table 2.2   Probability of Exceeding Design Wind Speed During Design Life of Building

Annual probability P_a	Mean recurrence interval ($1/P_a$) years	Exposure period (design life), n (years)					
		1	5	10	25	50	100
0.1	10	0.1	0.41	0.15	0.93	0.994	0.999
0.04	25	0.04	0.18	0.34	0.64	0.87	0.98
0.034	30	0.034	0.15	0.29	0.58	0.82	0.97
0.02	50	0.02	0.10	0.18	0.40	0.64	0.87
0.013	75	0.013	0.06	0.12	0.28	0.49	0.73
0.01	100	0.01	0.05	0.10	0.22	0.40	0.64
0.0067	150	0.0067	0.03	0.06	0.15	0.28	0.49
0.005	200	0.005	0.02	0.05	0.10	0.22	0.39

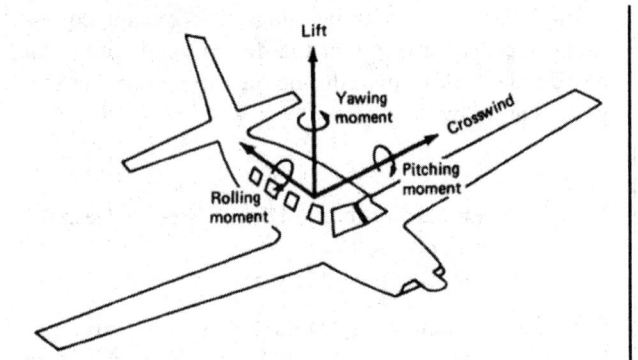

Figure 2.5 Six components of wind.

aeronautical work, in civil and structural engineering, the force and moment corresponding to the vertical axis (lift and yawing moment) are of little significance. Therefore, the flow of wind is considered two-dimensional, as shown in Fig. 2.6. When uplift forces need to be considered, the wind flow is three-dimensional as shown in Fig. 2.7; note that the more streamlined the wind flow is, the less the buffeting force on the building are as shown in Fig. 2.8.

Along wind or simply *wind* is the term used to refer to drag forces and *transverse wind* is used to refer to crosswind. The crosswind response, that is, motion in a plane perpendicular to the direction of wind, dominates over the along-wind response for most tall buildings. This complex nature of wake-excited response is a result of interaction of turbulence, building motion, and the dynamics of wake formation.

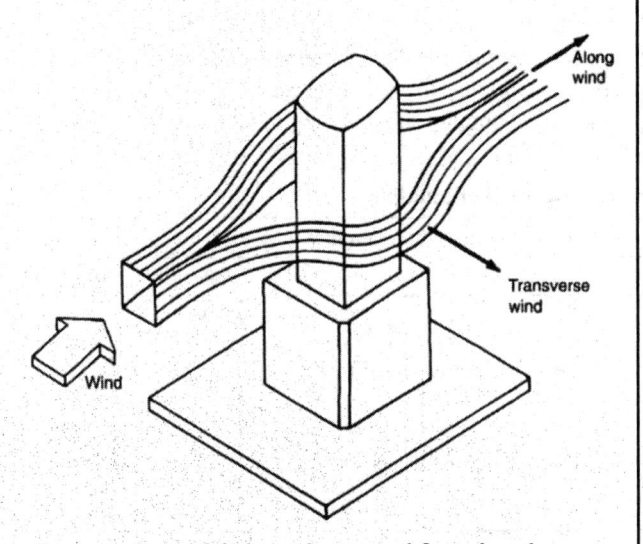

Figure 2.6 Simplified two-dimensional flow of wind.

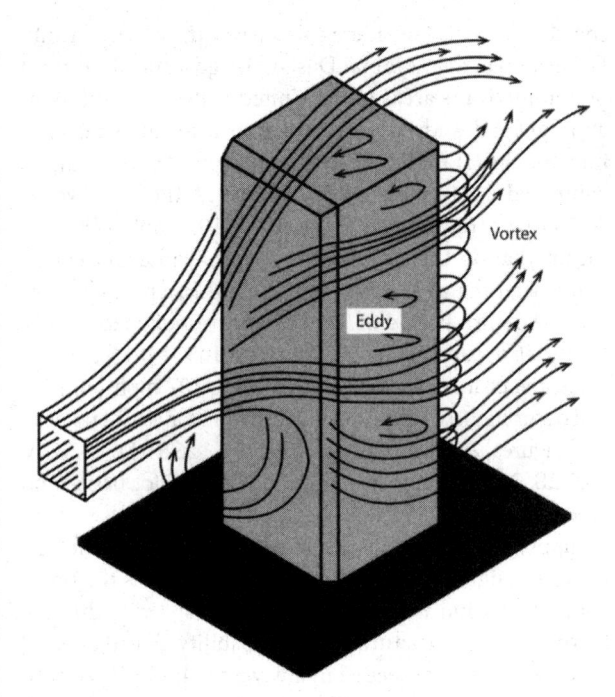

Figure 2.7 Three-dimensional wind flow.

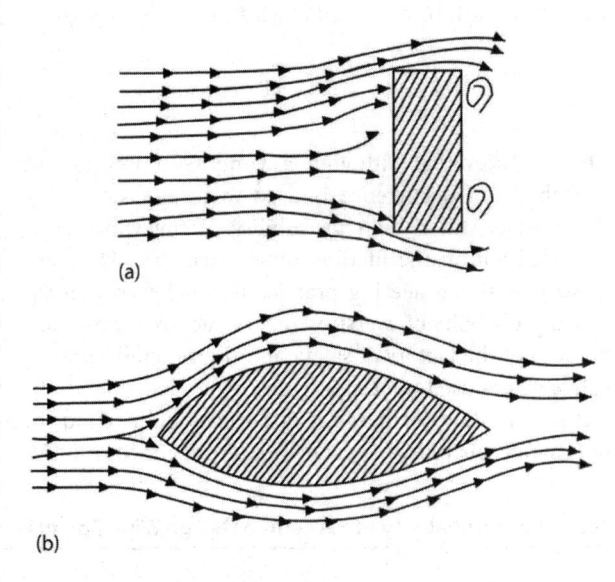

Figure 2.8 Wind flow around bluff and streamlined buildings: (a) bluff (rectangular) building and (b) streamlined building.

It is perhaps baffling to the inexperienced engineer to learn that a tall building is subject to wind excitations not only in a direction parallel to the wind but also in a direction perpendicular to it. Yet in many instances, the major criterion for design of very tall buildings is the crosswind response. While the maximum lateral wind-loading and deflection are generally in the direction parallel with the wind (along-wind direction), the maximum acceleration of a building leading to possible human perception of motion or even discomfort may occur in a direction perpendicular to the wind (crosswind direction).

There appear to be three distinctly different reasons why a building responds in a direction at right-angles to the applied wind forces; these are: (i) the biaxial displacement induced in the structure because of either asymmetry in geometry or in applied wind loading; (ii) the turbulence of wind; and (iii) the negative-pressure wake or trail on the building sides. For tall buildings it appears that the crosswind response is caused mainly by the wake.

Consider a cylindrically shaped building subjected to a smooth wind flow. The originally parallel stream lines are displaced on either side of the cylinder. This results in spiral vortices being shed periodically from the sides of the cylinder into the downstream flow of wind called the wake. At low wind speeds of say 50 to 60 mph (22.3 to 26.8 m/s), the vortices are shed symmetrically in pairs one from each side. These vortices can be thought of as imaginary projections attached to the cylinder that increase the drag force on the cylinder. When the vortices are shed, that is, break away from the surface of the cylinder, an impulse is applied to the cylinder in the transverse direction. This phenomenon of alternate shedding of vortices for a rectangular tall building is shown schematically in Fig. 2.9.

At low wind speeds, since the shedding occurs at the same instant on either side of the building, there is no tendency for the building to vibrate in the transverse direction. It is therefore subject to along-wind oscillations parallel to the wind direction. At higher speeds, the vortices are shed alternately first from one and then from the other side. When this occurs, there is an impulse in the along-wind direction as before, but in addition, there is an impulse in the transverse direction. The transverse impulses are, however, applied alternatively to the left and

then to the right. The frequency of transverse impulse is precisely half that of the along-wind impulse. This kind of shedding which gives rise to structural vibrations in the flow direction as well as in the transverse direction is called *vortex shedding* or the *Karman vortex street*, a phenomenon well known in the field of fluid mechanics.

There is a simple formula to calculate the frequency of the transverse pulsating forces caused by vortex shedding

$$f = \frac{V \times S}{D} \qquad (2.5)$$

where f = the frequency of vortex shedding in hertz
V = the mean wind speed at the top of the building
S = a dimensionless parameter called the Strouhal number for the shape
D = the diameter of the building

In Eq. (2.5), the parameters V and D are expressed in consistent units such as ft/s and ft, respectively.

The Strouhal number is not a constant but varies irregularly with the wind velocity. At low air velocities, S is also low and increases with the velocity up to a limit of 0.21 for a smooth cylinder. This limit is reached for a velocity of about 50 mph (22.4 m/s) and remains almost a constant at 0.20 for wind velocities between 50 and 115 mph (22.4 and 51 m/s).

Consider for illustration purposes, a circular prismatic-shaped high-rise building having a diameter equal to 110 ft (33.5 m) and a height-to-width ratio of 6 with a natural frequency of vibration equal to 0.16 Hz. Assuming a wind velocity of 60 mph (27 m/s), the vortex-shedding frequency is given by

$$f = \frac{V \times 0.2}{110} = 0.16\,\text{Hz}$$

where V is in ft/s.

If the wind velocity increases from 0 to 60 mph (27 m/s), the frequency of vortex excitation will rise from 0 to a maximum of 0.16 Hz. Since this frequency happens to

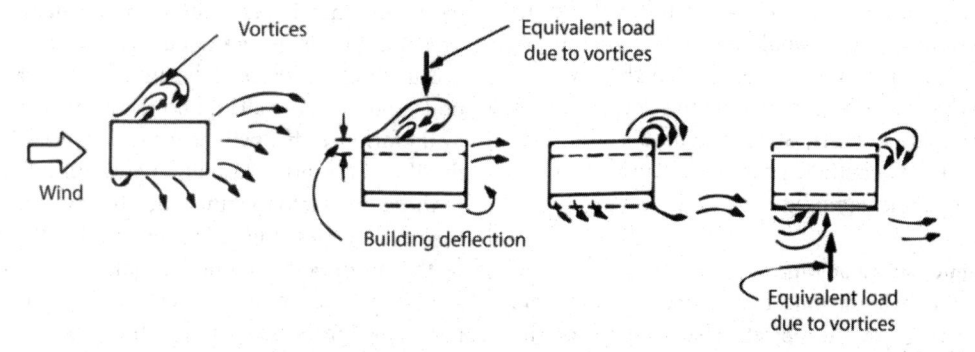

Figure 2.9 Vortex-shedding phenomenon.

be very close to the natural frequency of the building, and assuming a very low damping, the structure would pulsate as if its stiffness were zero at a wind speed somewhere around 60 mph (27 m/s). Note the similarity of this phenomenon with the ringing of church bells or the shaking of a tall lamppost whereby a small impulse added to the moving mass at each end of the cycle greatly increases the kinetic energy of the system. Similarly, during vortex shedding an increase in deflection occurs at the end of each swing. If the damping characteristics are small, the vortex shedding can cause building displacements far beyond those predicted on the basis of static analysis.

When the wind speed is such that the shedding frequency becomes approximately the same as the natural frequency of the building, a resonance condition is created. After the structure has begun to resonate, further increases in wind speed by a few percent will not change the shedding frequency, because the shedding is now controlled by the natural frequency of the structure. The vortex shedding frequency has, so to speak, locked-in with the natural frequency. When wind speed increases above that causing the lock-in phenomenon, the frequency of shedding is again controlled by the speed of the wind. The structure vibrates with the resonant frequency only in the lock-in range, and for wind speeds either below or above this range, the vortex shedding will not be critical.

Vortex shedding occurs for many building shapes. The value of S for different shapes is determined in wind tunnels by measuring the frequency of shedding for a range of wind velocities. One does not have to know the value of S very precisely because the lock-in phenomenon occurs within a range of about 10 percent of the exact frequency of the structure.

The crosswind response mechanism is very complex, and an exact analytical method that takes into account the variables of turbulence, building shape, structure stiffness, damping, and density has not been introduced into structural engineering practice. In cases where the crosswind response may become the controlling factor in the design, the only option would be wind tunnel investigation. However, it should be recognized that even in elaborate model tests it is not possible to simultaneously scale the Reynolds number, Strouhal number, the fluctuating lift and drag coefficients, the structural stiffness, and the aerodynamic damping.

2.4.7 Dynamic Nature of Wind

When wind hits a blunt object in its path, it transfers some of its energy to the object. The measure of the amount of energy transferred is called the *gust response factor*. As mentioned previously, wind turbulence (also called gustiness) is affected by terrain roughness and varies with the height above ground. A tall, slender, and flexible structure could have a significant dynamic response to wind because of buffeting. This dynamic amplification of response would depend on how the gust frequency correlates with the natural frequency of structure and also on the size of the gust in relation to the building size.

Unlike the mean flow of wind, which can be considered as static, wind loads associated with gustiness or turbulence change rapidly and even abruptly, creating effects much larger than if the same loads were applied gradually. Wind loads, therefore, need to be studied as if they were dynamic in nature. The intensity of a wind load depends on how fast it varies and also on the response of the structure. Therefore, whether the pressures on a building created by a wind gust, which may first increase and then decrease, are considered as dynamic or static depends to a large extent on the dynamic response of structure to which it is applied.

Buffering is a well-recognized phenomenon in wind engineering, in which a bluff cylinder is exposed to a uniform air flow sheds vortices or eddies from either side of the structure at a frequency n. This depends on the diameter of the structure, D, and on the velocity of the flow, V. The frequency is expressed in terms of the dimensionless Strouhal number, S, in Eq. (2.6).

$$S = \frac{n_0 D}{V} \qquad (2.6)$$

$S = 0.14$ for square cylinders.

Lateral force fluctuations with a frequency of n_e and longitudinal force fluctuations with a frequency of $2n_e$ coupled with the shedding of vortices mentioned above. The forces in a uniform steady flow are well organized and powerful at the Strouhal frequency. However, the steady flow results in large amplitudes oscillations that can be set up on lightly damped structures with Strouhal frequency that is resonant with the natural frequency. The structure is generally quiescent at other frequencies because of the absence of dynamic magnification. The natural wind effect on tall buildings is particularly represented by uniform steady flow; the velocity varies with height and the flow in a city is usually highly turbulent. The effect of these two factors is that the effective vortex shedding frequency varies along the length of the structure because of the mean velocity gradient and the regularity of the shedding is distributed by the turbulence in the flow. As a result, the forces associated with vortex shedding become random and less effective.

Let us consider the movement of a building 800 ft tall, designed to a drift index of $\frac{H}{400}$, acted upon by a wind gust. Under wind loads, the building bends slightly as its top moves. It first moves in the direction of wind, say with a magnitude of 2 ft (0.61 m), and then starts oscillating back and forth. After moving in the direction of wind, the top goes through its neutral position, then moves approximately 2 ft (0.61 m) in the opposite direction, and continues oscillating back and forth until it eventually stops. The time it takes a building to swing through a complete oscillation is known as a *period*. The period of oscillation for a tall steel building in the height range of 700 to 1400 ft (214 to 427 m) normally is in the range of 5 to 10 s, whereas for a 10-story concrete or masonry building it may be in the range of 0.5 to 1 s. The action of a wind gust depends not only on how long it takes the gust to reach its maximum intensity and decrease again, but on the period of the building itself. If the wind gust reaches its maximum value and vanishes in a time much shorter than the period of the building, its effects are dynamic. On the other hand, the gusts can be considered as static loads if the wind load increases and vanishes in a time much longer than the period for the building. For example, a wind gust that develops to its strongest intensity and decreases to zero in 2 s is a dynamic load for a tall building with a period of, say 5 to 10 s, but the same 2 s gust is a static load for a low-rise building with a period of less than 2 s.

2.4.7.1 AERODYNAMIC DAMPING

Aerodynamic damping is a factor that can influence the dynamic response of a structure. The damping force is generally induced by the motion of a body, and it is usually dissipative but not in all circumstances. The aerodynamic damping force is proportional to the ratio of air density to average building density and the frequency and could be positive opposing the motion or negative adding to the forces causing the motion, that is, reinforcing the motion. For a square-shaped building with wind normal to the face, the nature of the damping forces show that the longitudinal oscillation induces forces opposing the motion, while the transverse motion induce forces that reinforce the motion. The forces due to transverse motion are due to the square shape of the building, in which a small change in relative wind direction from the normal face will result in an into-wind component of transverse force.

Negative damping phenomenon has always been known as a source of instability of what is known galloping variety similar to that occurs in suspension bridges and ice-covered transmission cables. When the net damping in the system approaches zero, instability occurs; net damping includes both mechanical and aerodynamic damping. In cases like this, the oscillation amplitude increases significantly until the nonlinearities produced in the system restrain the motion. When nonlinearities are produced in the system, most likely the amplitudes are large and unacceptable.

For conventional shape buildings not exposed to wind speed much in excess of 100 mph, aerodynamic stability is not likely to occur with the stiffness, mass, and damping parameters of such buildings. On the other hand, structures with unconventional shaped of lighter weight, lower damping, lower flexibility, negative aerodynamic damping may be encountered which will reduce the overall available damping significantly. As a result, the sway amplitudes will be significantly high that may cause structural failure or significant damage to walls and partitions and nonconformable sway amplitudes. Current trends in tall buildings include the use of lightweight structural system, higher strength materials, more monolithic forms of construction and a wide range of architectural forms. Although these trends may be desirable from other points of view, the large negative aerodynamic damping should be monitored and considered.

2.4.8 Cladding Pressures

2.4.8.1 INTRODUCTION

The design of cladding for lateral loads is of major concern to architects and engineers. Although the failure of an exterior cladding resulting in broken glass may be of less consequence than collapse of a structure in an earthquake, the expense of replacement and hazards posed to pedestrians require that careful attention be given to its design. Cladding breakage in a windstorm is a complicated phenomenon as witnessed in hurricane Alicia, which hit Galveston and downtown Houston on August 18, 1983, causing breakage of glass in several tall buildings. Wind forces play a major role in glass breakage, which is also influenced by other factors, such as solar radiation, mullion and sealant details, tempering of the glass, double or single glazing of glass, and fatigue. It is known with certainty that glass failure starts at nicks and scratches which may be caused during its manufacturing and handling operations.

There appears to be no analytical approach available for a rational design of curtain walls of all shapes and sizes. Although most codes have tried to identify regions of high wind loads around building corners, the modern trend in architecture of using nonprismatic and curvilinear shapes combined with the unique topography of each site, has necessitated experimental determination of wind loads for each building.

It has become a routine nowadays to obtain design information concerning the distribution of wind pressures over a building's surface by conducting wind-tunnel studies. In the past, curtain wall has developed into an ornamental item and has emerged as a significant architectural element. Sizes of window panes have increased considerably, requiring that the glass lights be designed as structural elements for various combinations of forces due to wind, shadow effects, and temperature movement. Glass in curtain walls not only has to resist large forces, particularly in tall buildings, but must also be designed to accommodate the various distortions of the total building structure. Breaking of large panes of glass in tall buildings can cause serious damage to neighboring properties and can injure pedestrians.

2.4.8.2 DISTRIBUTION OF PRESSURES AND SUCTIONS

It has been known for some time that when air flows around the edges of a structure, the pressures produced at the corners are much in excess of the normal pressure on the center of elevation, as evidenced by damage caused to corner windows, eave and ridge tiles, etc., in a windstorm. Wind tunnel studies conducted on scale models of buildings have shown that three distinct pressure areas are developed. These are shown schematically in Fig. 2.10 and are listed below.

1. Positive-pressure zone on the upstream face (Region 1).

2. Negative-pressure zones at the upstream corners (Region 2).

3. Negative-pressure zone on the downstream face (Region 3).

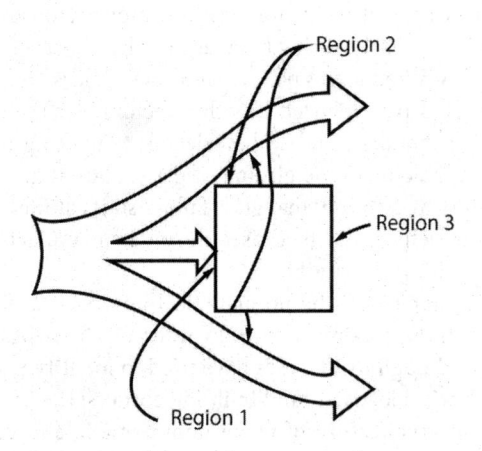

Figure 2.10 Distribution of pressures and suctions.

Highest negative pressures are created in the upstream corners designated as Region 2 in Fig. 2.10. These have been measured in wind-tunnel investigations and also have occurred in practice. Wind pressures on a buildings surface are not constant, but fluctuate continuously. The positive pressure on the upstream or the windward face fluctuates more than the negative pressure on the downstream or the leeward face. The negative-pressure region remains relatively steady as compared to the positive-pressure zone. The fluctuation of pressure is random and varies from point to point on the building surface. Therefore, the design of the cladding is strongly influenced by local pressures. As mentioned earlier, the design pressure can be thought of as a combination of the mean and the fluctuating velocity. As in the design of buildings, whether or not the pressure component arising from the fluctuating velocity of wind is treated as a dynamic or as a pseudo-static load is a function of the period of the cladding. The period of cladding on a building is usually on the order of 0.2 to 0.02 s, which is very much shorter than the period it takes for wind to fluctuate from a gust velocity to a mean velocity. Therefore, it appears that it is sufficiently accurate to consider both the static and the gust components of winds as equivalent static loads in the design of cladding.

Strength of glass, and indeed any other cladding material, is not known in the same manner as for steel or concrete. For example, it is not possible to buy glass based on yield strength criteria as for concrete or steel. Therefore, the selection, testing, and acceptance criteria for glass must necessarily be based on statistical probabilities rather than on absolute strength. The glass industry has addressed this problem, and commonly uses eight failures per thousand lights of glass as an acceptable probability of failure.

2.4.8.3 LOCAL CLADDING LOADS AND OVERALL DESIGN LOADS

The overall wind load, consisting of the composite effect of positive and negative pressures, is required to determine the required strength and stiffness of the building frame. The local wind loads which act on the various areas of the building enclosures are required for dictating the strength and stiffness of wall and roof elements and for the design of their fastenings. The two types of loads differ significantly, and it is important that these differences be understood. These are:

1. Local winds are more influenced by the configuration of the building surface on which they act than the overall loading.

2. The local load is the maximum load that may occur at any location at any time on any wall surface, whereas

the overall load is the summation of all loads (with due regard to their sign) occurring simultaneously over the building surfaces.

3. The intensity and character of local loading for any given wind direction and velocity differ substantially on various parts of the building surface, whereas the overall load is considered to have a specific intensity and direction.

4. The local loading is sensitive to the momentary nature of wind, but in determining the critical overall loading, only gusts of about 2 s or more are significant.

5. Generally, maximum local suctions are of greater intensity than the overall load.

6. Internal pressures caused by leakage of air through cladding systems have a significant effect on local cladding loads but usually are of no consequence in determining the overall load.

The relative importance of providing for these two types of wind loading is quite obvious. Although proper assessment of overall wind load is important, very few, if any, buildings have been toppled by winds. There are no classic examples of building failures comparable to the Tacoma bridge disaster. On the other hand, local failures of roofs, windows, and wall cladding are not uncommon, and in aggregate such failures continue to cost tens of millions of dollars each year.

The analytical determination of wind pressure or suction at a given surface of a building under varying wind direction and velocity is a very complex problem. Contributing to the complexity are the vagaries of wind action as influenced both by adjacent surroundings and the configuration of the wall surface itself. Much more research is needed on the micro effects of such architectural features as projecting mullions and column covers and deep window reveals. In the meantime, increasing use of model testing in boundary layer wind tunnels is providing vital information on wind loads on building surfaces.

Probably the most important fact established by these tests is that the negative or outward-acting wind loads on wall surfaces are greater and more critical than had formerly been assumed. It may be as much as twice the magnitude of positive loading. In most instances of local cladding failure, glass or panels have been blown off of the building, not into it, and the majority of such failures have occurred in areas near building corners. Therefore, it is important to give careful attention to the design of both anchorage and glazing details to resist outward-acting forces, especially near the corners of high-rise buildings.

Another feature that has come to light from model testing is that wind loads, both positive and negative, on

tall buildings do not vary in proportion to height above ground. Typically, the positive-pressure contours instead of being horizontal are usually found to follow a more concentric pattern as illustrated in Fig. 2.11, with the highest pressure being near the upper center of the façade and pressures at the very top being somewhat less than those a few stories below the roof.

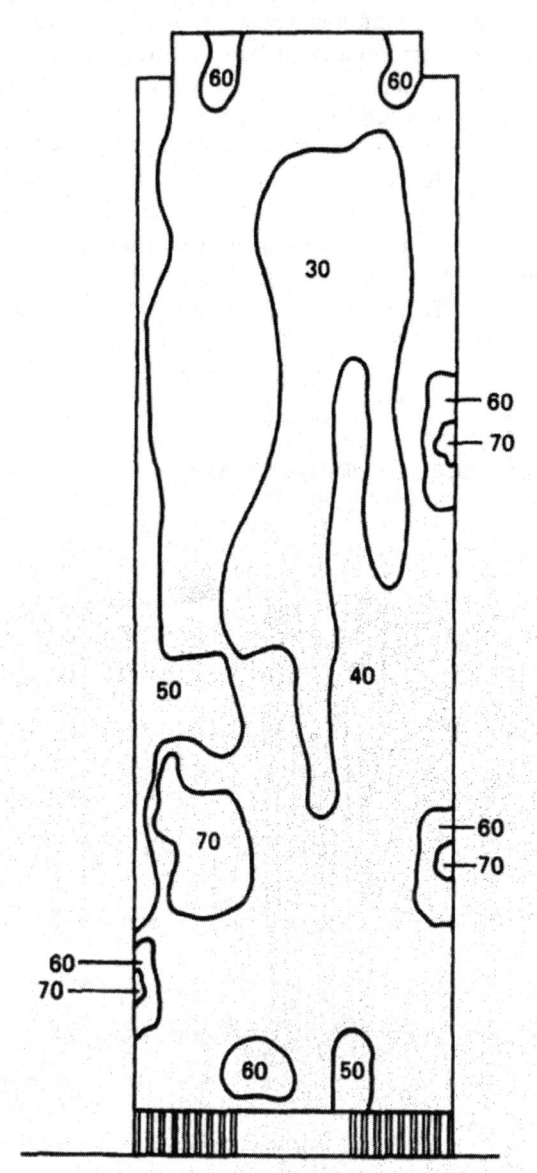

Figure 2.11   Distribution of positive pressures in psf.

2.5 CODE WIND LOADS

2.5.1 Introduction

Building codes and standards are documents which serve as compendiums for technical information and as sources for extracting minimum requirements of accepted design and construction practices. Codes and standards are, in fact, dynamic instruments that are revised periodically to reflect the state of the art. The international building code (IBC) issued by the International Code Council (ICC) is adopted in 50 states, the District of Columbia, the U.S. Virgin Islands, Guam, and the Northern Marianas Islands. Chapter 16 of the IBC contains the minimum magnitudes of some nominal loads and references ASCE/SEI 7 for others. For a specific project, the governing local building code should be consulted for any variances from the IBC or ASCE/SEI 7.

It is common for nominal loads to be referred to as service loads. These loads are multiplied by load factors in the strength design method. Exceptions are the wind load effect W and the earthquake load effect E: Both are defined to be strength-level loads where the load factor is equal to 1. These effects form the basis of the methodologies given in ASCE/SEI 7 for the determination of wind loads.

2.5.2   ASCE 7-16, Wind Load Provisions

2.5.2.1   OVERVIEW OF CODE REQUIREMENTS

According to IBC 2018 section 1609.1.1, wind loads on buildings and structures are to be determined by the provisions of Chapters 26 to 30 of ASCE/SEI 7-16. Five exceptions are given in IBC 1609.1.1 that permit wind loads to be determined on certain types of structures using industry standards other than ASCE/SEI 7 and one exception is given that permits the use of wind tunnel tests that conform to the provisions of Chapter 31 of ASCE/SEI 7-16.

Wind is assumed to come from any horizontal direction and its effects are applied in the form of pressures that act normal to the surfaces of a building or other structure (IBC 2018 1609.1.1). Positive wind pressure acts toward the surface and is commonly referred to as just pressure. Negative wind pressure, which is also called suction, acts away from the surface.

Positive pressure acts on the windward wall of a building and negative pressure acts on the leeward wall, the side walls, and the leeward portion of the roof (see Fig. 2.12).

In general, wind pressures must be considered on the *main wind-force resisting system* (MWFRS) and *components and cladding* (C&C) of a building or other structure. The MWFRS consists of structural elements that their function to resist the effects from the wind applied on the entire structure. Shear walls, moment frames, and braced frames are some examples of different types of MWFRSs.

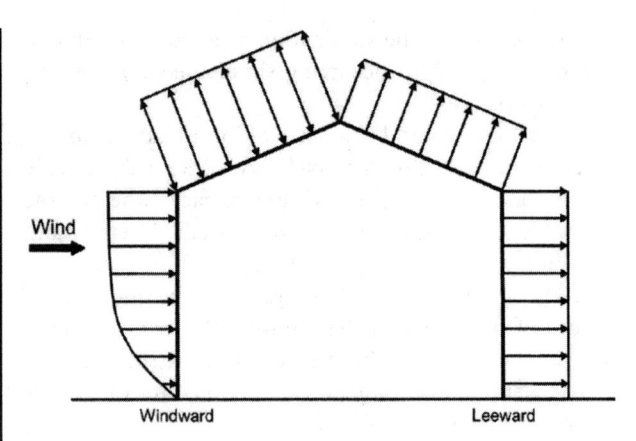

Figure 2.12   Application of wind pressures on a building with a gable or hip roof in accordance with IBC and ASCE/SEI 7.

Table 2.3 contains the procedures that are available in ASCE/SEI 7-16 to determine wind loads on MWFRSs and C&C. In the *directional procedure*, wind loads for specific wind directions are determined using external pressure coefficients that are based on wind tunnel tests of prototypical building models for the corresponding direction of wind. The provisions of this procedure cover a wide range of buildings and other structures.

In the *envelope procedure*, pseudo-external pressure coefficients derived from wind tunnel tests on prototypical building models that are rotated 360° in a wind tunnel are used to determine wind loads; the pressure coefficients envelope the maximum values that are obtained from all possible wind directions.

The alternate all-heights method of IBC 2018 section 1609.6, which is a simplification of the directional procedure in ASCE/SEI 7, is another method that is available to determine wind loads on buildings and other structures that meet the conditions of that section. Contrary to its name, this method is applicable to a certain class of buildings that have a height that is less than or equal to 75 ft.

With the exception of the wind tunnel procedure, all of these procedures and methods in Table 2.3 and the alternate all-heights method are static methods for estimating wind pressures. Static methods generally yield very accurate results for low-rise buildings.

Figure 26.1-1 of ASCE/SEI 7-16 provides an outline of the process that is required for determining wind loads. In-depth information on all of these procedures and methods, including the general requirements in ASCE/SEI 7-16 Chapter 26 and the alternate all-heights method of IBC 2018 section 1609.6, is given in the following sections.

Table 2.3   Summary of Wind Load Procedures in ASCE/SEI 7-16

System	ASCE/SEI 7-16 chapter	Description
MWFRS	27	Directional procedure for buildings of all heights
	28	Envelope procedure for low-rise buildings[1]
	29	Directional procedure for building appurtenances and other structures
	31	Wind tunnel procedure for any building or other structure
C&C	30	• Envelope procedure in parts 1 or 2, or • Directional procedure in parts 3, 4, and 5 • Building appurtenances in part 6
	31	Wind tunnel procedure for any building or other structure

[1]Low-rise buildings are defined in ASCE/SEI 7-16 26.2 as enclosed or partially enclosed buildings that have (1) a mean roof height less than or equal to 60 ft and (2) a mean roof height that is less than or equal to the least horizontal dimension of the building.

2.5.2.2   GENERAL REQUIREMENTS

Chapter 26 of ASCE/SEI 7-16 contains the following general requirements for determining wind loads on MWFRS and C&C:

• Basic wind speed (Figures 26.5-1 and 26.5-2 in ASCE 7-16)
• Wind directionality (26.6 in ASCE 7-16)
• Exposure (26.7 in ASCE 7-16)
• Topographic effects (26.8 in ASCE 7-16)
• Ground elevation effects (26.9 in ASCE 7-16)
• Velocity pressure (26.10 in ASCE 7-16)
• Gust effects (26.11 in ASCE 7-16)
• Enclosure classification (26.12 in ASCE 7-16)
• Internal pressure coefficients (26.13 in ASCE 7-16)

These requirements are used in conjunction with the methods and procedures contained in Chapters 27 through 31.

The following sections discuss these requirements and provide additional background information on fundamental concepts.

Wind Hazard Map   Regardless of the wind load procedure that is employed to determine wind pressures, the basic wind speed V must be determined at the location of the building or other structure (note that in the IBC 2018, the basic wind speed is designated V_{ult} and is defined as the ultimate design wind speed).

Figures 1609.3(1) through 1609.3(8) in the IBC and Figures 26.5-1A through 26.5-1D and Figures 26.5-2A through 26.5-2D in ASCE/SEI 7-16 are identical and provide basic wind speeds based on 3-s gusts at 33 ft above ground for Exposure C for different risk categories, which are defined in IBC 2018 Table 1604.5.

Since the wind speeds are ultimate design wind speeds, the wind load factors in the design load combinations are equal to 1 (see load combinations in Chapter 2 of ASCE/SEI 7-16). Table 2.4 provides a summary of the information associated with these maps.

The shaded areas on the wind speed maps are designated as special wind regions. These are areas where unusual wind conditions exist. The local authority having

Table 2.4   Summary of Basic Wind Speed Maps in the 2018 IBC and ASCE/SEI 7-16

Location	Figure no. IBC	Figure no. ASCE/SEI 7	Risk category*	Return period (years)
Conterminous	1609.3(4)	26.5-1A	I	300
United States	1609.3(1)	26.5-1B	II	700
Alaska	1609.3(2)	26.5-1C	III	1,700
Puerto Rico	1609.3(3)	26.5-1D	IV	3,000
Guam				
Virgin Islands				
American Samoa				
Hawaii	1609.3(8)	26.5-2A	I	300
	1609.3(5)	26.5-2B	II	700
	1609.3(6)	26.5-2C	III	1,700
	1609.3(7)	26.5-2D	IV	3,000

*See IBC 2018 Table 1604.5 for definitions of risk categories.

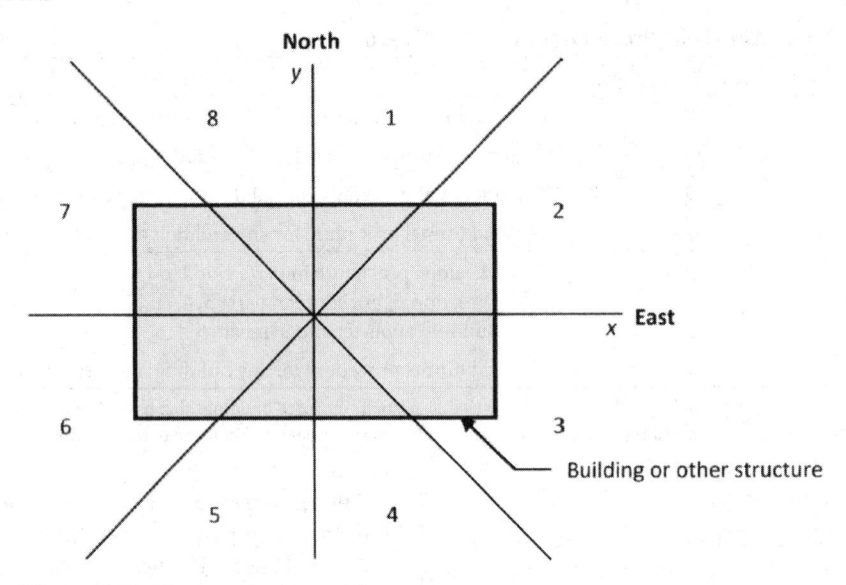

Figure 2.13 Sections for determining exposures.

jurisdiction over the project should be consulted to obtain the local design wind speed (ASCE/SEI 7-16 26.5.2).

The ultimate design wind speeds on the wind hazard maps do not include effects of tornadoes (ASCE/SEI 7-16 26.5.4). However, ASCE/SEI 7-16 C26.5.4 contains references and a tornadic gust wind speed map of the United States that corresponds to a return period of 100,000 years. The information presented in this section can be used as a guide in developing and designing buildings and other structures for the effects of tornadoes. Reference [4] also contains design guidance for tornadoes.

Where required, the basic design wind speeds in Figures 1609.3(1) through 1609.3(8) can be converted to allowable stress design wind speeds, V_{asd}, using Equation 16-33 in IBC 2018 1609.3.1:

$$V_{asd} = V_{ult} \sqrt{0.6} \qquad (2.7)$$

Values of V_{asd} are tabulated in IBC 2018 Table 1609.3.1 for various V_{ult}.

Wind Directionality The wind directionality factor K_d that is given in ASCE 7-16 section 26.6 accounts for the statistical nature of wind flow and the probability of the maximum effects occurring at any particular time for any given wind direction.

Table 26.6-1 OF ASCE 7-16 contains values of K_d as function of structure type. This factor is equal to 0.85 for the MWFRS and C&C of buildings.

Exposure

Wind direction and sectors According to IBC 1609.4 and ASCE/SEI 7-16 section 26.7, an exposure category must be determined upwind of a building or other structure for each wind direction that is considered in design. Wind must be assumed to come from any horizontal direction when determining wind loads (IBC 2018 section 1609.1.1 and ASCE/SEI 7-16 section 26.5.1). One rational way of satisfying this requirement is to assume that there are eight wind directions: four that are perpendicular to the main axes of the building or other structure and four that are at 45° angle to the main axes. Figure C26.7-8 of ASCE 7-16, which is reproduced here as Fig. 2.13, shows the sectors that are to be used to determine the exposure for a selected wind direction (IBC 2018 1609.4.1 and ASCE/SEI 7-16 26.7.1).

Surface roughness categories Surface roughness categories are defined in IBC 1609.4.2 and ASCE/SEI 7-16 section 26.7.2 and are summarized in Table 2.5.

Table 2.5   Surface Roughness Categories

Surface roughness category	Description
B	Urban and suburban areas, wooded areas or other terrain with numerous closely spaced obstructions having the size of single-family dwellings or larger
C	Open terrain with scattered obstructions having heights generally less than 30 ft; this category includes flat open country and grasslands
D	Flat, unobstructed areas and water surfaces; this category includes smooth mud flats, salt flats, and unbroken ice

Exposure Categories Exposure categories are based on the surface roughness categories defined above and essentially account for the boundary layer concept of surface roughness. Table 2.6 contains definitions of the three exposure categories given in IBC 2018 1609.4.3 and ASCE/SEI 7-16 section 26.7.3. Definitions of Exposure B and Exposure D are illustrated in Figures C26.7-1 and C26.7-2 of ASCE 7-16, respectively.

Table 2.6 Exposure Requirements

Exposure category	Definition

B	• Mean roof height $h \leq 30$ ft
	    Surface roughness category B prevails in the upwind direction for a distance > 1500 ft
	• Mean roof height $h > 30$ ft
	    Surface roughness category B prevails in the upwind direction for a distance > 2600 ft or 20 times the height of the building, whichever is greater

C	Applies for all cases where Exposure B or D does not apply
D	• Surface roughness category D prevails in the upwind direction for a distance > 5000 ft or 20 times the height of the building, whichever is greater • Surface roughness immediately upwind of the site is B or C, and the site is within a distance of 600 ft or 20 times the building height, whichever is greater, from an Exposure D condition as defined above

According to ASCE/SEI 7-16 section 26.7.3, the exposure category that results in the largest wind loads must be used for sites that are located in transition zones between exposure categories. However, the exception in ASCE/SEI 7-16 section 26.7.3 permits an intermediate exposure category to be used provided that it is determined by a rational analysis method defined in recognized literature. An example of such an analysis is given in C26.7 of ASCE/SEI 7-16.

Exposure requirements ASCE/SEI 7-16 section 26.7.4 contains exposure requirements that must be satisfied for all of the wind load procedures that are available in ASCE/SEI 7-16. A summary of these requirements is given in Table 2.7.

Topographic effects Buildings or other structures that are sited on the upper half of an isolated hill, ridge or escarpment can experience significantly higher wind velocities than those sited on relatively level ground. The *topographic factor* K_{zt} in section 26.8 accounts for this increase in wind speed, which is commonly referred to as wind speed-up.

Wind speed-up must be considered only when all of the five conditions in 26.8.1 are satisfied. When all of these conditions are met, K_{zt} is determined by Equation 26.8-1:

$$K_{zt} = (1 + K_1 K_2 K_3)^2 \qquad (2.8)$$

Values of K_1, K_2, and K_3 are obtained from Figure 26.8-1 of ASCE/SEI 7-16. K_1 accounts for the shape of the topographic feature and the maximum speed-up effect, K_2 accounts for the reduction in wind speed-up with respect to horizontal distance and K_3 accounts for the reduction of wind speed-up with respect to height above the local terrain.

Ground elevation effects The ground elevation factor, K_e, adjusts the velocity pressure, q_z, determined in accordance with ASCE/SEI 7-16 section 26.10 based on the reduced mass density of air at elevations above sea level.

Table 2.7 Exposure Requirements

Wind load procedure	Chapter	Requirements

Directional	27	• MWFRS of enclosed and partially enclosed buildings
		    Use an exposure category determined in accordance with 26.7.3 in each wind direction
		• Open buildings with monoslope, pitched or troughed free roofs
		    Use the exposure category determined in accordance with 26.7.3 from the eight sectors that results in the highest wind loads

Envelope	28	• MWFRS of all low-rise buildings designed using this procedure Use the exposure category determined in accordance with 26.7.3 from the eight sectors that results in the highest wind loads
Directional	29	• Building appurtenances and other structures Use an exposure category determined in accordance with 26.7.3 in each wind direction
C&C	30	• C&C Use the exposure category determined in accordance with 26.7.3 from the eight sectors that results in the highest wind loads

Table 26.9-1 of ASCE/SEI 7-16 contains values of K_e, which can be calculated using the equation in note 2 of the table where z_g is the ground elevation above sea level. A more complete version of Table 26.9-1 of ASCE/SEI 7-16 that includes air density values is provided in Table C26.9-1 of ASCE/SEI 7-16.

The constant 0.00256 in Equation 26.10-1 of ASCE/SEI 7-16 for q_z is used to convert a wind speed pressure based on the mass density of air for the standard atmosphere, which is defined as a temperature of 59°F and a sea level pressure of 29.92 inches of mercury. Values of air density other than the standard atmosphere values are adjusted using K_e. It is permitted to take $K_e = 1.0$ for all elevations, which is conservative except for elevations below sea level; however, using $K_e = 1.0$ to calculate q_z for all areas below sea level in the United States is not unconservative.

Velocity pressure The velocity pressure q_z at height z above the ground surface is determined by Equation 26.10-1 of ASCE/SEI 7-16; this is essentially Bernoulli's equation, and it converts the basic wind speed V to a velocity pressure:

$$q_z = 0.00256 K_z K_{zt} K_d K_e V^2 \qquad (2.9)$$

The terms in the equation above are discussed below. Note that at the mean roof height of the building, the velocity pressure is denoted q_h and the velocity pressure coefficient is denoted K_h; that is, the subscript changes from z to h.

Air density The constant 0.00256 in the equation is related to the mass density of air for the standard atmosphere (59°F and sea level pressure of 29.92 inches of mercury), and is obtained as follows (constant = one-half times the density of air times the velocity squared where the velocity is in miles per hour [mi/h] and the pressure is in pounds per square foot [lb/ft²]):

$$\text{Constant} = 0.5 \left[\frac{0.0765 \frac{\text{lb}}{\text{ft}^3}}{32.2 \frac{\text{ft}}{\text{s}^2}} \right] \times \left[\left(1 \frac{\text{mi}}{\text{h}}\right) \times 5280 \frac{\text{ft}}{\text{mi}} \times \frac{1 \text{ h}}{3600 \text{ s}} \right]^2$$

$$= 0.00256 \qquad (2.10)$$

The numerical constant of 0.00256 should be used except where sufficient weather data is available to justify a different value.

Velocity pressure exposure coefficient, K_z This coefficient modifies wind velocity (or pressure) with respect to exposure and height above ground. Values of K_z for Exposures B, C and D at various heights above ground level are given in Table 26.10-1 of ASCE/SEI 7-16. In lieu of linear interpolation and for heights greater than 500 ft above the surface, K_z may be calculated at any height z using the equations at the bottom of that table:

$$K_z = \begin{cases} 2.01 \left(\dfrac{15}{z_g}\right)^{\frac{2}{a}} & \text{for } z \le 15 \text{ ft} \\[12pt] 2.01 \left(\dfrac{z}{z_g}\right)^{\frac{2}{a}} & \text{for } 15 \text{ ft} \le z \le z_g \end{cases} \qquad (2.11)$$

The constant α is the 3-s gust speed power law exponent, which defines the approximately parabolic shape of the wind speed profile for each exposure (see Fig. 2.14). The nominal height of the atmospheric boundary layer, which is also referred to as the gradient height, is denoted as z_g. Values of α and z_g are given in Table 26.11-1 of ASCE/SEI 7-16 as a function of exposure.

The above discussion on the determination of K_z is valid for the case of a single roughness category (i.e., uniform terrain). Procedures on how to determine K_z for a single roughness change or multiple roughness changes are given in C27.3.1 of ASCE/SEI 7-16.

Topographic factor K_{zt} This factor modifies the velocity pressure exposure coefficients for buildings located on the upper half of an isolated hill or escarpment. See topographic effects section above.

Wind directionality factor K_d This factor accounts for the statistical nature of wind flow and the probability of the maximum effects occurring at any particular time for any given wind direction. See Figure 26.8-1 of ASCE/SEI 7-16 and wind directionality section above for more information on how to determine K_d.

Ground elevation factor K_e This factor adjusts the velocity pressure, q_z, determined in accordance with section 26.10 of ASCE/SEI 7-16 based on the reduced mass density of air at elevations above sea level.

Basic wind speed V The basic wind speed, V, is the 3-s gust speed at 33 ft above the ground in Exposure C (see IBC 1609.3 or ASCE/SEI 7-16 26.5.1).

Flowchart 1 (Fig. 2.15) can be used to determine the velocity pressures, q_z and q_h.

Gust-Effect Factor The effects of wind gusts must be included in the design of any building or other structure. The *gust-effect factor* defined in section 26.11 of ASCE/SEI 7-16 accounts for both atmospheric and aerodynamic effects in the along-wind direction.

The gust-effect factor depends on the natural frequency n_1 of the structure. In particular, the method in which the gust-effect is determined is contingent on whether the structure is rigid or flexible. By definition, a *rigid building*

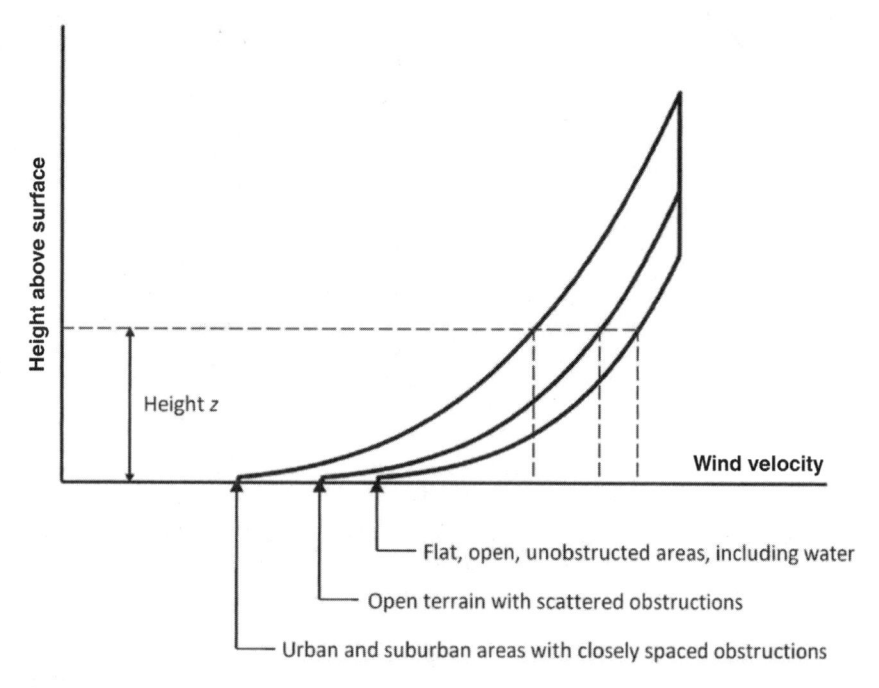

Figure 2.14   The variation of wind speed with respect to height and surface roughness.

or other structure is one where $n_1 \geq 1$ Hz, and a *flexible building or other structure* is one where $n_1 < 1$ Hz (ASCE/SEI 7-16 section 26.2).

Note that low-rise buildings that satisfy the definition in ASCE/SEI 7-16 section 26.2 (i.e., buildings with a mean roof height $h \leq 60$ ft and $h \leq$ least horizontal dimension of building) are permitted to be considered rigid (ASCE/SEI 7-16 section 26.11.2).

Flowchart 2 (Fig. 2.16) contains step-by-step procedures on how to determine the gust-effect factor for both rigid and flexible structures.

Approximate Natural Frequency   Many tools are available to determine the fundamental frequency n_1 of a structure. Most computer programs that are used to analyze structures can provide an estimate of n_1 based on member sizes and material properties that are used in the model. In the preliminary design stages, this information may not be known. Thus, section 26.11.3 provides equations to determine an approximate natural frequency n_a for concrete and steel buildings that meet the height and slenderness conditions of section 26.11.2.1 of ASCE/SEI 7-16.

1. Building height must be less than or equal to 300 ft
2. Building height must be less than four times its effective length L_{eff}, which is determined by Equation 26.11-1 of ASCE/SEI 7-16.

Enclosure Classification   The following discussion covers definitions for each type of classification and the requirements for protecting glazed openings in wind-borne debris regions.

Any building or other structure must be classified as enclosed, partially enclosed, or open based on the definitions in ASCE/SEI 7-16 section 26.2. A summary of these definitions is given in Table 2.8.

The quantities in Table 2.8 are as follows (see Fig. 2.17):

A_o = total area of openings in a wall that receives positive external pressure

A_g = gross area of wall in which A_o is identified

A_{oi} = sum of the areas of openings in the building envelope (walls and roof) not including A_o

A_{gi} = sum of the gross surface areas of the building envelope (walls and roof) not including A_g

Requirements for buildings that comply with more than one classification are given in ASCE/SEI 7-16 section 26.12.4. For a building that meets both open and partially enclosed definitions, the building is to be classified as open.

The situation is significantly different in certain regions where hurricanes can occur. *Hurricane-prone regions* are located along the Atlantic Ocean and Gulf of Mexico coasts where the basic wind speed for Risk Category II buildings is greater than 115 mph. Hawaii, Puerto Rico, Guam, Virgin Islands and American Samoa are also

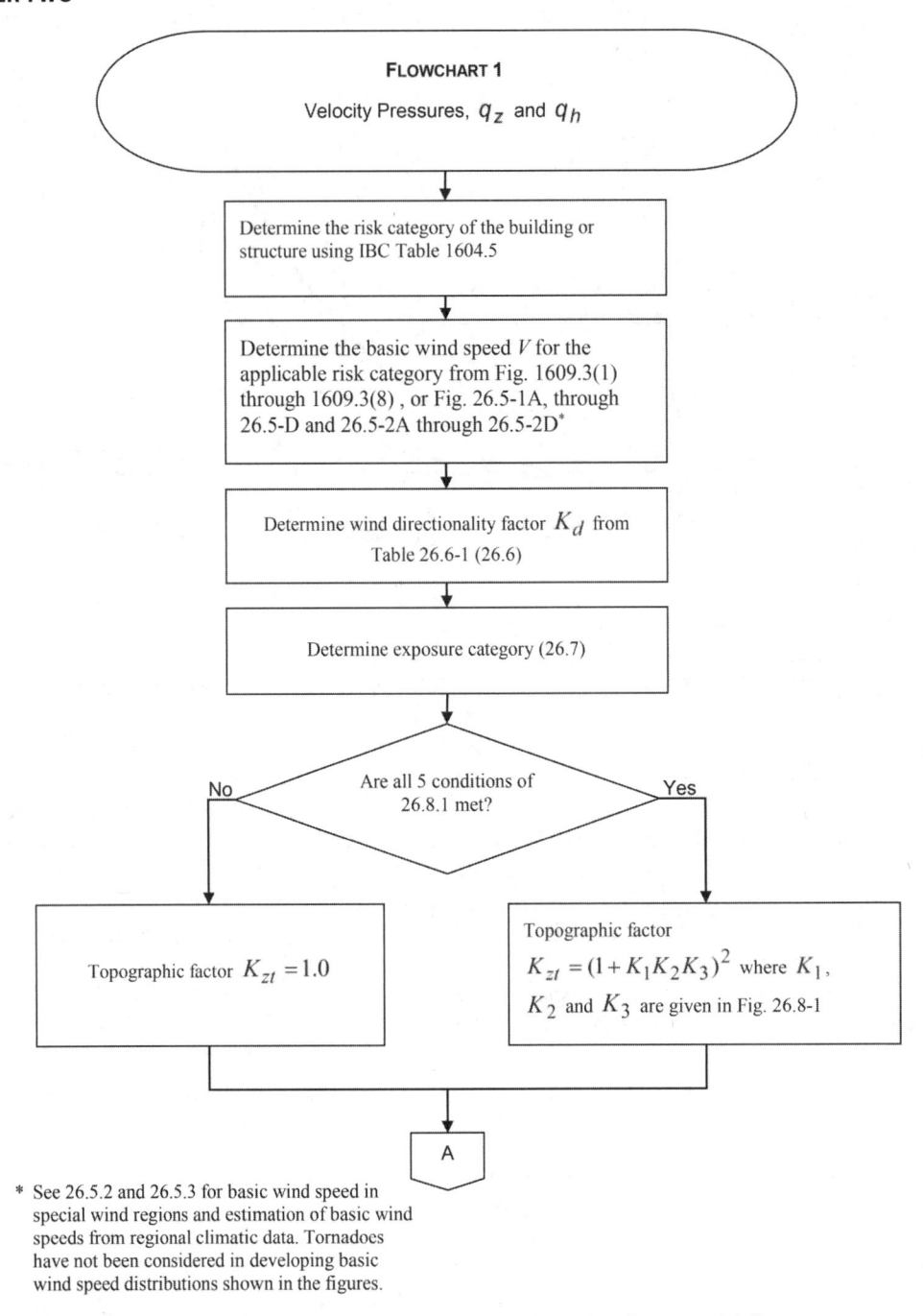

Figure 2.15 Determinations of velocity pressures, q_z and q_h (Flowchart 1, Ref. [1]).

classified as hurricane-prone regions. *Wind-borne debris regions* are in hurricane-prone regions and are located as follows (IBC 2018 section 202 and ASCE/SEI 7-16 section 26.12.3.1):

1. Within 1 mi of the coastal mean high-water line where the basic wind speed (ultimate design wind speed—IBC 2018) is greater than or equal to 130 mph, or

2. In areas where the basic wind speed (ultimate design wind speed—IBC 2018) is greater than or equal to 140 mph.

Actual locations of wind-borne debris regions are to be based on the wind speeds that are in the IBC and ASCE/SEI 7-16 figures, which are summarized in Table 2.9 (IBC 2018 section 202 and ASCE/SEI 7-16 section 26.12.3.1).

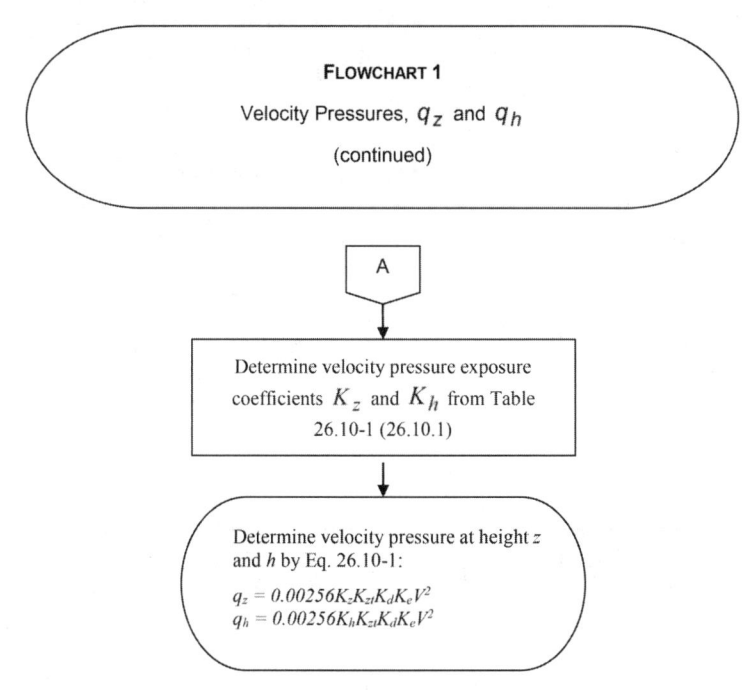

FLOWCHART 1

Velocity Pressures, q_z and q_h

(continued)

A

Determine velocity pressure exposure coefficients K_z and K_h from Table 26.10-1 (26.10.1)

Determine velocity pressure at height z and h by Eq. 26.10-1:

$q_z = 0.00256 K_z K_{zt} K_d K_e V^2$
$q_h = 0.00256 K_h K_{zt} K_d K_e V^2$

Figure 2.15   (*Continued*)

Special requirements are given in IBC 2018 1609.1.2 and ASCE/SEI 7-16 26.10.3.2 for the protection of glazed openings in wind-borne debris regions.

Internal Pressure Coefficient   Internal pressure coefficients (GC_{pi}) are given in ASCE/SEI 7-16 Table 26.13-1 and are based on the enclosure classifications defined in ASCE/SEI 7-16 26.12. These coefficients have been obtained from wind tunnel tests and full-scale data and are assumed to be valid for a building of any height even though the wind tunnel tests were conducted primarily for low-rise buildings. Gust and aerodynamic effects are combined into one factor (GC_{pi}); in accordance with ASCE/SEI 7-16 26.11.7, the gust-effect factor shall not be determined separately in the analysis.

For partially enclosed buildings that contain a single, relatively large volume without any partitions, the reduction factor R_i calculated by Equation 26.13-1 of ASCE/SEI 7-16 may be used to reduce the applicable internal pressure coefficient. This reduction factor is based on research that has shown that the response time of internal pressure increases as the volume of a building without partitions increases; as such, the gust factor associated with the internal pressure is reduced, resulting in lower internal pressure.

2.5.2.3   MAIN WIND-FORCE RESISTING SYSTEMS

Overview   Chapters 27, 28, 29, and 31 in ASCE/SEI 7-16 contain design requirements for determining wind pressures and loads on MWFRSs of buildings and other structures (see Table 2.10 summary of wind load procedures in ASCE/SEI 7-16). The provisions in Chapters 27 through 29 are discussed in the following sections. Chapter 31, which contains the requirements for wind tunnel procedures, is covered in this chapter.

Directional Procedure for Buildings (ASCE/SEI 7-16 Chapter 27)

Scope   The Directional Procedure of Chapter 27 applies to the determination of wind loads on the MWFRS of enclosed, partially enclosed and open buildings of all heights that meet the conditions and limitations given in 27.1.2 and 27.1.3, respectively.

A summary of the wind load procedures and their applicability for MWFRSs in accordance with Chapter 27 are given in Table 2.10.

Part 1 in this chapter is applicable to enclosed, partially enclosed, and open buildings of all heights; a wide range of buildings is covered by the provisions in this part. In general, wind pressures are determined as a function

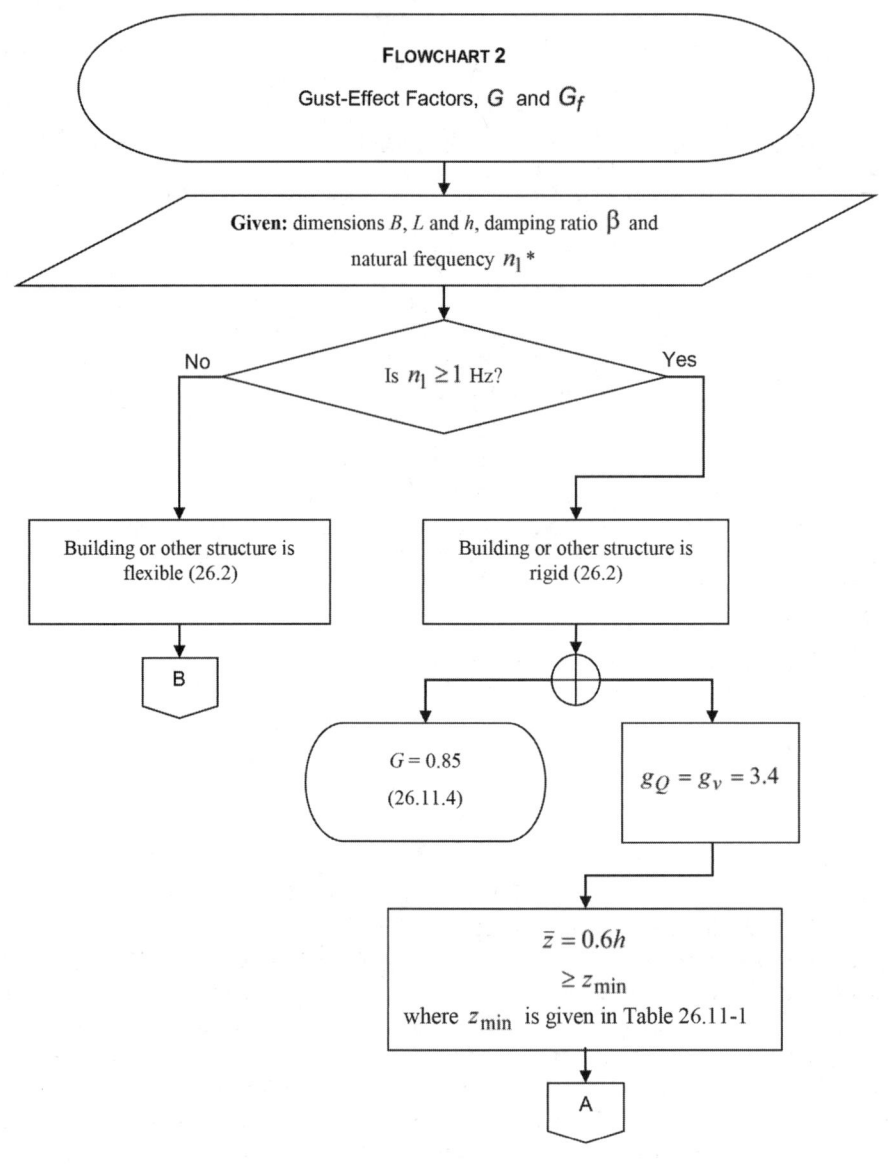

Figure 2.16   Gust-effect factors G and G_f (Flowchart 2, Ref. [1]).

of wind direction using equations that are appropriate for each surface of the building. A simplified method for a special class of buildings up to 160 ft in height is provided in Part 2, which is based on the provisions in Part 1.

In order to apply these provisions, buildings must be regularly shaped (i.e., must have no unusual geometrical irregularities in spatial form) and must exhibit essentially along-wind response characteristics. Buildings of unusual shape that do not meet these conditions must be designed by either recognized literature that documents such wind load effects or by the wind tunnel procedure in Chapter 31 (27.1.3).

Reduction in wind pressures due to apparent shielding of surrounding buildings, other structures or terrain is not permitted (27.1.4). Removal of such features around a building at a later date could result in wind pressures that are much higher than originally accounted for; as

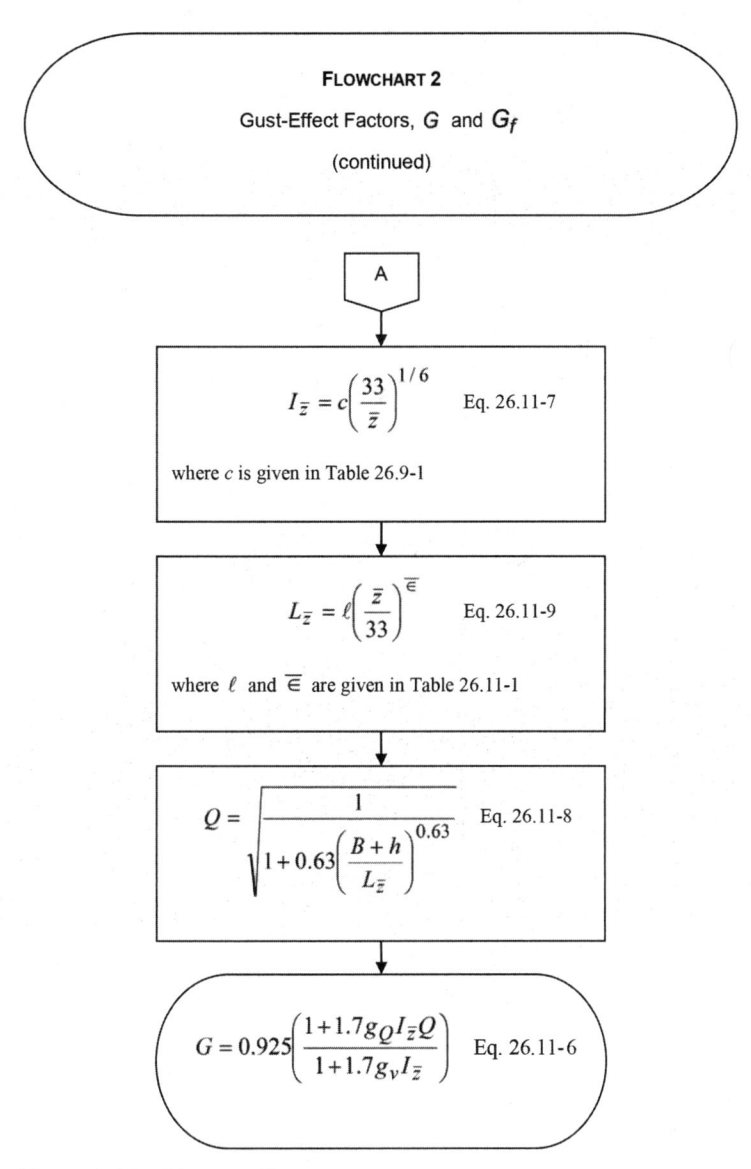

Figure 2.16   (*Continued*)

such, wind pressures must be calculated assuming that all shielding effects are not present.

Minimum design wind pressures and loads are given in section 27.1.5 of ASCE 7-16, which are applicable to buildings designed using Part 1 or Part 2. In the case of enclosed or partially enclosed buildings, wind pressures of 16 psf and 8 psf must be applied simultaneously to the vertical plane normal to the assumed wind direction over the wall and roof area of the building, respectively. Application of these minimum wind pressures are illustrated in Fig. 2.18 for wind along the two primary axes of the building. For open buildings, the minimum wind force is equal to 16 psf multiplied by the area of the open building either normal to the wind direction or projected on a plane normal to the wind direction. It is important to note that minimum design pressures or loads are load cases that must be considered separate from any other load cases that are specified in Parts 1 or 2.

Part 1—Enclosed, Partially Enclosed, and Open Buildings of All Heights

Overview   Part 1 of Chapter 27 is applicable to buildings with any general plan shape, height or roof geometry that matches the figures provided in this chapter. This procedure entails the determination of velocity pressures,

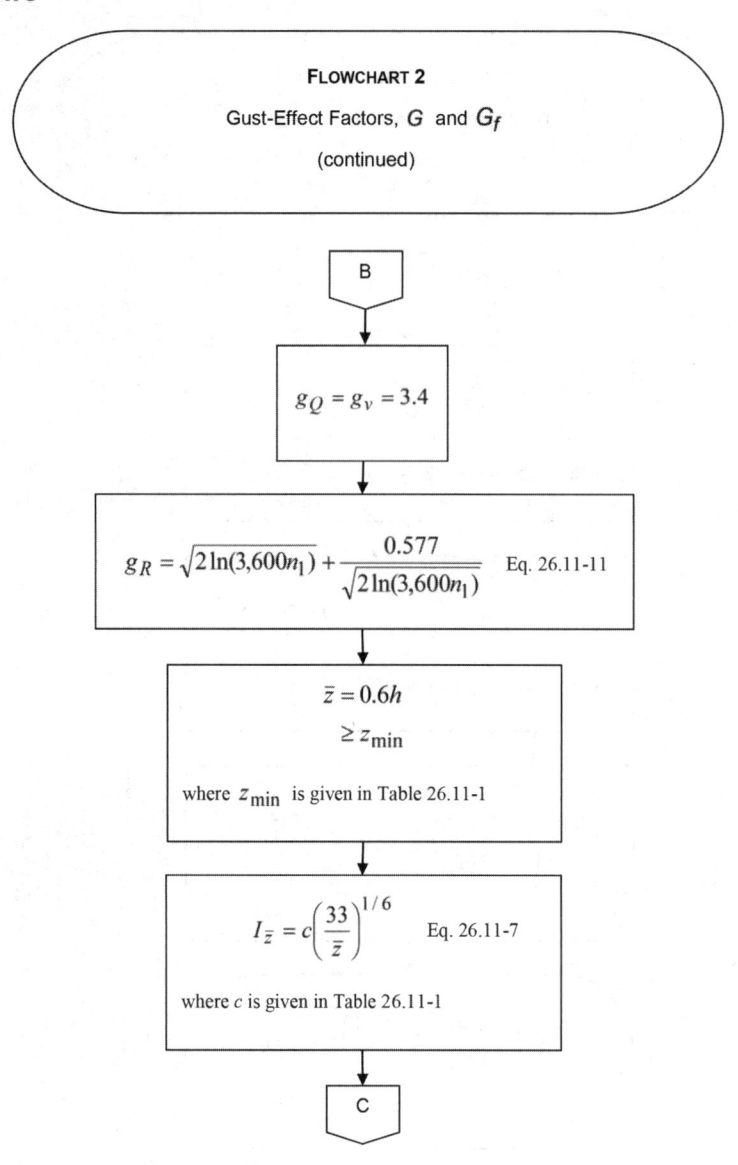

FLOWCHART 2

Gust-Effect Factors, G and G_f

(continued)

B

$$g_Q = g_v = 3.4$$

$$g_R = \sqrt{2\ln(3{,}600n_1)} + \frac{0.577}{\sqrt{2\ln(3{,}600n_1)}} \qquad \text{Eq. 26.11-11}$$

$$\bar{z} = 0.6h$$

$$\geq z_{\min}$$

where $z_{\min}$ is given in Table 26.11-1

$$I_{\bar{z}} = c\left(\frac{33}{\bar{z}}\right)^{1/6} \qquad \text{Eq. 26.11-7}$$

where c is given in Table 26.11-1

C

Figure 2.16 (*Continued*)

gust-effect factors, external pressure coefficients, and internal pressure coefficients for each surface of a rigid, flexible or open building. Table 27.2-1 contains the overall steps that can be used to determine wind pressures on such buildings.

Design Wind Pressures

Enclosed and partially enclosed rigid and flexible buildings
Design wind pressures p are calculated by Equation 27.3-1 for the MWFRS of enclosed and partially enclosed rigid buildings of all heights:

$$p = qGC_p - q_i(GC_{pi}) \qquad (2.12)$$

This equation is used to calculate the wind pressures on each surface of the building: windward wall, leeward wall, side walls, and roof. The pressures are applied simultaneously on the walls and roof, as depicted in Figure 27.3-1 (see also Fig. 2.12—Application of Wind Pressures on a Building with a Gable or Hip Roof in Accordance with IBC and ASCE/SEI 7). The first part of the equation is the external pressure contribution and the second part is the internal pressure contribution. External pressure varies with height above ground on the windward wall and is a constant on all of the other surfaces based on the mean roof height.

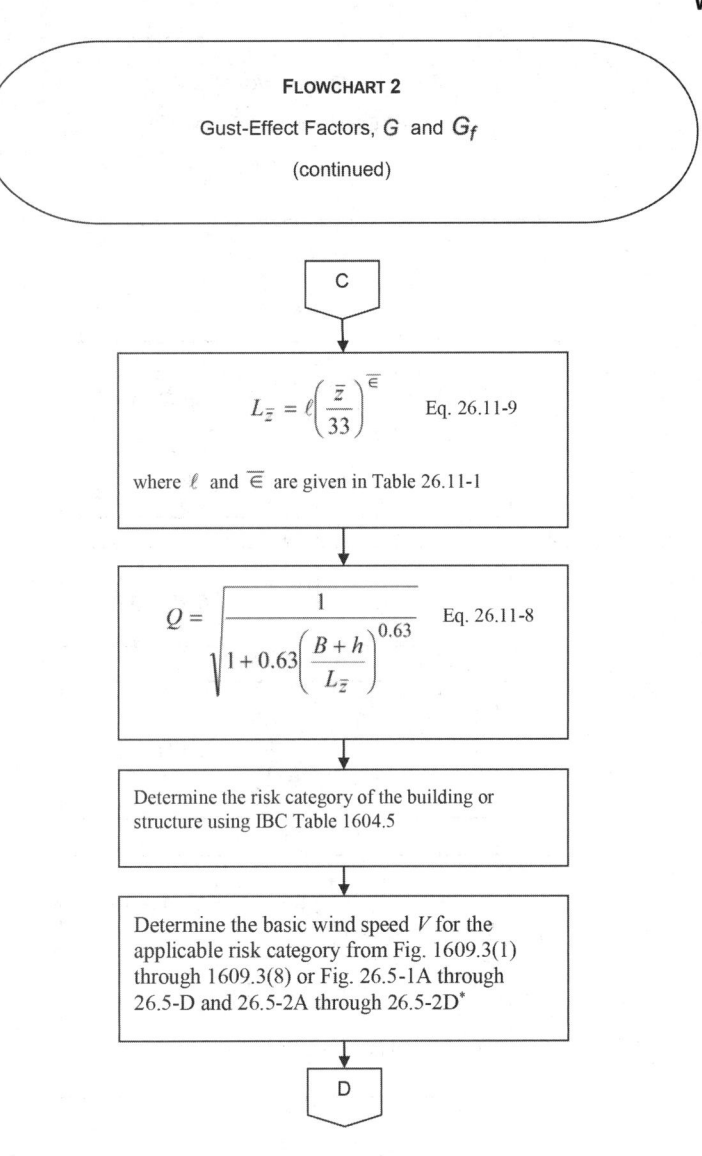

FLOWCHART 2

Gust-Effect Factors, G and G_f

(continued)

C

$$L_{\bar{z}} = \ell \left(\frac{\bar{z}}{33} \right)^{\bar{\in}} \qquad \text{Eq. 26.11-9}$$

where ℓ and $\bar{\in}$ are given in Table 26.11-1

$$Q = \sqrt{\frac{1}{1 + 0.63 \left(\dfrac{B + h}{L_{\bar{z}}} \right)^{0.63}}} \qquad \text{Eq. 26.11-8}$$

Determine the risk category of the building or structure using IBC Table 1604.5

Determine the basic wind speed V for the applicable risk category from Fig. 1609.3(1) through 1609.3(8) or Fig. 26.5-1A through 26.5-D and 26.5-2A through 26.5-2D*

D

* See 26.5.2 and 26.5.3 for basic wind speed in special wind regions and estimation of basic wind speeds from regional climatic data. Tornadoes have not been considered in developing basic wind speed distributions shown in the figures.

Figure 2.16 (*Continued*)

The gust-effect factor G for rigid buildings may be taken equal to 0.85 or may be calculated by Equation 26.11-6. For flexible buildings, G_f determined in accordance with 26.11.5 is to be used in Equation 27.3-1 instead of G.

External pressure coefficients C_p capture the aerodynamic effects, discussed above, and have been determined experimentally through wind tunnel tests on buildings of various shapes and sizes. These coefficients reflect the actual wind loading on each surface of a building as a function of wind direction.

Figure 27.3-1 contains C_p values for windward walls, leeward walls, side walls, and roofs for buildings with gable and hip roofs, monoslope roofs and mansard roofs. Wall pressure coefficients are constant on windward and side walls and vary with the plan dimensions of the building (i.e., vary with the aspect ratio of the building L/B) on the leeward wall. The table in the upper part of this figure also designates which velocity pressure to use—q_z or q_h—on a particular wall surface. Roof pressure coefficients vary with the ratio of the mean roof height to the plan

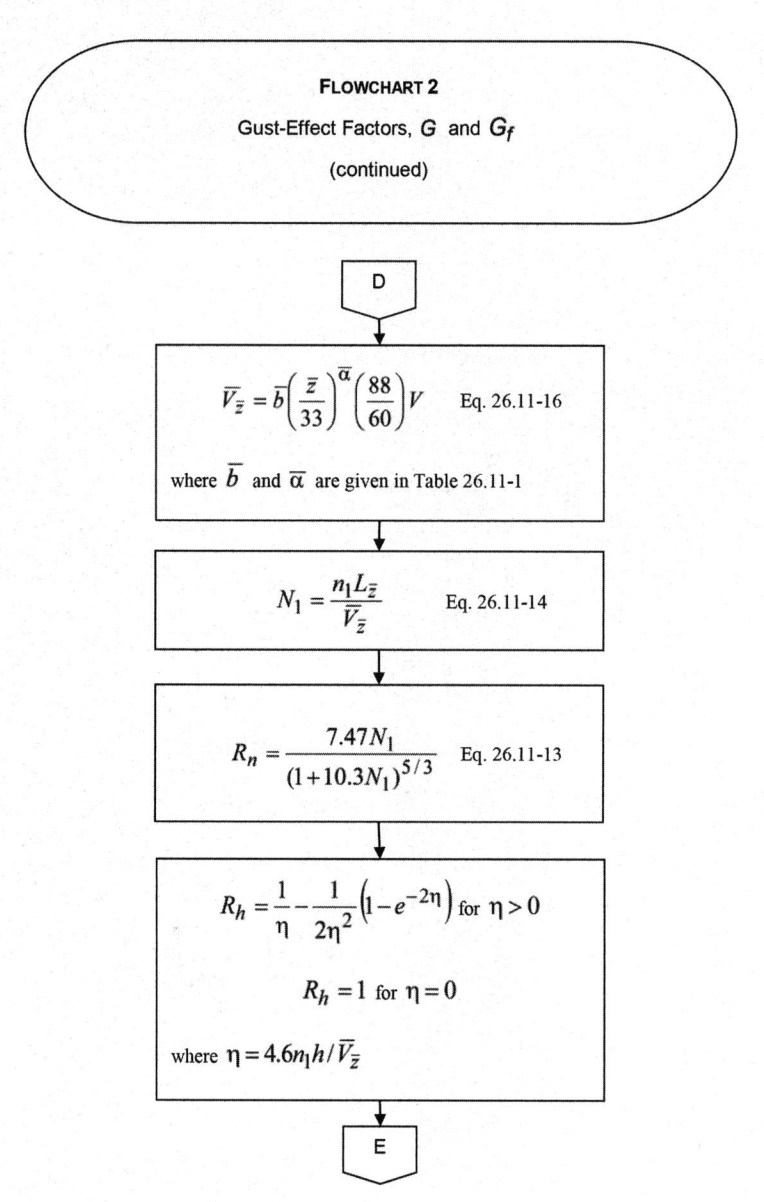

FLOWCHART 2

Gust-Effect Factors, G and G_f

(continued)

D

$$\overline{V}_{\bar{z}} = \overline{b}\left(\frac{\overline{z}}{33}\right)^{\overline{\alpha}}\left(\frac{88}{60}\right)V \qquad \text{Eq. 26.11-16}$$

where $\overline{b}$ and $\overline{\alpha}$ are given in Table 26.11-1

$$N_1 = \frac{n_1 L_{\bar{z}}}{\overline{V}_{\bar{z}}} \qquad \text{Eq. 26.11-14}$$

$$R_n = \frac{7.47N_1}{\left(1+10.3N_1\right)^{5/3}} \qquad \text{Eq. 26.11-13}$$

$$R_h = \frac{1}{\eta} - \frac{1}{2\eta^2}\left(1-e^{-2\eta}\right) \text{ for } \eta > 0$$

$$R_h = 1 \text{ for } \eta = 0$$

where $\eta = 4.6n_1h/\overline{V}_{\bar{z}}$

E

Figure 2.16 (*Continued*)

dimension of the building (h/L) and with the roof angle (θ) for a given wind direction (normal to ridge or parallel to ridge). All of these pressure coefficients are intended to be used with q_h, and the parallel to ridge wind direction is applicable for flat roofs. It is evident from the figure that negative roof pressures increase as the ratio h/L increases. Also, as θ increases, negative pressure decreases until a roof angle is reached where the pressure becomes positive; this is consistent with the aerodynamic effect of the separation zone (see Fig. 2.12 idealized wind flow around a gable roof building). Where two values of C_p are listed in the figure, the windward roof is subjected to either positive or negative pressure and the structure must be designed for both. Other important information on the use of this figure is given in the notes below the tabulated pressure coefficients.

The external pressure coefficients in Figure 27.3-2 for dome roofs are adapted from the 1995 edition of the Eurocode and are based on data obtained from a modeled atmospheric boundary layer flow that does not fully comply with the wind tunnel testing requirements given in Chapter 31. Two load cases must be considered. In Case A, pressure coefficients are determined between various locations on the dome by linear interpolation along arcs of the

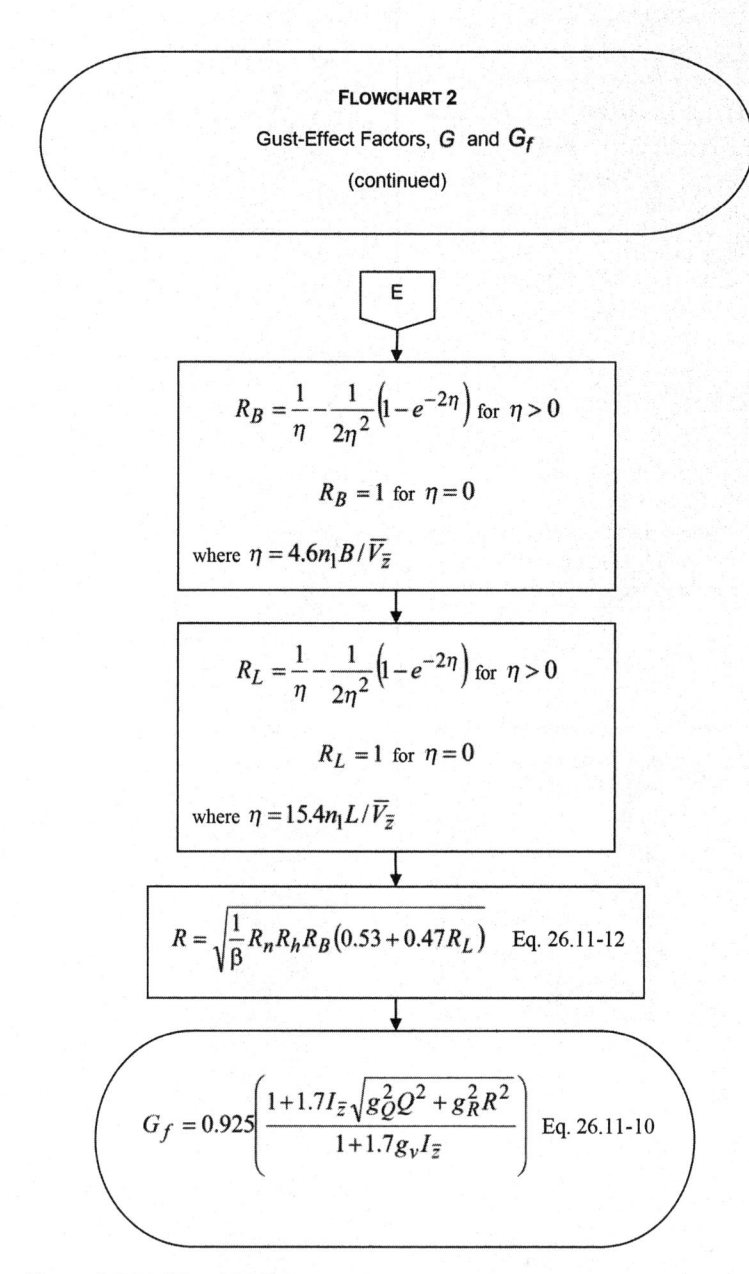

Figure 2.16   (*Continued*)

dome parallel to the direction of wind; this defines maximum uplift on the dome in many cases. In Case B, the pressure coefficient is assumed to be a constant value at a specific point on the dome for angles less than or equal to 25° and is determined by linear interpolation from 25° to other points on the dome; this properly defines positive pressures for some cases, which results in maximum base shear. Wind tunnel tests are recommended for domes that are larger than 200 ft in diameter and in cases where resonant response can be an issue (C27.3.1).

The pressure and force coefficients in Figure 27.3-3 for arched roofs are the same as those that were first introduced in 1972 (Ref. [5]). These coefficients were obtained from wind tunnel tests conducted under uniform flow and low turbulence.

The velocity pressure for internal pressure determination q_i is used in capturing the effects caused by internal pressure. On all of the surfaces of enclosed buildings and for negative internal pressure evaluation in partially enclosed buildings, q_i is to be taken as the velocity

Table 2.8   Enclosure Classifications

Classification	Definition
Open building	For each wall in the building, $$A_o \geq 0.8 A_g$$
Partially enclosed building	A building that complies with all of the following conditions: • $A_o \geq 1.1 A_{oi}$ • $A_o >$ lesser of 4 ft² or $0.01 A_g$ • $A_{oi}/A_{gi} \leq 0.2$
Enclosed building	A building that complies with the following conditions: $$A_o \leq \text{ smaller of } \begin{cases} 0.01 A_g \\ 4 \text{ ft}^2 \end{cases}$$
Partially open building	A building that does not comply with the requirements for open, partially enclosed or enclosed buildings

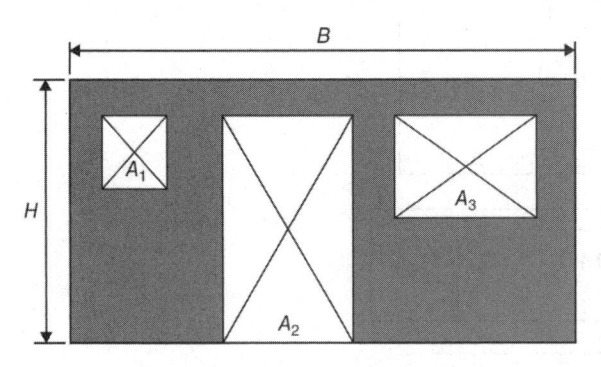

$$A_g = B \times H$$
$$A_o = A_1 + A_2 + A_3$$

Figure 2.17 Definition of wall openings for determination of enclosure classification.

Table 2.9   Wind-Borne Debris Region Wind Speed Figures

Classification	Figure no.
• Risk Category II buildings and other structures	Figures 1609.3(1) and 1609.3(5)
• Risk Category III buildings and other structures, except health care facilities	Figures 26.5-1B and 26.5-2B
• Risk Category III health facilities	Figures 1609.3(2) and 1609.3(6) Figures 26.5-1C and 26.5-2C
• Risk Category IV buildings and other structures	Figures 1609.3(3) and 1609.3(7) Figures 26.5-1D and 26.5-2D

pressure evaluated at the mean roof height q_h. For positive internal pressure evaluation, section 27.3.1 permits q_i to be set equal to q_z in partially enclosed buildings where q_z is the velocity pressure evaluated at the location of the highest opening in the building that could affect positive internal pressure. Note that it is conservative to set q_i equal to q_h in all cases where positive internal pressure is evaluated. In the case of low-rise buildings, the distance between the uppermost opening and the mean roof height is usually relatively small, and this approximation yields reasonable results. However, this approximation can be overly conservative in certain cases, especially for taller buildings where the distance between the uppermost opening and the mean roof height is relatively large. For buildings located in wind-borne debris regions with glazing that does not meet the protection requirements of section 26.12.3.2, q_i is to be determined assuming that the glazing will be breached.

The velocity pressure q_i is multiplied by the internal pressure coefficient (GC_{pi}). Both positive and negative values of (GC_{pi}) must be considered in order to establish the critical load effects.

Open buildings with monoslope, pitched, or troughed free roofs Design wind pressures p are calculated by Equation 27.3-2 for the MWFRS of open buildings with monoslope, pitched or troughed roofs:

$$p = q_h G C_N \tag{2.13}$$

In this equation, q_h is the velocity pressure at the mean roof height determined by Equation in section 2.5.2.1 and G is the gust-effect factor determined in accordance with 26.11. Net pressure coefficients C_N are given in Figures 27.3-4 through 27.3-7, which are based on the results from wind tunnel studies. Two load cases are identified in the figures, Load Case A and Load Case B. Both load cases must be considered in order to obtain the maximum load effects for a particular roof slope and blockage configuration.

For structures with free roofs that contain fascia panels where the angle of the plane of the roof from the horizontal is less than or equal to 5°, the fascia panels are to be considered as an inverted parapet. The contribution of the wind loads on the fascia panels to the wind loads on the MWFRS is to be determined using the provisions of ASCE/SEI 27.3.4 with q_p in Equation 27.3-3 taken as q_h.

Roof overhangs In the case of roof overhangs, the positive external pressure on the bottom surface of a windward roof overhang is determined using the external pressure coefficient for the windward wall ($C_p = 0.8$). This pressure is combined with the top surface pressures determined in accordance with Figure 27.3-1 (see Fig. 2.19).

Table 2.10 Summary of Wind Load Procedures in Chapter 27 of ASCE/SEI 7-16 for MWFRSs

| ASCE/SEI chapter | Part | Applicability | | Conditions |
		Building type	Height limit	
27	1	Enclosed Partially enclosed Open	None	• Regular-shaped building • Building does not have response characteristics making it subject to across-wind loading, vortex shedding, instability due to galloping or flutter • Building is not located at a site where channeling effects or buffeting in the wake of upwind obstructions warrant special consideration

	2	Enclosed, simple diaphragm	$h \leq 160$ ft	• Same conditions as in Part 1
				• Building must meet the conditions for either a Class 1 or Class 2 building:
				Class 1 Building:
				  1. $h \leq 60$ ft
				  2. $0.2 \leq L/B \leq 5.0$
				Class 2 building:
				  1. 60 ft $< h \leq 160$ ft
				  2. $0.5 \leq L/B \leq 2.0$
				  3. $n_1 \geq 75/h$
				• Building is an enclosed simple diaphragm building as defined in 26.2

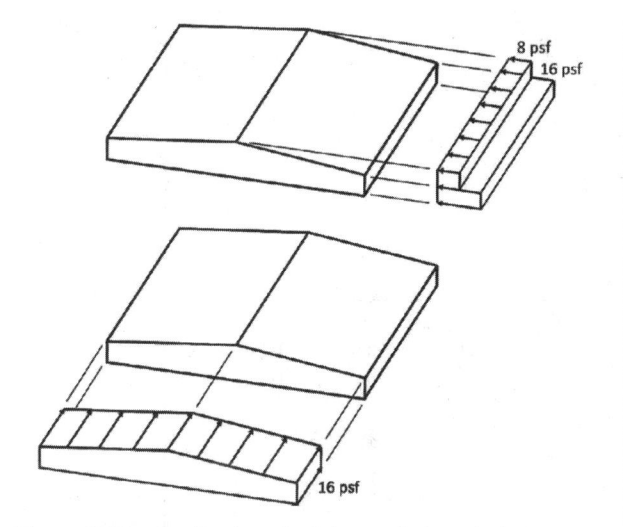

Figure 2.18 Application of minimum design wind pressures in accordance with section 27.1.5.

Parapets Design wind pressures p_p for the effects of parapets on the MWFRS of rigid or flexible buildings with flat, gable or hip roofs are calculated by Equation 27.3-3:

$$p_p = q_p(GC_{pn}) \qquad (2.14)$$

In this equation, q_p is the velocity pressure evaluated at the top of the parapet and (GC_{pn}) is the combined net pressure coefficient, which is equal to +1.5 for a windward parapet and −1.0 for a leeward parapet. It is important to note that p_p is the combined net pressure due to the combination of the net pressures from the front and back surfaces of the parapet.

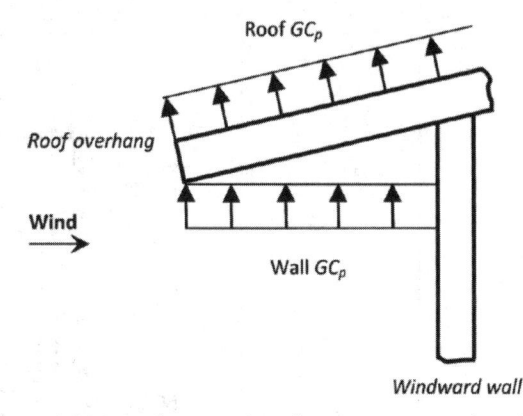

Figure 2.19 Application of wind pressures on a roof overhang, Part 1 of Chapter 27.

The pressures on the front and back of the parapet have been combined into one pressure, which is captured by the combined net pressure coefficients (GC_{pn}) for windward and leeward parapets. Since the wind can occur in any direction, a parapet must be designed for both sets of pressures. Note that the internal pressures inside the parapet cancel out in the determination of the combined pressure coefficient.

The pressures determined on the parapets are combined with the external pressures on the building to obtain the total wind pressures on the MWFRS.

Design wind load cases Buildings subjected to the wind pressures determined by Chapter 27 must be designed for the load cases depicted in Figure 27.3-8, which are reproduced here in Fig. 2.20 (in this figure, the subscripts x

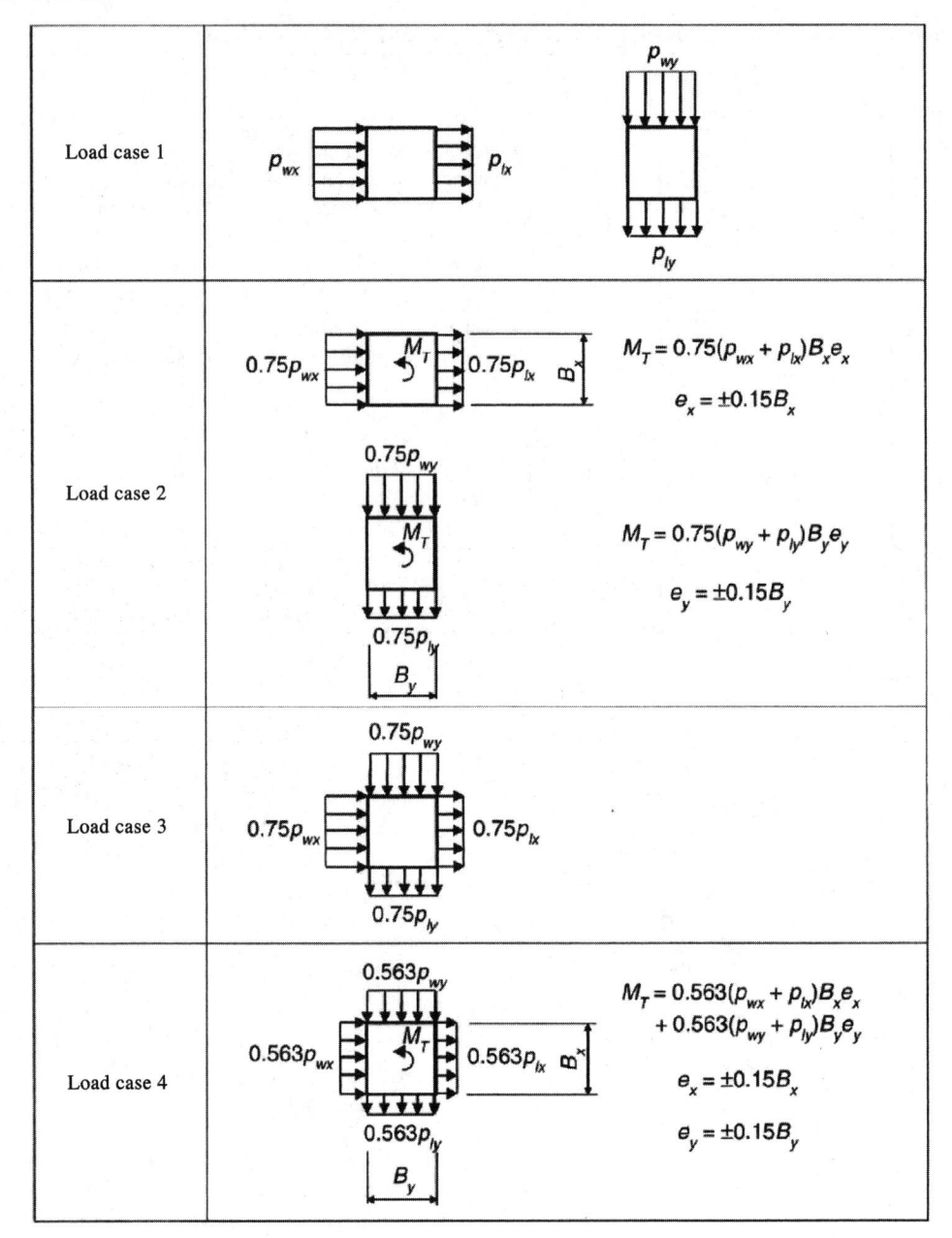

Figure 2.20 Design wind load cases in Part 1 of Chapter 27 (Ref. [1]).

and y refer to the principal axes of the building and w and l refer to the windward and leeward faces, respectively). In Load Case 1, design wind pressures are applied along the principal axes of a building separately.

Load Cases 2 accounts for the effects of nonuniform pressure on different faces of the building due to wind flow; these pressure distributions have been documented in wind tunnel tests. Nonuniform pressures introduce torsion on the building, and this is accounted for in design

by subjecting the building to 75 percent of the design wind pressures applied along the principal axis of the building plus a torsional moment M_T that is determined using an eccentricity equal to 15 percent of the appropriate plan dimension of the building. Torsional effects are determined in each principal direction separately.

A critical load case can occur when the design wind load acts diagonally to a building. This is accounted for in Load Case 3, where 75 percent of the maximum design wind pressures are applied along the principal axes of a building simultaneously.

Load Case 4 considers the effects due to diagonal wind loads and torsion. Seventy-five percent of the wind pressures in Load Case 2 are applied along the principal axes of a building simultaneously, and a torsional moment is applied, which is determined using 15 percent of the plan dimensions of the building.

In the case of flexible buildings, dynamic effects can increase the effects from torsion. Equation 27.3-4 accounts for these effects. The eccentricity e determined by this equation is to be used in the appropriate load cases in Fig. 2.18 in lieu of the eccentricities e_x and e_y that are given in that figure for rigid structures. An eccentricity must be considered for each principal axis of the building, and the sign of the eccentricity must be plus or minus, whichever causes the more severe load case.

Flowchart 3 (Fig. 2.21) can be used to determine design wind pressures on the MWFRS of buildings in accordance with Part 1 of Chapter 27.

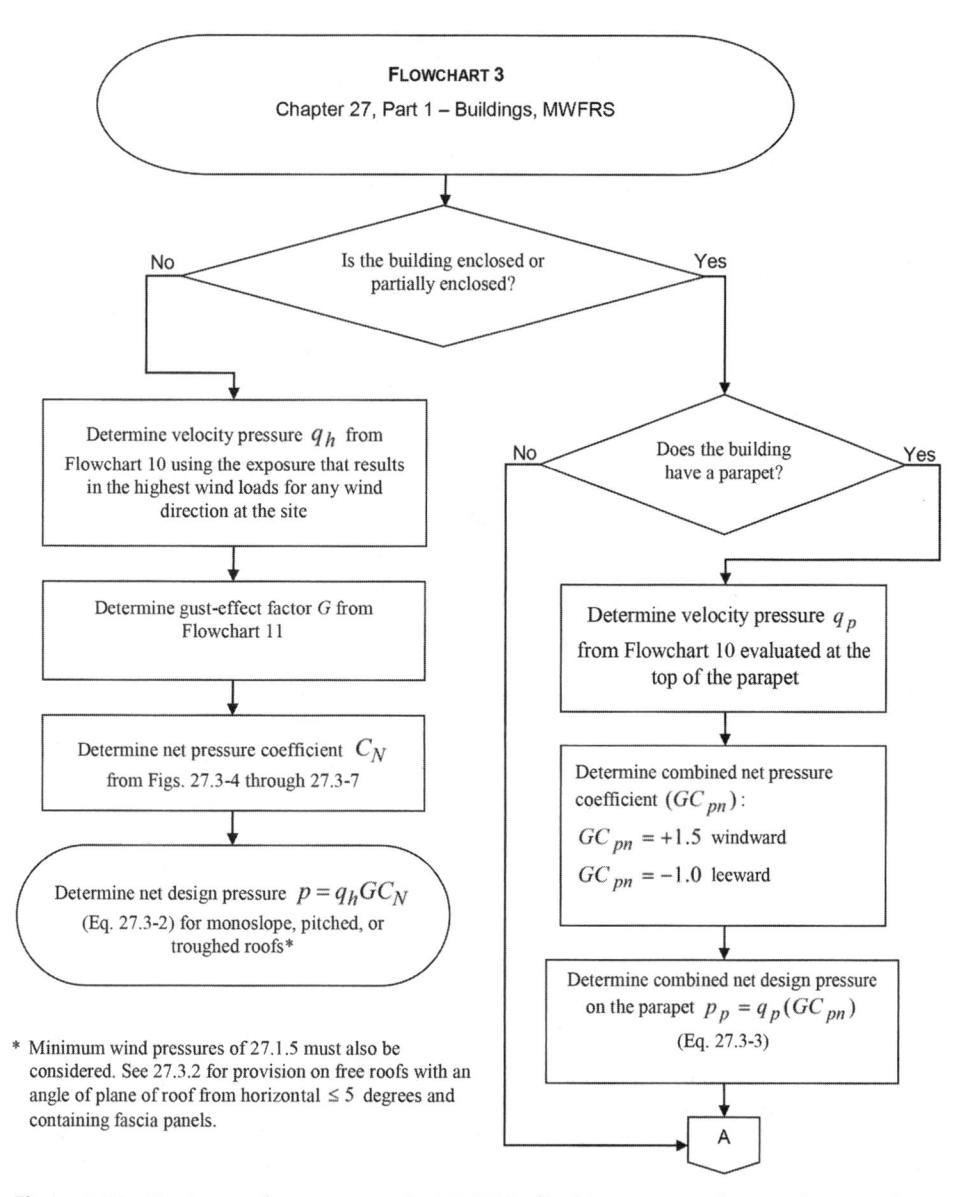

Figure 2.21 Design wind pressures on the MWFRS of buildings in accordance with Part 1 of Chapter 27 (Flowchart 3, Ref. [1]).

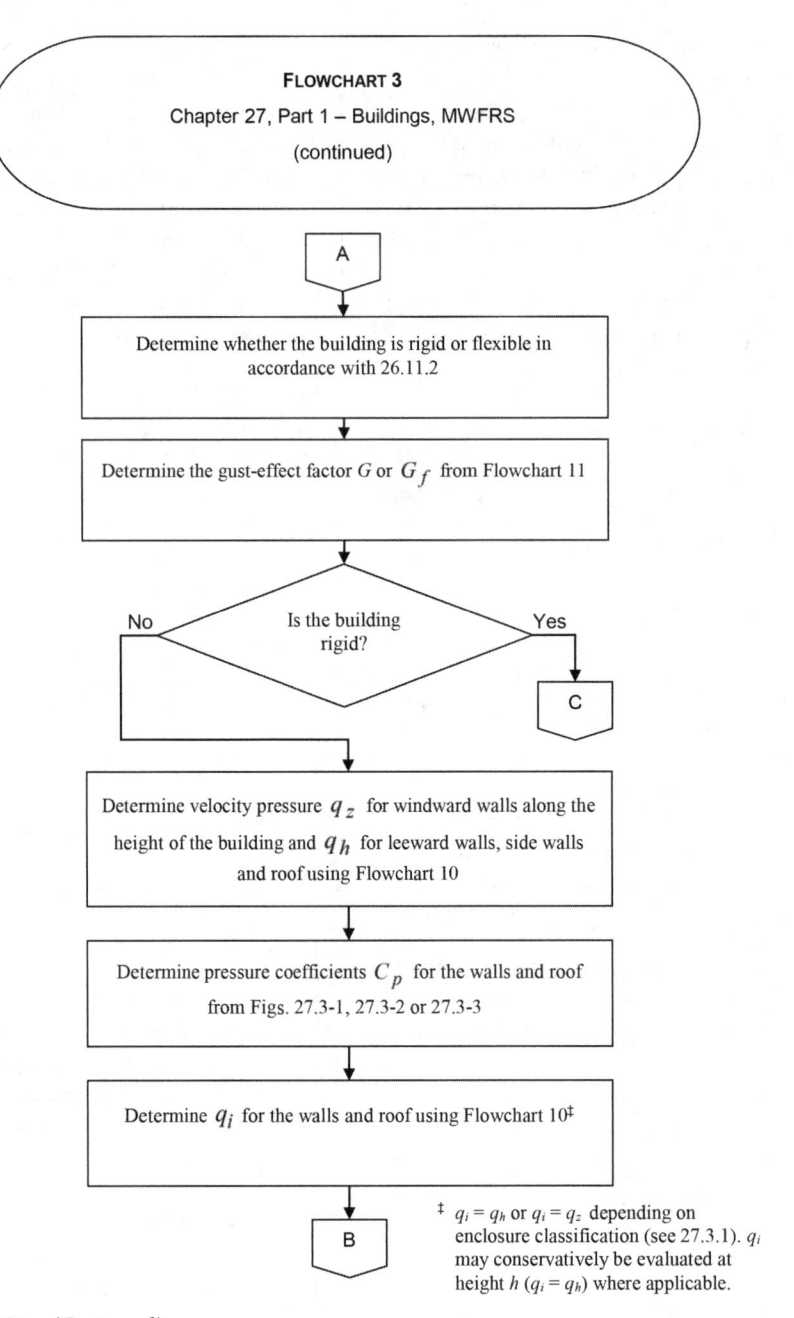

Figure 2.21   *(Continued)*

2.5.2.4   Components and Cladding

Overview   Chapter 30 contains design wind load provisions for C&C. These requirements may be used in the design of such elements provided the conditions and limitations of sections 30.1.2 and 30.1.3 are satisfied. The following sections discuss the six parts that are contained in this chapter.

A summary of the wind load procedures and their applicability for C&C in accordance with Chapter 30 are given in Table 2.11.

Similar to MWFRSs, reduction in wind pressures due to apparent shielding of surrounding buildings, other structures or terrain is not permitted in the design of C&C (section 30.1.4). Removal of such features around a building at a later date could result in wind pressures that are much higher than originally accounted for; as such, wind pressures must be calculated assuming that all shielding effects are not present.

Design wind pressures determined by Chapter 30 are permitted to be used in the design of air permeable roof or wall cladding. Examples of this type of cladding include siding, pressure-equalized rain screen walls, shingles, tiles, concrete roof pavers and aggregate roof surfacing. In general, this type of cladding allows partial air pressure equalization between their exterior and interior surfaces. Additional information can be found in C30.1.5.

General Requirements

Minimum design wind pressures According to section 30.2.2 of ASCE/SEI 7-16, the design wind pressure for C&C shall not be less than a net pressure of 16 psf

acting in either direction (positive or negative) normal to the surface. Like in the case of MWFRSs, this is a load case that needs to be considered in addition to the other required load cases in this chapter.

Tributary areas greater than 700 ft² C&C elements that support a tributary area greater than 700 ft² are permitted to be designed for wind pressures using the provisions for MWFRSs. The 700-ft² tributary area is deemed sufficiently large enough so that the localized wind effects are not pronounced as is the case of C&C; as such, the wind pressures on these elements are essentially equal to those determined by the method for MWFRSs.

External pressure coefficients Numerous figures are provided in this chapter that give values for the combined gust-effect factor and pressure coefficient (GC_p) for C&C. The gust-effect factor and pressure coefficients are not

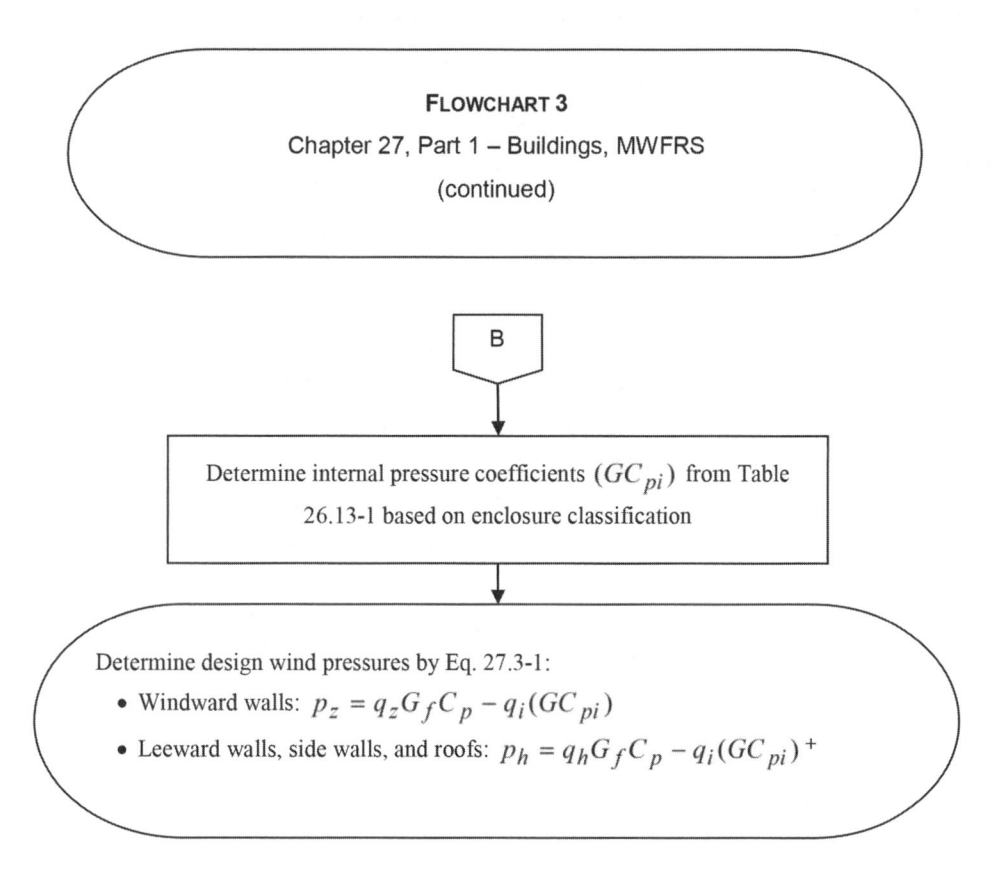

FLOWCHART 3

Chapter 27, Part 1 – Buildings, MWFRS

(continued)

B

Determine internal pressure coefficients (GC_{pi}) from Table 26.13-1 based on enclosure classification

Determine design wind pressures by Eq. 27.3-1:
- Windward walls: $p_z = q_z G_f C_p - q_i (GC_{pi})$
- Leeward walls, side walls, and roofs: $p_h = q_h G_f C_p - q_i (GC_{pi})^+$

$^+$ Notes:
1. See 27.3.5 and Fig. 27.3-8 for the load cases that must be considered.
2. Minimum wind pressures of 27.1.5 must also be considered.

Figure 2.21 *(Continued)*

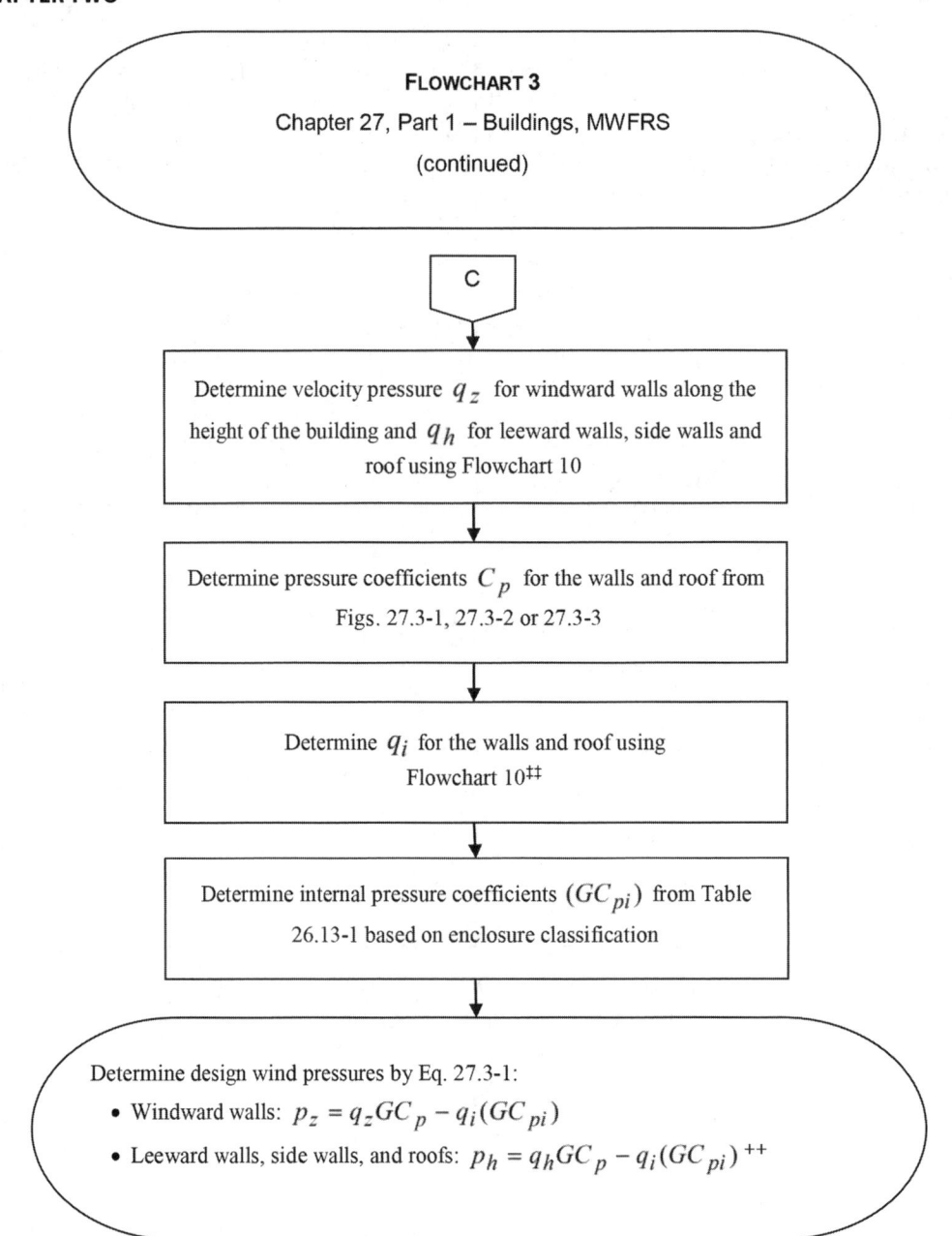

FLOWCHART 3

Chapter 27, Part 1 – Buildings, MWFRS

(continued)

C

↓

Determine velocity pressure q_z for windward walls along the height of the building and q_h for leeward walls, side walls and roof using Flowchart 10

↓

Determine pressure coefficients C_p for the walls and roof from Figs. 27.3-1, 27.3-2 or 27.3-3

↓

Determine q_i for the walls and roof using Flowchart 10‡‡

↓

Determine internal pressure coefficients (GC_{pi}) from Table 26.13-1 based on enclosure classification

↓

Determine design wind pressures by Eq. 27.3-1:

- Windward walls: $p_z = q_z GC_p - q_i(GC_{pi})$
- Leeward walls, side walls, and roofs: $p_h = q_h GC_p - q_i(GC_{pi})$ ++

‡‡ $q_i = q_h$ or $q_i = q_z$ depending on enclosure classification (see 27.3.1). q_i may conservatively be evaluated at height h ($q_i = q_h$) where applicable.

++ Notes:

1. See 27.3.5 and Fig. 27.3-8 for the load cases that must be considered.
2. Minimum wind pressures of 27.1.5 must also be considered.
3. See 27.3.3 for wind pressure on roof overhangs.

Figure 2.21 *(Continued)*

Table 2.11   Summary of Wind Load Procedures in ASCE/SEI 7-16 for C&C

ASCE/SEI chapter	Part	Building/element type	Height limit	Conditions
30	1	Enclosed, low-rise	$h \leq 60$ ft and $h \leq$ least horizontal dimension	• Regular-shaped building
		Partially enclosed, low-rise		• Building does not have response characteristics making it subject to across-wind loading, vortex shedding, instability due to galloping or flutter
		Enclosed with $h \leq 60$ ft	$h \leq 60$ ft	
		Partially enclosed with $h \leq 60$ ft		• Building is not located at a site where channeling effects or buffeting in the wake of upwind obstructions warrant special consideration
				• See additional conditions on selected figure(s) referenced in this part
	2	Enclosed, low-rise	$h \leq 60$ ft and $h \leq$ least horizontal dimension	• Same first three conditions as in Part 1
		Enclosed with $h \leq 60$ ft	$h \leq 60$ ft	• Building has with either a flat roof, a gable roof with $\theta \leq 45°$ or a hip roof with $\theta \leq 27°$
	3	Enclosed	$h > 60$ ft	• Same first three conditions as in Part 1
		Partially enclosed		• See additional conditions on selected figure(s) referenced in this Part
	4	Enclosed	$h \leq 160$ ft	• Same first three conditions as in Part 1
	5	Open	None	• Same first three conditions as in Part 1
				• See additional conditions on selected figure(s) referenced in this part
	6	Parapets	---	• Same first three conditions as in Part 1
				• All building types except enclosed buildings with $h \leq 160$ ft for which the provisions of Part 4 are used
		Roof overhangs	---	• Same first three conditions as in Part 1
				• All building types except enclosed buildings with $h \leq 160$ ft for which the provisions of Part 4 are used
		Roof structures and equipment	$h \leq 60$ ft	• Same first three conditions as in Part 1

permitted to be separated (section 30.2.4). Additional information on these coefficients is given in the following sections.

Velocity pressure   The velocity pressure q_z evaluated at height z is determined by Equation 26.10-1 of ASCE/SEI 7-16. Additional information on how to determine q_z is given in Section 2.5.2.2 of this chapter.

Envelope Procedures

Scope   Parts 1 and 2 in sections 30.3 and 30.4, respectively, contain methods to determine wind loads on C&C of low-rise buildings that meet the conditions of sections 30.3.1 and 30.4.1, respectively. Both of these parts are based on the Envelope Procedure in Chapter 28 of ASCE/SEI 7-16. Different methods were utilized in the development of these procedures than those that were developed for MWFRSs. Additional information is provided in the following sections.

Part 4—Buildings with $h \leq 160$ ft

Overview   Part 4 provides a simplified method of determining wind loads on C&C of enclosed buildings with a mean roof height less than or equal to 160 ft. Wind pressures on C&C located on various surfaces can be read directly from Table 30.7-2 for a building site classified as

Exposure C and an effective wind area of 10 ft². These pressures are modified by an effective area reduction factor, exposure adjustment factor and the topographic factor where applicable (see Table 30.7-2 and Equation 30.7-1 of ASCE/SEI 7-16).

Table 30.7-1 of ASCE 7-16 contains the overall steps that can be used to determine wind pressures on C&C of buildings designed by this method.

Design Wind Pressures

Wall and roof surfaces   Design wind pressures on designated zones of wall and roof surfaces are determined from Table 30.7-2 of ASCE 7-16 as a function of the basic wind speed V, mean roof height and roof angle for buildings that are located on primarily flat ground in Exposure C. These tabulated pressures are valid for an effective wind area of 10 ft² and have been determined using the applicable external pressure coefficients from Part 3 (namely, Figure 30.6-1 for flat roofs, ASCE 7-16 Figure 30.4-2A, 2B, and 2C for gable and hip roofs, and Figure 30.4-5A and 5B for monoslope roofs) and an internal pressure coefficient of ±0.18 for enclosed buildings. Modifications are made to these tabulated pressures based on the actual exposure and effective wind area.

Design wind pressures p are determined by Equation 30.7-1:

$$p = p_{\text{table}}(\text{EAF})(\text{RF})K_{ZT} \qquad (2.15)$$

In this equation, EAF is the exposure adjustment factor given in Table 30.7-2 of ASCE 7-16, which modifies the tabulated pressures in cases where the exposure at the site is different than Exposure C.

The effective area reduction factor (RF) is also given in Table 30.7-2 of ASCE 7-16 and modifies the tabulated

pressures for effective wind areas greater than 10 ft². Reduction factors, which are based on the graphs of the external pressure coefficients in the figures in Part 3, are provided for designated zones on walls and roofs for five different roof shapes and for roof overhangs.

Flowchart 4 (Fig. 2.22) can be used to determine design wind pressures on C&C of wall and roof surfaces in accordance with Part 4 of Chapter 30.

Parapets   Design wind pressures p for C&C elements of parapets for all building types and heights, except for

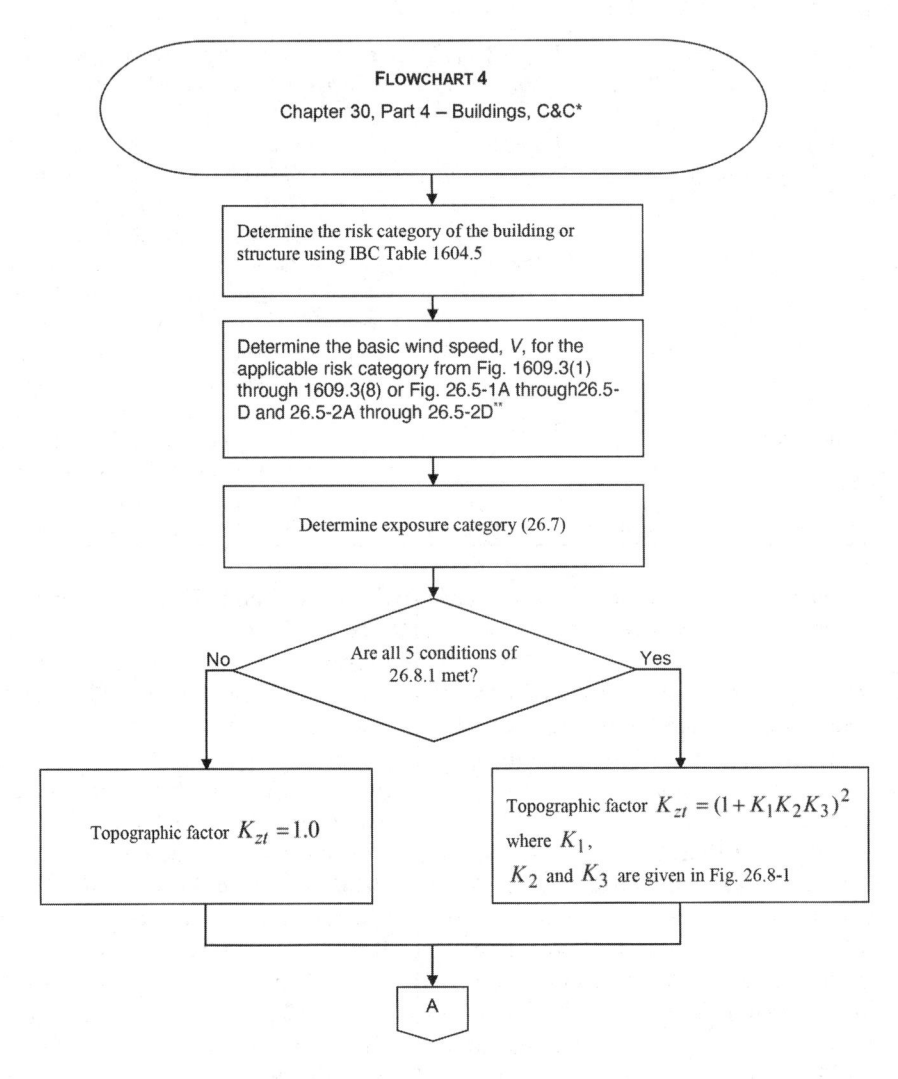

* C&C elements with tributary areas greater than 700 square feet may be designed using the provisions for MWFRSs (30.2.3). See 30.6.1.2 for parapets and 30.6.1.3 for roof overhangs.

** See 26.5.2 and 26.5.3 for basic wind speed in special wind regions and estimation of basic wind speeds from regional climatic data. Tornadoes have not been considered in developing basic wind speed distributions shown in the figures.

Figure 2.22   Design wind pressures on C&C of wall and roof surfaces in accordance with Part 4 of Chapter 30 (Flowchart 4, Ref. [1]).

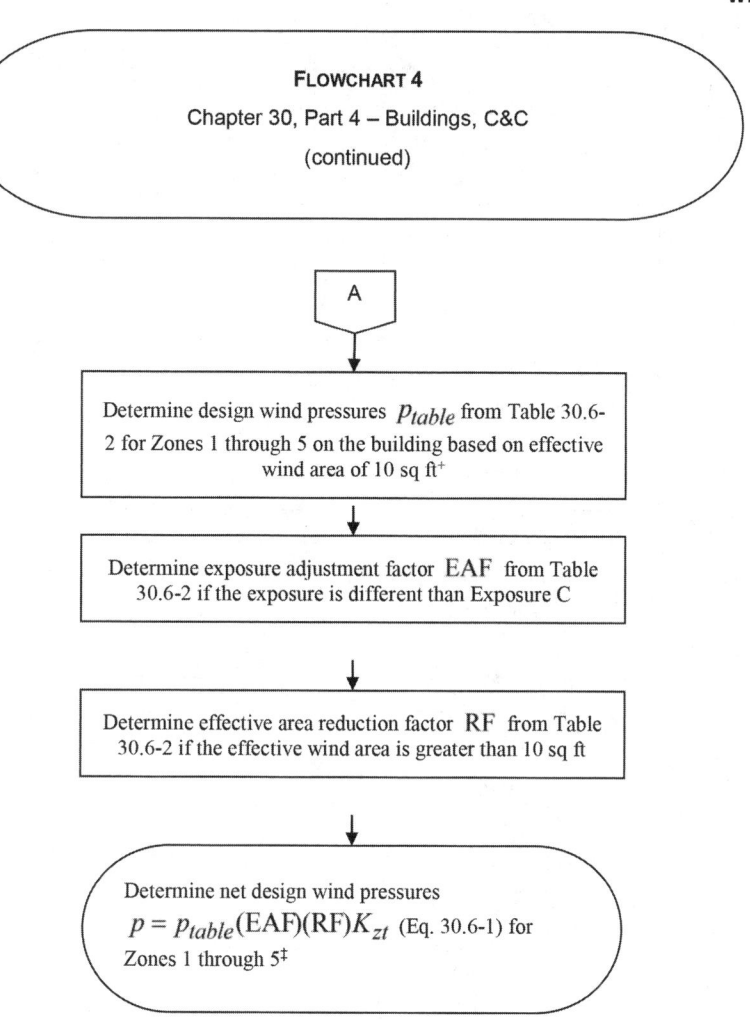

FLOWCHART 4

Chapter 30, Part 4 – Buildings, C&C

(continued)

A

Determine design wind pressures p_{table} from Table 30.6-2 for Zones 1 through 5 on the building based on effective wind area of 10 sq ft[+]

Determine exposure adjustment factor EAF from Table 30.6-2 if the exposure is different than Exposure C

Determine effective area reduction factor RF from Table 30.6-2 if the effective wind area is greater than 10 sq ft

Determine net design wind pressures

$$p = p_{table}(\text{EAF})(\text{RF})K_{zt} \text{ (Eq. 30.6-1) for}$$

Zones 1 through 5[‡]

[+] See 26.2 for definition of effective wind area.
[‡] Minimum wind pressures of 30.2.2 must also be considered.

Figure 2.22 (*Continued*)

enclosed buildings with a mean roof height less than or equal to 160 ft (see Part 4) are calculated by Equation 30.8-1:

$$\langle EQD \rangle \, p = q_p[(GC_p) - (GC_{pi})]$$

In this equation, q_p is the velocity pressure evaluated at the top of the parapet and (GC_{pi}) are the internal pressure coefficients from Table 26.11-1 based on the porosity of the envelope of the parapet. The external pressure coefficients (GC_p) are determined from the same figures as those in Parts 1 and 3 for walls and various roof configurations (see section 30.8 for a comprehensive list of applicable figures).

Similar to the requirements of Part 4, Load Case A and Load Case B must be considered for the windward and leeward parapets, respectively. Figure 30.8-1, which illustrates these load cases, is essentially the same as Figure 30.7-1 in Part 4 with the exception that the pressures in Part 6 must be determined from the applicable figures noted in section 30.8.

Table 30.8-1 contains the overall steps that can be used to determine wind pressures on C&C of parapets designed by this method.

Flowchart 5 (Fig. 2.23) can be used to determine design wind pressures on C&C of parapets in accordance with Part 6 of Chapter 30.

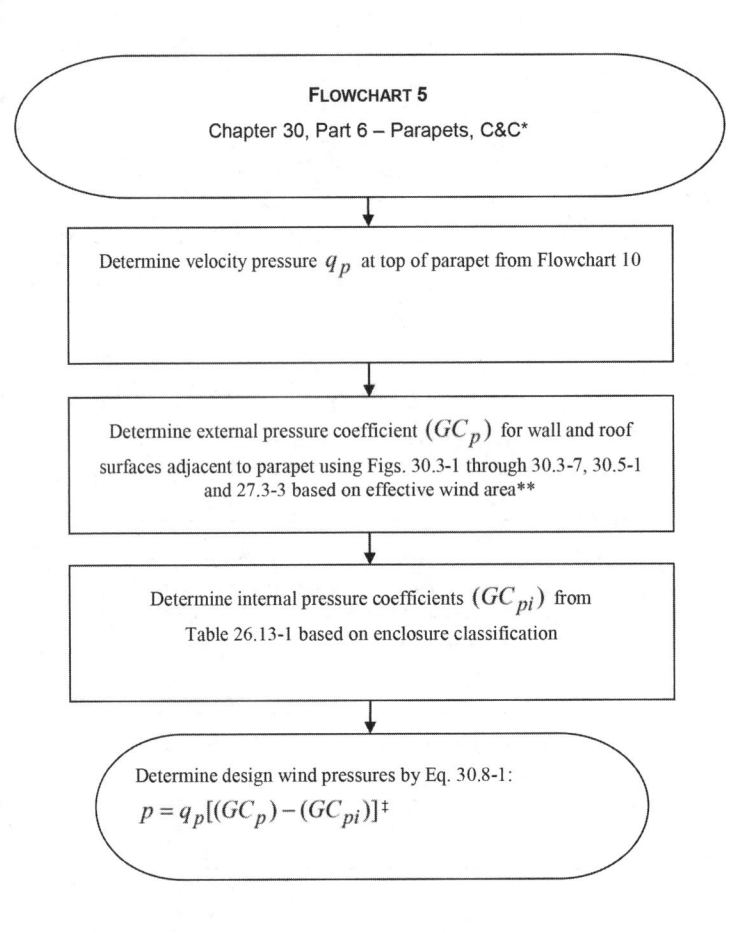

FLOWCHART 5

Chapter 30, Part 6 – Parapets, C&C*

Determine velocity pressure q_p at top of parapet from Flowchart 10

Determine external pressure coefficient (GC_p) for wall and roof surfaces adjacent to parapet using Figs. 30.3-1 through 30.3-7, 30.5-1 and 27.3-3 based on effective wind area**

Determine internal pressure coefficients (GC_{pi}) from Table 26.13-1 based on enclosure classification

Determine design wind pressures by Eq. 30.8-1:

$$p = q_p[(GC_p) - (GC_{pi})]^\ddagger$$

* C&C elements with tributary areas greater than 700 square feet may be designed using the provisions for MWFRSs (30.2.3).

** See 26.2 for definition of effective wind area.

$\ddagger$ Two load cases must be considered (see Fig. 30.8-1). Also, minimum wind pressures of 30.2.2 must also be considered.

Figure 2.23   Design wind pressures on C&C of parapets in accordance with Part 6 of Chapter 30 of ASCE/SEI 7-16 (Flowchart 5, Ref. [1]).

EXAMPLE 1: REINFORCED CONCRETE SHEAR WALL OFFICE BUILDING IN ORLANDO, FL PER ASCE 7-16

In this example, design wind pressures will be determined for the main force resisting for an office building located in Orlando, FL using the ASCE 7-16 procedure. The building is composed of 14 floors with a reinforced concrete shear wall lateral system.

Given

Plan dimensions:	60×120 ft (18.28×36.57 m), length N-S 60 ft, E-W 120 ft
Building height:	14 floors at 10 ft floor-to-floor = $10 \times 14 = 140$ ft (42.67 m)
Fundamental frequency:	1.1 Hz

Building classification:	Risk category II
Basic wind speed:	136 mph, 3 s gust speed for Orlando, FL, from wind speed map Figure 26.5-1B of ASCE 7-16 or ASCE 7 hazard tool (https://asce7hazardtool.online/)
Exposure category:	Exposure C
Enclosure classification:	Enclosed
Topographic factor K_{zt}:	The building is not situated on a hill, ridge or escarpment, $K_{zt} = 1.0$

Required

Wind pressures for the design of MWFRS.

Solution

Step-by-step solution is presented with reference to ASCE 7-16. Detailed calculations are shown for each step, then the results are presented in tables to show the variations over the height of the building. Table 2.12 presents the building dimensions, fundamental natural frequency, and enclosure classification.

Table 2.13 presents the velocity pressure exposure coefficient, K_z and wind velocity pressures q_z calculations.

Table 2.14 presents the gust-effect factors, G, calculations.

Table 2.15 presents external pressure coefficients, C_p, calculations.

Table 2.16 presents velocity pressure for internal pressure distribution, q_i.

Table 2.17 presents internal pressure coefficients, (GC_{pi}).

Table 2.18 presents design wind pressures, p (E-W wind) calculations.

Table 2.19 presents design wind pressures, p (N-S wind) calculations.

Table 2.20 presents design wind pressures on the parapet, p_p, calculations.

Step-by-step solution is presented as follows:

Step 1:

Check if the building meets all of the conditions and limitations of Sections 27.1.2 and 27.1.3 in order to use this procedure to determine the wind pressures on the MWFRS. For the building in question, the building is regularly shaped; that is, it does not have any unusual geometric irregularities in spatial form. Additionally, the building does not have response characteristics that make it subject to across-wind loading or other similar effects. Also, the building is not sited at a location where channeling effects or buffeting in the wake of upwind obstructions need to be considered.

The mean roof height of the building in question is greater than 60 ft; therefore, the provisions of Chapter 28 cannot be used.

Step 2:

Determine the enclosure classification of the building. The building is assumed to be enclosed.

Step 3:

Use Flowchart 3 to determine the design wind pressures, p, on the MWFRS.

1. Determine the design wind pressure effects of the parapet on the MWFRS.

    a. Determine the velocity pressure, q_p, at the top of the parapet from Flowchart 1.

    i. Determine the risk category of the building using IBC Table 1604.5. Due to the nature of its occupancy, this residential building falls under Risk Category II.

    ii. Determine basic wind speed, V, for the applicable risk category. The basic wind speed is given as 136 mph. For buildings with basic wind speed greater than 140 mph (see section 26.12.3.1), they are considered located in a wind-borne debris region. In this case, according to 26.12.3.2, glazing in buildings located in wind-borne debris regions must be protected with an impact-protective system or must be impact-resistant glazing. It is assumed that impact-resistant glazing is used over the entire height of the building. In the building in question, the wind speed is slightly less than 140 mph; it is assumed that the building is located in a wind-borne debris region. Therefore, the building is classified as enclosed.

    iii. Determine wind directionality factor, K_d, from Table 26.6-1. For the MWFRS of a building structure, $K_d = 0.85$.

    iv. Determine exposure category. Assume that Exposure C is applicable in all directions.

    v. Determine topographic factor, K_{zt}. The building is not situated on a hill, ridge or escarpment. Therefore, the topographic factor $K_{zt} = 1.0$ (see section 26.8.2).

    vi. Determine ground elevation factor, K_e. Ground elevation factor can be taken as 1.0 for all elevations (see section 26.9).

    vii. Determine velocity pressure exposure coefficient, K_z or K_h, from Table 26.10-1. For Exposure C at a height of 145 feet at the top of the parapet, K_z or $K_h = 1.37$ by linear interpolation.

    viii. Determine velocity pressure, q_p, evaluated at the top of the parapet by Equation 26.10-1.

$$q_z = 0.00256 K_z K_{zt} K_d K_e V^2 = 0.00256 \times 1.37 \\ \times 1.0 \times 0.85 \times 1.0 \times 136^2 = 55.1 \text{psf}$$

    b. Determine combined net pressure coefficient, (GC_{pn}), for the parapets. According to section 27.3.4, $(GC_{pn}) = 1.5$ for the windward parapet and $(GC_{pn}) = -1.0$ for the leeward parapet.

    c. Determine combined net design pressure, p_p, on the parapet by Equation 27.3-3.

$$p_p = q_p(GC_{pn}) \\ = 55.1 \times 1.5 = 82.7 \text{ psf on windward parapet} \\ = 55.1 \times (-1.0) = -55.1 \text{ psf on leeward parapet}$$

2. Determine whether the building is rigid or flexible according to section 26.11.2. In lieu of determining the natural frequency, n_1, of the building from a dynamic analysis, Equation 26.11-3 is used to compute

an approximate natural frequency, n_a, for concrete shear wall building, Equation 26.11-5 can be used, and for the building in question, it is given as 1.1 Hz. Because $n_a > 1.0$ Hz, the building is defined as a rigid building.

3. Determine gust-effect factor (G or G_f) from Flowchart 2. According to section 26.11.4, gust-effect factor, G, for rigid buildings may be taken as 0.85 or can be calculated by Equation 26.11-6. For simplicity, use $G = 0.85$.

4. Determine velocity pressure: q_z for windward walls along the height of the building, and q_h for leeward walls, side walls and roof using Flowchart 1. Most of the quantities needed to compute q_z and q_h were determined in this step above. The velocity exposure coefficients K_z and K_h are summarized in Table 2.13. Velocity pressures q_z and q_h are determined by Equation 26.10-1:

$$q_z = 0.00256 K_z K_{zt} K_d K_e V^2 = 0.00256 \times K_z \times 1.0$$
$$\times 0.85 \times 1.0 \times 136^2 = 40.25 K_z \text{ psf}$$

The velocity pressures at different heights are given in Table 2.14.

5. Determine pressure coefficients, C_p, for the walls and roof from Figure 27.3-1.

For wind in the E-W direction:

Windward wall: $C_p = 0.8$ for use with q_z
Leeward wall ($L/B = 120/60 = 2.0$): $C_p = -0.3$ for use with q_h
Side wall: $C_p = -0.7$ for use with q_h
Roof (normal to ridge with $q < 10$ degrees and parallel to ridge for all q with $h/L = 140/120 = 1.17 > 0.5$):

$C_p = -1.04, -0.18$ from windward edge to $h/2 = 70$ ft for use with q_h
$C_p = -0.7, -0.18$ from $h/2$ ft to $h = 120$ ft for use with q_h

For wind in the N-S direction:

Windward wall: $C_p = 0.8$ for use with q_z
Leeward wall ($L/B = 60/120 = 0.50$): $C_p = -0.5$ for use with q_h
Side wall: $C_p = -0.7$ for use with q_h
Roof (normal to ridge with $q < 10$ degrees and parallel to ridge for all q with $h/L = 140/60 = 2.33$):

$C_p = -1.04, -0.18$ from windward edge to $h/2 = 70 > 60$, since the length N-S is 60 ft, use 60 ft for use with q_h

6. Determine q_i for the walls and roof using Flowchart 1. According to section 27.3.1, $q_i = q_h = 54.7$ psf for windward walls, side walls, leeward walls and roofs of enclosed buildings.

7. Determine internal pressure coefficients, (GC_{pi}), from Table 26.13-1. For an enclosed building, (GC_{pi}) = +0.18, −0.18.

8. Determine design wind pressures p_z and p_h by Equation 27.3-1.

Windward walls:

$$p_z = q_z GC_p - q_h(GC_{pi})$$
$$= (q_z \times 0.85 \times 0.8) - 54.7(\pm 0.18)$$
$$= (0.68q_z \pm 9.85) \text{ psf (external} \pm \text{internal pressure)}$$

Leeward wall, side walls and roof:

$$p_h = q_h GC_p - q_h(GC_{pi})$$
$$= (54.7 \times 0.85 \times C_p) - 54.7(\pm 0.18)$$
$$= (46.5C_p \pm 9.85) \text{ psf (external} \pm \text{internal pressure)}$$

A summary of the maximum design wind pressures in the E-W and N-S directions is given in Tables 2.18 and 2.19, respectively.

The MWFRS of buildings whose wind loads have been determined by Chapter 27 must be designed for the wind load cases defined in Figure 27.3-8 (see section 27.3.5). In Case 1, the full design wind pressures act on the projected area perpendicular to each principal axis of the structure. These pressures are assumed to act separately along each principal axis. The wind pressures on the windward and leeward walls given in Tables 2.18 and 2.19 fall under Case 1.

In Case 2, 75 percent of the design wind pressures on the windward and leeward walls are applied on the projected area perpendicular to each principal axis of the building along with a torsional moment. The wind pressures and torsional moments, both of which vary over the height of the building, are applied separately for each principal axis. Wind pressures and torsional moments should be calculated over the height of the building. As an example of the calculations that need to be performed over the height of the building, the wind pressures and torsional moment at the mean roof height for Case 2 are as follows (see distribution in Case 2 below):

For E-W wind:

$$0.75p_{wx} = 0.75 \times 37.2 = 27.9 \text{ psf (windward wall)}$$
$$0.75p_{lx} = 0.75 \times 13.9 = 10.5 \text{ psf (leeward wall)}$$

$$M_T = 0.75(p_{wx} + p_{lx})B_x e_x$$
$$= 0.75(37.2 + 13.9) \times 60 \times (\pm 0.15 \times 60)$$
$$= \pm 20,696 \text{ ft-lb/ft}$$

For N-S wind:

$$0.75p_{wy} = 0.75 \times 37.2 = 27.9 \text{ psf (windward wall)}$$
$$0.75p_{ly} = 0.75 \times 23.2 = 17.4 \text{ psf (leeward wall)}$$

$$M_T = 0.75(p_{wy} + p_{ly})B_y e_y$$
$$= 0.75(37.2 + 23.2) \times 120 \times (\pm 0.15 \times 120)$$
$$= \pm 97,848 \text{ ft-lb/ft}$$

In Case 3, 75 percent of the wind pressures of Case 1 are applied to the building simultaneously. This accounts for wind along the diagonal of the building. In Case 4, 75 percent of the wind pressures and torsional moments defined in Case 2 act simultaneously on the building, see distribution in Case 4 below.

Table 2.12   MWFRS, ASCE/SEI Chapters 27 and 28

Length N-S =	60.0	ft
Length E-W =	120.0	ft
Number of levels below the eave =	14	

Level	Height above ground, z (ft)	
Parapet	145.0	Top of parapet, h_p (enter "-" if there is no parapet)
Ridge	140.0	Top of ridge, h_{ridge} (enter mean roof height where $\theta = 0$ deg)
h	140.0	Mean roof height, h
Eave	140.0	Eave height, h_{eave} (enter mean roof height where $\theta = 0$ deg)
14	140.0	
13	130.0	Leave the cells empty for z where there is no level
12	120.0	
11	110.0	
10	100.0	
9	90.0	
8	80.0	
7	70.0	
6	60.0	
5	50.0	
4	40.0	
3	30.0	
2	20.0	
1	10.0	

Roof angle, θ	0.00	deg
Mean roof height, h	140.0	ft
Basic wind speed, V	136	mi/h
Exposure category	C	ASCE/SEI 26.7.3
Risk category	II	ASCE/SEI Table 1.5-1
Fundamental natural frequency, n_1 (Hz)	1.10	N-S
	1.10	E-W
		See ASCE/SEI 26.11.3 for equations for approximate natural frequencies, n_a
Enclosure classification	Enclosed	

Table 2.13   Wind Velocity Pressures, q_z - ASCE/SEI Chapters 27 and 28

	Exposure category	C	
α		9.5	ASCE/SEI Table 26.11-1
z_g		900	ft

Velocity pressure exposure coefficient, K_z:

Level	z (ft)	K_z	ASCE/SEI Table 26.10-1
Parapet	145.0	1.37	
Ridge	140.0	1.36	
h	140.0	1.36	
Eave	140.0	1.36	
14	140.0	1.36	
13	130.0	1.34	
12	120.0	1.32	
11	110.0	1.29	
10	100.0	1.27	
9	90.0	1.24	
8	80.0	1.21	
7	70.0	1.17	
6	60.0	1.14	
5	50.0	1.09	
4	40.0	1.04	
3	30.0	0.98	
2	20.0	0.90	
1	10.0	0.85	

Table 2.14   Gust-Effect Factors, *G*—ASCE/SEI Chapters 27 and 28

Wind directionality factor, K_d		0.85	ASCE/SEI Table 26.6-1
Use ground elevation factor $K_e = 1.00$		Yes	ASCE/SEI 26.9
Ground elevation above sea level		0.0	ft
Ground elevation factor, K_e		1.00	

Wind velocity pressures, q_z			
Level	z (ft)	q_z (lb/ft^2)	ASCE/SEI Eq. (26.10-1)
Parapet	145.0	55.1	
Ridge	140.0	54.7	
h	140.0	54.7	
Eave	140.0	54.7	
14	140.0	54.7	

(*continued*)

Table 2.14 Gust-Effect Factors, *G*—ASCE/SEI Chapters 27 and 28 (*Continued*)

13	130.0	53.8
12	120.0	52.9
11	110.0	52.0
10	100.0	50.9
9	90.0	49.8
8	80.0	48.6
7	70.0	47.3
6	60.0	45.7
5	50.0	44.0
4	40.0	42.0
3	30.0	39.5
2	20.0	36.3
1	10.0	34.2

For detailed calculations of *G*, the following is used per the relevant sections of ASCE 7-16:

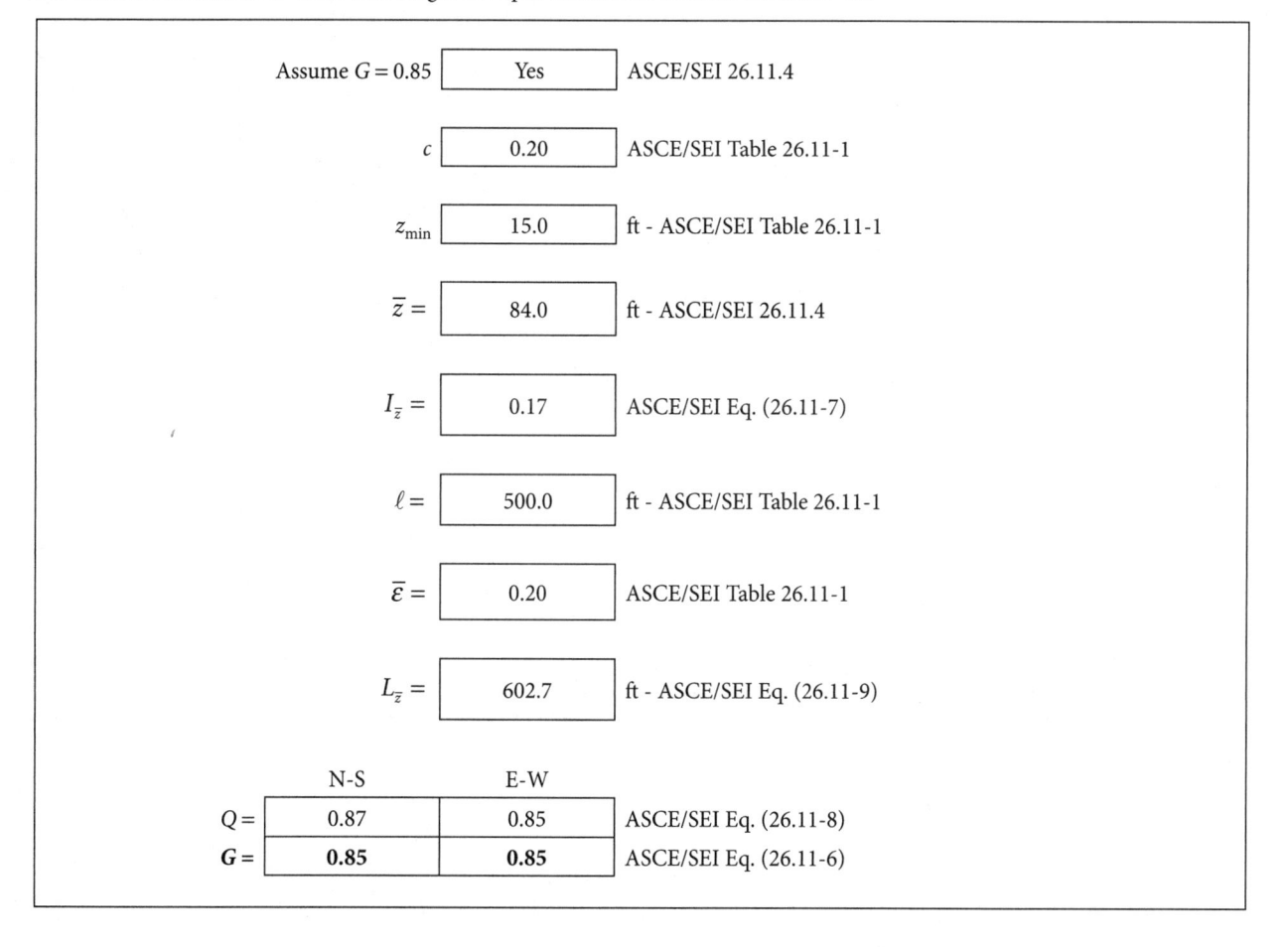

Assume $G = 0.85$	Yes	ASCE/SEI 26.11.4
c	0.20	ASCE/SEI Table 26.11-1
z_{min}	15.0	ft - ASCE/SEI Table 26.11-1
$\overline{z} =$	84.0	ft - ASCE/SEI 26.11.4
$I_{\overline{z}} =$	0.17	ASCE/SEI Eq. (26.11-7)
$\ell =$	500.0	ft - ASCE/SEI Table 26.11-1
$\overline{\varepsilon} =$	0.20	ASCE/SEI Table 26.11-1
$L_{\overline{z}} =$	602.7	ft - ASCE/SEI Eq. (26.11-9)

	N-S	E-W	
$Q =$	0.87	0.85	ASCE/SEI Eq. (26.11-8)
$G =$	**0.85**	**0.85**	ASCE/SEI Eq. (26.11-6)

Assume the gust factor $G = 0.85$ and a rigid building with $n_1 = 1.10 > 1.0$

Wind Loads: MWFRS, Chapter 27, Part 1—Enclosed and Partially Enclosed Rigid and Flexible Buildings (ASCE/SEI 27.3.1).

Table 2.15   External Pressure Coefficients, C_p (ASCE/SEI Figure 27.3-1)

			L/B	C_p
	Parapet	Windward		1.50
		Leeward		1.00
	Windward wall	N-S	All	0.80
		E-W	All	0.80
	Leeward wall	N-S	0.50	−0.50
		E-W	2.00	−0.30
	Side walls	N-S	All	−0.70
		E-W	All	−0.70
Roof		Normal to ridge for $\theta \geq 10$ deg (N-S)	Windward	0.00
				0.00
			Leeward	0.00
		Normal to ridge for $\theta < 10$ deg (N-S)	0 to $h/2$	−1.04
				−0.18
			$h/2$ to h	0.00
				0.00
			h to $2h$	0.00
				0.00
			$> 2h$	0.00
				0.00
		Parallel to ridge for all θ (E-W)	0 to $h/2$	−1.04
				−0.18
			$h/2$ to h	−0.70
				−0.18
			h to $2h$	0.00
				0.00
			$> 2h$	0.00
				0.00

Table 2.16   Velocity Pressure for Internal Pressure Distribution, q_i (ASCE/SEI 27.3-1)

$$q_i = \boxed{\quad 54.7 \quad} \text{ lb/ft}^2$$

q_i is conservatively taken as q_h for all internal pressure evaluations.

Table 2.17   Internal Pressure Coefficients, (GC_pi) [ASCE/SEI Table 26.13-1]

$$(GC_{pi}) = \boxed{\begin{array}{c} 0.18 \\ \hline -0.18 \end{array}} \qquad \text{Enclosed}$$

Table 2.18   Design Wind Pressures, p (E-W wind) [ASCE/SEI Equation (27.3-1)]

Building surface	Height above ground z (ft)	Velocity pressure q (lb/ft^2)	External pressure qGC_p (lb/ft^2)	Internal pressure $q_h(GC_{pi})$ (lb/ft^2)	Net pressure, p (lb/ft^2) $(+GC_{pi})$	Net pressure, p (lb/ft^2) $(-GC_{pi})$	Total horizontal pressure, p (lb/ft^2)
Windward wall	140.0	54.7	37.2	9.8	27.3	47.0	51.1
	—	—	—	—	—	—	—
	—	—	—	—	—	—	—
	140.0	54.7	37.2	9.8	27.3	47.0	51.1
	130.0	53.8	36.6	9.8	26.8	46.4	50.5
	120.0	52.9	36.0	9.8	26.2	45.8	49.9
	110.0	52.0	35.3	9.8	25.5	45.2	49.3
	100.0	50.9	34.6	9.8	24.8	44.5	48.6
	90.0	49.8	33.9	9.8	24.0	43.7	47.8
	80.0	48.6	33.0	9.8	23.2	42.9	47.0
	70.0	47.3	32.1	9.8	22.3	42.0	46.1
	60.0	45.7	31.1	9.8	21.3	40.9	45.0
	50.0	44.0	29.9	9.8	20.1	39.8	43.9
	40.0	42.0	28.6	9.8	18.7	38.4	42.5
	30.0	39.5	26.9	9.8	17.0	36.7	40.8
	20.0	36.3	24.7	9.8	14.8	34.5	38.6
	10.0	34.2	23.2	9.8	13.4	33.1	37.2
Leeward wall	All	54.7	−13.9	9.8	−23.8	−4.1	
Side walls	All	54.7	−32.5	9.8	−42.4	−22.7	

Roof, parallel to ridge for all θ

0	to	$h/2$	—	54.7	−48.3	9.8	−58.2	−38.5
0.0	to	70.0			−8.4	9.8	−18.2	1.5
$h/2$	to	h	—	54.7	−32.5	9.8	−42.4	−22.7
70.0	to	120.0			−8.4	9.8	−18.2	1.5
h	to	$2h$	—	54.7	0.0	—	—	—
—	to	—			0.0	—	—	—
	>2h		—	54.7	0.0	—	—	—
—	to	—			0.0	—	—	—

Table 2.19   Design Wind Pressures, *p* (N-S wind) [ASCE/SEI Eq. (27.3-1)]

Building surface	Height above ground z (ft)	Velocity pressure q (lb/ft²)	External pressure qGC_p (lb/ft²)	Internal pressure $q_h(GC_{pi})$ (lb/ft²)	Net pressure, p (lb/ft²) $(+GC_{pi})$	$(-GC_{pi})$	Total horizontal pressure, p (lb/ft²)
Windward wall	140.0	54.7	37.2	9.8	27.3	47.0	60.4
	140.0	54.7	37.2	9.8	27.3	47.0	60.4
	130.0	53.8	36.6	9.8	26.8	46.4	59.8
	120.0	52.9	36.0	9.8	26.2	45.8	59.2
	110.0	52.0	35.3	9.8	25.5	45.2	58.6
	100.0	50.9	34.6	9.8	24.8	44.5	57.9
	90.0	49.8	33.9	9.8	24.0	43.7	57.1
	80.0	48.6	33.0	9.8	23.2	42.9	56.3
	70.0	47.3	32.1	9.8	22.3	42.0	55.4
	60.0	45.7	31.1	9.8	21.3	40.9	54.3
	50.0	44.0	29.9	9.8	20.1	39.8	53.2
	40.0	42.0	28.6	9.8	18.7	38.4	51.8
	30.0	39.5	26.9	9.8	17.0	36.7	50.1
	20.0	36.3	24.7	9.8	14.8	34.5	47.9
	10.0	34.2	23.2	9.8	13.4	33.1	46.5
Leeward wall	All	54.7	−23.2	9.8	-33.1	−13.4	
Side walls	All	54.7	−32.5	9.8	-42.4	-22.7	

Roof, Normal to ridge with $\theta < 10$ deg

			Height	Velocity q	qGC_p	$q_h(GC_{pi})$	$(+GC_{pi})$	$(-GC_{pi})$	Total
0	to	$h/2$	—	54.7	−48.3	9.8	−58.2	−38.5	
0.0	to	60.0			−8.4	9.8	−18.2	1.5	
$h/2$	to	h	—	54.7	0.0	—	—	—	
—	to	—			0.0	—	—	—	
h	to	$2h$	—	54.7	0.0	—	—	—	
—	to	—			0.0	—	—	—	
	>2h		—	54.7	0.0	—	—	—	
—	to	—			0.0	—	—	—	

Table 2.20   Design Wind Pressures on the Parapet, p_p [ASCE/SEI Equation (27.3-3)]

Height above ground z (ft)	Velocity pressure q (lb/ft²)	Pressure p_p (lb/ft²) Windward	Leeward
145.0	55.1	82.7	−55.1

Wind Load Cases
Case 1:

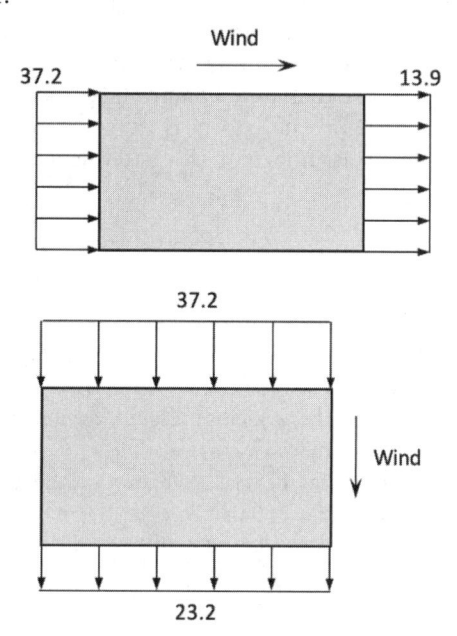

37.2   Wind   13.9

37.2

Wind

23.2

Case 2:

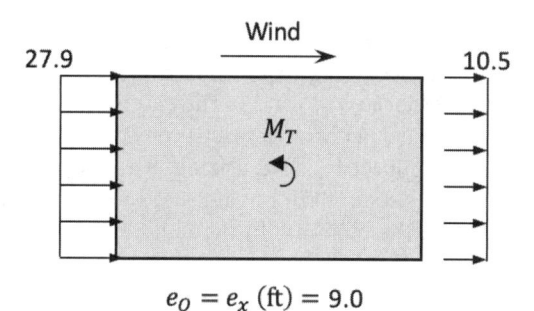

27.9   Wind   10.5

M_T

$$e_Q = e_x \text{ (ft)} = 9.0$$

$$M_T = \pm 20{,}696 \text{ ft-lb/ft}$$

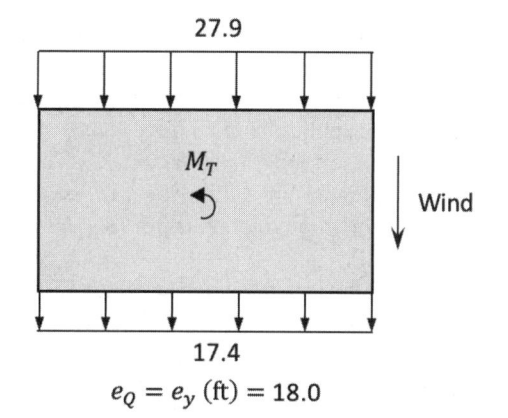

27.9

M_T

Wind

17.4

$$e_Q = e_y \text{ (ft)} = 18.0$$

$$M_T = \pm 97{,}848 \text{ ft-lb/ft}$$

Case 3:

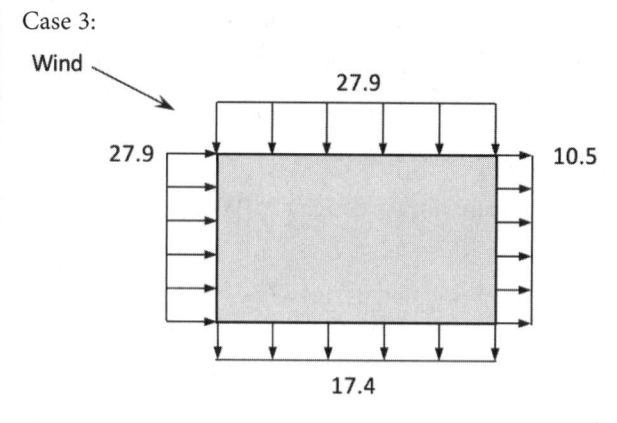

Wind   27.9

27.9   10.5

17.4

Case 4:

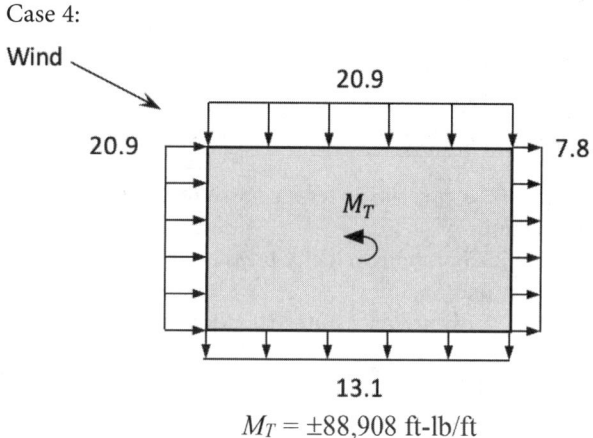

Wind   20.9

20.9   7.8

M_T

13.1

$$M_T = \pm 88{,}908 \text{ ft-lb/ft}$$

EXAMPLE 2: STRUCTURAL STEEL MOMENT-RESISTING FRAME OFFICE BUILDING IN CHICAGO, IL PER ASCE 7-16

In this example, design wind pressures will be determined for the main force resisting for an office building located in Chicago, IL using the ASCE 7-16 procedure. The building is composed of 18 floors with a structural steel moment-resisting frame lateral system.

Given

Plan dimensions:	145×125 ft (44.21 m $\times 38.11$ m), length N-S 125 ft, E-W 145 ft
Building height:	18 floors at 12 ft floor-to-floor $= 12 \times 18 = 216$ ft (65.85 m)
Fundamental frequency:	to be calculated.
Building classification:	Risk Category II
Basic wind speed:	115 mph, 3 s gust speed for Chicago, IL, from wind speed map Figure 26.5-1B of ASCE 7-16 or ASCE 7 hazard tool (https://asce7hazardtool.online/)
Exposure category:	Exposure B

Enclosure classification: Partially enclosed

Topographic factor K_{zt}:   The building is not situated on a hill, ridge, or escarpment, $K_{zt} = 1.0$

Required

Wind pressures for the design of MWFRS.

Solution

Step-by-step solution is presented with reference to ASCE 7-16. Detailed calculations are shown for each step, then the results are presented in tables to show the variations over the height of the building.

Table 2.21 presents the building dimensions, fundamental natural frequency, and enclosure classification.

Table 2.22 presents the Velocity pressure exposure coefficient, K_z and wind velocity pressures q_z calculations.

Table 2.23 presents the gust-effect factors, G, calculations.

Table 2.24 presents external pressure coefficients, C_p, calculations.

Table 2.25 presents velocity pressure for internal pressure distribution, q_i.

Table 2.26 presents internal pressure coefficients, (GC_{pi}).

Table 2.27 presents design wind pressures, p (E-W wind) calculations.

Table 2.28 presents design wind pressures, p (N-S wind) calculations.

Table 2.29 presents design wind pressures on the parapet, p_p, calculations.

Step-by-step solution is presented as follows:

Building dimensions, fundamental natural frequency, and enclosure classification are given and presented in Table 2.21.

Step 1:

Check if the building meets all of the conditions and limitations of sections 27.1.2 and 27.1.3 in order to use this procedure to determine the wind pressures on the MWFRS. For the building in question, the building is regularly shaped; that is, it does not have any unusual geometric irregularities in spatial form. Additionally, the building does not have response characteristics that make it subject to across-wind loading or other similar effects. Also, the building is not sited at a location where channeling effects or buffeting in the wake of upwind obstructions need to be considered.

The mean roof height of the building in question is greater than 60 ft, therefore, the provisions of Chapter 28 cannot be used.

Step 2:

Determine the enclosure classification of the building. For illustration, the building is assumed to be partially enclosed. Where a building has operable windows or where debris that may compromise some of the windows during a wind storm is expected, it is more appropriate to assume that the building is partially enclosed.

Step 3:

Use Flowchart 3 to determine the design wind pressures, p, on the MWFRS.

1. Determine the design wind pressure effects of the parapet on the MWFRS.

a. Determine the velocity pressure, q_p, at the top of the parapet from Flowchart 1.

i. Determine the risk category of the building using IBC Table 1604.5. Due to the nature of its occupancy, this residential building falls under Risk Category II.

ii. Determine basic wind speed, V, for the applicable risk category. The basic wind speed is given as 115 mph. For buildings with basic wind speed greater than 140 mph (see section 26.12.3.1), they are considered located in a wind-borne debris region. In this case, according to section 26.12.3.2, glazing in buildings located in wind-borne debris regions must be protected with an impact-protective system or must be impact-resistant glazing. It is assumed that impact-resistant glazing is not used over the entire height of the building. In the building in question, the wind speed is less than 140 mph; it is assumed that the building is not located in a wind-borne debris region. Therefore, the building is classified as partially enclosed.

iii. Determine wind directionality factor, K_d from Table 26.6-1. For the MWFRS of a building structure, $K_d = 0.85$.

iv. Determine exposure category. Assume that Exposure B is applicable in all directions.

v. Determine topographic factor, K_{zt}. The building is not situated on a hill, ridge, or escarpment. Therefore, the topographic factor $K_{zt} = 1.0$ (see section 26.8.2).

vi. Determine ground elevation factor, K_e. Ground elevation factor can be taken as 1.0 for all elevations (see section 26.9).

vii. Determine velocity pressure exposure coefficient, K_z or K_h, from Table 26.10-1. For Exposure B at a height of 221 ft at the top of the parapet, K_z or $K_h = 1.24$ by linear interpolation.

viii. Determine velocity pressure, q_p, evaluated at the top of the parapet by Equation 26.10-1.

$$q_z = 0.00256 K_z K_{zt} K_d K_e V^2 = 0.00256 \times 1.24$$
$$\times 1.0 \times 0.85 \times 1.0 \times 115^2 = 35.7 \text{ psf}$$

b. Determine combined net pressure coefficient, (GC_{pn}), for the parapets. According to section 27.3.4, $(GC_{pn}) = 1.5$ for the windward parapet and $(GC_{pn}) = -1.0$ for the leeward parapet.

c. Determine combined net design pressure, p_p, on the parapet by Equation 27.3-3.

$$p_p = q_p(GC_{pn})$$
$$= 35.7 \times 1.5 = 53.5 \text{ psf on windward parapet}$$
$$= 35.7 \times (-1.0) = -35.7 \text{ psf on leeward parapet}$$

Results for the wind parapet calculations are shown in Table 2.29.

2. Determine whether the building is rigid or flexible according to section 26.11.2. In lieu of determining the natural frequency, n_1, of the building from a dynamic analysis, Equation 26.11-3 is used to compute an approximate natural frequency, n_a.

Check if the approximate lower-bound natural frequencies given in 26.11.3 can be used for this building (26.9.2.1):

Building height = 216 ft < 300 ft
Building height = 225 ft < $4L_{eff} = 4 \times 125$
= 500 ft in both directions

Because both of these limitations are satisfied, Equation 26.11-2 can be used to determine an approximate value of n_a for steel moment resisting-frame systems:

$$n_a = \frac{22.2}{h^{0.8}} = \frac{22.2}{216^{0.8}} = 0.3 \text{ Hz}$$

Because $n_a < 1.0$ Hz, the building is defined as a flexible building.

Determine gust-effect factor (G or G_f) from Flowchart 2. The gust-effect factor, Gf, for flexible buildings is determined by Equation 26.11-10 in 26.9.5.

a) Determine g_Q and g_v

$$g_Q = g_v = 3.4$$

b) Determine g_R

Eq. 26.11-11:

$$g_R = \sqrt{2\ln(3600 n_1)} + \frac{0.577}{\sqrt{2\ln(3600 n_1)}}$$
$$= \sqrt{2\ln(3600(0.3))} + \frac{0.577}{\sqrt{2\ln(3600(0.3))}}$$
$$= 3.74 + 0.154 = 3.89$$

c) Determine $I_{\bar{z}}$

$$\bar{z} = 0.6h = 0.6(216) = 129.6 \text{ ft} > z_{min} = 30 \text{ ft}$$

Table 26.11-1 for Exposure B

$$I_{\bar{z}} = c\left(\frac{33}{\bar{z}}\right)^{1/6} = 0.3\left(\frac{33}{129.6}\right)^{1/6} = 0.24$$

Equation 26.11-7 and Table 26.11-1 for Exposure B

d) Determine Q

$$L_{\bar{z}} = \ell\left(\frac{\bar{z}}{33}\right)^{\bar{\epsilon}} = 320\left(\frac{129.6}{33}\right)^{1/3} = 504.9$$

Equation 26.11-9 and Table 26.11-9 for Exposure B

$$Q = \sqrt{\frac{1}{1 + 0.63\left(\frac{B+L}{L_{\bar{z}}}\right)^{0.63}}}$$

$$= \sqrt{\frac{1}{1 + 0.63\left(\frac{145 + 216}{504.9}\right)^{0.63}}} = 0.81$$

Equation 26.11-8

e) Determine R

The building falls under Risk Category II. The basic wind speed is 115 mph.

$$\bar{V}_{\bar{z}} = \bar{b}\left(\frac{\bar{z}}{33}\right)^{\bar{\alpha}}\left(\frac{88}{66}\right)V$$

$$= 0.45\left(\frac{129.6}{33}\right)^{\frac{1}{4}}\left(\frac{88}{60}\right)(115)$$

$$= 106.85$$

Equation 26.11-16 and Table 26.11-1 for Exposure B

$$N_1 = \frac{n_1 L_{\bar{z}}}{\bar{V}_{\bar{z}}} = \frac{(0.3)(504.9)}{106.85} = 1.42$$

Equation 26.11-14

$$R_n = \frac{7.47 N_1}{(1 + 10.3 N_1)^{5/3}} = \frac{7.47(1.42)}{(1 + 10.3(1.42))^{5/3}} = 0.11$$

Equation 26.11-13

$$\eta_h = \frac{4.6 n_1 h}{\bar{V}_{\bar{z}}} = \frac{4.6(0.3)(216)}{106.85} = 2.79$$

$$R_h = \frac{1}{\eta_h} - \frac{1}{2\eta_h^2}(1 - e^{-2\eta_h}) = \frac{1}{2.8} - \frac{1}{2(2.8^2)}(1 - e^{-2(2.8)}) = 0.29$$

Equation 26.11-15a

$$\eta_B = \frac{4.6n_1 B}{\overline{V}_{\bar{z}}} = \frac{4.6(0.3)(145)}{106.85} = 1.87$$

$$R_B = \frac{1}{\eta_B} - \frac{1}{2\eta_B^2}(1 - e^{-2\eta_B})$$

$$= \frac{1}{1.87} - \frac{1}{2(1.87^2)}(1 - e^{-2(1.87)}) = 0.39$$

Equation 26.11-15a

$$\eta_L = \frac{15.4n_1 L}{\overline{V}_{\bar{z}}} = \frac{15.4(0.3)(125)}{106.85} = 5.4$$

$$R_L = \frac{1}{\eta_L} - \frac{1}{2\eta_L^2}(1 - e^{-2\eta_L}) = \frac{1}{5.4} - \frac{1}{2(5.4^2)}(1 - e^{-2(5.4)}) = 0.17$$

Equation 26.11-15a

Suggested value for damping ration for steel buildings per C26.11 is $\beta = 0.01$.

$$R = \sqrt{\frac{1}{\beta} R_n R_h R_B (0.53 + 0.47 R_L)}$$

$$= \sqrt{\frac{1}{0.01}(0.11)(0.29)(0.40)(0.53 + 0.47(0.16))} = 0.88$$

Equation 26.11-12

f) Determine G_f

$$G_f = 0.925 \left(\frac{1 + 1.7 I_{\bar{z}} \sqrt{g_Q^2 Q^2 + g_R^2 R^2}}{1 + 1.7 g_v I_{\bar{z}}} \right)$$

$$= 0.925 \left(\frac{1 + 1.7(0.24)\sqrt{3.4^2(0.81^2) + 3.9^2(0.88^2)}}{1 + 1.7(3.4)(0.24)} \right)$$

$$= 1.083$$

Equation 26.11-10

A summary of parameters and results of the gust factor calculation is shown in Table 2.23.

3. Determine velocity pressure: q_z for windward walls along the height of the building, and q_h for leeward walls, side walls, and roof using Flowchart 1. Most of the quantities needed to compute q_z and q_h were determined in this step above. The velocity exposure coefficients K_z and K_h are summarized in Table 2.22. Velocity pressures q_z and q_h are determined by Equation 26.10-1:

$$q_z = 0.00256 K_z K_{zt} K_d K_e V^2 = 0.00256 \times K_z \times 1.0 \\ \times 0.85 \times 1.0 \times 115^2 = 28.78 K_z \text{ psf}$$

The velocity pressures at different heights are given in Table 2.22.

4. Determine pressure coefficients, C_p, for the walls and roof from Figure 27.3-1.

For wind in the E-W direction:
Windward wall: $C_p = 0.8$ for use with q_z
Leeward wall ($L/B = 145/125 = 1.16$): $C_p = -0.47$ for use with q_h
Side wall: $C_p = -0.7$ for use with q_h
Roof (normal to ridge with $q < 10$ degrees and parallel to ridge for all q with $h/L = 216/145 = 1.49 > 0.5$):

$C_p = -1.04, -0.18$ from windward edge to $h/2 = 108$ ft for use with q_h
$C_p = -0.7, -0.18$ from $h/2$ to $h = 216$ ft for use with q_h

For wind in the N-S direction:
Windward wall: $C_p = 0.8$ for use with q_z
Leeward wall ($L/B = 125/145 = 0.86$): $C_p = -0.5$ for use with q_h
Side wall: $C_p = -0.7$ for use with q_h
Roof (normal to ridge with $q < 10$ degrees and parallel to ridge for all q with $h/L = 216/125 = 1.73$):

$C_p = -1.04, -0.18$ from windward edge to $h/2 = 108$, for use with q_h
$C_p = -0.7, -0.18$ from $h/2$ ft to $h = 216$ ft for use with q_h

The results of the external pressure coefficients, C_p, calculations are shown in Table 2.24.

5. Determine q_i for the walls and roof using Flowchart 1. According to section 27.3.1, $q_i = q_h = 35.4$ psf for windward walls, side walls, leeward walls and roofs of partially enclosed buildings. The results are shown in Table 2.25.

6. Determine internal pressure coefficients, (GC_{pi}), from Table 26.13-1. For a partially enclosed building, $(GC_{pi}) = +0.55, -0.55$, shown in Table 2.26.

7. Determine design wind pressures p_z and p_h by Equation 27.3-1.

Windward walls:

$$p_z = q_z GC_p - q_h(GC_{pi})$$
$$= (q_z \times 1.083 \times 0.8) - 35.4(\pm 0.55)$$
$$= (0.866 q_z \pm 19.47) \text{ psf (external} \pm \text{internal pressure)}$$

Leeward wall, side walls, and roof:

$$p_h = q_h GC_p - q_h(GC_{pi})$$
$$= (35.4 \times 1.083 \times C_p) - 35.4(\pm 0.55)$$
$$= (38.34 C_p \pm 19.47) \text{ psf (external} \pm \text{internal pressure)}$$

A summary of the maximum design wind pressures in the E-W and N-S directions is given in Tables 2.27 and 2.28, respectively.

Table 2.21   Building Dimensions, Fundamental Natural Frequency, and Enclosure Classification

Length N-S = | 125.0 | ft

Length E-W = | 145.0 | ft

Number of levels below the eave = | 18 | Provide at least one level below the eave (max. 29)

Level	Height above ground, z (ft)	
Parapet	221.0	Top of parapet, h_p (enter "-" if there is no parapet)
Ridge	216.0	Top of ridge, h_{ridge} (enter mean roof height where $\theta = 0$ deg)
h	216.0	Mean roof height, h
Eave	216.0	Eave height, h_{eave} (enter mean roof height where $\theta = 0$ deg)
18	216.0	
17	204.0	Leave the cells empty for z where there is no level
16	192.0	
15	180.0	
14	168.0	
13	156.0	
12	144.0	
11	132.0	
10	120.0	
9	108.0	
8	96.0	
7	84.0	
6	72.0	
5	60.0	
4	48.0	
3	36.0	
2	24.0	
1	12.0	

Roof angle, $\theta =$ | 0.00 | deg

Mean roof height, $h =$ | 216.0 | ft

Basic wind speed, $V =$ | 115 | mi/h

(*continued*)

Table 2.21   Building Dimensions, Fundamental Natural Frequency, and Enclosure Classification (*Continued*)

Exposure category: | B |

Risk category: | II |

Fundamental natural frequency,
n_1 (Hz) = | 0.30 | N-S

| 0.30 | E-W

Enclosure classification: | Partially enclosed |

Table 2.22   Velocity Pressure Exposure Coefficient, K_z and Wind Velocity Pressures q_z Calculations

Exposure category: | B |

$\alpha =$ | 7 | ASCE/SEI Table 26.11-1

$z_g =$ | 1,200 | ft

Velocity pressure exposure coefficient, K_z:

Level	z (ft)	K_z	ASCE/SEI Table 26.10-1
Parapet	221.0	1.24	
Ridge	216.0	1.23	
h	216.0	1.23	
Eave	216.0	1.23	
18	216.0	1.23	
17	204.0	1.21	
16	192.0	1.19	
15	180.0	1.17	
14	168.0	1.15	
13	156.0	1.12	
12	144.0	1.10	
11	132.0	1.07	
10	120.0	1.04	
9	108.0	1.01	
8	96.0	0.98	

(*continued*)

Table 2.22 Velocity Pressure Exposure Coefficient, K_z and Wind Velocity Pressures q_z Calculations (*Continued*)

Level	z (ft)	K_z
7	84.0	0.94
6	72.0	0.90
5	60.0	0.85
4	48.0	0.80
3	36.0	0.74
2	24.0	0.66
1	12.0	0.57

Wind directionality factor, $K_d =$ | 0.85

Use ground elevation factor $K_e = 1.00$? | Yes

Ground elevation above sea level = | 0.0 | ft

Ground elevation factor, $K_e =$ | 1.00

Level	Wind velocity pressures, q_z	
	z (ft)	q_z (lb/ft^2)
Parapet	221.0	35.7
Ridge	216.0	35.4
h	216.0	35.4
Eave	216.0	35.4
18	216.0	35.4
17	204.0	34.9
16	192.0	34.3
15	180.0	33.6
14	168.0	33.0
13	156.0	32.3
12	144.0	31.6
11	132.0	30.8
10	120.0	30.0
9	108.0	29.1
8	96.0	28.1
7	84.0	27.1
6	72.0	25.9
5	60.0	24.6
4	48.0	23.1
3	36.0	21.2
2	24.0	18.9
1	12.0	16.5

Table 2.23   Gust-Effect Factors, G

Gust-Effect Factor for a Flexible Building, G_f ($n_1 < 1$ Hz)		

Damping ratio, $\beta =$ [0.010]

$\bar{b} =$ [0.45]

$\bar{\alpha} =$ [0.25]

$\bar{V}_{\bar{z}} =$ [106.8]   ft/s

	N-S wind	E-W wind
$g_r =$	3.89	3.89
$N_1 =$	1.42	1.42
$R_n =$	0.11	0.11
$\eta_h =$	2.79	2.79
$R_h =$	0.29	0.29
$\eta_B =$	1.87	1.61
$R_B =$	0.39	0.44
$\eta_L =$	5.40	6.27
$R_L =$	0.17	0.15
$R =$	0.88	0.91
$G_f =$	**1.08**	**1.10**

Building =   [**Flexible**]

Table 2.24 External Pressure Coefficients, C_p, Calculations

			L/B	C_p
	Parapet	Windward		1.50
		Leeward		1.00
	Windward wall	N-S	All	0.80
		E-W	All	0.80
	Leeward wall	N-S	0.86	−0.50
		E-W	1.16	−0.47
	Side walls	N-S	All	−0.70
		E-W	All	-0.70
Roof		Normal to ridge for $\theta \geq 10$ deg (N-S)	Windward	0.00
				0.00
			Leeward	0.00
		Normal to ridge for $\theta < 10$ deg (N-S)	0 to $h/2$	−1.04
				−0.18
			$h/2$ to h	−0.70
				−0.18
			h to $2h$	0.00
				0.00
			$> 2h$	0.00
				0.00
		Parallel to ridge for all θ (E-W)	0 to $h/2$	−1.04
				−0.18
			$h/2$ to h	−0.70
				−0.18
			h to $2h$	0.00
				0.00
			$> 2h$	0.00
				0.00

The top title row of the table reads: **External Pressure Coefficients, C_p (ASCE/SEI Fig. 27.3-1)**

Table 2.25 Velocity Pressure for Internal Pressure Distribution, q_i

Velocity Pressure for Internal Pressure Distribution, q_i (ASCE/SEI 27.3-1)

$$q_i = \boxed{\quad 35.4 \quad} \text{ lb/ft}^2$$

Table 2.26   Internal Pressure Coefficients, (GC_{pi})

Internal Pressure Coefficients, (GC_{pi}) [ASCE/SEI Table 26.13-1]		
$(GC_{pi}) =$	0.55	Partially enclosed
	−0.55	

Table 2.27   Design Wind Pressures, p (E-W wind)

					Net pressure, p (lb/ft²)		
Building surface	Height above ground z (ft)	Velocity pressure q (lb/ft²)	External pressure qGC_p (lb/ft²)	Internal pressure $q_h(GC_{pi})$ (lb/ft²)	$(-GC_{pi})$	$(-GC_{pi})$	Total horizontal pressure, p (lb/ft²)
Windward wall	216.0	35.4	31.2	19.5	11.7	50.7	49.5
	—	—	—	—	—	—	—
	—	—	—	—	—	—	—
	216.0	35.4	31.2	19.5	11.7	50.7	49.5
	204.0	34.9	30.7	19.5	11.2	50.2	49.0
	192.0	34.3	30.2	19.5	10.7	49.7	48.4
	180.0	33.6	29.6	19.5	10.1	49.1	47.9
	168.0	33.0	29.0	19.5	9.6	48.5	47.3
	156.0	32.3	28.4	19.5	9.0	47.9	46.7
	144.0	31.6	27.8	19.5	8.3	47.3	46.1
	132.0	30.8	27.1	19.5	7.6	46.6	45.4
	120.0	30.0	26.4	19.5	6.9	45.9	44.6
	108.0	29.1	25.6	19.5	6.1	45.1	43.9
	96.0	28.1	24.8	19.5	5.3	44.2	43.0
	84.0	27.1	23.8	19.5	4.3	43.3	42.1
	72.0	25.9	22.8	19.5	3.3	42.3	41.1
	60.0	24.6	21.6	19.5	2.2	41.1	39.9
	48.0	23.1	20.3	19.5	0.8	39.8	38.6
	36.0	21.2	18.7	19.5	−0.8	38.2	37.0
	24.0	18.9	16.7	19.5	−2.8	36.2	34.9
	12.0	16.5	14.6	19.5	−4.9	34.1	32.8
Leeward wall	All	35.4	−18.3	19.5	−37.8	1.2	
Side walls	All	35.4	−27.3	19.5	−46.8	−7.8	

Roof, parallel to ridge for all θ

0	to	$h/2$	—	35.4	−40.6	19.5	−60.1	−21.1
0.0	to	108.0			−7.0	19.5	−26.5	12.5
$h/2$	to	h	—	35.4	−27.3	19.5	−46.8	−7.8
108.0	to	145.0			−7.0	19.5	−26.5	12.5

(continued)

Table 2.27 Design Wind Pressures, *p* (E-W wind) (*Continued*)

			Height above ground z (ft)	Velocity pressure q (lb/ft^2)	External pressure qGC_p (lb/ft^2)	Internal pressure $q_h(GC_{pi})$ (lb/ft^2)	Net pressure, p (lb/ft^2)		Total horizontal pressure, p (lb/ft^2)
							$(-GC_{pi})$	$(-GC_{pi})$	
h	to	$2h$	—	35.4	0.0	—	—	—	
—	to	—			0.0	—	—	—	
	$>2h$		—	35.4	0.0	—	—	—	
—	to	—			0.0	—	—	—	

Table 2.28 Design Wind Pressures, *p* (N-S Wind)

Design Wind Pressures, *p* (N-S wind) [ASCE/SEI Eq. (27.3-1)]

Building surface	Height above ground z (ft)	Velocity pressure q (lb/ft^2)	External pressure q GC_p (lb/ft^2)	Internal pressure q_h (GC_{pi}) (lb/ft^2)	Net pressure, p (lb/ft^2)		Total horizontal pressure, p (lb/ft^2)
					$(+GC_{pi})$	$(-GC_{pi})$	
Windward wall	216.0	35.4	30.7	19.5	11.2	50.2	49.8
	216.0	35.4	30.7	19.5	11.2	50.2	49.8
	204.0	34.9	30.2	19.5	10.7	49.7	49.3
	192.0	34.3	29.7	19.5	10.2	49.1	48.8
	180.0	33.6	29.1	19.5	9.6	48.6	48.3
	168.0	33.0	28.5	19.5	9.1	48.0	47.7
	156.0	32.3	27.9	19.5	8.5	47.4	47.1
	144.0	31.6	27.3	19.5	7.8	46.8	46.5
	132.0	30.8	26.6	19.5	7.2	46.1	45.8
	120.0	30.0	25.9	19.5	6.4	45.4	45.1
	108.0	29.1	25.2	19.5	5.7	44.7	44.3
	96.0	28.1	24.3	19.5	4.8	43.8	43.5
	84.0	27.1	23.4	19.5	3.9	42.9	42.6
	72.0	25.9	22.4	19.5	2.9	41.9	41.6
	60.0	24.6	21.3	19.5	1.8	40.8	40.4
	48.0	23.1	20.0	19.5	0.5	39.4	39.1
	36.0	21.2	18.4	19.5	−1.1	37.9	37.6
	24.0	18.9	16.4	19.5	−3.1	35.9	35.5
	12.0	16.5	14.3	19.5	−5.2	33.8	33.5
Leeward wall	All	35.4	−19.2	19.5	−38.7	0.3	
Side walls	All	35.4	−26.8	19.5	−46.3	−7.3	
Roof, normal to ridge with $\theta \geq 10$ deg							
Windward	—	—	—	—	—	—	
			—	—	—	—	
Leeward	—	—	—	—	—	—	

(*continued*)

Table 2.28   Design Wind Pressures, *p* (N-S Wind) (*Continued*)

		Height above ground z (ft)	Velocity pressure q (lb/ft²)	External pressure q GC_p (lb/ft²)	Internal pressure q_h (GC_{pi}) (lb/ft²)	Net pressure, p (lb/ft²)		Total horizontal pressure, p (lb/ft²)
	Building surface					$(+GC_{pi})$	$(-GC_{pi})$	
	Roof, normal to ridge with $\theta < 10$ deg

0	to	$h/2$	—	35.4	−39.9	19.5	−59.4	−20.4	
0.0	to	108.0			−6.9	19.5	−26.4	12.6	
$h/2$	to	h	—	35.4	−26.8	19.5	−46.3	−7.3	
108.0	to	125.0			−6.9	19.5	−26.4	12.6	
h	to	$2h$	—	35.4	0.0	—	—	—	
—	to	—			0.0	—	—	—	
	>$2h$		—	35.4	0.0	—	—	—	
—	to	—			0.0	—	—	—	

Table 2.29   Design Wind Pressures on the Parapet, p_p

Design Wind Pressures on the Parapet, p_p [ASCE/SEI Equation (27.3-3)]

Height above ground	Velocity pressure	Pressure p_p (lb/ft²)	
z (ft)	q (lb/ft²)	Windward	Leeward
>221.0	35.7	53.5	−35.7

2.5.2.5   WIND TUNNEL PROCEDURE PER ASCE 7-16

Overview   The wind tunnel procedure in Chapter 31 can be utilized to determine wind loads on MWFRSs and C&C of any building or other structure in lieu of any of the procedures in the Chapters 27 through 30, and must be used where the conditions of these procedures are not satisfied (in particular, where a structure contains any of the characteristics defined in sections 27.1.3, 28.1.3, 29.1.3, or 30.1.3).

Wind tunnel tests should be seriously considered where buildings or other structures are not regularly shaped, are flexible and/or slender, have the potential to be buffeted by upwind buildings or other structures, or have the potential to be subjected to accelerated wind flow from channeling by buildings or topographic features. In the case of tall, slender buildings, only a wind tunnel test can properly capture any possible effects due to vortex shedding, galloping, or flutter. For buildings in the heart of a city, a wind tunnel test is mandatory since Exposures B through D cannot properly capture the conditions in such cases. Every project has its own unique characteristics and engineering judgment also plays a role in the decision-making process. When determining whether a wind tunnel test is required or not, it is always very important to keep in mind the limitations in Chapters 27 through 30, especially the general one related to along-wind response.

Of all of the methods that are contained in ASCE/SEI 7-16, the wind tunnel procedure is generally considered to produce the most accurate results. For certain types of buildings, the results from a wind tunnel test will be significantly smaller than those from any of the other methods. On the other hand, wind tunnel tests will yield results that are greater than those obtained from the other methods under certain conditions; as such, it is important to understand when such tests are required in order to adequately design the building or other structure for the effects of wind.

Information on the three basic types of wind tunnel test models that are commonly used is given in C31. Wind tunnel tests can also provide valuable information on snow loads, the effects of wind on pedestrians, and concentrations of air-pollutant emissions, to name a few. References are provided in C31 that provide detailed information and guidance for the determination of wind loads and other types of design data by wind tunnel tests.

Test Conditions   Basic requirements for test conditions of wind tunnel tests or any other tests that employ a fluid other than air are given in section 31.2. These seven conditions must be followed when any such test is conducted. Additional information on the basic procedures of conducting a wind tunnel test can be found in the references in C31.

Dynamic Response   The test conditions of section 31.2 are to be used when determining the dynamic response of a building or other structure. Mass distribution, stiffness, and damping must be properly accounted for in the model and in the subsequent analysis.

Load Effects   ASCE/SEI 31.4.2 prescribes limitations on the wind speed used in the tests, and section 31.4.3 gives lower limits on the magnitude of the principal loads that are to be applied to a building or structure for both the MWFRS and C&C. Two conditions are given that permit the limiting values to be reduced. C31 provides a comprehensive discussion on the reasons behind these limitations. See section 2.9 for wind tunnel testing and analysis.

2.6 PEDESTRIAN WIND STUDIES

A sheet of air moving over the earth's surface is reluctant to rise when it meets an obstacle such as a tall building. If the topography is suitable, it will prefer to flow around the building rather than over it. There are good physical reasons for this tendency, the predominant being that the wind will find the path of least resistance, that is, a path that requires minimum expenditure of energy. As a rule, it requires less energy for the wind to flow around the obstacle at the same level than for it to rise, which requires more potential energy. Also, if the wind has to go up or down, additional energy has to be expended in having to compress the column of air above or below it. Generally, wind will try to seek a gap at the same level. However, during high winds when the air stream is blocked by the broad side of a tall, flat building, its tendency is to drift in

a vertical direction rather than to go around the building at the same level; the circuitous path around the building would require expenditure of more energy. Thus the wind is driven in two directions. Some of the wind will be deflected upward, but most of it will spiral to the ground creating a so-called standing vortex or a minitornado at the sidewalk level.

Thus tall buildings and their smooth walls are not the only victims of wind buffeting. Pedestrians who walk past tall, smooth-skinned skyscrapers may be subjected to what someone has called the "Mary Poppins syndrome," referring to the tendency of the wind to lift the pedestrian literally off his or her feet. Another effect of this phenomenon has frequently been observed and is known as the "Marilyn Monroe effect," referring to the billowing action of ladies' skirts in the turbulence of wind around and in the vicinity of the building. Whatever the popular name may be, the point is that during windy days even a simple task such as crossing the plaza or taking an afternoon stroll becomes an extremely unpleasant experience to pedestrians, especially during winter months around buildings in the cold climates. Walking may become irregular, and the only way to keep walking in the direction of wind would be to bend the upper body windward.

In skyscrapers, the downward wind along the windward faces can cause strong ground-level winds, creating havoc for pedestrians and shops. The effect is often amplified when there are other tall buildings nearby lined up to form what is usually termed a street canyon.

A dramatic example of wind modification by buildings is arcades under tall buildings. Because of the high stagnation pressure on the upwind side of the building and the large suction on the downward side, a strong draft is generated through the arcade. Tall buildings supported by stilts have the same problem as arcades. Other examples are shown in Fig. 2.24.

For a successful design of a building at the pedestrian level, studies are therefore necessary to assist the building planners in overcoming the uncomfortable or dangerous wind conditions which often occur at the base of the buildings. Planners, designers, and developers are becoming increasingly aware of the potential for pedestrian-level problems and generally acknowledge that design assistance is required to predict the microclimate that will be adversely affected by a proposed design.

Although one can get some idea of wind flow patterns from the above examples, analytically it is impossible to estimate pedestrian-level wind conditions in outdoor areas of buildings and building complexes. There are innumerable variations to location, orientation, shapes, and topography, making it impossible to even attempt

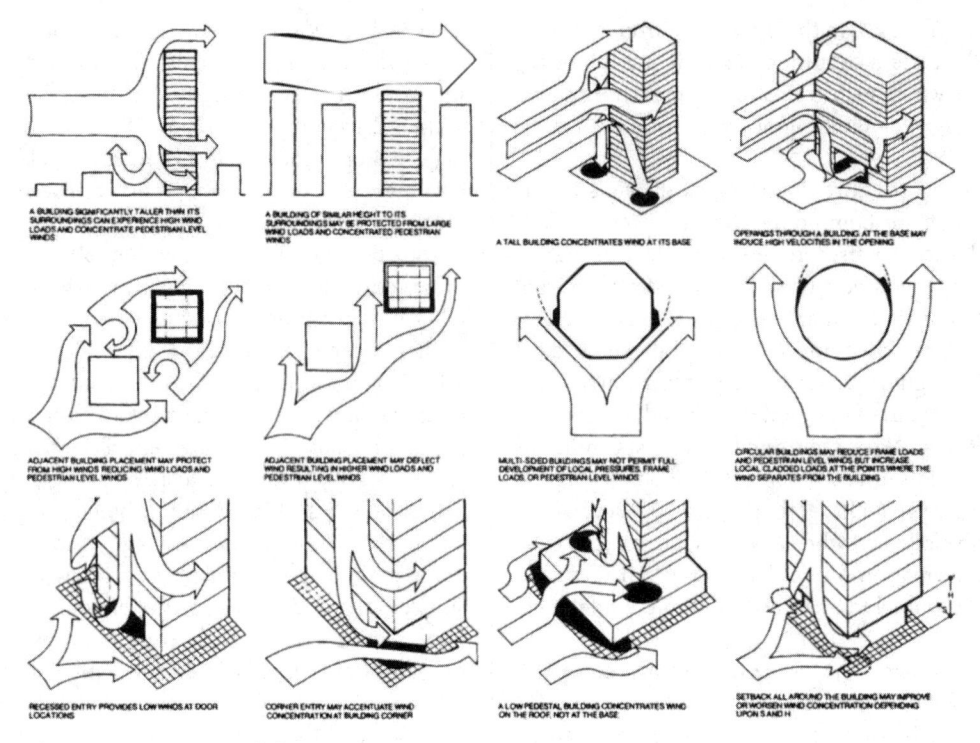

Figure 2.24   Near wind climate.

formulating an analytical solution to the problem. Based on actual field experience and results of wind tunnel studies, it is, however, possible to qualitatively recognize situations that adversely affect the pedestrian comfort within a building complex.

Model studies can provide reliable estimates of pedestrian-level wind conditions based on considerations of both safety and comfort. These studies require geometrically scaled models, including aerodynamically significant details in areas of interest. Simulation of characteristics of natural wind, including the simulation of the salient properties of the approach wind and the influence of nearby building structures and significant topographic features, is required for an accurate estimate of pedestrian wind conditions. Model studies are used to evaluate a building design and offer remedial suggestions where unacceptable pedestrian-level wind speeds are created. A model is constructed and pedestrian-level wind speed measurements are made at various locations for all the prevailing directions. The effects of adding a proposed building to an existing cluster of buildings are obtained by comparing measured wind speeds with and without the proposed building to a set of standard acceptance criteria. The acceptance criteria state how often a wind speed occurrence is permitted to occur for various levels

of activity. This is done for both the summer and winter seasons with acceptance criteria being more severe during the winter months. For example, an occurrence once a week of a mean speed of 15 mph (6.7 m/s) is considered acceptable for walking during the summertime, whereas only 10 mph (4.47 m/s) is considered acceptable during winter months.

2.7   FIELD MEASUREMENTS OF WIND LOADS

Gustave Eiffel, the noted engineer for the Eiffel Tower, was, perhaps, the earliest researcher to conduct studies and make measurements of the displacements of the top of his tower under the action of wind. His observations between 1893 and 1895 showed that under the effect of wind, the displacements at the top of the tower resembled an elliptical shape. The measured values, which ranged from 2.3 to 2.75 in (60 to 70 mm), were considerably less than those predicted by him. At about the same period, a series of tests were conducted on the 16-story Monadnock Building in Chicago to measure the movements of structure. The building's vibration characteristics were checked by measuring the oscillations of plumb bobs suspended in a stair shaft from the sixteenth floor. These observations were also cross-checked by transits and showed that the building was very stiff with a measured deflection of about $1\frac{1}{2}$ in. (12.7 mm). The next set of activities was spurred during the boom years of high-rise construction in the 1940s. New York's Empire State Building was the first to be instrumented for wind load measurements. Rathbun (1940) in his ASCE paper which describes the observations on the Empire State Building, compared the building oscillations to the tines of a tuning fork. His observation showed that the building's sway motion occurs primarily in the fundamental frequency of the building. In the next two decades no similar studies involving the measurement of the wind response of tall buildings were conducted. Since 1960, however, the flurry of high-rise building activity rekindled the interest in full-scale studies in a number of countries, the most important work coming from Canada, England, Australia, Hong Kong, and Holland. In Canada, Dalghish studied three buildings in the range of 34 to 58 stories. His work encompasses multiple aspects of wind engineering that include studies of cladding pressures, meteorological data for predicting overall structural loads, and measurements of mean and fluctuating pressures.

In England, Lee has carried out full-scale studies of the structural behavior of a 157 ft (48 m) tall concrete building by subjecting the building to forced vibrations. In comparison to the number of tall buildings completed in the United States, the number of buildings instrumented to validate wind tunnel tests has been very small. Some buildings have been instrumented, but because of the nature of the tests, it will be some time before enough data is gathered for results to be published.

Many difficulties arise in the field measurement of wind loads. Stack effects can create pressure differences as large as a moderate wind. Engineers and researchers alike have been interested in comparing the results of model testing with those of the prototype in order to determine what improvements are needed in testing, and to get a better picture of the actual physical phenomenon. Full-scale measurements boost confidence in model scale tests and improve our knowledge of the effects that we do not fully understand as yet such as Reynolds number and turbulence effects. Certain scaling inequalities inherent in the wind tunnel tests appear to promote the lingering doubt about the validity of wind tunnel tests.

The field-measured mean and fluctuating loads as compared to those predicted by the wind tunnel studies appear to show a moderate amount of agreement, but more often these studies expose the shortcomings of field studies. It is agreed among researchers that there simply are not enough field data on full-scale buildings to validate wind tunnel techniques. Therefore, it is unreasonable to expect the wind tunnel results to be any more precise than any of the other parameters encountered in structural engineering practice, such as the calculation of natural frequencies and damping of tall buildings. The level of structural damping is at best an educated guess. Thus, there are many uncertainties in the wind design of tall buildings, signifying once again that structural engineering is as much an art as it is a science.

Amidst this rather pessimistic observation, structural engineer Halvorson and wind engineer Isyumov have brought in evidence which appears to confirm the validity of wind tunnel tests. Their work consisted of a field-monitoring program to measure the dynamic response characteristics of Allied Bank Plaza Tower, a 71-story steel building in downtown Houston. The instrumentation consisted of two accelerometers located at the 71st floor of the tower. Wind measurements were made under two significant wind events: (i) an extratropical windstorm on April 1, 1983, with gusts speeds of up to 56 mph (25 m/s) and (ii) Hurricane Alicia on August 18, 1983, with fastest-mile wind speeds approaching the code-specified 50-year recurrence wind speed of 90 mph (40 m/s). Their observations showed that the sway response of the tower varied in a sinusoidal manner corresponding to the fundamental modes of vibration of the building. Full-scale measurements of accelerations showed good agreement with those predicted from the wind tunnel data. Based on the comparison between field measurements and wind tunnel results, the following conclusions are made by Halvorson and Isyumov.

1. The magnitude of structural loads experienced during Hurricane Alicia (which was determined as a 50-year recurrence event) are in good agreement with the loads predicted by the wind tunnel tests. The Houston code wind loads, on the other hand, overestimated the wind loads by approximately 100 percent.

2. The mean wind loads are only about 20 to 30 percent of the total structural loads. Dynamic effects brought about by wind gusts are 3 to 5 times as significant as the mean loads.

3. The lateral drift, which is a combination of mean and dynamic peak displacements, was evaluated from wind tunnel measurements and field records, respectively. Calculated drift indices in two directions were $2\frac{1}{2}$ and 2 ft (0.76 and 0.61 m) and compared well with the measured values of 2.58 and 1.25 ft (0.78 and 0.38 m).

4. Davenport's criterion of limiting the acceleration for a 10-year windstorm to approximately 20 mg appears to serve well the serviceability requirements of a tall building in regard to motion perception by building occupants.

5. Estimates from the field records indicate that the building exhibits a damping characteristic equivalent to 1.4 to 1.6 percent of the critical damping, which is somewhat higher than rule-of-thumb value of 1 percent used for steel buildings. The additional contribution appears to come from soil structure interaction provided by the foundation system, which consists of a $9\frac{1}{2}$ ft (289 m) thick mat founded on overconsolidated clay at a depth of 55 ft (16.77 m) below grade.

2.8   MOTION PERCEPTION: HUMAN RESPONSE TO BUILDING MOTIONS

Every building or other structure must satisfy a "strength" criteria, in which each member is sized to carry the design loads without buckling, yielding, fracturing, etc. It should also satisfy the intended function (serviceability) without excessive deflection and vibration. While strength requirements are traditionally specified by the building codes, the serviceability limit states for the most part are not included within the building codes. The reasons for not codifying the serviceability requirements are several: failure to meet serviceability limits are noncatastrophic, are a matter of judgement as to their application, involve the perceptions and expectations of the user or owner, and because the benefits themselves are often subjective and difficult to quantify. The fact that serviceability limits are usually not codified should not diminish their importance. A building which is correctly designed for code standards may nonetheless be too flexible for its occupants due to lack of deflection criteria. Excessive building drifts can cause safety-related frame stability problems because of excessive P-Δ effects. It can also cause portions of building cladding to fall, potentially injuring pedestrians below.

Perception of building motion under the action of wind is a serviceability issue. In locations where buildings are close together, the relative motion of an adjacent building may make the occupants more sensitive to otherwise imperceptible motion. Human response to building motions is a complex phenomenon involving many physiological and psychological factors. Some people are more sensitive than others to perceived building movements.

Although researchers have attempted to study this problem from motion simulators, there is no firm consensus on human comfort standards. Although building motion can be described by various physical quantities including maximum values of velocity, acceleration, and rate of change of acceleration, sometimes called jerk, it is generally agreed that acceleration, especially when associated with torsional rotations, is the best standard for evaluation of motion perception in tall buildings. A commonly used criterion is to limit the acceleration of upper floors to 2 percent of gravity (20 mg) for a 10-year return period.

2.9   WIND TUNNEL TESTING AND ANALYSIS

*****Authored by Dr. Denoon of CPP Wind Tunnel

2.9.1   Designing for Wind Effects

The various stages in designing buildings for wind effects can be described by the Davenport wind loading chain (Davenport, 1972) shown in Fig. 2.25. This describes the interaction, and interdependence, of the local wind climate and the individual building characteristics in the determination of wind loads and responses.

Each of the first four links of the chain can be estimated from codified approaches or more detailed site-specific analysis and/or boundary layer wind tunnel testing.

2.9.2   Wind Loading Mechanisms for Tall Buildings

There are several components to wind loading of tall buildings. Buildings may respond in a combination of along-wind, cross-wind, and torsional directions with these loads consisting of mean and fluctuating components.

Along-wind loads are caused by aerodynamic drag and buffeting from turbulence in the approach flow.

Cross-wind loads are caused by buffeting by lateral components of turbulence in the approach flow and also vortex-shedding whereby vortices shed alternately from the sides of the building excite the building orthogonal to the direction of the oncoming wind. The frequency at

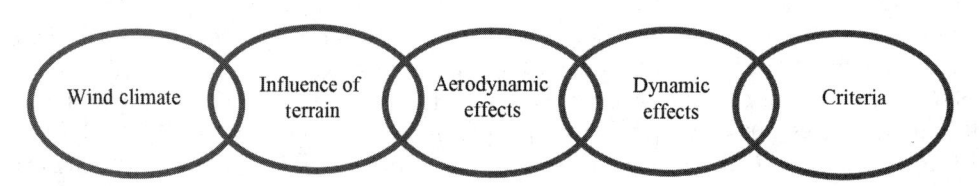

Figure 2.25   Davenport wind loading chain.

which vortices are shed from a tall building are a function of its shape, its width in the cross-wind direction, and the wind speed. When these combine to shed vortices at a frequency close to the natural frequency of vibration of the building then a phenomenon known as negative aerodynamic damping can occur. Effectively what happens is that a small excitation force can cause quite large resonant responses, and then the motion of the building increases the strength of the vortex shedding excitation.

Torsional responses can be caused by asymmetries in the aerodynamic shape of the building, the oncoming flow (such as partial shielding by an upwind building) or non-symmetric structural systems. Each of these components can be critical in tall building design.

Mean loads are due to the mean (or average) wind speeds but these are augmented by fluctuating loads caused by turbulence in the approach flow or generated by the building shape. The fluctuating loads comprise background and resonant components. The background component is due to the building reacting to the gust components of the approach wind, while the resonant component is due to the building vibrating at its own natural frequencies due to wind-induced excitation. Of these components, it is the resonant component of response that is affected by the structural dynamic properties of the building and hence subject to modification through structural alterations or the introduction of supplementary damping.

Codified approaches, such as ASCE 7-22 (ASCE, 2021), tend to concentrate on the estimation of along-wind loads with simpler estimates on the torsional components of response. ASCE 7 does not include any methods for even crudely estimating cross-wind responses, although a number of international codes and standards do. These estimates are, given the sensitivity of vortex-shedding to building shape, necessarily very approximate. They are intended to cover the majority of cases in building design with a degree of conservatism, but there are many situations where this will not be the case.

2.9.3   When to Use the Wind Tunnel

The appropriate times to use the wind tunnel are when the building being tested falls outside the scope of codes or standards, when there will be economies due to code-based conservatisms, more refined predictions of wind loads or accelerations are required, or when additional confidence in reliability is required. In all cases, a properly conducted wind tunnel test provides wind loading data that is specific to the building being designed for both its unique architecture and surroundings.

Examples of situations where added reliability or refined predictions include buildings with post-disaster functionality requirements such as hospital buildings or, buildings with mission critical contents such as data centers, or buildings with high occupant expectation such as top end residential buildings.

Cases where significant economies can be achieved include buildings situated within dense urban environments providing significant shielding, those that have favorable orientations relative to extreme wind directions, and those with very aerodynamic forms.

Buildings that fall outside the scope of codified design include those subject to significant cross-wind excitation, those that are subject to other strong resonant dynamic response, those with complex architecture, and those that may be buffeted or subjected to accelerated wind flows by surrounding buildings. Susceptibility to these effects is a function of the building architecture, location, and structural system. In simple terms, approximate guidelines for when wind tunnel testing may be advisable are when any of the four following conditions applies (Irwin, Denoon & Scott, 2013):

1. Height of the buildings is greater than 400 ft (120 m);
2. The height of the building is more than four times its average width (normal to the wind direction) over the top half of the building;
3. The lowest natural frequency of the buildings is below 0.25 Hz (period longer than 4 s); or
4. The reduced velocity U/fb at ultimate design wind speeds is greater than 5 where U is the mean hourly wind velocity at the top of the building, f is the lowest natural frequency, and b is the average width defined above.

2.9.4   Wind Climate

Wind climate analysis is the basis of all wind engineering design. Loads for a static structure vary proportionally to the square of the wind speed, and for dynamic structures undergoing cross-wind excitation this exponent can be as large as four. Consequently, an accurate understanding of the site wind climate is critical in the accurate determination of wind loads and response.

2.9.4.1   Storm Types

Depending on the project location, there are several types of wind storms that may affect the assessment of strength and serviceability design load effects. Storm types include synoptic gales, tropical cyclones (which also describes hurricanes and typhoons), thunderstorms, and tornadoes. Most wind engineering design is based on boundary-layer theory developed from synoptic storms. This provides the classic increase of wind speed with height above the ground with increasing turbulence associated with greater ground surface roughness (see Fig. 2.26).

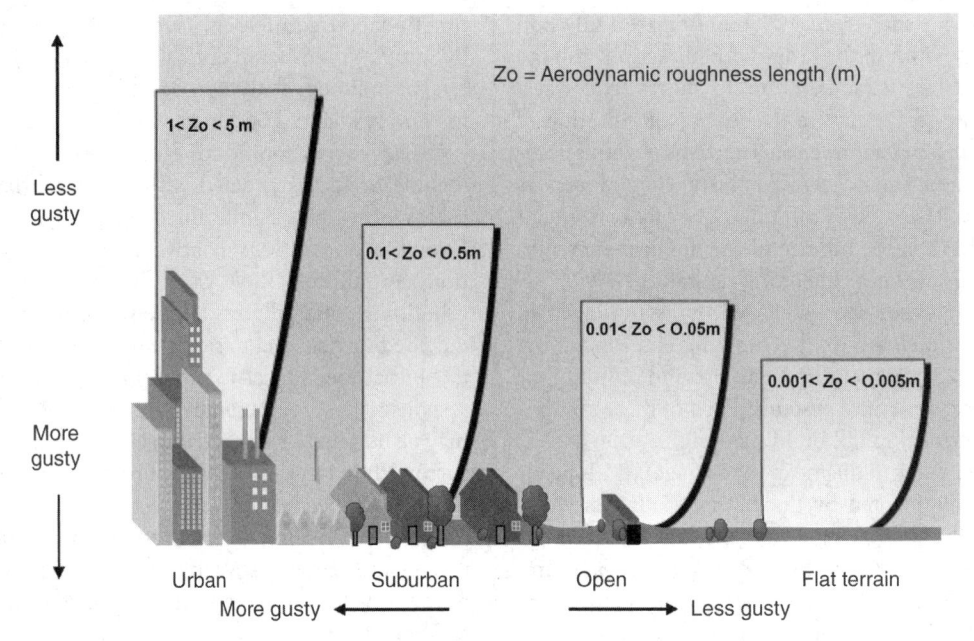

Figure 2.26 Classical boundary layer profiles.

Wind speed profiles within tropical cyclones also, largely, follow basic boundary layer characteristics. For the design all but the world's very tallest buildings, the boundary layer depth can be assumed to be greater than the building height, or at the very least retain relatively uniform characteristics over the upper portions of the building.

Traditionally it has been though that tornado risk at any location is significantly lower than for more traditional "straight-line" winds. Recent research has suggested that, at least in large parts of the inland US, this may not be the case for higher risk category buildings with large footprints such as hospitals or schools. ASCE 7-22 is the first standard to specifically address tornado loading in a manner related to, but independent from, wind loading requirements. Specific wind tunnel requirements have been drafted that use the traditional boundary layer wind tunnel testing to approximate the effects of tornadoes by modifying testing and analysis requirements. At the time of writing, it is not clear how many tall buildings these provisions will affect, or how ongoing research will affect the requirements in future.

Thunderstorms govern extreme wind speeds in many parts of the world, and they have fundamentally different characteristics. Peak wind speeds in a thunderstorm may peak in the range of 50 to 150 m above the ground, after which they can decrease again. They also have very different temporal characteristic with wind speeds increasing to a peak over a short period, and these peak wind speeds being maintained for only a minute or two, after which

they can rapidly decrease again. For buildings with long periods of vibration, this non-stationary storm behavior can limit the potential for the magnitude of resonant response to develop that would be expected during typical long duration storms, and the potentially lower wind speeds over the upper parts of tall buildings may also limit the base moments. For these reasons, the use of boundary-layer wind tunnel testing for tall buildings in regions where thunderstorms dominate will provide an extra level of reliability while research into physical modeling of non-stationary winds continues. In the meantime, the exclusion of thunderstorm events in the analysis of load and response effects, such as serviceability accelerations for occupant comfort, can sometimes be considered.

2.9.4.2 DETERMINATION OF DESIGN WIND SPEEDS

Site-specific wind speeds are a function of both the wind climate and the effects of local terrain. Terrain includes the effects of both upwind surface roughness and significant topography.

The most basic approach to determining site-specific wind speeds is the use of codified basic wind speeds with adjustment for terrain. In the US, ASCE 7 provides design wind speeds for strength and serviceability design mean recurrence intervals throughout US states and territories. These are adjusted for topography using simplified techniques and ground roughness modifications are introduced through the use of exposure categories.

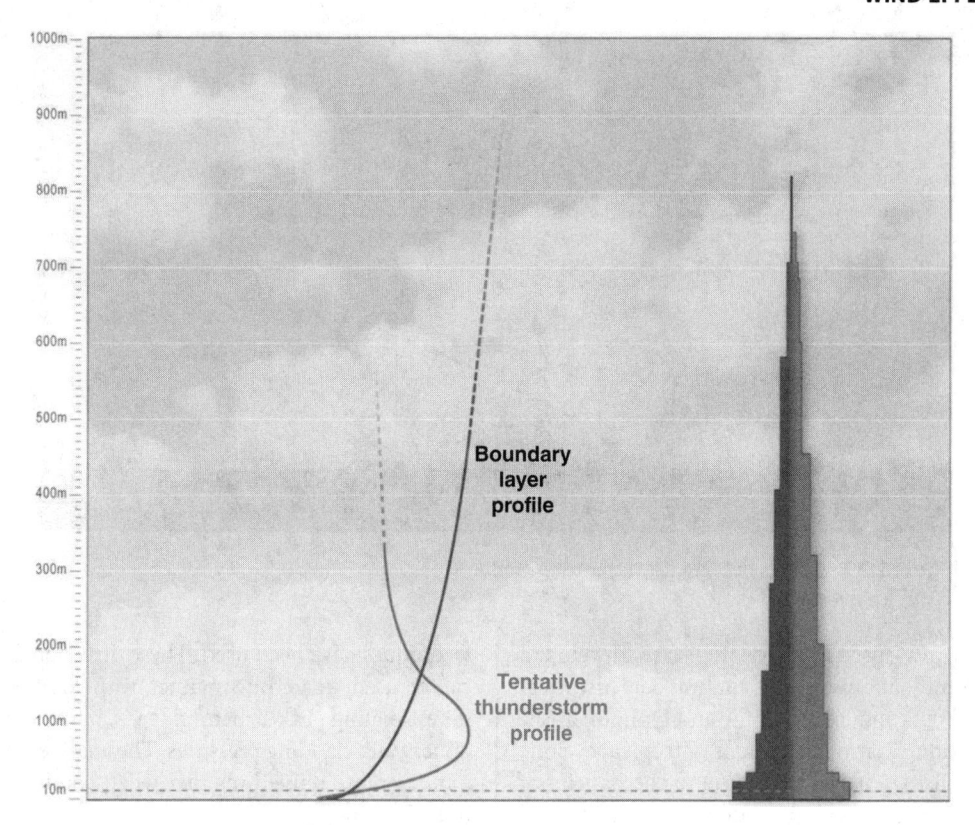

Figure 2.27 Thunderstorm profile.

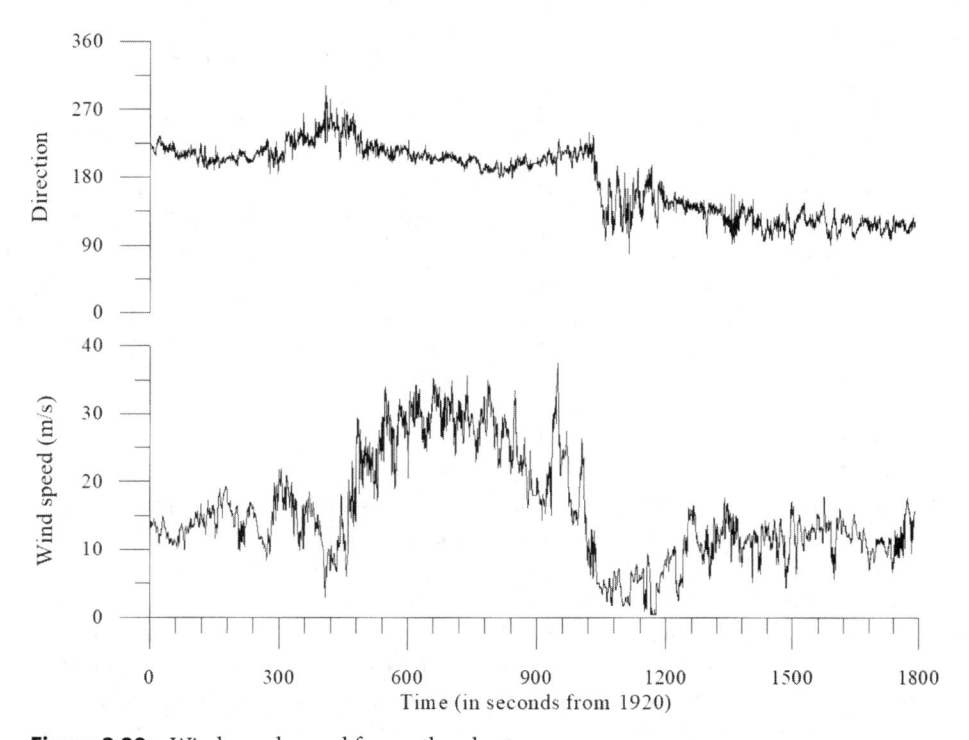

Figure 2.28 Wind speed record from a thunderstorm.

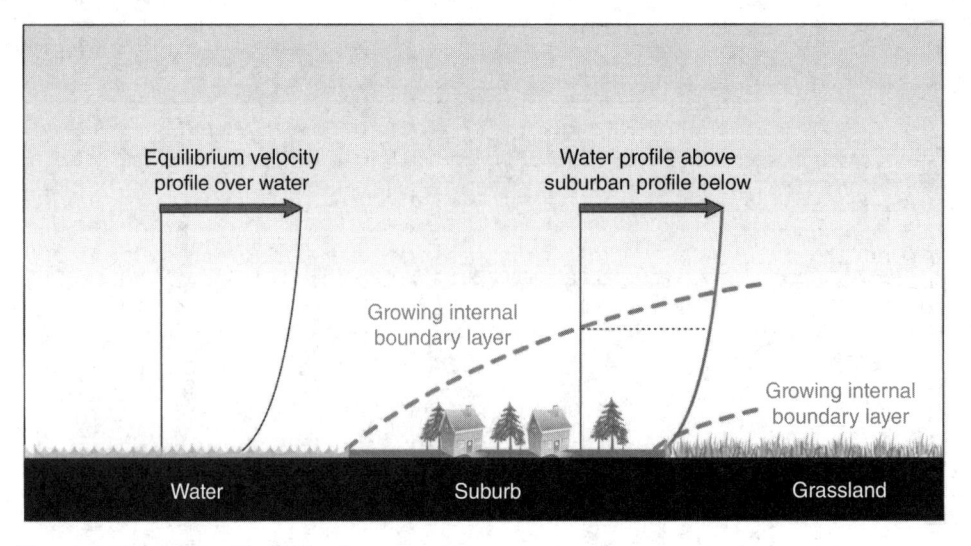

Figure 2.29 Effects of boundary layer transition.

The basic wind speeds in ASCE 7 are developed through a combination of the analysis of historical meteorological data and, in hurricane and typhoon areas, outputs from Monte Carlo simulations of tropical cyclone tracks and intensities are required due to the paucity of surface data. For historical measured data, these need to be corrected for the measurement devices used, their heights, and exposures to ensure that the predicted wind speeds have the correct gust duration at a standardized heigh (typically 10 m / 33 ft) in the equivalent of open country terrain. Each of these factors might change over the length of data record, requiring a range of correction factors. The data records also need to be filtered to remove erroneous data before extreme value analysis techniques are used to generate curves showing the relationship between wind speed and mean recurrence interval. In areas where multiple storm types contribute to the recurrence intervals of interest these analyses need to be conducted for each storm type individually and then the results combined to determine joint (combined) probabilities. When the in ASCE 7, as with many codes and standards, only one wind speed is provided for each MRI at any given location. This is an omni-directional wind speed. However, in most locations the wind climate will show marked directionality and directional wind speed analyses can be conducted by binning wind data into directional sectors and individual analyses conducted for each direction. This provides a much higher resolution of wind data for use in testing and analyses.

It is worth noting that ASCE 7 allows the use of site-specific wind climate analyses in place of the recommended values in the code as long as appropriate statistical techniques have been used. These directional wind speeds can be used, in a crude manner, with codified approaches for estimating peak structural loads, but not for local component and cladding pressures. The analyses can be used in a much more refined manner when combined with wind tunnel testing.

Once directional basic wind speeds have been calculated, they are adjusted to the site conditions taking into account the upwind terrain. There are accepted analytical techniques for this, which were also used in the generation of the exposure category profiles in ASCE 7. Directional site-specific analyses, however, take into account the effects of boundary layers that may not be fully developed due to multiple changes of upwind terrain.

Sometimes the effects of topography can be estimate analytically. In other cases, where the topography is more extreme then physical or numerical modeling may be required to determine the effects of speed-up or shielding from each wind direction as well as the effects on turbulence intensity.

2.9.5   The Boundary Layer Wind Tunnel

Boundary-layer wind tunnels come in a range of shapes and sizes, but what they all have is the ability simulate an atmospheric boundary layer to simulate the effects of far-field terrain. A typical wind tunnel with these capabilities is shown in Figs. 2.30 and 2.31. In this closed-return wind tunnel flow is expanded and straightened after the fan before being compressed again prior to entry to the working section of the wind tunnel. The boundary layer is kick—started by the combination of spires and trip

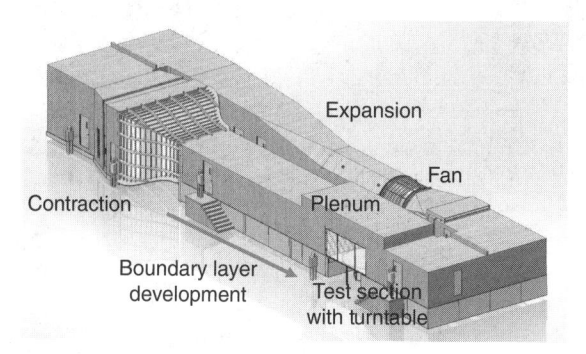

Figure 2.30 Boundary layer wind tunnel.

Figure 2.31 Interior of wind tunnel.

board and then allowed to grow naturally over the roughness blocks on the floor. By the start of the turntable there is a fully developed and stable boundary layer.

By altering the spire sizes, trip board height, and floor roughness different boundary layers can be simulated. In addition to modeling the correct variation of wind speed and turbulence intensity with height this arrangement allows the correct spectral distribution of turbulence (Fig. 2.32). This reflects the different scales of turbulence that exist in natural wind flow and without which wind tunnel modeling would not hold.

The turntable allows the test model to be rotated to simulate the effects of different wind directions, and has to be large enough to allow the inclusion of all the buildings that will have an individual impact on the project being tested. Overall, the test section has to have a large enough cross-sectional area to keep blockage by the models to an acceptable value (typically less than 8 percent of the cross-sectional area). Larger blockage ratios can be accepted if measures, such as a blockage tolerant section

in the wind tunnel show, have been incorporated. The effect of overly large blockage ratios is to create a "back-pressure" in the tunnel leading to slowing of the flow and an overestimate of loads and pressures. Conversely in wind tunnels with open test sections where the flow exits the tunnel onto an open test section without walls or ceiling there can be a negative blockage effect as the flow diverges around the test model leading to underestimates of loads and pressures.

For normal wind engineering applications, wind tunnels may require 15 to 20 m of floor roughness to generate stable boundary layers with cross-sectional widths of 2 to 5 m and heights of 2 to 3 m. The larger the cross-sectional area then the larger the models that can be used.

Typical wind tunnel scales that are used for buildings are in the range of 1:200 to 1:500. Most measurements are made in a non-dimensional coefficient format with pressures and/or applied loads non-dimensionalized by a reference velocity pressure measured in the approach flow in the wind tunnel. For sharp-edged (also known as bluff) bodies in turbulent flow these non-dimensional coefficients are the same at model scale as they are at full-scale as long as minimum Reynolds number requirements are met. In practice, this means that as long as the wind tunnel is run at a reasonable speed (in the order of 5 m/s or above) then the coefficients will be the same at any wind tunnel speed and can be applied to any design wind speed.

2.9.6 Wind Tunnel Test Techniques

Wind tunnel tests for wind-induced structural loads and responses fall into two categories: aerodynamic model tests and aeroelastic tests. Aerodynamic tests use a rigid model and responses are calculated through the analytical integration of dynamic properties whereas in an aeroelastic test model the dynamic properties are physically modeled.

2.9.6.1 Aeroelastic Model Testing

Aeroelastic models were the original approach to determining peak wind loads on tall buildings in wind tunnel testing. In an aeroelastic test model, dynamic properties including natural frequency, damping, and mass are modeled such that the test model will respond dynamically in the wind tunnel. This allows direct measurement of responses such as base moments and deflections in the time domain.

The benefit of aeroelastic testing is that the effects of aerodynamic damping are measured directly. Aerodynamic damping is normally slightly positive and additive to inherent structural damping. In most cases,

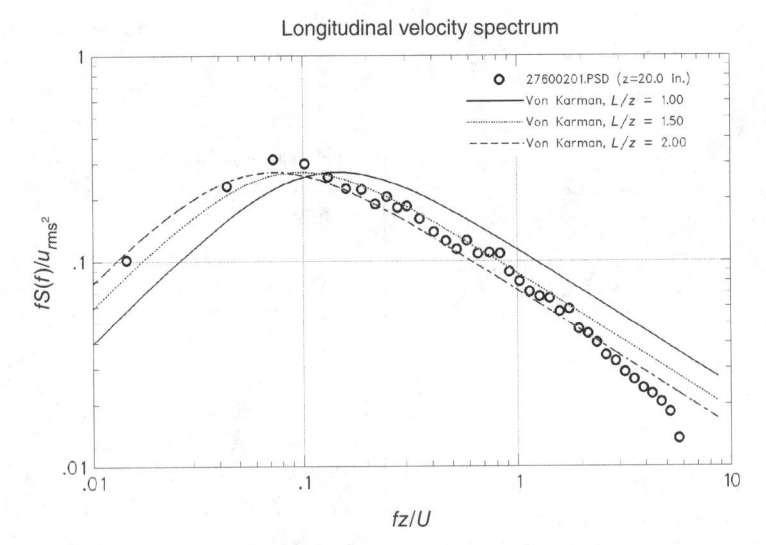

Figure 2.32   Longitudinal velocity spectrum.

as aerodynamic damping is small in comparison with structural damping, and potentially comparable to the uncertainties in the estimates of structural damping, it is ignored. However, this may not be the case where there are strong cross-wind (vortex shedding) responses. This can only be measured by aeroelastic testing.

There are several types of aeroelastic tests for buildings. The most appropriate technique depends on the size of the building and its structural characteristics. For buildings with symmetric structural systems where little torsional response is expected, and the frequencies are sufficiently high that significant responses are only expected in the primary translational directions then a simple two degree-of-freedom "stick" model may be appropriate. In a stick model, a rigid model of the tower is used with a sprung pivot at the base, or at an appropriate height to better approximate the primary mode shapes in a linear manner. Where torsional responses may be more important, but the mode shapes are largely uncoupled, then a torsional pivot can be added to this model type.

For taller building where higher modes may be significant or where mode shapes are highly coupled, then complex aeroelastic models may be required. These typically comprise a skeleton frame that mimics the structure of the building with a lightweight sectional shell.

Due to the scaling of the dynamic characteristics of the building in the test model, this means that wind speeds in the wind tunnel also need to be scaled to complete scaling requirements. For buildings subject to cross-wind excitation, this means testing at a range of wind speeds to ensure that peak responses are captured.

2.9.6.2   Aerodynamic Model Testing

More commonly used than aeroelastic testing is aerodynamic testing where applied wind forces are measured in the wind tunnel, and dynamic responses are calculated analytically, rather than being measured directly. The two most common approaches are high-frequency balance testing and high-frequency pressure integration. For both, testing at 36 wind directions at 10° intervals is industry best practice.

2.9.6.3   High-Frequency Balance

In the high-frequency balance (HFB) approach, a lightweight model is mounted on a very stiff balance, as shown in Fig. 2.33, that measures the applied moments and shears in translational and torsional directions. In this way, the model acts as a mechanical integrator. The model/balance combination should have natural frequencies of vibration that are higher than the scaled primary natural frequencies of vibration of the prototype building. This ensures that the measured time histories are not contaminated by model/balance resonance. This is normally achieved by a combination of using as stiff a balance as possible with the lightest possible model. Typically, natural frequencies of vibration of the combined system are in the range of 50 to 200 Hz. At lower natural frequencies, increased separation of reduced frequencies of the model/balance and the scaled prototype structure can be achieved by reducing the test wind speed in the wind tunnel. When doing this, though, it is necessary to ensure that minimum Reynolds Number requirements are still met while also noting that signal to noise ratios

Figure 2.33 High-frequency balance model of CAARC building.

in the instrumentation worsen as a result of the lower applied forces. For these reasons, it is desirable to use the stiffest balances and lightest models possible.

2.9.6.4 HIGH-FREQUENCY PRESSURE INTEGRATION

The high-frequency pressure integration (HFPI) method is an alternative to the HFB approach which used integration of pressures measured simultaneously over the entire building model to determine the total applied load. This has become a more common technique in recent years as pressure measurements systems have advanced and it is possible to measure pressures at many hundreds of locations simultaneously. An example of a pressure model of the CAARC building is shown in Fig. 2.34. Pressure taps (holes in the surface of the model) are connected by flexible, small diameter, vinyl tubes to banks of pressure transducers below the floor of the wind tunnel, as shown in Fig. 2.35. This system requires correction

Figure 2.34 Pressure model of CAARC building.

Figure 2.35 Tubing and pressure sensors below wind tunnel.

(physical or digital) for resonance in the pneumatic tubes and transducer volume to ensure that pressures are not being artificially amplified. Conversely, the tubes need to be short enough and of a narrow enough diameter to maintain frequency resolution. This technique relies on being able to physically extract a sufficient number of pressure tubes from a model at once. The total number of pressure taps needs to adequate to define the pressure fields acting on the building. While this can be achieved for the majority of buildings at reasonable test scales, for very architecturally complex buildings or those with reduced floor plates in lower parts of the building HFB testing may be more appropriate.

2.9.7 Calculating Design Wind Loads and Responses from Wind Tunnel Data

With the exception of aeroelastic testing, as discussed, wind loading data from the wind tunnel is extracted in non-dimensional form. This data then needs to be combined with the wind climate model in order to predict loads and responses. These will typically include strength-design loads, serviceability MRI loads for deflection assessments, and serviceability accelerations for the assessment of occupant comfort. The end goal of this is to produce strength design loads that meet the reliability intents of the code, serviceability loads that confirm building performance, and accelerations that won't result in excessive complaints from occupants. Each of these may require different wind climate models, or at least with different directionalities, to accurately predict the load effects of interest.

There are several approaches to combining wind tunnel data with wind climate models, each of which have pros and cons.

In the most simple approach, an omni-directional wind climate model is used which considers that the design wind speeds may come from any direction. This is clearly the most conservative method, and one that is rarely used except in locations where there is insufficient available wind data to be able to develop a directional model.

A more commonly used approach is the sector method where directional wind climate analyses are used to develop wind speeds with the appropriate MRI for each wind direction. This has the advantage of being an intuitive approach that is easy to understand, or explain, and scaling to code values at the most severe wind direction is straightforward. It is also consistent with codes where directional wind speeds are provided. It can, however, still be conservative as it is the recurrence interval of wind speeds, rather than load effect probabilities, that is being calculated.

The first probabilistic model used more widely to predict wind loads with a given probability of exceedance was the upcrossing, or outcrossing, technique (Lepage & Irwin, 1985). This approach uses mathematical models of wind statistics, and was developed in the days of limited computing power with a number of assumptions. It has the benefit of potentially increased precision in the prediction of load effects for a given probability, but has the disadvantages of several input assumptions and being conceptually complex to explain.

With more modern computing power the extreme load effect method uses measured wind records directly to develop wind loads on which statistical analyses can be conducted. The quality of this approach is entirely dependent on the quality of the input records and can be very computationally intensive. However, it is very intuitive and can be very effective with high quality wind storm time histories.

The final technique that is gaining increasing traction is multi-sector analysis (Holmes & Bekele, 2015). This is a statistical joint probability approach that is less computationally intensive than the extreme load effect method but does not have any of the approximations use in upcrossing methods.

For aerodynamic model test techniques, the initial analysis is normally done in the frequency domain. The use of frequency domain analysis provides more statistically reliable estimates of the loads based on a combination of mean plus a defined peak factor multiplied by the standard deviation of the dynamic components of response. The dynamic components include background and resonant responses. There are well accepted methods for calculating peak factors based on the frequencies of vibration and the time period over which the loads are being calculated.

2.9.8 Components and Cladding

AS well as overall structural wind loads and responses, an equally important part of the design of tall buildings for wind effects is the consideration of local pressures on the building envelope. Reliability of the building envelope is important from both the standpoints of life safety and protection of building contents. When high-frequency pressure integration is used for the determination of structural loads then the same data set can be used for this purpose, although sometimes with added pressure taps to capture more localized flow effects. Otherwise, this is a separate test to complement the loading studies. Typically, a minimum average number of pressure taps is around 1 per 1000 ft^2 of building surface area. For a typical tall building, five hundred or more pressure taps is not unusual. Results from this can be presented as simplified block zones, or interpolated pressure contours (see Fig. 2.36). Normally the peak 1-s pressures are presented with the zones and contours being based on an analysis of the effects of all directions, with 36 equally spaced wind azimuths at 10° intervals being standard.

2.9.9 Performance Objectives and Design Criteria

There are different performance objectives and design criteria in design for strength-design wind loads, serviceability deflections, and accelerations for occupant comfort.

Strength design loading criteria are generally set by local or national codes and form the basis of design either in terms of a probability of exceedance of strength design loads for a given purpose or occupancy of building, or by a defined design wind speed. In the US, these are described in ASCE 7 through means of Risk Category and associated mean recurrence intervals.

Traditionally, it has been assumed that the building will remain elastic in the design-level wind event. While seismic design has a long history of allowing limited inelastic, or plastic, behavior in extreme events when performance-based design (PBD) is employed to validate continuing structural integrity this is an approach only recently available for wind design. The ASCE/SEI Prestandard for Performance-Based Wind Design was published in 2019 and contains guidelines on application and performance standards to be met. At the time of writing, research is ongoing on sample and prototype buildings to assess the value of PBWD for different types of buildings in different wind climates.

For serviceability design, shorter MRIs are used in the determination of overall building deflections and story drifts. The drift limits do vary based on the building type

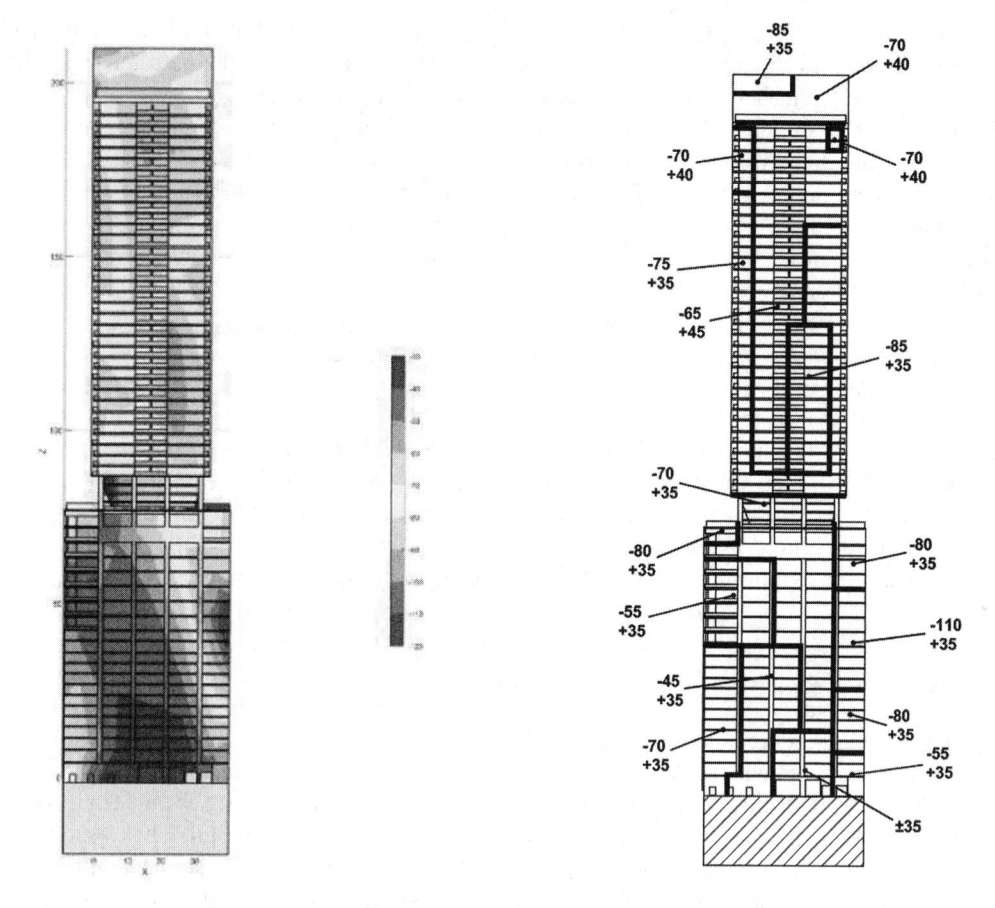

Figure 2.36 Examples of cladding pressure testing outputs: interpolated contours (L) and simplified block zones (R).

and local practice, and traditionally the MRIs have also varied. Industry consensus is moving toward the MRIs in Table 2.21 for strength and serviceability assessments (ASCE, 2019).

For building envelopes, a performance objective is to retain wind and watertight up to pressures associated with independently defined serviceability MRIs, while remaining attached to the building in even extreme events. Accelerations in tall buildings are also predicted by wind tunnel testing. These are important in limiting, or eliminating, occupant complaints as a result of perceptible wind-induced motion. Complaints can be generated by large accelerations causing distress to the occupants, or regularly perceptible motion causing annoyance among building occupants who are otherwise not doubtful of structural integrity as a result of the motion. Performance objectives with respect to accelerations should be agreed between all concerned parties including the building owners, tenants, and structural engineering team. Acceptable levels are subjective and should be judged on a number of factors including storm duration and contribution to the acceleration results and the likely reaction of occupants to motion. Traditionally, a 10-year MRI has been used in North America with simple guidelines based on occupancy type: residential or commercial. More recent guidelines use shorter MRIs and are frequency dependent, as motions become more perceptible at higher frequencies. The most commonly used modern guidelines are contained in ISO 10137 (ISO, 2007).

Table 2.30 Strength and Serviceability Assessments according to Risk Category

Risk category	Strength design MRI (years)	Serviceability assessment MRI (years)
I	300	—
II	700	10
III	1700	25
IV	3000	50

These use frequency-dependent guidelines at a one-year MRI with separate guidelines for residential and office properties. The residential guidelines are set at around 70 percent of those for office buildings. It is worth noting that many residential buildings have been constructed and occupied using accelerations that would not meet the current residential guidelines, and these values can be hard to meet in many locations.

Where building accelerations do not meet performance objectives, there are a number of approaches that can be taken to reduce the loads such as increasing building stiffness, increasing building mass, or altering the aerodynamic characteristics of the building. Another commonly used approach is the introduction of supplementary damping devices such as sloshing dampers, tuned mass dampers, or distributed damping. Increased damping is not normally relied on to reduced loads, unless distributed damping systems with significant redundancy have been employed.

2.9.10   Structural Dynamic Properties for Wind Engineering Analyses

In order to calculate wind-induced dynamic responses from wind tunnel test results, or conduct a properly scaled aeroelastic test, the structural engineer is required to provide input in the form of structural dynamic properties including natural frequencies of vibration, mode shapes, mass distributions, and structural damping ratios. Frequencies and mode shapes are normally generated from a finite element model of the primary structure, and the conduct of this modeling is critical to the accuracy of the results. As a general rule, increasing natural frequency of vibration for most buildings will result in decreased wind loads and therefore it is important to include as many elements that will contribute to the building stiffness as possible.

As natural frequencies are, to a degree, amplitude dependent in concrete structures depending on the degree of cracking assumed it is common to provide the wind engineer with two sets of structural properties for strength design and serviceability design. While the serviceability characteristics may assume uncracked properties, it should be noted that any degree of cracking assumed for strength design should be rather less than normally assumed for seismic design. It is also well known that frequencies measured in many as-built buildings are higher than those used during design, especially for stockier buildings. Again, this emphasizes the importance of the structural modeling in the prediction of wind loads.

The one approximation that does need to be made in the assumed inherent structural damping. For tall, steel buildings common values are in the range of 0.005 to 0.015, and for concrete buildings 0.010 to 0.020. The upper parts of the ranges are normally used for strength design, and values lower in the range for serviceability loads and accelerations. For particularly slender buildings, lower values may be applicable at time, and for stockier buildings higher values may apply. The selection of appropriate structural damping ratios is normally a collaborative effort between the structural and wind engineers.

2.9.11   Load Combinations and Design Load Cases

At any given wind direction, there will be building responses about all three primary axes whether due to direct excitation or modal coupling. Although the peak load in two or more directions may be largest at the same wind direction, these loads will probably not occur simultaneously due to lack of correlation in the response of different modes of vibration. An example of translational time histories, and resultant load traces is shown below in Fig. 2.37. This figure shows the time history traces and an ellipse designed to encompass these time histories. A+ shows the maximum moment about the x-axis with a companion load about the y-axis. B+ shows the peal load about the y-axis with a corresponding companion load about the x-axis. These are example of load combinations. In a building exhibiting torsional response, there would also be a companion torsional moment to apply. The number of load cases is dependent on the building and its response characteristics, but usually between 10 and 24 load cases are provided.

The load cases are usually determined for tall buildings on the basis of base moments, these acting as an approximation for maximizing load effects over the height of the building. Where there are significant changes in architecture and/or structural system in a tower (e.g., where there is a transfer level) it can be beneficial to also calculate load cases at intermediate heights.

Once the load cases have been determined, floor-by-floor translational shears and torsional moments are developed for each case. When applied by the structural engineer to their design model, these generate the same base moment combinations. This is an area where the peak loads need to be decomposed into mean, background, and resonant components. The distribution of resonant loads is a function of the mass and mode shape. When high-frequency pressure integration testing has been conducted the mean component can be derived directly from the mean pressures, and this can also be used as an approximation to distributing the background loading components. If high-frequency balance has been used then the mean and background load distributions

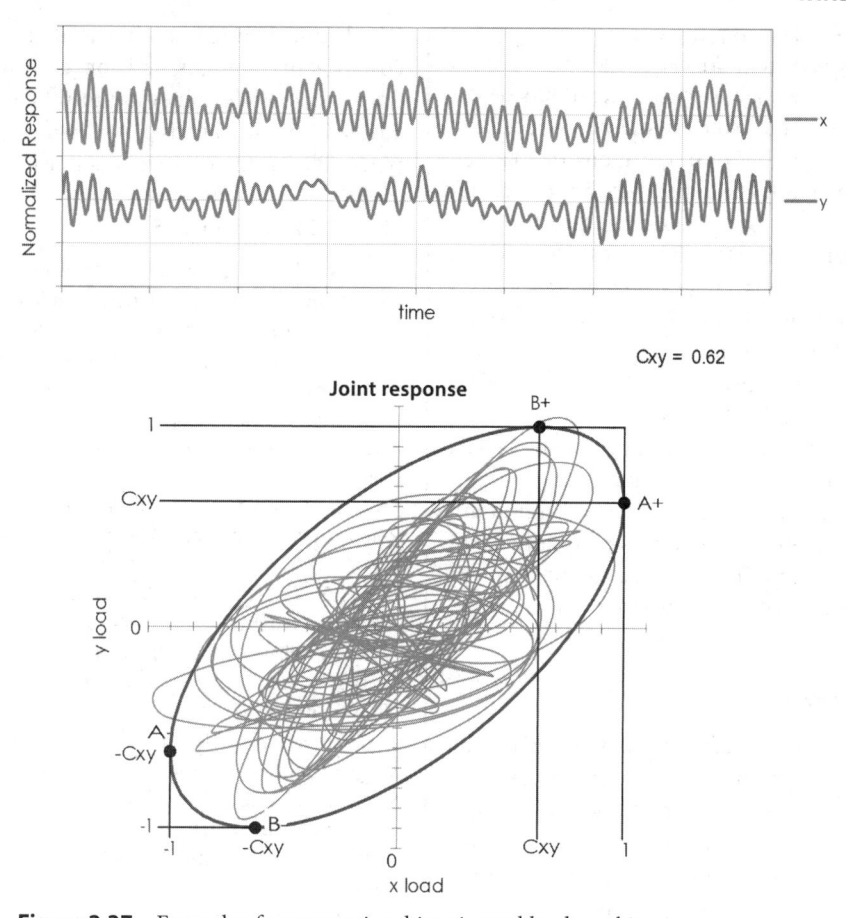

Figure 2.37   Example of response time histories and load combinations.

need to be approximated to ensure that base shears and moments match those calculated. Typically these distributions are based on the approach wind mean profile or a trapezoidal distribution.

The approximations inherent in the HFB approach limit the use of the technique in performance-based wind design where it is necessary to provide simultaneous loading time histories over the height of the building.

2.9.12   Other Uses of Wind Tunnels in Design

The wind tunnel is also used for a number of other design applications. These include outdoor wind comfort and safety, door operability, stack effect assessment, and exhaust dispersion.

Studies on outdoor wind comfort and safety provides information on wind speeds in public area such as at street level, outdoor dining, and recreational areas whether on the podium, terraces, or balconies. The wind speeds measured in the wind tunnel can be analyzed for comparison

with comfort and safety criteria. Some cities require this as part of the planning approval process and may have specific criteria to be met.

Door operability can be an issue on tall buildings with doors being difficult to open or close either as a result of wind speeds past them or pressures across them. Sometimes it is the external pressure that governs this, but in some other cases internal pressures (driven by stack effect) can drive these effects. The pressures measured on the building during the cladding study can be used to inform analyses on door operability by using both the external pressures and the pressures at openings that contribute to internal pressures and combine with thermal stack effect.

Wind tunnel testing can also be employed in the investigation of dispersion of building exhausts to ensure that there is adequate dilution, especially around fresh air intakes, public areas, or surrounding properties. The appropriate dilution is dependent on the exhaust composition from those that are hazardous such as laboratory fume cupboards to those that are merely undesirable such as cooking odors.

2.9.13 Specification of Wind Tunnel Tests and Quality Assessment Checks

Checking of wind tunnel data is complex and outside the realm of expertise of all but wind engineering specialists. There are, though, a number of publications that can be used to set a foundation for the specification of requirements for wind tunnel testing, assuring that the tests have been adequately conducted, and that the results are rational. These publications range from the fairly non-technical to the much more in-depth. The Guide to Wind Tunnel Testing of High-Rise Buildings (Irwin, Denoon, & Scott, 2013) is a good primer to the subject, while ASCE 49-21 (ASCE, 2021) is the US standard for how to conduct wind tunnel testing for buildings and structures. Elsewhere the Australasian Wind Engineering Society publish a Quality Assurance Manual with minimum basic requirements (AWES, 2019) and much more detail about the conduct of wind tunnel testing can be found in the dated, but still relevant, ASCE Manual of Practice 67 (ASCE, 1999).

2.10 CHAPTER SUMMARY

This Chapter addresses the different approaches for the evaluation of wind loading appropriate for the design of tall buildings. The reduction of mass, stiffness and damping, make contemporary tall buildings much more vulnerable to severe aerodynamic excitation requiring a careful examination of their static and dynamic wind response.

Issues affecting a tall building include: (i) The dynamic and static responses of the building as a whole due to wind acting on the primary system; (ii) Local pressures exerted on the exterior cladding, sometimes in association with internal pressures; and (iii) street-level wind conditions affecting pedestrian comfort and safety.

Wind-tunnel model studies are accepted globally as alternative procedures for providing information on the action of wind as required in the design of tall buildings. Although the International Building Code (IBC), ASCE 7-16 and other standards around the world attempt to quantify wind-induced building motions and related accelerations, it should be noted that the code procedures are mainly for isolated buildings located in homogeneous terrain. Such procedures may not be representative for contemporary buildings located in complex terrain or topography. For this reason, most codes alert the designer that their provisions may not be applicable to buildings with unusual properties, exterior geometry, response characteristics and siting. In addition to improved reliability of building performance there is also an economic motivation for undertaking wind-tunnel studies.

Wind-tunnel model studies lead to more tailored and cost-effective designs and, on occasions, expose situations where code values are insufficient, confirming the advisability of carrying out wind-tunnel studies for potentially wind-sensitive buildings.

SELECTED REFERENCES

1. Fanella, D. A., Structural load Determination 2018 IBC and ASCE/SEI 7-16, International Code Council (ICC), 2018.

2. American Society of Civil Engineers (2017), Minimum Design Loads and Associated Criteria for Buildings and Other Structures (ASCE 7-16), ASCE, Reston, VA, USA.

3. International Code Council. International Building Code (IBC 2018). Falls Church, VA, USA. International Code Council, 2018.

4. Coull and Stafford-Smith, Tall Buildings with Particular Reference to Shear Wall Structures, New York, Pergamon Press, 1967.

5. T.Y. Lin and S.D. Stotesbury, Design of Prestressed Concrete Structures, New York, John Wiley.

6. American Society of Civil Engineers (1999), *Wind Tunnel Studies of Buildings and Structures* (ASCE Manual of Practice Number 67), ASCE, Reston, VA, USA.

7. American Society of Civil Engineers (2021), *Minimum Design Loads and Associated Criteria for Buildings and Other Structures* (ASCE 7-22), ASCE, Reston, VA, USA.

8. American Society of Civil Engineers (2021), *Wind Tunnel Testing for Buildings and Other Structures* (ASCE 49-21), ASCE, Reston, VA, USA.

9. American Society of Civil Engineers (2019), *Prestandard for Performance-Based Wind Design*, ASCE, Reston, VA, USA.

10. Australasian Wind Engineering Society (2019), *Wind Engineering Studies of Buildings* (AWES-QAM-1-2019) Davenport, A.G. (1972), *Technical Committee No. 7, Wind Loading and Wind Effects, Theme Report*, American Society of Civil Engineers (ASCE) and International Association for Bridge and Structural Engineering (IABSE) International Conference on Design and Planning of Tall Buildings, Lehigh University, Bethlehem, PA, USA, 21–26 August 1972.

11. Holmes, J.D. and Bekele, S.A. (2015), *Directionality and wind induced response – calculation by sector methods*, 14th International Conference on Wind Engineering, ICWE14, Porto Alegre, Brazil.

12. Lepage, M.F. and Irwin, P.A. (1985*), A technique for combining historical wind data with wind tunnel tests to predict extreme wind loads*, Proceedings of the 5th US National Conference on Wind Engineering, Lubbock, TX, USA, 06–08 November 1985.

13. Irwin P., Denoon, R., and Scott, D. (2013), *Wind Tunnel testing of High-Rise Buildings: an output of the CTBUH Wind Engineering Working Group*, Council on Tall Buildings and Urban Habitat, Chicago, IL, USA.

14. ISO (2007), *Bases for design of structures – Serviceability of buildings and walkways against vibrations*, ISO 10137 International Organization for Standardization, Geneva.

Chapter 3
Seismic Effects

3.1 INTRODUCTION

3.1.1 Nature of Earthquakes

Catastrophic earthquakes appear in the headlines with discomforting frequency, causing thousands of lives to be lost and property damage running into hundreds of millions of dollars. This truly global phenomenon has begun to be understood, and considerable emphasis has been and is being placed on the analytical studies of earthquake response of buildings, supported by experimental studies both in the laboratory and in the field in an effort to prevent much of this destruction and loss of life.

Accounts of destructive earthquakes appear all through recorded history. Early humans, in their inability to comprehend such a strange and destructive phenomenon, attributed the whole mechanism of earthquakes to the angry work of gods. Although in ancient times it was tempting to think of earthquakes as somehow otherworldly, they are in fact, among the most common of the earth's phenomena.

We will limit our discussion to the class of earthquakes in which the energy release is both near the earth's surface and large enough to damage structures. Such shallow-focus earthquakes are related to the forces which bring about the gradual distortions of the earth's crust. According to a well-established theory known as the elastic rebound theory, the distortions and the associated strains and stresses in the outer layers of the earth build up with the passage of time until ultimately the stress at some location becomes high enough to fracture the rock or cause it to slip along some previously existing fault plane.

Slippage at one location causes an increase in the stress in the adjacent rock, resulting in a rapid propagation of slippage along the fault plane. The result is the sudden rebound of the elastic strain. The strain energy that has accumulated in the rock is suddenly released and propagates in all directions from the source in a series of shock waves. If the amount of energy released is small, or if the fault slippage occurs in an uninhabited region, the shock waves travel unnoticed except by sensitive seismographs. If this energy release is great, the effect at nearby locations is chaotic. The earth may experience violent motion in all directions, lasting for a few seconds in a moderate earthquake or for a few minutes in a very large earthquake.

Although most of the earthquakes of record have occurred in well-defined earthquake belts, an examination of earthquake records for the world reveals the truly global nature of this phenomenon. Thus, even in the seismically inactive parts of the world, some measure of earthquake resistance should be built into the design of all structures in which failure will be a major catastrophe.

The earth's crust is composed of a dozen or so large plates and several smaller ones ranging in thickness from 20 to 150 miles (32 to 241 km). The plates are in constant motion, riding on the molten mantle below and normally traveling at the rate of a millimeter a week, equivalent to the growth rate of a fingernail. The plates' travels result in continental drift and the formation of mountains, volcanoes, and earthquakes. If plates carrying two continental masses grind past each other, as the Pacific and the North American plates do under California's San Andreas

Fault, friction locks them together. When slippage occurs, the earth around the fault creates a so-called strike-slip earthquake. Still, another kind of tectonic phenomenon results when an oceanic and a continental plate meet each other. For example, the oceanic plate that forms part of the Pacific floor off Mexico is pushing north-eastward against the North American plate, which is creeping westward. The oceanic plate dips under the continental crust, but the relative movement between the two plates is resisted by friction. When frictional forces are overcome, the stuck section of the plate lurches forward, generating shock waves of a thrust quake similar to the one responsible for the Mexican disaster of 1985.

3.1.2   Some Recent Earthquakes

Scientists estimate that over one million earthquakes occur every year. Some are very small and cause no damage, while others are violent and cause severe damage. One of the largest and most violent earthquakes to hit North America occurred near Anchorage, Alaska, in 1964. In the twentieth century, there have been only three earthquakes of magnitude 8 or larger to affect a metropolitan area. The first one was the 1906 earthquake which destroyed much of San Francisco, California, estimated at 8.3 on the Richter scale (m_L). In 1923, Tokyo and Yokohama in Japan were badly damaged, and the third was the September 1985 quake which hit Mexico City. The most powerful earthquake ever recorded was off the coast of Chile in May, 1960. It reached 9.5 on the moment magnitude scale.

The devastating earthquake which hit Mexico City in September of 1985 measured 8.1 on the surface magnitude scale. In just 4 min an estimated 300 buildings collapsed in downtown Mexico City. Fifty more were later judged dangerously close to falling, and hundreds of others were regarded as unsafe. Just 36 h after the first tremor, the second earthquake, known as the aftershock, battered Mexico City. This tremor, which lasted for about a minute and was not as powerful as the first, toppled some already weakened buildings. The estimated death toll numbered over 8000 persons with property damage estimated at $8 billion. The strength of this earthquake set the skyscrapers swinging as far north as Houston, 1100 miles (1770 km) from the epicenter. Tidal waves rolled ashore on the coast of El Salvador more than 800 miles (1287 km) to the southeast. The widespread damage is a chilling reminder that the world's well-defined quake-prone areas can be struck at any time without warning and with deadly effect. The same region in which the September 1985 earthquakes occurred had experienced six earthquakes with a magnitude of at least

7.0 since 1911. Thus, the latest shocks came as no surprise to seismologists.

Mexico City is built on soft, moist sediment of an ancient lake bed, which when jolted shakes like a bowl of jelly producing a large effect with a predominant vibration period of about 2 s. In addition, the city is undergoing subsidence at the rate of up to 10 in. (25.4 mm) annually, creating uneven settlement in some building foundations. The unusual severity of the quake and the resonance of the soil structure are major factors behind the extent of the damage. It is estimated that because of the soft subsoils some buildings were subjected to acceleration equal to 1.0 g. Most seriously affected by the earthquakes were structures in the 5- to 15-story range which had natural frequencies close to that of the soil, causing them to resonate with greater vigor than other buildings.

During the 1985 Mexican earthquake, coastal towns only 50 miles (80 km) from the epicenter suffered less damage than Mexico City, which was 200 miles (320 km) from the epicenter, because the shoreline is made of solid rock and thus shakes less violently than the alluvial lake bed on which Mexico City is built. Thus even though the seismic waves were diminished in intensity in their travel from the epicenter, they were amplified by the city's foundation.

At 4:31 a.m., in the pre-dawn hours of Monday January 17, 1994, the San Fernando Valley (about 30 miles north-west of central Los Angeles) was shaken by its most devastating earthquake in 60 years. The so-called Northridge earthquake had a magnitude of 6.7 and peak ground accelerations of about 1.0 g in several locations. According to news accounts, the earthquake was responsible for more than 50 deaths (of which 22 were attributed to earthquake-induced heart attacks) and at least 5000 injuries. According to the City of Los Angeles, more than 10,000 buildings were red-tagged (prohibited entry) or yellow-tagged (restricted entry), and more than 25,000 dwelling units were vacated. Some areas flooded from broken water mains. Some areas were heavily damaged by fire. Damage estimates ranged from $50 to $100 billion.

The earthquake generated a large number of strong-motion recordings over a wide variety of geologic site conditions, including free-field stations on rock and soil as well as recordings of motions from instrumented structures of varying types of construction.

Although the epicenter was located in the suburban city of Reseda in the San Fernando Valley, peak horizontal accelerations approaching 0.5 g were recorded at sites as far as 22 miles (36 km) from the epicenter in downtown Los Angeles. Recordings at two stations in the epicentral area, Sylmar County Hospital (alluvium) and Tarzana Cedar Hills Nursery (alluvium over siltstone), yielded the largest free-field accelerations on soil sites and unusually high values of peak accelerations. The Tarzana station 4.4 miles (7 km) south of the epicenter, recorded peak horizontal and vertical accelerations of 1.82 and 1.18 g, respectively.

A magnitude 7.7 earthquake occurred in the Gilan Province between the towns of Rudbar and Manjil in northern Iran on Thursday, June 21, 1990, at 12:30 a.m. local time. The earthquake, the largest ever to be recorded in that part of the Caspian Sea region, may have been amplified by two or more closely spaced earthquakes occurring in rapid succession. The event, which was exceptionally close to the surface for this region, was unusually destructive causing widespread damage in areas within a 62-mile (100 km) radius of the epicenter near the city of Rasht about 124 miles (200 km) northwest of Tehran. One hundred thousand adobe houses sustained major damage or collapsed resulting in forty thousand fatalities, and sixty thousand injured. Five hundred thousand people were left homeless.

On October 17, 1989, the 7.1 magnitude Loma Prieta earthquake occurred in the Santa Cruz mountains due to movements occurring along a 25-mile (40 km) segment of the San Andreas fault. Measurements along the surface of the earth after the earthquake showed that the Pacific plate moved 6.25 ft (1.9 m) to the northwest and 4.25 ft (1.3 m) upward over the North American plate. The upward motion resulted from the deformation of the plate boundary at the bend in the San Andreas fault. At the surface the fault motion was evident as a complex series of cracks and fractures.

This earthquake, according to the geologists, was not unexpected. During the 1906 San Francisco earthquake, there was only about 1 m of movement on the Santa Cruz segment of the San Andreas fault, while farther north in the San Francisco area, there was more than 8 ft (2.5 m) of movement. This indicated that all of the strain had not been released in the Santa Cruz segment in the 1906 earthquake so this segment was likely to break before the northern segment.

At the Stanford University campus, 30 miles northwest of the epicenter, 60 buildings sustained varying degrees of damage, with an estimated repair cost of $160 million.

The most deadly structural failure occurred when the upper deck of the Interstate 880 (Nimitz Freeway) in Oakland fell onto the lower roadway causing 41 deaths. Another spectacular failure occurred on the Oakland Bay Bridge. Interstate 280, the Embarcadero Freeway, and the Highway at Fell Street were also damaged.

On July 1993, at 10:17 p.m. local time, a magnitude 7.8 earthquake occurred in the Japan Sea off southwest Hokkaido. The subduction zone event and subsequent tsunami resulted in widespread damage in Northern Japan, with losses estimated at 6 billion yen ($60 million) and 196 earthquake and tsunami-related deaths. The majority of the casualties and damage were tsunami-caused. Tsunami run-up heights on Okushiri were as high as 100 ft (30.6 m), the highest ever recorded in Japan. Ground shaking produced relatively little damage to engineered structures, but secondary effects from liquefaction and landslides caused substantial damage at several localities.

The January 17, 1995, Kobe earthquake was the most damaging to strike Japan since the great Kanto earthquake destroyed large areas of Tokyo and Yokohama that killed 143,000 people in 1923. As of January 30, 1995, the toll from the earthquake in Kobe and adjacent cities had reached 5096 dead, 13 missing, and 26,797 injured. One-fifth of the city's 1.5 million population was left homeless and more than 103,521 buildings were destroyed with an estimated cost of restoring basic functions to be about $100 billion dollars. The total losses, including losses of privately owned property and reduction in business activity, may be twice this amount, which would be ten times higher than losses resulting from the 1994 Northridge, California earthquake. The epicenter was located about 20 km south-west of downtown Kobe. The earthquake was assigned a Japan Meteorological Agency magnitude of 7.2.

California is the most earthquake-prone state in the United States with a number of active faults criss-crossing the region (Fig. 3.1). Among these, the San Andreas Fault is the longest passing under the Gulf of California through the San Joquin Valley and San Francisco. It continues under the Pacific Ocean off the coast of Northern California. The land west of the fault is slowly moving north, while the land east of the fault is moving south. In 1906, movement along this fault caused the famous San Francisco earthquake.

Some faults are deep and others are close to the surface of the earth. The point beneath the surface where the rocks break and move is called the focus of an earthquake. Directly above the focus, on the earth's surface, is the epicenter. The most violent shaking is often found near the epicenter of an earthquake.

Three types of waves are caused during an earthquake: (i) the primary, or P waves, which travel the fastest, pulling and pushing rock particles; (ii) secondary shear waves, or S waves, which are responsible for most of the earthquake damage: they are slower than P waves and cause the rock mass to move in a direction at right-angles

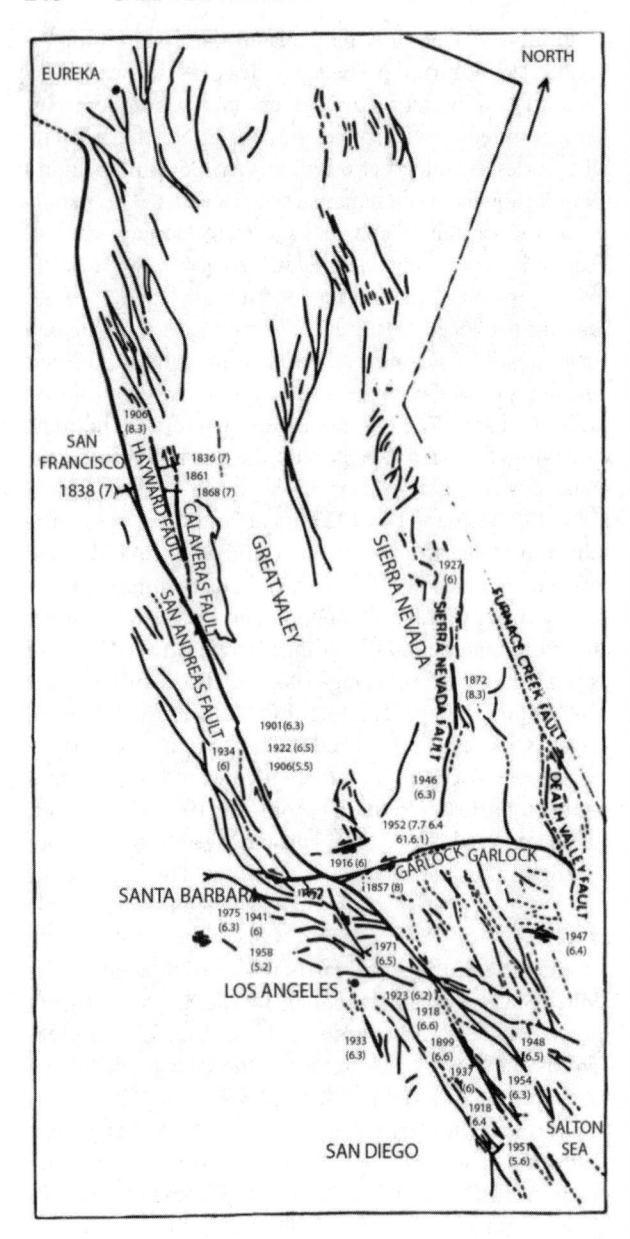

Figure 3.1 Simplified fault map of California. (Courtesy of Dr. Farzad Naeim.)

to the direction of propagation; (iii) the surface waves generated by the P and S waves which move along the earth's surface, similar to the way waves travel in an ocean.

Earthquakes often occur in the same three parts of the world. One major earthquake zone extends nearly all the way around the edge of the Pacific Ocean, going through New Zealand, the Philippines, Japan, Alaska, and along the western coasts of North and South America. The San Andreas Fault, where as many as 30 major earthquakes have occurred in historic times is part of this zone. A second major zone of earthquakes is found near the Mediterranean Sea, extending across Asia, encompassing Italy, Greece, Turkey, and part of India. The third zone is located from Iceland south, through the middle of the Atlantic Ocean.

Two large earthquakes with a moment magnitude of 7.7 and 7.6 hit eastern Türkiye and north Syria on February 6, 2023. The epicenter of the first earthquake was Kahramanmaraş-Pazarcık, whereas the second was in Kahramanmaraş-Elbistan. The two earthquakes were approximately 9 h apart and their effect extended to 11 provinces with a population exceeding 15 million in southern Türkiye causing massive collapses and damage to buildings, bridges, airports, ports, retaining structures, hydraulic structures, lifelines, etc. More than 100,000 buildings collapsed or heavily damaged, 15 bridges out of 1000 bridges were affected.

Türkiye surrounded by active faults from the north, south, and the east direction of the country. Historically, the country has been seismically active and sever earthquakes occurred in different parts of the country that caused severe damages to buildings and infrastructure and as well as tens of thousands of deaths and causalities. One of the active faults in Türkiye is the East Anatolian Fault which 450 km long extends along the south-west to the north-east of the country. At 4:17 local time (01:17 GMT) on February 6, 2023 an earthquake with a moment magnitude, M_w, of 7.7 according to AFAD, Disaster, and Emergency Management Presidency (www.afad.gov.tr) (AFAD 2023), occurred on the East Anatolian Fault. The epicenter of the Pazarcık-Kahramanmaraş, Earthquake is located at N37.288°, E37.043° at approximately 40 km north-west of Gaziantep, and 33 km south-east of Kahramanmaraş, with a focal depth of 8.6 km according to AFAD. Nine hours later, at 13:24 local time (10:24 GMT), another earthquake with an M_w of 7.6 occurred at Elbistan-Kahramanmaraş which is located at N38.089°, E37.239°, also on the East Anatolian Fault approximately 98 km north-west of Adıyaman, and 62 km north-east of Kahramanmaraş, with a focal depth of 7.0 km according to AFAD. Figure 3.2 shows the tectonic plates of Türkiye and the surrounding region as well as the direction of motion in arrows. The plate boundaries are indicated by red arrows, the black arrows indicate the relative motion across the plate boundaries. The gold star and the blue star indicate the epicenters of the Pazarcık-Kahramanmaraş and the Elbistan-Kahramanmaraş earthquakes, respectively.

The common type of construction in the affected region is multistory reinforced concrete, three to 15 stories, with

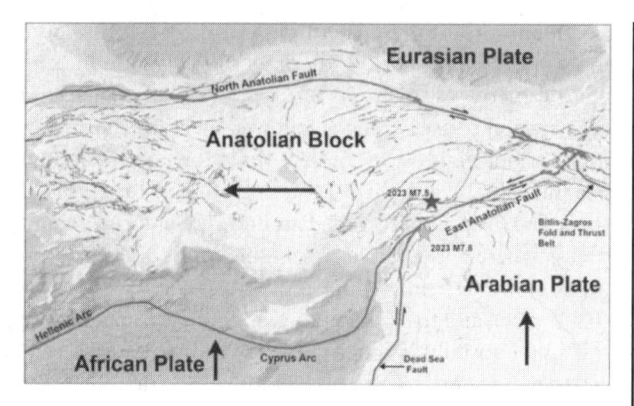

Figure 3.2   Tectonic map of Türkiye. (USGS)

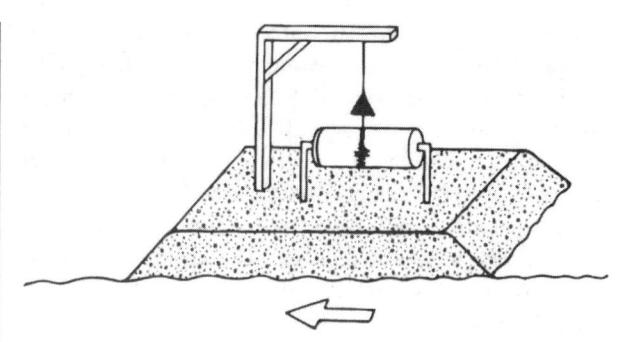

Figure 3.3   Conceptual model of displacement seismograph.

hollow-clay brick infill walls. Another type of construction that exists in the region is unreinforced masonry old structures.

3.1.3 Seismograph

For purposes of seismic engineering, earthquakes are recorded on a strong-motion accelerograph. The accelerograph is normally at rest and is triggered only when the ground acceleration exceeds a pre-set value. Earthquake records can be made for three components of ground acceleration, two horizontal and one vertical.

A basic seismograph consists of a recording device at the end of a heavy pendulum attached to a frame in such a way that it will remain nearly still even when the earth is subjected to seismic motions. A constantly moving drum, attached to the ground through a heavy foundation mass, is placed under the writing device. When the earth is still, the writing device records a straight line, and a zig-zag or wavy line during an earthquake. The higher the wavy lines on the seismograph, the stronger the earthquake.

When an earthquake wave travels through the ground underneath such a pendulum, the motion of the ground is traced by the pen. This is because the pendulum stays stationary due to its inertia, and the trace is actually the movement of the drum. In most seismographs used today, the ground motion is made to produce a small electric signal, and this signal records the motion.

In the conceptual seismograph shown in Fig. 3.3, the lengths of wavy lines are proportional to the ground displacement. This type of seismograph is called a displacement seismograph. It is possible, however, to record the ground acceleration by appropriately changing the period and damping of the pendulum. Such a device is called an acceleration seismograph.

The current method of recording earthquakes is by the use of accelerographs which produce an analogue trace of acceleration versus time in the form of either a photographic trace on film or paper or a scratch on waxed paper.

3.1.4 Measures of Earthquake Magnitude and Intensity

There are two commonly used earthquake parameters of interest to the structural engineer. They are an earthquake's magnitude and its intensity. The magnitude is a measure of the amount of energy released by an earthquake while the intensity is the apparent effect of the earthquake as experienced at a specific location. The magnitude is the easier of the two parameters to measure because, unlike the intensity which can vary with location, the magnitude of a particular earthquake is constant. The magnitude scale was first developed by Prof. Richter and then extended to cover world-wide events by others who invented the surface wave magnitude and moment wave magnitude scale. The most widely used scale to measure magnitude is the moment magnitude scale. Using the magnitude scale, the energy released, measured in Ergs can be estimated from the equation:

$$Log\ E = 11.4 + 1.5M \qquad (3.1)$$

where E is the energy released in Ergs, and M is the magnitude. Because the open-ended magnitude scale is logarithmic, each unit increment in the magnitude represents an increase in energy by a factor of 31.6. For example, a quake registering 6 on the magnitude is 31.6 times more powerful than one measuring 5. The empirical observation that earthquakes of magnitude less than 5 are not expected to cause structural damage, whereas for magnitudes greater than 5, potentially damaging ground motions will be produced.

Although an earthquake has only one magnitude it will have many different intensities. In the United States, intensity is measured according to the Modified Mercalli Index (MMI).

The intensity values of MMI describe the degree of shaking on a scale of 1 to 12 and are expressed in Roman numerals (I, II, …, XII). The successive intensities are not derived from a physical measurement, but merely represent ratings given to earthquake effects upon people and the amount of damage to buildings. The intensity levels are represented by a rather long description of earthquake effects ranking earthquake intensities from I (barely felt) to XII (total destruction). Destructive earthquakes of modified Mercalli intensity IX or X at epicentral regions generally register more than 6.5 on the Richter scale.

3.1.5 Seismic Design

For structural engineering purposes, an accelerograph record of the time-history of ground shaking gives the best measure of earthquake intensity. Figure 3.4 shows the recorded ground acceleration for the El Centro earthquake, California. From Fig. 3.4 it is seen that during a short initial period, the intensity of ground acceleration increases to strong shaking, followed by a strong acceleration phase, which is followed by gradual decreasing motion. Figures 3.5 and 3.6 show the corresponding ground velocity and displacement plotted as functions of time. The maximum recorded ground acceleration is about 0.33 g, where g is the acceleration due to gravity of 32 ft/s^2 (9.75 m/s^2). The maximum ground velocity is 13.7 in./s (0.348 m/s), and the maximum ground displacement from the initial position is about 8.3 in. (211 mm).

Shown in Fig. 3.7 are acceleration, velocity, and displacement records from the 1994 Northridge earthquake, California. The peak values of ground acceleration, velocity, and displacement are 0.843 g, 50.39 in./s (1.28 m/s), and 12.81 in. (325.5 mm).

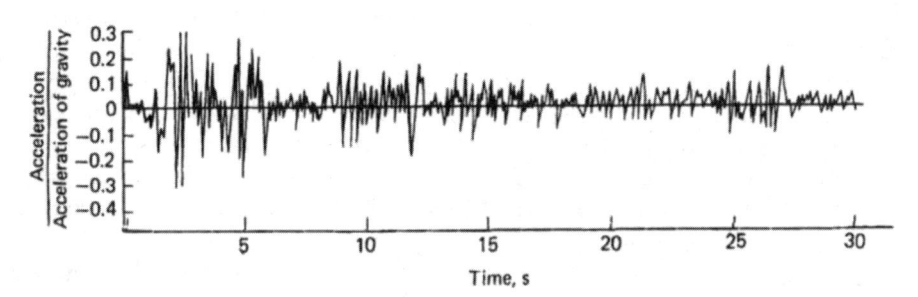

Figure 3.4   El Centro, California, earthquake ground acceleration record.

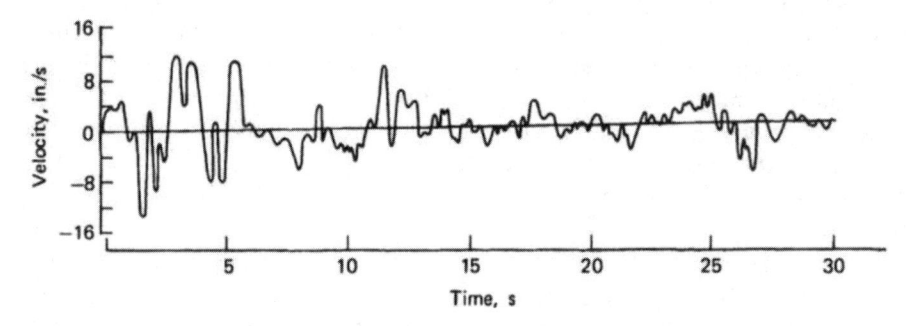

Figure 3.5   El Centro, California, earthquake ground velocity record.

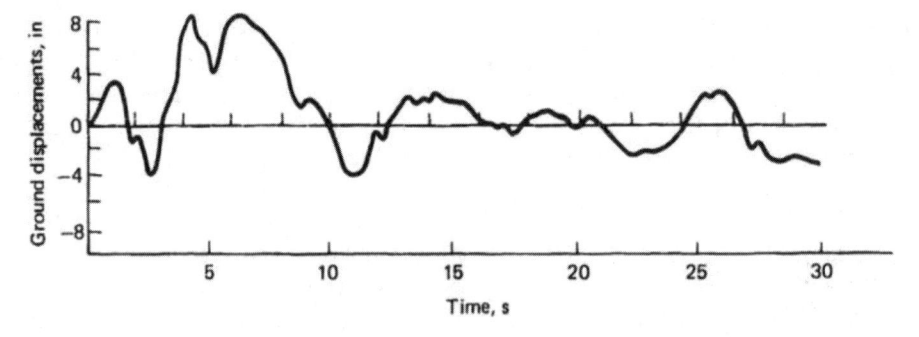

Figure 3.6   El Centro, California, earthquake ground displacement record.

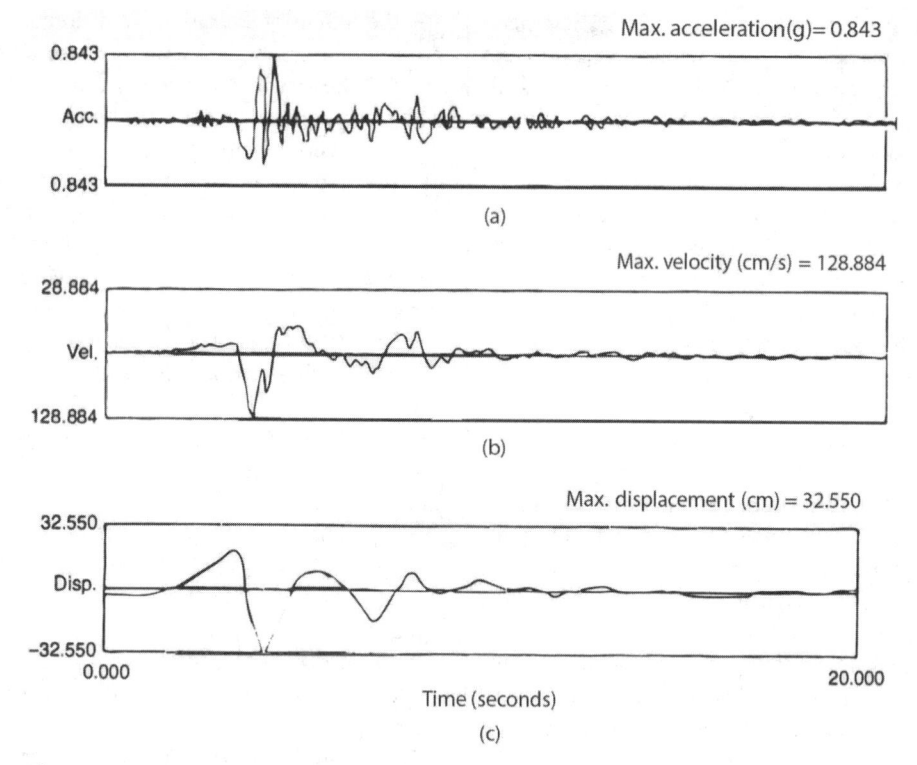

Figure 3.7   Northridge, California earthquake ground records, 1994. Sylmar county hospital parking lot (N–S direction): (a) acceleration; (b) velocity; (c) displacement. (Courtesy of Dr. Farzad Naeim.)

Although peak ground acceleration is considered most often as the single measure of damage potential, it is a combination of several characteristics of ground motion that are important to structural response. These are: (i) amplitude; (ii) duration; (iii) frequency content that are related to: (i) magnitude; (ii) site distances from the fault; (iii) site and intervening soil conditions.

It is not possible to predict with any certainty when and where earthquakes will occur, how strong they will be, and what characteristics the ground motions will have; therefore, the engineer must estimate the ground shaking. A simple method is to use a seismic zone map, such as those used in conjunction with building codes. The map is based on the seismic history and geological information from the area. A detailed method of estimating design earthquake is to conduct a site-specific seismic evaluation, which takes into account seismic history, active faults in the vicinity of the building site, and the stress-strain properties of materials through which the seismic waves travel.

Earthquake forces result from the vibratory motion of the ground on which the structure is supported. The vibratory motion of the ground sets up inertia forces both vertically and horizontally, but it is customary to neglect the vertical component except for cantilevers, since most members have adequate reserve strength for vertical loads because of safety factors specified in codes for gravity loads. The horizontal forces, equal to mass times acceleration represent the inertia forces occurring at the critical instant of maximum deflection and zero velocity during the largest cycle of vibration as the structure responds to the earthquake motion. In virtually all earthquake design practice, the structure is analyzed as an elastic system, although it is acknowledged that the elastic limit of the structural members will be exceeded during the violent shaking of a major earthquake.

3.1.6   Uncertainties in Seismic Design

In most structural engineering problems, we can evaluate with a fair degree of accuracy the dead loads and the live loads which the structure must be able to support. The strength and properties of materials used in the construction of a structure are well defined by many tests. With the available theories of mechanics and digital computers we can determine to a high degree of precision the moments, shears, and forces that the members will be subjected to. From the material and geometric properties, the amount of resistance the members can provide to these forces can be assessed. With this complete and adequate information, a realistic factor of safety against collapse can be established.

Consider, for example, the lateral design of a tall building subjected to wind loads. Unless the building is very slender, with a height-to-width ratio of greater than 6 or so, the building, if designed for code-designated wind forces, is more than likely to perform adequately throughout its life span. Even when the slenderness ratio is greater than 6 and the building shape is unusual, the engineer can fairly accurately determine the design lateral loads by either using correction factors to the basic wind loads to take into account the dynamic nature of wind or by resorting to wind tunnel tests to determine the worst possible wind load that can be expected over the life of the building. In other words, engineers can get fairly reasonable estimate of lateral loads either by analytical procedures or by wind-tunnel experiments. The measured wind loads on actual buildings, although the available data are limited, seem to indicate that the actual wind loads on buildings are somewhat smaller than those usually assumed in design, adding an additional degree of confidence.

The situation with regard to earthquake forces is entirely different. The seismic forces specified in the code are quite small, relative to the actual forces expected at least once in the life cycle of the building. It is important to understand the principles behind the code specifications and the justification for designing for lateral earthquake forces of 3 to 20 percent of gravity as compared to dynamic analysis requirements of over 50 percent of gravity. Accelerations derived from actual earthquakes are high when compared to the code forces used in design. A design based on manipulation of numbers to come up to code requirements without appreciation of the code intent will certainly not assure adequate earthquake resistance in case of a major earthquake. A better approach is to design on some reasonable basis, recognizing the uncertain nature of demands and to provide for all the reserve capacity that can be incorporated at little or no extra cost in initial construction or at only a slight sacrifice in architectural features. This in essence is the underlying philosophy of earthquake design.

The seismic loads on the structure during an earthquake result from inertia forces, which are created by ground accelerations. The magnitude of these loads is a function of the following factors:

- mass of the building;
- the dynamic properties of the building, such as its mode shapes, periods of vibration, and its damping;

- the intensity, duration, and frequency content of ground motion and soil structure interaction.

From the structural viewpoint, the intensity of vibration of the earth's surface at the building site is of interest. Such intensity of vibration is a function of: (i) amount of energy released; (ii) distance from center of earthquake to the structure; and (iii) character and thickness of foundation material. The magnitude of earthquake, which is a function of the energy released, can be predicted on a regional basis from probability theories. Mathematical theories are available to predict the effect of distance for various soil conditions underlying a site using assumed bedrock vibrations. However, there are too many unknowns to be able to predict quantitatively with any degree of certainty the ground vibration for some unknown future earthquake. Qualitatively, the following are apparent:

1. Ground shaking is strongest in the vicinity of the causative fault, and the intensity diminishes with distance from the fault.

2. Deep deposits of soft soils tend to produce ground surface motions having predominantly long-period characteristics.

3. Shallow deposits of stiff soils result in ground motions having predominantly short-period characteristics.

4. The soil amplification varies with frequency and intensity of the bedrock motions.

In spite of the great strides made in earthquake engineering during the last five decades, numerous uncertainties still exist. The traditional static approach of determining the force level for a given earthquake motion and designing the structure to withstand these forces with a considerable degree of safety has very serious limitations because of the following problems:

1. There simply are not enough empirical data available at the present time to make a reliable prediction as to the intensity and nature of future earthquakes at a given time.

2. Foundation and soil interaction and geological conditions have a profound effect on the structural performance, but at present there exists no clear-cut method which can correctly incorporate these effects.

3. Analysis by elastic assumptions does not take into account the change in properties of the building materials during the progress of an earthquake.

Because of these uncertainties, it is necessary when applying the static load criteria to evaluate the capabilities of the structure to perform satisfactorily beyond the elastic-code-stipulated stresses. Ductility, which involves deformations into the inelastic range is a necessity if structures designed for the static forces are to be capable of resisting earthquakes of the intensity of those which have been actually recorded.

Recent codes require that the function of buildings be taken into account in the earthquake design. This is a direct consequence of the lessons learned from the 1971 San Fernando and later earthquakes, in which several medical facilities were rendered useless and vacated, becoming liabilities rather than maintaining emergency service. Higher loads are specified for other vital public buildings whose functioning is considered indispensable in rescue and recovery efforts.

In earthquake-resistant design, it is not necessary to consider the simultaneous action of wind and earthquake loads, since the probability of this occurrence is quite low. There is no record of an extreme wind and earthquake load hitting a building at the same time. It is expected, as in the case of wind loads, that under the action of moderate earthquake loads, the building structure will remain within the elastic range. Under code-specified earthquake forces, which represent the action of a moderately large earthquake, it is reasonable to expect the structure to maintain elastic behavior. In the case of catastrophic earthquakes, however, a different philosophy exists that permits the building to venture into the plastic range. Certain portions of the building are permitted to suffer minor damage, provided the stability of the structure as a whole is not impaired. The occasional excursion of the building into the plastic range is accepted on the premise that the peak forces produced by earthquakes are of short duration and therefore can be more readily absorbed by the movement of the building than a sustained static load.

3.1.7 Design Ground Motion

It is important to recognize that the discipline of design ground motion specification is in a state of evolution. For typical building applications, even with the use of computers for probabilistic studies, the exact prediction of ground motions at a site is not possible.

The seismic hazard at a site is usually evaluated using probabilistic methods by considering all possible earthquakes in the area, estimating the associated shaking at the site and calculating the probabilities of these occurrences. Given the current limited knowledge and understanding of the earthquake process, all assessments of earthquake hazard are inherently uncertain. Probability methods are not used by design engineers on a regular basis and therefore the methods of generating response spectra may be difficult to comprehend. However, a brief discussion of earthquake hazard analysis techniques used in determining seismic ground motions is given below.

Available procedures for assessing seismic ground motions vary from fully deterministic procedures through hybrid (partly deterministic to partly probabilistic), to fully probabilistic procedures. In a deterministic approach, a single maximum earthquake is specified by magnitude and location with respect to a site of interest. By using an attenuation function, the reduction in the intensity of the seismic motion due to the distance of the site from the specified seismic source is calculated. The resulting reduced ground motion is used to design or evaluate the seismic vulnerability of a facility. The target earthquake, usually the maximum credible, also called the maximum considered earthquake, is selected by consideration of the historical seismic record and physical characteristics of the seismic source. The deterministic approach does not consider the likelihood of the occurrence of an earthquake nor does it consider the importance of the target earthquake relative to other possible seismic events, such as those due to larger but more distant earthquakes or smaller but closer earthquakes. For this reason, this approach is seldom used in practice.

The probabilistic approach, on the other hand, addresses the questions of how strongly and how often the ground will shake, by considering all possible earthquakes likely to affect the site. As in the deterministic procedure, an attenuation function, together with the distance from the seismic source, is used to estimate ground motion at the site resulting from a variety of seismic events. The rate of earthquake occurrence for each seismic source is also considered. Thus the probabilistic procedure combines information on earthquake size, location, probability of occurrence, and resulting ground motion to give results in terms of expected ground motion and associated annual probabilities of occurrence. The objective of the analysis is to provide an estimate of ground motion at a specific site. Typically the ground motion is expressed in terms of peak acceleration, velocity, displacement or a design spectrum at the site. Duration of strong motion is an important measure, but is not explicitly used in design criteria at the present time.

3.2 TALL BUILDING BEHAVIOR DURING EARTHQUAKES

3.2.1 Introduction

The behavior of a tall building during an earthquake is a vibration problem. The seismic motions of the ground do not damage a building by impact as does a wrecker's ball, or by externally applied pressure such as wind, but rather by internally generated inertial forces caused by vibration of the building mass. An increase in the mass has two undesirable effects on the earthquake design.

First, it results in an increase in the force, and second, it can cause buckling and crushing of vertical elements such as columns and walls when the mass pushing down exerts its force on a member bent or moved out of plumb by the lateral forces. This phenomenon is known as the p-Δ effect. The greater the vertical force, the greater the movement due to p-Δ. It is almost always the vertical load that causes buildings to collapse; in earthquakes, buildings very rarely fall over—they fall down. The seismic motions of the ground cause the structure to vibrate, and the amplitude and distribution of dynamic deformations and their duration are of concern to the engineer. Note that although duration of strong motion is an important measure, it is not explicitly used in design criteria at the present time.

3.2.2 Response of Tall Buildings

In general, tall buildings respond to seismic motion somewhat differently than low-rise buildings. The magnitude of inertia forces induced in an earthquake depends on the building mass, ground acceleration, the nature of foundation, and the dynamic characteristics of the structure (Fig. 3.8). If a building and its foundation were infinitely rigid, it would have the same acceleration as the ground: the inertia force F for a given ground acceleration a given by Newton's Law $F = Ma$, where M is the building mass. For a structure that deforms only slightly, thereby absorbing some energy, the force F tends to be less than the product of mass and ground acceleration. Tall buildings are invariably more flexible than low-rise buildings, and in general experience accelerations much less than low-rise buildings. But a flexible building subjected to ground motions for a prolonged period may experience much larger forces if its natural period is near that of the ground waves. Thus, the magnitude of lateral force in a building is not a function of the acceleration of the ground alone but is influenced to a great extent by the type of response of the structure and its foundation, as well. This interrelationship of building behavior and seismic ground motion also depends on the building period and is expressed in the so-called response spectrum, explained later in this chapter.

Consider, for example, the behavior of a 30-story building during an earthquake. Although the motion of the ground is erratic and three-dimensional, the horizontal components in two mutually perpendicular directions are of importance. These components typically have varying periods and can be considered as short-period

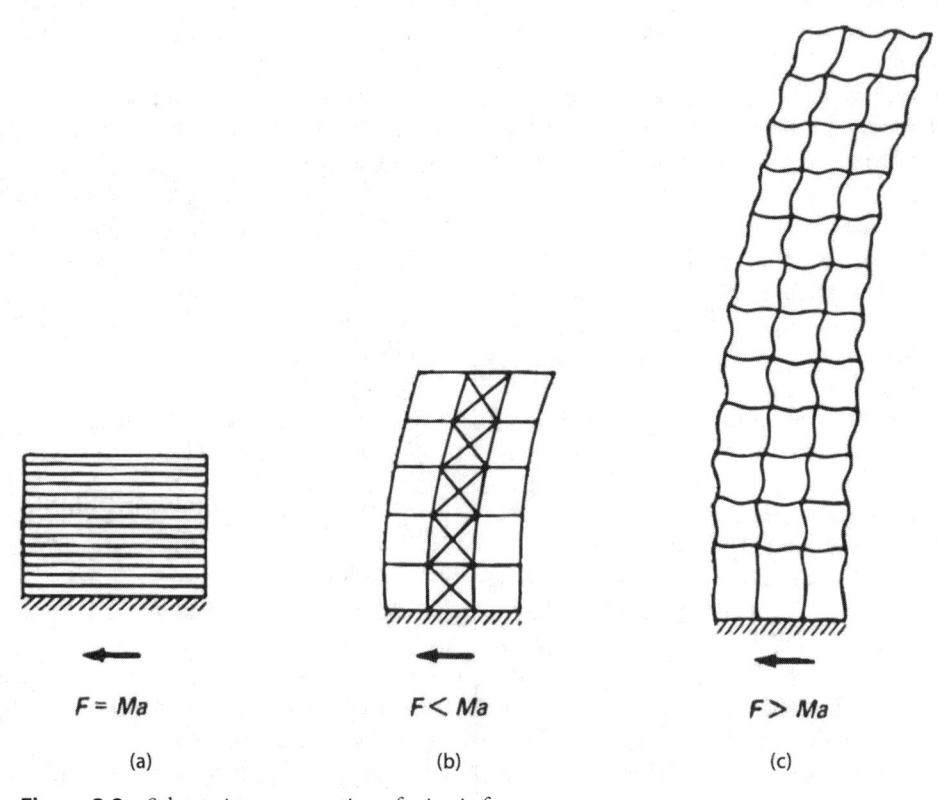

$F = Ma$ $F < Ma$ $F > Ma$

(a)   (b)   (c)

Figure 3.8 Schematic representation of seismic forces.

components when the period is less than 0.5 s, and long-period components for periods in excess of 0.5 s. The period of fundamental frequency T_1 of a tall building is a function of its stiffness, mass, and damping characteristics, and can vary over a broad range anywhere from 0.05 to 0.30 times the number of stories, depending upon the materials used in the construction and the structural system employed. As a preliminary approximation for steel-framed buildings, the period T_1 is approximately equal to $0.15N$, where N is the number of stories. A typical 30-story building would have a fundamental period of 4.5 s, with the periods of the next two higher modes, T_2 and T_3, approximately equal to one-third and one-fifth of T_1.

The second and third modes of vibration for the 30-story building are thus approximately equal to 1.5 and 0.9 s. During the first few seconds of earthquake, the acceleration of the ground reaches a peak value and is associated with relatively short-period components of the range 0 to 0.5 s, which have little influence on the fundamental response of the building. On the other hand, the long-period components that occur at the tail-end of the earthquakes, with periods closer to the fundamental period of the building have a profound influence on its behavior.

The intensity of ground motion reduces with the distance from the epicenter of the earthquake. The reduction of ground motion intensity, called attenuation, is at a faster rate for higher-frequency components than for lower-frequency components. The cause for the change in attenuation rate is not understood, but its existence is certain. This is a significant factor in the design of tall buildings, because a tall building, although situated farther from a causative fault than a low-rise building, may experience greater seismic loads because long-period components are not attenuated as fast as the short-period components. Therefore, the area influenced by ground shaking potentially damaging to, say, a 50-story building is much greater than to a one-story building.

Effective seismic design generally includes the following:

1. Selecting an overall structural concept including the layout of an LFRS that is appropriate to the anticipated level of ground shaking. This includes providing a redundant and continuous load path to ensure that the building responds as a unit when subject to ground motion.

2. Determining forces and deformations generated by the ground motion and distributing the same vertically to the LFRS with due consideration to the structural system, configuration, and site characteristics.

3. Analysis of the building for the combined effects of gravity and seismic loads to verify that adequate vertical and lateral strength and stiffness are achieved to satisfy the structural performance and acceptable deformation levels prescribed in the governing building code.

4. Providing details that allow for expected structural movements without damage to nonstructural elements such items as piping, window glass, plaster, veneer, and partitions. To minimize this type of damage, special care in detailing, either to isolate these elements or to accommodate the expected movement, is required. Breakage of glass windows can be minimized by providing adequate clearance at edges to allow for frame distortions. Damage to rigid nonstructural partitions can be largely eliminated by providing a detail, which will permit relative movement between the partitions and the adjacent structural elements.

5. In piping installations, use of expansion loops and flexible joints to accommodate relative seismic deflections between adjacent equipment items and the building floors.

6. Fasten freestanding shelving to walls to prevent toppling.

7. Stairways often suffer seismic damage because they tend to prevent drift between connected floors. This can be avoided by providing a slip joint at the lower end of each stairway to eliminate their bracing effect or by tying stairways to stairway shear walls.

3.2.3   Influence of Soil

The seismic motion that reaches a structure on the surface of the earth is influenced by the local soil conditions. The subsurface soil layers underlying the building foundation may amplify the response of the building to earthquake motions originating in the bedrock. Although it is difficult to visualize, it is possible that a number of underlying soils can have a period similar to the period of vibration of the structure. Low- to mid-rise buildings typically have periods in the 0.10 to 1.0 s range while taller, more flexible buildings have periods between 1 and 5 s or greater. Harder soils and bedrock will efficiently transmit short-period vibrations (caused by near-field earthquakes) while filtering out longer-period vibrations (caused by distant earthquakes) whereas softer soils will transmit longer-period vibrations.

As a building vibrates underground motion, its acceleration will be amplified if the fundamental period of the building coincides with the period of vibrations being transmitted through the soil. This amplified response is called resonance. Natural periods of soil are in the range of 0.5 to 1.0 s so that it is entirely possible for the building and ground to have the same fundamental period, and therefore, for the building to approach a state of resonance. This was the case for many 5- to 10-story buildings in the September 1985 earthquake in Mexico City. An obvious design strategy, if one can predict approximately the rate at which the ground will vibrate, is to ensure that buildings have a natural period different from that of the expected ground vibration to prevent amplification.

The intensity of ground motion reduces with the distance from the epicenter of the earthquake. The reduction, called attenuation, occurs at a faster rate for high-frequency (short-period) components than for lower-frequency (long-period) components. The cause of the change in attenuation rate is not understood, but its existence is certain. This is a significant factor in the design of tall buildings, because a tall building, although situated farther from a causative fault than a low-rise building, may experience greater seismic loads because long-period components are not attenuated as fast as the short-period components. Therefore, the area influenced by ground shaking potentially damaging to, say, a 50-story building is much greater than for a one-story building. As a building vibrates due to ground motion, its acceleration will be amplified if the fundamental period of the building coincides with that of the soil it rests upon. It is worth noting that natural periods of soil are typically in the range of 0.5 to 1.0 s. Therefore, it is entirely possible for a building and ground it rests upon to have the same fundamental period. This was the case for many 5- to 10-story buildings that were damaged in the September 1985 earthquake in Mexico City. Experience in several other earthquakes has confirmed that local soil conditions can have a significant effect on earthquake response. It is perhaps more challenging to picture, but the soil layers beneath a structure have a period of vibration T_{soil} similar to the period of vibration of a building T.Greater structural damage is likely to occur when the period of the underlying soil is close to the fundamental period of the structure. In these cases, a partial resonance effect may develop between the structure and the underlying soil. These conditions are addressed in ASCE.7-16. By classifying soil profiles, into Site Class A through F.

3.2.4 Damping

Buildings do not resonate with the purity of a tuning fork because they are damped; the extent of damping depends upon the construction materials, type of connections, and the presence of nonstructural elements. Damping is measured as a percentage of critical damping.

In a dynamic system, critical damping is defined as the minimum amount of damping necessary to prevent oscillation altogether. To get a mental picture of critical damping, imagine a tightly tensioned string immersed in a tank containing water. When the string is plucked, it oscillates about its mean position several times before coming to rest. If we replace water with a liquid of higher viscosity, the string will oscillate, but certainly not as many times as it did in water. By progressively increasing the viscosity of the liquid, it is easy to visualize that a state can be reached where the string, once plucked, will return to its neutral position without ever crossing it. The minimum viscosity of the liquid that prevents the vibration of the string altogether can be considered equivalent to critical damping.

The damping of structures under earthquake disturbances is influenced by a number of external and internal sources. Chief among them are:

1. External viscous damping caused by air surrounding the building. Since the viscosity of air is small, this effect is negligible in comparison to other types of damping.

2. Internal viscous damping associated with the material viscosity. This is proportional to the velocity and increases in proportion to the natural frequency of the structure.

3. Friction damping, also called Coulomb damping, occurring at connections and support points of the structure. It is a constant, irrespective of the velocity or amount of displacement. Steel buildings with bolted connections have more friction damping as compared to a fully welded construction. A prestressed concrete building has less damping as compared to a mild-steel-reinforced construction because in prestressed concrete cracking of concrete is relatively less.

4. Hysteresis damping that develops when the structure is subjected to load reversals in the inelastic range. The area inside of the hysteresis loop corresponds to the energy dissipated and is referred to as hysteretic damping. It increases with the level of displacement and is independent of the velocity of the structure.

5. Radiation damping resulting from energy dissipation through the ground on which the structure is built. It is a function of the characteristics of the ground such as density, Poisson's ratio, shear and elastic moduli, and the depth to which the structure is embedded in the ground.

6. Hysteresis damping around the foundation caused by the inelastic deformation of the ground adjacent to the foundation.

It is common practice in dynamic analysis of buildings to lump all of the various sources of damping into one type and to represent it as viscous damping. To arrive at a precise value of viscous damping that can effectively take into account all the aforementioned characteristics is impractical. Representative values of damping ratios used in practice vary anywhere from 0.02 to 0.10 depending upon the material used for the building and the level of design force used in the analysis. Vibration tests of existing buildings indicate that damping ratios vary from a low of 0.005 to a high of 0.075. For non-base-isolated buildings, analyzed for code-prescribed loads, the damping ratios used in practice vary anywhere from 1 to 10 percent of critical. The low-end values are for wind, while those for the upper end are for seismic design. The damping ratio used in the analysis of seismic base-isolated buildings is rather large compared to values used for nonisolated buildings and varies from about 0.20 to 0.35 (20 percent to 35 percent of critical damping). Base isolation, discussed elsewhere in this book, consists of mounting a building on an isolation system to prevent horizontal seismic ground motions from entering the building. This strategy results in significant reductions in interstory drifts and floor accelerations, thereby protecting the building and its contents from earthquake damage.

In earthquake design, the idea of critical damping is used to modify the ground response spectrum by assuming certain percentages of damping, generally of the order of 2 to 15 percent of critical. Low-end values are used in wind engineering, while upper-end values are used in earthquake engineering.

A level of ground acceleration generally at 0.1 g, where g is the acceleration due to gravity, is sufficient to produce some damage to weak construction. An acceleration of 1.0g, or 100 percent of gravity, is analytically equivalent, in the static sense, to a building that cantilevers horizontally from a vertical surface.

As stated previously, the process by which free vibration steadily diminishes in amplitude is called damping. In damping, the energy of the vibrating system is dissipated by various mechanisms, and often, more than one mechanism may be present at the same time. In simple laboratory models, most of the energy dissipation arises from the thermal effect of repeated elastic straining of the material and from the internal friction. In actual structures, however, many other mechanisms also contribute to the energy dissipation. In a concrete building, these include opening and closing of microcracks in concrete and friction between the structure itself and nonstructural

elements such as partition walls. Invariably, it is impossible to identify or describe mathematically each of these energy-dissipating mechanisms in an actual building Therefore, the damping in actual structures is usually represented in a highly idealized manner. For many purposes, the actual damping in structures can be idealized by a linear viscous damper or dashpot. The damping coefficient is selected so that the vibrational energy that dissipates is equivalent to the energy dissipated in all the damping mechanisms. This idealization is called equivalent viscous damping.

Figure 3.9 shows a linear viscous damper subjected to a force f_D. The damping forces f_D are related to the velocity u across the linear viscous damper by $f_D = cu$ where the constant c is the viscous damping coefficient; it has units of force x time/length.

Damping is defined as a force that resists dynamic motion which is considered a simple and realistic damping model for analysis purposes. The damping force, f_D, is proportional to the viscous friction of a fluid in a dashpot and therefore it is called viscous damping.

3.2.5 Building Motion and Deflections

Earthquake-induced motions, even when they are more violent than those induced by wind, evoke a totally different type of human response. First, because earthquakes occur much less frequently than windstorms, and second, because the duration of motion caused by an earthquake is generally short. People who experience earthquakes are grateful that they have survived the trauma and are less inclined to be critical of building motion. Earthquake-induced motions are, therefore, a safety rather than a human discomfort phenomenon.

Lateral deflections that occur during earthquakes should be limited to prevent distress in structural members and architectural components. Non-load-bearing in-fills, external wall panels and window glazing should be designed with sufficient clearance or with flexible supports to accommodate the anticipated movements.

3.2.6 Building Drift and Separation

Drift is the lateral displacement of one floor relative to the floor below. Buildings subjected to earthquakes need drift control to limit damage to interior partitions, elevator

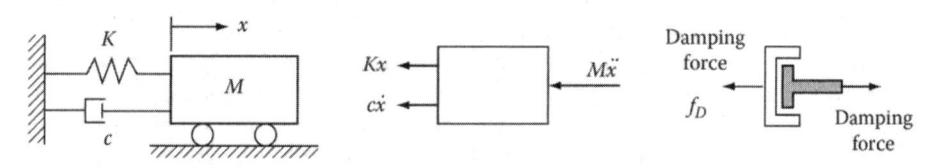

Figure 3.9 Linear viscous damper.

and stair enclosures, glass, and cladding systems. Stress or strength limitations do not always provide adequate drift control especially for tall buildings with relatively flexible moment-resisting frames (MRFs) and narrow shear walls.

Total building drift is the absolute displacement of any point relative to the base. Adjoining buildings or adjoining sections of the same building do not have identical modes of response and, therefore, have a tendency to pound against one another. Building separations or joints must be provided to permit adjoining buildings to respond independently to earthquake ground motion.

3.3   SEISMIC DESIGN CONCEPT

3.3.1   Determination of forces

There are two general approaches to determining seismic forces: an equivalent static force procedure and a dynamic analysis procedure. In this section, the equivalent static force procedure is illustrated. Dynamic analysis procedures are considered later, but some discussion of structural dynamics is included in this section in order to explain the rationale of the equivalent static force procedure.

3.3.2   Design of the Structure

The development of an adequate earthquake-resistant design for a structure includes the following: (i) selecting a workable overall structural concept; (ii) establishing member sizes; (iii) performing a structural analysis of the members to verify that stress and displacement requirements are satisfied; and (iv) providing structural and nonstructural details so that the building will accommodate the distortions and stresses that occur in the building. Elements that cannot accommodate these stresses and distortions such as rigid stairs, partitions, and irregular wings should be isolated to reduce detrimental effects on the lateral force-resisting system.

3.3.3   Structural Response

If the base of a structure is suddenly moved, as in the case of seismic ground motion, the upper part of the structure will not respond instantaneously but will lag because of the inertial resistance and flexibility of the structure. This concept is illustrated in Figs. 3.8a–c by showing the motion in one plane. The stresses and distortions in the building are the same as if the base of the structure were to remain stationary while time-varying horizontal forces are applied to the upper part of the building. These forces, called inertia forces, are equal to the product of the mass of the structure

times acceleration, or $F = ma$ (The mass m is equal to weight divided by the acceleration of gravity, i.e., $m = w/g$). Because the ground motion at a point on the earth's surface is three-dimensional (one vertical and two horizontal), the structures affected will deform in a three-dimensional manner. Generally, however, the inertia forces generated by the horizontal components of ground motion require greater consideration for seismic design since adequate resistance to vertical seismic loads is usually provided by the member capacities required for gravity load design. In the equivalent static procedure, the inertia forces are represented by equivalent static forces.

3.3.4   Path of Forces

Buildings are composed of vertical and horizontal structural elements which resist lateral forces. The vertical elements that are used to transfer lateral forces to the ground are: (i) shear walls; (ii) braced frames; and (iii) MRFs, or combinations. Horizontal elements that distribute lateral forces to vertical elements are: (i) most usually diaphragms, such as floor slabs; and (ii) horizontal bracing in special floors such as transfer floors. Horizontal forces produced by seismic motion are directly proportional to the masses of building elements and are considered to act at the center of the mass of these elements. All of the inertia forces originating from the masses on and off the structure must be transmitted to the lateral force-resisting elements, and then to the base of the structure and into the ground.

3.3.5   Demands of Earthquake Motion

The loads or forces that a structure sustains during an earthquake result directly from the distortions induced in the structure by the motion of the ground on which it rests. Base motion is characterized by displacements, velocities, and accelerations which are erratic in direction, magnitude, duration, and sequence. Earthquake loads are inertia forces related to the mass, stiffness, and energy-absorbing (e.g., damping and ductility) characteristics of the structure. During the life of a structure located in a seismically active zone, it is generally expected that the structure will be subjected to many small earthquakes, some moderate earthquakes, one or more large earthquakes, and possibly a very severe earthquake. In general, it is uneconomical or impractical to design buildings to resist the forces resulting from large or severe earthquakes within the elastic range of stress. If the earthquake motion is severe, most structures will experience yielding in some of their elements. The energy-absorption capacity of the yielding structure will limit the damage so that buildings that are properly designed and detailed can survive earthquake forces that are substantially greater than the design forces that are associated with allowable stresses in the elastic range. Seismic design concepts must consider building proportions and details for their ductility (capacity to yield) and reserve energy-absorption capacity for surviving the inelastic deformations that would result from a maximum expected earthquake. Special attention must be given to connections that hold the lateral force-resisting elements together.

3.3.6 Response of Buildings

A building is analyzed for its response to ground motion by representing the structural properties in an idealized mathematical model as an assembly of masses interconnected by springs and dampers. The tributary weight to each floor level is lumped into a single mass, and the force-deformation characteristics of the lateral force-resisting walls or frames between floor levels are transformed into equivalent story stiffnesses. Because of the complexity of the calculations, the use of a computer program is generally necessary, even when the design is by the equivalent static force procedure.

3.3.7 Response of Elements Attached to Buildings

Elements attached to the floors of buildings (e.g., mechanical equipment, ornamentation, piping, nonstructural partitions) respond to floor motion in much the same manner as the building responds to ground motion. However, the floor motion may vary substantially from the ground motion. The high-frequency components of the ground motion tend to be filtered out at the higher levels in the building while the components of ground motion that correspond to the natural periods of vibrations of the building tend to be magnified. If the elements are rigid and are rigidly attached to the structure, the forces on the elements will be in the same proportion to the mass as the forces on the structure. But elements that are flexible and have periods of vibration close to any of the predominant modes of the building vibration will experience forces in a proportion substantially greater than the forces on the structure.

3.3.8 Techniques of Seismic Design

For gravity loads, it has been a long-standing practice to design for strength and deflections within the elastic limits of the members. However, to control design within elastic behavior for the maximum expected horizontal seismic forces is impractical in high-seismicity areas. Therefore, several problems of building design should be recognized by the building owner, architect, and engineer as factors that may substantially increase the earthquake risk to their building. The solution lies in the design teams' understanding of seismic-resistant design rather than in the application of specific code provisions. A few of these problems are discussed below.

3.3.8.1 LAYOUT

A great deal of a building's resistance to lateral forces is determined by its plan layout. The objective in this regard is symmetry about both axes, not only of the building itself but of the arrangement of wall openings, columns, shear walls, etc. It is most desirable to consider the effect of lateral forces on the structural system from the start of the layout since this may save considerable time and money without detracting significantly from the usefulness or appearance of the building.

3.3.8.2 STRUCTURAL SYMMETRY

Experience has shown that buildings which are unsymmetrical in the plan have a greater susceptibility to earthquake damage than symmetrical structures. The effect of asymmetry will induce torsional oscillations of the structure and stress concentrations at re-entrant corners. Asymmetry in plan can be eliminated or improved by separating L-, T-, and U-shaped buildings into distinct units by use of seismic joints at junctions of the individual wings. Asymmetry caused by the eccentric location of lateral force-resisting structural elements, for example, a building that has a flexible front because of large openings and an essentially stiff (solid) rear wall, can usually be avoided by better conceptual planning, for example, by modifying the stiffness of the rear wall, or adding rigid structural elements to bring the center of rigidity (CR) of the lateral force-resisting elements close to the center of mass (CM).

3.3.8.3 IRREGULAR BUILDINGS

Those who have studied the performance of buildings in earthquakes generally agree that the building's form has a major influence. This is because the shape and proportions of the building have a major effect on the distribution of earthquake forces as they work their way through the building. Geometric configuration, type of structural members, details of connections, and materials of construction all have a profound effect on the structural-dynamic response of a building. When a building has irregular features, such as asymmetry in plan or vertical discontinuity, the assumptions used in developing seismic criteria for buildings with regular features may not apply. Therefore, it is best to avoid creating buildings with irregular features. For example, omitting exterior walls in the first story of a building to permit an open ground floor leaves the columns at the ground level as the only elements available to resist lateral forces, thus causing an abrupt change in rigidities at that level. This condition is undesirable. It is advisable to carry all shear walls down to the foundation. When irregular features are unavoidable, special design considerations are required to account for the unusual dynamic characteristics and the load transfer and stress concentrations that occur at abrupt changes in structural resistance. Examples of horizontal and vertical irregularities are illustrated in Figs. 3.10 and 3.11.

3.3.8.4 LATERAL FORCE-RESISTING SYSTEMS

There are several systems that can be used effectively for providing resistance to seismic lateral forces. Some of the more common systems are shown in Fig. 3.12. All of the systems rely basically on a complete, three-dimensional space frame; a coordinated system of moment frames, shear walls, or braced frames with horizontal diaphragms; or a combination of the systems.

1. In buildings where a space frame resists the earthquake forces, the columns and beams act in bending. During a large earthquake, story-to-story deflection

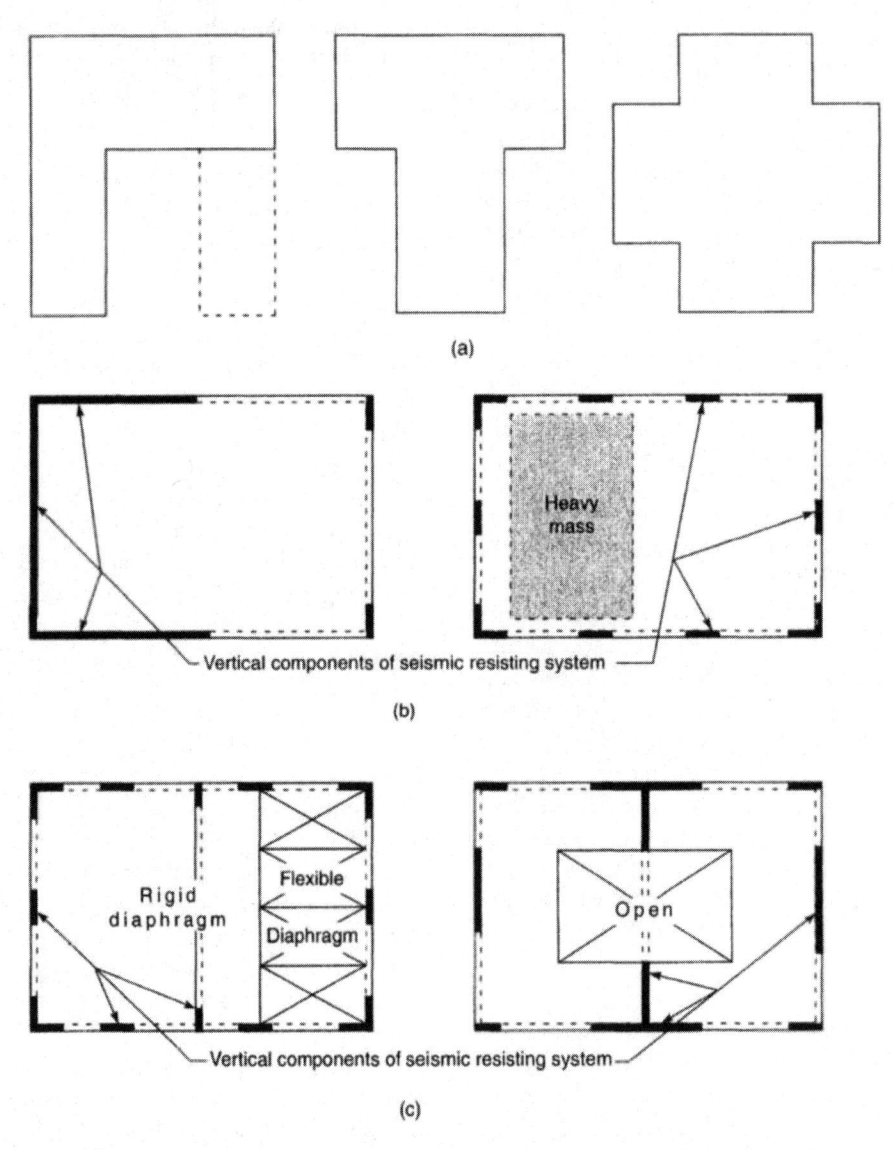

Figure 3.10 Horizontal irregularities: (a) geometric irregularities; (b) irregularity due to mass-resistance eccentricity; (c) irregularity due to discontinuity in diaphragm stiffness.

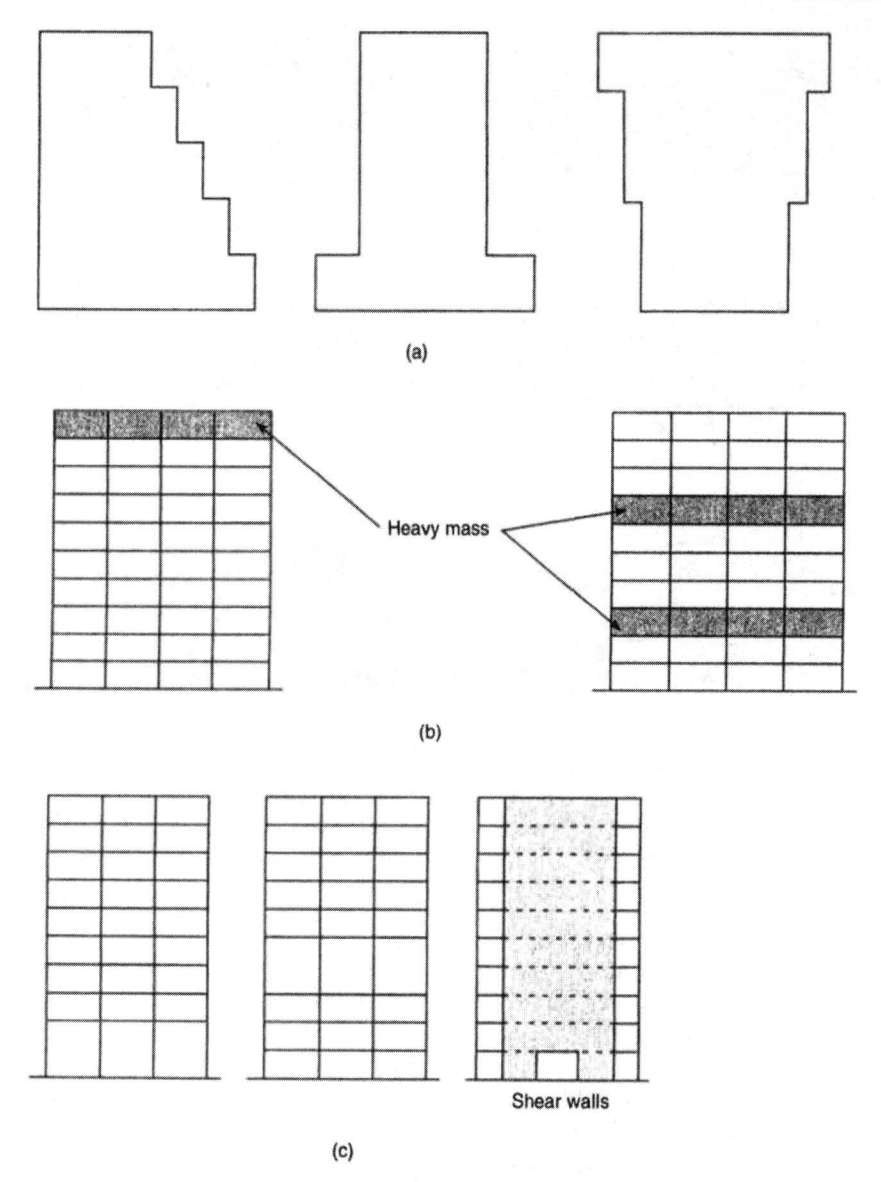

Figure 3.11 Vertical irregularities: (a) abrupt change in geometry; (b) large difference in floor masses; (c) large difference in story stiffnesses.

(story drift) may be accommodated within the structural system without causing failure of columns or beams. However, the drift may be sufficient to damage elements that are rigidly tied to the structural system such as brittle partitions, stairways, plumbing, exterior walls, and other elements that extend between floors. Therefore, buildings can have substantial interior and exterior nonstructural damage and still be considered as structurally safe. While there are excellent theoretical and economic reasons for resisting seismic forces by frame action, for particular buildings this system may be a poor economic risk unless special damage-control measures are taken.

2. A shear wall (or braced frame) building is normally more rigid than a framed structure. With low design stress limits in shear walls, deflection due to shear forces is relatively small. Shear wall construction is an economical method of bracing buildings to limit damage, and this type of construction is normally economically feasible up to about 15 stories. Notable exceptions to the excellent performance of shear walls occur when the height-to-width ratio becomes great enough to make overturning

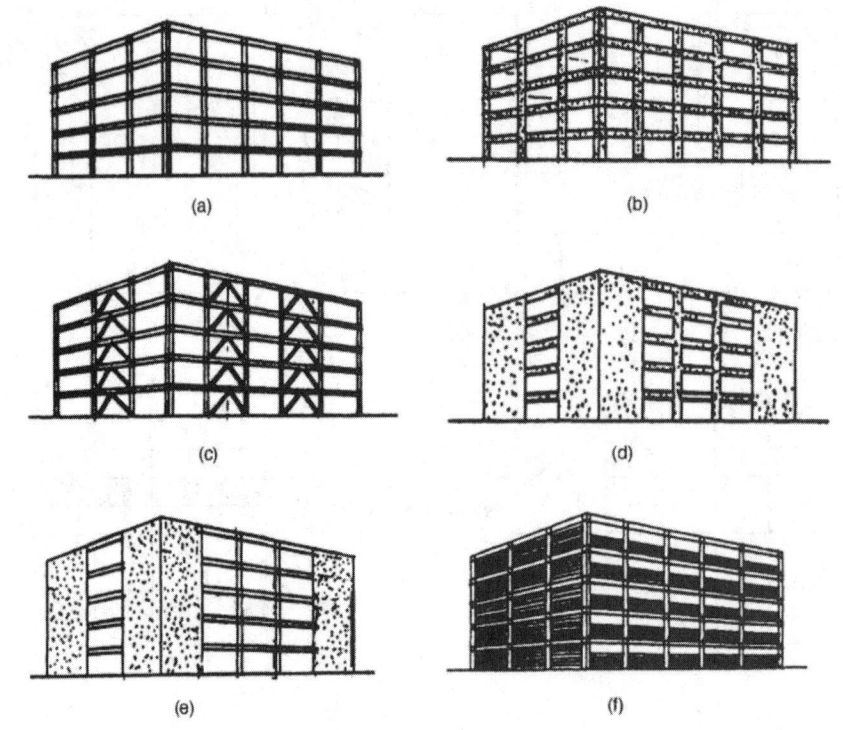

Figure 3.12 Lateral force-resisting systems: (a) steel moment-resisting frame building; (b) reinforced concrete moment-resisting frame building; (c) braced steel frame building; (d) reinforced concrete shear wall building; (e) steel frame building with cast-in-place concrete shear walls; (f) steel frame building with in-filled walls of nonreinforced masonry.

a problem and when there are excessive openings in the shear walls. Also, if the soil beneath its footings is relatively soft, the entire shear wall may rotate, causing localized damage around the wall.

3. The structural systems mentioned above may be used in combination. When frames and shear walls are combined, the system is called a dual bracing system. The type of structural system used, with specified details concerning the ductility and energy-absorbing capacity of its components, will establish the appropriate response modification factor, R, defined later, to be used for calculating the total base shear and to distribute the lateral seismic forces.

The design engineer must be aware that a building does not merely consist of a summation of parts such as walls, columns, trusses, and similar components but is a completely integrated system or unit that has its own properties with respect to lateral force response. The designer must follow the flow of forces through the structure into the ground and make sure that every connection along the path of stress is adequate to maintain the integrity of the system. It is necessary to visualize the

response of the complete structure and to keep in mind that the real forces involved are not static but dynamic, are usually erratically cyclic and repetitive, and can cause deformations well beyond those determined from the elastic design.

Structural engineers have the choice of three basic alternative types of lateral-force-resisting systems (LFRSs) as shown in Fig. 3.13.

These basic systems have a number of variations, related to the structural materials used and the connections between the members. Table 3.1 shows a summary of the seismic performance of structural systems in previous earthquakes.

Figures 3.14 and 3.15 show examples of structural systems suitable for different site conditions and occupancy types.

3.3.8.4.1 Shear Walls Shear walls are designed to support lateral forces distributed to the walls from diaphragms and transmit them to the ground. The forces in these walls are predominately shear forces. For an effective design, shear walls must run from the top of the building to the Foundation with no offsets and a minimum of openings.

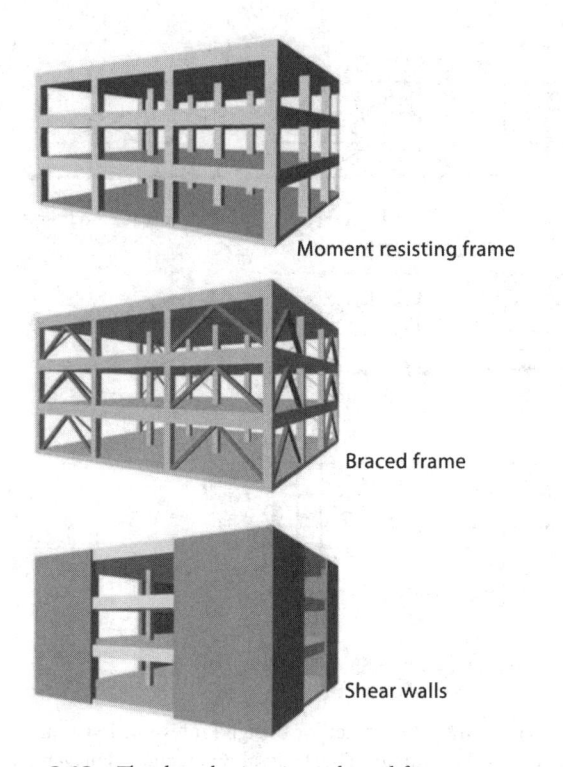

Moment resisting frame

Braced frame

Shear walls

Figure 3.13 The three basic seismic lateral force resisting systems (LFRSs). (From FEMA-454, Designing for Earthquakes: A Manual for Architects Washington, DC: Federal Emergency Management Agency, 2006.)

3.3.8.4.2 Braced Frames Braced frames are effective seismic systems with relatively high stiffness and act in the same way as shear walls; however, they generally provide less resistance but better ductility depending on their detailed design. They provide more architectural design Freedom than shear walls.

There are three general types of braced frames: conventional concentric braced frames (CBFs), concentric buckling restrained braced frames (BRBFs), and eccentric braced frames (EBFs). In the concentric frame, the center lines of the bracing members meet the horizontal beam at a single point. In the EBF, the braces are deliberately designed to meet the beam some distance apart from one another: the short piece of beam between the ends of the braces is called a link beam which is typically controlled by shear, bending, or combination depending on the length of the link beam. The purpose of the link beam is to provide ductility to the system: under heavy seismic forces, the link beam will yield, distort, and dissipate the energy of the earthquake in a controlled way, thus protecting the remainder of the structure.

3.3.8.4.3 Moment-resisting Frames An MRF is the engineering term for a frame structure with no diagonal bracing in which the lateral forces are resisted primarily by bending in the beams and columns mobilized by strong joints between columns and beams. MRFs provide

Table 3.1   Comparative Seismic Performance of Selected Structural Systems

Structural system	Earthquake performance	Specific building performance and energy absorption	General comments
Reinforced Concrete Wall	San Francisco, 1957 Alaska, 1964 Japan, 1966 Los Angeles, 1994 Variable to Poor	• Buildings in Alaska, San Francisco and Japan performed poorly with spandrel and pier failure • Brittle system	• Proportion of spandrel and piers is critical, detail for ductility and shear.
Steel Brace	San Francisco, 1906 Taft, 1952 Los Angeles, 1994 Variable	• Major braced systems performed well. • Minor bracing and tension braces performed poorly.	• Details and proportions are critical.
Steel Moment Frame	Los Angeles, 1971 Japan, 1978 Los Angeles, 1994 ? Good	• Los Angeles and Japanese buildings 1971/78 performed well • Energy absorption is excellent. • Los Angeles 1994, mixed performance.	• Both conventional and ductile frame have performed well if designed for drift.
Concrete Shear Wall	Caracas, 1965 Alaska, 1964 Los Angeles, 1971 Algeria, 1980 Variable	• Poor performance with discontinuous walls. • Uneven energy absorption.	• *Configuration is critical;* soft story or L-shape with torsion have produced failures.
Reinforced Concrete Ductile Moment Frame	Los Angeles, 1971| ? Good	• Good performance in 1971, Los Angeles • System will crack • Energy absorption is good. • Mixed performance in 1994 Los Angeles	• Details *critical*

Source: Federal Emergency Management Agency. *Primer for Design Professionals: Communicating with Owners and Managers of New Buildings on Earthquake Risk.* FEMA-389. Washington, DC: FEMA, 2004.

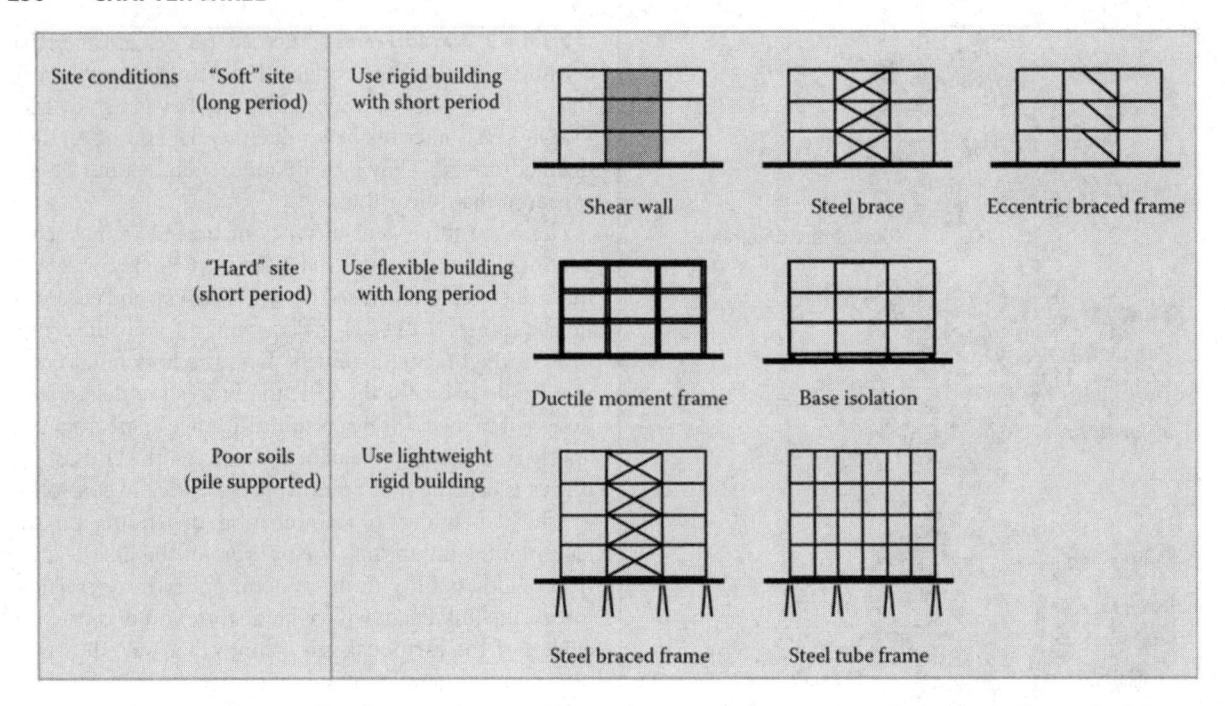

Figure 3.14   Examples of structural systems for site conditions. (Source: FEMA-389, Primer for Design Professionals: Communicating with Owners and Managers of New Buildings on Earthquake Risk. Washington DC: Federal Emergency Management Agency, 2004.)

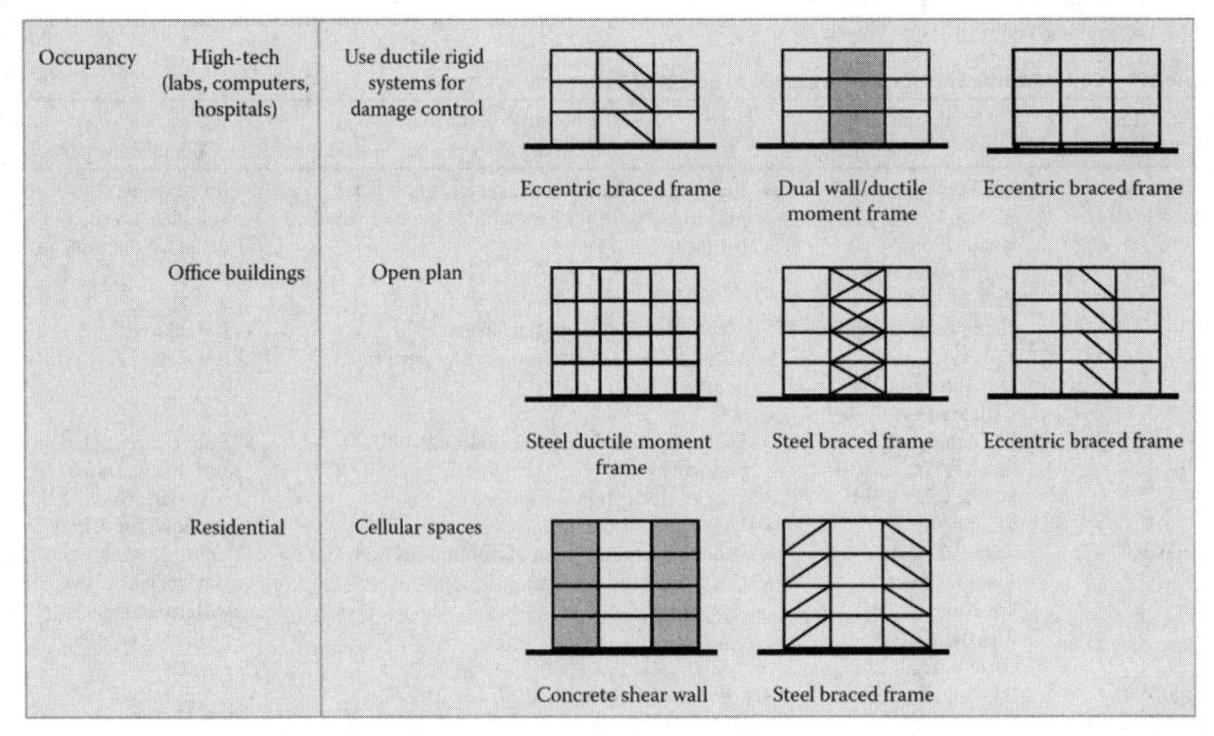

Figure 3.15   Structural systems for occupancy types. (Source: FEMA-389, Primer for Design Professionals: Communicating with Owners and Managers of New Buildings on Earthquake Risk. Washington DC: Federal Emergency Management Agency, 2004.)

the most architectural design freedom due to the elimination of diagonal braces and shear walls.

These systems are to some extent alternatives although designers often times mix systems using one type in one direction and another type in the other direction or combining them in one of the directions. This must be done with care, however, mainly because the different systems are of varying stiffness such as shear wall systems which are much stiffer than MRF systems and braced systems which falls in between. It is difficult to obtain balanced resistance when they are mixed. However, for high-performance structures, dual systems can be used, which is a combination of moment resisting frames and another system such as a shear wall or braced frame. Examples of effective mixed systems are the use of shear-wall core together with a perimeter MRF or a perimeter steel moment frame with interior eccentric-braced frames. Another variation is the use of shear walls combined with an MRF in which the frames are designed to act as a fail-safe backup case of shear-wall failure.

The framing system must be chosen at an early stage in the design because the seismic system plays the major role in determining the seismic performance of the building and more importantly has considerable influence on the architecture of design. For example, if shear walls are chosen as the seismic force-resisting system, the building planning must be able to accept a pattern of permanent structural walls with limited openings that run uninterrupted through every floor from roof to foundation. Moment frames can be designed with different ductility levels as required per the codes for height and other seismic requirements.

3.3.8.5 DIAPHRAGMS

The earthquake loads at any level of a building will be distributed to the vertical structural elements through the floor and roof diaphragms. The roof/floor deck or slab responds to loads like a deep beam. The deck or slab is the web of the beam carrying the shear and the perimeter spandrel or wall is the flange of the beam resisting bending.

Three factors are important in diaphragm design.

1. The diaphragm must be adequate to resist both the bending and shear stresses and be tied together to act as one unit.

2. The collectors and drag members, see Fig. 3.16, must be adequate to transfer loads from the diaphragm into the

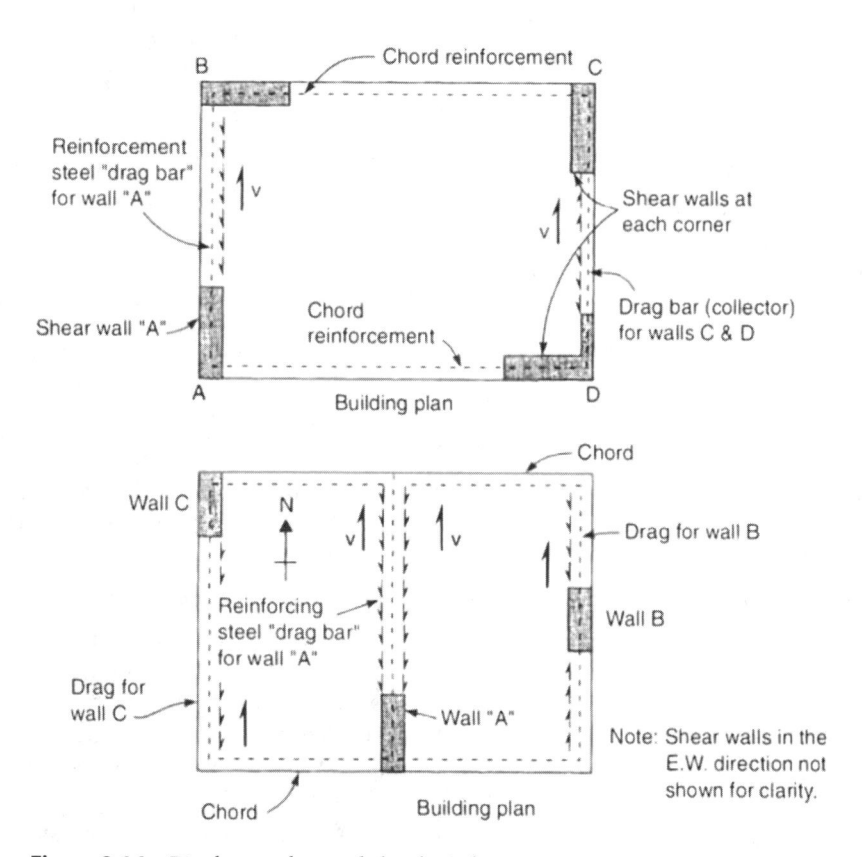

Figure 3.16 Diaphragm drag and chord reinforcement.

lateral load-resisting vertical elements such as shear walls and moment and braced frames.

3. Openings or re-entrant corners in the diaphragm must be properly placed and adequately reinforced.

Inappropriate location or large-size openings (stair or elevator cores, atriums, skylights) create problems similar to those related to cutting a hole in the web of a beam. This reduces the natural ability of the diaphragm to transfer the forces and may cause failure (Fig. 3.17).

3.3.8.6 DUCTILITY

Ductility is the capacity of building materials, systems, or structures to absorb energy by deforming into the inelastic range. The capability of a structure to absorb energy, with acceptable deformations and without failure, is a very desirable characteristic in any earthquake-resistant design. Brittle material such as concrete must be properly reinforced with steel to provide the ductility characteristics necessary to resist seismic forces. In concrete columns, for example, the combined effects of flexure (due to frame action) and compression (due to the action of the overturning moment of the structure as a whole) produces a common mode of failure; buckling of the vertical steel and spalling of the concrete cover near the floor levels. Columns must, therefore, be detailed with proper spiral reinforcing or hoops to have greater reserve strength and ductility.

Ductility is often measured by the hysteretic behavior of critical components such as a column-beam assembly of a moment frame. The hysteretic behavior is usually examined by observing the cyclic moment-rotation (or force deflection) behavior of the assembly as shown in Fig. 3.17b. The slope of the curves represents the stiffness

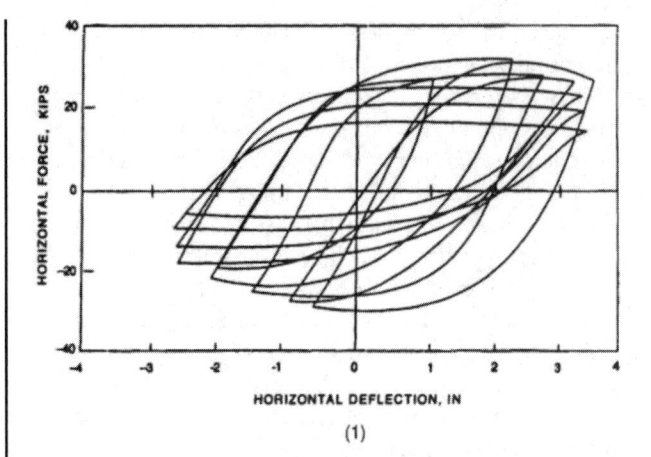

(1)

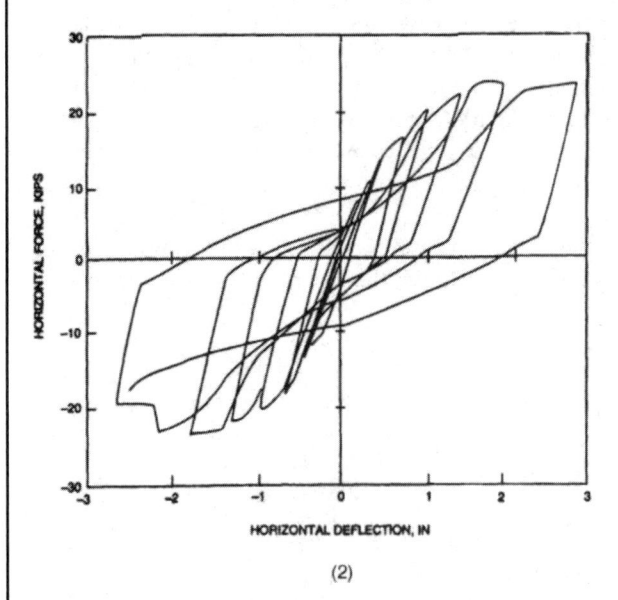

(2)

Figure 3.17b Hysteretic behavior: (1) curve representing large energy dissipation; (2) curve representing limited energy dissipation.

of the structure, and the enclosed areas are sometimes full and fat, or they may be lean and pinched. Structural assemblies with curves enclosing a large area representing large dissipated energy are regarded as superior systems for resisting seismic loading.

3.3.8.7 NONSTRUCTURAL COMPONENTS

For both analysis and detailing, the effects of nonstructural elements such as partitions, filler walls, and stairs must be considered. The nonstructural elements that are rigidly tied to the structural system can have a substantial influence on the magnitude and distribution of earthquake forces, causing a shear wall-like response with considerably higher lateral forces and overturning moments. Any element that is not strong enough to resist the forces

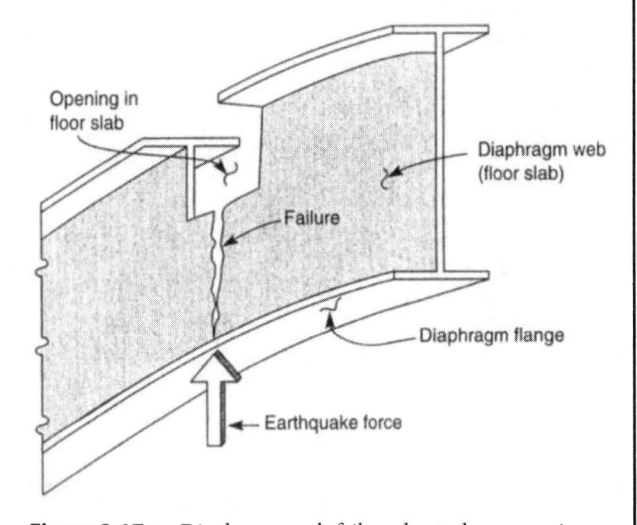

Figure 3.17a Diaphragm web failure due to large opening.

that it attracts will be damaged; therefore, it should be isolated from the lateral force-resisting system.

3.3.8.8 FOUNDATION

The differential movement of foundations due to seismic motions is an important cause of structural damage, especially in heavy, rigid structures that cannot accommodate these movements. Adequate design must minimize the possibility of relative displacement, both horizontal and vertical, between the various parts of the foundation and between the foundation and superstructure.

3.3.8.9 DAMAGE CONTROL FEATURES

The design of a structure in accordance with the seismic provisions will not fully ensure against earthquake damage because the horizontal deformations that can be expected during a major earthquake are several times larger than those calculated under design loads. A number of things can be done without increasing construction cost to limit damage, which otherwise would be expensive to repair following a strong earthquake. An important factor to keep in mind is the nature and geometry of the building when it responds to earthquake motions. As a rough guide, it should be assumed that deflections (story drift) may be four times that resulting from the required lateral forces.

Important things that can be detailed to minimize earthquake damage are as follows:

1. Provide details which allow structural movement without damage to nonstructural elements. Damage to such items as piping, glass, plaster, veneer, and partitions may constitute a major financial loss. To minimize this type of damage, special care in detailing, either to isolate these elements or to accommodate the movement, is required.

2. Breakage of glass windows can be minimized by providing adequate clearance at edges to allow for frame distortions.

3. Damage to rigid nonstructural partitions can be largely eliminated by providing a detail at the top and sides which will permit relative movement between the partitions and the adjacent structural elements.

4. In piping installations, the expansion loops and flexible joints used to accommodate temperature movement and are often adaptable to handling the relative seismic deflections between adjacent equipment items attached to floors.

5. Fasten free-standing shelving to walls to prevent toppling.

6. Concrete stairways often suffer seismic damage due to their inhibition of drift between connected floors. This can be avoided by providing a slip joint at the lower end of each stairway to eliminate the bracing effect of the stairway or by tying stairways to stairway shear walls.

7. If only cosmetic paint and plaster repairs are undertaken without regard to structural rehabilitation after damage from an earthquake, the structure may be left vulnerable to further damage and possible collapse in the event of a second strong earthquake.

3.3.8.10 REDUNDANCY

Redundancy is a highly desirable characteristic for earthquake-resistant design. When the primary element, connection or system fails, the lateral force can be redistributed to a secondary system without affecting the lateral stability of the structure and to prevent progressive failure.

3.4 IBC 2018 AND ASCE 7-16 CODE REQUIREMENTS EQUIVALENT LATERAL FORCE PROCEDURE

3.4.1 Introduction

The effects of earthquake motion on structures and their components should be determined according to ASCE 7-16 Chapters 11, 12, 13, 15, 17 and 18, as applicable, per the International Building Code (IBC) 2018 section 1613.1.

The earthquake provisions contained in IBC 2018 and in ASCE 7-16 are based on the requirements set forth in Reference [4], NEHRP Recommended Seismic Provisions for New Buildings and Other Structures (FEMA P-1050-1). The history of the creation and subsequent development of this document can be found in Reference [5], Earthquake-Resistant Design Concepts—An Introduction to the NEHRP Recommended Seismic Provisions for New Buildings and Other Structures (FEMA P-749). Table 3.2 shows a summary of the ASCE 7-16 chapters that contain seismic load provisions referenced by IBC 2018. Exceptions to these requirements are shown in [1, 2, 3]. These exceptions are not related to high-rise buildings.

3.4.2 Seismic analysis Procedure per IBC 2018 and ASCE 7-16

Detailed provisions and explanation for the seismic provisions according to ASCE 7-16 as well as the detailed procedure are presented in the "Structural Load Determination 2018 IBC and ASCE 7-16."[1] In this section, a brief description of the seismic provisions and procedure is presented, next the equivalent lateral force procedure (ELFP) is presented in detail.

3.4.2.1 SEISMIC DESIGN CRITERIA

• Mapped Acceleration Parameters: The mapped risk-targeted maximum considered earthquake (MCE$_R$) spectral response accelerations at periods of 0.2 second

Table 3.2   Summary of Chapters in ASCE/SEI 7 That Are Referenced by the 2018 IBC for Earthquake Load Provisions, Reference[1]

Chapter	Title
11	Seismic Design Criteria
12	Seismic Design Requirements for Building Structures
13	Seismic Design Requirements for Nonstructural Components
15	Seismic Design Requirements for Nonbuilding Structures
16	Nonlinear Response History Analysis
17	Seismic Design Requirements for Seismically Isolated Structures
18	Seismic Design Requirements for Structures with Damping Systems
19	Soil-Structure Interaction for Seismic Design
20	Site Classification Procedure for Seismic Design
21	Site-Specific Ground Motion Procedures for Seismic Design
22	Seismic Ground Motion, Long-Period Transition and Risk Coefficient Maps
23	Seismic Design Reference Documents

Table 3.3   Site Classification According ASCE 7-16 20.3[1]

Site Class	$\overline{v}_s$ (feet/sec)	$\overline{N}$ or $\overline{N}_{ch}$ (blows per foot)	$\overline{s}_u$(psf)
A–Hard rock	> 5,000	NA	NA
B–Rock	2,500 to 5,000	NA	NA
C–Very dense soil and soft rock	1,200 to 2,500	> 50	> 2,000
D–Stiff soil	600 to 1,200	15 to 50	1,000 to 2,000
	< 600	< 15	< 1,000
E–Soft clay soil	Any profile with more than 10 feet of soil with the following characteristics: • Plasticity index $PI > 20$ • Moisture content $w \geq 40\%$ • Undrained shear strength $\overline{s}_u < 500$ psf		
F–Soils requiring site response analysis in accordance with ASCE/SEI 21.1	See ASCE 20.3.1		

and 1.0 second are given in IBC 2018 Figures 1613.2.1(1) and 1613.2.1(2) and ASCE 7-16 Figures 22-1 to 22-8 for sites that have an effective average small-strain shear wave velocity of 2500 feet per second and 5-percent damping.

• Site Class: Six site classes are defined in Table 20.3-1 of ASCE 7-16, shown in Table 3.3. A site is to be classified as one of these six based on one of three soil properties measured over the top 100 feet of the site:

    • Effective average small-strain shear wave velocity, $\overline{v}_s$
    • Average field standard penetration, $\overline{N}$, or average standard penetration resistance for cohesionless soil layers, $\overline{N}_{ch}$
    • Average undrained shear strength, $\overline{s}_u$

Flowchart 2 (Fig. 3.31) below presents the procedure for site classification procedure for seismic design.

• Site Coefficients and Risk-Targeted MCE_R Spectral Response Acceleration Parameters

The risk-targeted MCE_R spectral response acceleration for short periods, S_{MS}, and at a 1-s period, S_{M1}, adjusted for site class effects, can be determined from IBC Equations 16-36 and 16-37, respectively, or ASCE/SEI Equations 11.4-1 and 11.4-2, respectively, Equations 3.2 and 3.3 below. Flowchart 3 (Fig. 3.32) below shows the procedure

of determining the site coefficients and risk-targeted MCE_R spectral response acceleration parameters.

$$S_{MS} = F_a S_S \tag{3.2}$$
$$S_{M1} = F_v S_1 \tag{3.3}$$

• Design Spectral Response Acceleration Parameters: The design earthquake spectral response accelerations at short periods, S_{DS}, and at 1-s period, S_{D1}, are determined by IBC 2018 Equations 16-38 and 16-39, respectively, or ASCE 7-16 Equations 11.4-3 and 11.4-4, respectively, Equations 3.4 and 3.5 below. Flowchart 3 (Fig. 3.32) shows the procedure for determining design spectral response acceleration parameters.

$$S_{DS} = (2/3)\, S_{MS} \tag{3.4}$$
$$S_{D1} = (2/3\, S_{M1} \tag{3.5}$$

• Design Response Spectrum: A generalized form of the design acceleration response spectrum based on the design spectral response acceleration parameters (ASCE 7-16 Figure 11.4-1, shown in Fig. 3.18). This spectrum should be used where required by the provisions in ASCE 7-16 but should not be used in cases where site-specific ground motion procedures are required per chapter 21 of ASCE 7-16.

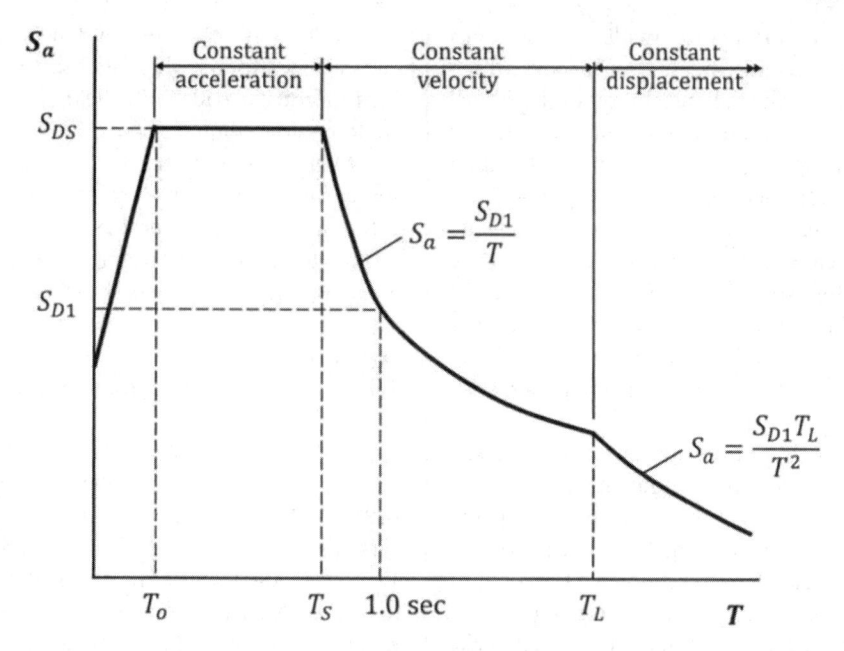

Figure 3.18   Design response spectrum [1].

- Site-Specific Ground Motion Procedures: Chapter 21 of ASCE 7-16 may be used for site-specific ground motion procedures given in ASCE 7-16 to determine ground motions for any structure and must be used for any structure that is located on a site that has been classified as Site Class F. Exceptions exist for cases as outlined in section 20.3.1 of ASCE 7-16 and as applicable.

- Importance Factor and Risk Category: According to section 1.5 of ASCE 7-16, the classification of buildings and other structures is based on "the risk to human life, health, and welfare associated with their damage or failure by nature of their occupancy or use, according to Table 1.5-1 for the purposes of applying flood, wind, snow, earthquake, and ice provisions." [2]. In seismic design, each building shall be assigned to the highest risk category and the minimum loads shall incorporate the importance factors given in Table 1.5-2 of ASCE 7-16 as applicable.

Risk categories are also defined in IBC 2018 Table 1604.5 and contain the provisions and definitions of risk categories. A concise summary of risk categories and a general description of the structures are shown in Table 3.4.

3.4.2.2   Seismic Design Category

A Seismic Design Category (SDC) must be assigned to all buildings and structures according to section 1613.2.5 of IBC 2018 and section 11.6 of ASCE 7-16. The SDC is a function of the risk category and the design spectral

accelerations at the site. Seismic design categories range from A to F. SDC A is assigned to buildings and structures with minimal risk and SDC F is assigned to buildings and structures with the highest risk. As the SDC of a structure increases, the strength and detailing requirements for a structure increase.

Flowchart 4 (Fig. 3.33) below shows the procedure of determining the seismic design category.

Table 3.4   Risk Categories and General Description of Structures

Risk category	General description
I	Low-hazard facility
II	Standard-occupancy building
III	High-occupancy building
IV	Essential or hazardous facility

3.4.2.3   Seismic Design Requirements for Building Structures

3.4.2.3.1 Structural Design Basis: Section 12.1 of ASCE 7-16 contains the basic requirements for seismic analysis and design of building structures, which states that the structure must have complete lateral and vertical force-resisting systems that are capable of providing adequate strength, stiffness and energy-dissipation capacity when subjected to the design ground motion, which is assumed to occur along any horizontal direction of a structure. Inadequate strength and stiffness, large displacements can occur, which could lead to local or overall instability, or both. All structural members in a building or a structure must be designed to resist all applicable axial, shear and bending moment forces determined according to ASCE 7-16, including the members that are not part of the seismic force-resisting system and all connections between the members. Also, the structure must have sufficient strength and stiffness so that the deflection limits per ASCE 7-16 are not exceeded.

Continuous load path with adequate strength and stiffness must be provided to allow the transfer of all forces from the point of application to the final point of resistance, generally to the ground. Without an adequate load path, individual members will move independently and may detach from one another which may lead to partial or total collapse. ASCE 7-16 section 12.1.3 contains minimum requirements for tying members together so that the structure can act as a unit in resisting earthquake loads.

Foundations must be designed to resist the forces caused by the ground motion and must be able to transfer these forces between the structure and the ground. Also, the foundations must be designed to accommodate ground movements without inducing large displacements into the structure, particularly at sites that maybe subject to liquefaction or lateral spreading. The selection of a foundation system for a site that maybe subjected to large displacement during a seismic event is very important to ensure that the structure remains connected and not torn apart by differential ground displacement. Examples of good foundation systems for such cases is mat foundation, and in the case of spread footing or piers are used, reinforced concrete grade beams should be used to tie the individual foundation elements together, which would allow the foundation system to move as one unit. Section 12.13 of ASCE 7-16 contains design requirements for foundations and Chap. 18 of IBC 2018 provides design and detailing requirements for different types of foundation systems organized with respect to seismic design categories.

3.4.2.3.2 Structural System Selection: Six general categories of Seismic Force Resisting Systems are presented and listed in Table 12.2-1. Each category contains a number of SFRSs. The six categories are:

1. Bearing wall system: a structural system in which the walls support all or a major part of the vertical/gravity loads and some of these walls support lateral loads as well.

2. Building frame system: a structural system that contains a space frame where its function is to support the vertical/gravity loads; and shear walls or braced frames are to resist earthquake loads.

3. MRF system: a structural system with a complete space frame that supports vertical/gravity loads. The entire space frame or part of is used as the SFRS. The seismic forces in moment frames are resisted by the flexure action in the members (beams and columns) that are connected by rigid (moment) connections. For seismic design category (SDC) D, E, or F deformation compatibility must be satisfied according to section 12.12.5 of ASCE 7-16.

4. Dual System: a structural system that is composed of a space frame that supports vertical/gravity loads. MRFs and shear walls or braced frames are to support earthquake loads based on their rigidities per section 12.2.5.1 of ASCE 7-16. The dual systems are also referred to as shear wall-frame interactive systems. The moment-resisting frames must act as a backup for the shear walls or braced frames and must resist at least 25 percent of the earthquake load.

5. Cantilever column system: a structural system in which the vertical/gravity loads as well as the earthquake loads are supported by the columns that act as cantilevers from their base. The system could be used in one-story buildings or in the top of multistory buildings. The system did not show good performance in past earthquakes, therefore, there are restrictions on its use as a SFRS.

6. Steel system not specifically designed for seismic resistance: a structural system that is permitted only in low seismic risk zones and does not conform to any other structural system above with the exception to cantilever column system.

Table 12.2-1 also includes the important parameters such as:

• *Response modification factor, R,* which is a measure of the system's ductility with a minimum value equals to 1.0 (no ductility, i.e., the system will behave in a linear-elastic manner), and a maximum value equals to 8.0 (highest ductility level that accounts for effective damping and energy dissipation). An R-value greater than 1.0 would decrease seismic base shear and would increase the level of design and detailing.

• *Overstrength factor, Ω_0* which accounts the actual seismic forces on certain members which could be significantly larger than forces obtained by analysis using the ASCE 7-16 prescribed provisions. The over strength factor, Ω_0, ranges from 2 to 3.

• *Deflection amplification factor, C_d* which is used to adjust the lateral displacements obtained from analysis using prescribed seismic forces based on ASCE 7-16 to expected lateral displacements during a design earthquake. The table shows that C_d is equal or slightly less than the corresponding R-values for a SFRS. For more ductile systems, the difference between the R-value and the C_d-value increases.

• Height limit with respect to SDC. Height limits are imposed on SFRS for certain SDCs that have No Limits "NL" for height. Ductile SFRS such as special reinforced concrete moment frames and special steel moment frames. Additionally, some SFRS are Not Permitted "NP" for a certain SDCs.

Other structural systems that are not in Table 12.2-1 can be used if they meet certain requirements according to section 12.2.1.1 and C12.2.1.1 of ASCE 7-16.

3.4.2.3.3 Diaphragm Flexibility, Configuration Irregularities and Redundancy Diaphragm Flexibility

Diaphragm Flexibility: lateral forces due to earthquake loads are distributed to the SFRS by diaphragms. The distribution of the forces to the SFRS depends on the relative flexibility of the diaphragm. In the case of flexible diaphragms, the lateral forces are distributed to the vertical elements of the SFRS using tributary areas. For diaphragms that are not flexible or for rigid diaphragms, the forces are distributed in proportion to the relative rigidities of the SFRSs in the direction of analysis based on the location of the CM and the CG.

In structural analysis, the relative flexibility of diaphragms must be considered with the stiffness of the SFRS, see section 12.3.1 of ASCE 7-16. Some diaphragms can be readily idealized as flexible or rigid diaphragms.

For example, untopped steel decking or wood structural panels can be idealized as flexible diaphragms, see section 12.3.1.1 per ASCE 7-16 for detailed provisions. Other diaphragms can be readily idealized as rigid diaphragms such as concrete slabs and concrete-filled metal deck with span-to-depth ratios less than or equal to 3. Section 12.3.2.1 indicates that structures must have no horizontal irregularity, see section 12.3.2.1 of ASCE 7-16.

In the cases where the idealized conditions for flexible or rigid diaphragms above are not met, they are permitted to be idealized as flexible diaphragm when the computed diaphragm maximum in-plane deflection under lateral loads is greater than two times the average story drift of adjoining vertical elements of the SFRS according to section 12.31.3 of ASCE 7-16. Figure 3.19 (Figure 12.3-1 of ASCE 7-16) explains this provision. Flowchart 6 (Fig. 3.35), adopted from [1] can be used to determine whether a diaphragm is flexible, rigid or semirigid.

Irregular and regular classifications: Discussion on structural symmetry and irregular structure is presented in sections 3.3.8.2 and 3.3.8.3 of this chapter. ASCE 7-16 provides provisions and classification for structural irregularity in section 12.3.2 and Tables 12.3-1 (Horizontal Structural Irregularities) and 12.3-2 (Vertical Structural Irregularities). These tables provide description of each type of horizontal and vertical irregularities, reference section and applicable SDC. Graphical representation of structural horizontal and vertical irregularities are shown in Tables 3.5 and 3.6. The graphical representation is obtained from Ref. 1. For a detailed explanation of the provisions and limitations, the reader is referred to Ref. 1.

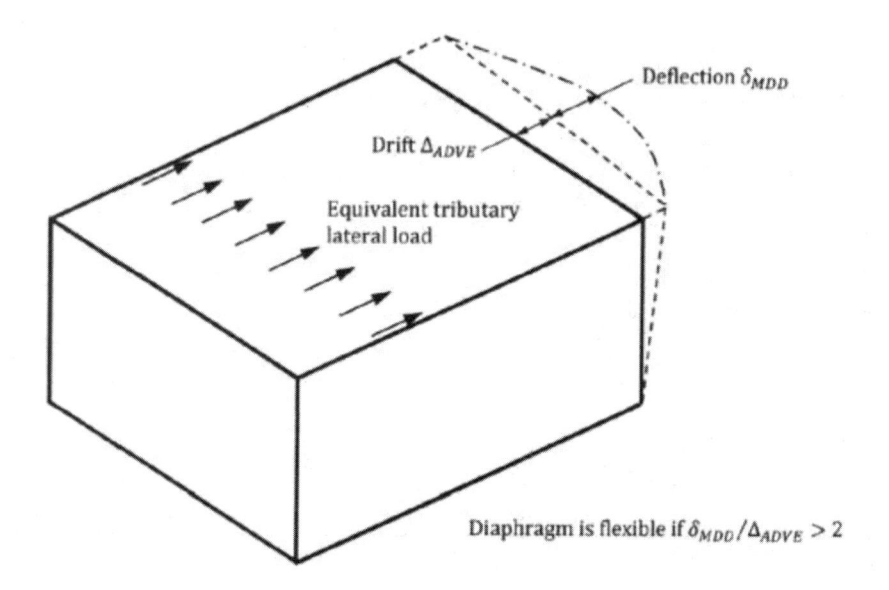

Figure 3.19   Definition of flexible diaphragm (ASCE/SEI 12.3.1.3).[1]

Table 3.5   Horizontal Structural Irregularities

Type of irregularity	Graphical explanation*
1a. Torsional Irregularity 1b. Extreme Torsional Irregularity	 $\Delta_{avg} = \dfrac{\Delta_{max} + \Delta_{min}}{2}$ Seismic force Torsional irregularity: $\Delta_{max} > 1.2\Delta_{avg}$ Extreme torsional irregularity: $\Delta_{max} > 1.4\Delta_{avg}$
2. Reentrant Corner Irregularity	 Reentrant corner irregularity: $b > 0.15a$ and $d > 0.15c$
3. Diaphragm Discontinuity Irregularity	 Diaphragm discontinuity irregularity: $a_1 b_1 > 0.5ab$
4. Out-of-Plane Offset Irregularity	 Wall below Wall above
5. Nonparallel System Irregularity	

*Figures obtained from [1].

Table 3.6   Vertical Structural Irregularities

Type of irregularity	Graphical explanation*
1a. Stiffness–Soft Story Irregularity 1b. Stiffness–Extreme Soft Story Irregularity	 Soft story • Stiffness-soft story irregularity: Soft story stiffness < 0.7(story stiffness above) or < 0.8(average stiffness of 3 stories above) • Stiffness-extreme soft story irregularity: Soft story stiffness < 0.6(story stiffness above) or < 0.7(average stiffness of 3 stories above)
2. Weight (Mass) Irregularity	 M_6 M_5 M_4 Weight (mass) irregularity: $M_5 > 1.5M_4$ or $M_5 > 1.5M_6$
3. Vertical Geometric Irregularity	 L_1 L_2 Vertical geometric irregularity: $L_1 > 1.3L_2$
4. In-Plane Discontinuity in Vertical Lateral Force-Resisting Element Irregularity	 L_1 Offset L_2 In-plane discontinuity irregularity: Offset > L_1 or Offset > L_2

(*Continued*)

Table 3.6 Vertical Structural Irregularities (*Continued*)

Type of irregularity	Graphical explanation*
5a. Discontinuity in Lateral Strength–Weak Story Irregularity 5b. Discontinuity in Lateral Strength–Extreme Weak Story Irregularity	 Lateral strength – weak story irregularity: $S_1 < 0.80S_2$ Lateral strength – extreme weak story irregularity: $S_1 < 0.65S_2$

*Figures obtained from [1].

Redundancy: Redundancy is a desirable feature to have in any structure where multiple load paths exist to resist the lateral loads. The redundancy factor is a measure of the redundancy inherent in any structure. For less redundant structures, the redundancy factor, ρ, reduces the response modification, R, which results in increasing the seismic demand on the structure. The redundancy factor, ρ, is either 1.0 or 1.3 according to section 12.3.4 of ASCE 7-16. Redundancy factor equals to 1.0 means that the structure is sufficiently redundant and no increase in seismic forces is required. Redundancy factor equals to 1.0 is used in the following conditions:

- Structures assigned to SDC B or C.
- Drift calculations and P-delta effects.
- Design of nonstructural components.
- Design of nonbuilding structures that are not similar to buildings.
- Design of collector elements, splices and their connections for which the seismic load effects including overstrength factor of 12.4.3 of ASCE 7-16 are used.
- Design of members or connections where the seismic load effects including overstrength factor of 12.4.3 of ASCE 7-16 are required for design.
- Diaphragm loads determined by Equation 12.10-1, including the limits imposed by Equations 12.10-2 and 12.10-3 of ASCE 7-16.
- Structures with damping systems designed in accordance with Chapter 18 of ASCE 7-16.
- Design of structural walls for out-of-plane forces, including their anchorage.

For conditions other than the listed above and for structures that are assigned to SDC D, E or F, ρ equal to 1.3 must be used unless one of the two conditions in 12.3.4.2 of ASCE 7-16 is met, in which case it is permitted to be set equal to 1.0. See Table 12.3-3 of ASCE 7-16 for detailed requirements of the first condition. A dual system is included under the "Other" types of elements and is considered to be inherently redundant.

Seismic Load Effects and Combinations: All structural members, including those that are not part SFRS, must be designed using the seismic load effect of section 12.4 of ASCE 7-16, unless exempted by ASCE 7-16. These effects include axial forces, shear forces and bending moments resulted from the application of horizontal and vertical forces, see section 12.4.2 of ASCE 7-16. Some structural members must be designed for the seismic load effects including the overstrength factor, Ω_0, see section 12.4.3 of ASCE 7-16.

The load combinations defined in sections 2.3 and 2.4 of ASCE 7-16 include the seismic load effect, E, which consists of the horizontal seismic effect, Eh, and the vertical seismic effect, Ev, determined from Equations 12.4-3 and 12.4-4a of ASCE 7-16, respectively:

$$E_h = \rho Q_E. \tag{3.6}$$
$$E_v = 0.2S_{DS}D \tag{3.7}$$

where,

Q_E is the effect (axial forces, shear forces, and bending moments) on the structural members obtained from the structural analysis.

ρ is the redundancy factor.

D is the Dead load.

S_{DS} is the design spectral acceleration for a short period.

The seismic load effects including the overstrength factor, E_m, consist of effects of horizontal seismic forces including overstrength, E_{mh}, and vertical seismic load effect, E_v. The horizontal seismic load effect with overstrength is determined by Equation 12.4-7 of ASCE 7-16:

$$E_{mh} = \Omega_0 Q_E. \tag{3.8}$$

The sign convention to use in the load combinations that contain E is important considering the dead load effect. If the dead load effects are additive to the seismic load effects, the strength design load combination 6 in 2.3.6 or the allowable stress load combinations 8 and 9 in 2.4.5, E is determined by Equation 12.4-1:

$$E = E_h + E_v \qquad (3.9)$$

In the case where the overstrength factor must be used, the strength design load combination 6 in 2.3.6 or the allowable stress load combinations 8 and 9 in 2.4.5, E must be taken equal to E_m, which is determined by Equation 12.4-5:

$$E_m = E_{mh} + E_v \qquad (3.10)$$

In the case where the seismic load effects, E, counteract the dead load effects, D, the strength design load combination 7 in 2.3.6 or the allowable stress load combination 10 in 2.4.5, E is determined by Equation 12.4-2:

$$E = E_h - E_v \qquad (3.11)$$

In the case where the overstrength factor must be used, the strength design load combination 7 in 2.3.6 or the allowable stress load combination 10 in 2.4.5, E must be taken equal to E_m, which is determined by Equation 12.4-6:

$$E_m = E_{mh} - E_v \qquad (3.12)$$

Refer to sections 2.3, 2.4, and 12.4 of ASCE 7-16 for the complete provisions on load combinations including seismic load effects.

Direction of Loading: Seismic loads must be applied in the direction that results in the most critical load effect on the structural members per section 12.5.1 of ASCE 7-16, which is based on the SDC. Table 3.7 reproduced from [1] summarizes these provisions.

Analysis Procedure Selection: According to the section 12.6 and Table 12.6-1 of ASCE 7-16 there are several parameters that affect the analysis procedure selection. These parameters are:
- Seismic design category,
- Risk category of the structure,
- Characteristics of the structure such as the structure's height and period, and
- The presence of any structural irregularities.

Flowchart 7 (Fig. 3.36) adopted from [1] can be used to determine the permitted analysis procedure.

Modeling Criteria Section 12.7 of ASCE 7-16 contains the requirements for creating an adequate model that can predict a structure's behavior subjected to seismic loads including modeling of foundations, effective seismic weight, structural modeling and interaction effects.

Foundation modeling For the purpose of determining the seismic loads on a structure, it is permitted to assume fixed supports at the base of a structure. According to section 11.2 of ASCE 7-16, the base of the structure is defined as the level at which the horizontal seismic ground motion are considered to be imparted by the structure. See section ASCE 7-16 Commentary section C11.2 for more information on where the base of the structure is considered for some of the common situations. If the foundation flexibility is considered, the requirements of section 12.13.3 of ASCE 7-16, foundation load-deformation characteristics, or Chapter 19 of ASCE 7-16, soil structure interaction for seismic design must be satisfied.

Table 3.7   Direction of Loading (Based on Section 12.5 of ASCE 7-16)

SDC	Requirements
B	Design seismic forces applied independently in each of two orthogonal directions and orthogonal interaction effects are permitted to be neglected.

C	• Conform to requirements of SDC B
	• Structures with horizontal irregularity Type 5:
	  • Orthogonal combination procedure: Apply 100 percent of the seismic forces in one direction and 30 percent of the seismic forces in the perpendicular direction on the structure simultaneously where the forces are computed in accordance with 12.8 (Equivalent Lateral Force Procedure), 12.9.1 (modal response spectrum analysis) or 12.9.2 (linear response history procedure), or
	  • Simultaneous application of orthogonal ground motion: Apply orthogonal pairs of ground motion acceleration histories simultaneously to the structure using 12.9.2 (linear response history procedure) or Chap. 16 (nonlinear response history procedure).

D-F	• Conform to requirements of SDC C • Any column or wall that forms part of two or more intersecting SFRSs that is subjected to axial load due to seismic forces along either principal plan axis greater than or equal to 20 percent of the axial design strength of the column or wall must be designed for the most critical load effect due to application of seismic forces in any direction.*

*Either of the procedures of 12.5.3.1a or 12.5.3.1b for SDC C are permitted to be used to satisfy this requirement.

Effective seismic weight   Adequate and accurate calculation of the effective seismic weight, W, is important as it affects the base shear and the distribution of seismic loads on the structure over its height. Section 12.7.2 of ASCE 7-16 defines the effective seismic weight, W, which includes the dead load of the structure (self-weight) in addition to:

1. in areas used for storage, a minimum of 25 percent of the floor live load;

2. where partitions must be included in accordance with 4.3.2, the actual partition weight or 10 psf of floor area, whichever is greater;

3. total operating weight of permanent equipment;

4. where the flat roof snow load, p_f, exceeds 30 psf, 20 percent of the uniform design snow load regardless of the roof slope; and

5. weight of landscaping and other materials at roof gardens and similar areas.

Structural modeling   To determine forces in the structural members and story drifts due to ground motion and to obtain a more realistic structural behavior, a mathematical model of the structure that includes member stiffness and strength must be created. For the analysis of concrete and masonry structure, the cracked section properties must be included and for the analysis of steel structures, panel zone deformation contribution to the story drift must be included.

For structures with horizontal irregularity Types 1a, 1b, 4 or 5, a three-dimensional analysis is required when a modal response spectrum or response history analysis is performed. In these analyzes, a minimum of three dynamic degrees of freedom, translation in two orthogonal directions and torsional rotation about the vertical axis, must be included at each level of the structure. If the structures have flexible diaphragms and Type 4 horizontal structural irregularities, three-dimensional model is not required.

The use of MRFs with rigid elements that are not part of the SFRS is common, therefore the interaction between the relatively flexible moment-resisting and the rigid elements must be considered in avoid any unexpected detrimental effects from occurring during an earthquake. A good and common example is infill masonry walls with reinforced concrete moment-resisting frame; if the masonry infill is fit tightly against the columns, the shear forces in the column will increase and will result in forming hinges and may lead to catastrophic failure.

Equivalent Lateral Force Procedure   The effect of ground motion on a structure is determined using a simplified design procedure known as the ELFP. This procedure is appropriate and valid for essentially regular structures without certain types of significant discontinuities where the primary response to ground motion is in the horizontal direction without substantial torsion in the first mode of vibration.

Linear static analysis is used to determine the effects of inelastic dynamic response by determining the design base shear V, which is distributed over the height of the structure at each floor level. The structure is analyzed for these static forces, which are distributed to the members of the SFRS considering the flexibility of the diaphragms.

The provisions of the ELFP are presented in section 12.8 of ASCE 7-16. This analysis procedure can be used for all structures assigned to SDC B and C as well as some types of structures assigned to SDC D, E and F (see Table 3.8).

3.4.2 Seismic Base Shear, V

The seismic base shear V is determined by multiplying the seismic response coefficient C_s by the effective seismic weight W (see Equation 12.8-1 of ASCE 7-16):

$$V = C_s W \tag{3.13}$$

Equations for C_s are given in 12.8.1, which form the design response spectrum:

For constant acceleration, $0 \le T \le T_S = S_{D1}/S_{DS}$ (Equation 12.8-2 of ASCE 7-16):

$$C_s = \frac{S_{DS}}{\left(\dfrac{R}{I_e}\right)} \tag{3.14}$$

For constant velocity, $T_S < T \le T_L$ (Equation 12.8-3 of ASCE 7-16):

$$C_s = \frac{S_{D1}}{T\left(\dfrac{R}{I_e}\right)} \tag{3.15}$$

For constant displacement, $T > T_L$ (Equation 12.8-4 of ASCE 7-16):

$$C_s = \frac{S_{D1}T_L}{T^2\left(\dfrac{R}{I_e}\right)} \tag{3.16}$$

A lower limit of C_s is defined in Equation 12.8-5 of ASCE 7-16:

$$C_s = 0.044 S_{DS} I_e \ge 0.01 \tag{3.17}$$

This equation provides a minimum base shear as a function of the short period design acceleration S_{DS} and the importance factor I_e. In no case is V permitted to be less than 1 percent of the effective seismic weight W.

For structures that are located where $S_1 \ge 0.6$, which applies to sites near active faults, Equation 12.8-6 provides an additional lower limit for C_s:

Table 3.8   Permitted Analytical Procedures (ASCE 7-16 section 12.6 and [1])

		Analysis methods		
SDC	Structural characteristics	Equivalent lateral force procedure (section 12.8)	Modal response spectrum analysis (section 12.9)	Seismic response history procedures (Chapter 16)
B, C	All structures	P	P	P
D, E, F	Risk Category I or II buildings not exceeding 2 stories above the base	P	P	P
	Structures of light-frame construction	P	P	P
	Structures with no structural irregularities and not exceeding 160 feet in structural height	P	P	P
	Structures exceeding 160 feet in structural height with no structural irregularities and with $T < 3.5T_S$	P	P	P
	Structures not exceeding 160 feet in structural height and having only horizontal irregularities of Type 2, 3, 4 or 5 or vertical irregularities of Type 4, 5a or 5b	P	P	P
	All other structures	NP	P	P

P = Permitted, NP = Not permitted.

$$C_s = \frac{0.5S_1}{\left(\dfrac{R}{I_e}\right)} \tag{3.18}$$

The design response spectrum in accordance with the ELFP is shown in Fig. 3.20.

The seismic response coefficient C_s is permitted to be calculated using a value of $S_S = 1.5$ for regular structures that are five stories or less above the base and that have a period T that is less than or equal to 0.5 s. This cap on S_S is based on the performance of buildings in past earthquakes that were designed using the applicable provisions of ASCE-16.

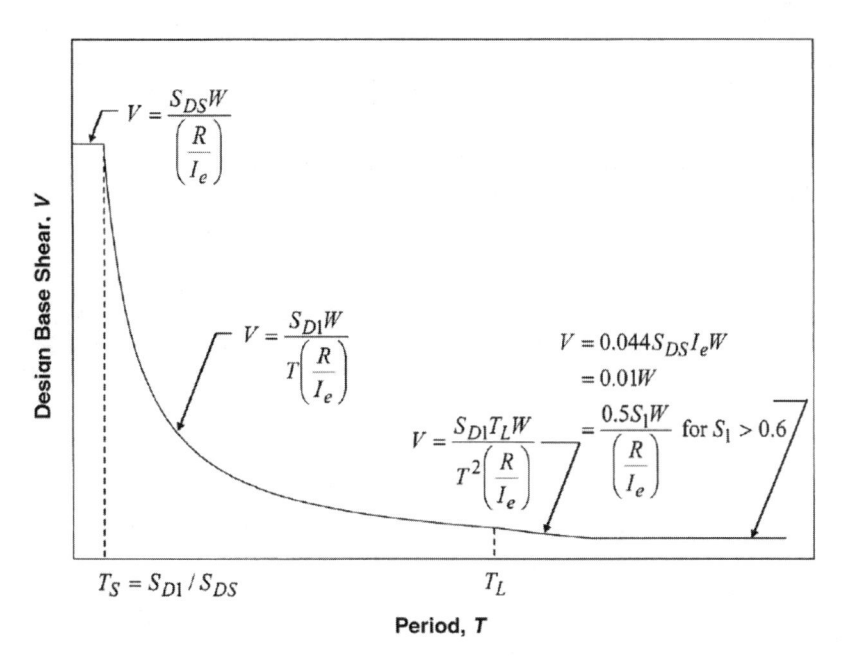

Figure 3.20   Design response spectrum according to the equivalent lateral force procedure (12.8).

3.4.3 Period Determination

The fundamental period of the structure T in the direction of analysis is determined based on the structural properties and deformational characteristics of the SFRS using fundamental principles of a dynamic analysis.

In the preliminary design stage, the member sizes may not be known to determine the period of the structure. Therefore in this case Equations 12.8-7 through 12.8-10 in 12.8.2.1 of ASCE 7-16 can be used to determine an approximate fundamental period T_a for a variety of structure types as shown in Table 3.3. These equations provide a lower-bound value for the approximate period, which, in turn, provides an upper-bound value of V.

The approximate period for masonry of concrete shear wall systems may be determined using the equation for all other structural systems in lieu of the equation given in Table 3.9.

If the fundamental period of a structure T is determined using a rational dynamic analysis, it must be taken less than or equal to the coefficient for upper limit on calculated period C_u in Table 12.8-1 of ASCE 7-16 times the appropriate approximate fundamental period T_a. The purpose of setting a limit on the calculated period is so that an unusually low base shear is not obtained for overly flexible structures.

The values of C_u in Table 12.8-1 of ASCE 7-16 remove the conservatism of the lower-bound equations that are used to determine T_a. As is evident from the values of C_u in the table, the limiting periods are larger in regions of lower seismicity since structures in these areas usually have longer periods (i.e., are more flexible) than those in regions of higher seismicity [1].

3.4.4 Vertical Distribution of Seismic Forces

When the seismic base shear V has been determined, it is distributed over the height of the building in accordance with Equations 12.8-11 and 12.8-12 of ASCE 7-16:

$$F_x = C_{vx}V \tag{3.19}$$

$$C_{vx} = \frac{w_x h_x^{k}}{\sum_{i=1}^{n} w_i h_i^{k}} \tag{3.20}$$

Table 3.9   Approximate Period T_a (ASCE 7-16 and [1])

Structure type		T_a
Moment-resisting frame systems in which the frames resist 100 percent of the required seismic forces and are not enclosed or adjoined by components that are more rigid and will prevent the frames from deflecting when subjected to seismic forces	Steel	$0.028h_n^{0.8}$
	Concrete	$0.016h_n^{0.9}$
Steel eccentrically braced frame in accordance with Table 12.2-1 lines B1 or D1		$0.030h_n^{0.75}$
Steel buckling-restrained braced frames		$0.030h_n^{0.75}$
Structures less than or equal to 12 stories in height where the SFRS consists entirely of concrete or steel moment-resisting frames where the average story height is greater than or equal to 10 feet		$0.1N$
Masonry or concrete shear wall structures		$\dfrac{0.0019h_n}{\sqrt{\dfrac{100}{A_B}\sum\limits_{i=1}^{x}\left(\dfrac{h_n}{h_i}\right)^2 \dfrac{A_i}{\left[1+0.83\left(\dfrac{h_i}{D_i}\right)^2\right]}}}$
All other structural systems		$0.020h_n^{0.75}$

Notes:
1. h_n = vertical distance from the base to the highest level of the SFRS of a structure
2. N = number of stories above the base of a structure
3. A_B = area of base of structure (square feet)
4. A_i = web area of shear wall i (square feet)
5. D_i = length of shear wall i (feet)
6. h_i = height of shear wall i (feet)
7. x = number of shear walls in the structure effective in resisting lateral forces in the direction of analysis

where,

F_x is the lateral seismic force located at level x above the base of the structure; w_i and w_x are the portions of the total effective seismic weight W located or assigned to level i or x; and h_i and h_x are the heights from the base of the structure to level i or x.

The exponent related to the structure period k is determined as follows:

$$k = \begin{cases} 1.0 & \text{for } T \leq 0.5 \text{ seconds} \\ 0.75 + 0.5T & \text{for } 0.5 \text{ seconds} < T < 2.5 \text{ seconds} \\ 2.0 & \text{for } T \geq 2.5 \text{ seconds} \end{cases}$$

The purpose of using the parameter k is to approximate the effects of higher modes, which are usually more dominant in structures with a longer fundamental period of vibration in structures that are more flexible. It is permitted to take $k = 2.0$ for structures with a period between 0.5 and 2.5 seconds instead of using the equation given above for that period range.

For structures with a fundamental period less than or equal to 0.5 seconds, V is distributed linearly over the height, varying from zero at the base to a maximum value at the top as shown in Fig. 3.21a. When T is greater than 2.5 seconds, a parabolic distribution is to be used as shown in Fig. 3.21b. As noted above, for a period between these two values, a linear interpolation between a linear and parabolic distribution is permitted or a parabolic distribution may be utilized.

It is important to note that the lateral forces F_x determined using the equations above are not the inertial forces that occur in the structure at any particular time during an actual earthquake. Rather, they provide design story shears that are consistent with enveloped results obtained from more refined analyses.

3.4.5 Horizontal Distribution of Forces [1]

The seismic design story shear V_x in story x is determined by summing the lateral forces acting at the floor or roof level supported by that story and all of the floor levels above, including the roof. The story shear is distributed to the vertical elements of the SFRS in the story based on the lateral stiffness of the diaphragm.

For diaphragms that rigid or semi-rigid (not flexible), V_x is distributed based on the relative stiffness of the vertical resisting elements and the diaphragm. Inherent and accidental torsion must also be included in the overall distribution (12.8.4.1 and 12.8.4.2 of ASCE 7-16).

The inherent torsional moment M_t is determined based on the distance between the CM and the CG. The accidental torsional moment M_{ta} is determined based on the assumption that the CM is displaced each way from its actual location by a distance equal to 5 percent of the dimension of the structure perpendicular to the direction of analysis. This is meant to account for any uncertainties in the actual locations of the CM and CR, this is due to tolerances in the constructed structure, that is, actual member sizes and actual locations of the members in the structure.

Illustrated in Fig. 3.22 is a floor or roof level with a diaphragm that is not flexible. The story shear V_x at this level acts through the CM. The inherent torsional moment M_t is equal to V_x times the eccentricity e_x between the CR and the CM. The accidental torsional moment M_{ta} is equal to $0.05b$ for V_x acting in the direction shown. The total torsional moment is the sum of M_t and M_{ta}. Similar

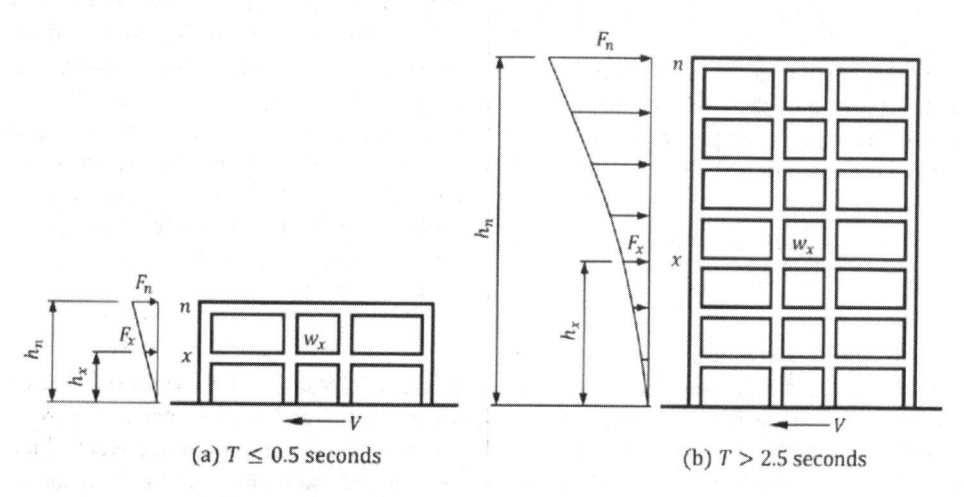

(a) $T \leq 0.5$ seconds

(b) $T > 2.5$ seconds

Figure 3.21 Vertical distribution of seismic forces (12.8.3) [1].

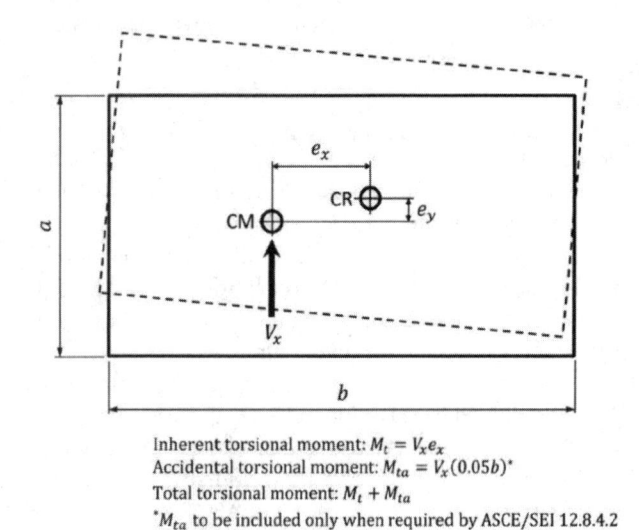

Inherent torsional moment: $M_t = V_x e_x$
Accidental torsional moment: $M_{ta} = V_x(0.05b)^*$
Total torsional moment: $M_t + M_{ta}$

$^*M_{ta}$ to be included only when required by ASCE/SEI 12.8.4.2

Figure 3.22   Inherent and accidental torsional moments (12.8.4.1 and 12.8.4.2 – ASCE 7-16) [1].

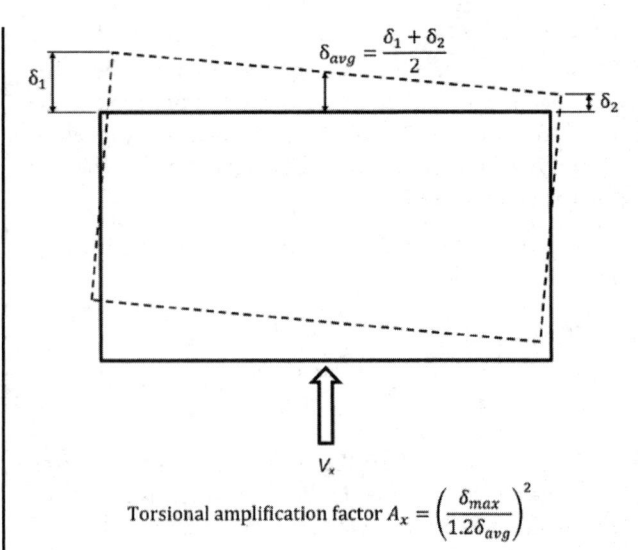

Torsional amplification factor $A_x = \left(\dfrac{\delta_{max}}{1.2\delta_{avg}}\right)^2$

Figure 3.23   Torsional amplification factor (12.8.4.3 of ASCE 7-16).

calculations can be performed to obtain the total torsional moment for V_x acting in the perpendicular direction.

In cases where earthquake forces are required to be applied concurrently in two orthogonal directions, the 5 percent displacement of the CM should be applied along a single orthogonal axis, which produces the greatest effects in the structural members. It is not required to apply the 5 percent displacement along two axes.

For flexible diaphragms, V_x is distributed to the vertical elements of the SFRS based on the mass that is tributary to the elements. Since the floor system usually has the greatest mass, V_x can be distributed based on the area of the diaphragm tributary to each line of resistance. Torsion need not be considered in the overall distribution.

If Type 1a or 1b torsional irregularity exists in structures assigned to SDC C, D, E or F, the accidental torsional moment is to be amplified in accordance with section 12.8.4.3 of ASCE 7-16. In particular, M_{ta} at each level is to be multiplied by the torsional amplification factor A_x, which is determined by Equation 12.8-14 of ASCE 7-16:

$$1.0 \le A_x = \left(\frac{\delta_{max}}{1.2\delta_{avg}}\right)^2 \le 3.0 \qquad (3.21)$$

where,

δ_{max} is the maximum displacement that occurs at level x in the structure computed assuming $A_x = 1$ and

δ_{avg} is the average of the displacements at the extreme points of the structure at level x computed assuming $A_x = 1$, as shown in Fig. 3.23. By setting $A_x = 1$ when determining the displacements,

the need for iterations in the calculation of A_x is eliminated.

3.4.6   Overturning

Overturning effects caused by seismic forces is another failure mode that must be designed for. For overturning, the critical load combinations are typically those where the effects from gravity and seismic loads counteract.

3.4.7   Story Drift Determination

Design story drift Δ is determined according with section 12.8.6 of ASCE 7-16. It is calculated as the difference of the deflections δ_x at the CM of the diaphragms at the top and bottom of the story under consideration as shown in Figure 12.8-2 of ASCE 7-16 and as shown in Fig. 3.24).

The design seismic forces F_x at each floor would result in elastic horizontal displacement from an elastic analysis of the structure, δ_{xe}, at each floor. These elastic deflections are multiplied by the deflection amplification factor C_d and divided by the importance factor I_e to obtain the deflections δ_x, which are estimates of the actual deflections that are likely to occur from the ground motion (see Equation 12.8-15 of ASCE 7-16):

$$\delta_x = \frac{C_d\delta_{xe}}{I_e}$$

Figure 3.25 shows the relationship between the lateral seismic forces and displacements. For a structure's elastic response to a ground motion, the deflection would be δ_E and the corresponding seismic shear force would be V_E, which is many times greater than the design force V that

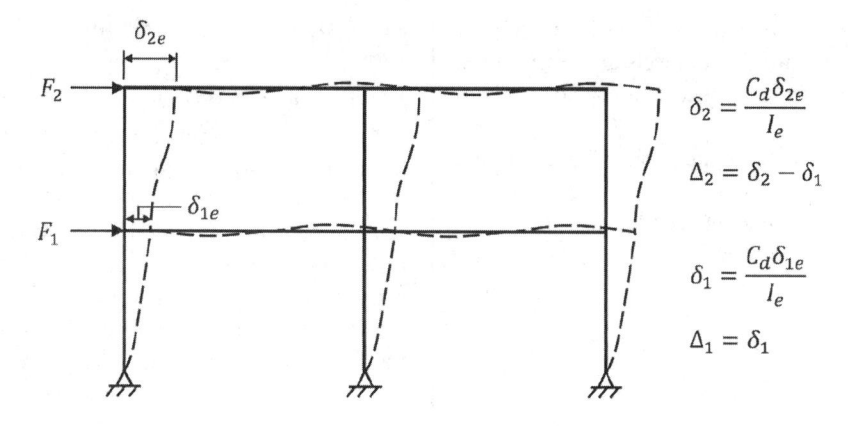

Figure 3.24 Story drift determination (12.8.6 – ASCE 7-16).

is determined by the provisions of 12.8 of ASCE 7-16. In particular, $V_E = RV$.

Figure 3.25 also shows the actual inelastic response of the structure to ground motion. If the SFRS is designed and detailed properly, it is expected to reach significant yield (the formation of the first plastic hinge) when subjected to seismic forces greater than the design seismic forces. Additional plastic hinges form as the seismic force increases until a maximum displacement is reached. Redundant structures that are designed and detailed properly can reach full yielding at force levels that are two to four times the prescribed design force levels.

As shown in Fig. 3.25, the actual inelastic response is different than the idealized one mainly due to over-strength and the related increase in stiffness. The actual displacement of the system could be less than $R\delta_{xe}$. To

consider this difference, the elastic displacement, δ_{xe}, is multiplied by the deflection amplification factor, C_d.

The importance factor, I_e, is included in the calculation of the inelastic displacement, δ_x. this is done to avoid over conservatism for structures with importance actors greater than 1.0.

The deflections δ_x at each floor level are used to calculate the design story drifts. These calculated deflections are compared with drift limits given in Table 12.12 of ASCE 7-16.

3.4.8 P-Delta Effects

When a structure deflects horizontally, the vertical loads are displaced from their original position. As a result, additional effects are introduced into the structural members, which cause the structure to deflect even further. The deflections will continue to increase, leading to overall instability of the structure if sufficient strength and stiffness are not provided.

Therefore, flexible structures are expected to be more prone to stability issues and P-delta effects more than rigid ones.

According to ASCE 7-16 section 12.8.7, member forces and story drifts induced by P-delta effects must be considered in member design and in the evaluation of overall stability of a structure where such effects are significant.

Instead of a more refined analysis, P-delta effects are not required to be considered when the stability coefficient θ is less than or equal to 0.10 where θ is determined by Equation 12.8-16 of ASCE 7-16:

$$\theta = \frac{P_x \Delta I_e}{V_x h_{sx} C_d} \qquad (3.22)$$

where P_x is the total vertical design load at and above level x:

$$P_x = \sum_{i=x}^{n} (P_D + P_L)_i \qquad (3.23)$$

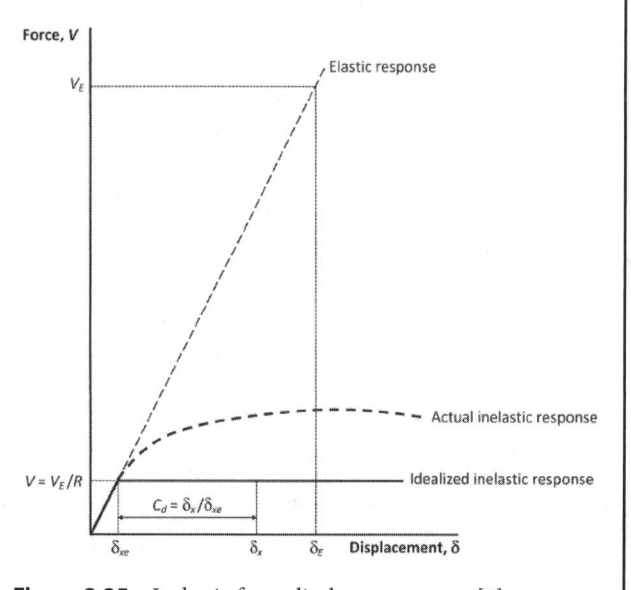

Figure 3.25 Inelastic force-displacement curve [1].

where P_D and P_L are the dead load and live loads, respectively, and n is the total number of levels in the structure.

The other quantities in the Equation are as follows and as shown in Fig. 3.26:
- Δ = design story drift defined in 12.8.6 occurring simultaneously with V_x
- V_x = seismic force acting between levels x and $x - 1$
- h_{sx} = story height below level x

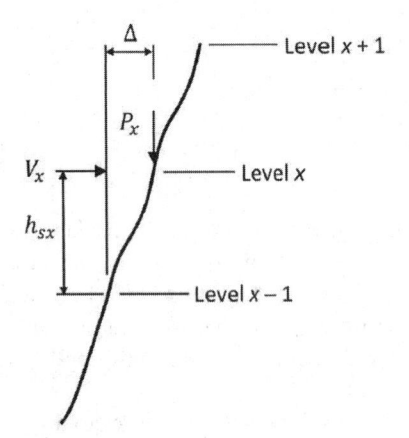

Figure 3.26   P-delta effects (12.8.7 – ASCE 7-16).

Where the stability coefficient θ is greater than θ_{max} determined by Equation 12.8-17 of ASCE 7-16, a structure is considered to be potentially unstable due to P-delta effects and must be redesigned.

$$\theta_{max} = \frac{0.5}{\beta C_d} \leq 0.25 \qquad (3.26)$$

Where β is the ratio of the shear demand to the shear capacity for the story under consideration, which is permitted to be conservatively taken as 1.0.

The θ_{max} equation must be satisfied even where computer software is utilized to determine second-order effects. However, the value of θ is permitted to be divided by $(1 + \theta)$ before checking the θ_{max} Equation.

P-delta effects must be considered when $0.10 < \theta \leq \theta_{max}$, and when considered, the member forces and displacements must be increased accordingly using a rational analysis or by multiplying them by $1.0/(1-\theta)$. Two types of rational analyzes are presented in [4].

Flowcharts 1 to 8 shown in Figs. 3.30 to 3.37 provide a step-by-step procedure on how to determine seismic design requirements, site classification, seismic ground motion values, SDC, diaphragm flexibility, analysis procedure, and the design seismic forces and their distribution based on the requirements of the ELFP.

3.4.9   Diaphragms, Chords and Collectors

3.4.9.1   DIAPHRAGM DESIGN

A Diaphragm is defined in IBC 2018 section 202. It is a horizontal or sloped system that transfer lateral forces to the vertical elements of the SFRS. Diaphragms are analyzed as deep beams in which the floor or the roof system resists shear forces and the chords (diaphragm edges perpendicular to the seismic forces) act as flanges that resist tension and compression forces. The forces distributed to the SFRS are based on the flexibility of the diaphragm. Another function of the diaphragm is the transfer of gravity loads that are perpendicular to its surface to framing members such as beams, joists, and columns that support the diaphragm. Diaphragm increases the flexural lateral stability of framing members if attached to its surface properly.

Figure 3.27 shows a schematic of a diaphragm with its elements (chords and collectors) for a seismic force applied in a specific direction. Diaphragms shears are transferred to the shear walls through the web of the floor or roof system. In the case where the roof shear wall does not extend over the full length of the shear wall, collector beams (parallel to seismic force) would be needed to collect the shear forces from the diaphragm to transfer it to shear wall.

Figure 3.28 shows the same diaphragm with an opening. In this case sub-diaphragms are created on each side of the opening as shown in the figure with sub-chord forces developed in the sub-diaphragm. Also, collector elements would be required on each side of the opening to collect the diaphragm shear into the sub-diaphragm.

Diaphragms, chords, and collectors are designed according to the provisions specified in section 12.10.1 and 12.10.2 of ASCE 7-16. Precast diaphragms in structures assigned to SDC C through F are to be designed using the provisions in section 12.10.3 of SCE 7-16.

Diaphragm Design Forces   Floor and roof diaphragms must be designed to resist seismic forces from base shear determined from the structural analysis or the force, F_{px} (below), whichever is greater for any structure assigned to seismic design categories B-F. The determination of F_{px} is according to Equation of 12.10-1 of ASCE 7-16.

$$0.4S_{DS}I_e w_{px} \geq F_{px} = \frac{\sum_{i=x}^{n} F_i}{\sum_{i=1}^{n} w_i} w_{px} \geq 0.2S_{DS}I_e w_{px}$$

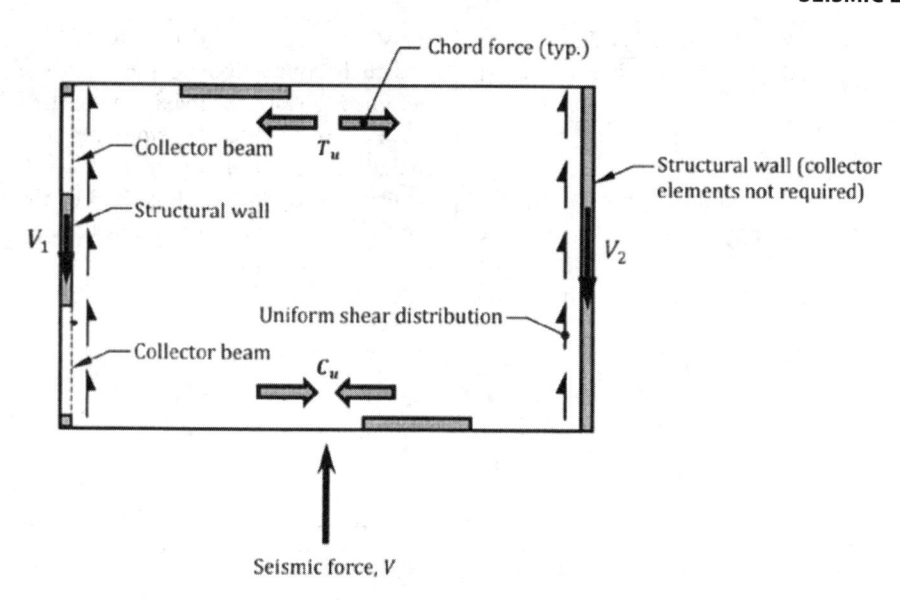

Figure 3.27   Diaphragm force distribution [1].

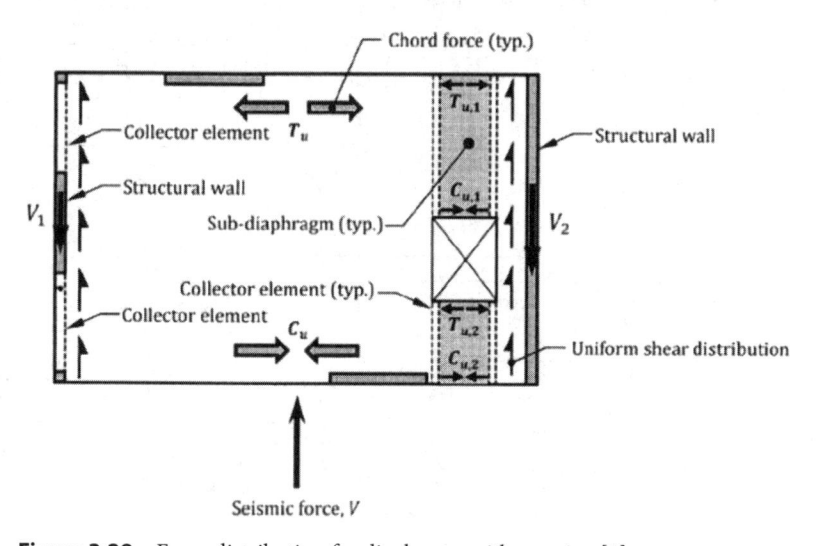

Figure 3.28   Force distribution for diaphragm with opening [1].

where,
F_i is the design seismic force applied to level i
w_i is the weight tributary to level i
w_{px} is the weight tributary to the diaphragm at level x

For structures with Type 4 structural irregularity, the transfer forces from the vertical seismic force-resisting elements above the diaphragm to the vertical seismic force-resisting elements below the diaphragm must be increased by the overstrength factor, Ωo, which should be added to the diaphragm inertial forces that originate on the diaphragm below, as shown in Fig. 3.29.

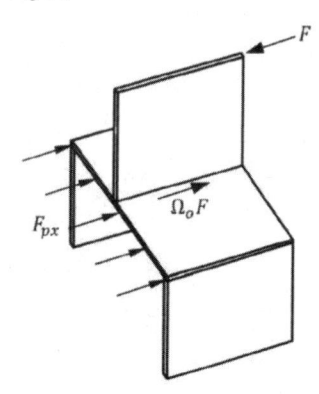

Figure 3.29   Example of vertical offsets in the seismic force-resisting system (12.10.1.1) [1].

Note that for structures assigned to SC D, E, or F, the redundancy factor, ρ, used with inertial forces is equal to 1.0.

3.4.9.2 COLLECTOR ELEMENTS

Collector elements or as also known "drag struts" are designed according to the provisions of section 12.10.2.1 for structures assigned to SDC C-F. Collector elements and their connections to the vertical elements must be designed to resist the maximum of the following:

1. Forces calculated using the seismic load effects including overstrength of 12.4.3 with seismic forces determined by the ELFP of 12.8 or the modal response spectrum analysis procedure of 12.9.1;

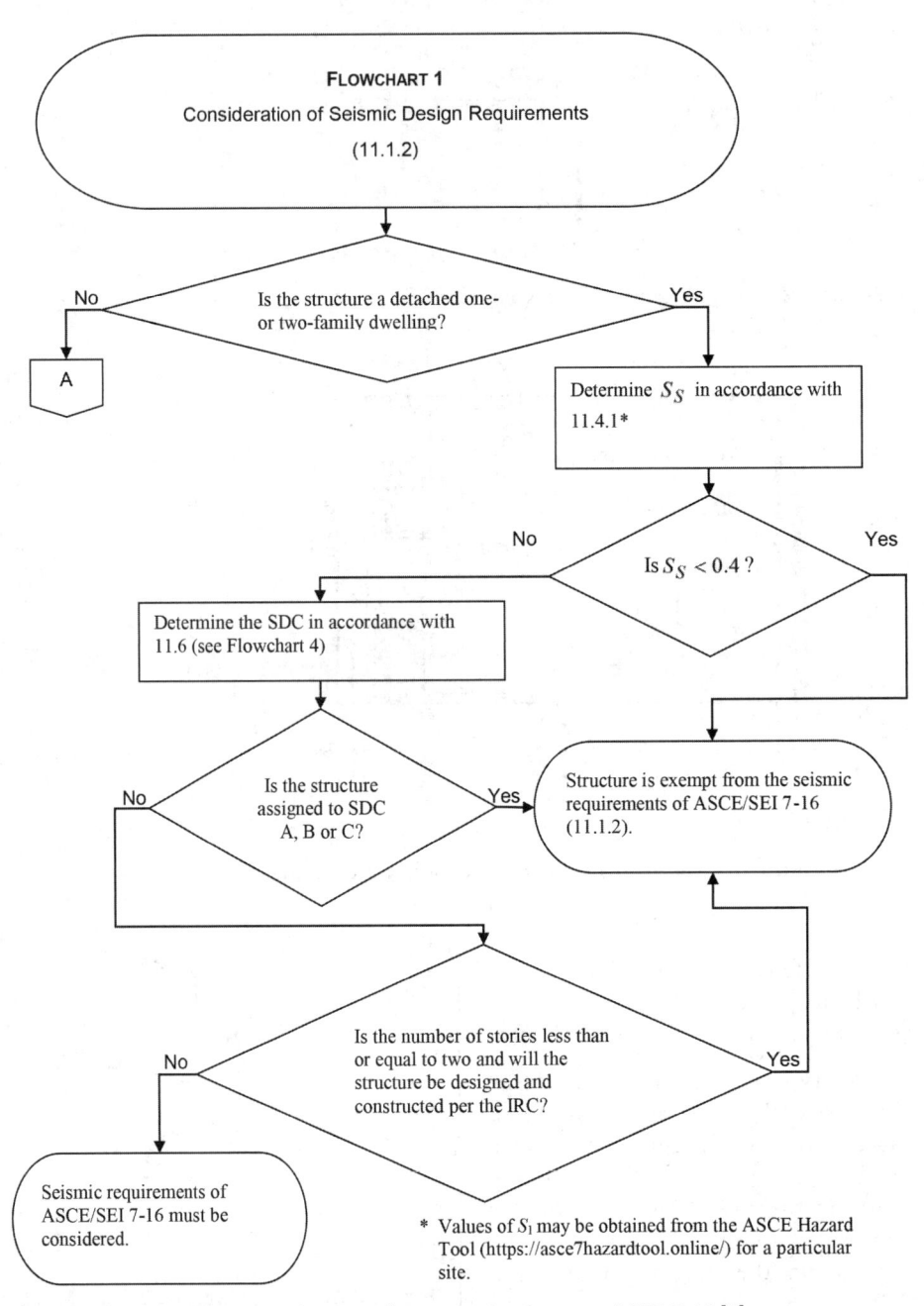

Figure 3.30 Consideration of seismic design requirements per ASCE 7-16 [1].

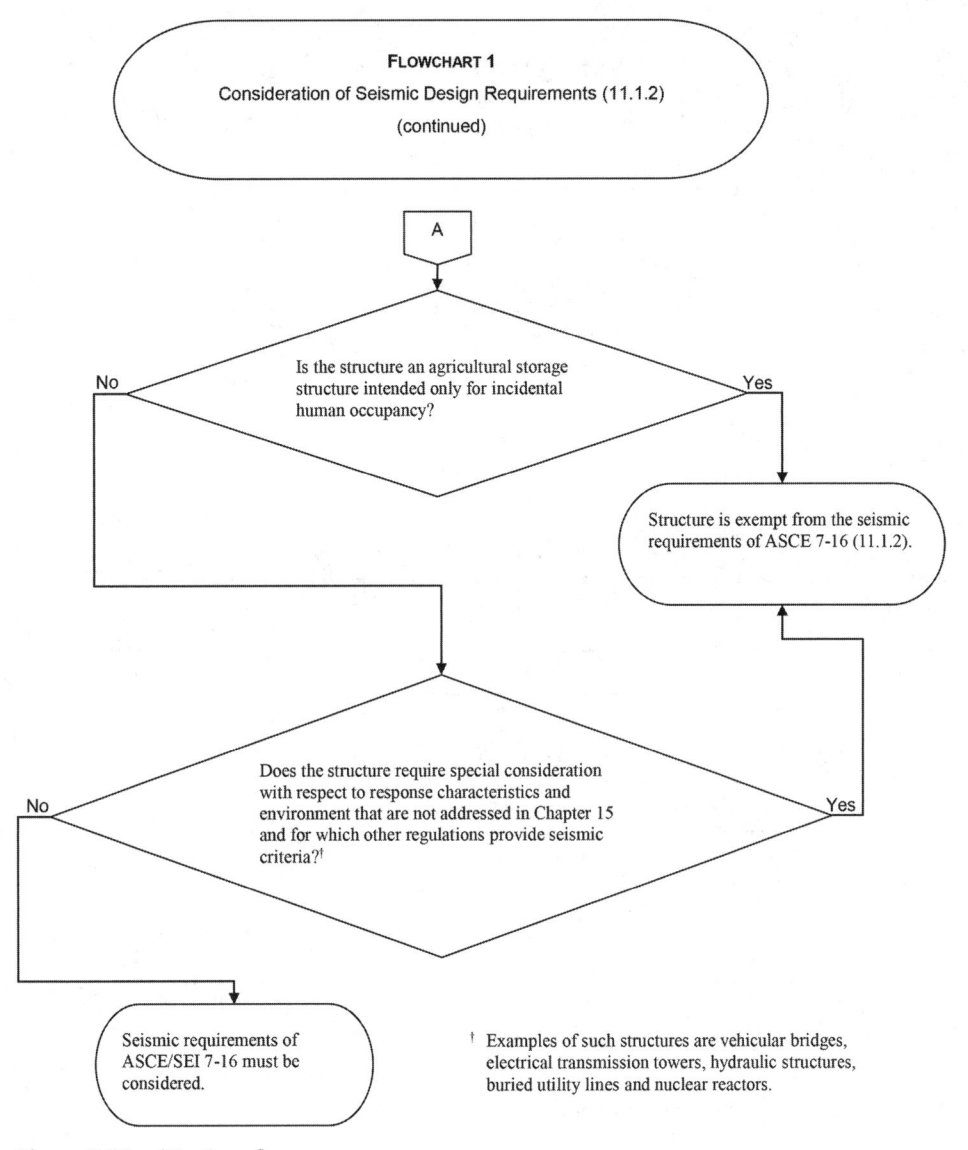

Figure 3.30 (*Continued*)

2. Forces calculated using the seismic load effects including overstrength of 12.4.3 with seismic forces determined by Equation 12.10-1 for diaphragms; and

3. Forces calculated using the load combinations of 2.3.6 with seismic forces determined by Equation 12.10-2, which is the lower-limit diaphragm force.

The use of the overstrength factor is to help keep inelastic behavior in the ductile elements of the SFRS and not in the collectors or their connections as it is critical that the collectors and their connections be able to perform as intended during a seismic event.

Additional information on the design and detailing of diaphragms, chords and collectors can be found in Ref. [11].

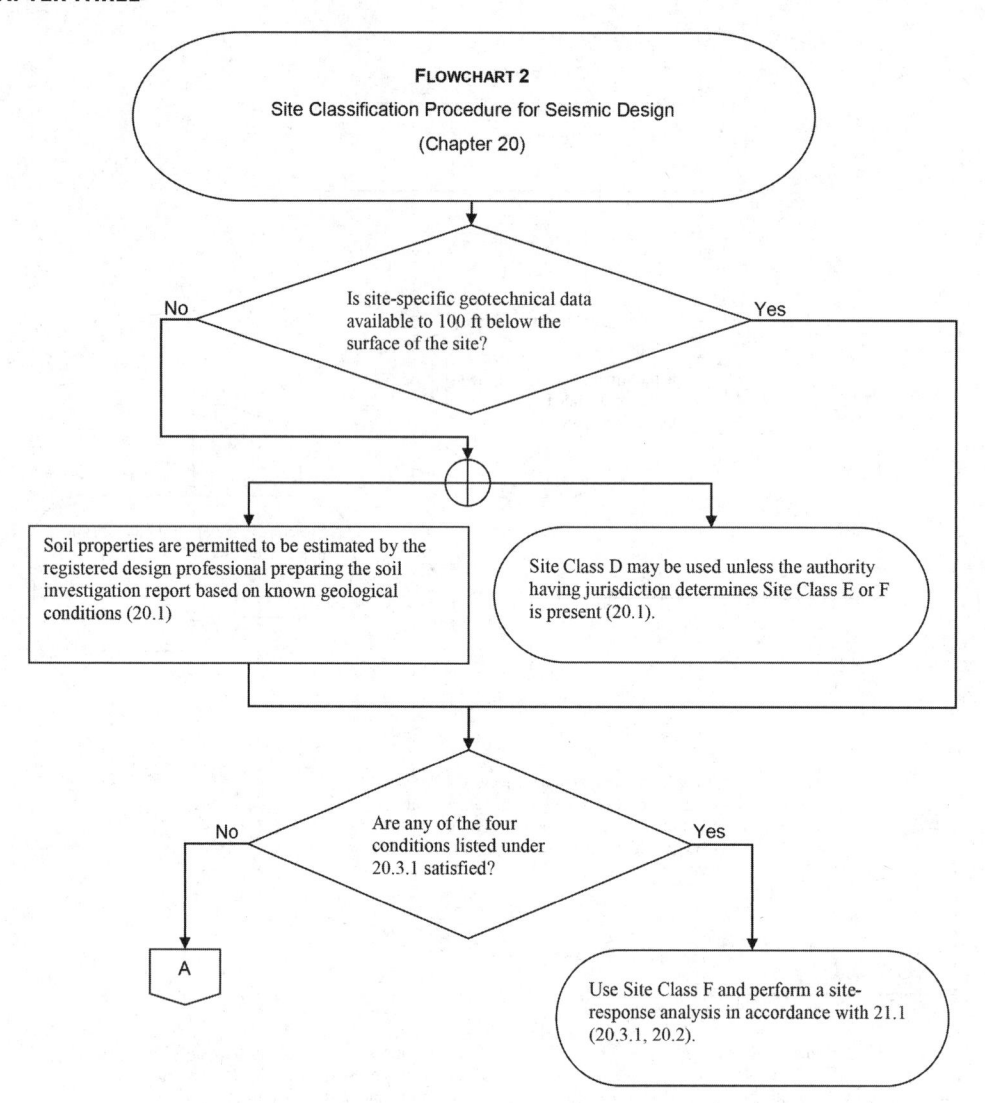

Figure 3.31 Site classification procedure for seismic design per ASCE 7-16 [1].

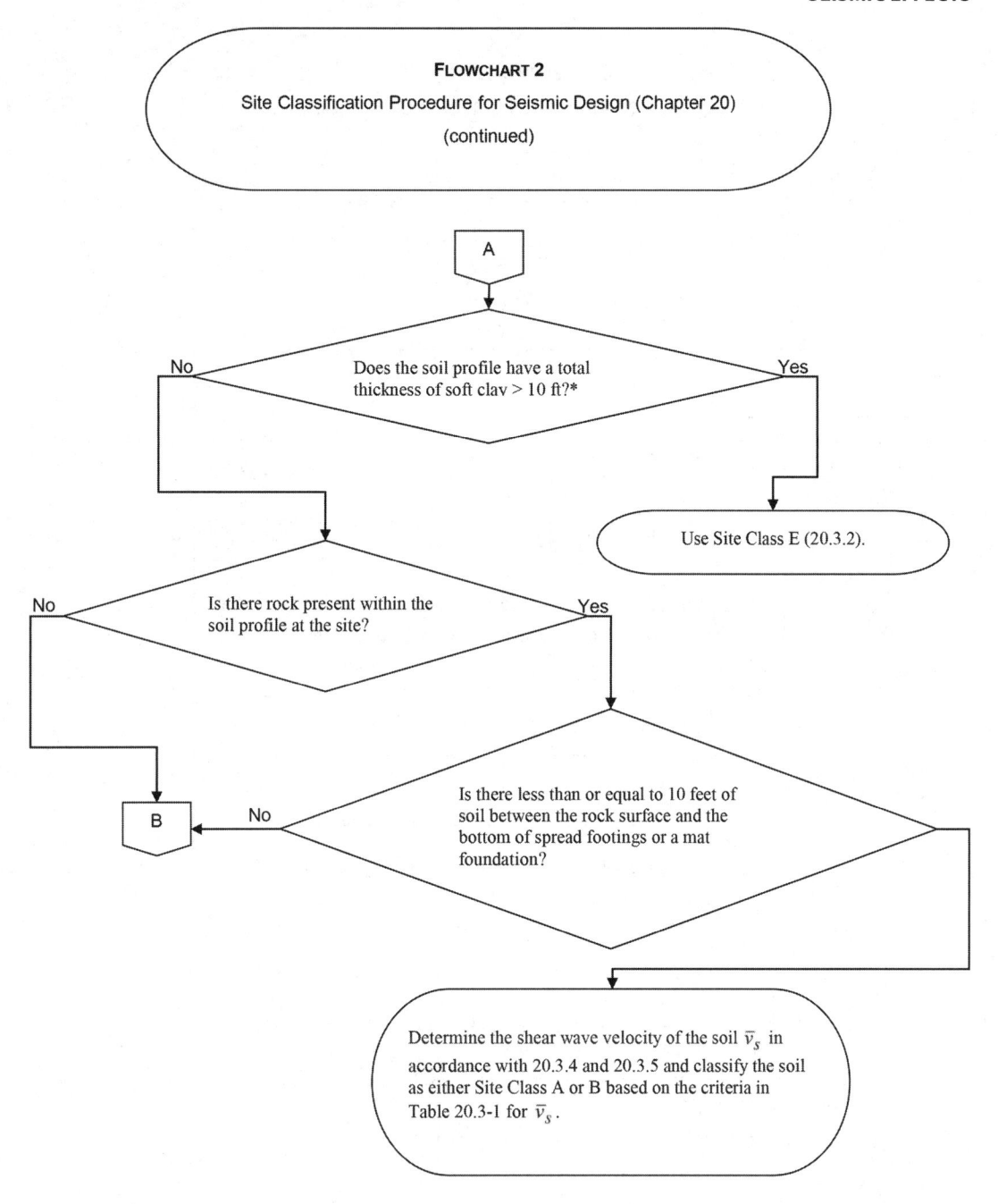

FLOWCHART 2
Site Classification Procedure for Seismic Design (Chapter 20)
(continued)

A

Does the soil profile have a total
thickness of soft clay > 10 ft?*

No   Yes

Use Site Class E (20.3.2).

Is there rock present within the
soil profile at the site?

No   Yes

B

Is there less than or equal to 10 feet of
soil between the rock surface and the
bottom of spread footings or a mat
foundation?

No

Determine the shear wave velocity of the soil $\overline{v}_s$ in
accordance with 20.3.4 and 20.3.5 and classify the soil
as either Site Class A or B based on the criteria in
Table 20.3-1 for $\overline{v}_s$.

* A soft clay layer is defined by $s_u < 500$ psf, $w \geq 40$ percent, and $PI > 20$ (20.3.2).

Figure 3.31   (*Continued*)

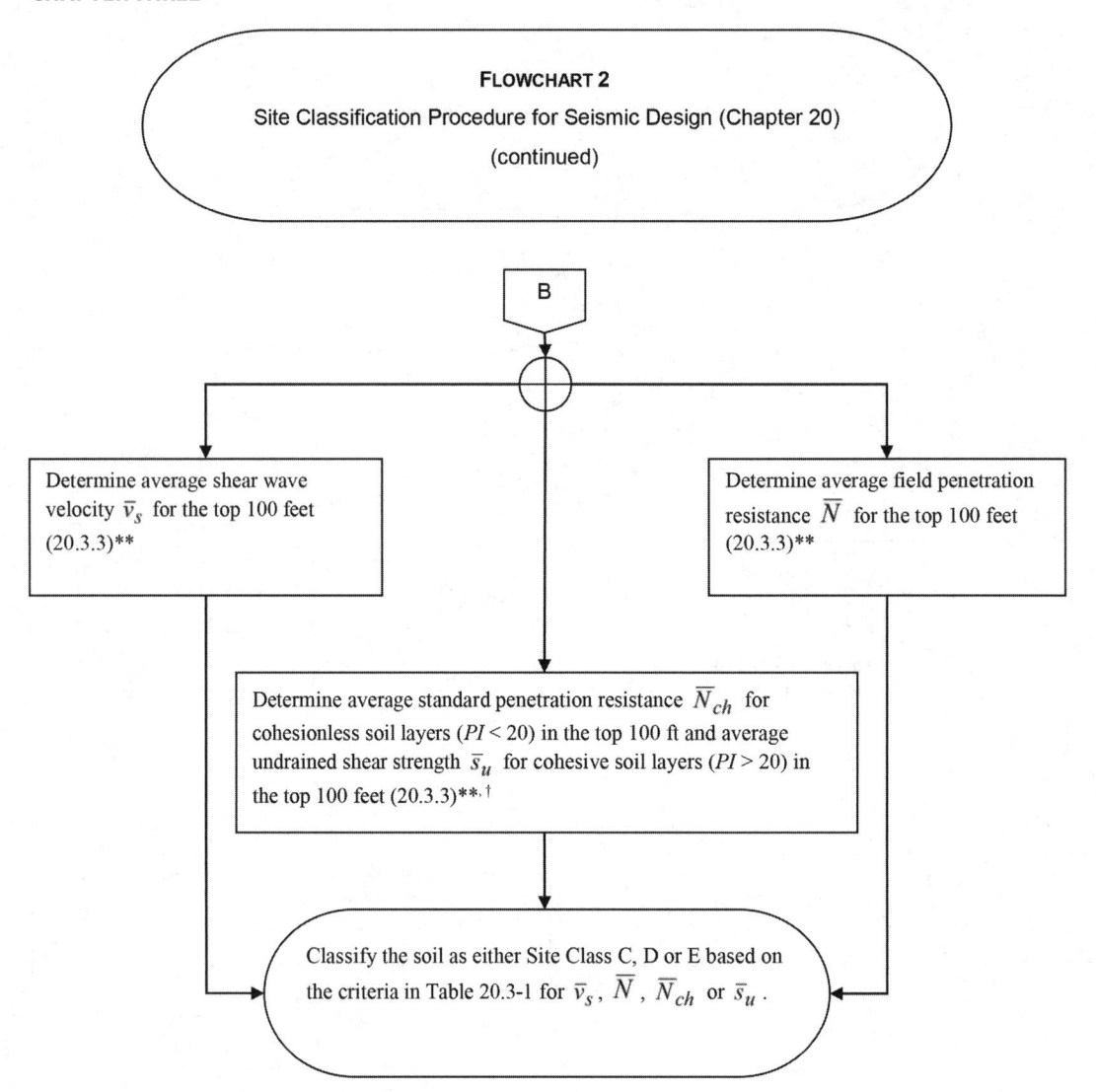

FLOWCHART 2

Site Classification Procedure for Seismic Design (Chapter 20)

(continued)

B

Determine average shear wave velocity $\bar{v}_s$ for the top 100 feet (20.3.3)**

Determine average field penetration resistance $\overline{N}$ for the top 100 feet (20.3.3)**

Determine average standard penetration resistance $\overline{N}_{ch}$ for cohesionless soil layers ($PI < 20$) in the top 100 ft and average undrained shear strength $\bar{s}_u$ for cohesive soil layers ($PI > 20$) in the top 100 feet (20.3.3)**,[†]

Classify the soil as either Site Class C, D or E based on the criteria in Table 20.3-1 for $\bar{v}_s$, $\overline{N}$, $\overline{N}_{ch}$ or $\bar{s}_u$.

** Values of $\bar{v}_s$, $\overline{N}$ and $\bar{s}_u$ are computed in accordance with 20.4 where soil profiles contain distinct soil and rock layers (20.3.3).

[†] Where the $\overline{N}_{ch}$ and $\bar{s}_u$ criteria differ, the site must be assigned to the category with the softer soil [20.3.3(3)].

Figure 3.31   (*Continued*)

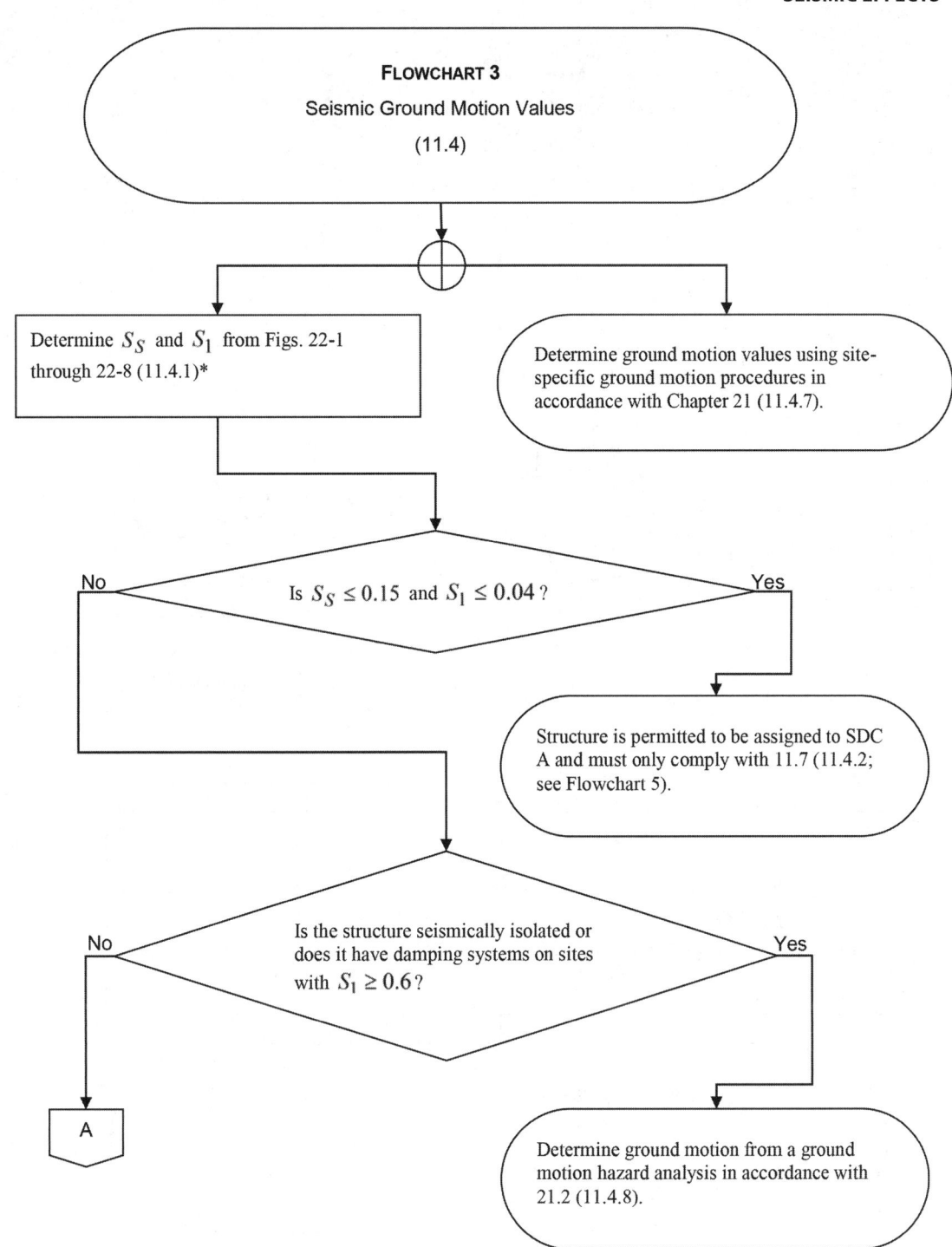

* Values of S_S and S_1 may be obtained from the from the ASCE
  Hazard Tool (https://asce7hazardtool.online /) for a particular site.

Figure 3.32 Seismic ground motion values per ASCE 7-16 [1].

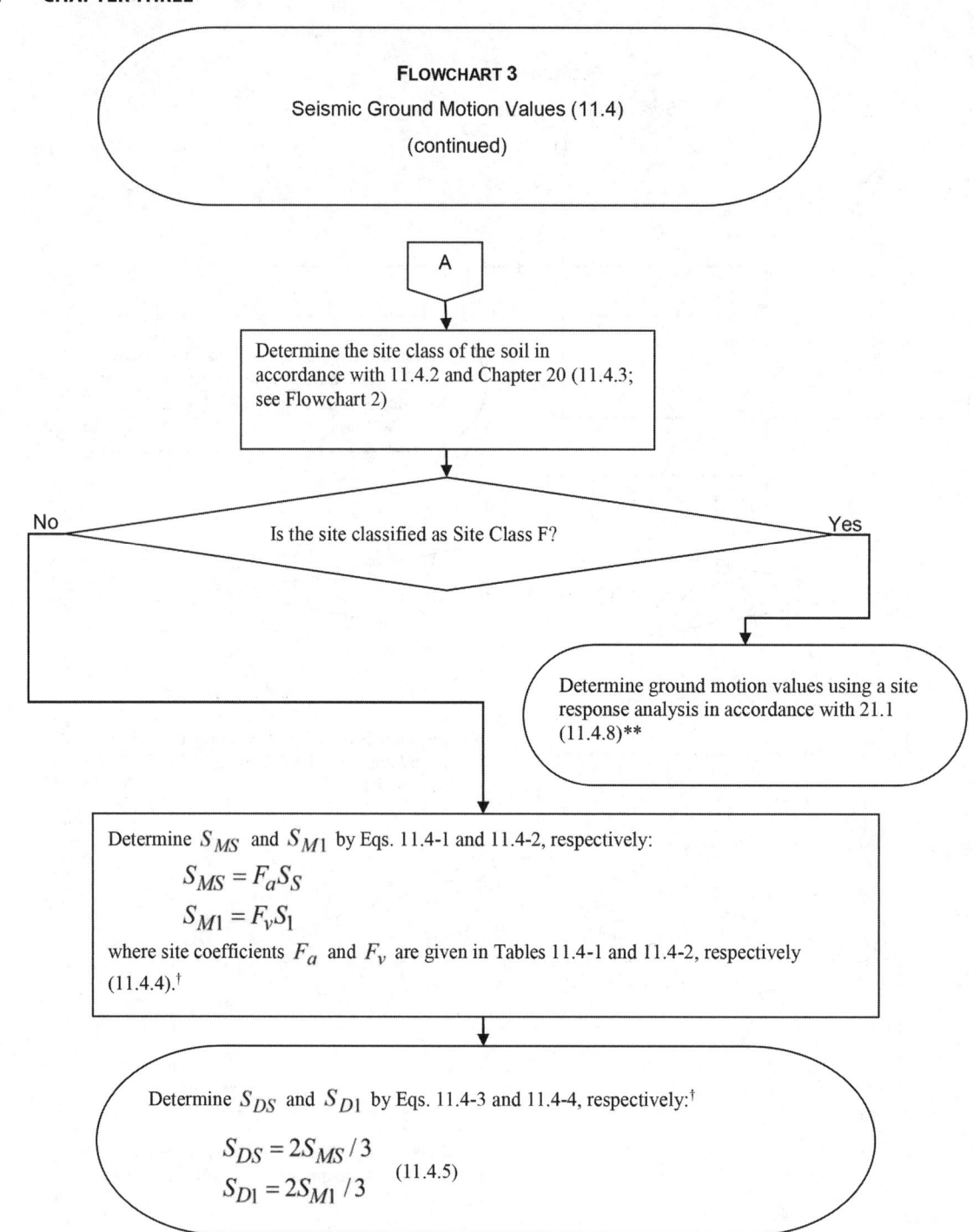

FLOWCHART 3

Seismic Ground Motion Values (11.4)

(continued)

A

Determine the site class of the soil in accordance with 11.4.2 and Chapter 20 (11.4.3; see Flowchart 2)

Is the site classified as Site Class F?

No

Yes

Determine ground motion values using a site response analysis in accordance with 21.1 (11.4.8)**

Determine S_{MS} and S_{M1} by Eqs. 11.4-1 and 11.4-2, respectively:

$$S_{MS} = F_a S_S$$
$$S_{M1} = F_v S_1$$

where site coefficients F_a and F_v are given in Tables 11.4-1 and 11.4-2, respectively (11.4.4).[†]

Determine S_{DS} and S_{D1} by Eqs. 11.4-3 and 11.4-4, respectively:[†]

$$S_{DS} = 2S_{MS}/3$$
$$S_{D1} = 2S_{M1}/3$$

(11.4.5)

** A site response analysis in accordance with 21.1 is required for structures on Site Class F sites unless the exception in 20.3.1(1) is satisfied for structures with periods $T \le 0.5$ sec.

[†] Where the simplified design procedure of 12.14 is used, only the values of F_a and S_{DS} must be determined in accordance with 12.14.8.1 (11.4.4, 11.4.5).

Figure 3.32   (*Continued*)

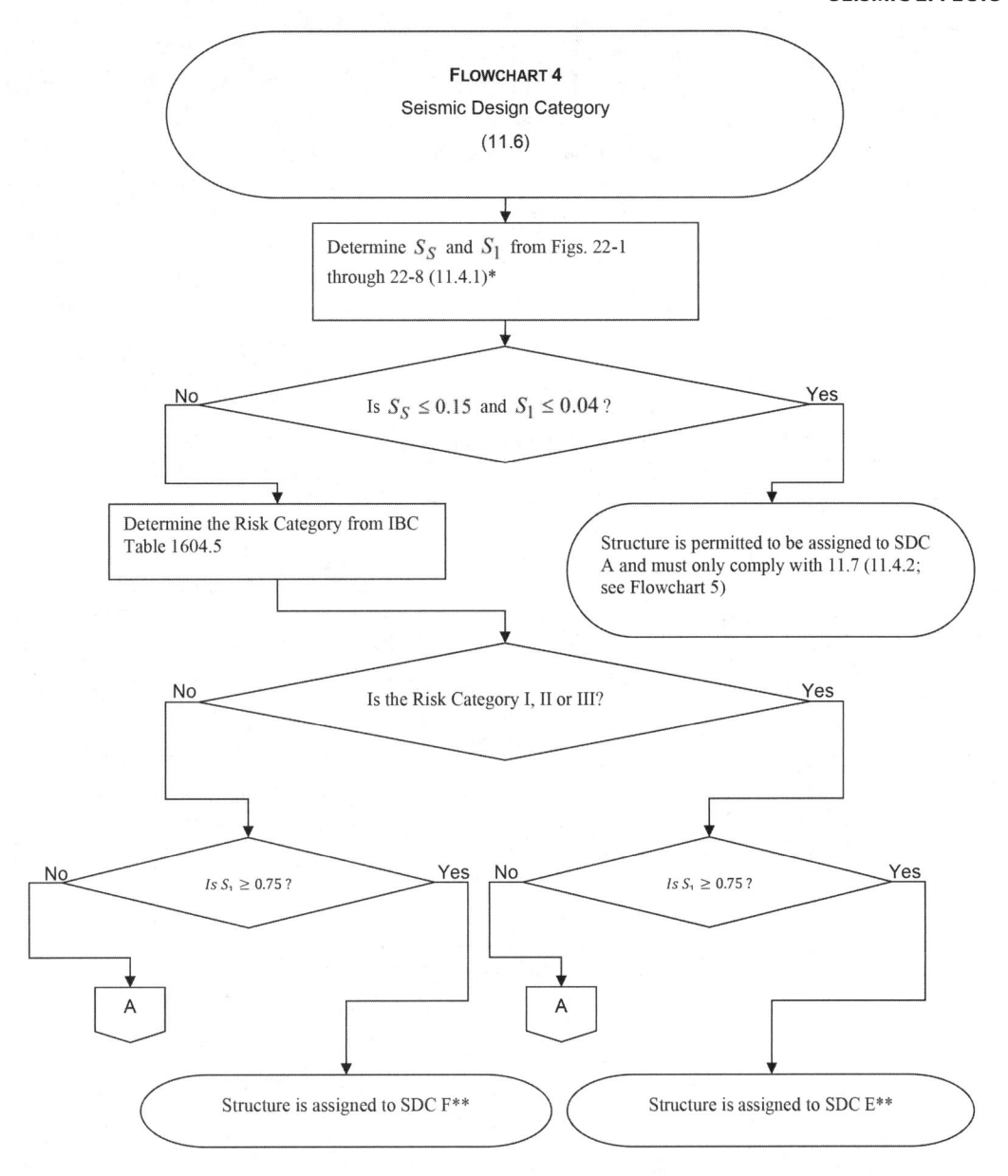

* Values of S_S and S_1 may be obtained from the from the ASCE Hazard Tool (https://asce7hazardtool.online /) for a particular site.

** A structure assigned to SDC E or F must not be located where there is a known potential for an active fault to cause rupture of the ground surface at the structure (11.8).

Figure 3.33 Diaphragm flexibility per ASCE 7-16 [1].

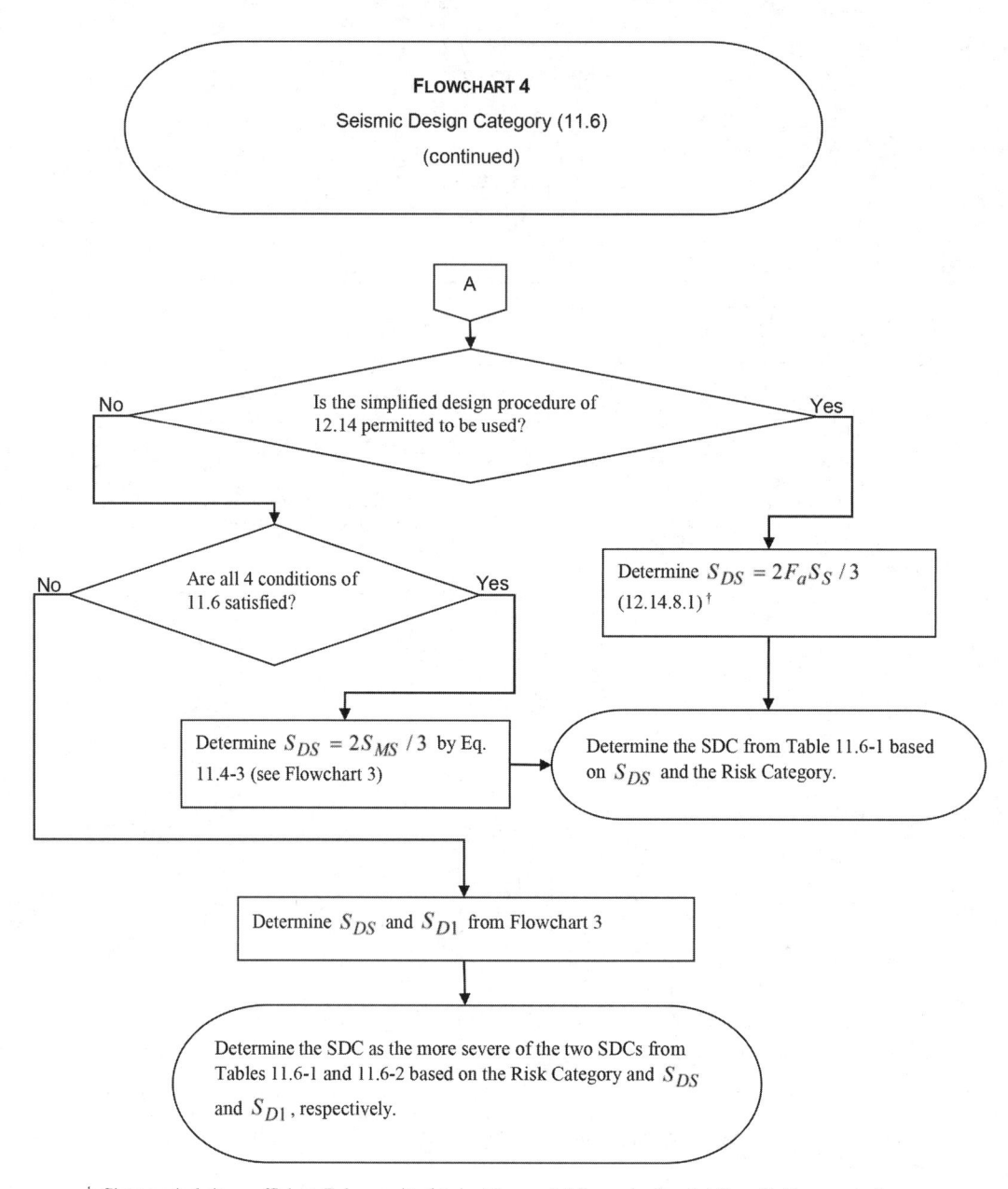

† Short-period site coefficient F_a is permitted to be taken as 1.0 for rock sites, 1.4 for soil sites or may be determined in accordance with 11.4.3. Rock sites have no more than 10 feet of soil between the rock surface and the bottom of spread footing or mat foundation. Mapped spectral response acceleration S_S is determined in accordance with 11.4.2 and need not be taken larger than 1.5 (12.14.8.1).

Figure 3.33   (*Continued*)

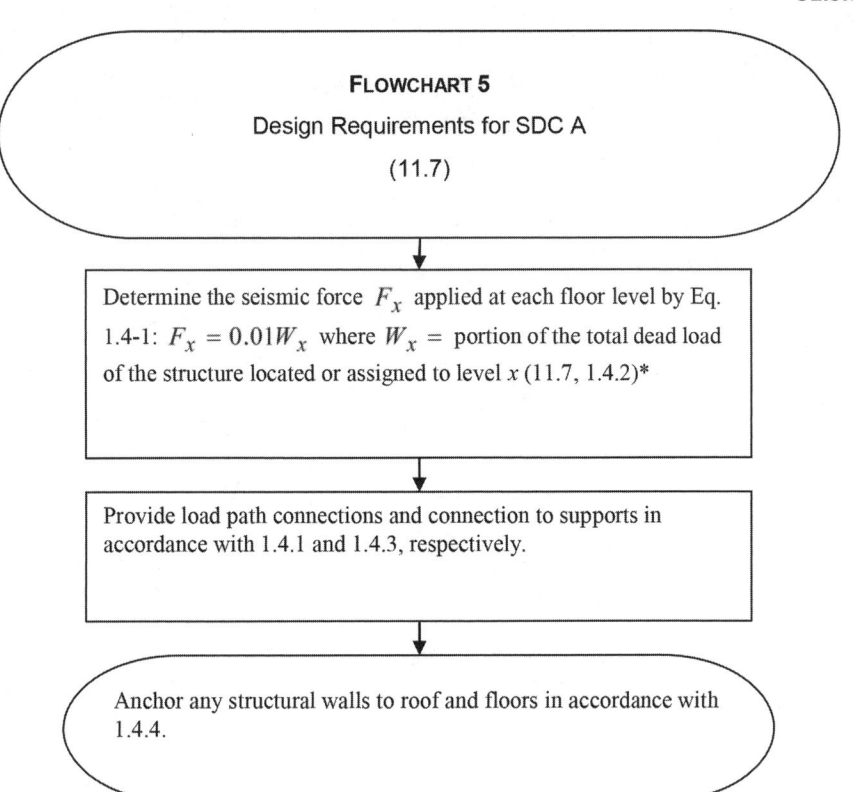

FLOWCHART 5

Design Requirements for SDC A

(11.7)

Determine the seismic force F_x applied at each floor level by Eq. 1.4-1: $F_x = 0.01W_x$ where $W_x =$ portion of the total dead load of the structure located or assigned to level x (11.7, 1.4.2)*

Provide load path connections and connection to supports in accordance with 1.4.1 and 1.4.3, respectively.

Anchor any structural walls to roof and floors in accordance with 1.4.4.

* These forces are applied simultaneously at all levels in one direction. The structure is analyzed for the effects of these forces applied independently in each of two orthogonal directions (1.4.2).

Figure 3.34   Design requirements for seismic design category (SDC) A per ASCE 7-16 [1].

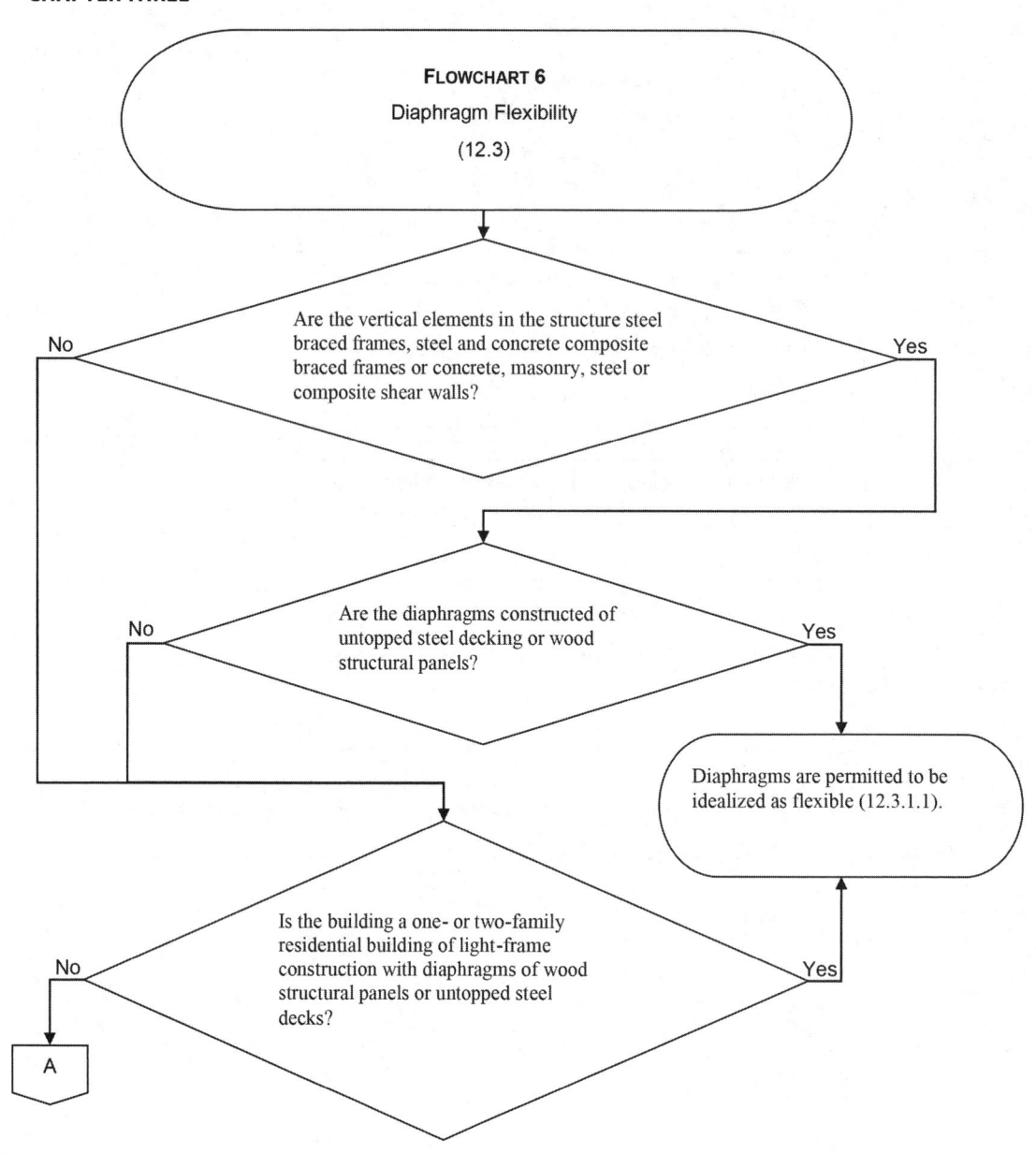

Figure 3.35   Diaphragm flexibility per ASCE 7-16 [1].

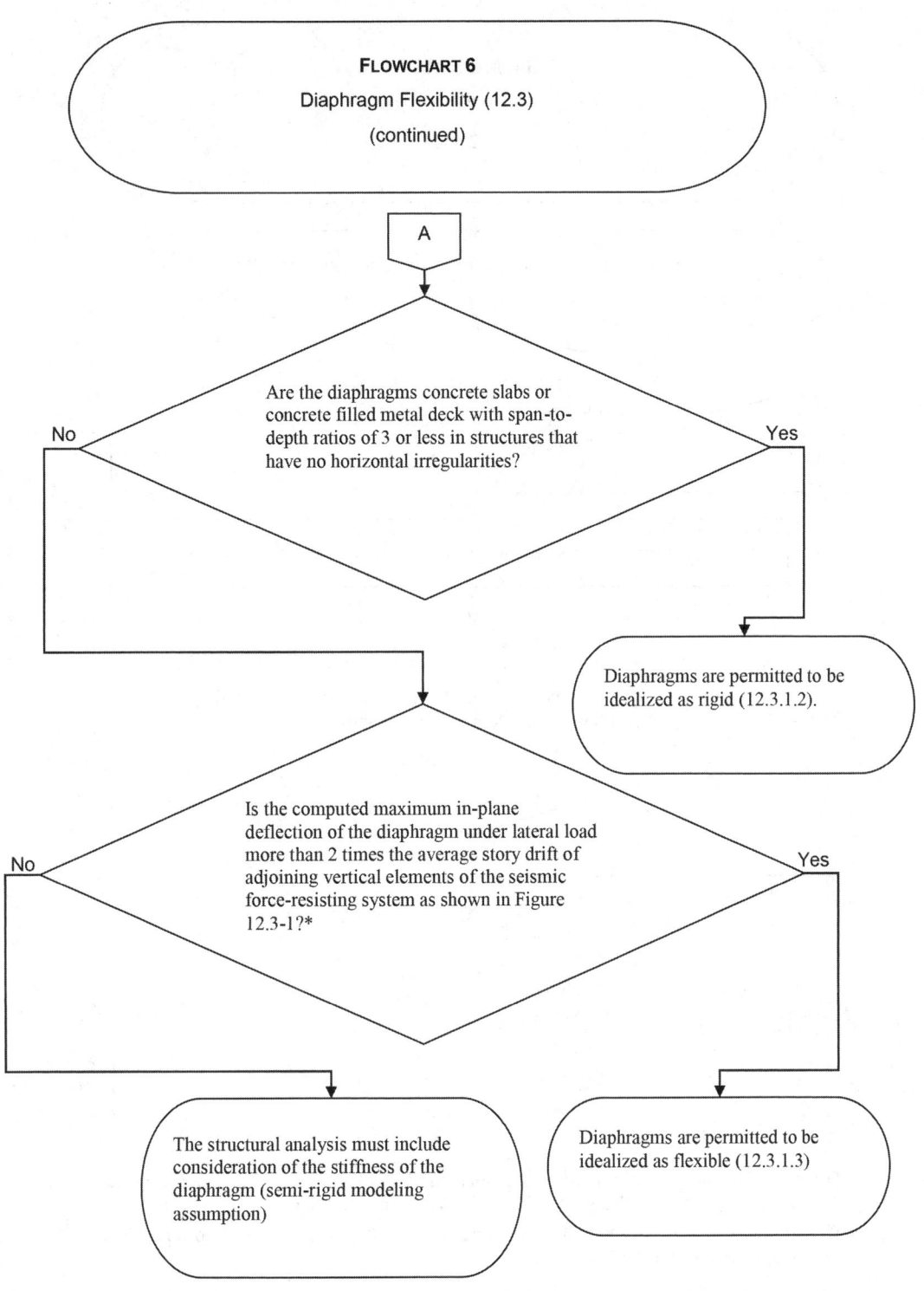

FLOWCHART 6
Diaphragm Flexibility (12.3)
(continued)

A

Are the diaphragms concrete slabs or concrete filled metal deck with span-to-depth ratios of 3 or less in structures that have no horizontal irregularities?

No   Yes

Diaphragms are permitted to be idealized as rigid (12.3.1.2).

Is the computed maximum in-plane deflection of the diaphragm under lateral load more than 2 times the average story drift of adjoining vertical elements of the seismic force-resisting system as shown in Figure 12.3-1?*

No   Yes

The structural analysis must include consideration of the stiffness of the diaphragm (semi-rigid modeling assumption)

Diaphragms are permitted to be idealized as flexible (12.3.1.3)

* The loading used for this calculation must be that prescribed in 12.8.

Figure 3.35   (*Continued*)

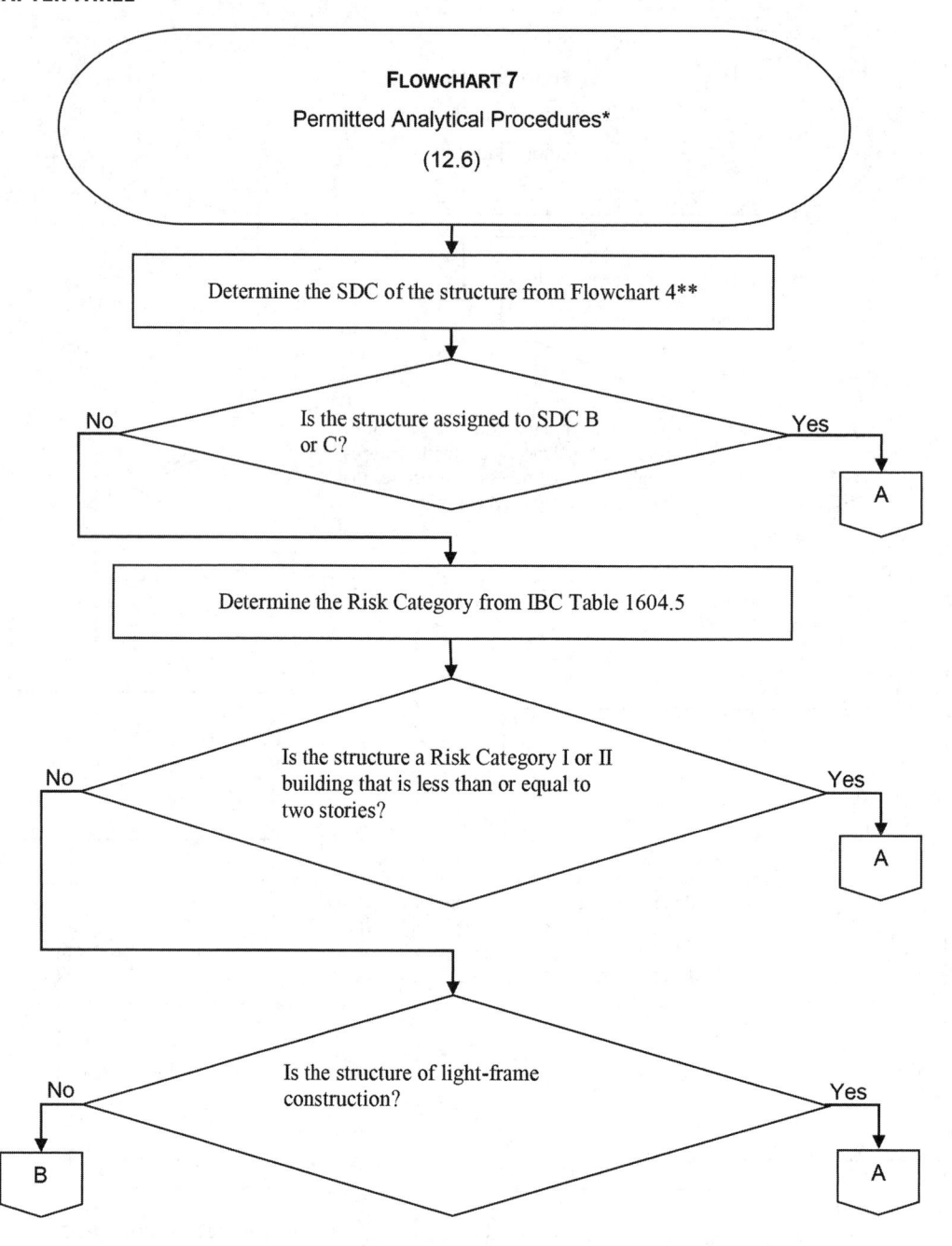

* The simplified alternative structural design method of 12.14 may be used for simple bearing wall or building frame systems that satisfy the 12 limitations in 12.14.1.1.

** This flowchart is applicable to buildings assigned to SDC B and higher. See Flowchart 5 for design requirements for SDC A.

Figure 3.36   Permitted analytical procedures A per ASCE 7-16.

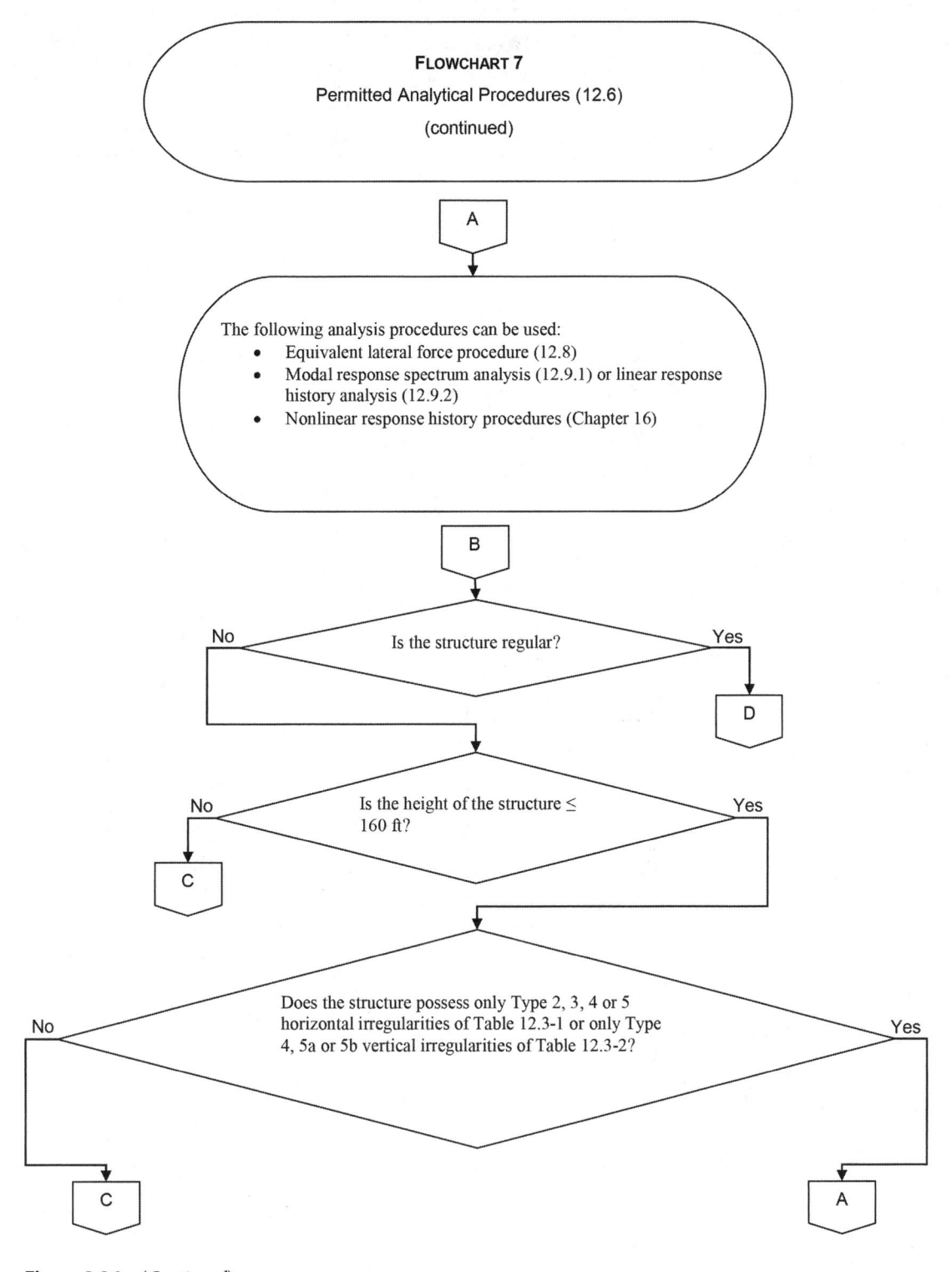

Figure 3.36 *(Continued)*

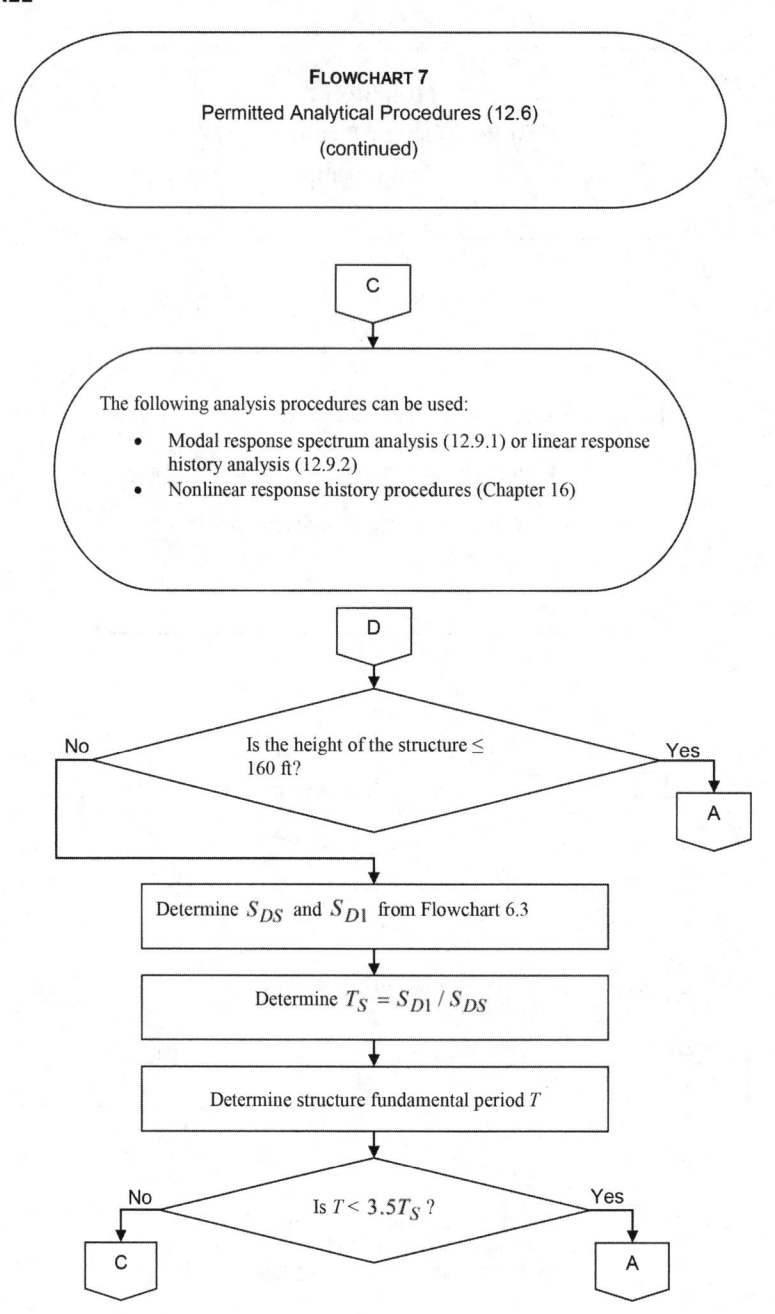

Figure 3.36   (*Continued*)

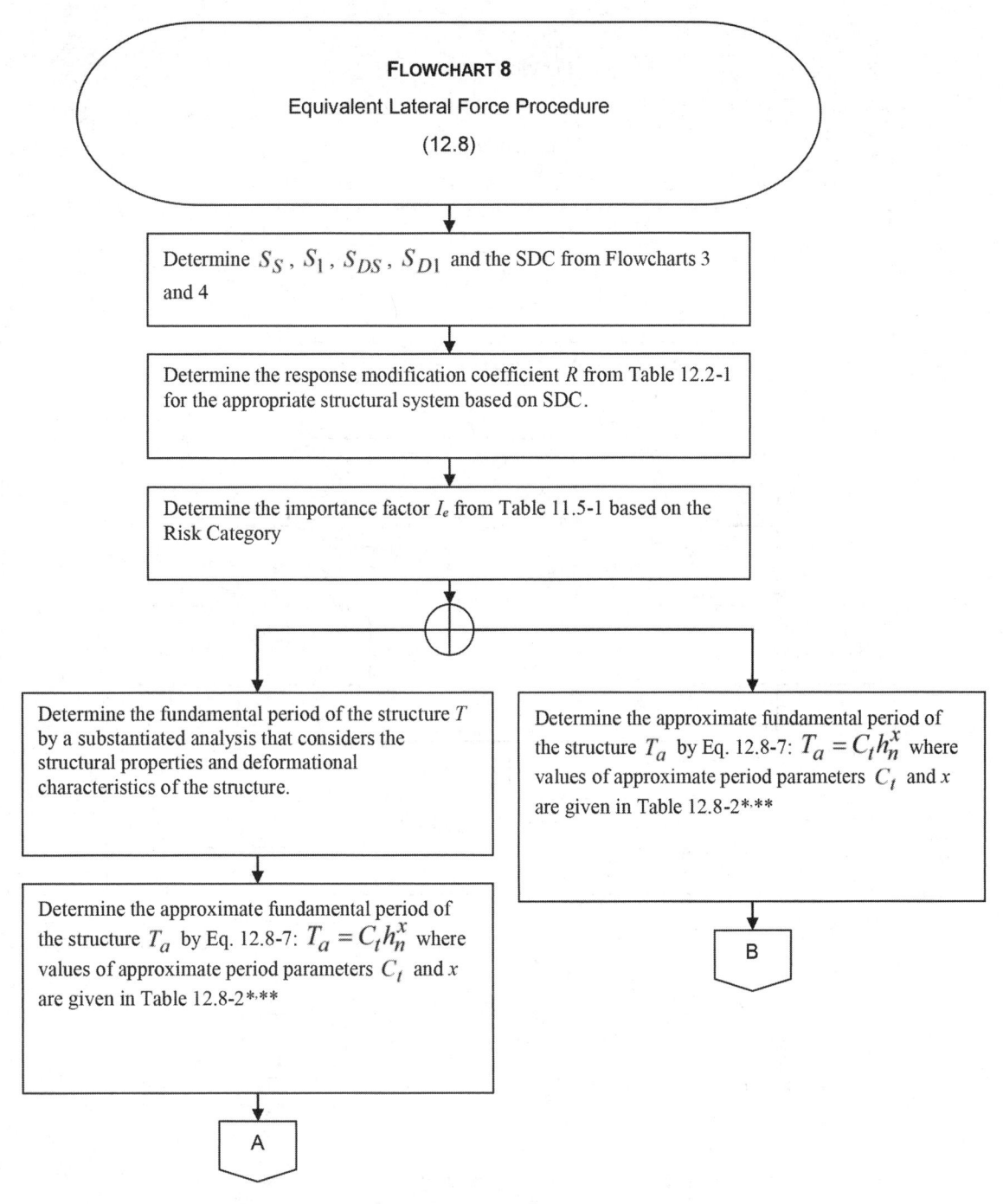

FLOWCHART 8

Equivalent Lateral Force Procedure

(12.8)

Determine S_S, S_1, S_{DS}, S_{D1} and the SDC from Flowcharts 3 and 4

Determine the response modification coefficient R from Table 12.2-1 for the appropriate structural system based on SDC.

Determine the importance factor I_e from Table 11.5-1 based on the Risk Category

Determine the fundamental period of the structure T by a substantiated analysis that considers the structural properties and deformational characteristics of the structure.

Determine the approximate fundamental period of the structure T_a by Eq. 12.8-7: $T_a = C_t h_n^x$ where values of approximate period parameters C_t and x are given in Table 12.8-2*,**

Determine the approximate fundamental period of the structure T_a by Eq. 12.8-7: $T_a = C_t h_n^x$ where values of approximate period parameters C_t and x are given in Table 12.8-2*,**

B

A

* h_n = height in feet above the base to the highest level of the structure.

** Alternate equations for T_a are given in 12.8.2.1 for concrete or steel moment resisting frames and masonry or concrete shear wall structures (see Table 3.9).

Figure 3.37 Equivalent lateral force procedure (ELFP) A per ASCE 7-16 [1].

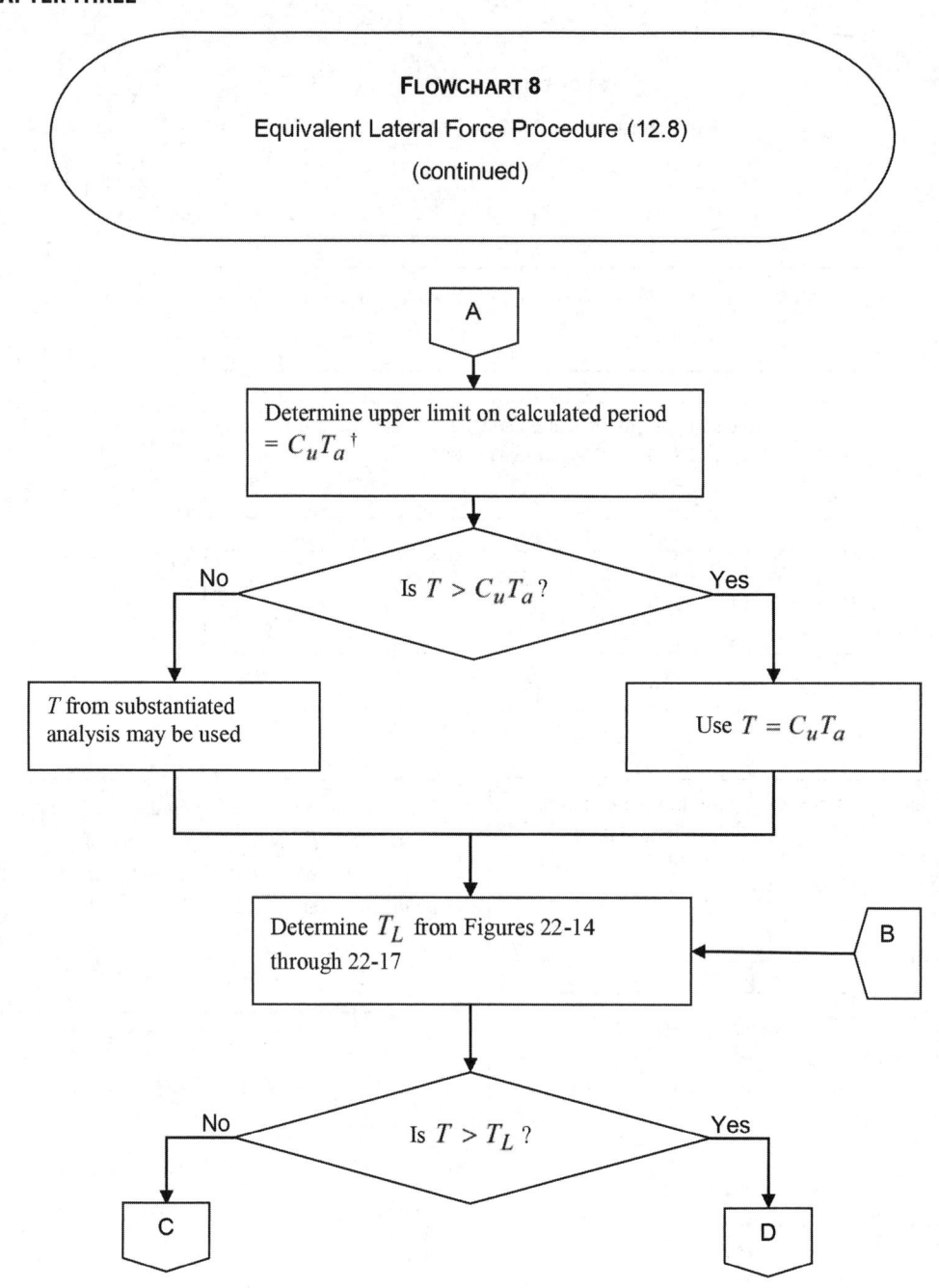

FLOWCHART 8

Equivalent Lateral Force Procedure (12.8)

(continued)

A

Determine upper limit on calculated period
$= C_u T_a$ †

Is $T > C_u T_a$?   No ← → Yes

T from substantiated analysis may be used

Use $T = C_u T_a$

Determine T_L from Figures 22-14 through 22-17 ← B

Is $T > T_L$?   No ← → Yes

C

D

† C_u = coefficient for upper limit on calculated period given in Table 12.8-1.

Figure 3.37   (*Continued*)

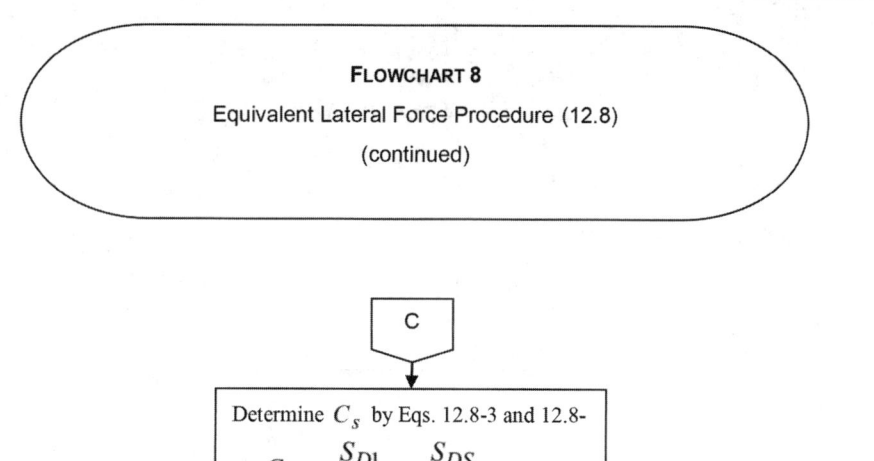

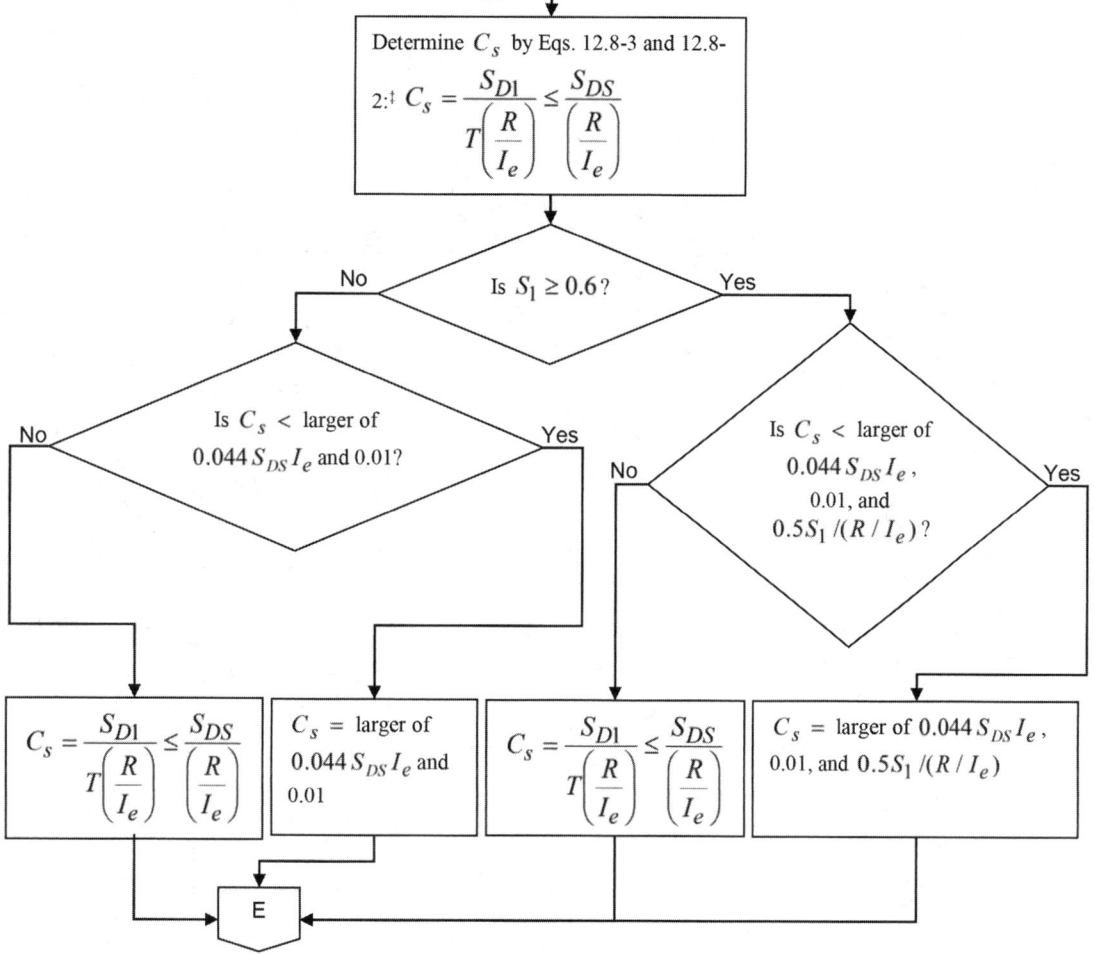

‡Value of C_s is permitted to be calculated using a value of 1.0 for S_{DS}, but not less than $0.7S_{DS}$, provided all the criteria of 12.8.1.3 are met.

Figure 3.37   (*Continued*)

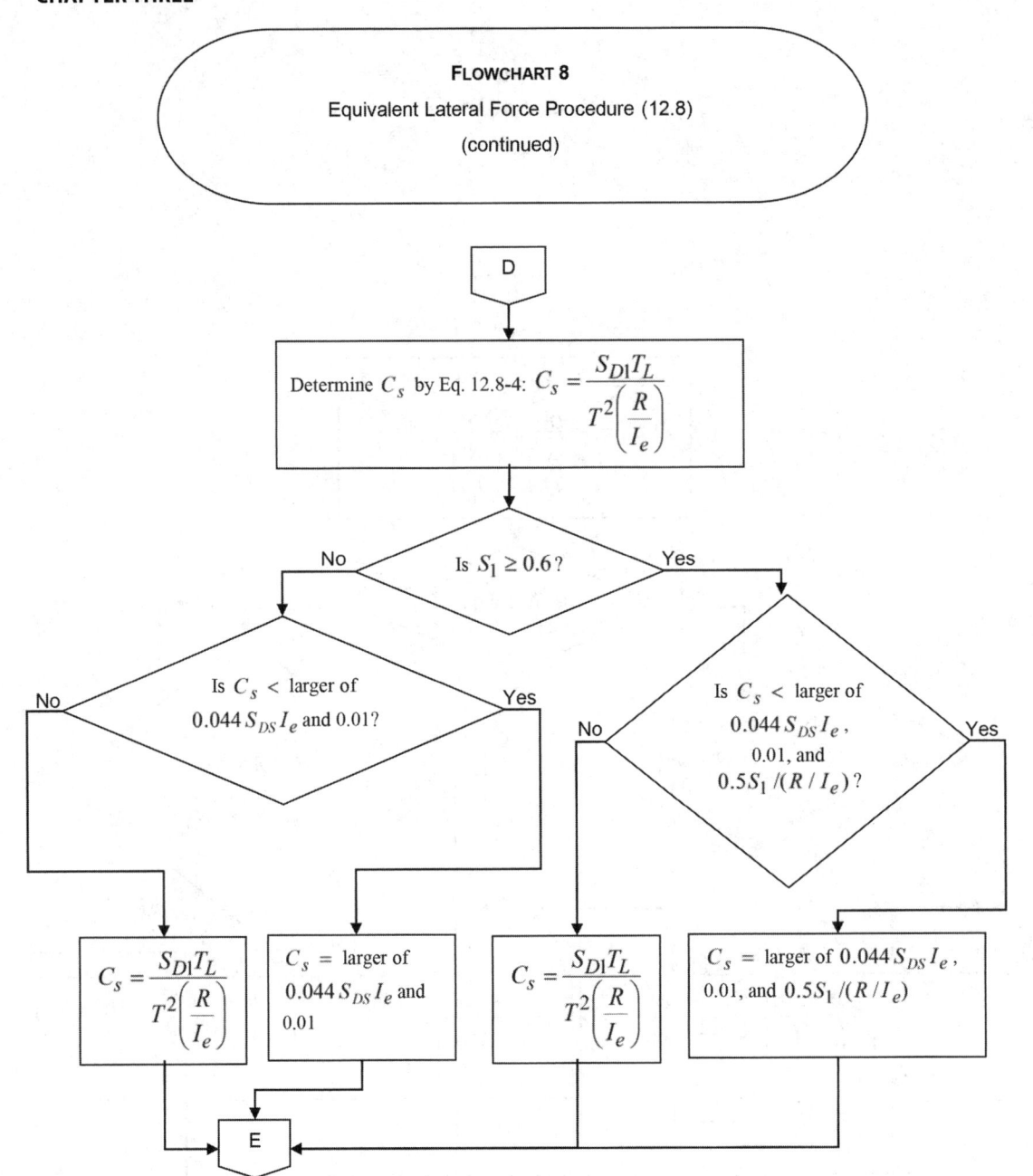

Figure 3.37   (*Continued*)

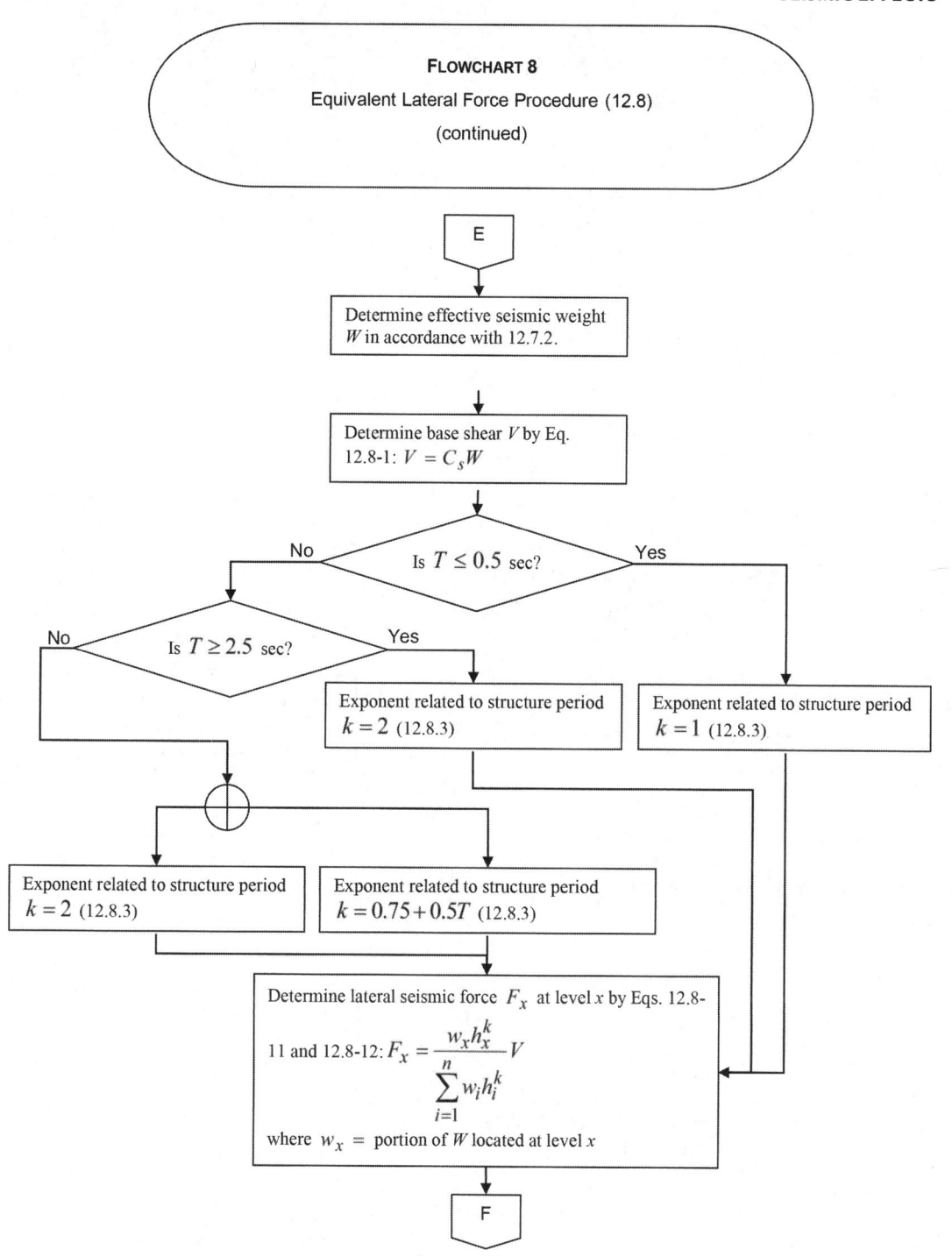

Figure 3.37 (*Continued*)

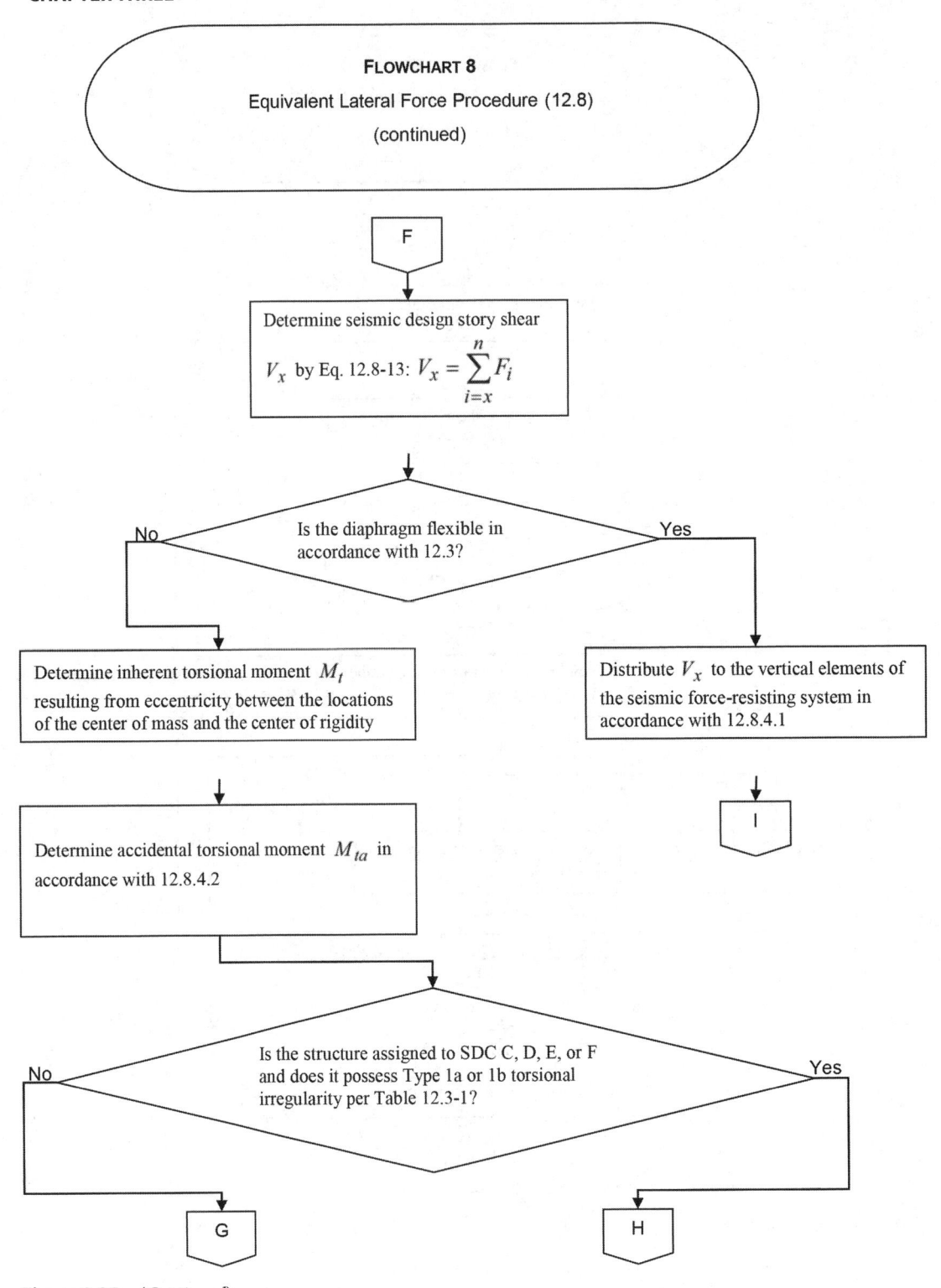

Figure 3.37   (*Continued*)

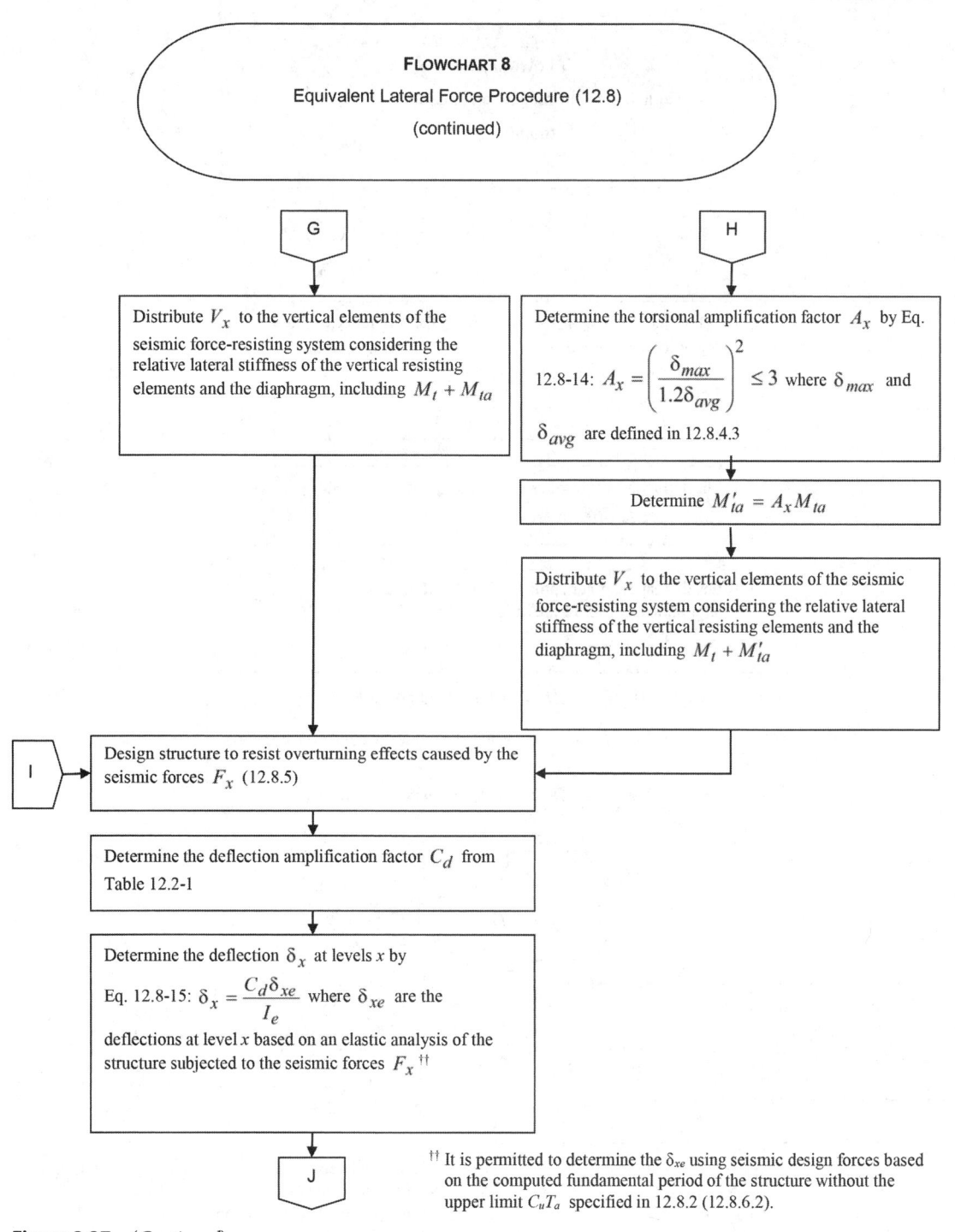

FLOWCHART 8

Equivalent Lateral Force Procedure (12.8)

(continued)

G

Distribute V_x to the vertical elements of the seismic force-resisting system considering the relative lateral stiffness of the vertical resisting elements and the diaphragm, including $M_t + M_{ta}$

H

Determine the torsional amplification factor A_x by Eq.

12.8-14: $A_x = \left(\dfrac{\delta_{max}}{1.2\delta_{avg}} \right)^2 \leq 3$ where δ_{max} and δ_{avg} are defined in 12.8.4.3

Determine $M'_{ta} = A_x M_{ta}$

Distribute V_x to the vertical elements of the seismic force-resisting system considering the relative lateral stiffness of the vertical resisting elements and the diaphragm, including $M_t + M'_{ta}$

I

Design structure to resist overturning effects caused by the seismic forces F_x (12.8.5)

Determine the deflection amplification factor C_d from Table 12.2-1

Determine the deflection δ_x at levels x by

Eq. 12.8-15: $\delta_x = \dfrac{C_d \delta_{xe}}{I_e}$ where δ_{xe} are the

deflections at level x based on an elastic analysis of the structure subjected to the seismic forces F_x [††]

J

[††] It is permitted to determine the δ_{xe} using seismic design forces based on the computed fundamental period of the structure without the upper limit $C_u T_a$ specified in 12.8.2 (12.8.6.2).

Figure 3.37   (*Continued*)

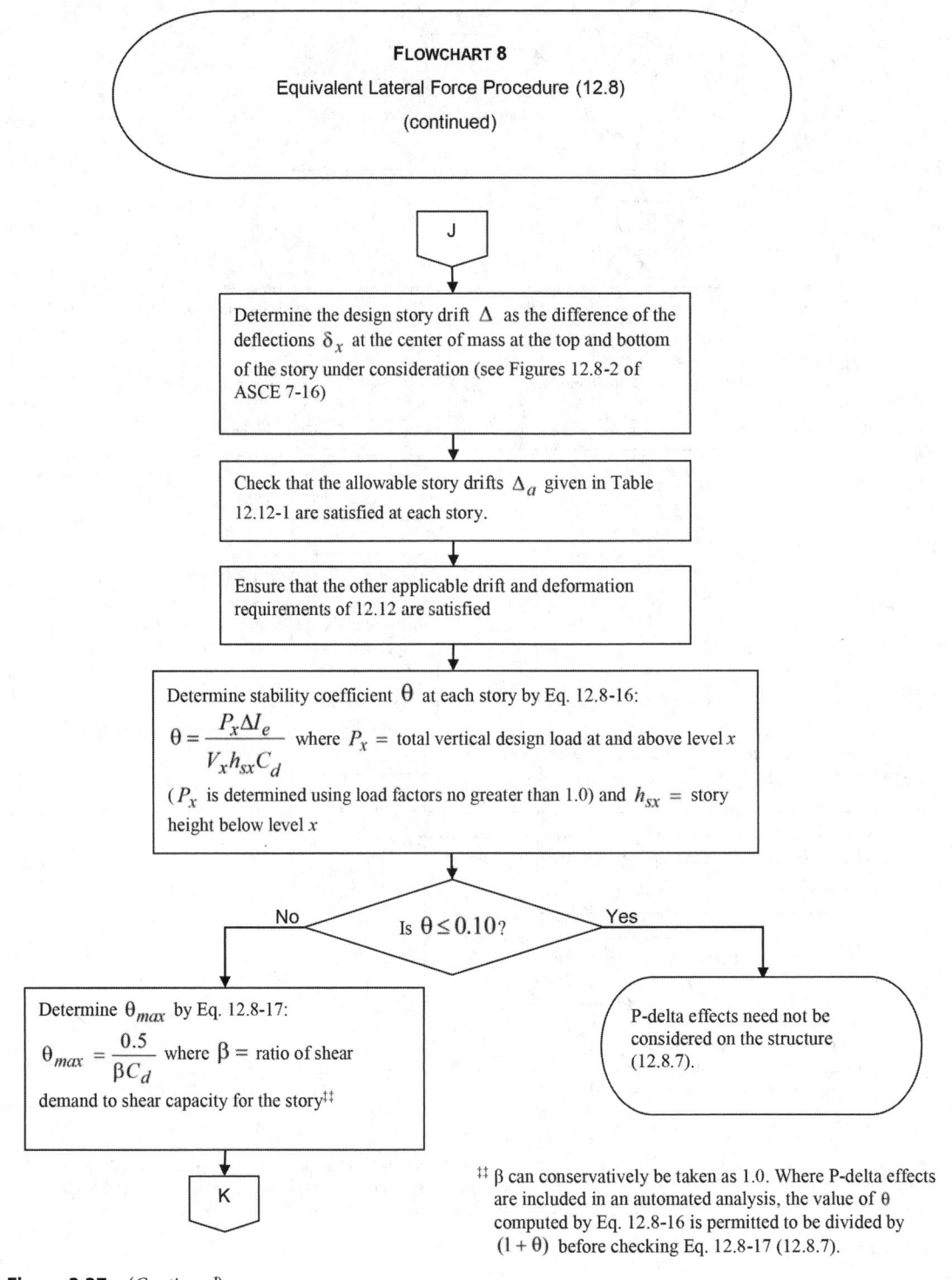

FLOWCHART 8
Equivalent Lateral Force Procedure (12.8)
(continued)

J

Determine the design story drift Δ as the difference of the deflections δ_x at the center of mass at the top and bottom of the story under consideration (see Figures 12.8-2 of ASCE 7-16)

Check that the allowable story drifts Δ_a given in Table 12.12-1 are satisfied at each story.

Ensure that the other applicable drift and deformation requirements of 12.12 are satisfied

Determine stability coefficient θ at each story by Eq. 12.8-16:

$$\theta = \frac{P_x \Delta I_e}{V_x h_{sx} C_d}$$ where P_x = total vertical design load at and above level x

(P_x is determined using load factors no greater than 1.0) and h_{sx} = story height below level x

Is $\theta \le 0.10$?   No   Yes

Determine θ_{max} by Eq. 12.8-17:

$$\theta_{max} = \frac{0.5}{\beta C_d}$$ where β = ratio of shear

demand to shear capacity for the story[‡‡]

K

P-delta effects need not be considered on the structure (12.8.7).

[‡‡] β can conservatively be taken as 1.0. Where P-delta effects are included in an automated analysis, the value of θ computed by Eq. 12.8-16 is permitted to be divided by $(1 + \theta)$ before checking Eq. 12.8-17 (12.8.7).

Figure 3.37 (*Continued*)

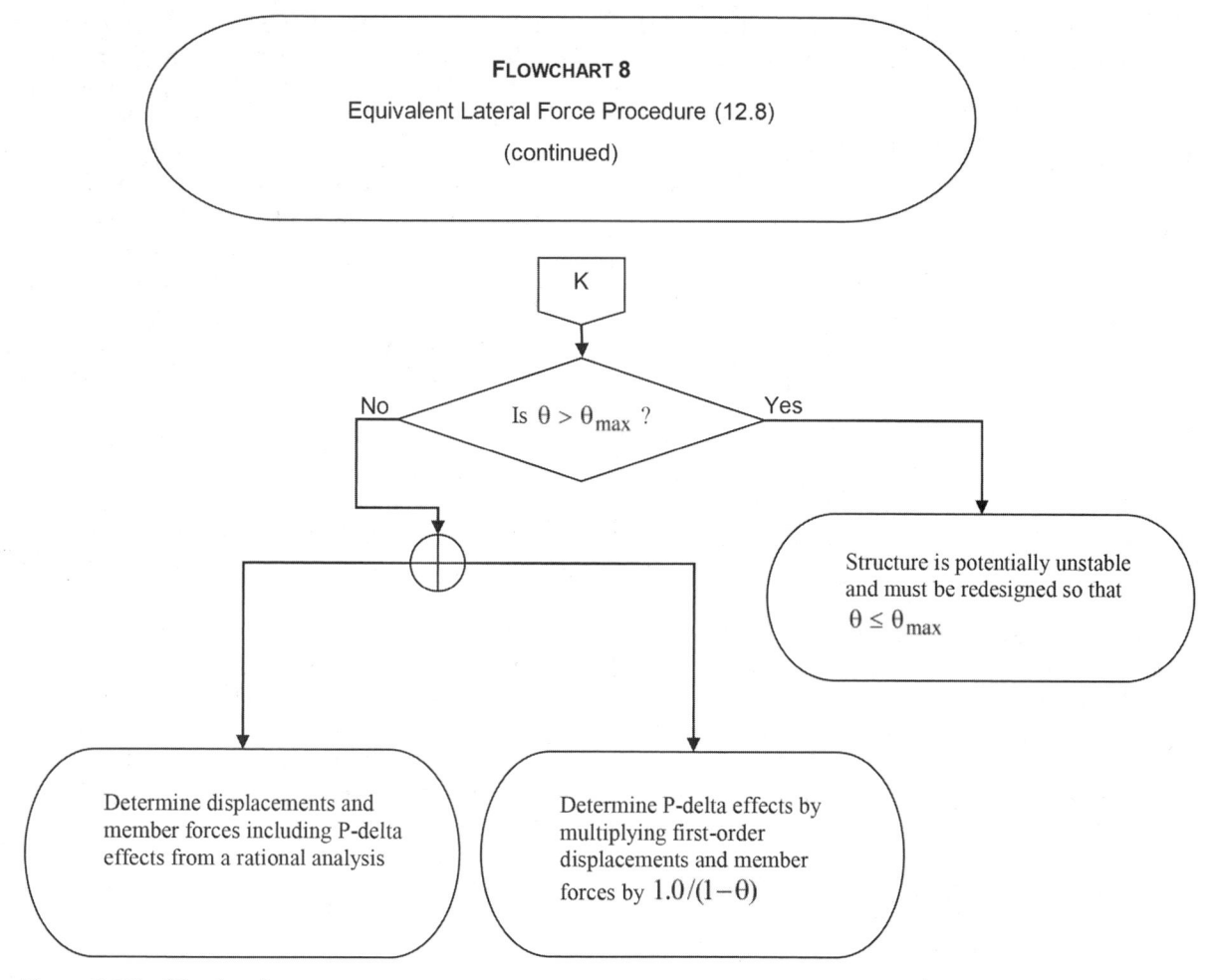

FLOWCHART 8

Equivalent Lateral Force Procedure (12.8)

(continued)

K

No — Is $\theta > \theta_{max}$? — Yes

Structure is potentially unstable and must be redesigned so that $\theta \leq \theta_{max}$

Determine displacements and member forces including P-delta effects from a rational analysis

Determine P-delta effects by multiplying first-order displacements and member forces by $1.0/(1-\theta)$

Figure 3.37   (*Continued*)

Example: Analysis of 20-story building using the ELFP of ASCE 7-16

A 20-story residential reinforced concrete building located in Los Angeles, CA (Latitude: 34.053, Longitude: −118.245) has a footprint of 120 ft × 150 ft and given the following information:

Material: Cast-in-place, reinforced concrete
Structural system: Moment-resisting frame
Risk Category: II
Site Class: C

Solution:

Example: Analysis of a 20-story building using the ELFP of ASCE 7-16

Step 1: Determine the seismic ground motion values from Flowchart 3.

1. Determine the mapped accelerations S_S and S_1. Instead of using Figures 22-1 and 22-2 of ASCE 7-16, the mapped accelerations are determined by inputting the latitude and longitude of the site into the ASCE Hazard Tool (https://asce7hazardtool. online/). The output is as follows: $S_S = 1.979$ and $S_1 = 0.705$.

2. Determine the site class of the soil. The site class of the soil is given as Site Class C.

3. Determine accelerations S_{MS} and $S_{M1.}$

Note: Because $S_1 > 0.2$, a ground motion hazard analysis in accordance with section 21.2 may be required (11.4.8(3)). For purposes of this example, assume such an analysis is not required. Site coefficients F_a and F_v are determined from Tables 11.4-1 and 11.4-2, respectively:

For Site Class C and $S_S > 1.50$: $F_a = 1.20$, For Site Class C and $S_1 > 0.6$: $F_v = 1.40$

$$S_{MS} = 1.20 \times 1.979 = 2.375$$
$$S_{M1} = 1.40 \times 0.705 = 0.987$$

4. Determine design accelerations S_{DS} and S_{D1}. From Equations 11.4-3 and 11.4-4:

$$S_{DS} = (2/3) \times 2.375 = 1.583$$
$$S_{D1} = (2/3) \times 0.987 = 0.658$$

The above results are shown in Tables 3.10, 3.11, and 3.12.

Step 2: Determine the SDC from Flowchart 4 (Fig. 3.33).

1. Determine if the building can be assigned to SDC A in accordance with 11.4.1. Because $S_S = 1.979 > 0.15$ and $S_1 = 0.705 > 0.04$, the building cannot be automatically assigned to SDC A.

2. Determine the risk category from IBC Table 1604.5. Given a residential building with a residential occupancy is less than 300 people congregate in one area, therefore, the Risk Category is II.

3. Because $S_1 < 0.75$, the building is not assigned to SDC E or F.

4. Check if all four conditions of 11.6 are satisfied.

Check if the approximate period, T_a, is less than $0.8 T_S$.

Use Equation 12.8-7 with approximate period parameters for "other structural systems:" $T_a = C_t h_n{}^x = 0.028(225)^{0.8} = 2.13$ s.

where C_t and x are given in Table 12.8-2.

$$T_S = S_{D1} / S_{DS} = 0.658 / 1.583 = 0.42 \text{ s}$$
$$2.13 \text{ s} > 0.8 \times 0.42 = 0.33 \text{ s}$$

Because this condition is not satisfied, the SDC cannot be determined by Table 11.6-1 alone (11.6).

5. Determine the SDC from Tables 11.6-1 and 11.6-2 and Flowchart 4.

From Table 11.6-1, with $S_{DS} > 0.50$ and Risk Category II, the SDC is D.

From Table 11.6-2, with $S_{D1} > 0.20$ and Risk Category II, the SDC is D.

The results for seismic design category selection are shown in Table 3.13. Therefore, the **SDC is D** for this building.

Step 3: Determine the permitted analytical procedure.

1. Check if $T < 3.5TS$.

$$T = 2.13 \text{ s} < 3.5 T_S = 3.5 \times 0.42 = 1.68 \text{ s}$$

2. Determine if the structure is regular or not.

The structures are assumed to be geometrically regular, however, to determine Type 1a torsional irregularities and Type 1b extreme torsional irregularities, the story drifts are needed from analysis. The drifts are determined from code-prescribed forces. Since it is not evident that a higher-order analysis method is required, the equivalent lateral force procedure maybe used to determine the lateral seismic forces.

Use Flowchart 8 (Fig. 3.37) to determine the lateral seismic forces from the equivalent lateral force procedure.

a. Determine the SDC. SDC was determined above as SDC D.

b. Determine the response modification coefficient, R, from Table 12.2-1 of ASCE 7-16. For moment-resisting frame in SDC D, intermediate reinforced concrete moment-resisting frame is not permitted, therefore, special reinforced concrete moment-resisting frame is selected. The response modification coefficient, R, for this system is 8.

c. Determine the importance factor, I_e, from Table 1.5-2 of ASCE 7-16. For Risk Category II, $I_e = 1.0$.

d. Determine the period of the structure, T. The approximate period of the structure was determined above in step 2 as 2.13 s.

e. Determine the long-period transition period, T_L, from Figure 22-14 or from ASCE Hazard Tool (https://asce7hazardtool.online/), $T_L = 8 \text{ s} > T = 2.13 \text{ s}$.

f. Determine seismic response coefficient, C_S. The seismic response coefficient, C_S, is determined by Equation 12.8-3:

$$C_s = \frac{S_{D1}}{T\left(\dfrac{R}{I_e}\right)} = \frac{0.658}{2.13\left(\dfrac{8}{1.0}\right)} = 0.039$$

Eq. 12.8-3—ASCE 7-16

The values of C_s should not exceed the value obtained from Equation 12.8-2:

$$C_s = \frac{S_{DS}}{R/I_e} = \frac{1.583}{8/1.0} = 0.198$$

Eq. 12.8-2—ASCE 7-16

Also, C_S must not be less than the larger of $0.044 S_{DS} I_e = 0.069$ (governs) and 0.01 (Equation 12.8-5).

Thus, the value of C_S from Equation 12.8-3 governs, $C_S = 0.069$.

    g. Determine effective seismic weight, W, according to 12.7.2. The seismic weight which includes the weight of the member sizes and the superimposed dead load at each level. The seismic weight for each level is given and shown in Table 3.14. The total effective seismic weight, W, is also shown in Table 3.14.

    h. Determine seismic base shear, V.

        Seismic base shear is determined by Equation 12.8-1:

    i. Determine seismic base shear, V.

        Seismic base shear is determined by Equation 12.8-1:

$$V = C_S W = 0.07 \times 30{,}900 = 2{,}152.3 \text{ kips}$$

    j. Determine exponent related to structure period, k. Because $0.5 \text{ s} < T = 2.13 \text{ s} < 2.5 \text{ s}$, k is determined as follows:

$$k = 0.75 + 0.5T = 1.82.$$

    k. Determine lateral seismic force, F_x, at each level, x.

        F_x is determined by Equations 12.8-11 and 12.8-12 of ASCE 7-16. A summary of the lateral forces, F_x, and the story shears, V_x, is given in Table 3.15.

        Three-dimensional analysis was performed independently in both directions, N-S & E-W, for the seismic forces in Table 3.15 using a commercial structural analysis program. Rigid diaphragms were assigned in the model at each level and reduced stiffnesses were assigned to the structural members assuming cracked sections, according to section 12.7.3 of ASCE 7-16. According to section 12.8.4.2 of ASCE 7-16, the center of mass was displaced each way from its actual location at a distance equal to 5 percent of the building dimension perpendicular to the applied forces to account for accidental torsion when determining whether a horizontal irregularity exists.

        The design and MCE_R horizontal response spectrum are plotted and shown in Figs. 3.38 and 3.39 per sections 11.4.6 and 11.4.7 respectively. The design and MCER vertical response spectrum are plotted and shown in Figs. 3.40 and 3.41 per sections 11.9.2 and 11.9.3, respectively. The equivalent lateral force design procedure is shown in Fig. 3.42.

Step 4: Determine the elastic and the inelastic displacements in the N-S direction.

From this model, elastic displacements at the center of mass are obtained at each level in the N-S direction, the elastic displacements are shown in Table 3.16. The inelastic displacement is calculated according to Equation 12.8-15 with deflection amplification factor $C_d = 5.5$ for reinforced concrete special moment-resisting (Table 12.2-1 of ASCE 7-16).

The allowable drifts have been calculated according to section 12.12 of ASCE 7-16. The allowable drift for Risk Category II is given in Table 12.12-1 of ASCE 7-16 as $\Delta_{a/hsx} = 0.20$. The results are shown in Table 3.16 which shows that all drifts are less than the allowable drift limits. According to section 12.12.1.1, for moment-resisting frames assigned to SDC D, E, or F, the allowable drift should be divided by the redundancy factor, ρ. In this example the redundancy factor is equal to 1.3. The results in Table 3.16 include $\rho = 1.3$ in the allowable drift results.

Step 5: P-delta effects in the N-S direction.

P-delta effects can be obtained from computer analysis, the procedure in Flowchart 8 (Fig. 3.37). P-delta effects need not be considered when the stability coefficient, θ, determined by Equation 12.8-16 is less than or equal to 0.10 per the requirements of section 12.8.7.

$$\theta = \frac{P_x \Delta I_e}{V_x h_{sx} C_d} \qquad \text{Equation 12.8-16—ASCE 7-16}$$

The P-delta effects calculations for each floor are shown in Table 3.17.

Step 6: Determine the design seismic forces in the diaphragm in both directions by Equation 12.10-1 and compare it with the lower and upper limits given in Equations 12.10-2 and 12.10-3, respectively.

$$0.2 S_{DS} I_e w_{px} \leq F_{px} = \frac{\sum_{i=x}^{n} F_i}{\sum_{i=x}^{n} w_i} w_{px} \leq 0.4 S_{DS} I_e w_{px}$$

The diaphragm forces according to the equation above with check with lower and upper limits are shown in Table 3.18.

Table 3.10   Seismic Ground Motion Values (ASCE/SEI 11.4)

	Input	
Risk category	II	ASCE/SEI Table 1.5-1
Mapped MCE_R, 5% damped, spectral response acceleration parameter at short periods, S_S	1.979	ASCE/SEI 11.4.2
Mapped MCE_R, 5% damped, spectral response acceleration parameter at a period of 1 s, S_1	0.705	
Site class	C	ASCE/SEI Table 11.4.3
Long-period transition period, T_L	8	s

Table 3.11   Site Coefficients and MCER Spectral Response Acceleration Parameters (ASCE/SEI 11.4.4)

Short-period site coefficient, F_a	1.200	—
Long-period site coefficient, F_v	1.400	—
MCE_R spectral response acceleration parameter for short periods, adjusted for site class effects, S_{MS}	2.375	ASCE/SEI Eq. (11.4-1)
MCE_R spectral response acceleration parameter at a period of 1 s, adjusted for site class effects, S_{M1}	0.987	ASCE/SEI Eq. (11.4-2)

Table 3.12   Design Spectral Response Acceleration Parameters (ASCE/SEI 11.4.5)

Design spectral acceleration parameter at short periods, S_{DS}	1.583	ASCE/SEI Eq. (11.4-3)
Design spectral acceleration parameter at a period of 1 s, S_{D1}	0.658	ASCE/SEI Eq. (11.4-4)

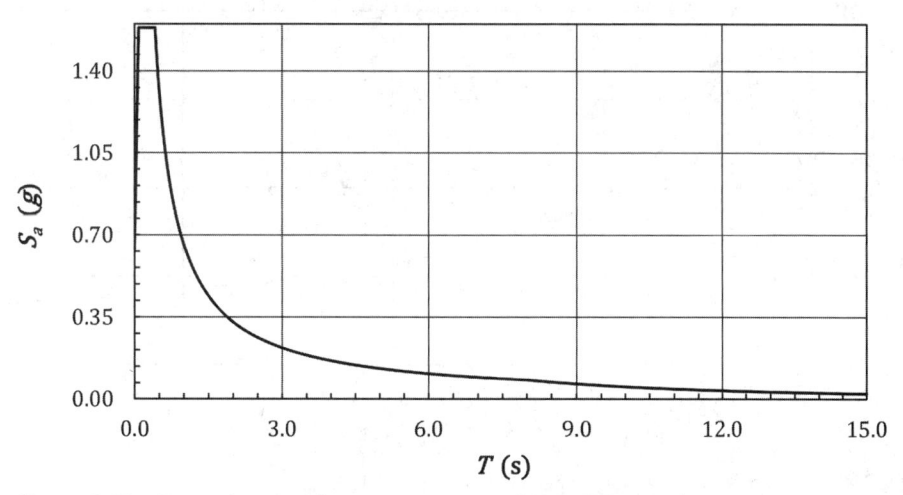

Figure 3.38   Design horizontal response spectrum (ASCE/SEI 11.4.6).

$T_0 = 0.08\ s$
$T_s = 0.42\ s$

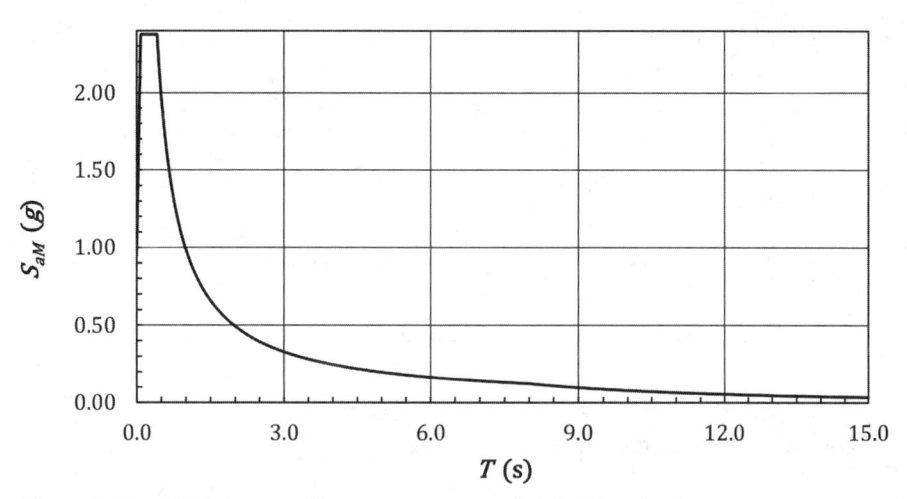

Figure 3.39   MCE$_R$ horizontal response spectrum (ASCE/SEI 11.4.7).

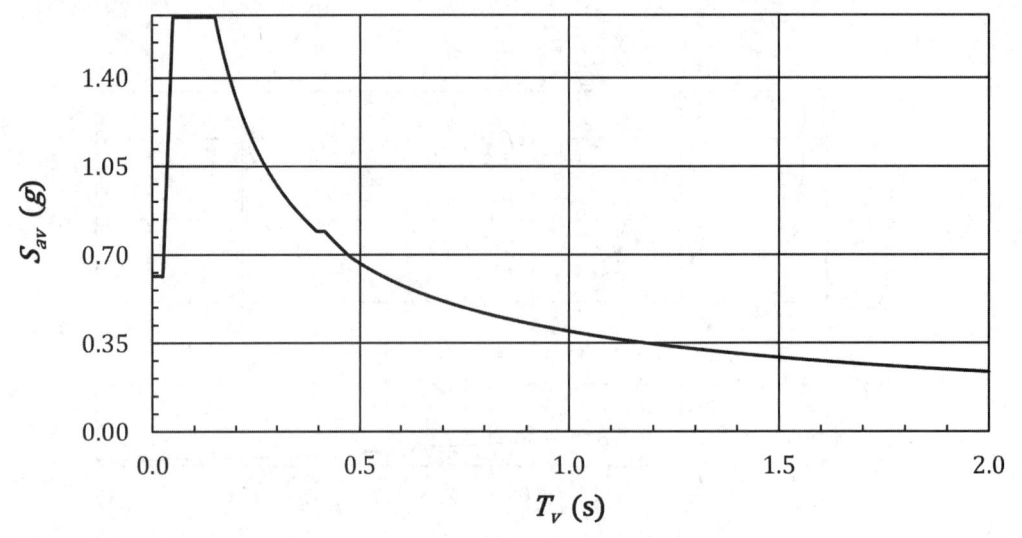

Figure 3.40 Design vertical response spectrum (ASCE/SEI 11.9.2).

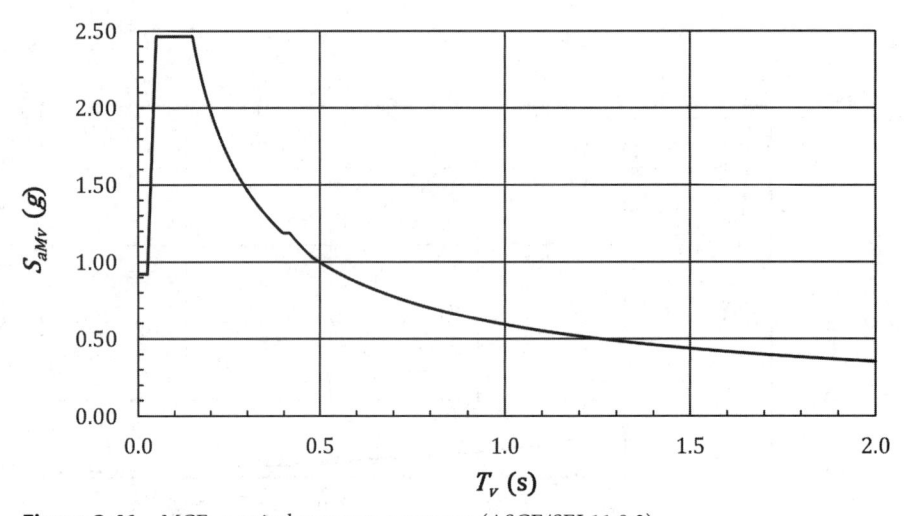

Figure 3.41 MCE$_R$ vertical response spectrum (ASCE/SEI 11.9.3).

Table 3.13   Seismic Design Category (ASCE/SEI 11.6)

SDC based on risk category and S_1:	—
SDC based on short-period response acceleration parameter:	D
SDC based on long-period response acceleration parameter:	D
SDC:	D

Equivalent Lateral Force (ELF) Procedure (ASCE/SEI 12.8):

Table 3.14   Story Height and Story Weight

Level	Weight tributary to level, w_x (kips)	Height, h_x (ft)	Story height (ft)
20	1,450	225.0	11.0
19	1,550	214.0	11.0
18	1,550	203.0	11.0
17	1,550	192.0	11.0
16	1,550	181.0	11.0
15	1,550	170.0	11.0
14	1,550	159.0	11.0
13	1,550	148.0	11.0
12	1,550	137.0	11.0
11	1,550	126.0	11.0
10	1,550	115.0	11.0
9	1,550	104.0	11.0
8	1,550	93.0	11.0
7	1,550	82.0	11.0
6	1,550	71.0	11.0
5	1,550	60.0	11.0
4	1,550	49.0	11.0
3	1,550	38.0	11.0
2	1,550	27.0	11.0
1	1,550	16.0	16.0

$W = 30,900$

Vertical Distribution of Seismic Forces (ASCE/SEI 12.8.3)

Exponent related to the period of the structure, $k = 1.82$

Table 3.15   Vertical Distribution of Seismic Forces in the N-S Direction

Level	Weight tributary to level, w_x (kips)	Height, h_x (ft)	$w_x h_x^{\,k}$	Lateral force, F_x (kips)	Story shear, V_x (kips)
20	1,450	225.0	26,951,272	261.2	261.2
19	1,550	214.0	26,304,660	255.0	516.2
18	1,550	203.0	23,902,152	231.7	747.9
17	1,550	192.0	21,603,466	209.4	957.3
16	1,550	181.0	19,409,677	188.1	1,145.4
15	1,550	170.0	17,321,938	167.9	1,313.3
14	1,550	159.0	15,341,488	148.7	1,462.0
13	1,550	148.0	13,469,662	130.6	1,592.6
12	1,550	137.0	11,707,914	113.5	1,706.1
11	1,550	126.0	10,057,829	97.5	1,803.6
10	1,550	115.0	8,521,151	82.6	1,886.2
9	1,550	104.0	7,099,818	68.8	1,955.0
8	1,550	93.0	5,795,998	56.2	2,011.2
7	1,550	82.0	4,612,154	44.7	2,055.9
6	1,550	71.0	3,551,122	34.4	2,090.3
5	1,550	60.0	2,616,232	25.4	2,115.6
4	1,550	49.0	1,811,497	17.6	2,133.2
3	1,550	38.0	1,141,929	11.1	2,144.3
2	1,550	27.0	614,125	6.0	2,150.2
1	1,550	16.0	237,580	2.3	2,152.5
	30,900		222,071,665	2,152.5	

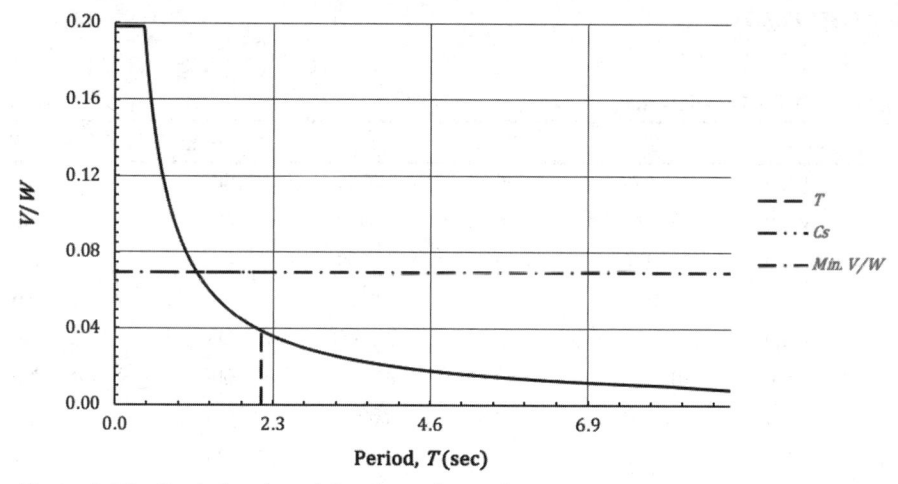

Figure 3.42 Equivalent lateral force procedure—design spectrum.

Story Drift Determination
(ASCE/SEI 12.8.6)

Allowable story drift, $\Delta_a/h_{sx} = 0.20$ (ASCE/SEI Table 12.12-1)

Table 3.16 Story Drift Determination in the N-S Direction

Level	δ_{xe} (in.)	δ_x (in.)	Δ (in.)	Δ_a (in.)	Status
20	5.90	32.45	1.10	2.03	OK
19	5.70	31.35	1.65	2.03	OK
18	5.40	29.70	1.38	2.03	OK
17	5.15	28.33	1.27	2.03	OK
16	4.92	27.06	1.49	2.03	OK
15	4.65	25.58	1.82	2.03	OK
14	4.32	23.76	1.16	2.03	OK
13	4.11	22.61	1.10	2.03	OK
12	3.91	21.51	1.10	2.03	OK
11	3.71	20.41	1.87	2.03	OK
10	3.37	18.54	1.49	2.03	OK
9	3.10	17.05	1.93	2.03	OK
8	2.75	15.13	1.65	2.03	OK
7	2.45	13.48	1.93	2.03	OK
6	2.10	11.55	1.38	2.03	OK
5	1.85	10.18	1.93	2.03	OK
4	1.50	8.25	1.65	2.03	OK
3	1.20	6.60	1.93	2.03	OK
2	0.85	4.68	1.93	2.03	OK
1	0.50	2.75	2.75	2.95	OK

P-Delta Effects (ASCE/SEI 12.8.7)

Table 3.17 *P*-Delta Calculations

Level	P_x (kips)	θ	β	θ_{max}	θ_{max} Status	Consider *P*-delta?
20	1,240.0	0.0072	1.00	0.0909	OK	No
19	3,390.0	0.0149	1.00	0.0909	OK	No
18	5,540.0	0.0140	1.00	0.0909	OK	No
17	7,690.0	0.0140	1.00	0.0909	OK	No
16	9,840.0	0.0176	1.00	0.0909	OK	No
15	11,990.0	0.0228	1.00	0.0909	OK	No
14	14,140.0	0.0154	1.00	0.0909	OK	No
13	16,290.0	0.0155	1.00	0.0909	OK	No
12	18,440.0	0.0164	1.00	0.0909	OK	No
11	20,590.0	0.0294	1.00	0.0909	OK	No
10	22,740.0	0.0247	1.00	0.0909	OK	No
9	24,890.0	0.0338	1.00	0.0909	OK	No
8	27,040.0	0.0306	1.00	0.0909	OK	No
7	29,190.0	0.0376	1.00	0.0909	OK	No
6	31,340.0	0.0284	1.00	0.0909	OK	No
5	33,490.0	0.0420	1.00	0.0909	OK	No
4	35,640.0	0.0380	1.00	0.0909	OK	No
3	37,790.0	0.0467	1.00	0.0909	OK	No
2	39,940.0	0.0493	1.00	0.0909	OK	No
1	42,090.0	0.0509	1.00	0.0909	OK	No

Table 3.18 Diaphragm Force Calculations in the N-S Direction

Level	Weight tributary to level i, w_i (kips)	Weight tributary to the diaphragm, w_{px} (kips)	F_i (kips)	Sw_i (kips)	ΣF_i (kips)	F_{px} (kips)	Min. F_{px} (kips)	Max. F_{px} (kips)	Design force (kips)
20	1,450	1,450	261.2	1,450	261.2	261.2	459.1	918.3	459.1
19	1,550	1,500	255.0	3,000	516.2	258.1	475.0	949.9	475.0
18	1,550	1,500	231.7	4,550	747.9	246.6	475.0	949.9	475.0
17	1,550	1,500	209.4	6,100	957.3	235.4	475.0	949.9	475.0
16	1,550	1,500	188.1	7,650	1,145.4	224.6	475.0	949.9	475.0
15	1,550	1,500	167.9	9,200	1,313.3	214.1	475.0	949.9	475.0
14	1,550	1,500	148.7	10,750	1,462.0	204.0	475.0	949.9	475.0
13	1,550	1,500	130.6	12,300	1,592.6	194.2	475.0	949.9	475.0
12	1,550	1,500	113.5	13,850	1,706.1	184.8	475.0	949.9	475.0
11	1,550	1,500	97.5	15,400	1,803.6	175.7	475.0	949.9	475.0
10	1,550	1,500	82.6	16,950	1,886.2	166.9	475.0	949.9	475.0
9	1,550	1,500	68.8	18,500	1,955.0	158.5	475.0	949.9	475.0
8	1,550	1,500	56.2	20,050	2,011.2	150.5	475.0	949.9	475.0
7	1,550	1,500	44.7	21,600	2,055.9	142.8	475.0	949.9	475.0
6	1,550	1,500	34.4	23,150	2,090.3	135.4	475.0	949.9	475.0
5	1,550	1,500	25.4	24,700	2,115.6	128.5	475.0	949.9	475.0
4	1,550	1,500	17.6	26,250	2,133.2	121.9	475.0	949.9	475.0
3	1,550	1,500	11.1	27,800	2,144.3	115.7	475.0	949.9	475.0
2	1,550	1,500	6.0	29,350	2,150.2	109.9	475.0	949.9	475.0
1	1,550	1,500	2.3	30,900	2,152.5	104.5	475.0	949.9	475.0

3.5 DYNAMIC ANALYSIS PROCEDURE

3.5.1 Introduction

Buildings with symmetrical shape, stiffness and mass distribution and with vertical continuity and uniformity behave in a fairly predictable manner whereas when buildings are eccentric or have areas of discontinuity or irregularity, the behavioral characteristics are very complex. In such instances dynamic analysis can be helpful in determining important seismic response characteristics that may not be evident from the static procedure, such as: (i) the effects of the structure's dynamic characteristics on the vertical distribution of lateral force; (ii) increase in the dynamic loads in the structure's lateral force resisting system due to torsional motions; and (iii) the effects of higher modes that could substantially increase story shears and deformations.

The ASCE 7-16 and IBC 2018 static method, ELFP, is based on a single mode response with approximate load distributions and corrections for higher mode response. These simplifications are appropriate for simple regular structures. However, they do not consider the full range of seismic behavior in complex structures.

Therefore, dynamic analysis is required for buildings with unusual or irregular geometry, since it results in distributions of seismic design forces that are in agreement with the actual distribution of mass and stiffness of the building.

A structure is considered irregular if it has any of the characteristics given in Tables 3.5 and 3.6 exist.

According to ASCE 7-16, the permitted analysis procedure is described above in Table 3.8 and

• ELFP, which is a static method (ASCE 7-16 Section 12.8)

• Modal response spectrum analysis (ASCE 7-16 section 12.9.1) or linear response history analysis (ASCE 7-16 section 12.9.2)

• Nonlinear response history procedures (ASCE 7-16 Chapter 16)

Determining the behavior of a structure in an earthquake is basically a vibration problem. Using dynamic analysis, calculations can be made of the earthquake-induced vibrations which will indicate the general nature and amplitude of deformations expected.

Structures that are built into the ground and extended vertically some distance above the ground respond either as simple or complex oscillators when subjected to seismic ground motions. Simple oscillators are represented by single-degrees-of-freedom systems (SDOF), and complex oscillators are represented by multi-degree-of-freedom (MDOF) system.

A simple oscillator is represented by a single lump of mass on the upper end of a vertically cantilevered pole or by a mass supported by two columns as shown in Fig. 3.43. The idealized system represents two kinds of structures: (i) a single-column structure with a relatively large mass at its top; and (ii) a single-story frame with flexible columns and a rigid beam. The mass M is the weight W of the system divided by the acceleration of gravity g. That is, $M = W/g$.

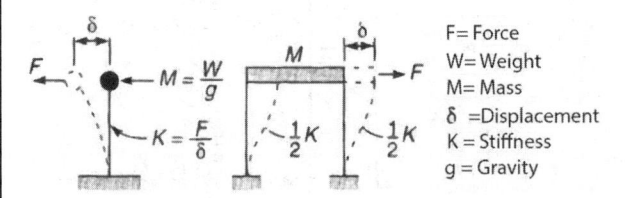

Figure 3.43 Idealized single degree-of-freedom system.

The stiffness K of the system, which is a ratio equal to a horizontal force F applied to the mass divided by the corresponding displacement δ. If the mass is deflected and then suddenly released, it will vibrate at a certain frequency, which is called its natural or fundamental frequency of vibration. The reciprocal of frequency is called the period of vibration. It represents the time for the mass to move through one complete cycle. The period T is given by the relation:

$$T = 2\pi \sqrt{\frac{M}{K}} \qquad (3.27)$$

In an ideal system having no damping ($\beta = 0$), the system would vibrate forever (Fig. 3.44). In a real system where there is always some damping, the amplitude of motion will gradually decrease for each cycle until the structure comes to a complete stop (Fig. 3.45). The system responds in a similar manner if, instead of displacing the mass at the top, a sudden impulse is applied to the base of the system.

Tall buildings may be analyzed as MDOF systems by lumping story-masses at intervals along the length of a vertically cantilevered pole. During vibration, each mass

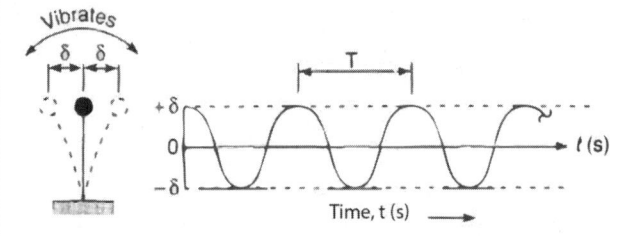

Figure 3.44 Undamped free vibrations of a single degree-of-freedom system.

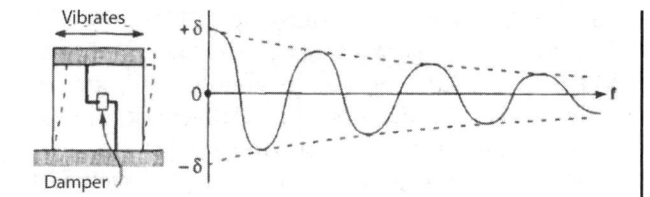

Figure 3.45 Damped free vibration of a single degree-of-freedom system.

will be deflected in one direction or another. For example, for higher modes of vibration, some masses may go to the right while others are going to the left. Or all masses may simultaneously deflect in the same direction as in the fundamental mode of vibration. An idealized MDOF system has a number of modes equal to the number of masses. Each mode has its own natural period of vibration with a unique mode shape formed by a line connecting the deflected masses. When ground motion is applied to the base of the multi-mass system, the deflected shape of the system is a combination of all mode shapes; but modes having periods near predominant periods of the base motion will be excited more than the other modes. Each mode of a multi-mass system can be represented by an equivalent single-mass system having generalized values M and K for mass and stiffness. The generalized values represent the equivalent combined effects of story

masses m_1, m_2,... and stiffness k_1 k_2 ... . This concept, shown in Fig. 3.46, provides a computational basis for using response spectra based on single-mass systems for analyzing multi-storied buildings. Given the period, mode shape, and mass distribution of a multi-storied building, we can use the response spectra of SDOF system for computing the deflected shape, story accelerations, forces and overturning moments. Each predominant mode is analyzed separately and the results are combined statistically to compute the multi-mode response.

Buildings with symmetrical shape, stiffness and mass distribution and with vertical continuity and uniformity behave in a fairly predictable manner whereas when buildings are eccentric or have areas of discontinuity or irregularity, the behavioral characteristics are very complex. The predominant response of the building may be skewed from the apparent principal axes of the building. The torsional response as well as the coupling or interaction of two translational directions of response must be considered. This is similar to the Mohr's circle theory of principal stresses.

Thus, three-dimensional methods of analysis are required as each mode shape is defined in three dimensions by the longitudinal and transverse displacement and the rotation about a vertical axis. Thus building irregularities complicate not only the method of dynamic analysis but also the methods used to combine modes.

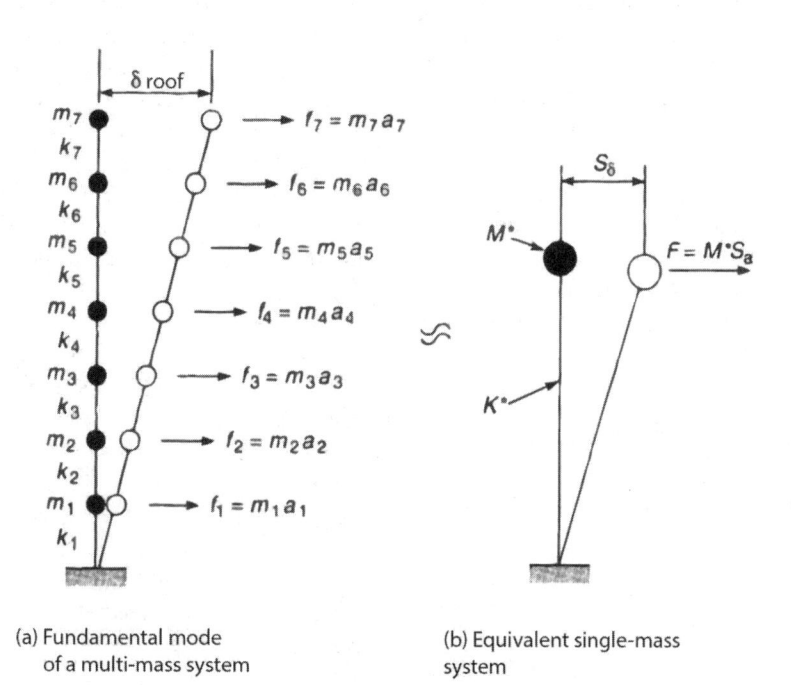

(a) Fundamental mode of a multi-mass system

(b) Equivalent single-mass system

Figure 3.46 Mathematical representation of a multi-mass system by a single-mass system.

For a building that is regular and essentially symmetrical, a two-dimensional model is generally sufficient for the modal analysis of the structure subject to ground motions in each of its principal axis. Note that when the aspect ratio (length-to-width) of the diaphragm is large, torsional response may be predominant thus requiring a 3-D analysis in an otherwise symmetrical and regular building.

The building is modeled as a system of masses lumped at each floor level, each mass having one-degree-of-freedom, that of lateral displacement in the direction under consideration. The weights used in computing the masses are those prescribed in the static procedure. The analysis will include, for each principal axis, all significant modes of vibration. The relative significance of higher modes will be determined by the values of modal participation factors and modal spectral accelerations.

When a structure is unsymmetrical in plan, has discontinuities in the vertical or horizontal planes, large plan aspect ratios, flexible horizontal diaphragms or other irregularities, a three-dimensional model is required. In a three-dimensional analysis, at each floor level there will be three-degrees-of-freedom. The primary displacement generally occurs in a direction parallel to the direction of the ground motion. There will also be a displacement component normal to this direction and rotation about the vertical axis of the building.

For moderate to high-rise buildings, the effects of higher modes may be significant. For a fairly uniform building, the dynamic characteristics can be approximated by using the general modal relationship shown in Table 3.19. The fundamental period of vibration of the buildings may be estimated by using the code formulas. Approximate periods for the second through fifth modes of vibration, and the roof and base participation factors, defined presently, may be estimated by using the relationship shown in Table 3.19. For example, the second and third mode periods are equal to 0.327 and 0.186 times the fundamental mode.

Table 3.19   General Modal Relationships

Mode	1	2	3	4	5
Ratio of period to 1st mode period	1.000	0.327	0.186	0.121	0.083
Participation factor at roof	1.31	−0.47	0.24	−0.11	0.05
Base shear participation factor	0.828	0.120	0.038	0.010	0.000

For most buildings, an inelastic response can be expected to occur during a major earthquake. Although nonlinear inelastic programs are available, they are not representative of typical design practice because: (i) their proper use requires special background; (ii) results produced are difficult to interpret and apply to traditional design criteria; and (iii) the necessary computations are expensive. Therefore, analyzes used in practice are essentially based on linear elastic analysis. One such method called the response spectrum method is the most widely used approach.

3.5.2   Response Spectrum Method

The word "spectrum" expresses the idea that a broad range of quantities is summarized in one graph. For a given earthquake and percentage of critical damping, the graph shows related quantities such as acceleration, velocity, or deflection for a complete range or spectrum of building periods.

The plot of a response spectrum (Figs. 3.47 and 3.48) may be visualized as a response of a series of progressively longer cantilever pendulums with increasing natural periods subjected to a common lateral motion of the base. Imagine the common base being moved through a ground motion corresponding to that occurring in a given earthquake. A plot of maximum response, such as acceleration versus the period of the pendulums will provide the acceleration response spectrum. The absolute value of the peak acceleration response occurring during the excitation for each pendulum is represented by a point on the acceleration spectrum curve. In Fig. 3.49, the response spectra for the 1940 El Centro earthquake are illustrated. Using the ground acceleration as input, a family of response spectrum curves can be generated for various levels of damping, where higher values of damping result in lower spectral response.

In order to establish the concept of the response spectrum method, let us consider a SDOF structure such as an elevated water tank supported on columns or a revolving restaurant supported at the top of a tall concrete core. These structures can be adequately modeled as SDOF structures by considering the columns and the core as flexible cantilevers and the tank or the restaurant as the only mass at the tip of the cantilever; the mass of the columns or core is ignored.

A multistory building will have as many modes of vibration as it has degrees of freedom. The use of lumped mass models to represent the actual distributed mass of a structure is a convenient tool for reducing degrees of freedom to a manageable few. In multistory buildings it is generally sufficient to assume the masses as concentrated

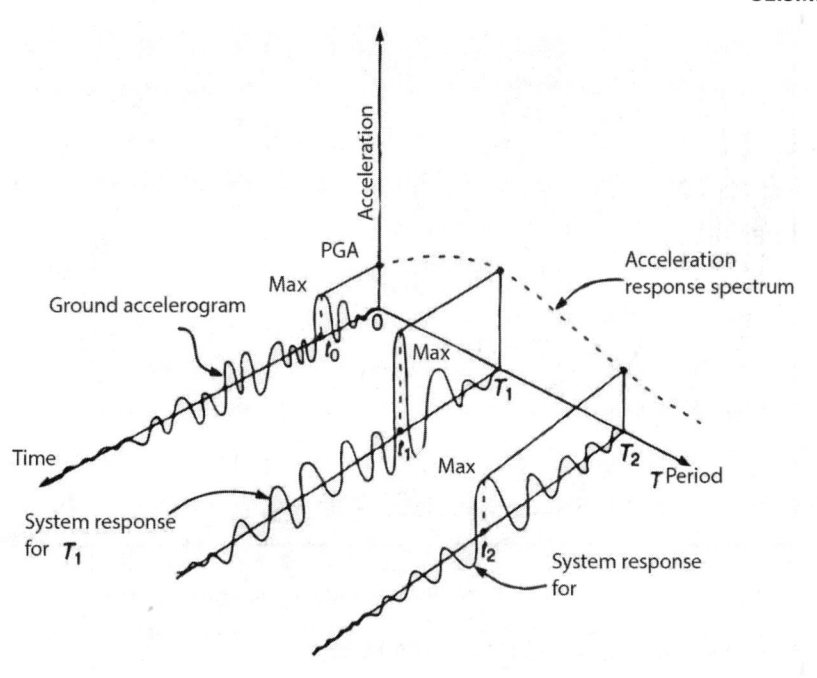

Figure 3.47 Graphical description of response spectrum.

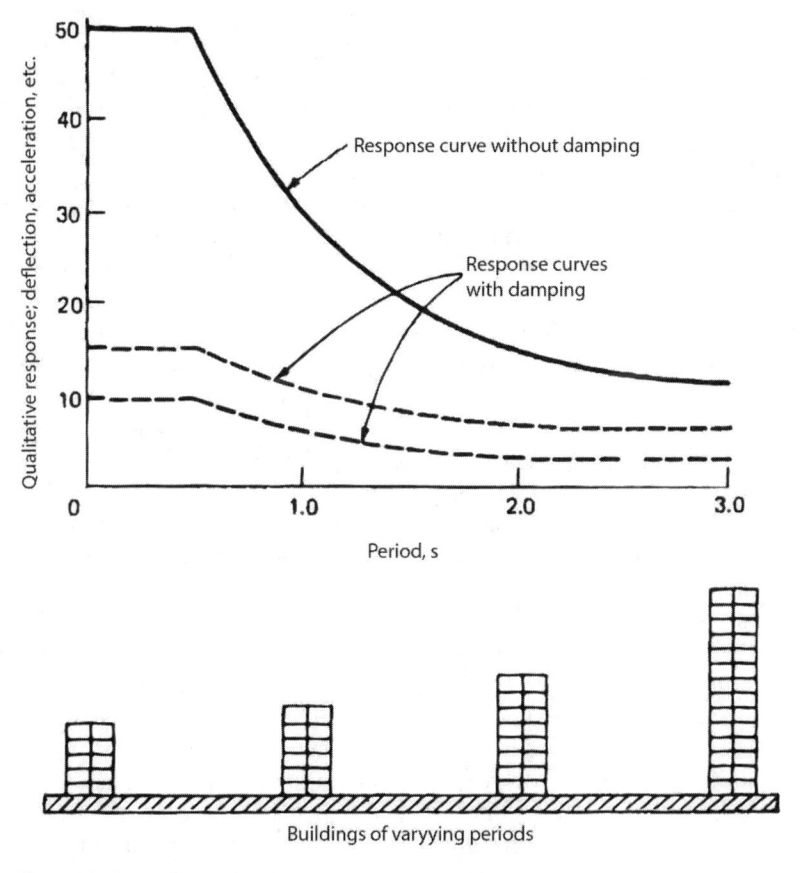

Figure 3.48 Response spectrum.

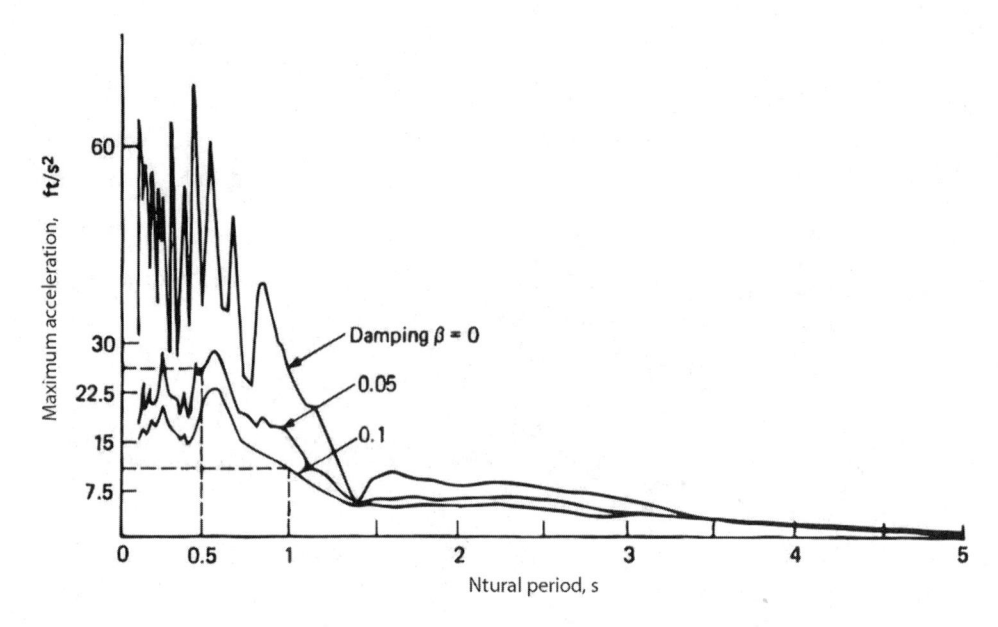

Figure 3.49 Acceleration spectrum: El Centro earthquake.

at the floor levels and to formulate the problem in terms of these masses.

Assuming that masses are concentrated at each level, for a planar analysis the number of modes of vibration corresponds to the number of levels in the multistory structure. Each mode of vibration has its own characteristic frequency or period of vibration. The actual motion of a tall building at any instant is a unique linear combination of its natural modes of vibration. During vibration, the masses of the structure vibrate in phase with displacements as measured from its initial position, always having the same relationship to each other. All masses vibrating in one of the natural modes pass the equilibrium position at the same time and reach their extreme positions at the same instant.

Using certain simplifying assumptions, it can be shown that each mode of vibration behaves as an independent SDOF system with a characteristic frequency. The assumptions required for and the proof of the proposition will not be attempted in this section because it is explained in subsequent sections. For now, suffice it to note that this method called the *modal superposition method*, consists of obtaining the total response of the building by appropriately combining the appropriate modes of vibration.

Since a multistory building has several degrees of freedom, in general it vibrates with as many different mode shapes and periods as it has degrees of freedom. Each mode of vibration contributes to the base shear,

and for elastic action of the structure, this modal base shear can be determined by multiplying an effective mass by an acceleration read from the response spectrum for the period of that mode and for the assumed damping. Therefore, the procedure for determining the base shear for each mode of a MDOF structure is the same as that for determining the base shear for a SDOF structure except that an effective mass is used instead of the total mass. The effective mass is a function of the actual mass at each floor and the deflection at each floor and is greatest for the fundamental mode and becomes progressively less for higher modes. The mode shape must therefore be known in order to compute the effective mass.

Since the actual deflected shape of the building consists of linear combinations of the modal shapes, higher modes of vibration also contribute, though to a lesser degree, to the structural response. These can be taken into account by using the concept of a participation factor. Further mathematical explanation of this concept is deferred to a later section, but suffice it to note that the base shear, for each mode is determined as the summation of products of effective mass and spectral acceleration. The force at each level for each mode is obtained by distributing the base shear in proportion to the product of the floor weight and displacement. The design values are then computed by using modal combination methods, such as complete quadratic combination (CQC) or the square root of the sum of the squares.

3.5.3   Development of Design Response Spectrum

In practice it is rare that a structural engineer is called upon to develop a design response spectrum. However, it is important to understand the various assumptions used in their development. Three basic types of response spectrum are used in practice. These are:

1. Response spectrum from actual earthquake records.
2. Smoothed design response spectrum.
3. Site-specific response spectrum.

Response spectrum from actual earthquake records. A response spectrum curve can be generated by subjecting a series of damped SDOF mass-spring systems to a given ground excitation. Response spectrum graphs are generated by numerical integration of actual earthquake records to determine maximum values for each period of vibration.

Spectral curves developed from actual earthquake records are quite jagged, being characterized by sharp peaks and troughs. Because the magnitude of these troughs and peaks can vary significantly for different earthquakes, and because of the uncertainties of future earthquakes, several possible earthquake spectra are used in the evaluation of the structural response.

Smooth response spectrum. As an alternative to the use of several earthquake spectra, a smooth spectrum representing an upper-bound response to ground motions may be generated. The sharp peaks in earthquake records indicate the resonant behavior of the system when the natural period of the system approaches the period of forcing function, especially for systems with little or no damping. However, even a moderate amount of damping, shown as β in Figs. 3.50, 3.51 and 3.52, has a tendency to smooth out the peaks and reduce the spectral response.

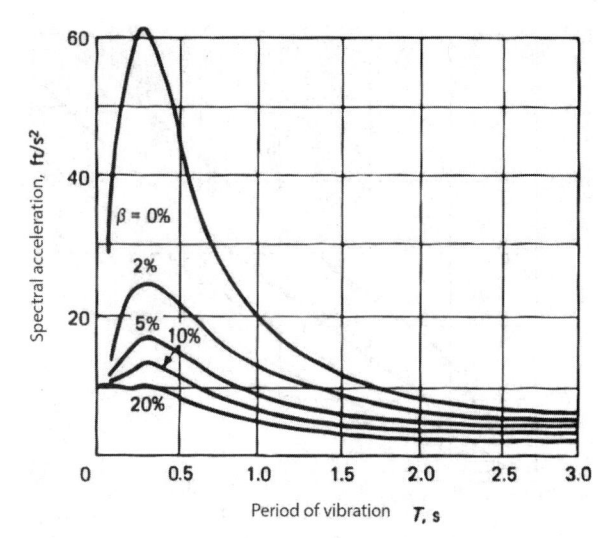

Figure 3.50 Smoothed acceleration spectra for the El Centro earthquake.

Because most buildings in practice have at least some degree smooth out the peaks and reduce the spectral response. Because most buildings in practice have at least some degree of damping, the peaks in response spectra are of little significance. Figure 3.49 shows the smoothed acceleration spectra for the El Centro, California, earthquake. The other two response spectra for velocity and displacement, shown in Figs. 3.51 and 3.52, are obtained from the acceleration spectrum, since they are related to one another. The three spectra can be represented in one graph, as shown in Fig. 3.53, in which the horizontal axis denotes the natural period and the ordinate the spectrum velocity, both on a logarithmic scale. The acceleration and displacement are represented on diagonal axes inclined at 45° to the horizontal. The plot,

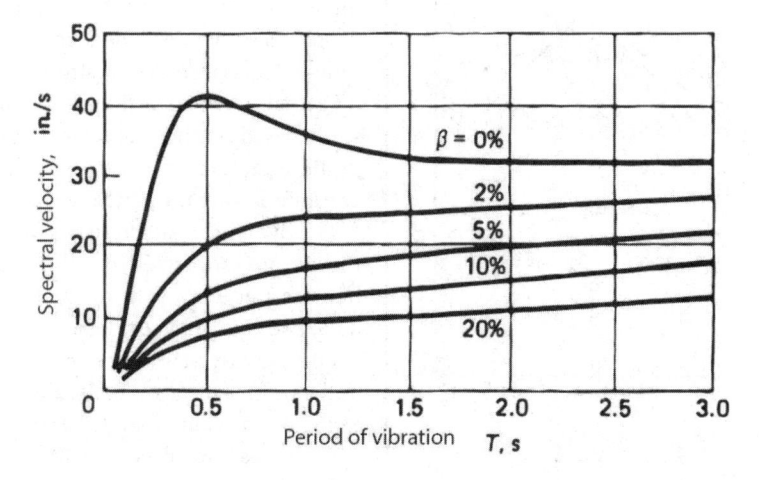

Figure 3.51 Smoothed velocity spectra for the El Centro earthquake.

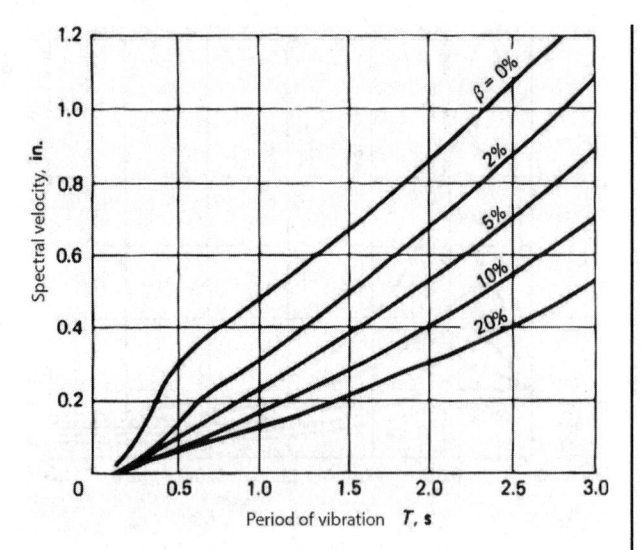

Figure 3.52 Smoothed displacement spectra for El Centro earthquake.

which encompasses all the spectral parameters, is called a *tripartite response spectrum*. From this plot, the following observations can be made.

1. For very stiff, structures, the spectral acceleration approaches the maximum ground acceleration. Structures in this period range would behave like rigid bodies attached to the ground.

2. For moderately short periods of the order of 0.1 to 0.3 s, the spectral accelerations are about twice as large as the maximum ground acceleration.

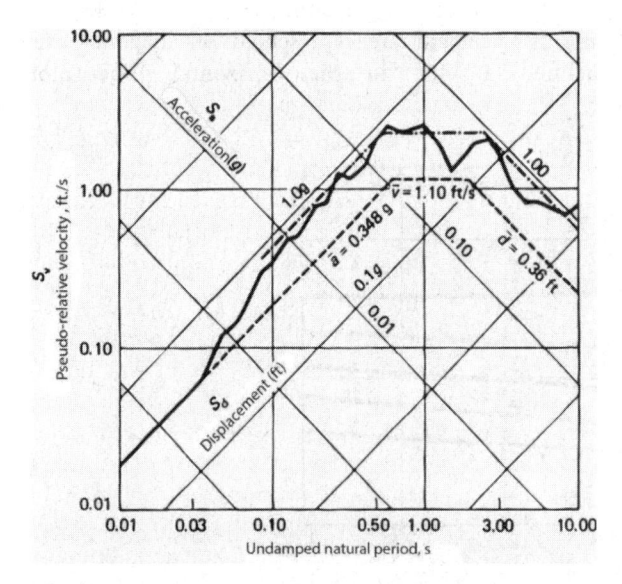

Figure 3.53 Tripartite response spectra for El Centro earthquake (5 percent damping, north-south component).

3. For long period buildings, the maximum spectral displacements approach the maximum ground displacements.

4. For intermediate values of period, the maximum spectral velocity is several times the input velocity.

Thus, in the short-period range, the variation of the spectrum curve tends to show a correlation with the line of maximum ground acceleration. In the medium-period range, the correlation is with maximum ground velocity while in the higher-period range, the correlation is with the displacement.

Because of the above characteristics, it is possible to represent an idealized upper-bound response spectrum by a set of three straight lines (Fig. 3.46). The values of ground acceleration ($\bar{a} = 0.348g$), maximum velocity ($\bar{v} = 1.10$ ft/s), and displacement ($\bar{d} = 0.36$ ft) for the El Centro earthquake are also shown in Fig. 3.46.

Unique design spectra. For especially important structures or where local soil conditions are not amenable to simple classification, the use of recommended smooth spectrum curves is inadequate for final design purposes. In such cases, site-specific studies are performed to determine more precisely the expected intensity and character of seismic motion. The development of site-specific ground motions is generally the responsibility of geotechnical consultants working in concert with the structural engineer. However, it is important for the structural engineer to be aware of the procedure used in the generation of a site-specific response spectrum. This is considered next.

The seismicity of the region surrounding the site is determined from a search of an earthquake database. A list of active, potentially active, and inactive faults is compiled from the database along with their nearest distance from the site.

The predicted response of the deposits underlying the site and the influence of local soil and geologic conditions during earthquakes are determined based on statistical results of studies of site-dependent spectra developed from actual time-histories recorded by strong motion instruments.

Several postulated design earthquakes are selected for study based on the characteristics of the faults. The peak ground motions generated at the site by the selected earthquakes are estimated from empirical relationships.

The dynamic characteristics of the deposits underlying the site are estimated from the results of a nearby downhole seismic survey, from the logs of borings, static test data, and dynamic test data.

The causative faults are selected from the list of faults as the most significant faults along which earthquakes are expected to generate motions affecting the site.

Several earthquakes with different probability of occurrence that may be generated along the causative faults

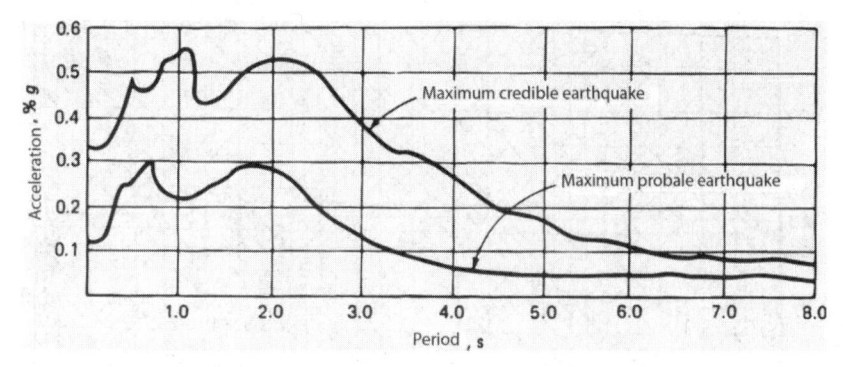

Figure 3.54   Unique site-specific design spectra.

are selected. The maximum credible earthquake (MCE) for example, constitutes the largest earthquake that appears to be reasonably likely to occur. Since the probability of such an earthquake occurring during the lifetime of the subject development is low, the ground motions associated with the MCE events are estimated to have 10 percent probability of being exceeded in 250 years, for example.

Several other design-basis earthquakes are also postulated from the data. For example, in Fig. 3.55d, earthquake "C" is considered to be an event having a 50 percent probability of being exceeded in 50 years.

The slip rates of the faults are estimated from published data. Using the slip rates, the accumulated slip over an approximate 475-year period (corresponding to 10 percent probability of being exceeded in 50 years) and over an approximate 72-year period (corresponding to 50 percent probability of being exceeded in 50 years) are determined. Using the surface displacement versus magnitude relationships, the magnitudes for each significant fault are determined.

Using a statistical analysis approach, the peak ground motion values (acceleration, velocity, and displacement) anticipated at the site are estimated. By applying structural amplification factors to these values, the spectral bounds for acceleration, velocity, and displacement are obtained for each desired value of structural damping, most usually 2, 5, and 10 percent of critical damping. The ground motion values vary with the magnitude of the earthquake and the distance of the site from the source of energy release.

The peak ground motion values for velocity and displacement are developed by relating peak ground velocity and displacement to the peak ground acceleration for four site classifications: rock, stiff soil, deep cohesionless soil, and soft to medium soil.

The ground motion values obtained above provide a basis by which site-dependent response spectra are computed. For each of the four site classes, spectral bounds are obtained by multiplying the ground motion values by damping-dependent amplification factors.

A schematic representation of acceleration spectra is shown in Fig. 3.54 for maximum credible and maximum probable events. Tripartite response spectra for four seismic events characterized as earthquakes A, B, C, and D for a downtown Los Angeles site are shown in Figs. 3.55a-d. Response spectra A is for an MCE of magnitude 8.25 occurring at San Andreas fault at a distance of 34 miles while B is for a magnitude 6.8 earthquake occurring at Santa Monica—Hollywood fault at a distance of 3.7 miles from the site. Response spectra C and D are for earthquakes with a 10 and 50 percent probability of being exceeded in 50 years.

3.5.4   Time-History Analysis

3.5.4.1   INTRODUCTION

The mode superposition or the spectrum method outlined in the previous section is a useful technique for the elastic analysis of structures. It is not directly transferable to inelastic analysis because the principle of superposition is no longer applicable. Also, the analysis is subject to uncertainties inherent in the modal superimposition method. The actual process of combining the different modal contributions is, after all, a probabilistic technique and in certain cases may lead to results not entirely representative of the actual behavior of the structure. Time-history analysis overcomes these two uncertainties, but it requires a large computational effort. It is not normally employed as an analysis tool in the practical design of buildings. The method consists of a step-by-step direct integration in which the time domain is discretized into a number of small increments δt; and for each time interval the equations of motion are solved with the displacements and velocities of the previous step serving as initial functions. The method is applicable to both elastic and inelastic analyzes. In elastic analysis, the stiffness characteristics of

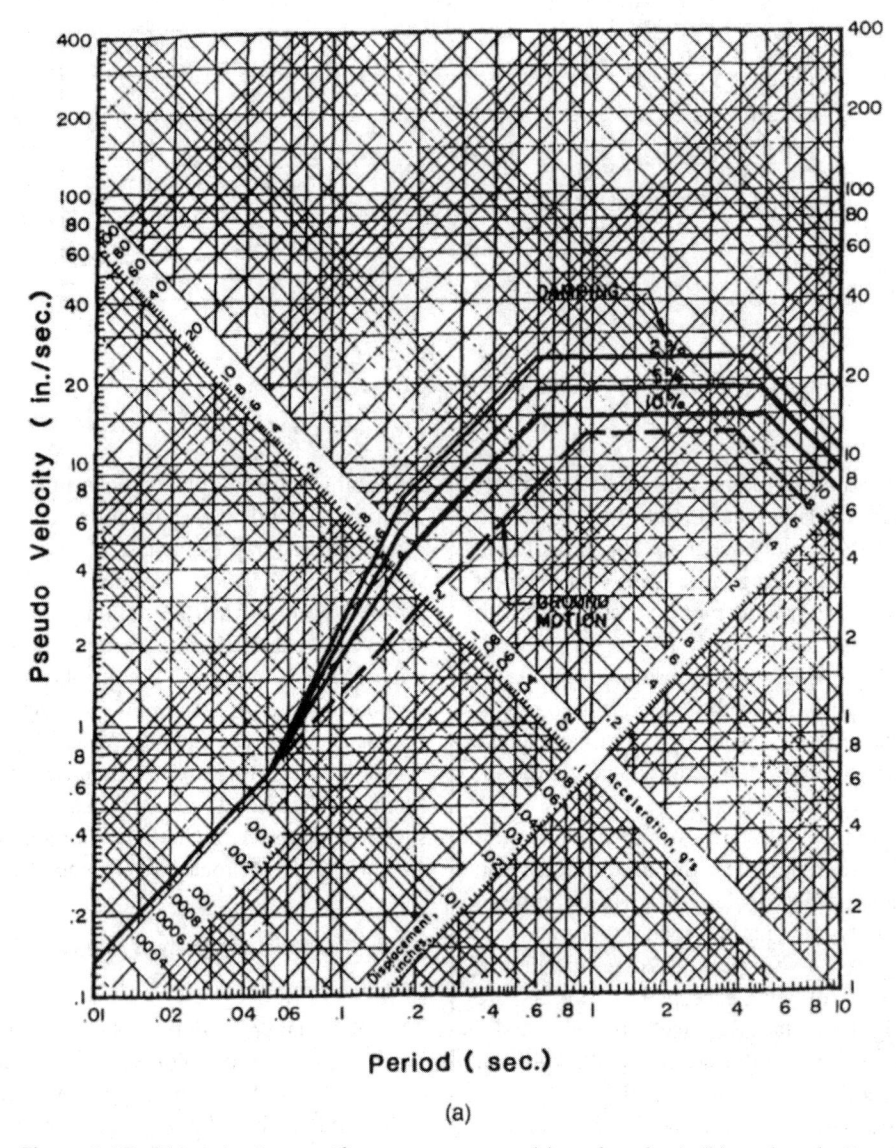

Figure 3.55 Tripartite site-specific response spectra: (a) earthquake A; (b) earthquake B; (c) earthquake C; (d) earthquake D.

the structure are assumed to be constant for the whole duration of the earthquake. In the inelastic analysis, however, the stiffness is assumed to be constant through the incremental time δt only. Modifications to structural stiffness caused by cracking, formation of plastic hinges, etc., are incorporated between the incremental solutions. A brief outline of the method, which is thus applicable to both elastic and inelastic analysis, is given below.

3.5.4.2 Analysis Procedure

In this method, earthquake motions are applied directly to the base of the computer model of a given structure.

Instantaneous stresses throughout the structure are calculated at small intervals of time for the duration of the earthquake or a significant portion of it. The maximum stresses that occur during the earthquake are found by scanning the computer output.

The procedure usually includes the following steps:

1. An earthquake record representing the design earthquake is selected.

2. The record is digitized as a series of small-time intervals of about 1/40 to 1/25 of a second.

3. A mathematical model of the building is set up. Usually consisting of a lumped mass on each floor.

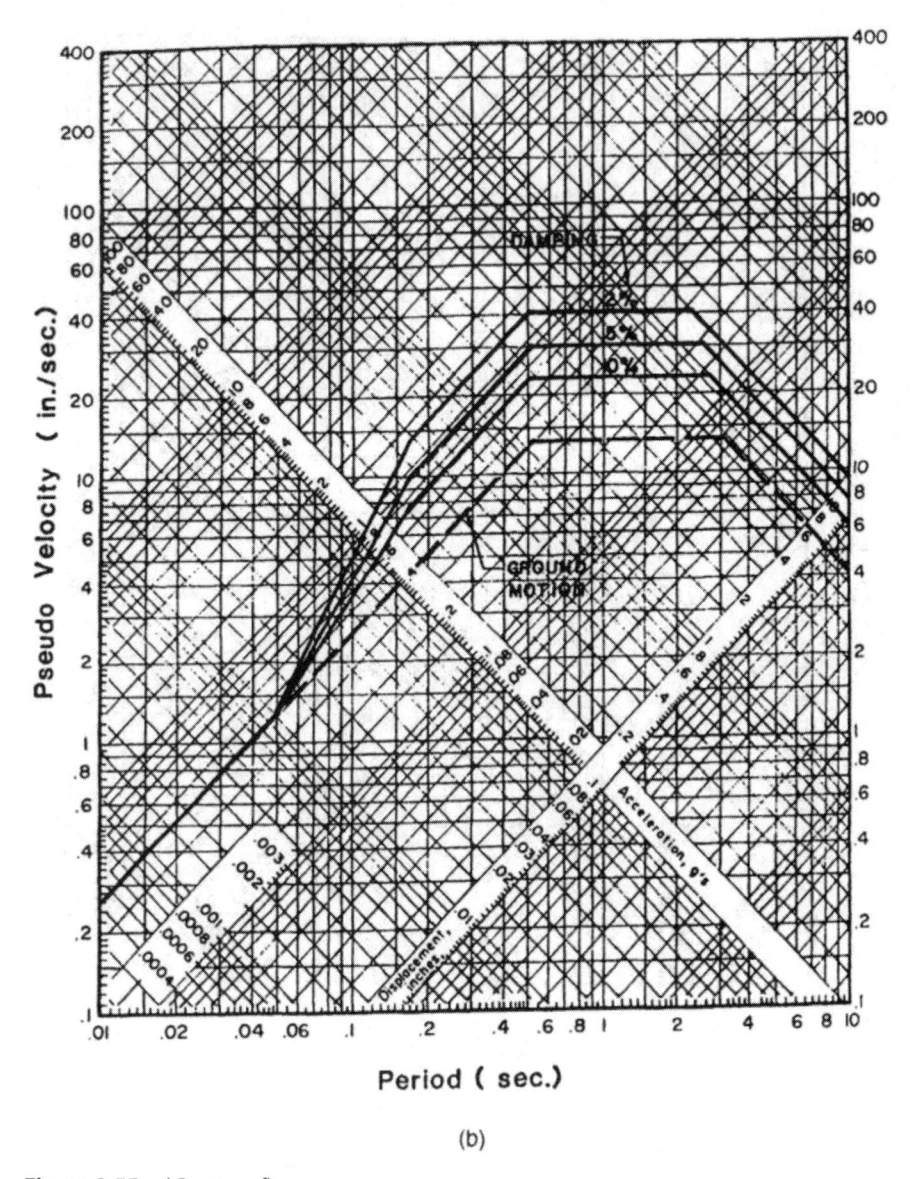

(b)

Figure 3.55 *(Continued)*

Damping is considered proportional to the velocity in the computer formulation.

4. The digitized record is applied to the model as accelerations at the base of the structure.

5. The computer integrates the equations of motion and gives a complete record of the acceleration, velocity, and displacement of each lumped mass at each interval.

The accelerations and relative displacements of the lumped masses are translated into member stresses. The maximum values are found by scanning the output record.

This procedure automatically includes various modes of vibration by combining their effect as they occur, thus eliminating the uncertainties associated with modal combination methods.

The time-history technique represents one of the most sophisticated method of analysis used in building design; however, it has the following sources of uncertainty:

1. The design earthquake must still be assumed.

2. If the analysis uses unchanging values for stiffness and damping, it will not reflect the cumulative effects of stiffness variation and progressive damage.

3. There are uncertainties related to the erratic nature of earthquakes. By pure coincidence, the maximum response of the calculated time history could fall at either a peak or a valley of the digitized spectrum.

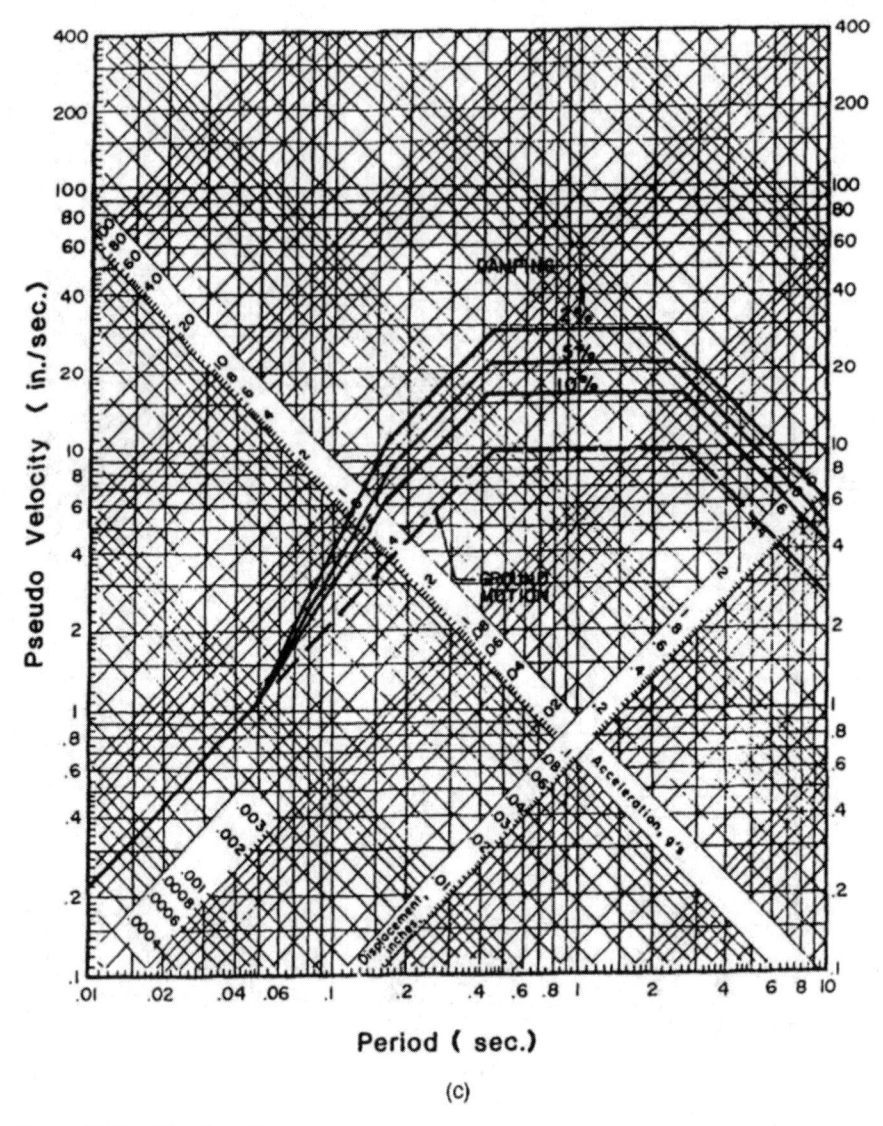

Figure 3.55   *(Continued)*

4. Small inaccuracies in estimating the properties of the structure will have considerable effect on the maximum response.

5. Errors latent in the magnitude of the time step chosen are difficult to assess unless the solution is repeated with several smaller time steps.

3.5.5   Overview Of ASCE 7-16 Requirements for Dynamic Analysis

3.5.5.1   MODAL RESPONSE SPECTRUM ANALYSIS

A modal response spectral analysis is permitted for any structure assigned to SDC B through F and is required

for regular and irregular structures where $T \geq \_3.5TS$ and certain types of irregular structures assigned to SDC D through F (see Table 12.6-1). The following requirements must be satisfied for this method of analysis.

Number of Modes   The distribution of mass and stiffness of a structure affects the forces and displacements due to ground shaking is the basis for performing modal response spectral analyzes. According to 12.9.1.1 of ASCE 7-16, a combined modal mass participation of 100 percent of the structure's mass must be achieved by including a sufficient number of modes in the analysis.

When performing such an analysis, it is permitted to represent all modes with periods less than 0.05 s in a

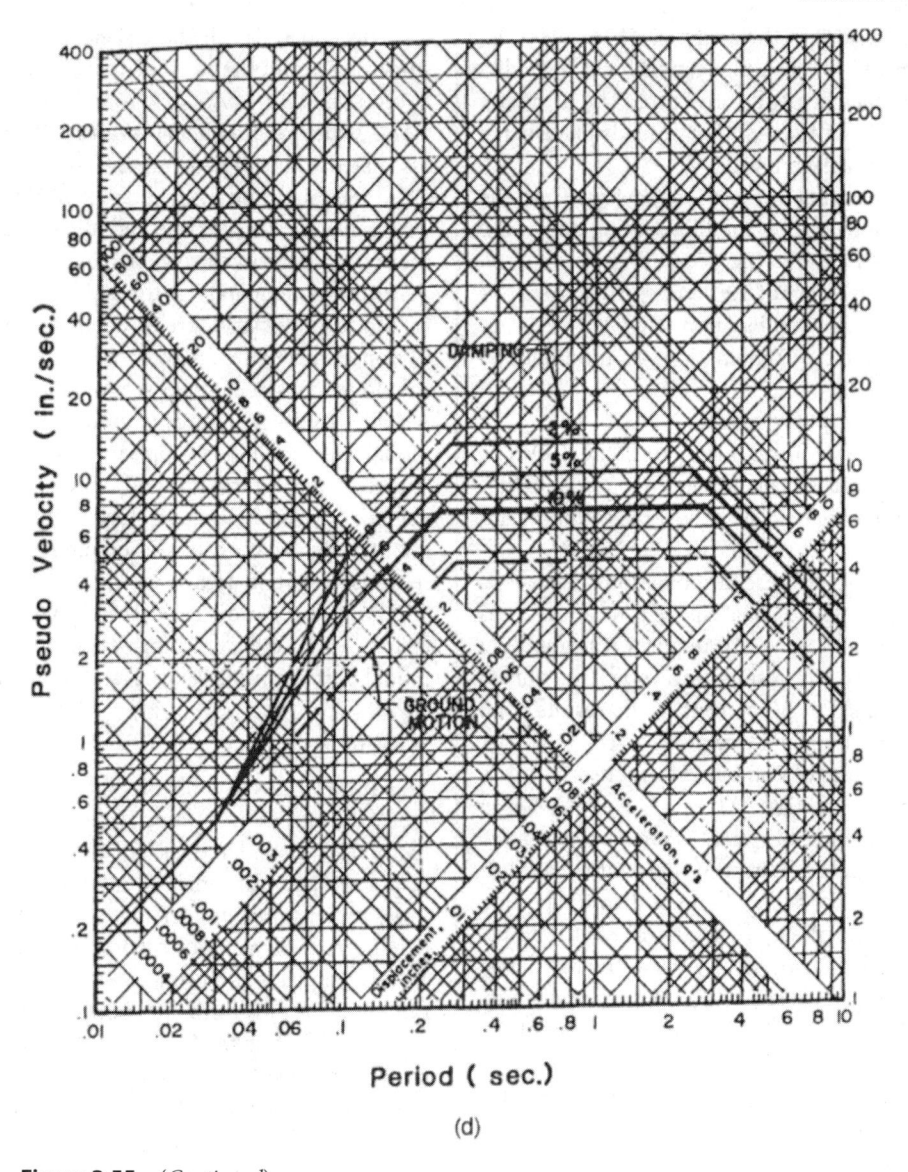

Figure 3.55 (*Continued*)

single rigid body mode that has a period of 0.05 s. The exception in this section permits an analysis that includes a minimum number of modes to obtain a combined modal mass participation of at least 90 percent in each orthogonal horizontal direction of response considered in the model. In general, 90 percent modal mass participation is usually a small fraction of the total number of modes, even in the case of high-rise buildings. For example, it is possible that the first three modes of a twenty-story building may achieve 90 percent modal mass participation.

Modal Response Parameters The forces and deflections can be calculated for each mode using the properties of each mode once the required number of modes has been determined as discussed in section 3.5.5.1.1. The calculations of forces and deflections are performed using the properties of each mode and either the general response spectrum defined in 11.4.6 of ASCE 7-16 or the site-specific response spectrum defined in 21.2. The spectral ordinates must be divided by (R/I_e) regardless of the spectrum that is used. The division by R accounts for inelastic behavior and multiplication by I_e provides the additional strength needed for important structures. Additionally, to obtain the expected inelastic displacements, the displacements obtained using the response spectrum modified by (R/I_e) must be multiplied by (C_d/I_e). At this stage of the analysis, a base shear, V_t, has been determined for each mode, which has been distributed over the height of the structure. The structure has been analyzed for these forces, and displacements have been obtained at each level.

Combined Response Parameters   The results from the analysis of each mode can be combined using one of the following methods:

- Square root of the sum of the squares (SRSS).
- Complete quadratic combination (CQC).
- Complete quadratic combination as modified by ASCE 4 (CQC-4) [10].
- Approved equivalent approach.

ASCE/SEI 12.9.1.3 requires that either the CQC or CQC-4 methods be used where closely spaced modes have significant cross-correlation of translational and torsional response.

In general, any one of these combination methods is applied to one direction of analysis at a time. In cases where section 12.5 of ASCE 7-16 requires consideration of orthogonal effects, the results from one direction of loading may be added to 30 percent of the results from the loading in the orthogonal direction.

Scaling Design Values of Combined Response

Scaling of forces   Section 12.9.1.4 of ASCE 7-16 requires scaling the forces in cases where the combined response for the modal base shear, V_t, determined according to section 12.9.1.3 of ASCE 7-16 is less than 100 percent of the calculated base shear, V, by the ELFP, the forces must be scaled by the factor V/V_t. Also, if the calculated fundamental period, T, exceeds C_uT_a in a given direction, C_uT_a should be used instead of T in the direction of the analysis. Basically, this scaling requirement provides a minimum base shear for design in cases where the computed T has been based on an incorrect, such as an excessively flexible, analytical model.

Scaling of drifts   According to section 12.9.1.4.2 of ASCE 7-16, drifts must be scaled by a factor of C_SW/V_t when V_t is less than C_SW where the governing seismic response coefficient, C_S, is determined by Equation 12.8-6 of ASCE 7-16. This provision is intended to provide a minimum drift for structures that are located in proximity to an active fault.

Horizontal shear distribution   The distribution of the horizontal forces to the SFRSs should be performed according to section 12.8.4 of ASCE 7-16. The amplification of torsion required in section 12.8.4.3 of ASCE 7-16 is not required when the accidental torsion effects are included in the model.

P-delta effects   Section 12.8.7 of ASCE 7-16, P-delta requirements, are also applicable to modal response spectrum analysis. The base shear used to determine the story shears and the story drifts is determined according to section 12.8.6 of ASCE 7-16.

Soil structure interaction effects   Reduction in forces associated with soil structure interaction is permitted provided the provisions of Chapter 19 of ASCE 7-16 are used. Any generally accepted procedures that are approved by the authority having jurisdiction are also permitted.

Structural modeling   A three-dimensional mathematical model of the structure is required to perform a modal response spectrum analysis. This mathematical model should be constructed per sections 12.7.3 and 12.9.1.8 of ASCE 7-16. In structures without rigid diaphragms, the model must include a representation of the diaphragm's stiffness characteristics and any additional dynamic degrees of freedom that are needed to accurately account for the participation of the diaphragm in the dynamic response of the structure.

Examples on modal spectrum analysis are presented below. See Reference [9] for details on dynamic analysis, modal periods, mode shapes, and modal participation.

3.5.5.2 Linear Response History Analysis

The linear response history analysis method requirements are given in section 12.9.2 of ASCE 7-16. This method is considered as an alternative to the modal response spectrum analysis method. In this method, a three-dimensional model of a structure is constructed per the requirements of section 12.9.2.2 of ASCE 7-16, and the model is subjected to spectrum-matched ground motions. The only difference between the modal response spectrum analysis method and this method is that in the modal response spectrum analysis method, the system response is determined by statistical combinations (SRSS or CQC) of the modal responses, whereas in the linear response history analysis method, the system response is obtained by simultaneous solution of the full set of equations of motion. The commentary of section 12.9.2, C12.9.2 of ASCE 7-16 gives additional information on the proper application of this method.

3.5.6 Modal Analysis: Hand Calculation Procedure

Two examples are presented in the following to illustrate the modal analysis method. In the first part of each example, modal analysis is performed to determine base shear for each mode using given building characteristics and ground motion spectra. In the second part, story forces, accelerations and displacements are calculated for each mode, and are combined statistically using the SRSS combination.

The formulas for obtaining the modal story participation factor, modal story lateral forces, modal base shear, modal deflections, and drifts are given within the framework of the second example.

3.5.6.1 EXAMPLE 1: THREE-STORY BUILDING

Given

Masses $\dfrac{W}{g}$, mode shapes $\phi_s{}'$, periods T_1, T_2, and T_3, and correspond accelerations a_1, a_2, and a_3. $T_1 = 0.9645$s, $T_2 = 0.3565$ s, and $T_3 = 0.1825$ s corresponding spectral accelerations from Fig. 3.56 are 0.251 g for mode 1, and 0.41 for modes 2 and 3.

Required

(i) Modal analysis to determine base shears.

(ii) Story forces, overturning moments, accelerations and displacements for each mode.

(iii) SRSS combinations.

Modal Analysis: Influence of Higher Modes   The results of modal analysis for determining base shears, and story forces, accelerations, and displacements are shown in Figs. 3.57 and 3.58. It should be noted that higher modes of response become increasingly important for taller or irregular buildings. For this regular three-story structure, the first mode dominates the lateral response. A comparison of the modal story shears and the SRSS

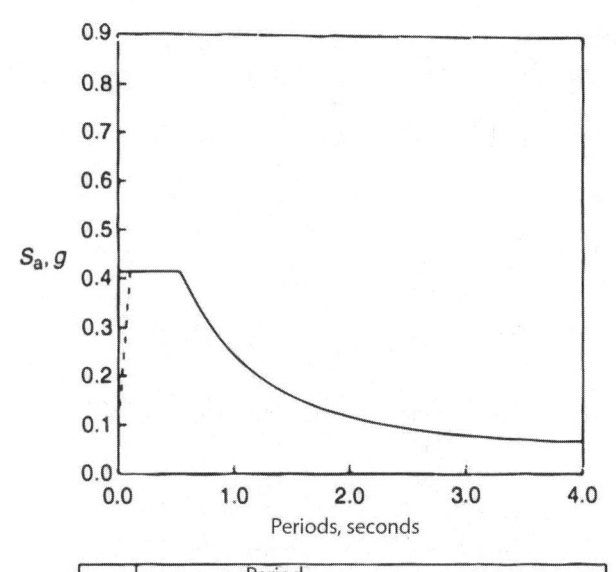

	Period							
	0.0	.586	.80	1.0	1.5	2.0	3.0	4.0
S_a, g	.14	.41	.300	.240	.160	.120	.080	.060

Figure 3.56   Three-story building: response spectrum.

story shears is shown in Fig. 3.59. For example, if only the first mode shears had been used for analysis, this represents 89 percent of the SRSS shear at the roof, 99 percent at the third floor, and 95 percent at the second floor. While the second mode shear at the roof is 50 percent of the first mode shear, when combined on an SRSS basis the first mode accounts for 79 percent of the SRSS response with 20 percent for the second mode and 0.6 percent for the third mode. These percentages are 91 percent, 8 percent, and 1 percent at the base.

The effective modal weight factor, a_m, also shows the relative importance of each mode. In this example, ($a_1 = 0.804$, $a_2 = 0.149$, $a_3 = 0.048$). 80.4 percent of the building mass participates in the first mode, 14.9 percent in the second mode, and 4.8 percent in the third mode.

3.5.6.2 EXAMPLE 2: SEVEN-STORY BUILDING

The results are shown in a format similar to the format used in the static force procedure such that a comparison of static force and dynamic analysis procedures can be made. A seven-story reinforced concrete moment resisting space frame building adapted from the Tri Services Manual, is chosen to illustrate the procedure. The modal analyzes are performed on the basis of the response spectrum shown in Figs. 3.60a-c. Observe that these three figures contain the same response information, only the format is different. Figure 3.60a shows spectral accelerations and periods on the vertical and horizontal axes respectively, t. Figure 3.60b shows the same information on the tripartite diagram, while Fig. 3.60c is a numerical representation of the same response spectra.

The first step is to develop a mathematical model of the building. Story masses are obtained from the calculated weights of the building tributary to each story. Although three-dimensional computer models are more common and required in some cases in seismic design practice, we will assume, for ease of presenting the sample modal analysis, that the seven-story building is analyzed as a series of two-dimensional frames. The periods and mode shapes are determined for the first three modes of vibration by a two-dimensional computer model.

In this program, each mode is normalized for $\Sigma(W/g)$ $\phi^2 = 1.0$. The mode shapes are shown in Fig. 3.61 with a normalized value of $\dfrac{1}{2}$ in. at the top story.

Modal analysis to determine total base shear and story accelerations   Figure 3.61 illustrates a hand-calculation procedure to determine the total base shear and the story accelerations using mass, mode shape, period, and response spectrum data. The following equations are used to determine the participation factors.

T_{m1} sec.		0.964			0.356			0.182		

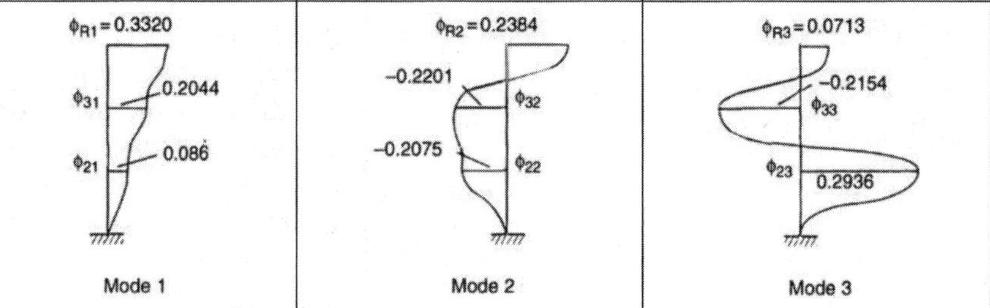

Mode 1   Mode 2   Mode 3

Level	Mass $\left(\dfrac{k \cdot \sec^2}{\text{ft.}}\right)$	Mode 1			Mode 2			Mode 3		
		ϕ_{x1}	$m_x\phi_{x1}$	$m_x\phi_{x1}^2$	ϕ_{x2}	$m_x\phi_{x2}$	$m_x\phi_{x2}^2$	ϕ_{x3}	$m_x\phi_{x3}$	$m_x\phi_{x3}^2$
R	5.81	0.3320	1.929	0.640	0.2384	1.385	0.330	0.0713	0.4143	0.030
3	7.32	0.2044	1.496	0.306	−0.2201	−1.611	0.355	−0.2154	−1.577	0.340
2	7.32	0.0860	0.630	0.054	−0.2075	−1.519	0.315	0.2936	2.149	0.631
				1.000*						
Σ	20.45		4.055	*		−1.745	1.000		0.9863	1.001

PF*$_{Rm}$	$\dfrac{\Sigma m\phi}{\Sigma m\phi^2}\phi_{R1} = 1.346$			−0.416		0.070
PF$_{3m}$		0.829		0.384		−0.212
PF$_{2m}$		0.349		0.362		0.289
α_m	$\dfrac{(\Sigma m\phi)^2}{\Sigma m(\Sigma m\phi^2)} = 0.8040$			0.149		0.048
S_a		0.251 g		0.41 g		0.41 g
$v = \alpha_m S_a W$		132.7 Kips		40.2 Kips		13.0 Kips

* Note that the sum of the modal participation factors $\displaystyle\sum_{m=1}^{3} PF_{xm} = 1.0$ and the sum of modal base shear participation factors

$\displaystyle\sum_{m=1}^{3} \alpha_m = 1.0.$

** The mode shapes have been normalized by the computer program so that $\sum m\phi^2 = 1.0$.

Figure 3.57 Three-story building: modal analysis to determine base shears.

Modal story participation factor The story modal participation factor will be calculated for each mode by using the equation

$$PF_{xm} = \left(\frac{\displaystyle\sum_{i=1}^{n} \frac{w_i}{g}\phi_{im}}{\displaystyle\sum_{i=1}^{n} \frac{w_i}{g}\phi_{im}^2} \right) \phi_{xm} \qquad (3.28)$$

where PF_{xm} = modal participation factor at level x for mode m

w_i/g = mass assigned to level i

ϕ_{im} = amplitude of mode m at level i

ϕ_{xm} = amplitude of mode m at level x

n = level n under consideration

It should be noted that some references define the modal participation factor as the quantity within the bracket in Eq. (3.28). Also, in some references, ϕ is normalized to 1.0 at the uppermost mass level.

Level	PF_{xm}	$\dfrac{m_x \phi_{xm}}{\Sigma m_x \phi_{xm}}$	F_{xm} (k)	V_{xm} (k)	ΔOTM_{xm} (ft·k)	OTM_{xm} (ft·k)	$a_{xm} = \dfrac{F_{xm}}{W_x}$	δ_{xm} (in.)	Δ_{xm} (in.)
R	1.346	0.476	63.2	63.2	772	0	0.337	3.065	1.182
3	0.829	0.369	48.9	112.1	1233	772	0.208	1.892	1.101
2	0.349	<u>0.155</u>	20.6	132.7	1416	2005	0.087	0.791	0.791
		1.000				3421			

(a) Mode 1

Level	PF_{xm}	$\dfrac{m_x \phi_{xm}}{\Sigma m_x \phi_{xm}}$	F_{xm} (k)	V_{xm} (k)	ΔOTM_{xm} (ft·k)	OTM_{xm} (ft·k)	$a_{xm} = \dfrac{F_{xm}}{W_x}$	δ_{xm} (in.)	Δ_{xm} (in.)
R	−0.416	−0.793	−31.9	−31.9	−389	0	−0.171	−0.212	0.407
3	0.384	0.923	37.1	5.2	57	−389	−0.157	0.195	0.011
2	0.362	<u>0.870</u>	35.0	40.2	429	−332	−0.148	0.184	0.184
		1.000				97			

(b) Mode 2

Level	PF_{xm}	$\dfrac{m_x \phi_{xm}}{\Sigma m_x \phi_{xm}}$	F_{xm} (k)	V_{xm} (k)	ΔOTM_{xm} (ft·k)	OTM_{xm} (ft·k)	$a_{xm} = \dfrac{F_{xm}}{W_x}$	δ_{xm} (in.)	Δ_{xm} (in.)
R	0.070	0.420	5.5	5.5	67	0	−0.029	0.0094	0.037
3	−0.212	−1.599	−20.8	−15.3	−168	67	−0.087	−0.028	0.066
2	0.289	<u>2.179</u>	28.3	13.0	139	−101	0.118	0.038	0.038
		1.000				38			

(c) Mode 3

Level	PF_{xm}	$\dfrac{m_x \phi_{xm}}{\Sigma m_x \phi_{xm}}$	F_{xm} (k)	V_{xm} (k)	ΔOTM_{xm} (ft·k)	OTM_{xm} (ft·k)	$a_{xm} = \dfrac{F_{xm}}{W_x}$	δ_{xm} (in.)	Δ_{xm} (in.)
R			71.0	71.0	867	0	0.379	3.072	1.251
3			64.8	113.3	1246	867	0.275	1.893	1.094
2			49.5	139.3	1486	2035	0.208	0.812	0.813
						3423			

(d) SRSS combination

Level	V_{SRSS}	Mode 1			Mode 2		Mode 3	
		V_1	V_1/V_{SRSS}	$(V_1/V_{SRSS})^2$	V_2	$(V_2/V_{SRSS})^2$	V_1	$(V_3/V_{SRSS})^2$
R	71.0	63.2	0.89	0.79	−31.9	0.202	5.5	0.006
3	119.3	112.1	0.989	0.98	5.2	0.002	−15.3	0.018
2	139.3	132.7	0.953	0.91	40.2	0.083	13.0	0.009

Figure 3.58 Three-story building: modal analysis to determine story forces, accelerations and displacements.

Level	V_{SRSS}	Mode 1			Mode 2		Mode 3	
		V_1	V_1/V_{SRSS}	$(V_1/V_{SRSS})^2$	V_2	$(V_2/V_{SRSS})^2$	V_1	$(V_3/V_{SRSS})^2$
R	71.0	63.2	0.89	0.79	−31.9	0.202	5.5	0.006
3	119.3	112.1	0.989	0.98	5.2	0.002	−15.3	0.018
2	139.3	132.7	0.953	0.91	40.2	0.083	13.0	0.009

Figure 3.59 Three-story building: comparison of modal story shears and the SRSS story shears.

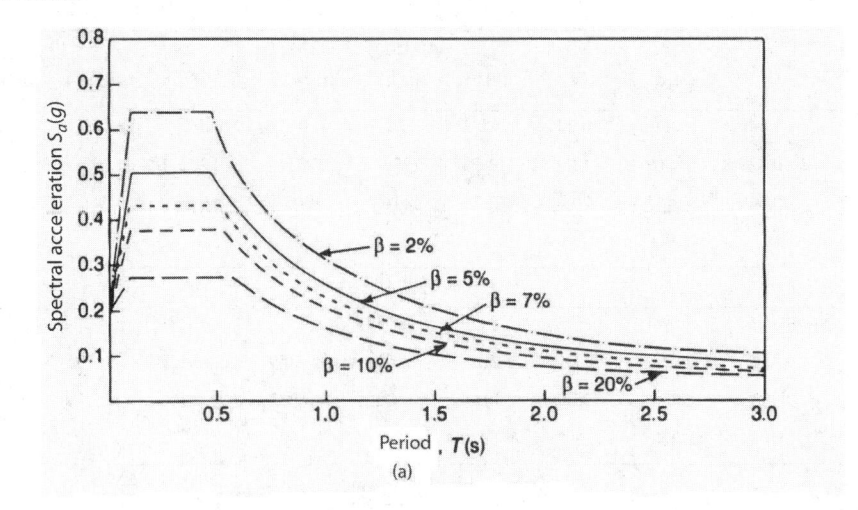

(a)

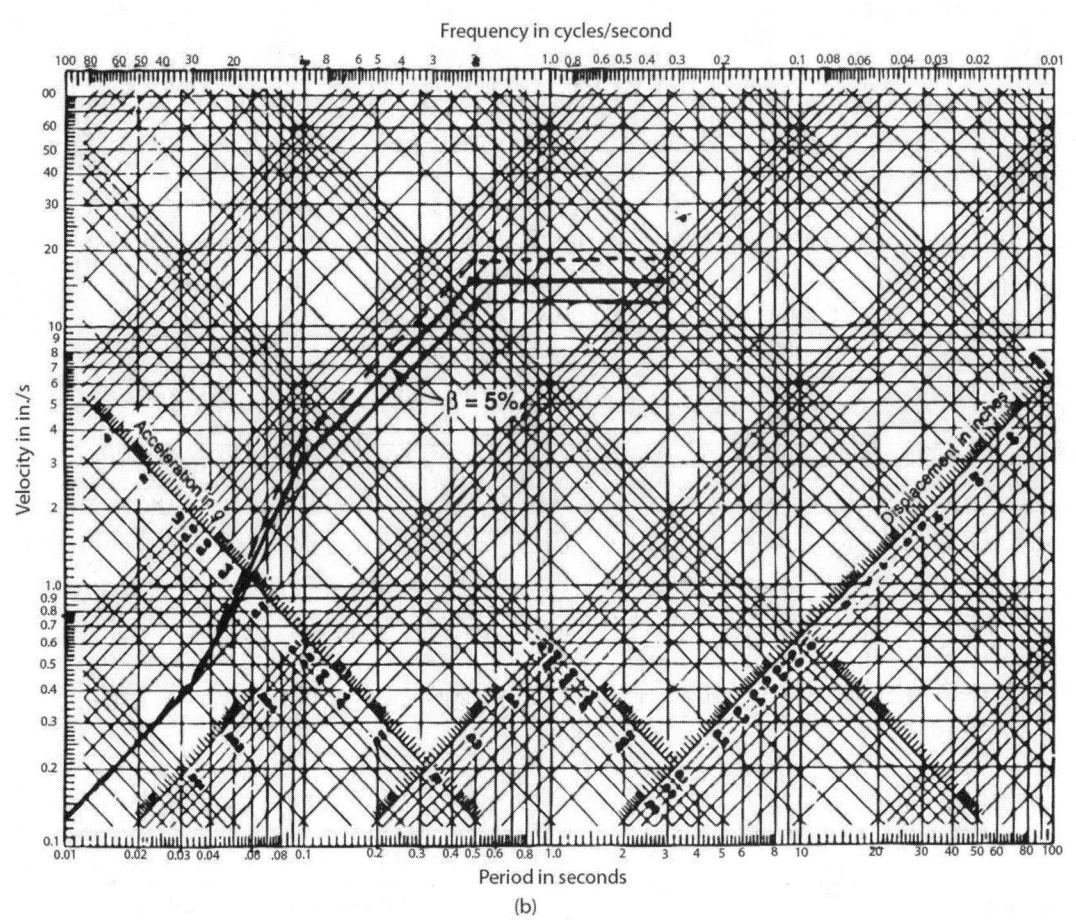

(b)

Figure 3.60   Response spectrum for 7-story building example: (a) acceleration spectrum; (b) tripartite diagram; (c) response spectra numerical representation.

Spectral acceleration, $S_a(g)$

T / β	0.1	0.48	0.50	0.80	1.0	1.75	1.5	1.75	2.0	2.25	2.5	3.0
2%	0.64	0.64	0.59	0.37	0.30	0.24	0.20	0.17	0.15	0.13	0.17	0.10
5%	0.50	0.50	0.48	0.30	0.24	0.192	0.16	0.137	0.12	0.107	0.096	0.08
7%	0.44	0.44	0.44	0.28	0.22	0.18	0.15	0.13	0.11	0.10	0.09	0.07
10%	0.38	0.38	0.38	0.25	0.20	0.16	0.13	0.11	0.10	0.09	0.08	0.066
20%	0.27	0.27	0.27	0.20	0.16	0.12	0.10	0.09	0.08	0.07	0.06	0.05

(c)

Figure 3.60 (*Continued*)

Level	$\dfrac{w}{g}$ $\left(\dfrac{\text{k}-\text{s}^2}{\text{ft}}\right)$	Mode 1				Mode 2				Mode 3				SRSS
		ϕ_1	$\frac{w}{g}\phi_1$	$\frac{w}{g}\phi_1^2$	$a_1(g)$	ϕ_2	$\frac{w}{g}\phi_2$	$\frac{w}{g}\phi_2^2$	$a_2(g)$	ϕ_3	$\frac{w}{g}\phi_3$	$\frac{w}{g}\phi_3^2$	$a_3(g)$	$a_x(g)$
Roof	43.78	0.0794	3.48	0.276	0.362	0.0747	3.27	0.744	−0.235	0.0684	2.99	0.205	0.120	0.448
7	45.34	0.0745	3.38	0.252	0.340	0.0411	1.86	0.076	−0.129	−0.0040	−0.18	0.001	−0.007	0.364
6	45.34	0.0666	3.02	0.201	0.304	−0.0042	−0.19	0.001	0.013	−0.0644	−2.92	0.188	−0.113	0.325
5	45.34	0.0558	2.53	0.141	0.254	−0.0471	−2.14	0.101	0.148	−0.0630	−2.86	0.180	−0.111	0.314
4	45.34	0.0425	1.93	0.082	0.194	−0.0718	−3.26	0.234	0.226	−0.0023	−0.10	0.000	−0.004	0.298
3	45.34	0.0279	1.27	0.035	0.127	−0.0697	−3.16	0.220	0.219	0.0604	2.74	0.166	0.106	0.275
2	56.83	0.0149	0.85	0.013	0.068	−0.0467	−2.65	0.124	0.147	0.0677	3.85	0.261	0.119	0.201
1	—	0	0	0	0	0		0	0	0	0	0	0	0
Σ	327.31		16.46	1.000		−6.27		1.000			3.52	1.001		
PF_{roof} Eq. (3.28)		$\frac{16.46}{1.000}(0.0794)=1.31$				$\frac{-6.37}{1.000}(0.0747)=-0.47$				$\frac{3.52}{1.001}(0.0684)=0.24$				Σ = 1.08
α Eq. (3.29)		$\frac{(16.46)^2}{(327.31)(1.000)}=0.828$				$\frac{(-6.27)^2}{(327.31)(1.000)}=0.120$				$\frac{(3.52)^2}{(927.31)(1.001)}=0.038$				Σ = 0.986
T		0.880 sec				0.288 sec				0.164 sec				
S_a		0.276g				0.500g				0.500g				
α_{roof} Eq. (3.30)		(1.31)(0.276) = 0.362g				(−0.47)(0.500) = −0.235g				(0.24)(0.500) = 0.120g				0.448
V Eq. (3.31)		(0.828)(0.276)(10,539) = 2408 kips				(0.12)(0.500)(10,539) = 632 kips				(0.038)(0.500)(10,539) = 200 kips				2498 kips
V/W		0.229				0.060				0.019				0.237

$W = \Sigma\left(\dfrac{w}{g}\right) \times g = 327.31 \times 32.2 = 10,539 \text{ kips} = \text{Building Weight.}$

$A_G = 0.20g$ Site PGA.

$\beta = 0.05$ Damping Factor.

Figure 3.61 Seven-story building: modal analysis to determine base shears.

Modal base shear participation factor The effective modal weight (or modal base shear participation factor) is calculated for each mode using

$$\alpha_m = \frac{\left(\sum_{i=1}^{n} \frac{w_i}{g} \phi_{im} \right)^2}{\sum_{i=1}^{n} \frac{w_i}{g} \sum_{i=1}^{n} \frac{w_i}{g} \phi_{im}^2} \tag{3.29}$$

where α_m = the modal base shear participation factor for mode m.

Next, the spectral acceleration for the period T of each mode is determined from the response spectrum. The story accelerations "a" are determined from

$$a_{xm} = PF_{xm} S_{am} \tag{3.30}$$

where α_{xm} = modal story acceleration at level x for mode m
PF_{xm} = modal participation as determined by Eq. (3.28)
S_{am} = spectral acceleration for mode m

Next, the base shears "V" are determined from

$$V_m = \alpha_m S_{am} W \tag{3.31}$$

where V_m = total lateral force for mode m
W = total seismic dead load of the building which includes the dead load plus applicable portions of other loads

For the example problem, the sum of the participation factors, PF_{xm} and α_m, add up to 1.08 and 0.986, respectively. These values being close to 1.0 indicate that most of the modal participation is included in the three modes considered in the example. The story accelerations and the base shears are combined by the SRSS. The modal base shears are 2408 kips, 632 kips, and 200 kips for the first, second, and third modes, respectively. These are used in Fig. 3.64 to determine story forces. The SRSS base shear is 2498 kips.

Story forces, accelerations, and displacements Figures 3.62 to 3.64 are set up in a manner similar to the static design procedure of section 3.4. In the static lateral procedure, $\dfrac{Wh}{\sum Wh}$, is used to distribute the force on the

$$T = 0.880 \text{ sec}$$

Modal base shear $V = 2408$ kips

(1)	(2)	(3)	(4)	(5)	(6)	(7)	(8)	(9)	(10)	(11)		
Story	ϕ	h ft	Δh ft	w kips	$\dfrac{w\phi}{\sum w\phi}$	F kips $(V_1) \times (6)$	V kips Σ (7)	ΔOTM K-ft (4)–(8)	OTM K-ft Σ (9)	Accel. g (7) ÷ (5)	δ^* ft	$\Delta\delta$ ft
Roof	0.0794	65.7		1410	0.211	508			0	0.360	0.228	
			8.7				508	4420				0.014
7	0.7450	57.0		1460	0.205	494			4420	0.338	0.214	
			8.7				1002	8717				0.022
6	0.0666	48.3		1460	0.184	443			13,137	0.303	0.192	
			8.7				1445	12,572				0.031
5	0.0558	59.6		1460	0.154	371			25,709	0.254	0.161	
			8.7				1816	15,799				0.039
4	0.0425	30.9		1460	0.117	282			41,508	0.193	0.122	
			8.7				2098	10,253				0.042
3	0.0279	22.2		1460	0.077	185			59,761	0.127	0.080	
			8.7				2283	19,862				0.057
2	0.0149	13.5		1830	0.052	125			79,623	0.068	0.043	
			13.5				2408	32,508				0.043
Grd.	0	0		0	0	0			112,131	0	0	
					Σ 1.000	2408		112,191				

$$* \text{Displacement } \delta_{x1} = \frac{g}{4\pi^2} \times T_1^2 \times \frac{F}{W}$$

$$= \frac{32}{4\pi^2} \times 0.88^2 \times \text{acceleration}$$

$$= 0.632 \times \text{acceleration}$$

Figure 3.62 Seven-story building: first mode forces and displacements.

$T_2 = 0.288$ sec

Modal base shear $V_2 = 632$ kips

(1)	(2)	(3)	(4)	(5)	(6)	(7)	(8)	(9)	(10)	(11)		
Story	ϕ	h ft	Δh ft	w kips	$\dfrac{w\phi}{\Sigma\, w\phi}$	F kips $(V_2) \times (6)$	V kips $\Sigma\,(7)$	ΔOTM K-ft $(4)\times(8)$	OTM K-ft $Z(9)$	Accel. g $(7)\div(5)$	δ^* ft	$\Delta\delta$ ft
Roof	0.0747	65.7		1410	0.522	−330			0	−0.234	−0.016	
			8.7				−330	−2871				0.007
7	0.0411	57.0		1460	0.297	−188			−2871	−0.129	−0.009	
			8.7				−518	−4507				0.010
6	−0.0042	48.3		1460	0.030	19			−7378	0.013	0.001	
			8.7				−499	−4341				0.009
5	−0.0471	39.6		1460	0.341	216			−11,719	0.148	0.010	
			8.7				−283	−2462				0.005
4	−0.0718	30.9		1460	0.520	329			−14,181	0.225	0.015	
			8.7				46	400				0.000
3	−0.0697	22.2		1460	0.504	319			−13,781	0.219	0.015	
			8.7				365	3176				0.005
2	−0.0467	13.5		1830	0.423	267			−10,605	0.146	0.010	
			13.5				632	8532				0.010
Grd.	0	0							−2073	0	0	
				Σ	0.999	632		−2073				

$$* \text{ Displacement } \delta_{x2} = \frac{g}{4\pi^2} \times T_2^2 \times \frac{F}{w}$$

$$= \frac{32}{4\pi^2} \times 0.288^2 \times \text{acceleration}$$

$$= 0.068 \times \text{acceleration}$$

Figure 3.63 Seven-story building: second mode forces and displacements.

assumption of a straight-line mode shape. In the dynamic analysis, the more representative $\dfrac{W\phi}{\Sigma\,W\phi}$ is used to distribute the forces for each mode. Story shears and overturning moments are determined in the same manner for each method. Modal story accelerations are determined by dividing the story force by the story weight. Modal story displacements are calculated from the accelerations and the period by using the following equations:

$$\delta_{xm} = PF_{xm}S_{am} = PF_{xm}S_{am}\left(\frac{T_m}{2\pi}\right)^2 g \qquad (3.32)$$

where δ_{xm} = lateral displacement at level x for mode m
S_{am} = spectral displacement for mode m calculated from the response spectrum
T_m = modal period of vibration

Modal interstory drifts Δ are calculated by taking the difference between the δ values of adjacent stories. The values shown in Figs. 3.62 to 3.64 are summarized in Fig. 3.65.

The fundamental period of vibration as determined from a computer analysis is 0.88 s. The periods of the second and third modes of vibration are 0.288 s and 0.164 s respectively. From Figs. 3.60a–c, using a response curve with 5 percent of critical damping ($\beta = 0.05$), it is determined that the second and third mode spectral accelerations (0.500 g) are 80 percent greater than the first mode spectral acceleration (0.276 g). On the basis of mode shapes and modal participation factors, modal story forces, shears, overturning moments, acceleration, and displacements are determined.

Figure 3.64a shows story forces obtained by multiplying the story acceleration by the story mass. The shapes of story force curves (Fig. 3.65a) are quite similar to the shapes of the acceleration curves (Fig. 3.65d), because the building mass is essentially uniform.

Figure 3.65b shows story shears which are a summation of the modal story forces in Fig. 3.65a. The higher modes become less significant in relation to the first mode because the forces tend to cancel each other due to the reversal of direction. The SRSS values do not differ substantially from the first mode values.

$T_3 = 0.164$ sec

Modal base shear $V_3 = 200$ kips

(1)	(2)	(3)	(4)	(5)	(6)	(7)	(8)	(9)	(10)	(11)		
Story	ϕ	h ft	Δh ft	w kips	$\dfrac{w\phi}{\Sigma w\phi}$	F kips $(V_3)\times(6)$	V kips $\Sigma(7)$	ΔOTM K-ft $(4)\times(8)$	OTM K-ft $\Sigma(9)$	Accel. g $(7)\div(5)$	δ^* ft	$\Delta\delta$ ft
Roof	0.0684	65.7		1410	0.849	170			0	0.121	0.003	
			8.7				170	1479				0.003
7	−0.0040	57.0		1460	−0.051	−10			1479	−0.007	0.000	
			8.7				160	1392				0.003
6	−0.0644	48.3		1460	−0.830	−166			2871	−0.114	−0.003	
			8.7				−6	−52				0.000
5	−0.0630	39.6		1460	−0.813	−163			2819	−0.112	−0.003	
			8.7				−169	−1470				0.003
4	−0.0023	30.9		1460	−0.028	−6			1349	−0.004	0.000	
			8.7				−175	−1523				0.002
3	0.0604	22.2		1460	0.778	156			−174	0.107	0.002	
			8.7				−19	−165				0.001
2	0.0677	13.5		1830	1.094	219			−339	0.120	0.003	
			13.5				200	2700				0.003
Grd.	0	0							2361	0	0	
				Σ	0.999	200		2361				

* Displacement $\delta_{x3} = \dfrac{g}{4\pi^2}\times T_3^2 \times \dfrac{F}{W}$

$d \qquad = \dfrac{32}{4\pi^2}\times 0.64^2 \times$ acceleration

$d \qquad = 0.022 \times$ acceleration

Figure 3.64   Seven-story building: third-mode forces and displacements.

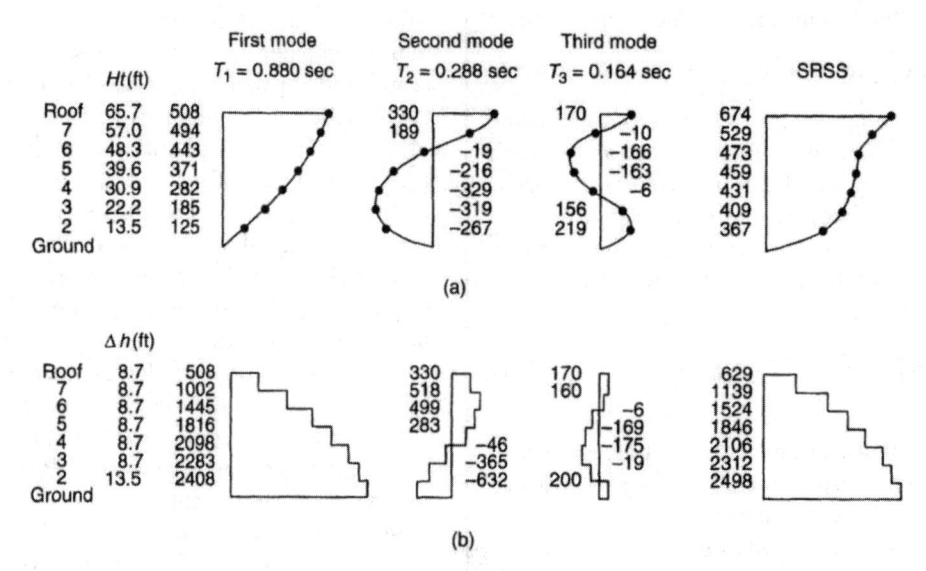

Figure 3.65   Seven story building. Modal analysis summary: (a) modal story forces (kips); (b) modal story shears (kips); (c) modal story overturning moments (kip-ft); (d) modal story accelerations (g´s); (e) modal lateral displacements (in.).

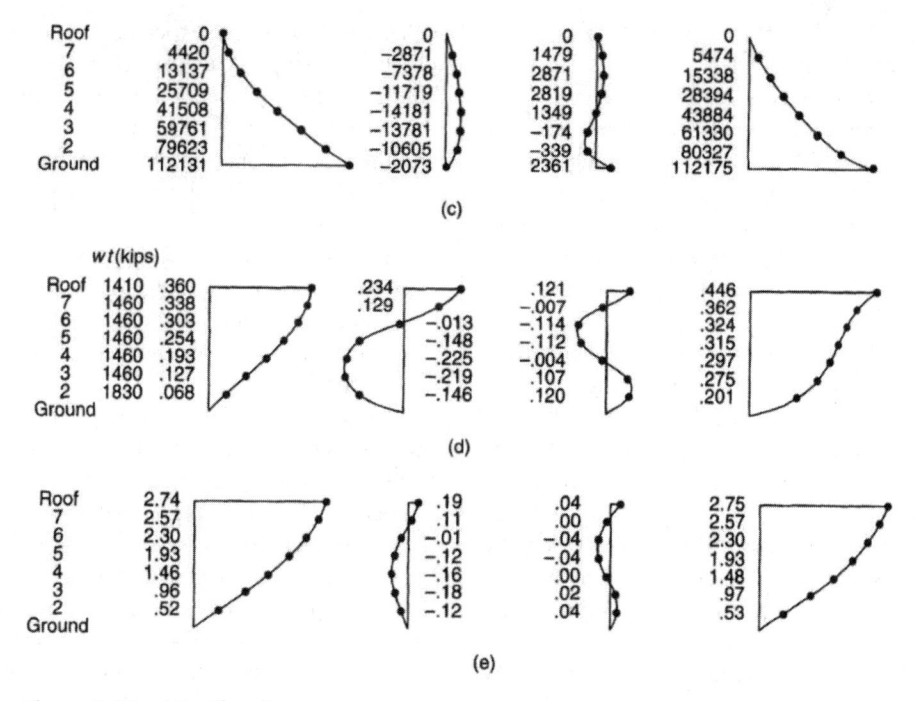

Figure 3.65 *(Continued)*

Figure 3.65c shows the building overturning moments. Again, the higher modes become somewhat less significant because of the reversal of force direction. The SRSS curve is essentially equal to the first mode curve.

Figure 3.65d shows story accelerations. Observe that the second and third modes do play a significant role in the structure's maximum response. While the shape of an individual mode is the same for displacements and accelerations, accelerations are proportional to displacements divided by the squared value of the modal period, which accounts for the greater accelerations in the higher modes. The shape of the SRSS combination of the accelerations is substantially different from the shapes of any of the individual modes because it accounts for the predominance of the various modes at different story levels.

Figure 3.65e shows the modal displacements. Observe that the fundamental mode predominates, while the second and third mode displacements are relatively insignificant. The SRSS combination does not differ greatly from the fundamental mode. It should be noted that for taller and irregular buildings the influence of the higher modes becomes larger. Consideration of higher modes of vibration may become necessary to demonstrate that at least 90 percent of the participating mass of the structure is included in the analysis.

3.6 SEISMIC VULNERABILITY STUDY AND RETROFIT DESIGN

3.6.1 Introduction

Earthquakes create havoc. Even minor earthquakes which take no lives can cause severe disruptions to utilities, business, transportation, and individuals whose homes and livelihoods are drastically changed within minutes. They happen without warning, and they can happen in many places, from Maine to Alaska, along the New Madrid fault in the Midwest, in areas we usually don't think of as vulnerable.

The most visible damage from an earthquake is to man-made structures; to buildings, bridges, and roads. Older buildings in particular suffer damage from earthquakes. Many buildings in almost all parts of the world were constructed before earthquake-resistant technology was developed. And building damage, whether it is total or partial building collapse, is generally considered the most dangerous aspect of an earthquake, because that's how most lives are lost.

Most Americans are aware of California's vulnerability to earthquakes. But California is not the only state facing earthquake risks. At least 40 states are seismically vulnerable to some degree. The vulnerability is obviously less severe, in terms of people's safety, in some geographic regions than others. Yet the possibility of functional disruption and significant property damage from an earthquake, even if no lives are lost, should be a matter of concern for every community.

Seismic rehabilitation entails costs as well as disruption. In fact, the effects of a rehabilitation program are similar to those of an earthquake because strengthening, in terms of cost and the need to vacate while strengthening is underway, is analogous to building repair after an earthquake. The crucial difference is that strengthening occurs at a specified time and no deaths or injuries will occur during the process.

Before selecting an approach to seismic rehabilitation, building owners must first get the facts on the nature of its seismic hazard, and on the risk to buildings likely to be affected by an earthquake and the people who live and work in the affected buildings. Seismic rehabilitation involves six basic steps.

Step one is seismic hazard assessment, usually done by geologists and seismologists. This is needed to identify the level of risk associated with the seismic vulnerability of buildings.

Step two is to develop an overall damage estimate by determining the probable extent of damage to various types of buildings and structures, as well as the disruptions to the use of the facilities.

Step three is determining building priorities, that is, categorize the buildings to be strengthened and, within each category, set the priorities for attention. Again, building owners will exercise their judgment on what needs to be done in what sequence.

Step four involves decisions on engineering methods, construction costs, indirect costs, and overall effectiveness of the methods employed.

Step five covers the economic impact of rehabilitation. Once the data have been gathered on vulnerability, loss, damage, and impact, the emphasis in the seismic rehabilitation program development is on the process of analyzing data, assessing impacts, considering trade-offs and evaluating options. The intention is to reach a consensus on acceptable levels of risk, given the economic resources available.

The final step is implementing the seismic strengthening program appropriate for the building and its usage.

In a seismic vulnerability study, it is convenient to classify the damage within a building into two categories, structural and nonstructural.

Structural damage refers to the degradation of the building's support system, such as frames and walls while nonstructural damage is any damage that does not affect the integrity of the building's physical support system. Examples of nonstructural damage are chimneys that collapse, broken windows or ornamental features, and a collapsed ceiling. The expected type of damage depends on the building's structural characteristics and age, its configuration, construction materials, the site conditions, the proximity of the building to neighboring buildings, and the type of nonstructural elements.

An earthquake can cause a building to experience four types of damage:

1. the entire building collapses;
2. portions of the building collapse;
3. components of the building fail and fall;
4. entry-exit routes are blocked, preventing evacuation and rescue.

Any of the above, may result in unacceptable risk to human lives. It can also mean loss of property, and interruptions of use and normal function.

Another type of damage that should be included in the vulnerability study is the structural damage from the "pounding" action that results when two buildings, insufficiently separated, collide. This condition is particularly severe when the floor levels of the two buildings do not match, and the stiff floor framing of one building can badly damage the more fragile walls or columns of its neighbor.

There are two basic ways to improve a building's structural performance in an earthquake: strengthening load-resisting components and decreasing demand on load-resisting components.

The strengthening of load-resisting members may range from upgrading a single element to replacing the entire system including the diaphragms and chord members because after all, a seismic strengthening program is a response to a unique set of demands.

The second method of improving a building's earthquake resistance is by decreasing demand on existing systems. There are four basic ways to reduce the demand:

1. reduce the weight of the building;
2. increase the fundamental period of vibration;
3. increase the response modification factor, R;
4. provide alternative methods such as base isolation and supplemental damping techniques.

Nonstructural architectural elements can also create life-threatening hazards. For example, windows may break or architectural cladding such as granite veneer with insufficient anchorage can cause damage. Consequently, a seismic retrofit program should explore techniques for dealing with nonstructural components, such as veneers, lighting fixtures, glass doors, and windows, raised computer access floors, as well as ceilings. Similarly, because damage to mechanical and electrical components can impair building functions that may be essential to life safety, seismic strengthening should be considered for components such as mechanical and electrical equipment, ductwork and piping, elevators, emergency power systems, communication systems, and computer equipment.

3.6.2. SEAOC's Vision 2000: Performance-Based Engineering

Vision 2000 is a SAEOC project specially devoted to the development of a framework for performance-based engineering of buildings—buildings that are engineered to avoid economic losses associated with damage and post-earthquake loss of function.

It is a radical departure from typical and current structural engineering practice in that it seeks to provide the structural engineering profession with tools to explicitly, rather than implicitly, design for multiple, specifically defined, levels of performance. These performance levels are defined in terms of specific limiting damage states, against which a structure's performance can be objectively measured. Recommendations have been developed as to which performance levels should be attained, by buildings of different occupancy and use, under several levels of earthquake loading. This tiered specification of performance levels to be achieved at predetermined earthquake hazard levels becomes the design performance objective and a basis for design. It recognizes the importance of the performance of all the various component systems to overall building performance and defines both responsibility and methodology for design to obtain the desired performance.

Performance-based engineering is defined as "selection of design criteria, appropriate structural systems, layout, proportioning, and detailing for a structure and its nonstructural components and contents and the assurance of construction quality control such that at specified levels of ground motion and with defined levels of reliability, the structure will not be damaged beyond certain limiting states."

Vision 2000 has identified five performance levels (Fig. 3.66). These are: Fully Operational, Operational, Life Safe, Near Collapse, and Collapse. Each of these performance levels has associated with it defined levels of damage to structural, architectural, mechanical, and electrical building components as well as tenant furnishings. Figure 3.67 provides a broad overview of where each performance level falls within the overall spectrum of possible damage states.

As with performance levels, there are infinite possible hazard levels that could be used in the development of design performance objectives. Among these four hazard levels have been selected. As shown in Fig. 3.64, these are: Frequent earthquakes, having a 50 percent chance of exceedance in 30 years (43-year mean return period); Occasional earthquakes, having a 50 percent chance of exceedance in 50 years (72-year mean return period); Rare earthquakes, having a 10 percent change of exceedance in 50 years (475-year mean return period) and Very Rare earthquakes, having a 10 percent chance of exceedance in 100 years (950-year return period).

In order to execute performance-based engineering, it is necessary to have a series of design parameters and acceptance criteria for each performance level for the various structural and nonstructural components that comprise the building. Design parameters are calculable response measures such as element forces, interstory drifts, plastic rotations, etc. that can be derived from a structural analysis of building response to a particular design earthquake. Acceptance criteria are the limiting values for design parameters in order to attain a given performance level. As an example, if interstory drift ratio is a design parameter used for certain classes of building, acceptance criteria would be certain defined drift ratios for each performance level such as 0.020 for the Near Collapse level, 0.015 for the Life Safe level, 0.01 for the Operational level, and 0.005 for the Fully Operational level. A wide variety of potential design parameters may need to be defined including deformation, strength, and energy-based parameters.

Vision 2000 also notes that since current seismic resistant design relies on the inelastic response of the structure to dissipate most of the input energy of the earthquake, it is imperative that the post-elastic behavior of the structure be addressed at the conceptual state of design. Application of capacity design principals is recommended to provide for the inelastic response by designating the ductile links or "fuses" in the lateral force-resisting system. The designated "fuses" will be counted upon to yield and to dissipate the input energy of the earthquake while the rest of the system remains elastic. This concept gives a clear understanding of the inelastic response of the structure such that the design and quality assurance programs may be focused on the critical links in the system. At the final design and detailing stage, these critical links will be rigorously detailed to provide the ductility required.

At each stage of design, an acceptability check is performed to verify that the selected performance objectives are being met. The specific extent and methodology of the analysis will vary with the performance objectives and the design approach used, however, the general concept of the acceptability analysis remains the same. Several acceptability analysis approaches are summarized in the

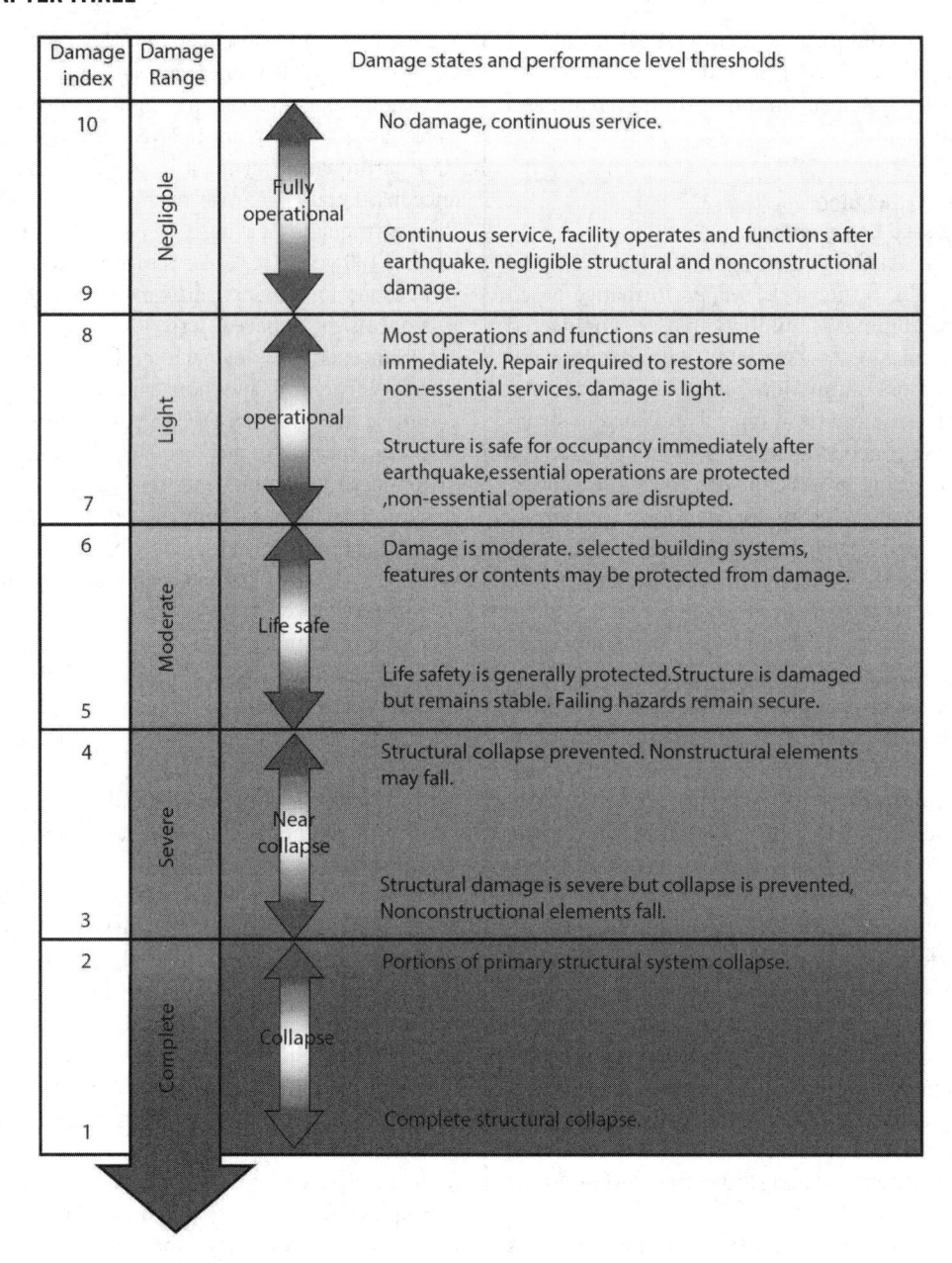

Figure 3.66   Vision 2000: damage states and performance levels.

Vision 2000 report including general elastic procedures, component based elastic analysis procedures, the capacity spectrum procedure, pushover analysis methods, dynamic nonlinear time-history procedures, and the drift demand method.

Vision 2000 design methodology is included in Appendix B of the 1996 Blue Book, and is expected to go through a consensus review and trial design process. It is anticipated that in about ten years all design approaches and acceptability checks will become part of the code.

3.7   DYNAMIC ANALYSIS: THEORY

3.7.1   Introduction

A good portion of the loads that occur in multistory buildings can be considered as static loads requiring static analysis only. Although almost all loads except

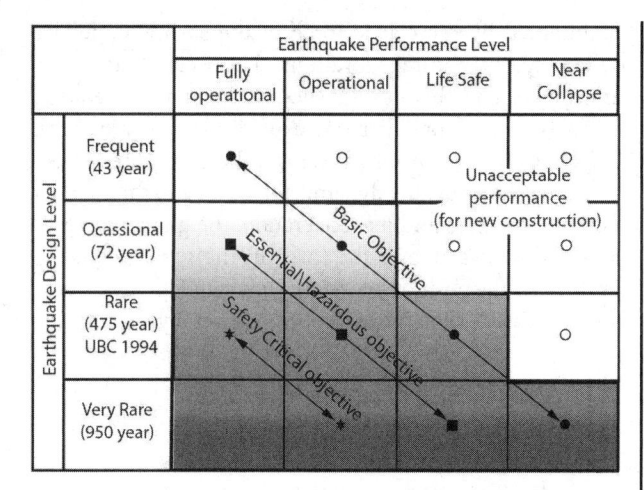

		Earthquake Performance Level			
		Fully operational	Operational	Life Safe	Near Collapse
Earthquake Design Level	Frequent (43 year)	●	○	○	○ Unacceptable performance (for new construction)
	Ocassional (72 year)	■		○	○
	Rare (475 year) UBC 1994	★	■	●	○
	Very Rare (950 year)	★	★	■	●

(diagonal arrows labeled: Basic Objective, Essential\Hazardous objective, Safety Critical objective)

Figure 3.67 Vision 2000: performance objectives for building design.

dead loads are transient, it is customary in most designs to treat these loads as static. For example, lateral loads imposed by transient wind pulses are usually treated as static loads. Even in earthquake design, which is clearly a dynamic problem, one of the acceptable methods of design is to use the so-called equivalent force system that is supposed to represent the static equivalent of dynamic forces. Under such assumptions, the analysis of a multistory building reduces to a single solution of the problem under static loads. The magnitude, nature, and origin of loads can be clearly visualized, and with the availability of computer programs the analysis can be performed without undue complexities. For example, once the wind load, building geometry, and member properties are known, the analysis for forces and deflections becomes a trivial computing task.

Although the equivalent static load approach is a recognized method of earthquake analysis, the state-of-the-art method for high-rise buildings uses a dynamic solution. Indeed, most building codes make the dynamic analysis mandatory for buildings whose configurations violate the assumptions made in the derivation of code equivalent forces. In today's architectural environment, it is more than likely that a tall building will not meet the spirit of the code assumptions either because the building has vertical steps, has a nonuniform configuration, or other idiosyncrasies. It is, therefore, necessary to have a thorough understanding and mastery of the concepts and methods of solutions used in dynamic analysis.

Consider a multistory, multi-bay plane frame structure subjected to lateral wind loads. Although wind load is dynamic in nature, normal design practice, except in the case of very slender buildings, is to treat the wind load as

an equivalent static load. The variation of wind load with respect to time is not considered: the load is assumed to be applied gradually to the structure. Under these conditions, the analysis of the structure becomes trivial. For a given set of loads, properties of the members and boundary conditions, there is but one unique solution.

Assume that the structure, instead of being subjected to wind, is subjected to seismic loads. During an earthquake, the structure is subjected to rapid ground displacements and experiences a number of different forces that include inertia forces, damping forces, and elastic forces. Although there is no external force per se, the mass of the structure generates an equivalent forcing function when subjected to acceleration.

It is easy to visualize this force by considering a simplified response of the structure during an earthquake. Before the start of an earthquake, the structure is in static equilibrium and would remain so if the movement of the ground due to the earthquake takes place very slowly; the structure would simply ride to the new displaced position. When the ground moves suddenly, the inertia of mass distributed in the structure attempts to prevent the displacement of the structure, thus creating seismic loads analogous to an externally applied lateral force.

Earthquake forces are considered dynamic, because they vary with time. Since the load is time-varying, the response of the structure, including deflections, forces, and bending moments, is also time-dependent. Instead of a single solution as in a static case, a separate solution is required at each instant of time for the entire duration of an earthquake. The resulting inertia forces are a function of deflections, which are themselves related to the inertia forces. It is therefore necessary to formulate the problem in terms of differential equations by relating the inertia forces to the second derivative of structural displacements.

In the following sections, a brief mathematical treatment is presented for SDOF systems with and without damping forces, followed by an analysis of MDOF systems. The modal superposition method and orthogonality conditions which form the backbone of dynamic analysis are explained with reference to a system with two-degrees-of-freedom to show how coupled equations of motions for a MDOF system are transformed into a set of independent SDOF systems. The section concludes with a summary highlighting the practical aspects of dynamic analysis.

3.7.2 Systems with Single-Degree-of-Freedom

Consider a portal frame shown in Fig. 3.68 consisting of an infinitely stiff beam supported by flexible columns.

Assuming that the beam is completely rigid and that columns have negligible mass as compared to the beam, the structure can be visualized as a spring-supported mass for the horizontal motion of the beam.

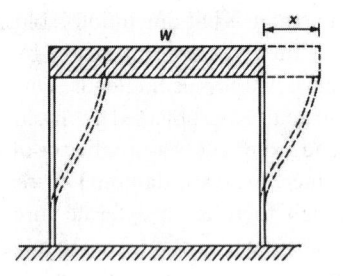

Figure 3.68   Single-bay single-story portal frame.

An analytical model for the system is shown in Fig. 3.69. Under the action of gravity force W, the spring will be extended by a certain amount. If the spring is very stiff, the extension is small, and vice versa.

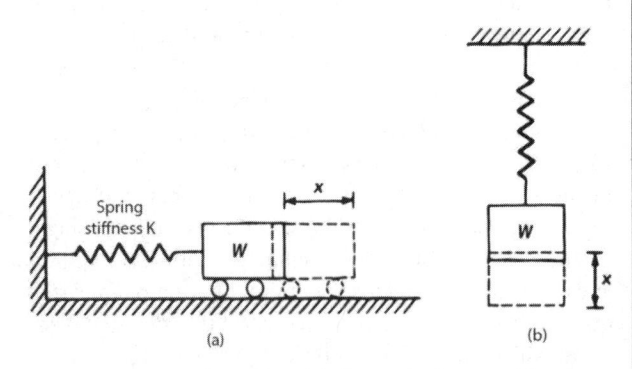

Figure 3.69   Analytical models for single degree-of-freedom system: (a) model in horizontal position; (b) model in vertical position.

The extension x experienced by the spring can be related to the stiffness of the spring k by the relation

$$x = \frac{W}{k} \qquad (3.33)$$

k is called the spring constant or spring stiffness and denotes the load required to produce a unit extension of the spring. If W is measured in kips and the extension in inches, the spring stiffness will have a dimension of kips per inch. The weight W comes to rest after the spring has extended by the length x. Equation (3.33) expresses the familiar static equilibrium condition between the internal force in the spring and the externally applied force W.

If a vertical force is applied or removed suddenly, vibrations of the system are produced. Such vibrations, maintained by the elastic force in the spring alone, are called free or natural vibrations. The weight moves up and down, and therefore is subjected to an acceleration given by the second derivative of displacement x, with respect to time, $\ddot{x}$. At any instant t, there are three forces acting on the body: the dynamic force equal to the product of the body mass and its acceleration, the gravity force W acting downward, and the force in the spring equal to $W + kx$ for the position of weight shown in Fig. 3.70. These are in a state of dynamic equilibrium given by the relation

$$\frac{W}{g}\ddot{x} = W - (W + kx) = -kx \qquad (3.34)$$

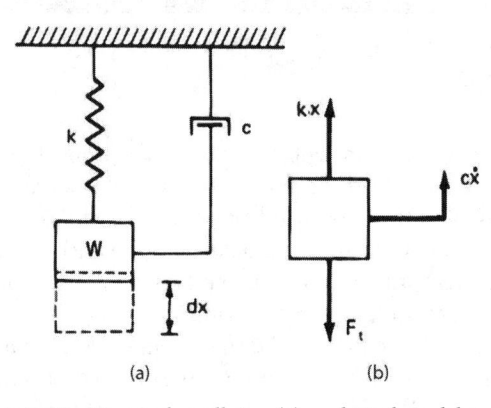

Figure 3.70   Damped oscillator: (a) analytical model; (b) forces in equilibrium.

The above equation of motion it is called Newton's law of motion and is governed by the equilibrium of inertia force that is a product of the mass W/g, and acceleration $\ddot{x}$, and the resisting forces that are a function of the stiffness of the spring.

The principle of virtual work can be used as an alternative to Newton's law of motion. Although the method was first developed for static problems, it can readily be applied to dynamic problems by using D'Alembert's principle. The method establishes dynamic equilibrium by including inertial forces in the system.

The principle of virtual work can be stated as follows: For a system that is in equilibrium, the work done by all the forces during a virtual displacement is equal to zero. Consider a damped oscillator subjected to a time-dependent force $F_{(t)}$ as shown in Fig. 3.70a. The free-body diagram of the oscillator subjected to various forces is shown in Fig. 3.70a.

Let δx be the virtual displacement. The total work done by the system is zero and is given by

$$m\ddot{x}\,\delta\dot{x} + c\dot{x}\,\delta x + kx\,\delta x - F_{(t)}\,\delta x = 0 \qquad (3.35)$$

$$(m\ddot{x} + c\dot{x} + kx - F_{(t)})\,\delta x = 0 \qquad (3.36)$$

since δx is arbitrarily selected,

$$m\ddot{x} + c\dot{x} + kx - F_{(t)} = 0 \qquad (3.37)$$

This is the differential equation of motion of the damped oscillator.

The equation of motion for an undamped system can also be obtained from the principle of conservation of energy, which states that if no external forces are acting on the system and there is no dissipation of energy due to damping, then the total energy of the system must remain constant during motion and consequently its derivative with respect to time must be equal to zero.

Consider again the oscillator shown in Fig. 3.70a without the damper. The two energies associated with this system are the kinetic energy of the mass and the potential energy of the spring.

The kinetic energy of the spring

$$T = \frac{1}{2}m\dot{x}^2 \qquad (3.38)$$

where $\dot{x}$ is the instantaneous velocity of the mass.

The force in the spring is kx; work done by the spring is $kx\,\delta x$. The potential energy is the work done by this force and is given by

$$V = \int_0^x kx\,\delta x = \frac{1}{2}kx^2 \qquad (3.39)$$

The total energy in the system is a constant. Thus

$$\frac{1}{2}m\dot{x}^2 + \frac{1}{2}kx^2 = \text{constant } c_0 \qquad (3.40)$$

Differentiating with respect to x, we get

$$m\dot{x}\ddot{x} + kx\dot{x} = 0 \qquad (3.41)$$

Since $\dot{x}$ cannot be zero for all values of t, we get

$$m\ddot{x} + kx = 0 \qquad (3.42)$$

which has the same form as Eq. (3.33). This differential equation has a solution of the form:

$$x = A\sin(\omega t + \alpha) \qquad (3.43)$$
$$\dot{x} = \omega A\cos(\omega t + \alpha) \qquad (3.44)$$

where A is the maximum displacement and ωA is the maximum velocity. Maximum kinetic energy is given by

$$T_{\text{max}} = \frac{1}{2}m(\omega A)^2 \qquad (3.45)$$

Maximum potential energy is

$$V_{\text{max}} = \frac{1}{2}kA^2 \qquad (3.46)$$

Since $T = V$,

$$\frac{1}{2}m(\omega A)^2 = \frac{1}{2}kA^2$$

or

$$\omega = \sqrt{\frac{k}{m}} \qquad (3.47)$$

which is the natural frequency of the simple oscillator. This method, in which the natural frequency is obtained by equating maximum kinetic energy and maximum potential energy, is known as *Rayleigh's method*.

3.7.3 Multi-Degree-of-Freedom Systems

In these systems, the displacement configuration is determined by a finite number of displacement coordinates. The true response of a multi-degree system can be determined only by evaluating the inertia effects at each mass particle because structures are continuous systems with an infinite number of degrees-of-freedom. Although analytical methods are available to describe the behavior of such systems, the methods are limited to structures with uniform material properties and regular geometry. The methods are complex, requiring the formulation of partial differential equations. The analysis is greatly simplified by replacing the entire displacement of the structure with a limited number of displacement components, and assuming the entire mass of the structure is concentrated in a number of discrete points.

Consider a multistory building with n degrees-of-freedom as shown in Fig. 3.71. The dynamic equilibrium

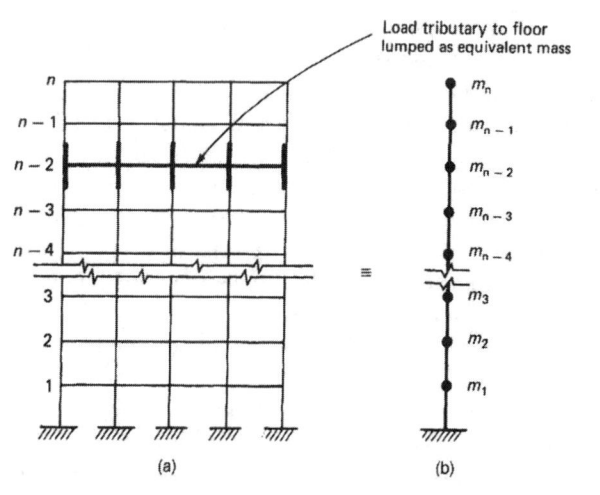

Figure 3.71 Multi degree-of-freedom system: (a) multistory frame; (b) analytical model with lumped masses.

equations for undamped free vibration can be written in the general form.

$$\begin{bmatrix} m_{11} & m_{12} & m_{13} & \cdots & m_{1m} \\ m_{21} & m_{22} & m_{23} & \cdots & m_{2m} \\ m_{31} & m_{32} & m_{33} & \cdots & m_{3m} \\ \cdots & \cdots & \cdots & \cdots & \cdots \\ m_{n1} & m_{n2} & m_{n3} & \cdots & m_{nm} \end{bmatrix} \begin{bmatrix} \ddot{x}_1 \\ \ddot{x}_2 \\ \ddot{x}_3 \\ \vdots \\ \ddot{x}_n \end{bmatrix}$$

$$+ \begin{bmatrix} k_{11} & k_{12} & k_{13} & \cdots & k_{1n} \\ k_{21} & k_{22} & k_{23} & \cdots & k_{2n} \\ k_{31} & k_{32} & k_{33} & \cdots & k_{3n} \\ \cdots & \cdots & \cdots & \cdots & \cdots \\ k_{n1} & k_{42} & k_{n3} & \cdots & k_{nm} \end{bmatrix} \begin{bmatrix} x_1 \\ x_2 \\ x_3 \\ \vdots \\ x_n \end{bmatrix} = 0 \quad (3.48)$$

The above system of equations can be written in matrix form:

$$[M]\{\ddot{x}\} + [K]\{x\} = 0 \quad (3.49)$$

where $[M]$ = the mass or inertia matrix
$\{\ddot{x}\}$ = the column vector of accelerations
$[K]$ = the structure stiffness matrix
$\{x\}$ = the column vector of displacements of the structure

If the effect of damping is included, the equations of motion would be of the form

$$[M]\{\ddot{x}\} + [C]\{\dot{x}\} + [K]\{x\} = \{P\} \quad (3.49a)$$

where $[C]$ = the damping matrix
$\{\dot{x}\}$ = the column vector of velocity
$\{P\}$ = the column vector of external forces

General methods of solutions of these equations are available, but tend to be cumbersome. Therefore, in solving seismic problems simplified methods are used; the problem is first solved by neglecting damping. Its effects are later included by modifying the design spectrum to account for damping. The absence of precise data on damping does not usually justify a more rigorous treatment. Neglecting damping results in dropping the second term, and limiting the problem to free-vibrations results in dropping the right-hand side of Eq. (3.49a). The resulting equations of motion will become identical to Eq. (3.49).

During free vibration the motions of the system are simple harmonic, which means that the system oscillates about the stationary position in a sinusoidal manner; all masses follow the same harmonic function, having similar angular frequency ω. Thus

$$x_1 = a_1 \sin \omega_1 t$$
$$x_2 = a_2 \sin \omega_2 t$$
$$\vdots$$
$$x_n = a_n \sin \omega_n t$$

or in matrix notation

$$\{x\} = \{a_n\} \sin \omega_n t$$

where $\{a_n\}$ represents the column vector of modal amplitudes for the nth mode, and ω_n the corresponding frequency. Substituting for $\{x\}$ and its second derivative $\{\ddot{x}\}$ in Eq. (3.49) results in a set of algebraic expressions.

$$-\omega_n^2[M]\{a_n\} + [K]\{a_n\} = 0 \quad (3.50)$$

Using a procedure known as Cramer's rule, the above expressions can be solved for determining the frequencies of vibrations and relative values of amplitudes of motion a_{11}, a_{12}, ..., a_n. The rule states that nontrivial values of amplitudes exist only if the determinant of the coefficients of a is equal to zero because the equations are homogeneous, meaning that the right-hand side of Eq. (3.50) is zero. Setting the determinant of Eq. (3.50) equal to zero, we get

$$\begin{bmatrix} k_{11} - \omega_1^2 m_{11} & k_{12} - \omega_1^2 m_{12} & k_{13} - \omega_1^2 m_{13} & \cdots & k_{1n} - \omega_n m_{1n} \\ k_{21} - \omega_2^2 m_{21} & k_{22} - \omega_2^2 m_{12} & k_{23} - \omega_2^2 m_{23} & \cdots & k_{2n} - w_n m_{2n} \\ k_{31} - \omega_3^2 m_{31} & k_{32} - \omega_3^2 m_{32} & k_{33} - \omega_3^2 m_{33} & \cdots & k_{3n} - \omega_n m_{3n} \\ \cdots - \cdots & \cdots - \cdots & \cdots - \cdots & \cdots & \cdots - \cdots \\ k_{n1} - \omega_n^2 m_{n1} & k_{n2} - \omega_n^2 m_{n2} & k_{n3} - \omega_n^2 m_{n3} & \cdots & k_{nn} - \omega_n m_{nn} \end{bmatrix} = 0$$

$$(3.51)$$

With the understanding that the values for all the stiffness coefficients k_{11}, k_{12}, etc., and the masses m_1, m_2, etc., are known, the determinant of the equation can be expanded, leading to a polynomial expression in ω^2. Solution of the polynomial gives one real root for each mode of vibration. Hence for a system with n degrees of freedom, n natural frequencies are obtained. The smallest of the values obtained is called the fundamental frequency and the corresponding mode the fundamental or first mode.

In mathematical terms the vibration problem is similar to those encountered in stability analyses. The determination of frequency of vibrations can be considered similar to the determination of critical loads, while the modes of vibration can be likened to evaluation of buckling modes. Such types of problems are known as *eigenvalue* or *characteristic value* problems. The quantities ω^2 which are analogous to critical loads are called eigenvalues or characteristic values, and in a broad sense can be looked upon as unique properties of the structure similar to geometric properties such as area or moment of inertia of individual elements.

Unique values for characteristic shapes, on the other hand, cannot be determined because substitution of ω^2 for a particular mode into the dynamic equilibrium equation [Eq. (3.50)] results in exactly n unknowns for the characteristic amplitudes $x_1 \ldots x_n$ for that mode. However, it is possible to obtain relative values for all amplitudes in terms of any particular amplitude. We are, therefore, able to obtain the pattern or the shape of the vibrating mode but not its absolute magnitude. The set of modal amplitudes that describe the vibrating pattern is called *eigenvector* or *characteristic vector*.

3.7.4 Modal Superposition Method

Modal superposition is a method in which the equations of motions are transformed from a set of n simultaneous differential equations to a set of n independent equations by the use of so-called normal coordinates. These equations are solved for the response of each mode, and the total response of the system is obtained by superposing individual solutions. Two concepts which are necessary for the understanding of the modal superposition method are: (i) normal coordinates and; (ii) property of orthogonality. These will be explained first, followed by application of the method to a 2-story structure. The treatments are necessarily incomplete in the mathematical sense but are sufficiently thorough to provide a sound conceptual understanding.

3.7.4.1 NORMAL COORDINATES

In a static analysis it is common and convenient to represent the displacements of a structure by a system of geometric coordinates such as the cartesian system of coordinates that indicate the linear and angular positions of the elements with respect to a static position. For example, in a planar system, coordinates x and y and rotation θ are used to describe the position of the displaced structure. If the structure is restrained to move only in the horizontal direction and if rotations are of no consequence, only one coordinate is sufficient to describe the displacement. The displacements in general can also be obtained indirectly by any independent system of coordinates which are sufficient in number to specify the deflected position of all elements of the system. These coordinates are called generalized coordinates and their number is equal to the number of degrees-of-freedom of the system. In dynamic analysis, however, it is more advantageous to use free-vibration mode shapes to represent the displacements because of their orthogonality properties, which will be explained shortly, and because a close approximation of the displacement can be made by considering only the first few modes.

The dynamic analysis of multi-degree-of-freedom systems becomes extremely difficult if a system of direct coordinates is employed to describe their motion. A static analysis, on the other hand, can be handled without undue complications by using any set of consistent coordinates, and the resulting forces and displacements can be converted from one set of coordinates to another without much difficulty. To avoid the computational problems, in structural dynamics the normal modes of vibration are employed as generalized coordinates to describe the motion. Using this system, the undamped motion equations become uncoupled, greatly simplifying their solution. While the mathematical description of normal modes and their properties may be intriguing, there is nothing complicated about the concept. Let us indulge in some analogies to bring home the concept. For example, we can consider a set of normal modes as being similar to the primary colors, red, blue, and yellow. None of the primary colors can be constructed as a combination of the others, but any secondary color such as green or pink can be created by combining the primary colors, each with a distinct proportion. The proportions can be looked upon as scale factors, while the primary colors themselves can be considered similar to normal modes. To further reinforce the concept of generalized coordinates, it is helpful to recall its application in the solution of beam bending problems in which the deflection curve of the beam is represented in the form of trigonometric series. Considering the case of a simply supported beam subjected to vertical loads as shown in Fig. 3.72a, the deflection at any point can be represented by the following series:

$$y = a_1 \frac{\sin \pi x}{l} + a_2 \frac{\sin 2\pi x}{l} + a_3 \frac{\sin 3\pi x}{l} \quad (3.52)$$

Geometrically, this means that the deflection curve can be obtained by superposing simple sinusoidal curves such as shown in Fig. 3.72b–e.

The first term in Eq. (3.52) represents the full sine curve, the second term the half-sine curve, etc. The coefficients a_1, a_2, a_3, etc., represent the maximum ordinates of the sine curves, and the numbers 1, 2, and 3 the number of waves or mode shapes. By determining the coefficients a_1, a_2, a_3 etc., the trigonometric series can be made to represent any deflection curve with a degree of accuracy that depends on the number of terms considered in the series.

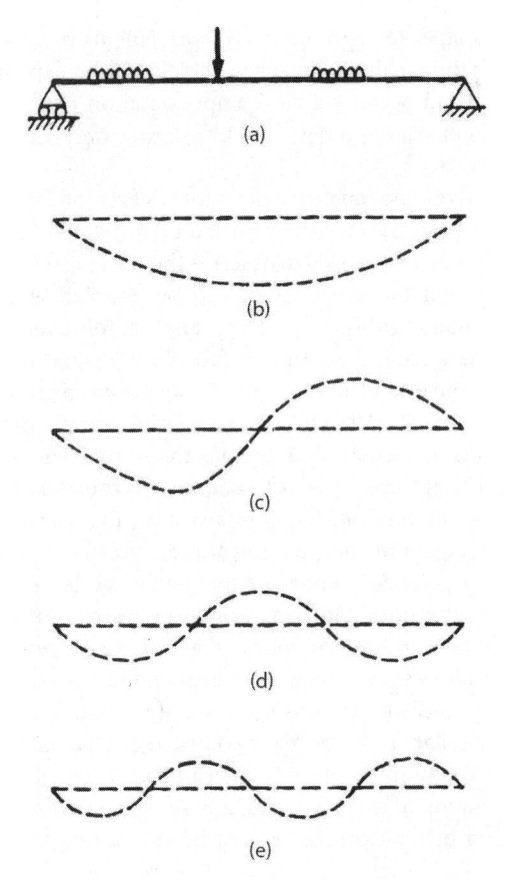

(a)

(b)

(c)

(d)

(e)

Figure 3.72   Generalized displacement of simply supported beam: (a) loading; (b) full sine curve; (c) half-sine curve; (d) one-third sine curve; (e) one-fourth sine curve.

3.7.4.2   ORTHOGONALITY

An important property of force-displacement relationship rarely used in static problems but very important in structural dynamics is the so-called orthogonal property. This property is best explained with reference to an example shown in Fig. 3.73.

Consider a 2-story lumped-mass system subjected to free-vibrations. The two modes of vibration can thus be considered as elastic displacements of the system due to two different loading conditions. A theorem known in structural mechanics as Betti's reciprocal theorem is used to derive the orthogonality conditions. This law states that the work done by one set of loads on the deflection due to a second set of loads is equal to the work done by the second set of loads acting on the deflections due to the first. Using this theorem with reference to Fig. 3.73, we get

$$\omega_1^2 m_1 x_{1b} + \omega_1^2 m_2 x_{2b} = \omega_2^2 m_1 x_{1a} + \omega_1^2 m_2 x_{2a} \quad (3.53)$$

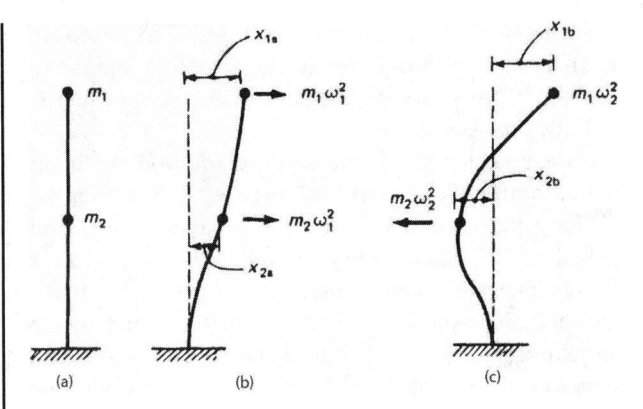

Figure 3.73   Two-story lumped-mass system illustrating Betti's reciprocal theorem: (a) lumped model; (b) forces acting during first mode of vibration; (c) forces acting during second mode of vibration.

This can be written in matrix form

$$\omega_1^2 \begin{bmatrix} m_1 & 0 \\ 0 & m_2 \end{bmatrix} \begin{bmatrix} x_{1b} \\ x_{2b} \end{bmatrix} = \omega_2^2 \begin{bmatrix} m_1 & 0 \\ 0 & m_2 \end{bmatrix} \begin{bmatrix} x_{1a} \\ x_{2a} \end{bmatrix}$$

or

$$(\omega_1^2 - \omega_2^2)\{x_b\}^T [M]\{x_a\} = 0 \quad (3.54)$$

If the two frequencies are not the same, i.e., $\omega_1 \neq \omega_2$ we get

$$\{x_b\}^T [M]\{x_a\} = 0 \quad (3.55)$$

This condition is called the orthogonality condition, and the vibrating shapes $\{x_a\}$ and $\{x_b\}$ are said to be orthogonal with respect to the mass matrix $[M]$. By using a similar procedure, it can be shown that

$$\{x_a\}^T [k]\{x_b\} = 0 \quad (3.56)$$

The vibrating shapes are therefore orthogonal with respect to stiffness matrix as they are with respect to the mass matrix. In the general case of the structures with damping, it is necessary to make a further assumption in the modal analysis that the orthogonality condition also applies for the damping matrix. This is for mathematical convenience only and has no theoretical basis. Therefore, in addition to the two orthogonality conditions mentioned previously, a third orthogonality condition of the form

$$\{x_a^T\} c \{x_b\} = 0 \quad (3.57)$$

is used in the modal analysis.

To bring out the essentials of the normal mode method, it is convenient to consider the dynamic analysis of a two-degree-of-freedom system. We will first analyze the

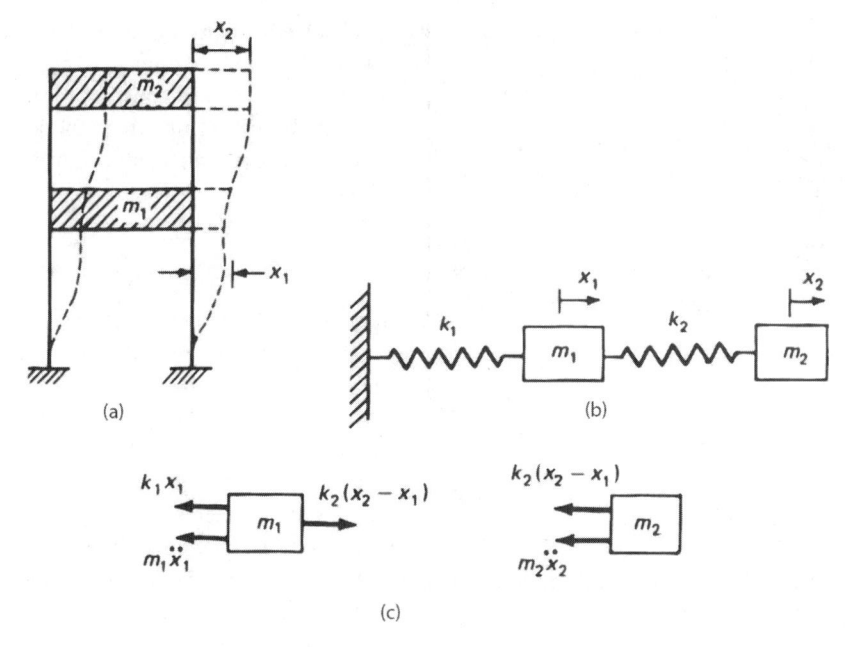

Figure 3.74 Two-story shear building: free vibrations: (a) building with lumped masses; (b) mathematical model; (c) free-body diagram with lumped masses.

system by a direct method and then show how the analysis can be simplified by the modal superposition method.

Consider a 2-story dynamic model of a shear building shown in Fig. 3.74 subject to free vibrations. The masses m_1 and m_2 at levels 1 and 2 can be considered connected to each other and to the ground by two springs having stiffnesses k_1 and k_2. The stiffness coefficients are mathematically equivalent to the forces required at levels 1 and 2 to produce unit horizontal displacements relative to each level.

It is assumed that the floors and therefore the masses m_1 and m_2 are restrained to move in the direction x and that there is no damping in the system. Using Newton's second law of motion, the equations of dynamic equilibrium for masses m_1 and m_2 are given by

$$m_1\ddot{x}_1 = -k_1x + k_2(x_2 - x_1) \tag{3.58}$$

$$m_2\ddot{x}_2 = -k_2(x_2 - x_1) \tag{3.59}$$

Rearranging terms in these equations gives

$$m_1\ddot{x}_1 + (k_1 + k_2)x_1 - k_2x_2 = 0 \tag{3.60}$$

$$m_2\ddot{x}_2 - k_2x_1 + k_2x_2 = 0 \tag{3.61}$$

The solutions for the displacements x_1 and x_2 can be assumed to be of the form

$$x_1 = A\sin(\omega t + \alpha) \tag{3.62}$$

$$x_2 = B\sin(\omega t + \alpha) \tag{3.63}$$

where ω represents the angular frequency and α represents the phase angle of the harmonic motion of the two masses. A and B represent the maximum amplitudes of the vibratory motion. Substitution of Eqs (3.62) and (3.63) into Eqs (3.60) and (3.61) gives the following equations

$$(k_1 + k_2 - \omega^2m_1)A - k_2B = 0 \tag{3.64}$$

$$k_2A + \left(k_2 - \omega^2m_2\right)B = 0 \tag{3.65}$$

To obtain solution for the nontrivial case of A and $B \neq 0$, the determinant of the coefficients of A and B must be equal to zero, thus

$$\begin{bmatrix} (k_1 + k_2 - \omega^2m_1) & -k_2 \\ -k_2 & (k_2 - \omega^2m_2) \end{bmatrix} = 0 \tag{3.66}$$

Expansion of the determinant gives the relation;

$$(k_1 + k_2 - \omega^2m_1)(k_2 - \omega^2m_2) - k_2^2 = 0 \tag{3.67}$$

or

$$m_1m_2\omega^4 - m_1k_2 + m_2(k_1 + k_2)\omega^2 + k_1k_2 = 0 \tag{3.68}$$

Solution of this quadratic equation yields two values for ω^2 of the form

$$\omega_1^2 = \frac{-b + \sqrt{b^2 - 4ac}}{2a} \tag{3.69}$$

$$\omega_2^2 = \frac{-b - \sqrt{b^2 - 4ac}}{2a} \tag{3.70}$$

where $a = m_1 m_2$
$$b = - [m_1 k_2 + m_2(k_1 + k_2)]$$
$$c = k_1 k_2$$

As mentioned previously, the two frequencies ω_1 and ω_2 which can be considered as intrinsic properties of the system are uniquely determined.

The magnitudes of the amplitudes A and B cannot be determined uniquely but can be obtained in terms of ratios $r_1 = A_1/B_1$ and $r_2 = A_2/B_2$ corresponding to ω^2_1 and ω^2_2, respectively. Thus

$$r_1 = \frac{A_1}{B_1} = \frac{k_2}{k_1 + k_2 - \omega_1^2 m_1} \tag{3.71}$$

$$r_2 = \frac{A_2}{B_2} = \frac{k_2}{k_1 + k_2 - \omega_2^2 m_1} \tag{3.72}$$

The ratios r_1 and r_2 are called the amplitude ratios and represent the shapes of the two natural modes of vibration of the system.

Substituting the angular frequency ω_1 and the corresponding ratio r_1 in Eqs (3.62) and (3.63), we get

$$x_1' = r_1 B_1 \sin(\omega_1 t + \alpha_1) \tag{3.73}$$

$$x_2' = B_1 \sin(\omega_1 t + \alpha_1) \tag{3.74}$$

These expressions describe the first mode of vibration, also called the fundamental mode. Substituting the larger angular frequency ω_2 and the corresponding ratio r_2 in Eqs (3.62) and (3.63), we get

$$x_1'' = r_2 B_2 \sin(\omega_2 t + \alpha_2) \tag{3.75}$$

$$x_2'' = B_2 \sin(\omega_2 t + \alpha_2) \tag{3.76}$$

The displacements x_1'' and x_2'' describe the second mode of vibration. The general displacement of the system is obtained by summing the modal displacements, thus

$$x_1 = x_1' + x_1''$$
$$x_2 = x_2' + x_2''$$

Thus for systems having two degrees of freedom, we are able to determine the frequencies and mode shapes without undue mathematical difficulties. Although the equations of motions for multidegree systems have similar mathematical form, solutions for modal amplitudes in terms of geometrical coordinates become unwieldy. Use of orthogonal properties of mode shapes makes this laborious process unnecessary. We will demonstrate how the analysis can be simplified by using the modal superposition method. Consider again the equations of motion for the idealized 2-story building discussed in the previous section. As before, damping is neglected, but instead of free vibrations we will consider the analysis of the system subject to time-varying force functions F_1 and F_2 at levels 1 and 2. The dynamic equilibrium for masses m_1 and m_2 is given by

$$m_1 \ddot{x}_1 + (k_1 + k_2)x_1 - k_2 x_2 = F_1 \tag{3.77}$$

$$m_2 \ddot{x}_2 - k_2 x_1 + k_2 x_2 = F_2 \tag{3.78}$$

These two equations are interdependent because they contain both the unknowns x_1 and x_2. These can be solved simultaneously to get the response of the system, which was indeed the method used in the previous section to obtain the values for frequencies and mode shapes. Modal superposition method offers an alternate procedure for solving such problems. Instead of requiring simultaneous solution of the equations, we seek to transform the system of interdependent or coupled equations into a system of independent or uncoupled equations. Since the resulting equations contain only one unknown function of time, solutions are greatly simplified. Let us assume that solution for the above dynamic equations is of the form:

$$x_1 = a_{11} z_1 + a_{12} z_2 \tag{3.79}$$

$$x_2 = a_{21} z_1 + a_{22} z_2 \tag{3.80}$$

What we have done in the above equations is to express displacement x_1 and x_2 at levels 1 and 2 as a linear combination of properly scaled values of two independent modes. For example, a_{11} and a_{12}, which are the mode shapes at level 1, are combined linearly to give the displacement x_1. z_1, and z_2 can be looked upon as scaling functions. *Substituting for x_1 and x_2 and their derivatives $\ddot{x}_1$ and $\ddot{x}_2$ in the equilibrium Eqs (3.77) and (3.78) we get*

$$m_1 a_{11} \ddot{z}_1 + (k_1 + k_2)a_{11} z_1 - k_2 a_{21} z_1 + m_1 a_{12} \ddot{z}_2$$
$$+ (k_1 + k_2)a_{12} z_2 - k_2 a_{22} z_2 = F_1 \tag{3.81}$$

$$m_2 a_{21} \ddot{z}_1 - k_2 a_{11} z_1 + k_2 a_{21} z_1 + m_2 a_{24} z_2$$
$$- k_2 a_{12} z_2 + k_2 a_{22} z_2 = F_2 \tag{3.82}$$

We seek to uncouple Eqs (3.81) and (3.82) by using the orthogonality conditions. Multiplying Eqs (3.81) by a_{11} and Eqs (3.82) by a_{21} we get

$$m_1 a_{11}^2 \ddot{z}_1 + (k_1 + k_2)a_{11}^2 z_1 - k_2 a_{11} a_{21} z_1 + m_1 a_{11} a_{12} \ddot{z}_2$$
$$+ (k_1 + k_2)a_{11} a_{12} z_2 - k_2 a_{11} a_{22} z_2 = a_{11} F_1 \tag{3.83}$$

$$m_1 a_{21}^2 \ddot{z}_1 - k_2 a_{11} a_{21} z_1 + k_2 a_{21}^2 z_1 + m_2 a_{21} a_{22} \ddot{z}_2$$
$$- k_2 a_{12} a_{21} z_2 + k_2 a_{21} a_{22} z_2 = a_{21} F_2 \tag{3.84}$$

Adding the above two equations, we get

$$(m_1 a_{11}^2 + m_2 a_{21}^2)\ddot{z}_1 + \omega_1^2(m_1 a_{11}^2 + m_2 a_{21}^2)z_1 = a_{11}F_1 + a_{21}F_2 \tag{3.85}$$

Similarly, multiplying Eqs (3.81) and (3.82) by a_{12} and a_{22} and adding we obtain

$$(m_1 a_{12}^2 + m_2 a_{22}^2)\ddot{z}_2 + \omega_2^2(m_1 a_{12}^2 + m_2 a_{22}^2)z_2 = a_{12}F_1 - a_{22}F_2 \tag{3.86}$$

Equations (3.85) and (3.86) are independent of each other and are the uncoupled form of the original system of coupled differential equations. These can be further written in a simplified form by making use of the following abbreviations:

$$\begin{aligned} M_1 &= m_1 a_{11}^2 + m_2 a_{21}^2 \\ M_2 &= m_1 a_{12}^2 + m_2 a_{22}^2 \end{aligned} \tag{3.87}$$

$$\begin{aligned} K_1 &= \omega_1^2 M_1 \\ K_2 &= \omega_2^2 M_2 \end{aligned} \tag{3.88}$$

$$\begin{aligned} P_1 &= a_{11}F_1 + a_{21}F_2 \\ P_2 &= a_{12}F_1 + a_{22}F_2 \end{aligned} \tag{3.89}$$

M_1 and M_2 are called the generalized masses, K_1 and K_2 the generalized stiffnesses, and P_1 and P_2 the generalized forces.

Using these notations, each of the Eqs (3.85) and (3.86) takes the form similar to the equations of motion of a single-degree-of-freedom system, thus

$$M_1 \ddot{z}_1 + k_1 z_1 = P_1 \tag{3.90}$$

$$M_2 \ddot{z}_2 + k_2 z_2 = P_2 \tag{3.91}$$

The solution of these uncoupled differential equations can be found by any of the standard procedures given in textbooks on vibration analysis. In particular, Duhamel's integral provides a general method of solving these equations irrespective of the complexity of the loading function. However, in seismic analysis usually a response spectrum is available for the forcing function. Therefore, the maximum values of the response corresponding to each modal equation is obtained from the response spectrum. Direct superposition of modal maximum would, however, give only an upper limit for the total system which, in many engineering problems, would be too conservative. To alleviate this problem approximations based on probability considerations are generally employed. One method employs the so-called root mean square procedure, also called the square root of sum of the squares (SRSS) method. As the name implies, a probable maximum value is obtained by combining the square

root of the sum of the squares of the modal quantities. Although this method is simple and widely used, it is not always a conservative predictor of earthquake response because more severe combinations of modal quantities can occur, for example, when two modes have nearly the same natural period. In such cases a more appropriate combination of modal quantities, such as the Complete Quadratic Combination (CQC) is more appropriate.

An attempt has been made in this section to bring out the essentials of structural dynamics as related to seismic design of buildings. A certain amount of mathematical presentation has been unavoidable. Lest the reader lose the physical meaning of the various steps, it is worthwhile to summarize the essential features of dynamic analysis.

Dynamic analysis of high-rise buildings is accomplished by idealizing them as systems with multiple degrees-of-freedom. The dead load of the building together with a percentage of live load (estimated to be present during an earthquake) is modeled as a system of masses lumped at floor levels. In a planar analysis, each mass has one degree-of-freedom corresponding to lateral displacement in the direction under consideration, while in a three-dimensional analysis it has three degrees-of-freedom corresponding to two translational and one torsional displacement. Free-vibrations of the buildings are evaluated, without including the effect of damping. Damping is taken into account by modifying the design response spectrum. The dynamic model representing the building has a number of mode shapes equal to the number of degrees-of-freedom of the model. Mode shapes have the property of orthogonality, which means that no given mode shape can be constructed as a combination of others, yet any deformation of the dynamic model can be described as a combination of its mode shapes, each multiplied by a scale factor. Each mode shape has a natural frequency of vibration. The mode shapes and frequencies are determined by solving an eigenvalue problem. The total response of the building to a given response spectrum is obtained by summing a number of modal responses. The number of modes required to adequately determine the forces for design is a function of the dynamic characteristics of the building. Generally for tall, regular buildings, six to ten modes in each direction are considered sufficient. Since each mass responds to earthquakes in more than one mode, it is necessary to evaluate effective modal mass values. These values indicate the percentage of the total mass that is mobilized in each mode. The acceleration experienced by each mass undergoing various modal deformations is determined from the response spectrum, that has been adjusted for damping. Product of acceleration for a particular frequency multiplied by the effective modal mass gives the static equivalent of forces at each discrete level. Since these forces do not reach their maximum values simultaneously during an earthquake event, statistical methods are used to achieve the combinations. The resulting forces are used as design static forces.

3.8 SUMMARY

Since earthquakes can occur almost anywhere, some measure of earthquake resistance in the form of reserve ductility and redundancy should be built into the design of all structures to prevent catastrophic failures. The magnitude of inertial forces induced by earthquakes depends on the building mass, ground acceleration and the dynamic response of the structure. The shape and proportion of a building have a major effect on the distribution of earthquake forces as they work their way through the building. If irregular features are unavoidable, special design considerations are required to account for load transfer at abrupt changes in structural resistance.

Two approaches are recognized in modern codes for estimating the magnitude of seismic loads. The first approach, termed the ELFP uses a simple method to take into account the properties of the structure and the foundation material. The second is a dynamic analysis procedure in which the modal responses are combined in a statistical manner to find the maximum values of the building response. Note that the level of force experienced by a structure during a major earthquake is much larger than the forces usually employed in the design. By prescribing detailing requirements the structure is relied upon to sustain post-yield displacements without collapse.

The complex random nature of an accelerogram makes it necessary to employ a more general characterization of ground motion. The most practical representation is by earthquake response spectra to postulate the intensity and vibration content future ground motion at a given site. Duration of ground motion, although important, is not used explicitly in design criteria at the present time (2024).

Multistory buildings are analyzed as a MDOF system. They are represented by lumped masses attached at story intervals along the height of a vertically cantilevered pole. Each mode of the building system is represented by an equivalent SDOF system using the concept of generalized mass and stiffness. With the known period, mode shape, mass distribution and response spectrum, one can compute the deflected shape, story accelerations, forces and overturning moments. Each predominant mode is analyzed separately, and by using either the SRSS or CQC method, the peak modal responses are combined to give a reasonable value between an upper bound as the absolute sum of the modes and the lower bound as the maximum value of a single mode.

The time-history analysis technique represents the most sophisticated method of dynamic analysis for buildings. In this method, the mathematical model of the building is subjected to accelerations from earthquake records that represent the expected earthquake at the base of the structure. The equations of motion are integrated by using computers to obtain a complete record of acceleration, velocity, and displacement of each lumped mass. The maximum value is found by scanning the output record. Even with the availability of sophisticated computers, use of this method is restricted to the design of special structures such as nuclear facilities, military installations, and base-isolated structures.

In seismic design, nearly elastic behavior is interpreted as allowing some structural elements to slightly exceed specified yield stress on the condition that the elastic linear behavior of the overall structure is not substantially altered. For a structure with a multiplicity of structural elements forming the LFRS, the yielding of a small number of elements will generally not affect the overall elastic behavior of the structure if excess load can be distributed to other structural elements that have not exceeded their yield strength. For ductile framing systems the maximum number of moment-frame-beams with flexural overstresses should be limited to 20 percent of the beams in the direction of force on any story. The number of frame-columns with flexural overstress should be limited to 10 percent of the frame-columns in any story.

Although for new buildings, the ductile design approach is quite routine, seismic retrofitting of existing non-ductile buildings with poor confinement details is generally extremely expensive. Therefore, it is necessary to formulate an alternative method which attempts at a realistic assessment of damage resistance of the building. The method is based on the concept of "trade-off" between ductility and strength. In other words, structural systems of limited ductility may be considered valid in seismic design, provided they can resist correspondingly higher forces. The concept of an "inelastic demand ratio" is used to describe the ability of the structural elements to resist stresses beyond yield stress.

SELECTED REFERENCES

[1] Fanella, D. A. (2018), *Structural load Determination 2018 IBC and ASCE/SEI 7-16*, International Code Council (ICC).
[2] American Society of Civil Engineers (2017), *Minimum Design Loads and Associated Criteria for Buildings and Other Structures* (ASCE 7-16), ASCE, Reston, VA, USA.

[3] International Code Council. International Building Code (IBC 2018). Falls Church, VA, USA. International Code Council, 2018.

[4] Building Seismic Safety Council (BSSC) (2015), *NEHRP Recommended Seismic Provisions for New Buildings and Other Structures*, FEMA P-1050-1. Federal Emergency Management Agency, Washington, DC.

[5] Building Seismic Safety Council (BSSC) (2010), *Earthquake-Resistant Design Concepts—An Introduction to the NEHRP Recommended Seismic Provisions for New Buildings and Other Structures*, FEMA P-749. Federal Emergency Management Agency, Washington, DC.

[6] International Code Council (ICC) (2017), 2018 International Existing Building Code, Washington, D.C.

[7] Federal Emergency Management Agency (FEMA) (2009), *Quantification of Building Seismic Performance Factors*, P-695, Applied Technology Council, Washington, DC.

[8] Federal Emergency Management Agency (FEMA) (2011), *Quantification of Building Seismic Performance Factors: Component Equivalency Methodology*, P-795, Washington, DC.

[9] Chopra, Anil K. (2001), *Dynamics of Structures—Theory and Application to Earthquake Engineering*, 2nd Ed. Prentice Hall, NJ.

[10] American Society of Civil Engineers (ASCE) (1998), *Seismic Analysis of Safety-Related Nuclear Structures*, ASCE 4-8. Reston, VA.

[11] National Council of Structural Engineering Associations (NCSEA) (2009), *Guide to the Design of Diaphragms, Chords and Collectors Based on the 2006 IBC and ASCE/SEI 7-05*. International Code Council, Washington, DC.

[12] Mays, Timothy W. (2010). *Guide to the Design of Out-of-Plane Wall Anchorage Based on the 2006/2009 IBC and ASCE/SEI 7-05*. International Code Council, Washington, DC.

Chapter **4**

Lateral Systems: Steel Buildings

4.1 INTRODUCTION

4.1.1 Steel in High-Rise Buildings

Although the use of steel in structures can be traced back to 1856 when Bessemer's steel-making process was first introduced, its application to tall buildings received stimulus from the 984 ft (300 m) Eifel Tower, which was constructed in 1889. After the turn of the 19th century, several tall buildings, from the 286 ft (87 m) Flatiron Building in 1902 to the 1046 ft (319 m) Chrysler Building in 1929, were constructed in downtown Chicago and Manhattan. The height record was broken by the 1250 ft (381 m) Empire State Building in 1931, the old twin towers of the World Trade Center buildings at 1350 ft (412 m) in 1972 (collapsed in 2001), followed almost immediately by the 1450 ft (442 m) Willis Tower (older name Sears Tower) in Chicago in 1974.

The role of steel members, which in the early structures were relegated to carrying gravity loads only, has been completely upgraded to include wind and seismic resistance in systems ranging from the modest portal frame at one end of the spectrum, to innovative systems involving outrigger systems, mega frames, interior super-diagonally braced frames etc., at the other.

Today there are innumerable structural steel systems that can be used for the lateral bracing of tall buildings. It would be an exercise in futility to try to classify all these systems into distinct categories because there is no single criterion that can be used for a comprehensive cataloging of all systems. However, for purposes of presentation, the different structural systems currently used in the design of tall steel buildings are broadly divided into the following categories roughly based on their relative effectiveness in resisting the lateral loads.

- Frames with semirigid connections (4.2).
- Rigid frames (4.3).
- Braced frames (4.4).
- Staggered truss system (4.5).
- Eccentric bracing systems (4.6).
- Interacting system of braced and rigid frames (4.7).
- Outrigger and belt truss systems (4.8).
- Framed tube systems (4.9).
- Trussed tube system (4.10).
- The bundled tube (4.11).
- Ultimate High-Efficiency Structures (4.12).
- Diagrid (4.13)
- Superframe (4.14)

4.2 FRAMES WITH SEMIRIGID CONNECTIONS

4.2.1 Introduction

Semirigid connections, as the name implies, are those with rotational characteristics intermediate in degree between fully rigid and simple connections. These connections offer known rotational restraint at the beam ends resulting in a significant reduction in midspan gravity moments. However, they are not sufficiently rigid to prevent the entire rotation between the beam and the intersecting column.

Although several specifications such as the AISC, the British, and the Australian codes permit semirigid connections it has rarely been used because of the difficulty in predicting the rather complex response of these connections. However, reasonable success has been obtained by another type of partially rigid connection which the AISC designates as Type 2 wind connection, with similar provisions found in the British and Australian codes.

In the following sub-sections, a brief description of the behavior of three basic types of connections is given with particular emphasis on the design of Type 2 wind connection.

4.2.2 Review of Connection Behavior

Three basic types of construction and associated assumptions are permissible under the AISC specifications, and each governs in a specific manner the size of members and the types and strength of their connections.

Type 1. Commonly designated as "rigid-frame" (continuous frame), assumes that beam-to-column connections have sufficient rigidity to hold virtually unchanged the original angles between intersecting members.

Type 2. Commonly designated as "simple framing" (unrestrained, free-ended), assumes that, insofar as gravity loading is concerned, ends of beams and girders are connected for shear only and are free to rotate under gravity load.

Type 3. Commonly designated as "semirigid framing" (partially restrained), assumes that the connections of beams and girders possess a dependable and known moment capacity intermediate in degree between the rigidity of Type 1 and the flexibility of Type 2.

Connections in simple frames are typically designed to transfer vertical shear only. They are also designed for axial loads if they transfer chord and drag forces due to seismic loads. In either case, it is assumed that there is no bending moment at the connection. Connections in fully rigid frames on the other hand are called upon to develop resistance to both shear and bending moments. They are assumed to have sufficient rigidity to hold virtually unchanged the original angles between connecting members. Semirigid connection behavior is intermediate between the simple and rigid condition, its fixity varying anywhere from a low 5 percent to a high 90 percent of full-fifty.

Completely simple and completely rigid behavior are, of course, ideal conditions. Practically, it is necessary to accept something less than ideal, since real frames perform in the broad range between fully a rigid and simple

support condition. For example, consider the typical beam-to-column connection consisting of a double-angle web connection as shown in Fig. 4.1. The angles fastened to the beam web are usually considered completely flexible and are assumed to carry only shear. Actually, they offer some restraint to the moment and thus oppose the rotation at the beam end. The relationship between the applied moment and rotation is complex and can only be determined by experiment. When rotation under gravity load takes place, the upper part of the connection is in tension while the lower part is compressed against the column. The rotation is accommodated by the deformation of the angles. Therefore, to minimize the rotational restraint, the angles are made as thin as possible.

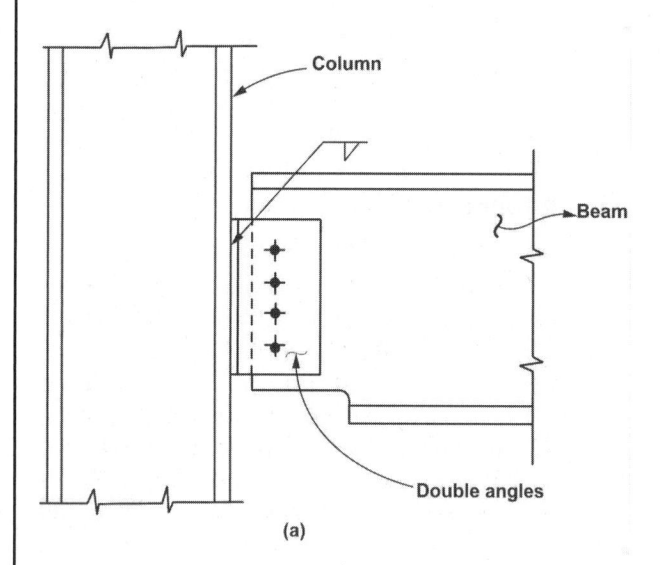

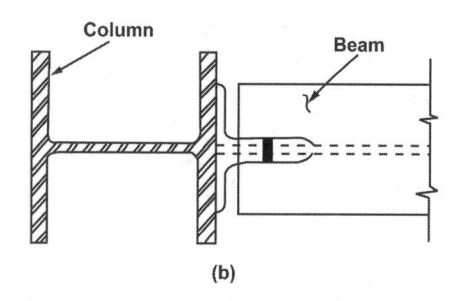

Figure 4.1 Beam-column bolted shear connection: (a) elevation view; (b) plan view.

Unstiffened seated beam connection shown in Fig. 4.2 is an example of a Type 2 connection. The behavior of the seat angle is shown schematically in Fig. 4.3. The bottom angle bends as a cantilever, except its bending is partially restrained by the bottom flange of the beam. The moment-rotation characteristics of a seat angle connection

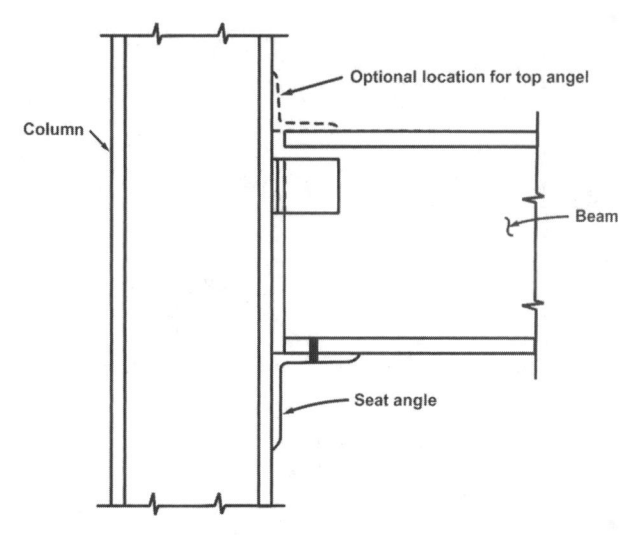

Figure 4.2   Unstiffened seated beam connection.

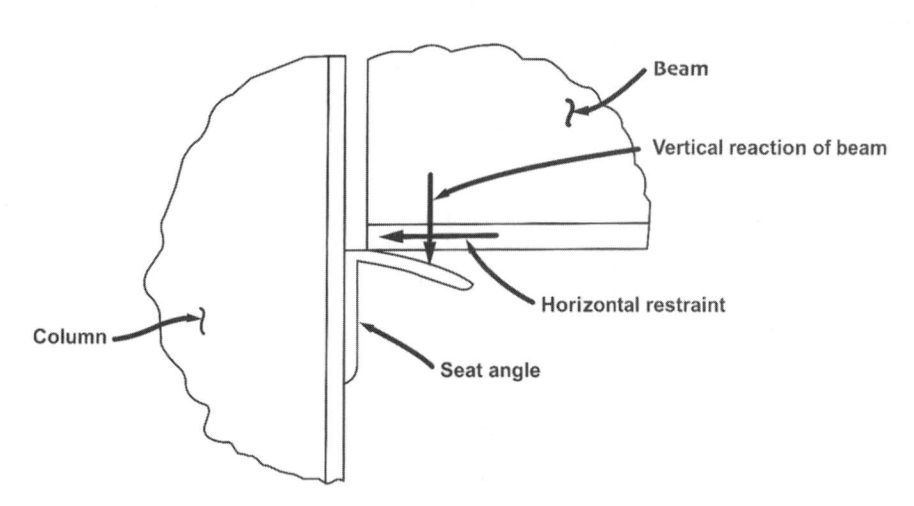

Figure 4.3   Cantilever bending of seat angle.

primarily depend on the beam depth, the thickness of top angle, the diameter of bolts, and the thickness of column flanges. The connection is typically stiffer than web angle connection but still considered to be a simple flexible connection. By adding top and bottom angles to the double-angle web connection, it is possible to develop a connection that has greater moment resistance than either of the previously described connections.

The top and bottom angles are assumed to carry the moment, and web angles the shear. Although the load distribution may appear to be arbitrary, such a division of function produces adequately proportioned connections. Structural tees used in place of top and bottom

angles increase the rotational restraint considerably. The increase occurs because the top tee is loaded in tension without any eccentricity, whereas the top angles are loaded eccentrically, resulting in large deformations.

4.2.3   Beam Line Concept
One of the methods for understanding the behavior of beam-to-column connection is to study a plot of moment-rotation characteristics, as shown in Fig. 4.4. The vertical axis shows the end moment. The resulting rotation is plotted along the horizontal axis in radians. Superimposed upon this plot is the so-called beam line, which expresses the resulting end moment M and rotation θ for

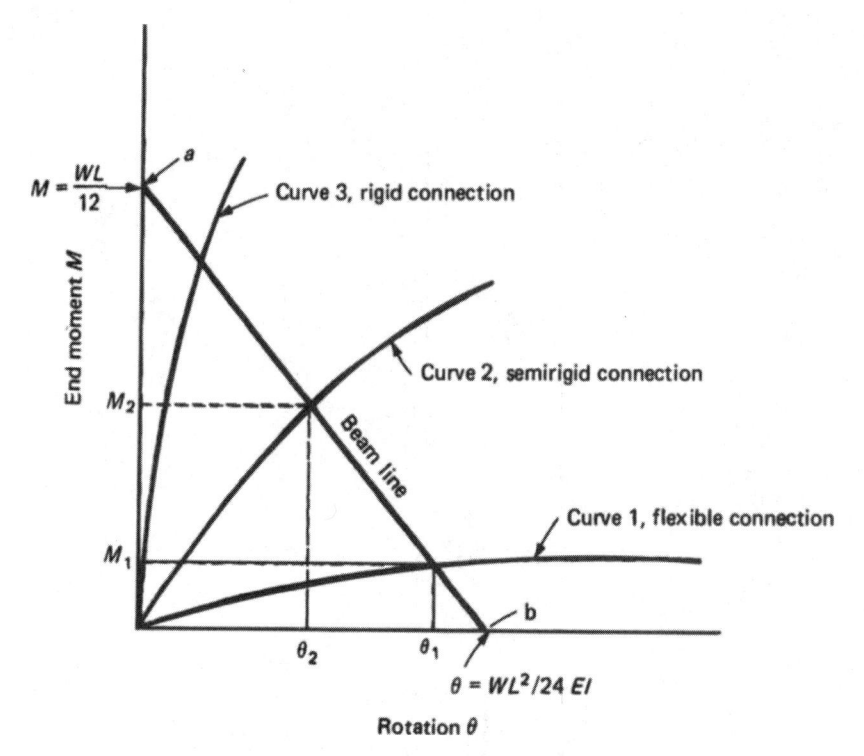

Figure 4.4 Beam line concept: moment-rotation (M–θ) curves.

a uniformly loaded beam for any end restraint, ranging from full fixity to simply supported condition.

The relation between end moment M and rotation θ can be expressed by the following equation:

$$M = -\frac{2EI\theta}{L} - \frac{WL}{12} \qquad (4.1)$$

This is a straight-line relationship and can be plotted by considering the rotation of a simply supported beam and the fixed end moment of a completely restrained beam. Point a on the beam line is the end moment when the connection is completely restrained. Thus, in Eq. (4.1), rotation $\theta = 0$, giving

$$M = -\frac{WL}{12} \qquad (4.2)$$

Point b is the rotation at the end of beam when the beam has zero restraint at the ends. In other words, the beam behaves as a simply supported beam. Substituting $M = 0$ in Eq. (4.1), we get

$$\theta = -\frac{WL^2}{24EI} \qquad (4.3)$$

The point at which the beam line intersects the connection line gives the resulting end moment and rotation

under the given load. The dependence of the beam behavior on the rigidity of the connection can be studied by using this diagram. In developing the M-θ relationship, it is assumed that the behavior of the two end connections is the same and that the beam is subjected to loads placed symmetrically on the beam. The behavior of all three types of connections, namely the flexible, the semirigid, and the rigid connection, can be studied by using the beam line diagram. Curve 1 represents a flexible connection which is typical of a double-angle web connection. Under a uniform load W, the beam ends rotate through an angle θ_1, which is very nearly equal to the rotation θ of a completely unrestrained beam. Corresponding to this rotation, a moment M is generated at the ends signifying that even with the so-called flexible connection, some end moment is generated. Normally the bending moment developed is about 5 to 20 percent of the fully fixed moment.

Curve 2 represents a semirigid connection such as an end plate connection, Fig. 4.5, detailed in such a way that under working loads it elastically yields to provide the necessary rotation of the connection. Although the beam is detailed to undergo a rotation equal to θ_2, significant moment M_2 corresponding to the rotation θ_2 develops at the beam ends. The restraint offered by this type of connection can vary anywhere from a low of 20 percent to

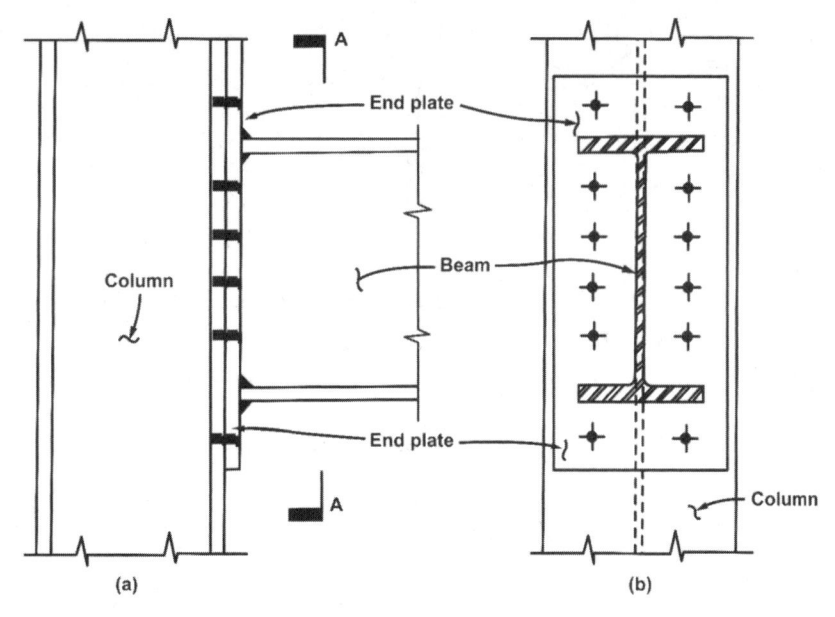

Figure 4.5 Beam-column end-plate connection: (a) elevation; (b) section A-A, side view.

a high of 90 percent of the full fixity. The resulting end moments could be 20 to 90 percent of the moment generated in a fully fixed beam.

Curve 3 represents the moment-rotation characteristics of a rigid connection such as a welded beam-to-column connection shown in Fig. 4.6. The beam develops end moments which are about 90 to 95 percent of fully fixed condition, especially when column flange stiffeners are used as in Fig. 4.6.

Wind moments applied to flexible and semirigid connections present some very intriguing problems because some means of transferring these moments must be provided in these connections, which are supposed to be flexible. Additional restraint provided for carrying the wind load will result in an increase in the end moment due to gravity loads. A rigorous mathematical solution of flexible and semirigid connections is not possible, but based on the performance of buildings using these types of connections, AISC, in older versions of the specifications, provided for two approximate solutions. In the first method, the connection is designed for the moment caused by the combination of gravity and wind loads using a one-third increase in the allowable stresses. In the second method, the connection is designed for the moment induced by wind loads only, using a one-third increase in the stress allowances. The connection must, however, be designed to yield plastically for any combination of gravity and wind moments. Any additional moment that could occur at the ends beyond the

wind moments is relieved because of the yielding of the connection. This type of connection necessitates some inelastic but self-limiting deformation of the connection components without overstressing the fasteners. The *self-limiting deformation criteria* are imposed to prevent the use of semirigid connections for cantilever beams. This is because cantilever beam deflections are not limited by the rotation of the connections but may continue to progress, resulting in failure of the connection, and even the beam itself.

Although the AISC specification permitted the designer to take advantage of the reduction in the midspan moment of a beam with semirigid connections, in practice this procedure has not found wide acceptance primarily because of a lack of reliable analytical techniques. The Type 2 wind connection, which basically ignores the beam restraint for gravity loads, has found relatively greater acceptance. The behavior of the Type 2 wind connection is considered next.

4.2.4 Type 2 Wind Connections

Although the design of Type 2 wind connections is empirical in its approach, many significant tall buildings have been built with this technique, the most notable example being the Empire State Building, for many years the world's tallest. It must be pointed out; however, that significant stability and stiffness are incorporated into the structure by the exterior stone cladding and interior braces. Other major buildings that have used Type 2

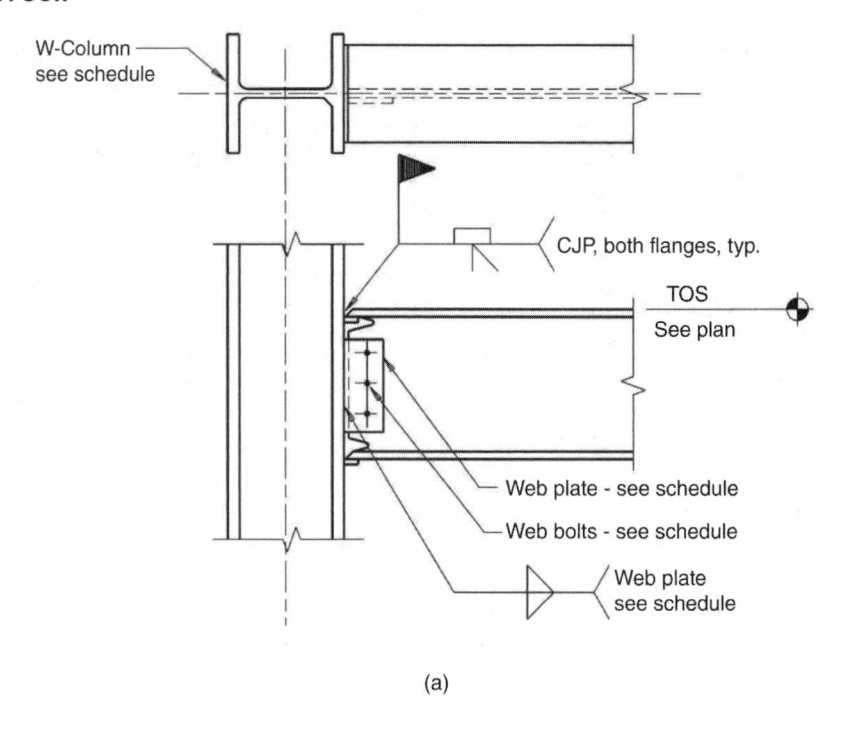

(a)

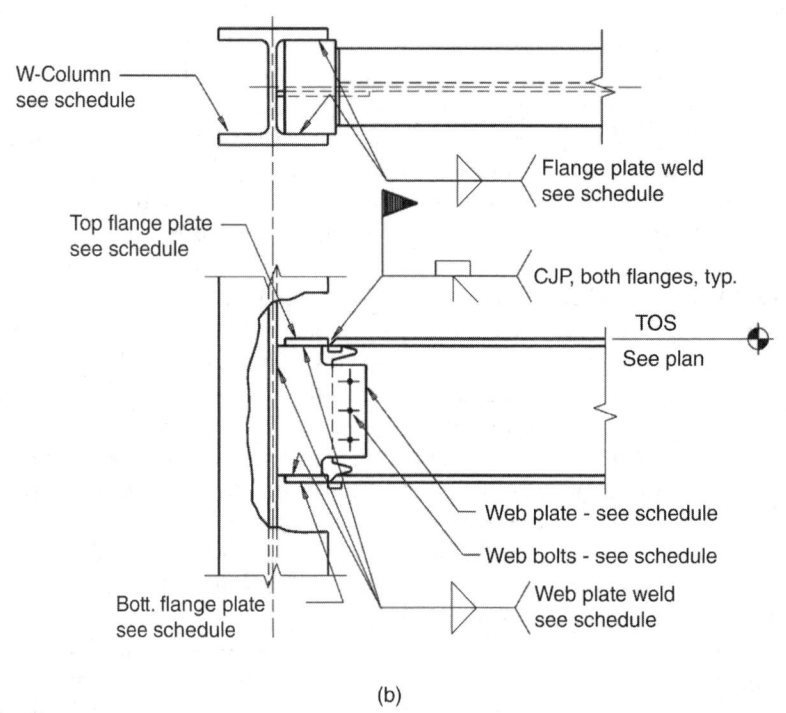

(b)

Figure 4.6 Beam-column welded moment connections: (a) column strong axis connection; (b) column weak axis connection. (Source: AISC Detailing Manual, 3rd Edition.)

construction are the United Nations Secretariat Building and the Chrysler Building, both in New York, and the Alcoa Building in Pittsburgh.

Design of Type 2 wind connections for tall buildings is based on the practice of ignoring beam end moments generated by a connection's resistance to gravity load while counting on the same connection to resist wind moments calculated on the assumption of fully rigid behavior. This is a time-tested procedure and it is safe provided the actual end moment, which can be higher than the design moment, does not overstress the fasteners. Connections designed under this procedure are generally semirigid with components which, by deforming inelastically, prevent fastener distress. Since it is difficult to calculate the true combined moment at the connections, reliance must be placed on joint configurations of demonstrable ductility. Thus the Type 2 wind connection can be defined as a type of connection that develops lateral resistance through special wind connections which provide some restraint to the ends of a simple beam designed for gravity loads only. Basic to the Type 2 wind design is the requirement that the connection should have adequate inelastic deformation capacity to avoid connector overstress under full loading.

To understand the Type 2 wind connection, it is instructive to trace its behavior through a complete loading sequence, starting from the gravity loads to reversible wind loadings. Figure 4.4 shows a moment-rotation curve for a beam with a known moment-rotation characteristic. The shape of the curve depends on the design of the connection and may range from almost full fixity of Type 1 rigid frame to Type 2 simple framing. In Fig. 4.4, point 1 at the intersection of the connection curve and the beam line ab represents the gravity moment M_1 and the corresponding rotation θ_1 at the ends due to a uniformly distributed load w. The end moment at the left-hand side of the girder is counterclockwise, while the end moment on the right-hand side is clockwise as shown in the free-body diagram of the beam in Fig. 4.7. The beam can be considered as a typical beam of a moment frame as shown in Fig. 4.7. The intersection points L_1 and R_1 in Fig. 4.9 represent the application of vertical load only to the beam; there is no wind moment acting on the connection or the beam. Assume that the frame is subject to wind loads acting from left to right, as represented in Fig. 4.8a. The beam is subjected to end moments which act in a clockwise direction as shown in the free-body diagram of the beam in Fig. 4.8b. The windward or the left-end moment acts in a direction opposite to the gravity moment, while the right-end or the leeward moment acts in the same sense as the gravity moment. The gravity

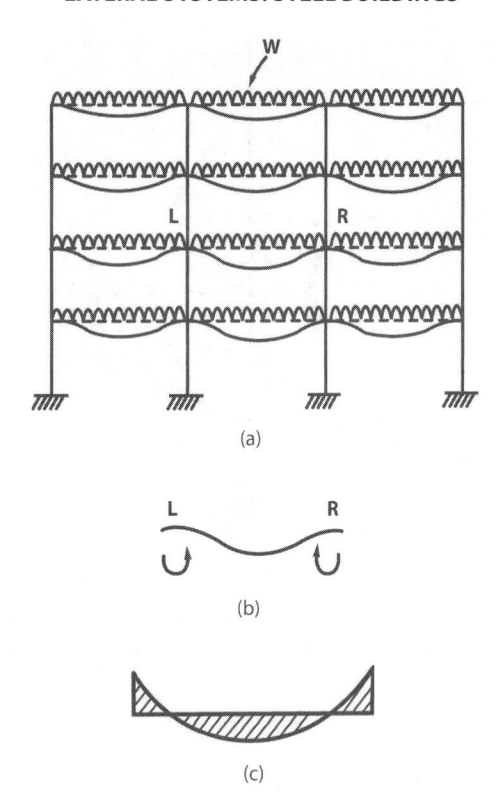

Figure 4.7 (a) Portal frame subjected to gravity loads; (b) end moments; (c) moment diagram.

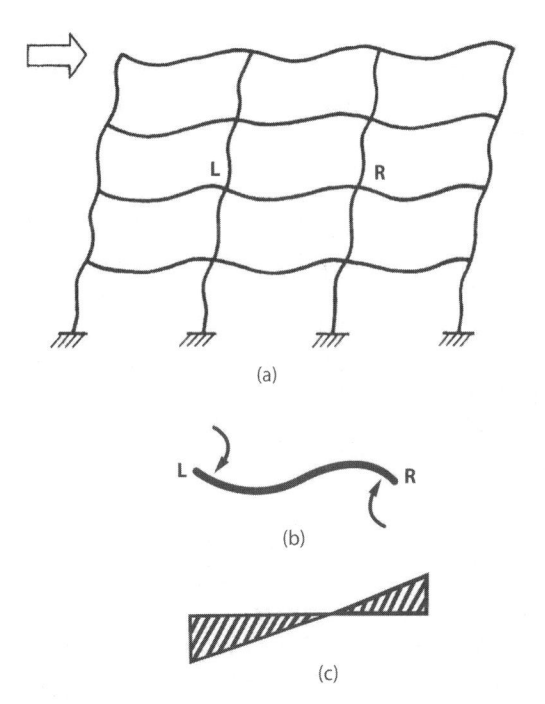

Figure 4.8 (a) Portal frame subjected to lateral loads from left; (b) end moments; (c) moment diagram.

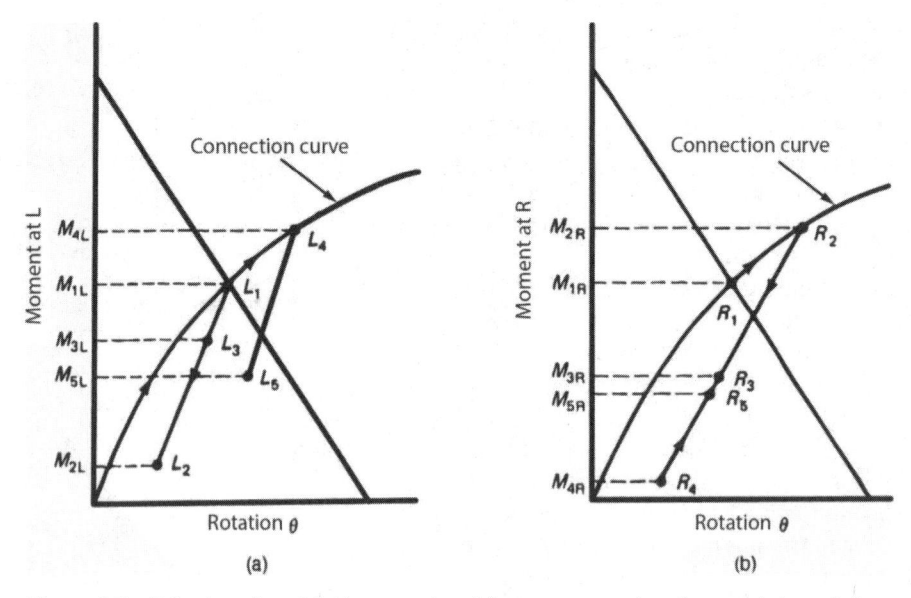

Figure 4.9   Behavior of semirigid connection: (a) moment-rotation characteristics at left support; (b) moment-rotation characteristics at right support.

moment at the left end is relieved by the wind action, while the right-end gravity moment becomes additive to the wind moment. Because of the additional moment, the right-end connection moves from the original point R_1 to R_2 along the connection curve. The left-end moment moves downward from L_1 to point L_2 because of the reduction in gravity moment. The windward moment does not retrace its path along the M-θ curve but travels on a line parallel to the slope of the curve. Recall that this characteristic is similar to the inelastic stress-strain diagram of a material such as steel subjected to load reversals. The decrease in moment occurs along a straight line because the moment is entirely elastic in this region. The rotations at the ends of the beam are, however, the same. The beam rotates by the same amount until the entire wind moment is developed between points R and L as illustrated in Fig. 4.9.

Since wind is a transient load, we have to consider the condition when it stops acting on the structure. The only loads on the beam are gravity loads similar to the condition that we started with, except the moment at the ends will not revert back to M_1 because during the loading cycle the connection has undergone inelastic rotations. The left-end connection goes from L_2 to L_3 and the right-end connection goes from R_2 to R_3. Both the connections move on an elastic line parallel to the initial slope. Static equilibrium requires that the moment at each end be the same. The resulting moment is, however, less than M_1.

Consider the wind now acting in the right-to-left direction as shown in Fig. 4.10a. The left-end connection

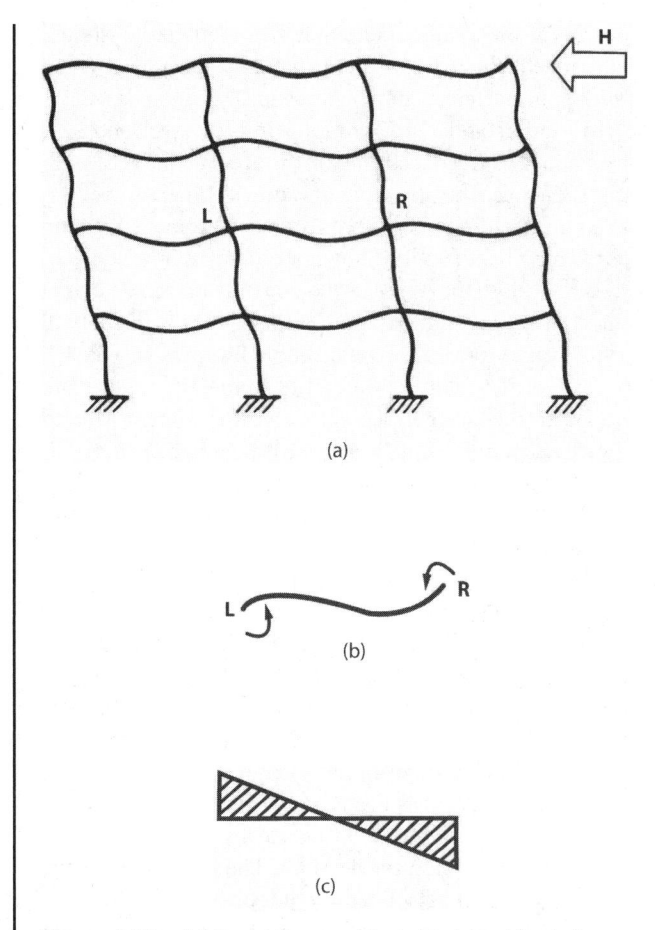

Figure 4.10   (a) Portal frame subjected to lateral loads from the right; (b) end moments; (c) moment diagram.

receives additional moment and travels from L_3 to L_4. Part of this rotation is inelastic and part is elastic as shown in Fig. 4.9a. The right-end connection is partially relieved of the moment and moves down to point R_4 as shown in Fig. 4.9b. Both the connections rotate through the same angle until the entire moment is developed. Figure 4.9a, b shows the behavior when the wind stops and the only loads are gravity loads. The connections move from R_4 to R_5 and L_4 to L_5. Both ends have the same moment in order to satisfy static equilibrium. The moments corresponding to R_5 and L_5 are considerably less than the gravity moment at points R_1 and L_1 under gravity loads only. This phenomenon of reduction in moment due to cyclic loading is referred to as "shake-down." The gravity moment, therefore, has shaken down considerably. From this point on, the connection behaves entirely as an elastic connection. Regardless of the direction of the wind, the maximum moment on the connection can never exceed points L_5 and R_5. The connection is said to have undergone a complete shakedown. After shakedown, the maximum moment occurring at the connection is considerably smaller than the original gravity movement. A brief summary of the connection behavior is given in Table 4.1.

One of the drawbacks for the not-so-wide use of semi-rigid frames is the lack of sufficient information available to the designers about the moment-rotation relationship. The moment-rotation characteristics of a connection depend upon many physical parameters such as type of connection, the size of angles, end plates, top and bottom angles, and gauge for bolt location. In short, an exact relationship can only be obtained by conducting experiments. However, Lothers in his text *Advanced Design in Structural Steel*, has presented formulas for a parameter Z to define the relationship between moment and rotation for the connections.

The factor Z for a connection is analogous to the flexibility factor $L/4EI$ which corresponds to the rotation at the end of a beam upon the application of a unit moment. The slope of the moment-rotation relation defines the parameter Z. Although the moment-rotation characteristics for most types of connections are nonlinear for the full spectrum of elastic and inelastic deformations, their behavior in the design range can be considered elastic. The reciprocal of the initial tangent to the moment-rotation curve can be considered as sufficiently accurate for determining the value of Z.

Approximate values of Z are given for four types of connections by Defalco and Marino in the 1960 AISC Journal. These are reproduced here, by permission, in Figs. 4.11 to 4.14. Type A connection consists of a double-angle connection as shown in Fig. 4.11, while Type B,

Table 4.1   Summary of Connection Behavior

Load case		Moment at L		Moment at R		Connection behavior
1. Gravity	W		M_{1L}		M_{1R}	Equal gravity moments at each end
2. Gravity plus wind from left	$W + H$		M_{2L}		M_{2R}	L unloads elastically while R loads along connection curve
3. Remove wind from left	W		M_{3L}		M_{3R}	Elastic recovery of $M_{H/2}$ at both connections
4. Gravity plus wind from right	$W - H$		M_{4L}		M_{4R}	L loads elastically up to and then along the connection curve. R unloads elastically
5. Remove wind from right	W		M_{5L}		M_{5R}	Elastic recovery of $M_{H/2}$ at both connections
6. Gravity plus wind from left	$W + H$		$M_{6L} = M_{5L} - \dfrac{M_H}{2}$		$M_{6R} = M_{5R} + \dfrac{M_H}{2}$	L unloads elastically by $M_{H/2}$ while R loads elastically by $M_{H/2}$
7. Remove wind from left	W		M_{5L}		M_{5R}	Elastic response at both ends. The connections have "shaken down" with the gravity moments considerably smaller than the initial gravity moments
8. Gravity plus wind from right	$W - H$		$M_{5L} + \dfrac{M_H}{2}$		$M_{5R} - \dfrac{M_H}{2}$	
9. Gravity plus wind from left	$W + H$		$M_{5L} - \dfrac{M_H}{2}$		$M_{5R} + \dfrac{M_H}{2}$	

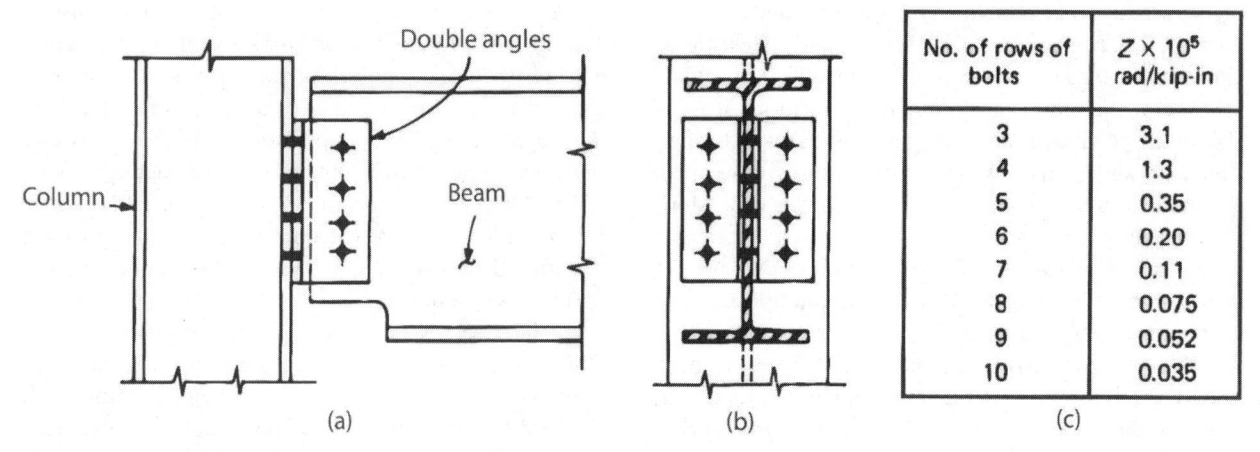

No. of rows of bolts	$Z \times 10^5$ rad/kip-in
3	3.1
4	1.3
5	0.35
6	0.20
7	0.11
8	0.075
9	0.052
10	0.035

Figure 4.11   Double-angle connection (Type A): (a) elevation; (b) side view; (c) values of stiffness factor Z.

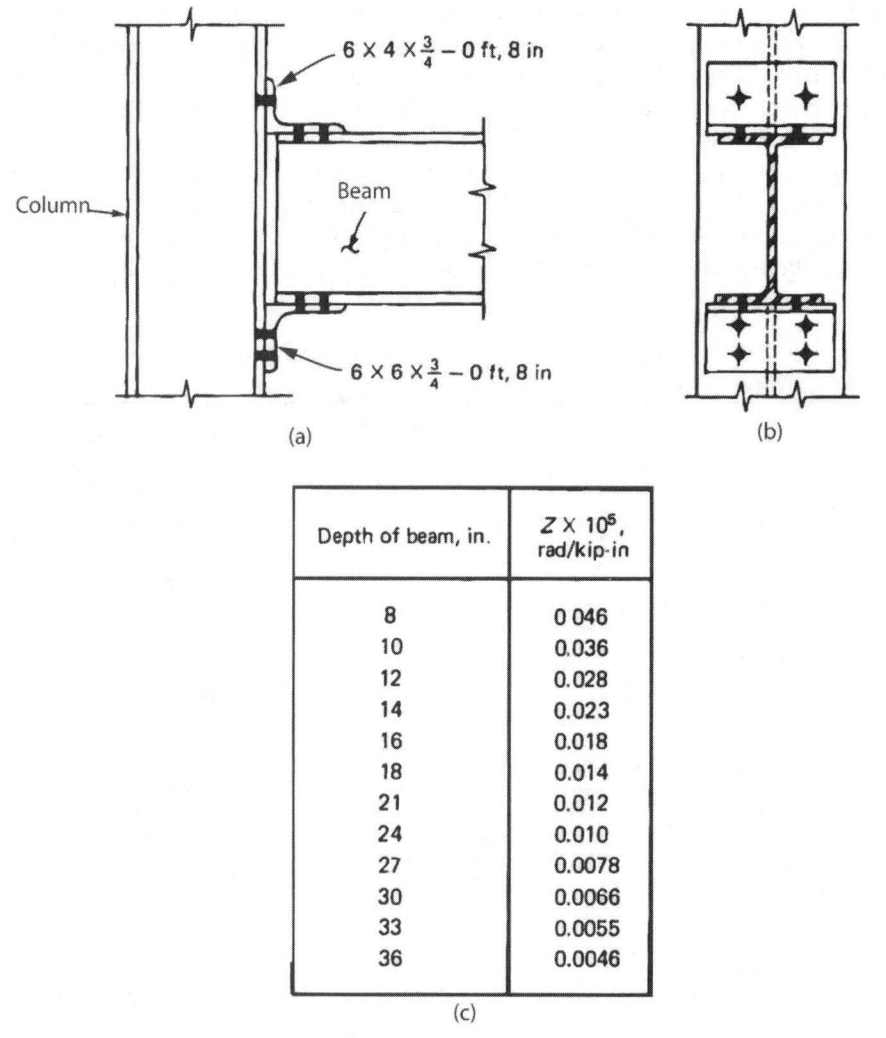

Depth of beam, in.	$Z \times 10^5$, rad/kip-in
8	0 046
10	0.036
12	0.028
14	0.023
16	0.018
18	0.014
21	0.012
24	0.010
27	0.0078
30	0.0066
33	0.0055
36	0.0046

(c)

Figure 4.12   Top and bottom clip angle connection (Type B): (a) elevation; (b) side view; (c) values of stiffness factor Z.

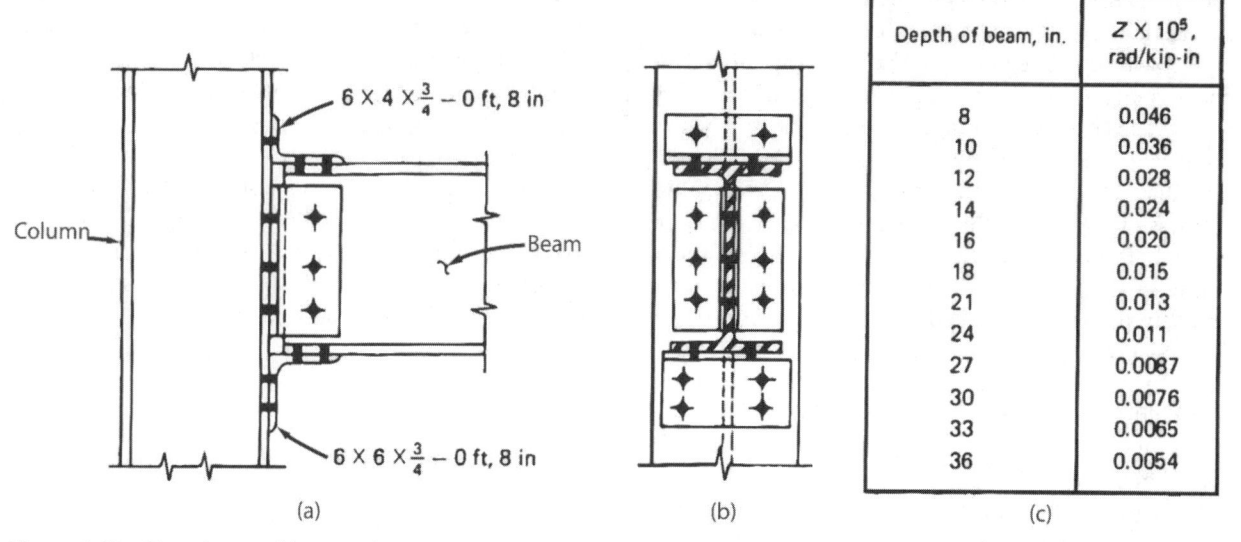

Depth of beam, in.	$Z \times 10^5$, rad/kip-in
8	0.046
10	0.036
12	0.028
14	0.024
16	0.020
18	0.015
21	0.013
24	0.011
27	0.0087
30	0.0076
33	0.0065
36	0.0054

(a)   (b)   (c)

Figure 4.13   Type C top and bottom clip angle connection with shear plate: (a) elevation; (b) side view; (c) values of stiffness factor Z.

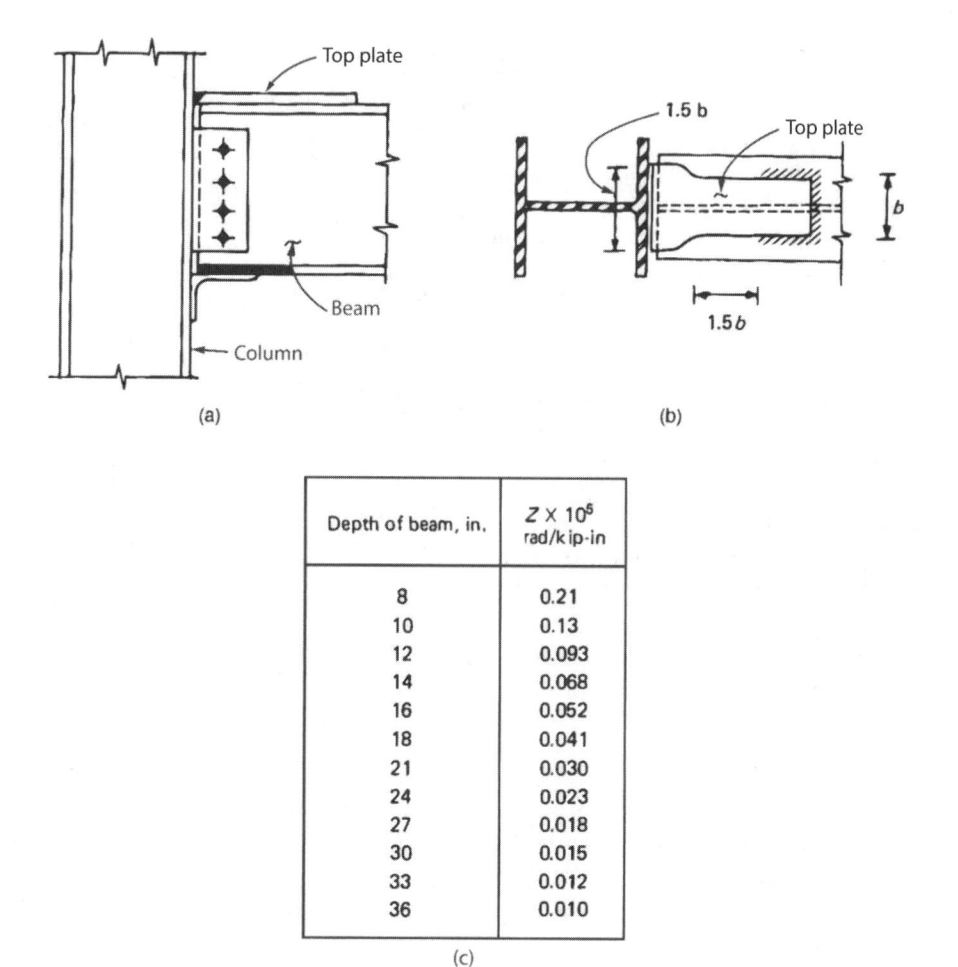

(a)   (b)

Depth of beam, in.	$Z \times 10^6$ rad/kip-in
8	0.21
10	0.13
12	0.093
14	0.068
16	0.052
18	0.041
21	0.030
24	0.023
27	0.018
30	0.015
33	0.012
36	0.010

(c)

Figure 4.14   Type D seat angle with top plate connection: (a) elevation; (b) plan; (c) values of stiffness factor Z.

shown in Fig. 4.12, consists of a top and bottom clip angle connection. Type C connection in Fig. 4.13 is similar to type B with the exception that the shear capacity of the beam is augmented by bolting two angles to the beam web. Type D connection is a seat angle connection with a top plate connection as shown in Fig. 4.14. More important than the ultimate moment capacity is the initial stiffness of the curve, which is represented by the slope of the M-θ curve at the origin. The connection rotational stiffness, designated as Z, represents a zero value for a perfectly pin-ended connection and infinity for a fully fixed beam.

Analyses of frames that incorporate Type 2 wind and semirigid (Type 3) connections must include considerations of:

1. Connection ductility.

2. Evaluation of the drift characteristics of frames with less than fully rigid connections.

3. Effect of partial restraints on column and frame stability.

4.2.4.1 Design Outline for Type 2 Wind Connections

1. Calculate the end reaction due to gravity loads.

2. Determine the magnitude of wind moment by an elastic analysis, assuming rigid connections.

3. Select the type of semirigid connection from an inventory of standard connections, such as double-angle, single-angle, shear tab, top-and-bottom angle, or header plate connection. Make an initial guess on the various dimensions and thicknesses for connection components.

4. Determine the moment-rotation characteristics of the connection to evaluate the Z value. Since the available data do not cover all types of connections, it is advisable to restrict connection designs to those for which M-θ curves have been well established. The AISC, however, allows the use of any connection as long as its behavior can be demonstrated by either tests or by a rational analysis.

5. With the known value of Z for the connection, plot the moment-rotation curve on the beam line.

6. Check the connection for wind moment by calculating the design values for each component such as bolts, connecting angles, welds, etc.

7. Check the connection for ductility by verifying that all connection materials such as bolts, welds, and plates are not stressed beyond their ultimate strengths (with proper safety factors) under the simultaneous action of gravity and wind moments.

8. Since the connections were assumed fully rigid in the wind analysis, Steps 1 to 7 yield a conservative

design. If required, the connection design can be modified by incorporating the effect of nonrigid connections in the wind analysis. This can be done by using a reduced bending rigidity for the beam to account for less than 100 percent rigidity of the connection.

Iteration of Steps 2 to 8 gives results that will converge to an optimum solution. Since semirigid connections provide less frame stiffness than fully rigid connections, it is necessary to use a modified girder stiffness, k_r which incorporates the Z factor as follows:

$$K_r = \frac{3(I/L)}{4(L'/L) - (L/L')} \qquad (4.4)$$

where K_r = modified relative stiffness of beam

$$L' = L + \frac{3EI}{Z}$$

L = beam span
I = moment of inertia of the beam

Although the use of reduced stiffness of the beam in the determination of wind moments is optional, it is important that reduction in stiffness be accounted for in determining the lateral drift and p-Δ effects.

4.2.5 Concluding Remarks

In spite of reported success of many buildings built by using Type 2 wind connections, there exists little unanimity of opinion about its applicability for buildings taller than five stories or so. Engineers who design buildings using Type 2 wind connections are automatically put in a defensive position of explaining the paradox of the joints acting as rigid for wind, and as pins for gravity loading. A straightforward application of the method for frames can result in structures in which the columns are overstressed and exhibit sway deflections much in excess of calculated values. Other cautious approaches that take into consideration the relative softness of connections have been proposed by several investigators but have not found general application in the design office.

4.3 RIGID FRAMES (MOMENT FRAMES)

4.3.1 Introduction

A frame is considered rigid when its beam-to-column connections have sufficient rigidity to hold virtually unchanged the original angles between intersecting members. A rigid-frame high-rise structure typically comprises parallel or orthogonally arranged bents consisting of columns and girders with moment-resistant joints. Resistance to horizontal loading is provided by the bending resistance of the columns, girders, and joints. The continuity of the frame also assists in resisting gravity loading more efficiently by reducing the positive moments in the center span of girders.

Typical deformations of a moment-resisting frame under lateral load are indicated in Fig. 4.15. The point of contraflexure is normally located near the midheight of the columns and midspan of the beams. The lateral deformation of a frame as will be seen shortly, is due partly to frame racking, which might be called shear sway, and partly to column shortening. The shear-sway component constitutes approximately 80 to 90 percent of the overall lateral deformation of the frame. The remaining component of deformation is due to column shortening, also called cantilever or chord drift component.

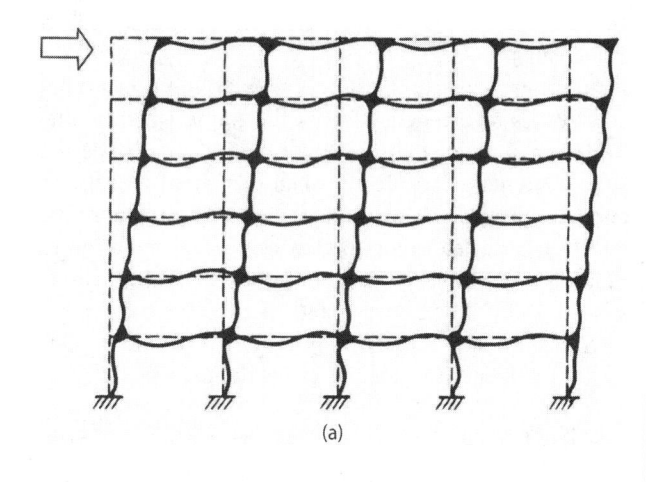

(a)

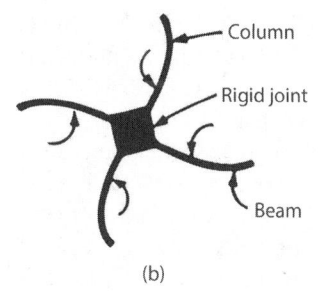

(b)

Figure 4.15   (a) Response of rigid frame to lateral loads; (b) flexural deformation of beams and columns due to non-deformability of connections.

Moment-resisting frames have advantages in high-rise construction due to their flexibility in architectural planning. They may be placed on the exterior, or throughout the interior of the building with minimal constraint on the architectural planning.

The size of members in a moment-resisting frame is often controlled by stiffness rather than strength to control drift under lateral loads. The lateral drift is a function of both the column stiffness and beam stiffness. In a typical application, the beam spans are 20 ft (6 m) to 30 ft (9 m) while the story heights are usually between 12 ft (3.65 m) to 14 ft (4.27 m). Since the beam spans are greater than the floor heights, the beam moment of inertia needs to be greater than the column inertia by the ratio of beam span to story height for an effective moment-resisting frame.

Moment-resisting frames are normally efficient for buildings up to about 30 stories in height. The lack of efficiency for taller buildings is due to the moment resistance derived primarily through flexure of its members.

The connections in steel moment-resisting frames are important design elements. Joint rotation can account for a significant portion of the lateral sway. The strength and ductility of the connection are also important considerations especially for frames designed to resist seismic loads.

Steel moment-resisting frames with welded connections have been regarded up until the January 1994 Northridge earthquake, as one of the safest for having the required strength, ductility, and reliability.

The Northridge, magnitude 6.7 earthquake which caused damage to over 200 steel moment-resisting frame buildings has shaken engineers' confidence in its use for seismic design. Almost without exception, the connections that failed were of the type with full penetration field weld of top and bottom flanges, and a high-strength bolted shear tab connection. The majority of the damage consists of fractures of the bottom flange weld between the column and girder flanges. There were also a large number of instances where top flange fractures occurred. Although many factors may have contributed to the poor performance it is believed that the basic joint configuration is not conducive to ductile behavior.

4.3.2   Deflection Characteristics

The lateral deflection components of a rigid frame can be thought of as being caused by two components similar to the deflection components of a prismatic cantilever beam. One component can be likened to the bending deflection and the other to the shear deflection. Normally for prismatic members when the span-to-depth ratio is greater than 10 or so, the bending deflection is by far the more predominant component. Shear deflections contribute a small portion to the overall deflection and are therefore generally neglected in calculating deflections.

The deflection characteristics of a rigid frame, on the other hand, are just the opposite; the component analogous to the beam shear deflection dominates the deflection picture and may amount to as much as 80 percent of the total deflection, while the remaining 20 percent comes from the bending component. The bending and the shear components of deflection are usually referred to as the cantilever bending and frame racking, each with its own distinct deflection mode.

4.3.2.1 CANTILEVER BENDING COMPONENT

This phenomenon is also known as *chord drift*. The wind load acting on the vertical face of the building causes an overall bending moment on any horizontal cross-section of the building. This moment, which reaches its maximum value at the base of the building, causes the building to rotate about the leeward column and is called the *overturning moment*. In resisting the overturning moment, the frame behaves as a vertical cantilever responding to bending through the axial deformation of columns resulting in compression in the leeward columns and tension or uplift in the windward columns. The columns lengthen on the windward face of the building and shorten on the leeward face. This column length change causes the building to rotate and results in the chord drift component of the lateral deflection, as shown in Fig. 4.16a.

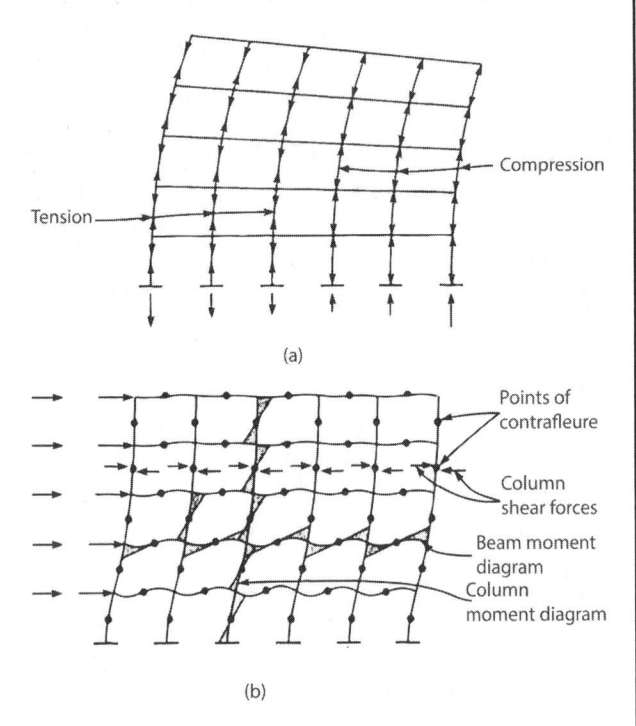

Figure 4.16 Rigid frame deflections: (a) forces and deformations caused by external overturning moment; (b) forces and deformations caused by external shear.

Because of the cumulative rotation up the height, the story drift due to overall bending increases with height, while that due to racking tends to decrease. Consequently the contribution to story drift from overall bending may, in the uppermost stories, exceed that from racking. The contribution of overall bending to the total drift, however, will usually not exceed 10 to 20 percent of that of racking, except in very tall, slender, rigid frames. Therefore, the overall deflected shape of medium-rise frame usually has a shear configuration.

For a normally proportioned rigid frame, as a first approximation, the total lateral deflection can be thought of as a combination of three factors:

1. Deflection due to axial deformation of columns (15 to 20 percent).

2. Frame racking due to beam rotations (50 to 60 percent).

3. Frame racking due to column rotations (15 to 20 percent).

In addition to the above, there is a fourth component that contributes to the deflection of the frame which is due to deformation of the joint. In a rigid frame, since the sizes of joints are relatively small compared to column and beam lengths, it is a common practice to ignore the effect of joint deformation. However, its contribution to building drift in very tall buildings consisting of closely spaced columns and deep spandrels could be substantial, warranting a closer study. This effect is called *panel zone deformation* and is discussed at length in Chap. 11.

4.3.2.2 SHEAR RACKING COMPONENT

This phenomenon is analogous to the shear deflection in a beam and is caused in a rigid frame by the bending of beams and columns. The accumulated horizontal shear above any story of a rigid frame is resisted by shear in the columns of that story (Fig. 4.16b). The shear causes the story-height columns to bend in double curvature with points of contraflexure at approximately mid-story-height levels. The moments applied to a joint from the columns above and below are resisted by the attached girders, which also bend in double curvature, with points of contraflexure at approximately mid-span. These deformations of the columns and girders allow racking of the frame and horizontal deflection in each story. The overall deflected shape of a rigid frame structure due to racking has a shear configuration with concavity upwind, a maximum inclination near the base, and a minimum inclination at the top, as shown in Fig. 4.16b. This mode of deformation accounts for about 80 percent of the total sway of the structure. In a normally proportioned rigid-frame building with column spacing at about 35 to 40 ft

(10.6 to 12.2 m) and a story height of 12 to 13 ft (3.65 to 4.0 m), beam flexure contributes about 50 to 65 percent of the total sway. The column rotation, on the other hand, contributes about 10 to 20 percent of the total deflection. This is because in most unbraced frames the ratio of column stiffness to girder stiffness is very high, resulting in larger joint rotations of girders. So generally when it is desired to reduce the deflection of unbraced frame, the place to start adding stiffness is in the girders. However, in nontypical frames, such as those that occur in framed tubes with column spacing approaching floor-to-floor height, it is necessary to study the relative girder and column stiffness before making adjustments in the member properties.

4.3.3   Methods of Analysis

Because of the large-scale availability of computers, the analysis of rigid-frame buildings, even in preliminary design stages, is accomplished most effectively by using stiffness analysis and finite element programs. Hand calculations are rarely undertaken except for very preliminary purposes because of the approximate nature of analysis and the longer time it takes to do hand computations.

Among the better-known approximate methods of analysis, the cantilever and portal methods are perhaps the most popular and considered by some engineers to be sufficiently accurate for use in the final analysis of buildings of intermediate height range. The portal method is considered reasonably valid for buildings less than 25 stories, and the cantilever method is assumed to be valid for buildings in the 25- to 30-story range. A brief description of each method is given in Chap. 10.

4.3.4   Calculation of Drift

Calculation of drift due to lateral loads is a major task in the analysis of tall building frames. Although it is convenient to consider the lateral displacements to be composed of two distinct components, whether or not the cantilever or the racking component dominates the deflection is dependent on factors such as height-to-width ratio of the building and the relative rigidity of the column-to-girder connection. Unless the building is very tall or very slender, it is usually the racking component that dominates the deflection picture. A simple method for determining the deflection of a tall building is to assume that the entire structure acts as a vertical cantilever in which the axial stress in each column is proportional to its distance from the centroidal axis of the frame. This approach assumes that the frame is infinitely stiff with respect to longitudinal shear and hence

underestimates the deflection. Methods of calculation which take into account the shear racking component are given in Chap. 10.

4.4   BRACED FRAMES

4.4.1   Introduction

Rigid frame systems are not efficient for buildings taller than about 30 stories because the shear racking component of deflection due to the bending of columns and girders causes the drift to be too large. A braced frame attempts to improve upon the efficiency of a rigid frame by virtually eliminating the bending of columns and girders. This is achieved by adding web members such as diagonals or chevron braces. The horizontal shear is now primarily absorbed by the web and not by the columns. The webs carry the lateral shear predominantly by the horizontal component of axial action allowing for nearly a pure cantilever behavior.

4.4.2   Behavior

In simple terms, braced frames may be considered as cantilevered vertical trusses resisting lateral loads primarily through the axial stiffness of columns and braces. The columns act as the chords in resisting the overturning moment, with tension in the windward column and compression in the leeward column. The diagonals and girders work as the web members in resisting the horizontal shear, with diagonals in axial compression or tension depending upon their direction of inclination. The girders act axially, when the system is a fully triangulated truss. They undergo bending also when the braces are eccentrically connected to them. Because the lateral load on the building is reversible, braces are subjected in turn, to both compression and tension; consequently, they are most often designed for the more stringent case of compression.

The effect of the chords' axial deformations on the lateral deflection of the frame is to tend to cause a "flexural" configuration of the structure, that is, with concavity downwind and a maximum slope at the top (Fig. 4.17a). The effect of the web member deformations, however, is to tend to cause a "shear" configuration of the structure (i.e., with concavity upwind, a maximum slope at the base, and a zero slope at the top; Fig. 4.17b). The resulting deflected shape (Fig. 4.17c) is a combination of the effects of the flexural and shear curves with a resultant configuration depending on their relative magnitudes, as determined mainly by the type of bracing. Nevertheless, it is the flexural deflection that most often dominates the deflection characteristics.

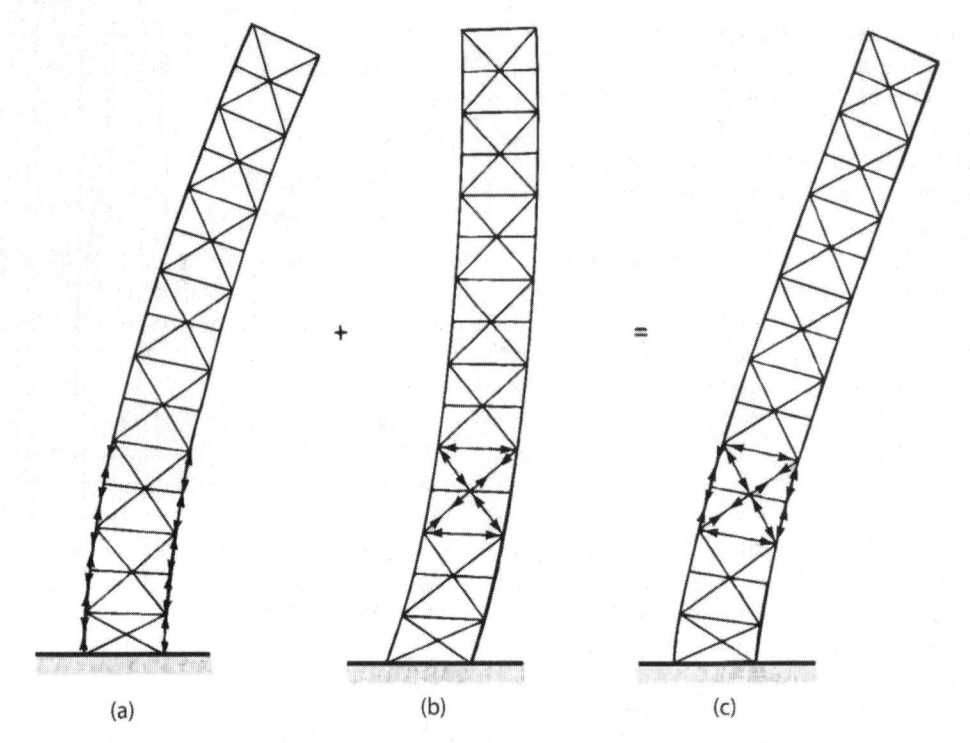

Figure 4.17   Braced frame deformation: (a) flexural deformation; (b) shear deformation; (c) combined configuration.

The role of web members in resisting shear can be demonstrated by following the path of the horizontal shear down the braced bent. Consider the typical braced frames, shown in Fig. 4.18a–e, subjected to an external shear force at the top level. In Fig. 4.18a, the diagonal in each story is in compression, causing the beams to be in axial tension; therefore, the shortening of the diagonal and extension of the beams gives rise

to the shear deformation of the bent. In Fig. 4.18b, the forces in the braces connecting to each beam end are in equilibrium horizontally with the beam carrying insignificant axial load.

In Fig. 4.18c, half of each beam is in compression while the other half is in tension. In Fig. 4.18d, the braces are alternately in compression and tension while the beams remain basically unstressed. And finally in Fig. 4.18e,

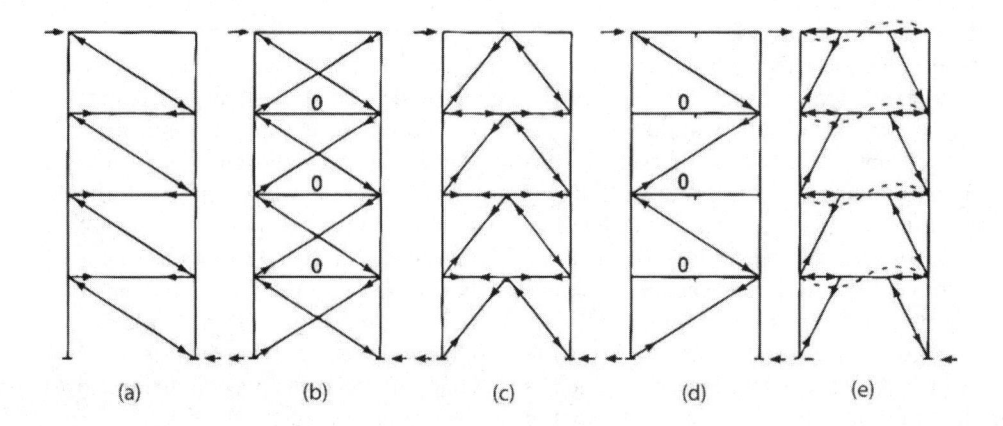

Figure 4.18   Load path for horizontal shear through web numbers: (a) single diagonal bracing; (b) *X*-bracing; (c) chevron bracing; (d) single-diagonal, alternate direction bracing; (e) knee bracing.

the end parts of the beam are in compression and tension with the entire beam subjected to double curvature bending. Observe that with a reversal in the direction of horizontal load, all actions and deformations in each member will also be reversed.

In a braced frame the principal function of web members is to resist the horizontal shear forces. However, depending upon the configuration of the bracing, the web members may pick up substantial compressive forces as the columns shorten vertically under gravity loads. Consider, for example, the typical bracing configurations shown in Fig. 4.19. As the columns in Fig. 4.19a,b shorten, the diagonals are subjected to compression forces because

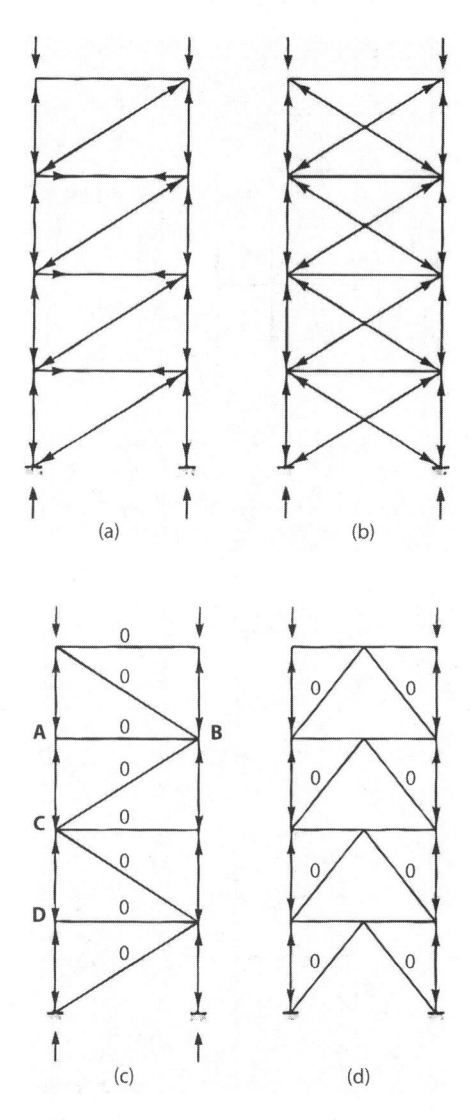

(a)   (b)

(c)   (d)

Figure 4.19 Gravity load path: (a) single diagonal single direction bracing; (b) X-bracing; (c) single diagonal alternate direction bracing; (d) chevron bracing.

the beams at each end of the braces are effective in resisting the horizontal component of the compressive forces in the diagonal. At a first glance, this may appear to be the case for the frame shown in Fig. 4.19c. However, the diagonals shown in Fig. 4.19c will not attract significant gravity forces because there is no triangulation at the ends of beams where the diagonals are not connected (nodes A and D, in Fig. 4.19c). The only horizontal restraint at the beam end is by the bending resistance of columns, which usually is of minor significance in the overall behavior. Similarly, in Fig. 4.19d the vertical restraint from the bending stiffness of the beam is not large; therefore as in the previous case, the diagonals experience only negligible gravity forces.

4.4.3 Types of Braces

Braced frames may be grouped into two categories as either concentric braced frames (CBFs) or eccentric braced frames (EBFs) depending upon their ductility characteristics. In CBFs the axes of all members, that is, columns, beams, and braces intersect at a common point such that the member forces are axial. EBFs, which will be discussed later in this chapter, utilize axis offsets to deliberately introduce flexure and shear into framing beams to increase ductility.

The CBFs can take many forms some of which are shown in Fig. 4.20. Depending upon the diagonal force, length, required stiffness, and clearances, the diagonal member can be made of double angles, channels, tees, tubes, or wide flange shapes. Figure 4.16 shows an inverted K bracing also called chevron bracing consisting of double angles. Besides performance, the shape of the diagonal is often based on connection considerations.

The least objectionable locations for braces are around service cores and elevators, where frame diagonals may be enclosed within permanent walls. The braces can be joined together to form a closed or partially closed three-dimensional cell for effectively resisting torsional loads.

Common types of interior bracing are shown in Fig. 4.20a–n. Figure 4.20e–n shows bracings across single bays in one-story increments. Figure 4.20a shows diagonal bracing in two-story increments. Shown in Fig. 4.20b,c is a K-braced frame, while Fig. 4.20d shows bracing for a three-bay frame. Any reasonable pattern of braces with single or multiple braced bays can be designed for resisting the lateral loads.

Finding an efficient and economical bracing system for a tall building presents the structural engineer with an excellent opportunity to use innovative design concepts. However, the availability of proper depth for bracing

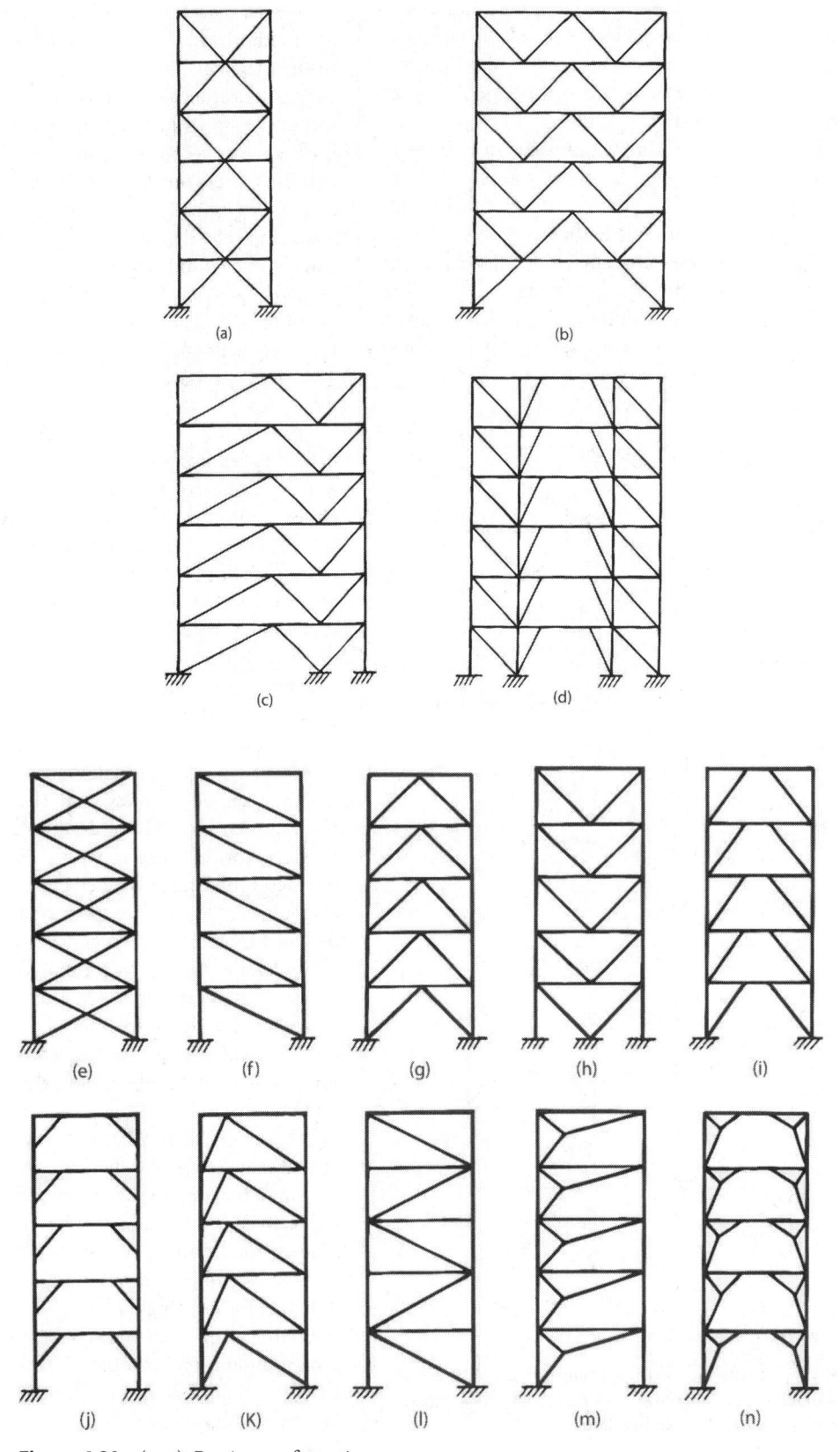

Figure 4.20   (a–n): Bracing configurations.

is often an overriding consideration. As a preliminary guide, a height-to-width ratio of 8 to 10 is considered proper for a reasonably efficient bracing system.

4.5 STAGGERED TRUSS SYSTEM

4.5.1 Introduction

Most high-rise residential-type structures such as apartments and hotels are generally in the neighborhood of 60 by 150 ft (18.3 × 45.75 m) to 200 ft (61 m) long. Their floor plans normally lend themselves to central double-loaded corridors of about 6 to 8 ft (1.83 to 2.43 m) in width. A study was done at the Massachusetts Institute

of Technology (MIT) in the mid-1960s under the sponsorship of U.S. Steel Corporation for the purpose of developing an economical framing system for such tall, narrow structures. The staggered truss system evolved as an out-growth of the research done by the departments of architecture and civil engineering at MIT. In this system, story-high trusses span in the transverse direction between the columns at the exterior of the building. The required flexibility in residential unit layouts is achieved by arranging the trusses in a staggered plan at alternate floors, as shown schematically in Fig. 4.21. The floor system acts as a diaphragm transferring lateral loads in the

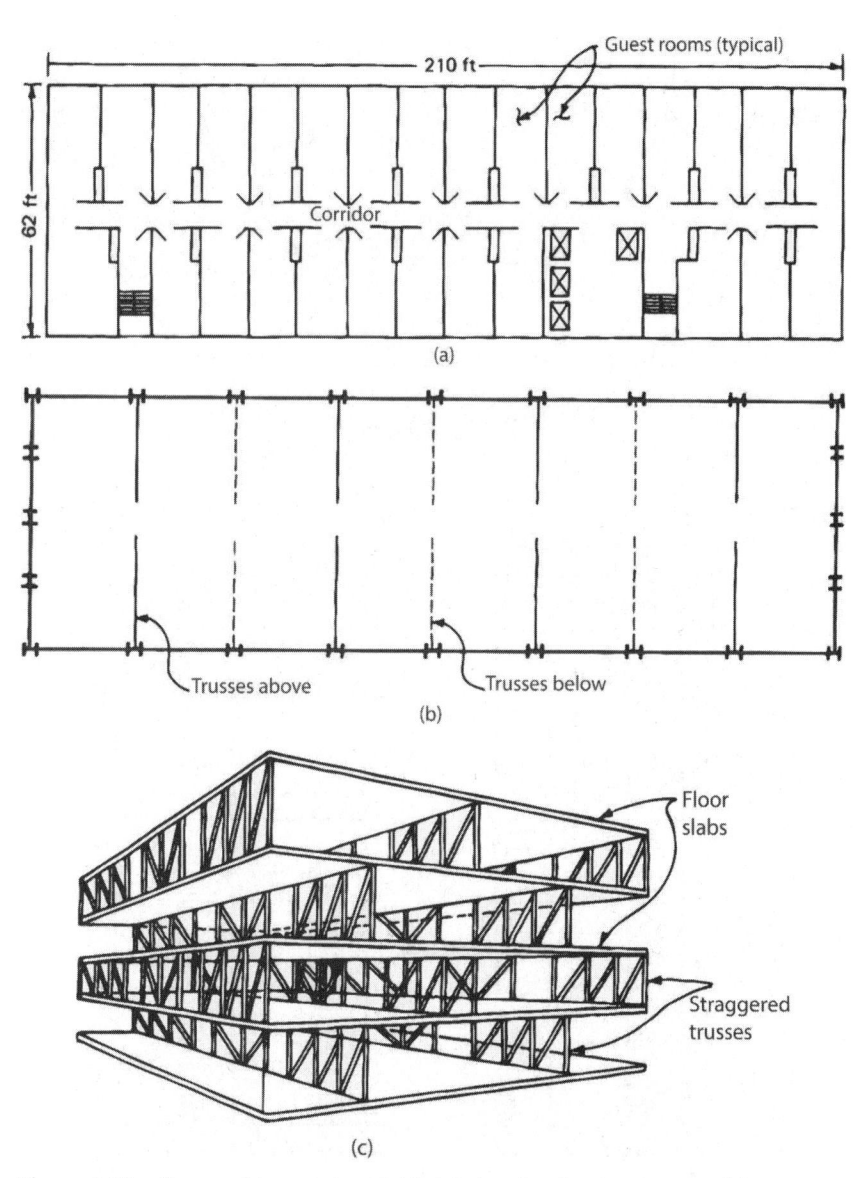

Figure 4.21 Staggered truss system: (a) hotel plan showing guest rooms; (b) arrangement of staggered trusses; (c) perspective view of truss arrangement.

short direction to the trusses. Lateral loads are thereby resisted by truss diagonals and are transferred into direct loads in the columns. The columns therefore receive no bending moments.

The truss diagonals are eliminated at the corridor location to allow for openings. Since the diagonal is eliminated, the shear is carried by the bending action of the top and bottom chord members. Similarly, other openings can be provided for in the truss to allow for additional openings at a slight structural premium when required by architectural layout. The system was first used for a housing project for the elderly in St. Paul, Minnesota, completed in 1967. Since then, a number of long, narrow, high-rise buildings for apartment houses, hotels, and in some cases for office buildings have been built using this concept.

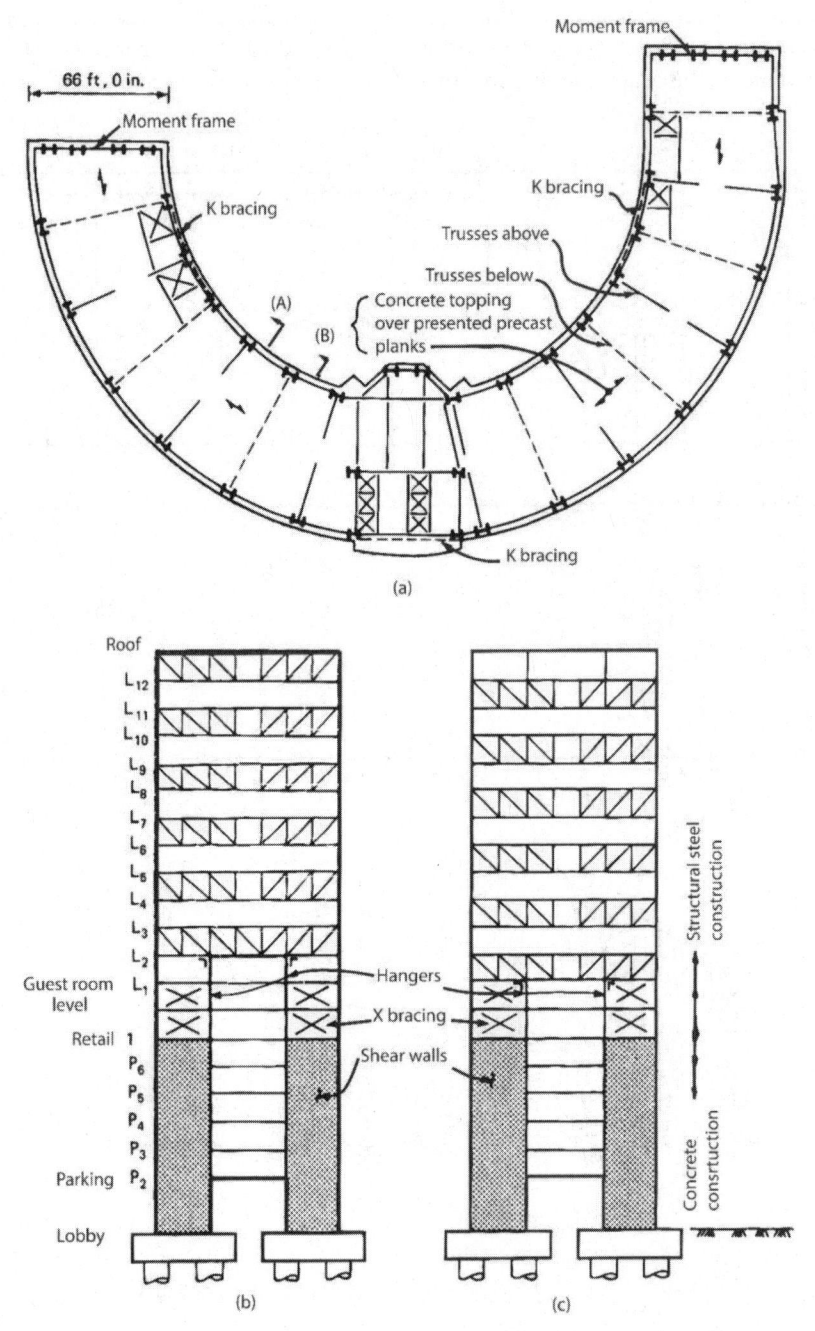

Figure 4.22 Staggered truss system for a semicircular building: (a) plan; (b) section A; (c) section B.

Because the staggered truss resists major gravity and lateral loads in direct stresses, the system is quite stiff. In general, no material needs to be added for drift control, and high-strength steels are conveniently used throughout the entire frame. The system has been used for 35- to 40-story buildings. Spans must be long enough to make the trusses efficient, with 45 ft (13.72 m) being considered as the minimum practical limit. In a typical hotel or residential building, a staggered truss system will normally reduce the steel requirement by as much as 30 to 40 percent as compared to a conventional moment-connected framing. Since the trusses are supported only by the perimeter columns, the need for interior columns and associated foundations is eliminated, contributing to the economy of the system.

An added advantage of the system is that it allows for public spaces free of interior columns on the lower levels. The most economical use of staggered trusses is achieved by placing the trusses between units, since in a normal hotel or housing these units are spaced uniformly across the length of the building. It is possible to extend these units through trusses by providing for additional openings. However, varying the spacings could create a variety of unit sizes that can be accommodated within the trusses. Thus, one-, two-, or three-bedroom apartments can be arranged on a single floor merely by varying the column spacings. The system is not limited to simple rectangular plans. It can be effectively used in curvilinear plans as shown in Fig. 4.22.

4.5.2 Physical Behavior

Consider the three-dimensional interaction between vertical bracings of a building interconnected through floor diaphragms, as shown in Fig. 4.23. Assume that for architectural reasons it is required to eliminate the bracing at column lines A and B below level 2. If there is no other

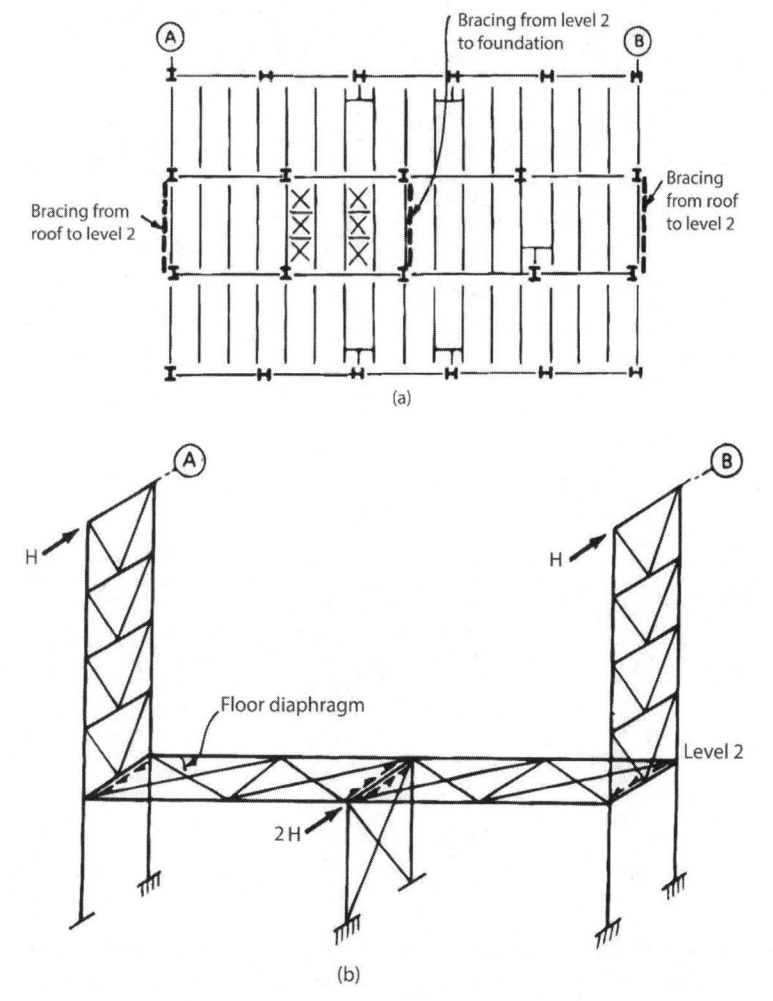

Figure 4.23 Conceptual model for staggered truss system: (a) building plan; (b) lateral load transfer through diaphragm action.

bracing below level 2, the columns at the extremities of the bracing must resist both the overturning moment and shear forces below level 2.

The overturning moment manifests itself as compressive and tensile forces in the columns, while the shear forces introduce bending moments in the columns resulting in a rather inefficient structural system. Assume that architecturally it is permissible to introduce a bracing at the center of the building below level 2 as shown in Fig. 4.23a,b. For purposes of lateral analysis, generally it is sufficiently accurate to assume that the slab diaphragm is rigid in its own plane. As a consequence of this assumption, most of the shear at the exterior braces at level 2 is transferred to the interior bracings through the diaphragm action of the floor slab. The columns under the exterior braces are therefore subjected to axial stresses only, while the shear is resisted by the interior bracing. This in essence is the structural action in a staggered truss system in which the lateral force is transmitted across the floor to the truss on the adjacent column line and continues down on the truss line across the next floor down the next truss, etc., as shown schematically in Figs. 4.24 and 4.25. Thus, between the floors, lateral forces are resisted by the truss diagonals, and at each floor these forces are transferred to the truss below by the floor system acting as a diaphragm. The columns between the floors receive no bending moments, resulting in a very efficient and stiff structure. Since the trusses are placed at alternate levels on adjacent column lines, two-bay-wide column-free interior floor space is created in the longitudinal direction.

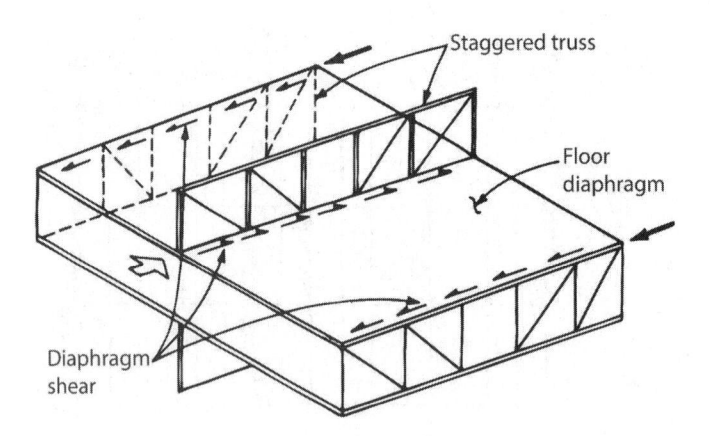

Figure 4.24 Load path in staggered truss system.

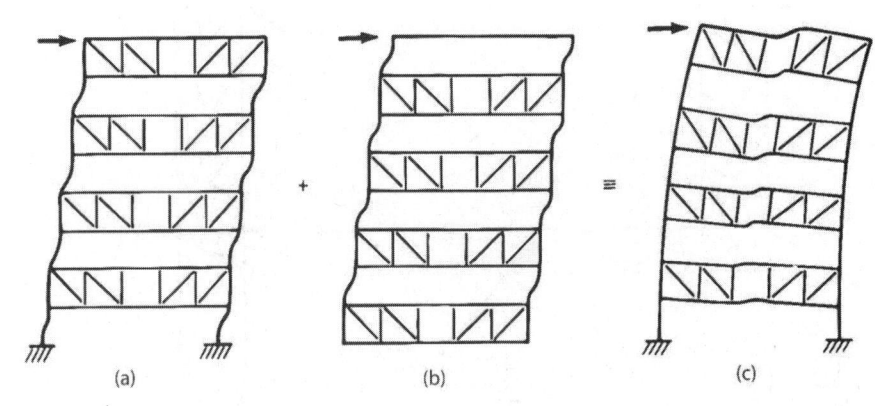

Figure 4.25 Conceptual two-dimensional model for staggered truss system: (a,b) lateral deformation of adjacent bays; (c) overall behavior. Note the absence of local bending of columns.

4.5.3 Design Considerations

4.5.3.1 FLOOR SYSTEM

As in other structural systems, the floor system in a staggered truss scheme needs to fulfill two primary requirements: (i) collect and transmit the gravity loads to the vertical elements; and (ii) resist the lateral loads as a shear diaphragm and provide a continuous path for transferring the lateral loads from the bottom chord of one truss to the top chord of the adjacent truss down through the structure. In addition to these structural requirements, the floor system must permit flexibility for apartment size and location, must provide fireproofing and an acceptable ceiling, and should be usable as temporary bracing during steel erection. Thus, one could use precast concrete planks, long-span composite steel decks, open-web joist system, or any other system consistent with the structural and architectural requirements. Precast planks and flat-bottomed steel decks are often used as exposed ceilings with minimum of finish. For spans up to 30 ft (9.15 m), 8 in. (203 mm) planks are required, while for spans less than 24 ft (7.3 m), 6 in. (152 mm) planks are adequate. In a composite steel deck system, a $7\frac{1}{2}$ in. (190 mm) deck is required for spans up to 30 ft (9.15 m), and for spans up to 24 ft (7.3 m), 6 in. (152 mm) deep steel deck is adequate. When precast planks are used, shear transfer is achieved by the use of welded plates cast in the planks or by welding shear connectors on the truss chord.

In the case of a metal deck system, generally adequate shear transfer is obtained by the connection of the steel deck to the trusses. Planks used for erection purposes should have connection weld plates, even when shear connectors are provided. The choice of the floor system depends on the geographical location as well as local conditions. In earthquake zones, the lighter floor produces smaller seismic forces. In cold climates, the cost of grouting between precast planks in winter is increased by the necessity for heating. The floor system may consist of either a series of simple or continuous spans over the chords of the trusses. Because of the large spacing between the trusses, the continuous spans are usually limited to a maximum of two bays. Generally, one end of each span is supported on the lower chord, while the other end is made continuous by simply running the floor slab across the top chord of the truss.

Since the trusses are staggered at alternate floors, the equivalent lateral load on each truss is equal to the lateral load acting on two bays. Hence floor panels on each side of the truss must transmit half of that load to the adjacent truss in the story immediately below. The floor system acts as a deep beam resisting both the in-plane shears and bending moments. Adequate provision should be made to transfer longitudinal shear between each span so that the whole floor system acts as a single unit.

Although any floor system that is capable of carrying gravity loads and diaphragm shear can be used in staggered truss systems, economic considerations generally favor the use of either precast concrete planks or long-span composite metal deck, both with a topping of concrete reinforced with welded-wire fabric.

The design of the floor components for gravity loads is identical to that of a conventional high-rise building. Although continuity may be developed at the support connections to the truss chords, simple span behavior is assumed for convenience. Because the staggered truss system depends upon the diaphragm action of the entire floor to transfer lateral loads from one truss to another, the floor is assumed to have chords to resist the in-plane shears and moments. The bending can be resisted by the floor slab or by flange action at the exterior walls. The floor system of two adjacent bays is considered to be a continuous beam for the in-plane forces. Since there are no published criteria for general use, for any specific project it is advisable to study the diaphragm behavior in somewhat greater detail to develop and decide upon design procedures and criteria.

4.5.3.2 COLUMNS

In the staggered truss system, the lateral loads are taken out of the system by the floor, which acts as a diaphragm, and the truss diagonals, which act as vertical braces. Therefore, lateral loads are transferred into direct loads in the columns. No shear exists in the columns to create bending in the transverse direction of the building. Thus the columns can be designed as braced members to resist minor axis buckling, and the strong direction of the column can be used to resist lateral loads in the longitudinal direction. Another aspect of column design for staggered truss systems that should be considered is the effect of truss deflections in causing excessive weak-axis bending moment. The axial compression of the top chord and elongation of the bottom chord introduce bending in columns. This problem can be solved by introducing a camber in the truss by deliberately making the truss bottom chord smaller than the top chord. As an alternate, the connection between the bottom chord and column can be designed to slip under dead load conditions. Torquing the bolts after the application of dead loads to the truss will limit the bending moments in the columns.

As a second alternative, the connection of the bottom chord can be designed to remain loose, increasing the effective length of the column in the weak axis to two stories between the top chords of the trusses at two levels. If none of the above procedures is applicable, then these moments should be provided for in the design of the columns.

4.5.3.3 TRUSSES

The design of the staggered truss system is quite conventional. Loading conditions and design methods are similar in principle to other framing systems. The floor system spans only from the bottom chord of one truss to the top chord of the next, and the resulting large floor area supported by each truss allows maximum live load reduction.

In the transverse direction the lateral loads are transferred from the bottom chord of a truss to the top chord of an adjacent truss through floor diaphragm action down through the truss diagonals to the bottom chord. The sequence of events starts over, thus causing the entire transverse lateral load to be transferred through the floor system by diaphragm action and through the truss by direct stresses. The only additional bending occurs in the truss chords at the corridor openings or in other places where diagonals are eliminated.

The span-to-depth ratio of trusses is usually in the range of 6:1, giving adequate depth for the efficient design of top and bottom chords. Usually the panel width of trusses is not a governing criterion. Larger panel lengths with fewer web members decrease the fabrication costs and may work out to be more economical.

For maximum efficiency, just as in any other structural system, it is preferable to maintain a uniform spacing of trusses. This allows for maximization of typical truss units and reduces fabrication costs. However, when required by architectural arrangement, it is possible to vary the column and thus the truss spacing. Vierendeel openings other than those required for corridors should be avoided in the interest of economy. Truss design is based on continuous chord and pinned members, preferably using a computer analysis. Generally W or S shapes are selected for the chord members since angles are not efficient in resisting the secondary bending. Also, when planks are used wide flanges offer good bearing areas. Since the staggered truss system resists loads primarily by direct stresses, deflections are generally not a problem and therefore high-strength steels can be economically employed. The truss as a member supports a very large area and it is likely that if reduced live loads are used, the maximum live load reduction will be permissible. If any chord member is considered to support an area equal to the truss panel length times the bay spacing, it would be prudent to base the chord moment design on a reduced live load based on the smaller tributary area.

The simplest method of stacking trusses is a configuration called the checkerboard pattern, in which the trusses are placed at alternate columns and floors. It is possible, however, to obtain a greater variety of spaces by using different layouts on alternate levels.

Longitudinally, the lateral forces in a staggered truss structure can be resisted by any conventional bracing system such as braced frames and core shear walls. However, many projects lend themselves to the design of rigid frames on the two broad faces: because (i) the main columns are oriented with webs parallel to the spandrels; and (ii) a large number of columns are generally available to participate in the moment frame. In some cases, deep precast fascia beams used for architectural reasons can be directly bolted to the columns to serve as stiff structural spandrels.

In the transverse direction, at the roof and at the bottom floors it is normally not possible to carry the rhythm of the staggered truss for the full height of the building. Posts and hangers are usually required to support these levels. At the bottom level, the lateral loads may need to be transferred to the foundation by diagonal bracing. These are shown schematically in Fig. 4.22b,c with respect to the curvilinear layout of a staggered truss system shown in Fig. 4.22a.

4.6 ECCENTRIC BRACING SYSTEMS

4.6.1 Introduction

Older CBFs were excellent from strength and stiffness considerations and are therefore used widely either by themselves or in conjunction with moment frames when the lateral loads are caused by wind. However, they were of questionable value in seismic regions because of their poor inelastic behavior. Moment-resistant frames possess considerable energy dissipation characteristics but are relatively flexible when sized from strength considerations alone. Eccentric bracing is a unique structural system that attempts to combine the strength and stiffness of a braced frame with the inelastic behavior and energy dissipation characteristics of a moment frame. The system is called eccentric because deliberate eccentricities are employed between beam-to-column and beam-to-brace connections. The eccentric beam element acts as a fuse by limiting large forces from entering and causing buckling of braces. The eccentric segment of the beam, called the link, undergoes flexural or shear yielding prior to formation of plastic hinges in the other bending members and well before buckling of any compression members. Thus the system maintains stability even under large inelastic deformations. The required stiffness during wind or minor earthquakes is maintained because no plastic hinges are formed under these loads and all behavior is elastic. Although the deformation is larger than in a concentrically braced frame because of bending and shear deformation of the "fuse," its contribution to deflection is not significant because of the relatively small length of the fuse. Thus the elastic stiffness of the eccentrically braced frames can be considered the same as the concentrically braced frame for all practical purposes.

4.6.2 Ductility

The ductile behavior is highly desirable when the structure is called upon to absorb energy such as when subjected to strong ground motions. Steel's capacity for deformation without fracture combined with its high strength makes it an ideal material for use in eccentric bracing systems. In a properly designed and executed connection, steel continues to resist loads even after the maximum load is reached. This property by virtue of which steel sustains the load without fracture is called ductility. A brittle material, on the other hand, does not undergo large deformations at the onset of yielding. It fractures prior to, or just when it reaches the maximum load.

4.6.3 Behavior of Frame

Eccentrically braced frames can be configured in various forms as long as the brace is connected to at least one link. The underlying principle is to prevent buckling of the brace from large overloads that may occur during major earthquakes. This is achieved by designing the link to yield.

The shear yielding of beams is a relatively well-defined phenomenon; the load required for shear yielding of a beam of given dimensions can be calculated fairly accurately. The corresponding axial load and moments in columns and braces connected to the link can also be assessed fairly accurately. Using certain overload factors, which will be explained shortly, the braces and columns are designed to carry more load than could be imposed by the shear yielding of link. This assures that in the event of a large earthquake, it is the link that blows the fuse and not the columns or braces connected to it.

Consider the bracing shown in Fig. 4.26 subjected to horizontal loads. Note that the connections between the

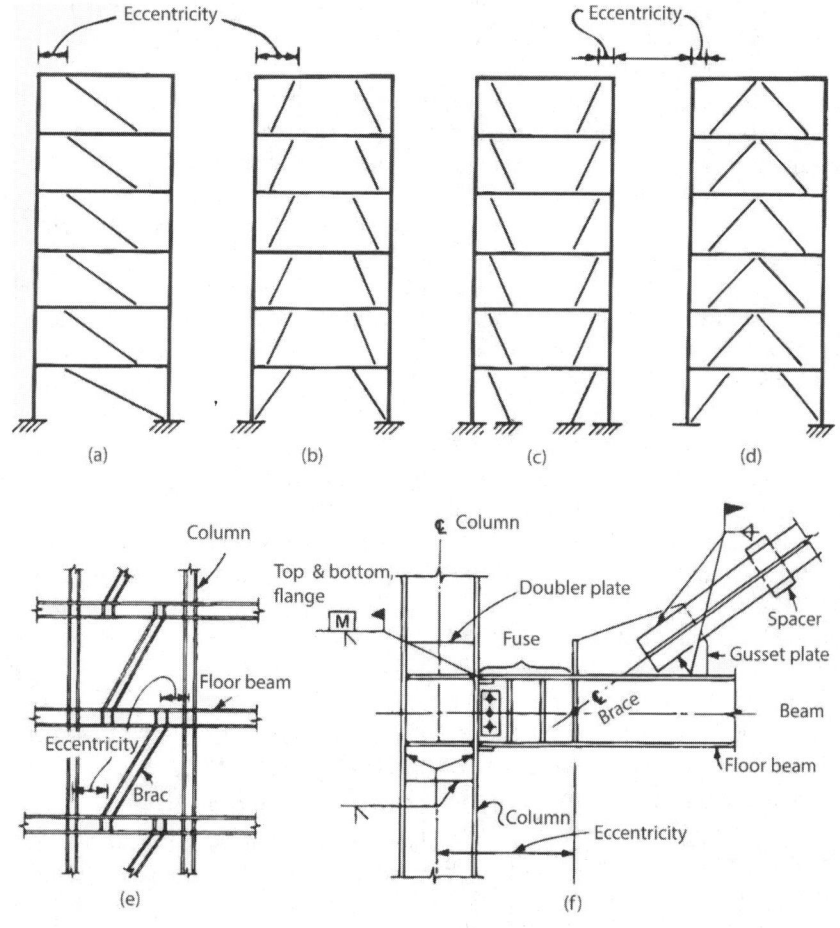

Figure 4.26 Eccentric bracing system: (a–d) common types of bracing; (e) elevation; (f) detail.

column and beams are moment-connected to achieve brace action. The force in the brace is transmitted to the beam as a horizontal force inducing axial stresses, and as a vertical force inducing shear stresses in the beam web. Of more concern in the design of the link are the cyclic shear forces induced in the beam. Assuming the link and its moment connection to the column are adequate in bending, the mechanism of failure is by shear yielding of the beam web provided web buckling is prevented. This is achieved by providing adequate stiffeners in the link.

4.6.4   Essential Features of the Link
Whether the link develops plastic hinges or yields in shear is a function of its length. Links longer than twice the depth tend to develop plastic hinges while shorter links tend to yield in shear. Links can be identified either as short or long. The short link experiences moderate rotation, while the long one a relatively larger rotation.

The cyclic shear yielding is an excellent energy dissipation mechanism because large cyclic deflections can take place without failure or deterioration in the hysteretic behavior. This is because yielding occurs over a large segment of the beam web and is followed by a cyclic diagonal field. The web buckles after yielding in shear, but the tension field takes over the load-carrying mechanism to prevent failure, resulting in a hysteretic loop having a large area representing good energy dissipation.

4.6.5   Analysis and Design Considerations
To force the formation of a hinge in the beam web, the plastic moment capacity of the beam should exceed the beam shear yield capacity. In calculating the plastic moment capacity of the beam, the contribution of the web is neglected because the web is assumed to have yielded. In design, the beam is first selected for the required shear capacity and then the plastic moment capacity is checked to be slightly larger than the shear yield capacity. As in ductile frame design, the column is selected by using the weak beam-strong column concept to assure that plastic hinges are formed in the beams and not in the columns. If the plastic moment of the beam selected is larger than that required by design, the column is designed in an equally conservative manner. To assure that the braces are prevented from buckling, they are designed to withstand forces somewhat larger than those given by the analysis.

This conservatism is necessary to take into account the fact that the actual beam designed is likely to have additional capacity due to factors such as: (i) beam strain hardening; (ii) actual yield stress being more than the theoretical value; (iii) interaction of floor slabs with beams tending to increase the plastic moment capacity. The brace-to-beam connection can be designed either as a welded or bolted connection. The bolts are designed as friction bolts and checked for bearing capacity because of the likelihood of slippage in the event of a large earthquake. The beam-to-column connection is designed as a moment connection by welding the beam flanges to the column with full-penetration welds. Single-side shear plate connection with fillet welds is used to develop the high shear forces in the link. Lateral support is provided at the top and bottom flanges of the beam to prevent lateral torsional buckling and weak axis bending.

4.6.6   Deflection Considerations
The lateral deflection of an eccentrically braced frame can be estimated as the sum of three components: (i) deflection due to elongation of the brace; (ii) deflection due to axial strain in the columns, usually referred to as the chord drift; and (iii) the deflection due to deformation of the eccentric element. Because the braces and columns are designed to remain elastic even under a severe earthquake, their deflection contributions are very nearly constant even after the shear yielding of the link. The beams in eccentric bracing are much heavier than in a corresponding concentrically braced frame, therefore are likely to contribute little to the deflection under elastic condition. Therefore, an eccentrically braced frame is not an unreasonably flexible system as compared to a concentric frame.

4.6.7   Conclusions
Buildings using eccentric bracing are lighter than moment-resisting frames and, while retaining the elastic stiffness of concentrically braced frames, are more ductile. The eccentric bracing system has the following characteristics:

1. It provides a stiff structural system without imposing undue penalty on the steel tonnage.

2. Eccentric beam elements yielding in shear, act as fuses to dissipate excess energy during severe earthquakes.

3. Premature failure of the link does not cause the structure to collapse because the structure continues to retain its vertical load-carrying capacity and stiffness.

4.7   INTERACTING SYSTEM OF BRACED AND RIGID FRAMES

4.7.1   Introduction
Even for buildings in the range of 10 to 15 stories, unreasonably heavy columns result if lateral bracing is confined to the building service core because the available depth for bracing is usually limited. In addition, high uplift forces that may occur at the bottom of core columns can present

foundation problems. In such instances, an economical structural solution can be arrived at by using rigid frames in conjunction with the core bracing system. Although deep girders and moment connections are required for frame action, rigid frames are often preferred because they are least objectionable from the interior space planning considerations. Oftentimes, architecturally, it may be permissible to use deep spandrels and closely spaced columns on the building façade because usually the columns will not interfere with the space planning and the depth of spandrels need not be shallow for passage of air conditioning ducts. A schematic floor plan of a building using this concept is shown in Fig. 4.27a.

As an alternative to perimeter frames, a set of interior frames can be used with the core bracing. Such an arrangement is shown in Fig. 4.27b in which frames on grid lines 1, 2, 6, and 7 participate with core bracing on lines 3, 4, and 5. Yet another option is to moment-connect the girders between the braced core and perimeter columns as shown in Fig. 4.27c in which the frame beams act as outriggers by engaging the exterior columns to resist the bending moments.

For slender buildings with height-to-width ratios in excess of 5, an interacting system of moment frames and braces becomes uneconomical if braces are placed only within the building core. In such situations, a good

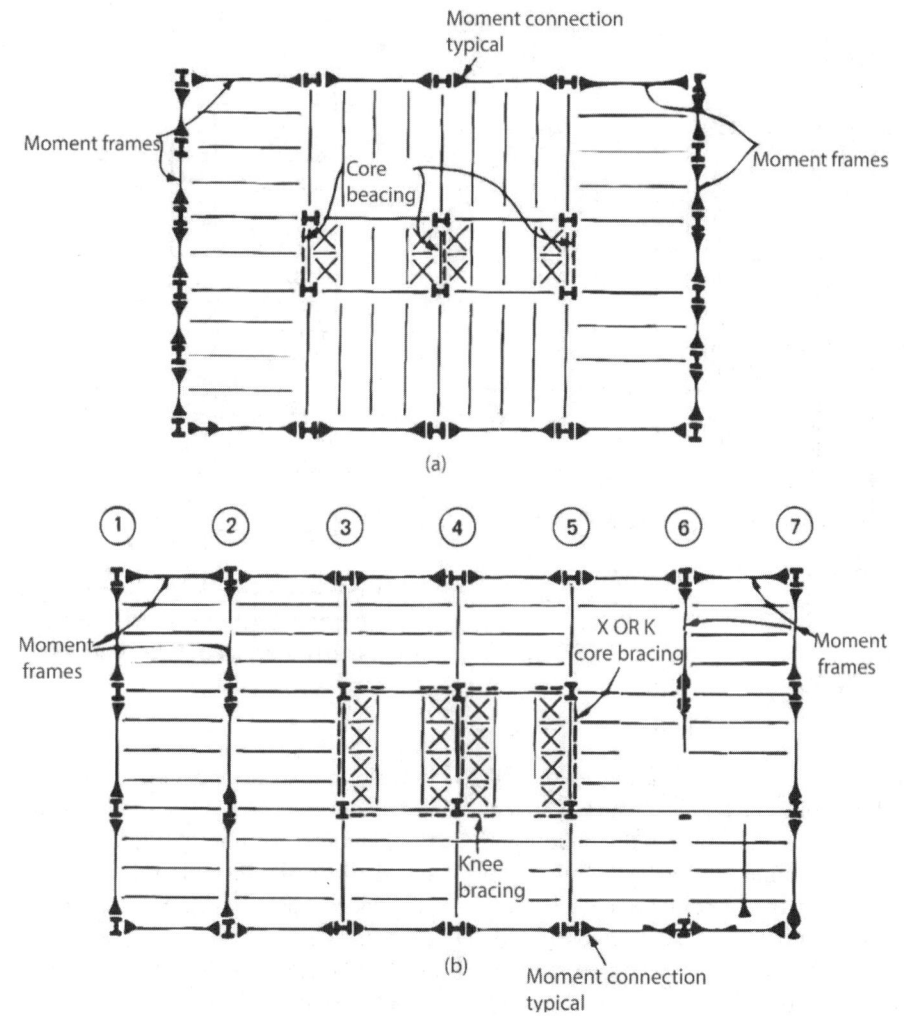

Figure 4.27 Schematic plans showing interacting braced and rigid frames: (a) braced core and perimeter frames; (b) braced core and interior and exterior frames; (c) braced core and interior frames; (d) full depth interior bracing and exterior frames; (e) transverse cross-section showing primary interior bracing, secondary bracing, and basement construction.

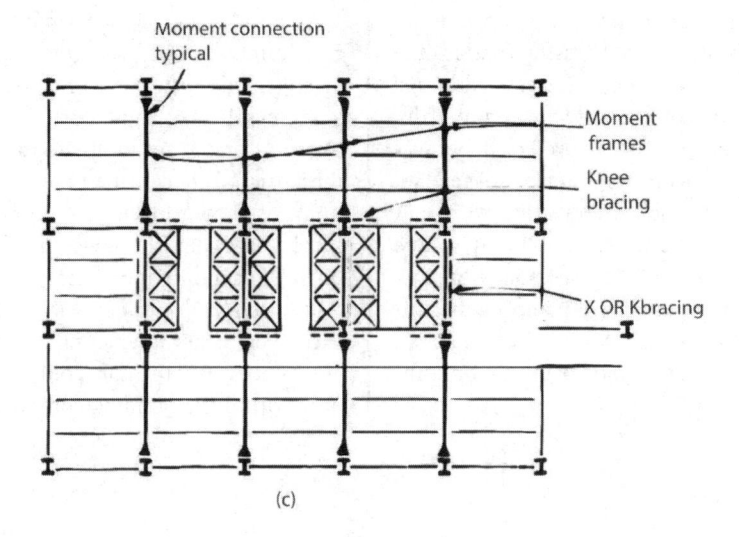

(c)

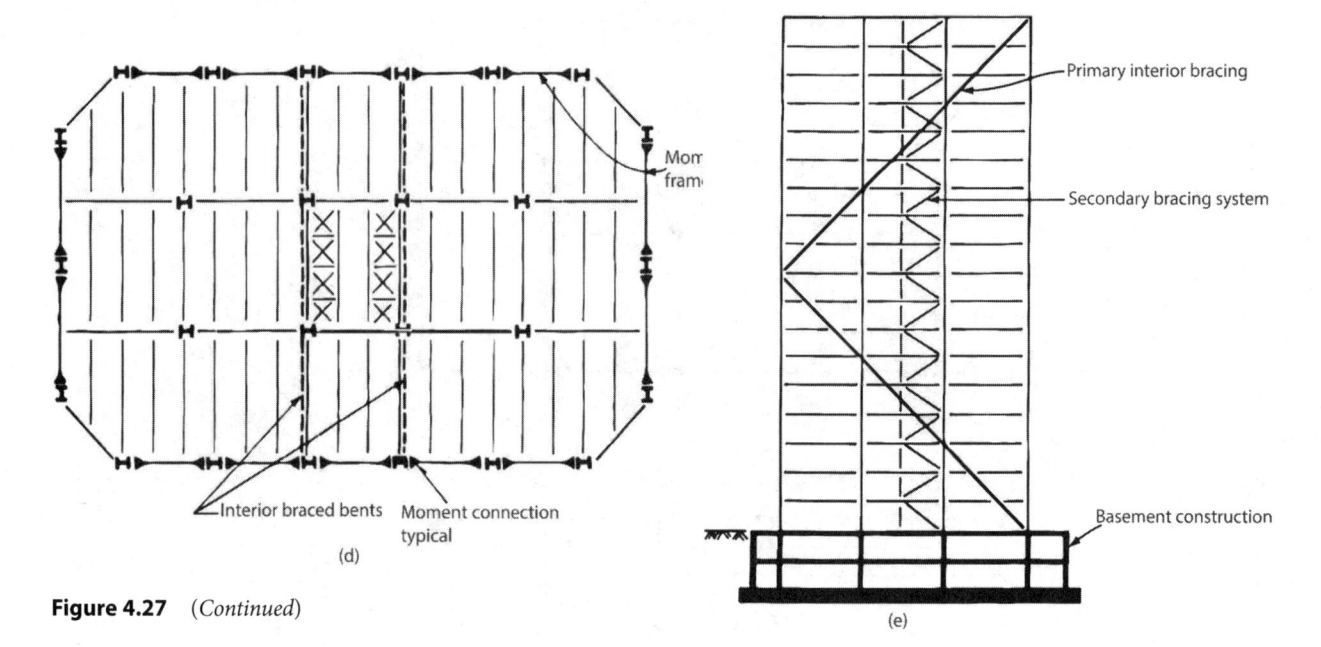

Figure 4.27 (*Continued*)

(d)

(e)

structural solution is to spread the bracing to the full width of the building along the façades if such a system does not compromise the architecture of the building. If it does, then a possible solution is to move the full-depth bracing to the interior of the building. Such a bracing concept is shown in Fig. 4.27d, in which moment frames located at the building façade interact with two interior-braced bents. These bents stretched out for the full width of the building form giant K braces, resisting overturning and shear forces by developing predominantly axial forces. A transverse cross-section of the building is shown in Fig. 4.27e, wherein a secondary system of braces required to transfer the lateral loads to the panel points of the K braces is also indicated. The diagonals of the K braces running through the interior of the building result in sloping columns whose presence has to be acknowledged architecturally as a trade-off for structural efficiency.

All of the above bracing systems and any number of their variations can be used either singly or in combination and can be made to interact with moment-connected frames. The magnitude of their interaction can be controlled by varying the relative stiffness of various structural elements to achieve an economical structural system.

4.7.2 Physical Behavior

If the lateral deflection patterns of braced and unbraced frames are similar, the lateral loads can be distributed between the two systems according to their relative stiffness. However, in normally proportioned buildings the unbraced and braced frames deform with their own characteristic shapes, necessitating that we study their behavior as a unit.

Insofar as the lateral-load-resistance is concerned, rigid and braced frames can be considered as two distinct units. The basis of classification is the mode of deformation of the unit when subjected to lateral loading. The deflection characteristics of a braced frame are similar to those of a cantilever beam. Near the bottom, the vertical truss is very stiff, and therefore the floor-to-floor deflections will be less than half the values near the top. Near the top, the floor-to-floor deflections increase rapidly mainly due to the cumulative effect of chord drift. The column strains at the bottom of the building produce a deflection at the top; and since this same effect occurs at every floor, the resulting drift at the top is cumulative. This type of deflection often referred as chord drift is difficult to control requiring material quantities well in excess required for gravity needs.

Rigid frames deform predominantly in a shear mode. The relative story deflections depend primarily on the magnitude of shear applied at each story level. Although near the bottom the deflections are larger, and near the top smaller as compared to the braced frames, the floor-to-floor deflections can be considered more nearly uniform. When the two systems, the braced and rigid frames are connected by rigid floor diaphragms, a nonuniform shear force develops between the two. The resulting interaction helps in extending the range of application of the two systems to buildings up to about 40 stories in height.

Figure 4.28 shows the individual deformation patterns of a braced and unbraced frames subjected to lateral loads. Also shown are the horizontal shear forces between the two frames connected by rigid floor slabs. Observe that the braced frame acts as a vertical cantilever beam, with the slope of the deflection greatest at the top of the building, indicating that in this region the braced frame contributes the least to the lateral stiffness.

The rigid frame has a shear mode deformation, with the slope of deformation greater at the base of the structure where the maximum shear is acting. Because of the different lateral deflection characteristics of the two elements, the frame tends to pull back the brace in the upper portion of the building while pushing it forward in the lower portion. As a result, the frame participates more effectively in the upper portion of the building where lateral shears are relatively less. The braced frame carries most of the shear in the lower portion of the building. Thus, because of the distinct difference in the deflection characteristics, the two systems help each other a great deal. The frame tends to reduce the lateral deflection of the trussed core at the top, while the trussed core supports the frame near the base. A typical variation of horizontal shear carried by each frame is shown in Fig. 4.28b in which the length of arrows conceptually indicates the magnitude of interacting shear forces.

Although the framed part of a high-rise structure is usually more flexible in comparison to the braced part, as the number of stories increases, its interaction with the braced frame becomes more significant, contributing greatly to the lateral resistance of the building. Therefore, when the frame part is fairly rigid by itself, its interaction with the braced portion of the building can result in a considerably more rigid and efficient design.

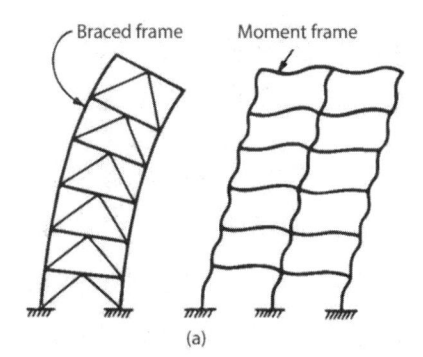
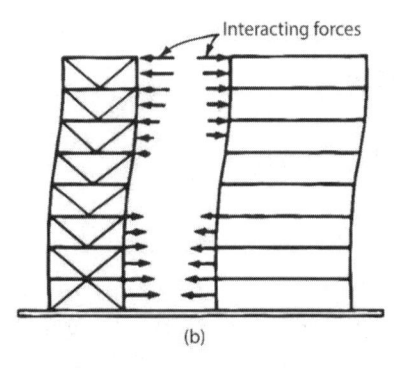

Figure 4.28 Interaction between braced and unbraced frames: (a) characteristic deformation shapes; (b) variation of shear forces resulting from the interaction.

4.8 OUTRIGGER AND BELT TRUSS SYSTEMS

4.8.1 Introduction

Innovative structural schemes are continuously being sought in the design of high-rise structures with the intention of limiting the wind drift to acceptable limits without paying a high premium in steel tonnage. The savings in steel tonnage and cost can be dramatic if certain techniques are employed to utilize the full capacities of the structural elements. Various wind-bracing techniques have been developed to this end; this section deals with one such system, namely, the belt truss system, also known as the core-outrigger system in which the axial stiffness of the perimeter columns is invoked for increasing the resistance to overturning moments.

This efficient structural form consists of a central core, comprising either braced frames of shear walls, with horizontal cantilever "outrigger" trusses or girders connecting the core to the outer columns. The core may be centrally located with outriggers extending on both sides (Fig. 4.29a) or it may be located on one side of the building with outriggers extending to the building columns on one side (Fig. 4.29b).

When horizontal loading acts on the building, the column-restrained outriggers resist the rotation of the core, causing the lateral deflections and moments in the core to

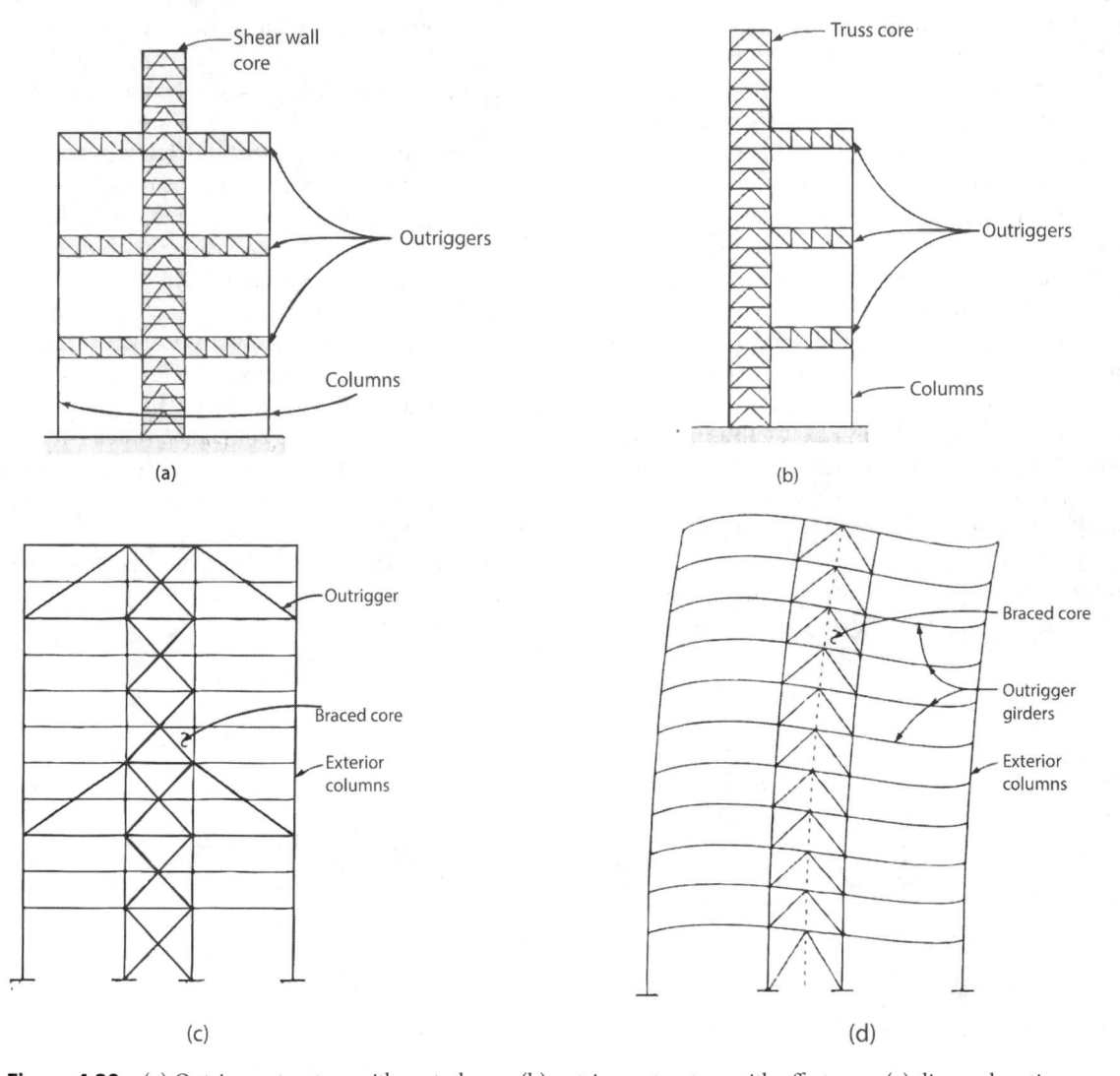

Figure 4.29   (a) Outrigger structure with central core; (b) outrigger structure with offset core; (c) diagonals acting as outriggers; (d) floor girders acting as outriggers; (e) plan with cap truss; (f,g) behavior of tied cantilever; (h) restraining spring at $X = 0$; (i) spring at $X = 0.25L$; (j) spring at $X = 0.50L$; (k) spring at $X = 0.75L$; (l) simplified analytical model for single outrigger system.

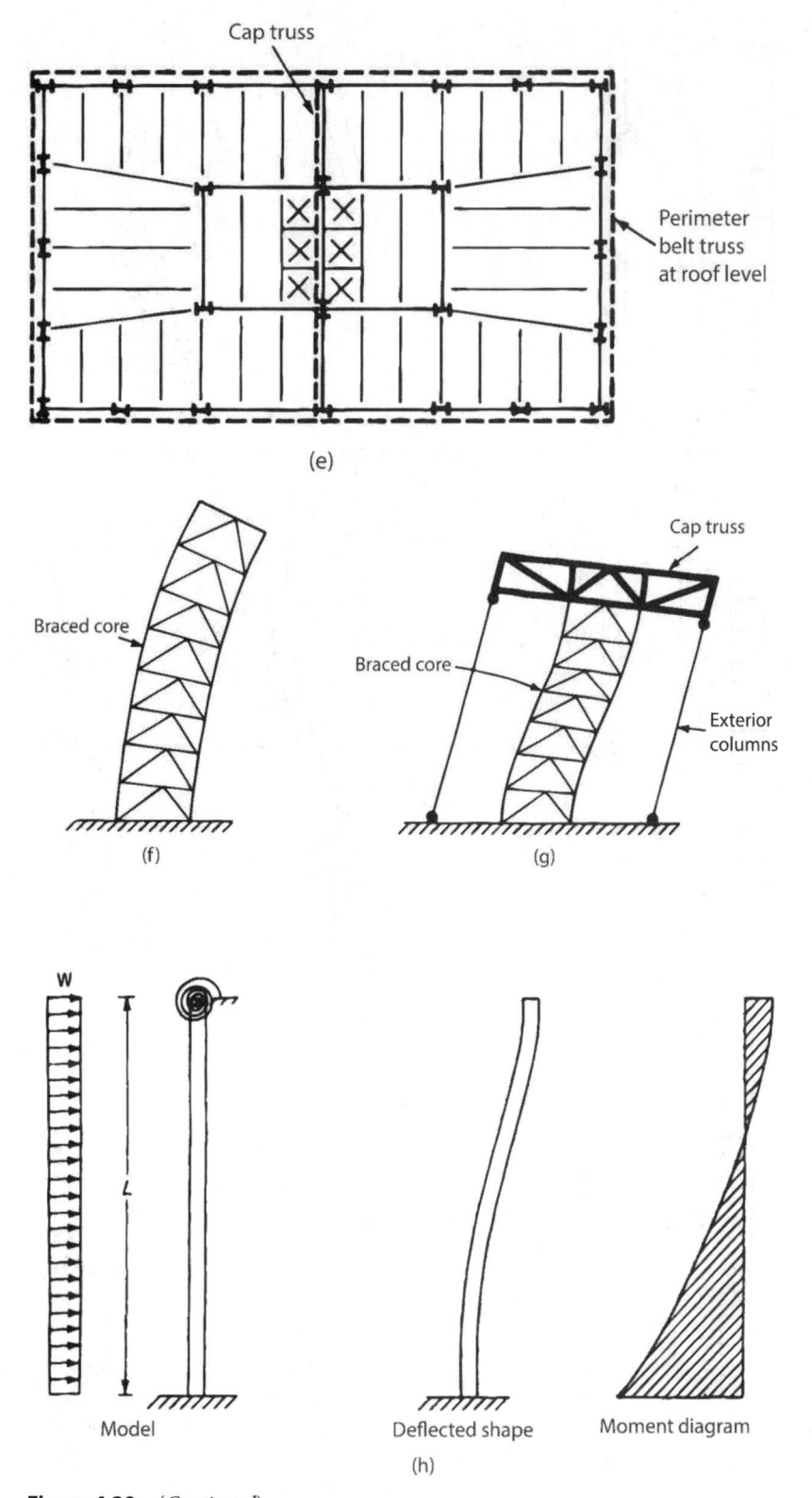

Figure 4.29 (*Continued*)

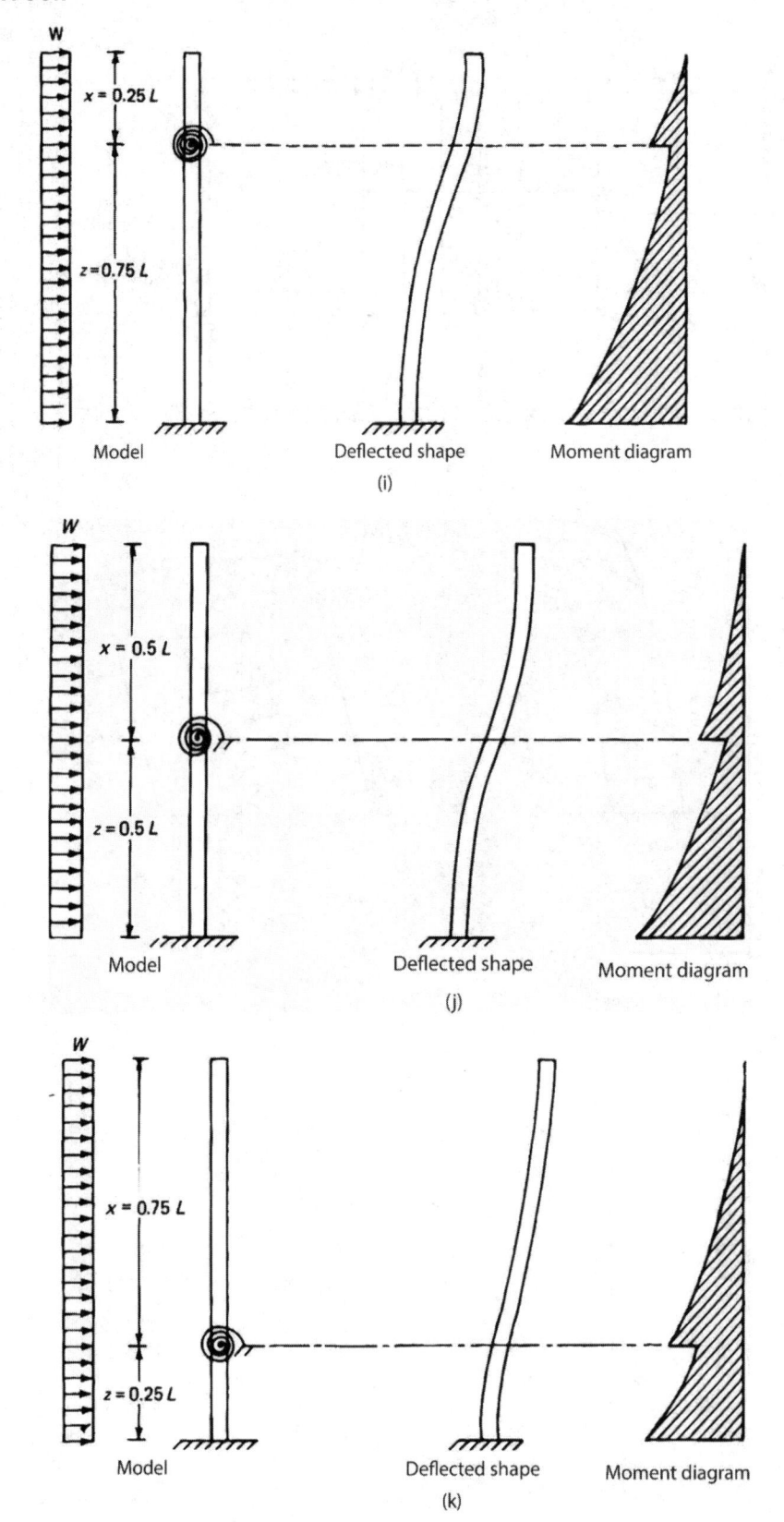

Figure 4.29 (*Continued*)

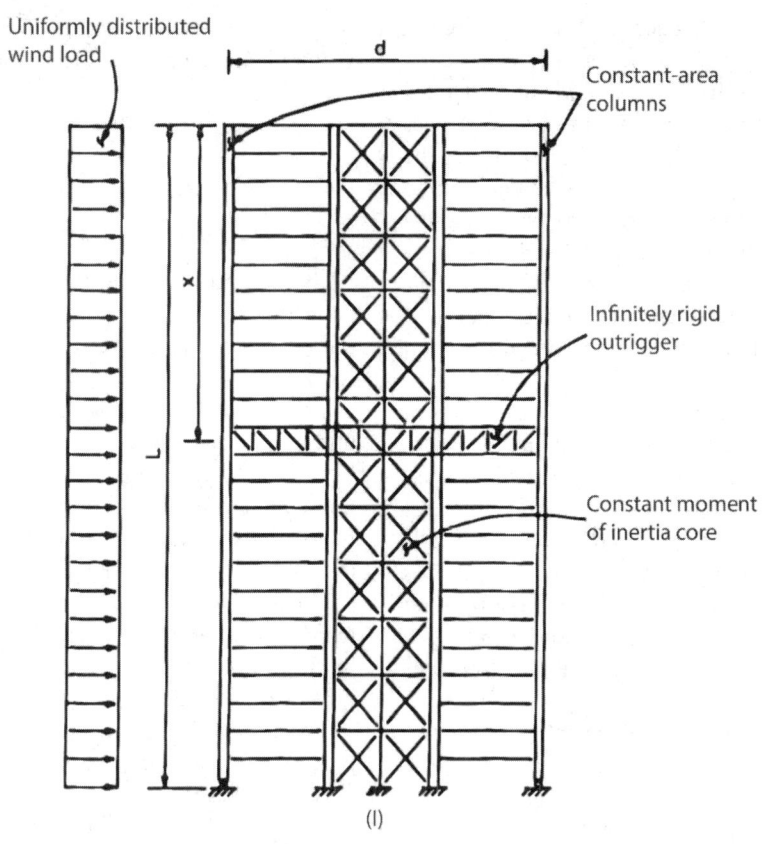

Figure 4.29 (*Continued*)

be smaller than if the free-standing core alone resisted the loading. The result is to increase the effective depth of the structure when it flexes as a vertical cantilever, by inducing tension in the windward columns and compression in the leeward columns.

In addition to those columns located at the ends of the outriggers, it is usual to also mobilize other peripheral columns to assist in restraining the outriggers. This is achieved by including a deep spandrel girder, or a "belt truss," around the structure at the levels of the outriggers.

To make the outriggers and belt truss adequately stiff in flexure and shear, they are made at least one, and often two-story deep. It is also possible to use diagonals extending through several floors to act as outriggers as shown in Fig. 4.29c. And finally, girders at each floor may be transformed into outriggers by moment connections to the core and, if desired, to the exterior columns as well (Fig. 4.29d). In all these cases it should be noted that while the outrigger system is very effective in increasing the structure's flexural stiffness, it does not increase its resistance to shear, which has to be carried mainly by the core.

In the following sub-section, the stiffening effect of a single outrigger located at the top of the structure is examined first. Next, the effect of lowering the truss along the height is studied with the object of finding the most optimum location for minimizing the building drift. Finally, a compatibility analysis for the optimum location of a two-outrigger structure is presented.

4.8.2   Physical Behavior

A traditional approach to lateral bracing is to provide braced frames around the building core with a system of moment connected frames at the exterior. However, for buildings taller than 40 stories or so, the core, if kept consistent with the vertical transportation, does not provide adequate stiffness to keep the drift down to acceptable limits.

A method of increasing the efficiency of the system is to use a "cap" or "hat" truss to tie the braced core to the exterior columns. The tied columns, in addition to the usual function of supporting the gravity loads, also assist in resisting the overturning moments. The tie-down action of the cap truss creates a restoring couple resulting in a point of inflection in its deflection curve. This reversal in curvature reduces the bending moment in the core, and hence the building drift at the top. The belt truss functioning as a horizontal fascia stiffener mobilizes other exterior columns to take part in restraining the rotation of the cap truss. A general improvement of up to 25 to 30 percent in stiffness can be realized by using a sufficiently stiff belt truss to tie all the peripheral columns.

The cap and belt truss system is also beneficial in equalizing the differential movement between interior and exterior columns resulting from temperature effects, and unequal shortening due to axial load imbalance.

The behavior of the system is explained with reference to Fig. 4.29e which shows a schematic plan of a core braced building. Although the building has a belt truss connecting all the exterior columns, we will assume that their effect may be represented by two equivalent columns tied to each end of the cap truss (Fig. 4.29f,g). This idealization is not necessary for evaluating the restraining effect of columns, but keeps the explanation simple.

The structure in Fig. 4.29f,g is basically a cantilever with a single degree of redundancy because its rotation is restrained only at the top by the stretching and shortening of windward and leeward columns. The resultant of the tensile and compressive forces is equivalent to a restoring couple opposing the rotation of the core.

In terms of its restraining action, the cap truss may be conceptualized as an equivalent spring located at the top. Its stiffness, defined as the restoring couple due to a unit rotation of the core, may be calculated as follows.

Since the cap truss is assumed rigid, the exterior columns undergo axial compression and tension equal in magnitude to the product of the rotation of core and their distance from the center of the core. Assuming that the columns are located at a distance of $d/2$ from the center of the core, the axial deformation of the columns is equal to $\theta \times \dfrac{d}{2}$ where θ is the rotation of the core. Since by our definition, $\theta = 1$, the axial deformation is equal to $d/2$ units. The axial load P in the columns corresponding to this deformation is given by

$$P = \frac{AEd}{2L}$$

where P = axial load in the columns
   A = area of columns
   E = modulus of elasticity
   d = distance between the exterior columns
   L = height of the building

The restoring couple which is the rotational stiffness of the cap truss is given by the product of the axial load in

the columns and the distance from the center of core. Using the notation K for the stiffness we get

$$K = \sum P_i d_i$$
$$= P \times \frac{d}{2} \times 2$$
$$= Pd$$

The equivalent spring located at the top reduces the deflection of the core by inducing a reversal in its bending curvature as shown in Fig. 4.29. The magnitude of reduction in drift due to the cap truss depends on the given geometry of the building and the area of the columns. The stiffnesses of the cap and belt truss which in a practical building are not infinitely large, also influence the reduction in drift.

Now let us consider the interaction of the core with the outrigger located not at top but somewhere along the height. How does the location influence its effectiveness? Is top location the best? What are the structural implications if it is moved toward the bottom, say to the mid-height of the building? Before we seek a general solution for these rather intriguing questions, it is useful to study its restraining action for a few specific locations, for example at the top, three-quarters of the height, mid-height, and quarter height.

A compatibility method is chosen for the analysis in which the rotation of the core at the outrigger level is matched with the corresponding rotation of the outrigger. Since the primary action of the outrigger is to reduce the rotation of the core, conceptually it can be replaced by an equivalent spring. The stiffness of the spring, defined as the moment required to induce unit rotation of the core, however, it not a constant but depends on its location. Its derivation will be explained shortly but for now we will proceed with the idea that the outrigger is mathematically equivalent to a spring with a stiffness inversely proportional to its distance from the bottom. In other words, its stiffness is minimum when at the top, and maximum at the bottom.

4.8.3   Deflection Calculations

Case 1: Outrigger located at top, $x = 0$, $z = L$ (Fig 4.29 h)

The rotation compatibility condition at $Z = L$ can be written as

$$\theta_W - \theta_S = \theta_L \tag{4.5}$$

where θ_w = rotation of the cantilever at $Z = L$ due to a uniform lateral load W, in radians

θ_S = rotation due to spring restraint located at $Z = L$, in radians. The negative sign indicates that the rotation of the cantilever due to the spring stiffness acts in a direction opposite to the rotation due to external load

θ_L = final rotation of the cantilever at $Z = L$, in radians

For a cantilever with uniform moment of inertia I and modulus of elasticity E subjected to uniform horizontal load W,

$$\theta_W = \frac{WL^3}{6EI} \qquad (4.6)$$

If M_1 and K_1 represent the moment and stiffness of the spring located at $Z = L$, Eq. (4.5) can be rewritten thus

$$\frac{WL^3}{6EI} - \frac{M_1 L}{EI} = \frac{M_1}{K_1} \qquad (4.7)$$

and

$$M_1 = \frac{WL^3}{1/K_1 + L/EI} \qquad (4.8)$$

The resulting drift Δ_1 at the building top can be obtained by superposing the deflection of the cantilever due to external uniform load W and the deflection due to the moment induced by the spring, thus

$$\Delta_1 = \Delta_{load} = \Delta_{spring}$$
$$= \frac{WL^4}{8EI} - \frac{M_1 L^2}{2EI}. \qquad (4.9)$$
$$= \frac{L^2}{2EI}\left(\frac{WL^2}{4} - M_1\right) \qquad (4.10)$$

Case 2: Outrigger located at $z = \dfrac{3L}{4}$ (Fig. 4.29i)

The general expression for lateral deflection y for a cantilever subjected to a uniform lateral load is given by

$$y = \frac{W}{24EI}(x^4 - 4L^3 x + 3L^4) \qquad (4.11)$$

Note that x is measured from the top.

Differentiating with respect to x, the general expression for slope of the cantilever is given by

$$\frac{dy}{dx} = \frac{W}{6EI}(x^3 - L^3) \qquad (4.12)$$

The slope at the spring location is given by substituting $Z = 3L/4$, that is, $x = L/4$ in Eq. (4.12). Thus

$$\frac{dy}{dx}\left(\text{at } z = \frac{3L}{4}\right) = \frac{W}{6EI}\left(\frac{L^3}{64} - L^3\right)$$
$$= \frac{WL^3}{6EI} \times \frac{63}{64} \qquad (4.13)$$

Using the notation M_2 and K_2 to represent the moment and stiffness of spring at $Z = 3L/4$, the compatibility equation at location 2 can be written thus

$$\frac{WL^3}{6EI}\left(\frac{63}{64}\right) - \frac{M_2}{EI}\left(\frac{3L}{4}\right) = \frac{M_2}{K_2} \qquad (4.14)$$

Noting that $K_2 = 4K_1/3$, the expression for M_2 can be written thus

$$M_2 = \left(\frac{WL^3/6EI}{1/K_1 + L/EI}\right)\frac{63/64}{3/4} = \left(\frac{WL^3/6EI}{1/K_1 + L/EI}\right)1.31 \qquad (4.15)$$

Noting that the terms in the parenthesis represent M_1, Eq. (4.15) can be expressed in terms of M_1

$$M_2 = 1.31 M_1$$

The drift is given by the relation:

$$\Delta_2 = \frac{WL^4}{8EI} - \frac{M_2 3L}{4EI}\left(L - \frac{3L}{8}\right) \qquad (4.16)$$

or

$$\Delta_2 = \frac{L^2}{2EI}\left(\frac{WL^2}{4} - 1.23 M_1\right) \qquad (4.17)$$

Case 3: Outrigger at mid-height, $z = \dfrac{L}{2}$ (Fig. 4.29j)

The rotation at $Z = L/2$ due to external load W can be shown to be equal to $7WL^3/48EI$, giving the rotation compatibility equation:

$$\frac{7WL^3}{48EI} - \frac{M_3 L}{2EI} = \frac{M_3}{K_3} \qquad (4.18)$$

where M_3 and K_3 represent the moment and stiffness of the spring at $Z = L/2$. Noting that $K_3 = 2K_1$, the expression for M_3 works out as

$$M_3 = \left(\frac{WL^3/6EI}{1/K_1 + L/EI}\right) \times \frac{7}{4} \qquad (4.19)$$

Since the expression in the parentheses is equal to M_1, M_3 can be expressed in terms of M_1

$$M_3 = 1.75 M_1 \qquad (4.20)$$

The drift is given by the equation:

$$\Delta_3 = \frac{WL^4}{8EI} - \frac{M_3 L}{2EI}\left(L - \frac{L}{4}\right) \tag{4.21}$$

or

$$\Delta_3 = \frac{L^2}{2EI}\left(\frac{WL^2}{4} - 1.31 M_1\right) \tag{4.22}$$

Case 4: Outriggers at quarter-height, $z = \dfrac{L}{4}$ (Fig. 4.29k)

The rotation at $Z = L/4$ due to uniform lateral load can be shown to be equal to $WL^3/6EI[(37/64)]$, giving the rotation compatibility equation:

$$\frac{WL^3}{6EI}\left(\frac{37}{64}\right) - \frac{M_4 L}{4EI} = \frac{M_4}{K_4} \tag{4.23}$$

where M_4 and K_4 represent the moment and stiffness of the spring at $Z = L/4$. Noting that $K_4 = 4K_1$, M_4 in Eq. (4.23) can be expressed in terms of M_1.

$$M_4 = 2.3 M_1 \tag{4.24}$$

The drift for this case is given by the expression:

$$\Delta_4 = \frac{WL^4}{8EI} - \frac{M_4 L}{4EI}\left(L - \frac{L}{8}\right) \tag{4.25}$$

or

$$\Delta_4 = \frac{L^2}{2EI}\left(\frac{WL^2}{4} - M_1\right) \tag{4.26}$$

Equations (4.10), (4.14), (4.22), and (4.26) give the building drift for four different locations of the belt and outrigger trusses.

The value of K_1 which corresponds to stiffness of the spring when it is located at $Z = L$ can be derived as follows. A unit rotation given to the core at the top results in extension and compression of all perimeter columns, the magnitudes of which are given by their respective distances from the center of gravity of the core. The resulting force multiplied by the lever arm gives the value for stiffness K_1. Thus, if p_1 is measured in kips and the distance in feet, K_1 has units of kip feet. The force p in each exterior column is given by the relation $p = AE\,\delta/L$; since by definition δ corresponds to column extension or compression due to unit rotation of the core, $\delta = d/2$, where d is the distance between the exterior columns. Therefore,

$$p = \frac{AE}{L}\left(\frac{d}{2}\right) \tag{4.27}$$

and its contribution to the stiffness K_1 is given by the relation:

$$\begin{aligned} K_i &= p_i d \\ &= \frac{A_i E}{L}\frac{d^2}{2} \end{aligned} \tag{4.28}$$

The total contribution of all exterior columns on the long faces is given by the summation relation:

$$K_1 = \sum_{i=1}^{n} K_i = \frac{d^2 E}{2L}\sum_{i=1}^{n} A_i \tag{4.29}$$

The contribution of the columns on the short faces can be worked out in a similar manner.

4.8.4   Optimum Location of Single Truss

The preceding analysis has indicated that the beneficial action of the outrigger is a function of two distinct characteristics, the stiffness of the equivalent spring and the magnitude of the rotation of cantilever at the spring location due to external loads. The stiffness varies inversely as the distance of the outrigger from the base. For example, it is at a minimum when located at the top and a maximum when at the bottom. The rotation of the free cantilever for a uniformly distributed horizontal load varies parabolically from a maximum value at the top to zero at the bottom. Therefore, from the point of view of spring stiffness it is desirable to locate the outrigger at the bottom, whereas from a consideration of rotation, the converse is true. It is obvious that the optimum location is somewhere in between.

The method of analysis is based on the following assumptions.

1. The area of the perimeter columns and the moment of inertia of the core are uniform throughout the height.

2. The outrigger and the belt trusses are flexurally rigid and induce only axial forces in the columns.

3. The lateral resistance is provided only by the bending resistance of the core and the tiedown action of the exterior columns.

4. The core is rigidly fixed at the base.

5. The rotation of the core due to shear deformation is negligible.

6. The intensity of lateral load remains constant for the whole height.

7. The structure is linearly elastic.

Figure 4.29l shows the analytical model for the single out-rigger truss located at a distance x from the top. As before, a compatibility method is used by matching the rotations. From the compatibility relation, the restoring moment M_x at the location of the outrigger is evaluated. Next, the deflection of the core at top due to the restoring couple is calculated and maximized using the principles of calculus. Solution of the resulting third-degree polynomial yields an optimum value of x for which the deflection of the core due to external load is minimum. The detail calculations are as follows.

The rotation θ of the cantilever at a distance x from the top, due to a uniformly distributed load w is given by the relation:

$$\theta = \frac{W}{EI}(x^3 - L^3)$$

The rotation at top due to the restoring couple M_x is given by the relation.

$$\theta = \frac{M_x}{EI}(L - x)$$

The compatibility relation at x is given by

$$\frac{W}{6EI}(x^3 - L^3) - \frac{M_x}{EI}(L - x) = \frac{M_x}{K_x} \quad (4.30)$$

where W = intensity of the wind load per unit height of the structure

M_x = the restoring moment due to outrigger restraint

K_x = spring stiffness at x equal to $\dfrac{AE}{(L-x)} \times \dfrac{d^2}{2}$

E = modulus of elasticity of the core
I = moment of inertia of the core
A = area of the perimeter columns
L = height of the building
x = location of truss measured from the top
d = distance out-to-out of columns

Next, obtain the deflection at the top of the structure due to M_x:

$$Y_M = \frac{M_x(L - x)(L + x)}{2EI} \quad (4.31)$$

From our definition, the optimum location of the belt truss is that location for which the deflection Y_M is a maximum. This is obtained by differentiating Eq. (4.31) with respect to x and equating to zero. Thus,

$$\frac{d}{dx}\left[\frac{W(x^3 - L^3)(L + x)}{12(EI)^2(1/AE + 1/EI)} \right] = 0 \quad (4.32)$$

$$4x^2 + 3x^2 L - L^3 = 0 \quad (4.33)$$

giving the optimum location at $x = 0.455L$. If the flexibility of the outrigger is taken into account, even for the overly simplified model, the corresponding equation for the solution of x becomes too involved for hand calculations. Extension of the solution to two or more outrigger trusses further complicates the solution, thus necessitating a formulation suitable for a computer. This is considered next.

4.8.5 Optimum Location for a Two-Outrigger System

In the preceding analyses only one compatibility equation was necessary because the one-outrigger structure is once redundant. On the other hand a two-outrigger structure is twice redundant requiring two compatibility equations corresponding to the degree of redundancy.

The analytical model is shown schematically in Fig. 4.30a. Since the formulation is for computer analysis, the column and core properties, and the distribution of lateral

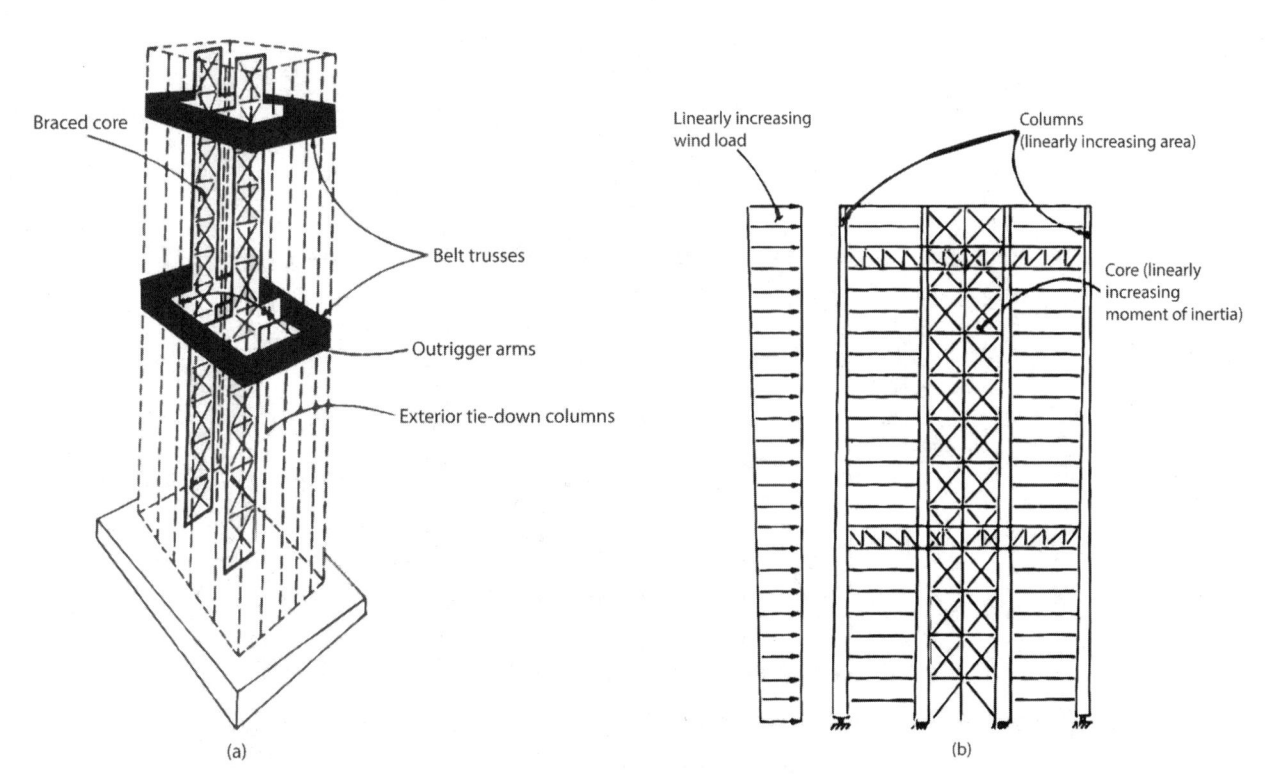

(a)   (b)

Figure 4.30 (a) Conceptual model for two outriggers; (b) analytical model.

load need not be uniform. A linear variation which is a reasonable representation of a real structure is used along the building height as shown in Fig. 4.30b.

4.8.5.1 COMPUTER SOLUTION

The analytical model is shown in Fig. 4.31. A flexible approach has been employed for the solution. The method is briefly explained with reference to the example problem. The moments at the outrigger locations are chosen as the unknown arbitrary constants M_1 and M_2, and the structure is released by removing the rotational restraints, making it statically determinate, so that the effect of any loading can be easily calculated. The flexibility coefficients f_{11} and f_{22} are calculated by using integrals of the form

$$\int \frac{m_i m_j}{EI} ds + k \int \frac{s_i s_j}{GA} ds + \int \frac{n_i n_j\, ds}{EA} \qquad (4.34)$$

where m, s, and n represent the moment, shear force, and the axial load distribution on the statically determinate

system due to the application of a unit moment at the location and in the direction of the arbitrary constants. E, G, I, and A are the familiar notations for material and member properties of the element of the structure for which the integral is being calculated. Note that different forms of energy are significant in different members.

Next, the compatibility equations for the rotations at the truss locations are set up and the magnitudes of the arbitrary constants M_1 and M_2 obtained. The tip deflection for the structure is obtained by superposition of the solutions for the external load and for the moments M_1 and M_2. A single solution to the problem is trivial and may easily be carried out by hand calculations. A computer solution is necessary, however, since the object of the exercise is to seek an optimum combination of the truss locations to minimize lateral drift, requiring many solutions for different truss locations. A computer program was written for this purpose and computations were carried out for a 46-story example structure shown in Fig. 4.31. The results of the analysis are given in the form of graphs in Fig. 4.32.

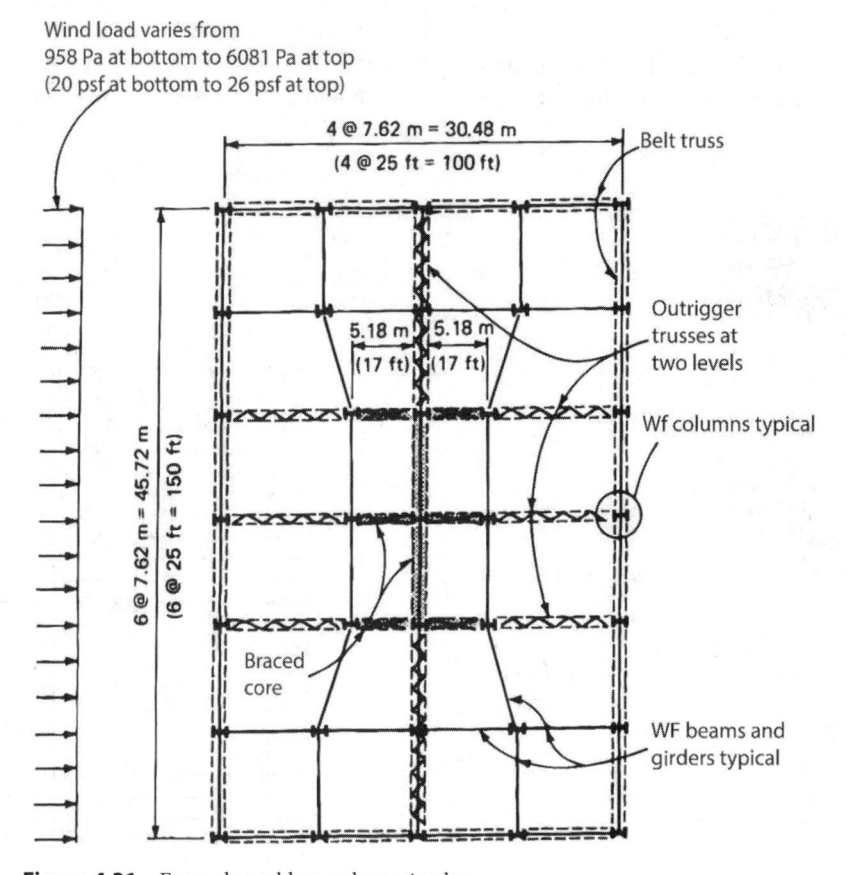

Figure 4.31 Example problem: schematic plan.

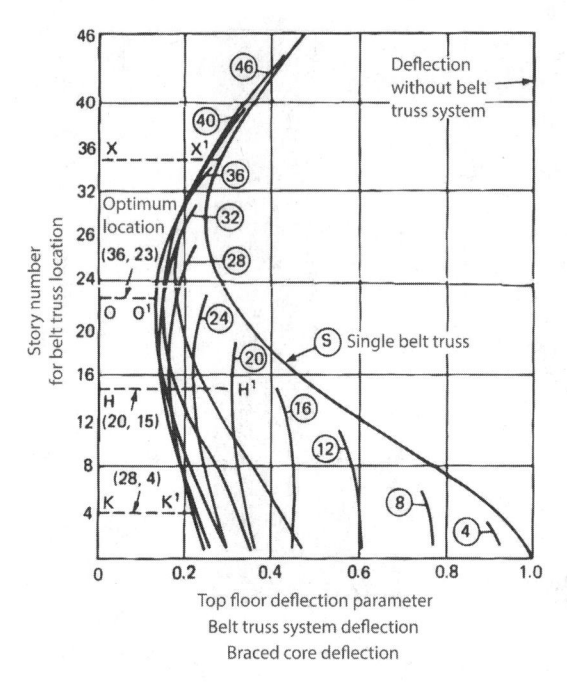

Figure 4.32 Graph for optimum belt truss location.

4.8.5.2 EXPLANATION OF GRAPHS

The magnitudes of the top floor deflection of the structure for three assumed modes of resistance have been presented in a nondimensional form in Fig. 4.32. The vertical ordinate with the value of the deflection parameter equal to 1 represents the top floor deflection, obtained by assuming that there are no belt trusses. The resistance is provided by the cantilever action of the braced core alone. The curve designated as S represents the deflection, assuming that a single belt truss located anywhere along the height of the structure is acting in conjunction with the braced core. The deflection for a particular location of the truss is obtained by the horizontal distance between the curve S and the vertical axis measured at the floor level (e.g., distance XX' multiplied by the cantilever deflection gives the top floor deflection for the location of the belt truss at floor 35). It is seen that the drift is quite sensitive to the truss location. The most favorable location is at floor 27; the resulting deflection is reduced to less than a third of the pure cantilever deflection.

The curves designated as 4, 8, …, 46 represent the top floor deflections obtained by assuming that there are two belt trusses located anywhere along the height of the structure. To obtain each curve, the location of the upper outrigger was considered fixed in relation to the building height, while the location of the lower outrigger was moved in single-story increments, starting from the first floor to the floor immediately below the top outrigger.

The number designations of the curves represent the floor number at which the upper outrigger is located. The second outrigger location is given on the vertical axis. The horizontal distance between the curve and the vertical axis is the top deflection parameter for the particular combination of truss locations given by the curve designation and the vertical ordinate. For example, let us assume that the deflection at the top is desired for the combination (20, 15), the numbers 20 and 15 being the floors at which the upper and lower outriggers are located. The procedure is to select the curve with the designation 20 and to draw a horizontal line from the vertical ordinate at 15 to this curve. The required top deflection parameter is the horizontal distance between the vertical axis and the curve 20 (distance HH' in Fig. 4.32). Similarly, distance KK' gives the deflection parameter for the combination (28, 4). It is seen from Fig. 4.32 that the relative location of the trusses has a significant effect on controlling the drift. Furthermore, it is evident that a deflection very nearly equal to the optimum solution can be obtained for a number of combinations. For the example problem, a deflection parameter of 0.15, which differs negligibly from the optimum value of 0.13, is achieved by the combinations (40, 23), (32, 23), etc. The effectiveness of the belt truss system is self-evident from the figure.

4.8.6 Example Projects

Figures 4.33a,b shows photographs of The First Wisconsin Center, a 42-story, 1.3 million square foot (120,770 m²) bank and office building which utilizes the concept of belt and outrigger system of lateral bracing. The building rises 601 ft (183 m) from a two-level glass-enclosed plaza and sports three belt trusses located at the bottom, middle, and top of the building. The belt truss at the bottom serves as a transfer truss, eliminating every other exterior column. Outrigger trusses are used at the top and middle of the building to engage the braced core to the exterior columns. Mechanical levels are located at the trussed floors. The architecture and engineering for the project is by the Chicago Office of Skidmore, Owings & Merrill.

As a second example, Fig. 4.33c shows a schematic partial elevation of a wind bracing system for a building in Houston, Texas called One Houston Center. The building, consisting of 48 stories, rises to a height of 681 ft (207.5 m) above grade. To minimize the building drift, two-story-deep outrigger trusses tie the perimeter columns to the K-braced core between the 33rd and 35th levels. Three outrigger trusses on each side of the core run through the building space. Use of these outrigger trusses helped in reducing the drift to less than 1/460 of the building height. The structural engineering for the project was by

(a)

(b)

Figure 4.33a,b   First Wisconsin Center, Milwaukee.

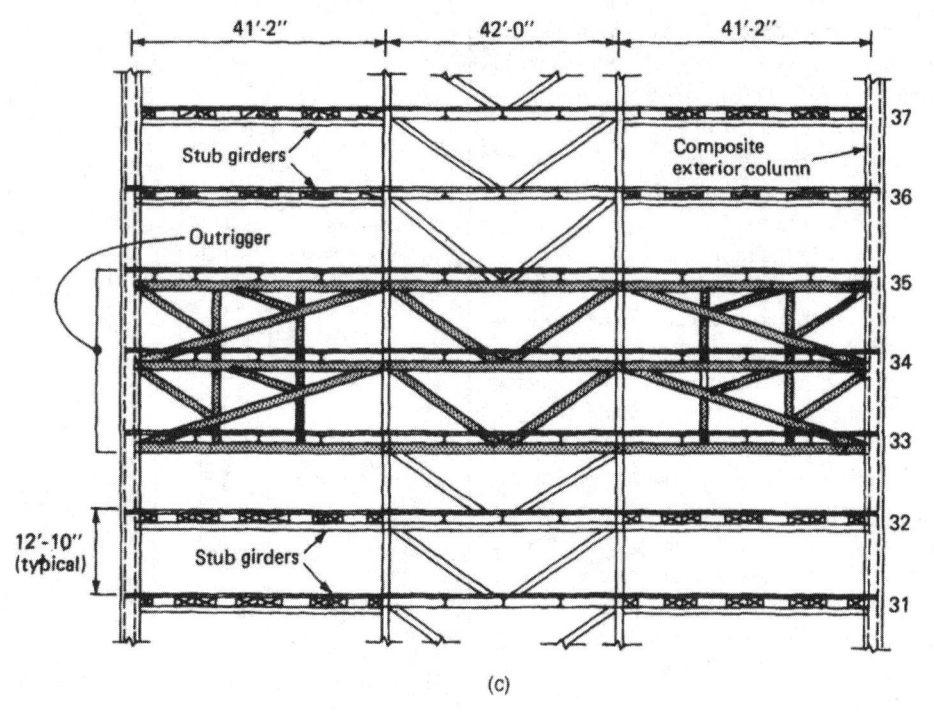

Figure 4.33c   One Houston Center.

the Houston office of Walter P. Moore & Associates. A novel system of floor framing, called a stub girder system, was used to frame typical bays measuring 41 ft 2 in. by 30 ft (12.54 m by 9.14 m). This unique system of floor framing, which is discussed in Chap. 9, attempts to reduce the building cost by simultaneously minimizing structural steel tonnage and floor-to-floor height.

4.8.7   Concluding Remarks

Although the analysis presented herein is based on certain simplifying assumptions, it is believed that the results do provide sufficiently accurate information for determining the optimum location of belt trusses in high-rise structures. Significant reductions in building drift can be obtained by judiciously selecting the locations. Furthermore, since solutions very nearly equal to the optimum solution are obtained for various combinations of truss locations, it should be relatively easy to choose a combination that satisfies simultaneously the structural, mechanical, and architectural requirements.

4.9   FRAMED TUBE SYSTEM

4.9.1   Introduction

In its simplest terms, the tube system can be defined as a fully three-dimensional system that utilizes the entire building perimeter to resist lateral loads.

Several buildings that were considered among the world's tallest buildings are tubular systems. They are the 110-story Sears Tower, the 100-story John Hancock Building, and the 83-story Standard Oil Building, all in Chicago, and the 110-story World Trade Center towers in New York. The earliest application of the tubular concept is credited to the late Dr. Fazlur Khan of the architectural engineering firm of Skidmore, Owings & Merill, who first introduced the system in a 43-story apartment building in Chicago.

The introduction of the tubular system for resisting lateral loads has brought about a revolution in the design of high-rise buildings. All recent high-rise buildings in excess of 50 to 60 stories employ the tubular concept in one form or another. In essence, the system strives to create a three-dimensional wall-like structure around the building exterior. In a framed tube this is achieved by arranging closely spaced columns and deep spandrels around the entire perimeter of the building. Because the entire lateral load is resisted by the perimeter frame, the interior floor plan is kept relatively free of core bracing and large columns, thus increasing the net leasable area of the building. As a trade-off, views from the interior of the building may be hindered by closely spaced exterior columns.

The necessary requirement to create a wall-like structure is to place columns on the exterior relatively close to each other and to use deep spandrel beams to tie the columns. The structural optimization reduces to examining different column spacings and member proportions. In practice, the framed tubular behavior is achieved by placing columns at 10 ft (3.05 m) to as much as 20 ft (6.1 m) apart, with spandrel depth varying from 3 to 5 ft (0.90 to 1.52 m).

The tube system can be constructed of reinforced concrete, structural steel, or a combination of the two, termed composite construction.

The method of achieving the tubular behavior by using columns on close centers connected by a deep spandrel is by far the most used system because rectangular windows can be accommodated in this design. A somewhat different system for steel buildings that permits larger spacing of columns is called the braced tube which has diagonal or K-type bracing at the building exterior. A similar concept for concrete buildings is to infill window penetrations in a systematic pattern to achieve the same effect as a diagonal or K-type bracing. Yet another is to use two or more tubes tied together to form a bundled tube. In this sub-section, a description of the framed tube system is given; other systems are explained in later sections.

4.9.2   Framed Tube Behavior

To understand the behavior of a framed tube, consider a square-shaped 50-story building as shown in Fig. 4.34a consisting of closely spaced exterior columns and deep spandrel beams. Assuming that the interior columns are designed for gravity loading only, their contribution to lateral load resistance is negligible. The floor system, as in other types of lateral bracing systems, is considered a rigid diaphragm and is assumed to distribute the wind load to various elements according to their stiffness. Its contribution to lateral resistance in terms of its out-of-plane action is considered negligible. The system resisting the lateral load thus comprises of four orthogonal rigidly jointed frame panels forming a tube in plan as shown in Fig. 4.34b.

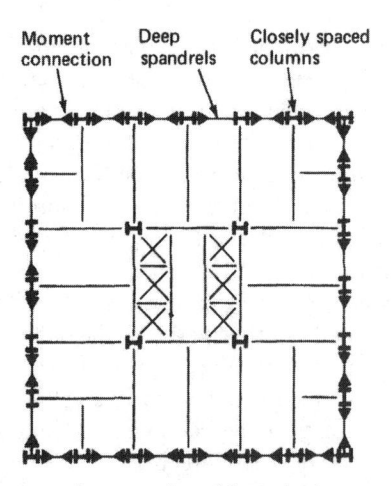

Figure 4.34a   Schematic plan of framed tube.

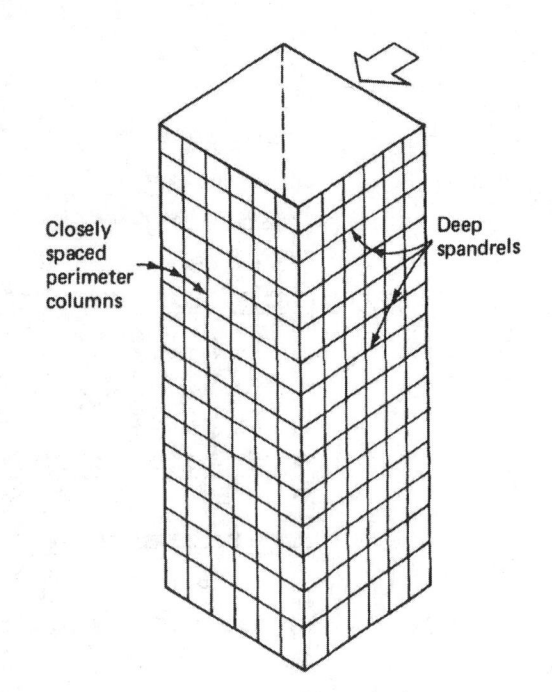

Figure 4.34b   Isometric view of framed tube.

The frame panels are formed by closely spaced perimeter columns that are connected by deep spandrel beams at each floor level. In such structures, the "strong" bending direction of the columns is aligned along the face of the building, in contrast to the typical rigid frame bent structure where it is aligned perpendicular to the face. The basic requirement has been to place as much of the lateral load-carrying material at the extreme edges of the building to maximize the inertia of the building's cross-section. Consequently, in many structures of this form, the external tube is designed to resist the entire lateral loading. The frames parallel to the lateral load act as "web" of the perforated tube, while the frames normal to the loads act as "flanges." Vertical gravity forces are resisted partly by the exterior frames and partly by some interior columns or an interior core. When subjected to bending under the action of lateral forces, the primary mode of action is that of a conventional vertical cantilevered tube, in which the columns on opposite sides of the neutral axis are subjected to tensile and compressive forces. In addition, the frames parallel to the direction of the lateral load are subjected to the usual in-plane bending, and the shearing or racking action associated with an independent rigid frame.

The discrete columns and spandrels may be considered, in a conceptual sense, equivalent to a continuous three-dimensional wall. The model becomes a hollow tube cantilevering from the ground with a basic stress distribution as shown in Fig. 4.35.

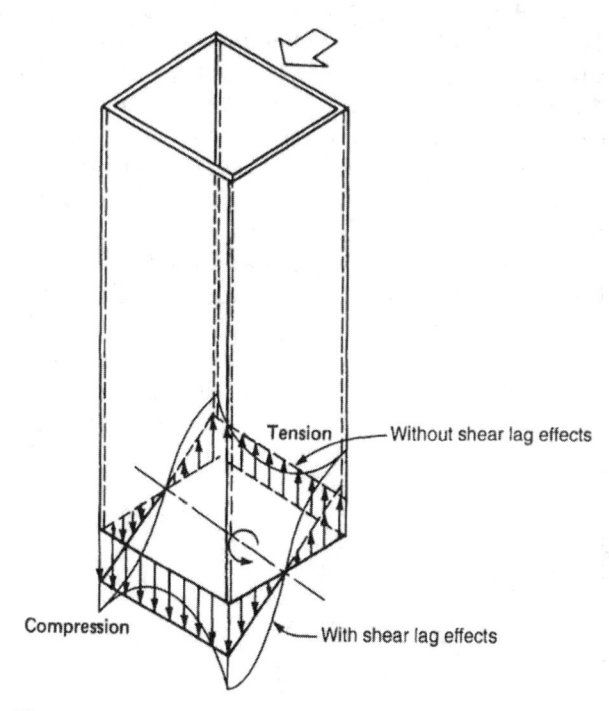

Figure 4.35 Axial stress distribution in square hollow tube with and without shear lag.

Although the structure has a tube-like form, its behavior is much more complex than that of a solid tube; unlike a solid tube, it is subjected to shear lag effects. The influence of shear lag is to increase the axial stresses in the corner columns and reduce those in the inner columns of both the flange and the web panels as shown by the dotted lines, in Fig. 4.35. Ignoring the shear lag consequences for now, the analogy of the hollow tube can be used to visualize the axial stress distribution in buildings with other plan forms such as rectangular, circular, and triangular as shown in Fig. 4.36a–c. This philosophy of creating a fully three-dimensional structural system utilizing the entire building footprint to resist lateral loads has allowed for considerable freedom in manipulating building plans. The rigorous organization of orthogonal bay spacing required with the previous types of bracing is no longer necessary. The only requirements are for the structure to be continuous around the exterior to invoke a three-dimensional response, and be of a closed-cell form, to resist torsional loads. Depending upon the height and dimensions of the building, the exterior column spacing is usually of the order of 10 to 15 ft (3 to 4.6 m), although a spacing as close as 3.8 ft (1.0 m) has been used for the 110-story World Trade Center twin towers, New York

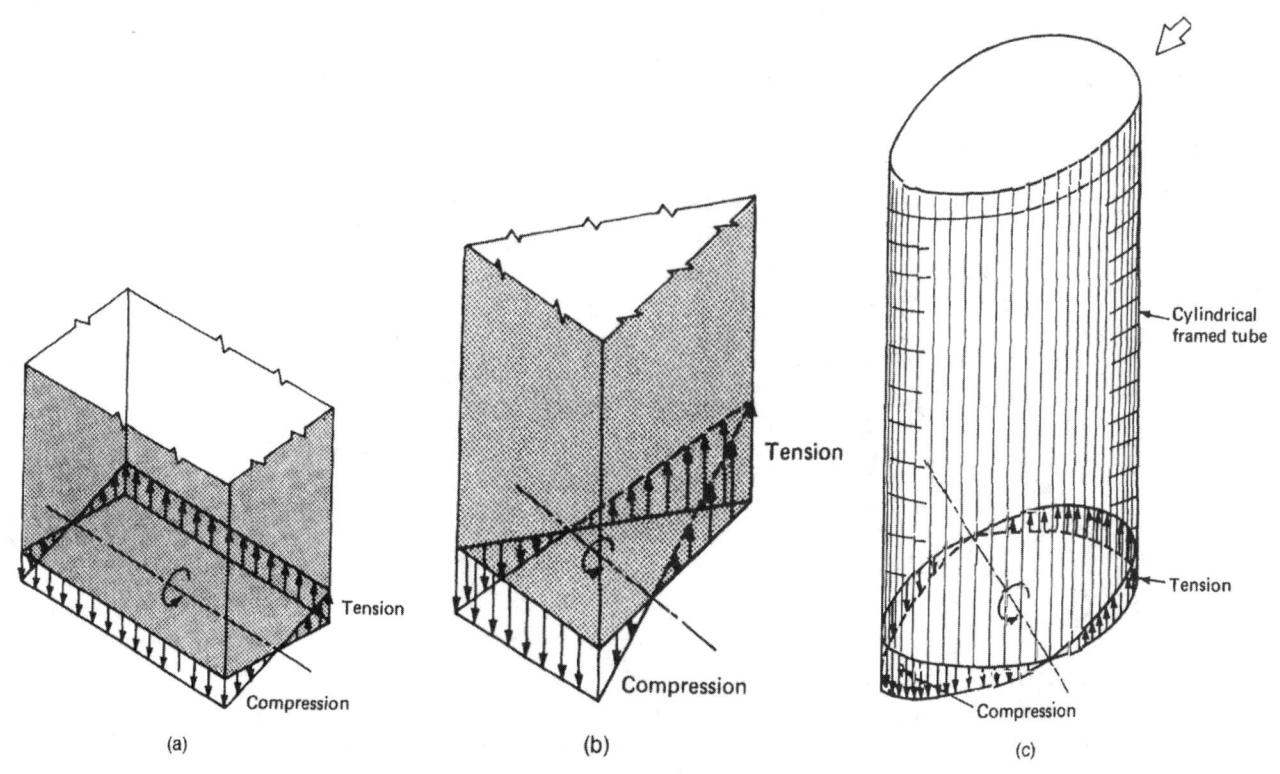

Figure 4.36 Axial stress distribution in tube structures: (a) rectangular tube; (b) triangular tube; (c) circular tube.

(Fig. 4.37a) (collapsed in 2001 during the September 11 attack). The efficiency of the system is directly related to the building height-to-width ratio, plan dimensions, spacing, and size of columns and spandrels.

Figure 4.37 shows examples of free-form tubular configurations. Although in simplistic terms the tube is similar to a hollow cantilever, in reality its response to lateral loads is in a combined bending and shear mode. The overall bending of the tube is due to axial shortening and elongation of the columns while the shear deformation is due to bending of individual columns and spandrels. The underlying principle for an efficient design is to eliminate or minimize the shear deformation so that the tower as a whole bends essentially as a cantilever.

4.9.3   Shear Lag Phenomenon

The primary action of the hollow tube as discussed previously is complicated by the fact that the flexibility of the spandrel beams increases the stresses in the corner columns and reduces the same in the inner columns of both the flange and the web panels. Contrary to what one may expect, even for a solid-wall tube, the distribution of axial forces is not uniform over the windward and leeward walls and linear over the side walls. The behavior can be readily appreciated by considering the shear deformations of the tube walls which are relatively "thin" as compared to the height and plan dimensions of the building. The extent to which the actual axial load distribution departs from the ideal is termed "shear lag

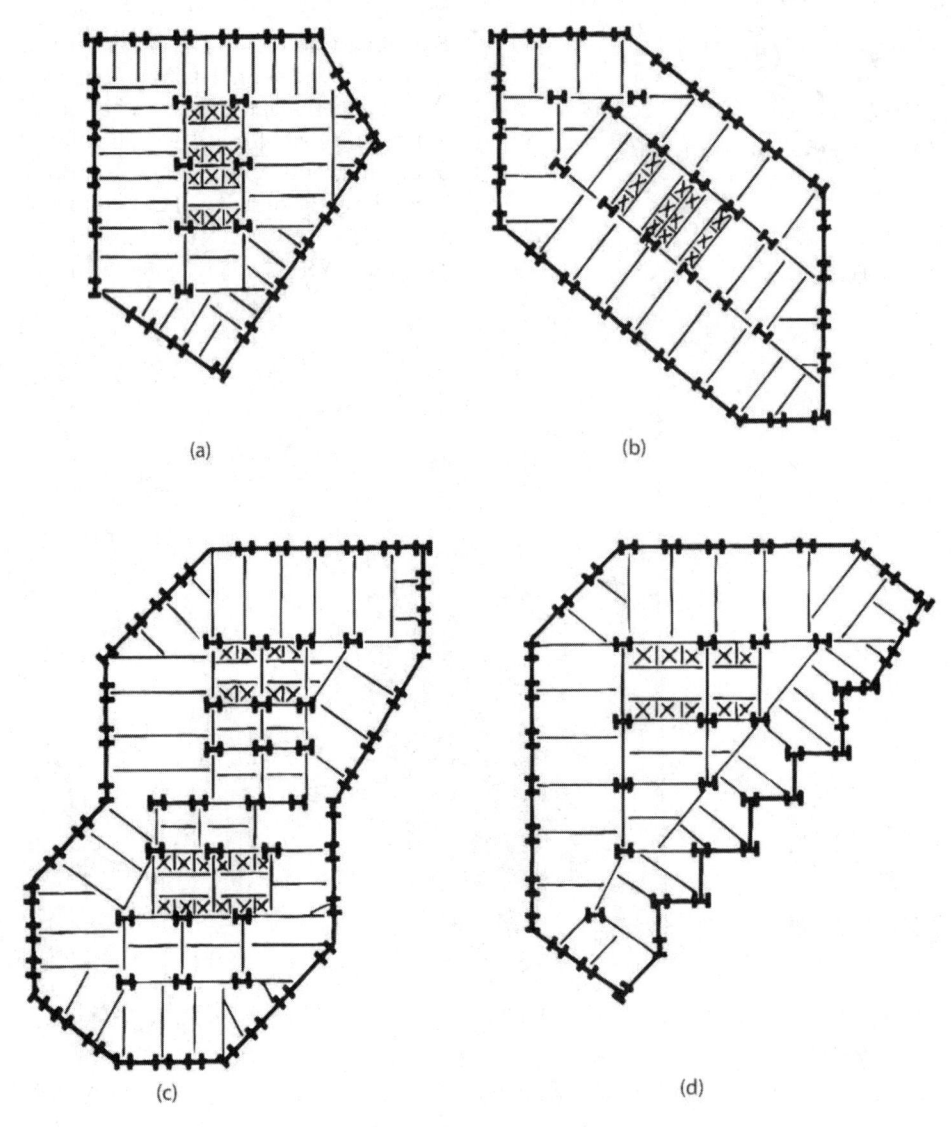

(a)

(b)

(c)

(d)

Figure 4.37   Free-form tubular configurations.

effect." An understanding of this phenomenon is essential in developing optimal tubular systems.

The closest structure to a perfect tube is a system of continuous perimeter walls without any discontinuities. A hollow box represents such a system, as shown in Fig. 4.38a. In order to fix ideas, assume the hollow box represents a 50-story steel building with a typical floor-to-floor height of 13 ft (3.94 m), giving a total height of 650 ft (198 m) for 50 stories. Assume the building is square with a plan dimension of 110 by 110 ft (33.5 by 33.3 m). For a normally proportioned 50-story building, in a seismically benign zone, such as zone 1 per the Uniform Building Code (UBC), or seismic design category (SDC) A or B per the International Building Code (IBC), the unit quantity of structural steel, including that required for gravity, is in the range of 22 to 24 psf (1053 to 1149 Pa) of the gross area of the building. Conservatively, assume that all of the 24 psf (1149 Psa) of structural steel is available for the lateral bracing of the building. The most efficient manner of using this material, in an academic sense, is to convert the 24 psf of steel into an equivalent wall located at the perimeter of the tube as shown below.

Total steel available for bracing = 24 × 110 × 1150 × 50 = 14,520,000 lb (64,585 kN)
Area of perimeter wall = 4 × 110 × 650 = 286,000 ft² (26,569 m²)
Equivalent thickness of wall = 14,520,000/286,000 × 3.4 = 15 in. (381 mm)

In comparison to the plan dimensions of the building, the calculated wall thickness of 15 in., (381 mm) is relatively small, giving a length-to-thickness ratio of 1:88. Because of this characteristic, the structure has a tendency to behave like a thin-walled beam. In a thin-walled beam, the shear stresses and strains are much larger than those in a solid beam and often result in large shearing deformations with a significant effect on the distribution of bending stresses. Because of the resulting large shear strains, the usual assumption used in engineers' bending theory is violated. This assumption, which states that plane sections before bending remain plane after bending, is known as the *Bernoulli hypothesis* and forms the basis for mathematical relations used in engineering mechanics. However, in thin-walled structures, the large shearing strains cause the plane of bending to distort. For the hollow box structure, the element E on the flange face distorts as shown in Fig. 4.38c. The final outcome due to the cumulative effect of distortion of all such elements is that under lateral load, the originally flat plane of the cross-section distorts as shown in Fig. 4.39. Because of these distortions, the simple stress distribution given by the engineers' theory of bending is no longer applicable. The bending stresses will not be proportional to the distance from the neutral axis of the section; the stress at the center of the flanges "lags" behind the stress near the web because of the lack of shear stiffness of the wall panel. This phenomenon is known as shear lag and plays an important role in the design of tubular high-rise structures. The bending stresses in the webs are also affected in a similar manner.

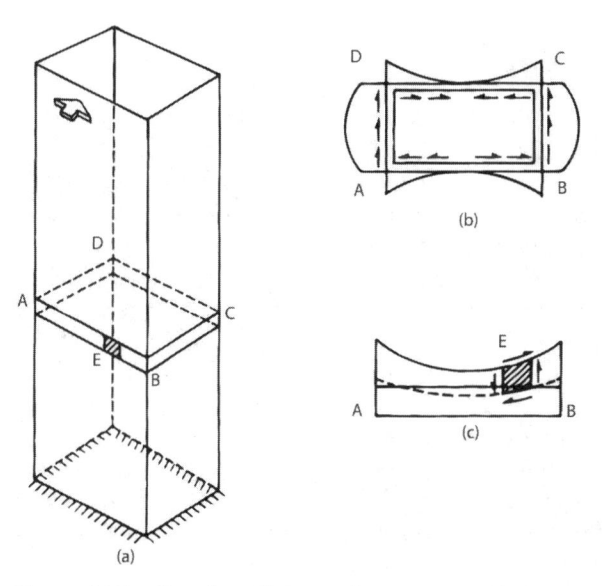

Figure 4.38 Shear lag effects in tube structures: (a) cantilever tube subjected to lateral loads; (b) shear stress distribution; (c) distortion of flange element caused by shear stresses.

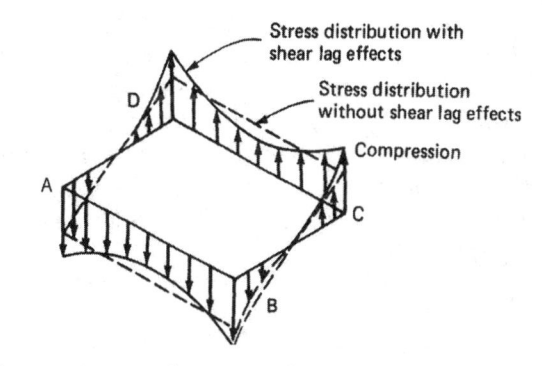

Figure 4.39 Axial stress distribution.

To better appreciate the shear lag phenomenon, let us take a closer look at our hypothetical high-rise building. In a normally proportioned steel building, the usual method of framing for gravity loads necessitates that interior columns be located in and around the core to maintain the span of floor beams in the economical range of 35 to 45 ft (10.6 to 13.64 m). The interior gravity columns and floor beams amount to about one-third of the total steel required for the building. For the example, the effective unit quantity of steel available for the bearing wall is therefore equal to 16 psf (766 Pa), effectively reducing the equivalent thickness of the tube wall to 10 in. (254 mm), with a proportional increase in the shear strain. The departure of bending stress distribution from those predicted on the basis of plane sections becomes even more severe. A high-rise building in practice has to accommodate penetrations in the exterior wall for obvious reasons, which means that the bending efficiency of the tube is further reduced because of these penetrations. Therefore, even with the most efficient distribution of material, shear lag is still present. It cannot be completely eliminated but can be minimized. Thus the structural optimization in a framed tube design reduces to an examination of different column spacings, and corresponding sizes with spandrels, which result in the least shear lag effects.

The shear lag effects in tubular buildings consisting of discrete columns and spandrels may readily be appreciated by considering the basic mode of action involved in resisting lateral forces. The primary resistance comes from the web frames which deform so that the columns T are in tension and C are in compression (Fig. 4.40). The web frames are subjected to the usual in-plane bending and racking action associated with an independent rigid frame. The primary action is modified by the flexibility of the spandrel beams which causes the axial stresses in the corner columns to increase and those in the interior columns to decrease.

The principal interaction between the web and flange frames occurs through the axial displacements of the corner columns. When column C, for example, is under compression, it will tend to compress the adjacent column C1 (Fig. 4.40) since the two are connected by the spandrel beams. The compressive deformations of C1 will not be identical to that of corner column C since the flexible connecting spandrel beam will bend. The axial deformation of C1 will be less, by an amount depending on the stiffness of the connecting beam. The deformation of column C1 will in turn induce compressive deformations of the next inner column C2, but the deformation will again be less. Thus each successive interior column will experience a smaller deformation and hence a lower stress than the outer ones. The stresses in the corner column will be greater than those from pure tubular action, and those in the inner columns will be less. The stresses in the inner columns lag behind those in the corner columns. Hence the term shear lag is used to describe this phenomenon.

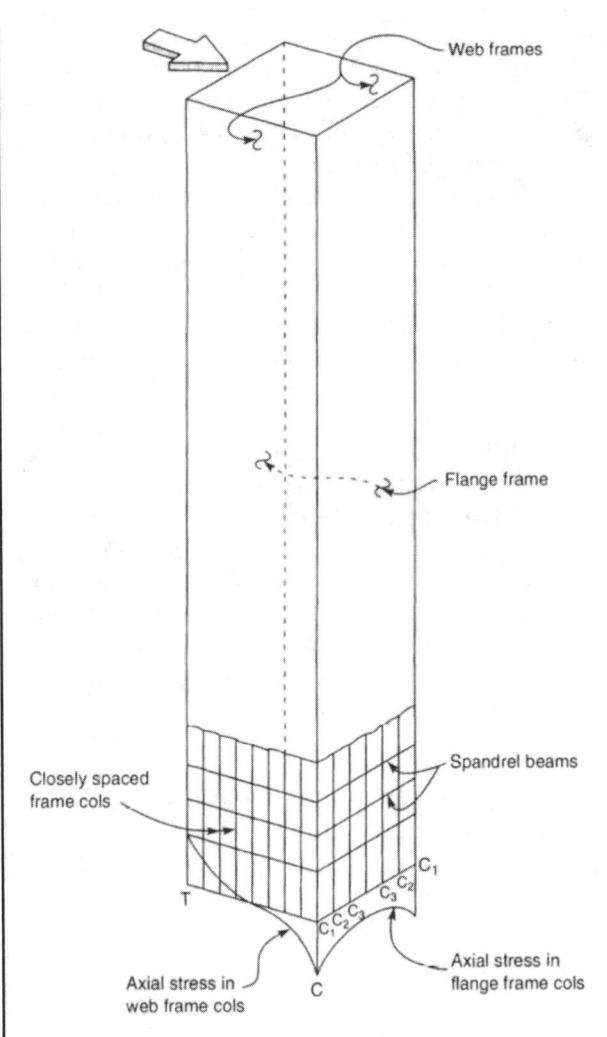

Figure 4.40   Shear lag in framed tube.

The difference between stress distribution, as predicted by ordinary engineer's beam theory and the actual distribution is illustrated in Fig. 4.40. Because the column stresses are distributed less effectively than in an ideal tube, the moment resistance and the flexural rigidity are reduced. Thus, although a framed tube is highly efficient for tall buildings, it does not fully utilize the potential stiffness and strength of the structure because of the effects of shear lag in the perimeter frames.

4.9.4   Irregularly Shaped Tubes

The framed tube concept can be executed with any reasonable arrangement of column and spandrels around the building perimeter. However, non-compact plans, and plans with re-entrant corners considerably reduce the efficiency of the system. For framed tubes, a compact plan may be defined as one with an aspect ratio of not greater than 1.5 or so. Elongated plans with longer aspect ratios impose considerable premium on the system because: (i) in wind-controlled design, the elongated building elevation acts like a sail collecting large wind loads; (ii) the resulting shear forces most usually require closer spacing and or larger size columns and spandrels parallel to the wind; (iii) shear leg effects are more pronounced especially for columns orientated perpendicular to the direction of the wind.

In a similar manner, a sharp change in the tubular form results in a less efficient system because the shear flow must pass around the corners solely through axial shortening of the columns. Also, a secondary frame action at these locations alters the load distribution in the framed tube columns as explained below.

Consider the framed tubes shown in Figs. 4.41a,b. To keep the explanation simple, let us assume that shear lag

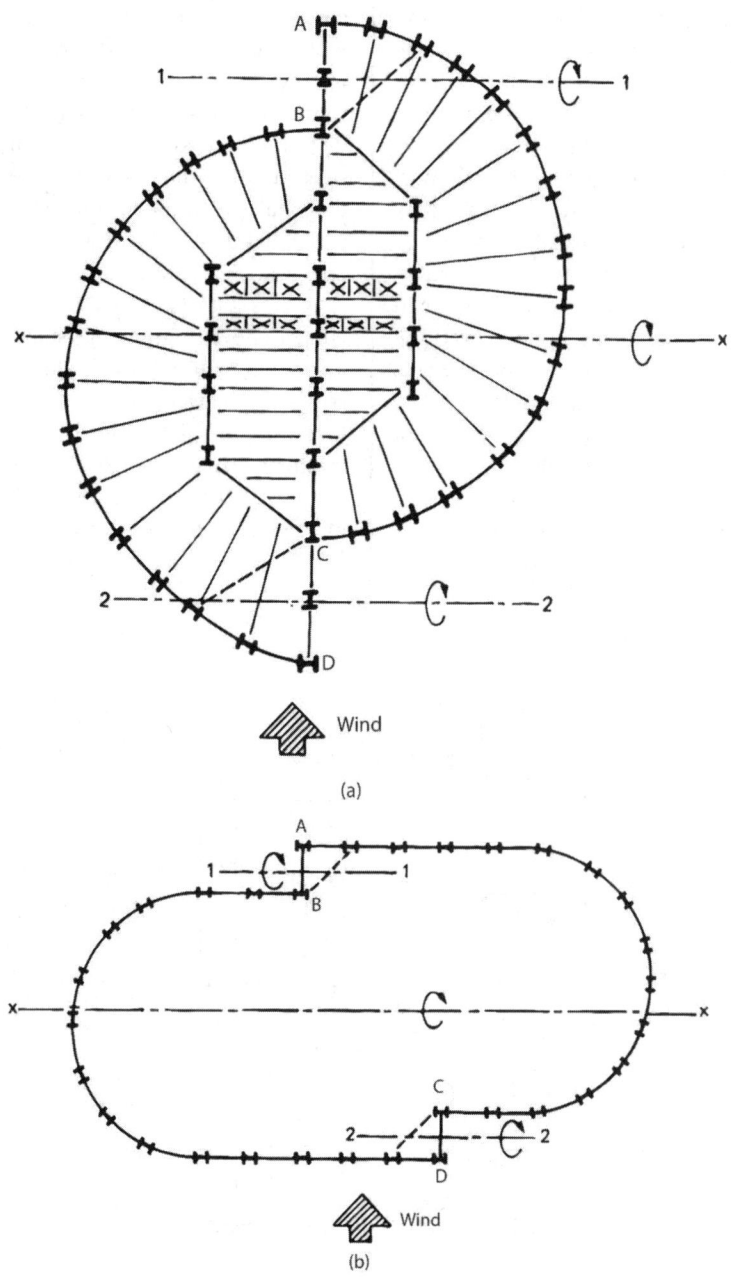

Figure 4.41 Offset tubes: (a) semicircular tubes; (b) rectangular tube with semicircular sides.

effects are negligible. For lateral loads in the N–S direction, the leeward columns A and B are subjected to compression while the windward columns C and B to tension. In addition to the primary action, the frames AB and CD experience a secondary bending action about their own local axes 1-1 and 2-2, as shown in Fig. 4.41. The resulting axial forces in columns A and B are compression and tension respectively. The effect of local bending is to increase the compressive force in A while decreasing the same for column B. The final force is the summation of the two; the primary tube action and the secondary local frame action.

In framed tubes, a limited number of columns can be transferred with little, if any, structural premium because the vierendeel action of the façade frame is generally sufficient to transfer the load. However, if the transfer is too severe requiring removal of a large number of columns, a one- or two-story deep transfer girder of truss may be necessary. Temporary shoring is usually required to support the dead and construction loads until a sufficient number of vierendeel frames are constructed, or in concrete construction, until the girder has achieved the design strength. A schematic view of a shoring system is shown in Fig. 4.42 consisting of steel columns braced in two perpendicular directions. Steel plates at the base of columns are supported on a sand bed contained in a steel box. Shims are used between the girder and shores to compensate for any construction irregularities or settling of sand. When the shoring is no longer required, sand is

released from the box to transfer the load to the girder. The rate of loading is controlled by manipulating the quantity of sand removed from the box.

Generally in a transfer system, be it a vierendeel frame, steel truss, plate or concrete girder, the vertical load is collected from tube columns, and channeled in to a limited number of columns below. Removal of a large number of columns usually requires an additional lateral system to carry the horizontal shears. Generally braced frames or concrete shear walls are provided within the building core to resist the shear forces. Occasionally steel plates with welded shear studs are used to increase the shear resistance of concrete walls. The shear forces from the frame columns are transferred to the core through the diaphragm action of the slab at the transfer level. In a steel building, this is sometimes achieved by providing diagonal steel bracing under the transfer floor and, in concrete building by increasing the thickness of the concrete slab and adding additional reinforcement.

Figures 4.43 to 4.45 show conceptually the ideas discussed above. Figure 4.43 shows the vertical transfer of a frame tube to a set of eight columns. A transfer girder spanning between the columns supports the column loads. Figure 4.44 shows the transfer of overturning moment, that is, the axial compressive and tensile forces in the tube columns. And finally, the horizontal transfer of shear forces through the floor diaphragm is shown in Fig. 4.45.

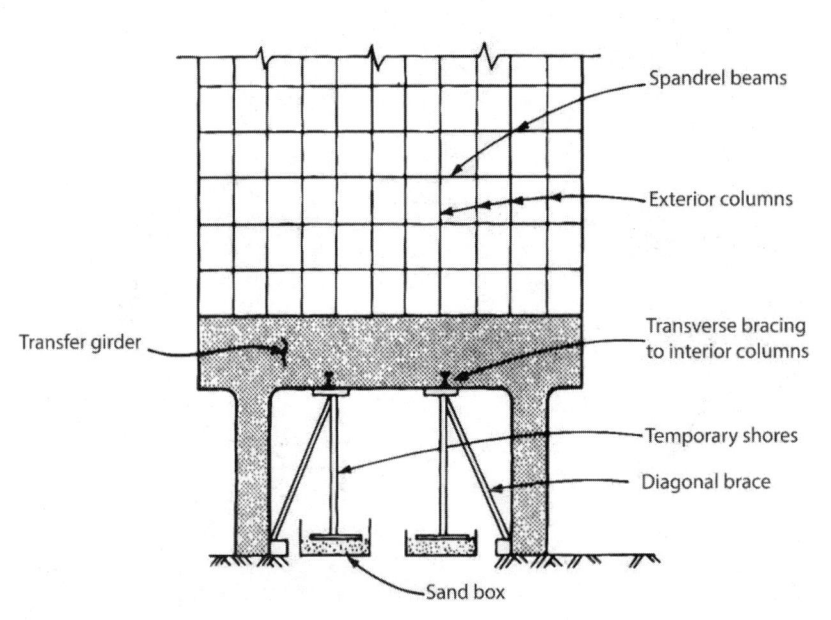

Figure 4.42   Shoring system for a tube structure.

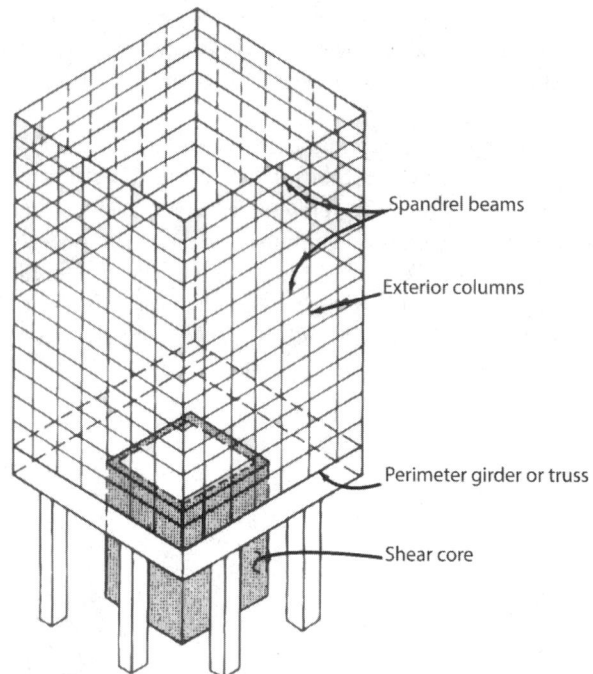

Figure 4.43 Vertical transfer.

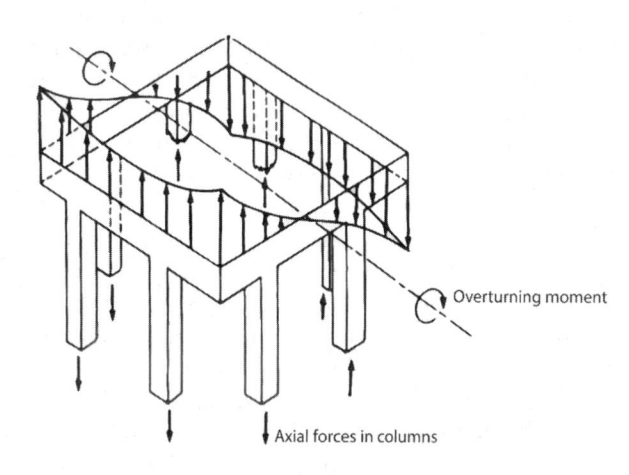

Figure 4.44 Column axial forces due to overturning moment.

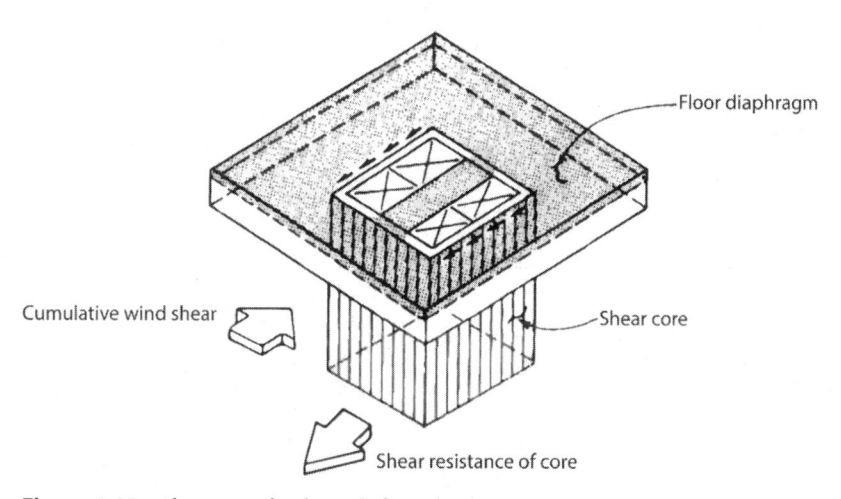

Figure 4.45 Shear transfer through floor diaphragm.

4.10 TRUSSED TUBE SYSTEM

A trussed tube system represents a classic solution for improving the efficiency of the framed tube by increasing its potential for use to even greater heights as well as allowing greater spacing between the columns. This is achieved by adding diagonal bracing to the face of the tube to virtually eliminate the shear lag effects in both the flange and web frames.

The framed tube, as discussed previously, even with its close spacing of columns is somewhat flexible because the high shear stresses in the frames parallel to the wind cannot be transferred effectively around the corners of the tube. For maximum efficiency, the tube should respond to lateral loads with the purity of a cantilever with compression and tension forces spread uniformly across the windward and leeward faces. The framed tube, however, behaves like a thin-walled tube with openings. The axial forces tend to diminish as they travel around the corners, with the result that the columns in the middle of the windward and leeward faces may not sustain their fair share of compressive and tensile forces. This effect referred to previously as the shear lag effect limits the framed tube application to 50- or 60-story buildings unless the column spacing is very small as with the 109-story World Trade Center Towers, New York which has columns at 3.8 ft (1.0 m). For taller buildings with their usual column spacing of 10 to 15 ft (3.0 to 4.58 m), the frames parallel to the lateral loads act as multibay rigid frames. Consequently, column and beam designs are controlled by their bending action rather than axial, resulting in unacceptably large sizes. Furthermore, out of the total sway, only about 25 percent is from the cantilever component, while the remainder is from the frame shear racking component. Because of the shear racking, the corner columns take more than their share of axial load, while those in between less, as compared to an ideal tube. A trussed tube overcomes this problem by stiffening the exterior frames.

The most effective trussed tube action may be obtained by replacing vertical columns with closely spaced diagonals in both directions (Fig. 4.46). However, this system

is not popular because it creates problems in the detailing of curtain walls. However, this system has evolved to what is known as "Diagrid System," which is discussed in section 4.13.

The diagonally braced tube, shown in Fig. 4.47a,b is by far the most usual method of increasing the efficiency of the framed tube. It represents an elegant solution by introducing a minimum number of diagonals on each façade that intersect at the same point as the corner columns. The system is tubular in that the fascia diagonals interact with the trusses on the perpendicular faces to achieve three-dimensional behavior.

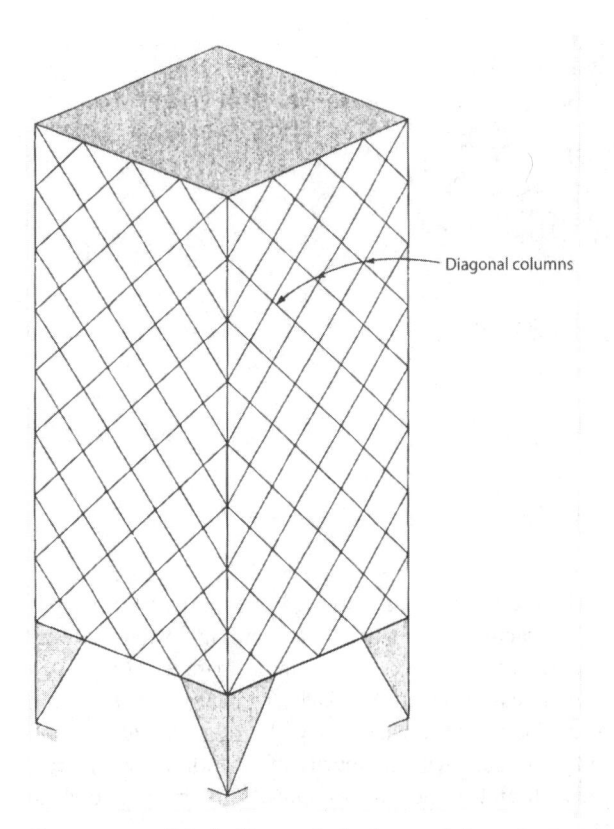

Figure 4.46   Tube building with closely spaced diagonal columns.

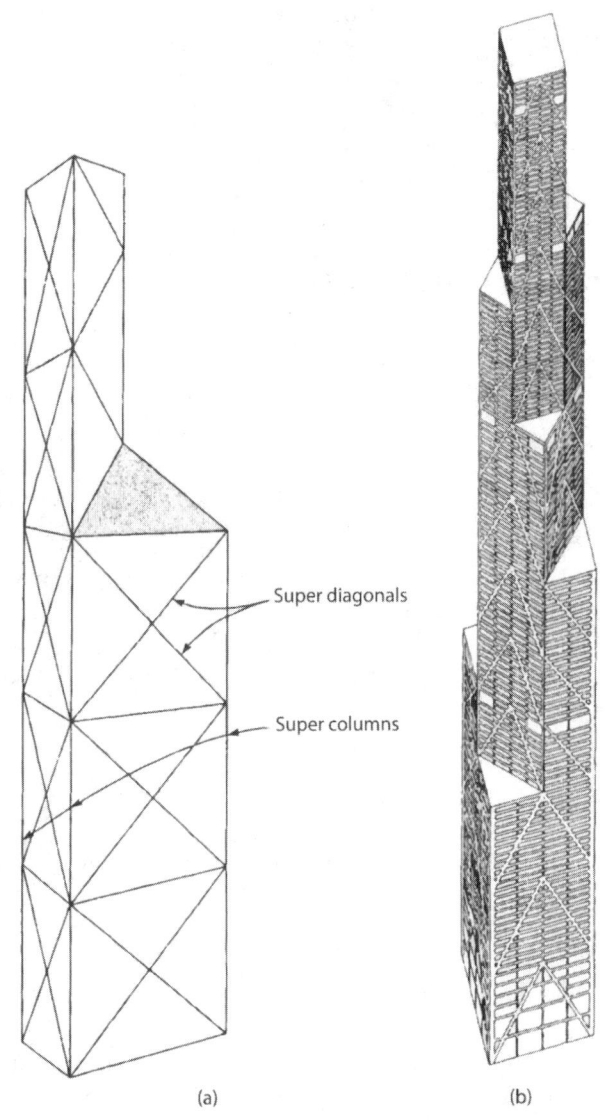

Figure 4.47   (a) Tube building with multistory diagonal bracing; (b) rotated square tube with super diagonals. (Adapted from an article by Mahjoub Elnimeiri, in Civil Engineering Journal.)

The diagonals of a braced tube connected to the columns at each intersection, virtually eliminate the effects of shear lag in both the flange and web frames. As a result, the structure behaves under lateral loading more like a braced frame, with greatly diminished bending of frames. Consequently, the spacing of the columns can be larger and the size of the columns and spandrels less, thereby allowing larger size windows than in the conventional tube structure. In the braced-tube structure the bracing contributes also to the improved performance of the tube in carrying gravity loading. Differences between gravity load stresses in the columns are evened out by the braces transferring axial loading from the more highly to the less stressed columns.

The principle of façade diagonalization can readily be used for partial tubular concepts. For example, in long rectangular buildings, the end frames along the short face may be diagonalized with moment-resisting frames on the long faces. The end diagonal frame may be in the form of a channel or C shape to provide lateral resistance in both directions. Many variations are possible, each having an impact on the exterior architecture.

To use the idea of a trussed tubular system in reinforced concrete construction, a diagonal pattern of window perforations in an otherwise framed tube construction is filled in between adjacent columns and girders. The result is a reduction in shear lag for the system under lateral loads. As with the steel-framed trussed tube, the façade diagonalization offers the additional benefit of equalizing the gravity loads in the exterior columns.

EXAMPLES

1. First International Plaza, Dallas

An example of a trussed tube structural system is shown in Fig. 4.48a, a 56-story office building in downtown Dallas. Designed by structural engineers Ellisor & Tanner, Inc., the building consists of 1.9 million square feet (174,437 m²) of office space and rises to a height of 710 ft (216 m) above grade. Exterior columns are spaced at 25 ft (7.62 m) on center with a floor-to-floor height of 12 ft, 6 in. (3.81 m) to facilitate intersection of diagonals with columns and spandrels. There are two X braces, each 28 stories tall, on each of the four sides, as shown in Fig. 4.48b. The diagonal bracing, in addition to carrying the wind loads, helps to distribute gravity loads along the exterior columns on each side of the building. The vertical bracing is located within the glass line to minimize problems associated with fireproofing the exterior structure and the effects of temperature on exposed steel. All members in the primary exterior framing—beams, columns, and diagonals—were fabricated from high-strength W14

shapes except for a few near the bottom which were built-up shapes. W14 shapes were selected because of the wide range of available sections and because all shapes had the same nominal inside dimension between flanges,

(a)

Figure 4.48 First International Plaza: (a) photograph; (b) schematic bracing.

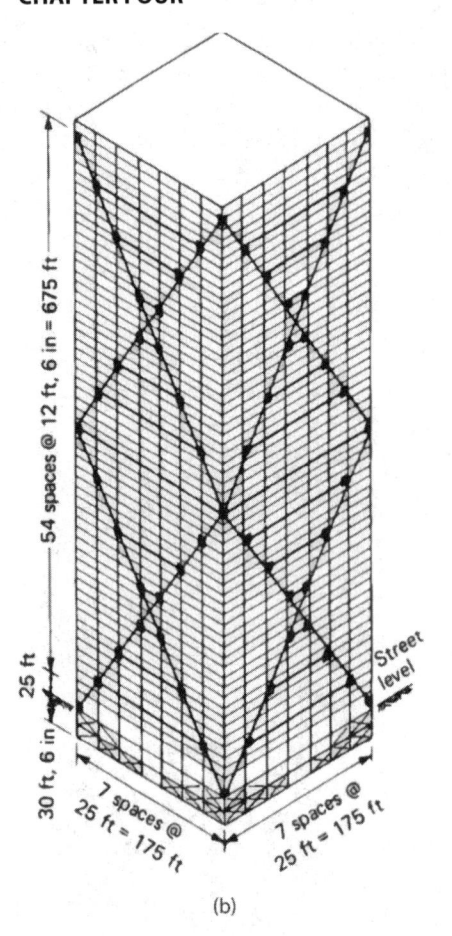

(b)

Figure 4.48 (*Continued*)

Figure 4.49 John Hancock Tower, Chicago.

thus facilitating connection details. Gusset plates used to connect the diagonals, columns, and beams are approximately 10 ft (3.0 m) wide and 12 ft (3.65 m) tall. The diagonals, which were fabricated four stories in length, were field-welded with full-penetration welds at one end and were high-strength-bolted through splice plates at the other end. The welded connection at one end reduced gusset and splice material, while the bolted connection at the other end provided the necessary tolerances for erection. Corner gusset assemblies were required where diagonal bracing met at the corners of the building. These were fabricated from four plates, two in each of the two directions. The four plates were joined by electroslag welds and were stress-relieved after fabrication.

2. John Hancock Center, Chicago

Perhaps the most notable example of this structural system is the John Hancock Center in Chicago (Fig. 4.49) designed by the Chicago office of Skidmore, Owings & Merrill. The building is 100 stories with a rectangular plan that tapers from the ground level to the top. The

diagonals are placed at 45° angles to each other, forming enormous X braces on each side. The diagonals serve multiple functions, acting as inclined columns to resist some of the gravity load, absorbing most of the wind shear, and stiffening the tube so that it mimics the behavior of a solid tube. This unique design, combined with the use of high-strength steel, enabled the engineers to achieve a 100-story building with only 29.7 psf (1422 Pa) of steel as compared to 42.2 psf (2020.5 Pa) for the 102-story Empire State Building.

The building is for multiuse involving commercial, parking, office, and apartment-type space in one building. The ground floor plan measures 164 ft (50 m) by 262 ft (80 m) and the clear span from the central core is approximately 60 ft (18 m). The building is tapered to the top to a dimension of 100 ft (30 m) by 160 ft (49 m), and the clear span reduces to 30 ft (9 m). The floor height is 12 ft 6 in. (3.8 m) in the office sector and 9 ft 4 in. (2.8 m) in the apartment sector. The structural system consists of diagonally braced exterior frames which act together as a tube. The floors are 5 in. (127 mm) thick including lightweight concrete topping and metal deck. The columns, diagonals, and ties are I-sections fabricated from three plates with a maximum thickness of 6 in. (150 mm). The maximum column dimension is 36 in. (920 mm). Floor framing, fabricated from rolled beams with simple connections, are designed for gravity loading only. The interior columns are designed for gravity loads using rolled and built-up sections. Almost all the steel is ASTM A-36. Connections were shop welded and field bolted except that field welding was used in spandrels, main ties, and column splices. The building was completed in 1968 reaching 1127 ft (344 m).

3. CitiCorp Center, New York

For the CitiCorp Center in New York, structural engineer William J. LeMessurier incorporated giant triangular trusses into the exterior façade of the building. These façade trusses collect about half the gravity loads and resist the entire wind loads on the building. The loads collected on the façade are channeled into four massive columns at the base. Because the shear resistance of the giant trusses is no longer available below the transfer level, a central core is designed to resist wind shears.

Diagonal bracing is employed in the plane of the transfer floor to achieve shear load transmission from the tubular frame to the core. The structural system is shown schematically in Fig. 4.50.

4.11 THE BUNDLED TUBE

The previous sections discussed tubular systems which are generally applicable to prismatic vertical profiles including a variety of nonrectilinear, closed-plan forms, such as circular, hexagonal, triangular, and other polygonal shapes. The most efficient shape is a square, whereas a triangular shape has the least inherent efficiency. The high torsional stiffness characteristic of the exterior tubular system has advantages in structuring unsymmetrical shapes. However, for buildings with significant vertical offsets, the discontinuity in the tubular frame introduces some serious inefficiencies. A bundled tube configured with many cells, on the other hand has the ability to offer vertical offsets in buildings without loss in efficiency.

It allows for wider column spacings in the tubular walls than would be possible with only an exterior framed tube.

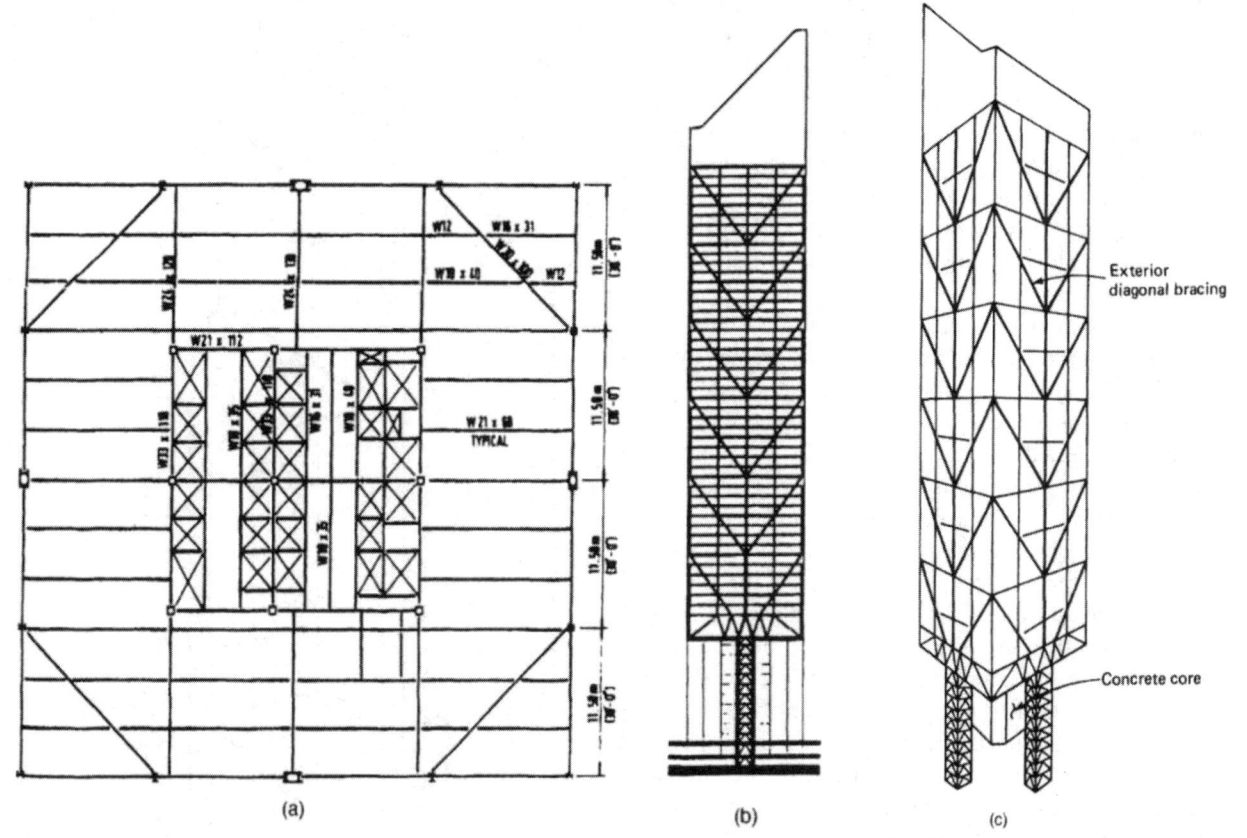

Figure 4.50 CitiCorp Center (Structural Engineers, Le Messurier consultants Inc.): (a) typical floor framing plan; (b) elevation; (c) lateral bracing system.

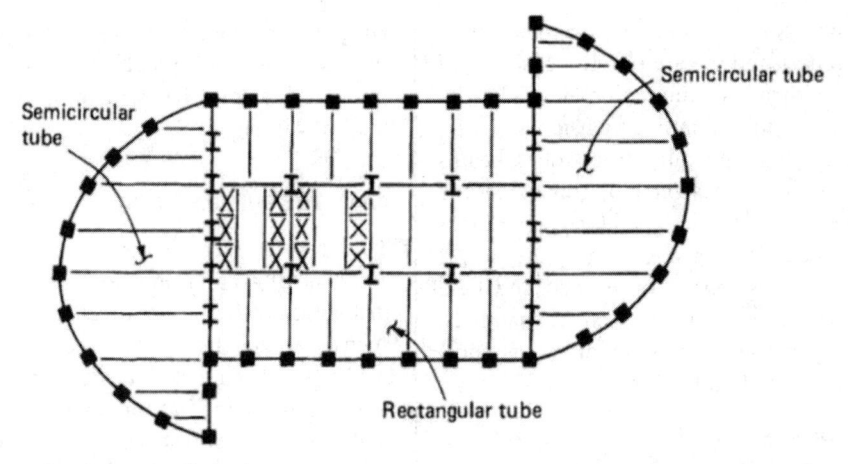

Figure 4.51 Bundled tube.

It is this spacing which makes it possible to place interior frame lines without seriously compromising interior space planning. In principle, any closed-form shape may be used to create the bundled form (Fig. 4.51).

4.11.1 Behavior

The bundled tube structure may be regarded as a set of tubes that are interconnected with common interior panels to form a perforated multicell tube, in which the frames in the lateral load direction resist the shears, while the flange frames carry most of the overturning moments. The cells can be curtailed at different heights without diminishing structural integrity. The torsional loads are readily resisted by the closed form of the modules. The greater spacing of the columns, and shallower spandrels, permitted by the more efficient bundled tube structure, provides for larger window openings than are allowed in the single-tube structure.

The shear lag experienced by conventional framed tubes is greatly reduced by the addition of interior framed "web" panels across the entire width of the building. When the building is subjected to bending under the action of lateral forces, in high in-plane rigidity of the floor slabs constrains the interior web frames to deflect equally with external web frames, and the shears carried by each are proportional to their lateral stiffness. Since the end columns of the interior webs are activated directly by the webs, they are more highly stressed than in a single tube where they are activated indirectly by the exterior web through the flange frame spandrels. Consequently, the presence of the interior webs reduces substantially the nonuniformity of column forces caused by shear lag. The vertical stresses in the normal panels are more nearly uniform, and the structural behavior is much closer to the proper tube than the framed tube. Any interior transverse frame panels will act as flanges in a similar manner to the external normal frames.

Because the bundled tube design is derived from the layout of individual tubes, it is possible to achieve a variety of floor configurations by simply terminating a tube at any desired level. In the simple case of two tubes, the corresponding plan shapes are shown schematically in Fig. 4.52. Figure 4.53 shows the diversity in plans that can

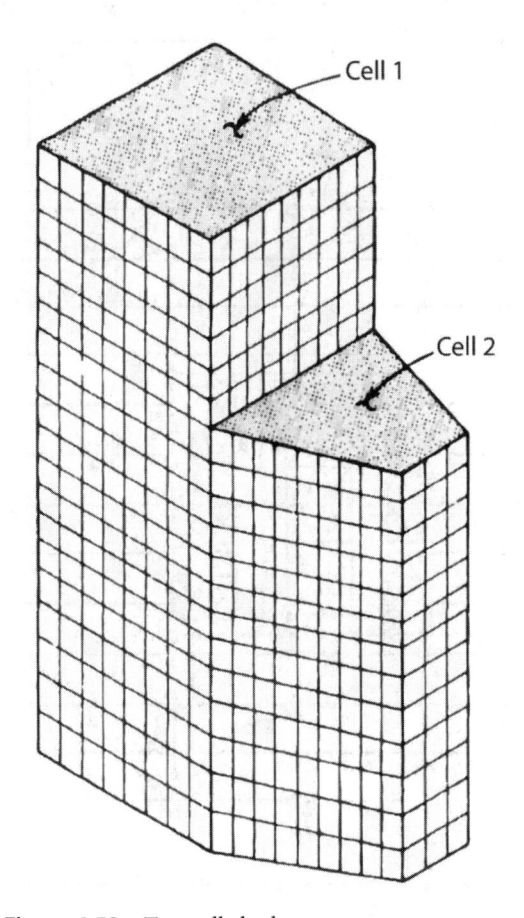

Figure 4.52 Two-celled tube.

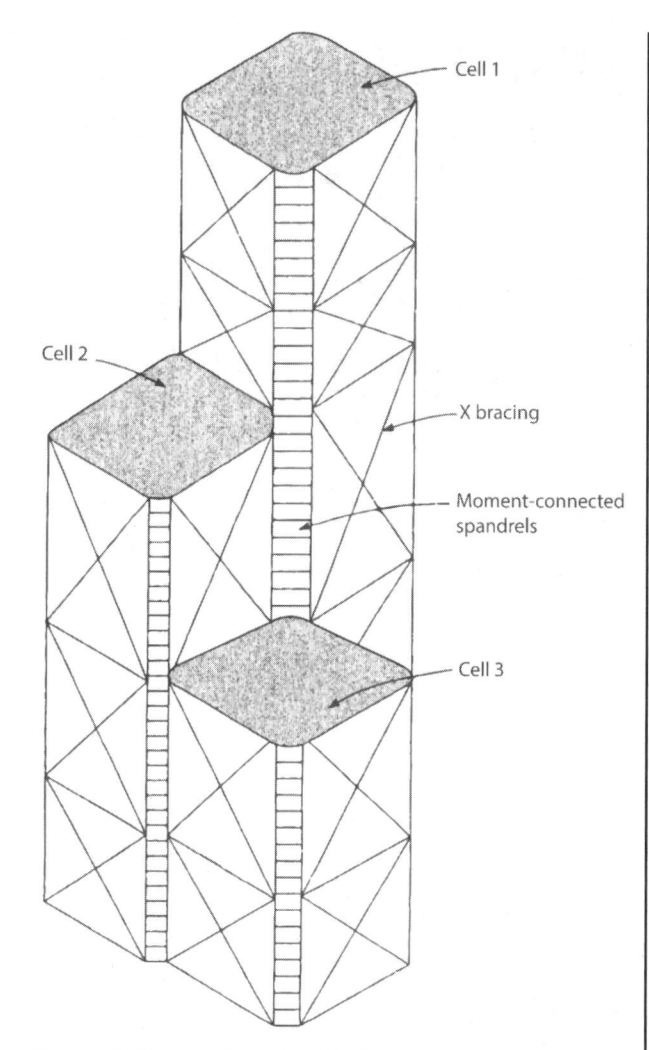

Figure 4.53 Bundled tube with diagonals.

be achieved in a three-cell bundled tube. For a structure to behave as a bundled tube it is not necessary that the adjacent tubes be of similar shape. A bundled tube consisting of a square and a triangular cell would respond, in a conceptual sense, in a manner similar to two square or two triangular cells.

A distinct advantage of the bundled tube is that the individual tubes can be assembled in any configuration and terminated at any level without loss of structural integrity. This feature enables creation of setbacks with a variety of shapes and sizes. The disadvantage, however, is that the floors are divided into tight cells by a series of columns that run across the building width.

The structural principle behind the modular concept is that the interior rows of columns and spandrels act as internal webs of a huge tubular cantilever in resisting shear forces, thus minimizing the shear lag effects. Without the beneficial effect of these internal diaphragms,

most of the exterior columns in a framed tube toward the center of the building would be of little use in resisting the overturning moment. The modular system can be seen as an extension of the perimeter tubular system with stiffened interior frames in both directions. The interior frames of the cell parallel to the wind resist shear forces, in a manner similar to the two end frames generating peak axial stresses at points of intersection of the web frames with the flange frames. The internal diaphragms tend to distribute the axial stresses equally along the flange frames by minimizing shear lag effects.

EXAMPLES

1. The Willis Tower (previously known as Sears Tower), Chicago

Reaching to a height of 1454 ft (443 m) above street level, the Sears Tower contains 109 floors and encloses 3.9 million gross sq. ft (362,000 m²) of office space. The basic shape is composed of nine areas 75 ft (22.9 m) square, for an overall floor dimension 225 ft (68.6 m) square. Two of the nine constituent tubes are terminated at the 50th floor, two more at the 66th and three at the 90th, creating a variety of floor shapes ranging from 41,000 sq. ft (3,800 m²) to 12,000 sq. ft (1,100 m²) in gross area (Fig. 4.54a). The structure acts as a vertical cantilever fixed at the ground in resisting wind loads. The walls of the nine tubes bundled

(a)

Figure 4.54 (a) Willis (old Sears) Tower, Chicago; (b) Four Allen Center, Houston, plan; (c) building section.

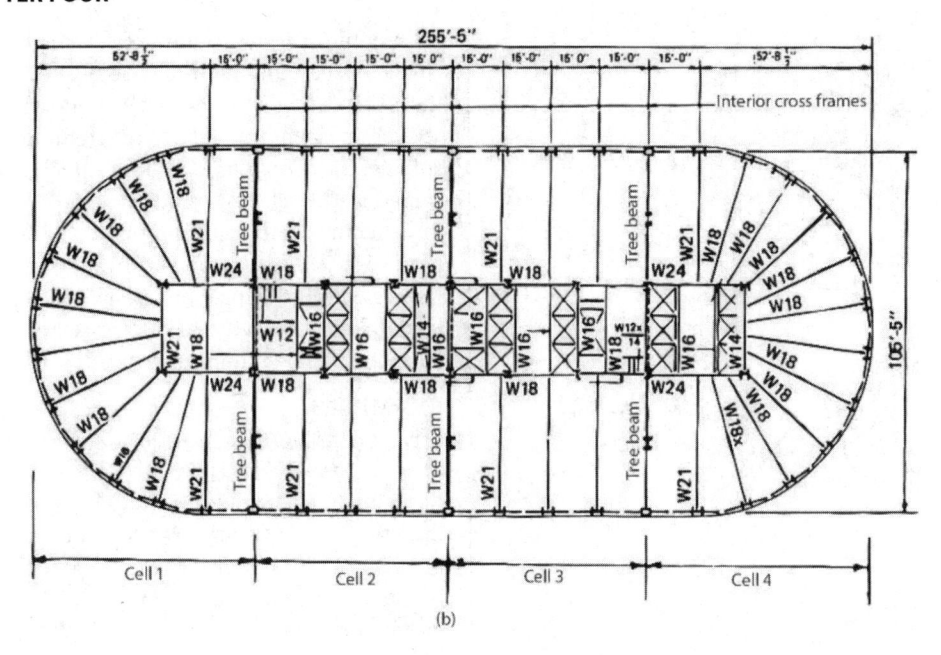

(b)

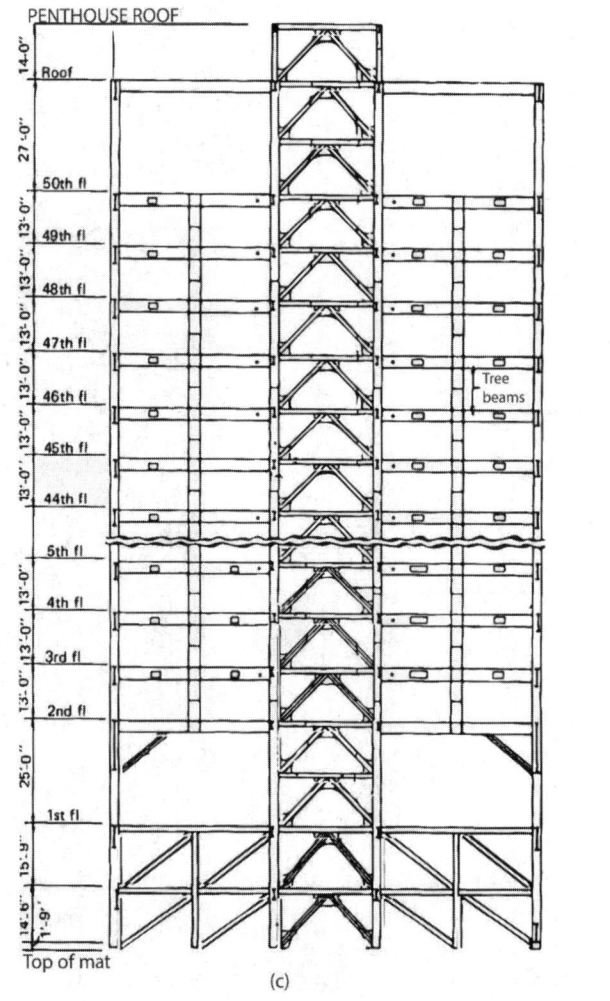

(c)

Figure 4.54   *(Continued)*

together are composed of columns at 15 ft (4.6 m) on centers and deep beams at each floor. Two adjacent tubes share one set of columns and beams. All beam-column connections are welded. Trussed levels with diagonals between columns are provided at three intermediate mechanical levels, two of them immediately below the setbacks at the 66th and 90th floors. The beams are 42 in. (1.07 m) and columns 39 in. (0.99 m) deep built-up I-sections with the flange width and thickness decreasing with increasing height. Except for column splices, all field connections are bolted. The floors are supported on one-way 40 in. (1.0 m) deep trusses spanning 75 ft (23 m) and spaced at 15 ft (4.6 m). Each truss frames directly into a column. The span direction is alternated every six floors to equalize the gravity loads on columns. A floor slab of 2.5 in. (63 mm) lightweight concrete cast on a 3 in. (76 mm) steel deck spans 15 ft (4.6 m) between the trusses.

Beams and columns of built-up I sections are 42 and 39 in. (1070 and 490 mm) depth, respectively. Column flanges vary from 24 by 4 in. (609 by 102 mm) at the bottom to 12 by 2.75 in. (305 by 19 mm) at the top, and beam flanges from 16 by 2.75 in. (16 by 70 mm) to 10 by 1 in. (254 by 25 mm). A total of 76,000 tonnes (69,000 tonnes) of structural steel was used in the project, consisting of grades A588, A572, and A36.

The steel-tube structure was shop-fabricated into units of two-story-high columns and half-span beams on each side, typically weighing 14 tonnes (15 tons). The shop fabrication eliminated 95 percent of field welding. Automated electroslag welding was used for the butt welds of beams to columns. The continuity plates across columns at the joints were fillet-welded by the innershield welding process.

Because site storage space was unavailable, the frame units were delivered exactly when needed and lifted off the truck into place. Except for column splices, all field connections were grade A490 high-strength friction-grip bolts in shear connections. The building designed by the Chicago Office of Skidmore, Owings & Merrill was completed in 1974.

2. Four Allen Center, Houston, Texas

To achieve the structural action of a bundled tube it is not necessary to have closely spaced columns dividing the building plan into secondary cells. It is possible to achieve an equivalent action, though not as efficiently as that with closely spaced columns, by inserting a minimum number of columns between the windward and leeward faces of the tube to reduce the shear lag. An example of this application can be seen in the 695 ft (211.83 m) office building called Four Allen Center in downtown Houston. In plan, the building is an elongated rectangle with semicircular ends with overall dimensions of approximately 110 by 260 ft (33.5 by 79.25 m) as shown in Fig. 4.54b,c. The tower is remarkably slender, with a height-to-width ratio in excess of 6.0. The elongated shape of the tower creates a striking knife-edged silhouette on the skyline. Although the columns are spaced at a reasonably close interval of 15 ft (4.57 m) along the entire perimeter, which is normally sufficient to achieve a framed tube solution for a building of this height, it was clear from the analysis that a pure framed tube was impractical because of the high plan-aspect ratio. The extent of shear lag was too severe to allow a single perimeter tube solution to be economical. Structural engineers Ellisor & Tanner devised a modified bundled tube by introducing interior cross frames, effectively subdividing the plan into a four-celled grid as shown in Fig. 4.54b. These cross frames were formed by horizontal tree beams interacting with diagonal trusses. A treecolumn element consists of short vertical stub columns attached to the girders at middistance between the exterior and core columns. This in effect creates a vierendeel truss action between the core and exterior columns, greatly increasing the shear stiffness of the system. Connections between the stub columns are designed to carry no axial forces and therefore the columns are not required to go all the way to the foundation. A schematic elevation of the framing is shown in Fig. 4.54c.

4.12 ULTIMATE HIGH-EFFICIENCY STRUCTURES

It is interesting to recall that there have been no quantum leaps in the height of tall buildings that are being built today as compared to the buildings built during the 1930s and 1940s. For example, the Empire State Building scraped the sky at 1250 ft (381 m), which is not a small feat by any stretch of the imagination. The success of this awe-inspiring building and other similar buildings of the era is credited to the holding-down power of the heavy exterior stone cladding and the masonry partitions that were in vogue during that period. Modern high-rise technology has largely replaced the heavy cladding and interior masonry systems with relatively lightweight counterparts. The holding-down power of these systems is no longer present in modern construction. Let us examine how we can employ the relatively lightweight materials to help in providing resistance to the lateral loads. The tube system, with its characteristic deployment of the columns at the building perimeter, certainly provides the much-required separation between the windward and leeward faces of the building for resisting the overturning moments. However, since the exterior columns, especially in a framed tube, are placed relatively close to each other, their tributary areas for collection of gravity loads are rather small. Therefore, the beneficial effect of gravity load in counteracting the tensile forces of the columns is somewhat limited, first because of the relatively light materials used in current construction practice, and second by the limited tributary area for the exterior columns. If somehow we could induce more gravity loads into these columns, would not the efficiency of the system be improved? This is precisely the idea behind the ultimate high-efficiency structures first envisioned by the master builder Dr. Fazlur Khan. The main premise behind the ultimate high-efficiency structure is to transfer as much gravity load as practicable to the columns resisting the overturning moments. Note that this idea is routinely used by engineers when they are confronted with high uplift forces while designing interior-core-braced buildings with limited separation between the leeward and windward columns. They would normally rearrange the floor framing to achieve more flow of gravity loads into the wind columns or may choose to eliminate certain interior columns altogether in order to collect more gravity loads on the wind columns. A similar approach is possible in a tube building—eliminate as many interior columns as possible, perhaps all the interior columns. This way the holding-down power of gravity loads is put to use in the most efficient manner. It must be recognized that there is a certain amount of trade-off in the floor framing system because it is economically prohibitive to clear-span the floor members using traditional approaches. Accommodation has to be made to achieve the transfer of entire building load into the exterior columns without paying a significant premium. This is considered next.

Consider a tube building with closely spaced exterior columns and deep spandrels. Imagine that all the interior columns are eliminated completely, leaving a column-free volume inside. Within this basic configuration it is possible to provide a system of transfer floor trusses at approximately every 15th floor, corresponding to the levels at which the low, low-mid, high-mid, and high-rise elevators terminate. The trusses could be designed as one- to two-story-deep vierendeel trusses spanning in two directions for the full width and length of the building. Where appropriate, the transfer levels could be made into skylobbies or other forms of common areas. The trusses then can support interior columns within the zones between two trusses. Any type of conventional structural steel framing such as composite rolled beams,

haunch girders, or stub girders can be employed to span the distance between the core columns and the exterior of the building. Since the interior columns carry gravity loads from a limited number of floors, their sizes will be substantially smaller than in a conventional system in which they would rise from the foundation level to the building top. Another advantage is that the columns within any particular zone bounded by two transfer levels may be located at will to suit the interior space planning desired for specific occupancies. This in itself may be reason enough to consider this system because of the tremendous flexibility offered in the layout of columns for mixed development use. For instance, in the space planned for office use, the interior columns may be located at 60 ft (18.28 m), and so on. For major tenants willing to lease the full block of floors between two transfer levels, columns can be arranged to suit their particular needs. Column transfers within the zones can be achieved with little structural difficulty because: (i) the loads to be transferred are relatively small; and (ii) it is possible to hang the upper six floors or so in one zone from the transfer girder, creating opportunities to have some floors entirely column-free. In fact, the floor framing options are limited only by the imagination of the designer. The principal structural advantage is, of course, that total dead and live load from every floor is transferred only to the exterior columns, thereby increasing the structural capacity of the system to withstand lateral loading. Undoubtedly there would be some premium in the tonnage of steel for the trusses and their associated fabrication and erection costs. However, the premium spread over the total square footage of the building is likely to be small.

It is, of course, necessary to tie the windward and the leeward columns of the tube with a structural system capable of resisting the shear forces caused by the lateral loads. This can be achieved with a system of deep spandrel beams when the perimeter columns are closely spaced, or with a system of diagonal bracing when the columns are spaced apart as in a trussed tube system. Dr. Fazlur Khan has shown that by progressively shifting the exterior columns to the corners of a rectangular trussed tube, the efficiency of the system can be greatly improved. An ultimate structure for a rectangular building, then, will have just four corner columns interconnected by massive diagonals as shown in Fig. 4.55.

The efficiency of a building to resist lateral loads can be increased further by using interior bracing with a structural system in which the total gravity load of the building is made to bear on a limited number of exterior columns. To increase the uplift capacity, interior columns

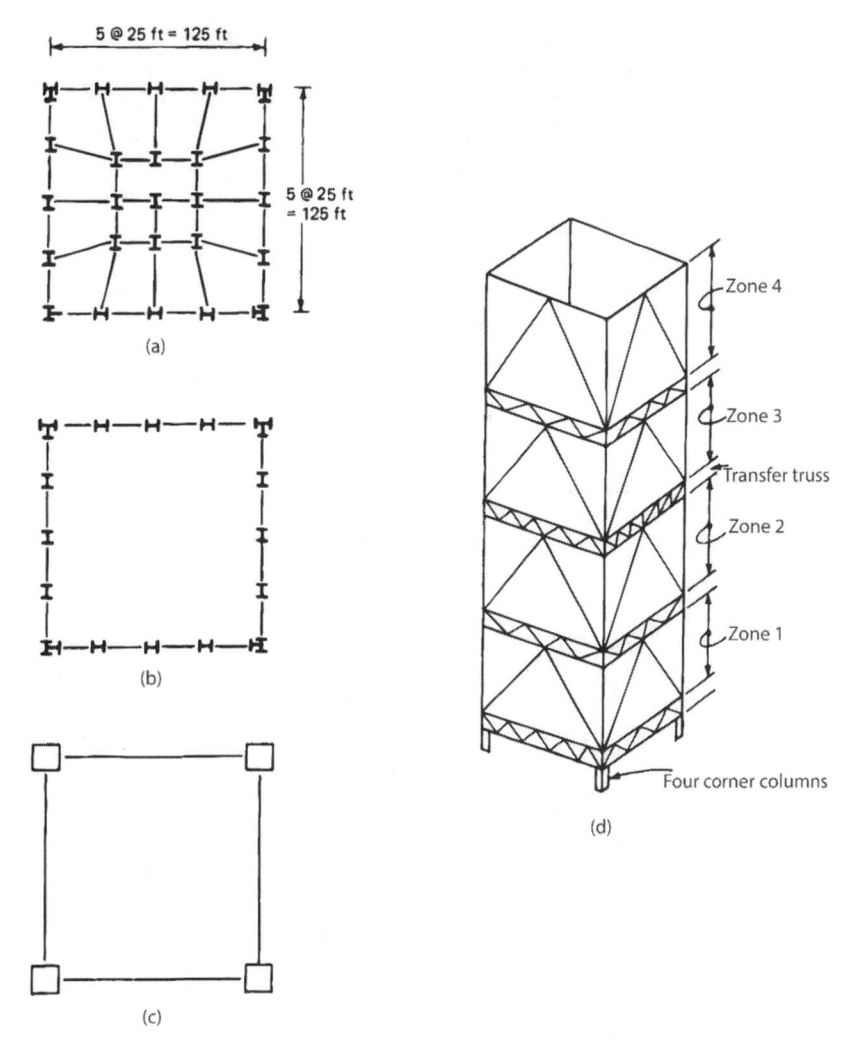

Figure 4.55 Floor plans: (a) conventional framing with interior and exterior columns extending to the foundation; (b) exterior columns extend to foundation; (c) interior columns are supported by transfer trusses; (d) Khan's concept of the ultimate high-rise building.

are completely eliminated within the building envelope (Figs. 4.55a–d). However, to achieve an economical floor system, it is necessary to use columns in the core area without unduly interfering with the leasing requirements. Therefore, a structural system is needed for transferring the loads from the interior columns to the building envelope. If this system can simultaneously work as a shear-resisting element, then the need for closely spaced columns or diagonal bracing on the perimeter can be eliminated. In other words, a system of interior bracing that performs the dual function of channeling the loads from the interior columns to the exterior while at the same time functions as a shear element between the windward and leeward columns is likely to be the optimum system. From a pure structural point of view, such a system is highly desirable. Examples of buildings using this concept are given in Chap. 6.

4.13 DIAGRID SYSTEM

The diagrid is a three-dimensional exterior bracing system which offers an alternative to the outriggers systems for buildings of comparable heights. Buildings with outrigger systems are typically designed with reinforced concrete cores or steel-braced cores. With the use of the outriggers in a building, the moments and the drifts are reduced below the location of the outrigger but without providing shear rigidity; therefore, cores with high shear rigidity are used with the outrigger system. In the diagrid structure, however, bending and shear rigidities are provided by the diagonals located at the perimeter of the building. The diagrid system could be considered as an innovative version of the conventional braced tube system with all vertical perimeter columns eliminated. With the elimination of the vertical perimeter columns, the diagonals will carry the gravity loads as well as the lateral loads, unlike the conventional braced tube system in which the diagonal braces carry lateral loads only.

The diagrid structure is more effective in resisting and significantly reducing shear deformation because the shear is carried by axial action of the diagonal members, which makes the system more effective and more economical when compared with a framed tube system without diagonals which carries shear by bending of the vertical perimeter columns. Figure 4.56 shows the braced tube system and the diagrid system.

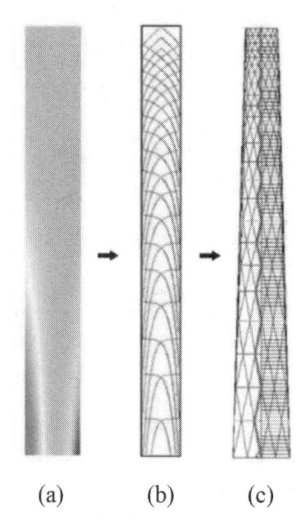

(a)   (b)   (c)

Figure 4.58   (a) Continuum stress in cantilever beam; (b) principal stress directions; (c) optimized diagrid shape.

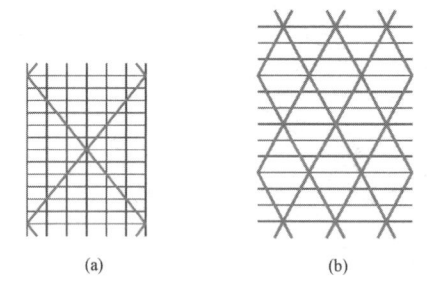

(a)   (b)

Figure 4.56   (a) Braced tube; (b) diagrid.

As shown in Fig. 4.57, regular grid and repeatable node geometry of the exterior diagonal members could be the simplest and the easiest for analysis, fabrication, and erection. Figure 4.58 shows that the diagonals can be arranged with varying inclination along the height of the building. The inclination angle of the braces is mirrored

to the angle of the internal principal stresses of a continuous cantilever subjected to a uniform bending load. The stress distribution of the cantilever beam shows that the arrangement of the members, following principal stress distribution, would require to be more vertical at the base where axial stresses and flexural deformation are dominant, and would require to be more horizontal at the top of the building (free end) where the shear stresses are dominant.

One of the challenges of the diagrid system is that the diagonal members are typically long, slender and unbraced and required to be capable of resisting rare wind and seismic loads without buckling. Performance-based design evaluation of this system is typically required to assure safety and code equivalence. As a

Figure 4.57   Evolution of bracing systems in tall buildings.

necessary consequence, buildings with diagrid system tend to perform more elastically and with less energy dissipation in large seismic events than more conventional structural systems. Another challenge that is present in the design of the diagrid system is the node design; the performance of the diagrid system connections is integral to the performance of the building as a whole, which requires intensive analysis. Scale modeling and testing are often required. If the nodes are modularized, great savings can be achieved. The case studies Lotte Super Tower, Seoul, South Korea; Shenzhen Rural Commercial Bank HQ, Shenzhen, China; KAUST Solar Chimney presented in Chap. 1 are examples of buildings designed with the diagrid system.

Additionally, Poly International Plaza in China's capital, Beijing, is another example of a building designed with the diagrid system. The elliptical shaped building is 162 m (532 ft) tall, shown in Fig. 4.59. The exterior diagrid system is composed of concrete-filled steel tubes creating a column-free interior space. The external diagrid system is combined with a concrete shear wall core. Extensive finite element analysis of the nodes was performed; the accuracy of the node analysis was confirmed by testing to confirm the necessity of the concrete fill, Fig. 4.60.

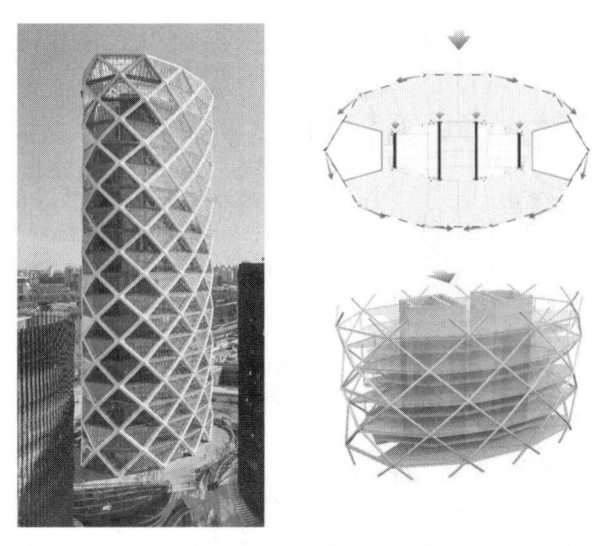

Figure 4.59 (a) Construction photo of Poly International Plaza, Beijing; (b) diagrid analysis and typical floor plan.

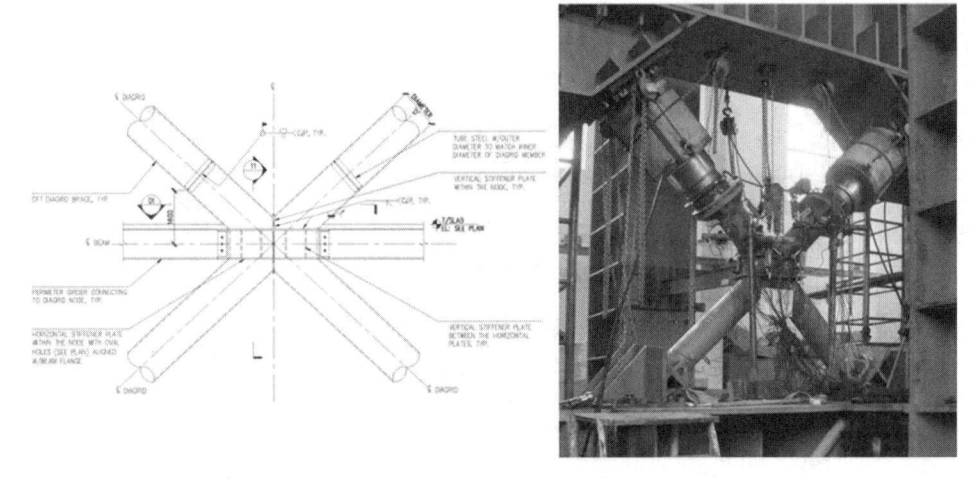

Figure 4.60 (a) Typical diagrid node detail; (b) laboratory test of typical diagrid node.

4.14   SUPERFRAME

The concept of the superframe structure was initiated by Fazlur Khan but was not realized in his lifetime. The superframe structure is a tall building topology that utilizes large perimeter elements to allow the full perimeter footprint to be utilized for lateral load resistance. Horizontal trusses at distinct levels to distribute the gravity loads to the corner frames to maximize the overturning moment resistance, minimize the net tension under wind loads, and to allow for modular building form. This would also allow for unique interior column layouts. The superframe concept was originated for a proposed 640 m (2100 ft) building in Chicago consisting of three concentric tube structures. The concept has been realized in the design of two supertall towers in China. The superframe concept is shown schematically in Fig. 4.61.

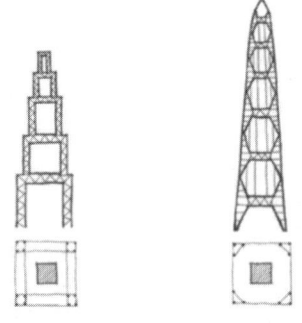

Figure 4.61   (a) Superframe structural system; (b) articulated superframe structural system (Luo, Y., 2015).

A good example of a building that utilized the superframe system is the 521 m (1709 ft) Guizhou Culture Plaza Tower in Guiyang, shown in Fig. 4.62. The tower captures the use of both the superframe and the tapered diagonal braced tube. Modularization has also been considered; the tower consists of eight 16-level structural modules. The gravity loads are transferred to the four-corner superframe columns by the superframe girders which maximize the overturning moment resistance and minimizes the net tension under wind loads. The superframe columns are open, built-up lattice structures with chamfered corners which with the overall tapered form of the tower minimize the dynamic effects of the wind on the tower. The superframe girders are used to dissipate energy during a seismic event through frame action. Additionally, the perimeter structure is connected to a reinforced concrete floor through thick slabs acting as virtual outriggers on the superframe girder floors.

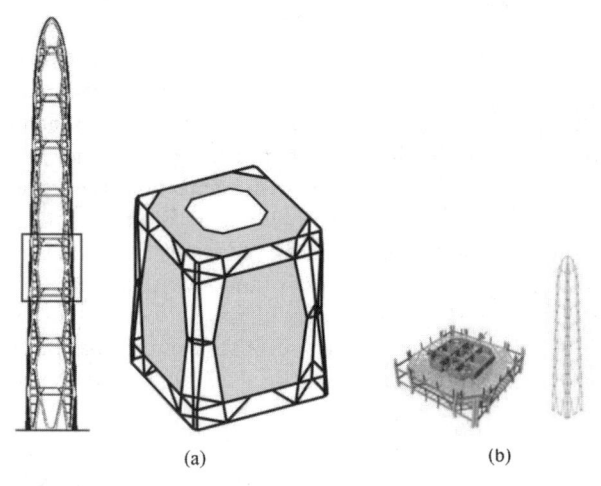

Figure 4.62   (a) Guizhou Cultural Plaza Tower; (b) 3D view of superframe module.

Another example is the Tianjin Goldin Finance 117 tower located in Tianjin, China. The tower is 597 m (1959′) tall with 117 stories. The lateral system consists of reinforced concrete megacolumns and a braced perimeter megaframe structure. Additionally, belt trusses are also used at intermediate locations to enhance the torsional stiffness of the tower and transfer gravity loading to the corner megacolumns. The structural system is shown in Fig. 4.63.

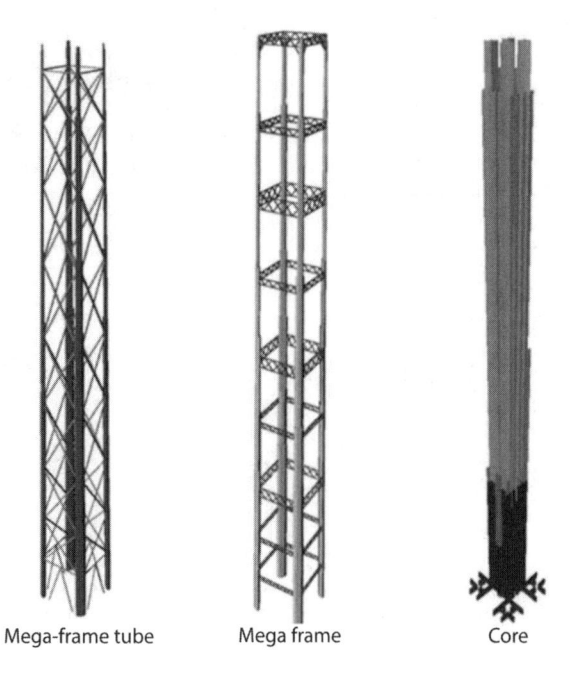

Mega-frame tube   Mega frame   Core

Figure 4.63   Tianjin Goldin Finance 117 Tower structural system (Lee, 2012).

Chapter **5**

Lateral Bracing Systems for Concrete Buildings

Analogous to steel or composite construction, concrete offers a wide range of structural systems suitable for high-rise buildings. There are perhaps as many structural concepts as there are engineers, making it awkward if not impossible to classify all the concepts into distinct categories. However, for purposes of presentation, it is convenient to group the most common systems into separate categories, each with an applicable height range as shown in Fig. 5.1. Although the height range for each group is logical for normally proportioned buildings, the appropriateness of each system to a particular building can only be judged when all other factors influencing the lateral load behavior are taken into account. Such factors include building geometry, severity of exposure to wind, seismicity of the region, ductility of the frame, and limits imposed on the size of the structural members.

Oftentimes, systems combining the characteristics of two or more can be employed to fulfill the specific project requirement. The multitude of systems available presents an opportunity for an experienced engineer to come up with a structural system that will serve its optimum function in the overall sense of the project. Although the selection of a system requires knowledge of both horizontal and lateral systems, the material presented in this chapter emphasizes the requirements of lateral systems only. Gravity systems are covered separately in Chap. 8.

Figure 5.1 shows 15 different categories of structural systems, starting with the most elementary system consisting of floor slabs and columns. At the other end of the spectrum is the bundled tube system, which is appropriate for very tall buildings and for buildings with a large plan aspect ratio. Almost all of the systems described in

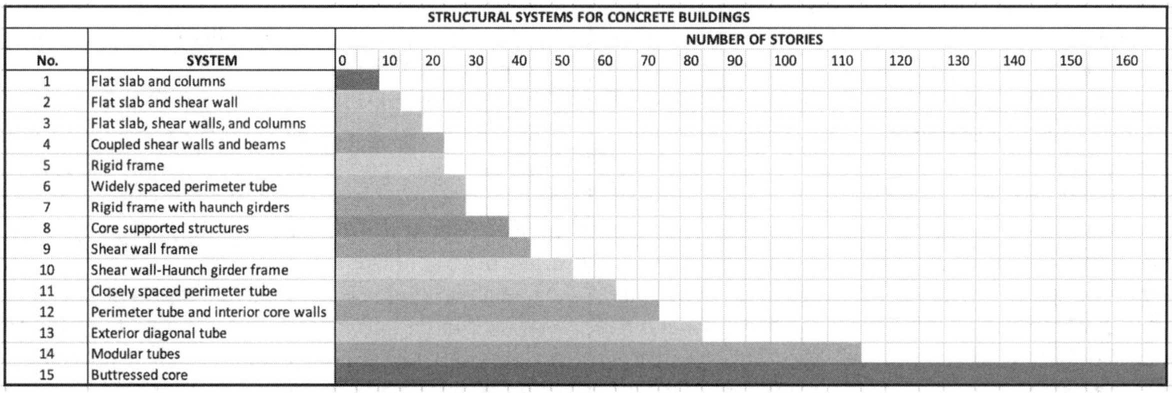

STRUCTURAL SYSTEMS FOR CONCRETE BUILDINGS																			
		NUMBER OF STORIES																	
No.	SYSTEM	0	10	20	30	40	50	60	70	80	90	100	110	120	130	140	150	160	
1	Flat slab and columns																		
2	Flat slab and shear wall																		
3	Flat slab, shear walls, and columns																		
4	Coupled shear walls and beams																		
5	Rigid frame																		
6	Widely spaced perimeter tube																		
7	Rigid frame with haunch girders																		
8	Core supported structures																		
9	Shear wall frame																		
10	Shear wall-Haunch girder frame																		
11	Closely spaced perimeter tube																		
12	Perimeter tube and interior core walls																		
13	Exterior diagonal tube																		
14	Modular tubes																		
15	Buttressed core																		

Figure 5.1 Structural systems for concrete buildings.

Chap. 4, and shortly in Chap. 6, are equally applicable to concrete buildings. Therefore, proper selection can only be made when the engineer has become familiar with the systems described not only in this chapter but elsewhere throughout this work.

5.1   FRAME ACTION OF COLUMN AND TWO-WAY SLAB SYSTEMS

Concrete floors in tall buildings often consist of a two-way floor system such as a flat plate, flat slab, or a waffle system. In a flat-plate system, the floor consists of a concrete slab of uniform thickness which frames directly into columns. The flat slab system makes use of either column capitals, drop panels or both to increase the shear and moment resistance of the system at the columns where the shears and moments are greatest. The waffle slab consists of two rows of joists at right angles to each other; commonly formed by using square domes. The domes are omitted around the columns to increase the moment and shear capacity of the slab. Any of the three systems can be used to function as an integral part of the wind-resisting systems for buildings in the 10-story range. Two-way slab systems with columns can be used as a seismic system in seismic design categories up to Seismic Design Category (SDC) C up to a specific height according to ASCE 7-16 Table 12.2-1.

The slab system shown in Fig. 5.2 has two distinct actions in resisting lateral loads. First, because of its high in-plane stiffness, it distributes the lateral loads to various vertical elements in proportion to their bending stiffness. Second, because of its significant out-of-plane stiffness, it restrains the vertical displacements and rotations of the columns as if they were interconnected by a shallow wide beam.

The concept of "effective width" as explained below can be used to determine the equivalent width of the slab. Although physically no beam exists between the columns, for analytical purposes it is convenient to consider a certain width of the slab as a beam framing between the columns. The effective width is however, dependent on various parameters, such as column aspect ratios, distance between the columns, thickness of the slab, etc. Research has shown that values less than, equal to, and greater than full width are all valid depending upon the parameters mentioned above.

Note that the American Concrete Institute, ACI code permits the full width of slab between adjacent panel center lines for both gravity and lateral loads. The only stipulation is that the two-way systems analysis should take into account the effect of slab cracking in evaluating the

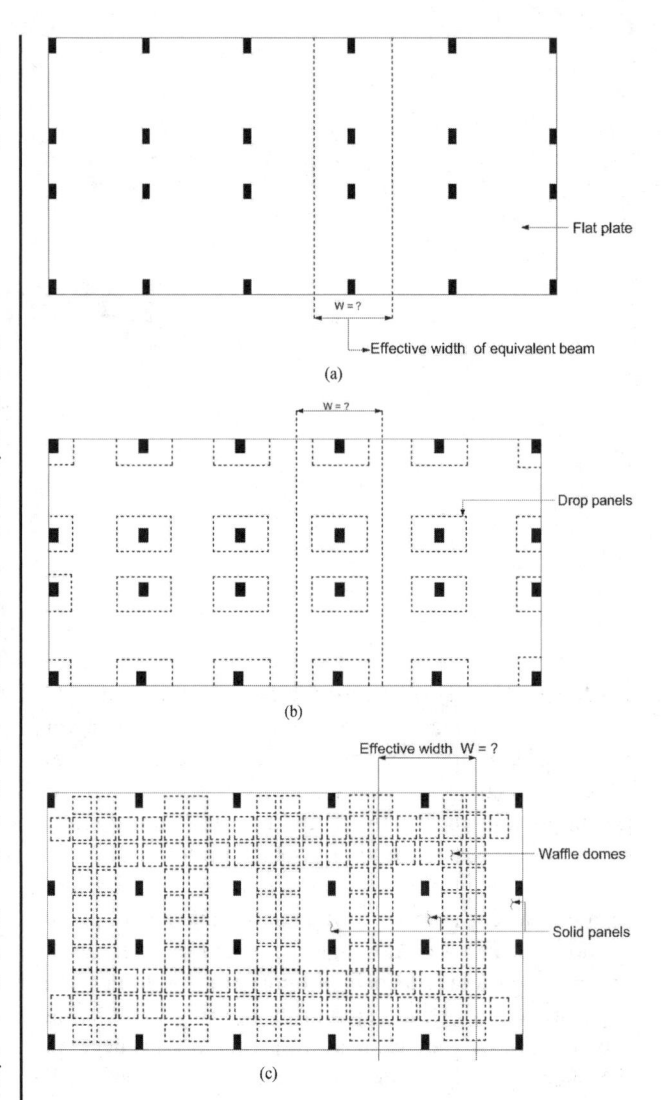

Figure 5.2   Lateral systems using slab and columns: (a) flat plate; (b) flat slab with drop panels; (c) two-way waffle system.

stiffness of frame members. Use of full width is explicit for gravity analysis and implicit (because it is not specifically prohibited) for the lateral loads.

However, engineers generally agree that using full width of the slab gives unconservative results for lateral load analysis. The method tends to overestimate the slab stiffness and underestimate the column stiffness, compounding the error in estimating the distribution of moments due to lateral loads.

The shortcomings of using full width in lateral load analysis can be overcome by determining the equivalent stiffness on the basis of a finite-element analysis. The stiffness thus obtained is appropriate for both gravity and lateral analysis.

Of particular concern in the detailing of two-way systems is the problem of stress concentration at the column-slab joint where, especially under lateral load, nonlinear behavior is initiated through concrete cracking and steel yielding. Shear reinforcement at the column-slab joint is necessary to improve the joint behavior and avoid early stiffness deterioration under lateral cyclic loading. Note that two-way slab systems without beams are not permitted by the ACI code in regions of high seismic risk (SDC D, E, and F). Their use in regions of low-to-moderate seismic risk (SDC A, B, and C) is permitted subject to certain requirements, mainly relating to reinforcement placement in the column strip. Since the requirements are too detailed to be the reader is referred to the International Building Code (IBC) and ACI codes for further details.

5.2 FLAT SLAB AND SHEAR WALLS

Frame action obtained by the interaction of slabs and columns is not adequate to give the required lateral stiffness for buildings taller than about 10 stories. The advantages of beamless flat ceilings could still be achieved in taller buildings by strategically locating shear walls to function as the main lateral-load-resisting element. The walls can be either planar, open, closed, or any combination of the three. Figure 5.3 shows an example in which planar shear walls are located in a simple orthogonal orientation. Skewed or irregular layouts require a three-dimensional analysis to include the effect of torsional loads. Planar shear walls in essence behave as slender cantilevers. When no major openings are present, stresses and deflections can be determined using simple bending theory. Complicated open-section shapes need special modeling and analyses techniques as outlined in Chap. 10.

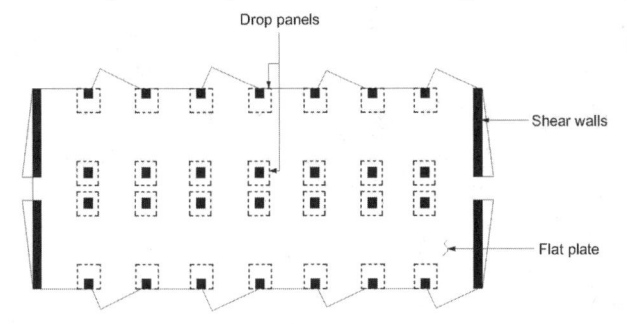

Figure 5.3   Shear wall flat slab system.

5.3 FLAT SLAB, SHEAR WALLS, AND COLUMNS

The applicable height range of slab and shear wall systems can be increased marginally by including the frame action of columns and slabs. The system is best suited for apartments, condominiums, and hotels, and is identical to the system described in the previous section. The difference is only in the analysis in that the frame action of columns and slabs is also taken into account in the lateral load analysis. Whether this action is significant or not is a function of relative stiffness of various elements. In most apartment or hotel layouts, the frame resistance to overturning moments is no more than 10 to 20 percent of the resistance offered by the shear walls. Many engineers, therefore, ignore the frame action altogether by designing the shear walls to carry the total lateral loads. However, in keeping with the current trend of taking advantage of all available structural actions, it is advisable to include the frame action in the analysis.

5.4 COUPLED SHEAR WALLS

When two or more shear walls are interconnected by a system of beams or slabs, the total stiffness of the system exceeds the summation of the individual wall stiffnesses because the connecting slab or beam restrains the individual cantilever action by forcing the system to work as a composite unit.

Such an interacting shear wall system can be used economically to resist lateral loads in buildings up to about 40 stories. However, planar shear walls are efficient lateral load carriers only in their plane. Therefore, it is necessary to provide walls in two orthogonal directions. However, in long and narrow buildings sometimes it may be possible to resist wind loads in the long direction by the frame action of columns and slabs because first, the area of the building exposed to the wind is small, and second, the number of columns available for frame action in this direction is usually large. The layout of walls and columns should take into consideration the torsional effects.

Walls around elevators, stairs, and utility shafts offer an excellent means of resisting both lateral and gravity loads without requiring undue compromises in the leasability of buildings. Closed- and partially closed-section shear walls are efficient in resisting torsion, bending moments, and shear forces in all directions, especially when sufficient strength and stiffness are provided around door openings and other penetrations. This is discussed in greater detail in Chap. 10.

5.5 RIGID FRAME

Cast-in-place concrete buildings have the inherent advantage of continuity at joints. Girders framing directly into columns, can be considered rigid with the columns; such a girder-column arrangement can be thought of as a portal frame. However, girders that carry shear and bending moments due to lateral loads often require additional construction depth, necessitating increases in the overall height of the building.

The design and detailing of joints where girders frame into building columns should be given particular attention, especially when buildings are designed to resist seismic forces. The column region within the depth of the girder is subjected to large shear forces. Horizontal ties must be included to avoid uncontrolled diagonal cracking and disintegration of concrete. Specific detailing provisions are given in the IBC and ACI codes to promote ductile behavior in SDC D, E, and F, and somewhat less stringent requirements in SDC B and C, as required by Table 12.2-1 of ASCE 7-16. The underlying philosophy is to design a system that can respond to overloads without loss in gravity-load carrying capacity.

Rigid-frame systems for resisting lateral and vertical loads have long been accepted as a standard means of designing buildings because they make use of the stiffness in the beams and columns that are required in any case to carry the gravity loads. In general, rigid frames are not as stiff as shear wall construction and are considered more ductile and less susceptible to catastrophic earthquake failures when compared to shear wall structures.

As discussed in Chap. 4, a rigid frame is characterized by its flexibility due to flexure of individual beams and columns and rotation at their joints. The strength and stiffness of the frame is proportional to the beam and column size and inversely proportional to the column spacing. Internally located frames are not very popular in tall buildings because the leasing requirements of most buildings limit the number of interior columns available for frame action. The floor beams are generally of long spans and are of limited depth. However, frames located at the building exterior do not necessarily have these disadvantages. An efficient frame action can thus be developed by providing closely spaced columns and deep beams at the building exterior.

5.6 WIDELY SPACED PERIMETER TUBE

The term tube, in the usual building terminology suggests a system of closely spaced columns 8 to 15 ft in center (2.43 to 4.57 m) tied together with a relatively deep spandrel. However, for buildings with compact plans it is possible to achieve tube action with relatively widely spaced columns interconnected with deep spandrels. An example of such a layout for a 28-story building is shown in Fig. 5.4.

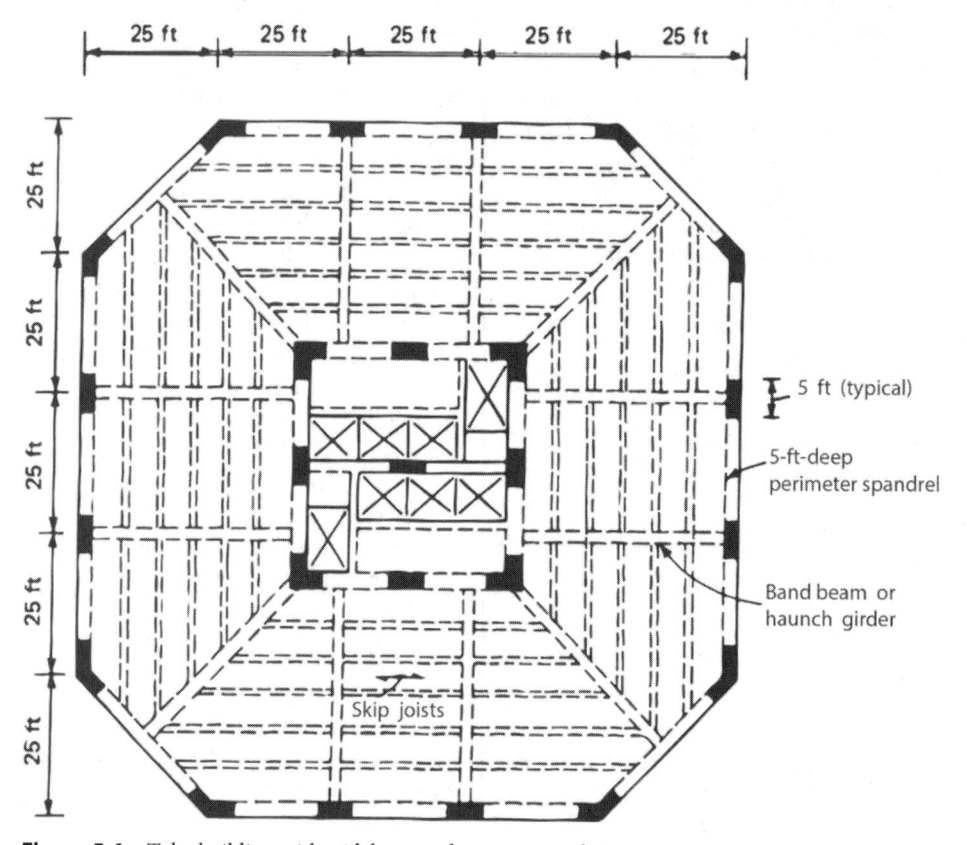

Figure 5.4 Tube building with widely spaced perimeter columns.

Lateral resistance is provided by the perimeter frame, consisting of 5 ft (1.5 m) wide columns, spaced at 25 ft (7.62 m) centers, and tied to a 5 ft (1.53 m) deep spandrel.

5.7 RIGID FRAME WITH HAUNCH GIRDERS

One of the drawbacks of a rigid frame system is the excessive depth of girder required to make the rigid frame economical. Rigid frames, when located at the perimeter, often can use deep spandrels without adversely affecting the floor-to-floor height of the building. Also, it may be architecturally acceptable to use relatively closely spaced columns at the exterior.

Office buildings usually are laid out with a span of about 40 ft (12.19 m) without any interior columns between the core and the exterior. To economically frame such a large span would require a girder depth of about 3 ft (0.91 m) unless the beams are post-tensioned. This requirement clearly impacts the floor-to-floor height and often is unacceptable because of the additional cost of curtain wall and the extra heating and cooling loads due to the increased volume of the building. A haunch girder, which consists of a girder of variable depth, gives the required stiffness for lateral loads without having to increase the floor-to-floor height. This is achieved by making the mid-section of the girder flush with the floor

system, thus providing ample beamless space for the passage of mechanical ducts.

Girders with haunches on either end are ideal for resisting lateral loads, but certain types of column layouts may limit the haunches to one end of the girder. Such an arrangement with single-end haunch girders is shown for a 28-story building in Fig. 5.5. To keep the skip-joist framing simple, the girders are framed into the spandrel without haunches at the exterior end. Girders that match haunch girder depth are used between the core columns, as shown in Fig. 5.5. The depth of these girders did not affect the mechanical distribution and therefore did not require an increase in floor-to-floor height. A 9 ft (2.74 m) high ceiling was accomplished with a 12 ft 10 in. (3.91 m) floor-to-floor height.

5.8 CORE-SUPPORTED STRUCTURES

One of the most frequent uses of shear walls is in the form of box-shaped cores around stairs and elevators, because this arrangement makes structural use of vertical enclosures required around the cores. Arrangement of internal cores is especially suitable for office buildings because it frees the lease space outside of the core from massive vertical elements. The walls around the core can be considered as a spatial system capable of transmitting lateral loads in both directions. Additional advantage

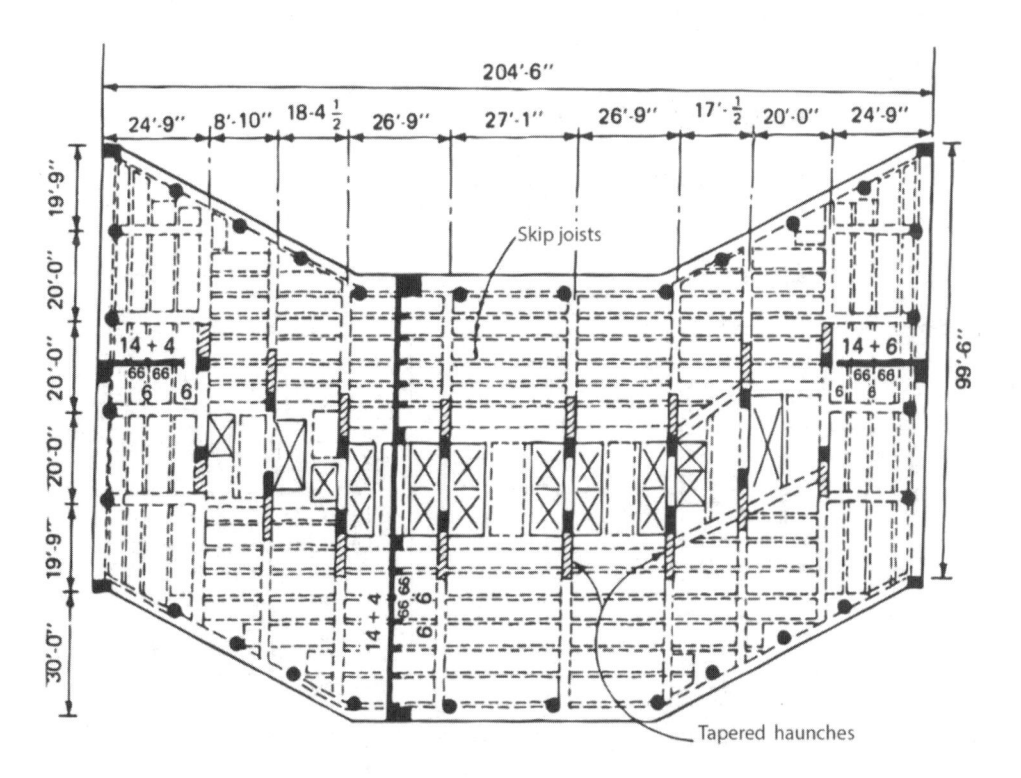

Figure 5.5 A 28-story haunch girder building.

of core structures is that being spatial structures, they have the ability to resist all types of loads: vertical loads, shear forces, and bending moments in two directions, as well as torsion, especially when adequate stiffness and strength are provided between flanges of open sections. The shape of the core to a large extent is governed by the elevator and stair requirements. Variations could occur from a single rectangular core to complicated arrangements of planar shear walls. Other structural elements surrounding the core may consist of either a cast-in-place or precast concrete or structural steel construction. In a precast or steel surround system, it is more than likely that the stiffness of the core will overwhelm the stiffness of other vertical elements. Even in a cast-in-place system, unless the exterior frame consists of relatively closely spaced columns and a deep spandrel, it is justifiable to ignore the resistance of other vertical members and to design the core system for the entire lateral load. Figures 5.6a–c show some examples of core arrangements.

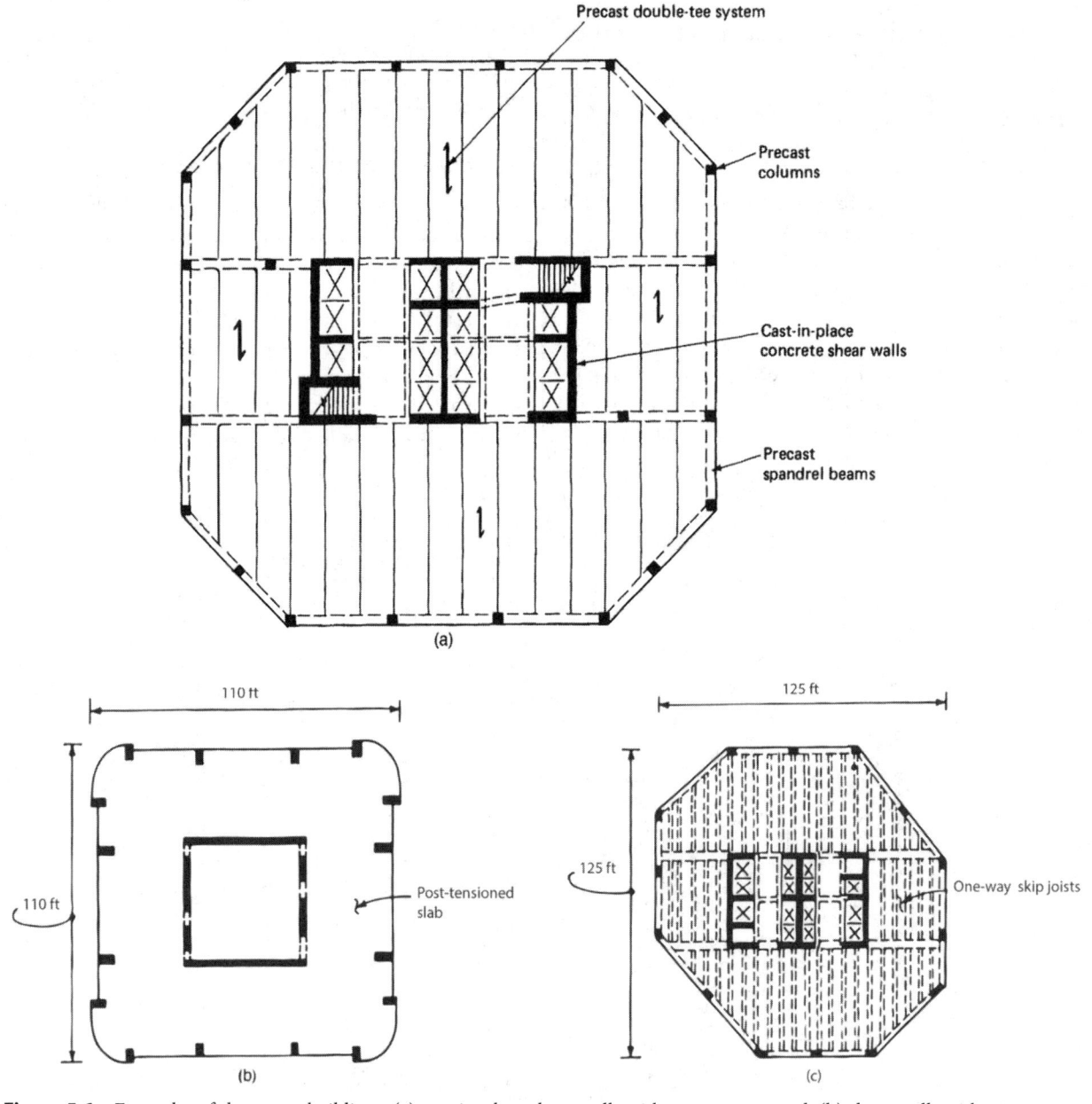

Figure 5.6   Examples of shear core buildings: (a) cast-in-place shear walls with precast surround; (b) shear walls with post-tensioned flat plate; (c) shear walls with one-way joist system.

5.9   SHEAR WALL-FRAME INTERACTION

Without question, this system is one of the most, if not the most, popular system for resisting lateral loads. The system has a broad range of applications and has been used for buildings as low as 10 stories to as high as 50 stories or even taller buildings. With the use of haunch girders, the applicability of the system is easily extended to buildings in the 70- to 80-story range.

The interaction of frame and shear walls has been understood for quite some time; the classical mode of the interaction between a prismatic shear wall and a moment frame is shown in Fig. 5.7; the frame basically deflects in a so-called shear mode while the shear wall predominantly responds by bending as a cantilever. Compatibility of horizontal deflection produces interaction between the two. The linear sway of the moment frame, when combined with the parabolic sway of the shear wall results in an enhanced stiffness because the wall is restrained by the frame at the upper levels while at the lower levels the shear wall is restrained by the frame. However, it is not always easy to differentiate between the two modes because a frame consisting of closely spaced columns and deep beams tends to behave more like a shear wall responding pre-dominantly in a bending mode. And similarly, a shear wall weakened by large openings may tend to act more like a frame by deflecting in a shear mode. The combined structural action, therefore, depends on the relative rigidity of the two, and their modes of deformation. Furthermore, the simple interaction diagram given in Fig. 5.7 is valid only if

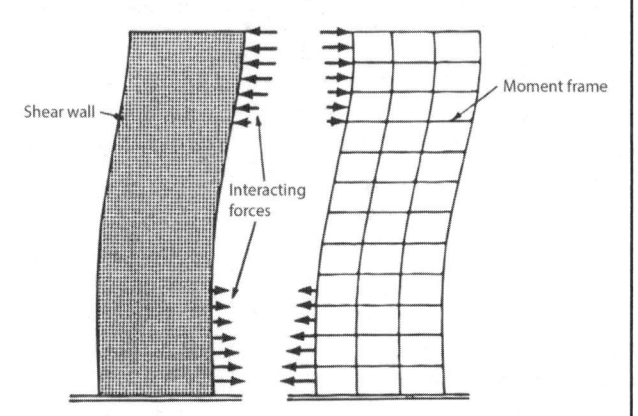

Figure 5.7   Shear wall-frame interaction.

• the shear wall and frame have constant stiffness throughout the height or;
• if stiffnesses vary, the relative stiffness of the wall and frame remains unchanged throughout the height.

Since architectural and other functional requirements frequently influence the configuration of structural elements, the above conditions are rarely met in a practical building. In a contemporary high-rise building, very rarely can the geometry of walls and frames be the same over the full height. For example, walls around the elevators are routinely stopped at levels corresponding to the elevator drop-offs, columns are made smaller as they go up, and the building geometry is very rarely the same for the full height. Because of the abrupt changes in the stiffness of walls and frames combined with the variation in the geometry of the building, the simple interaction shown in Fig. 5.7 does not even come close to predicting the actual behavior of the building structures. However, with the availability of two- and three-dimensional computer programs, capturing the essential behavior of the shear wall frame system is within the reach of everyday engineering practice.

To understand qualitatively the nature of interaction between shear walls and frames, consider the framing plan shown in Fig. 5.8a. The building is unusual in that it exhibits almost perfect symmetry in two directions and maintains a reasonably constant stiffness throughout its height. Therefore, in a qualitative sense, the interaction between the frames and shear walls is expected to be similar to that shown in Fig. 5.7.

The building is 25 stories and consists of four levels of basement below grade. The floor framing consists of 6 in. wide by 20 in. deep (152.4 by 508 mm) skip joist framing between haunch girders which span the distance of 35 ft 6 in. (10.82 m) between the shear walls and the exterior of the building. The girders are 42 in. wide by 20 in. deep (1.06 by 0.5 m) for the exterior 28 ft 6 in. (8.67 m) length, with a haunch at the interior tapering from a pan depth of 20 to 33 in. (0.5 to 0.84 m). Four shear walls of dimensions 1 ft 6 in. by 19 ft 6 in. (0.45 by 5.96 m) rise for the full height from a 5 ft (1.52 m) deep mat foundation. The exterior columns vary from 38 by 34 in. (965 by 864 mm) at the bottom to 38 by 24 in. (965 by 610 mm) at the top. Note that the haunch girder is made deliberately wider than the exterior column to simplify removal of flying forms. However, by using "hinges," it is also possible to simplify removal of formwork. Therefore, width of haunch girders can be consistent with the structural requirements; they need not be wider than exterior columns.

The lateral load resistance in the short direction of the building is provided by a combination of three types of frames: (i) two exterior frames along grids 1 and 8; (ii) two haunch girder frames along grids 2 and 7; and (iii) four shear wall-haunch girder frames along grids 3 through 6. The lateral load resistance in the long

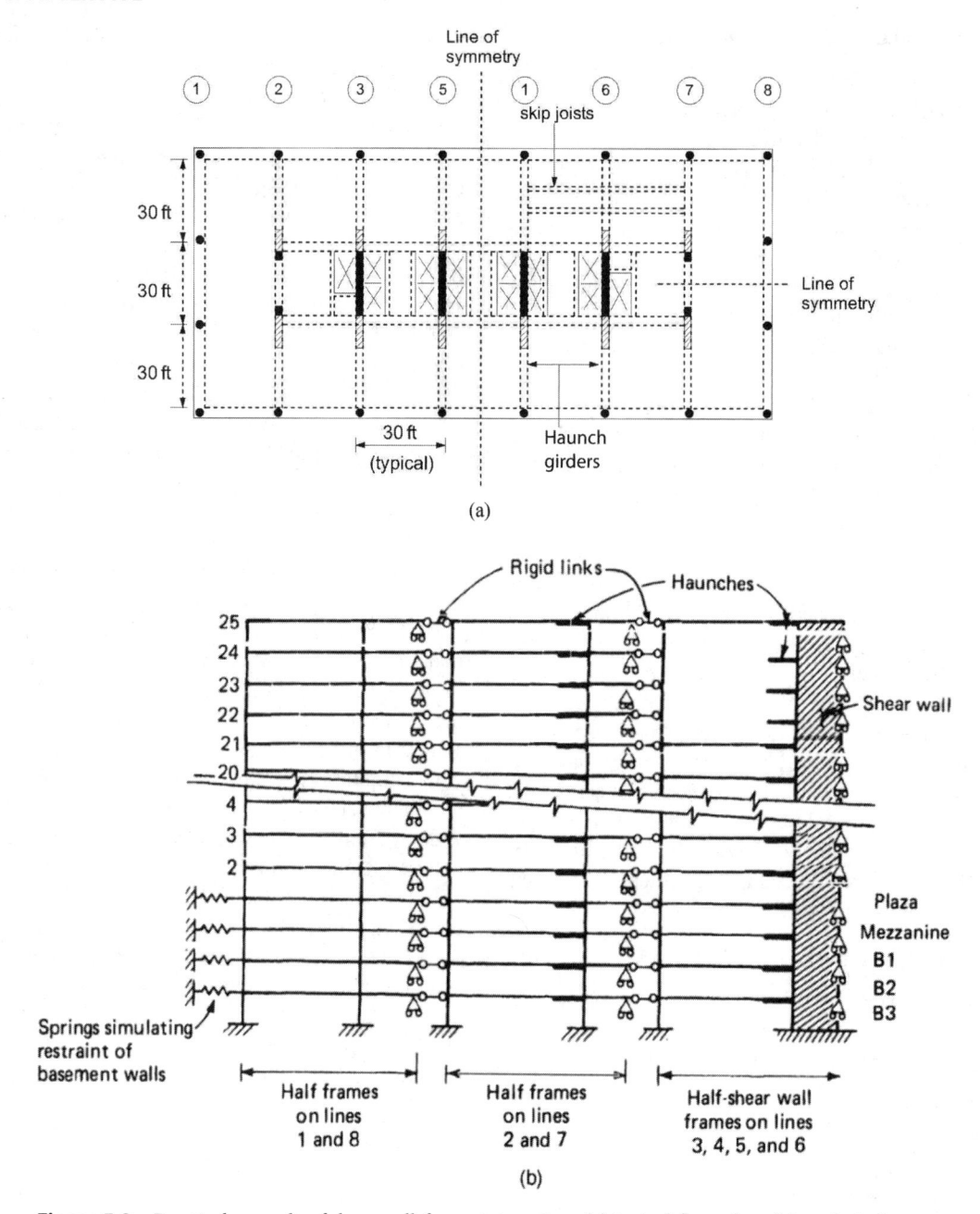

Figure 5.8   Practical example of shear wall-frame interaction: (a) typical floor plan; (b) analytical two-dimensional model for wind on long face; (c) distribution of lateral loads, (i) applied lateral loads, (ii) lateral loads resisted by end frames, (iii) lateral loads resisted by interior frames, (iv) lateral loads resisted by shear wall frames.

direction is provided primarily by frame action of the exterior columns and spandrels along the broad faces.

For purposes of structural analysis, the building is considered symmetrical about the two centerlines as shown in Fig. 5.8a. The lateral load analysis can be carried out by lumping together similar frames and using only one-half the building in the computer model; however, with today's computing capabilities, this is not necessary. This is shown in Fig. 5.8a for a two-dimensional computer model for analysis of wind on the broad face. In the model only, three equivalent frames are used to represent the lateral load resistance of eight frames,

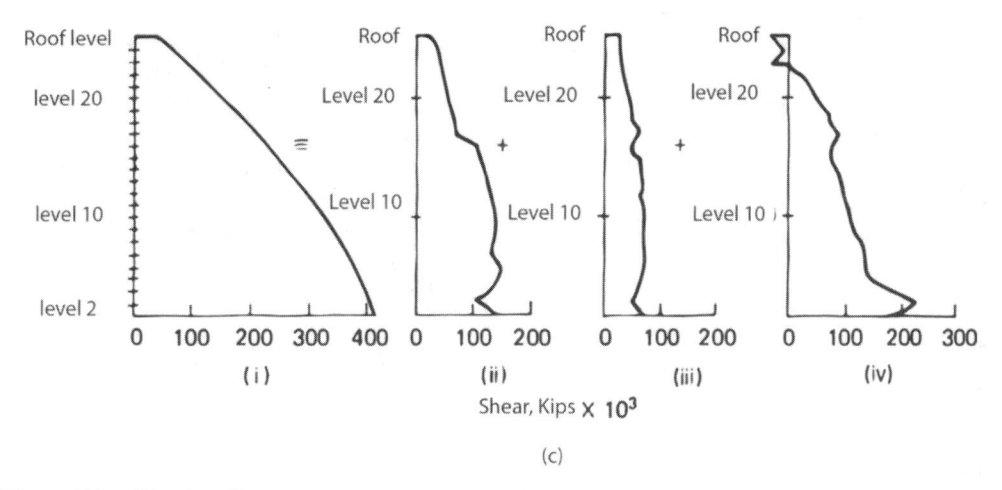

Figure 5.8   (*Continued*)

and only one-half of each frame is used to simulate the structural action of a full frame. The latter simplification is achieved by restraining the vertical displacement at the end of each frame as shown in Fig. 5.8b. Note spring restraints at the basement levels B1, B2, B3, and at the plaza are used to simulate the lateral restraint of the basement walls and soil structure interaction. The rigid links shown in Fig. 5.8b between the individual frames simulate the diaphragm action of the floor slab by maintaining the lateral displacements of each frame the same at each level.

The purpose of the example is to show qualitatively the nature of interaction that exists between shear walls and frames in a building without abrupt changes in the stiffness of either element. We will not burden the reader with an avalanche of computer results, but will limit the presentation to the distribution of horizontal shear in various frames. Part 1 of Fig. 5.8c shows the diagram of cumulative shear forces applied along the broad face of the building. The distribution of the shear forces among the three types of frames is indicated in parts 2 to 4 of Fig. 5.8c. Note the reversal in the direction of the shear force at the top of shear wall-frame in part 4 of Fig. 5.8c, which indicates that the shear wall-frame combination has a tendency to behave as a propped cantilever, not unlike the behavior observed in the simplified shear wall-frame interaction model shown in Fig. 5.7. This is not surprising because although the practical example consists of a combination of three different types of frames, in essence the structural system can be considered as a single shear wall acting in combination with a frame. There are no sudden changes in the stiffness of walls or frames along the building height

contributing to its departure from the fundamental behavior.

The example shown in Fig. 5.8 (although taken from an actual building in Houston, Texas) has very little in common with the usual types of structural systems that structural engineers are called upon to design. It is a rare event indeed to be commissioned to design a building that is symmetrical and has no significant structural discontinuity over its height. More usually, there is asymmetry because either some shear walls or frames drop off, or the building shape is architecturally modulated. The reality of present-day architecture precludes the generalization of even qualitative comments regarding the interaction between shear walls and frames. The structural engineer has very little choice, other than using computer analysis for determining the distribution of loads to various elements. Standard interaction diagrams that were helpful in assigning the lateral loads to simple shear wall and frame systems have limited application and are therefore generally not used in practice.

As a second example, let us consider a contemporary high-rise building that has asymmetrical floor plans and abrupt variation in stiffness of shear walls and frames throughout its height. Figure 5.9 shows a perspective of a twin-tower high-rise office development called The Lone Star Towers in Dallas, Texas. The first phase of the complex consists of a four-level, 1600-car below-grade parking structure and a 1,000,000 ft² (92,903 m²) office space. The building is 655 ft tall (200 m) and consists of 50 stories above grade. A variety of floor plans are accommodated between the second and the roof levels, resulting in a number of setbacks as shown in Fig. 5.10, with a major transfer of columns at level 40 (Fig. 5.10d). The floor plan, which is essentially rectangular at the second floor,

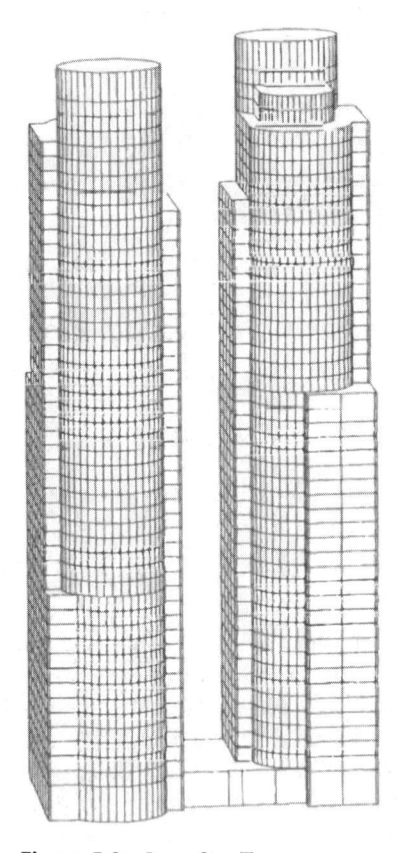

Figure 5.9   Lone Star Towers.

progressively transforms into a circular shape at the upper levels. Figure 5.10a–f shows the framing plans at various levels. The resistance to lateral load is provided by a system of I- and C-shaped shear walls interacting with haunch girder frames. The floor framing is of lightweight concrete consisting of 6 in. (152.4 mm) wide skip joists at 6 ft 6 in. (1.98 m) centers. The 20 in. (508 mm) depth of haunch girders at midspan matches that of pan joist construction. Tapered haunches are used at both ends of girders and vary from a depth of 20 in. (508 mm) at mid-bay to a depth of 2 ft 9 in. (0.84 m) at the face of columns and shear walls. High-strength normal-weight concrete of up to 10 ksi (68.95 mPa) is used in the design of columns and shear walls. The shear walls around low, mid, mid-low, mid-high, and high-rise elevators are either terminated or made smaller in their dimensions at various levels corresponding to the zoning of elevators. The analytical model necessarily resulted in a system of shear walls and frames which were asymmetrical not only with respect to the plan dimensions but also varied in stiffness over the building height. The final lateral load analysis was accomplished by using a three-dimensional computer model which included each and every structural member. The results were, however, verified by comparing the results with those of: (i) a relatively simpler three-dimensional model in which every other floor was lumped; and (ii) an equivalent shear wall and frame which represented the lateral stiffness of the building in the short direction. The agreement between the various computer results was within acceptable limits.

Figure 5.10g shows the distribution of horizontal shear forces among various lateral-load-resisting elements for wind loads acting on the broad face of the building. The lateral loads in the shear wall frames are shown by the curves designated as 12, 13, 14, and 15, which correspond to their locations on the grid lines shown in Fig. 5.10a–f. The shear forces in the frames are shown by curves designated as 10, 11, 16, 17, and 18. The results shown are from an earlier version of the tower, which consisted of 42 floors with slight modifications to the floor plans shown in Fig. 5.10.

The purpose of presenting limited results of the analysis is to show qualitatively how the distribution of transverse shear in a practical building is considerably different than in a structure with regularly placed, full-height shear walls and frames of fairly uniform stiffness. The large difference in the pattern of transverse shear distribution occurs for two reasons. First, the structure is complex, with stiffness varying significantly over the height. Second, the mathematical assumption of a rigid diaphragm that is commonly used in the modeling of the floor slab tends to bring about sharp shear transfers at levels where the stiffness of the building changes abruptly. Although it is possible to smooth out the harsh distribution and sudden reversals of transverse shear by modeling the floor slab as a flexible diaphragm by using a finite element model, such a complex modeling technique will have little effect on the final shear wall design.

5.10   FRAME TUBE STRUCTURES

The tube concept is an efficient framing system for tall slender buildings. In this system, the perimeter of the building consists of closely spaced columns connected by a relatively deep spandrel. The resulting system works as a giant vertical cantilever and is very efficient because of the large separation between the windward and leeward columns. The tube concept in itself does not guarantee that the system satisfies stiffness and vibration limitations. The "chord" drift caused by the axial displacement of the columns and the "web" drift brought about by the shear and bending deformations of the spandrels and columns may vary considerably depending upon the geometric and elastic properties of the tube. For example, if the plan aspect ratio is large, say much in excess of 1:1.5, it is likely that a supplemental lateral bracing is necessary to satisfy drift limitations. The number of stories that can be achieved economically by using the tube

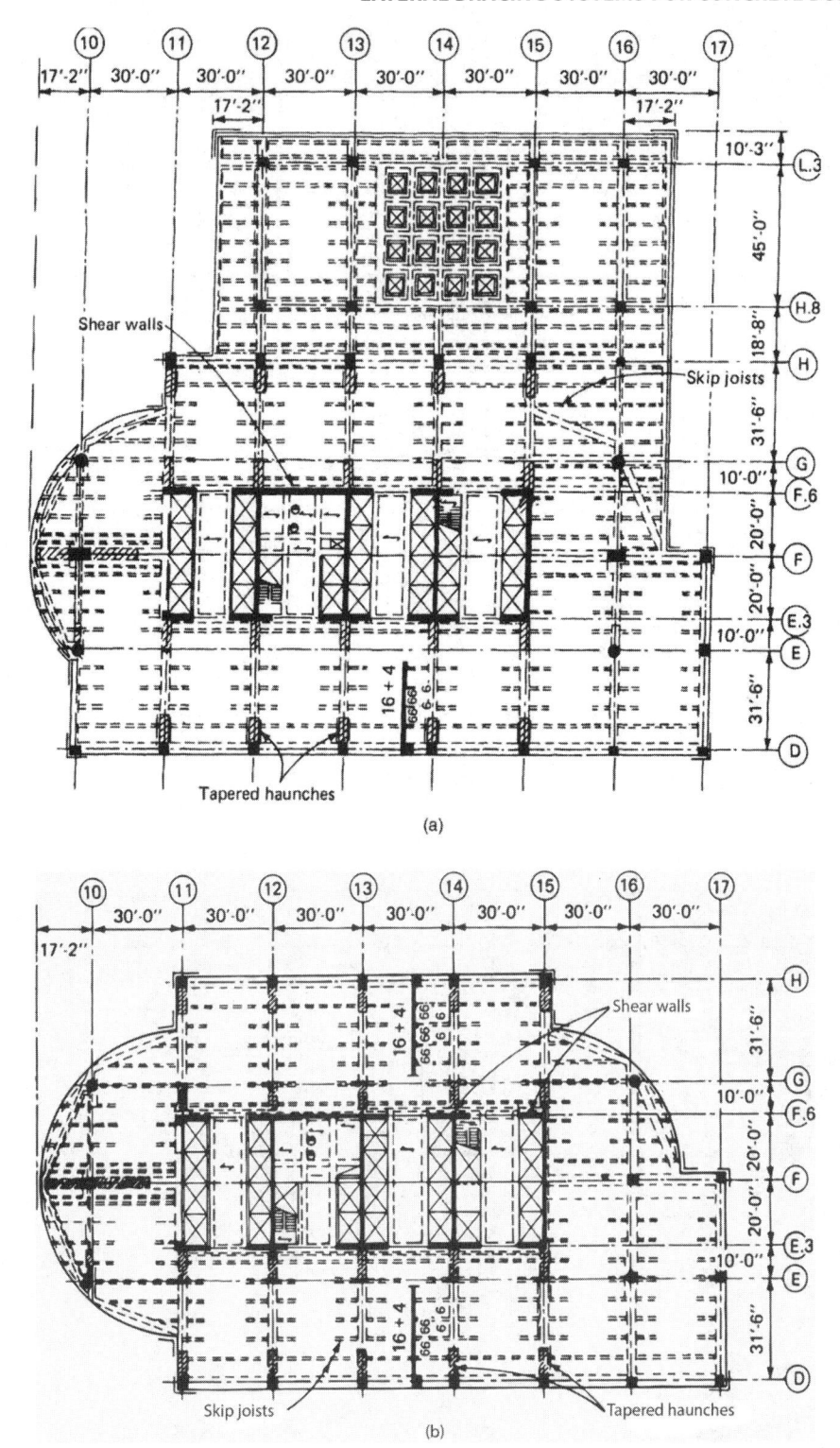

Figure 5.10 Example of shear wall-frame interaction in a 50-story building: (a) levels 2 through 14 floor plan; (b) levels 15 through 26 floor plan; (c) levels 27 through 39 floor plan; (d) levels 40 through 47 floor plan; (e) levels 48 and 49 floor plan; (f) levels 50 and 51 floor plan; (g) distribution of lateral loads.

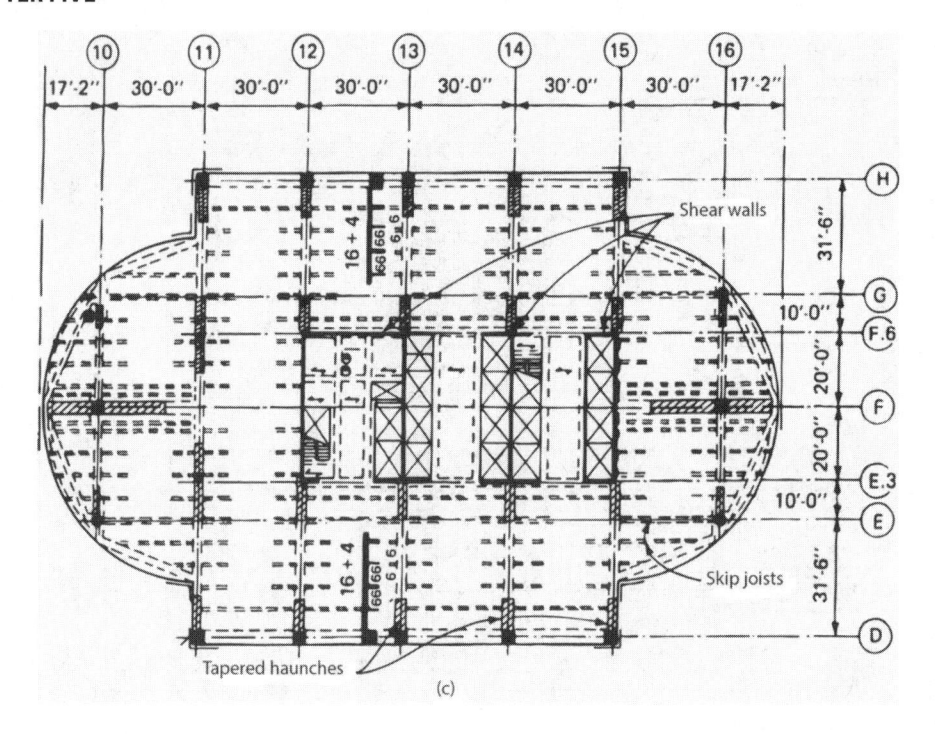

(c)

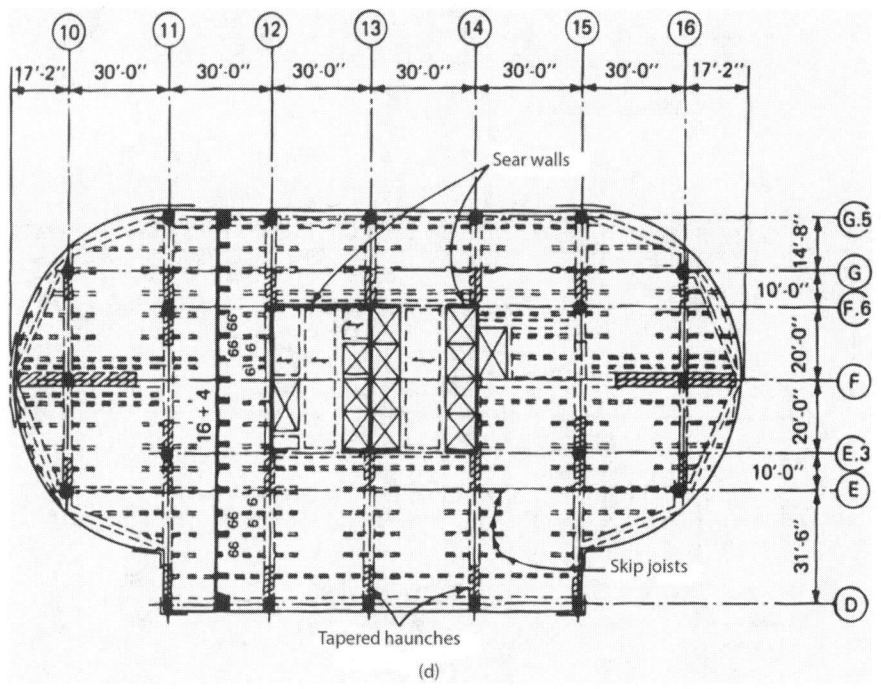

(d)

Figure 5.10   *(Continued)*

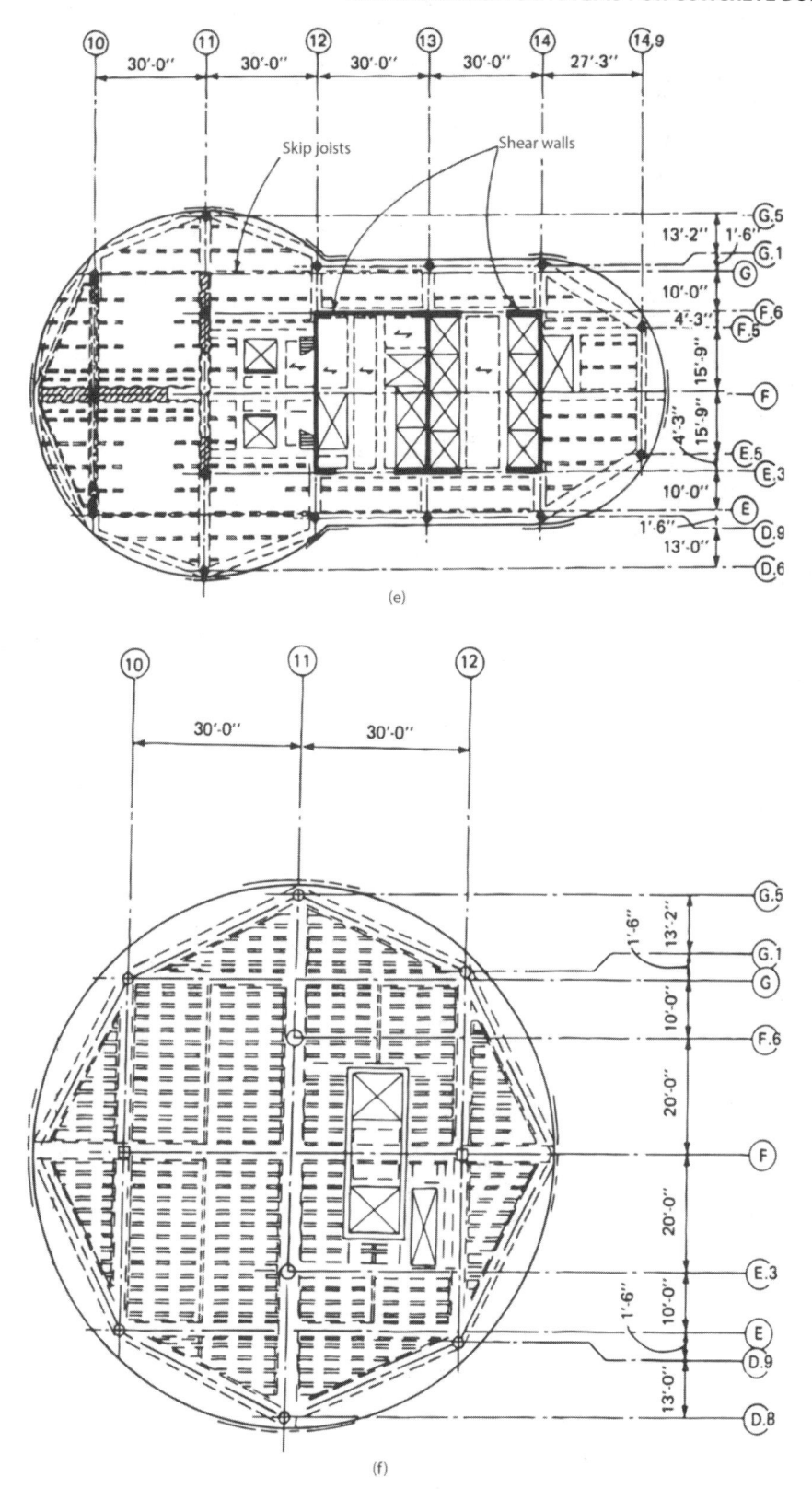

Figure 5.10 (*Continued*)

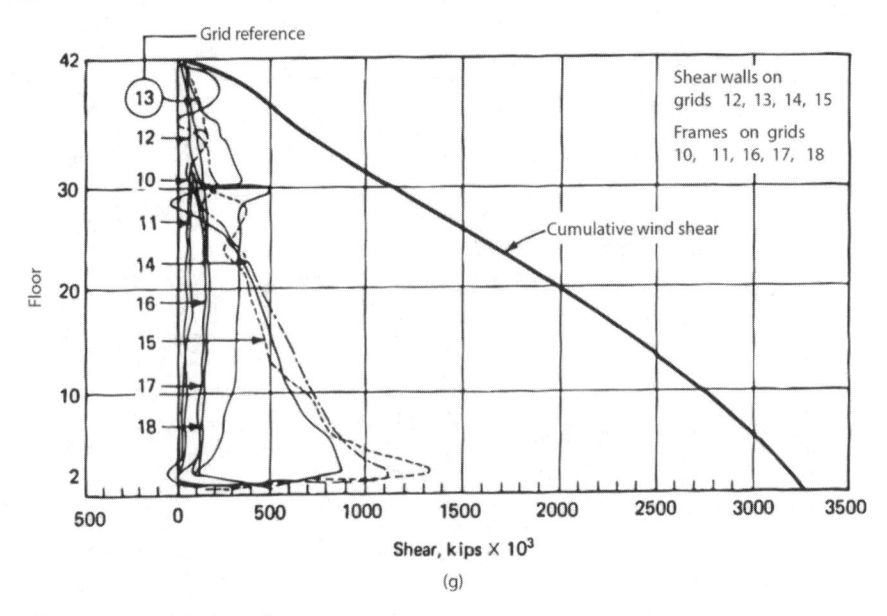

(g)

Figure 5.10 (*Continued*)

system depends on a number of factors such as spacing and size of columns, depth of perimeter spandrels, and plan aspect ratio of the building. The system should be given serious consideration for buildings taller than about 40 stories.

5.11 EXTERIOR DIAGONAL TUBE

Master builder Fazlur Khan of Skidmore, Owings & Merrill envisioned as early as 1972 that it was possible to build high rises in concrete rivaling those in structural steel. His quest to find a structural solution for eliminating the shear lag phenomenon led him to the diagonal tube concept. A brilliant manifestation of this principle in steel construction is seen in the John Hancock Tower in Chicago. Applying similar principles, Khan visualized a concrete version of the diagonal truss tube consisting of exterior columns spaced at about 10 ft (3.04 m) centers with blocked out windows at each floor to create a diagonal pattern on the façade. The diagonals could then be designed to carry the shear forces, thus eliminating bending in the tube columns and girders. Although Khan enunciated the principle in the 1970s, the idea had to wait almost 15 years to find its way to a real building. Currently, two high rises have been built using this approach. The first is a 50-story office structure located on Third Ave. in New York and the second is a mixed-use building located on Michigan Ave. in Chicago. The structural system for the building in

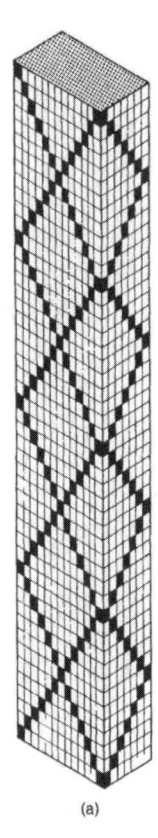

(a)

Figure 5.11 Exterior braced tube: (a) schematic elevation; (b) plan.

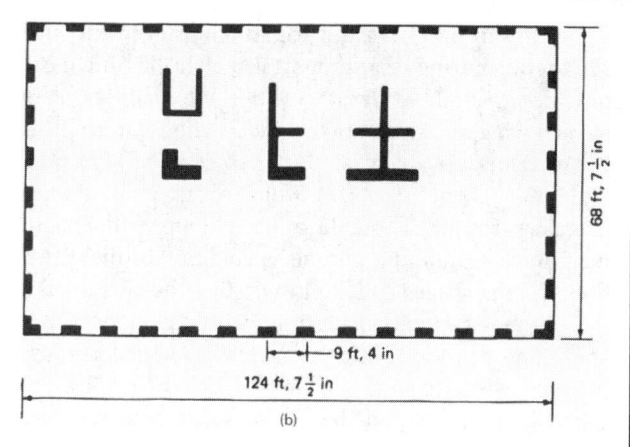

9 ft, 4 in
124 ft, 7½ in
68 ft, 7½ in
(b)

Figure 5.11   (*Continued*)

New York consists of a combination of a framed and trussed tube interacting with a system of interior core walls. All the three subsystems, namely, the framed tube, trussed tube, and shear walls, are designed to carry both lateral and vertical loads. The building is 570 ft (173.73 m) high with an unusually high height-to-width ratio of 8:1. The diagonals created by filling in the windows serve a dual function. First, they increase the efficiency of the tube by diminishing the shear lag, and second they reduce the differential shortening of exterior columns by redistributing the gravity loads. A stiffer, much more efficient structure is realized with the addition of diagonals. The idea of diagonally bracing this structure was suggested by Fazlur Khan to the firm of Robert Rosenwasser Associates, who executed the structural design for the building. Schematic elevation and floor plan of the building are shown in Fig. 5.11.

The Chicago version of the braced concrete tube is a 60-story multiuse project. The building rises in two tubular segments above a flared base. According to the designers, diagonal bracing was used primarily to allow maximum flexibility in the interior layout needed for mixed uses. In contrast to the building in New York, which has polished granite as cladding, the Chicago building sports exposed concrete framing and bracing.

5.12   MODULAR OR BUNDLED TUBE

The concept of bundled tube in concrete high rises is similar to that discussed in Chap. 4. The underlying principle is to connect two or more individual tubes in a bundle with the object of decreasing the shear lag effects. Figure 5.12a shows a schematic plan of a bundled tube structure. Two versions are possible using either framed or diagonally braced tubes as shown in Fig. 5.12b,c. A mixture of the two is, of course, possible.

5.13   BUTTRESSED CORE SYSTEM

The use of the Buttressed core structural system led to the dramatic increase in height of buildings. Over a period of more than three decades, the period between the completion of the World Trade Center (1972) and Taipei 101 (2004), the increase in building height was only 22 percent in the height of the world's tallest building (Baker et al., 2012). With the development of the buttressed core system and its use for Burj Khalifa in 2010, the increase in the world's tallest building (Burj Khalifa) was 60 percent from its predecessor, which showed a

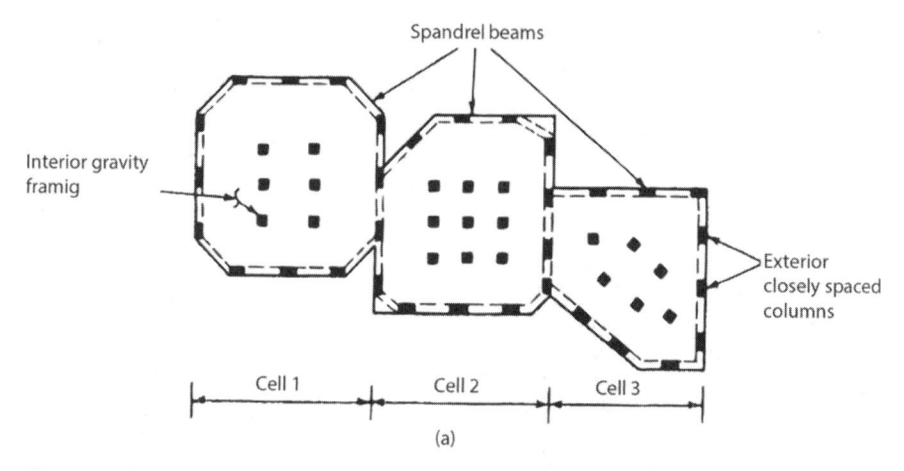

Spandrel beams
Interior gravity framig
Exterior closely spaced columns
Cell 1   Cell 2   Cell 3
(a)

Figure 5.12   Bundled tube: (a) schematic plan; (b) framed bundled tube; (c) diagonally braced bundled tube.

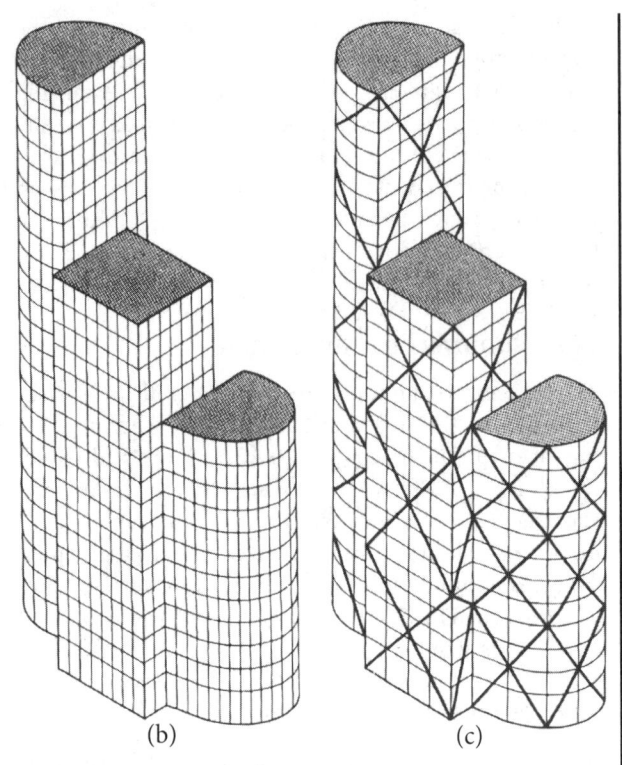

(b)   (c)

Figure 5.12   (*Continued*)

major change in the approach of the design of the tall buildings. This change is significant historically compared to the change from the Empire State Building, NY to the tube concept by Fazlur Khan in the 1970s; and from the development of the core plus outrigger system which helped in providing larger heights in tall buildings.

In addition to the dramatic increase in building heights, the buttressed core design employs conventional material and conventional construction technologies, that is, no innovative material or construction technologies were introduced to achieve this dramatic increase in building heights. The buttressed core is composed of a closed strong central core that anchors three building wings arranged in a tripod shape. The system is inherently stable in which each wing is anchored by the two other wings. In terms of resistance, the central core provides torsional resistance, and the wings provide shear resistance and larger moment of inertia. Examples of this system started from the 264 m (866 ft) tall Tower Palace III in Seoul, South Korea (designed by Skidmore Owings & Merrill LLP and completed in 2004) and then the 828 m (2717 ft) tall Burj Khalifa (designed by Skidmore Owings & Merrill LLP and completed in 2010) as shown in Figs. 5.13 and 5.14, respectively.

The Tower place III has a tripartite arrangement with 120° between wings that support its height. The architectural design required elevators within the oval floor plate of each wing, the structural engineers choose to connect the elevators via a central cluster of cores as shown in Fig. 5.13 (Baker et al., 2012), which became the primary lateral system for the building. The two upper floors are mechanical floors, the perimeter columns in these two floors were engaged to help in resisting the lateral load; the perimeter columns together with the floor plates above and below and the perimeter belt wall form virtual outriggers. In Burj Khalifa also, the three-wing "Y" shape floor plan, tripod-shaped, is used to brace the core structure. It is intended to reduce the torsional forces on the core regardless of the wind direction. There are no column transfers; the columns are in line to the top of the structure. Even for the floor plans that decrease in size, the columns stop at the walls below.

There are 27 floor size reductions along the height of the structure to control the wind shear, confuse the wind, and reduce the vortex shedding effect. Additionally, direct concrete outriggers, not virtual outrigger like in the case of Tower Palace III, were used every 30 floors to increase the lateral resistance of the system by engaging the perimeter columns in the lateral resistance. The multiple outriggers used allow the columns and walls redistribute of the load several times throughout the height of the building which will also control the differential shortening between the columns and the core. Unlike many tall buildings, where the columns at the base are massive, the columns in Burj Khalifa are relatively small and slightly bigger than the columns at the top of the building.

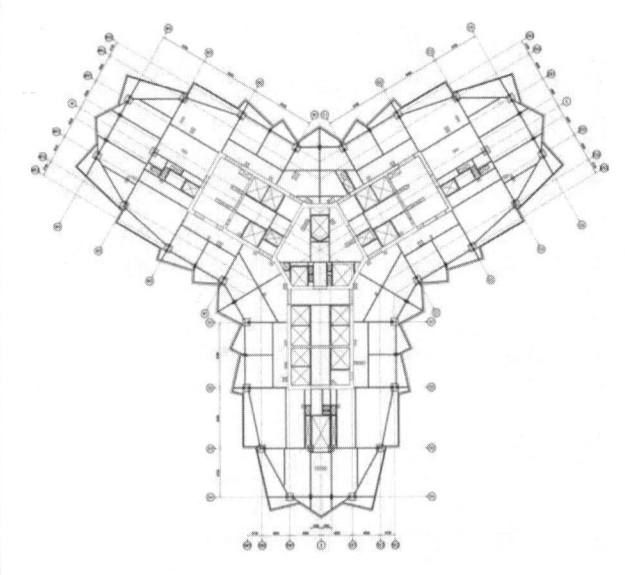

Figure 5.13   Plan view of Tower Palace III, Seoul, South Korea.

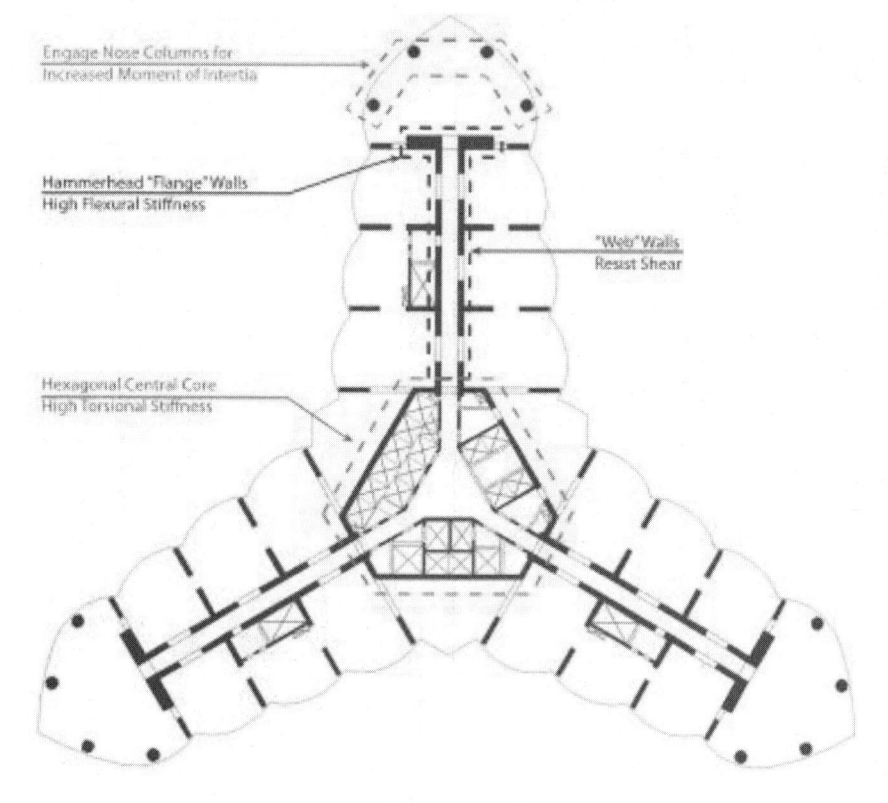

Figure 5.14 Plan view of Burj Khalifa, Dubai, UAE.

5.14 MISCELLANEOUS SYSTEMS

Figure 5.15 shows a schematic plan of a building with a cap truss. The cap truss takes on the form of a one- or two-story-high outrigger wall connecting the core to the perimeter columns. A one- or two-story wall at the perimeter acting as a belt truss may be used to tie the exterior columns together. As in steel systems, the introduction of cap truss results in a reversal of curvature in the bending mode of the shear core. A substantial portion of moment in the core is transferred to the perimeter columns by inducing tension in the windward columns and compression in the leeward columns (Fig. 5.15a).

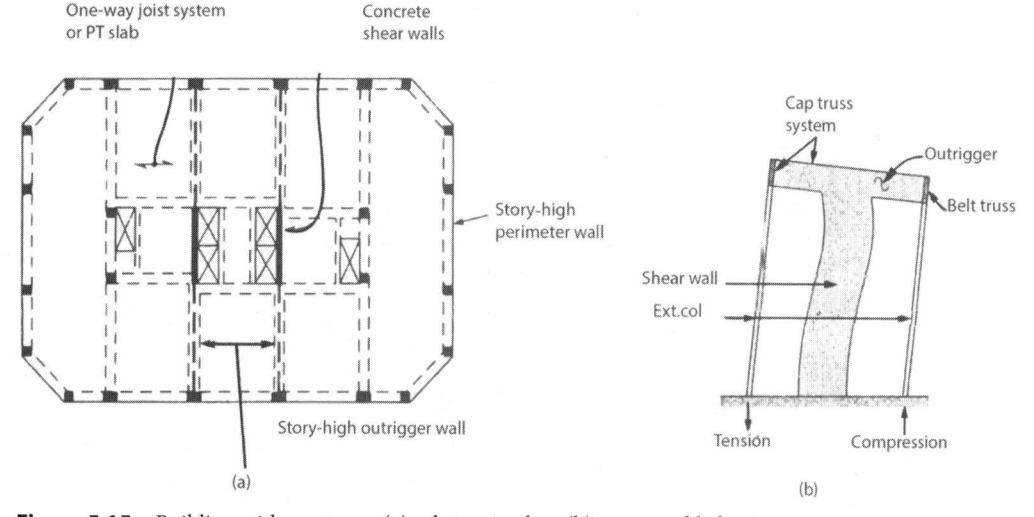

Figure 5.15 Building with cap truss: (a) schematic plan; (b) structural behavior.

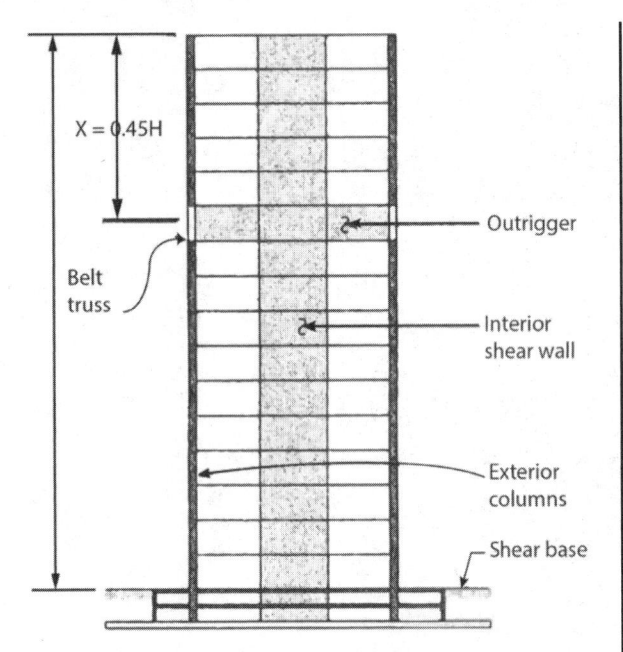

Figure 5.16 Single outrigger system: optimum location.

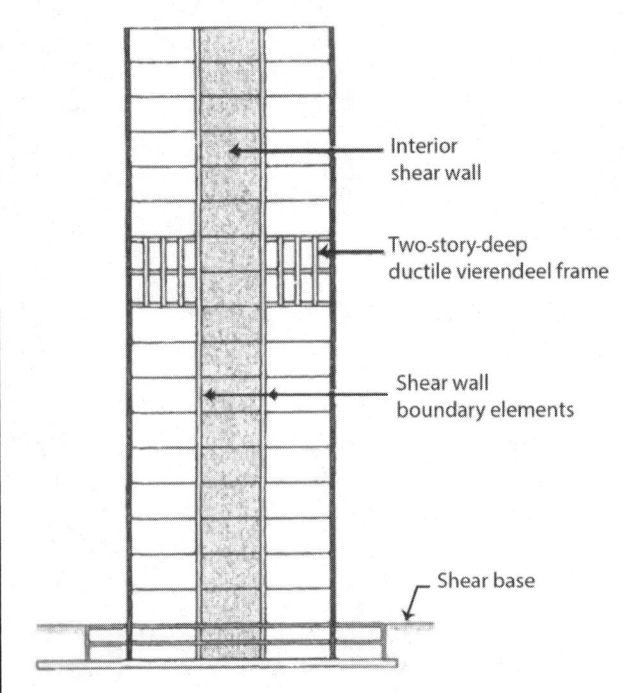

Figure 5.17 Outrigger system: seismic version.

The optimum location for a single outrigger system, as noted in Chap. 3 is at a height $x = 0.45H$ measured from the building top (Fig. 5.16).

In high seismic zones, it is prudent to use a one- or two-story-deep Vierendeel ductile frame functioning as outriggers and belt trusses (Fig. 5.17).

A cellular tube in which a building with a high plan aspect ratio is divided into four cells is shown in Fig. 5.18. By introducing a minimum number of interior columns, three on every other floor in the example building, it is possible to reduce the effect of shear lag on the long faces of the building. A two-story haunch girder

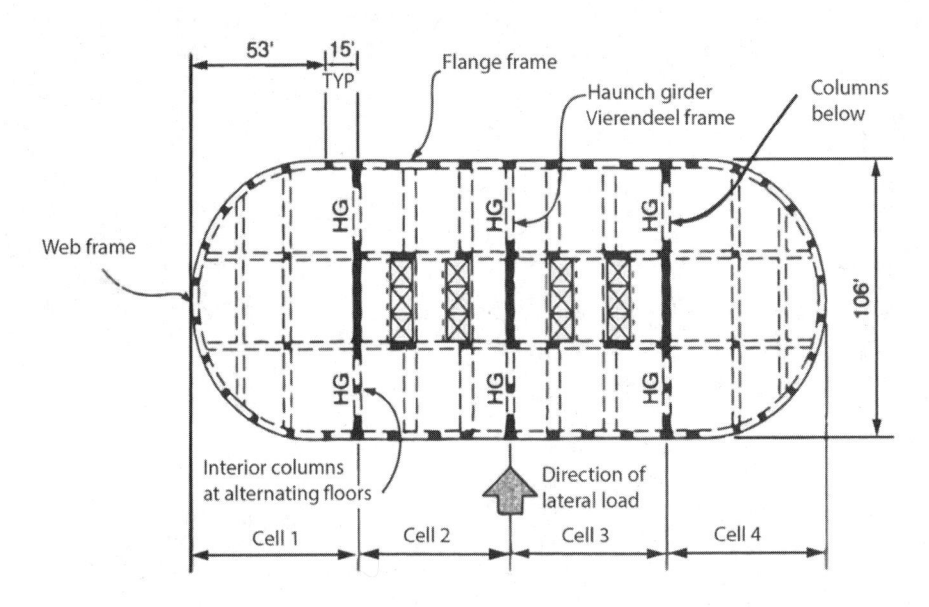

Figure 5.18 Cellular tube with interior Vierendeel frames.

Vierendeel frame effectively ties the building exterior columns to the interior shear walls thus mobilizing the entire flange frame in resisting the overturning moments.

The concept of full-depth interior bracing interacting with the building perimeter frame is shown in Fig. 5.19. The interior diagonal bracing consists of a series of wall panels interconnected between interior columns to form a giant K brace stretched out for the full width of the building (Fig. 5.19b).

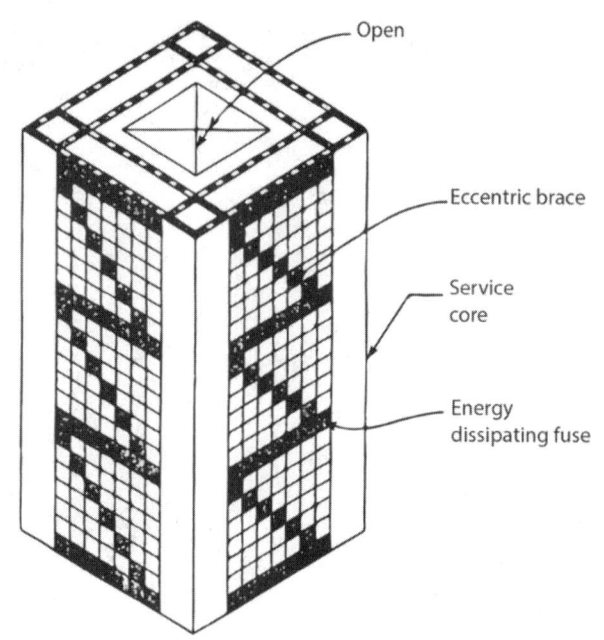

Figure 5.20   Eccentric bracing system for super tall buildings.

A system suitable for super tall buildings, consisting of a service core located at each corner of the building and interconnected by a diagonal in-fill bracing system, is shown in Fig. 5.20. The service core at each corner acts as a giant column carrying most of the gravity load and overturning moments. The diagonal braces at the building exterior resist the shear forces. Because of the deliberate eccentricity at one end, the resulting fuse helps in dissipating energy due to high seismic loads.

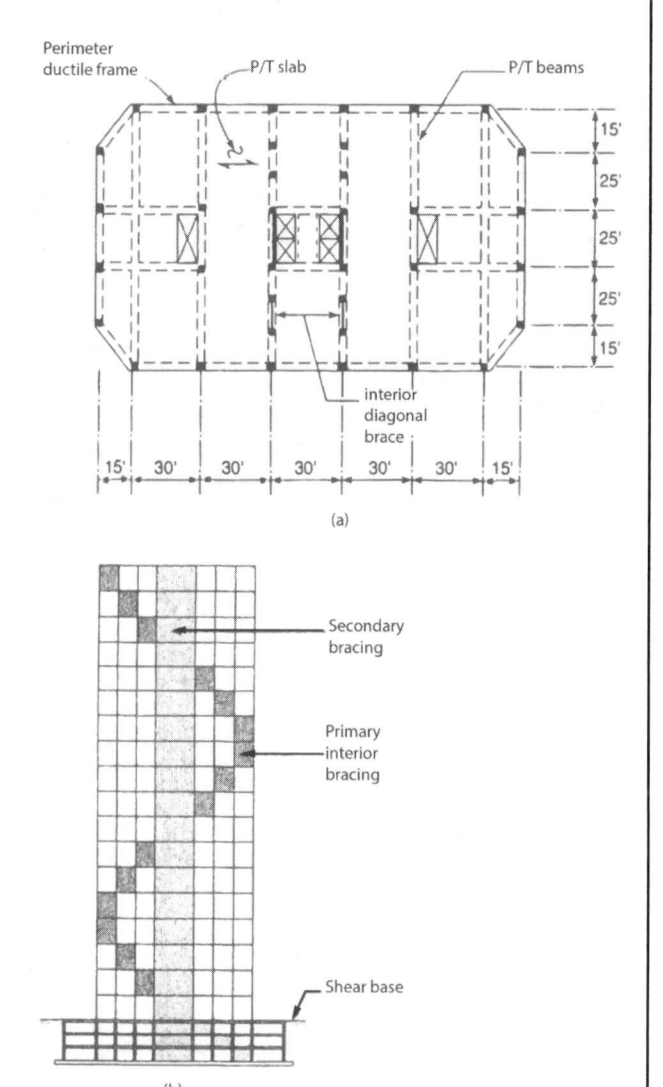

Figure 5.19   Full-depth interior brace: (a) plan; (b) schematic section.

Chapter **6**

Lateral Systems for Composite Construction

6.1 INTRODUCTION

In a broad sense, all high-rises are composite buildings because a functional building cannot be built by using only steel or concrete. For example, concrete is invariably used for floor slabs in an otherwise all-steel building. Similarly, in a critical sense, the use of mild steel reinforcement transforms A concrete building into a composite building. Without being too pragmatic, we will settle for the definition of composite buildings as those with a blend of structural steel and reinforced concrete.

The earliest composite construction consisted of structural steel beams and reinforced concrete slabs, with shear connectors in-between. The system, called the composite floor system, first developed for bridge construction, was readily adopted to buildings. Its phenomenal success inspired engineers to develop composite building systems by combining structural steel and reinforced concrete in a variety of vertical building systems. Since 1960, with the advent of high- and ultra-high-strength concrete, that is, concretes with compressive strengths in excess of 7000 psi to 19,000 psi (48.3 to 131 MPa), there has been a growing realization that a concrete column is more economical than a pure steel column. In fact, studies in North America indicate that concrete or composite columns are four to five times less expensive than all-steel columns. The favorable economics of concrete combined with its high stiffness and fireproofing characteristics has led to its marriage with steel, which has its own advantages namely strength, speed of construction, long-span capability, and lightness.

Depending upon regional preferences, especially in areas of low-seismic risk, the economics of concrete versus steel could go either way, except perhaps, for apartments, condominiums, and hotels. Concrete is favored for these buildings because the underside of the floor slab is often used as finished ceiling. Unlike office buildings, the air-conditioning duct work required in these buildings is relatively simple and thus there is no need for hung ceilings. On the other hand, an office building can be of either steel or reinforced concrete; the choice, although, depends to a great extent on the in-place cost of the frame, is nevertheless influenced by the speed of construction. If the building can be constructed faster and can bring a return on the investment sooner, the speed of construction necessarily enters the cost equation. The material choice, in other words, is based not on cost alone but on the speed of construction as well. Both steel and concrete possess advantages and disadvantages. It follows, therefore, that an ideal structural system is one that overcomes the disadvantages and exploits the advantages of both materials in a unified structural system.

Structural steel is well suited for providing generally column-free lease spaces required in contemporary high-rise office buildings. Because of its lightweight, it imposes less severe foundation requirements and goes up faster. And its lightness is quite often a major consideration in seismic design. Another advantage of steel construction is the use of steel decking for floor construction which offers a more flexible system for wiring a building than a solid concrete slab construction. Also, a steel frame is simpler to modify, to meet the changing needs of building tenants. With steel framing, it is less expensive to increase the load capacity of the floor, or cut holes in the floor to install stairways, atriums, etc., which may be required by changes in tenancy. Therefore, the adaptability to renovation and rehabilitation is an important factor in the selection process. Similarly, concrete buildings have their own advantages too. The advent of superplasticizers and high-strength concrete have made possible construction of reinforced concrete buildings without columns becoming cumbersomely large. Floor-framing techniques have progressed from flat slab construction to skip-joist and haunch girder systems, resulting in an increase in the span of concrete floor systems. Lateral bracing systems have been developed that rival those of steel systems. The advantages of concrete framing are low material costs, moldability, insulating and fire-resisting quality, and most of all, inherent stiffness. However, in relation to steel, concrete construction is generally slow.

The two building systems, concrete and steel, evolved independently of each other, and up until the 1960s, engineers were trained to think of tall buildings either in steel or concrete. Dr. Fazlur Khan, of Skidmore, Owings & Merrill, broke this barrier in 1969 by blending steel and concrete into a composite system for use in a relatively short, 20-story building, in which the exterior columns and spandrels were encased in concrete to provide the required lateral resistance. The system was basically a steel frame stabilized by reinforced concrete. However, today the advent of high-strength concrete has ushered in the era of super columns and mega frames where the economy, stiffness, and damping characteristics of large concrete elements are combined with the lightness and constructability of steel frames. Without this type of framing, many of our contemporary tall buildings may never have been built in their present form.

The term composite system has taken on numerous meanings in recent years to describe many combinations of steel and concrete. As used here, the term means any and all combinations of steel and reinforced concrete elements and is considered synonymous with other definitions such as mixed systems, hybrid systems, etc. The term is used to encompass both gravity- and lateral-load-resisting elements.

Since the advent of the first composite system, engineers have not hesitated to use a whole range of combinations to capitalize on the advantages of each material. This has resulted in numerous systems making distinct categorization a nearly impossible task. The systems can be best described by citing examples from the buildings built within the last two decades.

6.2   COMPOSITE ELEMENTS

To get an insight into different composite building schemes, it is instructive to study the various techniques of compositing both the horizontal and vertical elements. These are the following:

1. Composite slabs
2. Composite girders
3. Composite columns
4. Composite diagonals
5. Composite shear walls

The behavior of horizontal members such as slab, beam, girder, and spandrels subjected to gravity loads is covered in Chap. 9. In this section, their behavior only under lateral loads is considered.

6.2.1   Composite Slabs

In high-rise steel buildings and the use of high-strength, light-gauge (16 to 20 gauge) metal deck with concrete topping has become the standard floor-framing method. The metal deck usually has embossments pressed into the sheet metal to achieve composite action with the concrete topping. Once the concrete hardens, the metal deck acts as bottom tension reinforcement while the concrete acts as the compression component. The resulting composite slab acts as a horizontal diaphragm interconnecting all vertical elements at each level providing for the horizontal transfer of shear forces to bracing elements. Furthermore, it acts as a stability bracing for the compression flange of steel beams.

The shear stresses induced due to diaphragm shear forces are mostly in the concrete topping because the in-plane stiffness of concrete slab is significantly more than that of the metal deck. Thus the horizontal forces must transfer from the slab to the beam top flange through the welded studs. In addition to supporting the gravity loads, the slab serves as a load path for transferring tributary floor lateral loads to the lateral load-resisting elements. Diaphragm behavior, an important aspect of seismic design, is discussed later in Chap. 12.

6.2.2   Composite Girders

Consider a typical steel moment frame consisting of steel beams rigidly connected to columns. As discussed previously in Chap. 4, the stiffness of the frame most usually depends on the stiffness of the girder. This is because in a frame with its typical column spacing of 25 to 35 ft (7.6 to 10.67 m), and a floor-to-floor height of $12\frac{1}{2}$ to $13\frac{1}{2}$ ft (3.81 to 4.12 m), the columns are much stiffer than the beams. To limit sway under lateral loads it is more prudent to increase the girder stiffness rather than the column stiffness. Although frame beams are designed as non-composite beams, it is usual practice to use shear connectors at a nominal spacing of say, 12 in., especially in areas of high-seismic risk. The shear connectors primarily provided for the transfer of diaphragm shear also increase the moment of inertia of the girder. However, the moment of inertia does not increase for the full length of the girder because the girder responds by bending in a reversed curvature. Since concrete is ineffective in tension, the composite moment of inertia can be counted on only in the positive moment region. Although design rules are not well established, a rational method may be used to take advantage of the increased moment of inertia. Occasionally engineers have used a dual approach in wind design by using bare steel beam properties for strength calculations, and composite properties in the positive regions for drift calculations.

6.2.3 Composite Columns

Two types of columns are used in composite building systems. The first type consists of a steel core surrounded by a high-strength reinforced concrete envelope. The second type consists of a large steel pipe or tube filled with high-strength concrete. In the first type the steel section, most usually a wide flange section, is either a light section designed to carry construction dead loads only, or may be of a substantial weight to part take in resisting axial, shear, and bending moments together with the mild steel reinforcement and high-strength concrete envelope. Conceptually the behavior of a composite column is similar to a reinforced concrete column, if the steel section is replaced with an equivalent mild steel reinforcement. In fact, this concept provides the basis for generating the interaction diagram for the axial load and moment capacities of composite columns.

Compositing of only exterior columns by encircling steel sections with concrete is by far the most frequent application of composite columns. The reasons are entirely economic, because concrete forming around interior columns is quite involved and is not readily applicable to jump forms. Exterior columns, on the other hand, are relatively open-faced: concrete forms can be "folded" around the steel columns for placement of concrete, then unfolded and jumped to the next floor repeating the cycle without having to dismantle the entire framework.

However, in Japanese construction it is common practice to composite the interior columns as well. Their construction makes extensive use of welding for vertical as well as transverse reinforcement (Fig. 6.1).

Prior to 1986, the design of composite columns was covered only by the ACI code. The ACI design rules are based on the same principles as the design of reinforced concrete columns. The strength of the column approaches that of concrete columns as the percentage of steel column decreases. The ACI code permits the use of any shape inside of a composite column, with a stipulation that the yield strength of steel cannot be greater than 50 ksi. For tied columns, the minimum reinforcement ratio is one percent while the maximum ratio is limited to 8 and 6 percent for wind and seismic designs, respectively. The lateral spacing of longitudinal bars cannot exceed half the length of the shorter side of the section, 16 diameters of the longitudinal bars, or 48 diameters of lateral ties.

On the other hand, the AISC rules are similar to steel column design and requires that the cross-sectional area of the steel shape be at least 4 percent of the total composite cross-section. The strength of composite columns approaches that of steel columns as the percentage of steel section increases.

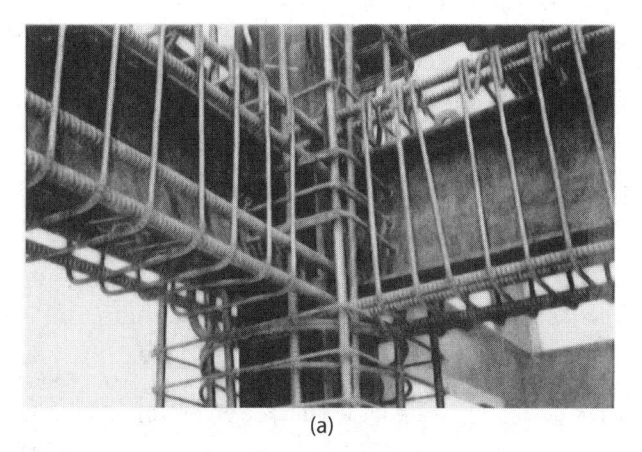

(a)   (b)

Figure 6.1 Japanese composite construction details: (a) beam-column intersection; (b, c) composite column with welded ties; (d) general view.

(c)

(d)

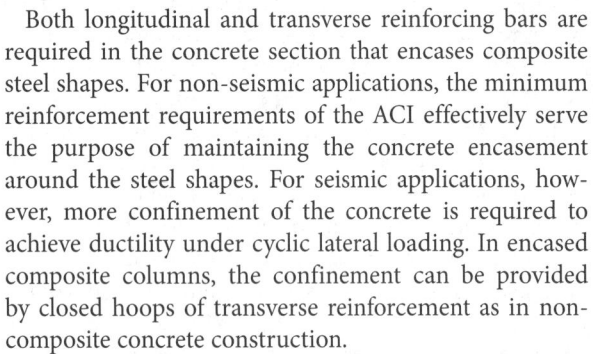

Figure 6.1   (*Continued*)

Both longitudinal and transverse reinforcing bars are required in the concrete section that encases composite steel shapes. For non-seismic applications, the minimum reinforcement requirements of the ACI effectively serve the purpose of maintaining the concrete encasement around the steel shapes. For seismic applications, however, more confinement of the concrete is required to achieve ductility under cyclic lateral loading. In encased composite columns, the confinement can be provided by closed hoops of transverse reinforcement as in non-composite concrete construction.

The second type of composite column consists of a steel pipe or tube filled with high-strength concrete. Typically neither vertical nor transverse reinforcement is used in the columns. However, as in the previous type, shear connectors are welded to the inner face of the pipe to provide for the interaction between the concrete and the outer shell. Since the compositing of column does not require any formwork this type is applicable to both the interior and exterior building columns.

The novel structural concept of concrete-filled steel columns for multistory construction was first proposed by Dr. A. G. Tarics of Reid & Tarics Associates, San Francisco, California, in 1972. This idea has been used in several buildings by the Seattle engineers Skilling, Ward, Magnusson, Berkshire Inc. (now known as Magnusson Klemencic Associates, MKA). Two Union Square building is one such application of huge pipe columns filled with concrete in composite construction. The 58-story building includes a core that carries 40 percent of the gravity loads and all of the lateral loads. The core framing consists of four 10 ft (3 m) diameter pipe columns filled with 19,000 psi (130 MPa) concrete.

The usual method of attaching steel members to pipe columns is to use a welded connection to the outside of the pipes in which case the flange forces are carried directly by the pipe. Sometimes plates are welded to the inside of the pipes acting as stiffeners to minimize local stresses in the thin-walled pipes.

Although the concept of concrete-filled steel columns has been used in several tall buildings, including some in high-seismic zones, there are still some nagging questions about the system. The post-yield behavior, the performance of the bond between concrete and steel under cyclic loading, the potential for local buckling of the pipe, and the heat of hydration of the concrete are some of the unanswered questions that need further investigation.

6.2.4   Composite Diagonals

Diagonals in a braced frame resist lateral forces primarily through axial stresses acting as part of a vertical truss. As a result, braced frames are more economical than moment-resisting frames on a material quantity basis. However, their use is often limited, because of potential interference of braces with architectural planning concerns.

The majority of braced frame applications is in structural steel. However, braced frames of composite construction have started finding application since the mid-1980s. The majority of these have been applied in concert with composite columns, in the form of "super columns" composed of large-diameter circular pipes filled with high-strength concrete.

6.2.5   Composite Shear Walls

In this system, as in conventional concrete systems, concrete shear walls are placed around the building core. The walls are usually C and I shapes with webs placed parallel to the elevator banks as shown in Fig. 6.2. The flanges of adjacent shear walls are invariably connected with link beams

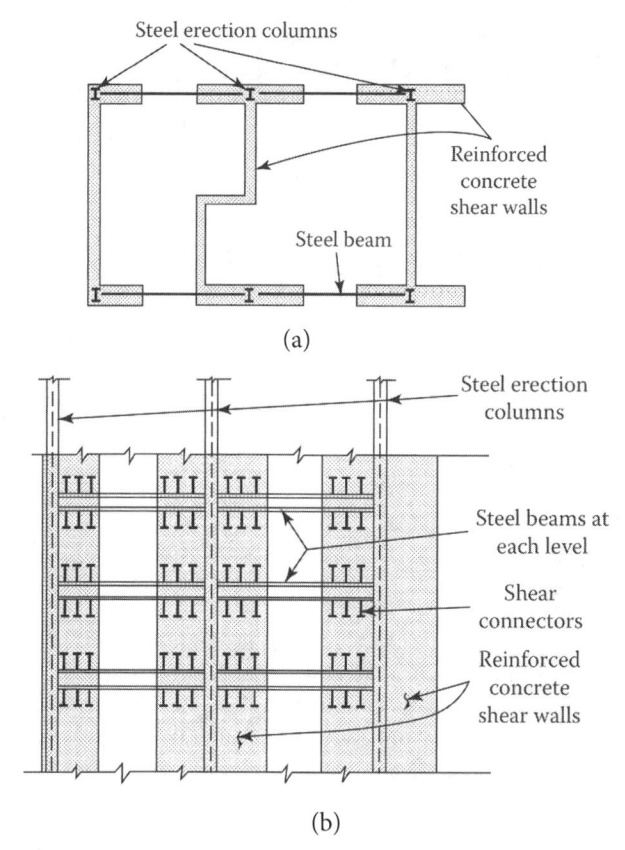

(a)

(b)

Figure 6.2   Composite shear wall with steel beams: (a) plan; (b) elevation.

to increase bending stiffness about an axis parallel to the web. The high shear forces in the link beams can result in brittle fracture unless the beam is properly detailed with diagonal reinforcement. In one version of composite shear walls, mostly used in areas of low-seismic risk, structural steel beams are used to link the shear walls. In this type of construction, steel columns within the shear walls are erected with the steel frame including floor beams.

The most usual method of connecting the steel beam is to extend it through to steel columns within the shear walls. The moment capacity is achieved by welding shear connectors to the top and bottom flanges of the beam as shown in Fig. 6.3. This type of construction has not been used in areas of high-seismic activity. Much research work needs to be done before seismic design guidelines are established. For resisting high in-plane shear forces, a full-length steel plate compositely attached to a concrete shear wall may be used. An example of such a construction is the core wall of Bank of China Building, Hong Kong. In this building, all the lateral forces are transferred to the core at the base. To resist the high shear forces, steel plates are attached to the concrete core through shear studs welded to the steel plates as shown in Fig. 6.4.

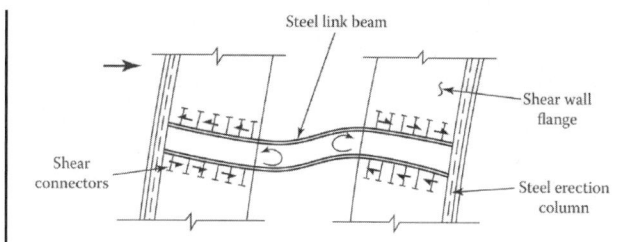

Figure 6.3   Moment transfer between steel beam and concrete wall.

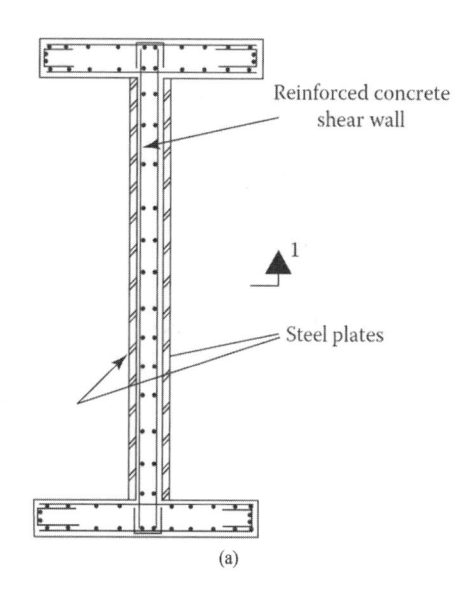

(a)

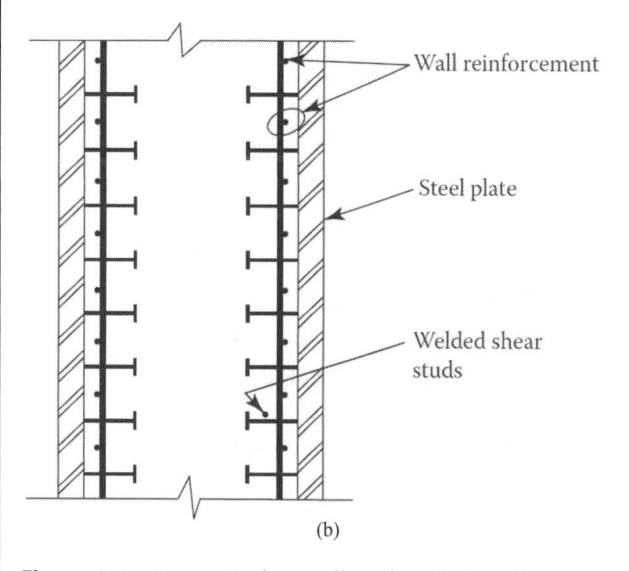

(b)

Figure 6.4   Composite shear walls with steel plates: (a) plan; (b) section; (c) and (d) shear wall with single steel plate; (e) steel plates on both sides of the wall; (f) and (g) reinforced concrete shear wall with steel boundary elements.

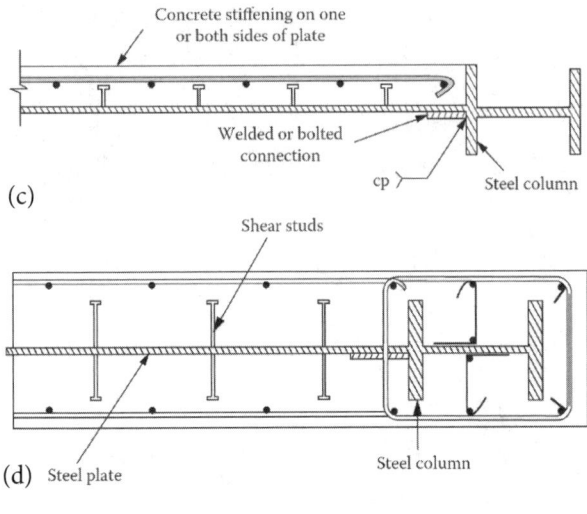

(c)

(d) Steel plate

(e)

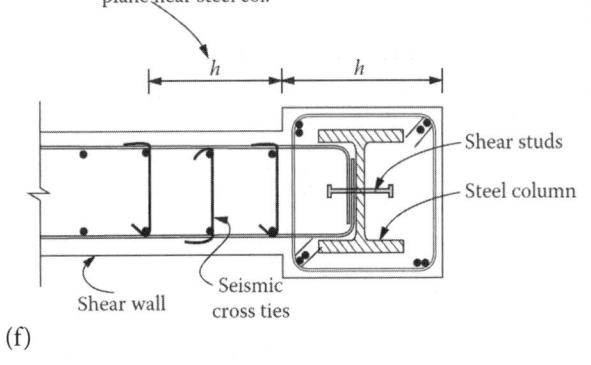

(f)

Figure 6.4   (*Continued*)

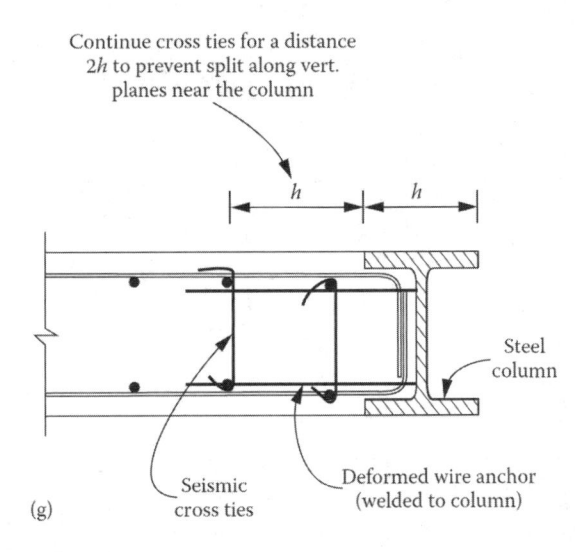

(g)

Figure 6.4   (*Continued*)

6.3   COMPOSITE BUILDING SYSTEMS

Composite building systems in use today may be conveniently classified into the following categories.

1. Shear wall systems
2. Shear wall-frame interacting systems
3. Tube systems
4. Vertically mixed systems
5. Mega frames with super columns

6.3.1   Shear Wall Systems

Core walls enclosing building services such as elevators, mechanical and electric rooms, and stairs have been used extensively to resist lateral loads in tall concrete buildings. Simple forms such as C and I shapes around elevators interconnected with coupling or link beams are used extensively. Their popularity as a lateral-load-resisting element has once again come to the forefront because of the current trend in architecture which appears to favor lean vertical elements around the perimeter. Earlier applications of this system were limited to buildings in the 30- to 40-story range, but with the advent of superplasticizers and high-strength concrete, it is now possible to use this system for taller buildings in the 50- to 60-story range. The range of application is mostly a function of available depth of core. Buildings using four-deep elevators provide a core depth of approximately 40 ft (12.2 m), which is sufficiently large to give an economical system for buildings in the 50- to 60-story range. The shear core basically responds to lateral loads in a bending mode and does not present undue complications in its analysis.

In the core-only system, the total lateral load is resisted by the shear walls and hence the remainder of the building can be conveniently framed in structural steel. Whether or not concrete or steel comes first in the construction is not always a known priority and is often influenced by the choice of construction method. In one version, concrete core is cast first by using jump or slip forms, followed by erection of the steel surround as shown in Fig. 6.5. Although the structural steel framing may not proceed as quickly as in a conventional steel building, the overall construction time is likely to be reduced because elevators, mechanical and electrical services can be installed rapidly in the core while construction outside the core proceeds. In another version, steel erection columns are

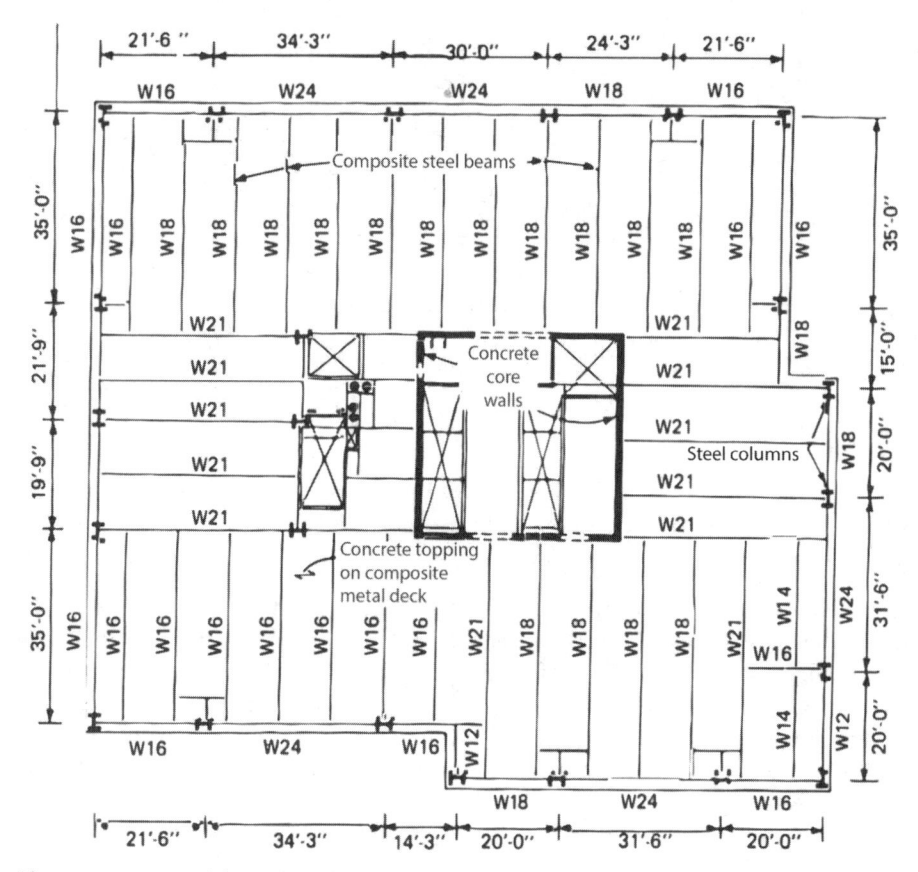

Figure 6.5 Typical floor plan of core structure with steel framing.

used within the shear walls to serve as erection columns, and steel erection proceeds as in a conventional steel building. After the steel erection has reached a reasonable level, concreting of core starts using conventional forming techniques. In order to facilitate faster jumping of forms to the next higher level, temporary openings are left in the floor slab around the shear walls.

The structural behavior of this system is no different than a concrete building with shear walls designed to take all the lateral forces. However, it behooves the engineer to recognize the absence of torsional stiffness. It is advisable to provide moment frames or other types of bracing around the building perimeter to counteract the torsional effects.

If all the lateral loads are resisted by concrete shear walls, the steel surround is designed as a simple framing for gravity loads only. Since there are no moment connections with welding or heavy bolting, the erection of steel proceeds much faster. The only nonstandard connection is between the shear walls and floor beams. Various techniques have been developed for this connection, chief among them, the embedded plate and pocket details, as shown in Fig. 6.6a. The floor construction invariably consists of a composite

metal deck with a structural concrete topping. This system has the advantage of keeping the steel fabrication and erection simple. Since columns carry only gravity loads, high-strength steel can be employed with the attendant savings. Interior as well as exterior columns can be made small, increasing the space-planning potential.

The floor within the core can be constructed either with cast-in-place beams and slabs or structural steel framing, steel decking, and concrete fill. At the connection between the floor slabs, both within and outside of the core shear keys are provided in the core walls to transmit lateral diaphragm forces from the floor system to the core. Although several options are available for connecting the steel beams to the concrete core, a weld plate detail shown in Fig. 6.6a is most popular, especially so in a slip-formed construction. During slip-form operation, the weld plates are set at the required locations, with the outer surface of the plate set flush with the wall surface. Anchorage of the weld plate is achieved by shear connectors welded to the inner surfaces, sometimes supplemented with a top-bent steel bar to resist high tensile forces. The shear connectors and steel bar ultimately

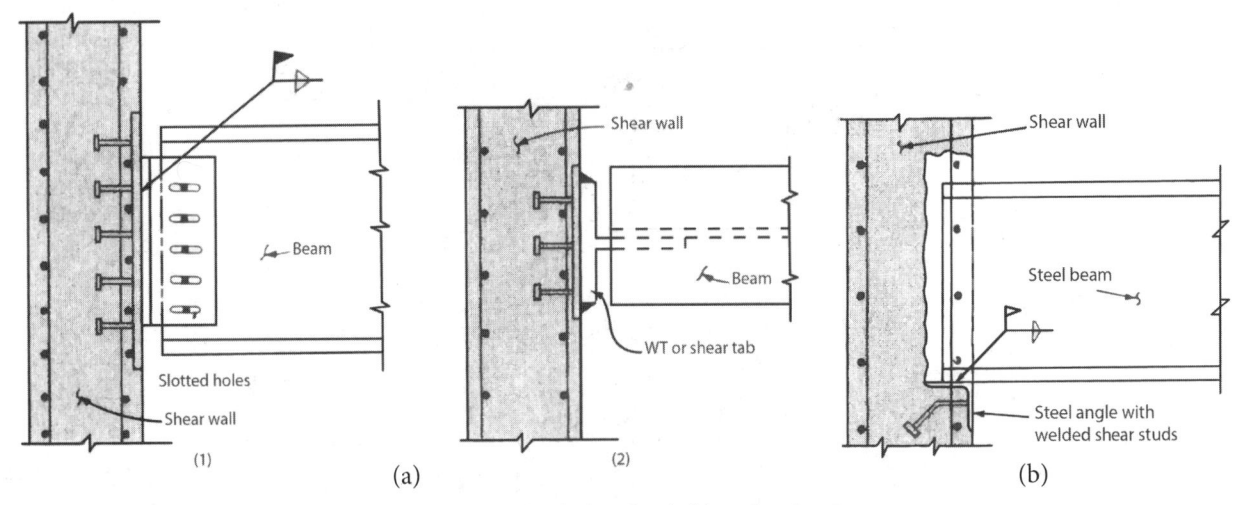

Figure 6.6   Beam to shear wall connection: (a) embedded plate detail; (b) pocket detail.

become embedded in the core wall. Experience with slip-formed composite buildings indicates that it is prudent to overdesign these connections to compensate for the misalignment of cores. Subsequent to the installation of weld plates, structural tee or shear tab connections with slotted holes are field-welded to the plate. Slotted holes are used for bolting of steel beams to provide additional tolerance for erection.

Slip forming is a special construction technique that uses a mechanized moving platform system. The process of slip forming is similar to an extrusion process. The difference is that whereas in an extrusion process the extrusion moves; in a slip-forming process the die moves while the extrusion remains fixed.

Although traditionally the core has been used to resist lateral forces, its contribution in supporting the vertical load is limited because a relatively small floor area is supported by the core. For a very tall building, this can result in an unfavorable stability condition. A method of overcoming this limitation is to apply external prestressing forces to the core to relieve the tensile stresses. An equivalent passive prestressing effect can be obtained by channeling increased vertical load to the core. Extending this fundamental idea to its limit results in the concept of a building totally supported on a single-core element. In practice, depending upon the floor area and the number of levels, several options are available for supporting the floors from the core. For example: (i) floors can be hung from the top of the center core; (ii) story-deep cantilever trusses located at one or two intermediate levels, such as at top and midheight of the building, can be used, the advantage being reduced length of hangers with fewer floor-leveling problems; or (iii) the floor system can be

cantilevered at each level. The selection of a suitable system depends on the economic consequence of each method. In addition to providing views unobstructed by exterior columns at each floor, the absence of columns provides for the commonly sought column-free space at the building entrances. Also, the undulations on the building exterior common in today's architecture are easy to accommodate in a core-only structural scheme. Galvanized bridge-strand cables can be used as hangers to support the structural steel framing which normally consists of composite beams, metal deck, and concrete topping. The floor beams are attached to the hanger with simple supports, while at the core pockets or anchor plates cast into the core walls provide for the support. It is common practice to slip-form the center core with an average concrete growth rate of 6 to 18 in./h (152 to 457 mm/h). After completion of the core, the second stage of construction in the hung-floor system is the erection of roof girders and draping of the floor-supporting cables. Erection of typical floor members between the core and the perimeter cables proceeds similarly to any other building. Erection of steel floor decks and welding of beams for composite action follows by placement of concrete topping. Because the elongation of the cable due to cumulative floor loads can be substantial, it is necessary to compensate for this effect properly.

6.3.2   Shear Wall-Frame Interacting Systems

This system has applications in buildings that do not have cores sufficiently large to resist the total lateral loads; interaction of shear walls with other moment frames located in the interior or at the exterior is called upon to

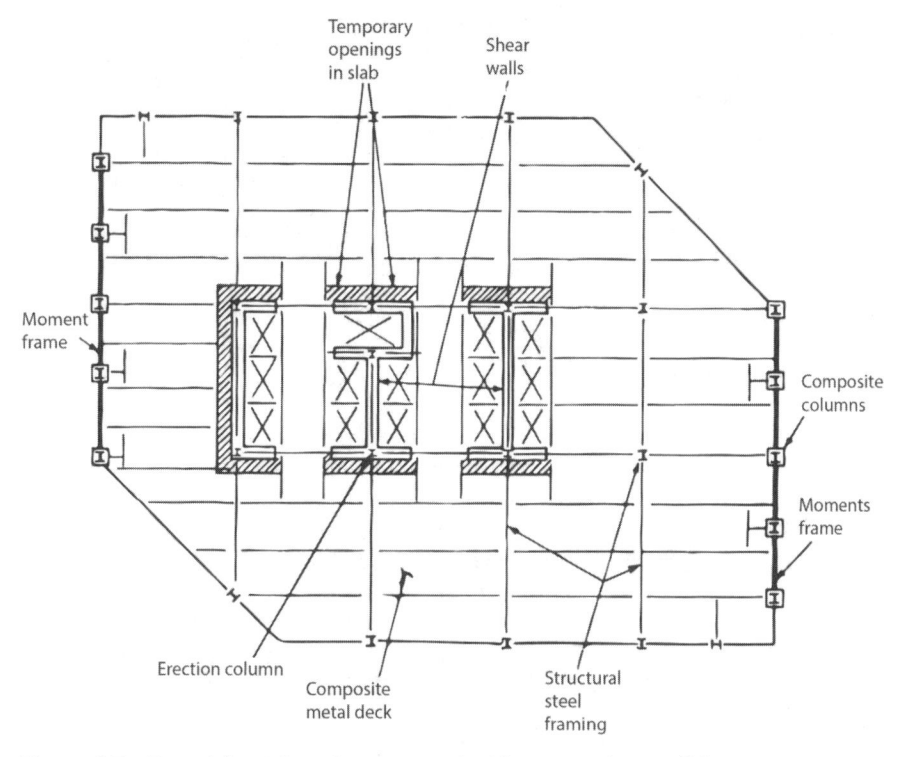

Figure 6.7 Typical floor plan of a composite building using shear wall frame interaction.

supplement the lateral stiffness of the shear walls. When frames are located on the interior, usually the columns and girders are of steel because the cost of formwork for enclosing interior columns and girders for composite construction far outstrips the advantages gained by additional stiffness of the frame. Interior columns typically have beams framing in four directions, making placement of mild steel reinforcement and formwork around them extremely cumbersome. On the other hand, it is relatively easy to form around exterior columns, and if desired, even the exterior spandrels can be made out of concrete without creating undue complexity. If steel erection precedes concrete construction, it is usually more cost-effective to use steel beams as interconnecting link beams between the shear walls. A schematic plan of a building using this system is shown in Fig. 6.7.

6.3.3 Tube Systems

A framing system used extensively in Louisiana and Texas is the composite concrete tube. It makes use of the well-known virtues of the tube system with the speed of steel construction. As in concrete and steel systems, closely spaced columns around the perimeter and deep spandrels form the backbone of the system. Two versions are currently popular: one system uses composite columns and concrete spandrels and the other uses structural steel

spandrels in place of concrete spandrels. A small steel section can be used as a steel spandrel in the former scheme to stabilize the steel columns. However, in the design of the concrete spandrel, its strength and stiffness contribution is generally neglected because of its relatively small size. Schematic plan and sections for the two versions of tubular system are shown in Figs. 6.8 and 6.9.

In either system, the speed of construction, rivaling that of an all-steel building, is maintained by erecting a steel skeleton first with interior steel columns, steel floor framing, and light exterior columns. Usually, the steel frame is erected some 10- to 12-stories ahead of the perimeter concrete tube. The key to the success of this type of construction lies in the rigidity of closely spaced exterior columns, which, together with deep spandrels, results in an exterior façade that behaves more like a bearing wall with punched windows than a moment frame.

6.3.4 Vertically Mixed Systems

Mixed-use buildings provide for two or more types of occupancies in a single building by vertically stacking different amenities. For example, lower levels of a building may house parking; middle levels, office floors; and top levels, residential units, such as apartments and hotel rooms. Since different types of occupancies economically favor different types of construction, it seems logical to

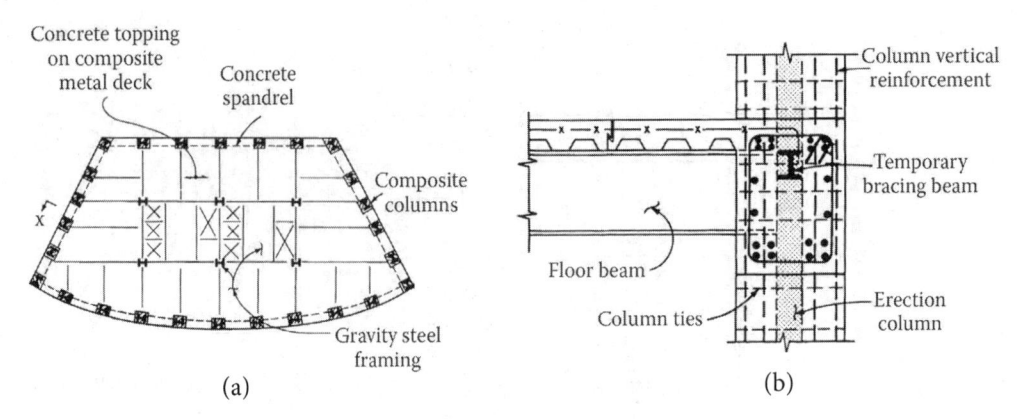

Figure 6.8 Composite tube with concrete spandrels: (a) typical floor plan; (b) typical exterior cross-section.

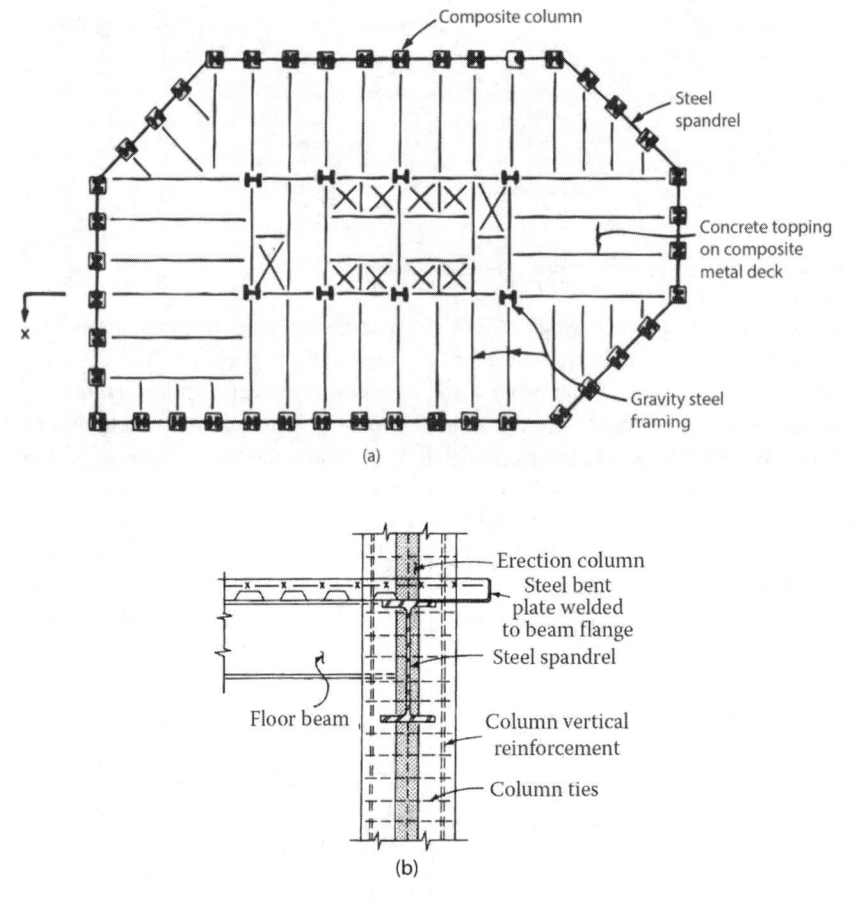

Figure 6.9 Composite tube with steel spandrels: (a) typical floor plan; (b) typical exterior cross-section.

mix construction vertically up the building height. As mentioned earlier, beamless flat ceilings are preferred in residential occupancies because of minimum finish required underneath the slab. Also, large spans of the order of 40 ft (12.2 m) required for optimum lease space for office buildings are too large for apartments. Additional columns can be introduced without unduly affecting the architectural layout of residential units.

The decrease in span combined with the requirement of flat-plate construction gives the engineer an opportunity to use concrete in the upper levels for apartments and hotels.

In certain types of buildings, use of concrete for the lower levels and structural steel for the upper levels may provide an optimum solution as shown in Fig. 6.10. The bracing for the concrete portion of the building is provided by the shear walls, while a braced steel core provides the lateral stability for the upper levels. A suggested technique of transferring steel columns onto a concrete wall consists of embedding steel columns for one or two levels below the transfer level. Shear studs shop-welded to the embedded steel column provide for the transfer of axial loads from the steel column to the concrete walls.

6.3.5 Mega Frames With Super Columns

As mentioned in Chap. 4, the most efficient method of resisting lateral loads in tall buildings is to provide "super

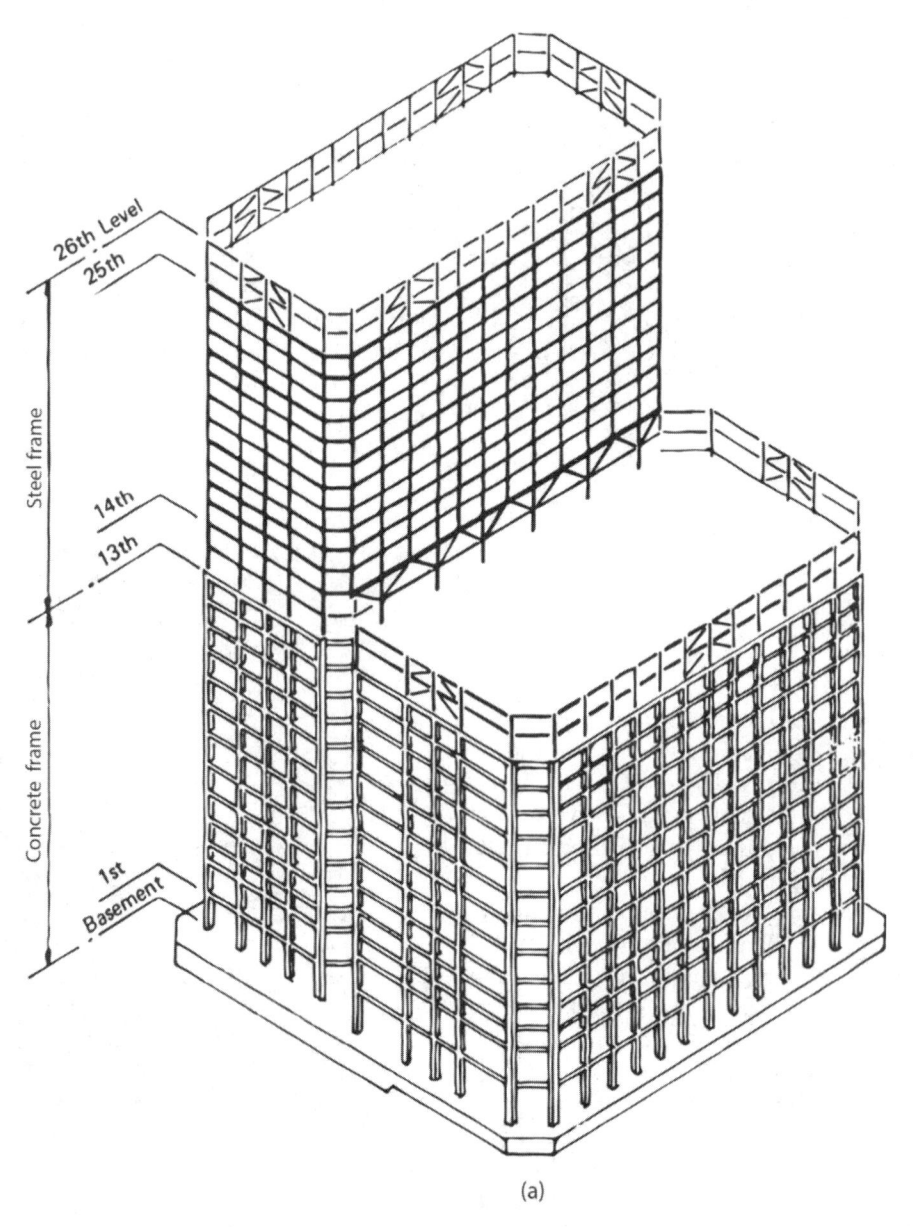

(a)

Figure 6.10 Vertically mixed system: (a) schematic perimeter framing; (b) schematic bracing concept.

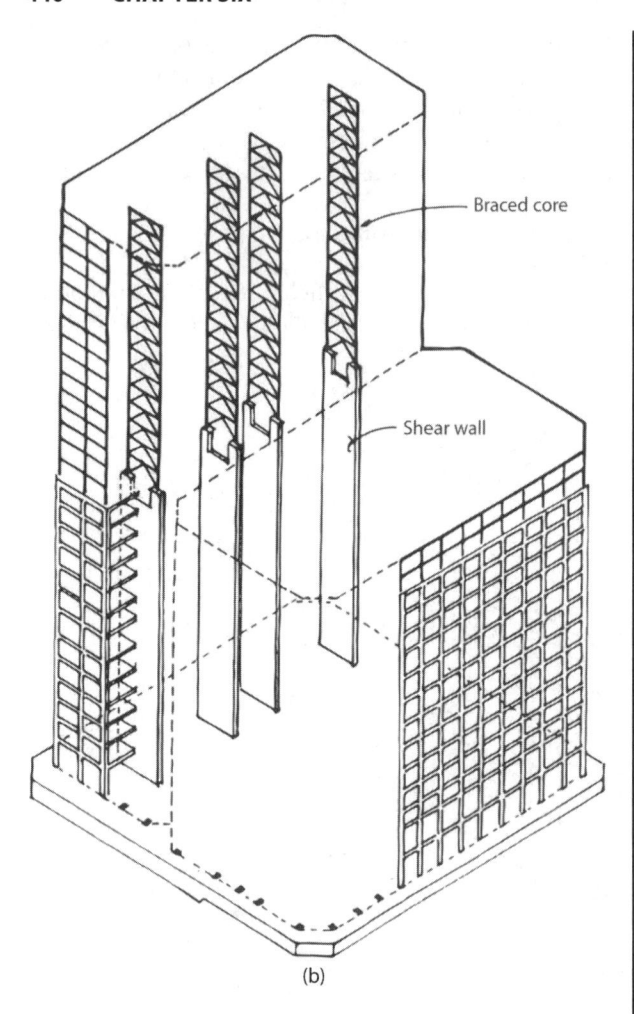

(b)

Figure 6.10 (*Continued*)

columns," placed as far apart as possible within the foot-print of the building. The columns are connected with a shear-resisting system such as welded steel girders, Vierendeel frames, or diagonals. This idea has given rise to a whole new category of composite systems characterized by their use of super columns.

The construction of super columns can take on many forms. One system uses large-diameter thin-walled pipes or tubes filled with high-strength concrete generally in the range of 6 to 20 ksi (41 to 138 MPa) compressive strength. Generally, neither longitudinal nor transverse reinforcement is used thereby simplifying construction. Another method is to form the composite column using conventional forming techniques; the only difference is that steel columns embedded in the super column are used for the erection of steel framing and also for additional axial and shear strength.

6.4   EXAMPLE PROJECTS

6.4.1   Composite Steel Pipe Columns

The 44-story Pacific First Center designed by Skilling Ward Magnusson, Inc., is an example of a building with large-diameter composite pipes. It has eight 7.5 ft (2.3 m) diameter pipe columns at the core, and perimeter columns with a maximum diameter of 2.5 ft (0.76 m), both filled with 19 ksi (131 MPa) concrete. Another example is the 62-story Gateway Tower in which 9 ft (2.7 m) diameter pipe columns exposed at the inner square of the hexagon are tied together with 10-story high X-braces.

The Union Square designed by the same firm is a 58-story building with four 10 ft (3 m) diameter pipe columns in the core. The pipes are filled with 19,000 psi (131 MPa) concrete. The building has fourteen more composite steel pipe columns of smaller diameter placed along the building perimeter to carry gravity loads. The steel pipes provided erection steel and replaced forms as well as vertical bars and horizontal ties for the high-strength concrete. There are no reinforcing bars in the pipe columns. The pipes are connected to the concrete with studs welded to the pipe's interior surfaces.

As a diversion from high-rise buildings, it may be interesting to look at composite columns in a non-high-rise application. Figure 6.11 is a photograph of a project called The Fremont Street Experience

(a)

Figure 6.11   Fremont Street Experience: (a) general view; (b) typical section through vault; (c) composite column; (d) tie-beam reinforcement detail.

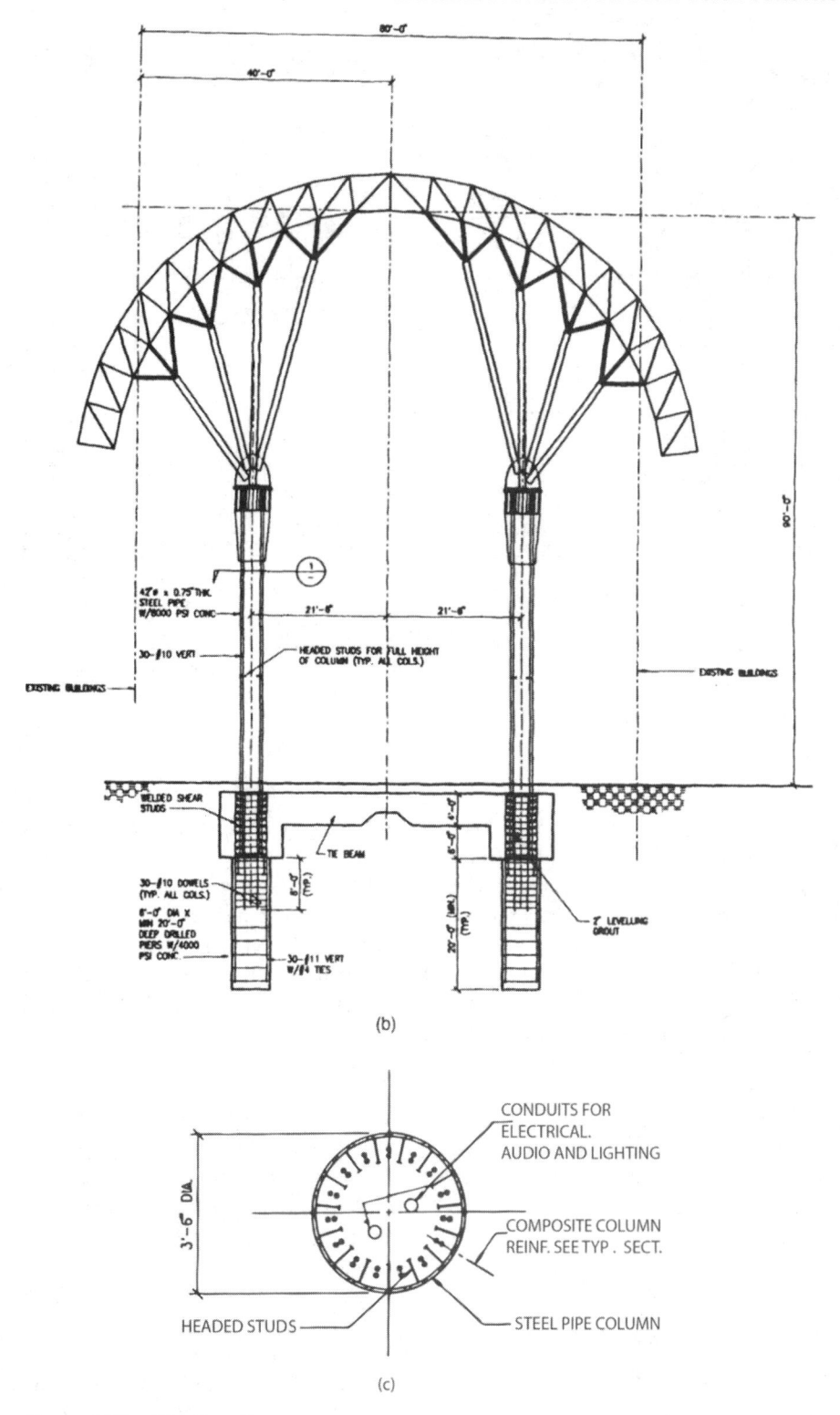

(b)

CONDUITS FOR
ELECTRICAL.
AUDIO AND LIGHTING

COMPOSITE COLUMN
REINF. SEE TYP . SECT.

HEADED STUDS

STEEL PIPE COLUMN

(c)

Figure 6.11 (*Continued*)

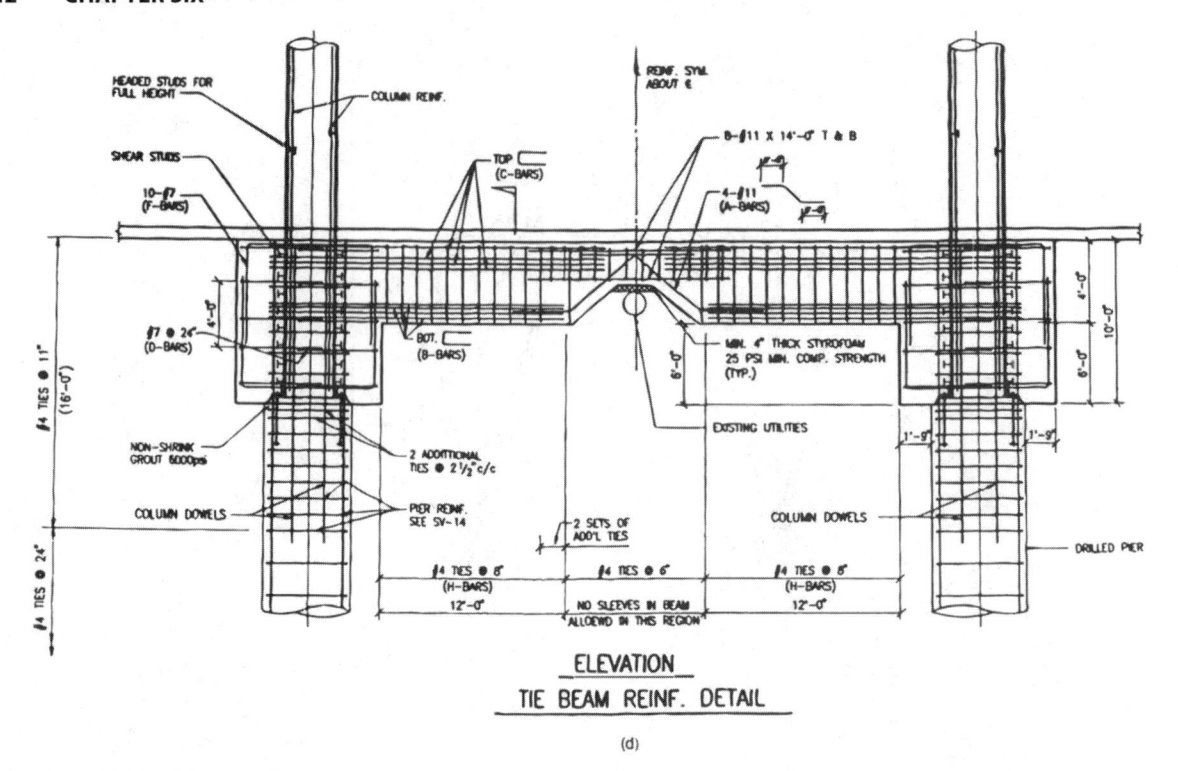

ELEVATION

TIE BEAM REINF. DETAIL

(d)

Figure 6.11 (*Continued*)

in Las Vegas, Nevada. By covering the street with a space frame vault, five downtown blocks have been transformed into a pedestrian mall. The space frame provides partial shade from the sun and supports a graphic display system to entertain visitors to Fremont Street. The overall dimensions of the frame are 1387 ft long (422.75 m) and 100 ft (30.5 m) wide, with a 50 ft (15.25 m) radius forming a semi-circle in cross-section. The space frame with a depth of 5.77 ft (1.76 m) is supported on composite columns at intervals of between 180 and 200 ft (54.87 and 61 m) along the lengthwise span of the vault. Typical space frame steel components are 3 in. (76 mm) diameter round steel tubing, with a wall thickness of 0.120 in. (3 mm). Struts with heavier wall thickness are also used where additional strength is required.

The composite columns consist of 42 in. diameter by 0.75 in. thick (1067 mm × 19 mm) steel pipes with 8000 psi (55.16 MPa) concrete. Headed studs $\frac{1}{2}$ in. (12.7 mm) diameter by 8 in. (203 mm) are welded to the inside face of the tube at a vertical spacing of 12 in. (305 mm) and a radial spacing of 9 in. (228 mm). The bending capacity of the pipe column is developed below the street level by using: (i) welded shear studs around the outer surface

of the column imbedded in the foundation; and (ii) by extending mild steel reinforcement inside of the column into the foundation. Figures 6.11*b–d* show schematic section through vault, composite column, and grade beam reinforcement details. The architectural design is by the Jerde Partnership Inc., Venice, California while the structural engineering of the support system for the vault is by John A. Martin & Associates, Inc., Los Angeles. The space frame design is by Pearce Systems International, Inc.

6.4.2 Formed Composite Columns

6.4.2.1 INTERFIRST PLAZA, DALLAS

A variation of the tube concept using formed composite columns was developed by LeMessurier Consultants, Inc for the 73-story 921 ft (281 m) tall Interfirst Plaza, Dallas, completed in 1985. To satisfy the request for offices with uninterrupted views, the weight of the entire building is placed on sixteen composite columns located up to 20 ft (6 m) in-board from the building exterior. The 20 ft distance between the glass and columns allowed for a continuous band of offices with uninterrupted views. To compensate for loss of bending rigidity, all loads are transferred to the ground through the composite columns interconnected with a system of seven-story two-way Vierendeel trusses, beginning at the 5th level and spanning 120 ft (36.6 m) and 150 ft (45.7 m). The composite columns vary in size from 5 to 7 feet (1.5 to 2.1 m) square, are made with 10 ksi (69 MPa) concrete, and are reinforced with 75 ksi (517 MPa) reinforcing bars and 50 ksi (345 MPa) W36 shapes imbedded in concrete. The concrete encasement of wide flange shapes ends at the 62nd level. A schematic floor framing plan and discussion of the structural system are given in Chap. 1.

6.4.2.2 BANK OF CHINA TOWER, HONG KONG

The prism-shaped building, designed by the architectural firm of I. M. Pei and Partners, and structural engineer Leslie E. Robertson, rises to a height of 76-stories. Each of the four quadrants of the building rises to a different height, and only one out of the four reaches the full 76-stories. The bracing system uses a system of space trusses to resist both lateral loads and almost the entire weight of the building. From the top quadrant down, the gravity load is systematically transferred out to the building corner columns. Transverse trusses wrap around the building at various levels and help in transferring the load to the corner columns. At the 25th floor, the column at the center of four quadrants is transferred to the four corners by the space truss system, providing an uninterrupted 158 ft (48 m) clear span for the banking lobby. To achieve continuity between different truss members of the space frame, instead of complex three-dimensional connections requiring expensive weldments, the members are made to act as a single unit by encasing them in reinforced concrete columns. The concrete encasement surrounding the steel columns acts as a shear transfer mechanism and also counteracts eccentricities in the truss system. The lateral loads are thus carried down to the fourth story through the space truss system and corner columns. At the fourth floor, shear forces are transferred to a system of interior composite core walls through 1/2 in. (12 mm) thick steel plate diaphragms acting compositely with the concrete slab. Although much of the shear force collected by the interior core walls is transferred to the 3 ft (0.9 m) thick slurry walls at the perimeter, the core walls are continued to the foundation to serve the dual function of resisting shear and of forming walls for the bank vault. The corner columns, which continue to the foundation, resist the overturning moments. The foundation for the building consists of caissons at bedrock. Some of the caissons are as large as 30 ft (9.1 m) in diameter. A schematic representation of floor plans, bracing concept, and photograph of the building are shown in Figs. 6.12 and 6.13.

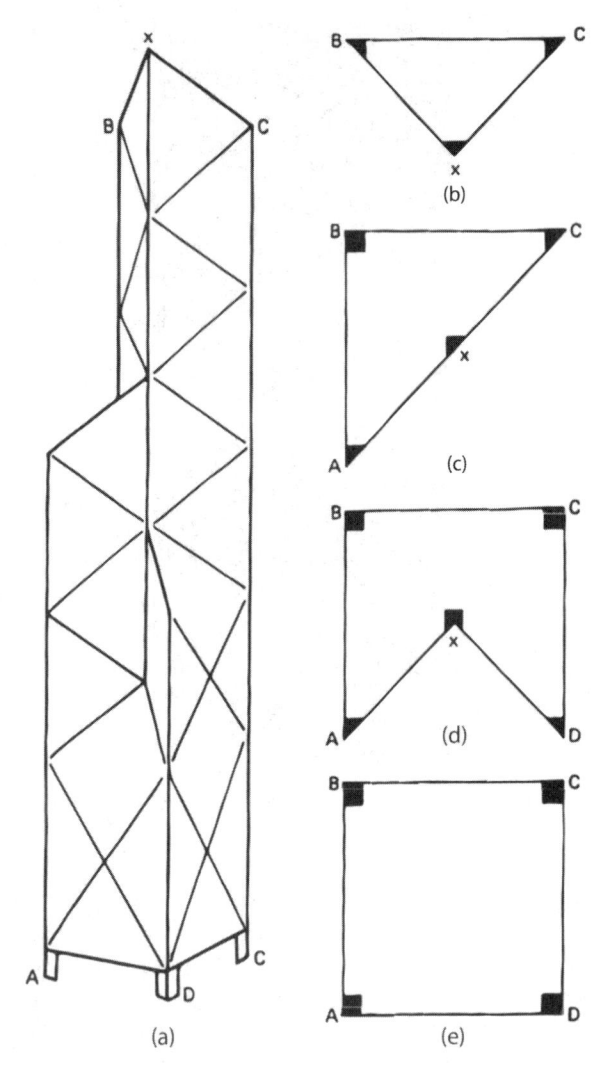

Figure 6.12 Bank of China Tower: (a) schematic elevation; (b)–(e) schematic plans.

6.4.2.3 BANK OF SOUTHWEST TOWER, HOUSTON, TEXAS

The Bank of Southwest Tower, an 82-story, 1220 ft (372 m) building, proposed for downtown Houston, Texas uses the unique concept of composite columns with interior steel diagonal bracing. The diagonals are used to transfer both the gravity and lateral loads into eight composite super columns. The building has a base of only 165 ft (50.32 m), giving it a height-to-width ratio of 7.4. The characteristic feature of the design consists of a system of internal braces that extend through the service core and span the entire width of the building in two directions. A typical bracing consists of an inverted K-type brace rising for nine floors, there being two such braces in each direction. Eight of these nine-story trusses are

Figure 6.13   Bank of China Tower: photograph.

a framed tube, the eight-column structural system frees the perimeter from view-obstructing columns, especially at the building corners. The structural engineering is by LeMessurier Consultants, Inc., and Walter P. Moore & Associates, Inc.

6.4.3   Composite Shear Walls and Frames

6.4.3.1   FIRST CITY TOWER

The building designed by structural engineers Walter P. Moore & Associates, Inc., consists of a base structure extending one level below the plaza, covering an entire downtown Houston block and tower extending through 49-stories giving 1.4 million square feet (130,000 m²) of office space. The building is a parallelogram in plan and is positioned on the site to create views (Fig. 6.14). Among the architecturally distinctive features of First City Tower are the four 11-story-high indentations on the façade where the glass panels open the building for viewing. These "vision strips" rise in staggered formation along the height of the two long faces.

Structural Anatomy   In common with most other projects, First City Tower went through a design metamorphosis. The final scheme adopted has the following composite components, as shown schematically in Fig. 6.15.

1. Composite floor-framing system, consisting of steel beams and concrete slab on formed steel deck interconnected with shear studs.

2. Composite stub girder system, consisting of rolled steel beams connected to the floor slab with a series of stubs welded to the beam. This system is discussed in Chap. 9, minimizes floor-to-floor height by combining the space required for mechanical ductwork within the design of the structural system.

assembled one on top of another within the tower. All the gravity loads are transferred to eight massive composite columns located at the building perimeter. Note that because the transverse diagonals are interconnected, a three-dimensional behavior is invoked; all eight columns participate in resisting the overturning moment. As compared to

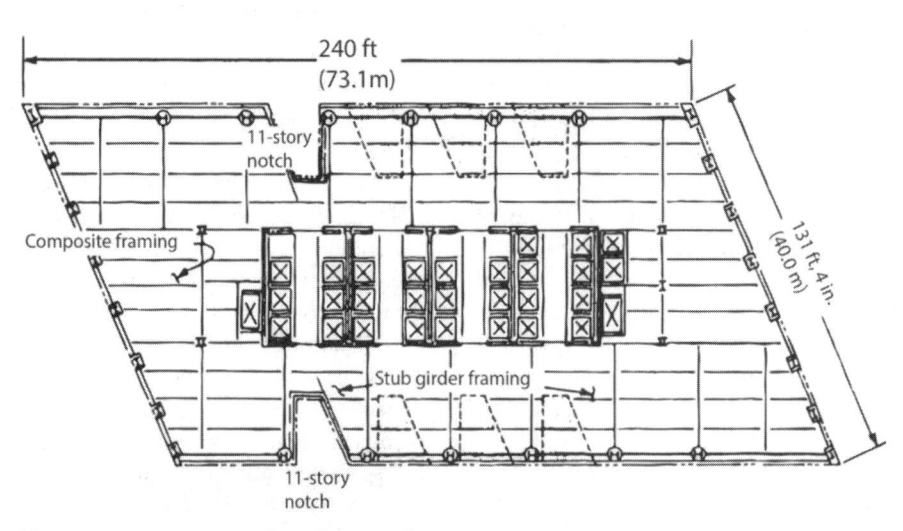

Figure 6.14   Composite floor-framing plan.

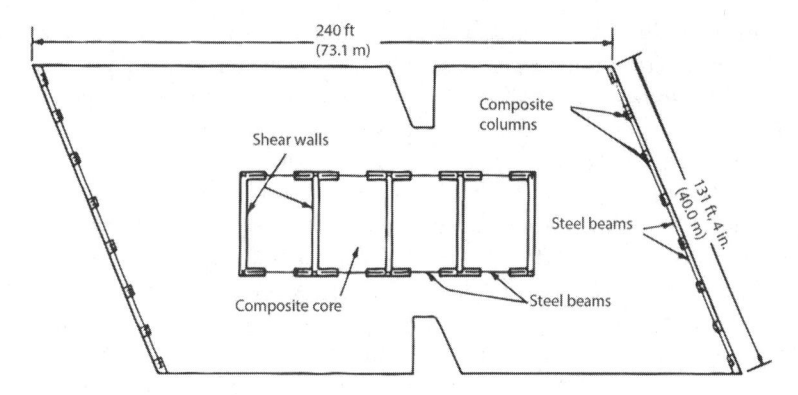

Figure 6.15   Composite elements.

3. Composite columns consisting of steel rolled shapes embedded in reinforced concrete columns (Fig. 6.16).

4. Composite frame consisting of composite columns and moment-connected steel beams (Fig. 6.16).

5. Composite shear walls consisting of a series of I- and C-shaped reinforced concrete shear walls interconnected with steel link beams (Fig. 6.16).

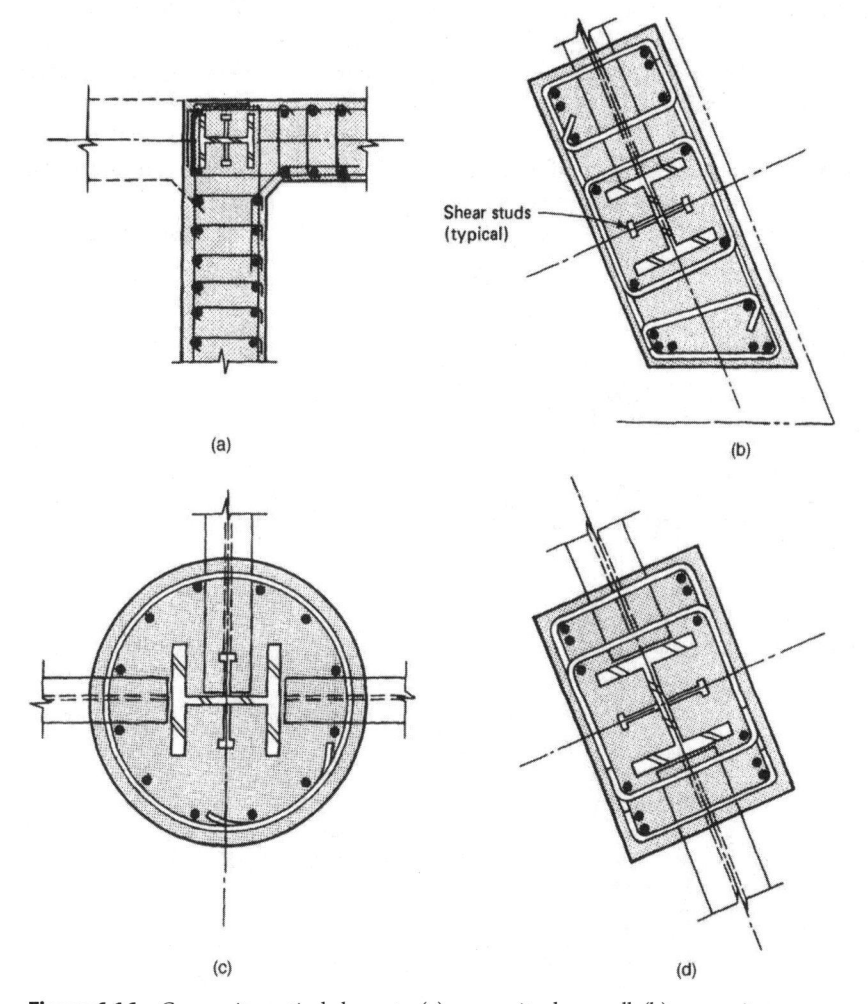

Figure 6.16   Composite vertical elements: (a) composite shear wall; (b) composite corner column; (c) typical circular column on long faces; (d) typical exterior column on short faces.

6. Composite construction that allows the initial growth of steel followed by placing of concrete to form composite columns and shear walls as shown schematically in Fig. 6.18.

Design   Figure 6.16 shows the arrangement of vertical reinforcement and structural steel core column for composite vertical framing elements. Typically, the embedded steel columns vary from a W14 by 370 lb/ft (455 by 418 mm, 551 kg/m) at the bottom to a W14 by 68 lb/ft (356 by 254 mm, 101 kg/m) member at the top. For rectangular columns, width of the concrete envelope is maintained for full height, while the depth of column into the building is reduced in the upper floors.

The vertical reinforcement in the columns varied from #18 bars (57 mm in diameter) at the bottom to #7 bars (22 mm in diameter) at the top. Open ties permitted in low-seismic zones are used throughout to facilitate the placement of column reinforcement. Mechanical tension splices are used for frame columns on the short faces, and compression mechanical splices are used for the gravity columns on the long faces. Shear studs shop-welded to the webs of steel columns are used to achieve load transfer from concrete to steel.

Figure 6.16a shows the arrangement of reinforcement and erection column embedded in the shear wall. Typically, a W10 by 72 lb/ft (267 by 257 mm, 107 kg/m) steel column is embedded at each shear wall corner. The ties enclosing the vertical reinforcement are needed if 1 percent or more reinforcement is required for compression in the walls. Typically, the ties are used in the shear walls up to level 6 only.

Figure 6.17 shows a typical connection detail between concrete shear wall and a typical interconnecting beam. The beam to embedded steel column connection is a typical bolted shear connection. Moment capacity required at the face of the shear wall is developed by means of shear transfer mechanism between the concrete and shop-welded studs at the top and bottom flanges of the beam. The stiffener plate, set flush with the wall face, has no structural purpose but helps in simplifying shear wall forming around the beam. A conventional moment connection was used between the core columns and beams at levels where the shear walls were dropped.

Composite Construction   The basic idea of composite construction is to let the construction of the steel frame advance to a predetermined number of stories first and then to envelop the column with concrete, as shown schematically in Fig. 6.18. The step-by-step process is as follows. After completion of the foundation system, the steel frame erection is started using standard procedures and AISC tolerances. Since the general idea is not to wait

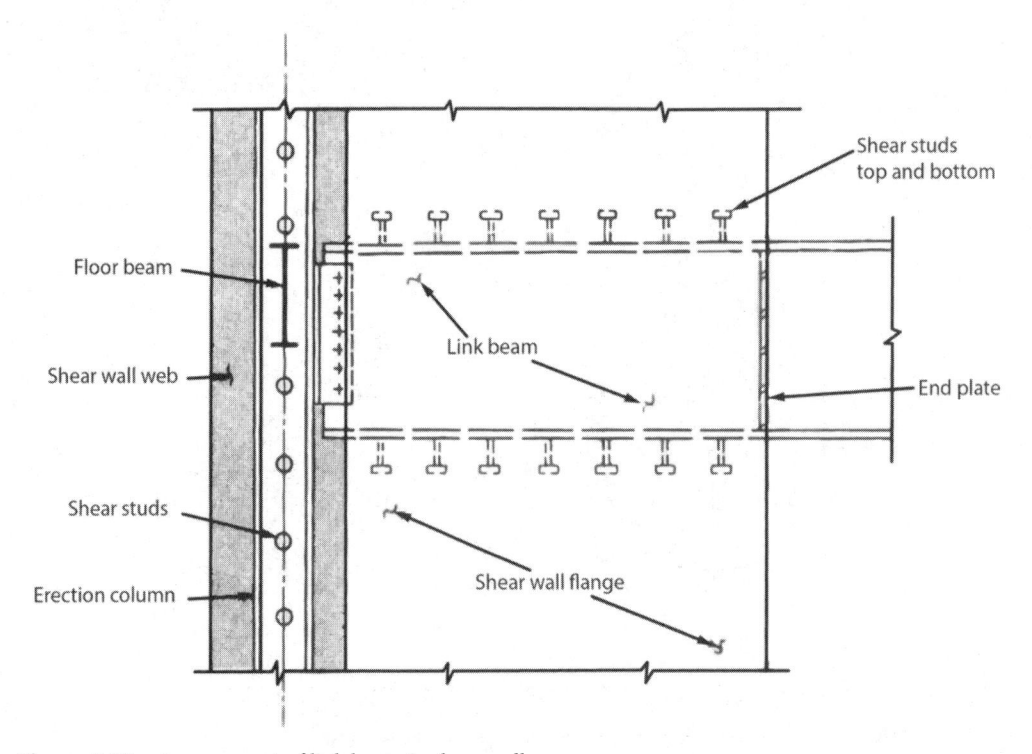

Figure 6.17   Arrangement of link beam in shear wall.

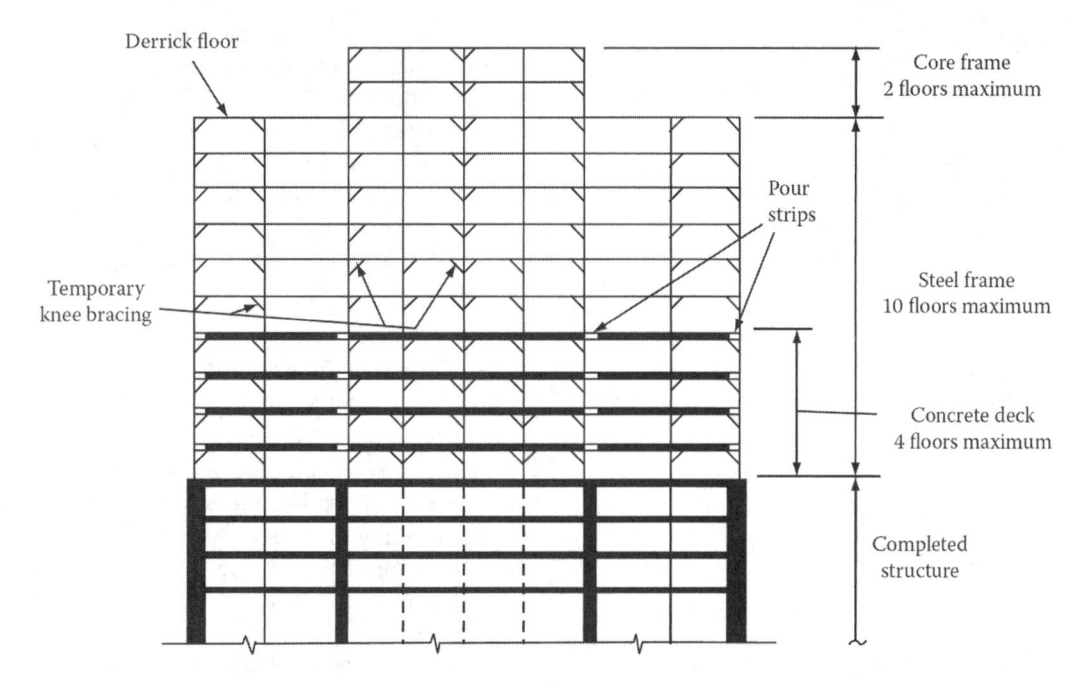

Figure 6.18   General construction sequence in composite structures.

for the concreting of shear walls before steel erection, small steel columns are utilized for vertical support at the intersection of shear wall web and flanges. In order to limit the size, heavy steel sections are embedded in the exterior columns. The erection of exterior spandrels, stub girder, and purlins proceeds as in a conventional steel frame.

The size of the interior steel columns embedded within the shear walls depends on the maximum separation of the derrick floor from the level to which concrete shear walls have been completed. For this project, the criterion was set at a maximum of 10 floors. This was established during the early phase of design with proper coordination from the steel erector and the general contractor. Metal deck installation and welding of spandrels on the short faces follow closely behind steel erection. Diaphragm action of the deck is established by welding metal deck to beam flanges and by the installation of shear studs welded through the deck. Concrete topping is placed on metal deck several floors above the concreted portion, with the exception of pour strips around the shear walls and exterior columns.

The completed floor slab serves as a platform to transport concrete from a material hoist to shear walls and exterior columns. The reinforcing steel for the exterior columns and shear walls is tied into position several floors high with story-high bar lengths. The formwork is then placed around the reinforcing cage.

Concrete is hoisted through the material hoist at the exterior of the building and then bugged over the concrete slab to the required location. Columns and walls below the floor are concreted by pouring concrete through the pour strips left around the shear walls and columns. Metal deck in the pour strip around the shear walls is cut to provide room for hoisting of shear wall forms. The forms are hoisted to the floor above through these temporary openings.

As a next step, the shear walls above the floors are poured and integration between floor slab diaphragm and shear walls is completed by concreting the pour strips. The various stages of the shear wall construction sequence are shown schematically in Fig. 6.19.

With proper sequencing of different trades, the construction of concrete shear walls and columns can proceed at a pace equal to the overall speed of a conventional steel building. Typically for First City Tower, steel erection proceeded at the rate of two floors per week, with concreting of shear walls and columns following closely behind at the same rate.

Temporary Bracing   In a pure steel or concrete building, there is no measurable lag between the levels at which the construction is proceeding and the level at which the building lateral resistance is established. For example, in a steel building moment connections or braces are connected between steel columns immediately behind the erection of the steel frame in order to provide

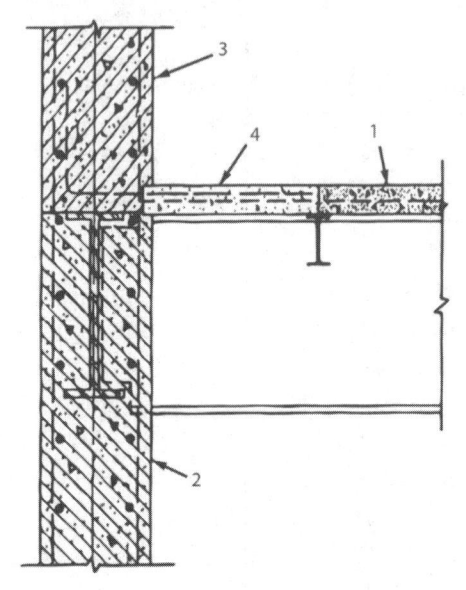

Figure 6.19 Shear wall construction sequence.

resistance and stability to lateral loads. Monolithic casting of concrete beams and columns and other elements, such as shear walls, automatically provides the required stability. However, for a composite building, stability is not achieved until after the concrete is placed and cured. Since the level at which the steel framing is erected is deliberately made sufficiently high above the concreted level, there exists in composite construction a distinct need for considering bracing during the erection process. Steel cables traditionally used in steel buildings to plumb

and stabilize the structure are often inadequate, requiring a more positive method of lateral stability. This was provided for the First City Tower by supplementary knee bracing around the building perimeter and also in the building core. The braces at the perimeter were later removed, while those in the core were left in place to become embedded in the composite shear walls since they did not interfere with the architecture of the building.

6.4.4 Composite Tube System

6.4.4.1 THE AMERICA TOWER, HOUSTON, TEXAS

In this system, a concrete exterior tube consisting of closely spaced columns and deep spandrels forms the essential wind-bracing system. Fast erection of the steel frame is achieved by using light structural columns around the perimeter. Interior columns and floor framing are identical to gravity-formed steel construction.

A somewhat new technique, as mentioned earlier, adapts a variation of this concept. This system does away with the concreting of the spandrels; instead, steel spandrels are employed in conjunction with composite columns to resist the lateral loads. This is the system used on the 42-story America Tower building, designed by structural engineers Walter P. Moore and Associates, Inc., Houston, Texas.

The America Tower project consists of 1 million square feet (92,858 m^2) of office space and is elliptical in

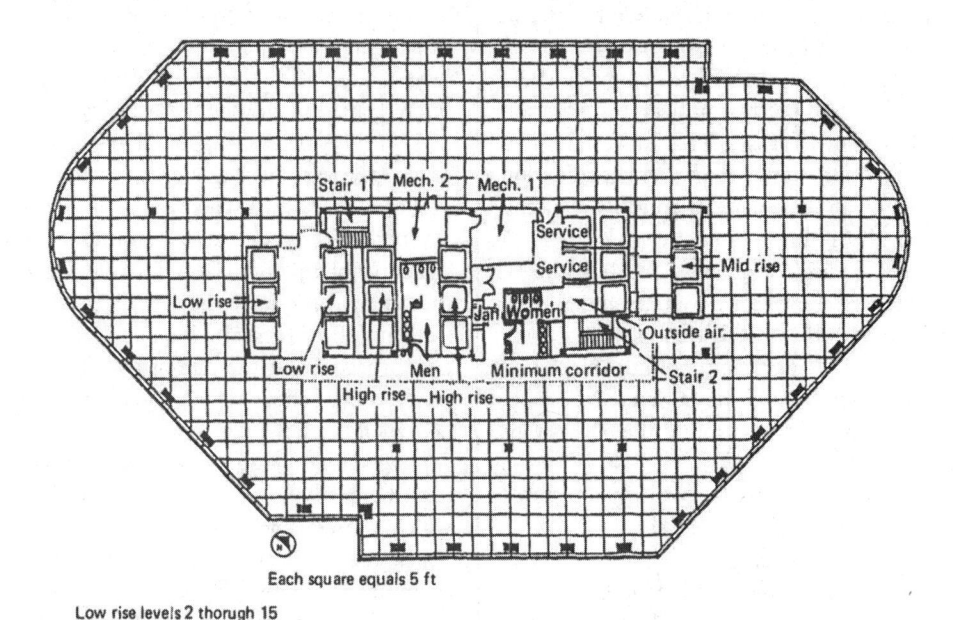

Each square equals 5 ft

Low rise levels 2 thorugh 15

Figure 6.20 America Tower architectural floor plan.

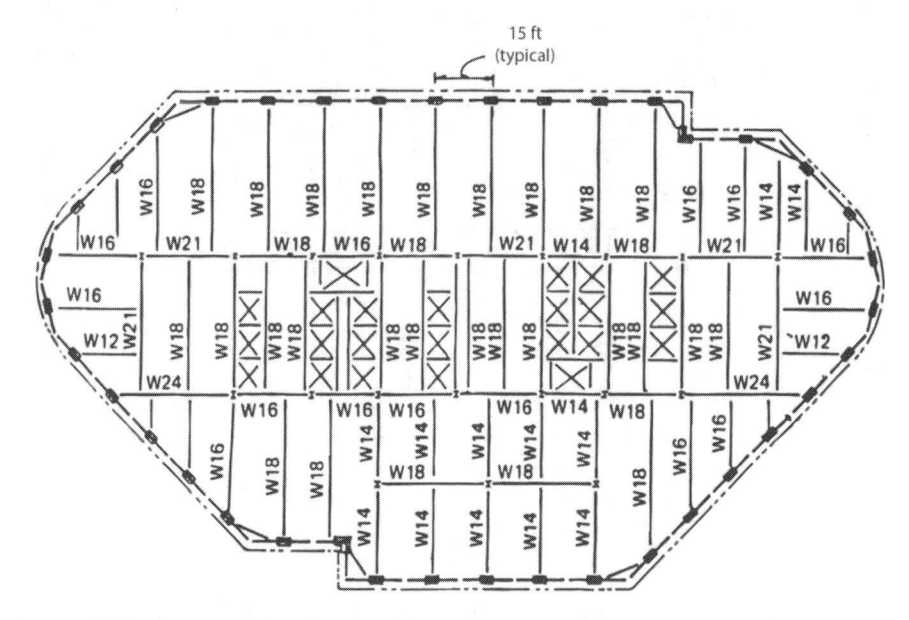

Figure 6.21 Structural floor framing plan.

plan with offsets on the long faces to highlight the verticality of the building and provide more corner offices. The exterior facade is designed with a curtain wall of alternating glass and precast panels. The architectural and structural framing plans are shown in Figs. 6.20 and 6.21.

Structural Anatomy The salient structural features are given in Table 6.1. A framed tube consisting of closely spaced exterior columns and spandrels was chosen as the proper structural solution.

Table 6.1 America Tower: Structural Features

Height above mat	619 ft 7 in. (188.85 m)
Width	125 ft 4 in. (38.2 m)
Height/width ratio	4.94
Interior columns	W14 rolled steel sections
Floor framing	W18 composite beams span 41 ft (12.45 m)
Exterior columns	42 by 27 in. (1.07 by 0.88 m) composite columns
Spandrels	W36, W33, W30 steel shapes
Mat	6 ft (1.83 m) thick

Construction As is common with most tube buildings, the system used for America Tower is the so-called prefabricated tree column system (Fig. 6.22). A typical tree column consists of a two-story column and two levels of spandrels. The major difference in the mixed tube construction is that the column sizes are small compared to an all-steel scheme, resulting in savings in the fabrication and erection of steel members.

Another feature that has resulted in considerable savings is the use of a hybrid tubular frame on the perimeter of the building. Composite columns used in the typical levels (floors 3 through the roof) are integrated into pure structural steel columns at the second level. This marriage of steel columns to composite columns simplified the construction of the exterior columns at nontypical lower levels.

Although the savings in the structural cost for this scheme appeared to be only marginal from the comparative study, the contractor elected to use this scheme because his recent job experience on another similar high-rise project had proved that a hybrid frame is more economical than a full-height composite scheme. Figure 6.23 shows a detail of the intersection of composite column with a steel column. The resistance to lateral loads during the erection of the tower was provided by the tube action of the erection columns and spandrel beams. The columns, which were sized for 12 floors of construction gravity loads, were found to be adequate for providing resistance to lateral loads. Additional welds between the erection column and the spandrel and stiffeners were necessary in order to develop the bending capacity of the columns.

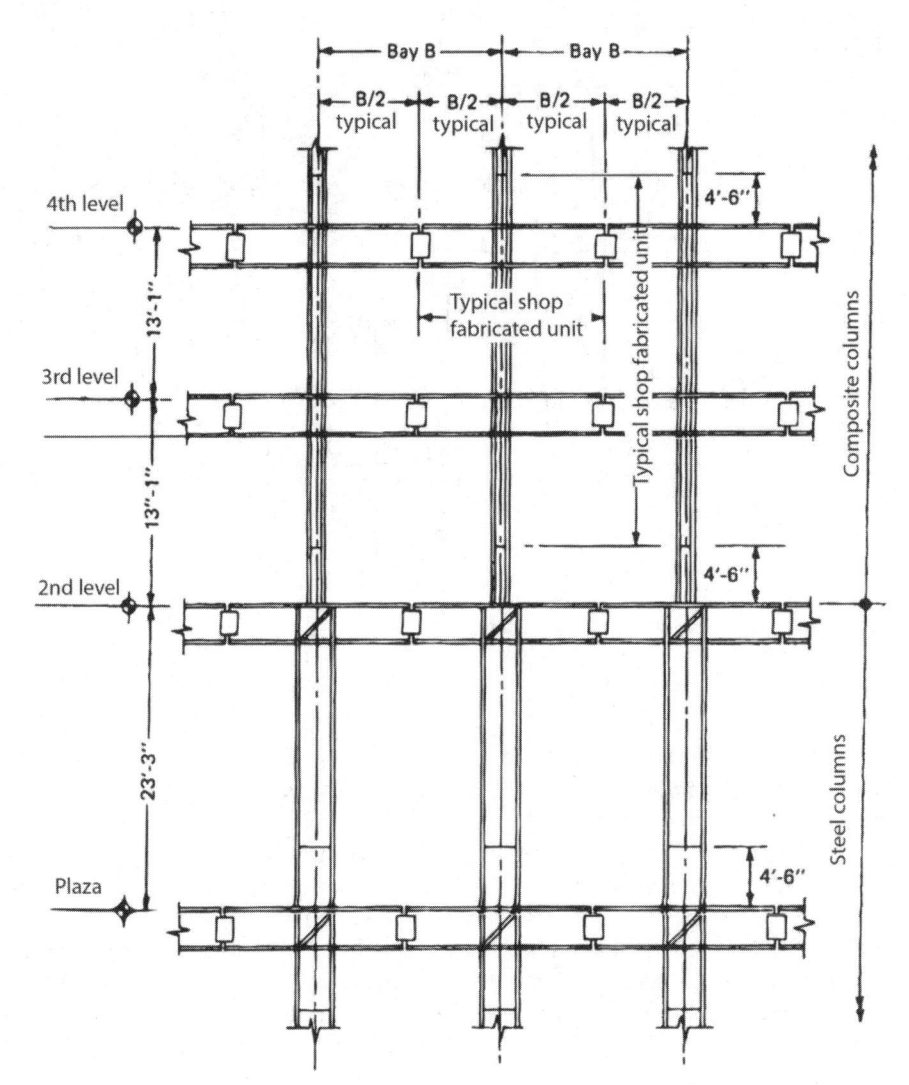

Figure 6.22 Typical frame fabrication unit.

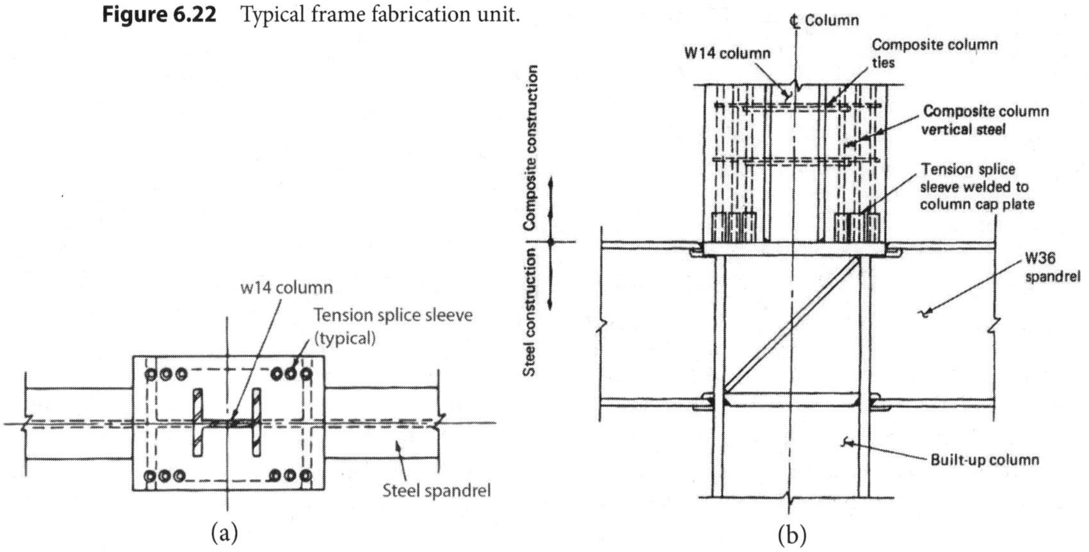

Figure 6.23 Composite column to steel column connection detail: (a) plan; (b) elevation.

6.4.5 Conventional Concrete System With Partial Steel Floor Framing

6.4.5.1 THE HUNTINGDON, HOUSTON, TEXAS

The Huntingdon in River Oaks, Houston, designed by structural engineers Walter P. Moore Associates, Inc. is a 34-story condominium project with living units in the tower, resident parking on three floors under the building, and a landscaped plaza featuring amenities such as a heated swimming pool, spa, pool-side room, etc. Construction consists of a poured-in-place concrete structure with a white precast exterior. Windows are dual-pane tinted glass with operable vents.

Structural Anatomy The building is essentially symmetrical in plan with a rectangular dimension of 142 ft 6 in. by 106 ft 6 in. (43.5 by 32.5 m) and is 490 ft (149.5 m) tall. The architectural design called for column-free space between the core and exterior of the building. The desired elevation dictated that exterior columns be spaced at 20 ft (6.1 m) centers around the building except at the corners. Columns were not permissible at the corners because each home has a balcony offering views in two directions. With this layout of exterior columns it was not possible to develop the tube action to resist the lateral loads because of lack of continuity of structural resistance around the corners. A preliminary analysis was performed with 24 in.

(610 mm) deep pan joists for floor system and an interacting system of shear wall and end frames for lateral loads. The analysis indicated that the system could be designed for strength without too much of a premium in material quantities, but the calculated lateral deflection under Houston hurricane wind code was too large for the occupancy of the building. Addition of stiffening elements without introducing more columns was necessary to maintain reasonable size of structural members and to reduce deflection.

The addition of girders between the core and exterior at each column line as shown in Fig. 6.24 was the answer. In comparison to the pan joist floor system, these girders are very stiff and help in mobilizing the exterior columns to resist the wind loads. The resulting structural system consisting of the core, end frames, and exterior columns reduced the deflection to well within the drift limitations.

Haunch Girders A framing system employing girders of constant depth crisscrossing the interior space between the core and the exterior often presents nonstructural problems because it limits the space available for passage of air-conditioning ducts. A system known as the haunch girder system, with wide acceptance in many parts of North America, alleviates nonstructural problems without making undue compromises in the structure. The

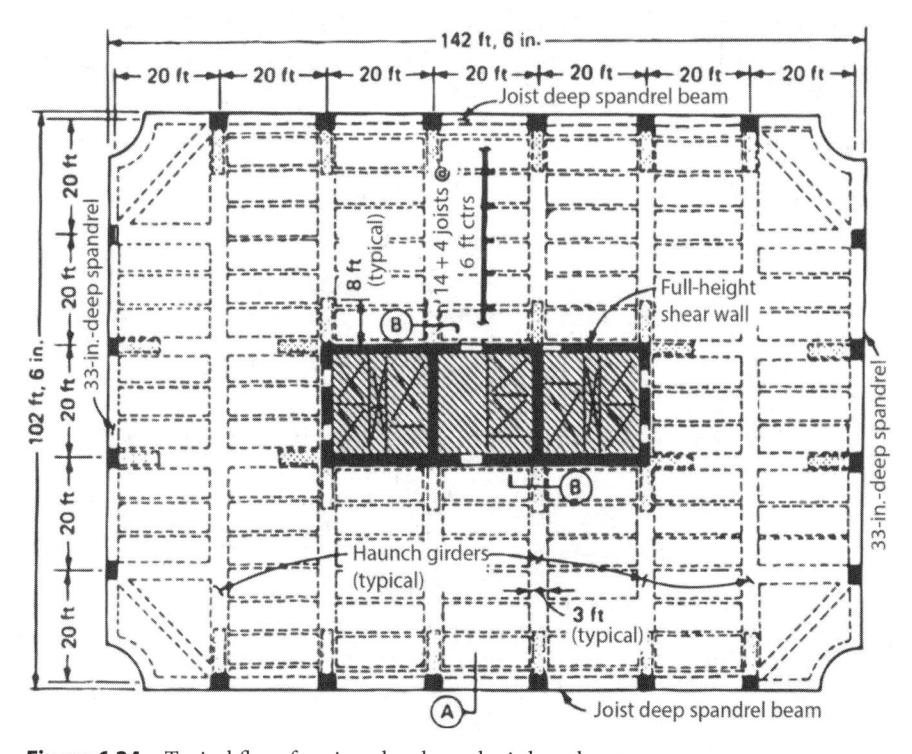

Figure 6.24 Typical floor framing plan: haunch girder scheme.

basic system consists of a girder of variable depth as shown in Figs. 6.25 and 6.26. The use of shallow depth at the center facilitates the passage of mechanical ducts and eliminates the need to raise the floor-to-floor height. Material quantity comparisons for gravity loads indicated that haunch girder framing is very economical. Additional comparison for lateral load design indicated even more savings from the reduced floor-to-floor height.

Flying Forms For the *in situ* construction to be competitive with other forms of construction, such as precast concrete or steel, it is imperative that the forming for the floor system should lend itself to repeated uses in a multistory frame. One-way joist system using flange-type steel pans is a reusable system that is currently popular in almost all high-rise buildings in North America. Wide pans in two common sizes of 53 and 66 in. (1346 and 1676 mm) widths have become a standard, replacing the 30 in. (762 mm) pans.

A conventional method of concrete joist construction is to nail the flanges of pans on a formed solid decking built on adjustable shores. When concrete has gained sufficient strength, shores are removed and pans stripped and stored in an area to await reuse until shoring and wooden decking are erected for the next floor. This method of forming is labor-intensive and also time-consuming, particularly when joists of different lengths are used.

A method of forming that fully utilizes the opportunity for repetition offered in multistory construction is a system known as flying form system. In this system joist pan forms are attached to slab, creating a system of deck surface, adjustable jack, and supporting framework. For stripping, the form is lowered by jack, slipped out of the slab, and lifted to the next floor for reuse as one rigid structure, resulting in economy of handling and placing.

The floor framing used in the Huntingdon project lent itself to the use of flying forms. To further simplify the formwork, the width of haunch girders was kept slightly larger than the maximum width of the exterior column.

Mixed-Floor Construction The architectural requirement for the project dictated that typical floors be constructed at two different elevations with a 2 ft 6 in. (762 mm) drop occurring at the outline of the interior core. This design allows flexibility to the buyers in locating wet areas, whereas the common areas such as the service and passenger elevator lobbies and stair landings

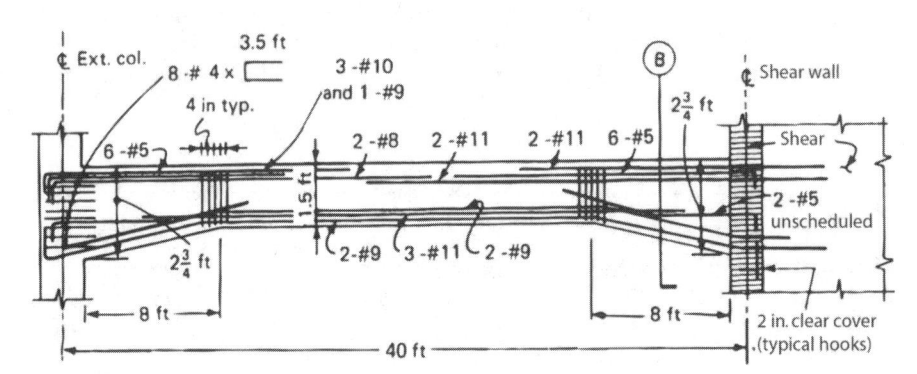

Figure 6.25 Haunch girder elevation and reinforcement.

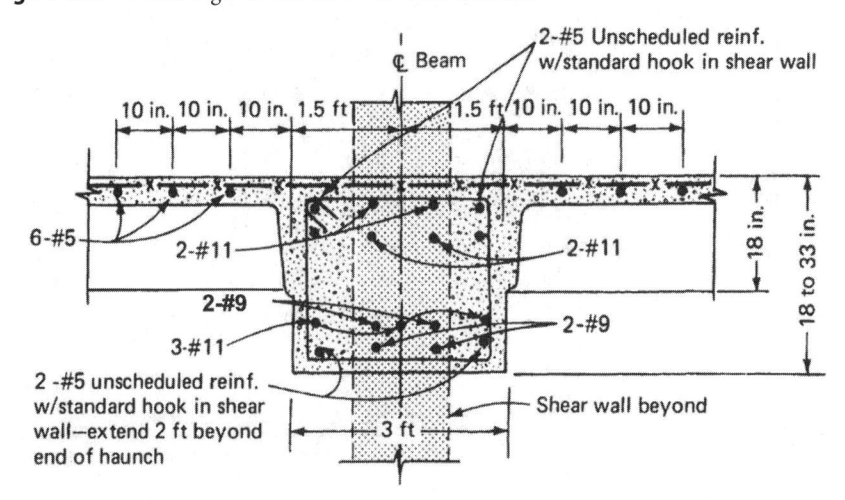

Figure 6.26 Haunch girder section.

are kept higher because of predetermined finish. Forming a poured-in-place concrete slab at two different elevations on either side of the core would have been somewhat complicated. This required taking a fresh look at the framing of slab inside the core.

Compared to an office building, the vertical transportation requirements for a condominium are relatively small. The core size dictated by the elevator and mechanical requirements for this project worked out to 50 by 20 ft (15.2 by 6.1 m). The elevator lobbies normally required in an office building are eliminated in the core arrangement because the elevators open on each floor directly to a private foyer serving a maximum of two homes, making it possible to have a very small core. The floor area that is enveloped within the core is very small. This, combined with the problem of a raised floor inside the core, made possible the use of a steel floor framing with composite deck, unwittingly creating yet another variation to the many types of mixed construction currently in vogue for tall buildings.

Pockets were provided in the shear wall to receive the steel beams in the core. Shelf angles were quick-bolted to the walls in areas where supports for the metal deck were required. This idea of using conventional steel framing in the core facilitated the construction of shear walls and the floor framing independent of the framing inside the core. Typically, the construction of shear walls and concrete framing proceeded at the rate of one floor a week. The steel framing in the core was deliberately left to lag behind the concrete framing so that the ironworkers could work without interfering with other trades.

6.5   HIGH-EFFICIENCY STRUCTURE: STRUCTURAL CONCEPT

A super-tall building is generally defined in architectural terms as a skyscraper when its silhouette has a slender form with a height-to-width ratio well in excess of 6. The slender proportion imposes engineering demands far greater than those of gravity, requiring the engineer to come up with innovative structural framing schemes. The ideal structural form is one that can at once resist the effect of bending, torsion, shear, and vibration in a unified manner. A perfect form is a chimney with its walls located at the farthest extremity from the horizontal center but as an architectural form it is less than inspiring for a building application. The next best and a more practical form is a skeletal structure with vertical stiffness, that is, columns located at the farthest extremity from the building center. Two additional requirements need to be incorporated within this basic concept to achieve high efficiency: (i) transfer as much gravity load, preferably all the gravity load into these columns to enhance their capacity for resisting overturning effects due to lateral loads; and (ii) connect columns on opposite faces of the building with a system capable of resisting the external shear forces. The ultimate structure for a rectangular building, then, will just have four corner columns interconnected with a shear-resisting system. Such a concept, proposed by the author for a super tall building, is shown in Fig. 6.27, in which the total gravity loads and overturning moments are resisted by four composite columns. The columns are located in-board

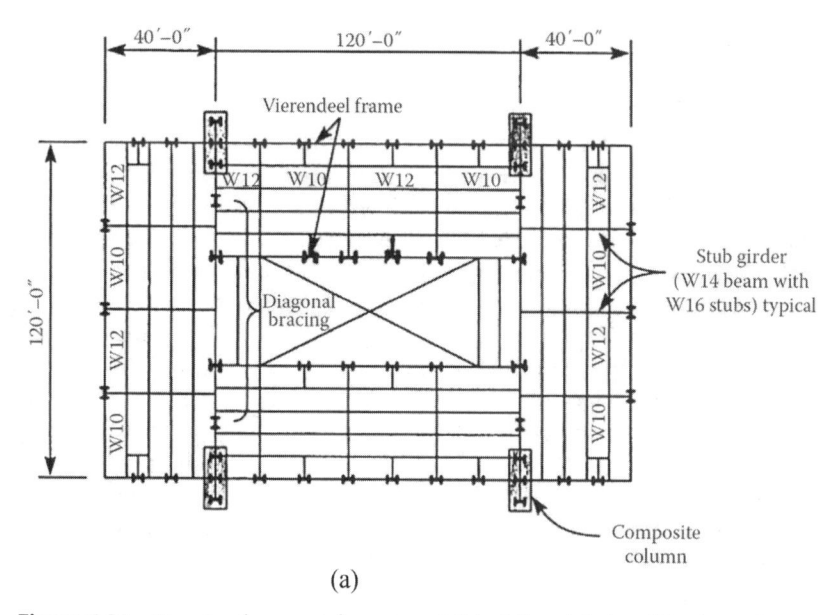

(a)

Figure 6.27   Structural concept for a super tall building: (a) plan; (b) schematic elevation; (c) interior view of mega module; (d) exterior view of mega module.

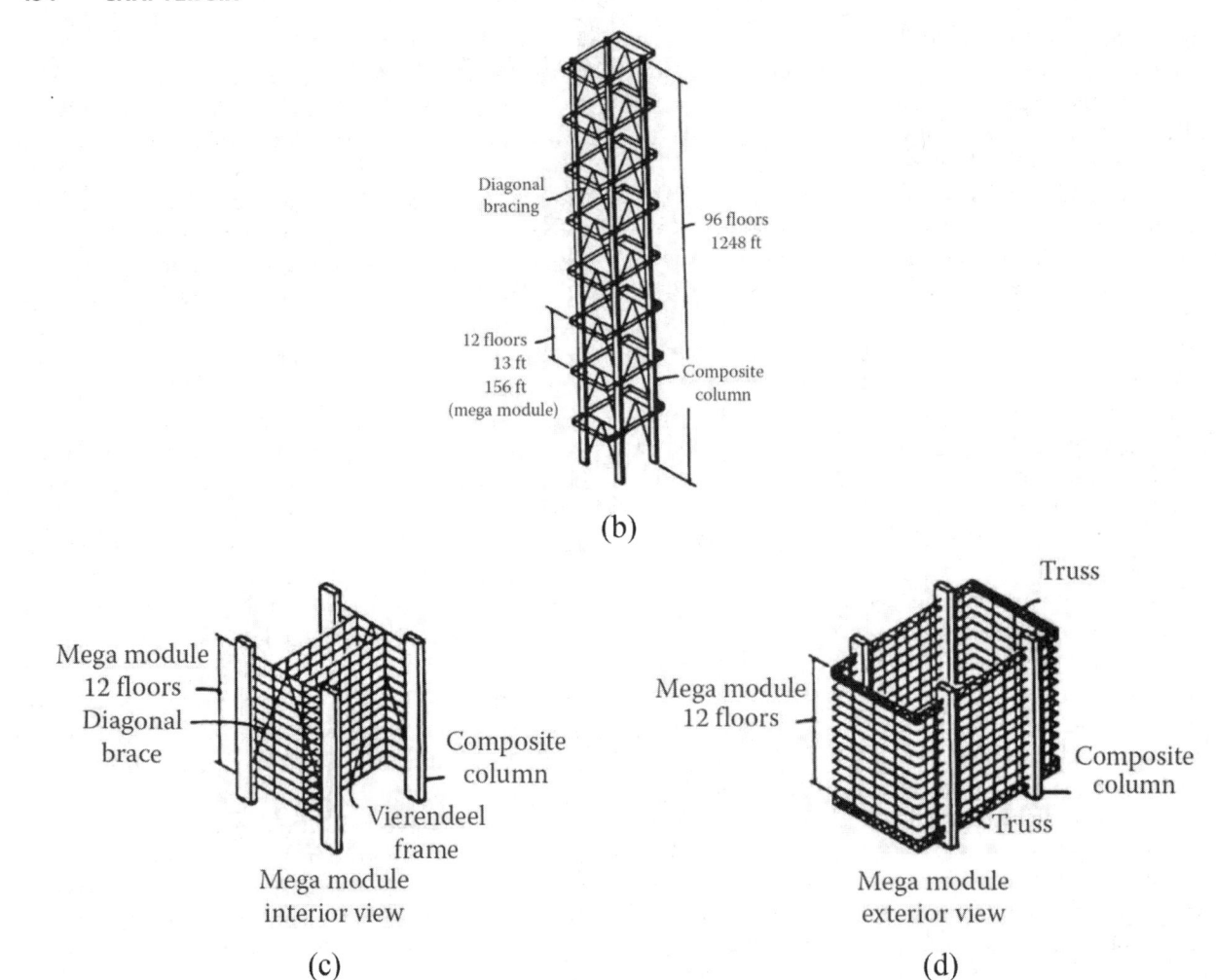

Figure 6.27 (*Continued*)

from the corners to allow for architectural freedom in modulating the short sides of the building. The shear in the transverse direction is resisted by a system of 12-story-high chevron braces while in the longitudinal direction the shear resistance is primarily provided by the full-height Vierendeel frames located on the long faces. The story-high longitudinal trusses located at every 12th floor permit cantilevering of the floor system. The primary function of the interior Vierendeel frame is to transfer the gravity loads of the interior columns to the composite columns via chevron braces. However, because of its geometry, it also resists external shear forces in the long direction.

Any number of conventional framing using composite beams, haunch girders, or stub girders may be used to span the distance from the core to the building perimeter. The scheme in Fig. 6.27 shows stub girders consisting of a W14 wide flange beam with W16 stubs welded at intervals to the top flange of W14 beam. This system is discussed in Chap. 9 has the advantage of reducing simultaneously the floor-to-floor height and the unit quantity of steel required for floor framing.

The author believes that the scheme shown in Fig. 6.27 can be modified readily for a variety of architectural building shapes. Any desired slicing and dicing of the building on the short faces may be accommodated without inflicting an undue penalty on systems efficiency. The structural concept is complete. Like all other schemes, it has to await its time before finding an application to a real building.

Chapter 7

Gravity Systems for Steel Buildings

7.1 INTRODUCTION

There are basically three groups of structural steel available for use in bridges and buildings per AISC Manual 15th edition, Table 2-4, reproduced in Table 7.1:

1. Carbon steel: American Society for Testing and Materials (ASTM) A36, A53, A500, A501 and A529, A709, A1043, A1085.

2. High-strength, low-alloy steels: ASTM-A572, A618, A709, A913, and A992, A1065.

3. Corrosion-resistant, high-strength, low-alloy steels: ASTM-A242, A588, and A847, A1065.

In the A572 category five grades of steel: 42, 45, 50, 60, and 65 are available for structural use. The grade numbers correspond to the minimum yield point in ksi, kilopounds per square inch of the specified steel. Carbon steel is available in 35 to 55 grades.

Steel buildings in the United States are designed as per the American Institute of Steel Construction (AISC) specifications, which were first published in 1923. The specifications are revised periodically to keep pace with new research findings and the availability of new materials. Steel construction for buildings is commonly referred to as steel skeleton framing, signifying that a majority of the members consist of linear structural elements such as beams and columns.

Skeleton framing is normally erected in two-story increments, each increment being called a tier. Light-gauge steel decking serving as a permanent form is the most common method of slab construction and could be used as positive reinforcement for the concrete topping if needed.

The rules for the design of structural steel members subject to any one, or combination of stress conditions due to bending, shearing, axial tension, axial compression, and web crippling are given in the AISC specifications. Members can be designed by the Load Resistance Factor Design (LRFD) or Allowable Strength Design (ASD) while using the appropriate load combinations for each method, sections 2.3 and 2.4 of ASCE 7-16 respectively. Only the LRFD method is used in this book using AISC 360-16, Specification for Structural Steel Buildings, dated July 7, 2016.

The functional needs of occupancy invariably dictate that floors be relatively flat. In a steel building, this is most often achieved by horizontal subsystems consisting of beams, girders, spandrels, and trusses over which spans a light-gauge metal deck. Concrete topping over the metal deck completes the floor system.

In this chapter, a brief description of some of the elements normally employed in the framing of steel buildings is given. We first describe gravity loads, followed by a description of metal deck, steel beams, joists, and finally, columns. The description of the metal deck is somewhat limited in this chapter because most usually metal deck is employed as a composite member, which we cover in greater detail in Chap. 9.

Table 7.1   Applicable ASTM Specifications for Various Structural Shapes (AISC Manual, 15th Edition)

Table 2-4
Applicable ASTM Specifications for Various Structural Shapes

Steel Type	ASTM Designation		F_y Yield Stress[a] (ksi)	F_u Tensile Stress[a] (ksi)	W	M	S	HP	C	MC	L	HSS Rect.	HSS Round	Pipe
Carbon	A36		36	58–80[b]	░	■	■	░	■	■	■			
	A53 Gr. B		35	60										■
	A500	Gr. B	42	58									░	
		Gr. B	46	58								░		
		Gr. C	46	62									■	
		Gr. C	50	62								■		
	A501	Gr. A	36	58								░	░	
		Gr. B	50	70								░	░	
	A529[c]	Gr. 50	50	65–100	░	░								
		Gr. 55	55	70–100	░	░								
	A709	36	36	58–80[b]	░									
	A1043[d,k]	36	36–52	58	░									
		50	50–65	65	░									
	A1085	Gr. A	50	65								░		
High-Strength Low-Alloy	A572	Gr. 42	42	60	░	░	░		░	░	░			
		Gr. 50	50	65	░	░	░	■	░	░	░			
		Gr. 55	55	70	░	░	░		░	░	░			
		Gr. 60[e]	60	75	░	░	░		░	░	░			
		Gr. 65[e]	65	80	░	░	░		░	░	░			
	A618[f]	Gr. Ia[k], Ib & II	50[g]	70[g]								░	░	
		Gr. III	50	65								░	░	
	A709	50	50	65	░	░	░	░	░	░	░			
		50S	50–65	65	░									
		50W	50	70	░	░	░	░	░	░	░			
	A913	50	50[h]	65[h]	░	░	░	░	░	░	░			
		60	60	75	░	░	░	░	░	░	░			
		65	65	80	░	░	░	░	░	░	░			
		70	70	90	░	░	░	░	░	░	░			
	A992		50[i]	65[i]	■									
	A1065[k]	Gr. 50[j]	50	60								░		

■ = Preferred material specification.

░ = Other applicable material specification, the availability of which should be confirmed prior to specification.

☐ = Material specification does not apply.

Table 2-4 (continued)
Applicable ASTM Specifications for Various Structural Shapes

Steel Type	ASTM Designation		F_y Yield Stress[a] (ksi)	F_u Tensile Stress[a] (ksi)	Applicable Shape Series							HSS		
					W	M	S	HP	C	MC	L	Rect.	Round	Pipe
Corrosion-Resistant High-Strength Low-Alloy	A588		50	70										
	A847[k]		50	70										
	A1065[k]	Gr. 50W[j]	50	70										

■ = Preferred material specification.
▨ = Other applicable material specification, the availability of which should be confirmed prior to specification.
□ = Material specification does not apply.

[a] Minimum, unless a range is shown.
[b] For wide-flange shapes with flange thicknesses over 3 in., only the minimum of 58 ksi applies.
[c] For shapes with a flange or leg thickness less than or equal to 1½ in. only. To improve weldability, a maximum carbon equivalent can be specified (per ASTM A529 Supplementary Requirement S78). If desired, maximum tensile stress of 90 ksi can be specified (per ASTM A529 Supplementary Requirement S79).
[d] For shape profiles with a flange width of 6 in. or greater.
[e] For shapes with a flange thickness less than or equal to 2 in. only.
[f] ASTM A618 can also be specified as corrosion-resistant; see ASTM A618.
[g] Minimum applies for walls nominally ¾ in. thick and under. For wall thickness over ¾ in., $F_y = 46$ ksi and $F_u = 67$ ksi.
[h] If desired, maximum yield stress of 65 ksi and maximum yield-to-tensile strength ratio of 0.85 can be specified (per ASTM A913 Supplementary Requirement S75).
[i] A maximum yield-to-tensile strength ratio of 0.85 and carbon equivalent formula are included as mandatory, and some variation is allowed, including for shapes tested with coupons cut from the web; see ASTM A992. If desired, maximum tensile stress of 90 ksi can be specified (per ASTM A992 Supplementary Requirement S79).
[j] The grades of ASTM A1065 may not be interchanged without approval of the purchaser.
[k] This specification is not a prequalified base metal per AWS D1.1/D1.1M:2015.

7.2 DESIGN LOADS

Gravity loads on buildings are of two kinds: (i) static and (ii) dynamic. Static load is considered permanent, whereas the dynamic load is time-dependent. The weight of every element within the structure is a static load and includes weights of load-bearing elements, beams, slabs, columns, walls, ceiling, floor and wall finishes, sprinkler systems, light fixtures, sheet metal ducts, permanent partitions, exterior cladding, cooling towers, central plants, pump rooms, thermal storage tanks, and other mechanical equipment.

Although their effect is similar to dead loads and essentially static, live loads are less accurately predictable because they are subject to greater variation. Most often in North America, a minimum value of 50 psf (2394 Pa) for live load plus 20 psf (958 Pa) for partition allowance is used in the design of speculative office buildings. Sometimes if tenants' requirements are known prior to the design of the building, allowance is made for expected usage of heavier loads such as book stacks, filing cabinets, computers, and business machines. It is becoming increasingly common in certain cities of the United States to design an area 20 ft (6.0 m) deep adjacent to the exterior of the building for the minimum 50 plus 20 psf (2394 plus 958 Pa) partition load, while the interior space is designed for a heavier load of 100 psf (4788 Pa). The rationale is that the 20 ft (60 m) deep space adjacent to the building exterior invariably consists of office space with light furniture, whereas heavily loaded areas such as storage and computer rooms are tucked away from the exterior closer to the building core.

The loads given in the codes have built-in empirical safety factors to account for the maximum possible loading conditions. They are given in the form of equivalent uniform and concentrated loads. Although there is a common understanding among engineers that the values given in the codes are rather conservative, invariably the design is done to the standards given in the codes applicable to the particular occupancy. However, when the actual layout and height of partitions are known, an attempt is made to justify a lower partition load. This is done more to justify the load-carrying capacity of an existing framing than to take advantage of load reduction in the design of a new facility.

Concentrated loads indicate possible single-load action at critical locations and are in addition to the uniform distributed load. It is perhaps obvious that the chances of having the full occupancy load simultaneously on every square foot of a large area in a building are next to zero. The larger the area under consideration, the less is the potential for having the full occupancy load. This probability aspect is taken into consideration in the codes by allowing the use of live load reduction factors. However recent failures of hung, long-span structures have brought to light the necessity of taking extra precautions, especially in the detailing of connections to account for extraordinary situations, such as people crowding because of ceremonies, parties, fire drills, etc.

Construction loads are caused by building construction activities requiring stockpiling of materials on relatively small areas. Construction materials, such as dry walls and glass lights for curtain wall, are usually stacked on each floor.

7.3   DESIGN LOAD COMBINATIONS

In accordance with IBC 1605.1, structural members of buildings and other structures must be designed to resist the load combinations of IBC 1605.2, 1605.3.1, or 1605.3.2. Load combinations that are specified in Chaps. 18 through 23 of the IBC, which contain provisions for soils and foundations, concrete, aluminum, masonry, steel, and wood, must also be considered. The structural elements identified in ASCE/SEI Chaps. 12, 13, and 15 must be designed for the load combinations with overstrength of ASCE/SEI 2.3.6 or 2.4.5.

IBC 1605.2 contains the load combinations that are to be used when strength design or load and resistance factor design is utilized. Load combinations using allowable stress design are given in IBC 1605.3. Both sets of combinations are covered in this section of this chapter. The combinations of IBC 1605.2 or 1605.3 can also be

used to check overall structural stability, including stability against overturning, sliding, and buoyancy (IBC 1605.1.1).

In ASCE/SEI 7-16, the load combinations with seismic load effects have been removed from ASCE/SEI Chap. 12 and placed in ASCE/SEI Chap. 2 in sections separate from the basic load combinations.

LOAD EFFECTS

The load effects that are included in the IBC and ASCE/SEI 7 load combinations are summarized in Table 7.2. More

Table 7.2   Load Effects

Notation	Load effect	Notes
D	Dead load	See IBC 1606
D_i	Weight of ice	See IBC 1614 and Chapter 10 of ASCE/SEI 7
E	Seismic load effect defined in ASCE/SEI 12.4.2	See IBC 1613, ASCE/SEI 12.4.2
E_m	Seismic load effect including overstrength defined in ASCE/SEI 12.4.3	See IBC 1613, ASCE/SEI 12.4.3
F	Load due to fluids with well-defined pressures and maximum heights	-
F_a	Flood load	See IBC 1612, Chapter 5 of ASCE/SEI 7
H	Load due to lateral earth pressures, groundwater pressure or pressure of bulk materials	See IBC 1610
L	Roof live load greater than 20 psf or less	See IBC 1607
L_r	Roof live load of 20 psf or less	See IBC 1607
R	Rain load	See IBC 1611
S	Snow load	See IBC 1608, Chapter 7 of ASCE/SEI 7
T	Cumulative effects of self-straining forces and effects	See ASCE/SEI 2.3.4 and 2.4.4
W	Load due to wind pressure	See IBC 1609, Chapters 26 to 31 of ASCE/SEI 7
W_i	Wind-on-ice load	See IBC 1614, Chapters 10 of ASCE/SEI 7

details on these load effects can be found in those documents, as well as in subsequent sections of this chapter.

LOAD COMBINATIONS USING STRENGTH DESIGN OR LOAD AND RESISTANCE FACTOR DESIGN

The basic load combinations where strength design or, equivalently, load and resistance factor design is used are given in IBC 1605.2 and summarized in Table 7.3. These equations establish the minimum required strength that needs to be provided in the members of a building or structure.

These load combinations apply only to strength limit states; serviceability limit states for deflection, vibration, drift, camber, expansion and contraction, and durability are given in Appendix C of ASCE/SEI 7.

Table 7.3   Summary of Load Combinations Using Strength Design or Load and Resistance Factor Design (IBC 1605.2)

IBC equation no.	Load combination
16-1	$1.4 (D+F)$
16-2	$1.2 (D+F) + 1.6 (L+H) + 0.5 (L_r$ or S or $R)$
16-3	$1.2 (D+F) + 1.6 (L_r$ or S or $R) + 1.6H + (f_1L$ or $0.5W)$
16-4	$1.2 (D+F) + 1.0W + f_1L + 1.6H + 0.5 (L_r$ or S or $R)$
16-5	$1.2 (D+F) + 1.0E + f_1L + 1.6H + f_2S$
16-6	$0.9D + 1.0W + 1.6H$
16-7	$0.9(D+F) + 1.0E + 1.6H$

$f_1 = 1$ for places of public assembly live loads in excess of 100 psf and for parking garages

$\quad = 0.5$ for other live loads

$f_2 = 0.7$ for roof configurations (such as sawtooth) that do not shed snow off the structure

$\quad = 0.2$ for other roof configurations

The seismic load effect, E, that is to be used in IBC Equation 16-5 (ASCE/SEI load combination 6) is equal to the following (see ASCE/SEI 12.4.2):

$$E = E_h + E_v$$

where $E_h =$ horizontal seismic load effect defined in ASCE/SEI 12.4.2.1 = Q_E

$\quad E_v =$ vertical seismic load effect defined in ASCE/SEI 12.4.2.2 = $0.2SDS$

$\quad \rho =$ redundancy factor defined in ASCE/SEI 12.3.4

$Q_E =$ effects of horizontal seismic forces applied to the structure

$S_{DS} =$ design spectral response acceleration parameter at short periods

Thus, IBC Equation 16-5 (ASCE/SEI load combination 6) can be written as follows:

$$(1.2 + 0.2 S_{DS})D + 1.2F + \rho Q_E + f_1L + 1.6H + f_2S$$

In IBC Equation 16-7 (ASCE/SEI load combination 7), the seismic load effect that is to be used is $E = E_h - E_v$ (see ASCE/SEI 12.4.2). Therefore, this equation can be written as follows:

$$(0.9 - 0.2 S_{DS})D + 0.9F + \rho Q_E + 1.6H$$

Refer to ASCE/SEI 7 12.4.2.2 for exceptions related to vertical seismic load effects.

Fluid load effects, F, occur in tanks and other storage containers due to stored liquid products. The stored liquid is generally considered to have characteristics of both a dead load and a live load. It is not a purely permanent load because the tank or storage container can go through cycles of being emptied and refilled. The fluid load effect is included in IBC Equations 16-1 through 16-5, where it adds to the effects from the other loads. It is also included in IBC Equation 16-7, where it counteracts the effects from uplift due to seismic load effects, E. Because the wind load effects, W, can be present when the tank is either full or empty, F is not incorporated in IBC Equation 16-6; that is, the maximum effects occur when F is set equal to zero.

The load combinations given in IBC 1605.2 are the same as those in ASCE/SEI 2.3.1 with some indicated in these sections.

According to IBC 1605.2.1, the load combinations of ASCE/SEI 2.3.2 are to be used where flood loads, F_a, must be considered in design (flood loads are determined by Chapter 5 of ASCE/SEI 7). In particular, the following modifications are to be made:

- V Zones or Coastal A Zones

$1.0W$ in IBC Equations 16-4 and 16-6 must be replaced by $1.0W + 2.0F_a$

- Noncoastal A Zones

$1.0W$ in IBC Equations 16-4 and 16-6 must be replaced by $0.5W + 1.0F_a$

Definitions of Coastal High Hazard Areas (V Zones) and Coastal A Zones are given in ASCE/SEI 5.2.

The load factors on F_a are based on a statistical analysis of flood loads associated with hydrostatic pressures, pressures due to steady overland flow, and hydrodynamic pressures due to waves, all of which are specified in ASCE/SEI 5.4.

In cases where self-straining loads, T, must be considered, their effects in combination with other loads are to be determined by ASCE/SEI 2.3.4 (IBC 1605.2.1). Instead of calculating self-straining effects based on upper bound values of this variable like other load effects, the most probable effect expected at any arbitrary point in time is used. More information, including load combinations that should be considered in design, is given in ASCE/SEI C2.3.4.

IBC 1605.2.1 requires that the load combinations of ASCE/SEI 2.3.3 be used where atmospheric ice loads must be considered in design. The following modifications to the load combinations must be made when a structure is subjected to atmospheric ice and wind-on-ice loads (atmospheric and wind-on-ice loads are determined by Chapter 10 of ASCE/SEI 7; see IBC 1614):

- $0.5(L_r$ or S or $R)$ in ASCE/SEI combination 2 (IBC Equation 16-2) must be replaced by $0.2D_i + 0.5S$
- $1.0W + 0.5(L_r$ or S or $R)$ in ASCE/SEI combination 4 (IBC Equation 16-4) must be replaced by $D_i + W_i + 0.5S$
- $1.0W$ in ASCE/SEI combination 5 (IBC Equation 16-6) must be replaced by $D_i + W_i$
- $1.0W + L + 0.5(L_r$ or S or $R)$ in ASCE/SEI combination 4 (IBC Equation 16-4) must be replaced by D_i

See ASCE/SEI C2.3.3 for more information on the load factors used in these equations. ASCE/SEI 2.3.5 provides information on how to develop strength design load criteria where no information on loads or load combinations is given in ASCE/SEI 7 or where performance-based design in accordance with ASCE/SEI 1.3.1.3 is being utilized. Detailed information on how to develop such load criteria that is consistent with the methodology used in ASCE/SEI 7 can be found in ASCE/SEI C2.3.5.

LOAD COMBINATIONS USING ALLOWABLE STRESS DESIGN

OVERVIEW

The basic load combinations where allowable stress design (working stress design) is used are given in IBC 1605.3. A set of basic load combinations is given in IBC 1605.3.1, and a set of alternative basic load combinations is given in IBC 1605.3.2. Both sets are examined below.

BASIC LOAD COMBINATIONS

The basic load combinations of IBC 1605.3.1 are summarized in Table 7.4.

Table 7.4   Summary of Basic Load Combinations Using Allowable Stress Design (IBC 1605.3.1)

Equation no.	Load combination
16-8	$D + F$
16-9	$D + H + F + L$
16-10	$D + H + F + (L_r$ or S or $R)$
16-11	$D + H + F + 0.75L + 0.75(L_r$ or S or $R)$
16-12	$D + H + F + (0.6W$ or $0.7E)$
16-13	$D + H + F + 0.75(0.6W) + 0.75L + 0.75(L_r$ or S or $R)$
16-14	$D + H + F + 0.75(0.7E) + 0.75L + 0.75S$
16-15	$0.6D + 0.6W + H$
16-16	$0.6(D + F) + 0.7E + H$

These load combinations apply to the design of all members in a structure and also provide for overall stability of a structure.

The seismic load effect, E, is a strength-level load. A factor of 0.7, which is approximately equal to 1/1.4, is applied to E in IBC Equations 16-12, 16-14, and 16-16 to convert the strength-level effects to service-level effects. Similarly, a factor of 0.6 is applied to W in IBC Equations 16-12, 16-13, and 16-15. The seismic load effect, E, that is to be used in IBC Equations 16-12 and 16-14 (ASCE/SEI load combinations 8 and 9) is equal to $E = E_h + E_v$.

Thus, IBC Equations 16-12 and 16-14 (ASCE/SEI load combinations 8 and 9) can be written as follows:

IBC Equation 16-12:

$$(1 + 0.14\, S_{DS})D + H + F + 0.7\rho Q_E$$

IBC Equation 16-14:

$$(1 + 0.105\, S_{DS})D + H + F + 0.525\rho Q_E + 0.75L + 0.75S$$

In IBC Equation 16-16 (ASCE/SEI load combination 10), the seismic load effect that is to be used is $E = E_h - E_v$. Therefore, this equation can be written as follows:

$$(0.6 - 0.14\, S_{DS})D + 0.6F + 0.7\rho Q_E + H$$

The exceptions applicable to these load combinations are given in ASCE/SEI 12.4.2.2 and in IBC 1605.3.1. See Reference 3 details of these exceptions.

Increases in allowable stresses that are given in the materials chapters of the IBC or in referenced standards are not permitted when the load combinations of IBC 1605.3.1 are used (IBC 1605.3.1.1).

According to IBC 1605.3.1.2, the load combinations of ASCE/SEI 2.4.2 are to be used where flood loads, F_a, must be considered in design. In particular, the following modifications are to be made:

- V Zones or Coastal A Zones

$1.5F_a$ must be added to the other loads in IBC Equations 16-12, 16-13, 16-14, and 16-15, and E is set equal to zero in IBC Equations 16-12 and 16-14.

- Noncoastal A Zones

$0.75F$ must be added to the other loads in IBC Equations 16-12, 16-13, 16-14, and 16-15, equal to zero in IBC Equations 16-12 and 16-14.

Where self-straining loads, T, must be considered in design, the provisions of ASCE/SEI 2.4.4 are to be used to determine the proper combination of T with other loads (IBC 1605.3.1.2). ASCE/SEI C2.4.4 provides load combinations for typical situations.

IBC 1605.3.1.2 requires that the load combinations of ASCE/SEI 2.4.3 be used where atmospheric ice loads must be considered in design. The following modifications to the load combinations must be made when a structure is subjected to atmospheric ice and wind-on-ice loads:

- $0.7D_i$ must be added to ASCE/SEI combination 2 (IBC Equation 16-9)
- (L_r or S or R) in ASCE/SEI combination 3 (IBC Equation 16-10) is to be replaced by $0.7D_i + 0.7W_i + S$
- $0.6W$ in ASCE/SEI combination 7 (IBC Equation 16-15) is to be replaced by $0.7D_i + 0.7W_i$
- $0.7D_i$ must be added to ASCE/SEI combination 1 (IBC Equation 16-8)

7.4   REQUIRED STRENGTH

According to the AISC specifications, the required strength must less or equal to the available strength for any element which is a function of the nominal strength and the corresponding resistance factor, if LRFD is used, or a safety factor, if ASD is used. The required strength is determined either with LRFD or ASD load combinations per the previous section. According the LRFD design philosophy, the available strength is calculated as the product of the resistance factor, φ, with the corresponding nominal strength such as φP_n, φM_n, φV_n, etc. According the ASD design philosophy, the available strength is calculated as the quotient of the nominal strength and the corresponding safety factor, Ω, such as P_n/Ω, M_n/Ω, V_n/Ω, etc. In some cases like in the proportioning of stability of bracing members that do not carry calculated forces, the specifications provide the required strength for such members.

7.5   METAL DECK SYSTEMS

The primary function of a floor system is to collect and distribute gravity loads to vertical elements such as columns and walls and occasionally to tension members such as hangers. This is accomplished by the out-of-plane bending action of the floor system. Another structural function of the floor system is to transmit the lateral loads to various lateral-load-resisting systems. This action is called the diaphragm action. In addition to these, the nonstructural functions of the floor system are: (i) to support nonstructural components such as finish materials in ceiling and floor, piping, ducts, wiring, lighting, and sprinklers; (ii) to provide protection from damage that may be caused by fire; and (iii) to provide resistance to transmission of sound.

7.6   OPEN-WEB JOIST SYSTEM

Open-web joists have been in use as floor and roof framing members since the early 1920s. The first joist used in 1923 was a warren-truss type with top and bottom chords of round bars and a web formed from single continuous bent bar. Since then, many types of joists have been developed, primarily to provide an economical floor and roof system. Their capacities and sizes are standardized and they are delivered to the site completely fabricated. They are made in standard depths from 8 to 30 in. (0.2 to 0.76 m) for the H series and 18 to 48 in. (0.46 to 1.2 m) for LH series, and 52 to 72 in. (1.32 to 1.84 m) for the DLH series. K series have been introduced which are more economical than H series. Open-web joists are manufactured utilizing hot-rolled or cold-formed steel. They are designed according to the standards set by Steel Joist Institute (SJI), primarily as simply supported uniformly loaded trusses. The top chord is assumed to be continuously braced by a floor or roof deck. The bottom chord is designed as an axially loaded tension member. The web members are designed to resist both the vertical and horizontal shear. Either a diagonal or horizontal bridging is used to stabilize the joist. The number of rows of bridging required is a function of the clear span of the joist and the chord size. Typically three to five rows may be required for office buildings.

The reasons for providing bridging in open-web joist construction are: (i) to provide stability for the joist during construction; (ii) to maintain alignment of the joist at the specified locations; (iii) to control the slenderness ratio of tension bottom chord to below 240, although this limitation is not essential for the structural integrity of tension members, it is adhered to in practice to maintain a certain minimum stiffness of the members to prevent undesirable lateral movements; and (iv) to provide lateral stability for compression diagonals.

Although other types of floor decks, such as precast concrete, gypsum, wood, or other materials capable of supporting the load can be used, in high-rise buildings concrete topping on metal deck is typical. The load tables published by the SJI are based on uniform load conditions. They can be used in selecting joists for gravity loads that can be expressed in terms of load per unit length of joist. Partitions, heavy pipes, and other elements running perpendicular to the joist, and mechanical units mounted on the joists should be treated as concentrated loads. In such cases the joist should be designed for the full combination uniform load and concentrated load.

In everyday engineering practice the engineer, relieved of the burden of designing the joists per se, selects the standard type and size required for the span and load conditions. Joists are manufactured with 50 ksi (344.75 MPa) steel for the chords and 36- or 50 ksi (248.2 and 344.75 MPa) steel for web members at the supplier's option. Chords of joists used for floor members are essentially parallel. Top chords are designed for uniform loads assuming either simple or continuous supports over the panel points.

It is a common practice to design the compression diagonals of an open-web joist with an effective length factor of $K = 1$ on the premise that the transverse flexural stiffness of the bottom tension chord coupled with the resistance of tension diagonals and bottom tension chord braces provides adequate stiffness. It behooves the engineer to verify that adequate bracing exists in unusual conditions with high stresses in the compression diagonals.

The roof joists should be checked for uplift capacity due to vertical suction forces induced by the wind loads. The net uplift force on roof joists is the gross uplift force due to wind, less the dead load of the roof system tributary to the joist, including the weight of the joist. If an adequate dead load exists in excess of the gross uplift load, no further consideration is necessary. If not, the design of the joist should be reviewed for reversal of stresses, such as compression of the bottom chord. Lateral bracing required to stabilize the bottom chord can be provided by: (i) additional bridging; (ii) rearranging the bridging provided for normal loading; or (iii) increasing the size of the bottom chord. The web system may also undergo stress reversal. Therefore, all components of the joist must be checked for stress reversal.

7.7   WIDE-FLANGE BEAMS

Wide-flange rolled structural shapes constitute the most common type of flexural members used in steel buildings to support transverse loads. These are used as beams, girders, and spandrels and can be designed as simply supported, fixed-ended, or partially fixed. They can be proportioned for shears and moments determined either by LRFD or ASD as discussed above.

7.7.1   Bending

Under the usual bracing conditions the plastic bending moment is $M_p = Z_x F_y$, where Z_x is the plastic section modulus and F_y is the yield strength of steel. Almost all beams are designed to bend about their major axis and are required to be laterally braced within certain intervals to resist lateral torsional buckling. Lateral supports for flexural members are required because the compression flange behaves in a manner similar to a column and tends to buckle in the absence of lateral supports. Metal decking welded to the beam top flange at sufficiently close intervals constitutes lateral bracing, as do cross beams framing into the sides of the beam if adequate connection is made to the beam top flange. With adequate lateral bracing, the design of a steel beam boils down to the selection of a bending member having a section modulus equal to or slightly larger than the calculated value, the design of bending of beams is performed according to chapter F of the AISC 360 specifications.

7.7.2   Shear

Steel beams used in floor framing are most often wide-flange shapes with an axis of symmetry about the plane of bending. They are designed for shear assuming: (i) the contribution of flanges to the shear capacity is negligible; and (ii) the parabolic variation of shear stress in the web is replaced by an average stress on the gross area of the web. With these assumptions, for purposes of calculating the shear stresses, symmetrical shapes can be reduced to an equivalent rectangle with dimensions $t_w d$, where t_w is the thickness of web and d is the beam depth. The nominal shear strength is calculated as $V_n = (0.6 F_y A_w) C_{v1}$ Chapter G of the AISC 360 specifications; where A_w is the shear area, $d t_w$, and C_{v1} is the web shear strength coefficient (shear buckling reduction coefficient). The design for beam shear is performed according to Chapter G of the AISC 360 specifications.

7.7.3   Serviceability

Serviceability requirements in steel and reinforced concrete buildings are based on human comfort criteria and are classified into two categories: deflection and floor vibration. The serviceability requirements were initially used in the early 1900s for brittle finishes and plaster. Although plaster has been replaced by other materials, the same serviceability limits are still being used in today's practice. Their impact is more on nonstructural components than structural elements. Typical gravity framing in steel structures is simply supported; the maximum deflection for a simply supported beam subjected to uniformly distributed loading is shown in Fig. 7.1.

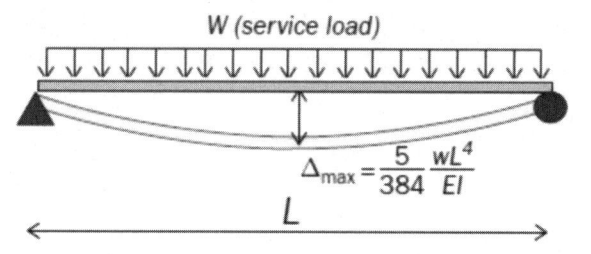

Figure 7.1   Maximum deflection of a simply supported beam subjected to distributed load.

To satisfy the deflection limit, steel beams are usually cambered upward to counteract dead load. For dead loads when the calculated deflections exceed 0.75 in. (19 mm). The maximum camber should be limited to about 2.5 in. (63.5 mm) because of floor-leveling problems during concreting operations. The allowable live-load deflection is a function of the type of ceiling suspended from the beams, and for plastered ceilings it should not exceed 1/360 of the span. Serviceability limits are based on service loads and are independent of the selected design method (ASD or LRFD). Serviceability may govern the design for long-span beams. In practice, the live load deflection is limited to $L/360$ and the dead + live load deflection is limited to $L/240$ for typical loading conditions and for shored construction; where L is the span of the beam and is twice the length for a cantilever. Table 7.5 shows the deflection limits for floor 7.

Table 7.5   Deflection Limits per ASCE 7-16

Member	Live load	Dead + live load	Snow or wind load
Floor members	L/360	L/240	—
Roof members not supporting ceiling	L/180	L/120	L/180
Roof members supporting nonplaster ceiling	L/240	L/180	L/240
Roof members supporting plaster ceiling	L/360	L/240	L/360

7.7.4   Concentrated Forces

Beams maybe subjected to concentrated loads that may cause the beam web to buckle. Several failure modes need to be checked per the AISC 360 specifications; these failure modes include:

a. Flange local bending
b. Web local yielding
c. Web local crippling
d. Web sidesway buckling
e. Web compression buckling
f. Web panel zone shear

7.8   COLUMNS

The design of a gravity column primarily carrying vertical loads reverts to selecting a steel section with a calculated design compressive strength, $\phi_c P_n$, given in Chapter E of the AISC 360 specifications, greater or equal to the required load, P_u. For buildings in excess of 40-stories or so, it becomes necessary to use built-up columns or cover-plated rolled-shapes. A few examples of these are given in Fig. 7.2.

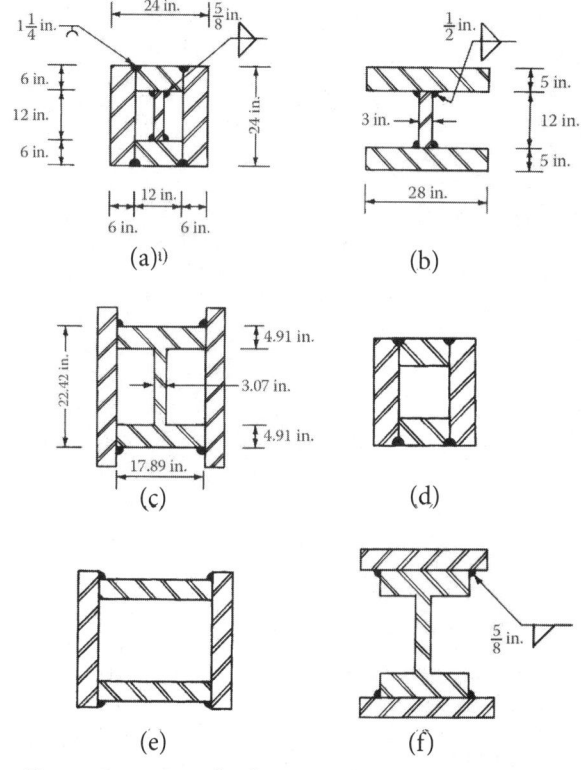

Figure 7.2   Example of gravity columns.

Chapter **8**

Gravity Systems in Concrete Buildings

Concrete floor systems are cast on temporary formwork or centering of lumber, plywood, or metal panels that are removed when the concrete has reached sufficient strength to support its own weight and construction loads. This procedure dictates that formwork be simple to erect and remove and be repetitive to achieve maximum economy. Although, in general, floor systems for high-rise buildings are the same as for their lower brethren, there are several characteristics which are unique to high-rise buildings. Floor systems in high-rise buildings are duplicated many times over, necessitating optimum solutions in their design because

1. savings that might otherwise be insignificant for a single floor may add up to a considerable sum because of large number of floors; and

2. dead load of floor system has a major impact on the design of vertical load-bearing elements such as walls and columns.

The desire to minimize dead loads is not unique to concrete floor systems but is of greater significance because the weight of concrete floor system tends to be heavier than steel floors and therefore has a greater impact on the design of vertical elements and foundation system. Another consideration is the impact of floor depth on the floor-to-floor height. Thus, it is important to design a floor system that is relatively lightweight without being too deep.

One of the necessary features of cast-in-place concrete construction is the large demand for job-site labor. Formwork, reinforcing steel, and the placing of concrete are the three aspects that demand the most labor. Repetition of formwork is a necessity for the economical construction of

cast-in-place high-rise buildings. Forms that can be used repetitively are "ganged" together and carried forward, or up the building, in large units, often combining column, beam, and slab elements in a large piece of formwork. Where the layout of the building frame is maintained constant for several stories, these result in economy of handling and placing costs. "Flying" form for flat work is another type that is used to place concrete for large floor areas. In a conventional construction method sometimes called the stick method, plywood sheets are nailed to a formed solid decking built on adjustable shores. When concrete has attained sufficient strength, the shores are removed and the plywood sheets are stripped, and if undamaged, they are stored for reuse for the next floor. This method of forming is labor-intensive and also time-consuming. If the floor-to-floor cycle is delayed one day because of formwork, the construction time of a tall building can be lengthened significantly. The flying form system typically shortens the construction time. In this system, floor forms are attached to a unit consisting of deck surface, adjustable jack, and supporting framework. For stripping, the form is lowered, slipped out of the slab, and shifted to the next floor as one rigid structure, resulting in economy of handling and placing.

8.1 FLOOR SYSTEMS

8.1.1 Flat Plates

Flat plates are two-way reinforced concrete floor systems with panels that have a long-to-short span ratio less than or equal to two (2) and the reactions from supported members are transferred in primarily two directions. The main flexural reinforcement runs predominantly in two orthogonal directions. Concrete slabs are often used to carry vertical loads directly to walls and columns without the use of beams or girders. Such a system called a flat plate (Fig. 8.1) is used where spans are not large and loads are not heavy as in apartment and hotel buildings.

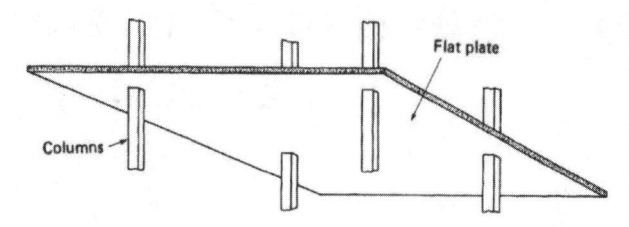

Figure 8.1   Flat plate system.

Flat plate is the term used for a slab system without any column flares or drop panels. Although column patterns are usually on a rectangular grid, flat plates can be used with irregularly spaced column layouts. They have been successfully built using columns on triangular grids and other variations. The minimum thickness of flat plates is determined according to ACI 318-19 section 8.3.1. The requirements for flat plates are summarized in Table 8.1 for Grade 60 reinforcement where ℓ_n is the clear span length in the long direction of the panel face-to-face of supports. Unless the calculated deflection limits of ACI 318-19 section 8.3.2 are satisfied, flat plates are not permitted to have a thickness less than 5 in. (ACI 318-19 section 8.3.1.1).

Table 8.1   Minimum Thickness of Flat Plates

Panel Location	Minimum Thickness, h
Exterior without edge beams	$\ell_n/30$
Exterior with edge beam*	$\ell_n/33$
Interior	$\ell_n/33$

*Edge beams with $\alpha_f \geq 0.8$, where α_f is calculated in accordance with Chapter 2 of ACI 318-19.

The term α_f in the footnote of Table 8.1 is the ratio of the flexural stiffness of the edge beam to the flexural stiffness of a width of slab bounded laterally by the centerline of the adjacent panel of the beam.

For 50 psf live load or less with relatively short spans, the thickness of flat plate will generally be controlled by the serviceability requirements in Table 8.1. For such spans and loads, the flexural reinforcement at the critical sections in the column and middle strips will usually be about the minimum amount specified in ACI 318-19

section 8.6.1 in such cases. Specifying the minimum slab thickness per the ACI provisions is recommended because specifying thicker slabs greater than the minimum needed for serviceability is economical; thicker slabs require more concrete without reduction in reinforcement. The serviceability requirements are independent of the concrete compressive strength, therefore, as 4000 psi concrete is the most economical, specifying concrete strength more than 4000 psi increases the cost without a reduction in slab thickness. For greater live load, in the range of 100 psf, and/or with relatively large spans, two-way shear requirements typically control the thickness of the flat plate instead of the serviceability requirements. Edge and corner columns are typically critical location due to the large unbalanced moments that can occur at these locations. The design of two-way slab systems is according to Chapter 8 of ACI 318-19, other section referred to in this chapter, and ACI 421 design guides.

8.1.2   Flat Slabs

Flat slab (Fig. 8.2) is also a two-way system of beamless construction but incorporates a thickened slab in the region of columns and walls. In addition to the thickened slab, the system can have flared columns. The thickened slab and column flares, referred to as drop panels, reduce shear and negative bending stresses around the columns. The dimensions of the drop panels must conform to ACI 318-19 section 8.2.4. The dimensions of drop panels per ACI are shown in Fig. 8.3; where ℓ_A and ℓ_B are the adjoining center-to-center span lengths. If these dimensional requirements are met, the shear strength around the column is reduced, the overall slab thickness, and the negative flexural reinforcement can be reduced.

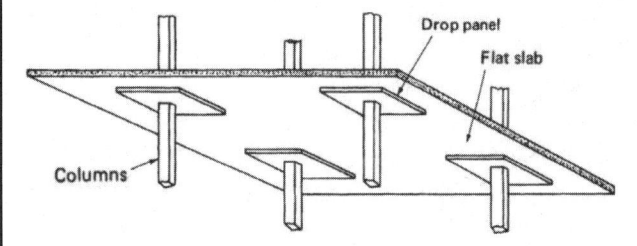

Figure 8.2   Flat slab system.

The minimum thickness of flat slabs is determined according to ACI 318-19 section 8.3.1. The requirements for flat slabs are summarized in Table 8.2 for Grade 60 reinforcement where ℓ_n is the clear span length in the long direction of the panel face-to-face of supports. Unless the calculated deflection limits of ACI 318-19 section 8.3.2 are satisfied, flat plates are not permitted to have a thickness

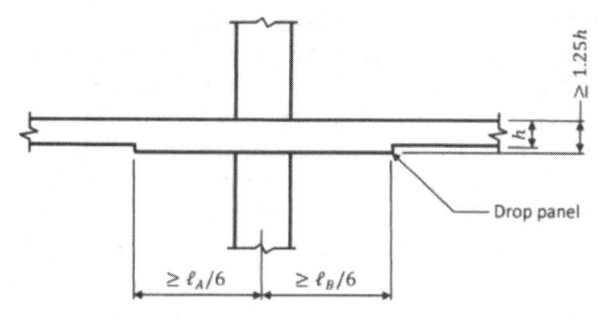

Figure 8.3 Dimensional requirements of drop panels per ACI 318-19.

less than 4 in. (ACI 318-19 section 8.3.1.1). Comparing Tables 8.1 and 8.2, it is evident that the minimum slab thickness for flat slabs is 10 percent less than that required for flat plates.

Table 8.2   Minimum Thickness of Flat Slabs

Panel Location	Minimum Thickness, h
Exterior without edge beams	$\ell_n/33$
Exterior with edge beam*	$\ell_n/36$
Interior	$\ell_n/36$

*Edge beams with $\alpha_f \geq 0.8$ where α_f is calculated in accordance with Chapter 2 of ACI 318-19

A flat slab system with a beamless ceiling has minimum structural depth, and allows for maximum flexibility in the arrangement of air-conditioning ducts and light fixtures. For apartments and hotels, the slab can serve as a finished ceiling for the floor below and therefore is more economical. Since there are no beams, the slab itself replaces the action of the beams by bending in two orthogonal directions. Therefore, the slab is designed to transmit the full load in each direction, carrying the entire load in shear and in bending.

The limitations of span are dependent upon the use of column capitals or drop panels. The criterion for the thickness of the slab is usually the punching shear around columns and the long-term deflection of the slab. In high-rise buildings, the slabs are generally 5 to 10 in. (127 to 254 mm) thick for spans of 15 to 25 ft (4.56 to 7.6 m). The design of two-way slab systems is according to Chapter 8 of ACI 318-19, other section referred to in this chapter, and ACI 421 design guides.

The term α_f in the footnote of Table 8.2 is the ratio of the flexural stiffness of the edge beam to the flexural stiffness of a width of slab bounded laterally by the centerline of the adjacent panel of the beam.

8.1.3   Two-Way Beam-Supported Slab System

In the two-way beam-supported slab system, the column-line beams are present on all four sides of a panel, as shown in Fig. 8.4. Similar to other two-way systems, the load is transferred in both orthogonal directions. The minimum thickness of this system is determined according to ACI 318-19 section 8.3.1. The requirements for two-way beam-supported slab are given in Table 8.3.1.2 of ACI 318-19, where ℓ_n is the clear span length in the long direction of the panel face-to-face of supports. In this table, α_{fm} is the average value of α_f for all beams on the edges of the panel. α_f is the ratio of the flexural stiffness of a beam to the flexural stiffness of a width of slab bounded laterally by the centerlines of adjacent panels, if any, on each side of the beam; and β is the ratio of the long-to-short dimensions of the clear spans in a panel.

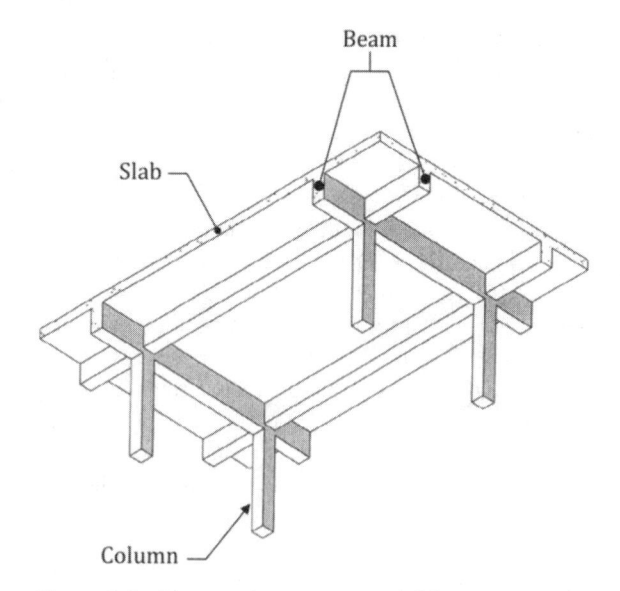

Figure 8.4   Two-way beam-supported slab system.

8.1.4   Waffle System

This system also called a two-way joist system (Fig. 8.5) is closely related to the flat slab system. To reduce the dead load of a solid slab construction, metal or fiberglass domes are used in the formwork in a rectilinear pattern as shown in Fig. 8.5. Domes are omitted near columns resulting in solid slabs to resist the high bending and shear stresses in these critical areas. The floor system is formed with domes that are 30, 41, and 52 in. wide, resulting in 3-, 4-, and 5-ft modules, respectively.

In contrast to a joist which carries loads in a one-way action, a waffle system carries the loads simultaneously in

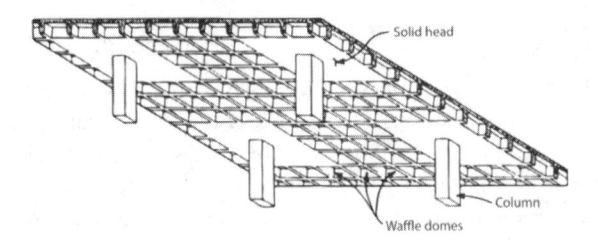

Figure 8.5 Waffle system.

two directions with a reinforced concrete slab cast integrally with the joists. The system is therefore more suitable for square bays than rectangular bays. The overall behavior of the system is similar to that of a flat slab. However, the waffle is more efficient for spans in the 30 to 40 ft (9.1 to 12.2 m) range because it has greater overall depth than a fat slab without the penalty of added dead weight.

The thickness of the slab is typically controlled by structural requirements or by fire resistance. Fire resistance controls the design in many cases. Normal-weight concrete slab with 4.5-in. thickness is commonly specified. Lightweight aggregate maybe advantageous in some cases. Since slab thickness is determined by fire resistance, and the joist width is set by industry standards, the only geometric variable is the joist depth. The standard joist width is 6 in. for a 3-ft module. Standard dome depths for the 3-ft module are 8, 10, 12, 14, 16, and 20 in. For design purposes, the two-way joist systems are considered as flat slabs with the solid heads around the columns acting as drop panels; therefore, the minimum thickness requirements in Table 8.2 are applicable. Since the system does not have a uniform cross-section, the deflection requirements can be determined by transforming the cross-section of the actual floor system into an equivalent section of uniform thickness. This is accomplished by determining a slab thickness that provides the same moment of inertia as the two-way joist section. Table 8.3 shows the equivalent thickness for the standard domes depth of a two-way joist system with 3-ft modules, a 6-in.-wide rib, and a 4.5-in.-thick slab [x].

Table 8.3 Equivalent Slab Thickness for Two-Way Joist Systems with 3-ft Module [x]

Dome Depth (in.)	Equivalent Slab Thickness (in.)
8	8.8
10	10.3
12	11.7
14	13.1
16	14.6
20	17.4

8.1.5 One-Way Concrete Ribbed Slabs

This system also referred to as a one-way joist system is one of the most popular systems for high-rise building construction in North America. The system is based on the well-founded premise that since concrete in a solid slab below the neutral axis is well in excess of that required for shear, much of it can be eliminated by forming voids. The resulting system shown in Fig. 8.6, has voids between the joists made with removable forms of steel, wood, plastic, or other material. The joists are designed as one-way T beams for the full-moment tributary to its width. However, in calculating the shear capacity, the ACI 318-19 section 9.8 allows for a 10 percent increase in the allowable shear stress of concrete. It is standard practice to use distribution ribs at approximately 10 ft (3.0 m) centers for spans greater than 20 ft (6 m). For maximum economy of formwork, the depth of beams and girders should be made the same as for joists.

8.1.6 Skip Joist System

In this system, instead of a standard 3 ft (0.91 m) spacing, joists are spaced at 5 ft or 6 ft 6 in. (1.52 and 1.98 m) spacings using 53 and 66 in. (1346 and 1676 mm) wide pans. The joists are designed as beams without using 10 percent increase in the shear capacity allowed for standard joists per section 9.8 ACI 318-19. Also the system is designed without distribution ribs thus requiring even less concrete. The spacing of vertical shores can be larger than for standard pan layout and consequently the formwork is more economical. Figure 8.7 shows a typical layout.

The fire rating requirement for floor systems is normally specified in the governing building codes. The most usual method of obtaining the rating is to provide a slab that will meet the code requirement without the use of sprayed-on fireproofing. In the United States, building codes, normally the slab thickness required for 2-h fire rating is 4 in. (101.6 mm) for lightweight concrete and $4\frac{1}{2}$ to 5 in. (114.3 to 127 mm) for normal-weight concrete, depending upon the type of aggregate. The thickness of slab required for fire rating is much in excess of that required by structural design. Therefore, the use of special pan forms with joists at 8 to 10 ft centers (2.43 to 3.04 m) should be investigated for large projects.

8.1.7 Band Beam System

This system shown in Fig. 8.8 uses wide shallow beams and should be investigated for buildings in which the floor-to-floor height is critical. Note that if lateral loads are resisted by parameter framing, it is not necessary to line band beams with either the exterior or interior columns. The slab in between the band beams is usually

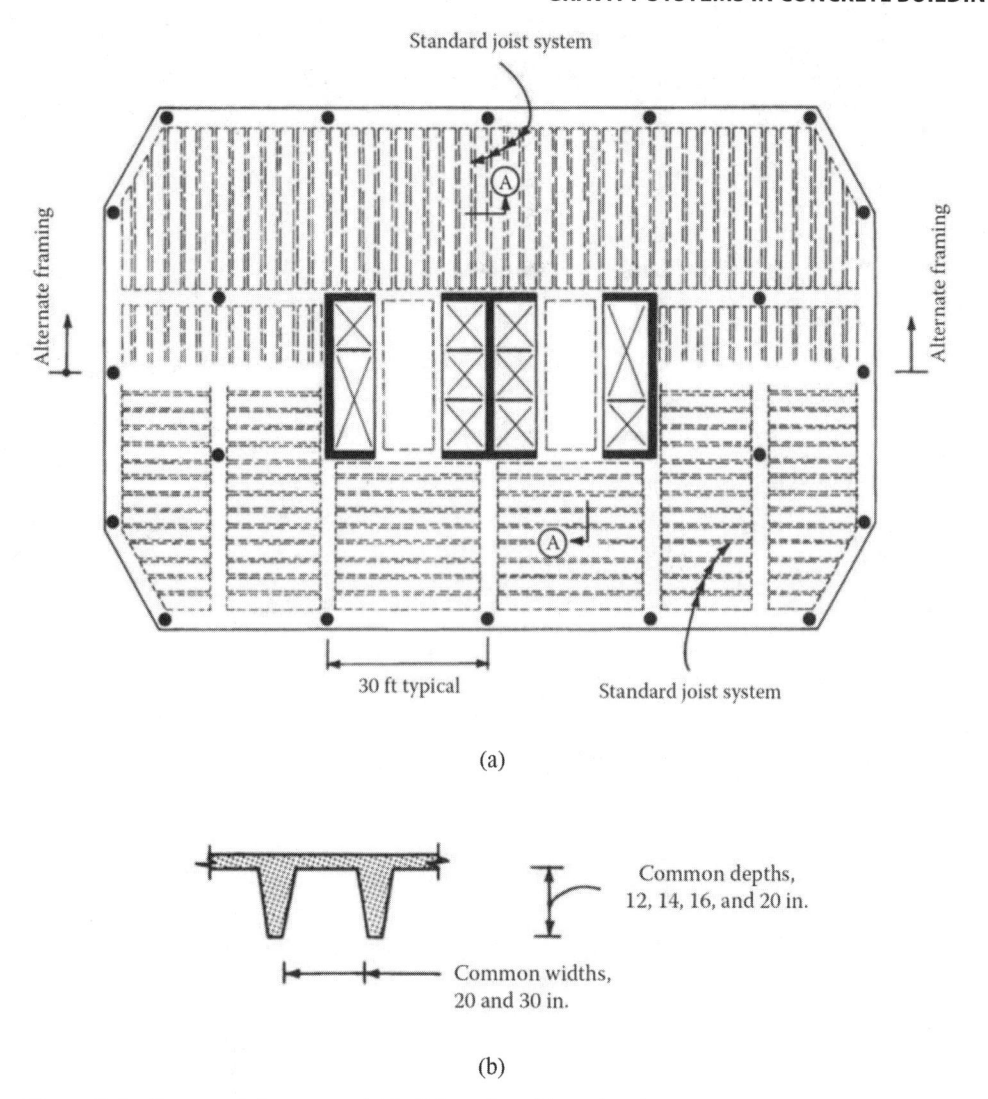

Figure 8.6   One-way joist system: (a) building plan; (b) section A.

designed as a bending-member with varying moment of inertia to take into account the increased thickness at the beams. A variation of the scheme uses standard or skip joists to span between the band beams.

8.1.8   Haunch Girder and Joist System
A floor-framing system with girders of constant depth crisscrossing the interior space between the core and the exterior often presents nonstructural problems because it limits the space available for the passage of air conditioning ducts. The haunch girder system widely accepted in certain parts of North America, achieves more headroom without making undue compromises in the structure. The basic system shown in Fig. 8.9 consists of a girder of

variable depth. The shallow depth at the center facilitates the passage of mechanical ducts and reduces the need to raise the floor-to-floor height. Two types of haunch girders are in vogue. One uses a tapered haunch (Fig. 8.9b) and the other a square haunch (Fig. 8.9c).

8.1.9   Beam and Slab System
This system consists of a continuous slab supported by beams generally spaced at 10 to 20 ft (3.04 to 6.08 m) on the center. The thickness of the slab is selected from structural considerations and is usually much in excess of that required for fire rating. The system has broad application and is generally limited by the depth available in the ceiling space for the beam stem. This system considered a

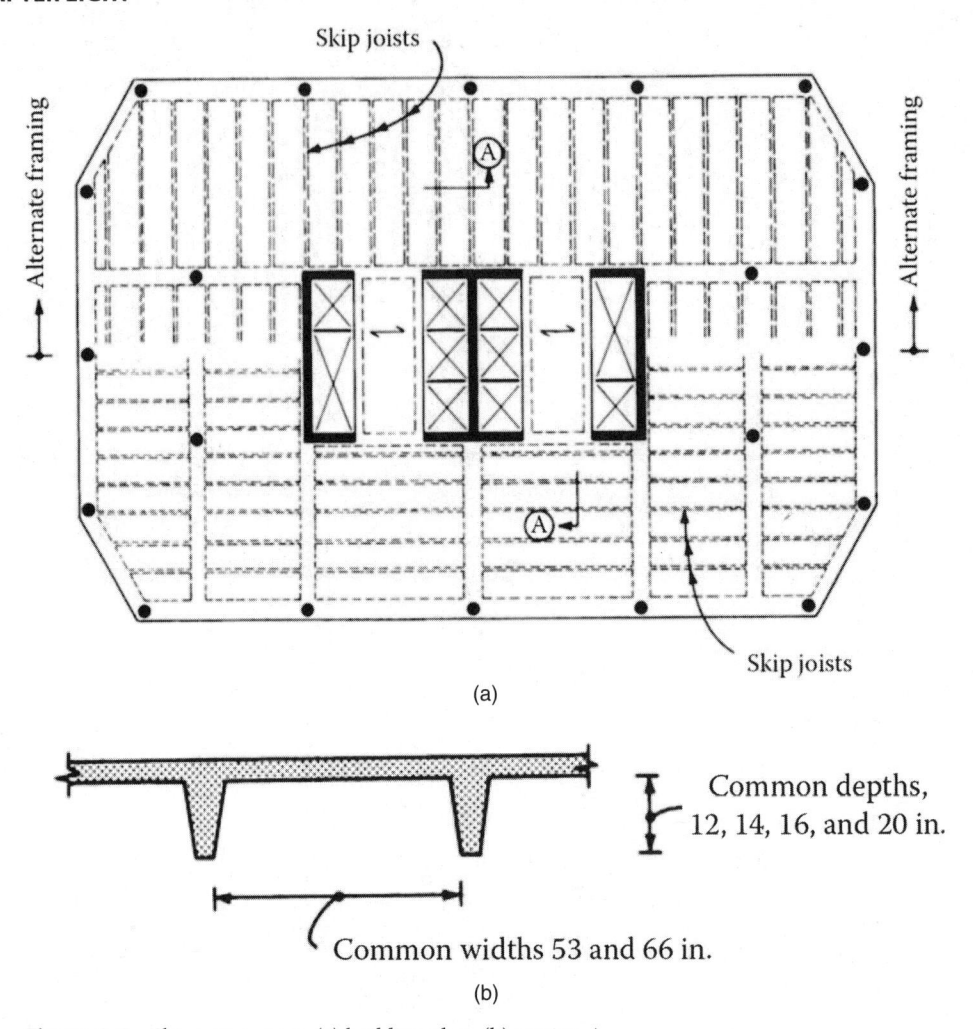

(a)

(b)

Figure 8.7   Skip joist system: (a) building plan; (b) section A.

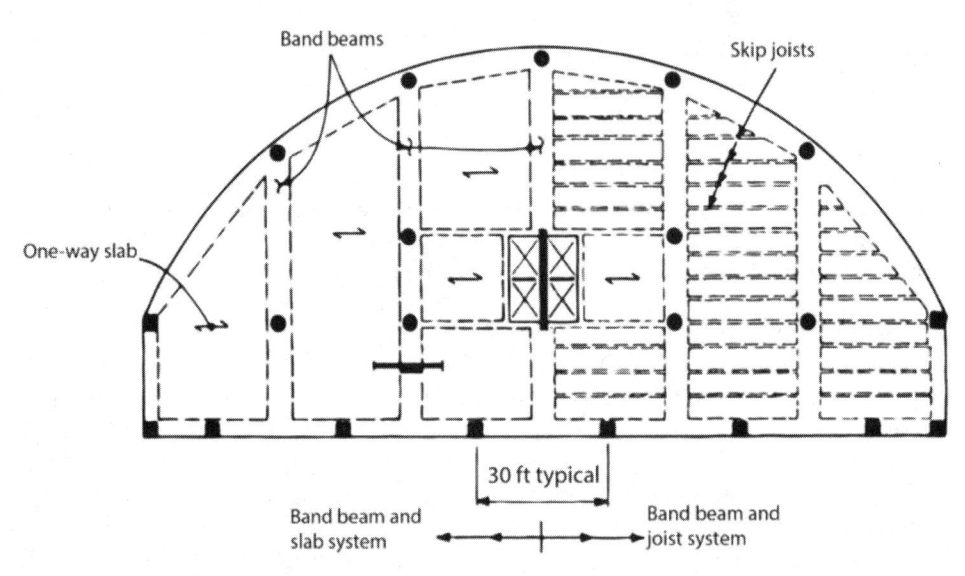

Figure 8.8   Band beam system.

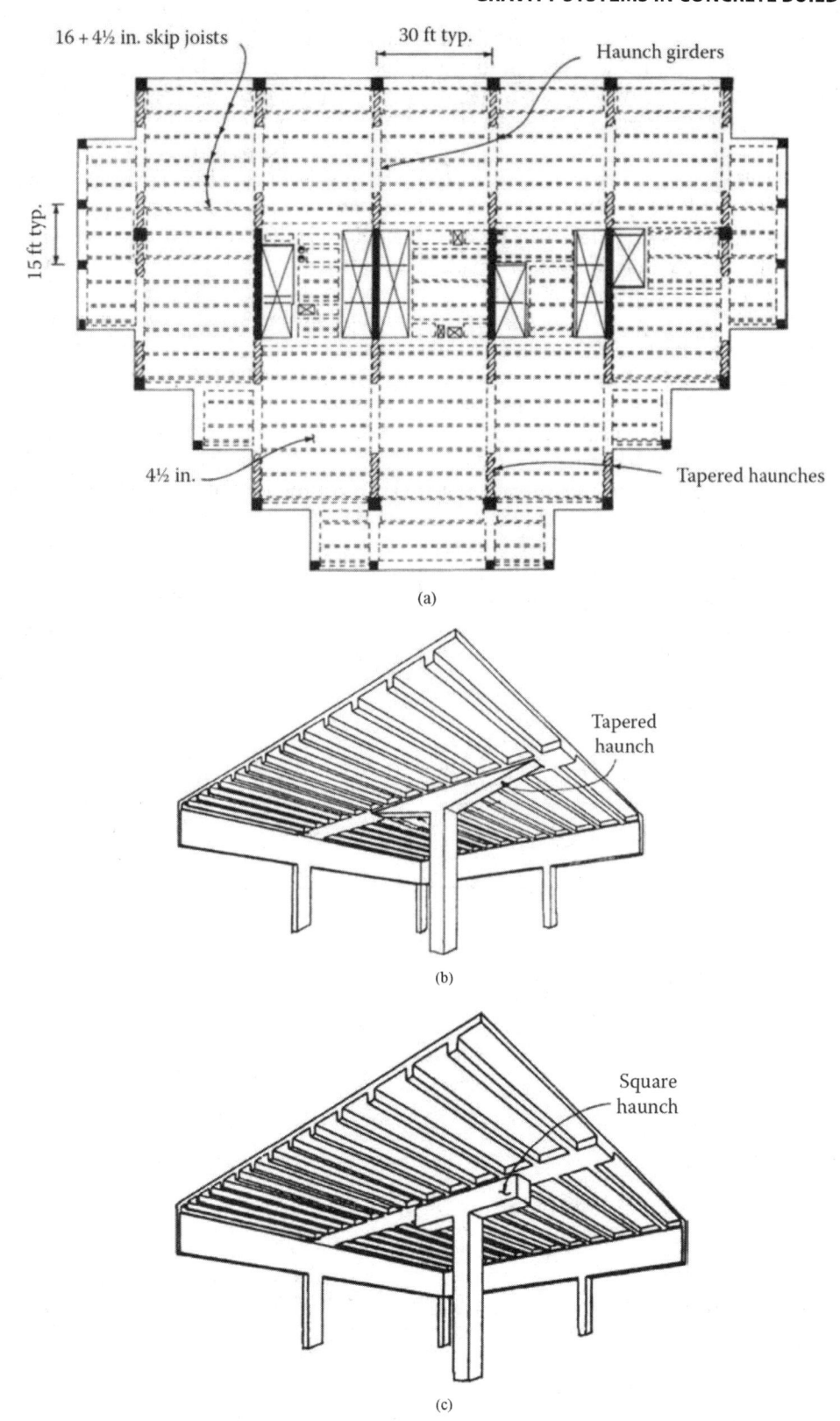

Figure 8.9   (a) Haunch girder and joist system. (b) Tapered haunch girder with skip joist system; (c) square haunch girder with skip joist system.

"heavy-duty" system is often used for framing nontypical floors such as ground floor and plaza levels, which are typically subjected to heavier superimposed loads due to landscape and other architectural features.

8.1.10   Design Examples

8.1.10.1   ONE-WAY SLAB-AND-BEAM SYSTEM

The analysis of the one-way slab system will be discussed to illustrate the simplifications normally made in a design office in analyzing the slabs and beams for gravity loads.

Figure 8.10 shows a uniformly loaded floor slab where the intermediate beams divide the floor slab into a series of one-way slabs. If a typical 1 ft width of the slab is cut out as a free-body in the longitudinal direction, it is evident that the slab will bend with a positive curvature between the beam stems, and a negative curvature at the supports. The deflected shape is similar to that of a continuous beam spanning between transverse girders, which act as simple supports. The assumption of simple support neglects the torsional stiffness of the beams supporting the slab. If the distance between the beams is the same, and if the slabs carry approximately the same load,

the torsional stiffness of the beams has little influence on the moments in the slab.

However, the exterior beams loaded from one side only, are twisted by the slab. The resistance to the end rotation of the slab offered by the exterior beam is dependent on the torsional stiffness of the beam. If the beam is small and its torsional stiffness low, a pin support may be assumed at the exterior edge of slab. On the other hand, if the exterior beam is large with a high torsional rigidity, it will apply a restraining moment to the slab. The beam, in turn, will be subjected to a torque requiring design for torsion.

Analysis by ACI Coefficients   Analysis by this method is limited to structures in which span lengths are approximately the same (with the maximum span difference between adjacent spans no more than 20 percent), the loads are uniformly distributed, and the live load does not exceed three times the dead load.

ACI values for positive and negative design moments are illustrated in Figs. 8.11, 8.12, and 8.13. In all expressions, l_n equals clear span for positive moment and shear, and the average of adjacent clear spans for a negative moment.

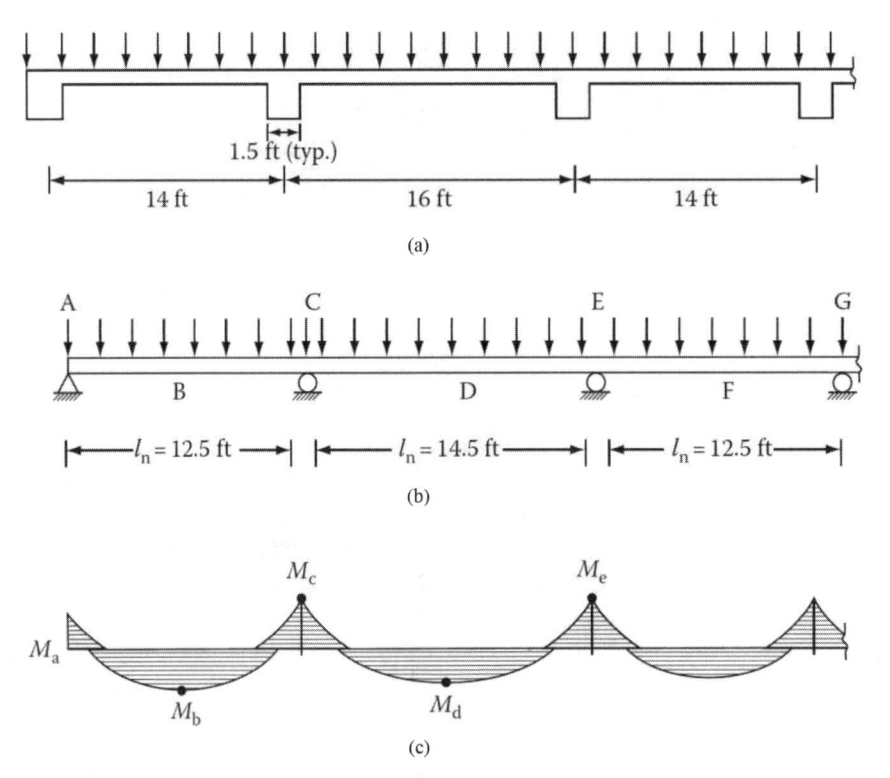

(a)

(b)

(c)

Figure 8.10   One-way slab example: (a) typical 1 ft strip; (b) slab modeled as a continuous beam; (c) design moments.

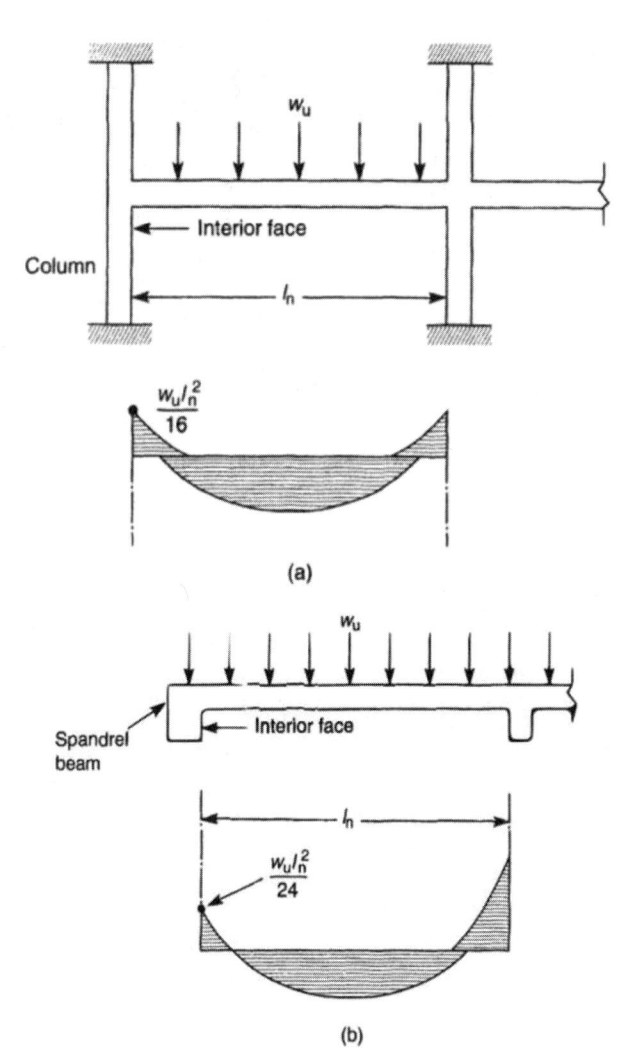

Figure 8.11 ACI negative moment coefficients at exterior supports: (a) slab built integrally with column; (b) slab built integrally with spandrel beam.

EXAMPLE

One-way mild steel reinforced slab.

Given

A one-way continuous slab as shown in Fig. 8.14.

$$f_c' = 4 \text{ ksi}, \quad f_y = 60 \text{ ksi}$$

$$\text{Ultimate load} = 0.32 \text{ kip/ft}$$

Solution

Use Table 8.4 to determine the minimum thickness of slab required to satisfy deflection limitations.

$$h_{\min} = \frac{l}{28} = \frac{12 \times 16}{28} = 6.86 \text{ in.} \quad \text{Use } 6\frac{1}{2} \text{ in. (165 mm)}$$

Analyze a 1 ft width of slab as a continuous beam using ACI coefficients to establish design moments for positive and negative steel (Fig. 8.14b).

$$M_a = \frac{w_u l_n^2}{24} = \frac{0.32 \times 12.5^2}{24} = 2.08 \text{ ft kips } (-\text{ve})$$

$$M_b = \frac{w_u l_n^2}{11} = \frac{0.32 \times 12.5^2}{11} = 4.55 \text{ ft kips } (+\text{ve})$$

At C, for negative moment, l_n is the average of adjacent clear spans: $l_n = (12.5 + 14.5)/2 = 13.5$ ft

$$M_c = \frac{w_u l_n^2}{10} = \frac{0.32 \times 13.5^2}{10} = 5.83 \text{ ft kips } (-\text{ve})$$

$$M_d = \frac{w_u l_n^2}{16} = \frac{0.32 \times 14.5^2}{16} = 4.21 \text{ ft kips } (+\text{ve})$$

$$M_e = \frac{w_u l_n^2}{11} = \frac{0.32 \times 14.5^2}{11} = 6.12 \text{ ft kips } (-\text{ve})$$

Compute reinforcement A_s per foot width of slab at critical sections. For example, at the second interior support, top steel must carry $M_e = 6.12$ kip ft. Note ACI code requires a minimum of $\frac{3}{4}$ in. cover for slab steel not exposed to weather or in contact with the ground.

We will use the "trial method" for determining the area of steel. In this method, the length of arm between the internal couple is estimated. Next, the tension force T is evaluated by using the basic relationship that applied moment equals the design strength; that is,

$$M_u = \phi T \times \text{arm}$$

$$T = \frac{M_u}{\phi \times \text{arm}}$$

where $\phi = 0.9$ for flexure, and $M_u =$ factored moment.

To start the procedure, the arm is estimated as $d - a/2$ by giving a value of $a = 0.15d$ where d is the effective depth. The appropriate area of steel A_s is computed by dividing T by f_y.

To get a more accurate value of A_s, the components of the internal couple are equated to provide a close estimate of the area A_c of the stress block. The compressive force C in the stress block is equated to the tension force T.

$$C = T$$
$$0.85 f_c' A_c = T$$
$$A_c = \frac{T}{0.85 f_c'}$$

Once A_c has been evaluated, locate the position of C, which is the centroid of A_c and recompute the arm between C and T. Using the improved value, find the

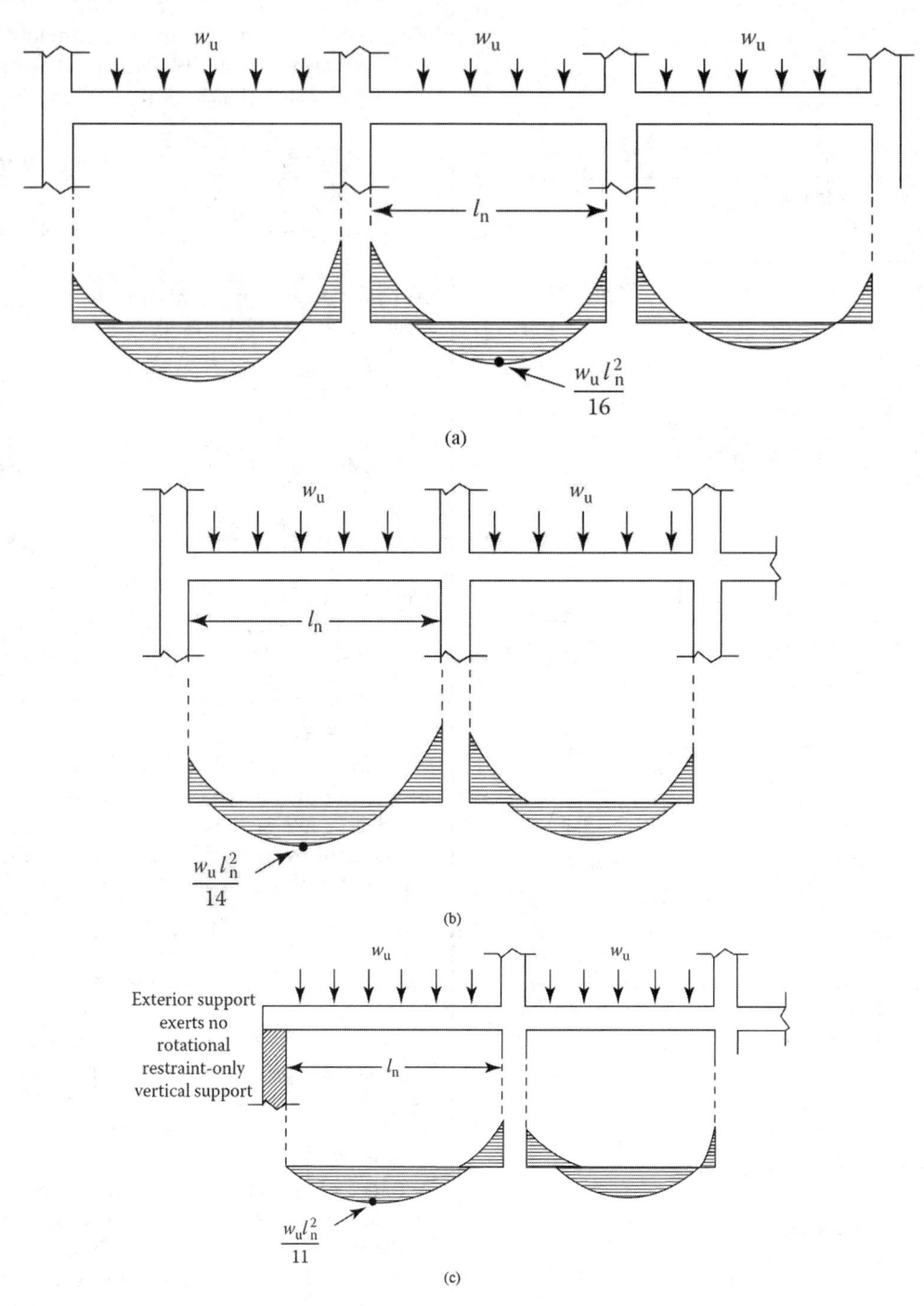

Figure 8.12 ACI positive moment coefficients: (a) interior span; (b) exterior span, discontinuous end integral with supports; (c) exterior span, discontinuous end unrestrained.

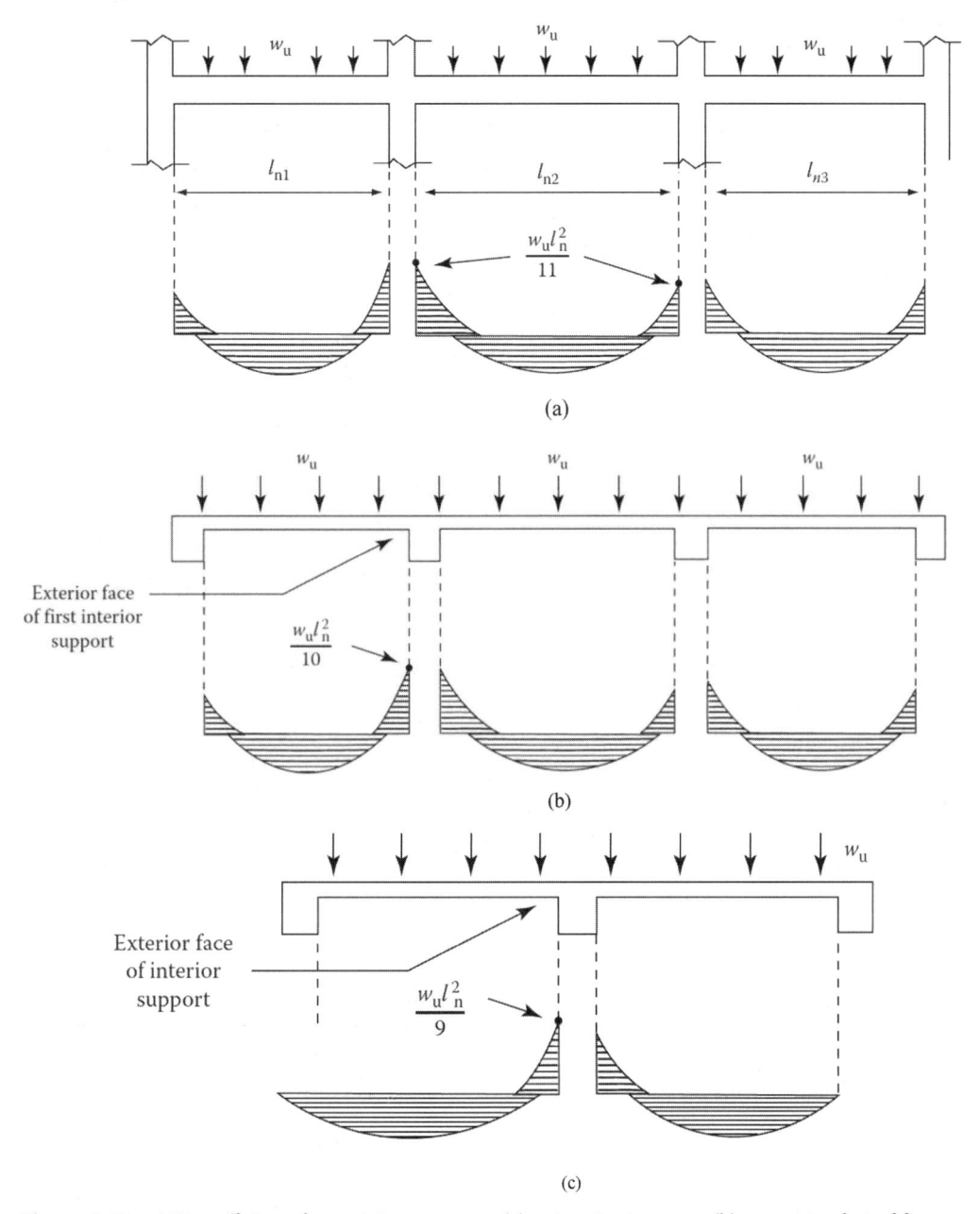

Figure 8.13 ACI coefficients for negative moments: (a) at interior supports; (b) at exterior face of first interior support, more than two spans; (c) at exterior face of first interior support, two spans.

second estimates of T and A_s. Regardless of the initial assumption for the arm, two cycles should be adequate for determining the required steel area.

For the example problem, the effective depth d for the slab is given by:

$$d = h - \left(0.75 - \frac{d_b}{2}\right) = 6.5 - (0.75 + 0.25) = 5.5 \text{ in.}$$

$$M_u = \phi T(d - a/2)$$

As a first trail, guess $a = 0.15d = 0.15 \times 5.5 = 0.83$ in.

$$6.12 \times 12 = 0.9T\left(5.5 - \frac{0.82}{2}\right) = 4.58T$$

$$T = 16.03 \text{ kips}$$

$$A_s = \frac{T}{f_y} = \frac{16.03}{60} = 0.27 \text{ in.}^2/\text{ft}$$

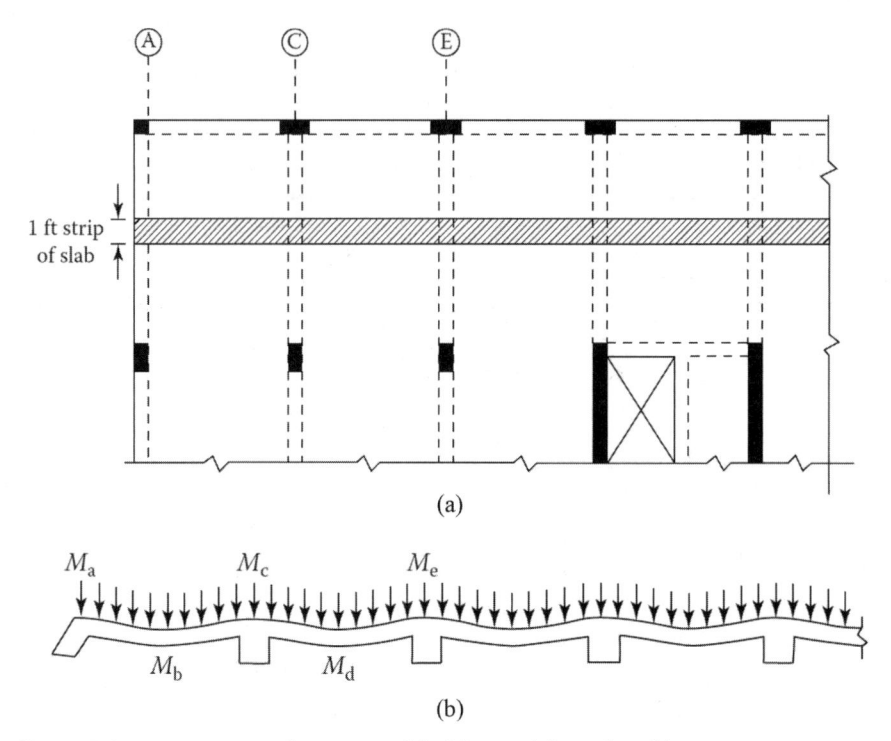

Figure 8.14 Design example, one-way slab: (a) partial floor plan; (b) section.

Table 8.4 Minimum Thickness of Beams or One-Way Slabs Unless Deflections Are Computed (ACI 318-19 Code Table 7.3.1.1 (One-Way Slabs) Table 9.3.1.1 (Beams))[†]
Members not supporting or attached to partitions or other construction likely to be damaged by large deflections

	Minimum Thickness h			
	Simply Supported	One End Continuous	Both Ends Continuous	Cantilever
Solid one-way slabs	$l/20$	$l/24$	$l/28$	$l/10$
Beams or ribbed one-way slabs	$l/16$	$l/18.5$	$l/21$	$l/8$

[†]Span length l in inches. Values in the table apply to normal-weight concrete reinforced with steel of $f_y = 60{,}000$ lb/in.[2]. For structural lightweight concrete with a unit weight between 90 and 120 lb/ft[3] multiply the table values by $1.65 - 0.005w_c$ respectively, but not less than 1.09; the unit weight w is in lb/ft[3]. For reinforcement having a yield point other than 60,000 lb/in.[2], multiply the table values by $0.4 + f_y/100{,}000$ with f_y in lb/in.[2]

Repeat the procedure using an arm based on the improved value of a. Equate $T = C$

$$16.03 = 0.85 f_c' A_c = 0.85 \times 4 \times a \times 12$$

$$a = 0.39 \text{ in.}$$

$$\text{Arm} = d - \frac{a}{2} = 5.5 - \frac{0.39}{2} = 5.31 \text{ in.}$$

$$T = \frac{M_u}{\phi\left(d - \dfrac{a}{2}\right)} = \frac{6.12 \times 12}{0.9 \times 5.31} = 15.37 \text{ kips}$$

$$A_s = \frac{15.37}{60} = 0.26 \text{ in.}^2$$

Check for temperature steel $= 0.0018 A_g$

$$= 0.0018 \times 6.5 \times 12 = 0.14 \text{ in.}^2/\text{ft}$$

Determine spacing of slab reinforcement to supply 0.26 in.²/ft

Using #4 rebars, $s = \dfrac{0.20}{0.26} \times 12 = 9.23$ in. Say 9 in.

Using #5 rebars, $s = \dfrac{0.31}{0.2} \times 12 = 14.31$ in. Say 14 in.

Use #4 @ 9 top at support e. Also by ACI code, the maximum spacing of flexural reinforcement should not exceed

18 in. or three times the slab thickness.

9 in. < 3 (6.5 in.) = 19.5 in. 9″ spacing is OK.

A schematic placement diagram of top steel is shown in Fig. 8.15.

8.1.10.2 T-Beam Design

Design for Flexure Design reinforcement for a simply supported T-beam spanning 30 ft as shown in Fig. 8.16.

$$W_u = 0.32 \times \frac{(16 \times 14)}{2} = 4.80 \text{ kip/ft}$$

Use $W_u = 5.0$ k/ft including the self-weight of beam. The minimum depth of beam to control deflections from Table 8.1 is

$$h_{min} = \frac{l}{16} = \frac{30 \times 12}{16} = 22.5 \text{ in.} \quad \text{Use 22.5 in.}$$

Try $b_w = 18$ in. The width must be adequate to carry shear and allow for proper spacing between reinforcing bars.

The effective width of T beam b_{eff} is the smallest of:
1. One-fourth of the beam span:

$$\frac{30}{4} = 7.5 \text{ ft} = 90 \text{ in.} \quad \text{(controls)}$$

2. Eight times the slab thickness on each side of the stem plus the stem thickness:

$$8 \times 6.5 \times 2 + 18 = 122 \text{ in.}$$

3. Center-to-center spacing of panel:

$$\frac{(16 \times 14)}{2} \times 12 = 180 \text{ in.}$$

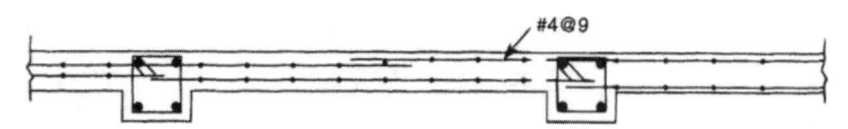

Figure 8.15 Slab reinforcement.

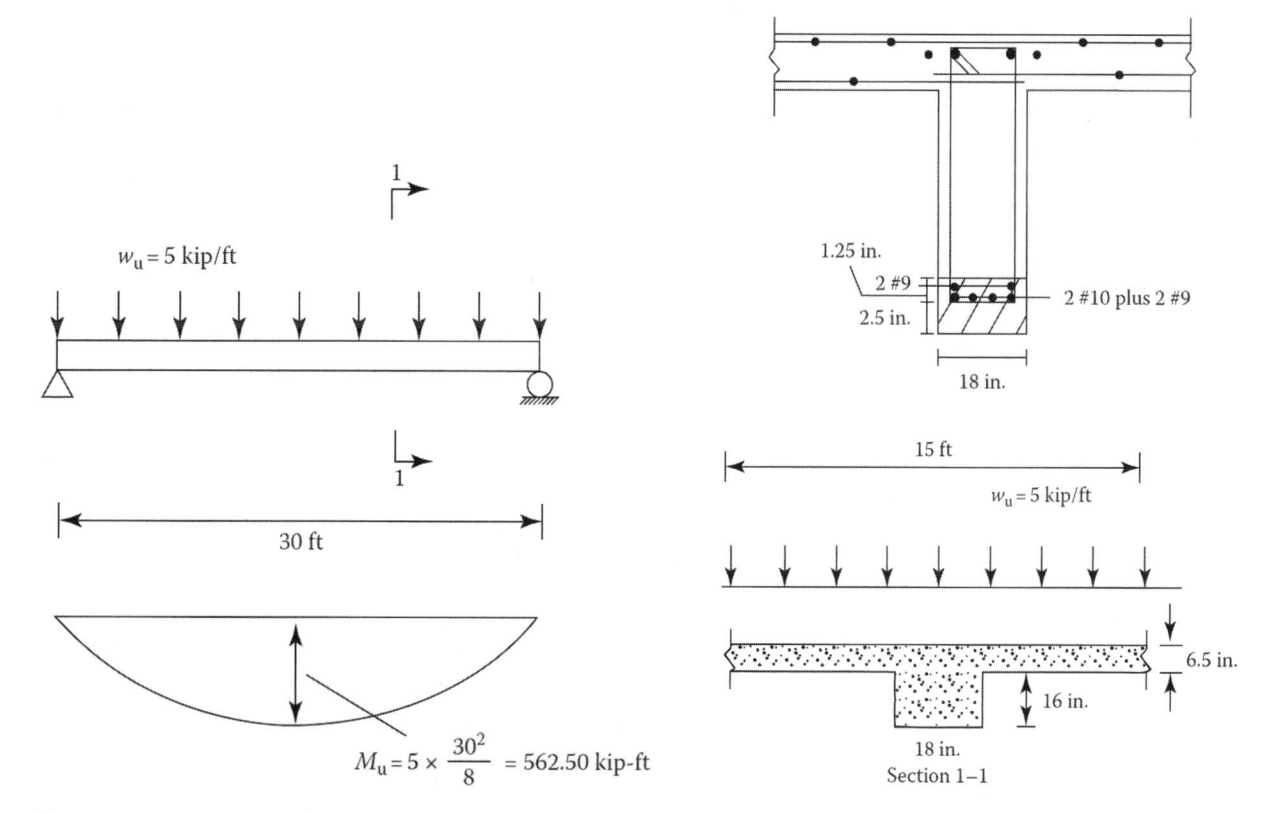

$$M_u = 5 \times \frac{30^2}{8} = 562.50 \text{ kip-ft}$$

Figure 8.16 Design example, simple beam.

Select the flexural steel A_s for $M_u = 562.50$ ft/kips using the trial method.

$$\text{Estimate } d = h - 2.6 = 22.5 - 2.6 = 19.9 \text{ in.}$$

$$M_u = \phi T \left(d - \frac{a}{2} \right) \quad \text{Guess } a = 0.8 \text{ in.}$$

$$562.50 \times 12 = 0.9 T \left(19.9 - \frac{0.8}{2} \right) = 17.557$$

$$T = 384.62$$

$$A_s = \frac{T}{f_y} = \frac{384.62}{60} = 6.41 \text{ in.}^2$$

Check value of a

$$384.62 = T = C = a b_{\text{eff}} (0.85 f_c')$$
$$= a(90)(0.85)4$$
$$a = 1.26$$

Repeat the procedure using an arm based on the improved value of a.

$$M_u = 562.50 \times 12 = 0.9 T \left(19.9 - \frac{1.26}{2} \right) = 17.34 T$$

$$T = \frac{562.50 \times 12}{17.34} = 389.2 \text{ kips}$$

Check value of a

$$389.2 = T = C = a \times 90 \times 0.85 \times 4$$

$$a = 1.27 \text{ in.}$$

$$A_s = \frac{T}{f_y} = \frac{389.2}{60} = 6.49 \text{ in.}^2$$

$$A_{s, \min} = \frac{200 \, b_w d}{f_y} = \frac{200 \times 18 \times 19.9}{60,000} = 1.19 \text{ in.}^2 < 6.49 \text{ in.}^2$$

Since 6.49 in.^{2} controls, use two #10 and four #9 bars

$$A_{s, \text{supplied}} = 6.54 \text{ in.}^2$$

The spacing of the reinforcement, s, closest to a tension face must not exceed:

$$s = 15 \left(\frac{40,000}{f_s} \right) - 2.5 C_s \quad \text{(ACI 318-19 equation x)}$$

And may not be greater than $12 \left(\dfrac{40,000}{f_s} \right)$

Where C_c is the least distance from surface of reinforcement or presenting steel to the tension face (in.). using

$f_s = 36$ ksi and $C_c = 2.0$ in., the minimum spacing is given by:

$$s = 15 \left(\frac{40,000}{36,000} \right) - 2.5(2) = 11.67 \text{ in. Controls}$$

$$s \leq 12 \left(\frac{40,000}{36,000} \right) = 13.3 \text{ in.}$$

The reinforcement provided is:

$$s = \frac{1}{3} \left[18 - 2 \left(2 + 0.5 + 1.\frac{128}{2} \right) \right] \cong 5 \text{ in.} < 11.67 \text{ in.}$$

T-Beam Design for Shear   The ACI procedure for shear design is an empirical method based on the assumption that a shear failure occurs on a vertical plane when shear force at that section due to factored service loads exceeds the concrete's fictitious vertical shear strength. The shear stress equation by strength of materials is given by:

$$\upsilon = \frac{VQ}{Ib}$$

where υ = shear stress at a cross-section under consideration
  V = shear force on the member
  I = moment of inertia of the cross-section about centroidal axis
  b = thickness of member at which v is computed
  Q = moment about centroidal axis of area between section at which v is computed and outside surface of member

This expression is not directly applicable to reinforced concrete beams. The ACI, therefore, uses a simple equation to calculate the average stress on the cross-section:

$$\upsilon_c = \frac{V}{b_w d}$$

where υ_c = nominal shear stress
  V = shear force
  b_w = width of beam web
  d = distance between the centroid of tension steel and compression surface

To emphasize that υ_c is not an actual stress but merely a measure of the shear stress intensity, it is termed a nominal shear stress.

For nonseismic design, the ACI 318-19 section 22.5 code assumes that concrete can carry some shear regardless of the magnitude of the external shearing force and that shear reinforcement must carry the remainder. Thus

$$V_u = \phi V_n = \phi (V_c + V_s)$$

where V_u = factored or ultimate shear force

V_n = nominal shear strength provided by concrete and reinforcement

V_c = nominal shear strength provided by concrete

V_s = nominal reinforcement provided by shear reinforcement

ϕ = strength reduction factor = 0.75 for shear and torsion, ACI 318-19 Table 21.2.1

Shear design computations can be made in terms of shear force V or in terms of unit shear stress.

We now calculate the stirrups using the strength equation in terms of shear forces.

$$V_u = \phi(V_c + V_s)$$

For the example problem,

$$V_u = 70 \text{ kips}$$

$$V_c = 2\sqrt{f_c'}b_w d$$

$$= 2\sqrt{4000}(18)(19.9) = 45 \text{ kips}$$

$$\phi\frac{V_c}{2} = 0.75\left(\frac{45}{2}\right) = 16.88 \text{ kips}$$

Since $V_u = 70$ kips exceeds $\phi\frac{V_c}{2}$ stirrups are required.

$$V_s = \frac{V_u}{\phi} - V_c$$

$$= \frac{70}{0.75} - 45 = 48.33 \text{ kips}$$

Spacing for two-legged #4 stirrups,

$$s = \frac{A_v f_y d}{V_s}$$

$$= \frac{2(0.2)(60)(19.9)}{48.33} = 9.9 \text{ in.}$$

Since V_s is less than $4\sqrt{f_c'}b_w d = 90$ kips

$$s = \frac{d}{2} = \frac{19.9}{2} = 9.95 \text{ say } 10 \text{ in.}$$

If $V_s \geq 4\sqrt{f_c'}b_w d$ the maximum spacing would have been $\frac{d}{4}$ but not to exceed 12 in.

The spacing s should not be less than $\frac{d}{2} = \frac{19.9}{2} = 9.95$ say 9 in., and the minimum area

$$A_{v,\min} = \frac{50b_w s}{f_y} = \frac{50(18)(9)}{60,000} = 0.135 \text{ in.}^2$$

$$A_{v,\text{provided}} = 0.4 \text{ in.}^2 > 0.135 \text{ in.}^2$$

Use #4 stirrups at 9 in. near the supports. A reduced spacing of stirrups equal to d may be used within the span where the calculated shear force is:

$$V_u \leq \frac{\phi V_c}{2} = 0.75\left(\frac{45}{2}\right) = 16.88 \text{ kips}$$

8.1.10.3 ANALYSIS OF TWO-WAY SLABS

Although two-way slabs may be designed by any method that satisfies the strength and serviceability requirements of the ACI code and design guides, most usually they are designed by the "equivalent-frame method" using computers. In this section, however only the Direct Design Method is discussed. The provisions of the Direct Design Method are given in ACI 421.3R-15 design guide.

In this method the simple beam moment in each span of a two-way system is distributed as positive and negative moments at midspan and at supports. Since stiffness considerations, except at the exterior supports, are not required, computations are simple and can be carried out rapidly.

Three steps are required for the determination of positive and negative design moments.

1. Determine simple beam moment:

$$M_0 = \frac{w_u l_2 l_n^2}{8}$$

where M_o = simple beam moment

w_u = ultimate uniform load

l_2 = slab width between columns transverse to the span under consideration

l_n = clear span between face of columns or capitals

2. For interior spans divide M_o into M_c and M_s, midspan and support moments as shown in Fig. 8.17. For exterior spans use Fig. 8.18 to divide M_0 into moments M_1, M_2, and M_3.

3. Distribute M_c and M_s in the transverse direction across the width between column and middle strips by using Tables 8.5 and 8.6 which give portion of moment in the column strips. The remainder is assigned to the middle strip.

Observe in Fig. 8.17, that for an interior span, the positive moment M_c at midspan equals $0.35M_0$, and the negative moment M_s at each support equal $0.65M_0$, values which are approximately the same as for a uniformly loaded fixed-end beam. These values are based on the assumption that an interior joint undergoes no significant rotation, a condition that is assured by the ACI code restrictions that limit: (i) the difference between adjacent span lengths to one-third of the longer span; and (ii) the ratio of live load to the dead load to 3.

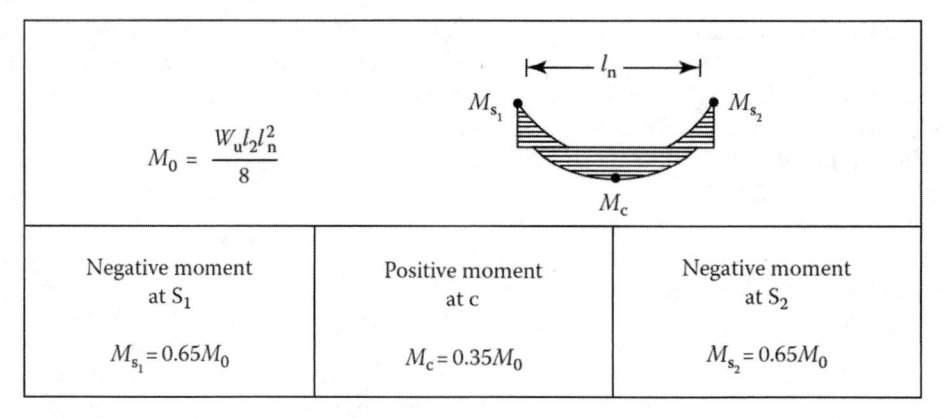

Figure 8.17   Assignment of moments at critical sections: interior span.

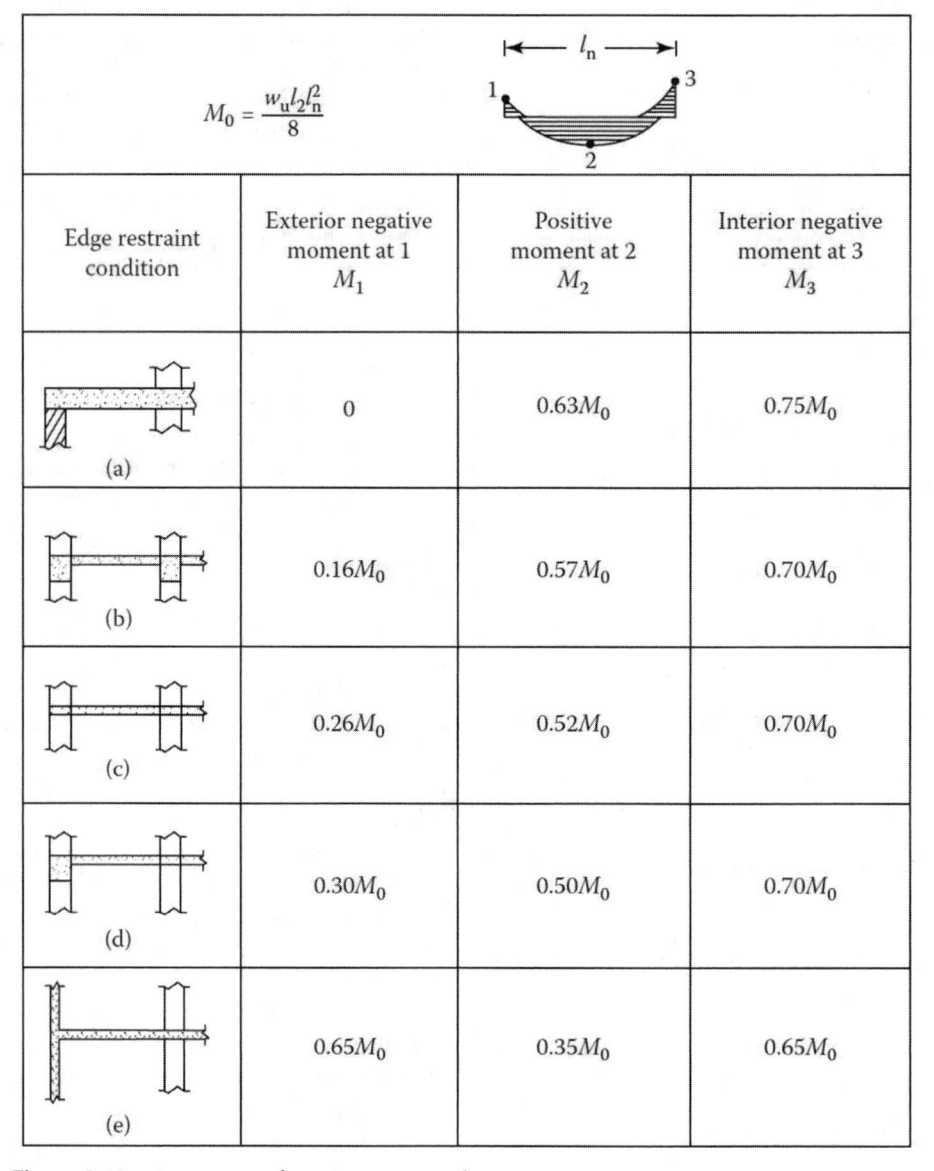

Figure 8.18   Assignment of moments to critical sections: exterior span.

Table 8.5   Portion of Positive Moment to Column Strip, Interior Span

$\alpha_1 \dfrac{l_2}{l_1}$	l_2/l_1		
	0.5	1.0	2.0
0	0.60	0.60	0.60
≥1.0	0.90	0.75	0.45

Table 8.6   Portion of Negative Moment to Column Strip at an Interior Support

$\alpha_1 \dfrac{l_2}{l_1}$	l_2/l_1		
	0.5	1.0	2.0
0	0.75	0.75	0.75
≥1	0.90	0.75	0.45

The final step is to distribute the positive and negative moments in the transverse direction between column strip and middle strips. The percentage of distribution factors are tabulated (Tables 8.5 and 8.6) for three values (0.5, 1, 2) of panel dimensions $\dfrac{l_2}{l_1}$, and two values (0 and 1) of $\alpha_1 \dfrac{l_1}{l_2}$. For intermediate values linear interpolation may be used. Table 8.5 is for interior spans while Table 8.6 is for exterior spans. For exterior spans the distribution of moment is influenced by the torsional stiffness of spandrel beam. Therefore, an additional parameter β_t, the ratio of the torsional stiffness of spandrel beam to flexural stiffness of slab is given in Table 8.7.

Table 8.7   Portion of Negative Moment to Column Strip at an Exterior Support

$\alpha_1 \dfrac{l_2}{l_1}$	β_t	l_2/l_1		
		0.5	1.0	2.0
0	0	1.0	1.0	1.0
0	≥2.5	0.75	0.75	0.75
≥1	0	1.0	1.0	1.0
≥1	≥2.5	0.90	0.75	0.45

For exterior spans the distribution of total negative and positive moments between columns strips and middle strips is given in terms of the ratio $\dfrac{l_2}{l_1}$, and the relative stiffness of the beam and slab, and the degree of torsional restraint provided by the edge beam. The parameter $\alpha = \dfrac{E_{cb} I_b}{E_{cs} I_s}$ is used to define the relative stiffness of the beam and slab spanning in either direction. E_{cb} and E_{cs} are the moduli of elasticity of the beam and slab and I_b and I_s are the moments of inertia. Subscripted parameters α_1 and α_2 are used to identify α for the directions of l_1 and l_2 respectively.

The parameter β_t in Table 8.7 defines the torsional restraint of the edge beam. If there is no edge beam, that is $\beta_t = 0$, all of the exterior moment at 1 (Fig. 8.18) is apportioned to the column strip. For $\beta_t \geq 2.5$, that is, for very stiff edge beams 75 percent of moment at 1 is assigned to the column strip. For values in-between, linear interpolation is permitted. In most practical designs, distributing 100 percent of the moment at 1 to the column strip while using minimum slab reinforcement in the middle strip yields acceptable results.

DESIGN EXAMPLE

Given

A two-way slab system as shown in Fig. 8.19.

$$w_d = 155 \text{ psf} \qquad w_d = 100 \text{ psf}$$

Determine the slab depth and design moments by the direct method at all critical sections in the exterior and interior span along column line B.

Solution

From Table 8.8, for $f_y = 60$ ksi, and for slabs without drop panels, the minimum thickness of slab is determined to be $\dfrac{l_n}{33}$ for the interior panels. The same thickness is used for the exterior panels since the system has beams between the columns along the exterior edges.

For example, l_n the clear span in the long direction $= 24 - 2 = 22$ ft. The minimum thickness

$$h = \frac{22(12)}{33} = 8 \text{ in.}$$

Interior span

$$w_u = 1.2(0.155) + 1.6(0.1) = 0.346 \text{ ksf}$$

$$M_0 = \frac{w_u l_2 l_n^2}{8} = \frac{0.346(20)(22)^2}{8} = 418.7 \text{ kip-ft}$$

Divide M_o between sections of positive and negative moments.

At midspan:

$$M_c = 0.35 M_o$$
$$= 0.35 \times 418.7 = 146.5 \text{ ft kips}$$

At supports:

$$M_s = 0.65 M_o$$
$$= 0.65 \times 418.7 = 272.2 \text{ ft kips}$$

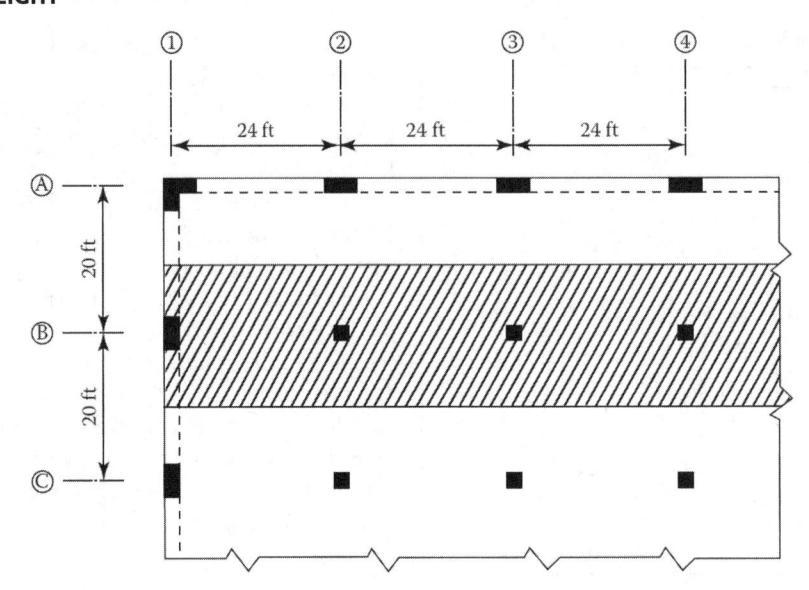

Figure 8.19 Design example, two-way slab.

Table 8.8 Minimum Thickness of Nonprestressed Two-Way Slabs Without Interior Beams (ACI 318-19, Table 8.3.1.1)[1]

| Yield Stress f_y^2, psi[†] | Without Drop Panels[3] | | | With Drop Panels[3] | | |
| | Exterior Panels | | | Exterior Panels | | |
	Without Edge Beams	With Edge Beams[4]	Interior Panels	Without Edge Beams	With Edge Beams[‡]	Interior Panels
40,000	$l_n/33$	$l_n/36$	$l_n/36$	$l_n/36$	$l_n/40$	$l_n/40$
60,000	$l_n/30$	$l_n/33$	$l_n/33$	$l_n/33$	$l_n/36$	$l_n/36$
80,000	$l_n/27$	$l_n/30$	$l_n/30$	$l_n/30$	$l_n/33$	$l_n/33$

[1] l_n is the clear span in the long direction, measured face-to-face of supports (in.).
[2] For f_y between the values given in the table, minimum thickness shall be calculated by linear interpolation.
[3] Drop panels as given in ACI 318-19 section 8.2.4.
[4] Slabs with beams between columns along exterior edges. Exterior panels shall be considered to be without edge beams if α_f is less than 0.8.

For the distribution of the midspan moment M_c between column and middle strips use Table 8.7. The value for α_1, the ratio of beam stiffness to slab stiffness for the example problem, is zero since there are no beams in the direction of spans under consideration. The ratio $\dfrac{l_2}{l_1} = \dfrac{20}{24} = 0.833$. From Table 8.5 column strip moment is 60 percent of total moment.

Moment to column strip = $0.60 \times 146.5 = 87.9$ ft kips
Moment to middle strip = $0.40 \times 146.5 = 58.6$ ft kips

For the distribution of support moment M_s between column and middle strips use Table 8.6. Since $\alpha_1 = 0$,

and $\dfrac{l_2}{l_1} = 0.833$, from Table 8.3 column strip moment is 75 percent of total moment.

Moment in column strip = $0.75 \times 272.2 = 204$ ft kips
Moment in middle strip = $0.25 \times 272.2 = 68$ ft kips

Exterior Span The magnitude of the moments at critical sections in the exterior span is a function of both M_o, the simple beam moment, and α_{ec}, the ratio of stiffness of exterior equivalent column to the sum of the stiffness of the slab and beam framing into the exterior joint. Instead computing α_{ec}, we use edge condition (d) given in Fig. 8.18 to evaluate the design moments at critical sections.

At the exterior column face

$$M_1 = 0.30 \times M_o = 0.30 \times 418.7 = 125.6 \text{ ft kips}$$

At midspan

$$M_2 = 0.50 \times M_o = 0.50 \times 418.7 = 209.4 \text{ ft kips}$$

At the interior column face

$$M_3 = 0.7 \times M_o = 0.7 \times 418.7 = 293 \text{ ft kips}$$

At the exterior edge of slab, the transverse distribution of the design moment to column strip is given in Table 8.7. Instead of calculating the value of β_t, we conservatively assign 100 percent of exterior moment to the column strip.

The moment to the column strip = 1×125.6 ft kips. The middle strip is assumed to be controlled by the minimum steel requirements, an assumption which is satisfactory in almost all practical designs.

8.2   PRESTRESSED CONCRETE SYSTEMS

Although mild-steel-reinforced concrete is well suited for the construction of high-rise floor systems, it requires considerable depth to work efficiently, especially in the 35 to 40 ft (10.66 to 12.2 m) span range normally required in modern office buildings. For example, to keep the long-term creep deflection to within acceptable limits, span-to-depth ratios of two-way flat slab and one-way beam or joist systems are limited to about $\frac{1}{30}$ to $\frac{1}{35}$ and $\frac{1}{15}$ to $\frac{1}{20}$, respectively. The relatively short span capability of flat slab system limits its application to high-rise multiple-unit residential buildings which can accommodate relatively closed-spaced columns within fixed and frequent partition layouts. When used for office layouts, the flat slab system becomes too heavy, imposing undue structural penalty, especially in difficult foundation conditions. The one-way joist system, although relatively lightweight, requires about 20 to 24 in. (508 to 610 mm) of structural depth for a 35 to 40 ft (10.66 to 12.2 m) span range. By using prestressing, it is possible to overcome the aforementioned shortcomings. Prestressing can thus be looked upon as an aid to boost the span range of conventionally reinforced floor systems by about 30 to 40 percent. This is the primary reason for the increase in the use of prestressed concrete. Some of the other advantages of prestressed concrete are:

1. Prestressed concrete is generally crack-free and therefore, more durable.

2. Shallower sections can be used because a larger depth of compression block is available in flexure.

3. Prestress concrete is resilient. Cracks due to overloading completely close and deformations are recovered soon after the removal of overload.

4. Fatigue strength (though not a design consideration in building design) is considerably more than that of conventionally reinforced concrete because tendons are subjected to smaller variations in stress due to repeated loadings.

5. Prestressed concrete members are generally crack-free and, therefore, are stiffer than conventional concrete members.

6. The structural members are self-tested for material and workmanship during stressing operations, thereby safeguarding against unexpected poor performance in service.

7. Prestress design can be controlled more since a predetermined force is introduced in the system; the magnitude, location, and technique of introduction of such an additional force is left to the designer, who can tailor the design according to project requirements.

A major motivation for the use of prestressed concrete comes from the reduced structural depth, which translates into lower floor-to-floor height, reduction in the area of curtain wall and building volume, with a consequent reduction in heating and cooling loads.

In prestressed concrete there is no savings in using high-strength strands instead of mild steel. This is because the savings in mild steel reinforcement quantities are just about offset by the higher unit cost of prestressing steel. The cost savings, however, come from the reduction in the quantity of concrete combined with indirect nonstructural savings resulting from reduced floor-to-floor height. Although from an initial cost consideration prestress concrete may be the least expensive, other costs associated with future tenant improvements such as providing large openings in slabs must be considered before selecting the final scheme.

Despite the advantages of prestressed concrete, there are some disadvantages such as fire, the difficulty in making penetrations in the slab due to the fear of cutting tendons, and the explosion resistance of the unbonded systems.

8.2.1   Methods of Prestressing

Centuries ago humans discovered the principle of prestressing by using metal bands or ropes around wooden planks to form barrels. They may not have had the dubious pleasure of figuring out the exact nature and magnitude of stresses, but they knew intuitively that the stronger the bands, the better the chance of containing liquids in the wooden barrels. It was not until the 1880s that a similar principle was applied to a reinforced concrete slab with the idea of counterbalancing the tensile stresses in concrete. Intentional compressive stress in concrete was induced to overcome the tensile stresses developed from external loads. These early attempts at prestressing did not meet with great success because the amount of prestressing that could be imposed via conventional steel was limited by the strength of the steel itself. Even at low stresses, it was difficult to maintain the prestress because creep and shrinkage of concrete would destroy the prestress in the course of time.

The eminent French engineer Eugene Freyssinet is credited as the forefather of prestressing as we know it today. He established the use of high-tensile-strength steel to ensure that even after creep and shrinkage losses, there remained adequate prestress to counteract the external loads.

Current methods of prestressing can be studied under two groups: (i) pre-tensioning and (ii) post-tensioning. In pre-tensioning, the tendons are stressed first and then, concrete is placed around the tendons. After the concrete has hardened, the tendons are released to impart prestress into the concrete member.

In post-tensioning, the tendons, which may consist of steel wires, strands, or bars, are tensioned and anchored against the concrete after it has hardened. The tensioning is accomplished by using hydraulic jacks. The tendons usually remain permanently unbonded to concrete and are placed directly in the forms, and stressed after the concrete has reached a minimum 75 percent of the design strength. The measured elongations are compared against the calculated values, and if satisfactory, the tendons projecting beyond the concrete are cut off. Formwork is removed after post-tensioning of tendons. However, the floor is back-shored to support shoring and construction loads from the floors above.

Post-tensioning is accomplished by using high-strength strands, wires, or bars as tendons. In North America, the use of strands by far leads to the other two types. The strands are either bonded or unbonded depending upon the project requirements. In bonded construction, the ducts are filled with a mortar grout after stressing the tendons while there is no grouting in unbonded construction.

For high-rise construction, unbonded construction is preferred because it eliminates the need for grouting. Post-tensioned members in multistory construction consist of slabs, joists, beams, and girders, with a large number of small tendons. Grouting each of the multitude of tendons is a time-consuming and expensive operation. Therefore, unbounded construction using strands is popular in North American high-rise construction.

8.2.2 Materials

Post-Tensioning Steel The basic requirement is the loss of tension in the steel due to shrinkage and creep of concrete, and the effects of stress relaxation should be a relatively small portion of the total prestress. In practice, the loss of prestress generally varies from a low of 15 ksi (103.4 MPa) to a high of 50 ksi (344.7 MPa). If mild steel having a yield of 60 ksi (413.7 MPa) is employed with an initial prestress of, say, 40 ksi, it is very likely that most of the prestress, if not the entire prestress, is lost because of shrinkage and creep losses. To limit the prestress losses to a small percentage of say, about 20 percent, the initial stress in the steel must be in excess of 200 ksi (1379 MPa). Therefore, high-strength steel is invariably used in prestressed concrete construction.

Although in general high-strength steel is produced by using alloys such as carbon, manganese, and silicon, prestressing steel achieves its high tensile strength by virtue of the process of cold-drawing, in which high-strength steel bars are drawn through a series of progressively smaller dyes. During this process, the crystallography of the steel is improved, because cold-drawing tends to realign the crystals.

High-strength steel used in North America is available in three basic forms: (i) uncoated stress-relieved wires, (ii) uncoated stress-relieved strands, and (iii) uncoated high-strength steel bars. Stress-relieved wires and high-strength steel bars are not generally used for post-tensioning and therefore are not considered here.

High-strength strands are fabricated in factories by helically twisting a group of six wires around a slightly larger center wire by a mechanical process called stranding. The resulting seven-wire strands are stress-relieved by a continuous heat treatment process to produce the required mechanical properties.

ASTM specification A416 specifies two grades of steel, 250 and 270 ksi (1724 and 1862 MPa), the higher strength being more common in the building industry. A modulus of elasticity of 27,500 ksi (189,610 MPa) is used for calculating the elongation of strands. To prevent the use of brittle steel which would result in a failure pattern similar to that of an over-reinforced beam, ASTM A-416 specifies a minimum elongation of 3.5 percent at rupture.

A special type of strand called the low-relaxation strand is increasingly used because it has a very low loss due to relaxation, usually about 20 to 25 percent of that for stress-relieved strand. With this strand less post-tensioning steel is required, but the cost is more because of the special process used in its manufacture.

Corrosion of unbonded strand is a possibility and can be prevented by using galvanized strands. This is not, however, popular in North America because: (i) various anchorage devices in use for posttensioned systems are not suitable for use with galvanized strand because of low coefficient of friction; (ii) damage can result to the strand because the heavy bite of the anchoring system can ruin the galvanizing; and (iii) galvanized strands are more expensive.

A little-understood, and infrequent occurrence of great concern in engineering is the so-called stress corrosion which occurs in highly stressed strands. The reason for the phenomenon is little known, but chemicals such as chlorides, sulfides, and nitrates are known to start this type of corrosion under certain conditions. It is also known that high-strength steels exposed to hydrogen ions are susceptible to failure because of loss in ductility and tensile strength. This phenomenon is called *hydrogen embrittlement* and is best counteracted by confining the strands in an environment having a pH value greater than 8.

Concrete   Concrete with compressive strengths of 5000 to 6000 psi (34 to 41 MPa) is commonly employed in the prestress industry. This relatively high-strength is desirable, for the following reasons. First, commercial anchorages are designed on the basis of high-strength concrete to prevent failure of concrete during the application of prestressing. Second, high-strength concrete has higher resistance in tension, shear, bond, and bearing and is desirable for prestressed structures which are typically under higher stresses than ordinary reinforced concrete. Third, its shrinkage is less and its higher modulus of elasticity and smaller creep result in smaller loss of prestress.

Post-tensioned concrete is considered as a self-testing system because if the concrete is not crushed under the application of prestress it should withstand subsequent loadings in view of the strength gain with age. In practice, it is not the 28-day strength that dictates the mix design but rather the strength of concrete at the transfer of prestress. Construction schedules on high-rise projects require post-tensioning as early as possible to facilitate early removal of forms for reuse in higher floors. Typically, the minimum strength of concrete at transfer is 70 to 75 percent of the 28-day strength. Assuming that stressing operation is on the fourth day or so, it is more than likely that the actual 28-day strength is much more than the specified strength. For example, assume that the design specifies a 28-day compressive strength of 5000 psi (34.47 MPa). The minimum strength required at transfer of prestress is 70 to 75 percent of 5000 psi, approximately equal to 4000 psi (27.6 MPa) at 4 days. This requirement would normally yield a concrete of 28-day strength of about 6000 psi (41.37 MPa), which is well in excess of the specified design strength. This rather wasted strength can be avoided by using the higher strength in the actual design.

Although high early strength (Type III) Portland cement is well-suited for post-tension work because of its ability to gain the required strength for stressing relatively early, it is not generally used because of higher cost. Invariably, Type I cement conforming to ASTM C-150 is employed in buildings.

The use of admixtures and fly ash is considered a good practice. However, the use of calcium chlorides or other chlorides is prohibited since the chloride ion may result in stress corrosion of prestressing tendons.

A slump of between 3 and 5 in. (76 to 127 mm) gives good results. The aggregate used in the normal production of concrete is usually satisfactory in prestressed concrete, including lightweight aggregates. However, care must be exercised in estimating the volumetric changes so that a reasonable prestress loss is calculated. Lightweight aggregates manufactured by using expanded clay or shale have been used in post-tensioned buildings. Lightweight aggregates that are not crushed after burning maintain their coating and therefore absorb less water. Such aggregates have drying and shrinkage characteristics similar to the normal-weight aggregates, although the available test reports are somewhat conflicting. The size of aggregate, whether lightweight or normal weight, has a more profound effect on shrinkage. Larger aggregates offer more resistance to shrinkage and also require less water to achieve the same consistency, resulting in as much as 40 percent reduction in shrinkage when the aggregate size is increased from say 3/4 to 1½ in. (19 to 38 mm). It is generally agreed that both shrinkage and creep are more a function of the cement paste than the type of aggregate. Lightweight aggregate has been gaining acceptance in prestressed construction since about 1955 and has a good track record.

8.2.3   PT Design

The design involves the following steps:

1. Determination of the size of the concrete member.

2. Establishing the tendon profile and prestressing force.

3. Calculating the prestressing force

4. Verifying the section for ultimate bending and shear capacity.

5. Verifying the serviceability characteristics, primarily in terms of stresses and long-term deflections.

It is well known that the depth of a member subjected to bending depends on many variables such as the magnitude of the design loads, shape of the cross-section, available clearance, span length, and allowable deflections. The deflections of prestressed members tend to be small because under service loads, they are usually uncracked and are much stiffer than nonprestressed members of the same cross-section. Also, the prestressing force induces deflections in an opposite direction to those produced by external loads. The final deflection, therefore, is a function of tendon profile and the magnitude of prestress. Appreciating this fact, the ACI 318-19 does not specify minimum depth requirements for prestressed members. However, as a rough guide, the suggested span-to-depth ratios given in Table 8.9 can be used to establish the depth of continuous flexural members. Another way of looking at the suggested span-to-depth ratios is to consider, in effect, that prestressing increases the span range by about 30 to 40 percent over and above the values normally used in nonprestressed concrete construction.

Table 8.9   Approximate Span Depth Ratios for Post-Tensioned Systems

Floor System	Simple Spans	Continuous Spans	Cantilever Spans
One-way solid slabs	40–48	42–50	14–16
Two-way flat slabs	36–45	40–48	13–15
Wide band beams	26–30	30–35	10–12
One-way joists	20–28	24–30	8–10
Beams	18–22	20–25	7–8
Girders	14–20	16–24	5–8

The above values are intended as a preliminary guide for the design of building floors subjected to a uniformly distributed superimposed live load of 50 to 100 psf (2394 to 4788 Pa). For the final design, it is necessary to investigate for possible effects of camber, deflections, vibrations, and damping. The designer should verify that adequate clearance exists for proper placement of post-tensioning anchors.

The tendon profile is established based on the type and distribution of load with due regard to clear cover required for fire resistance and corrosion protection. Clear spacing between tendons must be sufficient to permit easy placing of concrete. For maximum economy, the tendon should be located eccentric to the center of gravity of the concrete section to produce maximum counteracting effect to the external loads. For members subjected to uniformly distributed loads a simple parabolic profile is ideal, but in continuous structures parabolic segments forming a smooth reversed curve at the support are more practical. The effect is to shift the point of contraflexure away from the supports. This reverse curvature modifies the load imposed by post-tensioning from those assumed using a parabolic profile between tendon high points.

The design of a simple span is rather trivial and can be accomplished with hand calculations. In continuous and indeterminate structures, the induced moments are not directly proportional to the tendon eccentricity because the deflection due to post-tensioning is resisted at the supports. The support restraint introduces moments called the secondary moments. The name is a misnomer because it does not mean that its values are negligible or necessarily smaller than the primary moments.

The initial post-tension force immediately after transfer is less than the jacking force because of (i) slippage of anchors, (ii) frictional losses along tendon profile, and (iii) elastic shortening of concrete. The force is reduced further over a period of months or even years due to change in the length of concrete member resulting from shrinkage and creep of concrete and relaxation of the highly stressed steel. The effective prestress is the force in the tendon after all the losses have taken place. For routine designs, empirical expressions for estimating prestress losses yield sufficiently accurate results, but in cases with unusual member geometry, tendon profile, and construction methods it may be necessary to make refined calculations.

In North American practice, it is usually sufficient to specify effective force and tendon profile. The post-tension contractor submits calculations of prestress losses for the engineer's review. Therefore, the engineer is spared the drudgery of calculating the prestress losses. The post-tension design of a statically determinate structure is trivial and can be accomplished with little difficulty by hand. The floor framing systems normally encountered in practice are invariably statically indeterminate and are most usually designed by using computer programmes. Most programs use the concept of load balancing.

In this concept, prestressing is seen as a method to balance a certain portion of the external loads by inducing a counteracting load. This method, first developed by T. Y. Lin is very popular. Its application to statically indeterminate systems could be visualized just as easily as for statically determinate structures. Also, the procedure gives a simple method of calculating deflections by considering only that portion of the external load not balanced by the prestress. If the effective prestress completely balances the sustained loading, the post-tensioned member will not undergo any deflection and will remain horizontal irrespective of the modulus of rigidity or flexural creep of concrete.

In the load-balancing approach, the analysis of a prestressed member is reduced to the analysis of a nonprestressed member subjected to the load differential between externally applied loads and internally applied prestress. Since the analysis is performed with only the unbalanced portion of the external load, the inaccuracies in the method of analysis become relatively insignificant. Often, approximate method is all that is necessary for the final design. The load balancing method can be conveniently applied to multiple-span beams and slabs. The prestressing force need not be the same in all the spans. The load in each span can be balanced by choosing a suitable prestress and profile. For spans requiring higher prestressing, additional tendons can be added.

A question that usually arises is how much of the external load is to be balanced. The answer, however, is not simple. Balancing all the dead load often results in too much prestressing, leading to uneconomical design. On the other hand, there are situations in which the live load is significantly heavier than the dead load, making it more economical to prestress not only for full dead loads but also for a significant portion of the live load. However, in the design of typical floor framing systems, the prestressing force is normally selected to balance about 75 to 95 percent of the dead load and, occasionally, a small portion of the live load. This leads to an ideal condition with the structure having little or no deflection under dead loads.

Limiting the maximum tensile and compressive stresses permitted in concrete does not in itself assure that the prestressed member has an adequate factor of safety against flexural failure. Therefore, its nominal bending strength is computed in a procedure similar to that of a reinforced concrete beam. Under-reinforced beams are assumed to have reached the failure load when the concrete strain reaches a value of 0.003. Since the yield point of prestressing steel is not well defined, empirical relations based on tests are used in evaluating the strain and hence the stress in tendons.

The shear reinforcement in post-tensioned members is designed in a manner almost identical to that of nonprestressed concrete members, with due consideration for the longitudinal stresses induced by the post-tensioned tendons. Another feature unique to the design of post-tensioned members is the high stresses in the vicinity of anchors. Prestressing force is transferred to the concrete by anchoring the tendons with the aid of anchorages. Large stresses are developed at the anchorages, which have to be dealt with properly by providing well-positioned reinforcement in the region of high stresses. At a cross-section of a beam sufficiently far away (usually two to three times of the larger cross-sectional dimension of the beam) from the end zones, the axial and bending stresses in the beam due to an eccentric prestressing force are given by the usual P/A and MC/I relations. But in the vicinity of stress application, the stresses are distributed in a complex manner. Of importance are the transverse

tensile forces generated at the end blocks for which reinforcement is to be provided. The bursting tensile stress has a maximum value along the axis of the force. Its distribution depends on the location of bearing area and its relative proportion with respect to the areas of the end face.

Because of the indeterminate nature and intensity of the stresses, the design of reinforcement for the end block is primarily based on empirical expressions. Reinforcement is designed to carry the tensile stresses created in the end block by the tendon reactions and usually consists of closely spaced stirrups tied together with horizontal bars.

8.2.4 Practical Considerations

Condominiums and apartment buildings are economical if a structural system with a flat ceiling is used for floor construction. A post-tensioned flat plate is one such system for column spacings in the range of 20 to 30 ft (6.09 to 9.14 m). The formwork is simple and lends itself to quick construction. The resulting flat plate system has good acoustical characteristics while maintaining a minimum floor-to-floor height. A flat slab with drop panels, which can be considered as an extension of the flat plate system, is suitable for office buildings with clear spans in the range of 35 to 45 ft (10.6 to 13.7 m). In such systems, it may be economical to use long and narrow shear heads to accommodate flying forms. Post-tensioned joists clearly spanning between the core and building exterior offer an alternative method for framing office buildings.

As in reinforced concrete and structural steel construction, the use of post-tensioned concrete is only limited by the imagination and ingenuity of the engineer and the relative economics of various construction materials and labor at the bid time. Certain rules of thumb for span-to-depth ratios and the average value of posttensioning stresses in structural members are useful in conceptual design. The depth for slabs usually works out between $L/40$ and $L/50$, while for joists it is between $L/25$ and $L/35$. Beams can be much shallower, with a depth in the range of $L/20$ and $L/30$. Band beams offer perhaps the least depth without using as much concrete as flat slab construction. Although a span-to-depth ratio approaching 35 is adequate from strength and serviceability points of view, it is necessary to make sure that adequate space exists for proper detailing of anchorages. Detailing of beam-column intersection of shallow band beams should be carefully developed to avoid conflicts between the post-tensioning tendon anchorage, and main vertical column reinforcement. Bundling of column bars may be required to relieve congestion. Adequate clearance must be provided to permit access to stressing equipment.

Another thumb rule used in preliminary design is the compression stress level in the members due to post-tensioning. A minimum compression level of 125 and 150 psi (862 and 1034 KPa) is a practical and economical range for slabs while a range of 250 to 300 psi (1724 to 2068 KPa) has been found to be adequate for beams. Compression stresses as high as 500 psi (3447 KPa) have been used successfully in band-beam systems.

8.2.5 Building Examples

For the first example, refer to the two-way flat plate framing plan of the Museum Tower shown in Fig. 1.8c and Fig. 8.20. The primary tendons are 1/2 in. diameter (12.7 mm) strands that are banded in the north-south direction. Unbonded tendons run from left to right across the building width. Additional tendons are used in the end panels to allow for the increased moments due to lack of continuity at one end.

As a second example, Fig. 8.21 shows the framing plan for a post-tensioned band-beam-slab system. Shallow beams only 16 in. (0.40 m) deep span across two exterior bays of 40 ft (12.19 m) and an interior bay of 21 ft

(6.38 m). Post-tensioned slabs 8 in. (203 mm) deep, span between the band beams, typically spaced at 30 ft (9.14 m) on center. In the design of the slab additional beam depth is considered as a haunch at each end. Primary tendons for the slab run across the building width, while the temperature tendons are placed in the north-south direction between the band beams.

8.2.6 Cracking Problems in Post-Tensioned Floors

Cracking caused by restraint to shortening is the biggest problem associated with post-tensioned floor systems. The reason for this is that the restraint to shortening is a time-dependent complex phenomenon with only subjective empirical solution. Exact numerical solutions which prevent cracking altogether have not been developed yet.

Shrinkage of concrete is the biggest contributor to shortening in both prestressed and non-prestressed concrete. In prestressed concrete, out of the total shortening, only about 15 percent is due to elastic shortening and creep. Therefore, the problem is not that post-tensioned floors shorten that much more than non-prestressed concrete but it is the manner in which they shorten.

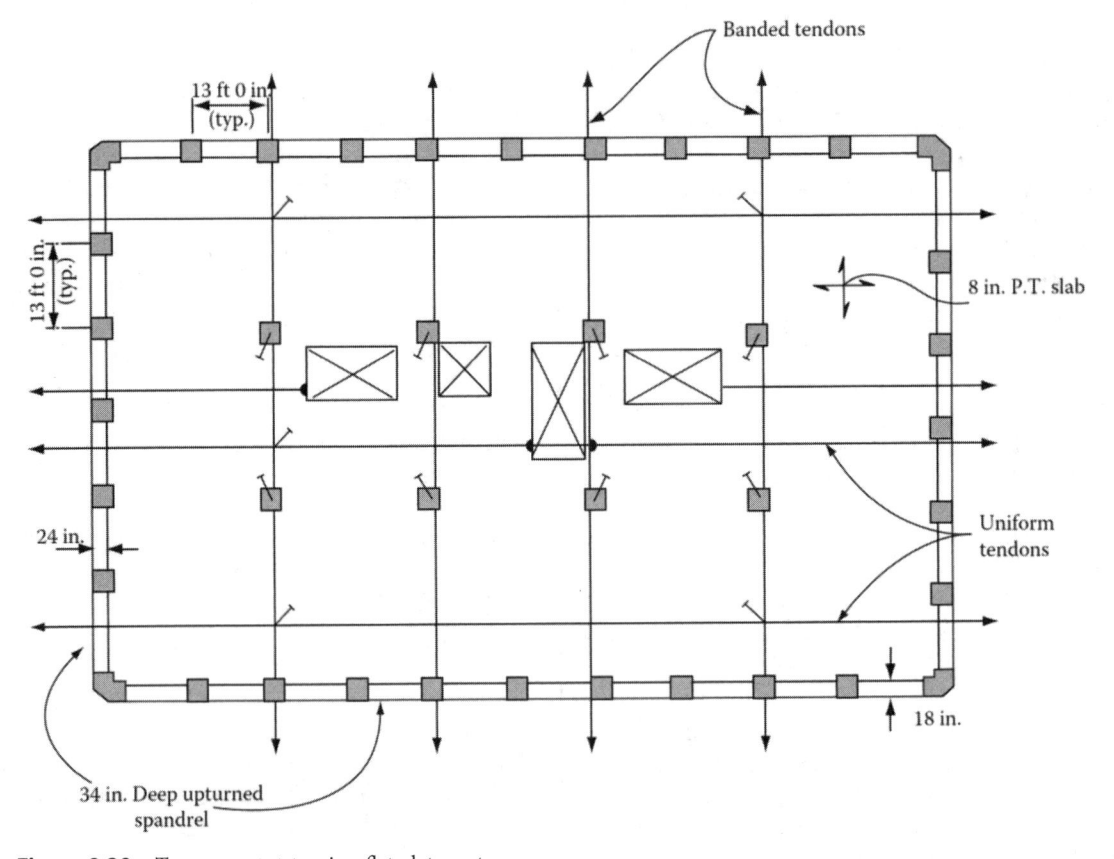

Figure 8.20 Two-way post-tension flat plate system.

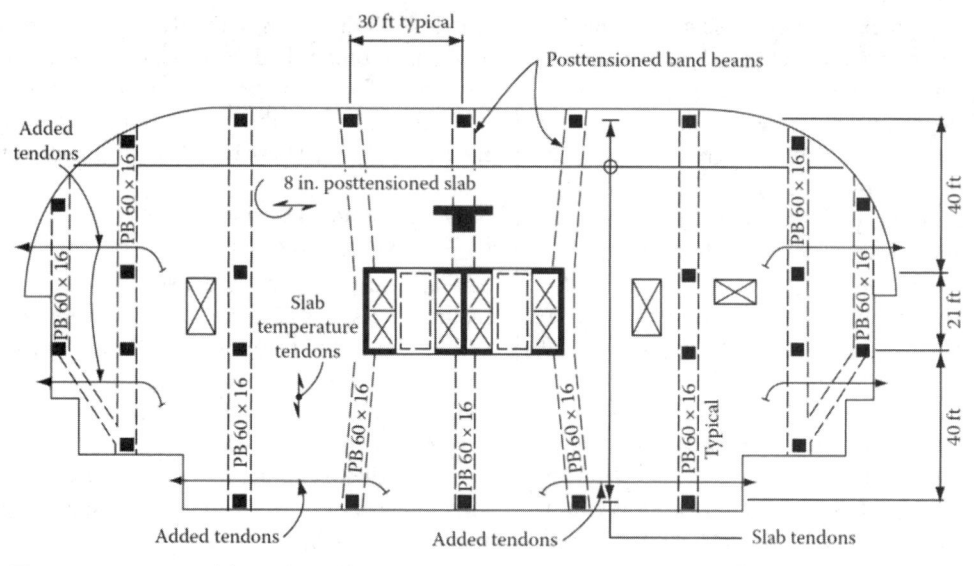

Figure 8.21 Typical floor plan of one-way post-tension slab system.

As a non-prestressed concrete slab tries to shorten, its movement is resisted internally by the bonded mild steel reinforcement. The reinforcement is put into compression and hence the concrete in tension. As the concrete tension builds up the slab cracks at fairly regular intervals allowing the ends of the slab to remain in the same position in which they were cast. In a manner of speaking, the concrete has shortened by about the same magnitude as a post-tensioned system, but not in overall dimensions. Instead of the total shortening occurring at the ends, the combined widths of many cracks which occur across the slab make-up for the total shortening. The reinforcement distributes the shortening throughout the length of the slab in the form of numerous cracks. Thus reinforced concrete tends to take care of its own shortening

problems internally by the formation of numerous small cracks, each small enough to be considered acceptable. Restraints provided by stiff vertical elements such as walls and columns tend to be of minor significance, since provision for total moment has been provided by the cracks in concrete.

This is not the case with post-tensioned systems in which shrinkage cracks, which would have formed otherwise, are closed by the post-tensioning force. Much less mild steel is present and consequently the restraint to the shortening provided is much less. The slab tends to shorten at each end generating large restraining forces in the walls and columns particularly at the ends where the movement is greatest (Fig. 8.22a). These restraining forces can produce severe cracking in the slab, walls,

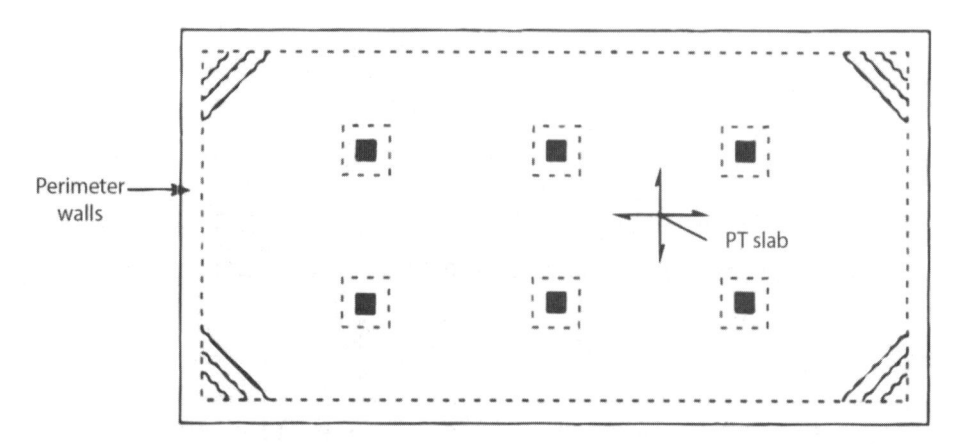

Figure 8.22a Cracking in post-tensioned slab caused by restraint of perimeter walls.

or columns at the slab extremities causing problems for engineers and building owners alike. The most serious consequence is perhaps water leakage through the cracks.

The solution to the problem lies in eliminating the restraint by separating the slab from the restraining vertical elements. If a permanent separation is not feasible, cracking can be minimized by using temporary separations to allow enough of the shortening to occur prior to making the connection.

Cracking in post-tensioned slab also tends to be proportional to initial pour size. Some general guidelines that have evolved over the years are as follows: (i) the maximum length between temporary pour strips (Fig. 8.22b) is 150 ft (200 ft if restraint due to vertical elements is minimal); and (ii) the maximum length of post-tensioned slab irrespective of the number of pour strips provided, is 300 ft. The length of time for leaving the pour strips open is critical and can range anywhere from 30 to 60 days. A 30-day period is considered adequate for average restraint conditions with relatively centered, modest-length walls while a 60-day period is more the norm for severe shortening conditions with large pour sizes and stiff walls at the ends.

To minimize cracking caused by restraint to shortening, it is a good idea to provide a continuous mat of reinforcing steel in both directions of the slab. As a minimum #4 bars at 36 in. on centers both ways, is recommended for typical conditions. For slab-pours in excess of 150 ft in length with relatively stiff walls at the ends, the minimum reinforcement should be increased to #4 bars at 24 in. on centers both ways.

8.2.7 Preliminary Design

8.2.7.1 INTRODUCTION

The aim of post-tensioned design is to determine the required prestressing force and hence the number, size, and profile of the tendons for behavior at service loads. The ultimate capacity must then be checked at critical sections to ensure that prestressed members have an adequate factor of safety against failure.

In statically determinate structures, as in simple beams, the moments induced by the post-tensioning are directly proportional to the eccentricity of the tendons with respect to the neutral axis of the beam. In indeterminate structures, as in continuous beams, the moments due to post-tensioning are usually not proportional to the tendon eccentricity. The difference is due to the restraint imposed by the supports to post-tensioning deformations. The moments resulting from the restraints to prestressing deformations are called "secondary moments."

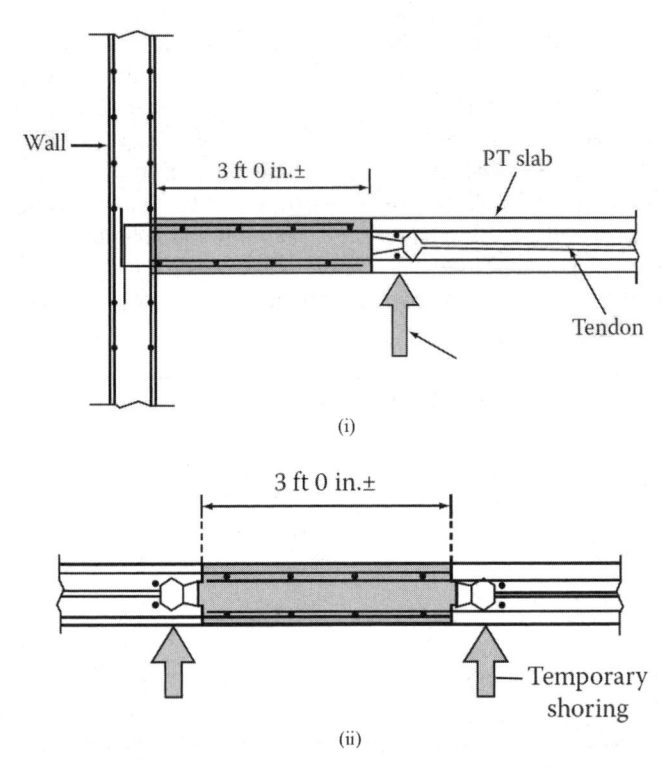

Figure 8.22b Temporary pour strip: (i) at perimeter of building; (ii) at interior of slab.

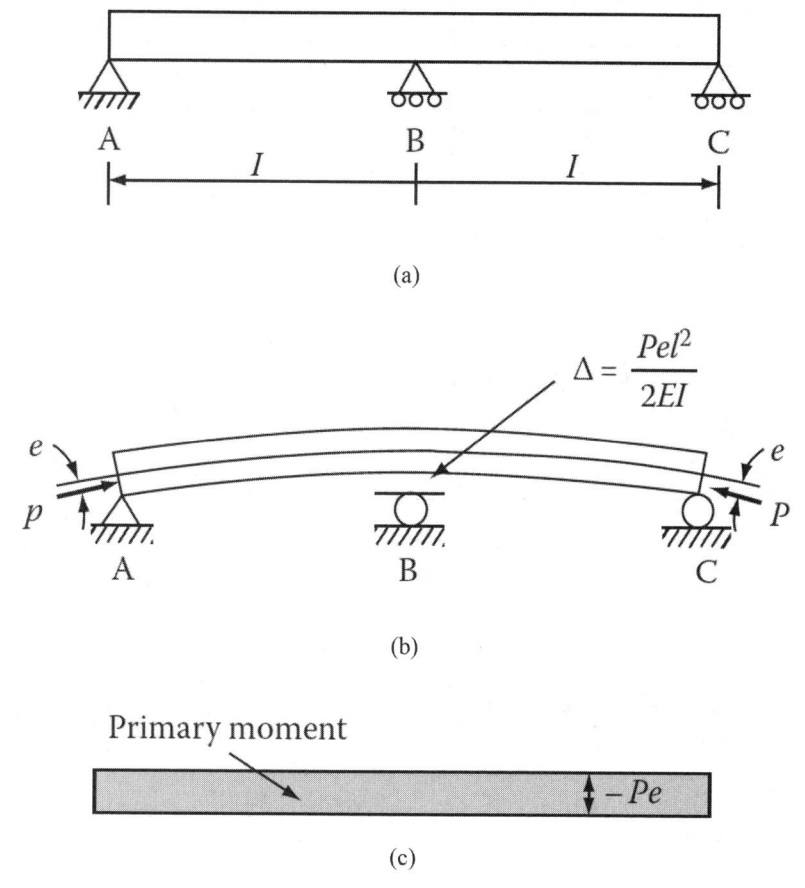

$$\Delta = \frac{Pel^2}{2EI}$$

Primary moment

$-Pe$

(c)

Figure 8.23 Concept of secondary moment: (a) two-span continuous beam; (b) vertical upward displacement due to PT; (c) primary moment; (d) reactions due to PT; (e) secondary moment; (f) final moments.

Primary and secondary moments, as well as the total moments due to post-tensioning are shown in Fig. 8.23 for a two span continuous beam. The beam has a post-tensioning force P, acting at a constant eccentricity e. Hence, the primary moment in the beam is Pe as shown in Fig. 8.23c. The primary moment will cause a theoretical upward deflection of $Pel^2/2EI$ at the center support. The reactions necessary to retain the support at its original position are shown in Fig. 8.23d. Observe the secondary moments are functions of the reactions, and for this reason, vary linearly between the supports. Also note that for this case, the secondary moment is 150 percent of the primary moment. The total moment due to post-tensioning may be expressed as the superposition of the primary and secondary moments as shown in Fig. 8.23f.

The preliminary design method presented in this section is based on an article published in Ref. 25. It uses the technique of load balancing in which the effect of prestressing is considered as an equivalent external load. For example, the parabolic profile in Fig. 8.25a exerts a horizontal force P_1 at the ends along with vertical components equal to $P \sin \theta$. The vertical component is neglected in design because it occurs directly over the supports. In addition to these loads, the parabolic tendon exerts a continuous upward force on the beam along its entire length. By neglecting friction between the tendon and concrete, we can assume that (i) the upward pressure exerted is normal to the plane of contact; and (ii) tension in tendon is constant. The upward pressure exerted is equal to the tension in the tendon divided by the radius of curvature. Due to the shallow nature of post-tensioned structures, the vertical component of the tendon force may be assumed constant. Considering one-half of the beam as a free body Fig. 8.25a, the vertical load exerted by the tendon may be derived by summing moments about support A. Thus $w_p = \dfrac{8pe}{L^2}$. Equivalent loads and moments

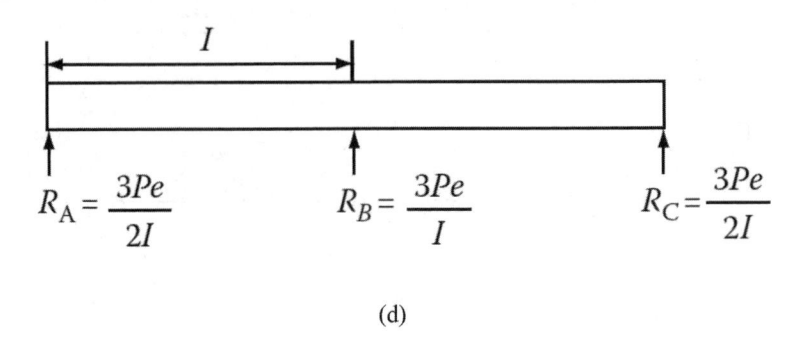

(d)

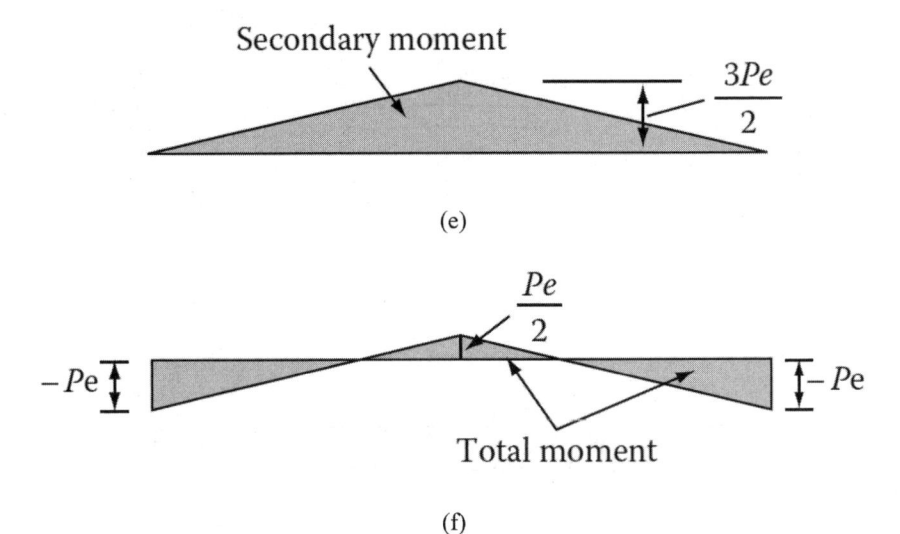

(f)

Figure 8.23   (*Continued*)

produced by other types of tendon profile are shown in Fig. 8.24.

8.2.7.2   GENERAL STEP-BY-STEP PROCEDURE

1. Determine preliminary size of members using the values given in Table 8.9 as a guide.

2. Determine section properties of the member such as the area A, moment of inertia I, and section modulus S_t and S_b.

3. Determine tendon profile with due regard to cover and location of mild steel reinforcement.

4. Determine effective span L_e by assuming $L_1 = \dfrac{1}{16}$ to $\dfrac{1}{19}$ of the span length for slabs, and $L_e = \dfrac{1}{10}$ to $\dfrac{1}{12}$ of the span length for beams (see Fig. 8.26).

5. Start with an assumed value for balanced load w_p equal to 0.7 to 0.9 times the total dead load.

6. Determine the elastic moments for the total dead plus live loads (working loads). For continuous beams and slabs use computer plane-frame analysis programs, moment distribution method or ACI coefficients if applicable, in the decreasing order of preference.

7. Reduce negative moments to the face of supports.

8. By proportioning the unbalanced load to the total load, determine the unbalanced moments at M_{ub} at critical sections such as at the supports and at the center of spans.

9. Calculate stress at bottom and top, f_b and f_t at critical sections. Typically, at supports the stresses f_t and f_b are in tension and compression. At center of spans, the stresses are in compression and tension.

10. Calculate the minimum required post-tension stress f_p by using the following equations.

For negative zones of one-way slabs and beams:

$$f_p = f_t - 6\sqrt{f_c'}$$

For positive moments in two-way slabs:

$$f_p = f_t - 2\sqrt{f_c'}$$

11. Find the post-tension force P by the relation $P = f_p \times A$ where A is the area of the section.

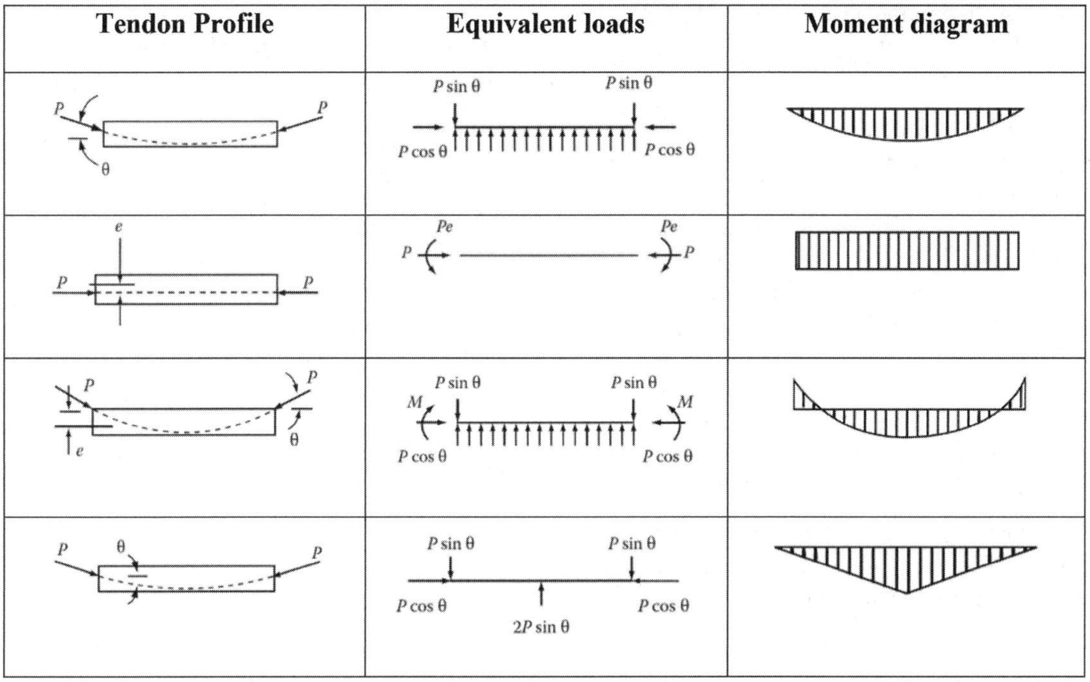

Tendon Profile	**Equivalent loads**	**Moment diagram**

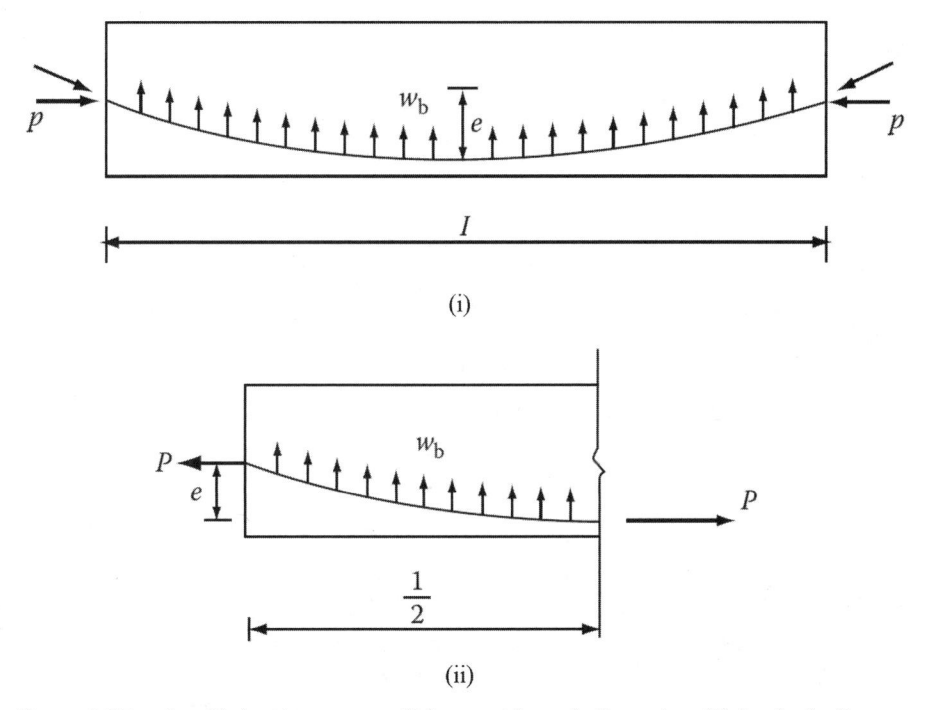

Figure 8.24 Equivalent loads and moments produced by prestressed tendons: (a) upward uniform load due to parabolic tendon; (b) constant moment due to straight tendon; (c) upward uniform load and end-moments due to parabolic tendon not passing through the centroid at the ends; (d) vertical point-load due to sloped tendon.

(i)

(ii)

Figure 8.25a Load balancing concept: (i) beam with parabolic tendon; (ii) free-body diagram.

12. Calculate the balanced load w_p due to P by the relation

$$w_p = \frac{8 \times Pe}{L_e^2}$$

where e = drape of the tendon
L_e = effective length of tendon between inflection points.

13. Compare the calculated value of w_p from step 12 with the value assumed in step 5. If they are about the same, the selection of post-tension force for the given loads and tendon profile is complete. If not, repeat steps 9 to 13 with a revised value of $w_p = 0.75 w_{p1} + 0.25 w_{p2}$. w_{p1} is the value of w_p assumed at the beginning of step 5 and w_{p2} is the derived value of w_p at the end of step 12. Convergence is fast requiring no more than three cycles in most cases.

8.2.7.3 SIMPLE SPANS

The concept of preliminary design discussed in this section is illustrated in Fig. 8.25b where a parabolic profile with an eccentricity of 12 in. is selected to counteract part of the imposed load consisting of a uniformly distributed dead load of 1.5 kip/ft and a live load of 0.5 kip/ft.

In practice, it is rarely necessary to provide a prestress force to fully balance the imposed loads.

A value of prestress, often used for building system is 75 to 95 percent of the dead load. For the illustrative problem, we begin with an assumed 80 percent of the dead load as the unbalanced load.

First Cycle The load being balanced is equal to $0.80 \times 1.5 = 1.20$ kip/ft. The total service dead plus live load = $1.5 + 0.5 = 2.0$ kip/ft of which 1.20 kip/ft is assumed in the first cycle to be balanced by the prestressing force in the tendon. The remainder of the load equal to $2.0 - 1.20 = 0.80$ kip/ft acts vertically downward producing a maximum unbalanced moment M_{ub} at center span given by

$$M_{ub} = 0.80 \times \frac{54^2}{8}$$

$$= 291.6 \text{ kip ft}$$

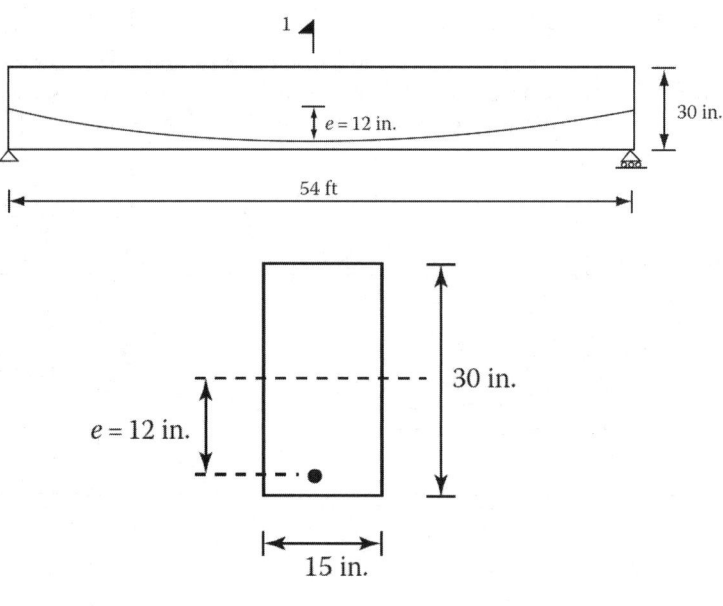

Section 1

Section properties

Area $A_g = 15 \times 30 = 450 \text{ in.}^2$

$y_b = y_t = 15 \text{ in.}$

$I_g = 15 \times \frac{30^3}{12} = 33750 \text{ in.}^4$

$S_b = S_t = \frac{33750}{15} = 2250 \text{ in.}^3$

$f_c' = 5000 \text{ psi}$

Allowable tension $= 6\sqrt{f_c'}$

$= 0.424 \text{ ksi}$

Figure 8.25b   Preliminary design: simple span beam.

The tension and compression in the section due to M_{ub} is given by

$$f_c = f_b = \frac{291.6 \times 12}{2250}$$

$$= 1.55 \text{ ksi}$$

The minimum prestress required to limit the tensile stress to $6\sqrt{f_c'} = 0.424$ is given by

$$f_p = 1.55 - 0.424 = 1.13 \text{ ksi}$$

Therefore, the required minimum prestressing force $P =$ Area of beam $\times 1.13 = 450 \times 1.13 = 509$ kips. The load balanced by this force is given by

$$w_p \times \frac{54^2}{8} = Pe = 509 \times 1$$

$\therefore w_p = 1.396$ kip/ft compared to the value of 1.20 used in the first cycle. Since these two values are not close to each other, we repeat the above calculations starting with a more precise value for w_p in the second cycle.

Second Cycle   We start with a new value of w_p by assuming the new value as 75 percent of the initial value + 25 percent of the derived value. For this example problem new value of

$$w_p = 0.75 \times 1.20 + 0.25 \times 1.396 = 1.25 \text{ kip/ft}$$

$$M_{ub} = (2 - 1.25) \times \frac{54^2}{8} = 273.3 \text{ kip ft}$$

$$f_b = f_t = \frac{273.3 \times 12}{2250} = 1.458 \text{ ksi}$$

The minimum stress required to limit the tensile stress to $6\sqrt{f_c'} = 6\sqrt{5000} = 0.424$ ksi is given by

$$f_p = 1.458 - 0.424 = 1.03 \text{ ksi}$$

Minimum prestressing force $P = 1.03 \times 450 = 465$ kips. The balanced load corresponding to the prestress value of 465 is given by

$$w_p = \frac{8Pe}{L^2} = \frac{8 \times 465 \times 1}{54^2}$$

$\therefore w_p = 1.27$ kip/ft which is nearly equal to the value assumed in the second cycle. Thus, the minimum prestress required to limit the tensile stress in concrete to $6\sqrt{f_c}$ is 465 kips.

To demonstrate how rapidly the method converges to the desired answer, we will rework the problem by assuming an initial value of $w_p = 1.0$ kip/ft in the first cycle.

First Cycle

$$w_p = 1.0 \text{ kip/ft}$$

$$M_{ub} = (2 - 1) \times \frac{54^2}{8} = 364.5 \text{ kip ft}$$

$$f_b = f_t = \frac{364.5 \times 12}{2250} = 1.9444 \text{ ksi}$$

$$f_p = 1.944 - 0.454 = 1.49 \text{ ksi}$$

$$P = 1.49 \times 450 = 670.5 \text{ kips}$$

$$w_p \times \frac{54^2}{8} = 670.5 \times 1$$

$$w_p = 1.84 \text{ kip/ft}$$

compared to 1.0 kip/ft used at the beginning of first cycle.

Second Cycle

$$w_p = 0.75 \times 1 + 0.25 \times 1.84 = 1.21 \text{ kip/ft}$$

$$M_{ub} = (2 - 1.21) \times \frac{54^2}{8} = 288 \text{ kip ft}$$

$$f_b = f_c = \frac{288 \times 12}{2250} = 1.536 \text{ ksi}$$

$$f_p = 1.536 - 0.454 = 1.082 \text{ ksi}$$

$$P = 1.082 \times 450 = 486.8 \text{ kips}$$

$$w_p = \frac{486.8 \times 1 \times 8}{54^2}$$

$$= 1.336 \text{ kip/ft}$$

compared to the value of 1.21 used at the beginning of second cycle.

Third Cycle

$$w_p = 0.75 \times 1.21 - 1.21 \times 0.25 \times 1.336 = 1.24 \text{ kip/ft}$$

$$M_{ub} = (2 - 1.24) \times \frac{54^2}{8} = 276.67 \text{ kip ft}$$

$$f_b = f_c = \frac{276.47 \times 12}{2250} = 1.475 \text{ ksi}$$

$$f_p = 1.475 - 0.454 = 1.021 \text{ ksi}$$

$$P = 1.021 \times 450 = 459.3 \text{ kips}$$

$$w_p = 459.3 \times \frac{1 \times 8}{54^2} = 1.26 \text{ kip/ft}$$

compared to 1.24 assumed at the beginning of third cycle. The value of 1.26 kip/ft is considered close enough for design purposes.

8.2.7.4   CONTINUOUS SPANS

The above example illustrates the salient features of load balancing. These are that generally the prestressing force is selected to counteract or balance a portion of dead load, and under this loading condition the net stress in the tension fibers is limited to a value $= 6\sqrt{f_c'}$. If it is desired to design the member for zero stress at the bottom fiber at center span (or any other value less than the code allowed maximum value of $6\sqrt{f_c'}$.) it is only necessary to adjust the amount of post-tensioning provided in the member.

There are some qualifications to the above procedure that should be kept in mind when using the technique to continuous beams. Chief among them is the fact that it is not usually practical to install tendons with sharp break in curvature over supports as shown in Fig. 8.26a. The stiffness of tendons requires a reverse curvature (Fig. 8.26b) in the tendon profile with a point of contraflexure some distance from the supports. Although this reverse curvature modifies the equivalent loads imposed by post-tensioning from those assumed for a pure parabolic profile between the supports, a simple revision to the effective length of tendon as will be seen shortly, yields results sufficiently accurate for preliminary designs.

Consider the tendon profiles shown in Fig. 8.27 for a typical exterior and an interior span. Observe three important features.

1. The effective span L_e, is the distance between the inflection points which is considerably shorter than the actual span.

2. The sag or drape of the tendon numerically equal to an average height of inflection points, less the height of the tendon midway between the inflection points.

3. The point midway between the inflection points is not necessarily the lowest point on the profile.

The upward equivalent uniform load produced by the tendon is given by

$$w_p = \frac{8Pe}{L_e^2}$$

where w_p = equivalent upward uniform load due to prestress

P = prestress force

e = cable drape between inflection points

L_e = effective length between inflection points

Note that relatively high loads acting downward over the supports result from the sharply curved tendon profiles located within these regions (Fig. 8.28).

Since the large downward loads are confined to a small region, typically $\frac{1}{10}$ to $\frac{1}{8}$ of the span, their effect is secondary as compared to the upward loads. Slight differences occur in the negative moment regions between the external load moments and the moment due to prestressing force. The differences are of minor significance and can be neglected in the design without losing meaningful accuracy.

As in simple spans the moments caused by the equivalent loads are subtracted from those due to external loads, to obtain the net unbalanced moment which produces the flexural stresses. To the flexural stresses, the axial compressive stresses from the prestress are added to obtain the final stress distribution in the members. The maximum compressive and tensile stresses are compared to the allowable values. If the comparisons are favorable, an acceptable design has been found. If not either the tendon profile or force (and very rarely the cross-sectional shape of the structure) is revised to arrive at an acceptable solution.

In this method, since the moments due to equivalent loads are linearly related to the moments due to external loads, the designer can by-pass the usual requirement of determining the primary and secondary moments.

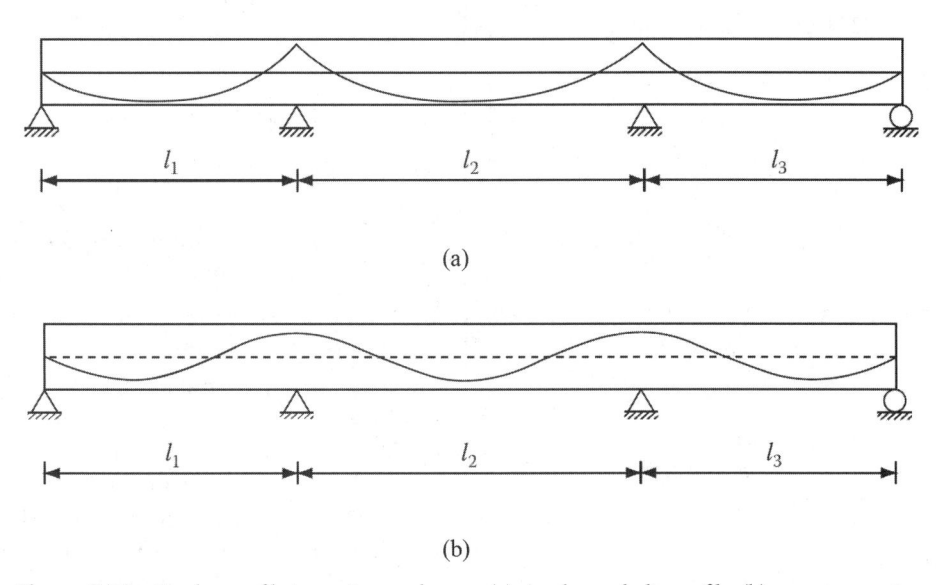

(a)

(b)

Figure 8.26   Tendon profile in continuous beams: (a) simple parabolic profile; (b) reverse curvature in tendon profile.

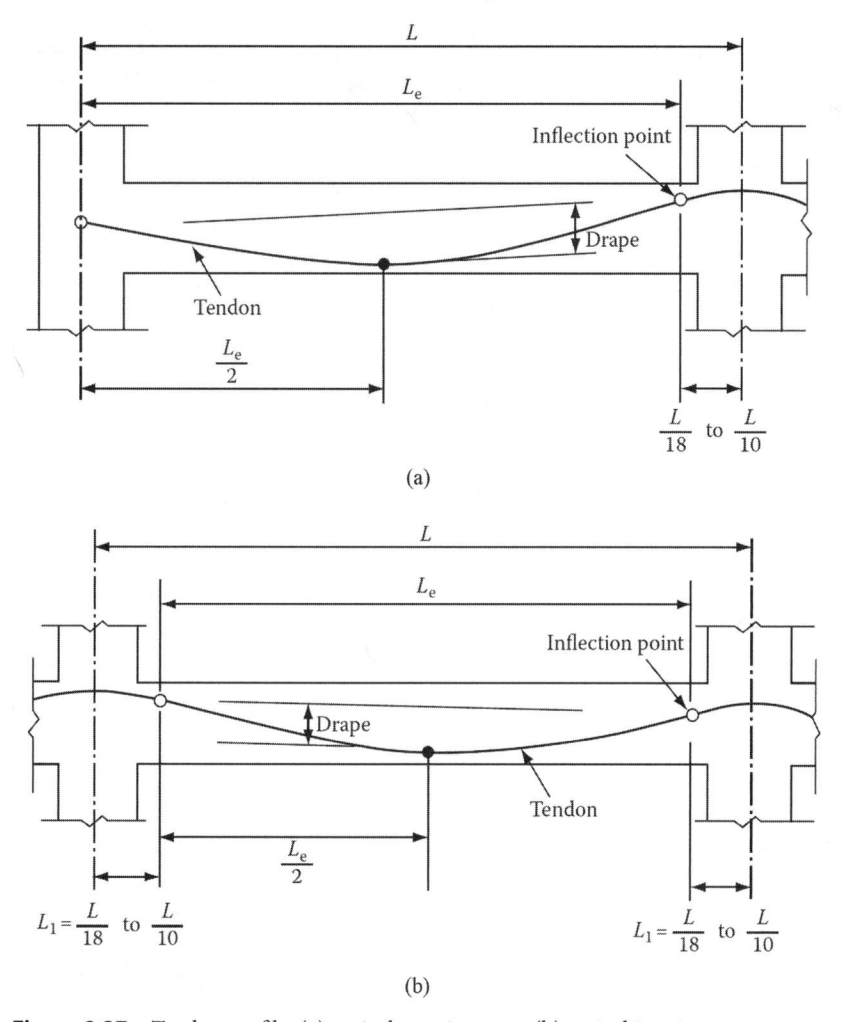

Figure 8.27 Tendon profile: (a) typical exterior span; (b) typical interior span.

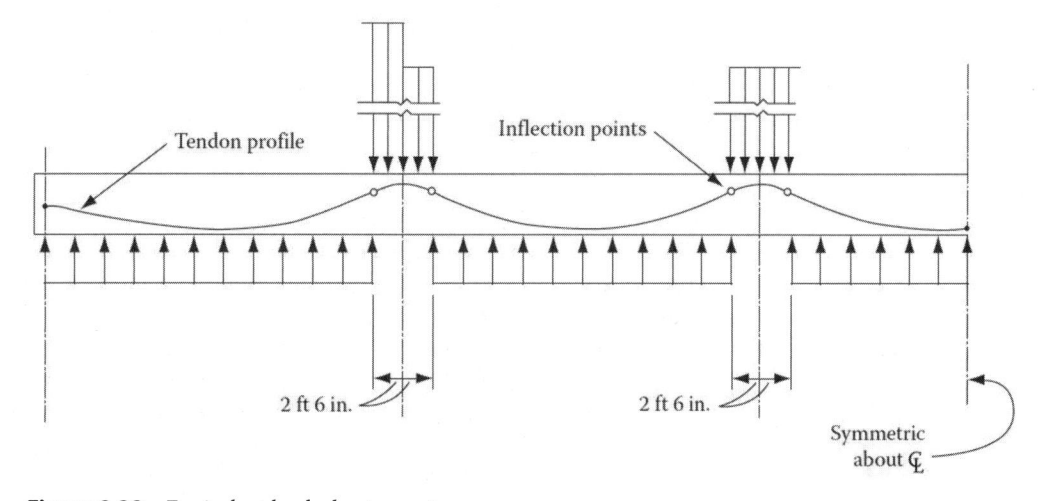

Figure 8.28 Equivalent loads due to prestress.

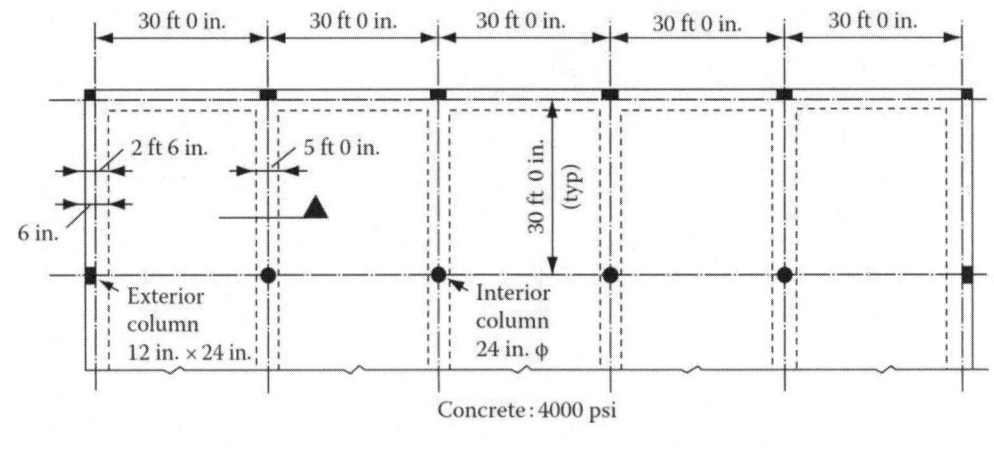

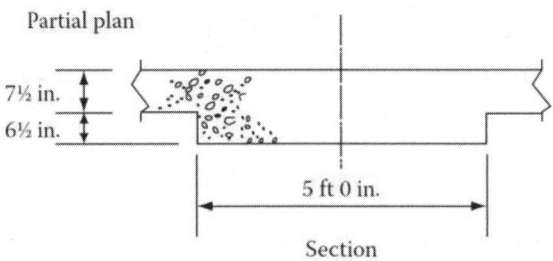

Figure 8.29   Example 1: one-way post-tensioned slab.

EXAMPLE 1: ONE-WAY SLAB   Given a 30′ 0″ column grid layout design a one-way slab spanning between beams is shown in Fig. 8.29.

Slab and beam depths:

Clear span of slab = 30 − 5 = 25 ft

Recommended slab depth = $\dfrac{\text{span}}{40} = \dfrac{25 \times 12}{40} = 7.5$ in.

Clear span for beams = 30 ft center-to-center span, less 2′ 0″ for column width = 30 − 2 = 28 ft

Recommended beam depth = $\dfrac{\text{span}}{25} = \dfrac{28 \times 12}{25} =$ 13.44 in. use 14 in.

Loading:

Dead load:
   7.5″ slab   = 94 psf
   Mech. & lights = 6 psf
   Ceiling   = 6 psf
   Partitions   = 20 psf
   Total dead load = <u>126 psf</u>

Live load:
   Office load = 100 psf
   Code minimum is 50 psf
   Use 100 psf per owner's request

   Total D + L = 226 psf

Slab design: Slab properties for 1′-0″ wide strip

$$I = \frac{bd^3}{12} = 12 \times \frac{7.5^3}{12} = 422 \text{ in.}^4$$

$$S_{\text{top}} = S_{\text{bot}} = \frac{422}{3.75} = 112.5 \text{ in.}^3$$

$$\text{Area} = 12 \text{ in.} \times 7.5 = 90 \text{ in.}^2$$

A 1 ft width of slab is analyzed as a continuous beam. The effect of column stiffness is ignored.

The moment diagram for a service load of 226 plf is shown in Fig. 8.30.

Moments at the face of supports have been used in the design instead of center line moments. Negative center line moments are reduced by a "$Va/3$" factor (V = shear at that support, a = total support width), and positive moments are reduced by $Va/6$ using average adjacent values for shear and support widths. A frame analysis may of course be used to obtain more accurate results.

The design of continuous strands will be based on the negative moment of 10.6 kip/ft. The additional prestressing required for the negative moment of 16.8 kip/ft will be provided by additional tendons in the end bays only.

Determination of Tendon Profile   Maximum tendon efficiency is obtained when the cable drape is as large as

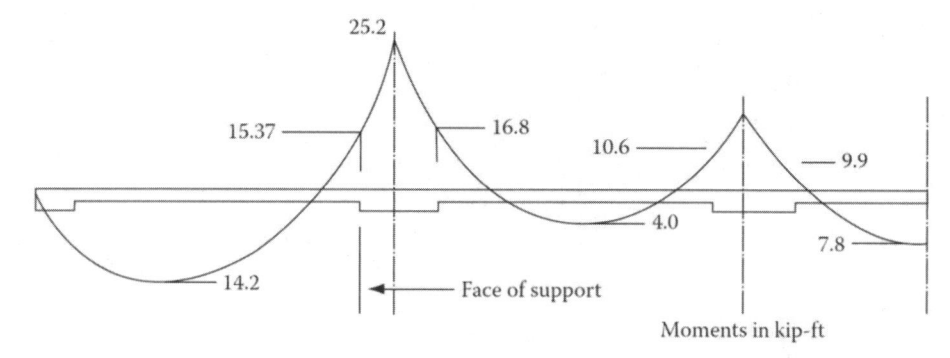

Figure 8.30 Example 1: one-way post-tensioned slab service load, $(D + L)$, moment diagram.

the structure will allow. Typically, the high points of the tendon over the supports and the low point within the span are dictated by concrete cover requirements and the placement of mild steel.

The high and low points of tendon in the interior bay of the example problem are shown in Fig. 8.31. Next, the location of inflection points are determined. For slabs, the inflection points usually range within $\frac{1}{16}$ to $\frac{1}{19}$ of the span. The fraction of span length used is a matter of judgment and is based on the type of structure. For this example, we choose $\frac{1}{16}$ of span which works out to 1′-10½″.

An interesting property useful in determining tendon profile shown in Fig. 8.32, is that, if a straight line (chord) is drawn connecting the tendon high point over the support, and the low point midway between, it intersects the tendon at the inflection point. Thus, the height of the tendon can be found by proportion. From the height, the bottom cover is subtracted to find the drape.

Referring to Fig. 8.32:

$$\text{Slope of the chord line} = \frac{h_1 - h_2}{(L_1 + L_2)}$$

$$h_3 = h_2 + L_2 \times (\text{Slope})$$

$$= h_2 + \frac{L_2(h_1 - h_2)}{(L_1 + L_2)}$$

$$\text{This simplifies to } h_3 = \frac{(h_1 L_2 + h_2 L_1)}{(L_1 + L_2)}$$

The drape h_d is obtained by subtracting h_2 from the above equation. Note, that notation e is also used in these examples to denote drape h_d.

The height of the inflection point as given above is exact for symmetrical layout of the tendon about the center span. If the tendon is not symmetrical, the value is approximate but sufficiently accurate for preliminary design.

Returning to our example problem we have $h_1 = 6.5$ in., $h_2 = 1$ in., $L_1 = 1.875$ ft, and $L_2 = 13.125$ ft.

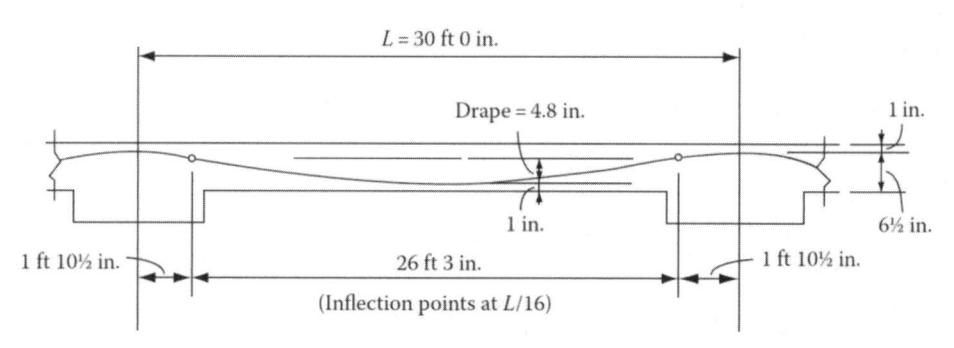

Figure 8.31 Example 1: one-way post-tensional slab tendon profile, interior bay.

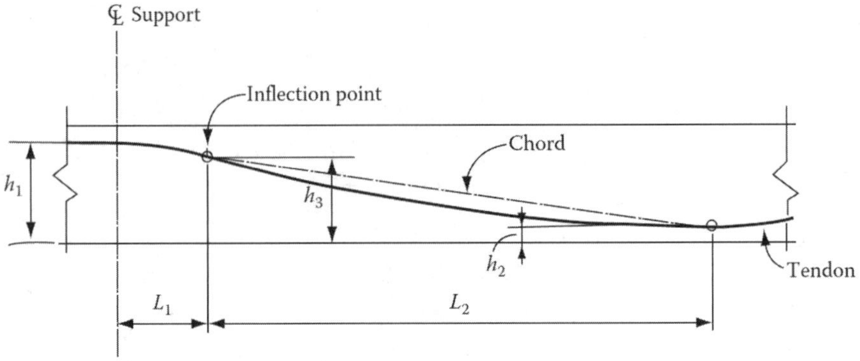

Figure 8.32 Dimensions for determining tendon drape.

Height of tendon at the inflection point:

$$h_3 = \frac{(h_1 L_2 + h_2 L_1)}{(L_1 + L_2)}$$

$$h_3 = \frac{6.5 \times 13.125 + 1 \times 1.875}{(1.875 + 13.125)} = 5.813''$$

Drape $h_d = e = 5.813 - 1 = 4.813''$ use 4.8″

Allowable stresses from ACI 318-19 are as follows.

f_t = Tensile stress = $6\sqrt{f_c'}$ (ACI 318-19, 8.3.4)

f_c = Compressive stress = $0.45 f_c'$ (ACI 318-19, Table 24.5.4.1)

For 4000 psi concrete:

$$f_t = 6\sqrt{4000} = 380 \text{ psi}$$
$$f_c = 0.45 \times 4000 = 1800 \text{ psi}$$

Design of Through Strands The design procedure is started by making an initial assumption of the equivalent load produced by the prestress. A first value of 65 percent of the total dead load is used.

First Cycle Assume

$$w_p = 0.65 \, w_d$$

where w_p = Equivalent upward load due to post-tensioning
w_d = Total dead load
∴ $w_p = 0.65 \times 126 = 82$ plf

The balancing moment caused by the equivalent load is calculated from

$$M_{pt} = M_s \frac{w_{pt}}{w_s}$$

where M_{pt} = balancing moment due to equivalent load (also indicated by notation M_b).
M_s = moment due to service load, $D + L$

In our example, $M_s = 10.6$ kip ft for the interior span.

$$M_{pt} = 10.6 \times \frac{82}{226} = 3.85 \text{ kip ft}$$

Next, M_{pt} is subtracted from M_s to give the unbalanced moment M_{ub}. The flexural stresses are then obtained by dividing M_{ub} by the section moduli of the structure's cross-section at the point where M_s is determined: Thus

$$f_t = \frac{M_{ub}}{S_t} \tag{8.1}$$

$$f_b = \frac{M_{ub}}{S_b} \tag{8.2}$$

In our case, $M_{ub} = 10.6 - 3.85 = 6.75$ kip ft. The flexural stress at the top of the section is found by

$$f_t = \frac{M_{ub}}{S_t} = \frac{6.75 \times 12}{112.5} = 0.72 \text{ ksi}$$

The minimum required compressive prestress is found by subtracting the maximum allowable tensile stress f_a given below, from the tensile stresses from Eqs. (8.1) and (8.2). Thus the smallest required compressive stress is:

$$f_p = f_{ts} - f_a$$

where f_{ts} = the tensile stress found from Eqs. (8.1) to (8.2) depending upon the sign of the moment
$f_a = 6\sqrt{f_c'}$ for one-way slabs or beams for the negative zones
$f_a = 2\sqrt{f_c'}$ for positive moments in two-way slabs

In our case,

$$f_p = 0.720 - 0.380 = 0.34 \text{ ksi}$$

and

$$P = 0.34 \times 7.5 \times 12 = 30.60 \text{ kip/ft}$$

Use the following equation to find the equivalent load due to prestress

$$w_p = \frac{8Pe}{L_e^2}$$

$$= 8 \times \frac{30.6 \times 4.81}{12 \times (26.25)^2} = 0.142 \text{ klf} = 142 \text{ plf}$$

This is more than 82 plf. N.G.

Since the derived value of w_p is not equal to the initial assumed value, the procedure is repeated until convergence is achieved. Convergence is rapid by using a new initial value for the subsequent cycle, equal to 75 percent of the previous initial value w_{p1}, plus 25 percent of the derived value w_{p2}, for that cycle.

Second Cycle Use the above criteria to find the new value of w_p for the second cycle

$$w_p = 0.75w_{p1} + 0.25w_{p2} = 0.75 \times 82 + 0.25 \times 142 = 97 \text{ plf}$$

$$M_b = \frac{97}{226} \times 10.6 = 4.55 \text{ kip ft}$$

$$M_{ub} = 10.6 - 4.55 = 6.05 \text{ kip ft}$$

$$f_t = f_b = \frac{6.05 \times 12}{112.5} = 0.645 \text{ ksi}$$

$$f_p = 0.645 - 0.380 = 0.265 \text{ ksi}$$

$$P = 0.265 \times 90 = 23.89 \text{ kips}$$

$$w_p = \frac{8 \times 23.89 \times 4.81}{12 \times (26.25)^2} = 0.111 \text{ klf} = 111 \text{ plf}$$

This is more than 97 psf. N.G.

Third Cycle

$$w_p = 0.75 \times 97 + 0.25 \times 111 = 100.5 \text{ plf}$$

$$M_b = \frac{100.5}{226} \times 10.6 = 4.71 \text{ kip ft}$$

$$M_{ub} = 10.6 - 4.71 = 5.89 \text{ kip ft}$$

$$f_t = f_b = \frac{5.89 \times 12}{112.5} = 0.629 \text{ ksi}$$

$$f_p = 0.629 - 0.380 = 0.248 \text{ ksi}$$

$$P = 0.248 \times 90 = 22.3 \text{ kips}$$

$$w_p = \frac{8 \times 22.3 \times 4.81}{12 \times (26.25)^2} = 0.104 \text{ klf} = 104 \text{ plf}$$

This is nearly equal to 100.5 plf, therefore, satisfactory.
Check compressive stress at the section.

Bottom flexural stress = 0.629 ksi.

Direct axial stress due to prestress $= \frac{22.3}{90} = 0.246$ ksi

Total compressive stress = 0.629 + 0.246 = 0.876 ksi is less than $0.45 f_c' = 1.8$ ksi. Therefore, satisfactory.

End Bay Design Design end bay prestressing using the same procedure for a negative moment of 15.37 kip ft.

Assume that at left support the tendon is anchored at the center of gravity of slab with a reversed curvature as shown in Fig. 8.36, profile 1. Assume further that the center of gravity of tendon is at a distance of 1.75 in. from the bottom of slab. With these assumptions we have: $h_1 = 3.75$ in., $h_2 = 1.75$ in., $L_1 = 1.875$ ft, and $L_2 = 13.125$ ft.

The height of the tendon inflection point at left end:

$$h_3 = \frac{3.75 \times 13.125 + 1.75 \times 1.875}{15} = 3.25 \text{ in.}$$

The height of the right end:

$$h_3 = \frac{6.5 \times 13.125 + 1.75 \times 1.875}{15} = 5.906 \text{ in.}$$

$$\text{Average height of tendon} = \frac{3.25 + 5.906}{2}$$

$$= 4.578'' \text{ use } 4.6 \text{ in.}$$

Drape $h_d = e = 4.6 - 1.75 = 2.85$ in.

First Cycle We start with the first cycle, as for the interior span, by assuming $w_{pt} = 82$ plf.

$$M_{pt} = 15.37 \times \frac{82}{226} = 5.58 \text{ kip ft}$$

$$M_{ub} = 15.37 - 5.58 = 9.79 \text{ kip ft}$$

$$f_t = f_b = \frac{9.79 \times 12}{112.5} = 1.04 \text{ ksi}$$

$$f_p = 1.04 - 0.380 = 0.664 \text{ ksi}$$

$$P = 0.664 \times 90 = 59.7 \text{ kip/ft}$$

$$w_p = \frac{8 \times 59.7 \times 2.85}{12 \times (26.25)^2} = 0.165 \text{ klf} = 165 \text{ plf}$$

This is more than 82 plf. N.G.

Second Cycle

$$w_p = 0.75 \times 82 + 0.25 \times 165 = 103 \text{ plf}$$

$$M_{pt} = 15.37 \times \frac{103}{226} = 7.0 \text{ kip ft}$$

$$M_{ub} = 15.37 - 7.0 = 8.37 \text{ kip ft}$$

$$f_t = f_b = \frac{8.37 \times 12}{112.5} = 0.893 \text{ ksi}$$

$$f_p = 0.893 - 0.380 = 0.513 \text{ ksi}$$

$$P = 0.513 \times 90 = 46.1 \text{ kips}$$

$$w_p = \frac{8 \times 46.1 \times 2.85}{12 \times (26.25)^2} = 0.127 \text{ klf} = 127 \text{ plf}$$

This is more than 103 plf. N.G.

Third Cycle

$$w_p = 0.75 \times 103 + 0.25 \times 127 = 109 \text{ plf}$$

$$M_{pt} = 15.37 \times \frac{109}{226} = 7.41 \text{ kip ft}$$

$$M_{ub} = 15.37 - 7.41 = 7.96 \text{ kip ft}$$

$$f_t = f_b = \frac{7.96 \times 12}{112.5} = 0.849 \text{ ksi}$$

$$f_p = 0.849 - 0.380 = 0.469 \text{ ksi}$$

$$P = 0.469 \times 90 = 42.21 \text{ kips}$$

$$w_p = \frac{8 \times 42.21 \times 2.85}{12 \times (26.25)^2} = 0.116 \text{ klf} = 116 \text{ plf}$$

This is nearly equal to 109 plf used at the start of third cycle.

Therefore, satisfactory.
Check compressive stress at the section:

$$f_b = 0.849 \text{ ksi}$$

$$\text{Axial stress due to prestress} = \frac{42.21}{90} = 0.469 \text{ ksi}$$

$$\text{Total compressive stress} = 0.849 + 0.469 = 1.318 \text{ ksi}$$

This is less than 1.8 ksi. Therefore, the design is O.K.
Check the design against the positive moment of 14.33 kip ft

$$w_p = 116 \text{ pfl}$$

$$M_b = 14.33 \times \frac{116}{226} = 7.36 \text{ kip ft}$$

$$M_{ub} = 14.33 - 7.36 = 6.97 \text{ kip ft}$$

$$\text{Bottom flexural stress} = \frac{6.97 \times 12}{112.5} = 0.744 \text{ ksi (tension)}$$

$$\text{Axial compression due to prestress} = \frac{42.21}{12 \times 7.5} = 0.469 \text{ ksi}$$

$$\text{Tensile stress at bottom} = 0.744 - 0.469 = 0.275 \text{ ksi}$$

This is less than 0.380 ksi. Therefore, the end bay design is O.K.

EXAMPLE 2: BEAM DESIGN, POSTTENSIONED CONTINUOUS BEAM DESIGN   Refer to Fig. 8.33 for dimensions and loading. Determine flange width of beam using the criteria given in the ACI 318-19.

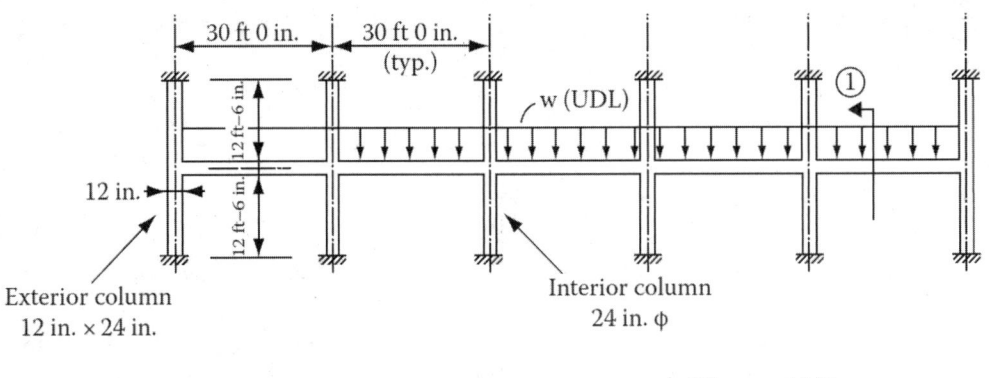

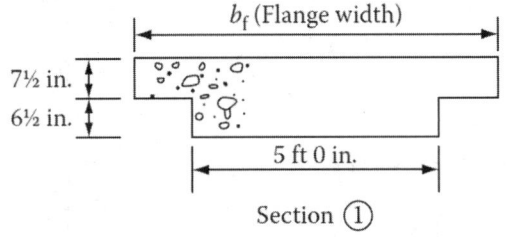

Figure 8.33   Example 2: post-tensioned beam, dimensions and loading.

The flange width b_f is the least of
1. Span/4
2. Web width + 16 × (flange thickness)
3. Web width + $\frac{1}{2}$ clear distance to next web

Therefore,

$$b_f = \frac{30}{4} = 7.5 \text{ ft (controls)}$$

$$= 5 + 16 \times \frac{7.5}{12} = 15 \text{ ft}$$

$$= 5 + \frac{25}{2} = 17.5 \text{ ft}$$

Section properties

$$I = 16{,}650 \text{ in.}^4 \quad Y = 7.69 \text{ in.}$$

$$S_t = 2637 \text{ in.}^3$$

$$S_b = 2166 \text{ in.}^3$$

$$A = 1065 \text{ in.}^2$$

Loading

$$\text{Dead load of } 7\frac{1}{2} \text{ in slab} = 94 \text{ psf}$$

$$\text{Mech. \& Elec.} = 6 \text{ psf}$$

$$\text{Ceiling} = 6 \text{ psf}$$

$$\text{Partitions} = 20 \text{ psf}$$

$$\text{Additional dead load due to beam self wt} = \frac{615 \times 60 \times 150}{144 \times 30} =$$

$$13.5 = 14 \text{ psf}$$

$$\text{Total dead load} = 140 \text{ psf}$$

$$\text{Live load at owner's request} = 80 \text{ psf}$$

$$D + L = 220 \text{ psf}$$

Uniform load per ft of beam = 0.220 × 30 = 6.6 klf. The resulting service load moments are shown in Fig. 8.34. As before we design for the moments at the face of supports.

Interior Span Calculate through tendons by using interior span moment of 427 kip-ft at the inside face of third column (Fig. 8.34).

Assume $h_1 = 11.5$ in., $h_2 = 2.5$ in., $L_1 = 2.5$ ft, $L_2 = 12.5$ ft. Refer to Fig. 8.32 for notations.

The height of inflection point

$$h_3 = \frac{11.5 \times 12.5 + 2.5 \times 2.5}{15} = 10 \text{ in.}$$

$$h_d = e = 10 - 2.5 = 7.5 \text{ in.}$$

First Cycle Assume $w_p = 3.5$ klf

$$w_p = 3.5 \text{ klf}$$

$$M_p = \frac{3.5}{6.6} \times 427 = 226 \text{ kip ft}$$

$$M_{ub} = 427 - 226 = 201 \text{ kip ft}$$

$$f_t = \frac{201 \times 12}{2637} = 0.915 \text{ ksi}$$

$$f_p = 0.915 - 0.380 = 0.535 \text{ ksi}$$

$$P = 0.535 \times 1065 = 570 \text{ kips}$$

$$w_p = \frac{8 \times 570 \times 7.5}{12 \times (27.5)^2} = 3.77 \text{ klf}$$

which is greater than 3.5 klf. N.G.

Second Cycle New value of

$$w_p = 0.75 \times 3.5 + 0.25 \times 3.77 = 3.57 \text{ klf}$$

$$M_p = \frac{3.57}{6.6} \times 427 = 231 \text{ kip ft}$$

$$M_{ub} = 427 - 231 = 196 \text{ kip ft}$$

$$f_t = \frac{196 \times 12}{2637} = 0.892 \text{ ksi}$$

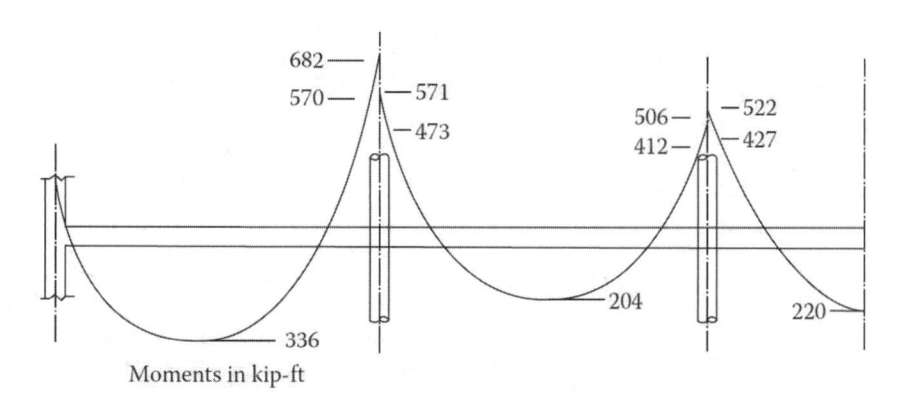

Figure 8.34 Example 2: post-tensioned beam service load moments.

$$f_p = 0.892 - 0.380 = 0.512 \text{ ksi}$$

$$P = 0.512 \times 1065 = 545 \text{ kips}$$

$$w_p = \frac{8 \times 545 \times 7.5}{12 \times (27.5)^2} = 3.60 \text{ klf}$$

which is nearly equal to 3.57 klf. Therefore, design is satisfactory.

Check design against positive moment of 220 kip/ft.

$$M_p = \frac{3.6 \times 220}{6.6} = 120 \text{ kip ft}$$

$$M_{ub} = 220 - 120 = 100 \text{ kip ft}$$

$$F_{\text{bot}} = \frac{100 \times 12}{2166} = 0.554 \text{ ksi (tension)}$$

$$\text{Axial comp. stress} = \frac{545}{1065} = 0.512 \text{ ksi (comp.)}$$

$$f_{\text{total}} = 0.554 - 0.512 = 0.042 \text{ ksi (tension)}$$

This is less than the allowable tensile stress of 0.380 ksi. Therefore, the design is satisfactory.

End Span Determine end bay prestressing for a negative moment of 570 kip-ft at the face of first interior column (Fig. 8.34).

First Cycle As before, assume $w_p = 3.5$ klf

$$w_p = 3.5 \text{ klf}$$

$$M_p = \frac{3.5}{6.6} \times 570 = 302 \text{ kip ft}$$

$$M_{ub} = 570 - 302 = 268 \text{ kip ft}$$

$$f_t = \frac{268 \times 12}{2637} = 1.22 \text{ ksi}$$

$$f_p = 1.22 - 0.380 = 0.84 \text{ ksi}$$

$$P = 0.84 \times 1065 = 894 \text{ kips}$$

$$w_p = \frac{8 \times 894 \times 7.5}{12 \times (27.5)^2} = 5.912 \text{ klf}$$

which is greater than 3.5 klf. N.G.

Second Cycle New value of

$$w_p = 0.74 \times 3.5 + 0.25 \times 5.912 = 4.1 \text{ klf}$$

$$M_p = \frac{4.1}{6.6} \times 570 = 354 \text{ kip ft}$$

$$M_{ub} = 570 - 354 = 216 \text{ kip ft}$$

$$f_t = \frac{216 \times 12}{2637} = 0.983 \text{ ksi}$$

$$f_p = 0.983 - 0.380 = 0.603 \text{ ksi}$$

$$P = 0.603 \times 1065 = 642 \text{ kips}$$

$$w_p = \frac{8 \times 642 \times 7.5}{12 \times (27.5)^2} = 4.24 \text{ klf}$$

This is nearly equal to 4.1 klf. However, a more accurate value is calculated as follows:

$$w_p = 0.75 \times 4.1 + 0.25 \times 4.24 = 4.13 \text{ klf}$$

Check the design against positive moment of 336 kip-ft

$$M_p = \frac{4.13}{6.6} \times 336 = 210 \text{ kip ft}$$

$$M_{ub} = 336 - 210 = 126 \text{ kip ft}$$

$$\text{Bottom flexural stress} = \frac{126 \times 12}{2166} = 0.698 \text{ ksi (tension)}$$

Axial compressive stress due to post-tension

$$= \frac{642}{1065} = 0.603 \text{ ksi (comp.)}$$

$$f_{\text{total}} = 0.698 - 0.603 = 0.095 \text{ ksi}$$

This is less than the allowable tensile stress of 0.380 ksi. Therefore, the design is O.K.

EXAMPLE 3: FLAT PLATE SYSTEM Figure 8.35 shows a schematic section of a two-way flat plate system. Design of post-tension slab for an office-type loading is required.

Given

Specified compressive strength of concrete, $f_c' = 4000$ psi

Modulus of elasticity of concrete, $E_c = 3834$ ksi

Allowable tensile stress in precompressed tensile zone = $6\sqrt{f_c'} = 380$ psi

Allowable fiber stress in compression = $0.45 f_c' = 0.45 \times 4000 = 1800$ psi

Tendon cover: Interior spans   Top 0.75 in.
                                 Bot. 0.75 in.

               Exterior spans   Top 0.75 in.
                                 Bot. 1.50 in.

Tendon diameter = ½ in.

   Minimum area of bonded reinforcement:
   In negative moment areas at column supports:

$$A_s = 0.00075 \, A_{cf}$$

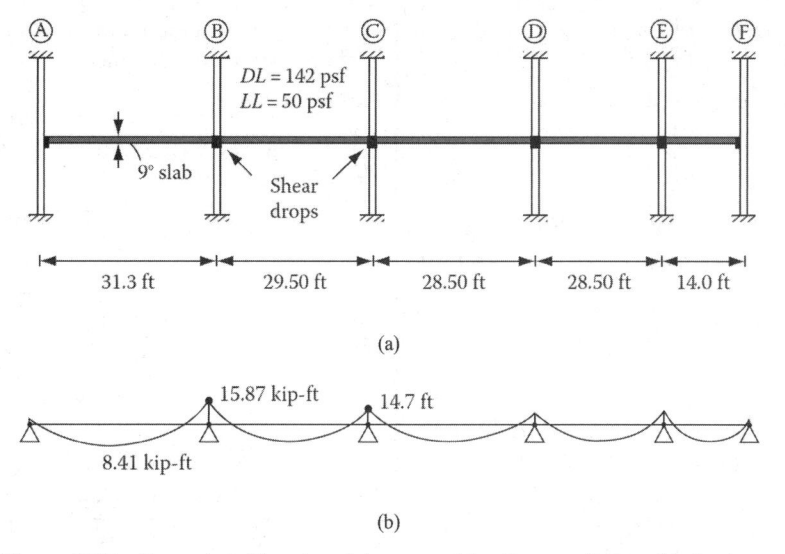

Figure 8.35 Example 3: Flat plate: (a) span and loading conditions; (b) elastic moments due to dead plus live loads.

where A_{cf} is the larger gross cross-sectional area of the slab beam strips of two orthogonal equivalent frames intersecting at a column of a two-way slab, in in.^{2}

In positive moment areas where computed concrete stress in tension exceeds $2\sqrt{f_c'}$:

$$A_s = \frac{N_c}{0.5 f_y}$$

Rebar yield: 60 ksi. Max bar size = #5

Rebar cover 1.63 in. at top and bottom

Post-tension requirements

Minimum post-tensioned stress = 125 psi
(see ACI 318-19, section 8.2.3)

Minimum balanced load = 65 percent of total dead load

Design

The flat plate is sized using the span-depth ratios given in Table 8.9.

The maximum span is 31 ft 4 in. between grids A and B. Using a span-depth ratio of 40, the slab thickness is $\frac{31.33 \times 12}{40} = 9.4$ in., rounded to 9 in.

The flat plate has "shear drops" intended to increase only the shear strength and flexural support width. The shear heads are smaller than a regular drop panel as defined in the ACI 318-19. Therefore, shear heads cannot be included in calculating the bending resistance.

Loading: Dead load of 9″ slab   112 psf
Partitions   20 psf

Ceiling and mechanical   10 psf
Reduced live load   50 psf

Total service load = 112 + 20 + 10 + 50 = 192 psf

Ultimate load = 1.4 × 142 + 1.7 × 50 = 285 psf

Slab properties (for a 1 ft-wide strip)

$$I = \frac{bh^3}{12} = 12 \times \frac{9^3}{12} = 729 \text{ in.}^4$$

$$S_{top} = S_{bot} = \frac{729}{4.5} = 162 \text{ in.}^3$$

$$\text{Area} = 12 \times 9 = 108 \text{ in.}^2$$

The moment diagram for a 1 ft-wide strip of slab subjected to a service load of 192 psf is shown in Fig. 8.35b.

The design of continuous strands will be based on a negative moment of 14.7 kip-ft at the second interior span. The end bay prestressing will be based on a negative moment of 15.87 kip-ft.

Interior Span Calculate the drape of tendon using procedure given for the previous problem. See Figs. 8.32 and 8.36.

$$h_3 = \frac{h_1 L_2 + h_2 L_1}{L_1 + L_2}$$

$L_1 = 1.84 \text{ ft}$ $h_1 = 8 \text{ in.}$

$L_2 = 12.90 \text{ ft}$ $h_2 = 1.25 \text{ in.}$

$L_e = 12.9 \times 2 = 25.8 \text{ ft}$

$$h_3 = \frac{8 \times 12.90 + 1.25 \times 1.84}{14.75} = 7.153$$

Tendon drape $= 7.153 - 1.25 = 5.90$ in.

First Cycle

Minimum balanced load $= 0.65 \times$ (total DL)

$$= 0.65(112 + 10 + 20) = 92 \text{ psf}$$

Moment due to balanced load $= \dfrac{92}{192} \times 14.7$

$$= 7.04 \text{ kip ft}$$

This is subtracted from the total service load moment of 14.7 kip-ft to obtain the unbalanced moment M_{ub}.

$$M_{ub} = 14.7 - 7.04 = 7.66 \text{ kip-ft}$$

The flexural stresses at top and bottom are obtained by dividing M_{ub} by the section moduli of the structure's cross-section.

$$f_t = \frac{7.66 \times 12}{162} = 0.567 \text{ ksi}$$

$$f_b = \frac{7.66 \times 12}{162} = 0.567 \text{ ksi}$$

The minimum required compressive prestress f_p is found by subtracting the maximum allowable tensile stress $f_a = 6\sqrt{f_c'}$ from the calculated tensile stress. Thus, the smallest required compressive stress is:

$$f_p = f_t - f_a$$

$$f_p = 0.567 - 0.380 = 0.187 \text{ ksi}$$

The prestress force is calculated by multiplying f_p by the cross-sectional area:

$$P = 0.187 \times 9 \times 12 = 20.20 \text{ kip/ft}$$

Determine the equivalent load due to prestress force P by the relation

$$w_p = \frac{8Pe}{L_e^2}$$

For the example problem,

$$P = 20.2 \text{ kip/ft}, \ e = 5.90 \text{ in.}$$

$$L_e = 2 \times 12.90 = 25.8 \text{ ft}$$

$$\therefore \ w_p = \frac{8 \times 20.20 \times 5.90}{25.8^2 \times 12} = 0.120 \text{ klf} = 120 \text{ plf}$$

Comparing this with the value of 92 plf assumed at the beginning of first cycle, we find the two values are not equal.

Therefore, we assume a new value and repeat the procedure until convergence is obtained.

Second Cycle

$$w_p = 0.75 \times 92 + 0.25(120) \ 99 \text{ plf}$$

$$M_b = \frac{99}{192} \times 14.7 = 7.58 \text{ kip ft}$$

$$M_{ub} = 14.7 - 7.58 = 7.12 \text{ kip ft}$$

$$f_t = f_b = \frac{7.12 \times 12}{162} = 0.527 \text{ ksi}$$

$$f_p = 0.527 - 0.380 = 0.147 \text{ ksi}$$

$$P = 0.147 \times 9 \times 12 = 15.92 \text{ kip/ft}$$

$$w_p = \frac{8 \times 15.92 \times 5.90}{25.8^2 \times 12} = 0.094 \text{ klf} = 94 \text{ plf}$$

This is less than 99 plf assumed at the beginning of second cycle. Therefore, we assume a new value and repeat the procedure.

Third Cycle

$$w_p = 0.75 \times 99 + 0.25(94) = 97.7 \text{ plf}$$

$$M_b = \frac{99.7 \times 14.7}{192} = 7.48 \text{ kip ft}$$

$$M_{ub} = 14.7 - 7.48 = 7.22 \text{ kip ft}$$

$$f_t = f_b = \frac{7.22 \times 12}{162} = 0.535 \text{ ksi}$$

$$f_p = 0.535 - 0.380 = 0.155 \text{ ksi}$$

$$P = 0.155 \times 9 \times 12 = 16.74 \text{ kip/ft}$$

$$w_p = \frac{8 \times 16.74 \times 5.90}{25.8^2 \times 12} = 0.99 \text{ klf} = 99 \text{ plf}$$

This is nearly equal to 97.7 plf assumed at the beginning of the third cycle. Therefore O.K.

Check compressive stress at the support:

$$M_p = \frac{99 \times 14.7}{192} = 7.58 \text{ kip ft}$$

$$M_{ub} = 14.7 - 7.58 = 7.12 \text{ kip ft}$$

$$f_b = \frac{7.12 \times 12}{162} = 0.527 \text{ ksi} = 527 \text{ psi}$$

Axial compressive stress due to post-tension $= \dfrac{16.74 \times 1000}{9 \times 12}$

$$= 155 \text{ psi}$$

Total compressive stress $= 527 + 155 = 682 \text{ psi}$

This is less than the allowable compressive stress of 1800 psi. Therefore, the design is satisfactory.

End Bay Design   The placement of tendon within the end bay presents a few problems. The first problem is in determining the location of the tendon over the exterior support. Placing the tendon above the neutral axis of the member results in an increase in the total tendon drape allowing the designer to use less prestress than would otherwise be required. Raising the tendon, however, introduces an extra moment that effectively cancels out some of the benefits from the increased drape. For this reason, the tendon is usually placed at neutral axis at exterior supports.

The second problem is in making a choice in the tendon profile: whether to use a profile with a reverse curvature over each support (profile 1), or over the first interior support only (profile 2). See Fig. 8.36. A profile with the reversed curvature over the first interior support only, gives a greater cable drape than the first profile suggesting a larger equivalent load with the same amount of prestress. On the other hand, the effective length L_e between inflection points of profile 1 is less than that of profile 2 which suggests the opposite. To determine which profile is in fact more efficient, it is necessary to evaluate the amount of prestress for both profiles. More usually, a tendon profile with reverse curvature over both supports is 5 to 10 percent more efficient since the equivalent load produced is a function of the square of the effective length.

The last item addresses the extra end bay prestressing required in most situations. The exterior span, in an equal span structure has the greatest moments due to support rotations. Because of this, extra prestressing is commonly added to end bays to allow efficient design of end spans. Although for design purposes, the extra end bay prestressing is considered to act within the end bay only, these tendons actually extend well into the adjacent span for anchorage as shown in Fig. 8.37. Advantage can be taken of this condition by designing the through tendons using the largest moment found within the interior

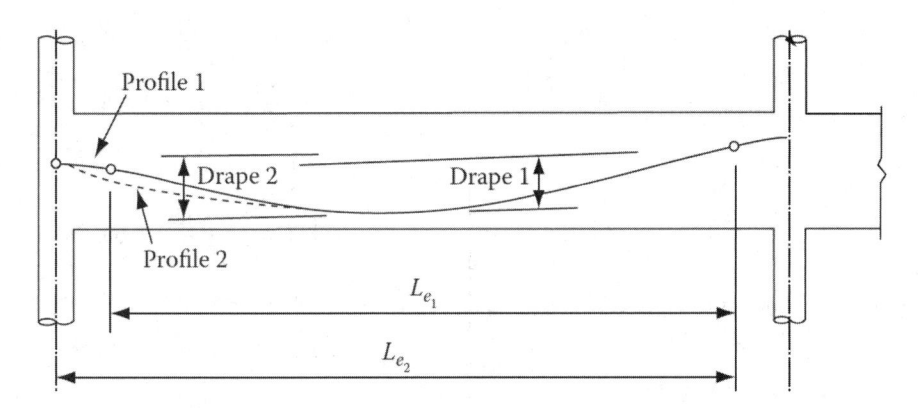

Figure 8.36   End bay tendon profiles.

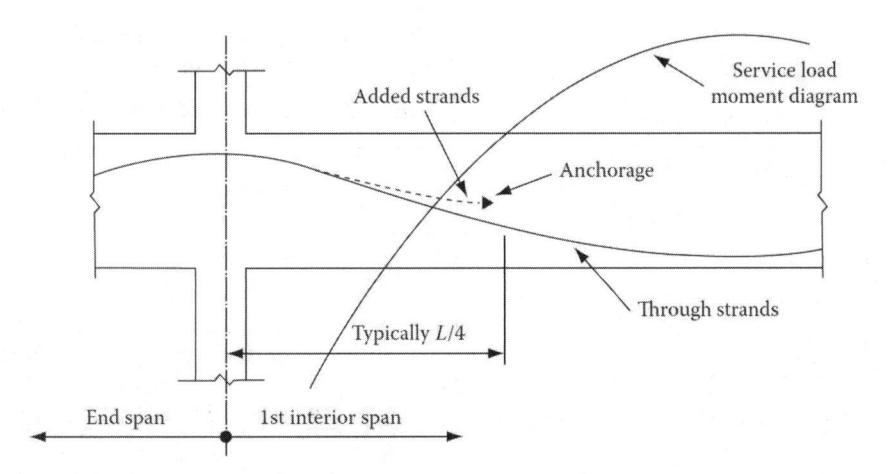

Figure 8.37   Anchorage of added tendons.

spans including the moment at the interior face of the first support. The end bay prestress force is determined using the largest moment within the exterior span. The stress at the inside face of the first support is checked using the equivalent loads produced by the through tendons and the axial compression provided by both the through and added tendons. If the calculated stresses are less than the allowable values, the design is complete. If not, more stress is provided either by through tendons or added tendons or both.

The design of end bay using profiles 1 and 2 follows.

Profile 1: Reverse curvature at the right support only (Fig. 8.38b). Observe the height of inflection point is exact if the tendon profile is symmetrical about the center of span. If it is not, as in span 1 of the example problem, a sufficiently accurate value can be obtained by taking the average of the tendon inflection point at each end as follows.

Left End

$$h_3 = \frac{4.5 \times 15.65 + 1.75 \times 0}{0 + 15.67} = 4.5 \text{ in.}$$

Right End

$$h_3 = \frac{8 \times 13.7 + 1.75 \times 1.95}{(1.95 + 13.70)} = 7.22 \text{ in.}$$

$$\text{Average } h_3 = \frac{4.5 + 7.22}{2} = 5.86 \text{ in.}$$

$$\text{Drape} = 5.86 - 1.75 = 4.11 \text{ in.}$$

First Cycle To show the quick convergence of the procedure, we start with a rather high value of

$$w_p = 0.75 \text{ DL} = 0.75 \times 142 = 106 \text{ plf}$$

$$M_b = \frac{106}{192} \times 15.87 = 8.76 \text{ kip ft}$$

$$M_{ub} = 15.87 - 8.76 = 7.11 \text{ kip ft}$$

$$f_t = f_b = \frac{7.11 \times 12}{162} = 0.527 \text{ ksi}$$

$$f_p = 0.527 - 0.380 = 0.147 \text{ ksi}$$

$$P = 0.147 \times 9 \times 12 = 15.87 \text{ kips}$$

$$w_p = \frac{8Pe}{L_e^2}$$

$$= \frac{8 \times 15.87 \times 4.11}{29.35^2 \times 12}$$

$$= 0.050 \text{ klf} = 50.0 \text{ plf}$$

This is less than 106 plf. N.G.

Second Cycle

$$w_p = 0.75(106) + 0.25(50.0) = 92 \text{ plf}$$

$$M_b = \frac{92}{192} \times \frac{92}{192} \times 15.87 = 7.60 \text{ kip ft}$$

$$M_{ub} = 15.87 - 7.60 = 8.27 \text{ kip ft}$$

$$f_t = f_b = \frac{8.27 \times 12}{162} = 0.612 \text{ ksi}$$

$$f_p = 0.612 - 0.380 = 0.233 \text{ ksi}$$

$$P = 0.233 \times 12 \times 9 = 25.16 \text{ kip/ft}$$

$$w_p = \frac{8 \times 25.16 \times 4.11}{29.35^2 \times 12}$$

$$= 0.080 \text{ klf} = 80 \text{ plf}$$

This is less than 91.5 psi used at the beginning of second cycle N.G.

Third Cycle

$$w_p = 0.75 \times 92 + 0.25 \times 80 = 89 \text{ plf}$$

$$M_b = \frac{89}{192} \times 15.87 = 7.356 \text{ kip ft}$$

$$M_{ub} = 15.87 - 7.356 = 8.5 \text{ kip ft}$$

$$f_t = f_b = \frac{8.5 \times 12}{162} = 0.631 \text{ ksi}$$

$$f_p = 0.631 - 0.380 = 0.251 \text{ ksi}$$

$$P = 0.251 \times 12 \times 9 = 27.10 \text{ kips}$$

$$w_p = \frac{8 \times 27.10 \times 4.11}{29.35^2 \times 12} = 0.086 \text{ klf} = 86 \text{ plf}$$

This is nearly equal to 89 plf used at the beginning of third cycle. Therefore O.K.

Profile 2. Reverse curvature over each support (Fig. 8.38c)
Left end

$$h_3 = \frac{4.5 \times 13.70 + 1.75 \times 1.95}{(13.70 + 1.95)} = 4.156 \text{ in.}$$

Right end

$$h_3 = \frac{8 \times 13.70 + 1.75 \times 1.95}{(13.70 + 1.95)} = 7.221 \text{ in.}$$

$$\text{Average } h_3 = \frac{4.516 + 7.221}{2} = 5.689 \text{ in.}$$

$$e = h_d = 5.689 - 1.75 = 3.939 \text{ in.}$$

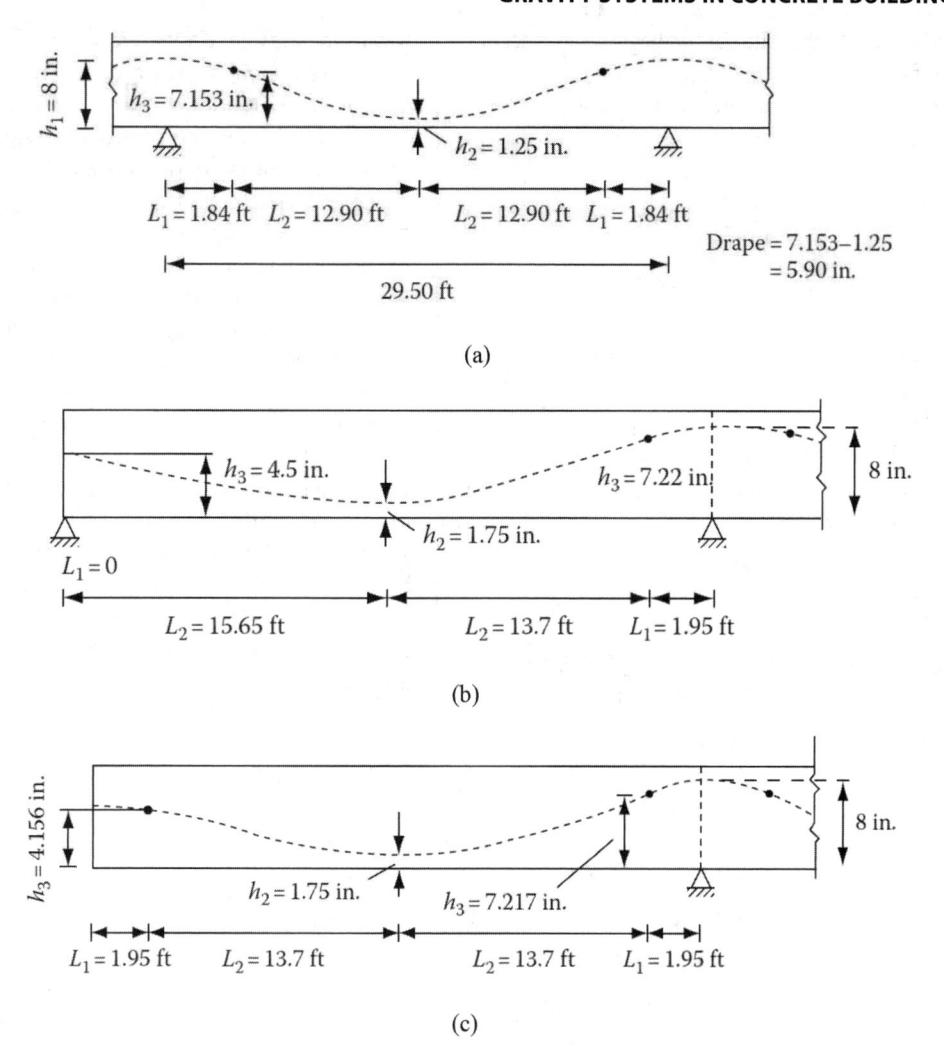

Figure 8.38 Example problem 3: flat plate, tendon profiles: (a) interior span; (b) exterior span, reverse curvature at right support; (c) exterior span, reverse curvature at both supports.

First Cycle   We start with an assumed balanced load of 0.65 DL = 92 plf

Balanced moment $M_b = 15.87 \times \dfrac{92}{192} = 7.60$ kip ft

$M_{ub} = 15.87 - 7.6 = 8.27$ kip ft

$f_t = f_b = \dfrac{8.27 \times 12}{162} = 0.613$

$f_p = 0.613 - 0.380 = 0.233$ ksi

$P = 0.233 \times 9 \times 12 = 25.12$ kips

$w_p = \dfrac{8 \times 25.12 \times 3.937}{(27.38)^2 \times 12} = 0.088$ klf $= 88$ plf

This is less than 92 plf. N.G.

Second Cycle

$w_p = 0.75 \times 92 + 0.25 \times 88 = 91$ plf

$M_b = 15.87 \times \dfrac{91}{192} = 7.52$ kip ft

$M_{ub} = 15.87 - 7.52 = 8.348$ kip ft

$f_t = f_b = \dfrac{8.348}{162} \times 12 = 0.618$ ksi

$f_b = 0.618 - 0.380 = 0.238$ ksi

$P = 0.238 \times 9 \times 12 = 25.75$ kips.

$w_p = \dfrac{8 \times 25.75 \times 3.937}{(27.38)^2 \times 12} = 0.090$ klf $= 90$ plf

This is nearly equal to the value at the beginning of second cycle. Therefore O.K.

Check the design against positive moment of 8.41 kip-ft.

$$w_p = 0.090 \text{ klf}$$

$$M_b = 8.41 \times \frac{0.090}{0.142} = 5.33 \text{ kip ft}$$

$$M_{ub} = 8.41 - 5.33 = 3.08 \text{ kip ft}$$

Bottom flexural stress $= \dfrac{3.08 \times 12}{162} = 0.228 \text{ ksi}$ (tension).

Axial compression due to post-tension $= \dfrac{25.75}{12 \times 9} = 0.238 \text{ ksi}$.

Total stress at bottom $= 0.228 - 0.238 = -0.10$ ksi (compression). This is less than allowable tension of 0.380 ksi. Therefore, design O.K.

8.2.7.5 MILD STEEL REINFORCEMENT DESIGN (STRENGTH DESIGN FOR FLEXURE)

In the design of prestress members, it is not enough to limit the maximum values of tensile and compressive stresses within the permitted values at various loading stages. This is because although such a design may limit deflections, control cracking, and prevent crushing of concrete, an elastic analysis offers no control over the ultimate behavior or the factor-of-safety of a prestressed member. To assure that prestressed members will be designed with an adequate factor-of-safety against failure, the ACI 318-19 code requires that M_u, the moment due to factored service loads, not exceed ϕM_n, the flexural design strength of the member.

The nominal bending strength is computed in nearly the same manner as that of a reinforced concrete beam. The only difference is in the method of stress calculation in the tendon at failure. This is because the stress-strain curves of high-yield-point steels used as tendons do not develop a horizontal yield range once the yield strength is reached. It continues to slope upward at a reduced slope. Therefore, the final stress in the tendon at failure f_{ps} must be predicted by an empirical relationship. The reader is referred to standard reinforced concrete design textbooks for further discussion of strength design of prestressed members.

Chapter 9

Composite Gravity Systems

Gravity systems in composite construction can be broadly classified into composite floor systems and composite columns. Composite floor systems can consist of simply supported prismatic or haunched structural steel beams, trusses, or stub girders linked via shear connectors to a concrete floor slab to form an effective T-beam flexural member. Formed metal deck supporting a concrete topping slab is an integral component in these floor systems used nearly exclusively in steel-framed buildings in North America.

9.1 COMPOSITE METAL DECKS

9.1.1 General Considerations

Metal deck is manufactured from steel sheets by a fully mechanized, high-speed cold-rolling process. Although it is possible to produce shapes up to ½ in. (12.7 mm) and even ¾ in. (19 mm) thick by cold forming, cold-formed steel construction is generally restricted to plates and sheets weighing from 0.5 psf (24 Pa) to a maximum of 9 psf (431 Pa).

Composite metal deck is manufactured with deformations specifically designed-in to produce composite action under flexure between the metal deck and concrete. The shear connection between the two is provided through lugs, corrugations, ridges, or embossments formed in the profile of the sheet to increase the chemical bond between the two materials. The steel deck profile is typically trapezoidal with relatively wide flutes suitable for through-deck welding of shear studs. Metal deck may also include closed cells to accommodate floor

electrification system. Non-cellular deck panels may be blended with cellular panels as part of the total floor system. Metal deck is commonly available in 1½ in., 2 in., 3 in. (38, 51, and 76 mm) depths with rib spacings of 6, 7½, 8, 9, and 12 in. (152, 190, 208, 228, and 305 mm).

A composite slab is usually designed as a simply supported reinforced concrete slab with the steel deck acting as positive reinforcement. Typical mesh used for control of temperature cracking does not provide enough negative reinforcement for typical beam spacing of 8 to 15 ft (2.44 to 4.57 m). Although the slab is designed as a simple span, it is a good practice to provide a nominal reinforcement of say #4 @ 18" c-c at the top to control excessive cracking of slab. It is generally believed that cracking of slab in the negative regions does not materially impair the composite beam strength.

The Steel Deck Institute (SDI), regarded as the industry standard by the metal deck manufacturers, has published a manual that encompasses the design of composite decks, form decks, and roof decks. A brief description of the SDI specifications is given in the following section.

9.1.2 SDI Specifications

The steel used for the fabrication of composite metal deck shall have a minimum yield point of 33 ksi (227.5 MPa). The specified yield point is the primary criterion for strength under static loading. The tensile strength is of secondary importance because fatigue strength and brittle fracture, which relate to tensile strength rather than yield point, are rarely of consequence; metal deck is rarely subjected to repetitive loads and the characteristic thinness invariably precludes the development of brittle fracture.

Considerable variations in the thickness of metal deck may occur because of rolling tolerances. Therefore, SDI stipulates that the delivered thickness of bare steel without the finish such as phosphatizing and galvanizing shall be not less than 95 percent of the specified thickness. The increase in the stiffness of deck due to galvanizing is not relied upon in the design of metal deck.

Opinions differ among engineers whether the metal deck used inside of a building which is not directly exposed to weather needs galvanizing or not. SDI does not mandate any particular type of finish. The appropriate finish is left to the discretion of the engineer with the recommendation that due consideration be given to the effects of environment to which the structure is subjected. However, SDI in its commentary recommends a galvanized coating conforming to ASTM A-525 G.60 requiring a minimum galvanizing of 0.75 ounce per square feet (2.24 Pa) of metal deck. Other salient features of SDI specifications are as follows:

1. Minimum compressive strength of concrete f'_c shall be 3.0 ksi (20.68 MPa). The compressive stress in concrete is limited to $0.4 f'_c$ under the applied load for unshored construction and under the total dead and live loads for shored construction. The flexural or shear bond is to be based on ultimate strength analysis with a minimum safety factor of 2. The minimum temperature and shrinkage reinforcement in a composite slab is a function of the area of concrete, as in ordinary reinforced concrete slab, but only the concrete area above the metal deck need be considered in calculating the area of concrete.

2. The use of admixtures containing chloride salts is prohibited because salts can corrode the steel deck.

3. When designing the section as a form in bending, the section properties are to be calculated as per AISC *Specification for the Design of Cold-formed Steel Structural Members*.

4. Bending stress is limited to 0.6 times the yield strength of steel. An upper limit of 36.0 ksi (248.2 MPa) is imposed on the allowable stress. In addition to the weight of wet concrete and deck, allowance should be made for construction live loads of 20 psf (958 Pa) of uniform load or a 150 lb (667 N) concentrated load. This is to account for the weight of one person working on a 1 ft (305 mm) width of deck. It is a common practice to allow for a 200 lb (890 N) point load as the equivalent load. This is because the loading is considered temporary with a 33 percent increase in the stress, which is equivalent to reducing the 200 lb (890 N) load by 25 percent. Clear spans are to be used in the moment calculations.

5. For calculating deflections, it is not necessary to consider the construction loads since the deck, which is designed to remain elastic, will rebound after the removal of construction loads. The calculated deflection based on the weight of concrete is limited to the smaller value $L/180$ or ¾ in. (19 mm), in which L is the clear span of the deck. Deflections of composite slabs due to live loads of 50 to 80 psf (2394 to 3830 Pa) are seldom a design concern because the deflections are usually less than $L/360$, where L is the span of deck. Because the slab is assumed to have cracked at the supports, the deflections are best predicted by using the average of the cracked and uncracked moment of inertia using transformed sections. Note that when slabs are cast level to compensate for the deflection of metal deck, a 10 to 15 percent of additional concrete is required for slab construction.

6. A minimum bearing of 1½ in. (38 mm) is specified for proper deck seating on supports.

7. A maximum average spacing of 12 in. (305 mm) for puddle welds is specified to obtain proper anchorage to supporting members. The maximum spacing between adjacent welds is limited to 18 in. (457 mm). Welding of decks with a thickness less than 0.028 in. (0.71 mm) is not practical because of the likelihood of burning off the sheet. Therefore, SDI stipulates use of welding washers for floor decks of less than 0.028 in. (0.71 mm) in thickness. Stud welding through the metal deck to the steel top flange can be used instead of puddle welds to satisfy the minimum spacing requirements. However, since it is possible to get uplift forces during wind storms, puddle welds should be used to prevent metal decks from blowing off buildings during construction.

8. Mechanical fasteners that satisfy the anchorage criteria can be used in lieu of puddle welds.

9. Side laps with proper fasteners are required between two longitudinal pieces of deck to: (i) prevent differential deflection, (ii) provide sufficient diaphragm strength, and (iii) sustain local construction loads without distortion or separation. The edges of the metal deck shall be connected with ¾ in. (19 mm) diameter fusion welds at a maximum spacing of 3 ft (0.9 m) throughout for simply supported spans. Button punching at 2 ft (0.61 m) on centers is an acceptable alternative to minimum fusion welding.

To function as formwork, decks supporting cantilevers should be proportioned to satisfy the following criteria: (i) dead load deflection should be limited to $L/90$ of overhang or 3/8 in. (9.5 mm), whichever is smaller; (ii) steel stress should be limited to 26.7 ksi (184 MPa) for dead load plus 200 lb (890 N) concentrated load at the outer edge of overhang, or steel stresses limited to 20.0 ksi

(138 MPa) for dead load plus 20 psf (958 Pa) of additional load, whichever is more severe; and (iii) the deck should receive one fusion weld at the cantilever end, and the spacing of welds throughout the cantilever span should not exceed 12 in. (0.30 m). Button punching can be used as an acceptable alternative to fusion welding.

9.2 COMPOSITE BEAMS

9.2.1 General Considerations

Three types of composite construction are recognized by the AISC specifications: (i) fully encased steel beams, (ii) concrete-filled HSS, and (iii) steel beams with shear connectors. In fully encased steel beams, the natural bond between concrete and steel interface is considered sufficient to provide the resistance to horizontal shear provided that: (1) the concrete thickness is 2 in. (50.8 mm) or more on the beam sides and soffit with the top of the beam at least 1 in. (25.4 mm) below the top and 2 in. (50.8 mm) above the bottom of the slab; and (2) the encasement is cast integrally with the slab and has adequate mesh or other reinforcing steel throughout the depth and across the soffit of the beam to prevent spalling of concrete. The third type consists of steel beams, metal deck, concrete topping, and shear connectors (mostly studs), by far the popular in the construction of buildings in North America. Invariably composite action is achieved by providing shear connectors between top flange of the steel beam and concrete topping. Fully encased steel beam construction is not used because encasing of beams with concrete requires expensive formwork.

Composite sections have greater stiffness than the summation of the individual stiffness of slab and beam and, therefore, can carry larger loads or similar loads with appreciably smaller deflection and are less prone to transient vibrations. Composite action results in an overall reduction of floor depth, and for high-rise buildings, the cumulative savings in curtain walls, electrical wiring, mechanical ductwork, interior walls, plumbing, etc., can be considerable.

Composite beams can be designed either for shored or unshored construction. For shored construction, the cost of shoring should be evaluated in relation to the savings achieved by the use of lighter beams. For unshored construction, steel is designed to support by itself the wet weight of concrete and construction loads. The steel section, therefore, is heavier than in shored construction.

In composite floor construction, the top flanges of the steel beams are attached to the concrete by the use of suitable shear connectors. The concrete slab becomes part of the compression flange. As a result, the neutral axis of the

section shifts upward, making the bottom flange of the beam more effective in tension.

Since concrete already serves as part of the floor system, the only additional cost is that of the shear connectors. In addition to transmitting horizontal shear forces from the slab into beam, the shear connector prevents any tendency for the slab to rotate independently of the beam.

The stud shear connector is a short length of round steel bar welded to the steel beam at one end and having an anchorage provided in the form of a round head at the other end. The most common diameters are ½, ⅝, and ¾ in. (12, 16, and 19 mm). The length is dependent on the depth of metal deck and should extend at least 1½ in. (38 mm) above the top of the deck. The welding process typically reduces their length by about 3/16 in. (5 mm). The upset head thickness of the studs is usually ⅜ or ½ in. (9 to 12 mm), and the diameter ½ in. (12 mm) larger than the stud diameter. The studs are normally welded to the beam with an automatic welding gun, and when properly executed, the welds are stronger than the steel studs. Studs located on the side of the trough toward the beam support are more effective than studs located toward the beam centerline. The larger volume of concrete between the stud and the pushing side of the trough helps in the development of a larger failure cone in concrete, thus increasing its horizontal shear resistance.

The length of stud has a definite effect on the shear resisted by it. As the length increases, so does the size of the shear cone, with a consequent increase in the shear value. The shear capacity of the stud also depends on the profile of the metal deck. To get a qualitative idea, consider the two types of metal decks shown in Figs. 9.1 and 9.2. The deck in Fig. 9.1 has a narrow hump compared to the one in Fig. 9.2. When subjected to a load V, the concrete and the metal deck tend to behave as a portal frame. The concrete in the troughs can be thought of as columns with the concrete over the humps acting as beams (Fig. 9.2b). A narrow hump of the portal frame, results in an equivalent

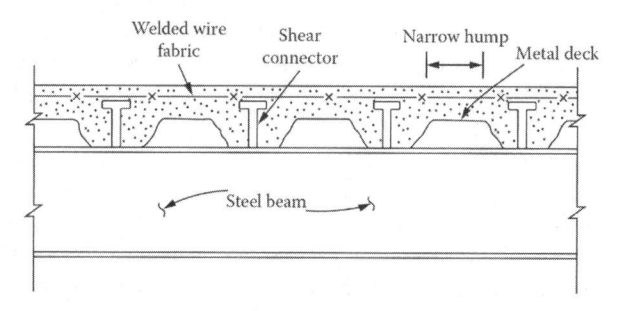

Figure 9.1 Composite beam with narrow hump metal deck.

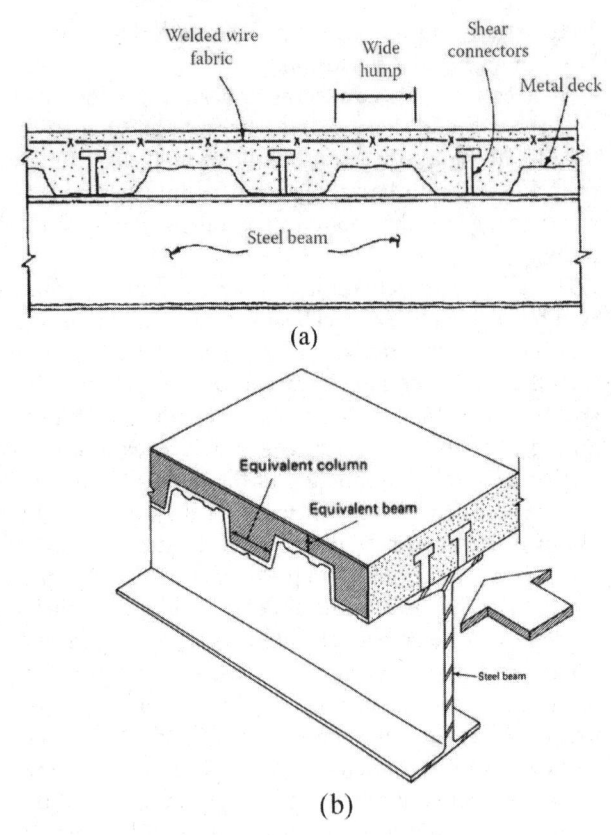

Figure 9.2 (a) Composite beam with wide hump metal deck; (b) simplified analytical model of composite metal deck subjected to horizontal shear.

beam of smaller span when compared to the one with a wider hump, meaning that a deck profile with the widest trough and narrowest hump will yield the highest connector strengths. However, other considerations such as volume of concrete, section modulus, and the stiffness of deck also influence the shear strength of the connector.

Metal decks for composite construction are available in the United States in three depths, (38 mm), 2 in. (51 mm), and 3 in. (76 mm). The earlier types of metal deck did not have embossments, and the interlocking between concrete and metal deck was achieved by welding reinforcement transverse to the beam. Later developments of metal deck introduced embossments to engage concrete and metal deck and dispensed with the transverse-welded reinforcement.

The spans utilizing composite metal deck are generally in the range of 8 to 15 ft (2.4 to 4.6 m).

In floor systems using 1½ in. (38 mm) decks, the electrical and telephone services are generally provided by the so-called poke-through system, which is simply punching through the slab at various locations and passing the under-floor ducts through them. A deeper deck is required if the power distribution system is integrated as part of the structural slab; as a result, 2 and 3 in. (51 and 76 mm) metal decks were developed. Experiments have shown that there is very little loss of composite beam stiffness due to the ribbed configuration of the metal deck in the depth range of 1½ to 3 in. (38 to 76 mm). As long as the ratio of width to depth of the metal deck is at least 1.75, the entire capacity of the shear stud can be developed similar to beams with solid slabs. However, with deeper deck a substantial decrease in shear strength of the stud occurs, which is attributed to a different type of failure mechanism. Instead of the failure of shear stud, the mode of failure is initiated by cracking of the concrete in the rib corners. Eventual failure takes place by separation of concrete from the metal deck. When more than one stud is used in a metal deck flute, a failure cone can develop over the shear stud group, resulting in lesser shear capacity per each stud. The shear stud strength is therefore closely related to the metal deck configuration and factors related to the surface area of the shear cone.

Often special details are required in composite design to achieve the optimum result. Openings interrupt slab continuity, affecting capacity of a composite beam. For example, beams adjacent to elevator and stair openings may have full effective width for part of their length and perhaps half that value adjacent to the openings. Elevator sill details normally require a recess in the slab for door installations, rendering the slab ineffective for part of the beam length. A similar problem occurs in the case of trench header ducts, which require the elimination of concrete, as opposed to the standard header duct, which is completely encased in concrete. When the trench is parallel to the composite beam, its effect can easily be incorporated into the design by suitably modifying the effective width of compression flange. The effect of the trench-oriented perpendicular to the composite beam could range from negligible to severe depending upon its location. If the trench can be located in the region of minimum bending moment, such as near the supports in a simply supported beam, and if the required number of connectors could be placed between the trench and the point of maximum bending moment, its effect on the composite beam design is minimum. If, on the other hand, the trench must be placed in an area of high bending moment, its effect may be so severe as to require that the beam be designed as a noncomposite beam.

The slab thickness normally employed in high-rise construction with composite metal deck is usually governed by fire-rating requirements rather than the thickness required by the bending capacity of the slab. In certain parts of the United States, it may be economical to use the minimum thickness required for strength and to use sprayed-on or some other method of fireproofing to obtain the required ratings. Some major projects have used 2½ in. (63.5 mm) thick concrete on 3 in. (76.2 mm) deep metal deck spanning as much as 15 ft (4.57 m). A comparative study which takes into consideration the vibration characteristics of the floor is necessary to zero in on the most economical scheme.

In continuous composite beams, the negative moment regions can be designed such that: (i) the steel beam alone resists the negative moment or (ii) it acts compositely with mild steel reinforcement placed in the slab parallel to the beam. In the latter case, shear connectors must be provided through the negative moment region.

Careful attention should be paid to the deflection characteristics of composite construction because the slender not-yet-composite shape deflects as wet concrete is placed on it. There are three ways to handle the deflection problem.

1. Use relatively heavy steel beams to limit the dead-load deflection and pour lens-shaped tapering slabs to obtain a nearly flat top. Although a reasonably flat surface results from this construction, the economic restraints of speculative office buildings do not usually permit the luxury of the added cost of additional concrete and heavier steel beams.

2. Camber the steel beam to compensate for the weight of steel beam and concrete. Place a constant thickness of slab by finishing the concrete to screeds set from the cambered steel. Continuous lateral bracing as provided by the metal deck is required to prevent the lateral torsion buckling of beam. If steel deck is not used, this system requires a substantial temporary bracing system to stabilize the beam during construction.

3. Camber and shore the steel beam. The beam is fabricated with a camber calculated to compensate for the deflection of the final cured composite section. Shores are placed to hold the steel at its curved position while the concrete is being poured. As in method 2, slab is finished to screeds set from cambered steel. Although methods 1 and 3 are occasionally used, the trend is to use method 2 because it is the least expensive.

9.2.2 AISC Design Specifications and Requirements (AISC 360-16, Chap. I)

The detailed AISC provisions are given in Chap. I of the AISC 360-16 provisions. Below outlined the general requirements that should be used together with the detailed requirements of Chap. I. Note that only Load Resistance Factor Design (LRFD) provisions are presented in this chapter; the allowable strength design (ASD) could be used as well.

Including provisions for solid slab, there are three categories of composite beams in the AISC specifications each with a differing effective concrete area.

1. *Solid slab.* The total slab depth is effective in compression unless the neutral axis is above the top of the steel beam. In high-rise floor systems with relatively thin slabs the neutral axis of steel beams is invariably below the slab, rendering the total slab depth effective in compression.

2. *Deck perpendicular to beam* (Fig. 9.3).

(a) Concrete below the top of steel decking shall be neglected in computations of section properties and in calculating the number of shear studs, but the concrete below the top flange of deck may be included for calculating the effective width.

(b) The maximum spacing of shear connectors shall not exceed 32 in. (813 mm) along the beam length.

(c) The steel deck shall be anchored to the beam either by welding or by other means at a spacing not exceeding 16 in. (406 mm).

(d) The shear capacity of the studs should be reduced using the shear stud reduction factors, R_g and R_p.

3. *Deck ribs parallel to beam* (Fig. 9.4).

(a) The major difference between perpendicular and parallel orientation of deck ribs is that when deck is parallel to beam, the concrete below the top of the decking can be included in the calculations of section properties and must be included when calculating the number of shear studs.

(b) If steel deck ribs occur on supporting beam flanges, it is permissible to cut high-hat to form a concrete haunch.

(c) When nominal rib height is 1½ in. (38.1 mm) or greater, minimum average width of deck flute should not be less than 2 in. for the first stud in the transverse row plus four stud diameters for each additional stud. This gives minimum average widths of 2 in. (51 mm) for one stud, 2 in. plus $4d$ for two studs, 2 in. plus $8d$ for three studs, etc., where d is the diameter of stud. Note that if a metal deck cannot accommodate this width requirement, the deck can be split over the girder to form a haunch.

(d) The shear capacity of the studs should be reduced using the shear stud reduction factors, R_g and R_p.

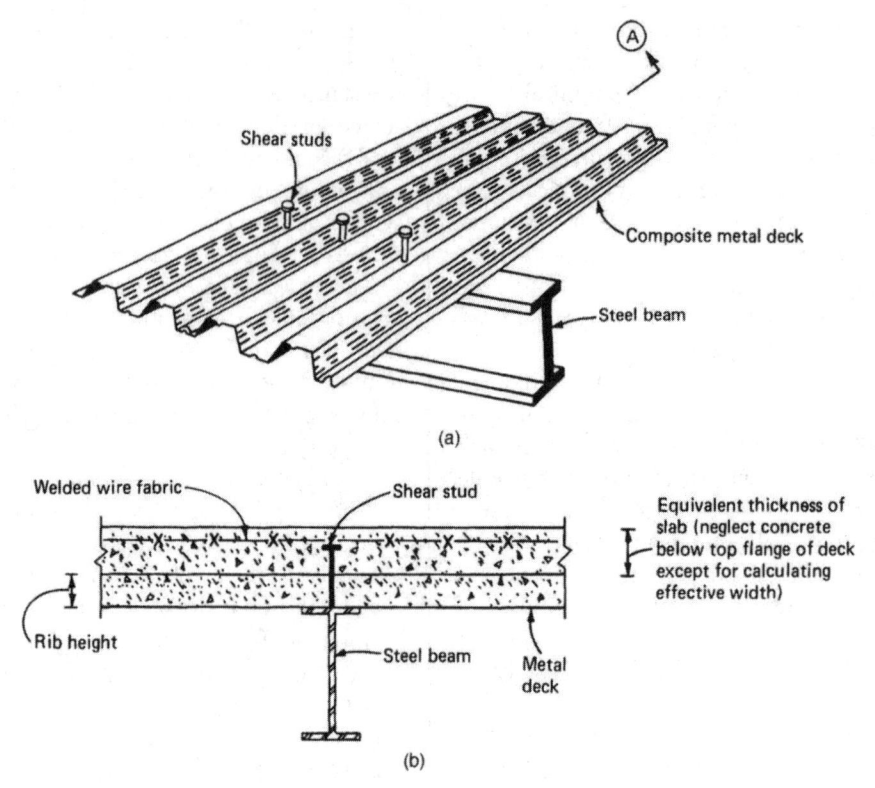

Figure 9.3   Composite beam with deck perpendicular to beam: (a) schematic view; (b) section A showing equivalent thickness of slab.

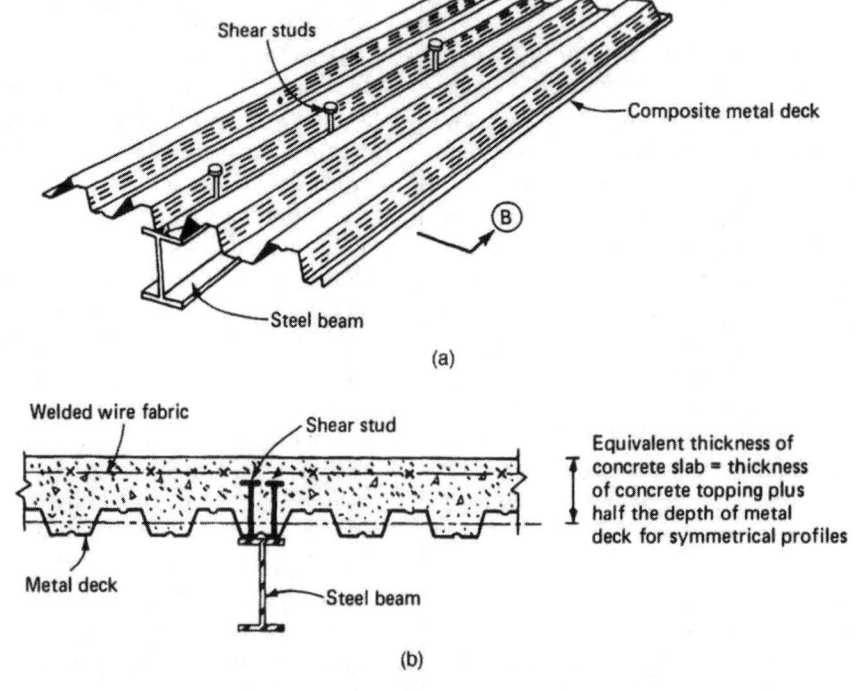

Figure 9.4   Composite beam with deck parallel to beam: (a) schematic view; (b) section A showing equivalent thickness of slab.

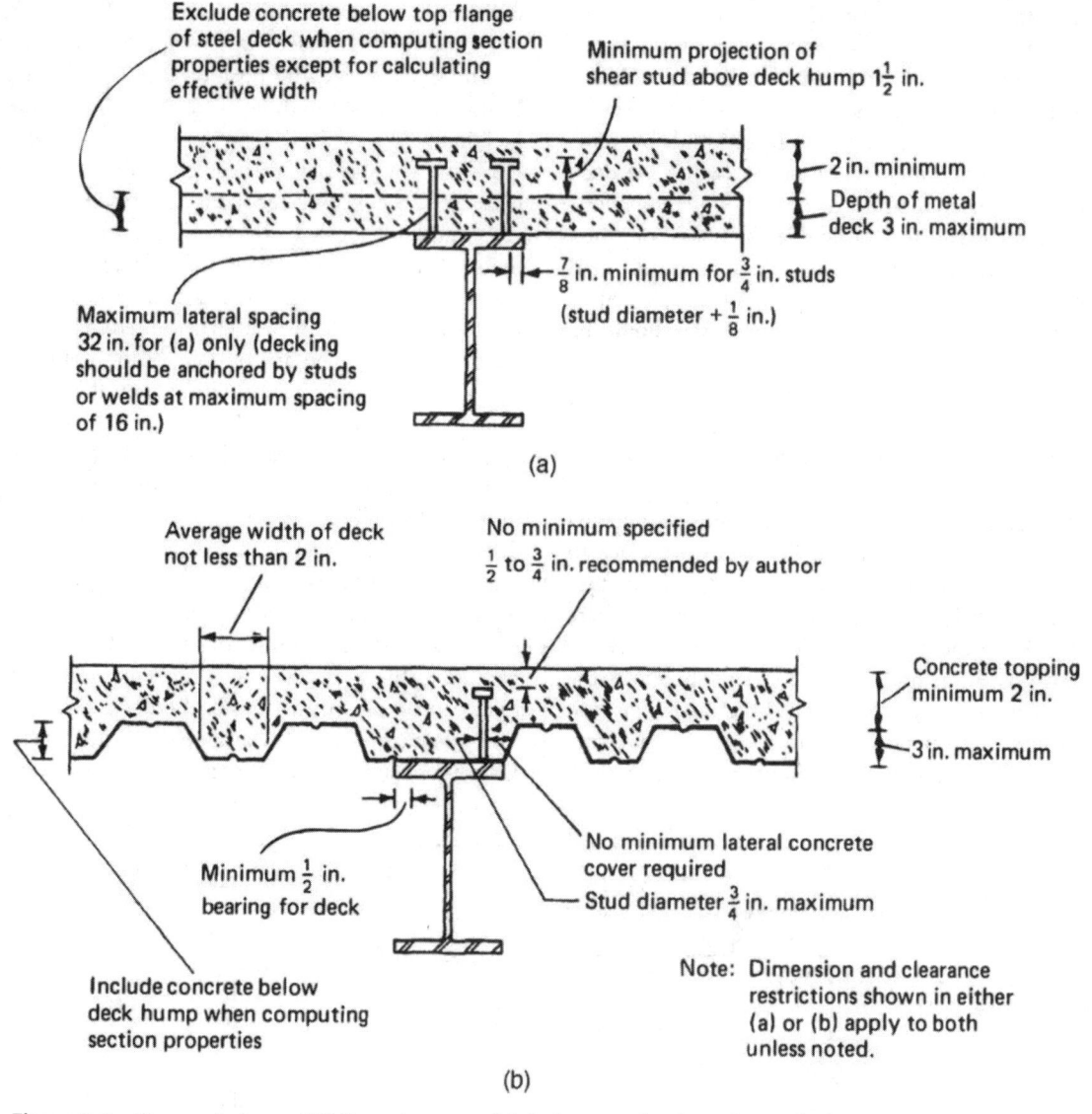

Figure 9.5 Composite beam, AISC requirements: (a) deck perpendicular to beam; (b) deck parallel to beam.

9.2.3 General AISC Requirements for Composite Beams and Explanation

1. The section properties do not change a much from deck running perpendicular or parallel to the beam, however, the change in the required number of studs can be significant.

2. Rib deck height shall not exceed 3 in.

3. Rib average width shall not be less than 2 in. If the deck profile is less than 2 in., the minimum clear width shall be used in the calculation.

4. The reduction formula for stud length is based on rib geometry, number of studs per rib, and embedment length of the studs.

5. After installation, the studs should extend a minimum of 1½ in. above the steel deck.

6. The concrete cover over the top of the stud shall be at least ½ in.

7. Studs can be placed as close to the web of deck as needed for installation and to maintain the necessary spacing.

8. Deck anchorages can be provided by the stud welds.

9. Maximum diameter of shear connectors is limited to ¾ in.

10. Total slab thickness including the ribs is used in determining the effective width without regard to the orientation of the deck with respect to the beam axis.

11. The slab thickness above the steel deck shall not be less than 2 in.

12. Shear studs may be welded directly through the deck or through pre-punched or cut-in-place holes in the deck. The usual procedure is to install studs by welding directly through the deck; however, when the deck thickness is greater than 16 gage for single thickness, or 18 gage for each sheet of double thickness, or when the total thickness of galvanized coating is greater than 1.25 ounces/sq. ft, special precautions and procedures as recommended by the study manufacturer should be followed.

13. The majority of composite steel floor decks have a stiffening rib in the middle of each deck flute. Because of the stiffener, studs must be welded off-center in the deck rib.

14. Although it is recommended that studs be detailed in the strong position, ensuring that studs are place in the strong position is not necessarily an easy task because it is not always easy for the installer to determine where along the beam the particular rib is located, relative to the end, midspan or point of zero shear. Therefore, the installer may not be clear on which is the strong position and which is the weak position. However, it is reassuring to note that even a large change in shear strength does not result in a proportional decrease in the flexural strength of composite beams (see Fig. 9.6a and b for the definition of strong and weak stud position). AISC sets the default value for shear strength equal to that for the stud weak position.

15. Uniform spacing of shear connectors is permitted, unless heavy concentrated loads are present.

16. Studs not located directly over the web of a beam tend to tear out of a thin flange before attaining full shear-resisting strength. To avoid this issue, the size of a stud not located over the beam web is limited to two times the flange thickness. The practical application of this limitation is to select only beams with flanges thicker than the stud diameter divided by 2.5.

17. The minimum spacing of connectors along the length of the beam, in both flat soffit concrete slabs and in formed steel deck with ribs parallel to the beam, is six diameters. If the steel beam flange is narrow, this spacing requirement may be achieved by staggering the studs with a minimum transverse spacing of three diameters between the staggered rows of studs. When deck ribs are parallel to the beam and the design requires more studs than can be placed in the rib, the deck may be split so that adequate spacing is available for stud installation. Figure 9.7a shows possible connector arrangements.

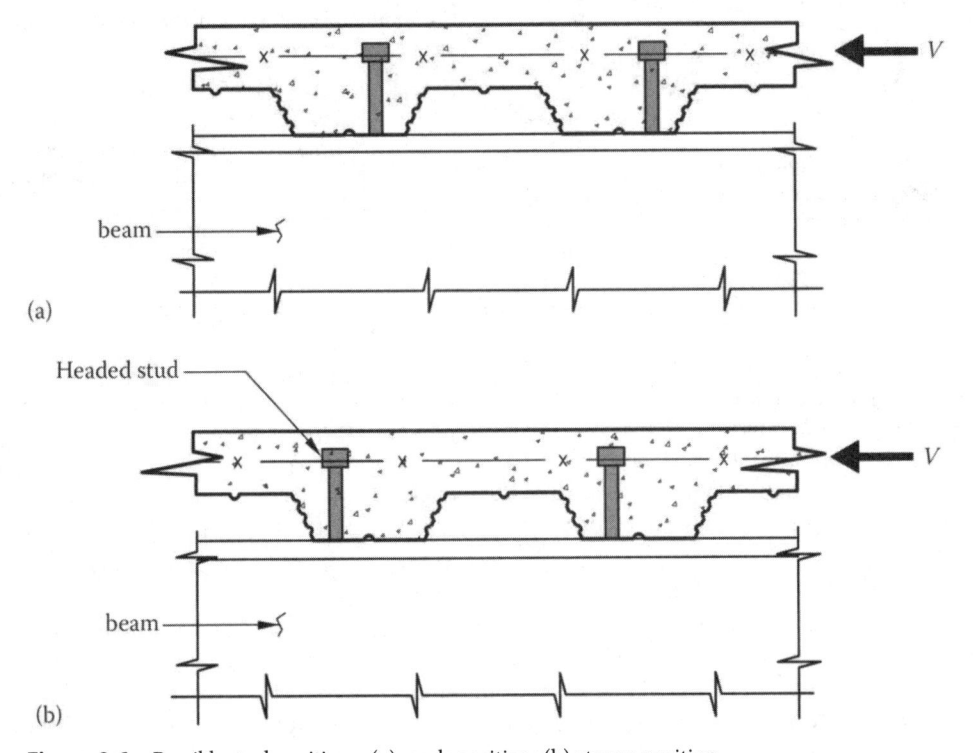

Figure 9.6   Possible stud positions: (a) weak position; (b) strong position.

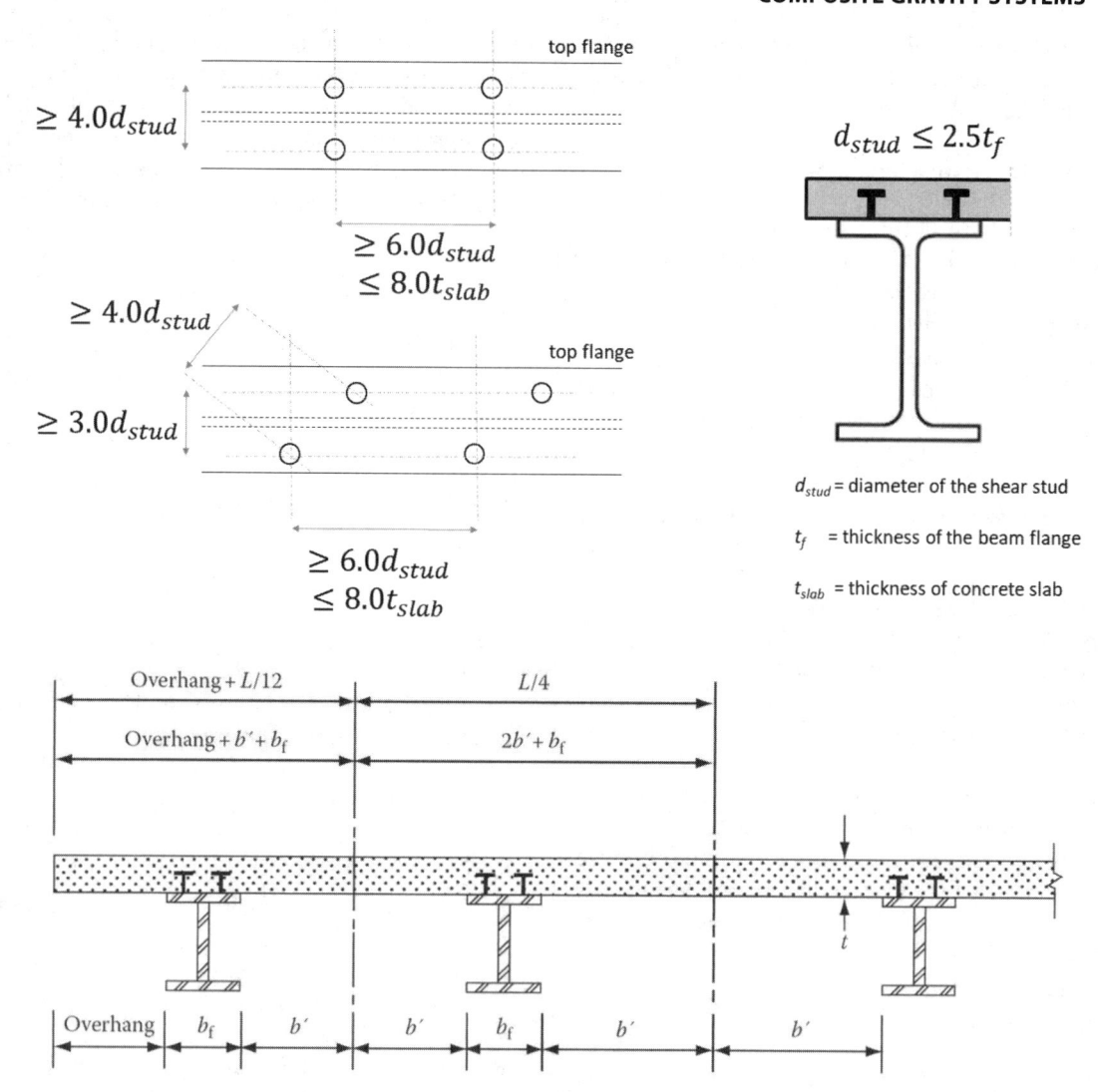

Figure 9.7 (a) Stud arrangement: provisions and limitations; (b) effective width concept as defined in the AISC 360-16 specifications.

9.2.4 Effective Width (AISC 360 Section I3.1a)

The effective width is based on the full span, center-to-center of supports, for both simple and continuous beams.

The effective width of the concrete slab shown in Fig. 9.7b is the sum of the effective width for each side of the beam centerline, each of which shall not exceed:

1. One-eighth of the beam span, center-to-center of supports

2. One-half the distance to the centerline of the adjacent beam

3. The distance to the edge of the slab

9.2.5 Positive Flexural Strength (AISC 360 Section I3.2a)

The design positive flexural strength using the LRFD procedure, $\phi_b M_n$, is determined using the resistance factor for flexure, $\phi_b = 0.9$ and M_n, the nominal flexural strength is determined as follows:

1. For $h/t_w \le 3.76\, E/F_y$, M_n is to be determined from the plastic stress distribution on the composite section for the limit state of yielding (plastic moment). AISC provides a user note indicating that all current ASTM A6, W, S, and HP shapes satisfy the above requirement for $F_y \le 70$ ksi.

2. For $h/t_w > 3.76\ E/F_y$, M_n is determined from the superposition of elastic stresses, considering the effects of shoring, for the limit state of yielding.

9.2.6 Negative Flexural Strength (AISC 360 Section I3.2b)

The design negative flexural strength using the LRFD procedure, $\phi_b M_n$, is determined for the steel section alone using $\phi_b = 0.9$. Alternatively, the available negative flexural strength may be determined from the plastic stress distribution of the composite section, provided that:

1. The steel beam is compact and is adequately braced.

2. Shear connectors connect the slab to the steel beam in the negative moment region.

3. The slab reinforcement parallel to the steel beam, within the effective width of the slab, is properly developed.

9.2.7 Load Transfer Between Steel Beam and Slab (AISC 360 Section I3.2d)

1. Load transfer for positive moment

The entire horizontal shear at the interface between the steel beam and the concrete slab shall be assumed to be transferred by shear connectors, except for concrete-encased beams. For composite action with concrete subject to flexural compression, the total horizontal shear force, V', between the point of maximum positive moment and the point of zero moment shall be taken as the lowest value according to the limit states of concrete crushing, tensile yielding of the steel section, or strength of the shear connectors:

a. Concrete crushing

$$V' = 0.85 f_c' A_c$$

b. Tensile yielding of the steel section

$$V' = F_y A_s$$

c. Strength of shear connectors

$$V' = \Sigma Q_n$$

where
A_c is the area of concrete slab with effective width, in.^{2}
A_s is the area of steel cross-section, in.^{2}
ΣQ_n is the sum of nominal strengths of shear connectors between the point of maximum positive moment and the point of zero moment, kip.

2. Load transfer for negative moment

In continuous composite beams where longitudinal reinforcing steel in the negative moment regions is considered

to act compositely with the steel beam, the total horizontal shear force between the point of maximum negative moment and the point of zero moment shall be taken as the lower value of tensile yielding of the steel reinforcement in the slab, or strength of the shear connectors:

a. Tensile yielding of the slab reinforcement

$$V' = A_r F_{ysr}$$

where
A_r is the area of adequately developed longitudinal reinforcing steel within the effective width of the concrete slab, in.^{2}
F_{ysr} is the specified minimum yield stress of the reinforcing steel, ksi
b. Strength of steel shear connectors (studs or channels)

$$V' = \Sigma Q_n$$

3. Strength of stud shear connectors (AISC 360 section I8.2a)

The nominal strength of one stud shear connector embedded in solid concrete or in a composite slab is:

$$Q_n = 0.5 A_{sa} \sqrt{f_c' E_c} \leq R_g R_p A_{sa} F_u$$

where
A_{sa} is the cross-sectional area of stud shear connector, in.^{2}
E_c is the modulus of elasticity of concrete $= w_c^{1.5} \sqrt{f_c'}$, ksi
F_u is the specified minimum tensile strength of a stud shear connector, ksi
R_g and R_p are the stud shear capacity reduction factors as described in the following.

$R_g = 1.0$—(a) for one stud welded in a steel deck rib with the deck oriented perpendicular to the steel shape; (b) for any number of studs welded in a row directly to the steel shape; (c) for any number of studs welded in a row through steel deck with the deck oriented parallel to the steel shape and the ratio of the average rib width to rib depth ≥ 1.5

$= 0.85$—(a) for two studs welded in a steel deck rib with the deck oriented perpendicular to the steel shape; (b) for one stud welded through steel deck with the deck oriented parallel to the steel shape and the ratio of the average rib width to rib depth ≤ 1.5

$= 0.7$—for three or more studs welded in a steel deck rib with the deck oriented perpendicular to the steel shape.

$R_p = 1.0$—for studs welded directly to the steel shape (in other words, not through steel deck or sheet) and having a haunch detail with not more than 50% of the top flange covered by deck or sheet steel closures.

$= 0.75$—(a) for studs welded in a composite slab with the deck oriented perpendicular to the beam and $e_{\text{mid-ht}} \geq 2$ in.; (b) for studs welded through steel deck, or steel sheet used as girder filler material, and embedded in a composite slab with the deck oriented parallel to the beam.

$= 0.6$—for studs welded in a composite slab with deck oriented perpendicular to the beam and $e_{\text{mid-ht}} \geq 2$ in.

$e_{\text{mid-ht}} \geq 2$ in. = distance from the edge of stud shank to the steel deck web, measured at mid-height of the deck rib, and in the load-bearing direction of the stud (in other words, in the direction of maximum moment for a simply supported beam), in.

w_c = weight of concrete per unit volume ($90 \leq$ wc ≤ 155 lb/ft^3).

Note the stud reduction factors R_g and R_p are equal to unity for composite steel beams with solid slab.

4. Required number of shear connectors

The number of shear connectors required between the section of maximum bending moment, positive or negative, and the adjacent section of zero moment shall be equal to the horizontal shear force divided by the nominal strength of one shear connector.

5. Shear connector placement and spacing

Shear connectors required on each side of the point of maximum bending moment, positive or negative, shall be distributed uniformly between that point and the adjacent points of zero moment. However, the number of shear connectors placed between any concentrated load and the nearest point of zero moment shall be sufficient to develop the maximum moment required at the concentrated load point.

9.2.8 Deflection Considerations

When a composite beam is controlled by deflection, the design should limit the behavior of the beam to the elastic range under serviceability load combinations. In other words, calculate deflections using elastic properties of composite beam, and working stress design (WSD). It is often not practical to make accurate stiffness calculations of composite flexural members. Therefore, for realistic deflection calculations, one may use a lower bound moment of inertia, I_{lb}, as defined below:

$$I_{\text{lb}} = I_s + A_s(Y_{ENA} - d_3)^2 + \left(\frac{\Sigma Q_n}{F_y}\right)(2d_3 + d_1 - Y_{ENA})^2$$

where

A_s is the area of steel cross-section, in.^{2}

d_1 is the distance from the compression force in the concrete to the top of the steel section, in.

d_3 is the distance from the resultant steel tension force for full section tension yield to the top of the steel, in.

I_{lb} is the lower bound moment of inertia, in.^{4}

The above equations should not be used for ratios, $\Sigma Q_n/C_f$, less than 0.25. This restriction is to prevent excessive slip, as well as substantial loss in beam stiffness.

I_s = the moment of inertia for the structural steel section, in.^{4}

ΣQ_n = the sum of the nominal strengths of shear connectors between the point of maximum positive moment and the point of zero moment to either side, kip.

$$Y_{ENA} = \left[\frac{A_s d_3 + \dfrac{\left(\dfrac{\Sigma Q_n}{F_y}\right)(2d_3 + d_1)}{A_s + \left(\dfrac{\Sigma Q_n}{F_y}\right)}}{}\right]$$

The use of constant stiffness in elastic analyses of continuous beams is analogous to the practice in reinforced concrete design. The stiffness calculated using a weighted average of moments of inertia in the positive moment region and negative moment regions may take the following form:

$$I_t = aI_{\text{pos}} + bI_{\text{neg}}$$

where

I_{pos} is the effective moment of inertia for a positive moment, in.^{4}

I_{neg} is the effective moment of inertia for a negative moment, in.^{4}

The effective moment of inertia is based on the cracked transformed section considering the degree of composite action. For continuous beams subjected to gravity loads only, the value of a may be taken as 0.6 and the value of b may be taken as 0.5 for calculations related to drift. For cases where elastic properties of partially composite beams are needed, the elastic moment of inertia may be approximated by:

$$I_{\text{equiv}} = I_s + \sqrt{\left(\frac{\Sigma Q_n}{C_f}\right)}(I_{tr} - I_s)$$

where

I_s is the moment of inertia for the structural steel section, in.[4]

I_{tr} is the moment of inertia for the fully composite uncracked transformed section, in.[4]

ΣQ_n is the strength of shear connectors between the point of maximum positive moment and the point of zero moment to either side, kip.

C_f is the compression force in concrete slab for fully composite beam; smaller of $A_s F_y$ and $0.85 f'_c A_c$, kip.

A_c is the area of concrete slab within the effective width, in.[2]

The effective section modulus, S_{eff}, referred to the tension flange of the steel section for a partially composite beam, may be approximated by:

$$S_{eff} = S_s + \sqrt{\left(\frac{\Sigma Q_n}{C_f}\right)}(S_{tr} - S_s)$$

where

S_s is the section modulus for the structural steel section, referred to the tension flange, in.[3]

S_{tr} is the section modulus for the fully composite uncracked transformed section, referred to the tension flange of the steel section, in.[3]

9.2.9 Shear Strength (AISC I4)

The shear strength of composite beams with steel-headed studs or steel channel anchors shall be determined per Chap. G of the AISC 360-16 specifications, which is based on the steel section alone.

9.2.10 Design Procedure for Composite Beam

Considering the typical day-to-day practice with computers, below outlined the design procedure to allow a better understanding of how a composite beam is designed.

The design steps are:

1. Limit material properties as follows:

Normal weight concrete: 3 ksi $\leq f'_c \leq$ 10 ksi

Lightweight concrete: 3 ksi $\leq f'_c \leq$ 6 ksi

Structural steel and reinforcing: $F_y \leq$ 75 ksi

2. Loads

Dead loads typically consist of slab and beam weight, miscellaneous of 10 psf for ceiling etc., and an allowance of 15 psf for partitions (when live loads are less than 100 psf). Live loads shall be as specified in the applicable building codes. Use higher loads if specified by the building user. Reduce live loads if permitted by the applicable building code.

3. Determine the required flexural strength

In the most typical case of simply supported beams subject to uniformly distributed load, w_u, and span l.

$$M_u = \frac{w_u l^2}{8}$$

4. Determine b_{eff}

The effective width of the concrete slab is the sum of the effective widths for each side of the beam centerline, which shall not exceed:

    a. One-eighth of the beam span, center to center of supports.

    b. One-half the distance to the center-line of the adjacent beam.

    c. The distance to the edge of the slab.

5. Calculate the moment arm for the concrete force measured from the top of the steel shape, $Y2$

Assume depth of compression block, $a = 1.0$ in. (Some assumption must be made to start the design process. An assumption of 1.0 in. has proven to be a reasonable starting point in many design problems.)

$$Y2 = t_{slab} - a/2$$

Enter AISC Steel Design Manual, Table 3-19 with the required strength and the calculated value of Y2. Select a beam and a plastic neutral axis location that indicates sufficient available strength. The trial beam is okay if $\phi_b M_n = 0.9 M_n \geq M_u$.

6. Check the beam deflections and available strength

Check the deflection of the steel beam under construction loads considering only the weight of steel beam and concrete topping, and an allowance of 20 psf for construction live loads as contributing to the construction load.

Limit deflection to a maximum of 2.5 in. to facilitate concrete placement. Camber the beam if the calculated deflection is greater than ¾ in. Provide camber equal to 75 percent of the calculated dead load deflection.

$$I_{req} = \frac{5}{384} \frac{w_{DL} l^4}{E\Delta}$$

Revise trial section if I_{req} is greater than that of the trial section.

7. Check selected member strength as an unshored beam under construction loads assuming adequate lateral bracing through the deck attachment to the beam flange. Also determine ΣQ_n. Calculate the required strength of the steel beam alone using the weight of the steel beam and concrete slab as dead loads, and an

allowance of 20 psf construction load. Label this load as $w_{(const)}$.

$$M_{u(\text{unshored})} = \frac{w_{(\text{const})}l^2}{8}$$

$$\phi M_{n(\text{steel})} \geq M_{u(\text{unshored})}$$

Using the assumed values of Y2 and the plastic neutral axis, PNA, determine the required horizontal shear ΣQ_n.

8. Check the value a, using the relation

$$a = \frac{\Sigma Q_n}{0.55 f_c' b}$$

9. Check live load deflection

$$\Delta_{LL} = \frac{5}{384} \frac{wl^4}{EI_{LB}}$$

10. Determine if the beam has sufficient available shear strength $\phi V_n \geq V_u$ using the provisions of Chapter G of AISC 360-16, which is typically not an issue for typical uniformly loaded beams.

11. Determine the required number of shear connectors = 2N

With the known direction and profile of metal deck, determine shear capacity of one ¾ in., diameter stud per rib for the given strength and weight of the concrete slab. Here, we identify ¾ in. diameter studs, because they are the most common type used in building construction, in North America.

$$N = \frac{\Sigma Q_n}{Q_n}$$

(on each side of the beam)

12. Check the spacing of shear connectors

Use one stud for every flute, starting at each support. If studs remain, double up studs on each end of the span. Check spacing requirement.

$$6 d_{stud} \leq \text{stud spacing} \leq 8 t_{slab}$$

Note that stud length shall extend a minimum of 1½ in. into slab after welding.

9.3 COMPOSITE COLUMNS

The term "composite column" in the building industry is taken to represent a unique form of construction in which structural steel is made to interact compositely with concrete. The structural steel section can be a tubular section filled with structural concrete or it could be a steel wide-flange section used as a core with a reinforced concrete surround.

Historically, composite columns evolved from the concrete encasement of structural steel shapes primarily intended as fire protection. Although the increase in strength and stiffness of the steel members due to concrete used as fireproofing was intuitively known, it was not until the 1940s that methods to actually incorporate the increases were developed. In fact, in the earlier days the design of the steel column was penalized by considering the weight of concrete as an additional dead load on the steel column. Later developments took account of the increased radius of gyration of the column because of the concrete encasement, and allowed for some reduction in the amount of structural steel. In some earlier high-rise designs the concrete encasement was ignored from the viewpoint of strength considerations, but the additional stiffness of concrete was included in calculating the lateral deflections.

After the development of sprayed-on contact fireproofing in the 1950s and 1960s, use of concrete for fireproofing of structural steel was no longer an economical proposition. The high formwork cost of concrete could not be justified for fireproofing.

Over the last 45 years or so, the use of encased structural steel columns has found applications in buildings varying from as low as 10 stories to as high as 70 stories or even taller buildings. These columns have been incorporated in an overall construction known as the composite system, which has successfully captured the essential advantages associated with steel and concrete construction: the speed of steel with the stiffness and moldability of concrete. Concrete columns with small steel-core columns used as erection columns were perhaps the earliest applications. Later much heavier columns were used, serving the dual purpose of both steel erection and load resistance. The heavier steel columns were used essentially to limit the size of composite vertical elements.

Another version consists of exterior concrete columns acting compositely with steel plate or precast cladding. Yet another version popular in some countries uses laced columns fabricated from light structural shapes such as angles, T sections, and channels. The concrete enclosure provides both fire-proofing qualities and also imparts additional stiffness to the light structural shapes, inhibiting their local buckling tendencies. Additional conventional reinforcement can be accommodated in the concrete encasement, as in conventionally reinforced concrete columns.

9.3.1 Encased Composite Columns (AISC I2.1)

9.3.1.1 Limitations (AISC I2.1a)

To qualify as an encased composite column, the following limitations shall be met:

1. The cross-sectional area of the steel core shall comprise at least 1% of the total composite cross-section.

2. Concrete encasement of the steel core shall be reinforced with continuous longitudinal bars and lateral ties or spirals. The minimum transverse reinforcement shall be at least 0.009 in.^{2} per in. of tie spacing.

3. The minimum reinforcement ratio for continuous longitudinal reinforcing, ρ_{sr}, shall be 0.004, where ρ_{sr} is given by

$$\rho_{sr} = \frac{A_{sr}}{A_g}$$

where

A_{sr} is the area of continuous reinforcing bars, in.^{2}
A_g is the gross area of composite member, in.^{2}

9.3.1.2 Compressive Strength (AISC I2.1b)

The design compressive strength, $\phi_c P_n$, for axially loaded encased composite columns shall be determined for the limit state of flexural buckling based on column slenderness as follows:

$$\phi_c = 0.75$$

a) When $P_{no}/P_e \leq 2.25$

$$P_n = P_{no} \left[0.658^{\left(\frac{P_{no}}{P_e}\right)} \right]$$

b) When $P_{no}/P_e > 0.25$

$$P_n = 0.877 P_e$$

where

$$P_{no} = F_y A_s + F_{ysr} A_{sr} + 0.85 A_c f_c'$$

$$P_e = \pi^2 (EI_{eff})/(KL)^2$$

where

A_s = area of steel section, in.^{2}
A_c = area of concrete, in.^{2}
A_{sr} = area of continuous reinforcing bars, in.^{2}
E_c = modulus of elasticity of concrete = $w_c^{1.5} \sqrt{f_c'}$, ksi
E_s = modulus of elasticity of steel = 29,000 ksi
f_c' = specified compressive strength of concrete, ksi
F_y = specified minimum yield stress of steel section, ksi

F_{ysr} = specified minimum yield stress of reinforcing bars, ksi
I_c = moment of inertia of the concrete section, in.^{4}
I_s = moment of inertia of steel shape, in.^{4}
I_{sr} = moment of inertia of reinforcing bars, in.^{4}
K = effective length factor
L = laterally unbraced length of the member, in.
w_c = weight of concrete per unit volume ($90 \leq w_c \leq 155$ lb/ft^3)

where

EI_{eff} is the effective stiffness of composite section, kip-in.^{2}

$$EI_{eff} = E_s I_s + E_s I_{sr} + C_1 E_c I_c$$

where

$$C_1 = 0.25 + 3 \left(\frac{A_s + A_{sr}}{A_g} \right) \leq 0.7$$

The available compressive strength should not be than that specified for a bare steel member per the requirements of Chapter E of AISC.

9.3.1.3 Tensile Strength (AISC I2.1c)

The tensile strength, P_n is given by:

$$P_n = A_s F_y + A_{sr} F_{ysr}$$

The design tensile strength, $\phi_t P_n$ is given by:

$$\phi_t = 0.90$$
$$\phi_t P_n = 0.90(A_s F_y + A_{sr} F_{ysr})$$

9.3.1.4 Shear Strength (AISC I4)

The design shear strength for encased composite members, $\phi_v V_n$, shall be determined based on one of the following:

(a) The available shear strength of the steel section alone as given in Chapter G of AISC 360-16.

(b) The available shear Strength of the reinforced concrete portion (concrete and steel reinforcement) alone as defined by ACI 318 with $\phi_v = 0.75$.

(c) The nominal shear strength of the steel section, as defined in Chapter G of AISC 360-16, plus the nominal strength of the reinforcing steel, as defined by ACI 318 with a combined resistance and a strength reduction factor of $\phi_v = 0.75$.

9.3.1.5 Load Transfer (AISC I6)

Loads applied to axially loaded encased composite columns shall be transferred between the steel and concrete in accordance with the following requirements:

(a) When the external force is applied directly to the steel section, shear connectors shall be provided to transfer the required shear force, V_r', as follows:

$$V_r' = P_r\left(1 - \frac{F_y A_s}{P_{no}}\right)$$

where

P_r is the required external force applied to column, kip
A_s is the area of steel cross-section, in.^{2}
P_{no} is the nominal axial compressive strength without consideration of length effects, kip

(b) When the external force is applied directly to the concrete encasement, shear connectors shall be provided to transfer the required shear force, V_r', as follows:

$$V_r' = P_r\left(\frac{F_y A_s}{P_{no}}\right)$$

(c) When the external force is applied concurrently to the steel section and concrete encasement, V_r' shall be determined as the force required to establish equilibrium of the cross-section.

(d) When load is applied to the concrete of an encased composite column by direct bearing the design bearing strength, R_n, of the concrete shall be:

$$R_n = 1.7 f_c' A_1$$

The design bearing strength, $\phi_B R_n$

$$\phi_B = 0.65$$
$$\phi_B R_n = 0.65(1.7 f_c' A_1)$$

where A_1 is the loaded area of concrete, in.^{2}

9.3.1.6   DETAILING REQUIREMENTS
(AISC I2.1E)

Detailing requirements for encased columns are as follows:

Clear spacing between the steel core and the steel reinforcement shall be 1.5 times the bar diameter at a minimum, but not less than 1.5 in.

For built-up composite section with two or more encased steel shapes, the steel shapes shall be interconnected with lacing, tie plates or equivalent to prevent buckling of the individual shapes before the concrete hardens.

Section I6.4 of AISC 360-16 requires that the force transfer shall be within the load introduction length not exceeding two times the minimum transverse dimension of the encased composite member. The anchors transferring the longitudinal shear shall be on at least two faces symmetrically arranged about the steel shape axes. The steel anchor spacing shall conform to Section I8.3e of AISC 360-16 specification both within and outside the load introduction length.

9.3.2   Filled Composite Columns (AISC I2.2)

9.3.2.1   LIMITATIONS (AISC I2.2A)

To qualify as a filled composite column the following limitations shall be met:

1. The cross-sectional area of the steel HSS shall comprise at least 1 percent of the total composite cross-section.

2. Filled composite members shall be classified for local buckling per section I1.4.

3. Minimum longitudinal reinforcement is not required and if provided, internal transverse reinforcement is not requirement for strength.

9.3.2.2   COMPRESSIVE STRENGTH
(AISC I2.2B)

The design compressive strength, $P_c = \phi_c P_n$, for axially loaded filled composite columns shall be determined for the limit state of flexural buckling based on the following formula:

a) for compact section:

$$P_{no} = P_p$$

$$P_p = F_y A_s + C_2 f_c'\left(A_c + A_{sr}\frac{E_s}{E_c}\right)$$

$C_2 = 0.85$ for rectangular sections and 0.95 for circular sections

b) for noncompact section:

$$P_{no} = P_p - \frac{P_p - P_y}{(\lambda_r - \lambda_p)^2}(\lambda - \lambda_p)^2$$

where

λ, λ_p, and λ_r are slenderness ratios determined from Table I1.1a.

P_p is determined from:

$$P_p = F_y A_s + 0.7 f_c'\left(A_c + A_{sr}\frac{E_s}{E_c}\right)$$

c) For slender sections

$$P_{no} = F_{cr} A_s + 0.7 f_c'\left(A_c + A_{sr}\frac{E_s}{E_c}\right)$$

where
For rectangular-filled sections

$$F_{cr} = \frac{9E_s}{\left(\dfrac{b}{t}\right)^2}$$

For round-filled sections

$$F_{cr} = \frac{0.72F_y}{\left[\left(\dfrac{D}{t}\right)\dfrac{F_y}{E_s}\right]^{0.2}}$$

where C_3 = coefficient for calculations of effective rigidity of filled composite compression member

$$= 0.45 + 3\left(\frac{A_s + A_{sr}}{A_g}\right) \leq 0.9$$

$$EI_{\text{eff}} = E_s I_s + E_s I_{sr} + C_3 E_c I_c$$

The available compressive strength should not be more than that specified for a bare steel member per the requirements of Chap. E of AISC.

9.3.2.3 Tensile Strength (AISC I2.2c)

The design tensile strength $P_t = \phi_t P_n$, for filled composite columns shall be determined as:

$$P_n = A_s F_y + A_{ysr} F_y$$

The design tensile strength

$$P_t = \phi_t P_n = 0.90(A_s F_y + A_{ysr} F_y)$$

$$\phi_t = 0.90$$

9.3.2.4 Shear Strength (AISC I4)

The design shear strength for filled composite members, $\phi_v V_n$, shall be determined based on one of the following:

(a) The available shear strength of the steel section alone as given in Chapter G of AISC 360-16.

(b) The available shear strength of the reinforced concrete portion (concrete and steel reinforcement) alone as defined by ACI 318 with $\phi_v = 0.75$.

(c) The nominal shear strength of the steel section, as defined in Chapter G of AISC 360-16, plus the nominal strength of the reinforcing steel, as defined by ACI 318 with a combined resistance and a strength reduction factor of $\phi_v = 0.75$.

9.3.2.5 Load Transfer (AISC I6)

Loads applied to axially loaded filled composite columns shall be transferred between the steel and concrete in accordance with the following requirements:

(a) When the external force is applied directly to the steel section, shear connectors shall be provided to transfer the required shear force, V'_r, as follows:

$$V'_r = P_r\left(1 - \frac{F_y A_s}{P_{no}}\right)$$

where
P_r is the required external force applied to column, kip
A_s is the area of steel cross section, in.^{2}
P_{no} is the nominal axial compressive strength without consideration of length effects, kip

(b) When the external force is applied directly to the concrete fill, shear connectors shall be provided to transfer the required shear force, V'_r, as follows:

$$V'_r = P_r\left(\frac{F_y A_s}{P_{no}}\right)$$

(c) When the external force is applied concurrently to the steel section and concrete encasement, V'_r shall be determined as the force required to establish equilibrium of the cross-section.

(d) When load is applied to the concrete of an encased composite column by direct bearing the design bearing strength, R_n, of the concrete shall be:

$$R_n = 1.7 f'_c A_1$$

The design bearing strength, $\phi_B R_n$

$$\phi_B = 0.65$$

$$\phi_B R_n = 0.65(1.7 f'_c A_1)$$

where A_1 is the loaded area of concrete, in.^{2}

9.3.2.6 Detailing Requirements (AISC I6.4b)

Section I6.4 of AISC 360-16 requires that the force transfer shall be within the load introduction length not exceeding two times the minimum transverse dimension of the rectangular steel member and two times the diameter of a round steel member both above and below the force transfer region. For the case of a filled composite section without steel reinforcement, the load introduction length shall extend beyond the force transfer region in the direction of the applied load. The steel anchor spacing shall conform to Section I8.3e of AISC 360-16 specification both within the load introduction length.

9.4 COMBINED AXIAL FORCE AND FLEXURE (AISC I5)

The available flexural and axial forces for a composite member are given in sections I2 and I3 of AISC 360-16, respectively. The interaction between flexure and axial loads shall account for stability per the requirements of Chapter C of AISC 360-16. To account for the length effect, the axial strength is determined according to section I2 of AISC 360-16 specifications.

For encased columns, the interaction between flexural and axial forces shall be based on section H1.1 or one of the methods presented in section I1.2 of AISC 360-16 specifications.

For filled composite members with noncompact or slender, the interaction between axial and flexural forces shall be based on section H1.1, defined in section I1.2d specifications or using the equations defined in section I5 of AISC 360-16.

9.5 COMPOSITE HAUNCH GIRDERS

Composite haunch girders, although not often used as a floor framing system, merit mention because they minimize the floor-to-floor height without requiring complicated fabrication. Figure 9.8 shows a schematic floor plan in which composite haunch girders frame between exterior columns and interior core framing. The haunch girder typically consists of a shallow steel beam, 10 to 12 in. (254 to 305 mm) deep for spans in the 35 to 40 ft (10.6 to 12.19 m) range. At each end of the beam, a triangular haunch is formed by welding a diagonally cut wide-flange beam usually 24 to 27 in. (610 or 686 mm) deep (Fig. 9.9). The haunch is welded to the shallow beam and to the columns at each end of the girder. In this manner, the last 8 or 9 ft (2.4 or 2.7 m) of the haunch girder at either end flares out toward the column with a depth varying from about 10 or 12 in. (254 or 305 mm) at the center to about 27 in. (686 mm) at the ends. The system uses less steel and provides greater flexibility for mechanical ducts, which can be placed anywhere under the shallow central span. The reduction in floor-to-floor height further cuts the costs of exterior cladding and heating and cooling loads. The system, however, is not common because of higher fabrication costs.

A variation of the same concept shown in Fig. 9.10 uses nontapered haunches at each end. The square haunch girder can be fabricated using a shallow-rolled section in the center and two deep-rolled sections, one at each end. Another method of fabricating the girder is to notch the bottom portion of the girder at midspan and reweld the flange to the web. The method requires more steel but comparatively less fabrication work.

In comparison to a shallow girder of constant depth, a haunch girder is significantly stiffer. Figure 9.10 shows the moment diagram in a haunch girder subjected to combined gravity and lateral loads. The corresponding stiffness properties including the effect of composite action in the positive moment regions are also shown in Fig. 9.10.

9.6 COMPOSITE TRUSSES

Figure 9.11 shows a typical floor-framing plan with composite trusses. To keep the fabrication simple, the top and bottom chords consist of T sections to which double-angle web members are welded directly without the use of gusset plates. The top chord is made to act compositely with the floor system by using welded shear studs. The space between the diagonals is used for the passage of mechanical and air-conditioning ducts. When the space between the diagonals is not sufficient, vertical members may be welded between the chords to form a Vierendeel panel.

9.7 COMPOSITE STUB GIRDERS

9.7.1 General Considerations

In high-rise design, maximum flexibility is achieved if structural, mechanical, electrical, and plumbing trades have their own designated space in the ceiling. This is achieved in a conventional system by placing HVAC ducts, lights, and other fixtures under the beams. Where deep girders are used, penetrations are made in the girder webs to accommodate the ducts. In an office building the typical span between the core and the exterior is about 40 ft (12.2 m), requiring 18 to 21 in. (457 to 533 mm) deep beams. Usual requirements of HVAC ducts, lights, sprinklers, and ceiling construction result in depths of 4 to 4.25 ft (1.21 to 1.3 m) between the ceiling and top of floor slab. The depth can, however, be decreased at a substantial penalty either by providing penetrations in relatively deep beams or by using shallower, less economical beam depths.

The stub girder system, Fig. 9.12, invented by engineer Dr. Joseph Caloco, attempts to eliminate some of these shortcomings while at the same time reduces the floor steel weight. The key components of the system are short stubs welded intermittently to the top flange of a shallow steel beam. Sufficient space is left between stubs to accommodate mechanical ducts. Floor beams are supported on top of, rather than framed into the shallow steel beam. Thus the floor beams are designed as continuous

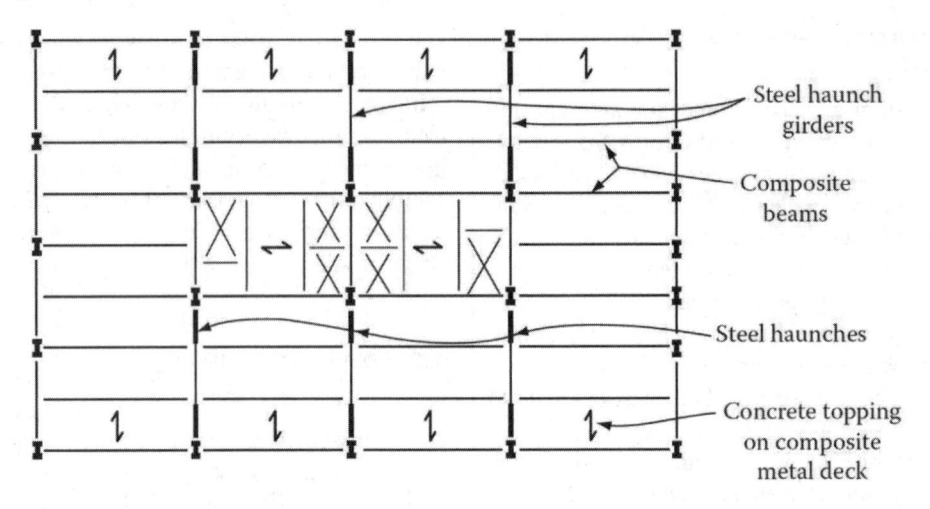

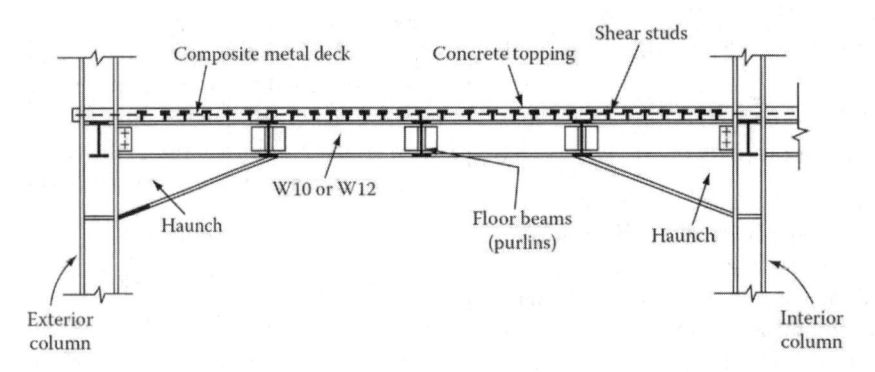

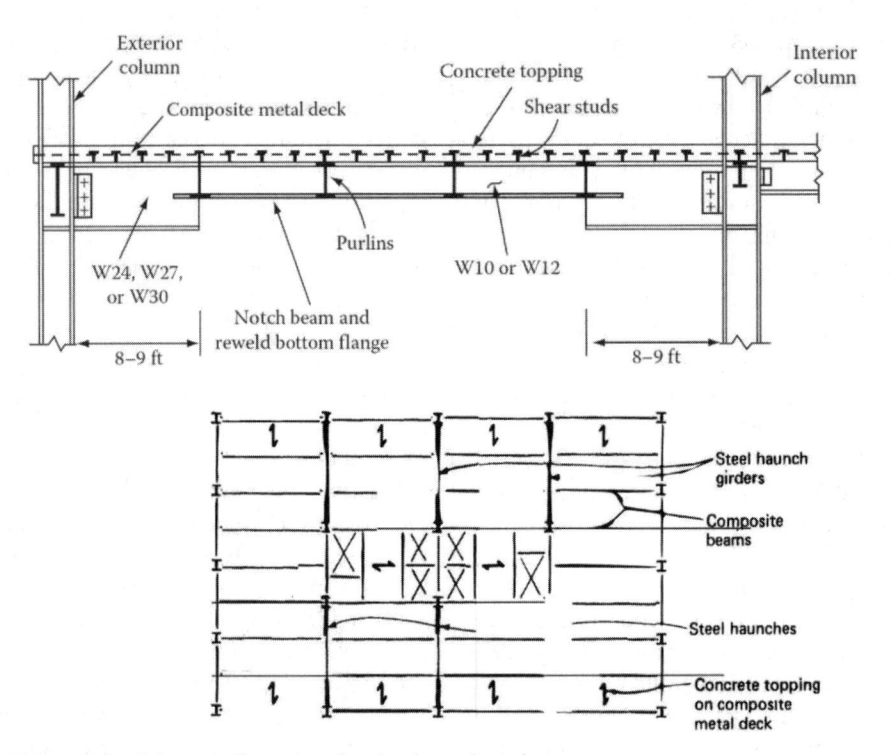

Figure 9.8   Schematic floor plan showing haunch girders.

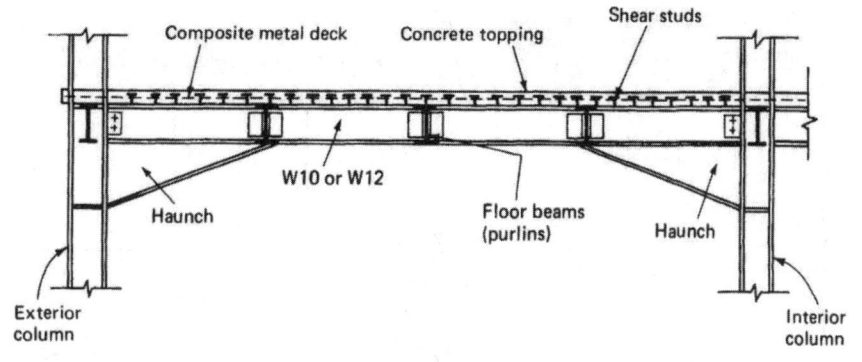

Figure 9.9   Composite girder with tapered haunch.

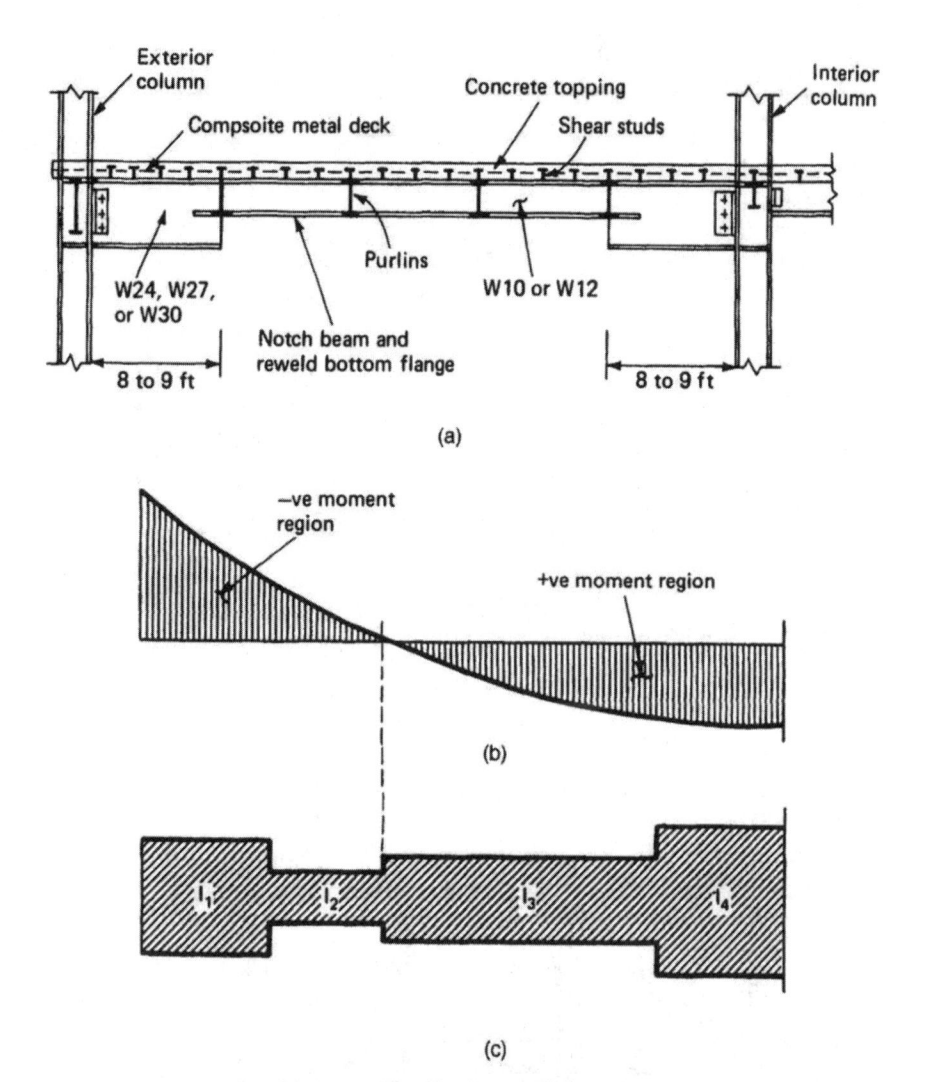

I_1 = Moment of inertia of unnotched steel section
I_2 = Moment of inertia of notched steel section
I_3 = Moment of inertia of composite notched section
I_4 = Moment of inertia of composite unnotched section

Figure 9.10   Composite girder with square haunch: (a) schematic elevation; (b) combined gravity and wind moment diagram; (c) schematic moment of inertia diagram.

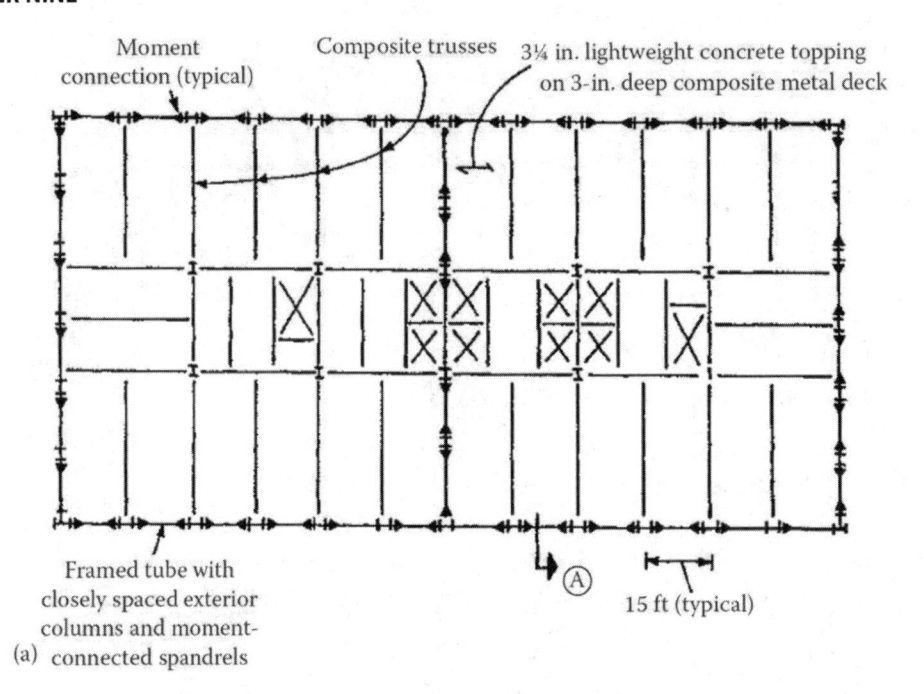

(a)

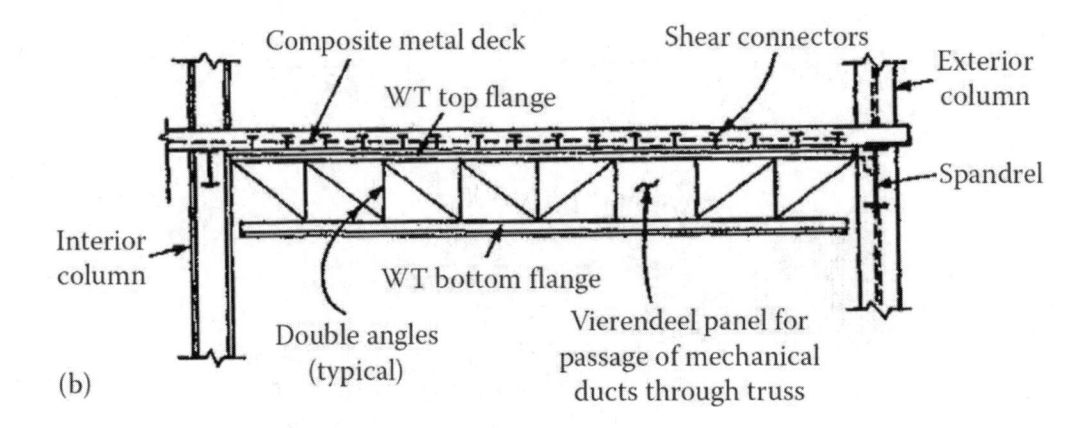

(b)

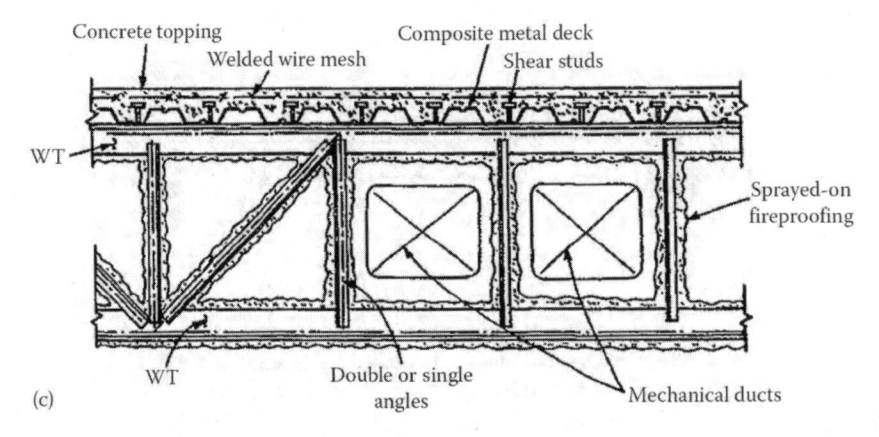

(c)

Figure 9.11 Composite truss: (a) framing plan; (b) elevation of truss, section A; (c) detail of truss.

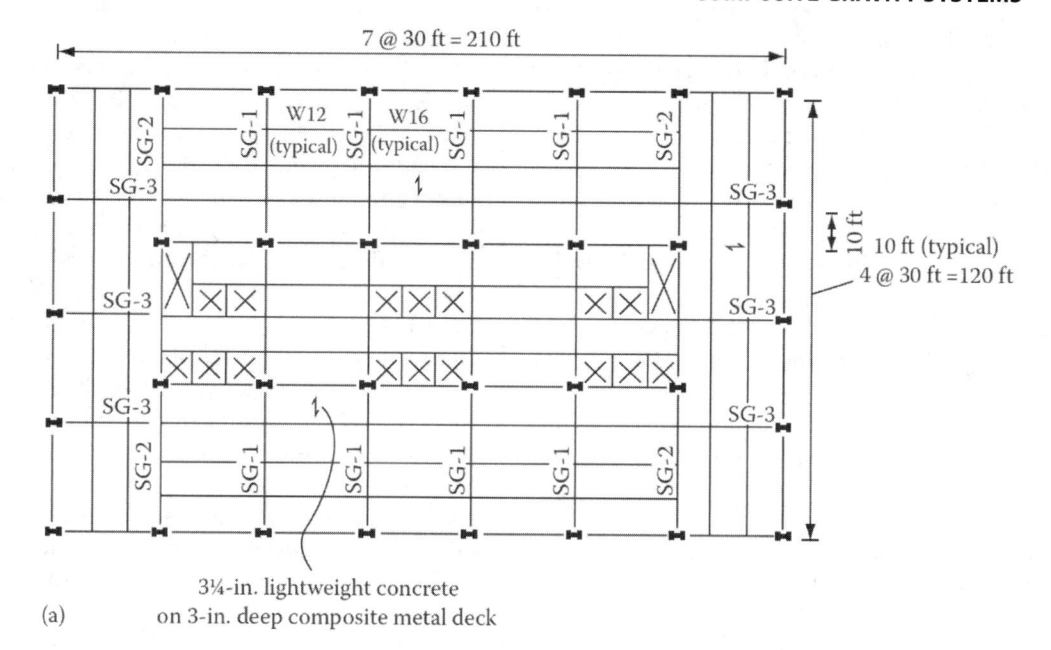

(a)

3¼-in. lightweight concrete
on 3-in. deep composite metal deck

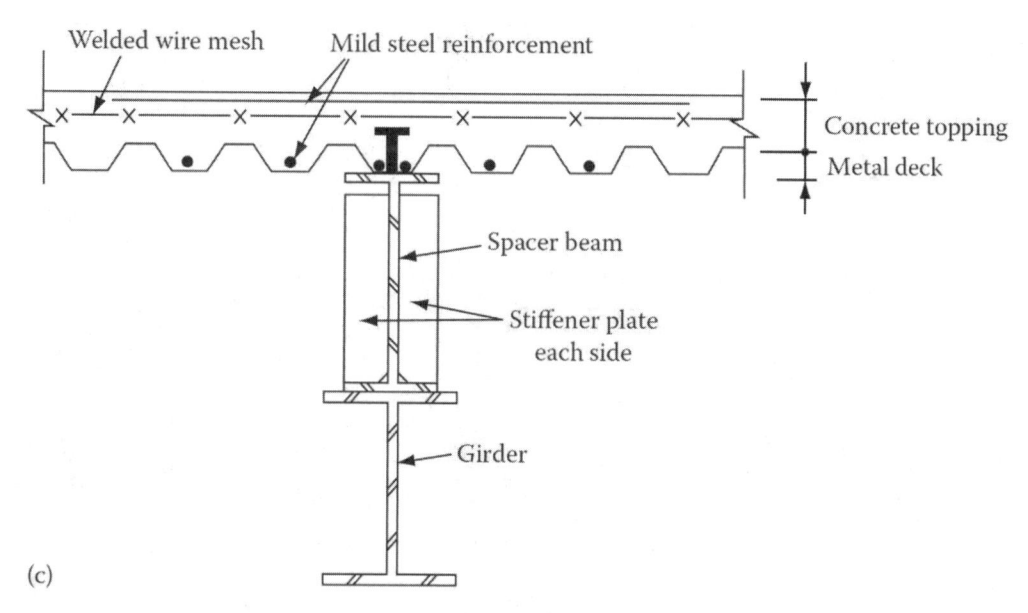

Figure 9.12 Stub girder framing: (a) framing plan; (b) elevation of stub girder SG-1; (c) section A through stub girder.

members which results in steel savings and reduced deflections. The stubs consist of short wide flange beams placed perpendicular to and between the floor beams. The floor system consists of concrete topping on steel decking connected to the top of stubs. The stub girders are spaced at 25 to 35 ft (7.62 to 10.7 m) on the center, spanning between the core and the exterior of the building.

The behavior of a stub girder is akin to a Vierendeel truss; the concrete slab serves as the compression chord, the full-length steel beam as the bottom tension chord, and the steel stubs as vertical web members. From an overall consideration, the structure allows installation of mechanical system within the structural envelope, thus reducing floor-to-floor height; the mechanical ducts run through and not under the floor.

9.7.2 Behavior and Analysis

The primary action of a stub girder is similar to that of a Vierendeel truss; the bending moments are resisted by tension and compression forces in the bottom and top chords of the truss and the shearing stresses by the stub pieces welded to the top of wide-flange beam. The bottom chord is the steel wide-flange which resists the tensile forces. The compression forces are carried by the concrete

slab. The effective width of concrete slab varies from 6 to 7 ft (1.83 to 2.13 m) requiring additional reinforcement to supplement the compression capacity of the concrete. The shear forces are resisted by the stub pieces, which are connected to the metal deck and concrete topping through shear connectors and to the steel beam by welding.

Because the truss is a Vierendeel truss as opposed to a diagonalized truss, bending of top and bottom chords constitutes a significant structural action. Therefore, it is necessary to consider the interaction between axial loads and bending stresses in the design.

Figure 9.12a shows a typical floor plan with stub girders SG1, SG2, etc. Consider stub girder SG1, spanning 40 ft (12.19 m) between the exterior and interior of the building (Fig. 9.12b). The deck consists of a 2 in. (51 mm) deep 19-gauge composite metal deck with a 3¼ in. (82.5 mm) lightweight structural concrete topping. A welded wire fabric is used as crack control reinforcement in the concrete slab.

The first step in the analysis is to model the stub girder as an equivalent Vierendeel truss. This is shown in Fig. 9.13a. A 14 in. (356 mm) wide flange beam is assumed as the continuous bottom chord of the truss. The slab and the steel beam are modeled as equivalent top

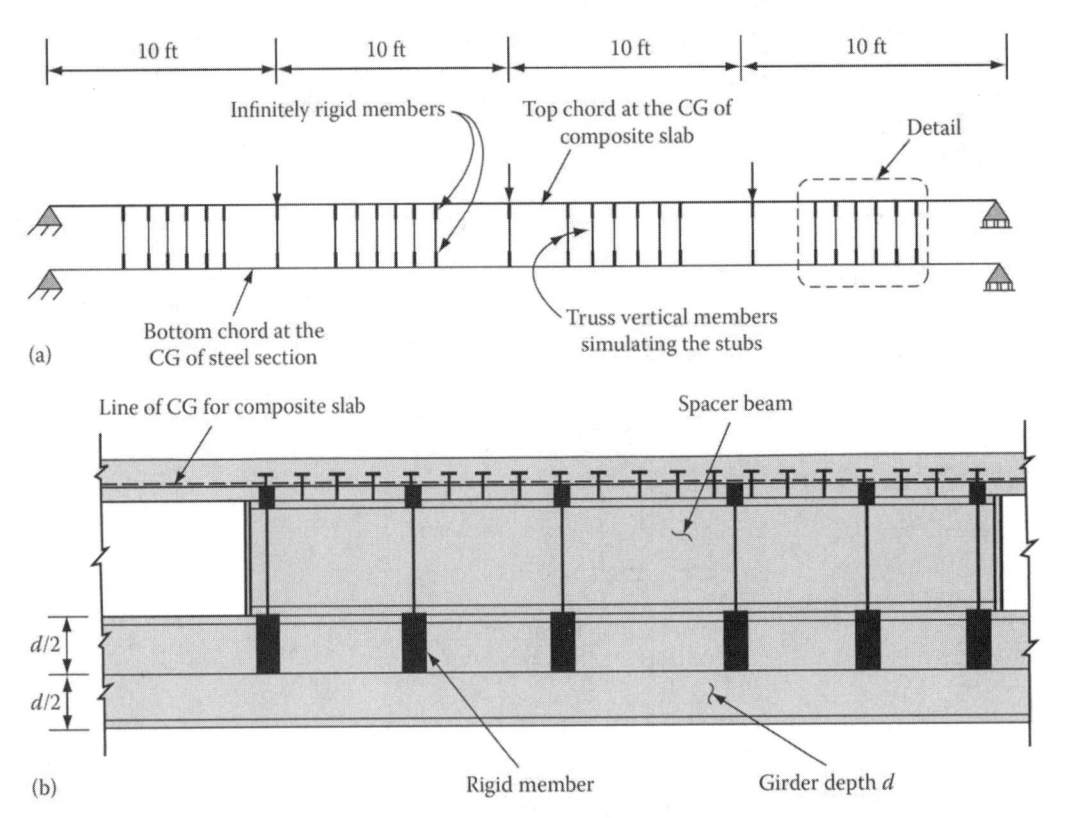

Figure 9.13   (a) Elevation of Vierendeel truss analytical model; (b) partial detail of analytical model.

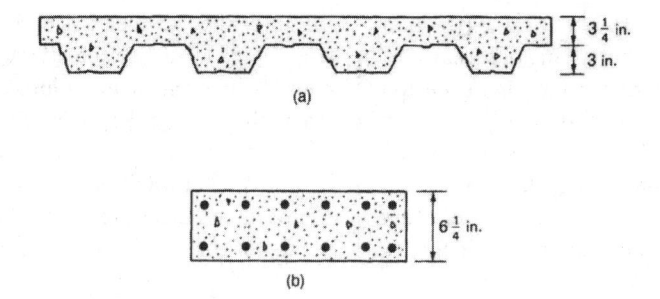

$3\frac{1}{4}$ in.

3 in.

(a)

$6\frac{1}{4}$ in.

(b)

Figure 9.14 Equivalent slab section.

and bottom chords. Note the beam elements representing these members are at the neutral axes of the slab and beam, as shown in Fig. 9.13b.

The stub pieces are modeled as a series of vertical beam elements between the top and the bottom chords of the truss with rigid panel zones at the top and bottom.

The various steps of modeling of stub girder are summarized as follows:

1. *Top chord of Vierendeel truss.* As shown in Fig. 9.13, the top chord consists of an equivalent transformed area of the concrete topping which is obtained by dividing the effective width of concrete slab by the modular ratio $n = E_s/E_c$. In calculating the transformed properties, advantage can be taken by including the mild steel reinforcement used in the slab. Although for strength calculations, modulus of elasticity of normal-weight concrete is used even for lightweight concrete slabs in composite beam design, in stub girder the lower value of n for lightweight concrete is used both for deflection and strength calculations. The moment of inertia I_t of the top chord is obtained by multiplying the unit value of I of composite slab, given in deck catalogs, by the effective width of slab such as shown in Fig. 9.14.

2. *Bottom chord.* The properties of the steel section are directly used for the bottom chord properties.

3. *Stub pieces.* The web area and moment of inertia of the stub in the plane of bending of the stub girder are calculated and apportioned to a finite number of vertical beam elements representing the stubs. The more elements employed to represent the stub pieces, the better will be the accuracy of the solution. As a minimum, the author recommends one vertical element for 1 ft (0.3 m) of stub width. The vertical segments between the top and bottom flanges of the stub and the neutral axes of the top and bottom chords is treated as an infinitely rigid member. Stiffener plates used at the ends of stubs can be incorporated in calculating the moment of inertia of the stubs.

9.7.3 Moment-Connected Stub Girder

The stub girder system, due to its large overall depth of approximately 3 ft (0.92 m), has a very large moment of inertia and can be used as part of a wind-resisting system. The model used for analysis is a Vierendeel truss, where the concrete slab and the bottom steel beam are simulated as linear elements and each stub piece is divided into seven elements. The gravity and wind load shear forces and moments are introduced as additional load cases in the computer analysis, and the combined axial forces and moments in each section of the stub girder are obtained. All parts of the stub girder are checked for combined axial forces, shear, and moments as shown above. The controlling section for the slab is generally at the end of the first stub piece furthermost from the column. Particular care is required to transfer the moment at the column girder interfaces. If wind moments are small, moment transfer can take place between the slab and the bottom steel beam. The slab needs to be attached to the column either by long deformed wire anchors or by welding reinforcing bars to the column. For relatively large moments, the solution for moment transfer is to extend the first stub piece to the column face. The top flange of the stub piece and the bottom flange of the W14 girder are welded to the column as in a typical moment connection. The design of the connection is, therefore, identical to welded beam-column moment connection. The girder should be checked along its full length for the critical combination of gravity and wind forces. Depending upon the magnitude of reversal of stresses due to wind load, bracing of the bottom chord may be necessary.

9.7.4 Strengthening of Stub Girder

Strengthening of existing stub girders for tenant-imposed higher loads is more expensive than in conventional composite construction. A speculative type of investment building is usually designed for imposed loads of

50 psf (2.4 kN/m^2) plus 20 psf (0.96 kN/m^2) as partition allowance. For heavier loads strengthening of local framing is required. The bottom girder, which is in tension and bending, is relatively easy to reinforce by welding additional plates or angles to the existing steel member. Reinforcing the top chord of the stub girder, which is in compression and bending, is somewhat tricky. Addition of structural steel angles by using expansion anchors to the underside of metal deck and welding of additional stub pieces to reduce the effective length of compression chord, which acts like a column, have been used in practice with good results. From the point of view of ultimate load behavior, it is acceptable to strengthen the bottom chord to resist total load without the truss action. However, it is important to check the lateral bracing requirements for the top flange of the bottom chord.

10.1 PRELIMINARY HAND CALCULATIONS

Even in today's high-tech computer-oriented world with all its sophisticated analysis capability, there still is a need for approximate analysis of structures. First, it provides a basis for selecting preliminary member sizes because the design of a structure, no matter how simple or complex, begins with a tentative selection of members. With the preliminary sizes, an analysis is made to determine if design criteria are met. If not, an analysis of the modified structure is made to improve its agreement with the requirements, and the process is continued until a design is obtained within the limits of acceptability. Starting the process with the best possible selection of members results in a rapid convergence of the iterative process to the desired solution.

Second, it is almost always necessary to compare several designs before choosing the one most likely to be the best from the points of view of structural economy and how well it fits in with other disciplines. Of the myriad structural systems only two or three schemes may be worthy of further refinement. Approximate methods are all that may be required to sort out the few final contenders from among the innumerable possibilities. Preliminary designs are therefore very useful in weeding out the weak solutions.

In the lateral load analysis of buildings, wind and earthquake forces are treated as equivalent loads and are reduced to a series of horizontal concentrated loads applied to the building at each floor. Portal and cantilever methods offer quick ways of analysis of rigid frames with unknown sizes. Both these methods are based on the well-observed characteristic of portal frames, namely that the points of contraflexure in beams and columns tend to form near the center of each column and girder segment. For purposes of analysis, the inflection points are assumed to occur exactly at the center of each member.

In the portal method, a rigid frame is treated as a series of consecutive single-bay portal frames. Interior columns are considered as part of two such portals, and the direct compression arising from the overturning effect on the leeward column of one portal is offset by the direct tension arising from the overturning effect on the windward column of the adjacent portal. If the widths of portals are unequal, the distribution of wind shear resisted by each portal can be assumed proportional to the aisle widths to maintain the interior column free of direct stress. Alternately, the column shears can be assumed to be unaffected by aisle widths resulting in axial stresses in the interior columns. With the shears in each column known and the points of contraflexure preestablished, the moments in beams and columns are determined. Simple statics yields axial and shear forces in beams and columns.

In the cantilever method, the building is analyzed as a cantilever standing on end fixed at the ground level. The overturning moment is assumed to be resisted by the axial compression of columns on the leeward side of the neutral axis and tension of columns on the windward side. The neutral axis for the frame is determined as the centroid of the areas of the columns in the bent. The axial forces in the columns due to overturning are assumed to be proportional to their distances from the neutral axis. As in the portal method, the points of inflection are assumed to occur at midheight of columns and midspan of girders. From the known axial forces in columns and the locations of the points of contraflexure, moments in columns and girders are obtained.

10.1.1 Portal Method

Consider the application of the portal method to a 30-story frame shown in Fig. 10.1, consisting of two equal exterior bays and a smaller interior bay. Table 10.1 lists the lateral loads assumed in the analysis. The procedure is as follows. Distribute the accumulated story shears to each column in proportion to the aisle widths such that

there are no direct stresses in the interior columns. Calculate the moments in the top and bottom of each column from the known shear in the column and the story height. Next, starting at the upper left corner of the frame, determine the girder moments where the column and girder moments are the same. Since the points of contraflexure are assumed at the center of the girder, the moments at each end are equal but opposite in sign. Determine the girder shears by the relation that shear multiplied by half of span length equals girder end moment. Next the axial stresses at the exterior columns are determined directly from girder shears. The results for the example frame are shown in Fig. 10.2.

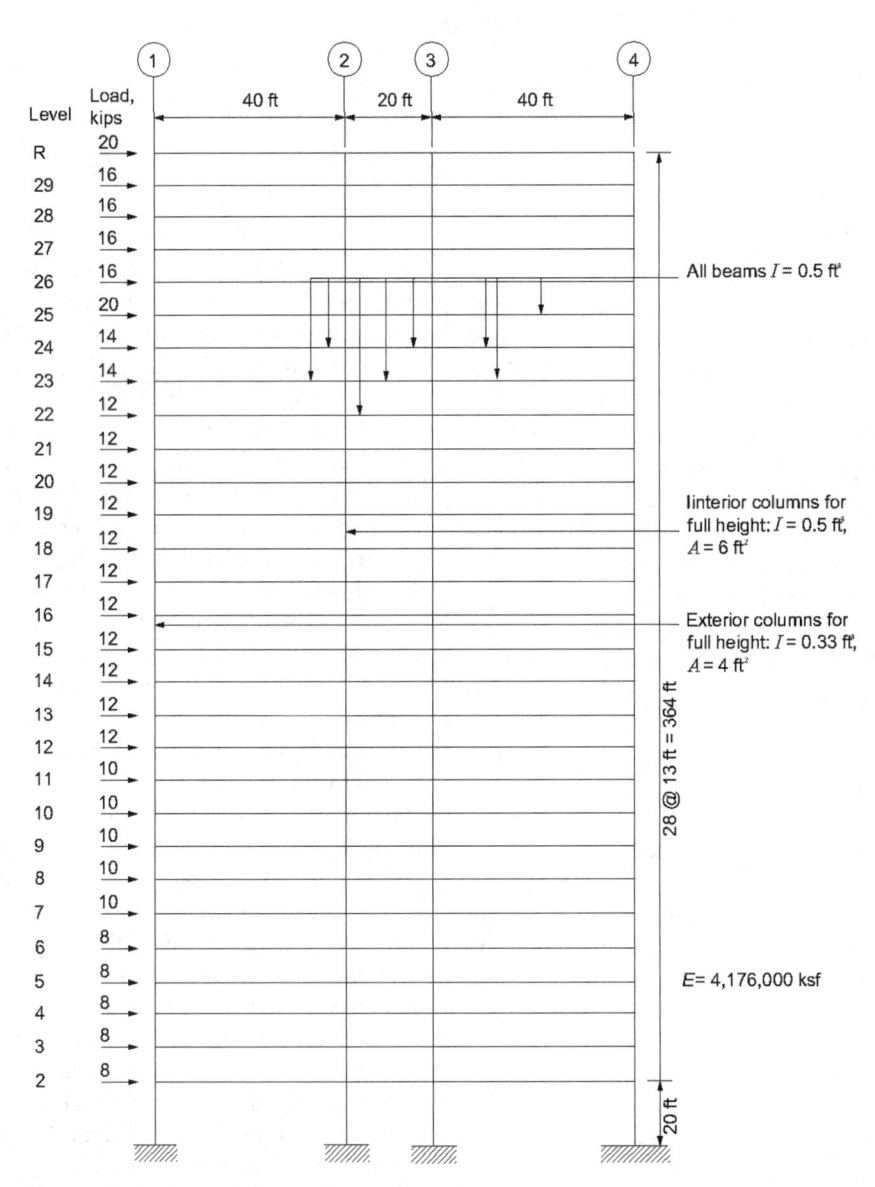

Figure 10.1 Example frame: dimensions and properties.

Table 10.1 Lateral Loads for 30-Story Building Shown in Fig. 10.1

Level	Story shear, kips	Accumulated shear, kips	Level	Story shear, kips	Accumulated shear, kips
R	20	20	15	12	222
29	16	36	14	12	234
28	16	52	13	12	246
27	16	68	12	12	258
26	16	84	11	10	268
25	14	98	10	10	278
24	14	112	9	10	288
23	14	126	8	10	298
22	12	138	7	10	308
21	12	150	6	8	316
20	12	162	5	8	324
19	12	174	4	8	332
18	12	186	3	8	340
17	12	198	2	8	348
16	12	210			

Note: 1 kip = 4.448 kN.

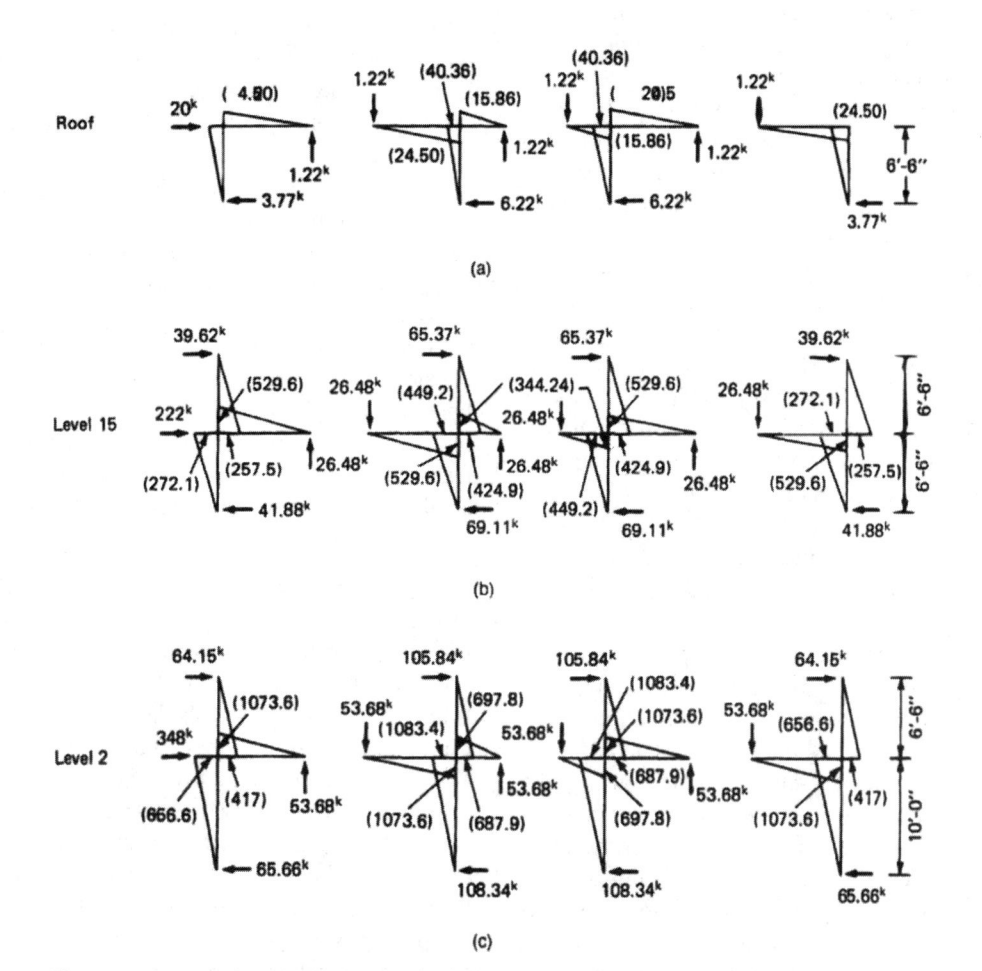

Figure 10.2 Portal method: (a) moment and forces at roof level; (b) moment and forces at level 15; (c) moment and forces at level 2. Note: all moments are in kip/ft and forces in kips.

10.1.2 Cantilever Method

Two assumptions are made in the analysis: (i) inflection points form at midspan of each beam and at midheight of each column; and (ii) the axial stresses in the columns vary as the distance from the frame centroidal axis. It is usually further assumed that all columns in a given story are of equal area resulting in column forces which vary as the distance from the center of gravity of the frame. To get a comparison with the portal method, we shall apply the cantilever method to the three-bay portal frame (Fig. 10.1) analyzed in the previous section.

The first assumption locates the points of contraflexure. Shown in Fig. 10.3a, part 1 is a free-body diagram of the top story above the points of contraflexure in the columns. The frame axis of rotation is located at the center of gravity of the columns, which for the example problem coincides with the line of symmetry of the frame. The column axial forces for the top story are obtained by equating the moment of the column reactions about the frame axis to the moment of the lateral forces taken about a horizontal plane through the assumed hinges of the top floor. These are also shown in Fig. 10.3a, part 1.

In a similar manner, the axial forces in the columns of other stories are computed by passing a section through the points of contraflexure of columns of each story and considering the moment equilibrium of the frame above the section (Fig. 10.3a, parts 2–4).

After the column axial forces are found, the girder shears are determined at once. For example, in Fig. 10.3a, part 3, the tension in the exterior windward column at the 15th level is 210.72 kips (937.28 kN). Tension in the same column at the 14th level is 187.08 kips (832.13 kN). Therefore, by the relation that the summation of the axial forces in the columns and the girder shear is equal to 0 at the joint where the 15th-story girder joins the exterior windward column, the girder shear is 210.72 − 187.08 = 23.64 kips (105.15 kN). Figure 10.3a, part 3, shows the method of obtaining this and the remaining shears for the 15th-level girder.

With the girder shears known, the girder moments follow directly. These equal the shear in the girder times one-half the span length. The study of the various joints will show that from the relation that $\Sigma M = 0$ at any joint, the sum of column moments must equal the sum of girder moments. Using this principle, the moments in the columns at the roof are obtained from roof girder moments (Fig. 10.3a, part 1). Since the points of contraflexure in the columns are at midheight, the column moments above the 29th level have the same value as at the roof level (Fig. 10.3a, part 2). Moments in the columns below the 29th level are obtained from the relation $\Sigma M = 0$, and in a similar manner column moments in other floors are

found. The column shears are obtained by dividing column moments by half the height of columns. As a check, observe that the shear in the columns of any level equals the sum of the horizontal external loads above that level. The moments and forces obtained by using the above procedure for the example problem are shown in Fig. 10.3a.

To get a feel for the accuracy of the procedures, the frame in Fig. 10.1 has been analyzed by a plane-frame computer analysis (Fig. 10.3b). The computer results vary considerably from either of the two methods. Chief among the reasons for the discrepancy are: (i) points of contraflexure in the lower stories are not at the midpoints; and (ii) the shears are greater in exterior girders than in the interior girders of that floor.

10.1.3 Lateral Stiffness of Frames

The lateral displacement of one floor relative to the floor below results from a combination of bending and shear deformation of the bent. The bending deformation or the chord drift, as it is sometimes called, is a consequence of axial deformation of the columns and is independent of the size, type, location, and arrangement of the web system. The shear deformation is due to the rotation of the joints in the frame, which causes bending of columns and girders of the frame. For relatively short frames with height-to-width ratios less than 3, the deflection due to axial shortening of columns can be neglected and the deflection of the frame can be assumed to be entirely due to joint rotations. Its contribution to deflection can, however, be obtained by considering the frame as a cantilever with an equivalent moment of inertial $I = 2ad^2$, where a is the area of exterior column and d is half the base of the portal frame. For taller frames, it is prudent to consider the axial deformation of the interior columns; the equivalent moment of inertia is determined by the relation $I = \sum_{1}^{n} a_1 d_1^2$, where a_1, a_2, . . ., a_n represent the areas of the columns and d_1, d_2, . . ., d_n represent their corresponding distances from the natural axis of the frame. To derive the shear deformation equations, consider the frame shown in Fig. 10.4. Isolate a typical floor and column segment between the points of contraflexure above and below the floor as shown in the figure.

Deflection Due to Column Rotations Consider the free-body diagram of a typical story bounded between the points of contraflexure in the columns above and below the ith level as shown in Fig. 10.5. When the number of stories is large, it is reasonable to assume that the shears in the columns above and below the floor do not differ appreciably. If the floor girders are rigid, the lateral deflection $\Delta_1/2$ of each column would be equal to the sum

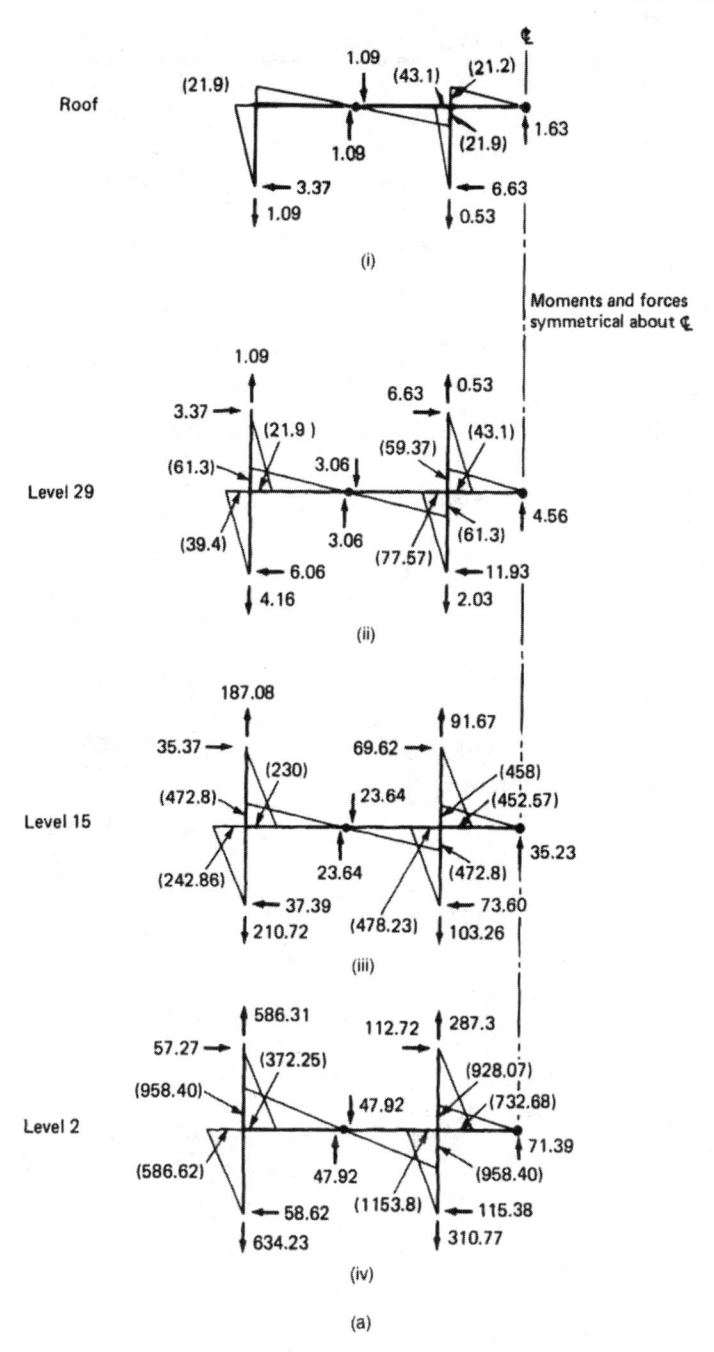

Figure 10.3a Cantilever method: (a) moment and forces at roof level; (b) moment and forces at level 29; (c) moment and forces at level 15; (d) moment and forces at level 2. Note: All moments are in kip/ft and forces in kips.

of the deflections of the two cantilevers of length $h/2$ under the action of wind shears V (Fig. 10.5).

$$\frac{\Delta_1}{2} = \frac{V\left(\dfrac{h}{2}\right)^3}{3EI_c} \quad \text{or} \quad \Delta_1 = \frac{Vh^3}{12EI_c} \tag{10.1}$$

giving for all columns $\Delta_1 = Vh^3/12E\,\Sigma I_c$.

Deflection Due to Girder Rotations Next consider the columns as rigid, giving rise to rotations of the girders as shown in Fig. 10.6a. Each girder undergoes a rotation equal to θ at each end giving rise to an internal moment of $12EI\theta/L$ for each girder. The total internal moment is given by the summation of such terms for each girder. Thus, the total internal moment due to girder rotation is $12E\theta\,\Sigma\,(I_{bi}/L_i)$. The external moment due to wind shears

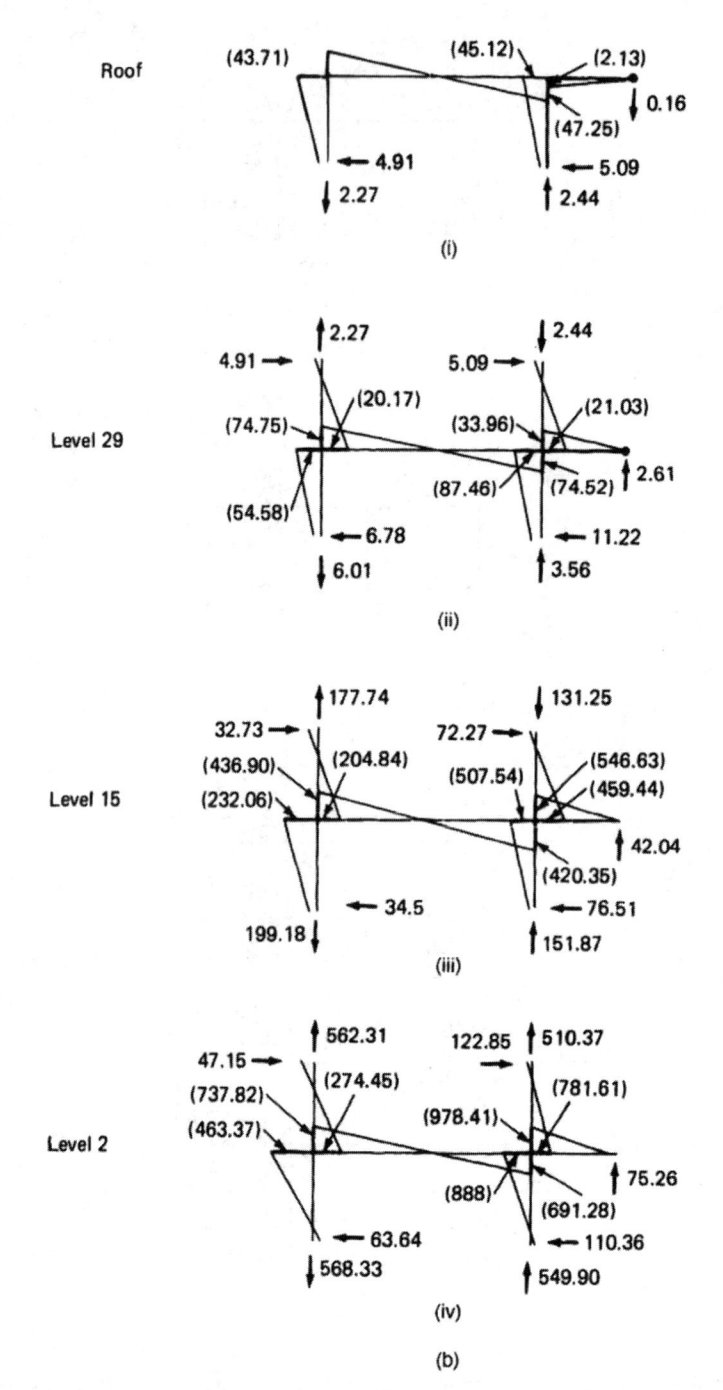

Figure 10.3b Cantilever method, moment and forces at: (a) roof level; (b) level 29; (c) level 15; (d) level 2. Note: all moments are in kip ft and forces in kips.

V is given by $V \times h$. Equating external moment to internal moment and noting that θ produces a displacement $\Delta_2 = \theta h$, we get

$$\Delta_2 = \frac{Vh^2}{12E \sum (I_{bi}/L_i)} \qquad (10.2)$$

The total frame shear deflection Δ_s is given by

$$\Delta_s = \Delta_1 + \Delta_2 = \frac{Vh^2}{12} \left\{ \frac{h}{(\sum EI)_{col}} + \frac{1}{\sum (EI/L)_{beam}} \right\} \qquad (10.3)$$

The deflection for the total number of stories is obtained by the summation of the deflections for each story.

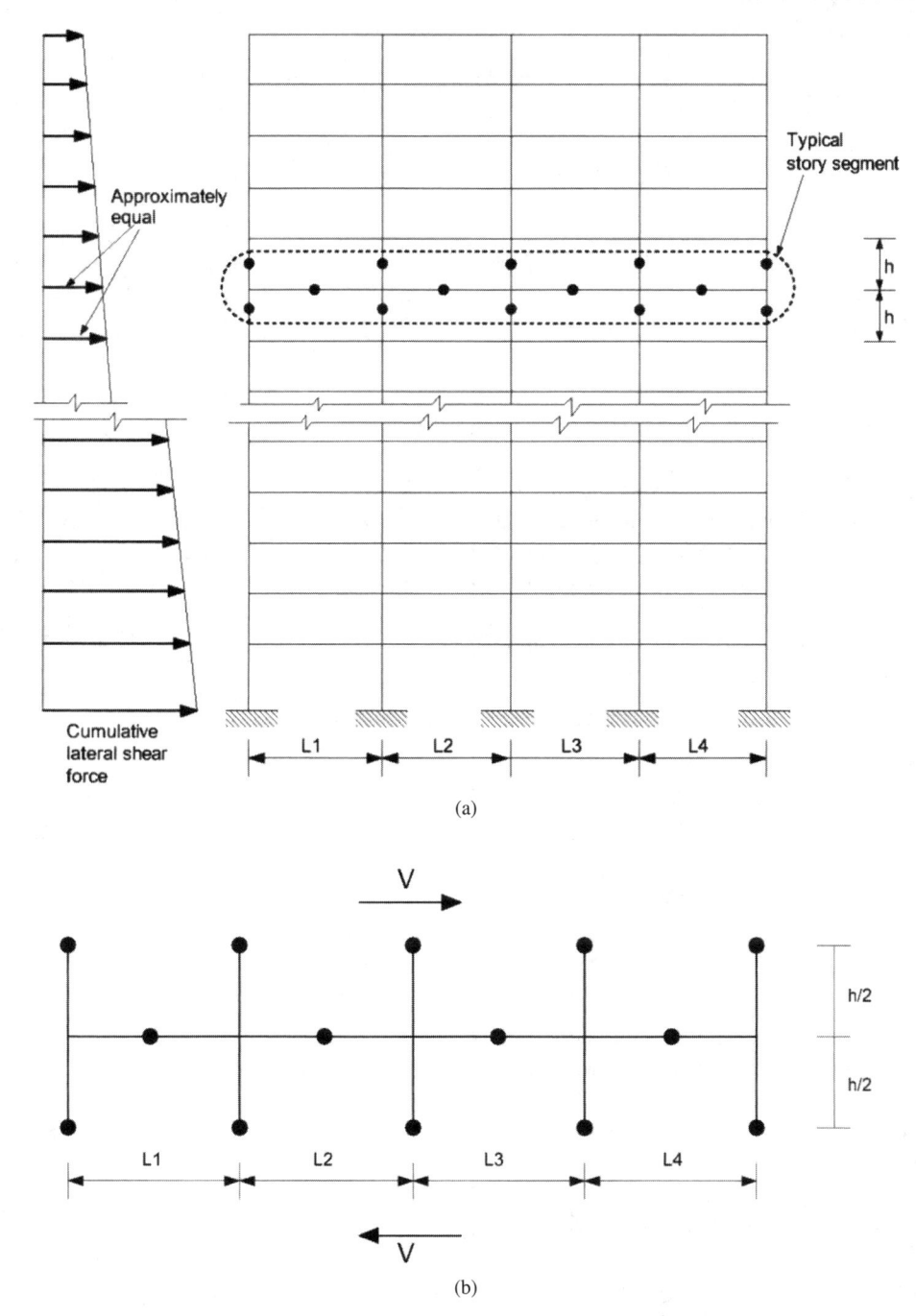

Figure 10.4 Deflections of portal frame: (a) frame subjected to lateral loads; (b) typical story segment.

An example of deflection calculations using the above procedure follows. To keep the presentation simple, we will consider the same example frame that was used for calculating moments and forces by the portal and cantilever methods (refer back to Fig. 10.1).

Deflection Calculations for Frame Shown in Fig. 10.1

Cantilever Deflection The neutral axis for the frame lies on the line of symmetry. The moment of inertia of the frame about the neutral axis is given by $I = 2(a_1 d_1^2 + a_2 d_2^2)$

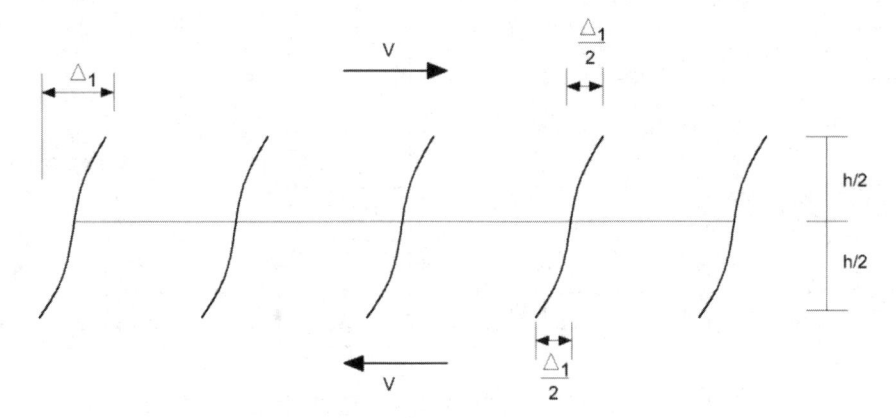

Figure 10.5 Lateral deflection of typical story due to bending of columns.

where a_1 and a_2 are the areas of the exterior and interior columns and d_1 and d_2 their distance from the neutral axis. Substituting $a_1 = 4$ ft^2 and $a_2 = 6$ ft^2, $d_1 = 53$ ft and $d_2 = 13$ ft, we get $I = 2(4 \times 53^2 + 6 \times 13^2) = 24{,}500$ ft^4 (211.46 m^4).

For purposes of deflection calculation, we can assume that the frame is subjected to a uniformly distributed horizontal load $= \dfrac{12}{13} = 0.9231$ kip/ft. The cantilever deflection at the top is given by

$$\Delta_{\text{cant}} = \frac{wl^4}{8EI} = \frac{0.9231 \times 384^4}{8 \times 4{,}176{,}000 \times 24{,}500} = 0.0245 \text{ ft} \ (7.47 \text{ mm})$$

Shear Deflection Due to Column Rotations This is given by

$$\Delta_1 = \frac{Vh^3}{12E \sum I_c}$$

For the example problem, the moments of inertia for the exterior and interior columns are, respectively, equal to 0.33 ft^4 and 0.5 ft^4, giving $\sum I_c = 2 \times 0.33 + 2 \times 0.5 = 1.66$ ft^4. Using an average cumulative shear value of $V = 210$ kips and $h = 13$ ft,

$$\Delta_1 = \frac{210 \times 13^3}{12 \times 4{,}176{,}000 \times 1.66} = 0.0056 \text{ ft} \ (1.70 \text{ mm})$$

Shear Deflection Due to Girder Rotations This is given by

$$\Delta_1 = \frac{Vh^2}{12E \sum (I/L)}$$

For the example problem, $\sum I/L$ of girders $= 0.5/40 + 0.5/26 + 0.5/40 = 0.0442$ ft, giving

$$\Delta_2 = \frac{210 \times 13^2}{12 \times 4{,}176{,}000 \times 0.0442}$$

$$= 0.016 \text{ ft/floor} \ (4.87 \text{ mm/floor})$$

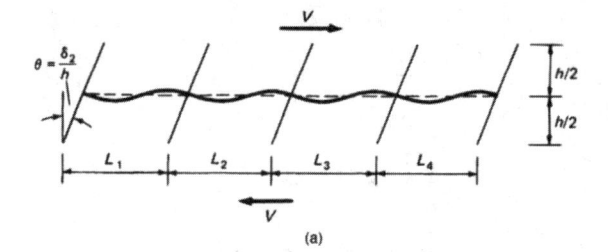

(a)

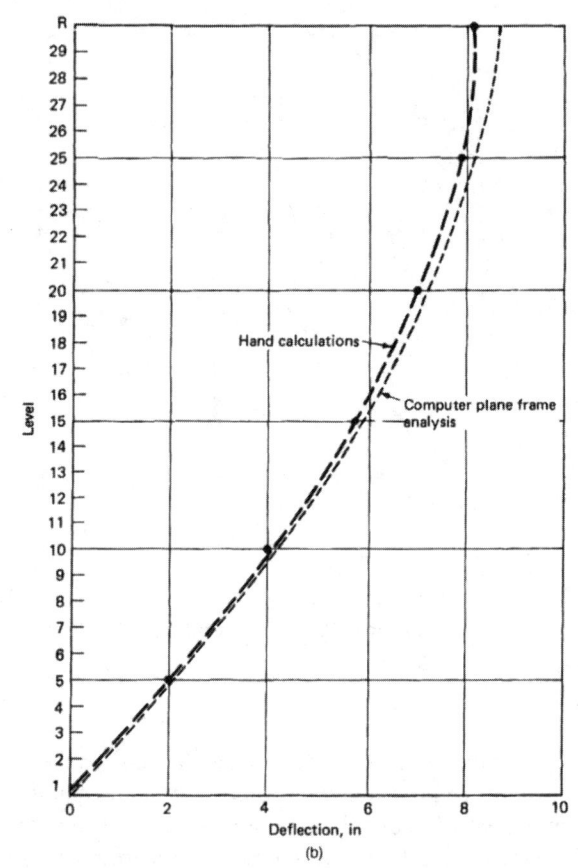

Figure 10.6 (a) Lateral deflection of typical story due to bending of girders; (b) deflection comparison (30-story frame).

The total shear deflection $\Delta_s = \Delta_1 + \Delta_2 = 0.0056 + 0.016 = 0.0216$ ft/floor (6.58 mm/floor). The shear deflection at top of 30 stories is given by $30 \times 0.0216 = 0.648$ ft. Therefore, total deflection at top due to chord drift and shear deformation is $0.0245 + 0.648 = 0.6725$ ft (204.97 mm). A comparison of floor-by-floor deflections obtained by using the above approach with those of a computer plane frame analysis is given in Fig. 10.6b. The appropriateness of the method for preliminary design is obvious.

Another method of calculation of frame deflection consists of representing the columns and beams as a single cantilever column with an equivalent flexural stiffness of I_e and shear stiffness of A_e to simulate the cantilever and shear modes of bending of the frame. The method is best explained with reference to Fig. 10.7 which shows a 19-story, three-bay unsymmetrical portal-frame with columns of varying moments of inertia. We first locate x, the distance of frame axis of bending from the windward column by equating moments of individual column areas to the moment of total area about the windward column. Using the values given in Fig. 10.7, we get,

$$4 \times 30 + 6 \times 50 + 6 \times 90 = (4 + 6 + 6 + 4)x$$

giving

$$x = 48 \text{ ft } (14.63 \text{ m})$$

from the windward column.

Calculate the moment of inertia of the frame about its axis of bending by the relation $I = \Sigma\, Ax^2$. Since the areas of the columns change at four locations, the corresponding four values of frame moment of inertia from the top work out equal to 21,120 ft^4, 42,240 ft^4, 63,360 ft^4, and 84,480 ft^4, respectively (182.3 m^4, 364.6 m^4, 546.86 m^4, 729.15 m^4).

Figure 10.8 shows the equivalent cantilever with varying moments of inertia. If the beams were infinitely rigid, the deflection calculated for the cantilever would have represented the total lateral deflection of the frame. Since in reality the beams are flexible, the deflection of the

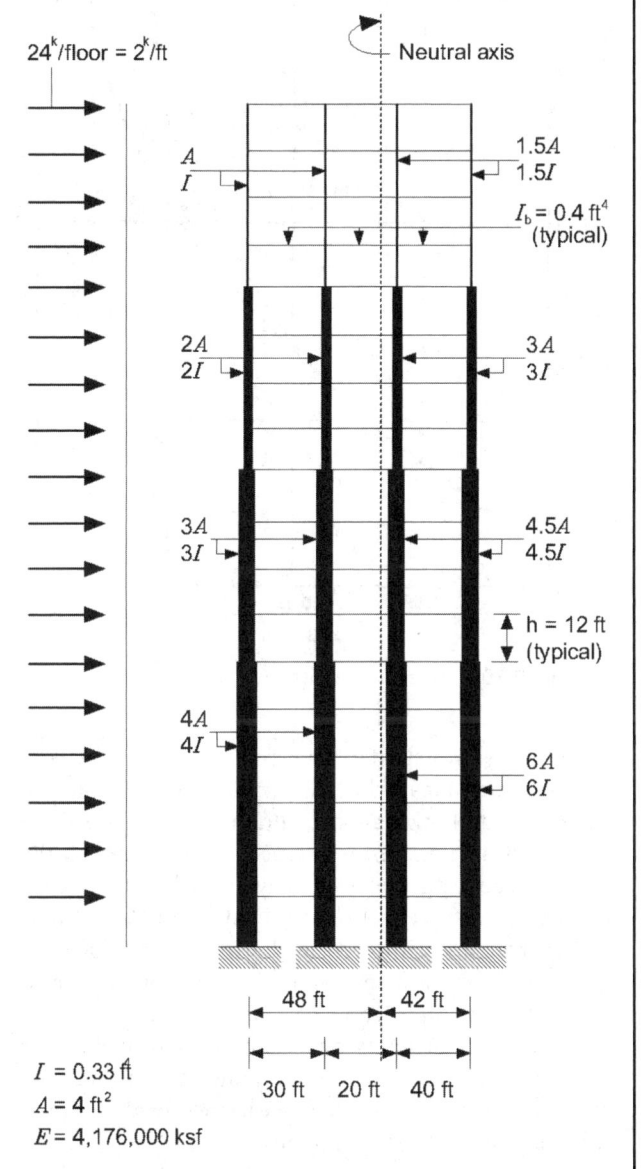

$I = 0.33$ ft
$A = 4$ ft^2
$E = 4,176,000$ ksf

Figure 10.7 Example portal frame for deflection calculations. Note the variation of column areas and moments of inertia at four locations.

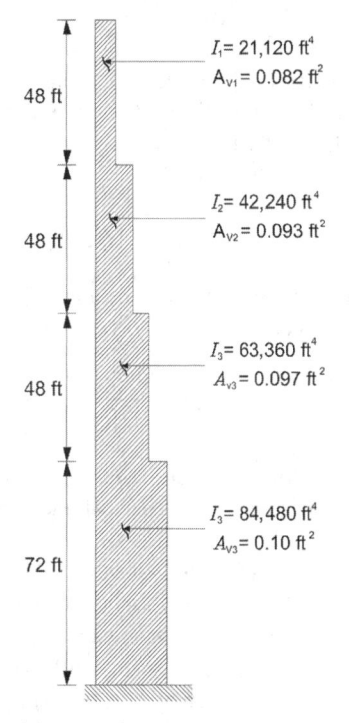

Figure 10.8 Equivalent cantilever representing the portal frame.

cantilever is increased by the racking component, which is equivalent to the shear deformation of the cantilever. This was shown equal to

$$\Delta_s = \frac{Vh^2}{12}\left[\frac{h}{(EI/h)_{\text{col}}} + \frac{1}{(EI/L)_{\text{beam}}}\right] \quad (10.4)$$

Defining story stiffness as the deflection per unit of horizontal shear, the equivalent story stiffness is given by the relation:

$$\frac{V}{\Delta_1} = \frac{12}{h^2\left\{1/(\sum EI)_{\text{col}} + 1/[\sum(EI/L)]_{\text{beam}}\right\}} \quad (10.5)$$

An equivalent shear area for the cantilever is worked out as follows. Consider the shear deformation of the cantilever per unit height h subjected to horizontal forces V as shown in Fig. 10.9. The shear deflection Δ_s is given by

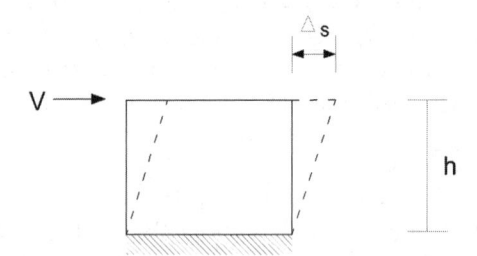

Figure 10.9 Shear deformations of cantilever of unit height.

$$\Delta_s = \frac{Vh}{GA_\upsilon} \quad (10.6)$$

The story stiffness Δ_s/h works out equal to $0.4EA_\upsilon/h$ in which it is assumed that $G = 0.4E$. Equating story stiffness relations of Eqs. (10.5) and (10.6), we get

$$\frac{0.4EA_\upsilon}{h} = \frac{12}{h^2\left\{1/(\sum E_cI)_{\text{col}} + 1/[\sum(E_bI/L)]_{\text{beam}}\right\}}$$

Assuming E is constant for beams and columns, that is, $E_c = E_b = E$, we get

$$A_\upsilon = \frac{30}{h\{1/(\sum I)_{\text{col}} + 1/[(IL)]_{\text{beam}}\}} \quad (10.7)$$

Using the numerical values shown in Fig. 10.7, the equivalent shear areas at four vertical locations work out, respectively, equal to 0.082 ft², 0.093 ft², 0.097 ft², and 0.1 ft² (0.0076 m², 0.0086 m², 0.0090 m², 0.0093 m²) from the top. These values are shown schematically in Fig. 10.8.

The deflection of the equivalent cantilever of varying moments of inertia can be obtained either by long-hand methods such as virtual work or by using a relatively simple stick computer model. Reasonable results can be obtained by assuming average properties for the equivalent cantilever. The average values for I and A for the example problem work out equal to 56,320 ft⁴ and 0.093 ft² (486 m⁴ and 0.0086 m²), respectively. Using a value of 216 kips for the average cumulative shear V, we get a total top deflection of 0.319 ft (94 mm) as compared to a value of 0.28 ft (82.3 mm) obtained from a stick computer model and a value of 0.24 ft (73 mm) as obtained from a plane frame analysis. Comparisons of deflections are shown in Fig. 10.10.

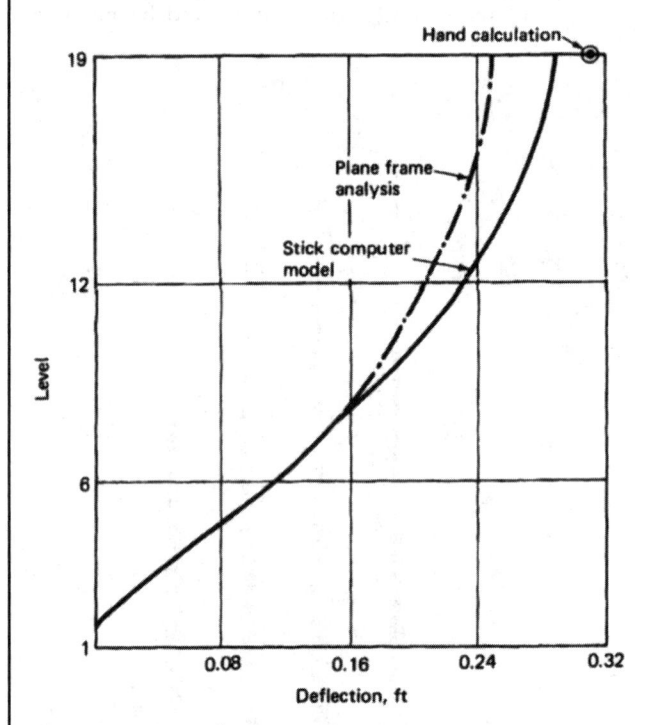

Figure 10.10 Deflection results.

The analysis presented thus far is based on the centerline dimensions, which in general overestimate the deflection. Although all structural members have finite widths, it is unnecessary, especially in view of the approximate nature of the analysis, to be overly concerned about the effect of joint widths on the stiffness of the structure. However, in those cases in which the dimensions of the members are large in comparison to story height and girder spans, it is possible to incorporate the effect of joints by assuming that no member deformation occurs within the joint. An approximate expression for the equivalent shear area for the equivalent column can be shown to be:

$$A_\upsilon = \frac{30}{h^2\{h\alpha_1^3/(\sum I)_{\text{col}} + \alpha_2^3/(\sum I/L)]_{\text{beam}}\}} \quad (10.8)$$

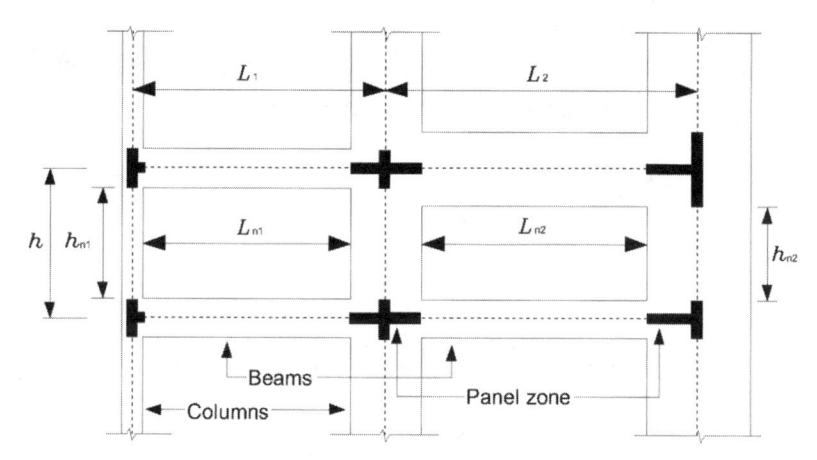

Figure 10.11 Typical beam-column joint; infinitely rigid panel zone.

where a_1 = the average ratio of clear height to center to center heights of columns (Fig. 10.11)

a_2 = the average of the ratio of the clear span to the centerline spans of girders (Fig. 10.11)

Analytical and experimental investigations have shown that an analysis based on rigid offset lengths to the outer face of supports overestimates the stiffness of the structure. The analysis should therefore include some method for compensating the deformations that do exist in the panel zones. A rigid zone reduction factor can be used to reduce the lengths of rigid offsets—a method similar to that employed in many commercial computer programs. Arbitrary reductions are assigned to joint sizes in an effort to compensate for the joint deformation.

The underlying principle in both the portal and cantilever methods is the assumption that the point of contraflexure is located at midheight and midspan of columns and girders. Rigorous computer analyses show that this assumption is violated in various degrees, especially at the top and bottom floors of a tall building. It is possible, however, to improve the results of the approximate analyses by refining the locations for points of contraflexure.

For example, the points of contraflexure at the lower floors, especially at the first floor may be assumed to, occur at a location closer to about $h/3$ below the second floor. Equivalent shear stiffness for the first story can be shown to be:

$$A_v = \frac{20}{h\{1/(\sum EI)_{col} + 1/5[\sum(EI/L)_{beam}]\}} \quad (10.9)$$

Further refinement of the analysis is generally considered unnecessary in view of the approximate nature of the analysis.

10.1.4 Framed Tube Structures

As mentioned earlier, the framed tube system in its simplest form consists of closely spaced exterior columns tied at each floor level by relatively deep spandrels. The behavior of the tube is in essence similar to that of a hollow perforated tube. The overturning moment under lateral load is resisted by compression and tension in the columns while the shear is resisted by bending of columns and beams primarily in the two sides of the building parallel to the direction of the lateral load. The bending moments in the beams and columns of these frames, called the web frames, can be evaluated using either the portal or the cantilever method. It is perhaps more accurate to use the cantilever method because tube systems are predominantly used for very tall buildings in the 40- to 80-story range in which the axial forces in the columns play a dominant role.

As mentioned earlier, because of the continuity of closely spaced columns and spandrels around the corners of the building, the flange frames are coaxed into resisting the overturning moment. Whether or not all the flange columns, or only a portion thereof, contribute to the bending resistance is a function of shear rigidity of the tube. A method for approximating the shear lag effects in a rectangular tube is to model it as two equivalent channels as shown in Fig. 10.12. The determination of width of the channel flange is subjected to engineering judgment and is usually taken as 15 to 20 percent of the width of the building.

Shown in Fig. 10.13 is the plan of a framed tube building delineating a limited number of columns in the leeward and windward faces as part of equivalent

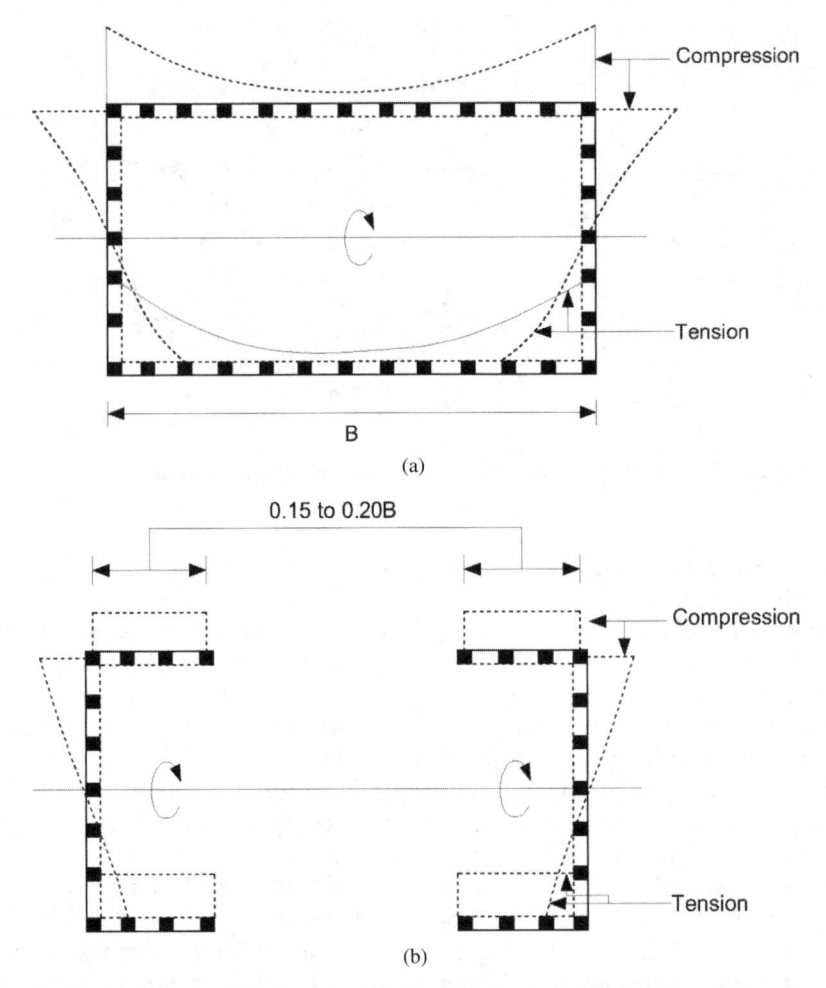

Figure 10.12   Framed tube: (a) axial stress distribution with shear lag effects; (b) axial stress distribution in equivalent channels without shear lag effects.

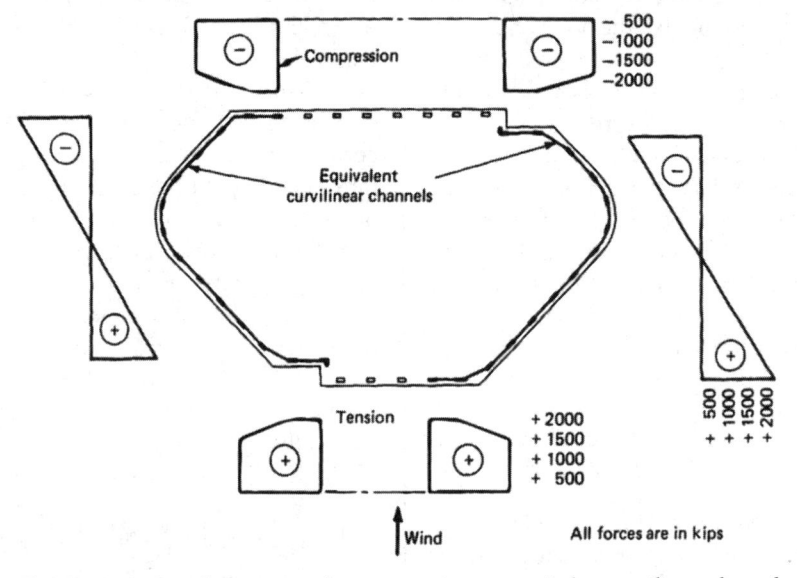

Figure 10.13   Axial stresses in columns assuming two equivalent curvilinear channels.

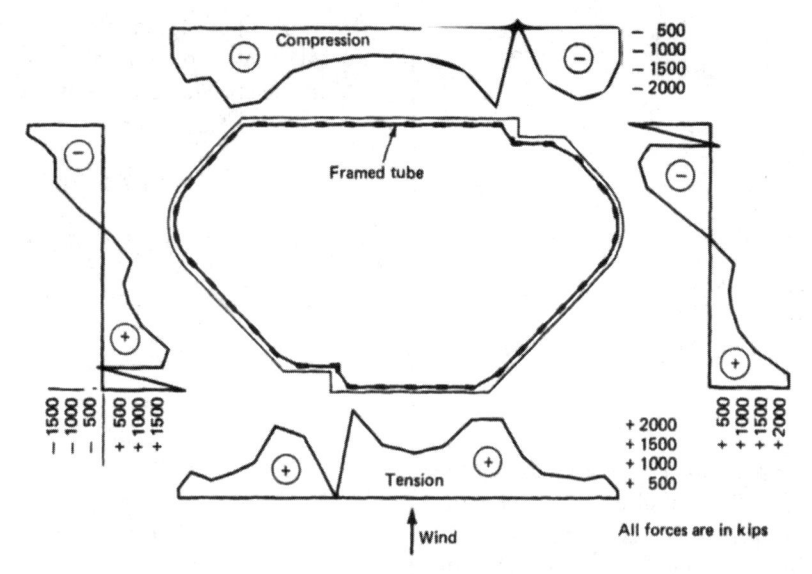

Figure 10.14 Axial forces in columns from three-dimensional analysis of framed tube.

channel flanges. The axial forces obtained on the basis of equivalent channels are, as shown in Fig. 10.13. Shown in Fig. 10.14 are the axial forces obtained from a three-dimensional computer analysis.

An equivalent column approach, as shown in the previous section, can be used to obtain approximate deflection values. In calculating the moment of inertia of the frame, it is only necessary to include the contribution of equivalent flange columns on the windward and leeward sides of the tube.

10.1.5 Coupled Shear Walls

Frequently, vertical rows of doors or windows occur within a continuous shear wall. When coupled by beams at each floor, the wall is usually referred to as a coupled shear wall. Another system popular in 20- to 30-story apartment and hotel buildings is the cross-wall system. This system (Fig. 10.21) consists of a continuous one-way slab spanning between load-bearing reinforced concrete walls, which resist both horizontal and vertical loads. The shear walls are either staggered in plan or placed,

in line with each other, as shown in Fig. 10.15. From the point of view of structural analysis, the behavior of both walls is very similar with a high degree of interaction between the horizontal and the vertical elements. Take, for example, the rotation θ at the center of gravity of the shear wall, as shown in Fig. 10.16. In having to comply with the deformed shape of the wall, the floor system undergoes not only a rotation, but a corresponding vertical displacement at all locations except at the center of gravity of the wall.

Continuous Medium Method Briefly, the analysis procedure is follows. The individual connecting beams of finite stiffness I_b are replaced by an imaginary continuous connection or laminae. The equivalent stiffness of the laminae for a story height $h = I_b/h$ giving a stiffness of $I_b\, dx/h$ for a height dx. When the wall is subjected to horizontal loading, the walls deflect, inducing vertical shear forces in the laminae. The system is made statically determinate by introducing a cut along the center of beams which is assumed to lie on the points of contraflexure. The displacement at each wall is determined and,

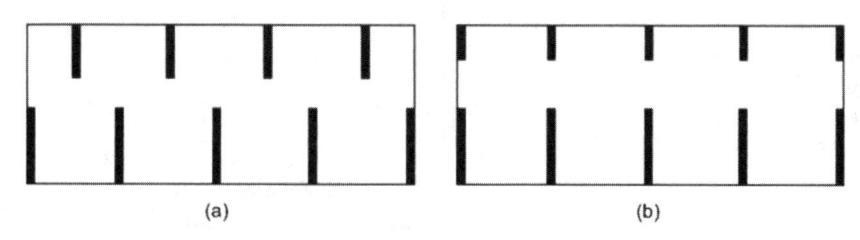

Figure 10.15 Coupled shear walls: (a) staggered shear walls; (b) walls in line with each other.

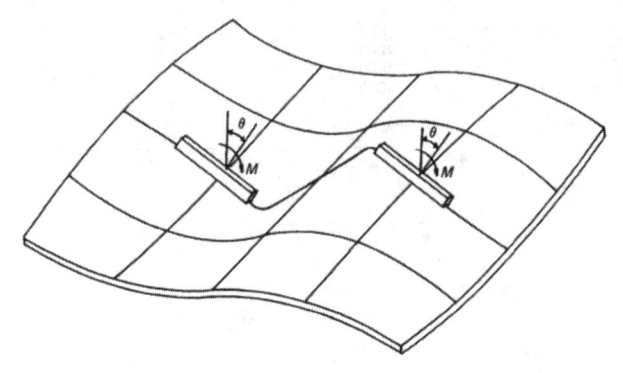

Figure 10.16   Displacement compatibility between slab and walls.

by considering the compatibility of deformation of the laminae, a second-order differential equation with the vertical shear force as a variable is established. The solution of this differential equation for the appropriate boundary conditions, of which fixed base is most common, leads to an equation for the integral shear T from which the moments and axial loads in the walls can be established. Once the distribution of T has been established, the shear force in the connecting beam at any level is obtained as the difference between the values of integral shear T at levels $h/2$ above and below the beam. At any level x, the bending moment in each wall can be established by superposition of the moment due to external lateral loads and a counteracting moment due to eccentricity of integral shear force from the center of gravity of the wall. The deflected form of the structure can then be established by integrating the moment curvature relationships.

The following notations are used in the development of the analysis.

A_1	Area of wall 1
A_2	Area of wall 2
I_2	Moment of inertia of wall 1
I_2	Moment of inertia of wall 2
H	Total height of wall
h	Story height
I_b'	Moment of inertia of interconnecting beam
l	Distance between centroids of walls 1 and 2
T	Integral shear force or sum of the laminae shears above a given level
q_n	Shear force in the laminae
w	Uniformly distributed lateral load
E	Modulus of elasticity assumed constant for the system
b	Width of opening
A_b	Area of connecting beam
I	$I_1 + I_2$
G	Shear modulus

I_b Moment of inertia of connecting beam reduced to take into account the effect of shear deformation in the beam. This is given by the relation:

$$I_b = \frac{I_b'}{1 + 2.4(d/b)^3(1+v)}$$

d depth of interconnecting beam
v Poisson's ratio
α, β, and μ Parameters given by the following relations

$$\alpha^2 = \frac{12I_b}{hb^3}\left[\frac{l^2}{I} + \frac{A}{A_1 A_2}\right]$$

$$\beta = \frac{6wlI_b}{Ib^3 h}$$

$$\mu = 1 + \frac{AI}{A_1 A_2 l^2}$$

Figure 10.17a shows a pierced shear wall subjected to a uniformly distributed horizontal load of intensity w. In Fig. 10.17b, the wall is imagined cut along the centerlines of the connecting beams. The structural action of the tie beams is replaced by an imaginary equivalent continuous laminae. The shear force T acting along the vertical axis of the wall is determined by considering the relative displacement of each wall.

Figure 10.17c shows an experimental setup for testing a coupled shear wall subjected to horizontal load.

Under lateral loads the two ends of beam at the cuts experience a vertical displacement consisting of contributions δ_1, δ_2, δ_3, and δ_4 as shown in Fig. 10.18.

The relative displacement δ_1 due to bending of each wall element (Fig. 10.18a) is given by:

$$\delta_1 = l\frac{dy}{dx}$$

The shear force $T = qh$ acting at each floor level at the center of connecting beams will cause a relative displacement δ_2 (Fig. 10.18b) due to bending of these beams. This is given by

$$\delta_2 = \frac{qb^3 h}{12EI_b'}$$

The same shear force causes a shear deformation δ_3 (Fig. 10.18c) in the beam given by

$$\delta_3 = \frac{qbh}{GA_b'}$$

For rectangular sections effective cross-sectional shear area A_b' can be considered $= A_b/1.2$. Therefore,

$$\delta_3 = \frac{1.2qbh}{GA_b}$$

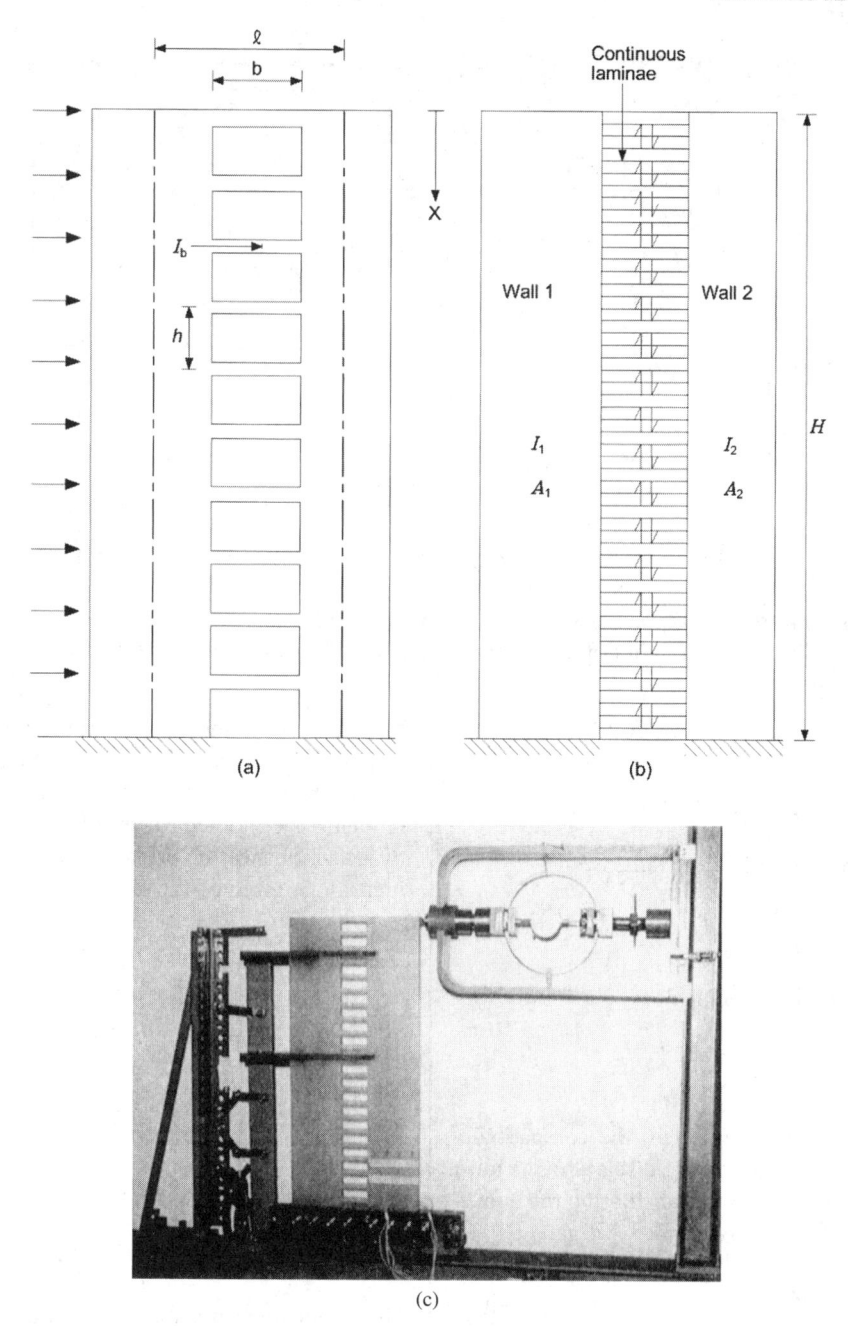

Figure 10.17   Analysis of coupled shear wall by continuous shear medium technique: (a) interconnected shear walls; (b) representation of discrete beams by a continuous shear medium; (c) experimental setup for a coupled shear wall model.

The displacement δ_4 (Fig. 10.18d) is the relative displacement of the two wall elements due to the axial deformation of the walls caused by T acting as a vertical load on the wall elements. This is determined as follows. The axial force in wall at any height is

$$T = \int_0^x q\, dx$$

and therefore

$$q = \frac{dT}{dx}$$

The strain in wall 1 due to axial loads is equal to

$$\frac{\text{stress}}{E} = \frac{\text{force}}{E \times \text{area}} = \frac{T}{EA_1}$$

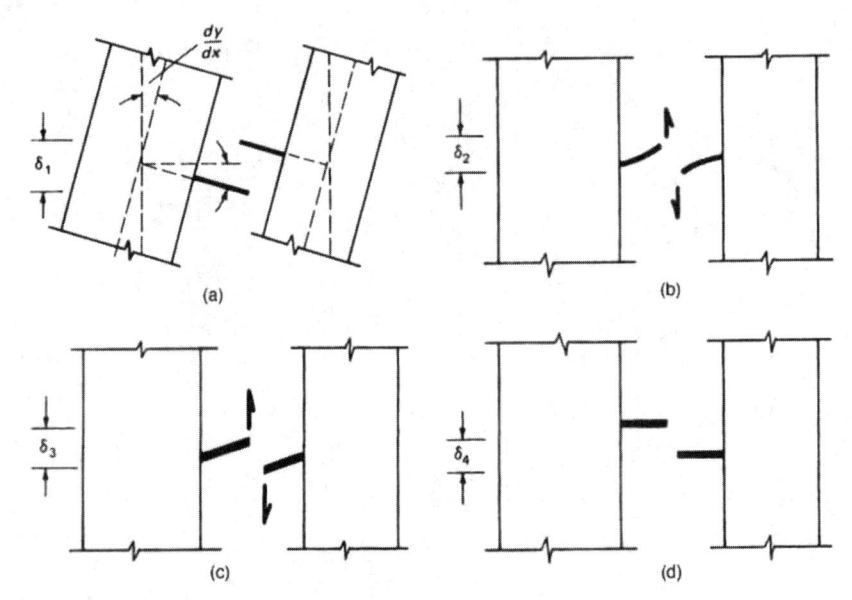

Figure 10.18   Displacement of connecting beam due to: (a) wall rotation; (b) beam bending; (c) beam shear deflection; (d) axial tension and compression in wall segment.

Therefore, total increase in length for wall 1 is

$$\int_x^H \frac{T}{A_1 E}\,dx$$

Similarly, total axial shortening for wall 2 is

$$\int_x^H \frac{T}{A_2 E}\,dx$$

Therefore, relative displacement is

$$\delta_4 = \frac{1}{E}\left(\frac{1}{A_1}+\frac{1}{A_2}\right)\int_x^H T\,dx$$

Since the two walls are connected, the compatibility condition stipulates that the relative displacements must vanish, i.e., $\delta_1 + \delta_2 + \delta_3 + \delta_4 = 0$. Substituting the above-derived expressions for δ_1, δ_2, δ_3, and δ_4, we get

$$l\frac{dy}{dx}+\frac{b^3hq}{12EI_b'}+\frac{qhb}{1.2GA_b}+\frac{1}{E}\left(\frac{1}{A_1}+\frac{1}{A_2}\right)\int_x^H T\,dx = 0$$

Substituting $q = dT/dx$ and $A = A_1 + A_2$, we get

$$\frac{dy}{dx}=-\frac{b^3h}{12lEI_b'}\frac{dT}{dx}-\frac{bh}{1.2GA_bl}\frac{dT}{dx}-\frac{A}{lEA_1A_2}\int_x^H T\,dx$$

Differentiating with respect to x, we get

$$EI\frac{d^2y}{dx^2}=-\frac{b^3hI}{12lI_b'}\frac{d^2T}{dx^2}-\frac{2bhI}{1.2lA_b}\frac{d^2T}{dx^2}-\frac{lAT}{lA_1A_2} \qquad (10.13a)$$

The first two terms on the right-hand side of the above equation, which pertain to the bending and shear deflection of the beam, can be combined into a single term by reducing the moment of inertia of the beam to include the effect of shear deformation. The reduced moment of inertia I_b is given by the relation

$$I_b = \frac{I_b'}{1+2.4(d/b)^3(1+v)}\sqrt{2}$$

Using this relation Eq. (10.3b) reduces to

$$EI\frac{d^2y}{dx^2}=-\frac{b^3hl}{12I_bl}\frac{d^2T}{dx^2}-\frac{AI}{lA_1A_2}T \qquad (10.13b)$$

The total applied moment M_x at any point x is given by

$$M_x = \frac{wx^2}{2}$$

Hence equation of statistical equilibrium is arrived at as follows. The applied moment less the moment due to T acting at an eccentricity l is

$$EI\frac{d^2y}{dx^2}=\frac{wx^2}{2}-Tl \qquad (10.13c)$$

From Eqs. (10.13b and c) the following governing second-order differential equation is derived:

$$\frac{d^2T}{dx^2}-\alpha^2 T = -\beta^2 x^2 \qquad (10.13d)$$

where

$$\alpha^2 = \frac{12I_b}{hb^3}\left[\frac{l^2}{I}+\frac{A}{A_1A_2}\right]$$

$$\beta = \frac{6wlI_b}{hb^3I}$$

The solution of Eq. (10.13d) gives

$$T = c_1 \sinh \alpha x + c_2 \cosh \alpha x + \frac{\beta}{\alpha^2}\left(x^2 + \frac{2}{\alpha^2}\right) \quad (10.14a)$$

where c_1 and c_2 are the constants of integration.

To eliminate these constants, we introduce boundary conditions. Most commonly, the boundary condition at the base of shear walls where $x = H$ is to assume a rigid foundation which permits no deformation. The deformation at the cut ends of the laminae is zero and hence $q = 0$ or $dT/dx = 0$ at $x = H$. At top of walls where $x = 0$ the wall is free; therefore, the integral of shear force must vanish, that is, $T = 0$ at $x = 0$. Substituting the boundary conditions in Eq. (10.14a) we get:

$$T = \frac{2\beta}{\alpha^4}\left\{1 + \frac{\sinh \alpha - \alpha H}{\cosh \alpha H}\sinh \alpha x - \cosh \alpha x + \frac{\alpha^2 x^2}{2}\right\} \quad (10.14b)$$

Once the distribution force T has been obtained, the shear force in the coupling beam may be determined as the difference in values of T at levels $h/2$ above and below that level. The bending moment in the beam is obtained by the product of shear force and half the clear span of beam. Since the walls are assumed to deflect equally, the bending moments are proportional to their stiffness. Therefore, the bending moments in walls 1 and 2 are given by

$$M_1 = \left(\frac{wx^2}{2}-Tl\right)\frac{I_1}{T}$$

$$M_2 = \left(\frac{wx^2}{2}-Tl\right)\frac{I_2}{I}$$

The general expression for deflection y at any point x can be obtained by integrating Eq. (10.13c) twice and substituting appropriate boundary conditions. Assuming the foundation for the walls to be rigid, the boundary conditions are

$$y = 0 \quad \text{and} \quad \frac{dy}{dx} = 0 \quad \text{at } x = H$$

Although interstory drifts are important, most usually in preliminary analysis the maximum deflection at top is of prime interest. This is given by the following expression:

$$y_{max} = \frac{wH^4}{2EI}\left[0.25\left(1-\frac{1}{\mu}\right)\right.$$

$$\left. -\frac{2}{\mu}\left\{\frac{\alpha H \sinh \alpha H - \cosh \alpha H + 1}{(\alpha H)^4 \cosh \alpha H}-\frac{1}{2(\alpha H)^2}\right\}\right]$$

$$(10.15)$$

where

$$\mu = 1 + \frac{AI}{A_1A_2l^2}$$

To analyze a system of coupled shear walls by this method requires laborious calculations. Several researchers have proposed simplifications of this procedure. Of particular interest is the one proposed by Coull and Choudhury (ref. 56). In this method, the stress distribution in coupled shear walls is obtained as a combination of two distinct actions: (i) walls acting together as a single composite cantilever with the neutral axis located at the centroid of the two elements; and (ii) walls acting as independent cantilevers bending about their own neutral axes. Semigraphical methods are presented for rapidly evaluating maximum stresses and deflections in coupled shear walls subjected to a variety of loading cases. The interested reader is referred to ref. 56 for further details of this method.

10.2 LUMPING TECHNIQUES

Analysis of tall buildings is a highly complex and indeterminate problem. Even with the availability of large-capacity high-speed computers, certain simplifying assumptions are necessary for all but the simplest of structures. One such assumption commonly made in rigid frame analysis is that the points of contraflexure occur at the midspan and midheight of beams and columns. This assumption makes it possible to lump a number of typical floors into a single floor. This is explained with reference to Fig. 10.19. The prototype frame, consisting of a 30-story, three-bay frame, is to be modeled as a lumped frame. Generally, in a high-rise building the floor-to-floor heights at the bottom and top few levels are different from typical floor-to-floor heights. Also as noted before, the points of contraflexure for beams and columns at these floors are not at their center. Therefore, it is appropriate to limit lumping to typical floors only, as shown in Fig. 10.20. For purposes of illustration let us assume that two floors of the prototype are considered equivalent to a single floor of the lumped model.

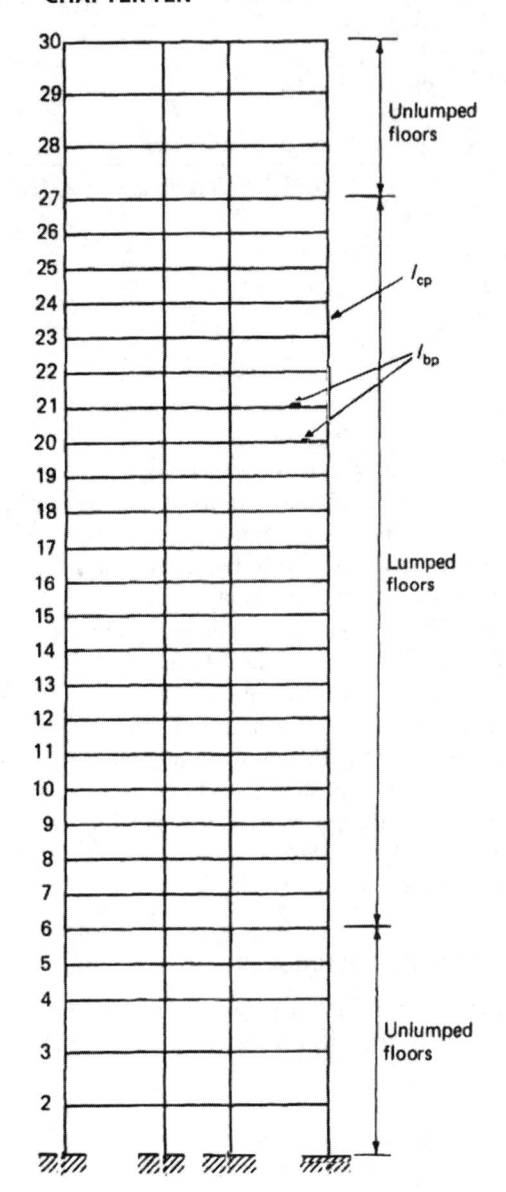

Figure 10.19   Prototype of unlumped frame.

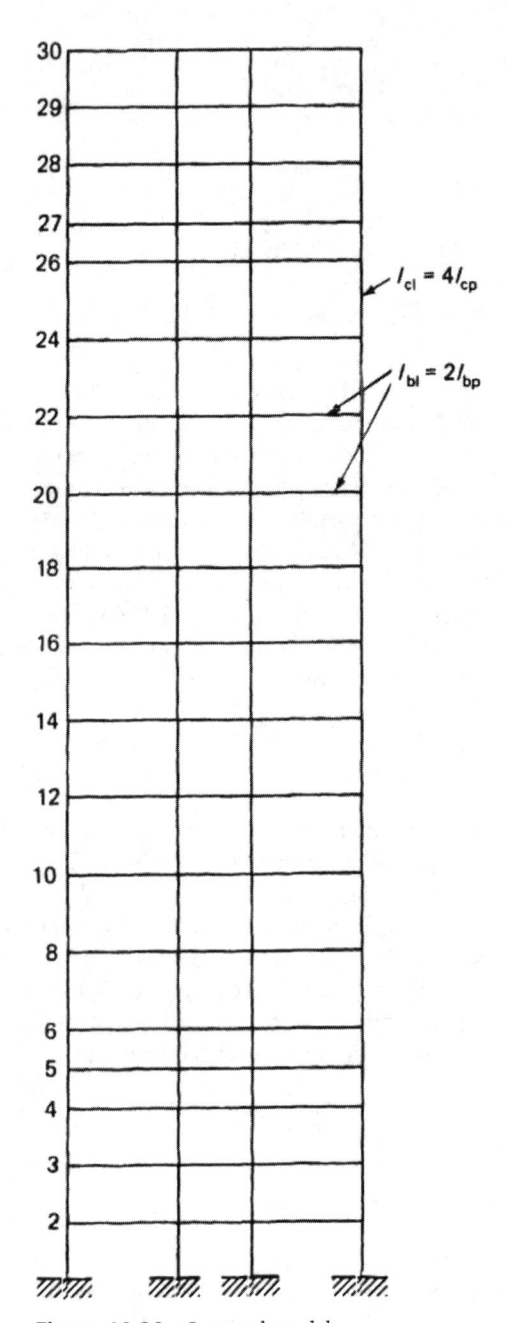

Figure 10.20   Lumped model.

The behavior of a rigid frame, as mentioned many times over, is a combination of the cantilever and shear racking modes. The cantilever behavior is a function of the location and axial stiffness of the columns only and does not depend on the arrangement of beams. Therefore, to maintain the cantilever behavior intact between the prototype and the lumped model, the areas and location of the columns are kept the same in both unlumped and lumped models; the lumped model column occupies the equivalent position of the prototype column with the actual prototype column areas.

The equivalence of the raking component between the two models is maintained by keeping the ratio of column and girder stiffness factors the same between the two. Since two floors of the actual model are lumped into one floor, the moment of inertia and area of the girder in the lumped model should be twice their values in the prototype model. If n floors are lumped into one floor, the corresponding properties will be n times the prototype values. To keep the explanation simple, it is useful to introduce the following notations:

I_{cp} = moment of inertia of column in the unlumped model (prototype)

I_{cl} = moment of inertia of the column in the lumped model

L = length of girder which is the same in both models

h_{cp} = height of column in the unlumped model (prototype)

h_{cl} = height of column in the lumped model

In the present example, two stories are lumped together. Therefore, ratio of $h_{cl}/h_{cp} = 2.0$. In general, this ratio can be considered as n, where n is the lumping ratio. Equating stiffness ratios of column and beams between the two models gives

$$\frac{I_{cp}/h_{cp}}{I_{cp}/L} = \frac{I_{cl}/h_{cl}}{I_{bl}/L} \qquad (10.16)$$

which simplifies to

$$I_{cl} = I_{cp}\left(\frac{h_{cl}}{h_{cp}}\right)\left(\frac{I_{bl}}{I_{bp}}\right) \qquad (10.17)$$

Since in the example the ratio of the heights of prototype and model columns is 2.0 and the moment of inertia of the model beam is twice that of the prototype, we get

$$I_{cl} = I_{cp}(2)^2 \qquad (10.18)$$

In the general case, the moment of inertia of the lumped model column works out to be n^2 times the prototype value. Lumping of nontypical floors can also be accomplished by assuming locations of point of contraflexure at, say, one-third the height for the lower stories and by using the principle of virtual work to equate the deflection properties. However, since the mathematical expressions get unwieldy, this procedure is not discussed here.

Discrepancies always exist in all but the simplest of structures between the unlumped and lumped models, especially at regions of abrupt change in stiffnesses and geometry. Although such deviations may be high locally, in general the overall behavior is kept unaltered.

10.3 PARTIAL COMPUTER MODELS

Consider a plane frame with an even number of bays subjected to lateral loads as shown in Fig. 10.21. The frame is symmetrical about column 4, and therefore computationally it is more efficient to analyze only one-half of the frame. It is only necessary to reduce the geometric properties of column 4 by 50 percent and to introduce appropriate kinematic boundary conditions at the centerline of the column. For the cantilever bending action of the frame due to lateral loads, the neutral axis can be considered to pass through the centerline of column 4. To reproduce this effect in the half model, it is only necessary to restrain the axial deformation of column 4 at each floor. Since only one-half the frame is analyzed, the model is subjected to half the horizontal loads as shown schematically in Fig. 10.21b.

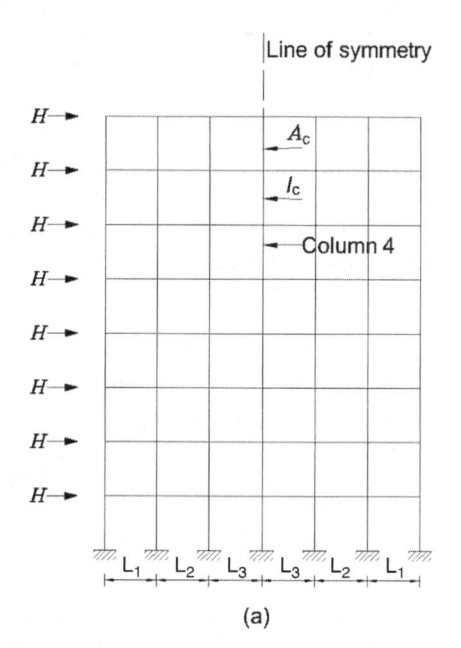
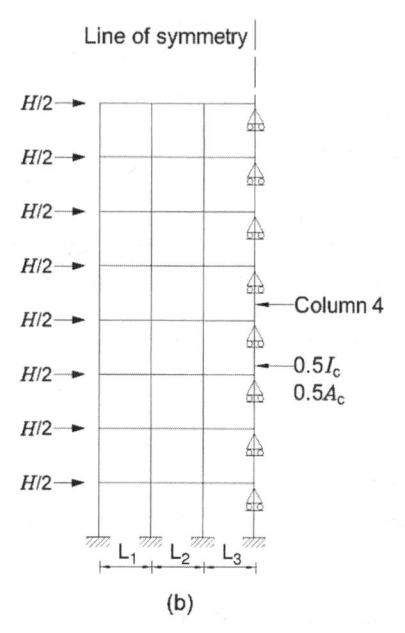

Figure 10.21 Symmetrical frame with even numbers of bays: (a) full model; (b) partial model.

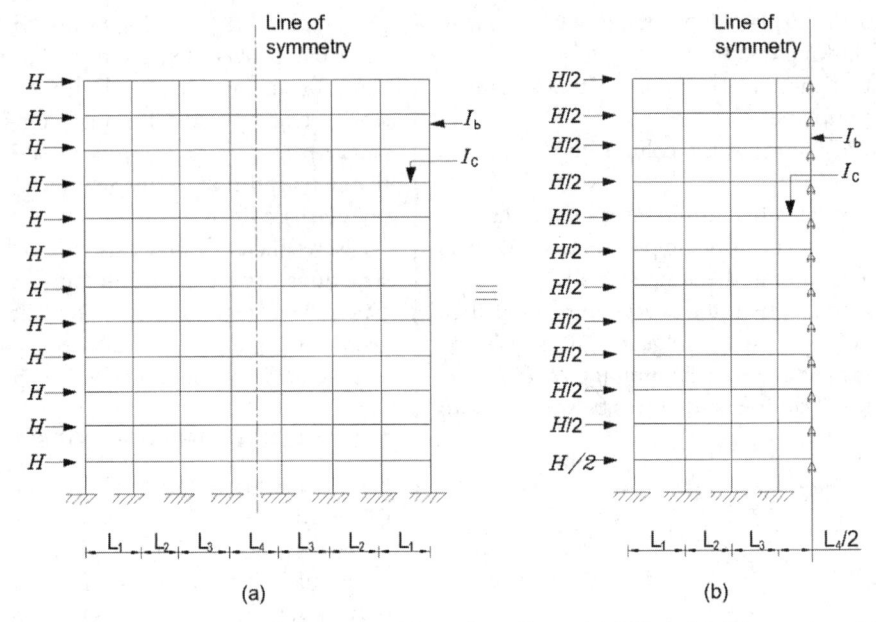

Figure 10.22 Symmetrical frame with odd number of bays: (a) full model; (b) partial model.

The procedure for analyzing a symmetrical structure with an odd number of bays is similar to the previous procedure as shown in Figs. 10.22 and 10.23. The only difference is that the neutral axis for frame bending action passes through the center of beam spans. This affect is duplicated in the half model by introducing fictitious vertical supports at the midspan of beams. Only one-half of the lateral loads is applied to the model as before.

A symmetrical tube can be analyzed by considering only a quarter of the model. In addition to the aforementioned boundary conditions, it is necessary to assure that the three-dimensional model does not twist due to the application of horizontal loads.

A tube is a three-dimensional structure, and as such responds by bending about both its principal axes and rotation about a vertical axis. In analyzing a quarter or half model, it is necessary to restrain the transverse bending and rotation of the tube. The kinematic restraints that preclude transverse movement and rotation of the model are shown in Fig. 10.23.

10.4   TORSION

10.4.1   Introduction

This section gives an overview of analysis of structural systems for torsion with particular emphasis on the torsion analysis of open-section cores. At first, we take a cursory look at the classical methods of torsion of elements such as circular, non-circular, and cellular sections,

and later on discuss warping torsion of structural systems consisting of open cores.

The terminology used in torsion analysis may be conveniently grouped under two headings: uniform or St Venant's torsion and warping torsion, often times referred to as constrained torsion or torsion-bending. The terms for uniform torsion are well established and given in most textbooks on structural mechanics. The purpose of recalling them here is to show how they relate to the warping theory.

The terms shown on the right-hand side of Table 10.2 relating to warping torsion have, in the past, been given little attention. Consequently, designers are generally not at ease with neither the concepts of warping behavior nor with its methods of analysis. The aim here is to introduce the concept of warping without indulging in an abundance of mathematics and to show, by numerical examples, the importance of considering warping in practical cases.

Table 10.2(a)   Torsion Terminology

Uniform (St Venant) torsion	Warping torsion
• Torsional shear stress	• Shear center
• Twist	• Open section
• Polar moment of inertia	• Warping deformation
• Membrane analogy	• Sectorial coordinate
• Shear flow	• Warping moment of inertia
• Cellular sections	• Bimoment
	• Normal stress
	• Tangential stress

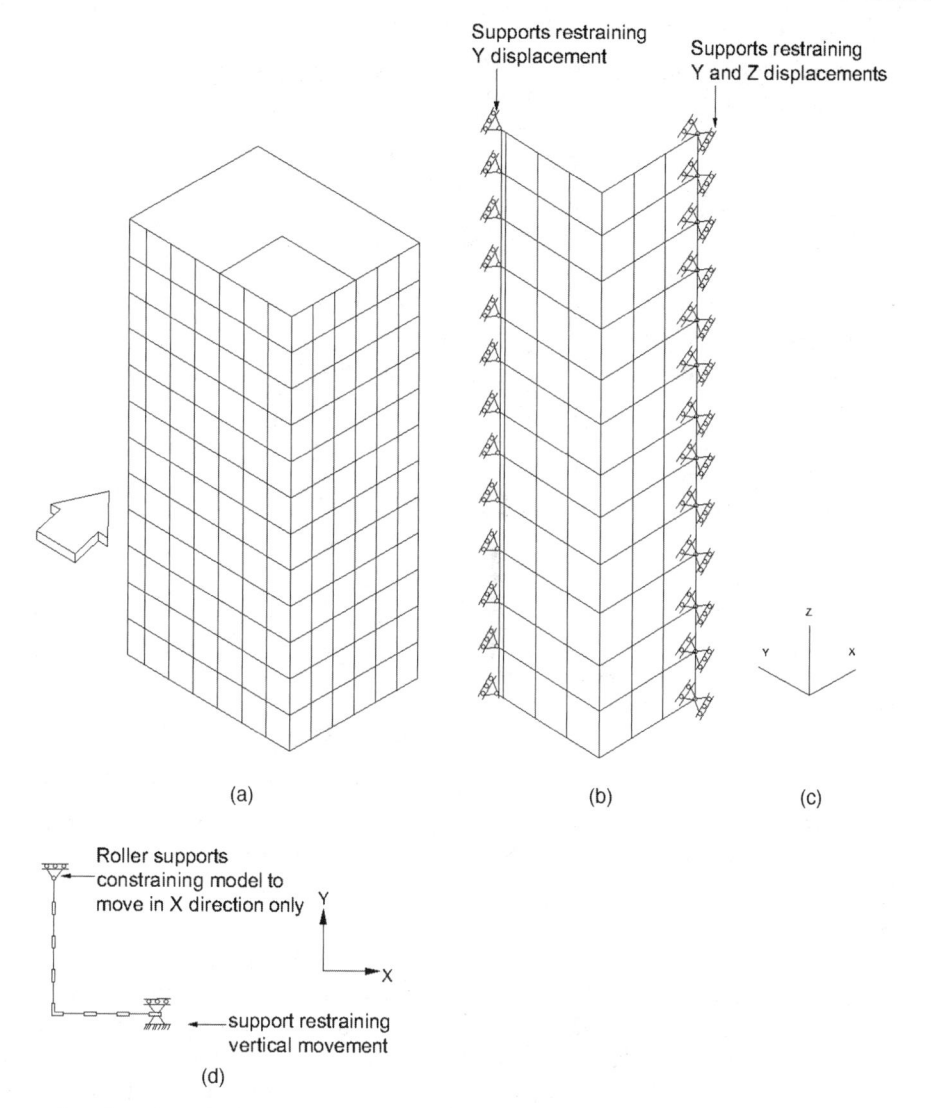

Figure 10.23 Partial analytical model for framed tube: (a) full model; (b) partial model; (c) coordinate axes system; (d) plan of partial model.

Table 10.2(b)   Analogy Between Bending and Warping Torsion

Elementary bending theory	Warping theory
Plane sections remain plane	Profile warps
$I_x = \int_A y^2 dA$	$I_\omega = \int_A \omega^2 dA$
Δ_x	θ_z
$M_x = -EI_x \dfrac{d^2x}{dz^2}$	$B = -EI_\omega \dfrac{d^2\theta}{dz^2}$
$\sigma_x = \dfrac{MC}{I_x}$	$\sigma_\omega = \dfrac{B\omega}{I_\omega}$
$\tau_x = -\dfrac{VQ_x}{I_x t}$	$\tau_\omega = -\dfrac{HQ_\omega}{I_\omega t}$

Terms such as sectorial coordinates, sectorial moment of inertia, bimoment, etc., will be introduced as and when the concepts are discussed instead of defining all of them at once at this stage.

Torsion, at an elemental level, occurs in various practical situations. One of the most common examples is a heavy curtain wall supported from a spandrel beam (Fig. 10.24a). This may not be much of a problem when the spandrel beam is part of the seismic or wind frame. However, when the building perimeter is not part of a lateral frame, the simple connections at the ends of spandrel do not offer adequate torsional restraint resulting in excessive rotations. Another example, with torsion in the reverse direction, is shown in Fig. 10.24b.

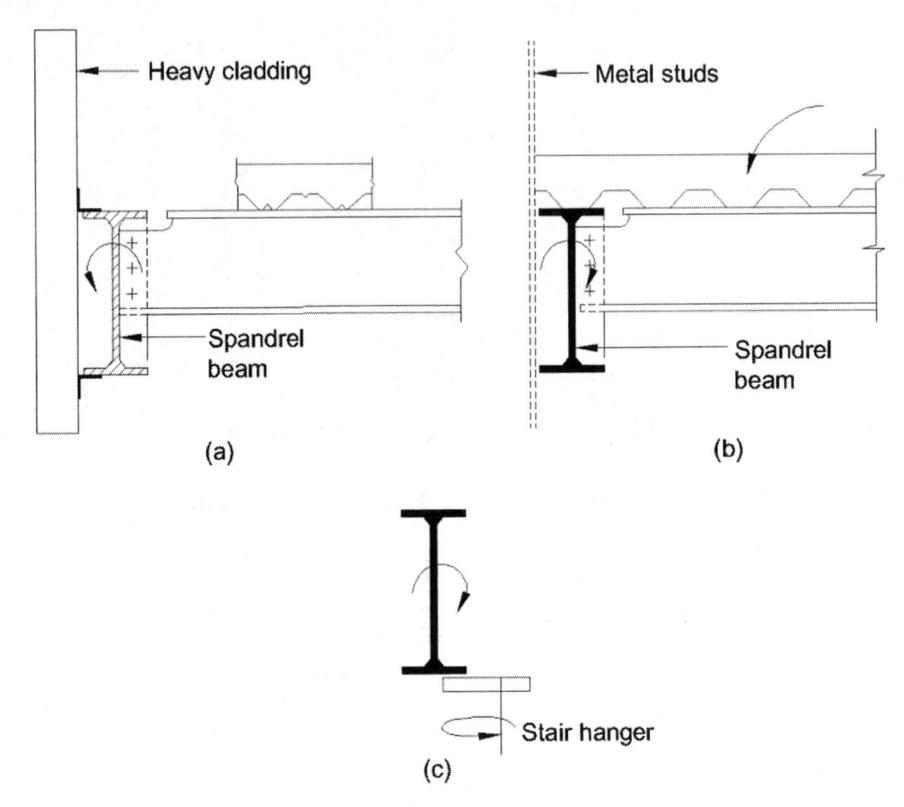

Figure 10.24 Examples of torsion: (a) anti-clockwise rotation due to heavy cladding; (b) clockwise rotation due to weight of concrete; (c) beam rotation due to eccentric hanger.

In this example, taken from author's experience, light gage metal studs were welded to a steel wide flange spandrel before concrete topping was placed on the metal deck. Subsequent concreting of deck resulted in an inward rotation of the spandrel with a consequent misalignment of the studs. A stair-support beam (Fig. 10.24c) located between stair and mechanical shafts is another example because the slab which prevents torsion is absent.

Torsional effects on buildings as a whole are enhanced when the center of twist is eccentric from the center of gravity for inertial loading, or from the center of area for wind loading. Minimum eccentricities are prescribed by building codes to account for accidental seismic torsion. And, to reflect the observed torsional behavior of buildings in turbulent wind, some building codes such as the NBC and the Houston building code require that all buildings be designed for partial as well as full wind loading.

Consider the twisting of a circular shaft as shown in Fig. 10.25a. The twisting of the shaft does not produce any longitudinal stress, that is, axial compression or tension, but only pure shear stresses. The shear stresses vary from zero at the center of the shaft, to maximum value at the perimeter. Because of the absence of axial deformation, a cylindrical layer peeled off of the shaft, changes its shape under the action of twist, from a rectangle to a parallelogram (Fig. 10.25b). The absence of longitudinal stresses indicates that the surfaces at the ends of the shafts remain plane. In other words, no warping will take place. The work done by the twisting moment is expended in developing shear stresses and only shear stresses as shown in Fig. 10.26.

Consider a rectangular section subjected to the action of a vertical load at the center of gravity of the section (Fig. 10.27). To find shear stress at any horizontal section, we introduce an imaginary horizontal cut at that section and obtain the shear stress by the relation VQ/It. By inspection, the resultant of the vertical shear stresses is at the center of gravity of the beam.

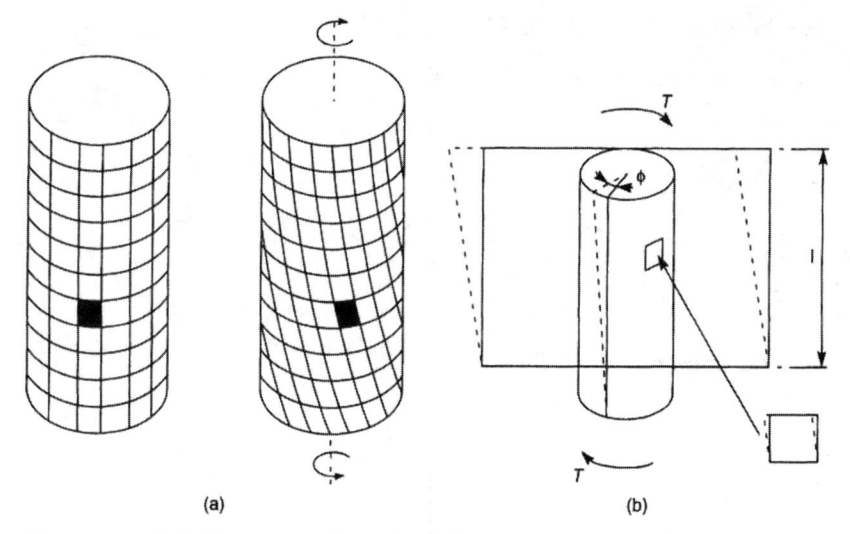

Figure 10.25   (a, b) Twisting of circular shaft.

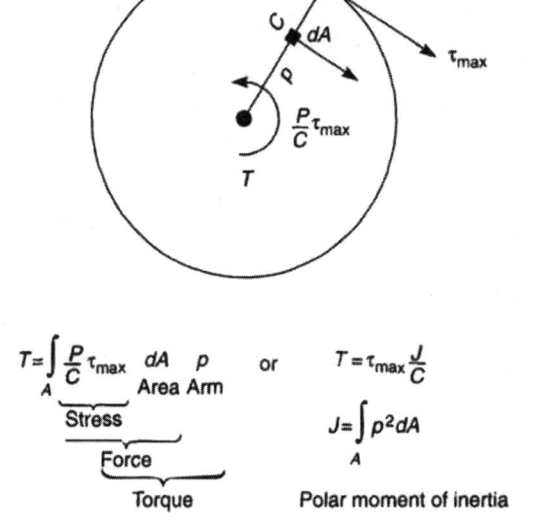

$$T = \int_A \frac{p}{c} \tau_{max}\ dA\ p \qquad \text{or} \qquad T = \tau_{max} \frac{J}{c}$$

Stress · Area · Arm

Force

Torque

$$J = \int_A p^2 dA$$

Polar moment of inertia

Figure 10.26   Variation of shear stresses in circular shaft.

Next we take a look at the torsional behavior of a thin-walled section. The main reason why a thin-walled section must be given special consideration is, the shear stresses and strains in it are much larger than those in solid sections. An examination of distribution of shear stresses through the cross-section shows that the shear stresses flow through the cross-section as if they were a fluid: hence the name, shear flow (Fig. 10.28).

Now consider a flanged section such as a C-shaped shear wall (Fig. 10.29). To find the shear flow, we abandon the idea of the horizontal cut. Instead, we consider a cut perpendicular to the profile and find the shear along the profile. The shear R_2 in web is in equilibrium with the vertical load V and while the horizontal shears R_1 and R_3 in the webs result in no net horizontal load, the resulting moment requires offsetting of the vertical load to a location left of the web. The resultant forces from shear stresses

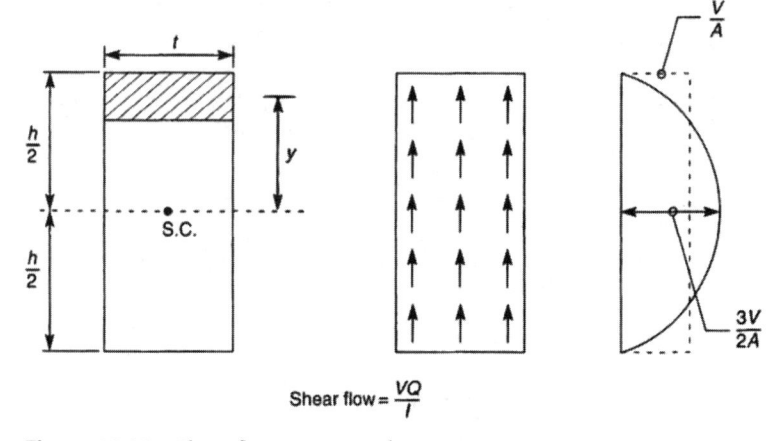

Shear flow $= \dfrac{VQ}{I}$

Figure 10.27   Shear flow in rectangular section.

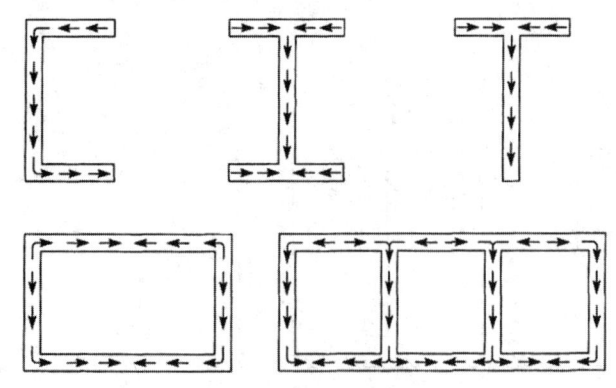

Figure 10.28 Shear flow in thin-walled sections: load at shear center.

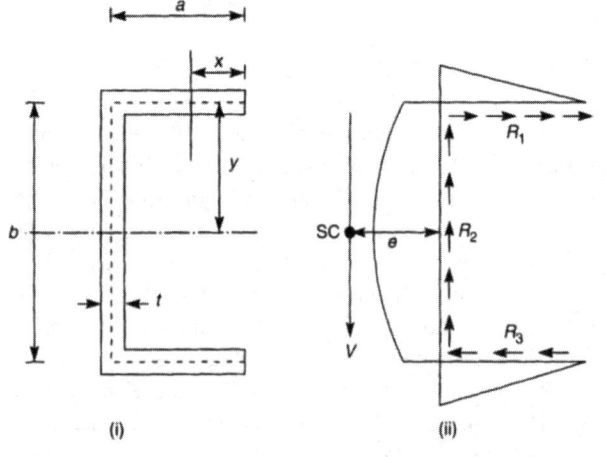

(i)

(ii)

Figure 10.29 Shear center in C-section.

$$R_1 = R_3 = \int_0^a \tau t \, dx_1$$

$$= \frac{Vbta^2}{4I_y}$$

$$= \frac{3Va^2}{b^2\left(1 + \frac{6a}{b}\right)}$$

For vertical equilibrium $\qquad R_2 = V,$

For zero rotational effect, $\qquad R_1 b = R_2 e$

Hence $\qquad e = \frac{R_1 b}{R_2} = \frac{3a^2}{b\left(1 + \frac{6a}{b}\right)}$ (10.19)

To find shear stresses in a cellular section, Fig. 10.30, a two-step approach is required because the problem is statically indeterminate. First, the section is rendered statically determinate by inserting a horizontal cut along the length of the section and the shear flow in the section is evaluated by the relation VQ/I. Next, the shear flow required to close the gap is evaluated. The final shear stress is evaluated by combining the two.

As an example, Fig. 10.31 shows schematically the final shear stresses in a hollow rectangular section. The section consists of webs of unequal thickness and is subjected to a vertical load at its shear center.

If we have a multiple cellular-section, the procedure is similar to that for a single-cell section. The only difference is the problem is statically indeterminate to the nth degree,

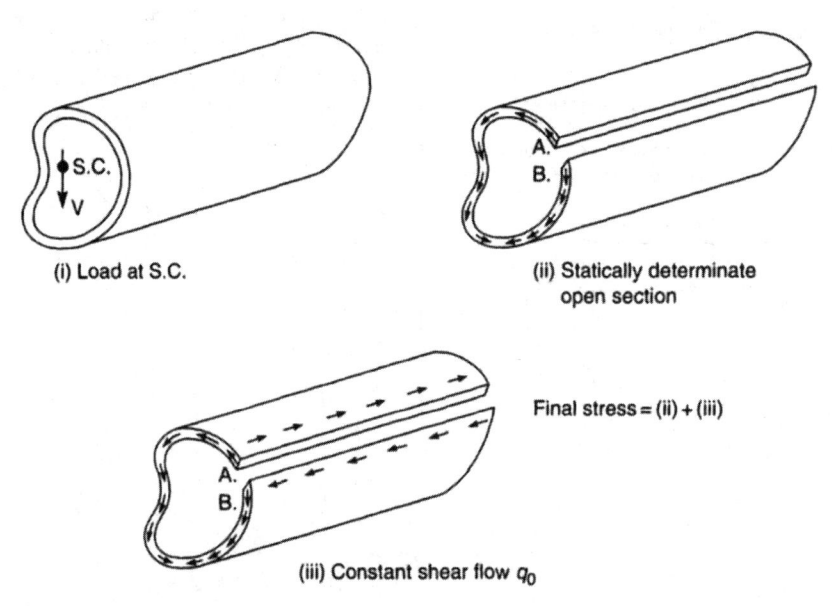

(i) Load at S.C.

(ii) Statically determinate open section

Final stress = (ii) + (iii)

(iii) Constant shear flow q_0

Figure 10.30 Shear stress in hallow section: load at shear center.

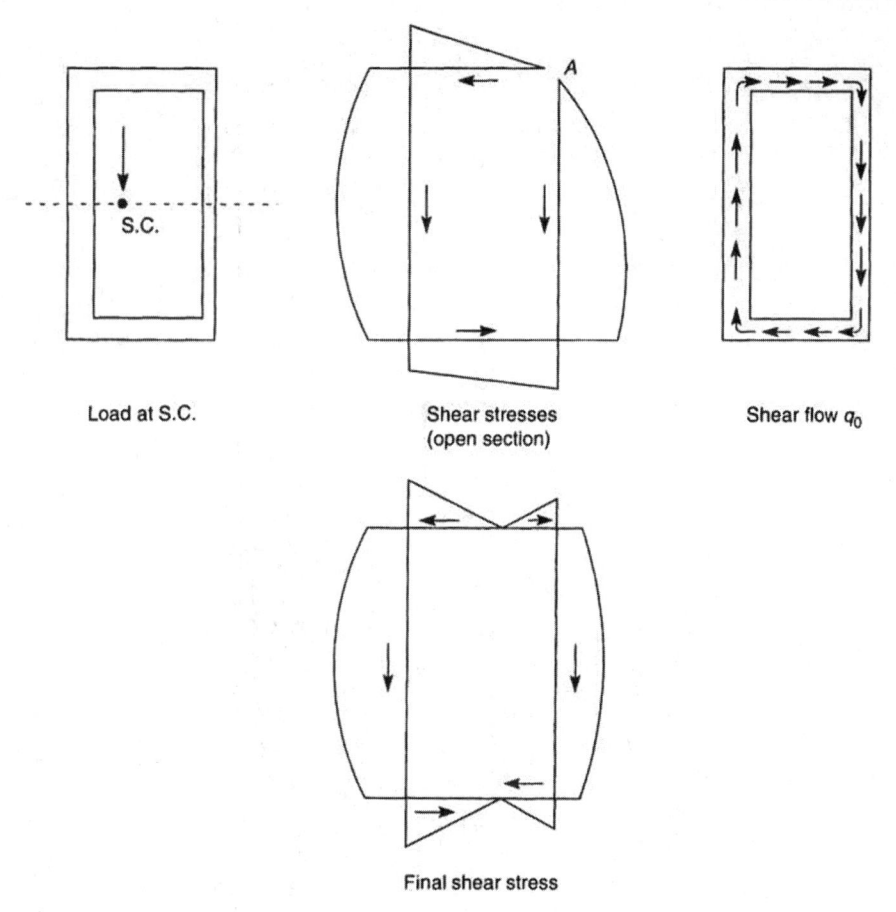

Figure 10.31 Shear stress in hollow rectangular section.

where n represents the number of cells. The example in Fig. 10.32 has two cells, hence, $n = 2$. Two cuts are made at A and B to render the section open. The shear flows q_1 and q_2 are evaluated by solving two simultaneous equations, and the final shear stress obtained by superposition.

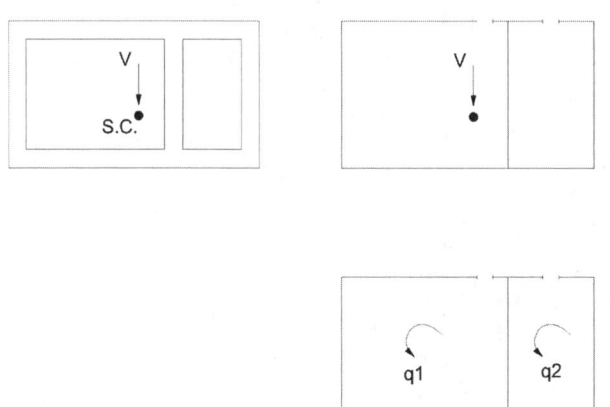

Figure 10.32 Shear flow in cellular sections: (a) load at shear center; (b) section rendered open with two cuts; (c) shear flows required for compatibility; (d) final shear flow = b + c.

The theory of torsion and related formulas discussed above are commonly referred to as St Venant torsion formulas, and are valid for beams of circular cross-sections. His formula can be accepted for non-circular sections only when the additional stress caused by warping deformation is ignored. Consider, for example a rectangular section shown in Fig. 10.33a. The vertical fibers of the section are moving up and down from their initial position in space due to torsion. The top and bottom of the beam do not remain plane, but become warped. However, no additional stresses are induced because the warping deformations are not restrained either at the ends or at any section along its length.

Let us examine the case when the bottom of the beam is fixed. The warping of the bottom surface of the beam is restrained resulting in longitudinal strains and stresses. If we separate an imaginary elemental beam, as shown in Fig. 10.33b, it can be seen that the deflected shape is similar to that of a laterally loaded cantilever. It is obvious that bending stresses manifest at the fixed end of the beam.

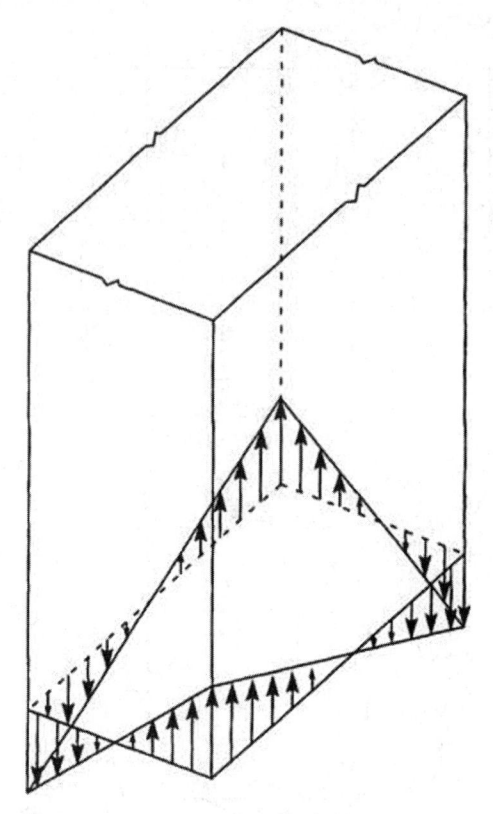

Figure 10.33a   Warping of solid beams.

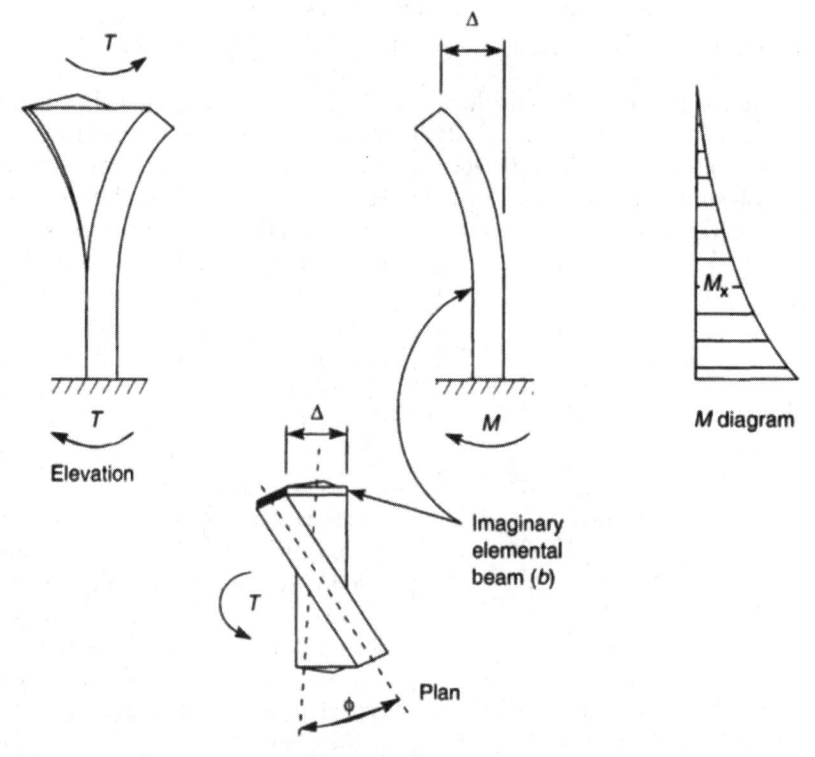

Figure 10.33b   Thin rectangular beam: bending moment due to warping restraint.

The presence of bending stresses implies that part of the work done by the twisting moment is used up in bending the beam and only the remainder will develop shear stresses associated with the St Venant twist. Hence the resistance to external twisting moment is offered as the sum of pure torsion plus some additional torsion which causes bending of the section. This second part is called "warping torsion, non-uniform torsion or flexural twist."

For the thin rectangular beam shown in Fig. 10.33b, very little energy is expended to cause elemental bending about the weak axis. For such beams we can safely neglect the warping component of the twisting moment because the effect of constraining warping is usually restricted to the vicinity of the restraint. This phenomenon is valid, to a lesser extent for thin-walled closed sections. On the other hand, the effect of constraining warping of thin-walled open sections does not diminish rapidly and has a considerable influence on the stress distribution over a greater portion of the section.

Flexural twist causes a pair of moments. Such a pair of moments, called "bimoment" although is a mathematical function, can be visualized in most practical cases. For example, consider a two-span continuous beam supported by an interior column as shown in Fig. 10.34. Since the two channels frame into opposite flanges of the column, a bimoment is introduced at the top of the column.

Restrained warping behavior involves a set of so-called sectorial parameters each of which has counterpart in the theory of bending of beams. Since the sectorial parameters are generally unfamiliar to practicing engineers it is perhaps appropriate to review them briefly here.

The sectorial coordinate, ω, at a point on the profile of a warping core is the parameter that expresses the axial response such as axial stress and strain at that point relative to other points around the core. The ω diagram can be constructed with the known location of the shear center and a point of zero warping deflection as an origin. The principal sectorial coordinate in warping theory is analogous to the distance c of a point from the neutral axis of a section in bending. Just as the parameter c is used in developing the well-known bending theory, the parameter ω is used in developing the warping theory.

A great advantage of the theory of bimoment is that internal strains and stresses can be found from formulas as simple as those used in the engineer's theory of bending. The bimoment and flexural twist can be used in a manner similar to bending moment and shearing forces. The procedure differs in that we use the sectorial coordinate ω instead of the linear coordinate c, to calculate the physical properties related to warping torsion.

To a beginner, the thin-walled beam theory with its differential equations presented later in this chapter, may

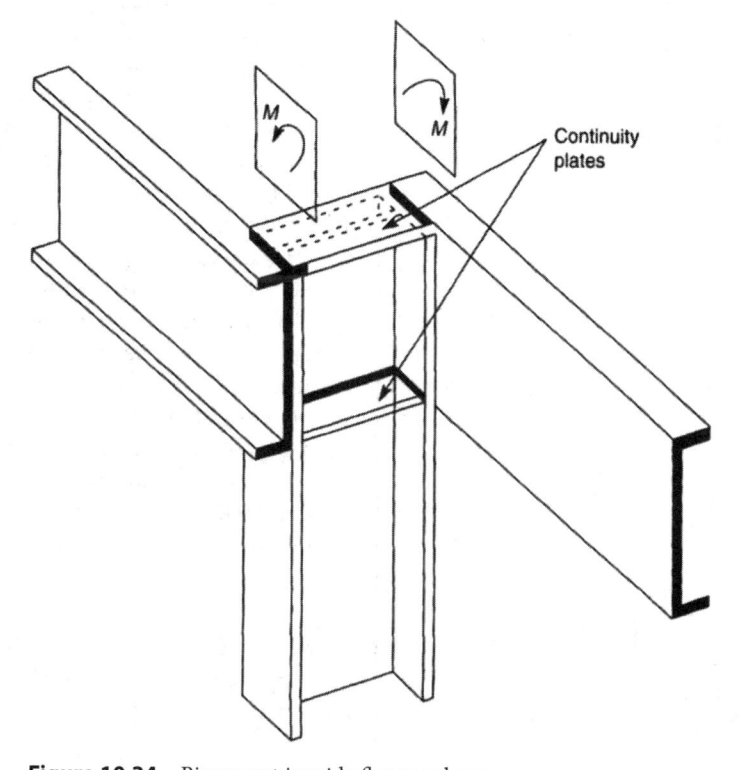

Figure 10.34 Bimoment in wide flange column.

look too academic for use in a down-to-earth practical design. In reality, once the idea of bimoment is assimilated, its use is not much more difficult than the use of bending moments or shear forces. It provides the engineer with a means for verifying the behavior tall shear wall buildings subjected to torsion.

10.4.2   Concept of Warping Behavior: I-Section Core

Perhaps the easiest model to describe the warping theory is an I-shaped shear wall with unequal flanges as shown in Figs. 10.35 and 10.36. In most shear wall buildings the core around elevators and stairs consists of a series of I and C-shaped shear walls. Therefore, the model chosen has practical significance. Since torsion is the subject of

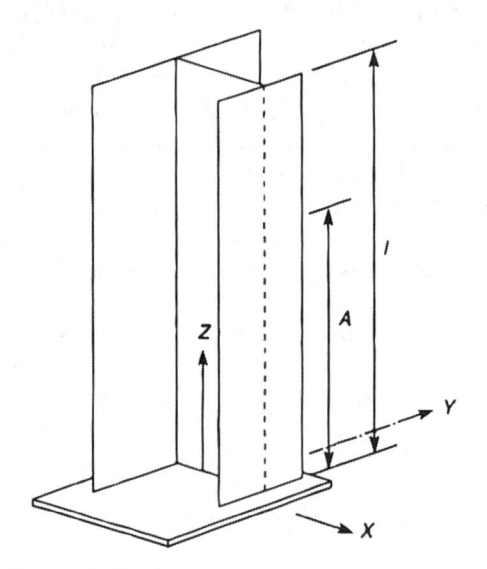

Figure 10.35   I-section core.

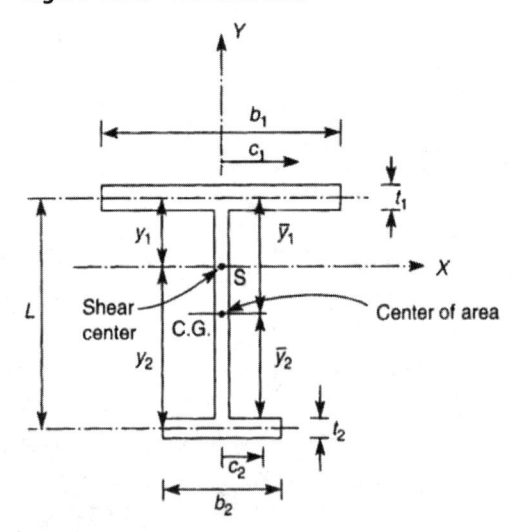

Figure 10.36   Core properties.

discussion, the location of shear center of the cross-section is of importance. Its location is determined in a manner similar to the location of the center of gravity of the section. The only difference is that instead of dealing with the areas of the segments, we use their moments of inertia.

If an axial force is applied to the center of area, only axial deformations and stresses will occur. If, however, the axial force is applied through a point other than the center of gravitybending about the transverse axes, and possibly warping, can also occur. Neglecting the web, the position of the center of gravity also called the center of area is given by

$$\bar{y}_1 = \frac{A_2 L}{A_1 + A_2} \quad \text{and} \quad \bar{y}_2 = \frac{A_1 L}{A_1 + A_2} \qquad (10.20)$$

The location of c.g. is important in relation to vertical axial forces. The shear center s on the other hand is important in relation to transverse forces. If a transverse force acts through s, the member will only bend. If, however, a transverse force acts elsewhere than through s, the member will twist and warp as well as bend. The shear center in this case is located along the y axis by

$$y_1 = \frac{I_2}{I_1 + I_2} L \quad \text{and} \quad y_2 = \frac{I_1}{I_1 + I_2} \qquad (10.21)$$

An inspection of Eqs. (10.20) and (10.21) indicates that the center of the area and the shear center generally will not coincide unless the section is doubly symmetric, in which case both points lie at the center of symmetry.

When a torque T is applied to the top of the member shown in Fig. 10.37a, it twists about the shear center axis causing the flanges to: (i) bend in opposite directions, about the y axis; and (ii) twist about their vertical axes. The effect of the flange-bending is to cause the flange sections to rotate in opposite directions about their y axes so that initially plane sections through the member become nonplanar, or warped. Diagonally opposite corners 1 and 4, in Fig. 10.37a displace downwards while 2 and 3 displace upward. At any level z up the height of the core, the torque $T = T_z$ is resisted internally by a couple $T_\omega(z)$ resulting from the shears in the flanges and associated with their in plane bending, and a couple $T_\upsilon(z)$ resulting from shear stresses circulating within the section and associated with the twisting of the flanges and the web. Then

$$T_\omega(z) + T_\upsilon(z) = T_z \qquad (10.22)$$

The rotation of the member about its shear center axis at a height z from the base is θ_z hence the horizontal displacement of flange #1 at that level is

$$x_1(z) = y_1 \theta_{(z)} \qquad (10.23)$$

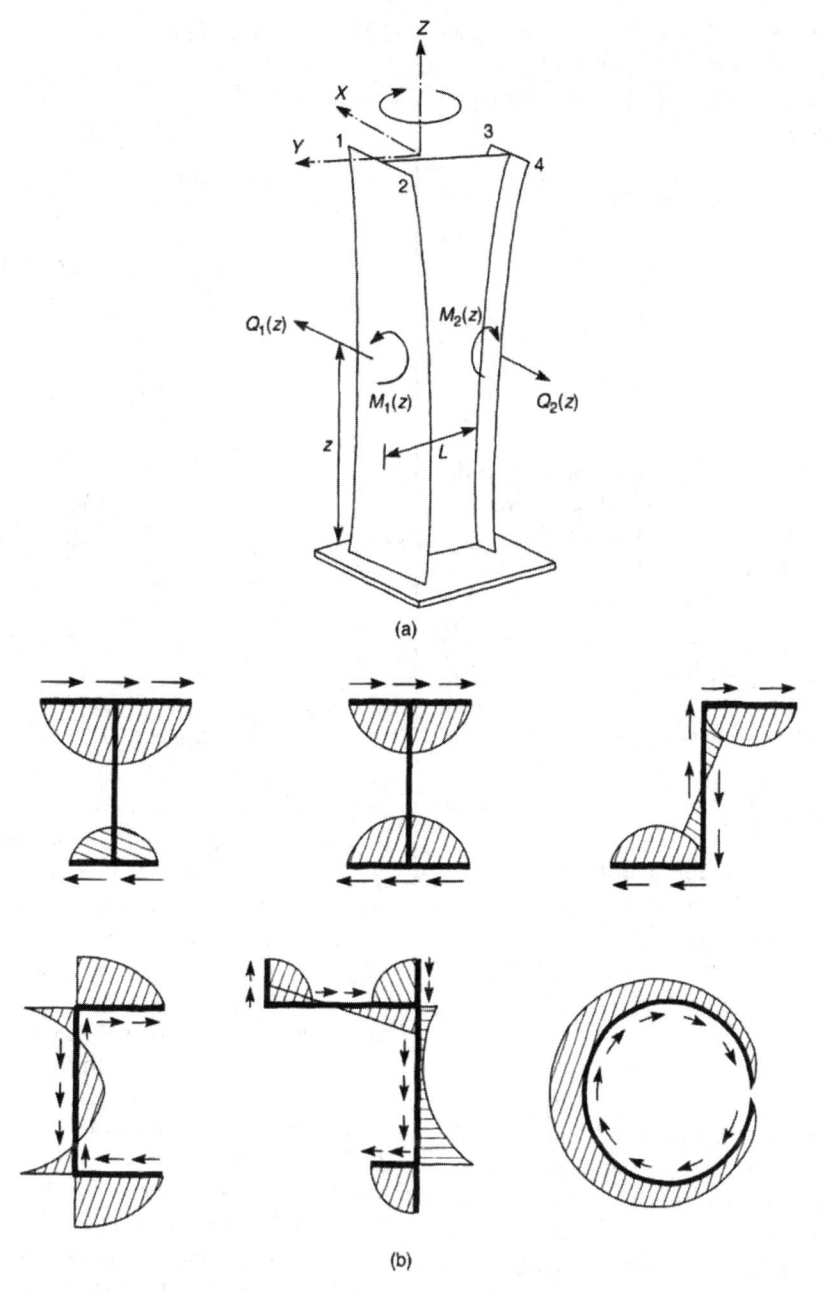

Figure 10.37 (a) Bending of flanges due to torque; (b) shear forces due to warping torsion.

and its derivatives are

$$\frac{dx_1}{dz}(z) = y_1 \frac{d\theta}{dz}(z)$$

$$\frac{d^2x_1}{dz^2}(z) = y_1 \frac{d^2\theta}{dz^2}(z) \qquad (10.24)$$

$$\frac{d^3x_1}{dz^2}(z) = y_1 \frac{d^3\theta}{dz^3}(z)$$

Similar expressions may be written for flange #2.

The shear associated with the bending in flanges #1 and #2 can be expressed by

$$Q_1(z) = -EI_1 \frac{d^3x}{dz^3}(z) = -EI_1 y_1 \frac{d^3\theta}{dz^3}(z) \quad (10.25)$$

and

$$Q_2(z) = -EI_2 \frac{d^3x}{dz^3}(z) = -EI_2 y_2 \frac{d^3\theta}{dz^3}(z) \quad (10.26)$$

Multiplying the shear forces Q_1 and Q_2 by their respective distances from the shear center we obtain the torque resisted by these forces. Therefore, the torque contributed by these shear forces is

$$T_\omega(z) = Q_1 y_2 + Q_2 y_2 = -(EI_1 y_1^2 + EI_2 y_2^2)\frac{d^3\theta}{dz^3}(z) \quad (10.27)$$

or

$$T_\omega(z) = -EI_\omega \frac{d^3\theta}{dz^3}(z) \quad (10.28)$$

where

$$I_\omega = I_1 y_1^2 + I_2 y_2^2 \quad (10.29)$$

I_ω is a geometric property of the section similar to the moments of inertia I_x and I_y, and is called the warping moment of inertia or warping constant. It expresses the capacity of the section to resist warping torsion. Neglecting the web, the torque resisted by the twisting of the section is

$$T_\upsilon(z) = GJ_1 \frac{d\theta}{dz}(z) \quad (10.30)$$

where J_1 is the torsion constant of the section given by

$$J_1 = \frac{b_1 t_1^3}{3} + \frac{b_2 t_2^3}{3} \quad (10.31)$$

in which b_1 and b_2 are the widths, and t_1 and t_2 are the thicknesses, of flanges #1, #2, respectively.

Summing the two internal torques, Eqs. (10.28) and (10.30), and equating the sum to external torques as in Eq. (10.22)

$$-EI_\omega \frac{d^3\theta}{dz^3}(z) + GJ_1 \frac{d\theta}{dz}(z) = T \quad (10.32)$$

Equation (10.32) is the fundamental equation for restrained warping torsion. It simply states that an external torque applied to an open core is resisted by a combination of internal torque due to St Venant shear stresses and a couple due to equal and opposite shear forces in the flanges. The distribution of shear forces due to torsion in typical shear wall profiles is shown in Fig. 10.37b.

Considering the stresses in the flanges due to bending, the compressive stress in flange #1 at c_1 from the y axis and z from the base is

$$\sigma_1(c_1, z) = \frac{M_1(z)c_1}{I_1} \quad (10.33)$$

The tensile stress in flange #2 at c_2 from the y axis is

$$\sigma_2(c_2, z) = \frac{M_2(z)c_2}{I_2} \quad (10.34)$$

Multiplying the right-hand side of Eq. (10.33), by the expression

$$\frac{L}{(y_1 + y_2)}\frac{y_1}{y_1}$$

which is equal to unity, and noting since $Q_1 = Q_2$, and the flange moments $M_1 = M_2 = M$, gives:

$$\sigma_1(c_1, z) = \frac{M_{(z)} L y_1 c_1}{I_1 y_1^2 + I_1 y_1 y_2} \quad (10.35)$$

and since, from Eq. (10.21)

$$I_1 y_1 y_2 = I_2 y_2^2 \quad (10.36)$$

Substituting Eq. (10.36) in Eq. (10.35)

$$\sigma_1(c_1, z) = \frac{M_{(z)} L y_1 c_1}{I_1 y_1^2 + I_2 y_2^2} \quad (10.37)$$

or

$$\sigma_1(c_1, z) = \frac{B_{(z)} \omega(c_1)}{I_\omega} \quad (10.38)$$

in which $B_{(z)} = M_{(z)} L$ is an action termed a bimoment, and $\omega(c) = y_1 c_1$, is a coordinate termed the sectorial area, or principal sectorial coordinate, for that point of the section. In its simplest form, as considered here, a bimoment consists of a pair of equal and opposite couples acting in parallel planes. Its magnitude is the product of the couple and the perpendicular distance between the planes.

The above simple treatment of torsion of an I section explains the concept of warping and how the equations of torsion bending, also called restrained warping, are related to simple bending theory. The analogy is perhaps even more obvious by comparing the terms given in Table 10.2b.

10.4.3 Sectorial Coordinate ω'

The sectorial coordinate also called the warping function at a point on the profile of a warping core is the parameter that expresses the axial response (i.e., displacement, strain, and stress) at that point, relative to the response at other points around the section. Conceptually, this is similar to the distance c we use in bending formula $f = \dfrac{Mc}{I}$ to find the bending stress f at a point in the cross section located at a distance c from the neutral axis.

The warping coordinate is defined in relation to two points: a pole $0'$ at an arbitrary position in the plane of the section, and an origin P_0 at an arbitrary location on the profile of the section (Fig. 10.38a,b). The value of the sectorial coordinate at any point P on the profile is then given by the area

$$\omega'_{(s)} = \int_0^s h\, ds \quad (10.39)$$

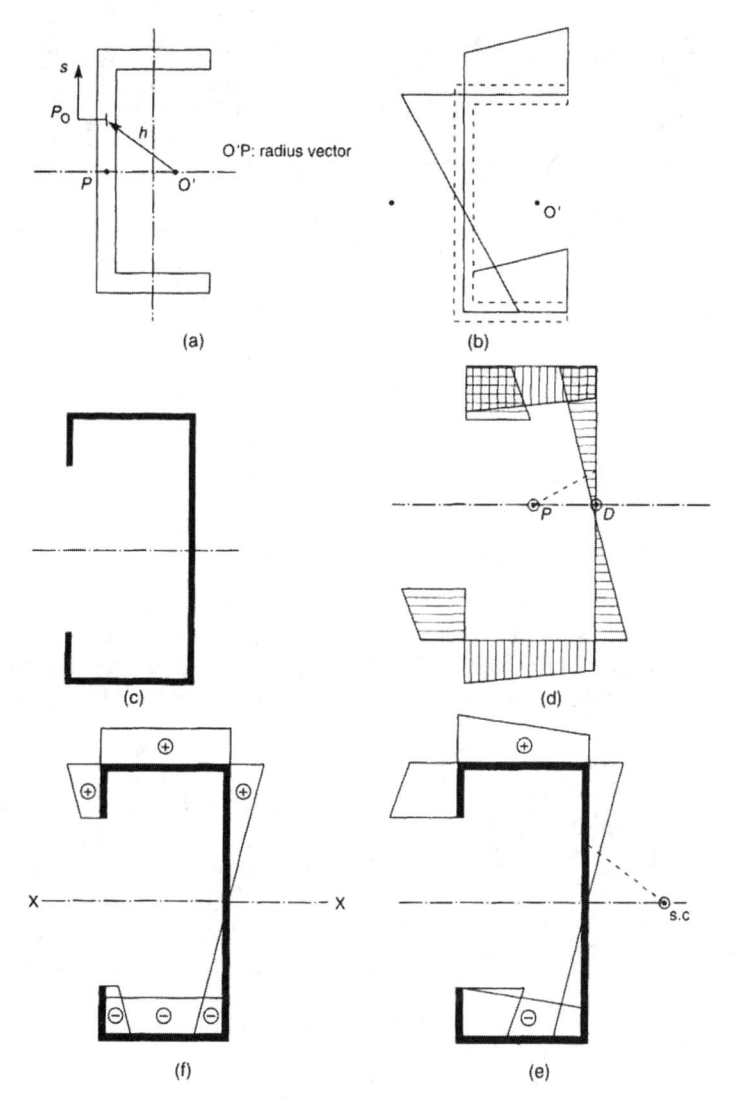

Figure 10.38 (a) Section profile; (b) sectorial coordinate ω_s diagram; (c) singly symmetric core; (d) ω' diagram; (e) Y coordinate diagram; (f) principal sectional coordinates; (g) sectorial coordinates for common profiles.

where h is the perpendicular distance from the pole $0'$ to the tangent to the profile at P and s is the distance of P along the profile P_0. It is evident that the warping function is an area and its magnitude depends on the location of the pole and of the point in the profile from which the integration is started.

In effect, the sectorial coordinate ω' is equal to twice the area swept out by the radius vector $0'P$ in moving from P_0 to P. The sectorial coordinate diagram (Fig. 10.38b) indicates the values of ω' around the profile. When the sectorial coordinates are related to the shear center as a pole, and to the origin of known zero warping displacement, Eq. (10.39) gives the principal sectorial coordinate

values, ω and their plot is the principal sectorial coordinate diagram. The principal sectorial coordinate of a section in warping theory is analogous to the distance c of a point from the neutral axis of a section in bending. The parameters ω and c are used in developing the corresponding warping and bending stiffness properties of the sections, and in determining the axial displacements and stresses. Sectorial coordinates for common profiles are shown in Fig. 10.38g.

10.4.4 Shear Center

The shear center of a section is a point in its plane through which a load transverse to the section must pass

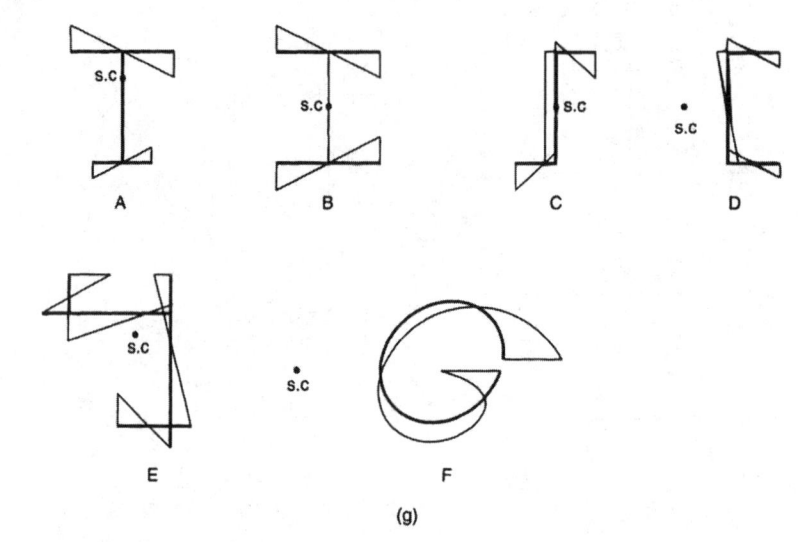

(g)

Figure 10.38   (*Continued*)

to avoid causing torque and twist. It is also the point to which warping properties of a section are related, in the way that bending properties of a section are related to the neutral axis.

Tall building cores are often singly or doubly symmetric in plan, which simplifies the location of the shear center. In doubly symmetric sections, the shear center lies at the center of symmetry while, in singly symmetric sections, it lies on the axis of symmetry.

The procedure for determining the location of shear for a singly symmetric section (Fig. 10.38c) is as follows.

1. Construct the ω_p diagram (Fig. 10.38d) by taking an arbitrary pole P on the line of symmetry, an origin D where the line of symmetry intersects the section, and by sweeping the ray PD around the profile.

2. Using the ω_p and the y diagrams for the section, Figs. 10.38d and 10.34e, respectively, calculate the product of inertia of the ω_p diagram about the X axis $I_{\omega_p x}$ using

$$I_{\omega_p x} = \int^A \omega_p y \, dA \qquad (10.40)$$

in which $dA = t \, ds$, the area of the segment of the profile of thickness t and lengths ds. The integral in Eq. (10.40), may be evaluated simply by using the product integral table, Table 10.3.

3. Calculate I_{xx}, the second moment of area of the section about the axis of symmetry.

4. Finally, calculate the distance a_x of the shear center 0 from 0′, along the axis of symmetry, using

$$\alpha_x = \frac{I_{\omega_p x}}{I_{xx}} \qquad (10.41)$$

Table 10.3   Product Integral Tables

	Linear M diagrams					Parabolic M diagrams		
	$M \overline{\Box}\ L$	$M_0 \overline{\triangle}\ L$	$\overline{\triangle} M_1\ L$	$M_0 \overline{\Box} M_1\ L$	$\overset{Origin}{\triangle}\ M_1\ L/2\ L/2$	$\overset{Origin}{\triangle} M_1\ L$	$M_0 \overset{Origin}{\triangle} M_1\ \vert\leftarrow L\rightarrow\vert$
$m \overline{\Box}\ L$	mML	$\frac{1}{2}m_0ML$	$\frac{1}{2}mM_1L$	$\frac{1}{2}mL(M_0+M_1)$	$\frac{2}{3}mM_1L$	$\frac{1}{3}mM_1L$	$\frac{1}{3}mL(2M_0-M_1)$
$m_0 \overline{\triangle}\ L$	$\frac{1}{2}m_0ML$	$\frac{1}{3}m_0M_0L$	$\frac{1}{6}m_0M_1L$	$\frac{1}{6}m_0L(2M_0+M_1)$	$\frac{1}{3}m_0M_1L$	$\frac{1}{12}m_0M_1L$	$\frac{1}{12}m_0L(5M_0-M_1)$
$\overline{\triangle} m_1\ L$	$\frac{1}{2}m_1ML$	$\frac{1}{6}m_1M_0L$	$\frac{1}{3}m_1M_1L$	$\frac{1}{6}m_1L(2M_1+M_0)$	$\frac{1}{3}m_1M_1L$	$\frac{1}{4}m_1M_1L$	$\frac{1}{4}m_1L(M_0-M_1)$
$m_0 \overline{\Box} m_1\ L$	$\frac{1}{2}ML(m_0+m_1)$	$\frac{1}{6}M_0L(2m_0+m_1)$	$\frac{1}{6}M_1L(m_0+2m_1)$	$\frac{L}{6}[m_0(2M_0+M_1)+m_1(2M_0+M_1)]$	$\frac{1}{3}M_1L(m_0+m_1)$	$\frac{1}{12}M_1L(m_0+3m_1)$	$\frac{L}{12}[m_0(5M_0-M_1)+3m_1(M_0-M_1)]$

10.4.5 Principal Sectorial Coordinate ω Diagram

The ω diagram is related to the shear center 0 as its pole and a point of zero warping deflection as an origin. In a symmetrical section the intersection of the axis of symmetry with the profile at D defines a point of antisymmetrical behavior, and hence of zero warping, therefore it may be used as the origin.

Values of ω can be found by sweeping the ray OD around the profile and taking twice the values of the swept areas.

For the section of Fig. 10.38c, the principal sectorial coordinate diagram is shown in Fig. 10.38f.

10.4.6 Sectorial Moment of Inertia I_ω

This geometric parameter expresses the warping torsional resistance of the core's sectional shape. It is analogous to the moment of inertia in bending.

The sectorial moment of inertia is derived from the principal sectorial coordinate distribution using the relation:

$$I_\omega = \int_0^A \omega^2 dA \qquad (10.42)$$

Note the similarity with the expression for the moment of inertia

$$I_{yy} = \int_0 x^2\, dA$$

10.4.7 Shear Torsion Constant J

When a beam is twisted, its fibers must undergo a shear strain to accommodate the twist. Associated with the strain are the shear stresses called St Venant shear stress. When an open-section core is subjected to torque (Fig. 10.38a) each wall twists developing St Venant shear stresses within the thickness of the wall. The stresses are distributed linearly across the thickness of the wall, acting in opposite directions on opposite sides of the wall's middle line. As the effective lever arm of these stresses is equal to only two-thirds of the wall thickness, the torsional resistance of these stresses is low. The torsion constant for this plate twisting action is

$$J = \frac{1}{3}k \sum_{}^{n} bt^3 \qquad (10.43)$$

in which b is the width and t the thickness of a wall. The summation includes the n walls that comprise the section. The plate twisting rigidity of an open section core is given by GJ.

k is a factor which makes allowance for small fillets within the cross-section. If there are no fillets, its value is equal to 1.00.

10.4.8 Calculation of Sectorial Properties: Worked Example

Consider again the shear core with unequal flanges as shown in Fig. 10.39. It is required to determine for the core,

1. The location of the shear center.
2. The principal sectorial coordinate, ω_s diagram.
3. The sectorial moment of inertia I_ω.
4. The St Venant torsion constant J.

1. Location of Shear Center. The axis of symmetry of the section is OY, therefore the shear center lies on the OY axis. We select an arbitrary pole P at the junction of the web and the upper flange of the core. The ω_p diagram is constructed as shown in Fig. 10.39 by taking an arbitrary point on the web as the sectorial origin. The sectorial areas for the section of the upper flange and the web are equal to zero while they are distributed skew symmetrically for the lower flange.

Using the ω_p and the Y coordinate diagrams, Fig. 10.39b and 10.39c, we calculate the integral $\omega_p\, dA$ by using the product integrals given in Table 10.3.

A summary of the calculations is given in the following Table 10.4.

Table 10.4   Calculations for Integral $\omega_p\, dA$

Segment	ω_p	x	$\int_0^s \omega_p x\, t\, ds$
DE			$\frac{1}{3} \times 5 \times 150 \times 5 \times 2 = 2500$ ft^5
EC			$\frac{1}{3} \times 5 \times 150 \times 5 \times 2 = 2500$ ft^5
AF	0		0
BF	0		0

For the whole section, $I_{\omega p} = 2500 \times 2 = 5000$ ft^5. The moment of inertia of the section about y axis

$$I_{yy} = \tfrac{1}{12}(2 \times 10^3 + 2 \times 20^3) = 1500\ \text{ft}^4$$

From Eq. (10.41), the distance of the shear center from the center of web is

$$\alpha_x = \frac{I_{\omega_p x}}{I_{yy}} = \frac{5000\ \text{ft}^5}{1500\ \text{ft}^4} = 3.33\ \text{ft}$$

2. Principal Sectorial Coordinate Diagram. This is constructed by using the shear center s.c. as the pole and sweeping the ray from the middle of the web, around the profile (Fig. 10.39d).

3. Sectorial Moment of Inertia I_ω. From Eq. (10.50)

$$I_\omega = \int \omega^2 dA = \int^S \omega^2 t\, ds$$

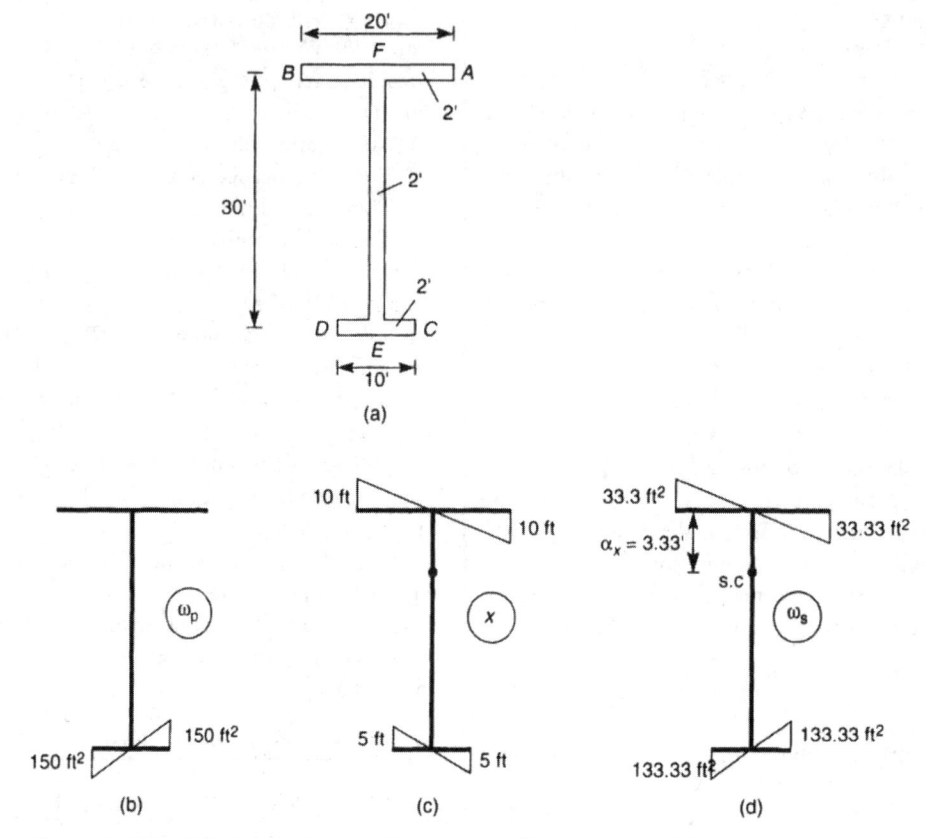

Figure 10.39   Calculation of sectorial properties: (a) cross-section; (b) ω_p diagram; (c) x-coordinate diagram; (d) principal sectorial coordinate ω_s diagram.

Using the ω diagram (Fig. 10.39) and the product integral table, Table 10.3, the calculations for evaluating I_ω are as shown in the following Table 10.5.

Table 10.5   Calculations for Sectorial Moment of Inertia

Segment	Variation of ω	$\int_0^s \omega^2 x t\, ds$
DE		$\frac{1}{3} \times 5 \times 133.33^2 \times 2 = 59{,}260$ ft^6
EC		$\frac{1}{3} \times 5 \times 133.33^2 \times 2 = 59{,}260$ ft^6
BF		$\frac{1}{3} \times 10 \times 33.33^2 \times 2 = 7406$ ft^6
AF		$\frac{1}{3} \times 10 \times 33.33^2 \times 2 = 7406$ ft^6

$\therefore I_\omega$ for the whole section $59260 \times 2 + 7406 \times 2 = 133{,}332$ ft^6

4. Torsion Constant J. For the I-section core, using Eq. (10.43)

$$J = \frac{1}{3}\sum_{1}^{n} bt^3 = \frac{1}{3} \times 2^3(20 + 10 + 30) = 160 \text{ ft}^4$$

10.4.9   General Theory of Warping Torsion

Before derivation of general warping torsion equations, it is instructive to consider qualitatively the difference between the behavior of thin-walled open sections and solid sections. A major difference lies in the manner in which the stresses attenuate along their length. Consider a square cantilever column loaded at top corner by a vertical load P as shown in Fig. 10.40a. The load can be replaced by four sets of loads acting at each corner, which together constitute a system of loads statically equivalent to the applied force P. The first set represents axial loading, the second and third sets represent bending about the x and y axes. The resulting axial and bending stresses can be computed by the usual engineer's theory of bending,

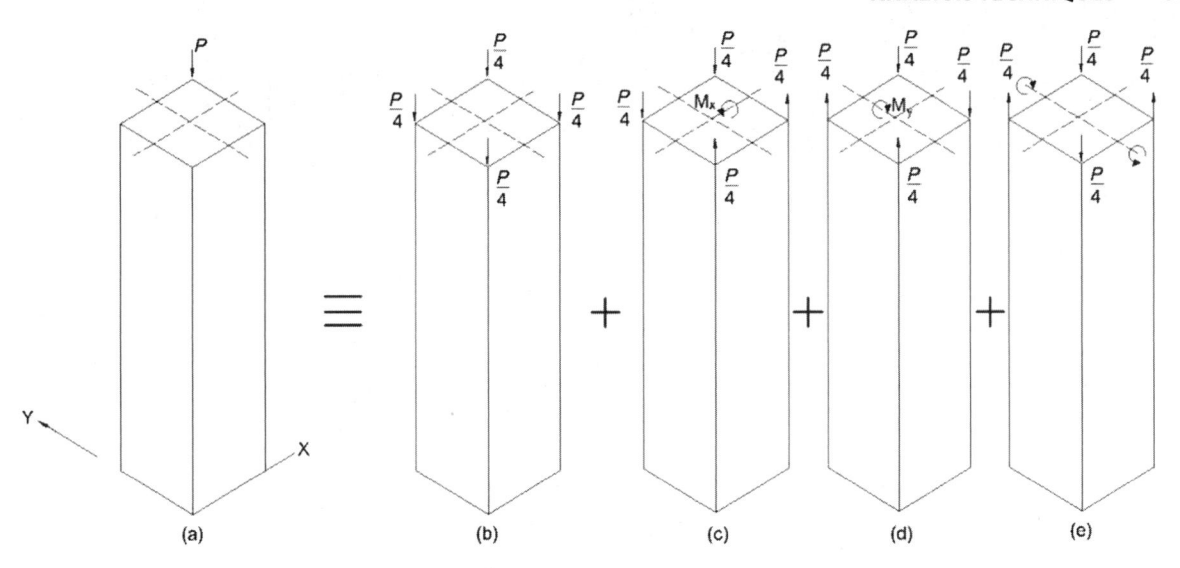

Figure 10.40   Cantilever column of solid cross section: (a) vertical load at corner; (b) symmetrical axial loading; (c) bending about x-axis; (d) bending about y-axis; (e) self-equilibrating loading producing bimoment. Note: the resulting warping stresses are negligible.

which assumes that Bernoulli's hypothesis is valid. In the last loading case, the cross-sections do not remain plane because the two pairs of loads on opposite faces of the column tend to twist the cross-section in opposing directions. This equal and opposite twisting results in warping of the cross-section. The last set of loads is, however, statically equivalent to zero and can be ignored by invoking St Venant's principle, which states that the perturbations imposed on a structure by a set of self-equilibrating system of forces affect the structure locally and will not appreciably affect parts of the structure away from the immediate region of applied forces. This statement simply means that the effect of self-equilibrating system of forces can be neglected in the analysis. The stresses caused by these forces attenuate rapidly toward zero and just about vanish at a distance equal to the characteristic dimension of the cross section. The stresses due to the self-equilibrating system of forces can be ignored throughout the whole length of the cantilever except at the very top region.

Now consider an I-shaped shear wall as shown in Fig. 10.41 which has the same overall dimensions as the column with the exception that it is composed of thin plates of thickness t. The first three sets of loads result in stress distributions which can be obtained as before by using the Bernoulli hypothesis. Although the fourth loading is self-equilibrating as before, its effect is far from local. The flanges, which are bending in opposite directions, do so as though they were independent of each other. The

web acts as a decoupler separating the self-equilibrating load into two subsets one in each flange. Each subset is not self-equilibrating and causes bending in each flange. The bending action of the flanges can be thought of as being brought about by equal and opposite horizontal forces parallel to the flanges. The compatibility condition between the web and flanges results in a twisting of the cross-section as shown in Fig. 10.42. Although the cross-section of each of the flanges remains plane, the wall as a whole is subjected to warping deformations. The restraint at the foundation prevents free warping at this end and sets up warping stresses.

The system of skew-symmetric loads which is equivalent to an internally balanced force system arising out of warping of cross-section is termed a *bimoment* in thin-walled beam theory. Mathematically it can be construed as a generalized force corresponding to the warping displacement, just as moment and torsion are associated with rotation and twisting deformation, respectively. In the present example bimoment can be visualized as a pair of equal and opposite moments acting at a distance e from each other. Its magnitude is equal to M times e and has units of force times the square of the distance (lb · in.^{2}, kip · ft^2, etc.).

Presently it will be shown that the warping stresses can be calculated by the relation

$$\sigma_\omega = \frac{B_\omega \omega_s}{I_\omega} \tag{10.44}$$

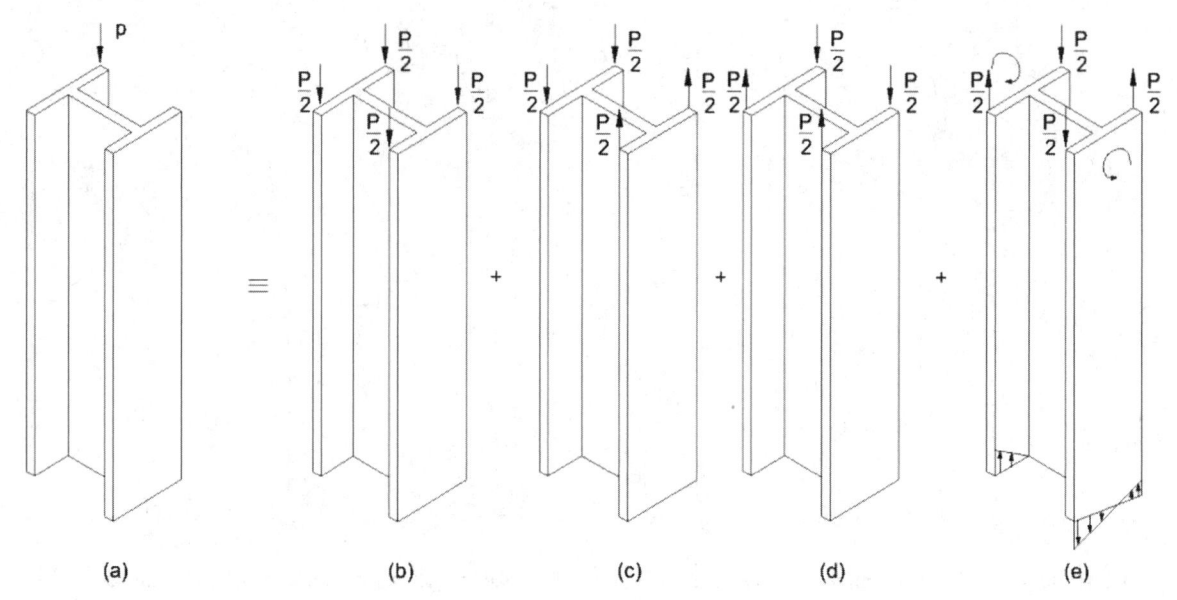

Figure 10.41 *I*-shaped cantilever beam: (a) vertical load at a corner; (b) symmetrical axial loading; (c) bending about *x*-axis; (d) bending about *y*-axis; (e) self-equilibrating loading producing bimoment. Note: the flanges bend in opposite directions producing significant warpings stresses.

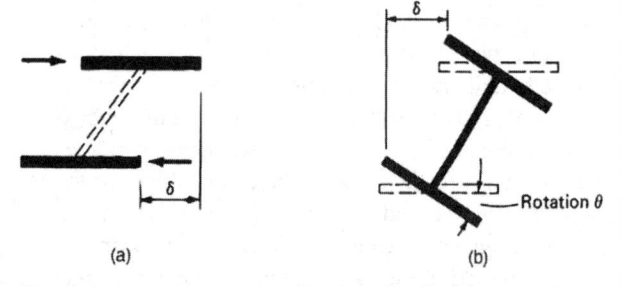

Figure 10.42   Plan section of *I*-shaped column: (a) displacement of flanges due to bimoment load; (b) rotation due to geometric compatibility between flanges and web.

where B_ω = the bimoment, a term that represents the action of a set of self-equilibrating forces
ω_s = the warping function
I_ω = the warping moment of inertia

The three terms B_ω, ω_s, and I_ω are conceptually equivalent moment M, linear coordinate or x or y, and moment of inertia I encountered in bending problems. Note the similarity between the bending stress as calculated by the familiar relation $\sigma_b = My/I$ and the warping stress formula given in Eq. (10.44).

In open section cores the magnitude of warping stresses can be very large and may even exceed the value of bending stress, depending upon the aspect ratio between the height, width, breadth, and thickness of the member.

A general theory which can account for warping was developed by Vlasov; in what follows a brief derivation of the fundamental equation and a description of the method of solution is given. The following notation is used in the derivation of warping torsion equation.

M_1	Torsional moment due to constrained torsion shear stresses
M_2	Torsional moment due to St Venant torsion shear stresses
M	Total torsional moment
z, s	Orthogonal system of coordinates
u	Longitudinal displacement due to warping along the z axis
υ	Transverse displacement along the tangent to the profile
θ_z	Torsional rotation at z
ω_s	Sectorial area
$\sigma_{z,s}$	Longitudinal stress due to warping
ε	Strain
E	Modulus of elasticity of the material
G	Modulus of rigidity of the material
μ	Poisson's ratio of the material
γ	Shear strain
τ	Shear stress
t_s	Thickness of thin-walled beam
I_ω	Warping moment of inertia
m_z	Intensity of applied torque
B_z	Bimoment
l	Length of beam
J	St Venant torsion constant
k	Characteristic parameter, $l\sqrt{(1-\mu^2)GJ/EI_\omega}$

Z_z Action matrix
G_0 Distribution matrix
Z_0 Initial boundary restraint matrix
C_1, C_2, C_3, C_4 Constants of integration

In Vlasov theory two fundamental assumptions are made:

1. The cross-section is completely rigid in the transverse direction.

2. The shear strain of the middle surface is negligible.

These two assumptions are almost completely satisfied in a practical core structure. The high in-plane stiffness of the floor slabs practically prevents distortion of the core at frequent intervals along its length, and the second assumption is valid for all but very low buildings.

Consider an open tube shown in Fig. 10.43. An orthogonal system of coordinates z, s is chosen, consisting of a generator and the middle line of the profile (Fig. 10.43a). The origin for the coordinate z is taken at the base, and any generator is taken as the origin for the curvilinear coordinate s. Let θ be the angle of rotation of the profile at a distance z from the base. This rotation is in the xy plane and is measured with respect to any arbitrary center of rotation R.

Consider the displacements of any point p on the middle surface of the tube. The transverse displacement υ in the direction of the tangent to the profile line is given by

$$\upsilon_{z,s} = \theta_z h_s \tag{10.45}$$

where h_s is the perpendicular distance from the tangent at p to the center of rotation R (Fig. 10.43). If $u_{z,s}$ is the longitudinal displacement along the generator, then considering the displacements at p, of an element $dz\, ds$

(Fig. 10.43b), lying on the middle surface, the condition of zero shear strain is given by the relation

$$\gamma = \frac{\partial u}{\partial s} + \frac{\partial \upsilon}{\partial z} = 0$$
$$\frac{\partial u}{\partial s} + h_s \frac{d\theta}{dz} = 0 \tag{10.46}$$

Integrating,

$$u_{z,s} = -\int^s h_s\, ds\, \frac{d\theta}{dz} \tag{10.47}$$

The integral s is taken along the profile from an arbitrary point to the point p for which the longitudinal displacement is required. The product $h_s\, ds$ is equal to twice the area of the elementary triangle whose base and height are equal to ds and h_s, respectively, and is usually given the symbol $d\omega$.

$$u_{z,s} = -\int^s \frac{d\theta}{dz}\, d\omega \tag{10.48}$$
$$= -\frac{d\theta}{dz}\, \omega_s \tag{10.49}$$

Since the displacement $u_{z,s}$ changes along the distance z, the strain ε_z is given by

$$\varepsilon_z = \frac{\partial u}{\partial z} = -\frac{d^2\theta}{dz^2}\, \omega_s$$

Hence, corresponding stress $\sigma_{z,s}$ is

$$\sigma_{z,s} = \frac{E}{1-\mu^2}\, \varepsilon_z$$
$$= \frac{E}{1-\mu^2}\, \frac{d^2\theta}{dz^2}\, \omega_s \tag{10.50}$$

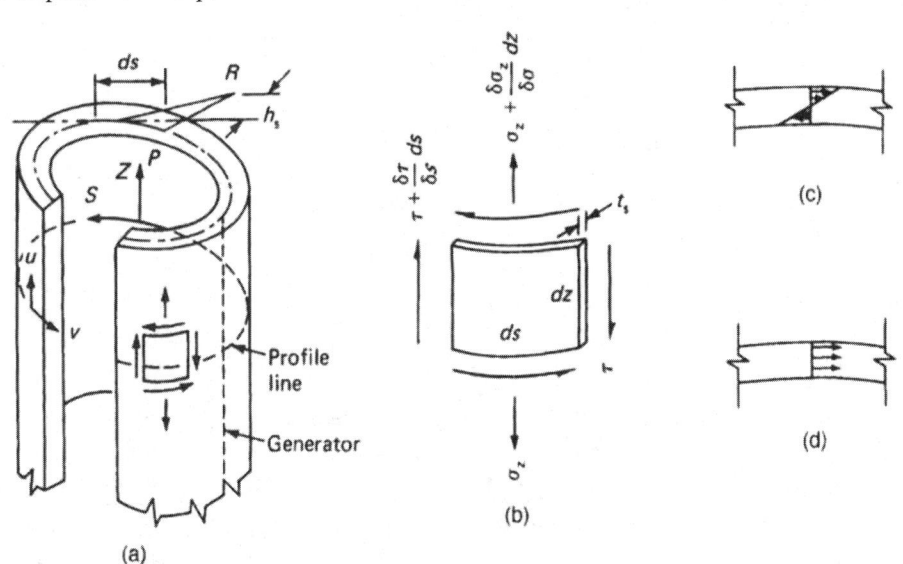

Figure 10.43   Open tube: (a) coordinate system; (b) equilibrium of element $dz\, ds$; (c) St Venant's torsion shear stresses; (d) constrained (warping) torsion shear stresses.

The origin of the coordinate s can now be found from the condition that there is no applied vertical load on the tube, that is,

$$\int_0^s \sigma_{z,s} \, t_s \, ds = 0 \qquad (10.51)$$

where t_s is the thickness of the tube.

The longitudinal stresses $\sigma_{z,s}$ are accompanied by shear stresses and are found from consideration of equilibrium of an element $t \, ds \, dz$ in the z direction (Fig. 10.43b)

$$t_s \frac{\partial \sigma}{\partial z} + \frac{\partial t_s \tau}{\partial s} = 0 \qquad (10.52)$$

$$\tau = \frac{E}{1-\mu^2} \frac{d^3\theta}{dz^3} \int_0^s \omega_s \, ds \qquad (10.53)$$

Using the condition of zero external shear forces, it may be deduced that the origin of the arbitrary center of rotation R is at the shear center. This determines completely the total stress distribution in terms of the derivatives of θ.

The torque M_1 carried by the membrane shear stresses (Fig. 10.43d) which accompany the longitudinal stresses is given by

$$M_1 = \int_0^s \tau t_s h_s \, ds = -\frac{E}{1-\mu^2} \frac{d^3\theta}{dz^3} \int_0^s \omega_s^2 t_s \, ds \qquad (10.54)$$

The quantity

$$\int_0^s \omega_s^2 t_s \, ds$$

is a structural constant, the so-called warping moment of inertia, and is usually denoted by I_ω. Hence

$$M_1 = -\frac{-E}{1-\mu^2} I_\omega \frac{d^3\theta}{dz^3} \qquad (10.55)$$

The torque M_2 carried by St Venant shear stresses (Fig. 10.43c) is given by

$$M_2 = GJ \frac{d\theta}{dz}$$

where GJ is the St Venant torsional rigidity of the section.

The total torque $M = M_1 + M_2$ is then given by

$$M = \frac{-EI_\omega}{1-\mu^2} \frac{d^3\theta}{dz^3} + GJ \frac{d\theta}{dz} \qquad (10.56)$$

Differentiating with respect to z, Eq. (10.56) becomes

$$\frac{EI_\omega}{1-\mu^2} \frac{d^4\theta}{dz^4} - GJ \frac{d^2\theta}{dz^2} = m_z \qquad (10.57)$$

Using notation

$$k = l \sqrt{\frac{(1-\mu^2)GJ}{EI_\omega}}$$

Equation (10.57) can be written

$$\frac{d^4\theta}{dz^4} - \frac{k^2}{l^2} \frac{d^2\theta}{dz^2} = m_z \qquad (10.58)$$

Equation (10.58) is the governing differential equation of constrained torsion. It can be used in the analysis of open section shear walls illustrated in Fig. 10.44.

Longitudinal Stresses and Bimoment Consider the relation between the longitudinal stresses and warping,

$$\sigma_{z,s} = -\frac{E}{1-\mu^2} \frac{d^2\theta}{dz^2} \omega_s$$

Multiplying both sides of this equation by $\omega_s t_s$ and integrating over the whole profile gives

$$\int \sigma_{z,s} \omega_s t_s ds = -\frac{E}{1-\mu^2} \frac{d^2\theta}{dz^2} \int \omega_s^2 t_s \, ds$$

and since

$$\int \omega_s^2 t_s \, ds = I_\omega$$

$$\int \sigma_{z,s} \omega_s t_s \, ds = -\frac{E}{1-\mu^2} \frac{d^2\theta}{dz^2} I_\omega \qquad (10.59)$$

The quantity

$$\int \sigma_{z,s} \omega_s t_s \, ds$$

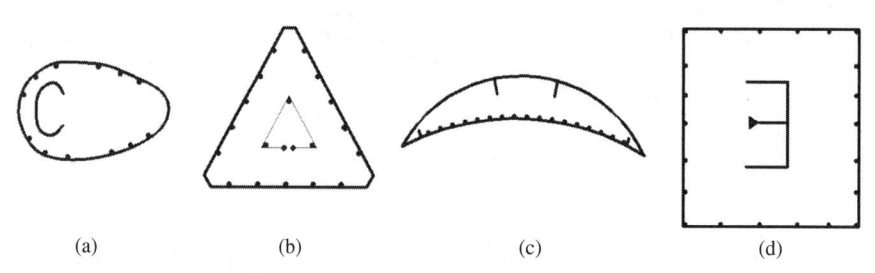

(a)   (b)   (c)   (d)

Figure 10.44 Examples of open section shear walls.

is a generalized force called the bimoment and is represented by B_z. Thus

$$B_z = -\frac{EI_\omega}{1-\mu^2}\frac{d^2\theta}{dz^2}$$

From Eq. (10.50)

$$\frac{E}{1-\mu^2}\frac{d^2\theta}{dz^2} = \frac{\sigma_{z,s}}{\omega_s} \tag{10.60}$$

$$\therefore \sigma_{z,s} = \frac{B_z\omega_s}{I_\omega}$$

Hence the magnitude of bimoment at z and the distribution of the sectorial area over the profile completely determine the longitudinal stresses.

Solution of the Differential Equation Using the notation $f_{(z)} = m\,(1-\mu^2)/EI_\omega$, Eq. (10.42) can be written as

$$\frac{d^4\theta}{dz^4} - \frac{k^2}{l^2}\frac{d^2\theta}{dz^2} - f_{(z)} = 0 \tag{10.61}$$

The solution is of the form,

$$\theta_{(z)} = C_1 + C_2 z + C_3 \sinh\frac{k}{l}z + C_4\cosh\frac{k}{l}z + \bar{\theta}_z \tag{10.62}$$

Differentiating Eq. (10.62) and using Eqs. (10.56) and (10.57), equations for the displacements θ_z and θ_z' and the two forces B_z and M_z can be written thus,

$$\theta_z' = C_1 + C_2 z + C_3\sinh\frac{k}{l}z + C_4\cosh\frac{k}{l}z + \bar{\theta}_z$$

$$\theta_z' = C_2 + C_3\frac{k}{l}\cosh\frac{k}{l}z + C_4\frac{k}{l}\sinh\frac{k}{l}z + \bar{\theta}_z'$$

$$B_z = -GJ\left[C_3\sinh\frac{k}{l}z + C_4\cosh\frac{k}{l}z + \frac{l^2}{k^2}\bar{\theta}_z''\right]$$

$$M_z = GJ\left[C_2 + \bar{\theta}_z' - \frac{l^2}{k^2}\bar{\theta}_z'''\right]$$

$$(10.63)$$

The constants C_1, C_2, C_3, and C_4 can be determined from the two boundary conditions at each end. However, calculations are greatly simplified if instead of the arbitrary constants C_1, C_2, C_3, and C_4, displacement and force boundary conditions, in terms of θ, θ', b, and M, are used in Eq. (10.63). If θ_0, θ_0', B_0, and M_0 are the two sets of displacements and forces at the section $z = 0$, and if there are no applied forces, the constants C_1, C_2, C_3, and C_4 from Eq. (10.47) are given by

$$C_1 = \theta_0 + \frac{1}{GJ}B_0$$

$$C_2 = \frac{1}{GJ}M_0$$

$$C_3 = \frac{l}{k}\theta_0' - \frac{1}{k}\frac{1}{GJ}M_0 \tag{10.64}$$

$$C_4 = \frac{-1}{GJ}B_0$$

Substituting these in Eq. (10.63) and writing in matrix form the general equations for the four quantities, θ, θ', B, and M will be of the form

$$\begin{bmatrix}\theta_z \\ \theta_z' \\ \dfrac{B_z}{GJ} \\ \dfrac{M_z}{GJ}\end{bmatrix} = \begin{bmatrix}1 & \dfrac{l}{k}\sinh\dfrac{k}{l}z & 1-\cosh\dfrac{k}{l}z & z-\dfrac{1}{k}\sinh\dfrac{k}{l}z \\ 0 & \cosh\dfrac{k}{l}z & \dfrac{-k}{l}\sinh\dfrac{k}{l}z & 1-\cosh\dfrac{k}{l}z \\ 0 & \dfrac{-1}{k}\sinh\dfrac{k}{l}z & \cosh\dfrac{k}{l}z & \dfrac{l}{k}\sinh\dfrac{k}{l}z \\ 0 & 0 & 0 & 1\end{bmatrix}\begin{bmatrix}\theta_0 \\ \theta_0' \\ \dfrac{B_0}{GJ} \\ \dfrac{M_0}{GJ}\end{bmatrix}$$

$$(10.65)$$

or in matrix notation, $Z_s = G_0 Z_0$, where Z_s is the action matrix, G_0 the distribution matrix, and Z_0 the initial boundary restraint matrix. If, in addition to the boundary restraints, concentrated forces and displacements are applied at any section $z = t$, then, using the principle of superposition, the expressions for the actions at any section $z(t \le z \le d)$ will be of the form

$$Z_{(z)} = G_{0(z)}Z_0 + G_{(z-t)}Z_t \tag{10.66}$$

where $G_{(z-t)}$ is of the same form as $G_{0(z)}$, except that the argument $(z-t)$ replaces z, and z_t refers to the restraint matrix at $z = t$. The solution represented by Eq. (10.66) can easily be extended to other loading cases, such as several loads applied at various sections and distributed loads, by simple superposition and integration, respectively.

Boundary Conditions The horizontal loading on the core is replaced by a statically equivalent system of loads parallel to the x axis and acting along the shear center axis and a uniform twisting moment. The simple beam theory is used to analyze the effects of loads through the shear center axis.

The effect of the uniformly distributed twisting moment m_z is accounted for by considering the restraint vector at $z = t$ as $[0, 0, 0, m_z]$ and integrating the last column of the distribution matrix G_{z-t} of Eq. (10.66) between the limits 0 and l.

Assuming the core to be completely rigid at the base, the boundary conditions at the ends of the core are

Bottom: $\theta_0 = 0$ and $\theta_0' = 0$ (10.67)
Top: $B_l = 0$ and $M_l = 0$

Using these boundary conditions in the general Eqs. (10.66), the expressions for the four quantities θ_z, θ_z', B_z, and M_z are written. The two quantities θ_z and B_z which are a measure of deflection and stresses will be

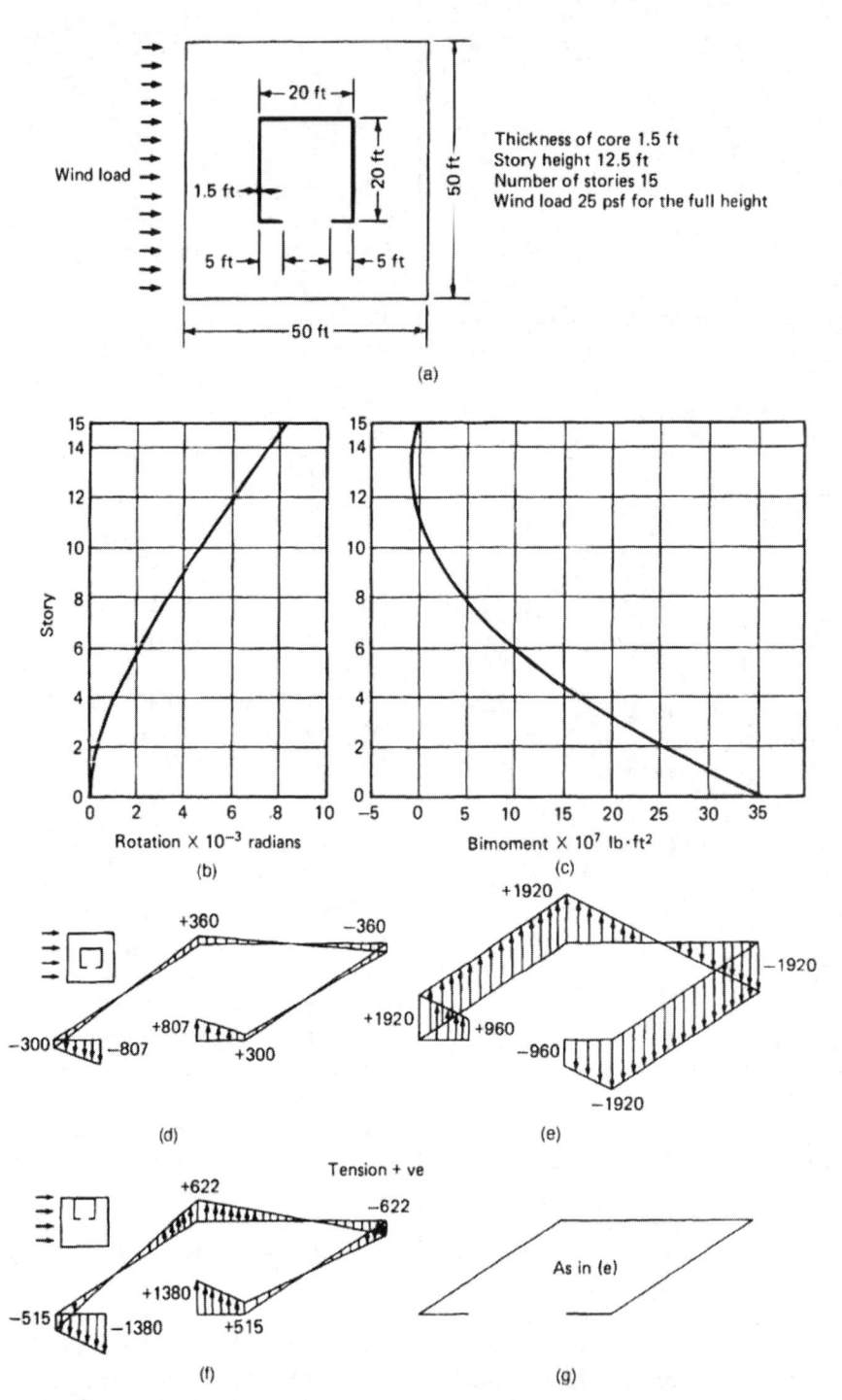

Figure 10.45 Example problem: (a) dimensions and loading; (b) rotations; (c) bimoment variation; (d) warping stresses (core at center), psi; (e) bending stresses, psi; (f) warping stresses (offset core), psi; (g) bending stresses.

$$\theta_z = \frac{-m}{GJ \cosh k}\left[\frac{-l^2}{k^2} - \frac{l^2}{k}\sinh k + z\left(1 - \frac{z}{2}\right)\cosh k\right.$$

$$\left. + \frac{l^2}{k^2}\cosh\frac{k}{l}z + \frac{l^2}{k^2}\cosh\frac{k}{l}z + \frac{l^2}{k}\sinh\frac{k}{l}(l-z)\right]$$

(10.68)

$$B_z = \frac{-ml^2}{k^2 \cosh k}\left[\cosh k - \cosh\frac{k}{l}z - \frac{k\sinh k}{l}(l-z)\right]$$

(10.69)

Equations (10.68) and (10.69) are the basic equations applicable in the analysis of a wide variety of complex-shaped shear walls. Using these equations, a back-of-the-envelope type of calculations may be performed to get an idea of the maximum rotation at the top, and the maximum axial and shear stresses at the base of shear wall structures subjected to torsion.

10.4.10 Torsion analysis of Shear Wall Structures: Worked Examples

The theory described previously, may be used to determine the stresses and rotations in shear wall structures subjected to torsion. To demonstrate the method, three buildings consisting of: (i) a single core; (ii) twin cores; and (iii) a randomly distributed shear walls will be considered.

EXAMPLE 1: As a first example consider a 15-story building consisting of a rather large core as shown in Fig. 10.45. The core is singly symmetric, therefore its shear center lies on the axis of symmetry. To keep the derivation of shear center location somewhat general we will consider the core with arbitrary dimensions as in Fig. 10.46. The procedure for calculating the location O of the shear center, the principal sectorial coordinates and the warping moment of inertia I_ω are as follows.

1. Construct the ω_p diagram (Fig. 10.46b) which is the diagram of sectorial areas with respect to an arbitrary pole P. Using the ω_p and x diagrams for the section, Fig. 10.46b and 10.46c, calculate the product of inertia of the ω_p diagram about the Y axis $I_{\omega p}x$ using

$$I_{\omega_p x} = \int^A \omega_p x \, dA$$

in which $dA = t \, ds$, the area of a segment of the profile of thickness t and length ds. The integral in Eq. (10.60) may be evaluated simply for straight-sided sections by using the product integrals given in Table 10.3.

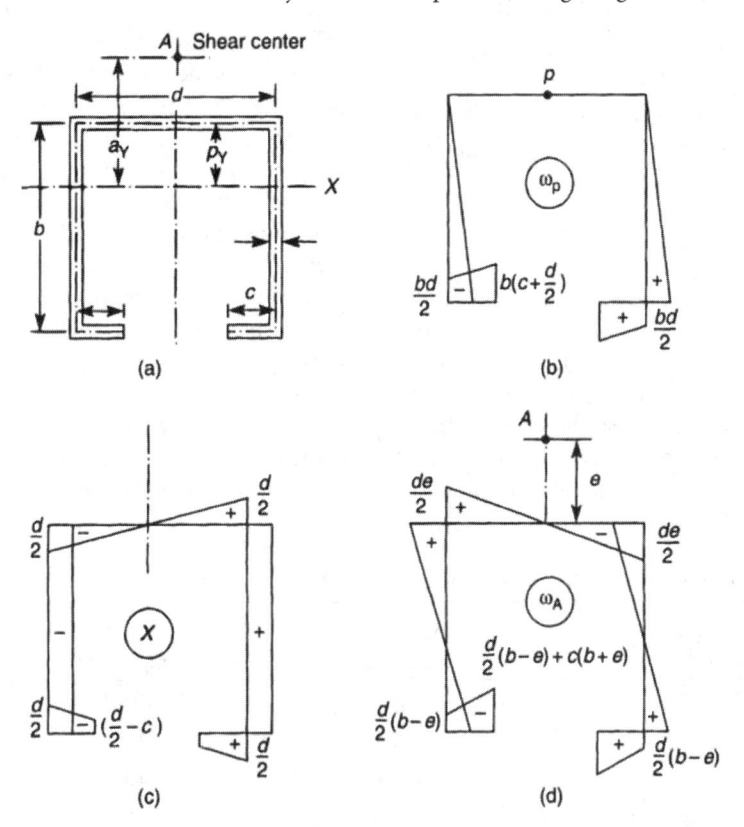

Figure 10.46 (a) Core dimensions; (b) diagram of sectorial area ω with respect to arbitrary point p; (c) diagram of x coordinate; (d) principal sectorial areas of ω.

2. Calculate I_{yy}, the second moment of area of the cross-section about the axis of symmetry.

3. As a final step, calculate the distance α_y of the shear center from p along the axis of symmetry, using

$$\alpha_x = \frac{I_{\omega_p x}}{I_{yy}}$$

4. The values the principal sectorial coordinates ω_p can be found by sweeping the ray AP around the profile and taking twice the values of the swept areas.

5. Sectorial moment of inertia I_ω. This geometric property expresses the core's warping resistance. It is analogous to the moment of inertia in bending, I_{xx} and I_{yy}.

The sectorial moment of inertia is derived from the principal sectorial coordinate distribution using

$$I_\omega = \int_0^A \omega^2 dA = \int_0^s \omega^2 t\, ds$$

For the assumed dimensions of the core shown in Fig. 10.46, the results for ω_s and I_ω are as indicated in Fig. 10.46b through d and as calculated in the following equation.

$$
\begin{aligned}
I_\omega = \int_0^s \omega^2 t\, ds = 2t & \left[\frac{d^3 e^2}{24} + \frac{d^2 e^3}{12} + \frac{d^2(b-e)^3}{12} \right. \\
& + \frac{c}{6} \left\{ \frac{d}{2}(b-e) \left[\frac{3d}{2}(b-e) + c(b+e) \right] \right. \\
& + \left[\frac{d}{2}(b-e) + c(b+e) \right] \\
& \left. \left. \times \left[\frac{3d}{2}(b-e) + 2c(b+e) \right] \right\} \right]
\end{aligned}
\tag{10.70}
$$

For the example problem shown in Fig. 10.45 the stresses at the base due to bending and twisting are compared for two cases. In the first case the core is located centrally, and in the second case it is offset with respect to the plan dimensions. The numerical calculations of the various sectorial properties are not given since these are given with excruciating details in the other two examples (Table 10.6).

A comparison of stresses for the example problem indicates that the warping stresses are of the same magnitude as the bending stresses. Observe that the warping stresses exceed the bending stress for the second case where the core is offset from the center of plan dimensions. The deflection at the building top, due to torsion bears a similar relation to the bending deflection, as will be demonstrated in the next example. All these examples, although simplified to make the numerical work less cumbersome, demonstrate the importance of considering warping stresses and deflection in practical building structures.

EXAMPLE 2. Consider a 25-story, 300 ft (91.44 m) building consisting of two cores as shown in Fig. 10.47a,b. To keep the analysis simple assume that the resistance to lateral loads and torque is provided solely by the core. The building is subjected to a uniform wind load of 25 psf (1.197 kN/m²) in the x direction.

It is required to determine the maximum deflection and rotation at the top, and the vertical stresses at the base due to bending and twisting. An elastic modulus $E = 3600$ ksi (24,822 MPa) and a shear modulus $G = 1565$ ksi (10,791 MPa) are assumed for the concrete properties. The procedure is first described and then illustrated numerically.

Step 1. *Determine the sectorial properties.* For the given structure, by inspection, the location of shear center O is determined at a point mid-way between the two cores. The ω diagram is related to the shear center O as its pole and a point of zero warping deflection as an origin. In a symmetrical section, as in the example problem, the intersection of the axis of symmetry with the profile at D defines a point of anti-symmetrical behavior, and hence of zero warping deflection: therefore, it may be used as the origin.

Values of ω are found from first principles, by sweeping the ray OD around the profile and taking twice the values of the swept areas. For the example problem, the principal sectorial coordinate diagram is shown in Fig. 10.47c.

Table 10.6   Example 3: Comparison of Bending Stresses at the Base

Wall no.	Section modulus in.³	Bending moment from comp. Analysis kip-in	Extreme bending stresses (ksi)	
			Comp. analysis	Warping theory
1	294,912	65,238	0.221	0.206
2	18,432	1038	0.056	0.04
3	165,888	64,191	0.387	0.360
4	73,728	6950	0.094	0.078
5	73,728	6950	0.094	0.078

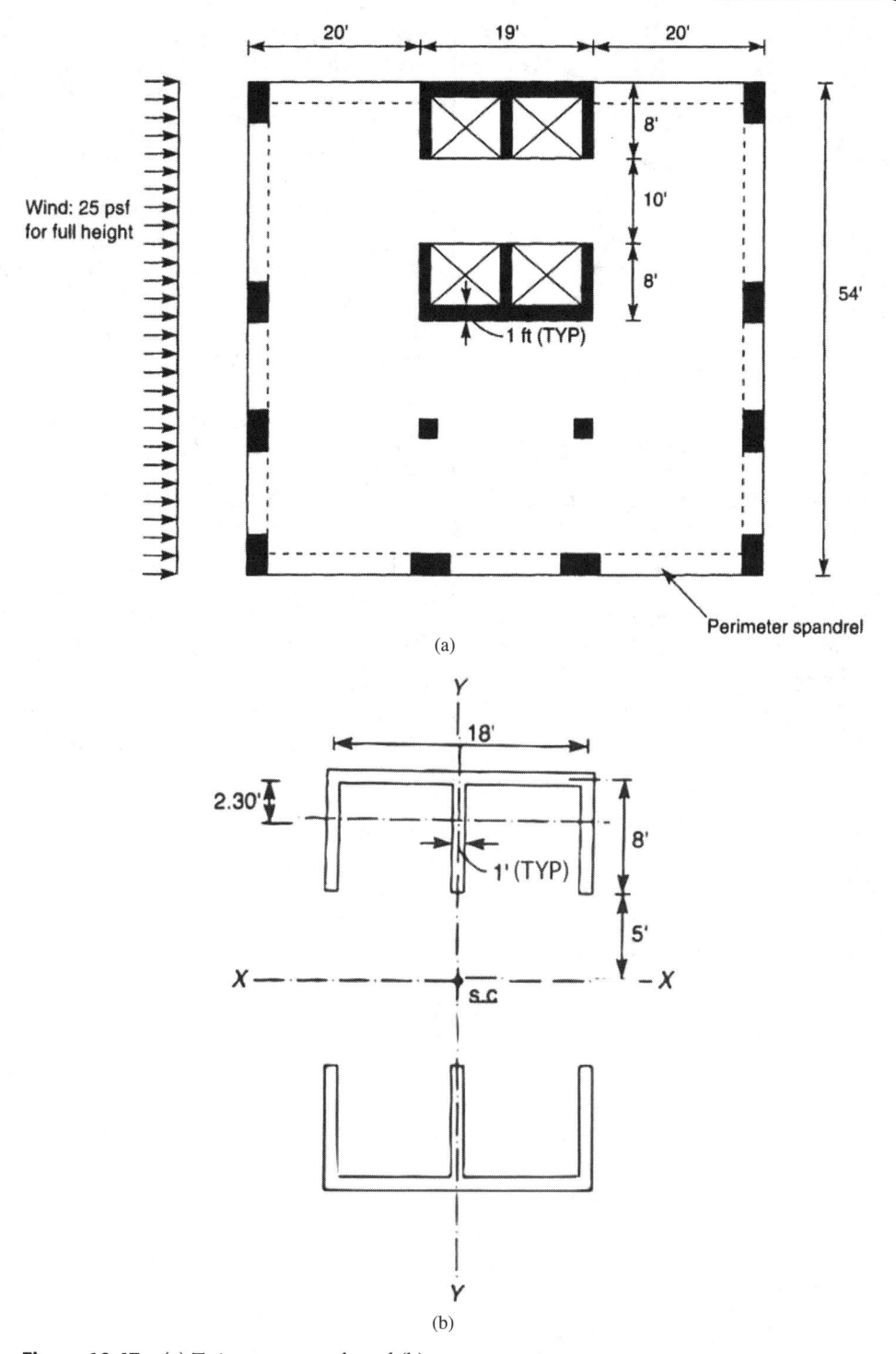

Figure 10.47   (a) Twin-core example and (b) core properties.

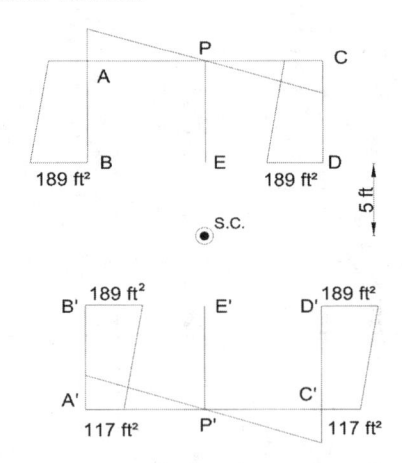

Figure 10.47c w_s diagram (sectorial coordinates).

Step 2. *Determine the sectorial moment of inertia I_ω from Eq. (10.42)*

$$I_\omega = \int^A \omega^2 dA = \int^s \omega^2 t\, ds$$

Using the ω_s diagram (Fig. 10.47c) and the product integral table (Table 10.3), the value for I_ω is evaluated as shown in the worksheets.

Step 3. *Torsion constant J for the core is determined from Eq. (10.43). For one core,*

$$J = \frac{1}{3}\sum bt^3 = \frac{1}{3}bt^3 = \frac{1}{3} \times 1^3 (3 \times 7.5 + 1 \times 19) = 13{,}834 \text{ ft}^4,$$

J for two cores = 13.834 × 2 = 27.671 ft⁴.

Step 4. *Determine eccentricity e of the line of action of wind resultant from the shear center.* The resultant wind force per unit height of the building is equal to 25 × 54 = 1.35 kip/ft (1.83 kN/m) acting at 13.5 ft (4.12 m) to the south of shear center. Therefore, the eccentricity, *e*, from the shear center is 13.5 ft (4.12 m). Since the external torque is the product of the horizontal loading and its eccentricity, the torsion due to wind is 1.35 × 13.5 = 18.225 kft/ft (24.70 kNm) per unit height, anticlockwise.

Step 5. *Determine bending deflection at the top and stresses at the base.* A bending analysis is now performed to determine the maximum lateral deflection at top and the bending stresses at the base. Deflection at top due to bending is calculated as follows.

$$\Delta_{y\,(\text{max})} = \frac{\omega l^4}{8 EI_{yy}} = 0.737 \text{ ft}$$

The deflection is $\frac{1}{406}$ of the height, as compared to the generally accepted limit of $\frac{1}{400}$ and therefore is acceptable. Observe that the building is very flexible in the *Y* direction because the moment of inertia of the core I_{xx} is about one-sixth of I_{yy}, indicating that supplemental bracing is required in the *y* direction. One solution is to add perimeter rigid frames as indicated in Fig. 10.47a. However, we continue the problem with the assumption made earlier, namely, that the core resists all the lateral loads.

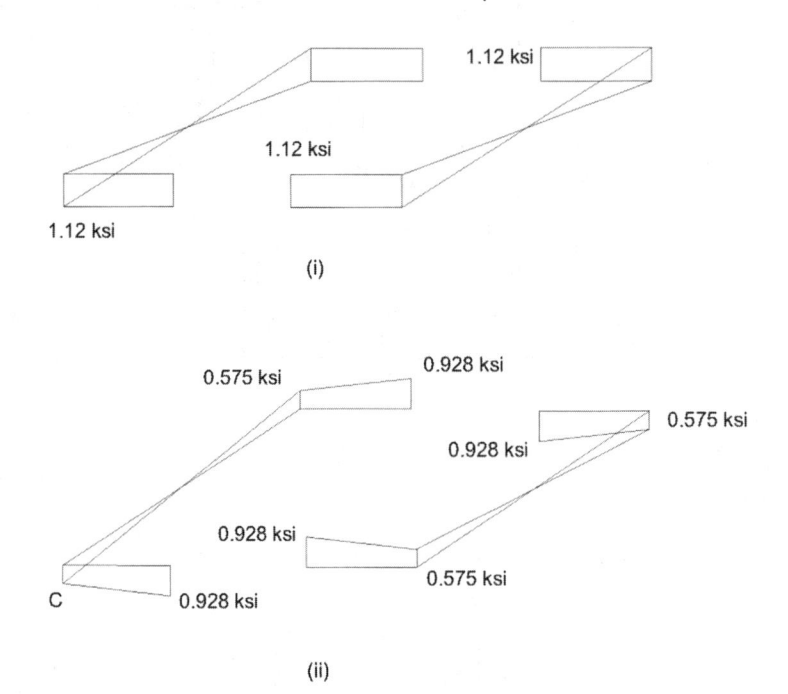

Figure 10.47d   Comparison of stresses: (i) bending stress σ_b; (ii) warping stress σ_ω.

$$\text{Maximum bending stress} = \frac{MC}{I} = \frac{60,750}{1788} \times 9.5$$

$$= 322.78 \text{ ksf}$$

$$= 2.24 \text{ ksi (16.78 MPa)}$$

The bending stress diagram is given in Fig. 10.47d.

Step 6. *Determine the parameter k using Eq. (10.26)*

$$k = l \sqrt{\frac{GJ}{EI_\omega}}$$

$$= 300 \sqrt{\frac{1565 \times 144 \times 28}{92,7180 \times 2.3}}$$

$$= 1.08$$

Step 7. *Determine the rotation and total deflection at the corner of top floor.* The rotation at any level of the building for a uniformly distributed torque "m" may be obtained from Eq. (10.69).

At the top, $Z = l$. Substituting $z = l$ in Eq. (10.69) we get

$$\theta = \frac{-ml^2}{GJ \cosh k} \left[-\frac{l}{k^2} - \frac{l}{k} \sinh k + \frac{1}{2} \cosh k + \frac{1}{k^2} \cosh k \right]$$

$$(10.71)$$

Substituting for the various parameters we get $\theta_H = 0.0196$ radians, anticlockwise. Therefore, the additional deflection at the south-east corner c of top floor due to torsion $\Delta_t = \theta_H \times$ distance of c from shear center $= 0.196 \times 50.10 = 0.9821$ ft (0.30 m) The total deflection at c due to bending and torsion $= \Delta_b + \Delta_t = 1.47 + 0.9821 = 2.45$ ft (0.75 m). This value represents $\frac{1}{122}$ of the building height, an unacceptably large value, confirming the earlier observation that the building is too flexible in the y direction.

Step 8. *Determine bimoments and warping stresses.* The warping stresses σ_ω at the base are determined from the bimoment B at that level. The bimoment is obtained for a uniformly distributed torque from Eq. (10.69). Then, at any point on the section where the principal sectorial coordinate is ω_s, the vertical warping stress is obtained from Eq. (10.44). The total axial stresses due to horizontal loading are obtained by combining the warping stresses with the bending stresses.

The vertical stresses at the base due warping are determined from the bimoment at the base. This is given by Eq. (10.69):

$$B_z = -\frac{ml^2}{k \cosh k} \left[\cosh k - \cosh \frac{k_z}{l} - k \sinh \frac{k}{l} (l - z) \right]$$

$$(10.72)$$

At the base

$$z = 0, \quad B_0 = -\frac{ml^2}{k^2 \cosh k} [\cosh k - l - k \sinh k] \quad (10.73)$$

$$k = 1.08, \sinh k = 1.3025, \cosh k = 1.642$$
$$\sinh 0 = 0, \cosh 0 = 1$$

Substituting the above values

$$B_0 = \frac{-ml^2}{1.08^2 \times 1642} [1.642 - 1 - 1.08 \times 1.3025]$$

$$= \frac{-ml^2}{2.504}$$

$$= -\frac{18.225 \times 300^2}{2.504}$$

$$= 656,100 \text{ kip/ft}^2$$

Warping stresses are given by $\sigma_\omega = \dfrac{B_0}{I_\omega} \omega_s$. At D,

$$\sigma_\omega = \frac{656,100}{927,180} \times 189$$

$$= 133.74 \text{ ksf}$$

$$= 0.928 \text{ ksi}$$

At C,

$$\sigma_\omega = \frac{656,100}{927,180} \times 117$$

$$= 82.79 \text{ ksf}$$

$$= 0.575 \text{ ksi}$$

Figure 10.47d shows a comparison of bending and warping stresses. The importance of warping torsion is obvious.

Example 2: Worksheet
Core bending properties

$$\text{Area of one core} = (1 + 8)1 + (8 - 0.5) \times 1 \times 3$$

$$= 41.5 \text{ ft}^2$$

Determine c.g.

$$3 \times 7.5 \times 1 \left(\frac{7.5}{2} + 0.5 \right) = 41.5 \, \bar{y}$$

$$\bar{y} = \frac{95.6}{41.5} = 2.30 \text{ ft}$$

$$I_{yy} = 19 \times \frac{1^3}{12} + 3 \left(\frac{1 \times 7.5^3}{12} \right) + 19 \times 1 \times 2.30^2 + 3.75 \times 1 \times 1.96^2$$

$$= 294 \text{ ft}^4$$

$$I_{xx} \text{ for two cores} = 2 \times 294 = 588 \text{ ft}^4$$

$$I_{xx} = 1 \times \frac{1^3}{12} + 7.5 \times 1 \times 9^2 \times 2 + 3 \times 7.5 \times \frac{1^3}{12}$$

$$= 571.58 + 1215 + 1.87$$

$$= 1788.45 \text{ ft}^4 \quad \text{for one core}$$

$$I_{xx} \text{ for two cores} = 2 \times 1788.45 = 3577 \text{ ft}^4$$

Core torsion properties

The shear center is located at the center of two cores. Using points P and P' as the principal poles, the ω_s is drawn by sweeping the radii OP and OP' around the profile

$$\omega_s \text{ Diagram (sectorial coordinates)}$$

Sectorial moment of inertia I_ω

This is obtained by using the product integral table as follows:

Segment	ω_s	$\int \omega_s^2 \, ds$
PA, PC, P'A', P'C'	117 / 9	$\frac{1}{3} \times 9 \times 117 \times 1 = 41{,}067 \text{ ft}^6$
AB, CD, A'B', C'D'	117 189 / 8	$\frac{8}{6}[117(2 \times 117 + 189) + 189(2 \times 189 \times 117)]$ $= 190{,}728 \text{ ft}^6$
	Total for half of one core = 41,067 + 190,728 = 231,795 ft^6	

Verify the integral $\int^A \omega^2 ds$ for segment AB by treating the area as a combination of a rectangle and a triangle thus:

$$\int \omega^2 ds = 117 \times 17 \times 8 + \tfrac{1}{3} \times 8\,72^2 + 2 \times \tfrac{1}{2} \times 117 \times 82 \times 8$$

$$= a^2 + b^2 + 2ab$$

$$= 109{,}512 + 13{,}824 + 67{,}392 = 190{,}728 \text{ ft}^6$$

$$I_w \text{ for two cores} = 4 \times 231{,}795 = 927{,}180 \text{ ft}^6$$

St Venant torsion constant GJ

$$J \text{ for one core} = \sum \frac{mt^3}{3} = \frac{1}{3} \times 1^3 (3 \times 7.5 + 1 \times 19)$$

$$= 13{,}834 \text{ ft}^4$$

$$J \text{ for two core} = 2 \times 13{,}834 = 27.67 \text{ ft}^4$$

$$G = \frac{E}{2(1+\mu)} = \frac{E}{2.3} \quad E = 57{,}000\sqrt{4000}$$

$$= 3600 \text{ ksi} = 518{,}400 \text{ ksf}$$

$$G = \frac{518{,}400}{2(1+0.15)} \quad \mu \text{ for concrete} = 0.15$$

$$= 225{,}391 \text{ ksf}$$

$$GJ = 225{,}391 \times 27.67$$

$$= 6{,}236{,}569 \text{ kft}^2$$

$$k = l\sqrt{\frac{GJ}{EI_\omega}}$$

$$= 300\sqrt{\frac{27.67}{927{,}180 \times 2.3}} = 1.08$$

Bending analysis
Deflection

The building deflection at top is given by the relation

$$\Delta_b = \frac{wl^4}{8EI} \qquad w = 25 \times 54 = 1.35 \text{ kip/ft}$$

$$= \frac{1.35 \times 300^4}{8 \times 518{,}400 \times 3677} \qquad l = 300 \text{ ft}$$

$$= 0.737 \text{ ft} \qquad E = 518{,}400 \text{ ksi}$$

$$\frac{\Delta_b}{l} = \frac{0.737}{3.00} = \frac{1}{406} \qquad I = 3577 \text{ ft}^4$$

Bending stresses

$$\text{Maximum bending moment at the base} = \frac{wl^2}{2}$$

$$= \frac{1.35 \times 300^2}{2}$$

$$= 60{,}750 \text{ kip ft}$$

$$\text{Bending stress } \sigma_b = \frac{MC}{I_{xx}}$$

$$= \frac{60{,}750 \times 9.5}{3577}$$

$$= 161.35 \text{ ksf}$$

$$= 1.120 \text{ ksi}$$

Torsion analysis

Rotation

The rotation θ_l at the top, for an open section subjected to a uniform torque of m is given by

$$\theta_l = -\frac{m}{GJ\cosh k}\left[-\frac{l^2}{k^2} - \frac{l^2}{k^2}\sinh k + l\left(l - \frac{l}{2}\right)\cosh k \right.$$
$$\left. + \frac{l^2}{k^2}\cosh k + \frac{l^2}{k}\sinh\frac{k}{l}(l-l) \right]$$

Substituting

$$m = 1.35 \times 13.5 = 18.225 \text{ kft/ft}$$
$$l = 300 \text{ ft}$$
$$k = 1.08, \sinh k = 1.3025, \cosh k = 1.642$$
$$\sinh 0 = 0, \cosh 0 = 1$$
$$\theta_l = 0.0196 \text{ radians.}$$

The distance R of shear center from the building corner

$$R = \sqrt{29.5^2 + 40.5^2} = 50.10 \text{ ft}$$

Deflection due to torsion $\Delta_t = \theta_l \times R$
$$= 0.0196 \times 50.10$$
$$= 0.9821 \text{ ft}$$

Total deflection due to bending and torsion $= \Delta_b + \Delta_t$
$$= 0.737 + 0.9821 = 1.72 \text{ ft}$$

The deflection index $\dfrac{\Delta}{l} = \dfrac{1.72}{300} = \dfrac{1}{174}$, an unacceptably large value indicating serious deficiency in the lateral load-resisting system.

Stresses due to torsion

The bimoment B_z due to a uniformly applied torque m is given by the relation

$$B_z = \frac{-ml^2}{k^2\cosh k}\left[\cosh k - \cosh\frac{kz}{l} - k\sinh\frac{k}{l}(l-z) \right]$$

At $z = 0$, the bimoment B_0 at the base is given by

$$B_0 = \frac{-ml^2}{k^2\cosh k}[\cosh k - 1 - k\sinh k]$$

Substituting the values for m, l, k, $\sinh k$ and $\cosh k$ as before,

$$B_0 = \frac{-ml^2}{2.504}$$
$$m = 1.35 \times 13.5 = 18.225 \text{ kft/ft} \quad l = 300 \text{ ft}$$
$$B_0 = \frac{18.225 \times 300^2}{2.504}$$
$$= 656,100 \text{ kft}^2$$

Warping stress at D,

$$\sigma_{\omega(D)} = \frac{B_0}{I_\omega}\omega_s$$
$$= \frac{656,100 \times 189}{927,180}$$
$$= 133.74 \text{ ksf}$$
$$= 0.928 \text{ ksi}$$

Warping stress at C,

$$\sigma_{\omega(C)} = \frac{656,100 \times 117}{927,180}$$
$$= 82.79 \text{ ksf}$$
$$= 0.575 \text{ ksi}$$

As noted previously, the analogy between the warping torsion and bending may be used to get an approximate idea of the rotation and stresses in core structures. Using this approach, for the example problem we get the rotation at top

$$\theta_l = \frac{ml^4}{8EI_\omega} = \frac{13.5 \times 300^4}{8 \times 518,400 \times 927,180}$$
$$= 0.0285 \text{ radians}$$

as compared to the value 0.0196 radians obtained by the more accurate analysis.

The biomoment B_0, at the base is given by $B_0 = \dfrac{ml^2}{2}$ as compared to the value of $\dfrac{ml^2}{2.504}$ obtained before which is pretty good for a five-minute back-of-the-envelope-type calculation.

EXAMPLE 3: The warping theory described with reference to single and twin-core structures may also be used to determine the stresses and rotations in a shear wall structure consisting of randomly distributed planar shear walls. To demonstrate the method of analysis, a 15-story building is analyzed, first by using the warping theory, and then by a commercially available three-dimensional computer program. The result for torsion which includes the rotations of the core, and the bending stresses at the base are compared and shown to give almost identical results. The warping analysis is given in the work sheet. The highlights of the analysis are as follows.

The building Fig. 10.48, consists of three walls, W_1, W_2, and W_3 in the transverse direction, and two walls W_4 and W_5 in the longitudinal direction. The uniform wind load of 25 psf (1.197 kN/m²) gives a resultant of $25 \times 70 = 1.75$ kip/ft (2.37 kNm) acting at the center of gravity of the plan in the Y direction. For the given structure, by

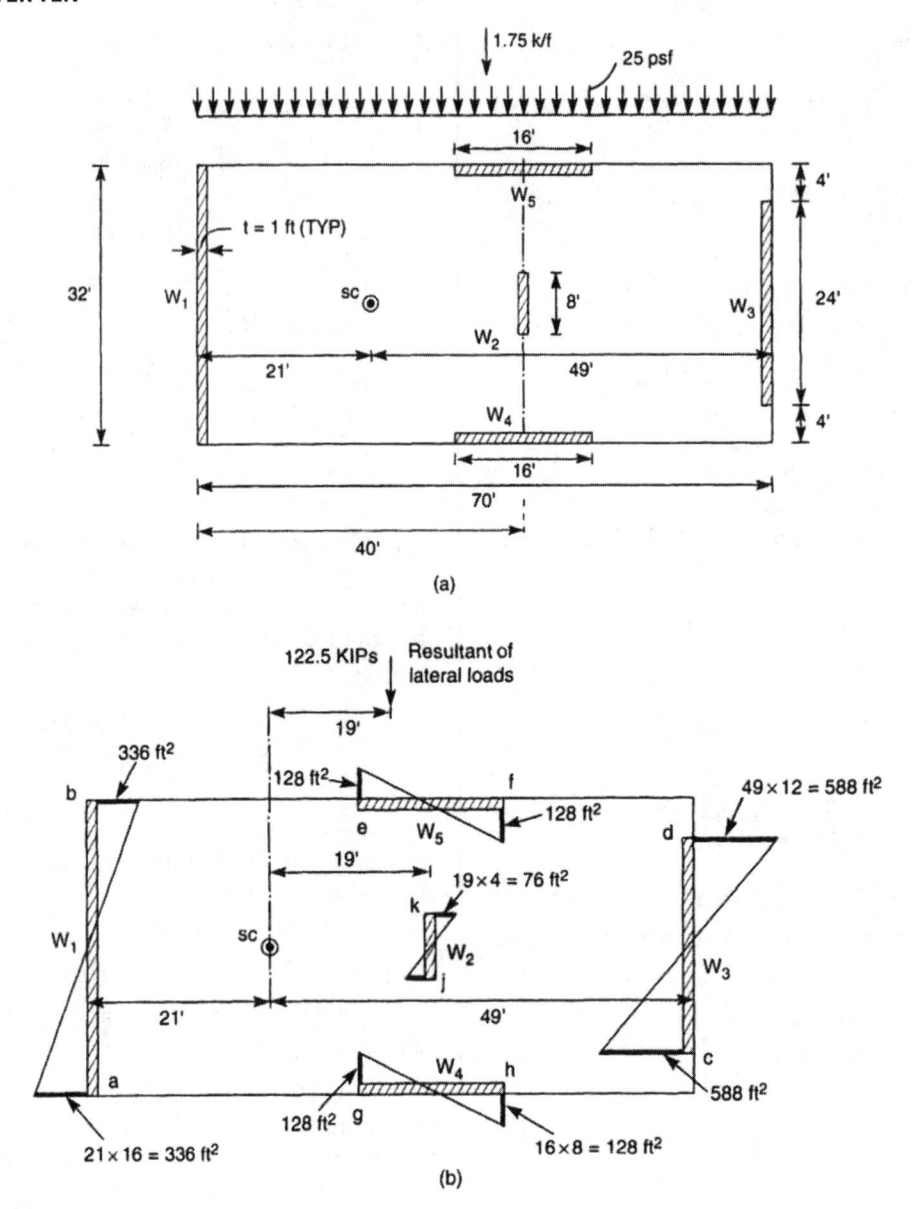

Figure 10.48 Randomly distributed shear walls, torsion example: (a) plan; (b) rotation comparison.

inspection, the location of the shear center O is determined to be on a line midway between the walls W_4 and W_5. The distance x of the shear center O from wall W_1, is obtained from the relation

$$\bar{x} = \frac{\sum I_{xx} x}{\sum I_{xx}}$$

Next the eccentricity e which is the distance from the line of action of wind resultant from the shear center is determined to be 14 ft (4.27 m). The external torque,

the product of the horizontal load and its eccentricity is evaluated to be equal to $1.75 \times 14 = 24.5$ kft/ft (10.12 kNm/m). To keep the comparison simple, both the computer and warping analyses are performed for only a uniformly applied torsion. For the example problem, the center of gravity of each wall defines a point of antisymmetrical behavior, and hence of zero warping deflection. Therefore, these points are used as the origins for determining the values of sectorial coordinate ω. A building analysis is then performed to determine the rotation at

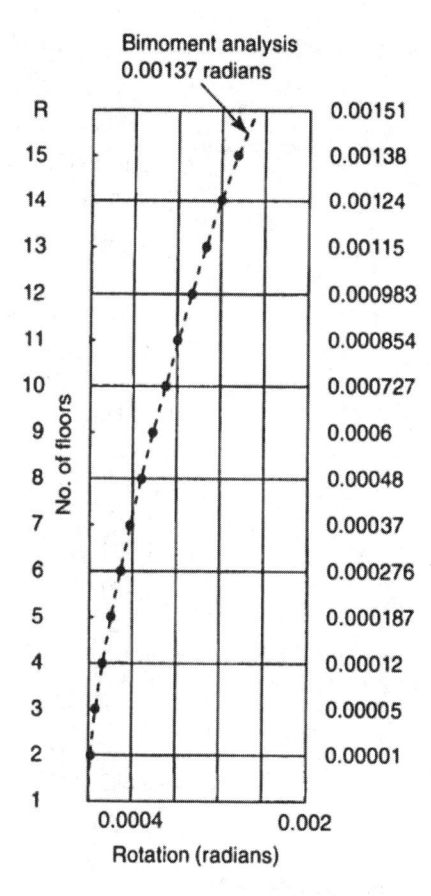

Bimoment analysis
0.00137 radians

No. of floors	Rotation value
R	0.00151
15	0.00138
14	0.00124
13	0.00115
12	0.000983
11	0.000854
10	0.000727
9	0.0006
8	0.00048
7	0.00037
6	0.000276
5	0.000187
4	0.00012
3	0.00005
2	0.00001

Rotation (radians)

Rotation comparison: etabs version 6 versus
warping analysis
(c)

Figure 10.48 (*Continued*)

the building top and the bending stresses at the base, as shown in the worksheets.

In the computer analysis, the building is analyzed with five walls assuming the slab to be infinitely rigid in its own plane. In the final part of the work sheet, the results of rotations and bending stresses from the two analyses are compared. From the close correlation between the two, the applicability of warping analysis to practical planar shear walls is obvious.

EXAMPLE 3: Worksheet
Location of shear center

By inspection, the location of shear center s.c. is determined to be on the common neutral axis x-x. The location of the other shear axis is given by the relation

$$e = \frac{\sum I_x x}{\sum I_x}$$

Notice the similarity between this and the following:

$$\bar{y} = \frac{\sum Ay}{\sum A}$$

which is used to find the neutral axis of a built-up section.

Just as the areas of individual parts are used to find the neutral axis, the moments of inertia of individual areas are used to find the shear axis of the building (Fig. 10.48). The procedure is the same; select a reference axis (y-y), determine I_x for each member section (about its own neutral axis x-x) and the distance x of the section from the reference axis (y-y). The resultant e is the distance from the chosen reference axis (y-y) to the parallel shear axis of the building.

$$e = \frac{I_{x_1} x_1 + I_{x_2} x_2 + I_{x_3} x_3 + I_{x_4} x_4}{I_{x_1} + I_{x_2} + I_{x_3} + I_{x_4}}$$

or:

$$e = \frac{\sum I_x x}{\sum I_x} \tag{10.74}$$

Calculate the location $\bar{x}_1$ of the shear center s.c. from the center line of wall W_1.

$$\sum I_x = 1 \times \frac{32^3}{12} + 1 \times \frac{8^3}{12} + 1 \times \frac{24^3}{12}$$
$$= 2730.67 + 42.67 + 1152$$
$$= 3925.34 \text{ ft}^4$$

$$\bar{x}_1 = \frac{\sum I_x x}{\sum I_x}$$
$$= (2736.67 \times 0 + 42.67 \times 40 + 1152 \times 70) / 3925.24$$
$$= \frac{82,346.8}{3925.24} = 20.979 \text{ft} \quad \text{Use 21 ft}$$

Verify location of s.c. from the center line of east wall W_3.

$$\bar{x}_2 = (1152 \times 0 + 42.67 \times 30 + 2730.67 \times 70) / 3925.24$$
$$= \frac{192,427}{3925.24} = 49 \text{ ft}$$

The eccentricity e of the line of action of wind resultant from the shear center

$$e = 35.21 = 14 \text{ ft}$$

Torsional moment m per foot height of the building

$$m = 1.75 \times 14 = 24.5 \text{ kf/f}$$

Torsion properties

Using the centers of each wall as the principal poles, the sectorial coordinate diagram representing the warping coordinates for the composite building is drawn by sweeping the radius vector passing through the shear center. The resulting ω_s diagram is shown in Fig. 10.48b.

The warping moment of inertia for the building is calculated as before by using the product integral table.

$$I_\omega = \frac{2}{3} \times 16 \times 336^2 + \frac{2}{3} \times 4 \times 76^2 + \frac{2}{3} \times 12 \times 588^2$$

$$+ \frac{2}{3} \times 8 \times 128^2 + \frac{2}{3} \times 8 \times 128^2 = 4{,}160{,}341 \text{ ft}^6$$

The St Venant's torsion J is calculated from the relation

$$J = \sum_{}^{n} bt^3$$

$$= 1^3 (32 + 8 + 24 + 16 + 16)$$

$$= 96 \text{ ft}^4$$

$$GJ = 225{,}360 \times 96$$

$$= 216{,}345{,}60 \times \text{kip/ft}^2$$

$$k = l\sqrt{\frac{GJ}{EI_\omega}}, \quad \frac{G}{E} = \frac{1}{2.3}$$

$$= 180 \sqrt{\frac{96}{4{,}160{,}341 \times 2.3}} = 0.57$$

Torsional rotation

The rotation θ_l at top due to a uniformly distributed torque of m per unit height is given by

$$\theta_l = \frac{-ml^2}{GJ \cosh k} \left[\frac{\cosh k - 1}{k^2} + \frac{\cosh k}{2} - \frac{\sinh k}{k} \right]$$

Substituting

$$k = 0.57, \sinh k = \sinh 0.57 = 0.601$$
$$\cosh k = \cosh 0.57 = 1.167, \ GJ = 216{,}345{,}60 \text{ kip/ft}^2$$

and

$$m = 1.75 \times 14 = 24.5 \text{ kf/f}$$
$$\theta_l = 0.00137 \text{ radians.}$$

Bending stresses due to torsion

The bimoment B_0 at the base is given by

$$B_0 = -\frac{ml^2}{k^2 \cosh k} [\cosh k - 1 - k \sinh k]$$

Substituting

$$k = 0.57, \sinh k = \sinh 0.57 = 0.601$$
$$\cosh k = \cosh 0.57 = 1.167, \ m = 1.75(35 - 21) = 24.5 \text{ kft/ft}$$
$$l = 15 \text{ stories @ 12 ft} = 180 \text{ ft}$$

$$B_0 = -\frac{24.5 \times 180^2}{0.57^2 \times 1.167} [1.167 - 1 - 0.57 + 0.601]$$

$$= 367{,}529 \text{ kip/ft}^2$$

The bending stresses at the extremities of walls are as follows.

Wall 1
$$\sigma_\omega \text{ at } a, b = \frac{367{,}529 \times 336}{4{,}160{,}341}$$
$$= 29.68 \text{ kip/ft}^2$$
$$= 0.206 \text{ ksi}$$

Wall 3
$$\sigma_\omega \text{ at } c, d = \frac{367{,}529 \times 588}{4{,}160{,}341}$$
$$= 51.941 \text{ kip/ft}^2$$
$$= 0.360 \text{ ksi}$$

Wall 5
$$\sigma_\omega \text{ at } e, f = \frac{367{,}524 \times 128}{4{,}160{,}341}$$
$$= 11.30 \text{ kip/ft}^2$$
$$= 0.078 \text{ ksi}$$

Wall 4
$$\sigma_\omega \text{ at } g, h\text{—similar to wall 5 at } e, f.$$

Wall 2
$$\sigma_\omega \text{ at } j, k = \frac{367{,}529 \times 76}{4{,}160{,}341}$$
$$= 671 \text{ kip/ft}^2$$
$$= 0.046 \text{ ksi}$$

10.4.11   Torsion Analysis of Steel Braced Core: Worked Example Torsional

The torsional behavior of braced cores in steel buildings can be predicted with sufficient accuracy by using the nonuniform torsion theory given in the preceding section. For example, consider the typical floor plan of a 30-story steel building as shown in Fig. 10.49a. Assume that wind resistance is provided by X-bracing of the core all around except between two corridor columns as shown in Fig. 10.49a. Without the luxury of bracing all around its perimeter, the core loses much of its torsional resistance and behaves more like an open core as shown in Fig. 10.49b. In analyzing the core, it will be assumed initially that the warping constraint of beam AB is negligible. These effects are considered later. In the analytical model, the braces can be considered as equivalent solid walls subjected only to shear forces. The normal forces can be assumed to be sustained by the columns alone, acting as vertical ribs. The idealized analytical model is shown in Fig. 10.49c. The differential equation of nonuniform torsion is of the form

$$\frac{d^4\theta}{dz^4} - \frac{k^2}{l^2} \frac{d^2\theta}{dz^2} = m_z \frac{1 - \mu^2}{EI_\omega} \qquad (10.75)$$

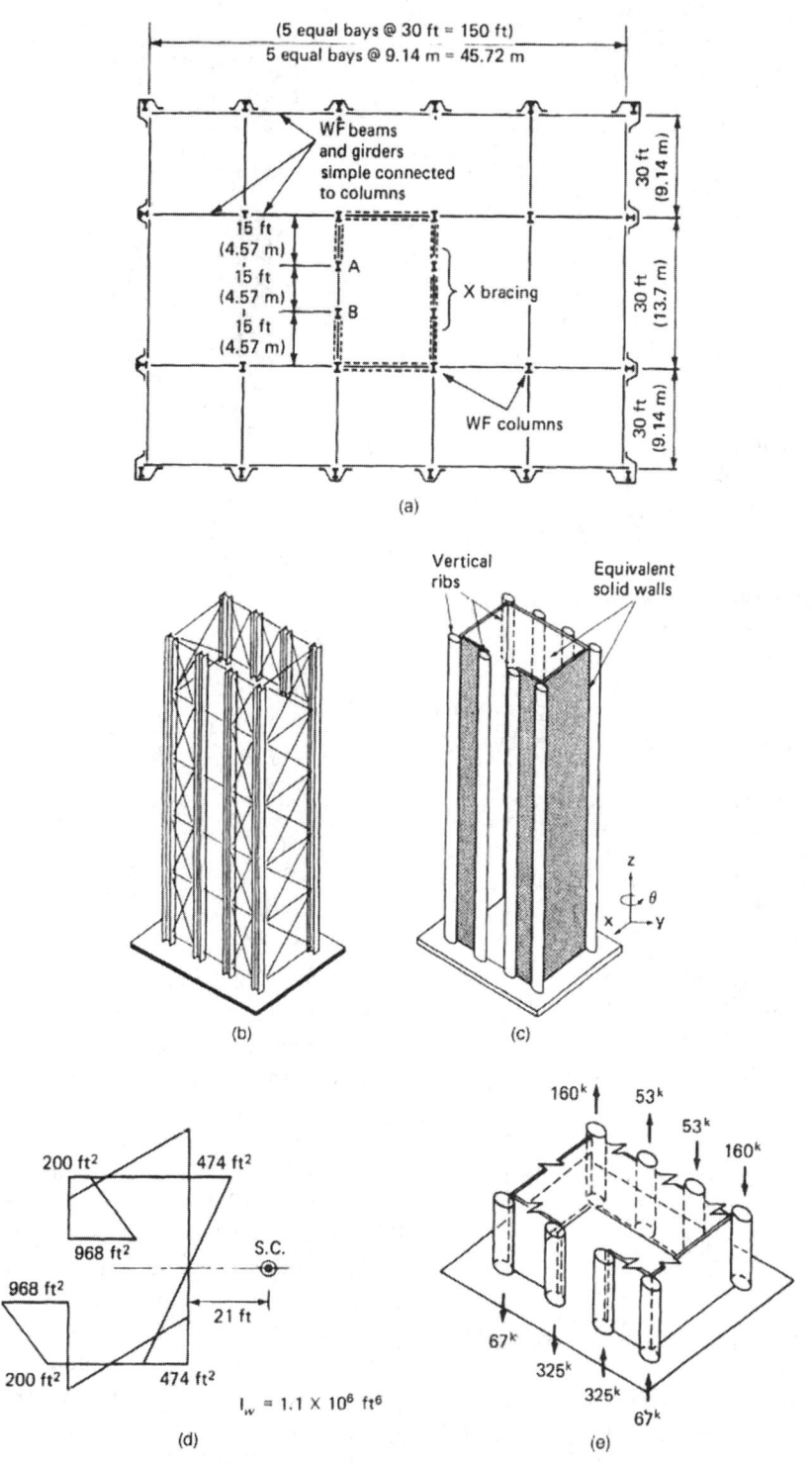

Figure 10.49 Example problem: (a) typical floor plan; (b) asymmetrical bracing resulting in an open section core; (c) analytical model for torsion analysis; (d) diagram of principal sectorial coordinates; (e) axial forces in core columns due to bending; (f) axial force in core columns due to bending; (g) core warping constrained by floor beams.

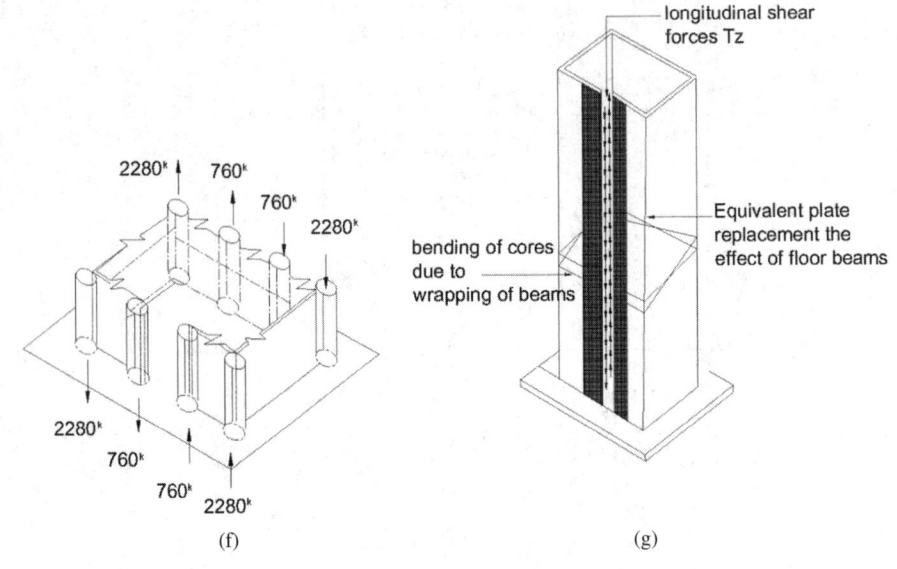

Figure 10.49 (*Continued*)

where θ = the rotation of the core at section z in the xy plane
k = the nondimensional parameter given by

$$k = l\sqrt{GJ/EI_\omega}$$

l = the height of the core
m_z = the intensity at z of the applied torque
μ = the Poisson's ratio
E = the modulus of elasticity
I_ω = the warping moment of inertia
J = the St Venant's torsional moment of inertia of the braced core

For the case of no applied external forces, Eq. (10.75) has a solution of the form:

$$\theta_z = \theta_0 + \theta_0' \frac{l}{k}\sinh\frac{k}{l}z + \frac{B_0}{GJ}\left(1 - \cosh\frac{k}{l}z\right)$$
$$+ \frac{M_0}{GJ}\left(z - \frac{1}{k}\sinh\frac{k}{l}z\right) \qquad (10.76)$$

where θ_0, θ_0', B_0, and M_0 are the rotation, warping, bimoment, and torque at the section $z = 0$.

In a high-rise core, the boundary conditions at the foundation level are $\theta_0 = 0$ and $\theta_0' = 0$. At the top, since there is no applied torque or bimoment, $M_l = 0$ and $B_l = 0$. Using these conditions in Eq. (10.76), we could write the expressions for the rotation θ_z and bimoment B_z at any section z. For the particular case of uniformly distributed twisting moment m_z, these expressions can be shown to be:

$$\theta_z = -\frac{m}{GJ\cosh k}\left[-\frac{l^2}{k^2} - \frac{l^2}{k}\sinh k + z\left(1 - \frac{z}{2}\right)\cosh k\right.$$
$$\left. + \frac{l^2}{k^2}\cosh\frac{k}{l}z + \frac{l^2}{k}\sinh\frac{k}{l}(l-z)\right] \qquad (10.77)$$

$$B_z = -\frac{ml^2}{k^2\cosh k}\left[\cosh k - \cosh\frac{k}{l}z - k\sin\frac{k}{l}(l-z)\right] \qquad (10.78)$$

Equations (10.77) and (10.78) are basic equations applicable to a wide variety of braced cores. These equations, together with the sectorial properties of the core, completely define the core rotations and axial stresses in the columns.

To show the order of the magnitude of axial forces induced in the columns, the results of torsion and bending analysis for the building shown in Fig. 10.49a are presented. A uniform wind load of 20 psf (97.64 kg/m^2) is assumed to act on the x face of the building for the full height. The diagram of principal sectorial coordinates and the sectorial moment of inertia for the core are given in Fig. 10.49d.

Conservatively, it will be assumed that the St Venant's torsional stiffness of the core is small compared to the warping stiffness. The parameter k is equal to zero, giving simple expressions for rotation and bimoment in the core. The bimoment B_0 at the base will be $ml^2/2$ giving a value of 5.5×10^6 kip/ft^2. The axial forces in the columns are calculated from the relation:

$$P_c = \frac{B_0\omega_c^2 A_c}{I_\omega} \qquad (10.79)$$

where A_c is the area of the column under consideration. The calculated values for the example problem are shown in Fig. 10.49e. To allow comparison, the axial forces induced in the columns due to bending of the core are shown in Fig. 10.49.

In practice, the unbraced face of a building core is not completely open; it is partially closed at each floor level by beams framing between the columns. The effect of the beams is to constrain the warping deformations of the core. At the same time, the beams themselves are subjected to large shear and bending moments in having to comply with the warping of the core as shown in Fig. 10.49g. One approach to the problem, which is explained later in this chapter, is to employ a stiffness method of analysis for the beam and core system by considering each story segment of the core as a thin-walled beam element and by adding the effect of beams by incorporating the warping stiffness of the beam into the total stiffness of the core. Herein a different method analogous to the continuous connection technique, which can be carried out without the aid of a computer, is presented. Briefly, the procedure consists of replacing the effect of individual beams by a continuous plate equivalent in mechanical properties to the connecting beams.

The structure is rendered statically determinate by introducing an imaginary cut along a line bisecting the imaginary plate. The equilibrium equation is written for the core in terms of external transverse forces and equivalent shear forces applied along the cut edges of the plate. The compatibility condition for the relative displacement at the cut section leads to the differential equation for the core beam assembly. The detailed steps are as follows.

The equilibrium equation for nonuniform torsion of a thin-walled beam subjected to the action of external transverse loads and longitudinal shear forces applied along the edges can be shown to be

$$EI_\omega \frac{d^4\theta}{dz^4} - GJ \frac{d^2\theta}{dz^2} = m + \frac{dT}{dz}\Omega \qquad (10.80)$$

where m = the external torsional moment at z
dT/dz = the derivative of the shear forces T_z at section z
Ω = twice the area of the enclosed contour between the core and the beam

The vertical displacement $u_{z,s}$ which results from warping of the thin-walled beam is expressed according to the equation

$$u_{z,s} = -\frac{d\theta}{dz}\omega_s \qquad (10.81)$$

where $d\theta/dz$ is the relative warping dependent upon the z coordinate and ω_s is the sectorial area which depends on the location of the point s on the contour.

The relative displacement at the cut section due to warping can be written thus

$$\delta_1 = -\frac{d\theta}{dz}(\omega_L - \omega_k)$$

or

$$\delta_1 = -\frac{\partial\theta}{\partial z}\Omega \qquad (10.82)$$

The relative displacement δ_2 at the cut due to the flexibility of the beam under the application of a shear force $T_z h$ is given by

$$\delta_2 = \frac{T_z h}{G}\left(\frac{\alpha^2 G}{12EI_b} + \frac{1.2}{A_b}\right) \qquad (10.83)$$

where h = the story height
α = the beam length
I_b = moment of inertia of beam about the y axis
A_b = the area of the beam
E and G = the familiar material properties of the beam
For compatibility of displacement, we should have $\delta_1 + \delta_2 = 0$; i.e.,

$$\frac{d\theta}{dz}\Omega + \frac{T_z h}{G}\left(\frac{\alpha^2 G}{12EI_b} + \frac{1.2}{A_b}\right) = 0 \qquad (10.84)$$

Differentiating with respect to z, we have

$$\frac{d^2\theta}{dz^2}\Omega + \frac{dT}{dz}\frac{h}{G}\left(\frac{\alpha^2 G}{12EI_b} + \frac{1.2}{A_b}\right) = 0 \qquad (10.85)$$

Substituting for dt/dz in the equilibrium equation and using the notation

$$J_b = \frac{\Omega}{\alpha h}\left\{\frac{\alpha^2 G}{12EI_b} + \frac{1.2}{A_b}\right\} \qquad (10.86)$$

we get

$$EI_\omega \frac{d^4\theta}{dz^4} - G(J + J_b)\frac{d^2\theta}{dz^2} = m_z \qquad (10.87)$$

This equation is identical to Eq. (10.75). Therefore, the two solutions given in Eqs. (10.75) and (10.78) for the rotation and bimoment of the open core can also be used for the solution of the core and beam assembly. For this purpose, it is only necessary to replace St Venant's torsional constant J by the sum $J + J_b$ as given by Eq. (10.87). The parameter k is now computed from the equation

$$k = l\sqrt{\frac{G(J - J_b)}{EI_\omega}} \qquad (10.88)$$

The calculation of bimoment and axial forces follows the procedure outlined earlier.

For purposes of illustration, it is assumed in the problem that a W16 × 26 beam is moment-connected across the corridor columns. The values for J_b and k are found to be respectively equal to 55.38 ft^4 (0.478 m^4) and 19.5. Substituting the value of k in Eq. (10.78), the bimoment at the base is found to be equal to 60 percent of the value obtained for the open core. The resulting values of the column loads are also 60 percent of the values shown in Fig. 10.49e.

10.4.12 Warping Torsion Constants for Open Sections

It is perhaps evident by now that although the concept of warping torsion is easy to assimilate, the calculation of sectorial properties are rather tedious. To alleviate this problem, formulas for the sectorial properties of open sections commonly used in shear wall structures, are given in Table 10.7.

Let us verify the value of I_ω derived previously in Example 2, by using the formula given in the table (ref. no. 7 in Table 10.7)

$$I_\omega = \frac{h^2 t_1 t_2 b_1^3 b_2^3}{12(t_1 b_1^3 + t_2 b_2^3)}$$

$$h = 30\ \text{ft}, \quad t_1 = t_2 = 2\ \text{ft}, \quad b_1 = 20\ \text{ft}, \quad b_2 = 10\ \text{ft}$$

$$I_\omega = \frac{30^2 \times 2 \times 2 \times 10^3 \times 20^3}{12(2 \times 10^3 + 2 \times 20^3)} = 133{,}336\ \text{ft}^6$$

This confirms the accuracy of the calculations done previously.

10.4.13 Computer Analysis

Modeling Techniques The classical warping torsion theory for open sections described in the previous sections is useful for the analysis of buildings with uniform single cores and for buildings with a random distribution

Table 10.7 Torsion Constants for Open Sections

Cross-section reference no.	Constants
1. Channel 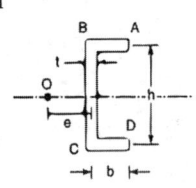	$e = \dfrac{3b^2}{h + 6b}$ $J = \dfrac{t^3}{3}(h + 2b)$ $I_\omega = \dfrac{h^2 b^2 t}{12}\dfrac{2h + 3b}{h + 6b}$
2. C section	$e = b\dfrac{3h^2 b + 6h^2 b_1 - 8b_1^3}{h^2 + 6h^2 b + 6h^2 b_1 + 8h_1^3 - 12hb_1^2}$ $J = \dfrac{t^3}{2}(h + 2b + 2b_1)$ $I_\omega = l\left[\dfrac{h^2 b^2}{2}\left(b_1 + \dfrac{b}{3} - e - \dfrac{2eb_1}{b} + \dfrac{2b_1^2}{k}\right) + \dfrac{h^2 e^2}{2}\left(b + b_1 + \dfrac{h}{6} - \dfrac{2b_1^2}{h}\right) + \dfrac{2b_1^3}{3}(b + e)^2\right]$
3. Hat section	$e = b\dfrac{3h^2 b + 6h^2 b_1 - 8b_1^2}{h^2 + 6h^2 b + 6h^2 b_1 + 8h_1^3 + 12hb_1^2}$ $J = \dfrac{t^3}{2}(h + 2b + 2b_1)$ $I_\omega = l\left[\dfrac{h^2 b^2}{2}\left(b_1 + \dfrac{b}{3} - e - \dfrac{2eb_1}{b} - \dfrac{2b_1^2}{h}\right) + \dfrac{h^2 c^2}{2}\left(b + b_1 + \dfrac{h}{6} + \dfrac{2b_1^2}{h}\right) + \dfrac{2b_1^2}{3}(b + e)^2\right]$
4. Twin channel with flanges inward 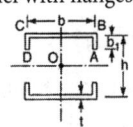	$J = \dfrac{t^2}{3}(2b + 4b_1)$ $I_\omega = \dfrac{tb^2}{24}(8b_1^2 + 6h^2 b_1 + h^2 b + 12b_1^2 h)$

(Continued)

Cross-section reference no.	Constants

5. Twin channel with flanges outward

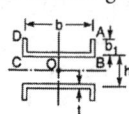

$$J = \frac{t^2}{3}(2b + 4b_1)$$

$$I_\omega = \frac{tb^2}{24}(8b_1^3 + 6h^2b_1 + h^2b - 12b_1^2h)$$

6. Wide flanged beam with equal flanges

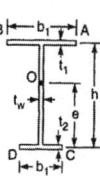

$$J = \tfrac{1}{3}(2t^3b + t_\omega^3h)$$

$$I_\omega = \frac{h^2tb^3}{24}$$

7. Wide flanged beam with unequal flanges

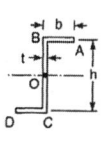

$$e = \frac{t_1b_1^2h}{t_1b_1^3 + t_2b_2^3}$$

$$J = \tfrac{1}{3}(t_1^3b_1 + t_2^3b_2 + t_\omega^3h)$$

$$I_\omega = \frac{h^2t_1t_2b_1^3b_2^3}{12(t_1b_1^3 + t_2b_2^3)}$$

8. Z section

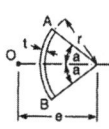

$$J = \frac{t^2}{3}(2b + h)$$

$$I_\omega = \frac{th^2b^3}{12}\left(\frac{b+h}{2b+h}\right)$$

9. Segment of a circular tube

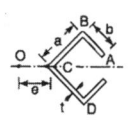

$$e = 2r\frac{\sin\alpha - \alpha\cos\alpha}{\alpha - \sin\alpha\cos\alpha}$$

$$J = \tfrac{2}{3}t^3r\alpha$$

$$I_\omega = \frac{2tr^5}{3}\left[\alpha^3 - 6\frac{(\sin\alpha - \alpha\cos\alpha)^2}{\alpha - \sin\alpha\cos\alpha}\right]$$

10.

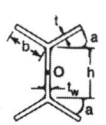

$$\text{For } \alpha = 45° \quad e = 1.06r$$
$$= 90° \quad e = 1.27r$$
$$= 180° \quad e = 2r$$
$$e = 0.707ab^2\frac{3a - 2b}{2a^3 - (a-b)^3}$$

11.

$$J = \tfrac{2}{3}t^3(a + b)$$

$$I_\omega = \frac{ta^4b^3}{6}\frac{3a + 2b}{2a^3 - (a-b)^2}$$

$$J = \tfrac{1}{3}(4t^3b + t_\omega^3a)$$

$$I_\omega = \frac{a^2b^2t}{3}\cos^2\alpha$$

of planar shear walls with uniform properties. It is also helpful in understanding warping behavior and in getting a feel for the magnitude of axial forces resulting from torsion. However, the properties of cores and shear walls usually vary with the height of buildings with the walls and cores often interconnected to other lateral load-resisting elements such as moment frames. Such complex assemblies cannot be analyzed by the classical theory and it is necessary to revert to a stiffness matrix computer analysis, which can consider the total structure as an assemblage of discrete elements.

A significant aspect of commercial computer programs most often used in engineering practice for modeling of shear walls is that they do not require any knowledge of warping theory, nor do they require the calculation of the sectorial properties. Building systems with shear walls in practice are analyzed by using special panel elements that combine the versatility of the finite element method with the design requirement that the wall output be in terms of total moments and forces, instead of the usual finite element output of direct stresses, shear stresses and principal stress values.

A simple cantilever wall as shown in Fig. 10.50 may be modeled either as an assemblage of panel elements between the floors, or as a single column placed at the center of gravity of the wall. The single column is given the properties representing the wall axial area, shear area, and moment of inertia. Rigid end zones on either side of the column are specified to capture the effects of finite dimensions of the wall on the stiffness of the system. In modeling the columns as panels, column lines C_1 and C_2 are used to define the extent of panels.

A discontinuous wall Fig. 10.51, in which the shear wall from above terminates on a series of columns below, requires four column lines C_1, C_2, C_3, and C_4, and three panels at each level.

When the width of a shear wall changes over the building height, as shown in Fig. 10.57, the model requires a column line corresponding to each width of wall. For example, in Fig. 10.52, seven columns lines are required to define the widths of the wall at various heights.

A wall with random openings may be modeled by using a combination of column lines and panel elements as shown in Fig. 10.53.

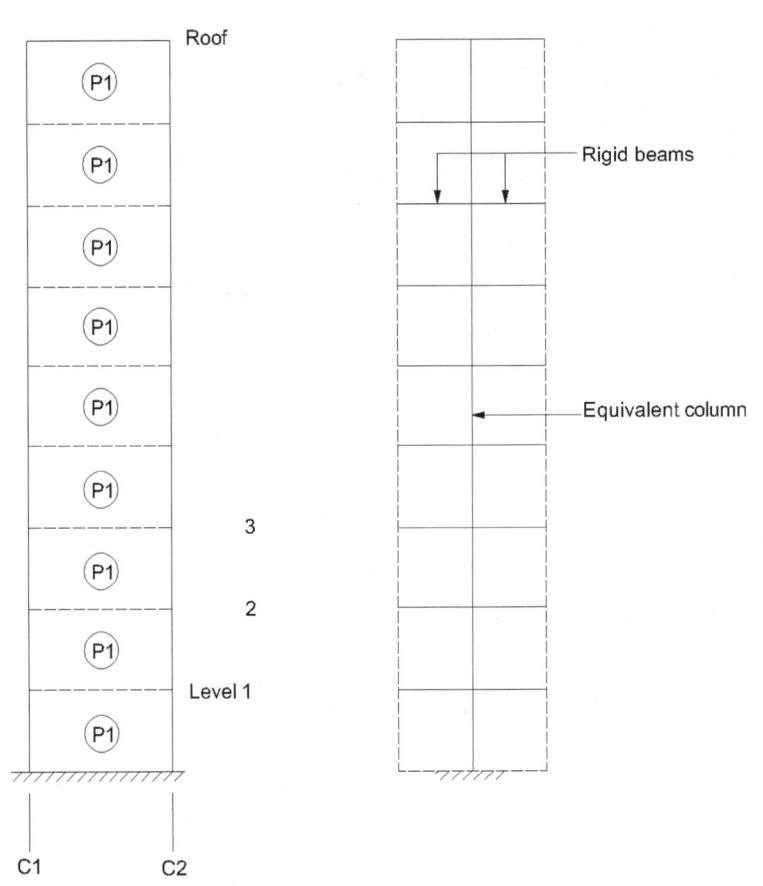

Figure 10.50 Modeling of simple shear walls: (a) panel elements; (b) equivalent column.

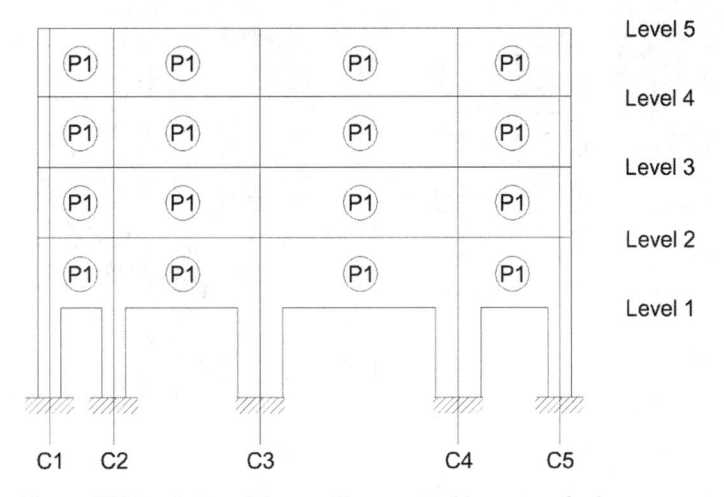

Figure 10.51 Series of shear walls supported by series of columns.

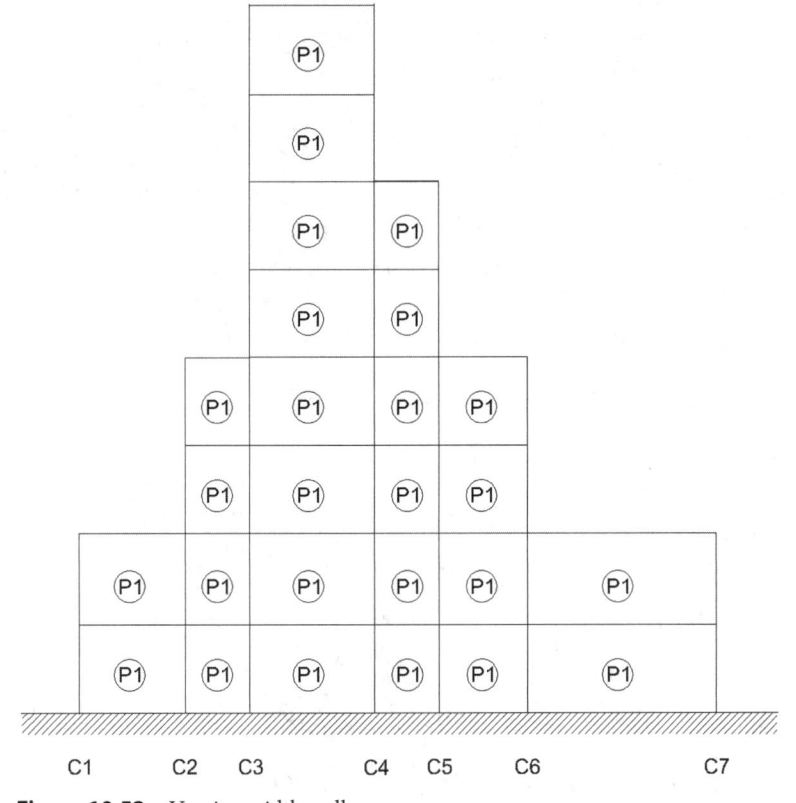

Figure 10.52 Varying width wall.

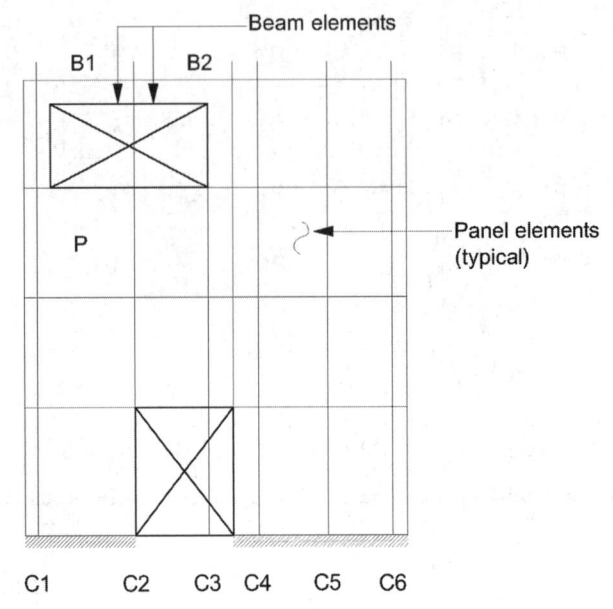

Figure 10.53   Shear wall with random openings.

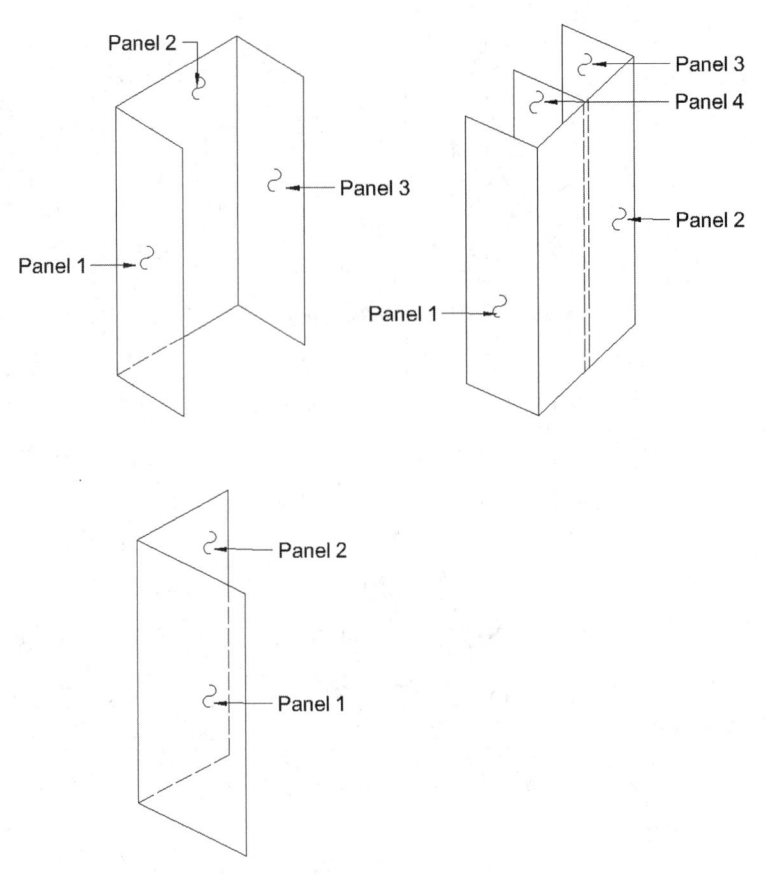

Figure 10.54   Modeling of three-dimensional shear walls.

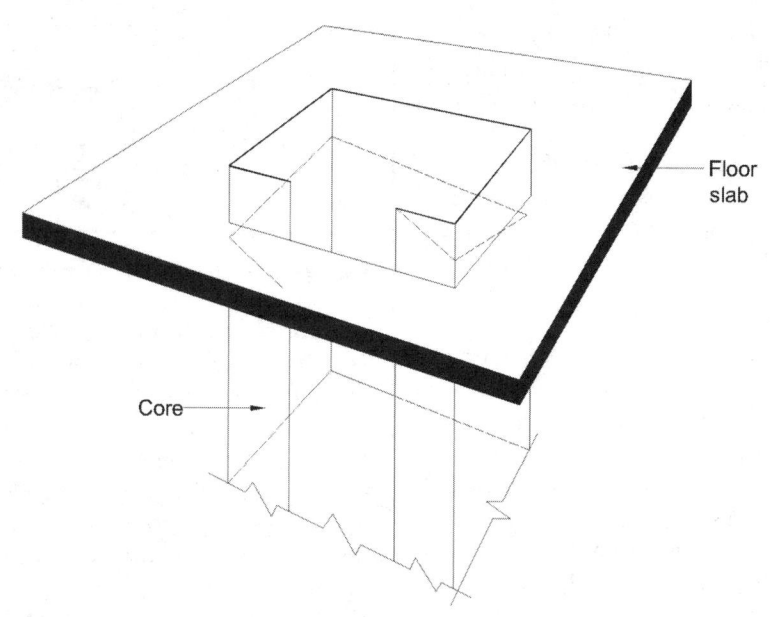

Figure 10.55   Warping deformation of core.

Modeling of three-dimensional shear walls such as C, E, and L-shapes is shown in Fig. 10.54 in which more than one panel element is used at each level.

Single column model

As an alternative to defining three-dimensional walls with panel elements, it is possible to use a single column model with an extra, seventh degree-of-freedom per node, to represent warping. The principal advantage of the single warping-column model is its extremely concise form of representing warping. In this formulation, the seventh, warping, degree of freedom is taken to be $d\theta/dz$ to express the magnitude of warping, while B, the bimoment, is taken as the corresponding generalized force. In other words, just as we associate the rotations θ_x and θ_y with their corresponding moments M_z and M_y, the rate of change of torsional rotation $\dfrac{d\theta}{dz}$ is associated with the bimoment B.

As part of his investigation into the torsional behavior of open section shear walls (ref. 20), the author developed a 14×14 stiffness matrix for defining the three-dimensional behavior of open sections. The purpose was to investigate the behavior of single and multiple open cores interconnected through rigid floor diaphragms. The investigation included the study of warping behavior of open sections as well as the "warping restraint" provided by the out-of-plane bending and twisting of the floor slab. A brief description of the methodology is given in the following sections.

Warping Stiffness of Floor Slab   A floor slab surrounding core may be considered to have two distinct actions. One is to hold the cross-sectional shape of the core intact, by preventing its in-plane distortion, and the other is to restrain its longitudinal deformation due to warping. The in-plane action of the slab is tacitly taken into consideration in the torsion theory by assuming the profile of the core to be rigid. The out-of-plane to be rigid. The out-of-plane stiffness of the slab, the warping stiffness, is the subject of this discussion.

Figure 10.55 shows the warping deformation of the cone while Fig. 10.56 shows the corresponding deformation of the slab. Observe that the slab is under "torture" due to the out-of-plane bending and twisting indicating the likelihood of very large interacting forces. The warping of the cross-section which is the longitudinal displacement of the open core is given by the relation:

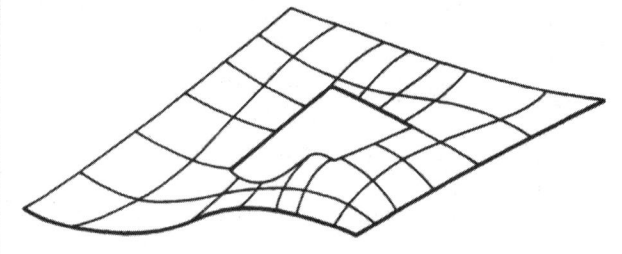

Figure 10.56   Warping deformation of floor slab.

$$u_{z,s} = -\theta_z' \omega_s$$

The cross-sectional distortion θ_z' can be considered equivalent to a generalized displacement similar to, say, the rotation of the core about the x-axis. Just as the axial displacement due to θ_x varies across the cross-section as a function of the coordinate y, similarly the warping displacement varies across the profile in a manner similar to the warping coordinate.

Thus the distortion of the cross-section θ'_z can be considered as a generalized displacement. A unit value of θ'_z, that is, a "unit warping displacement" introduces up-and-down longitudinal displacements which vary across the cross-section in a manner similar to the warping coordinate.

When the core undergoes a warping deformation, the floor slab rigidly connected to the core is forced to bend and twist. The out-of-plane displacements of the slab and the warping displacement of the core must be the same along the boundary common to the core and the slab.

The displacement of the slab at its common boundary with the wall is, therefore, known from the warping displacement of the core. For a unit warping displacement of the core, the slab is displaced along the contact boundary in a manner similar to the warping coordinate diagram ω_s of the core. This displacement gives rise to continuous interactive forces consisting of distributed axial forces and moments at the inner edges of the slab. The evaluation of the warping stiffness of the slab, therefore, reduces to the determination of these interactive forces and moments which can be mathematically converted into a bimoment function, as will be seen shortly.

To study the slab-core interaction, consider the slab isolated from the core structure and subjected to an as yet unknown distribution of reactive forces and moments corresponding to the warping deformation of the core.

The resulting force system, which consists of concentrated forces $P_1, P_2, P_3,\ldots, P_n$, and moments $M_1, M_2, M_3,\ldots,$ M_n applied at the points $k = 1, 2, 3,\ldots, n$ of the cross-section, can be expressed as a bimoment by the relation

$$B_\omega = \sum_{k=1}^{n} P_k \omega_k + \sum_{k=1}^{n} M_k \frac{\partial \omega_k}{\partial s} \qquad (10.89)$$

where B_ω = the warping stiffness of floor slab

$\displaystyle\sum_{k=1}^{n} P_k \omega_k$ = the summation of the product of concen-

trated forces and the warping displacement

$\displaystyle\sum_{k=1}^{m} M_k \frac{\partial \omega_k}{\partial s}$ = the summation of the concentrated transverse moments and the rate of change of warping function The forces and moments are diagrammatically shown in Fig. 10.57. These are evaluated by using a finite element analysis.

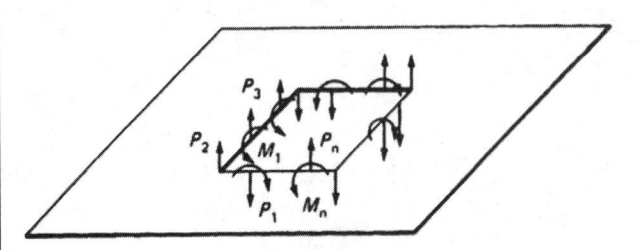

Figure 10.57   Forces and moments due to warping to floor slab.

Finite Element Analysis   A typical finite element idealization for the floor slab is shown in Fig. 10.58. To impose a unit θ'_z displacement at the inner boundary of the slab transverse displacements perpendicular to the plane of the slab and equal in magnitude and sense to the warping function are introduced at the nodes common to the core and the slab e.g., nodes $N_1, N_2,\ldots, N_n$ in Fig. 10.58. In addition to these vertical displacements, the slopes $\delta\omega/\delta x$ and $\delta\omega/\delta y$ are made equal in magnitude and sense to the slope of the ω_s diagram. Having thus given at the inner edge of the slab a displacement conforming to the warped outline of the core, the forces and transverse moments at the nodes are found from a finite element solution. The bimoment, which then corresponds to the required warping stiffness of the slab, is found from the relation given in Eq. (10.69). The effect of the floor system, which is mathematically equivalent to the bimoment, is incorporated into a stiffness type of analysis by adding the bimoment to the appropriate elements of the stiffness matrix of the core. A prerequisite for this operation is the derivation of stiffness coefficients for the open section. This is considered next.

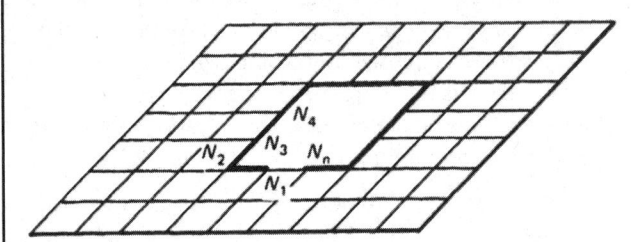

Figure 10.58   Typical finite element idealization of slab.

Twisting and Warping Stiffness of Open Sections   The stiffness matrix for the open section is derived by solving the governing differential equation of torsion by imposing appropriate boundary conditions at each end. Defining a member as the segment of the core between two floors, the stiffness matrix for the restrained member is established first. Elements of this matrix are the values of restraint exerted at the ends of the member when unit displacements are imposed one at a time at each end. Concentrating on the torsion analysis only, the number of degrees of freedom at each end will be two, the rotation θ_z and the warping θ_z'. Hence, the order of the restrained member stiffness matrix will be 4 by 4. Equation (10.49) is conveniently used for deriving the elements of the member stiffness matrix. This is reproduced here for convenience as Eq. (10.70).

$$\begin{bmatrix} \theta_z \\ \theta_z' \\ \dfrac{B_z}{GJ} \\ \dfrac{M_z}{GJ} \end{bmatrix} = \begin{bmatrix} 1 & \dfrac{l}{k}\sinh\dfrac{kz}{l} & 1-\cosh\dfrac{k}{l}z & z-\dfrac{l}{k}\sinh\dfrac{kz}{l} \\ 0 & \cosh\dfrac{kz}{l} & -\dfrac{k}{l}\sinh\dfrac{kz}{l} & 1-\cosh\dfrac{kz}{l} \\ 0 & -\dfrac{l}{k}\sinh\dfrac{kz}{l} & \cosh\dfrac{kz}{l} & \dfrac{l}{k}\sinh\dfrac{kz}{l} \\ 0 & 0 & 0 & 1 \end{bmatrix} \begin{bmatrix} \theta_0 \\ \theta_0' \\ \dfrac{B_0}{GJ} \\ \dfrac{M_0}{GJ} \end{bmatrix}$$

$$(10.90)$$

Now, to find the elements of the first row of the stiffness matrix, which correspond to the torque and bimoment at each end of the beam required to produce a unit rotation of $\theta = 1$ at $z = 0$ while all the other three displacements are zero, it is necessary to introduce the appropriate boundary conditions in Eq. (10.70). Thus

$$\begin{bmatrix} 0 \\ 0 \\ \dfrac{B_l}{GJ} \\ \dfrac{M_l}{GJ} \end{bmatrix} = \begin{bmatrix} 1 & \dfrac{l}{k}\sinh\dfrac{kz}{l} & 1-\cosh\dfrac{k}{l}z & z-\dfrac{l}{k}\sinh\dfrac{kz}{l} \\ 0 & \cosh\dfrac{kz}{l} & -\dfrac{k}{l}\sinh\dfrac{kz}{l} & 1-\cosh\dfrac{kz}{l} \\ 0 & -\dfrac{l}{k}\sinh\dfrac{kz}{l} & \cosh\dfrac{kz}{l} & \dfrac{l}{k}\sinh\dfrac{kz}{l} \\ 0 & 0 & 0 & 1 \end{bmatrix} \begin{bmatrix} 1 \\ 0 \\ \dfrac{B_0}{GJ} \\ \dfrac{M_0}{GJ} \end{bmatrix}$$

$$(10.91)$$

Solving these equations, the four forces M_0, B_0, M_l, and B_l at the two ends are obtained. In the same manner, the remaining elements of the stiffness matrix are obtained by introducing the appropriate boundary conditions. These are shown diagrammatically in Fig. 10.59. Adapting the sign convention shown therein, the member stiffness matrix for the thin-walled beam subjected to torsion will take the form,

$$\left[\dfrac{GJ}{2+k\sinh k - 2\cosh k}\right]$$

$$\times \begin{bmatrix} (1-\cosh k)\dfrac{l}{k}(k\cosh k-\sin k) & -(1-\cosh k)\dfrac{l}{k}(\sinh k-k) \\ \dfrac{-k}{l}\sinh k-(1-\cosh k) & \dfrac{k}{l}\sinh k & -(l-\cosh k) \\ (1-\cosh k)\dfrac{l}{k}(\sinh k-k) & -(1-\cosh h)\ \dfrac{l}{k}(k\cosh k-\sinh k) \end{bmatrix}$$

$$(10.92)$$

To the best of the author's knowledge, he was the first to derive θ, θ', B, and M in the form of stiffness coefficients. Such a matrix is very convenient for including the effect of beams and slabs, and can be extended to a generalized stiffness analysis by treating open cores as special columns with seven degrees of freedom at each end.

When the torsional rigidity GJ due to St Venant torsion becomes zero, the differential equation of constrained torsion will be analogous to the equation of bending of a beam. The elements of the $[S]$ matrix (Eq. (10.72)) can be written directly from the beam stiffness matrix if M, B, θ, θ', and I_ω are replaced by P, M, w, w', and I_{xx}, where P is the shear force, M is the bending moment, ω is the displacement, w' is the slope dw/dz, and I_{xx} is the moment of inertia. Further, this analogy can be used to check the elements of the matrix $[S]$ by writing the expanding forms for $\sinh k$ and $\cosh k$ in Eq. (10.72) and taking $k \to 0$ in the limit. For example, considering one element S_{11} of Eq. (10.72) we get

$$S_{11} = \dfrac{GJk}{l(2+k\sinh k-2\cosh k)}\sinh k$$

when $k \leftarrow 0$,

$$S_{11} = \dfrac{EI_\omega k^2\left(\dfrac{k}{l}\right)\left(k+\dfrac{k^3}{6}+\dfrac{k^5}{120}\right)}{l^2\left[2+h\left(k+\dfrac{k^3}{6}+\dfrac{k^5}{120}\right)-2\left(1+\dfrac{k^2}{2}+\dfrac{k^4}{24}\right)\right]}$$

$$= \dfrac{12EI_\omega}{l^3}$$

which is of the same form as the element corresponding to the shear force in the well-known beam stiffness matrix.

Stiffness Method Using Warping-Column Model Building structures are generally analyzed as three-dimensional frames, with the members oriented in any direction and subjected to axial force, shear and moment in two orthogonal planes, and torsion about their linear axes. Therefore, a general beam or column element in the analysis of three-dimensional frames, must include forces in three directions and moments about three axes. Such a beam element with six displacements at each end is shown in Fig. 10.60. The stiffness matrix which is the relationship between the end forces and displacements is a 12×12 matrix, corresponding to six degrees of freedom at each end. The stiffness coefficients depicting the force-displacement relation for the three-dimensional beam element is found by combining the stiffness terms

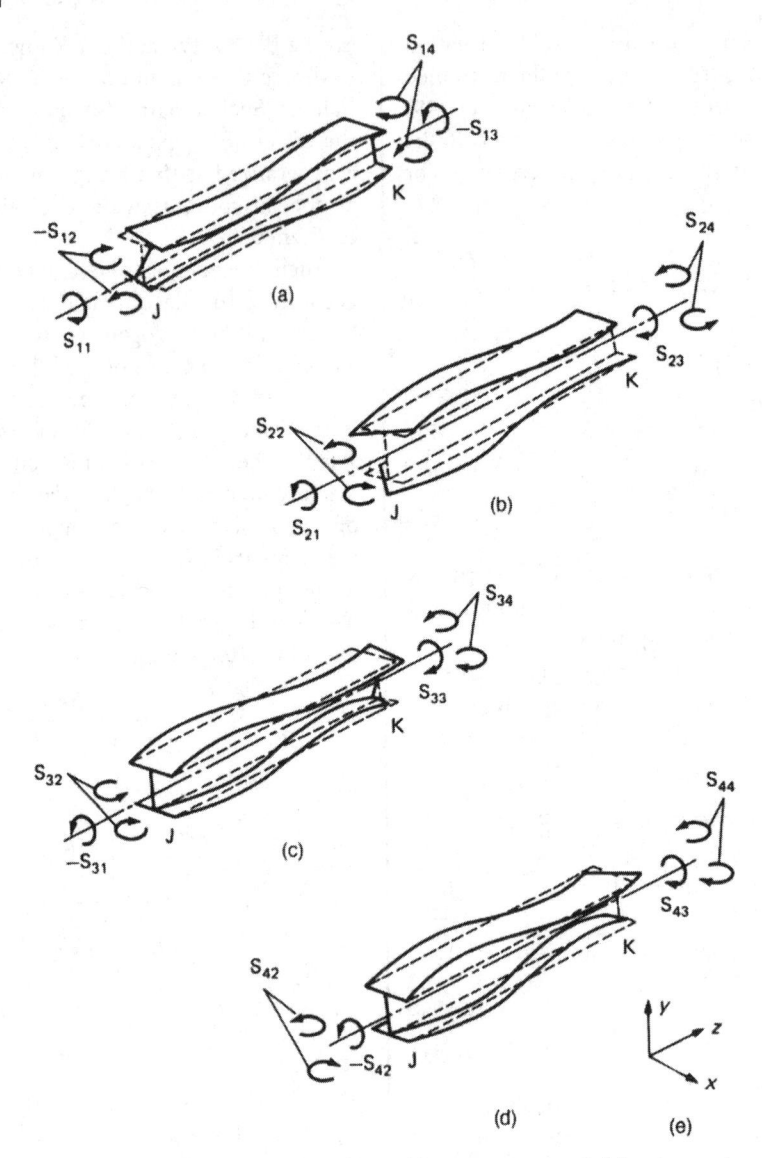

Figure 10.59 Stiffness coefficients for thin-walled open sections: (a) unit rotation at J; (b) unit warping displacement at J; (c) unit rotation at K; (d) unit warping displacement at K; (e) coordinate axes.

for axial deformation, bending about two axes, and torsion. The resulting 12×12 stiffness matrix is given in Fig. 10.60.

A non-planar shear wall such as an I- or a C-shaped wall is modeled in a three-dimensional analysis as an assemblage of floor-to-floor panel elements connected along their edges. The continuous connection between the panels provides for the principal interaction, the vertical shear, along their connecting edges.

As an alternative technique, a three-dimensional wall may be represented in all its aspects of behavior including warping, by a warping column element, with seven degrees of freedom at each floor level. Its assigned properties would include the warping moment of inertia I_ω, in addition to the familiar area A, to represent its resistance to axial load, inertias I_x and I_y to represent its resistance to bending about its principal axes, and J to represent its St Venant's resistance to torsion. Such a

		u	v	w	θ_x	θ_y	θ_z	u	v	w	θ_x	θ_y	θ_z
		1	2	3	4	5	6	7	8	9	10	11	12
u	1	$\frac{12EI_y}{L^3}$	S										
v	2	0	$\frac{12EI_x}{L^3}$	Y									
w	3	0	0	$\frac{EA}{L}$	M								
	4	0	$\frac{6EI_x}{L^2}$	0	$\frac{4EI_x}{L}$	M							
	5	$\frac{-6EI_y}{L^2}$	0	0	0	$\frac{4EI_y}{L}$	E						
	6	0	0	0	0	0	$\frac{GJ}{L}$	T					
u	7	$\frac{-12EI_y}{L^3}$	0	0	0	$\frac{6EI_y}{L^2}$	0	$\frac{12EI_y}{L^3}$	R				
v	8	0	$\frac{-12EI_x}{L^3}$	0	$\frac{-6EI_x}{L^2}$	0	0	0	$\frac{12EI_x}{L^3}$	I			
w	9	0	0	$\frac{-EA}{L}$	0	0	0	0	0	$\frac{EA}{L}$	C		
θ_x	10	0	$\frac{-6EI_x}{L^2}$	0	$\frac{2EI_x}{L}$	0	0	0	$\frac{-6EI_x}{L^2}$	0	$\frac{4EI_x}{L}$	A	
θ_y	11	$\frac{-6EI_y}{L^2}$	0	0	0	$\frac{2EI_y}{L}$	0	$\frac{6EI_y}{L^2}$	0	0	0	$\frac{4EI_y}{L}$	L
θ_z	12	0	0	0	0	0	$\frac{-GJ}{L}$	0	0	0	0	0	$\frac{GJ}{L}$

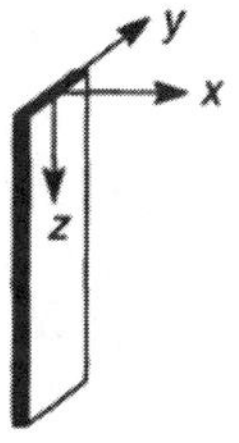

Figure 10.60   (a) Twelve-by-twelve stiffness matrix for prismatic three-dimensional element; (b) coordinate axes; (c) positive sign convention.

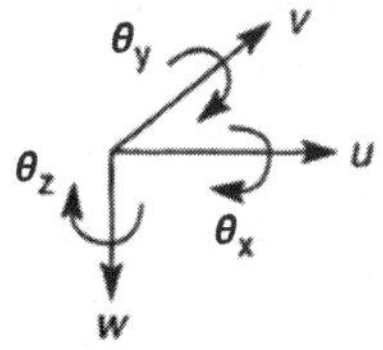

single-column model, with an extra, seventh degree of freedom, as developed by the author in (ref. 20), is particularly suitable for open-section walls that are uniform over the height. The seventh, warping, degree of freedom is the parameter $d\theta/dz$ which expresses the magnitude of warping. It is used as the warping degree of freedom, while B, the bimoment, becomes the corresponding generalized force. Thus, with seven degrees of freedom per node, the column element (Fig. 10.61) has a 14×14 stiffness matrix. A number of such story-height elements may be stacked vertically to represent a complete core. The interaction of slabs and beams at the floor levels may also be included in the stiffness matrix of the total structure by an appropriate combination of floor stiffness matrix with the stiffness matrix of the core, as given in the author's first publication (ref. 20). Engineers engaged in developing special-purpose computer programs may find this reference useful for including the additional warping degree of freedom for open-section shear wall buildings.

		u	v	w	θ_x	θ_y	θ_z	θ_z'	u	v	w	θ_x	θ_y	θ_z	θ_z'
		1	2	3	4	5	6	7	8	9	10	11	12	13	14
u	1	$\dfrac{12EI_y}{L^3}$													
v	2	0	$\dfrac{12EI_x}{L^3}$												
w	3	0	0	$\dfrac{EA}{L}$											
θ_x	4	0	$\dfrac{6EI_x}{L^2}$	0	$\dfrac{4EI_x}{L}$										
θ_y	5	$\dfrac{-6EI_y}{L^2}$	0	0	0	$\dfrac{4EI_y}{L}$									
θ_z	6	0	0	0	0	0	S_{11}								
θ_z'	7	0	0	0	0	0	S_{12}	S_{22}							
u	8	$\dfrac{-12EI_y}{L^3}$	0	0	0	$\dfrac{6EI_y}{L^2}$	0	0	$\dfrac{12EI_y}{L^3}$						
v	9	0	$\dfrac{-12EI_x}{L^3}$	0	$\dfrac{-6EI_x}{L^2}$	0	0	0	0	$\dfrac{12EI_x}{L^3}$					
w	10	0	0	$-\dfrac{EA}{L}$	0	0	0	0	0	0	$\dfrac{EA}{L}$				
θ_x	11	0	$\dfrac{6EI_x}{L^2}$	0	$\dfrac{2EI_x}{L}$	0	0	0	0	$\dfrac{-6EI_x}{L^2}$	0	$\dfrac{4EI_x}{L}$			
θ_y	12	$\dfrac{-6EI_y}{L^2}$	0	0	0	$\dfrac{2EI_y}{L}$	0	0	$\dfrac{6EI_y}{L^2}$	0	0	0	$\dfrac{4EI_y}{L}$		
θ_z	13	0	0	0	0	0	$-S_{11}$	$-S_{12}$	0	0	0	0	0	S_{11}	
θ_z'	14	0	0	0	0	0	S_{12}	S_{24}	0	0	0	0	0	$-S_{12}$	S_{22}

(Diagonal letters spell: S Y M M E T R I C A L)

(b) SC — coordinate axes

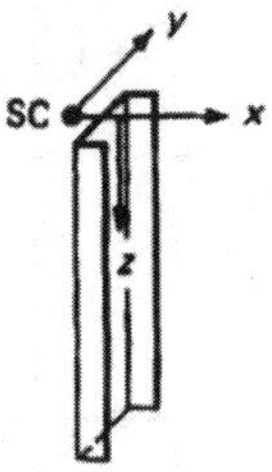

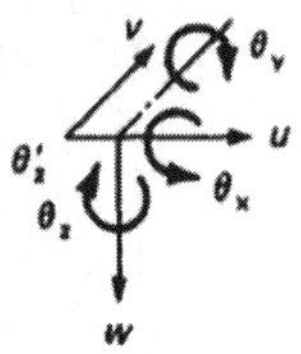

(c) positive sign convention

Torque bimoment relation as between P_x and θ_y.

$$S_{11} = \alpha \frac{k}{\ell}\sinh k \qquad S_{24} = \alpha \frac{\ell}{k}(\sinh k - k)$$

$$S_{12} = \alpha(1 - \cosh k)$$

$$S_{22} = \alpha(k \cosh k - \sinh k)$$

where

$$\alpha = \frac{GJ}{(2 + k \sinh k - 2 \cosh k)}$$

Figure 10.61   (a) Fourteen-by-fourteen stiffness matrix for thin-walled open section; (b) coordinate axes; (c) positive sign convention.

Chapter 11

Performance-Based Design

11.1 INTRODUCTION

Performance-based seismic design (PBSD) also known as PBD represents a shift, in engineering design methods building upon insights gained from studying how structures perform in earthquakes. This approach benefits from improvements in tools and computational resources. The evolution of PBSD principles is the result of collaboration, among experts and the financial support provided by the funded National Earthquake Hazards Reduction Program.

It is interesting to note that the primary use of PBSD initially centered around seismic rehabilitation of existing structures. These concepts were found to be just as effective, in the realm of new construction projects, which eventually led to their integration into building codes, for designing seismic-resistant new buildings.

PBSD intends to support code-based methods not to replace them. Although it is well developed for seismic design, conventional prescriptive methods are primarily used for addressing hazards such, as strong winds. An inclusive design approach that considers multiple hazards should promote the use of a performance-based approach to reduce risks linked to winds and flooding.

Building codes provide the minimum safety requirements by setting out prescriptive criteria. These guidelines determine the types of materials that can be used for approved nonstructural components, minimum strength and stiffness requirements, and construction details. Although these rules aim to guarantee that buildings meet performance criteria, they do not evaluate the performance of each design under this conventional method. As a result, structures constructed according to these guidelines may fail to meet the expected minimum standards in terms of performance.

PBSD assesses how a building would perform in situations considering uncertainties, in both hazard assessment and building response. This method helps in designing buildings or upgrading existing ones with a view of potential risks such as disruptions to occupancy and economic losses from future earthquakes. Moreover, PBSD establishes a language to enable conversations between stakeholders and design professionals on design options and decisions. It also sets the groundwork, for determining levels of safety and property protection that cater to the requirements of each project taking into account the viewpoints of different decision-makers.

In comparison with the prescriptive design approach, PBSD provides an approach to assess how well buildings, systems, and components perform. It allows for evaluating different designs to see if they can deliver results at a lower cost or determine if more robust performance is necessary, for critical facilities.

The PBSD procedures introduce the concept of performance based on objectives such as collapse prevention, life safety, immediate occupancy, and operational performance. They also introduced the concept of damage-related performance for both structural and non-structural components. While these procedures were initially intended for new buildings, they were first developed for seismic vulnerability studies and retrofit design of existing buildings.

11.2   DEFINING PERFORMANCE-BASED DESIGN

Performance-based design (PBD) is a concept that is constantly changing. The term, as it is currently defined has meanings, three of which are outlined below:

1. A design strategy that aligns with the safety requirements and performance standards of building codes while offering designers and building authorities a structured approach to assess different design possibilities currently outlined in codes. In this context, PBD encourages creativity and simplifies the process for designers to suggest building systems that are not addressed by code regulations.

2. A method that determines and chooses a performance objective, from performance options. These performance options are intended to meet design requirements that exceed the standards.

3. A design methodology that offers designers with tools and resources to meet performance objectives ensuring the prediction of a structure's performance.

11.3   PRESCRIPTIVE CODE APPROACH

The conventional methods used in building codes in the United States have been prescriptive-based codes. These codes are quantitative that depend on values provided by the codes to ensure reasonable safety levels against risks, like earthquakes. The requirements are formulated based on building and occupancy categories and are usually expressed using set values.

Prescriptive codes offer guidelines for addressing design and construction issues. Structures constructed following these codes are assumed to be safe. It is important to indicate that the standards in prescriptive-based codes represent the minimum requirements essential for protecting public health, safety, and overall, well-being. There are cases where it might be advantageous, suitable, or even essential to enhance safety measures beyond the mandated minimum thresholds.

Under this approach, all buildings assigned to a certain Seismic Design Category (SDC) are treated similarly. However, this method overlooks the building characteristics and systems needed to satisfy certain performance objectives beyond safety especially in upscale properties. How can we tackle these issues and others effectively? One emerging and progressive solution is the adoption of a performance-based approach to enhance or complement the existing code requirements.

11.4   PERFORMANCE-BASED DESIGN APPROACH

While detailed performance requirements are an addition, to the building codes in the United States, the concept is not entirely new. The different prescriptive life-safety codes include provisions for methods and materials or equivalencies. These code provisions allow for the use of methods, equipment, or materials that are not specifically mentioned in the code as long as they are approved by the code official. It is within these provisions that the traditional codes can be approached using the PBD method.

Under the concept of exploring approaches, materials, or equivalents the code official must grant approval, for the alternative or equivalent if it can be demonstrated to be on par in terms of quality, strength, durability, and safety. The professionals proposing the method or equivalence holds the responsibility of supplying all documentation to the code official. Given the code officials capacity to authorize methods and materials within existing codes performance-based codes serve as a framework for considering alternative designs based on performance. Essentially this process is not unfamiliar to the code official; it merely presents an approach, to evaluating and reviewing designs.

In the realm of building design considering performance has always been a practice. However, PBD offers an approach that is especially relevant when it comes to ensuring life-safety and mitigating risks from various hazards. For designers following performance-based codes offers a way to explore, record, and propose different materials, techniques, and equivalencies.

PBD focuses on the aspects and needs of each building than relying solely on a set code. Stakeholders, such as members of the design team builders, owners, and code officials who have an interest in the project's success are involved. The design team is a subset of stakeholders that consists of individuals, like architects representatives and relevant consultants.

It is crucial for all stakeholders to actively participate in the development, approval, and implementation of any PBD. Since stakeholders define the acceptable level of risk, their involvement from the early project stages is vital. The performance-based approach serves as the foundation for selecting design options for a project that needs to enhance occupancy requirements. This method facilitates a comparison of performance objectives across design alternatives as well as determining the stakeholder's acceptable performance levels and associated costs. PBD focuses on integrating property protection and life-safety strategies within systems than designing them in isolation.

11.4.1   Performance-Based Design for Natural Hazards

As indicated above, the performance-based design approach in building design is not new. The innovation lies in formalizing the decision-making process linked to anticipate performance, and ultimately establishing performance-based codes to govern building design and construction.

In the field of natural disasters, performance is used to indicate the extent of damage or stress. This in itself marks a shift in perspective because building owners or occupants typically assume that following building codes ensures an environment and expected levels of damage are not usually discussed between an architect and an owner or even an architect and their engineer. Recent earthquake incidents have highlighted the realization that damage, which could be severe, can occur in a building constructed according to prescriptive codes.

The study and application of PBD are currently most advanced, in the field of design. Experienced seismic engineering professionals have long acknowledged performance objectives that cater to the needs of property owners using them as a foundation, for setting design criteria. These objectives, also known as performance levels, can be described as follows:

Level 1: The building is mostly undamaged and intact and can be operational immediately (operational level).

Level 2: The building has some damage and requires repairs, but can still be occupied and functional after minor repairs of nonstructural components are done (immediate occupancy level).

Level 3: The building has both structural and nonstructural damage. Threat to life is low and any injuries from occupancy should be minor and rare (life-safety level).

Level 4: The building is severely damaged and may need to be demolished; not collapsed yet. There is a chance of injuries if occupied (near collapse level).

In this context, the building that complies with the codes is positioned low on the scale (at level 3) and many property owners, both private and public are willing to invest more to achieve a higher level of performance. For a hospital reaching at level 2 is essential with level 1 being the better target. Similarly, a high-tech manufacturing facility might aim for the same level due to the value of its assets and potential financial losses incurred from production shutdowns. On the other hand, a warehouse housing replaceable commercial inventory, with minimal occupants may choose to prioritize cost effectiveness at level 4.

In the last two decades or so, there has been a shift toward formalizing the PBSD approach. This has involved defining the performance levels in a manner. Through observations of damaged buildings advancements in materials science, experimental research, and analytical methods, engineers now possess a sophisticated understanding of how buildings respond to seismic activity. This enhanced knowledge enables engineers to predict with reliability how a structure will behave when subjected to different levels of shaking. While these predictions are not guaranteed, they are rooted in engineering principles that were nonexistent a few decades ago. Concurrently, comprehensive studies on aspects of PBSD are being conducted nationwide with a focus on California.

PBD is not suggested as a replacement for prescriptive building codes. Instead, it is viewed as an opportunity for improvement and customization of the design to align with the goals of building owners. Adhering to the code still stands as the requirement to guarantee safety.

Achieving a building code that focuses on performance than inspected design and construction methods is challenging. However, it is anticipated to see a rational balance between performance and prescriptive standards in the framework. This transition occurred in other industries such as the aerospace industry, like airplane design few decades ago, where airplanes are now consistently designed to meet performance criteria set by the military or airline companies.

Designers and building owners, in flood and earthquake-prone areas should consider a key objective:

• Can the real probabilities and frequencies of events be defined accurately the events during the building's lifespan?

• What level and types of damage can be tolerated, if any?

• Are there methods to achieve this acceptable standard, if any?

• What are the costs associated with performance levels over the building's lifetime?

• Do these performance levels meet exceed or fall below the required design codes?

Careful consideration of these questions by all parties marks the initial phase in designing, for optimal performance.

11.4.2 Performance-Based Seismic Design

In discussions, the methods for applying PBSD have been extensively developed. Nevertheless, these methods continue to progress, with research and development focusing on aspects, like terminology, analytical techniques, and ensuring accurate performance prediction. This section provides an overview of the approaches currently used in PBSD.

11.4.2.1   DETERMINING ACCEPTABLE RISK

The process for determining risk in building design begins with defining what level of performance can be achieved and discussing this with the owners' representatives. It is important to understand the damage and loss that may occur during earthquakes of varying magnitudes. The focus is on reaching an understanding that complete earthquake resistance's not feasible and compromises need to be made between seismic performance and costs. Acceptable risk involves considering the extent and types of damage and loss that building owners are willing to accept prioritizing the prevention of casualties while also addressing issues such as damage, nonstructural components, systems, and contents.

The conversation about the level of risk starts by figuring out the following: If a building is constructed based solely on the minimum code standards would the potential damage and loss in a design-level earthquake be considered acceptable? If the response to this query is yes, it implies a threshold of risk has been established allowing for continued design work. However, if the answer is no or uncertain further considerations should be taken into account:

• To what extent and types of damage can be considered acceptable?

• What are the consequences for long-term expenses and advantages throughout the building's lifespan?

• Can the desired level of performance be afforded within the expenses for the owners, ensuring minimum code requirements are always met?

It is important to clarify any uncertainties. The level of uncertainty in predicting performance depends on the design and conforming to building codes. While the design team can control this aspect for new buildings for a building retrofitting may present challenges due to existing building features that may not be ideal.

A new design that incorporates elements of seismic design such as a continuous load path, structural redundancy, plan and section symmetry, short spans and well-designed nonstructural connections, and bracing is likely to be more cost-effective and reliable in terms of performance compared to a design lacking these features.

The discussion of these matters should lead to a conclusion on performance objectives that will then be used as a target for the designers. However, it is ultimately the responsibility of the owner's representative to make the decision on performance objectives. It is crucial to grasp the implications of this decision. It falls upon the design team to provide all necessary information within their means.

Traditionally, architects have been the source of design information for building owners. With the complexity of PBSD, structural engineers are increasingly being consulted in matters. For large projects, key consultants like peer reviewers may participate in discussions in the early stages. In these cases, peer reviewers may be expected to engage in discussions with the design team in equal terms. The familiarity of all parties with PBSD terminology can significantly influence how effectively seismic performance issues are discussed and resolved.

When community representatives or committees with expertise are involved, the design team should make sure to clarify the issues. Discussing risk might be a concept for many building owners and understanding seismic performance language may also pose a challenge. Traditionally, starting a conversation about damage tolerance for a project has not been practice. Some owners may view it as a sign during meetings. However, reaching an agreement on design objectives and expectations can lead to achieving the desired performance level and prevent earthquake damage surprises in the future. Statements outlining performance objectives could be included in a project's building program and could serve as the basis for a PBD procedure.

ASCE/SEI 41-23 and older versions include tables that illustrate anticipated damage to elements such as architectural, mechanical, electrical, and plumbing components. These expectations apply to buildings designed using tools that offer the necessary analysis methods and detailing approaches to meet these performance standards across high, moderate, and low earthquake intensity regions.

11.4.3 Expected Performance of Buildings Designed to Current Seismic Codes

Current seismic design codes primarily focus on life safety and ensuring community safety. The suggested provisions set expectations without offering guarantees recognizing the possibility of building damage due to earthquakes. A general outline of performance expectations includes the following:

Structures designed according to guidelines should generally:

• Withstand minor earthquake ground motion without damage.

• Withstand moderate earthquake ground motion without structural damage though there may be nonstructural damage.

• Withstand significant earthquake ground motion equivalent to the strongest recorded or predicted intensity for the site without collapsing, but there is a potential for both structural and nonstructural damage.

It is anticipated that structural damage, even in major earthquakes, will likely be repairable for the majority of buildings meeting these standards. However, there may be cases where repairing the damage may not be cost-effective. Various factors such as the intensity and duration of ground shaking, building design, type of force resisting systems, construction materials, and workmanship influence the extent of damage.

Designers refer to codes as a basis since they establish the standards agreed upon by consensus. These codes do not offer guidance on material or system selection. Rather set criteria for their use once chosen. Additionally, they do not provide designers with insights into performance differences between systems. For example, comparing shear walls stiffness to frames and understanding its impact on resilience. Codes also overlook how certain structural systems may cause damage compared to others despite both being equally effective in resisting earthquakes. The subsequent sections address the anticipated performance of nonstructural elements.

11.4.4 Expected Performance of Structural Components

As mentioned above, according to the current seismic design provisions, noncritical facilities are designed for life safety, that is no damage in minor earthquakes, limited structural damage in moderate earthquakes, and providing resistance to collapse in a major earthquake. Resistance to collapse implies that while the structure may have lost a portion of its lateral stiffness and strength, the gravity load bearing elements still function and offer some safety margin against collapse. The structure might exhibit displacement with certain components of the seismic force resisting systems showing signs of cracking, spalling, yielding, buckling, and localized failure. After an earthquake, the structure should not be occupied until necessary repairs are completed. Strong aftershocks could potentially jeopardize the stability of the building. Repairing such structures maybe feasible but may not always be economically viable.

11.4.5 Expected Performance of Nonstructural Components

Current seismic design standards focus on ensuring integrity during earthquakes. Often overlook the performance of nonstructural elements such as room partitions, filing cabinets, lighting fixtures, and staircases. These standards do not account for the functionality of electrical or plumbing systems like fire sprinklers, HVAC equipment, or electrical panels. The majority of building damage in earthquakes has been attributed to components and systems failing. Building owners are frequently surprised when a structure remains standing after an earthquake, but not operational due to nonstructural damage. While current seismic regulations mandate securing components to prevent falling hazards, these elements can still suffer damage that compromises their functionality. Power outages, water supply disruptions, and HVAC failures can render a building uninhabitable. Additionally leaks from fire sprinkler breaks can lead to flooding and extensive damage within the building. Damage to elements, in buildings subjected to strong ground shaking may include cracked cladding and glazing broken light fixtures, misaligned doors, and dislodged ceiling tiles.

11.5 ENHANCING SEISMIC PERFORMANCE TO MITIGATE RISK

Enhancing performance to mitigate earthquake risk involves considering several factors. These include understanding seismic risk management and two key elements for performance: evaluating seismic hazards at the site and determining the desired seismic performance of structural and nonstructural elements for relevant earthquakes.

This section focuses on seismic design aspects for improved seismic performance regardless of building occupancy:
- Choosing structural materials and systems
- Selecting architectural/structural configurations
- Evaluating the anticipated performance of nonstructural components such as ceilings, partitions, HVAC systems, plumbing, and exterior cladding.

11.5.1 Selection of Structural Material and Structural System

An earthquake does not understand how a building would respond to ground shaking. It reveals weaknesses in the building that result from errors or deficiencies in its design and construction. However, variations in design and construction can influence its response significantly. The occupancy determines these variations leading each building type to have different design factors. A building with a moment frame structure will respond differently to ground motion compared to a building with shear walls. The moment frame structure is more flexible resulting in lower earthquake forces but more deflection than the shear wall structure. This increased motion may cause damage to components like partitions and ceilings. On the hand, the shear wall building is stiffer but will attract higher forces than a moment frame building; it deflects less but experiences higher accelerations that will affect acceleration-sensitive equipments such as air-conditioning equipment and heavy tanks. These structural and nonstructural system characteristics can be obtained from the seismic provisions; however, these provisions are not a design guide; these provisions do not offer direct guidance on the distinct performance characteristics of available systems or how to choose an appropriate structural system, for a specific site or building type.

11.5.2   The Selection of the Architectural Configuration

The design and layout of a building, including its size, proportions, and three-dimensional shape significantly influence its seismic performance. This is because the way a building is configured determines how earthquake forces are distributed throughout its structure. A thought-out configuration ensures that these forces are evenly spread out in both the horizontal and vertical directions allowing them to be directly and efficiently carried by the foundations. On the hand a designed configuration can lead to areas of high-stress concentration and torsion posing serious risks during earthquakes.

Over time engineers and researchers have identified configuration issues through observation of buildings performances during seismic events. Interestingly, many problematic configurations arise not out of negligence but due to requirements or site limitations that need to be addressed in the buildings design. Therefore, architects and engineers must collaborate closely from the beginning of the design process to find solutions that meet architectural needs but also prioritize safety and cost effectiveness. This teamwork involves selecting a system early on to fulfill the buildings requirements and then working together to refine design options that minimize or eliminate potential configuration issues. By approaching design decisions with both aesthetics and structural integrity in mind, professionals can create buildings that are both safe and aesthetically appealing.

The current seismic codes now include guidelines to address configuration issues. However, the approach of the code is to acknowledge these problems and try to resolve them by either increasing design forces or requiring advanced analysis. None of these methods is ideal to eliminate the problem. The key to solving the issue lies in design than relying solely on a prescriptive code. Potential design solutions, for addressing a soft first story condition that architects and engineers could consider exploring include:

• What would be the architectural implications if the soft story eliminated?

• Considering design approaches like increasing number of column or enhancing system stiffness to balance out differences between floors.

• Adding bracings or wall segments at the end of column lines if site conditions allow.

A broader challenge arises from the building's response becoming less predictable as the architectural and structural layout moves away from symmetry. This poses concerns for PBD.

11.5.3   Considerations for Nonstructural Components

It is known that majority of the damage that resulted in building closure after earthquakes is due to nonstructural components and systems. Even if a building meets regulations, it may not function properly due to nonstructural damage. Additionally, nonstructural components can affect performance during ground shaking. Structural analysis typically focuses on the structure only. Nonstructural elements attached to the building and heavy contents can introduce torsional forces depending on their placement. Some examples of structural/nonstructural interaction include:

• When heavy masonry partitions are rigidly attached to columns and under floor slabs, they can cause localized issues. If positioned asymmetrically, stress concentrations and torsional forces are generated. One common structural problem arises is short column conditions due to the insertion of partial masonry walls between columns. Adding these masonry walls after building completion is often viewed as a renovation that does not necessitate engineering analysis. As a result, the shortened columns become relatively stiff that attract high forces and eventually fail.

• Stairs in smaller buildings can function as bracing elements between floors leading to torsion issues; one solution is to disconnect the stair from the floor slab at one end to allow structural movement.

• In storage areas or library stacks, heavy storage items can introduce torsion into a structure. While the structure may have been designed to handle the load, it may not have accounted for nonsymmetrical loading effects over time such as when acquiring new library books.

11.6   CURRENT SPECIFICATIONS FOR PERFORMANCE-BASED SEISMIC DESIGN

The successful implementation of seismic risk management strategies relies on the application of PBD approaches. PBD aims to ensure that building designs can reliably perform as intended under seismic hazard conditions. This concept of defining levels of building performance and selecting performance objectives is relatively new in seismic design. It stems from observations made during earthquakes, where buildings that sustained damage still met the life-safety requirements outlined in the seismic code they were designed under as no fatalities or serious injuries occurred. These experiences underscore the importance for design professionals to clearly articulate what compliance to building codes entails and the limitations of design. Additionally, studying damaged buildings has enhanced the understanding of how structures respond to earthquake ground motions.

11.6.1 Building Performance Objectives

The core idea behind implementing PBSD is to establish a shared set of performance objectives. These objectives outline how a building should perform (e.g., ensuring life safety, tolerable damage levels, and post-earthquake functionality) when subjected to an earthquake hazard of defined intensity, such as maximum considered earthquake or an event with known return period. As the earthquake intensity increases, the buildings performance typically decreases. Setting performance objectives aims to provide a projection of how the building will perform in one or more earthquake scenarios.

11.6.2 Building Performance Levels

The buildings performance can be described in qualitative ways such as: the safety it provides to occupants during and after an earthquake; the cost and feasibility of restoring the building to its pre-earthquake state; the duration of time needed to conduct repairs and return the building to service; and the overall economic, architectural, or historic impacts on the community.

These performance aspects are closely linked to the extent of damage sustained by the building in an earthquake. Generally speaking, there are four levels of performance that can be defined based on these factors.

11.6.2.1 OPERATIONAL LEVEL

This represents the extent of damage to the building. The framework will preserve all of its strength and stiffness. Anticipated damage includes cracking in walls, partitions, and ceilings along with structural components. All mechanical, electrical, plumbing, and other systems essential for the functioning of the buildings are predicted to be operational from backup sources. There should be damage to elements. With levels of earthquake ground movement most buildings ought to meet or surpass this level of performance. Typically, though it may not be economically feasible to plan for this level of performance during ground shaking unless its for buildings that provide the essential services.

11.6.2.2 IMMEDIATE OCCUPANCY LEVEL

The building has suffered damage overall. While the structural systems are holding up well there may be some damage to elements like cladding, ceilings, and mechanical/electrical components that could require repair and cleanup. It is anticipated that utilities essential for operations may not be available although those crucial for life-safety systems would be functioning. Building owners aiming for this level of performance in earthquake scenarios or for important structures during severe shaking might find it beneficial. This level of protection offers benefits to building performance but without the added expenses of standby utilities and rigorous seismic testing to validate equipment performance.

11.6.2.3 LIFE-SAFETY LEVEL

The building has experienced damage impacting its strength and stiffness. However, gravity load bearing elements remain functional. While out of plane wall failures and parapet tipping are not anticipated, there will be some drift and certain elements of the force resisting system showing signs of cracking, spalling, yielding, and buckling. Nonstructural components are secure and do not pose a falling hazard. Various architectural, mechanical, and electrical systems have been affected. Occupancy should only resume once repairs are completed as the building may not be safe otherwise. Although feasible repairing, the structure may not be economically viable. This level of performance typically aligns with code compliance objectives.

11.6.2.4 COLLAPSE PREVENTION LEVEL OR NEAR-COLLAPSE LEVEL

The building has suffered damage with the lateral force resisting system losing much of its strength and stiffness from the earthquake. While load bearing columns and walls are still functioning, the building is on the brink of collapse. Structural elements have undergone degradation with cracking and spalling of masonry and concrete components as well as buckling and fracturing of steel elements. In addition, infills and unbraced parapets may fail and exits could be blocked. There are large permanent drifts in the building. Nonstructural components have also sustained damage posing falling hazards. Due to these issues, the building is deemed unsafe for occupancy making repair and restoration efforts impractical. This level of building performance serves as a basis for seismic rehabilitation ordinances in municipalities due to its effectiveness in reducing life-safety risks at a reasonable cost.

11.6.3 Some of the Current Performance-Based Design Guides

1. Tall Buildings Initiative: Guidelines for Performance-Based Seismic Design of Tall Buildings, 2017.

These Seismic Design Guidelines for Tall Buildings offer an alternative to the established seismic design methods found in the ASCE 7 standard and the International

Building Code (IBC). They are designed for engineers and building officials involved in the design and assessment of tall buildings. When implemented correctly, these guidelines aim to ensure that buildings can consistently meet the seismic performance goals outlined in ASCE 7 and sometimes even exceed them in aspects as indicated. Users have the flexibility to customize and adjust these guidelines to create designs that exceed the seismic performance requirements presented in this guide.

2. Performance-Based Seismic Design of Tall Buildings, 2017, CTBUH Technical Guide, 2017.

In 2008, the council of Tall Buildings and Urban Habitat (CTBUH) Seismic Working Group created a document called "Recommendations for the Seismic Design of High-Rise Buildings." This publication along with meetings of the working group highlighted that many experts prefer using PBSD over prescriptive code-based methods when designing tall buildings in high seismic areas. Since 75 percent of the world's tall buildings completed by 2016 were located in high seismic zones, nonstandard design and analysis approaches were necessary for building approval, it is clear that sharing the principles of PBSD design would benefit an international audience. Consequently, the CTBUH formed the PBSD working group to create a publication introducing PBSD concepts globally and showcasing real-world applications with examples.

3. An Alternative Procedure for Seismic Analysis and Design of Tall Buildings Located in the Los Angeles Region, 2017, 2020 edition and 2023.

The acronym LATBSDC stands for Los Angeles Tall Buildings Structural Design Council. The council has released a document called Alternate Design Criteria, which outlines a performance-based approach, for analysis and design of buildings. In this document, tall buildings are those exceeding 160 ft above the surrounding average ground surface. The council has utilized analysis and design methods that are commonly allowed in building standards while preparing this document.

The design methodology is based on capacity design principles along, with a series of PBD assessments. Therefore, for a structure to meet the criteria outlined in this process it must possess a ductile yielding mechanism of undergoing lateral deformations. For instance, a special moment resisting frame (SMRF) should exhibit zones characterized by flexural yielding at beam ends shear yielding in column beam panel zones and yielding at column bases.

Chapter 12
Special Topics

The purpose of this chapter is twofold: (i) to examine in detail a number of special subjects which we have met in several ways in earlier chapters; and (ii) to touch briefly on certain topics which are unique to the design of tall buildings. Our consideration of special topics opens with a discussion of differential shortening of columns.

A column in a tall building undergoes axial shortening considerably more than its lower brethren, requiring special attention in its design. A related problem, by no means unique to tall buildings, but one that gets aggravated to a greater extent, is the levelness of floors. Similarly, the problem of human response to transient vibration of floors is not unique to tall buildings but needs careful study because the cost of correcting the problem in a tall building with several floors is phenomenally more expensive than correcting the relatively fewer floors of a low-rise building. Next, we will consider in some detail the behavior of panel zones and their effect on lateral deflections of buildings.

Next, we move on to the gray area of the design of curtain wall systems that brings together several diversified disciplines. The section briefly presents the design and installation aspects of metal curtain walls, stone claddings, brick veneer systems, and glass fiber-reinforced concrete (GFRC) systems.

The next section is introductory in its presentation of mechanical devices that are used to increase the damping and thus reduce wind-induced sway acceleration and torsion-induced translation acceleration. Two devices, a tuned mass damper and a viscoelastic damper, are discussed. With the additional damping provided by these devices, the peak wind-induced resultant accelerations are designed to be within the benchmark limit of 20 milli-g (one-fiftieth of the acceleration due to gravity) for a recurrence interval of once in 10 years.

The next section presents a brief discussion of design aspects of two types of foundations: the drilled pier or caisson foundation, and the mat foundation. The next section presents a discussion of seismic design of floor diaphragms for horizontal forces including the effect of plan irregularities commonly encountered in practice. Next, earthquake mitigation technologies that include seismic isolation and energy dissipation are discussed, followed by an overview of SAC guidelines for repair, modification, and design of welded steel moment frames. The next section gives unit structural quantities for several types of concrete floor systems and high-rise steel and composite buildings.

12.1 DIFFERENTIAL SHORTENING OF COLUMNS

Columns in tall buildings are subjected to large axial displacements because they accumulate loads from a large number of floors and are also relatively long. A 60-story interior column in a steel building may shorten as much as 2 to 3 in. (50 to 76 mm) because of dead and live loads, while a concrete column may experience an additional 2 to 3 in. (50 to 76 mm) of shortening because of creep and shrinkage. If this shortening is not given due consideration, problems may develop in the performance of curtain walls and levelness of floor systems. Proper awareness of this problem is necessary on the part of the structural engineer, architect, and the curtain wall supplier to avoid lost time and money.

The maximum effect of column shortening is at the roof level reducing gradually toward the ground level. In a concrete frame, the axial shortening may take several years to complete because of the long-term effect of creep, although a major part of it occurs within the first few months of construction. There is very little the structural engineers can do to minimize frame shortening, but they should make the design team aware of the magnitude of frame shortening so that soft joints of appropriate widths are properly detailed between curtain wall joints to prevent load from being transferred into the building façade. Before fabrication of curtain wall connections, the in-place elevations of the structural frame should be measured, and in-place elevations of the structural frame should be measured and provided to the curtain wall contractor. The fabrication should be based on these, rather than the theoretical elevations. There must be sufficient space at the joints between the panels to allow for the expected movement of the structure as well as thermal expansion and contraction of the panels themselves. Insufficient space may result in bowed curtain wall panels, or in extreme cases, the panels may even pop off the building.

A similar problem occurs when mechanical and plumbing lines are attached rigidly to the structure. Frame shortening may force the pipes to act as structural columns resulting in their distress. A general remedy is to make sure that nonstructural elements are not brought in to bear the vertical loads. Sufficient compensation should be provided during design and construction to make sure that nonstructural elements are separated from structural elements.

The axial loads in all columns of a tall building are seldom the same, giving rise to the problem of so-called differential shortening. The problem is more acute in a composite structure because slender steel columns are subjected to large axial loads during construction. Determining the magnitude of axial shortening in a composite system is complicated because many variables that contribute to the shortening of columns cannot be predicted with sufficient accuracy. The lower part of the column, which is encased in concrete, is continually undergoing creep, and because the age and strength of concrete keep changing, their effect on creep is difficult to predict with any precision. The steel column at any given period during construction is partly enclosed in concrete at lower

floors, with the bare steel section projecting beyond the concreted levels by as many as 10 or 15 floors. Another factor difficult to predict is the gravity load redistribution due to continuity of beams attached to columns. If the building is founded on compressible material, foundation settlement is another factor that influences the relative changes in the elevations of the columns. The magnitude of load imbalance continually changes, making an accurate assessment of column shortening beyond the reach of day-to-day engineering practice. If all the variables are known, the prediction of differential shortening is no more complicated than a systematic evaluation of the PL/AE equation.

The routine method of analysis of high-rise structures is usually performed for the full frame without taking into account the sequential nature of construction. This approach for a 60-story concrete building may result in a calculated axial shortening of about 3 in. (76 mm) of immediate axial shortening and a mind-boggling 3 to 5 in. (76 to 254 mm) of additional displacement when creep effects are included in the computation. Fortunately, the method of construction in concrete buildings more or less takes care of the immediate shortening, and to a limited extent the creep effects on lower-level columns. This is because buildings are constructed one floor at a time and since each floor is leveled at the time of its construction, the column shortening which has occurred prior to the construction of that floor is of no consequence. Also, the lower-story columns of tall buildings undergo considerably smaller creep and shrinkage because the load is applied incrementally over a 15- to 24-month construction period.

Creep is difficult to quantify because it is time-dependent. Initially, the rate of creep is significant and diminishes as time progresses until it eventually reaches zero. Because of sustained loads the stress in concrete gradually gets transferred to the reinforcement with a simultaneous decrease in concrete stress.

Columns with different percentages of reinforcement and different volume-to-surface ratios creep and shrink differently. An increase in the percentage of reinforcement and volume-to-surface ratio reduces the strain due to creep and shrinkage under similar stresses. Differential shortening of columns induces moments in the connecting girders and spandrels resulting in gravity load transfers to adjacent columns. A column which has shortened less receives more load, thus compensating for the initial imbalance.

In this section, we will not address the foreshortening of concrete columns because attempts to quantify the shrinkage and creep effects are considered beyond the scope of this work. Leaving the problem to more theoretical minds, we will proceed in this section to a brief discussion of the differential shortening of steel columns. For this purpose, a column isolated from the remainder of the structural frame is studied as a cantilever. A closed-form solution for computing the axial shortening of columns is presented with a numerical example to demonstrate the practicality of the method. The section concludes with suggested details for field adjustment of column heights.

12.1.1 Calculations

Differential rather than the *absolute* shortening of column is more significant. If all columns shorten by the same amount, the floors would still be level. Relative displacement between columns occurs because of the difference between the P/A ratios of columns. If all columns in a building have the same slenderness ratio and are sized for gravity load requirement only, there will be no relative vertical movement between the columns. All columns will undergo the same displacement because the P/A ratio is nearly constant for all columns. In a real building, this condition is seldom present. Usually, the design of frame columns is governed by the combined gravity and lateral load, while non-frame columns are designed for gravity loads only.

For example, consider the tubular system used for buildings in the 50- to 80-story range. The system typically utilizes closely spaced exterior columns and widely spaced interior columns. Normally, in a steel system, high-strength steel columns up to 65 ksi are used in the interior of the building to collect gravity loads on large tributary areas resulting in a large P/A ratio. The exterior columns, on the other hand, usually have a small P/A ratio for two reasons; first, their tributary areas are small because of their close spacing of usually 8 to 12 ft (2.44 to 3.66 m); second, the columns are sized from lateral displacement considerations, resulting in areas much in excess of those required from the strength consideration alone. Because of this imbalance in the gravity stress level, these two groups of columns undergo different axial shortenings; the interior columns shorten much more than the exterior columns.

A somewhat reversed condition occurs in buildings utilizing interior-braced core columns and widely spaced exterior columns; the exterior columns experience more axial shortening than the interior columns. The behavior of columns of buildings utilizing other structural systems, such as interacting core and exterior frames, tends to be somewhere in between these two limiting cases.

In all cases, it is relatively easy to evaluate the shortening of columns. The procedure requires a step-by-step manipulation of the basic PL/AE equation as described below.

Consider a typical column of a 50-story building. Assume, for simplicity, that the variations in story heights, column areas, and the load increment at each floor are constant as shown in Fig. 12.1. The calculation of the axial shortening, in itself, at any floor is trivial. It is given by the summation equation:

$$\Delta_n = \frac{PL}{AE} \sum_{i=1}^{n} (NS + 1 - i) \qquad (12.1)$$

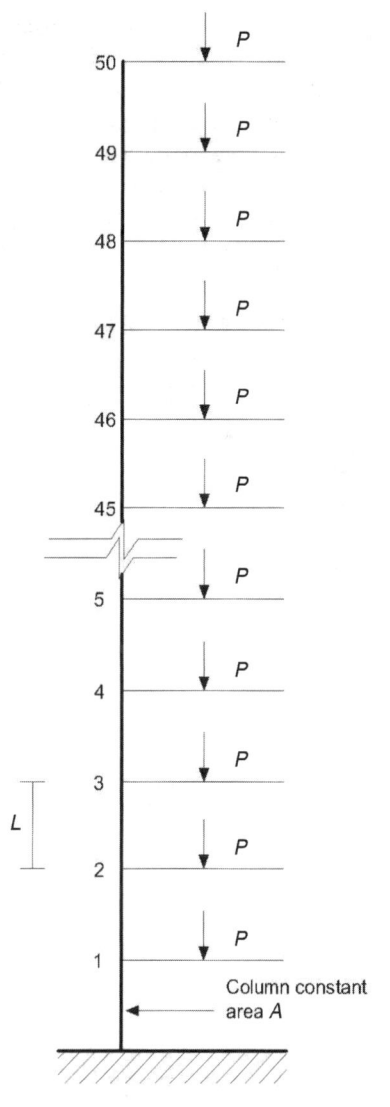

Figure 12.1   Simplified column for calculating column shortening.

where Δ_n = the axial shortening at level n

P = the load increment assumed constant at each level

L = the story height assumed constant for the full height

A = the column area assumed constant for the full height

NS = the number of stories

Using this equation, the axial shortening for the simplified column (Fig. 12.1) works out as given in Table 12.1. Having obtained these values, the next step is to evaluate the column length correction Δ_c at different levels. This is given by the difference in the axial shortenings at the level under consideration and the floor immediately below it. For instance, from Table 12.1 this value at level 30 for the example column is given by PL/AE (1065 – 1044) = 21 PL/AE. The magnitude of this correction in a normally proportioned building is rather small, perhaps $\frac{1}{8}$ in. (3.17 mm) at the most. Instead of specifying this value as a correction at each level, in practice it is usual to lump these corrections for a few floors, for example, every tenth floor or so. The lumped correction at the level is simply the difference between the values

of axial shortening at that level and at the lumped floor below it. For example, in Table 12.1 the lumped correction of 255 PL/AE units at the 30th floor is equal to the column shortening at the 30th level less the shortening at the 20th level.

$$\frac{PL}{AE}(1065 - 810) = 255\frac{PL}{AE}\text{ units}$$

Let us consider the practical case of a building column that has variations in story heights, load increments, and column areas as shown in Fig. 12.2. The summation equation for this general case takes the form:

$$\Delta_n = \frac{1}{E}\sum_{k=1}^{n}\frac{L_k}{A_k}\sum_{i=k}^{NS}P_i \qquad (12.2)$$

Table 12.2 shows in a tabular form the assumed dimensions and loading conditions for the column and the computations for obtaining the column length shortening values. The last column of the table shows the lumped corrections at levels 2, 10, 20, 30, 40, and the roof. Basically, these corrections represent the additional lengths over and above their theoretical lengths. For example,

Table 12.1   Axial Shortening Computations for Simplified Column

Level	Axial shortening	Column length correction at each level	Lumped column length correction	Level	Axial shortening	Column length correction at each level	Lumped column length correction
50	1275	1	55	25	950	26	
49	1274	2		24	924	27	
48	1272	3		23	897	28	
47	1269	4		22	869	29	
46	1265	5		21	840	30	
45	1260	6		20	810	31	355
44	1245	7		19	779	32	
43	1247	8		18	747	33	
42	1239	9		17	714	34	
41	1230	10		16	680	35	
40	1220	11	155	15	645	36	
39	1209	12		14	609	37	
38	1197	13		13	572	38	
37	1184	14		12	534	39	
36	1170	15		11	495	40	
35	1155	16		10	455	41	405
34	1139	17		9	414	42	
33	1122	18		8	372	43	
32	1104	19		7	329	44	
31	1085	20		6	285	45	
30	1065	21	255	5	240	46	
29	1044	22		4	194	47	
28	1022	23		3	147	48	
27	999	24		2	99	49	50
26	975	25		1	50	50	

Note: All *values* are in terms of *PL/AE*.

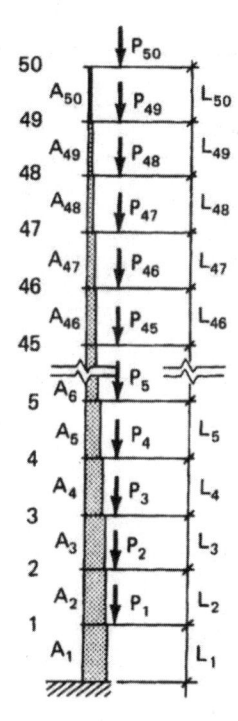

Figure 12.2 Axial shortening computations for a practical column.

$\Delta_c = 1\frac{1}{4}$ in. (31.75 mm) at the tenth level means that the actual fabricated length of column should be $1\frac{1}{4}$ in. longer than the theoretical length. This overlength could be achieved by increasing the length of column in each tier by $\frac{1}{4}$ in. (6.35 mm) (ten stories equal five tiers, therefore, $\frac{1}{4}$ in. times 5 gives $1\frac{1}{4}$ in.). The fabricator may elect to increase the length in each story by $\frac{1}{8}$ in. (3.2 mm) instead of $\frac{1}{4}$ in. per tier to achieve the same Δ_c at the tenth floor.

The value of $\Delta_c = 2$ in. (50.8 mm) at the 20th floor means the overlength of columns between levels 1 and 20 should be 2 in. However, an overlength of $1\frac{1}{4}$ in. (31.75 mm) up to the tenth level has already been achieved by specifying $\Delta_c = 1\frac{1}{4}$ in. at the tenth level. Therefore, the increment between the 10th and 20th levels should be 2 in. less $1\frac{1}{4}$ in. $= \frac{3}{4}$ in. (19.0 mm). A correction table incorporating the above information for the example column is shown in Table 12.2.

Table 12.2 Axial Shortening Computations for Practical Column

Level	Accumulated load, kips	Column section	Story height, in.	Column shortening Δn, in.	Column length correction at each level, in.	Lumped column length correction, in.	Column shortening from Eq. (11.4), in.
50	53	W14 × 43	156	5.14	0.023	0.73	5.11
49	106	43	210	5.12	0.061		5.08
48	159	53	168	5.05	0.051		5.02
47	212	53	156	5.00	0.073		4.95
46	265	68	156	4.93	0.071		4.89
45	318	68	156	4.86	0.086		4.82
44	371	84	156	4.77	0.081		4.75
43	424	84	156	4.69	0.092		4.67
42	477	95	156	4.60	0.092		4.59
41	530	95	156	4.51	0.102		4.50
40	583	111	156	4.41	0.09	1.02	4.42
39	636	111	156	4.32	0.105		4.33
38	689	127	156	4.21	0.09		4.24
37	742	127	156	4.12	0.107		4.15
36	795	142	156	4.01	0.103		4.04
35	848	142	156	3.91	0.109		3.96
34	901	167	156	3.80	0.09		3.86
33	954	167	156	3.71	0.105		3.76
32	1007	176	156	3.62	0.105		3.66
31	1060	176	156	3.50	0.110		3.56
30	1113	202	156	3.39	0.101	1.07	3.46
29	1166	202	156	3.29	0.106		3.36
28	1219	211	156	3.19	0.106		3.26

(Continued)

Table 12.2 Axial Shortening Computations for Practical Column (*Contiued*)

Level	Accumulated load, kips	Column section	Story height, in.	Column shortening Δn, in.	Column length correction at each level, in.	Lumped column length correction, in.	Column shortening from Eq. (11.4), in.
27	1272	211	156	3.08	0.110		3.16
26	1325	228	156	2.97	0.106		3.12
25	1378	228	156	2.86	0.111		2.95
24	1431	246	156	2.75	0.107		2.85
23	1484	246	156	2.65	0.111		2.74
22	1537	264	156	2.53	0.107		2.64
21	1590	264	156	2.43	0.110		2.53
20	1643	287	156	2.32	0.104	1.06	2.42
19	1696	287	156	2.20	0.108		2.32
18	1749	314	156	2.10	0.101		2.21
17	1802	314	156	2.00	0.105		2.10
16	1855	314	156	1.90	0.108		1.99
15	1908	314	156	1.79	0.111		1.88
14	1961	342	156	1.68	0.104		1.78
13	2014	342	156	1.58	0.107		1.67
12	2067	370	156	1.47	0.101		1.56
11	2120	370	156	1.37	0.104		1.45
10	2173	370	156	1.26	0.107	0.53	1.34
9	2226	370	156	1.16	0.109		1.23
8	2279	398	156	1.05	0.104		1.12
7	2332	398	156	0.94	0.107		1.01
6	2385	398	156	0.84	0.11		0.89
5	2438	398	210	0.73	0.11		0.78
4	2491	426	168	0.62	0.11		0.67
3	2544	426	156	0.51	0.11		0.56
2	2597	500	156	0.40	0.09		0.45
Mezzanine	2650	500	240	0.31	0.15		0.17
1	2770	W14 × 500	240	0.16	0.16		0.17

12.1.2 Simplified Approach

In a normally proportioned building the cross-sectional area of a column usually increases in a stepwise manner, from a minimum value at the roof to a maximum value at the base as shown in Fig. 12.3. The incremental steps are caused by the finite choice of column shapes. In tall buildings which merit column shortening investigations, the significance of these incremental steps diminishes rather quickly as compared to a low-rise building column. Therefore, it is possible without losing meaningful accuracy, to express the load and cross-sectional properties by continuous mathematical expressions as indicated in Fig. 12.4. The gravity load distribution may be assumed to vary linearly throughout the height (Fig. 12.4). A similar linear assumption for the column area overestimates the actual column areas as shown by the dashed curve in Fig. 12.5. Although mathematically it is possible to derive an equation to fit the curve, the author proposes a modified linear variation as indicated in Fig. 12.5 in which the equivalent column area at the bottom is taken as 0.9 × the

actual area. This simplification leads to less formidable expressions for a closed-form solution without any meaningful loss in accuracy.

Derivation of closed-form solution

The notations used in the derivation of closed-form solution (Fig. 12.5a) are as follows:

L = Height of the building (note previously in the longhand method, notion L was used to denote story height)

Δ_z = Axial shortening at a height x (also denoted as z), above foundation level

A_t = column area at top

A_b = Modified column area at bottom equal to 0.9 × actual area of column at bottom = 0.9 × A_B

A_x = Area of column at height x (also denoted as z), above foundation level

α = Rate of change of area of column

P_t = Axial load at top

P_b = Axial load at bottom

P_x = Axial load at height x above foundation

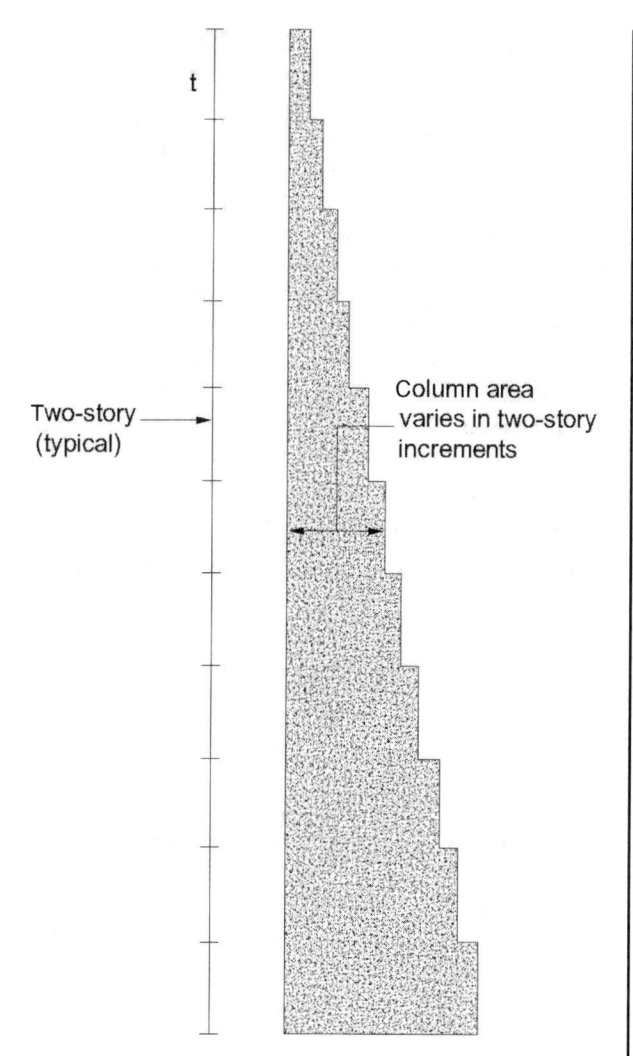

Figure 12.3 Variation of cross-sectional area of a high-rise column.

β = Rate of change of axial load
ε_x = Axial strain at height x
E = Modulus of elasticity

The area of column at height z is given by:

$$A_x = A_t \frac{x}{L} + A_b\left(1 - \frac{x}{L}\right)$$

$$= A_b - (A_b - A_t)\frac{x}{L}$$

$$= A_b - \alpha x$$

where $\alpha = \dfrac{A_b - A_t}{L}$

The axial load at height z above foundation is given by:

$$P_x = P_t \frac{x}{L} + P_b\left(1 - \frac{x}{L}\right)$$

$$= P_b - \beta x$$

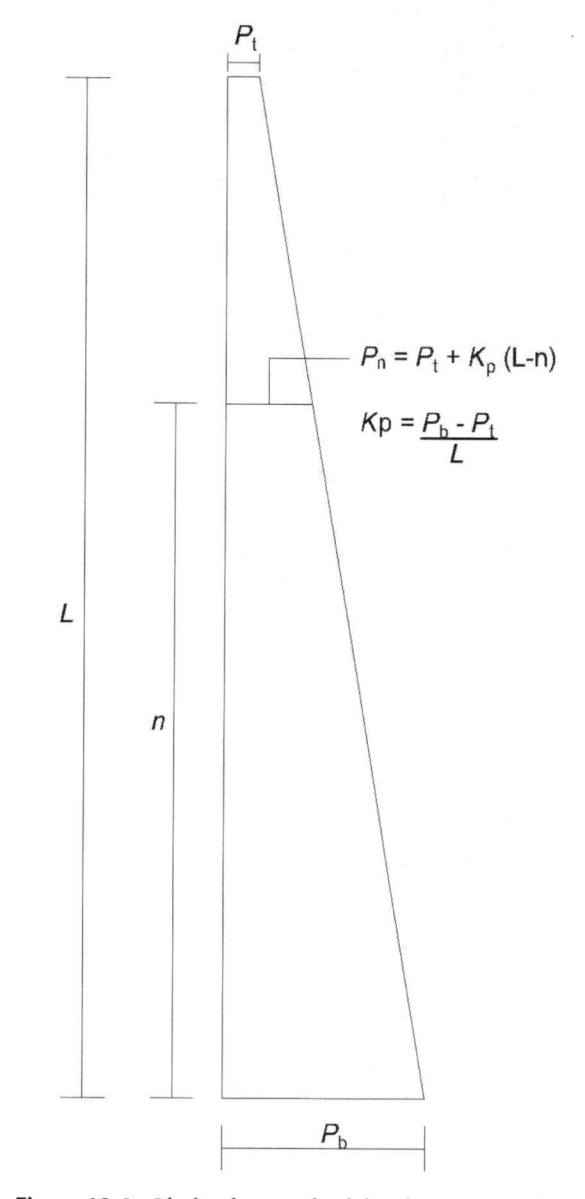

Figure 12.4 Idealized gravity load distribution on a column.

where $\beta = \dfrac{P_t - P_t}{L}$

The axial strain $\varepsilon_x = \dfrac{Px}{A_x E}$

Using vertical work: $P_z^1 \Delta_z = \displaystyle\int_0^z P_x^1 \varepsilon_x dx$

$$P_z^1 = 1 = P_x^1,$$

with $\Delta_z = \dfrac{1}{E}\displaystyle\int_0^z \dfrac{P_b - \beta x}{A_b - \alpha x}\,dx$

$$= \dfrac{P_b}{E}\int_0^z \dfrac{dx}{A_b - \alpha x} - \dfrac{\beta}{E}\int_0^z \dfrac{x\,dx}{A_b - \alpha x}$$

In Figure 12.4:

$$P_n = P_t + K_p\,(L\text{-}n)$$

$$K_p = \frac{P_b - P_t}{L}$$

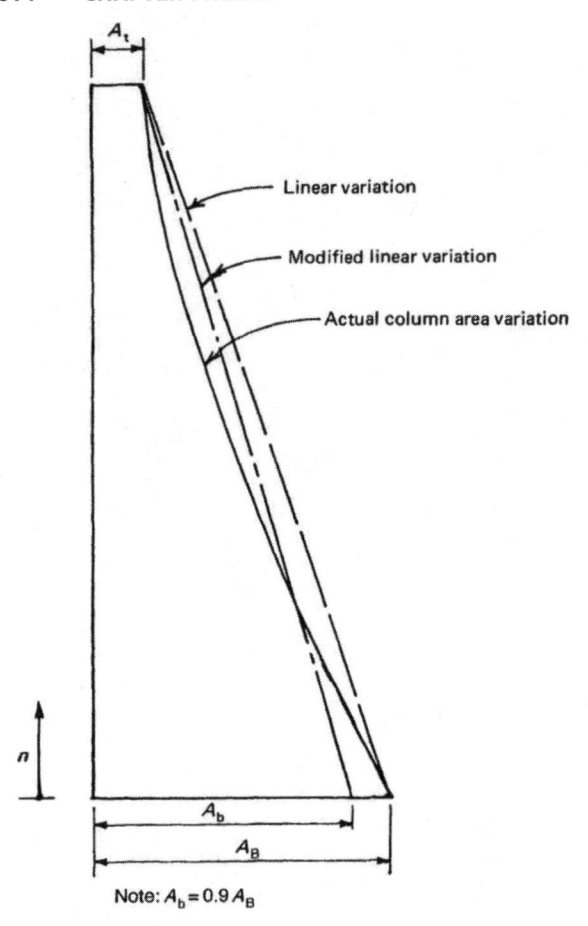

Figure 12.5a Idealized column cross-sectional areas.

Evaluating these integrals, we get the following final expression for Δ_z.

$$\Delta_z = \frac{P_b}{E}\left[-\frac{1}{\alpha}\ln\left(1-\frac{\alpha z}{A_b}\right)\right] - \frac{\beta}{E}\left[-\frac{1}{\alpha^2}\left\{\alpha z + A_b\ln\left(1-\frac{\alpha z}{A_b}\right)\right\}\right]$$

Example problem. Given:

Height of the building: $L = 682$ ft $= 8184$ in. (207.8 m)
Modulus of elasticity: $E = 29\,000$ ksi (200×10^3 MPa)
Axial load at top: $P_t = 53$ kips (237.5 kN)
Area of column at top: $A_t = 12.48$ in.^{2} (8052 mm^2)
Axial load at base: $P_b = 2770$ kips (12.32×10^3 kN)
Actual column area at base: $A_B = 147$ in.^{2} (94.84×10^3 mm^2)
Reduced column area at base: $A_b = 0.9 \times 147 = 133.3$ in.^{2} (86.0×10^3 mm^2)

Required. Axial shortening of column at top.

Solution Since column shortening is calculated at top, $z = L$.

$$\alpha = \frac{A_b - A_t}{L} = \frac{133 - 12.48}{8184} = 0.01476 \text{ in.}^2/\text{in.}$$

$$\beta = \frac{P_b - P_t}{L} = \frac{2770 - 53}{8184} = 0.332 \text{ kip/in.}$$

$$\ln\left(1 - \frac{\alpha L}{A_b}\right) = \ln\left(1 - \frac{0.01476 \times 8184}{133.3}\right)$$

$$= \ln(0.09362)$$

$$= -2.36847$$

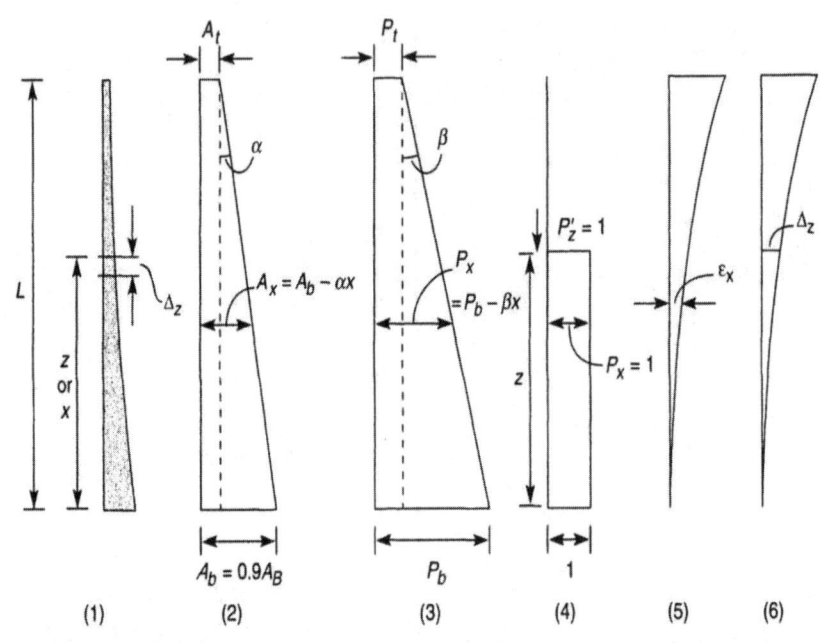

Figure 12.5b Axial shortening of columns: closed-form solution: (1) axial shortening Δ_z; (2) column area; (3) column axial load; (4) unit load at height z; (5) axial strain; (6) axial displacement.

$$\Delta_L \text{ at top} = \frac{2770}{29000}\left\{-\frac{1}{0.01476}\times(-2.36847)\right\} - \frac{0.332}{29000}$$

$$\times\{-4590.15(0.01476\times8184+133.3\times-2.36847\}$$

$$= 15.327 - 10.2$$

$$= 5.127$$

Similarly, the axial shortening is calculated at various heights by substituting appropriate values for z. The results given in column 8 of Table 12.2 agree closely with those from the longhand method. The appropriateness of the closed-form solution is obvious. A method of specifying overlengths for compensating differential shortening of columns is shown in Table 12.3.

Table 12.3   Column Length Correction Table

Level	Column length shortening, in. (mm)	Correction to scheduled column lengths, in. (mm)
50	$5\frac{1}{8}$ (13.75)	6 @ $\frac{1}{8} = \frac{3}{4}$ (19)
40	$4\frac{3}{8}$ (111)	8 @ $\frac{1}{8} = 1$ (25.4)
30	$3\frac{3}{8}$ (85.7)	8 @ $\frac{1}{8} = 1$ (25.4)
20	$2\frac{3}{8}$ (60.3)	8 @ $\frac{1}{8} = 1\frac{1}{8}$ (28.6)
10	$1\frac{1}{4}$ (31.75)	10 @ $\frac{1}{8} = 1\frac{1}{4}$ (31.75)

12.1.3   Column Shortening Verification During Construction

Assuming that the engineer has appropriately corrected column lengths to compensate for axial shortening, it becomes somewhat difficult during steel erection to determine whether the variations in the top elevations of columns are within allowable limits. For example, let us say the actual variation between an interior and exterior column at the time of erection is 2 in. (50.8 mm) when the columns are erected halfway up a 40-story building. Let us say 1 in. (25.4 mm) out of this 2 in. is in excess of the allowable erection tolerance. It is not immediately clear how much of this excess 1 in. is due to the overlength allowed for column shortening, because the column has undergone partial shortening due to already existing dead and construction loads. Therefore, there is a need for a second set of column shortening computations that gives the relative elevations of the columns "during erection." These values would indicate the amount by which the columns should be protruding

above their theoretical location at the time of erection and, therefore, serve as a benchmark for checking the relative elevations of columns during construction.

To illustrate the above idea, let us consider again the overly simplified column of Fig. 12.1. The total over-length specified at level 30, for example, is 1065 PL/AE units (Table 12.1). When loads P are applied starting at level 1, the overlength correspondingly starts decreasing by a factor PL/AE units for load P at each level. When erection is at the 30th level, the total shortening at the level would be 855 PL/AE units. The residual overlength at this level will be 210 PL/AE units. Physically, the residual overlength Δ_{R_n} at a given level represents the column shortening due to loads applied at and above that level as shown schematically in Fig. 12.6.

For the general case of column shown in Fig. 12.2, the residual overlength Δ_{Rm} at any level n works out to be

$$\Delta_{R_n} = \frac{1}{E}\sum_{i=n}^{NS} P_i \sum_{i=n}^{n}\frac{L_i}{A_i} \qquad (12.5)$$

Values of Δ_{R_n} calculated by using Eq. (12.5) are shown in Table 12.4.

Table 12.4   Residual Overlength of Column During Construction

Level	Residual overlength from Eq. (11.5), in.
50	0.33
45	1.87
40	2.15
35	2.20
30	2.12
25	1.96
20	1.72
15	1.49
10	1.12
5	0.68

12.1.4   Conclusions

Although the closed-form solution presented here is based on certain simplifying assumptions, the author believes that the accuracies obtained by the use of this simple method fully justify its application to normally proportioned tall buildings. Where the material and load distribution patterns are radically different from those shown in Fig. 12.3, the application of the more accurate long-hand procedure is recommended.

As mentioned previously, a steel frame does not have the luxury of built-in compensation of a cast-in-place concrete building because steel columns are prefabricated with predetermined lengths. Therefore, a suggested method for avoiding gross inaccuracies in predicting column length corrections would be to have

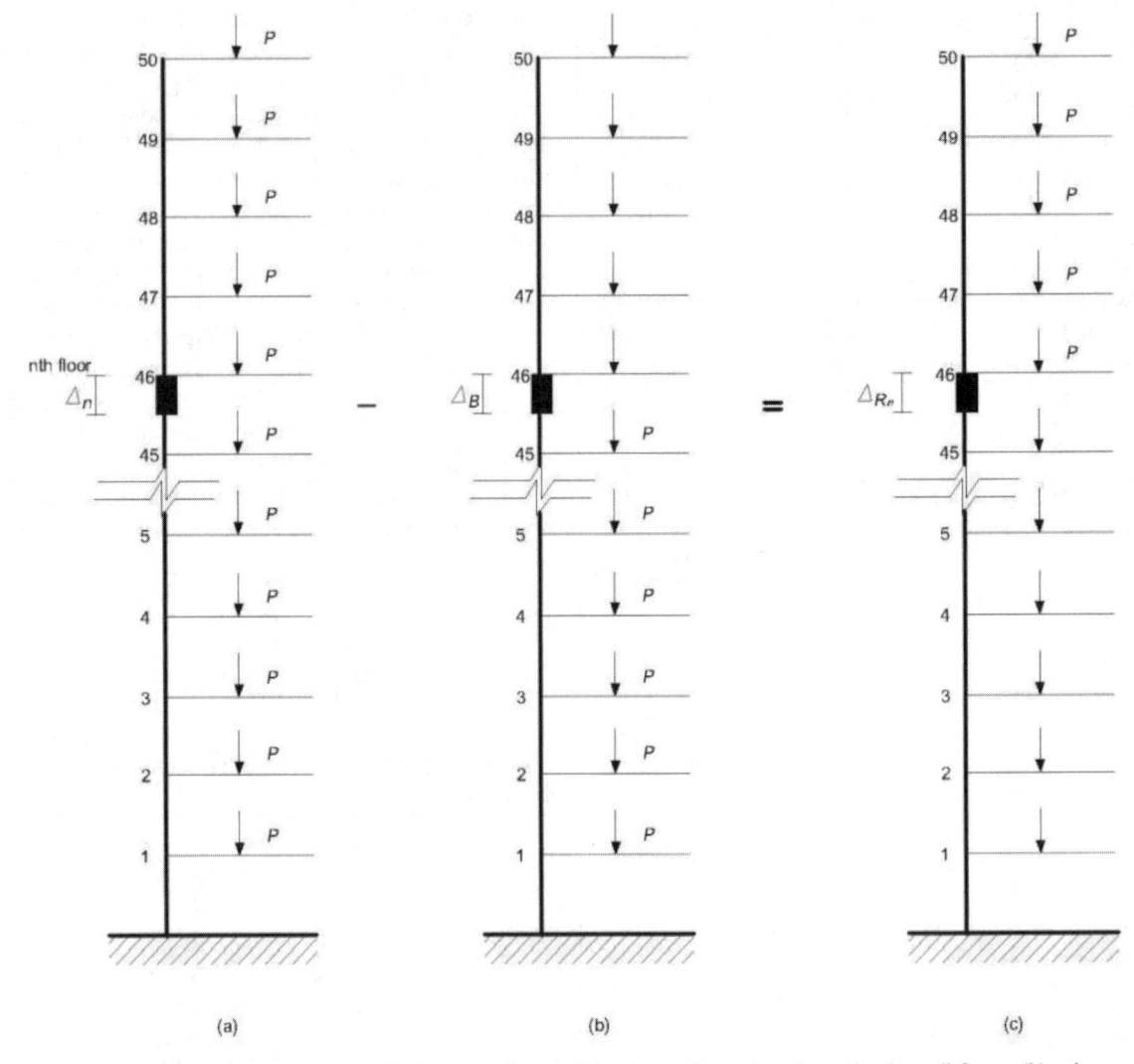

Figure 12.6 Physical interpretation of column overlength: (a) column shortening due to loads at all floors; (b) column shortening due to loads below the n the level; (c) column residual overlength $\Delta_{R_n} = \Delta_n - \Delta_B$.

the steel erector check the top elevation of columns at predetermined levels, say every fourth or sixth floor, and to make provisions for adjustments of column heights.

Some of the problems associated with differential shortening of columns can be eliminated in the fabrication stage. Certain columns can be made slightly longer than their nominal length to account for axial shortening. However, the uncertainties especially in composite structures are so many that some form of field adjustment is usually required.

For example, in steel buildings, it may become necessary to place removable shims between columns and their foundations in order to compensate for differential shortening. When the erection of structural steel reaches predetermined levels, the columns are temporarily unloaded

and steel shims removed. Another relatively simple method of adjusting column lengths which has been successfully used in composite construction is shown in Fig. 12.7.

12.2 FLOOR-LEVELING PROBLEMS

Floor-leveling problems are of increasing importance because stronger building materials and more refined designs have resulted in lighter construction more prone to deflections than earlier heavier buildings. In framed buildings, considerable trouble is encountered in trying to provide a level floor because of the many variable conditions that exist in practice. Concrete floors that are level at the time of construction may not be so at the time

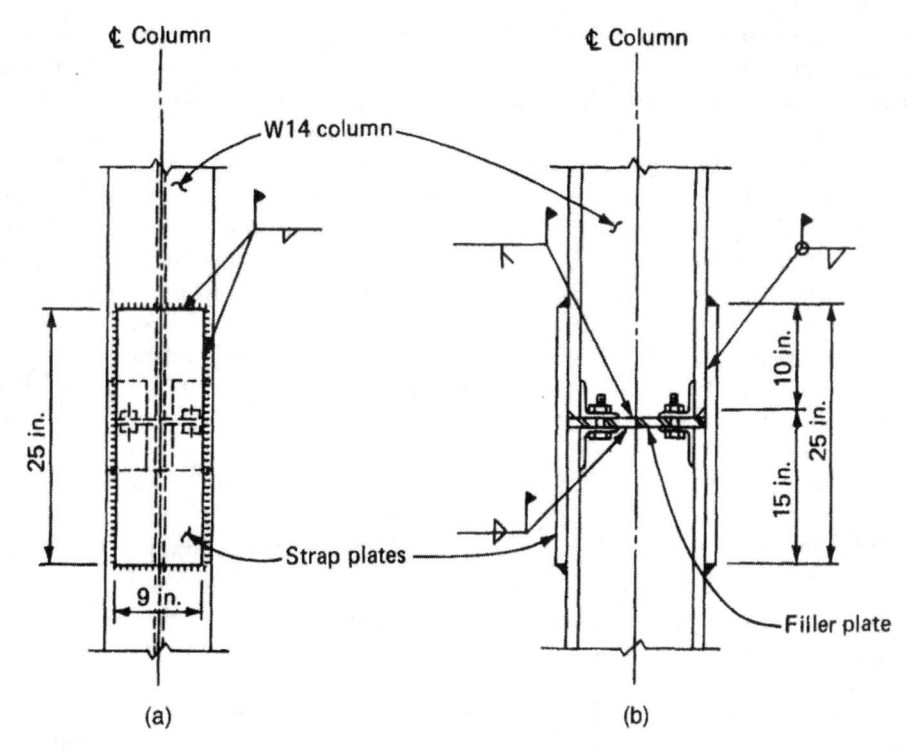

Figure 12.7 Splice detail for field correction of column length: (a) elevation; (b) section.

of occupancy. Many of the variable factors encountered are not mathematically determinable, even though an attempt is made by the engineer to compensate for such effects.

In steel buildings, floor beams for the normal 30–42 ft spans are furnished with a predetermined camber, while in concrete buildings the camber is built in the formwork. Usually, the specified camber ranges from a minimum of $\frac{1}{2}$ in. (12.7 mm) to a maximum of 2.5 in. (63.5 mm). Cambers smaller than $\frac{1}{2}$ in. (12.7 mm) are difficult to achieve, while a camber of substantially greater than 2.5 in. (63.5 mm) will result in other serviceability problems for beams in the 30- to 42-ft (9.14 to 12.80 m) range. Cambers are specified anticipating that the loading of floors will overcome the camber, resulting in a level floor. This is not always the case because: (i) rolling and construction tolerance combined with the long-term effect of creep of concrete is enough to affect the final result up or down in both steel and concrete construction; (ii) most usually, camber is calculated as if the beam were pin-connected or completely fixed. Actual conditions vary. For example, even with simple connections, steel beams experience partial fixity. Depending upon the degree of fixity the final result could again be up or down; (iii)

vertical members will shorten elastically during erection. The magnitude of elastic shortening between interior and exterior columns or between two adjacent exterior columns most usually is different, compounding floor-leveling problems.

Because of these variable factors, combined with the fact that none of these is mathematically determinable in the context of a practical design office, it makes it almost an accident if the floor turns out to be perfectly level. The problem comes to light at the time of interior finishing of the space when ceiling and partitions are being installed. One sure method of obtaining a level floor is to float the floor to remove the lumps and fill the low spots. Cement-based self-leveling underlayments are used for this purpose. In a floor built to commercially acceptable tolerance, the average fill over the entire floor area should not exceed $\frac{1}{2}$ in. (12.7 mm), which translates into an additional dead load of 6 psf (287.3 mm). Depending upon the type of construction, this additional load may represent an increase of 3 to 6 percent of the total working stress load. It is recommended that an allowance be made for this additional load in the design.

The most commonly specified tolerance for finished floor slab surfaces is $\frac{1}{8}$ in. (3.7 mm) in 10 ft (3.048 m), which is considered too stringent for most uses. The reasons for unlevelness are manyfold, including formwork sagging, deflection of members due to dead and live loads, finishing irregularities, or errors in setting of steel beams or formwork. As a result, the as-built surface of the floor always exhibits bumps and dips.

In recognition of this problem, the American Concrete Institute has revised its "Standard Tolerances for Concrete Construction and Materials" (ACI 117). The standard includes floor finish tolerances based on two measuring methods: the F-number system and the straight-edge method. F numbers describe floor flatness. The larger the F number, the flatter the floor. An F-60 floor is roughly twice as flat as an F-30 floor.

12.3   FLOOR VIBRATIONS

Earlier chapters have dealt with the subject of vibration induced in a building by outside sources of vibrational loads such as wind and earthquakes. In addition to these, a building is subjected to a variety of vibrational loads that come from within. Although almost all loads except dead loads are nonstatic, internal sources of vibration that might be a cause of concern in an office or a residential building are the oscillating machinery, passage of vehicles, and various types of impact loads such as those caused by dancing, athletic activities, and even by pedestrian traffic. The trend in the design of floor framing systems of high rises is for long spans using structural systems of minimum weight. To this end, high-strength steel with lightweight concrete topping is routinely employed. With the use of lightweight concrete, most building codes allow for a reduction in the thickness of slab required for fire rating. This results in a further reduction in the mass and stiffness of the structural system, thereby increasing the period of the structure, which at times may approach the period of the source causing the vibration. Resonance may occur, causing large forces and amplitudes of vibration.

The performance of such structures can be greatly improved by adding nonstructural elements such as partitions and ceilings which contribute greatly to the damping of vibrations. Nonstructural elements may also add to the mass and stiffness to produce the desired degree of solidity. Although the essential requirement in establishing the adequacy of a floor system is its strength, large deflections can be objectionable for several reasons: (i) excessive deflections and vibrations may give the user the negative impression that the building is not solid. In retail areas, for example, the China may rattle every time someone goes by, or mirrors in dressing rooms

of clothing stores may shake, giving the customer the somewhat nebulous but real feeling that the structure is not solid. In extreme cases, vibration may cause damage to the structure as a result of loosening of connections, brittle fracture of welds, etc. It is therefore important that the structure be able to absorb impact forces and vibrations without transmitting any humanly perceptible shaking or bouncing. Monolithic concrete buildings are more solid in this respect as compared to light-framed buildings with steel or precast concrete; (ii) excessive deflection may result in curvature or misalignments perceptible to the eye; (iii) large deflections may result in fracture of architectural elements such as plaster or masonry; and (iv) large deflections may result in the transfer of load to nonstructural elements such as curtain wall frames.

It is difficult to establish a general criterion related to the perception of vibrations. Feeling of bounciness varies from person to person, and what is objectionable to some may be barely noticeable to others. Among the criteria employed in the design of floor systems are limitations on the span-to-depth ratio and flexibility which normally lead to deeper sections than would be required from strength considerations alone. It is somewhat dubious that these limitations assure occupants' comfort.

Recognizing that there is no single scale by which the limit of tolerable deflection can be defined, the AISC specification does not specify any limit on the span-to-depth ratios for floor framing members. However, as a guide, the commentary on the specification recommends that the depth of fully stressed beams and girders in floors should not be less than $(F_y/800)$ times the span. If beams of lesser depth are used, it is recommended that the allowable bending stresses be decreased in the same ratio as the depth. Where human comfort is the criterion for limiting motion, the commentary recommends that the depth of steel beams supporting large open floor areas free of partitions and other sources of damping should not be less than one-twentieth of the span, to minimize perception of transient vibration due to pedestrian traffic.

Thus, there is no clear-cut requirement on the flexibility to limit the perception of vibration by occupants. Flexibility limits are given, however, from other considerations such as fracture of architectural elements like plaster ceilings. The rule-of-thumb limitations are $\frac{1}{150}$ to $\frac{1}{180}$ of the span for visibly perceptible curvature and $\frac{1}{240}$ to $\frac{1}{360}$ of the span for curvature likely to result in fracture of applied ceiling finishes.

In the design of floor systems, fatigue damage due to transient vibrations is not a consideration because it is tacitly assumed that the number of cycles to which the floor system is subjected is well within the fatigue limitations. However, damage due to fatigue can be a cause of concern in floors subjected to aerobic exercise activities.

Human response is directly related to the characteristics of the vertical motion of the floor system. Users perceive floor vibrations more strongly when standing or sitting on the floor than when walking across it. Human response to vibration seems to be a factor for consideration in design only when a significant proportion of the users will be standing, walking slowly, or seated.

Most of the experiments done on human response to vibrations are related to the physical safety and performance abilities of physically conditioned young subjects in a vibrating environment such as the research supported by NASA and various defense agencies. Very little information is available on the comfort of humans subjected to unexpected vibrations during the course of their normal duties such as slowly walking across a floor or sitting at a desk. Comfort is a subjective human response and defies scientific quantification. Different people report the same vibrations to be perceptible, unpleasant, or even intolerable. A measure for human response to steady sinusoidal vibration taken from Ref. 60 is shown in Fig. 12.8. Although there is no simple physical characteristic of vibration that completely defines the human response, there is enough evidence to suggest that acceleration associated in the frequency range of 1 to 10 Hz is the preferable criteria. This is the range for normally encountered natural frequencies of floor beams. Investigations have shown that human susceptibility to building floor vibrations is influenced by the rate at which the vibrations decay; people tend to be less sensitive to vibrations that decay rapidly. In fact, experiments have shown that people do not react to vibrations which persist for fewer than five cycles.

Human response ratings to a steady state of vibrations as originally documented by two researchers, Richer and Meister, have been found to be too severe for the design of building floors subject to transient vibrations caused by human activity. Lenzen (Ref. 61) has modified the Richer and Meister rating scale by multiplying the amplitude scale by 10 to account for the non-steady state of vibrations. The modified curves which account implicitly for damping are shown in Fig. 12.9. In this figure, the natural frequency f is plotted on the horizontal scale and the amplitude A_0 is plotted on the vertical scale.

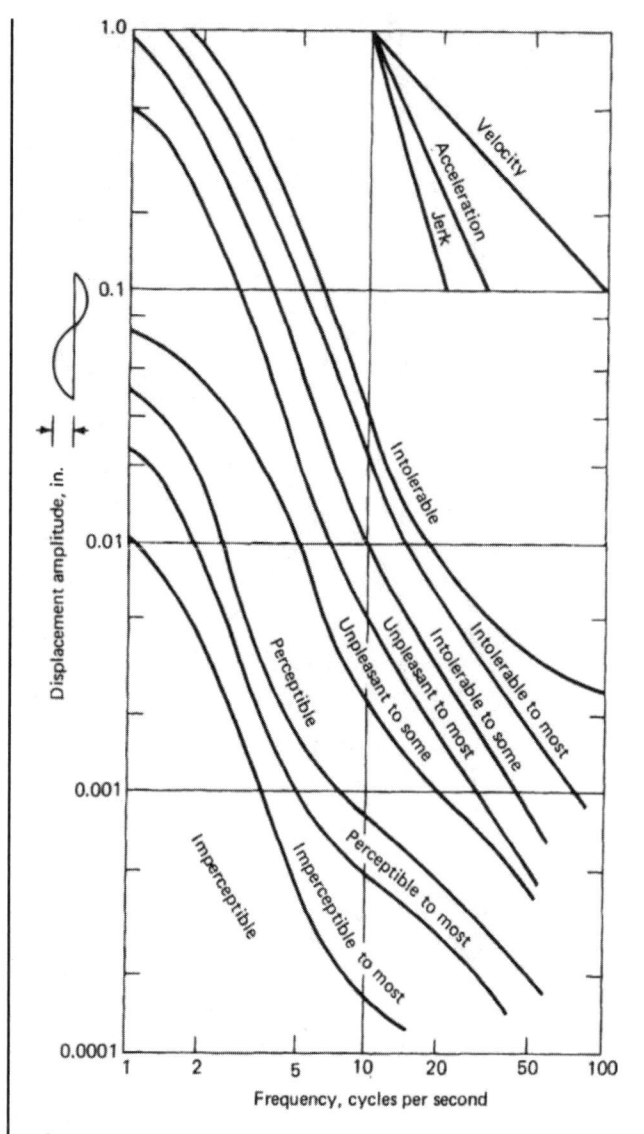

Figure 12.8 Response to sustained harmonic vibration.

The natural frequency f is given by the relation:

$$f = 1.57 \sqrt{\frac{EI_b g}{W_d l^4}}$$

where f = frequency in cycles per second

E = the modulus of elasticity of the system in ksi

I_b = transformed moment of inertia of the beam assuming full interaction with slab system in inches^4

g = acceleration due to gravity, 386.4 in./s

W_d = dead load tributary to beam in kips/in.

l = effective span of beam, in inches

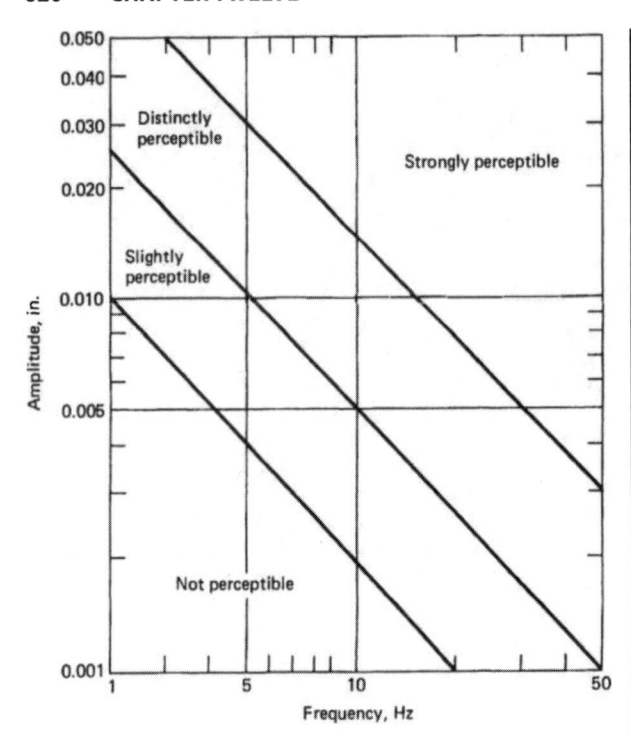

Figure 12.9   Richer, Meister vibration criteria.

The design amplitude A_0 is obtained by modifying the initial amplitude of vibration of a simply supported beam subjected to the impact load of a 190 lb person executing a heel drop. The initial amplitude for the most common value of E of 29,000 ksi is given by the relation:

$$A_{0t} = (DLF)_{max} \times \frac{l^3}{80EI_b}$$

where $(DLF)_{max}$ is the maximum dynamic factor which can be obtained from a graph given in Ref. 62.

Since a floor system usually consists of a number of parallel beams, Ref. 62 suggests that the design amplitude be obtained by dividing the initial amplitude by a factor N_{eff} to account for the action of multiple beams. Methods of estimating N_{eff} are given in Ref. 62.

The design procedure can thus be summarized as follows:

1. Compute the transformed moment of inertia of the beam under investigation. Use full composite action regardless of method of construction and assume an effective width equal to the sum of half the distances to adjacent beams. For composite beams on metal deck use an effective slab depth that is equal in weight to the actual slab including concrete in valleys of decking and the weight of decking itself.

2. Compute the frequency from the relation $f = 1.57\sqrt{EI_b g / W_d l^4}$.

3. Compute the heel drop amplitude of a single beam by using the relation $A_{0t} = (DLF)_{max} \times l^3/80EI_b$.

4. Estimate the effective number of beams, N_{eff} (Ref. 63) and compute the design amplitude by the relation $A_0 = A_{0t}/N_{eff}$.

5. Plot on the modified Richer–Meister scale (Fig. 12.9) the computed frequency f and the amplitude A_0.

6. Redesign if necessary.

Another response rating based on experimental data has been developed by Wiss and Parmelee (Ref. 47). In their method, the response rating R is given as a function of frequency, peak amplitude, and damping. Based upon the computed value of R, the expected human response is classified into one of the five following categories:

1. Imperceptible	$R < 1.5$
2. Barely perceptible	$1.5 < R < 2.5$
3. Distinctly perceptible	$2.5 < R < 3.5$
4. Strongly perceptible	$3.5 < R < 4.5$
5. Severe	$R > 4.5$

The response factor R is given by:

$$R = 5.08(FA_0/D^{0.217})^{0.265}$$

where R = response rating

F = frequency, in cycles per second

A_0 = Displacement in inches

D = Damping ratio expressed as a ratio of actual damping to critical damping

The damping coefficient D, among other things, depends on the inherent characteristics of the floor, such as ceiling, ductwork, flooring, furniture, and partitions. It should be noted that D cannot be determined theoretically but can only be estimated in relation to existing floors and their contents. For a rough estimate, the Canadian Standards Association suggests the following values:

Bare floors	$D = 0.03$
Finished floor with ceiling, mechanical ducts, flooring and furniture	$D = 0.06$
Finished floor with partitions	$D = 1.13$

Floor structures subjected to rhythmic activities such as dancing, aerobics, and other jumping exercises have been a source of annoyance to owners and engineers alike. Unlike vibration problems encountered in office occupancies, the vibrations due to rhythmic activities are continuous. These vibrations can be greatly amplified when periodic forces are synchronized with the floor frequency, a condition called resonance. Unlike transient vibrations, continuous vibrations may not decay. The National Building Code of Canada (NBC) in its commentary recommends that floor frequencies less than 5 Hz should be avoided for light residential floors, schools, auditoriums, gymnasiums, and other similar occupancies. It recommends a frequency of 10 Hz or more for very repetitive activities because of the possibility of getting resonance when the rhythmic beat is on every second cycle of vibration.

In a paper entitled "Vibration Criteria for Assembly Occupancies," Allen, Rainer, and Pernica have presented a procedure for designing floor structures subjected to rhythmic activities. Briefly the procedure is as follows:

1. Determine the density of occupancy based on type of activity. For example, if the floor area is 30 by 60 ft (9.15 by 18.3 m) and has an aerobic class of 50 people of average weight of 120 lb, the equivalent density of occupancy works out to be

$$\frac{50 \times 120}{30 \times 60} = 3.33 \text{ psf} \quad (159.6 \text{ Pa})$$

2. Choose an appropriate forcing frequency f and a dynamic load factor α. For aerobic exercises, the value of f suggested in the paper is between 1.5 and 3 Hz, while the value for α is given as 1.5.

3. Choose an acceptable limiting acceleration ratio, α_0/g at the center of the floor. The suggested value for physical exercise activity is 0.05.

4. Determine the lowest acceptable fundamental frequency f_0 of the floor system by the relation:

$$f_0 \geq f \sqrt{1 + \frac{1.3}{\alpha_0/g} \frac{\alpha W_p}{W_t}}$$

where W_p = weight per unit area of participants
$\quad W_t$ = total weight per unit area of structure, participants, furniture, etc.

5. Determine the natural frequency f_0 of the floor structure. In addition to the weight of the floor structure itself, weights of participants and furniture, if any, are to be included in the computation of f_0.

6. The frequency f_0 should be greater than or equal to the frequency obtained in step 4. If not, the options are to stiffen the floor system, relocate the activity, and convince the owner to accept a higher limiting acceleration by pointing out that no serious safety-related problems are known to have occurred for floors with frequencies higher than 6 Hz.

Increasing the frequency of the floor system by increasing the stiffness is usually cost prohibitive. The most prudent course is to make building owners aware of vibration-related problems during the early design phase.

12.4 PANEL ZONE EFFECTS

Structural engineers involved in the design of high-rise structures are confronted with many uncertainties when calculating lateral drifts. For example, they must decide the magnitude of appropriate wind loads and the limit of allowable lateral deflections and accelerations. Even assuming that these are well-defined, another question that often comes up in modeling of building frames is whether or not one should consider the panel zones at the beam-column intersections as rigid.

The panel zone can be defined as that portion of the frame whose boundaries are within the rigid connection of two or more members with webs lying in a common plane. It is the entire assemblage of the joint at the intersection of moment-connected beams and columns. It could consist of just two orthogonal members as at the intersection of a roof girder and an exterior column, or it may consist of several members coming together as at an interior joint, or any other valid combination. In all these cases, the panel zone can be looked upon as a link for transferring loads from horizontal members to vertical members and vice versa. For example, consider the free-body diagram of a frame element consisting of an assemblage of two identical beams and columns with points of zero moments at the ends (Fig. 12.10). These zero moment ends are, in fact, representative of points of inflection in the members.

Consider the frame element subjected to lateral loads. It is easy to see that because of these loads the columns are

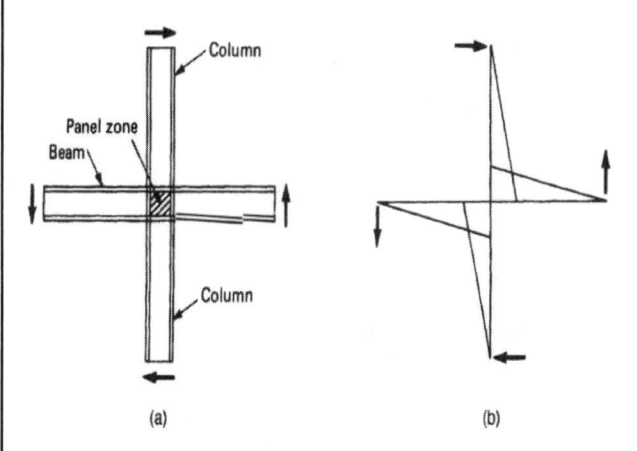

Figure 12.10 Typical frame element: (a) free-body diagram; (b) bending moments due to shear in beam and columns.

subjected to horizontal shear forces and corresponding bending moments as shown in Fig. 12.10b. Equilibrium considerations result in vertical shear forces in the beams at the inflection points and corresponding bending moments in the beams. The panel zone thus acts as a device for transferring the moments and forces between columns and beams. In providing for this mechanism the panel zone itself is subjected to large shear stresses.

The presence of high shear forces in a panel zone is best explained with references to the connection shown in Fig. 12.11a. The bending moment in the beam can be considered to be carried as tensile forces in the top flange and compressive forces in the bottom flange and the shear stresses can be assumed to be carried by the web. In the panel zone, the tensile force in the top flange is carried into the web by horizontal shear forces and, by a similar action, is converted back into a tensile force in the outer flange of the column. The distribution of the actual state of stress in the panel zone is highly indeterminate, but a reasonable approximation can be obtained by assuming that the tensile stresses are reduced linearly from a maximum at the edge of the corner B or D to zero at the external corner. If members AB and CD are assumed as stiffeners, a distinct load

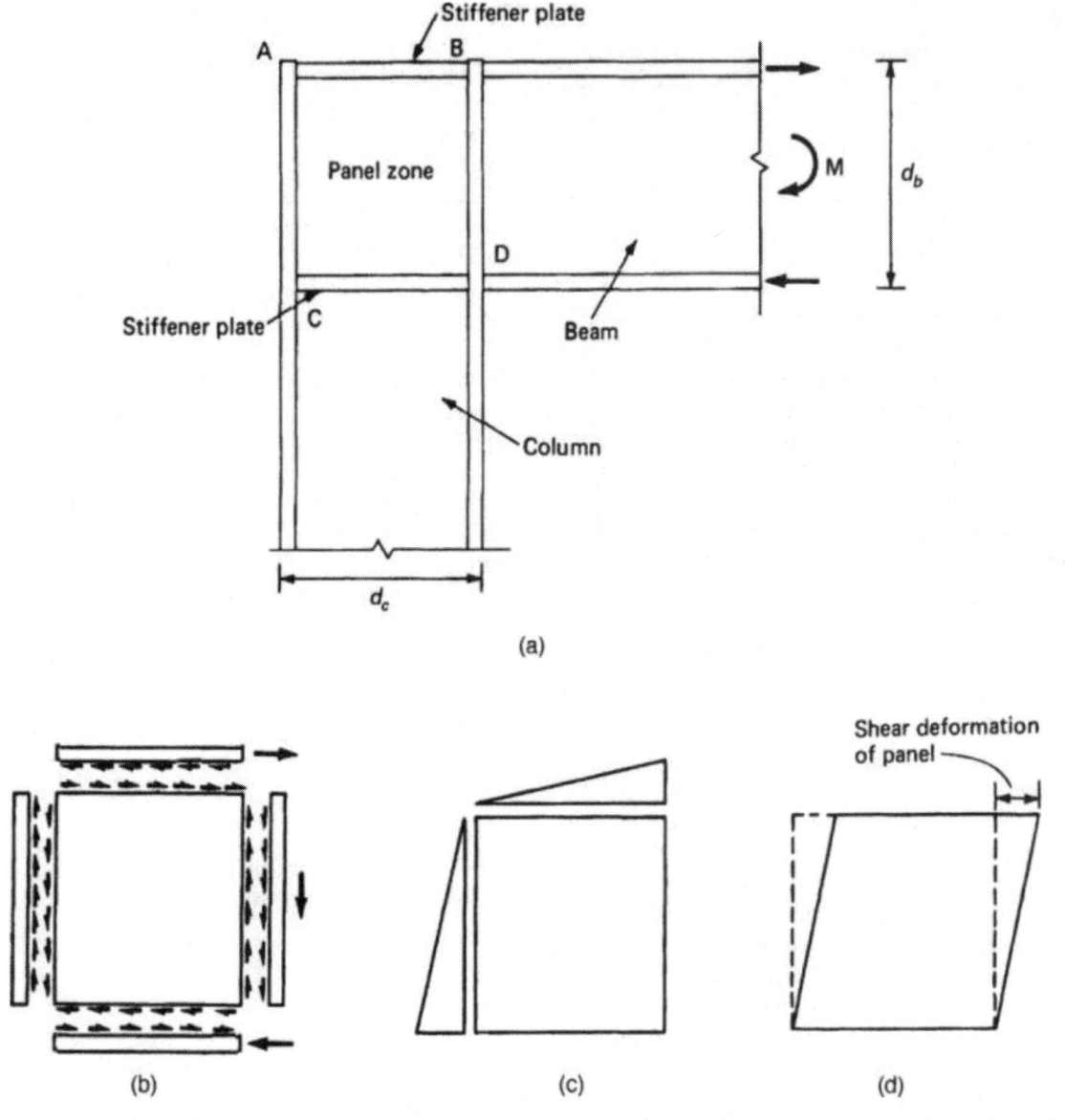

Figure 12.11   Panel zone behavior: (a) corner panel; (b) schematic representation of shear forces in panel zone; (c) linear distribution of tensile stresses; (d) shear deformation of panel zone.

path can be visualized for the compressive and tensile forces in the beam flange. Consideration of equilibrium of forces within the panel zone results in shear stress and a corresponding shear deformation as shown in Fig. 12.11d. It is this deformation that is of considerable interest in the calculation of drift of multistory buildings.

Before proceeding with a qualitative explanation of the behavior of panel zones and their influence on building drift, it is instructive to discuss some of the assumptions commonly made in the analysis of building frames. Prior to the availability of commercial analysis programs with built-in capability of treating panel zones as rigid joints, it was common practice to ignore their effects; the frame was usually modeled using actual properties along the centerlines of beams and columns.

If the size and number of joints in a frame were relatively large, an effort was made to include the effect of joint rigidity by artificially increasing the moments of inertia of beams and columns; the actual properties were usually multiplied by a square of the ratio of centerline dimensions to clear-span dimensions.

Nowadays, it is relatively easy to model the panel zone as a rigid element because of the availability of a large number of computer programs that include this feature. Flexibility of panel zones can also be considered in some of these programs, although somewhat awkwardly, by artificially decreasing the size of panel zones.

Computations of beam, column, and panel zone contributions to frame drift can be carried out by hand calculations by using virtual work method. For this purpose, consider again the typical frame element subjected to horizontal shear forces P_c and vertical shear forces P_b at the inflection points (Fig. 12.12a).

The notations used in the development of the method are as follows:

d_b = depth of panel zone
d_c = width of panel zone
h_c = clear height of column
L_c = clear span of beam
L = center-to-center span of beam
h = center-to-center height of column
I_c = moment of inertia of column
I_b = moment of inertia of beam
E = modulus of elasticity
G = shear modules
Δ_b = frame drift due to beam bending
Δ_c = frame drift due to column bending
Δ_p = frame drift due to panel zone shear deformation

The bending moment diagrams for the typical frame element can be obtained under three different assumptions.

1. The first assumption corresponds to ignoring the rigidity of panel zone; the bending moment diagrams for the external and unit loads can be assumed as shown in Figs. 12.12c and 12.12g. The bending moments increase linearly from the point of contraflexure to the centerline of the joint. By integrating the moment diagrams shown in Figs. 12.12c and 12.12g the column and beam bending contributions to the frame drift are given by:

$$\Delta_c = \frac{P_c h^3}{12EI_c}$$

$$\Delta_b = \frac{P_b L^3}{12EI_b}$$

2. In the second case, which corresponds to assuming that the panel zone is completely rigid, we get bending moment diagrams for external and unit loads as shown in Figs. 12.12b and 12.12f. The bending moments increase linearly from the points of contraflexure but stop at the face of beams and columns. Integration of moment diagrams gives the expressions for Δ_c and Δ_b as follows:

$$\Delta_c = \frac{P_c h_c^3}{12EI_c}$$

$$\Delta_b = \frac{P_b L_c^3}{12EI_b}$$

3. The third assumption, which attempts to account for the flexibility of panel zones, results in bending moment and shear force diagrams for external and unit loads as shown in Figs. 12.12d,h,e,i. Integration of bending moment and shear force diagrams leads to the following expressions in Δ_c, Δ_b, and Δ_p.

$$\Delta_c = \frac{1}{12EI_c}\left(P_c h_c^3 + P_c h_c^2 d_b\right)$$

$$\Delta_b = \frac{1}{12EI_b}\left(P_b L_c^3 + P_b L_c^2 d_c\right)$$

$$\Delta_P = \frac{P_c h_c^2}{d_b t_w d_c}$$

The effect of panel zone continuity plates may be determined by performing a finite element analysis of a typical frame unit as shown in Fig. 12.13. A series of finite element analyses can be performed to relate the effect of panel zone to basic section properties of beam and columns of the typical unit. Halvorson (Ref. 64) indicates that for a typical 13 ft, 1 in. high by 15 ft long (4 by 4.57 m) unit consisting of W 36 × 300 columns and W 36 × 230 beams, that the frame stiffness is approximately 8, 15, or

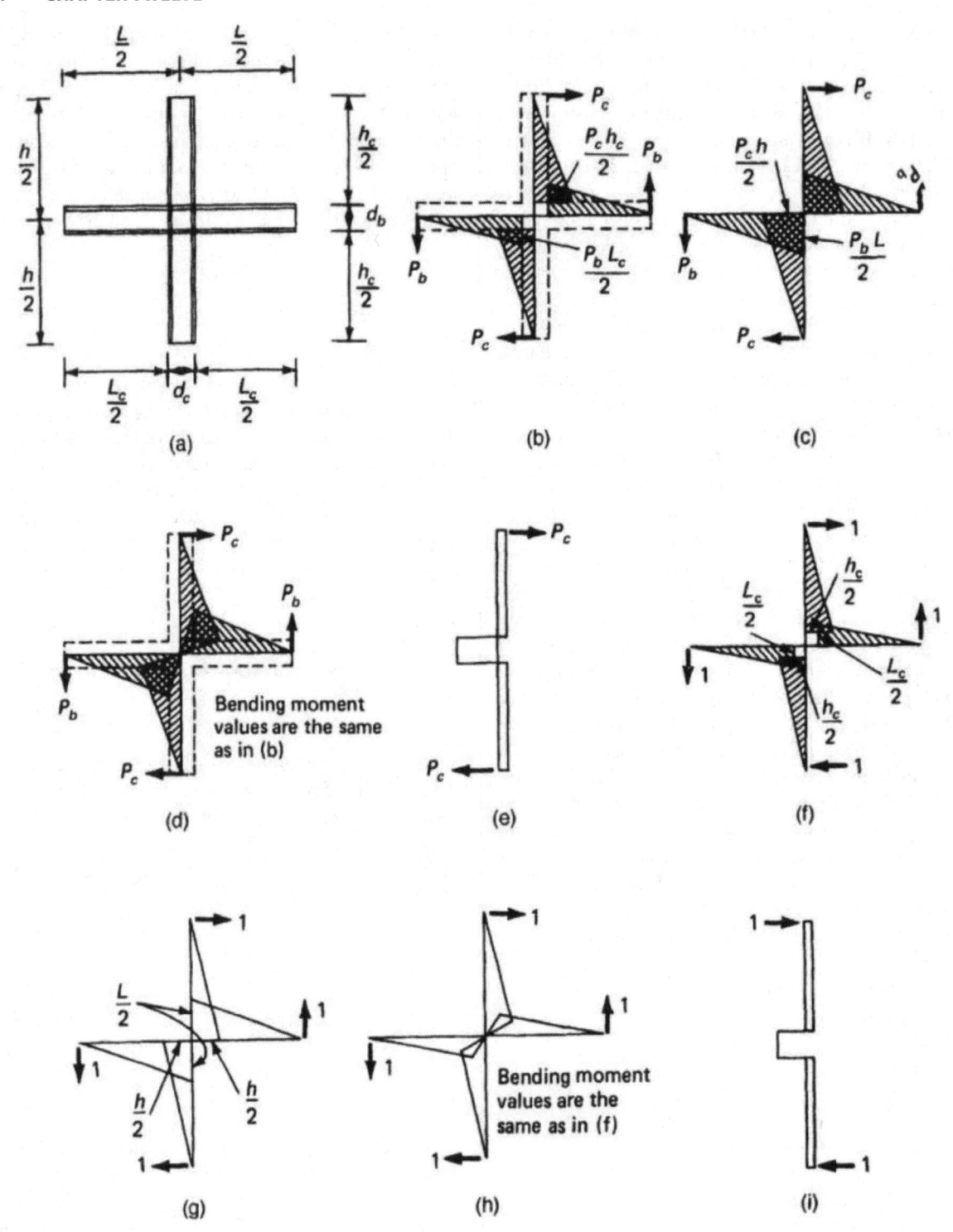

Figure 12.12 Typical frame segment: (a) geometry; (b) bending moment diagram with rigid panel zone; (c) bending moment diagram without panel zone; (d) bending moment diagram with flexible panel zone; (e) shear force diagram; (f-i) unit load diagrams.

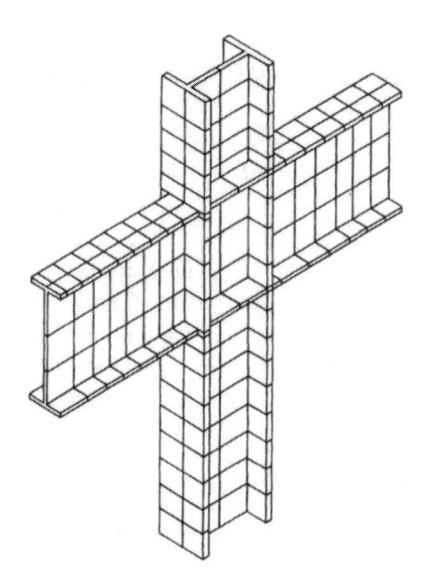

Figure 12.13 Finite element idealization of typical frame unit.

22 percent stiffer than a stick element model depending upon whether no, AISE minimum-, or full-continuity plates are provided, respectively.

Using the virtual work expressions given above, or by performing a finite element analysis, it is relatively easy to compute the contribution of panel zone deformation to frame drift. The author recommends that before undertaking the analysis of large tube-like frames, representative frame elements, say at one-fourth, one-half, and three-fourths the height of the building be analyzed to get a feel for the contribution of panel zone deformation to frame drift. Armed with the results, it is relatively easy to modify the properties of beam and columns such that the overall behavior of the frame is properly represented in the model.

12.5 CLADDING SYSTEMS

Cladding is an expensive part of the building, taking up as much as 10 to 20 percent of a building's initial cost. If the building façade is improperly designed or installed, repairs can cost many times more than the original cost because of difficulties of recladding a building that is already in use. Until the invention of structural steel building frame, about a hundred years ago, exterior building walls were load-bearing elements pierced by windows to provide light and ventilation. The walls served the dual purpose of providing support for the upper floors while simultaneously protecting the building against the elements. Today, the only function of building exterior is to act as a filter system.

The performance of a cladding system depends largely on the effect of natural forces on the cladding material. The natural forces are sunlight, temperature, water, wind, and seismic forces, in addition to the ever present gravity forces. Sunlight, particularly in the form of ultraviolet rays, produces chemical changes which cause fading and degradation of materials. Temperature creates expansion and contraction of materials in addition to the temperature differential between interior and exterior of the cladding. Water, in the form of wind-driven rain, can penetrate through the small openings and appear on the interior face of curtain walls. If trapped within the wall in the form of vapor, it can cause serious damage. Wind, acting either inward or outward, creates stresses that require the structural properties of the framing members, panels, and glass to be determined for its maximum effects. Gravity loads of cladding systems cause deflections in horizontal load-carrying members. Although gravity load itself is a one-time load, the members supporting the cladding system are subjected to variable floor and roof live loads, requiring that connections of the cladding to the frame be designed to provide sufficient relative movement to ensure that displacements do not impose vertical loads on the cladding itself. In addition, concrete buildings are subjected to creep, compounding the problem of cladding design.

In the design of curtain walls there are three matters of chief concern: (i) structural integrity, (ii) provision for movement, and (iii) weathertightness. For structural integrity of the wall, the stiffness rather than the requirement of strength is the primary concern, although anchorage failure due to inadequate strength should be given proper attention. Providing proper resistance to lateral loads is a routine procedure, especially in view of the fact that we now know more about the nature of wind loads than we did only a decade ago. The significance of negative wind pressure or suction forces acting on the wall augmented by the internal building pressure caused by air leakage is well understood, as reflected in most building codes.

In designing a curtain wall, it is important to provide for ample movement between the wall components themselves, relative movements between the components, and relative movement between the cladding and the building frame to which it is attached. Although the causes of movement are well known, it is not practical to predict the magnitude of the movements accurately. Since the movements manifest themselves at the joints of the cladding, the success of a cladding system lies in the proper detailing of the joints. Provision must be made to accommodate both the vertical and horizontal movement in the plane of the wall. Other design considerations such as weather-tightness, moisture control, thermal insulation, and sound transmissions are areas normally outside the sphere of structural engineering and therefore will not be discussed here.

Common types of curtain wall failures are: (i) structural failure of anchorages resulting in other wall elements disengaging from the building, creating serious hazard to the pedestrians; (ii) glass breakage, which in many instances has been a safety hazard; (iii) excessive air infiltration, which prevents proper conditioning of the interior space; and (iv) accumulation of condensation, causing significant damage to interior finishes.

Failure of cladding can arise because of differential vertical and lateral movement between building columns. Thermal movement, long-term creep of a concrete frame, and foundation settlements can also contribute to the failure. Structural engineers normally do not have a direct involvement in the design of building cladding systems. Typically, architect selects the cladding material for appearance and specifies requirements for the weatherproofing, performance, and durability of the cladding. From knowledge of the weight of the cladding and designated connections points, the engineer designs the structure to carry the anticipated loads. The anticipated movements of the structural frame should be indicated on the drawings in order for the curtain wall supplier to properly detail its connection to the structure. The effect of floor deflections due to the weight of cladding and live load, the magnitude of anticipated building drift, column shortening, foundation settlement, and thermal effects should all be taken into consideration.

The design of cladding systems in North America is traditionally by the architects. Many structural engineering firms specifically exclude this service by inserting a "no involvement with cladding" clause in their service contract. In spite of these attempts to keep out of conflict, if the building cladding turns out to be less than adequate, be it for watertightness, breakage, or any number of other problems, the structural engineer will be drawn into the dispute sooner or later. Therefore, it is important to communicate to the design team the expected behavior of the building frame both under gravity and lateral loads.

Building cladding is one place where a number of independent designs come together. The selection of cladding material for appearance and details for weatherproofing and performance characteristics are in the domain of the architect. The design of the structure for the weight of cladding is the responsibility of the structural engineer. It is likely that the exact weight and connection points of the cladding may not be known at the time the building frame is designed but may have to be conservatively assumed. Short- and long-term deflections of the structure due to gravity loads, column shortening due to creep and shrinkage in concrete and composite structures, expansion and contraction due to temperature, and lateral deflection of the building, especially the interstory drift, are some of the pieces of information required by the cladding supplier. The engineer should have a good understanding of tolerance required both horizontally and vertically for proper installation of curtain walls. In a core-braced steel frame, it is likely that the perimeter spandrel is designed as a simply supported beam for gravity loads and hence somewhat limber as compared to a wind-resisting moment-connected spandrel. Even when the gravity loads are heavy, typical simple connections at beam ends have very little stiffness to prevent the rotation of spandrel beams. This is, perhaps, the single most common cause for cladding problems.

The basic components that make up the exterior envelope consist of vision panels and spandrel panels. While vision panels are invariably of glass, the spectrum of materials available for spandrel panels is an ever increasing one. To name a few, spandrel panels can be made of aluminum, steel, glass, masonry, precast concrete, or GFRC. To reduce costs, many designs use a laminated thin-gauge metal sheet bonded to a metal honeycomb backup. Typically, framing employed for the mullions is aluminum because of the ease of producing complex custom shapes by extrusion.

The next sections discuss the design and installation aspects of exterior cladding systems. The proper design, installation, and operation of cladding in buildings is a culmination of the coordination activities among architects, engineers, cladding manufacturers, glazing contractors, building officials, and, of course, building owners. The following sections briefly explain how the glass commonly used in a building is made and discusses the strength and testing aspects of glass. Other cladding systems such as metal curtain wall, stone cladding, brick veneer systems, and GFRC systems are also briefly discussed.

12.5.1   Glass

Just as floor slabs are designed to carry dead load and occupancy loads, and roof decks to carry dead loads, snow loads, and uplift forces due to wind, so must windows and spandrel panels be designed to withstand the lateral loads due to wind. Glass panels in curtain walls are structural elements insofar as they resist wind pressure or suction and transfer wind forces to the building frame. In today's high-rise architecture, glass is employed in more imaginative ways than ever before. At present, the only way to construct a building 50- or 100-stories high is to frame it on a skeleton of steel or concrete and to cloak it with a skin of relatively lightweight cladding. Glass has always been the most suitable material in the eyes of modern architects as witnessed by scores of tall buildings built within the last two decades.

To counteract the fundamental drawback of glass, namely the inordinate amount of heat loss and gain in its pure state, the glass industry has invented a whole series of products intended to make glass buildings economical. There are many tinted glasses designed to reduce the blinding effect of the sun; there are tinted-glass sandwiches, double-glazed panels designed to reduce the blinding effect and to increase the insulating value of the glass wall. All these types of glass units may be used not only as vision panels but also as spandrel panels. There are reflective glasses coated with silver or gold mirror films and innumerable shading devices that reduce glare as well as heat gain and thus air-conditioning loads.

In laying the groundwork for a discussion of glass design, it may be beneficial to explain briefly how different types of glass commonly used in building are made.

Glass is an amorphous, organic, transparent, or translucent substance made of a mixture of silicates, borates, or phosphates and cooled from a liquid to a solid state. Archeological evidence from pre-Roman times indicates that glass making originated in the Near East, probably in the third millennium before Christ. Sand, flint, and quartz are the major sources of silica for glass manufacturing. Typical glass batches include, in addition to sand and other raw materials, up to 50 percent of broken glass of related composition, called waste. This waste promotes melting and homogenization of glass. Impurities, normally in the form of iron traces, cause the glass to be green or brown. To achieve a clear substance magnesium oxide is added to counteract the effect of iron traces. Glass can be colored by dissolving in it certain oxides and sulfides.

Carefully measured ingredients are mixed and allowed to undergo initial fusion before being subjected to full heat. In modern glass plants, the glass is melted in large tank furnaces heated by gas, oil, or electricity. Window glass, in use since the first century, was originally made by blowing hollow cylinders that were slit and flattened into sheets. In modern processes, molten glass can be directly drawn into sheets. Glass thus made is not entirely uniform in thickness because of the nature of the process by which it is made. The variations in thickness distort the appearance of objects viewed through panes of glass. The

traditional method of overcoming such defects has been to grind and polish the glass.

The modern procedure of making "plate glass," first introduced in France in 1668, consists of rolling the glass continuously between double rollers. After the rough sheet has been annealed, both sides of the glass sheet are finished continuously and simultaneously. *Annealing* is a process in which the glass is reheated to a temperature high enough to relieve internal stresses and then slowly cooled to avoid introducing new stresses.

Grinding and polishing are now being replaced by the more economical "float-glass" process. In this process, flat surfaces are formed on both sides by floating a continuous sheet of glass on a bath of molten tin. The temperature is high enough to allow the surface imperfections to be removed by fluid flow of glass. The temperature is gradually lowered as the glass moves along the tin bath and it passes through a long annealing oven at the end.

Annealed glass which has been reheated to a temperature near its softening point and forced to cool rapidly under carefully controlled conditions is described as "heat-treated" glass. Very often stresses are introduced intentionally in glass to impart strength to it. The objective is to introduce a surface compression because glass always breaks as a result of tensile stresses that generally originate across an infinitesimal surface scratch. Compression on the surface increases the amount of tensile stresses that can be endured before breakage occurs. One of the oldest methods of introducing surface compression is called "thermal tempering." It consists of heating the glass almost to the softening point and then cooling it rapidly with an air blast or by plunging it into a liquid bath. This process rapidly hardens the surface, and the subsequent contraction of the slower-cooling interior portions of the glass pulls the surface into compression. Surface compressions approaching 35 ksi (241.32 MPa) can be obtained in thick pieces by this method. Heat-treated glasses are classified as either "fully tempered" or "heat strengthened" according to the magnitude of compressive stresses induced during heat treatment. Federal specification DD-G-1403B calls for a minimum surface compression of 10 ksi (69 MPa) for a minimum edge compression of 9.7 ksi (66.8 MPa) for the glass to be classified as fully tempered glass. The corresponding minimum requirements for heat-strengthened glass are 3.5 and 5.5 ksi (24.1 and 38 MPa). Below this level, the glass is classified as annealed glass.

The fracture characteristics of heat-strengthened glass vary widely. Annealed glass fractures at about 3.5 ksi (24.1 MPa) while fully tempered glass does so at 10 ksi (69 MPa). The characteristic feature of tempered glass is that the glass fractures into small relatively harmless fragments, reducing the likelihood of injury to people. Therefore, many building regulations require the use of tempered glass for skylights, overhead glazing, sloped glazing, and other safety glazing applications.

"Laminated glass" units are made by bonding together two or more lights of glass with an elastomer interlayer. The interlayer is commonly a plastic film of 0.030 in. (0.76 mm) thickness. The laminated glass unit can be made either with annealed, heat-strengthened, or fully tempered glass sheets in any combination. The interlayer does not possess the strength or the stiffness necessary to render the composite unit as strong as an equivalent monolithic light of same thickness. The strength of a laminated unit is taken as 60 percent of the strength of a monolithic light of equal thickness. The actual strength could vary anywhere from 25 percent to full strength of an equivalent monolithic light depending upon the capacity of the interlayer to transmit horizontal shear loads.

"Insulating glass unit" consists of two lights of monolithic glass separated by a spacer and sealed around the perimeter. The sealed air space acts as a layer of insulation, greatly improving the heat-resisting properties of the unit. The edge seal may be fused glass or may be composed of elastomeric sealants and silicones capable of providing a moisture seal around the air space for the normal life of a building. The spacer contains a desiccant to absorb any moisture that may cause the fogging of the glass. Because the air space within an insulating unit is sealed, any pressure applied onto one face is effectively transferred to the other light, making the two lights share the external pressure.

"Wire glass," made by introducing wire mesh into the molten glass before it passes between the rollers, is used to prevent glass from shattering if it is struck. The presence of mesh does not increase the resistance of the light to breakage; it simply holds the pieces together should breakage occur. Building codes commonly consider the strength of wired glass as 50 percent the strength of an annealed monolithic light of the same thickness. Instead of casting a wire mesh within the glass light, an elastomeric film can be applied to the surface to improve the resistance to falling from the frame after breakage.

The strength of glass panels cannot be determined by the classical method of plate analysis because such an analysis is valid only for deflections considerably smaller than the plate thickness. Glass panels deflect many times their thickness and as a result develop membrane stresses which add significantly to their strength and stiffness. Moreover, glass is a brittle material exhibiting no observable yield strength. As such its behavior under loads is best described

by evaluating the strength under full-scale test results. Test results show a wide variation of strength for the same size and support conditions of glass panels. This is because the mechanism of glass failure is complex and is highly sensitive to different characteristics of flaws such as their size, orientation, and severity. The only practical approach is to evaluate the effects using an appropriate statistical model. To incorporate the scatter of statistical analysis, a so-called design factor is correlated with probability of failure at full design load. The glass industry uses the normal distribution as the standard model in recommending glass thickness for various design conditions. The published charts of recommended thickness are based on an expected glass breakage probability of 8 lights per 1000, resulting in a design factor of 2.5.

It should be noted that from the statistical point of view, it is virtually impossible to design a glass light without some probability of its failure under design load. Therefore, in conducting a full-scale mock-up test of curtain walls, glass breakage is not considered as a cause for test failure. Glass breakage may very well occur, since the test loads are usually much larger than the design load of glass. If breakage occurs, broken glass is replaced with plywood in order to complete mock-up tests. However, if the test indicates that premature failure of glass is due to inadequate stiffness of the glass supporting system, stricter deflection control should be imposed in the supporting system.

In designing cladding systems with double-glazed insulating sandwiches, care should be taken to prevent the buildup of ultraviolet light, which is known to play havoc with the strip of sealants used on the sides and edges.

Tempered or semitempered glass commonly used in spandrel areas appears to be susceptible to breakage from even minute inclusions of nickel sulfide. Tempered glass is used in these areas because of the large amount of heat that builds up in the space behind the spandrel between the ceiling and floor above. Heat-strengthened glass which is not fully tempered and does not have the same mechanical and thermal endurance qualities is considered much less susceptible to breakage from infiltration of nickel and carbon sulfide.

The performance of the glass supporting members such as mullions may have a bearing on the actual load-carrying capacity of the window if they are significantly more flexible than the support provided in strength tests. Moreover, relatively large lateral deflection may occur under design loads, resulting in in-plane movements and a tendency for the glass to slip out of its retaining frame.

The most commonly recommended deflection limitation for the glass supporting system is 1:175 of the span.

However, stiffer supports may be required for proper weathertightness, durability of sealants, or appearance.

Glass design from a structural point of view consists of selecting an appropriate thickness for a given area and design pressure from charts based on tests conducted by the glass industry.

12.5.2   Metal Curtain Wall

This is one of the most popular methods of cladding the exterior of a building. In simple terms it consists of a metal framework in which metal, glass, and other surface material are housed. Although aluminum curtain walls of today appear to present endless variety of design, the majority of these designs can be classified into the following five generally recognized systems.

1. Stick system.
2. Unit system.
3. Unit and mullion system.
4. Panel system.
5. Column and spandrel cover system.

In the stick system, the vertical mullions are usually attached first to the structural frame with anchors followed by installation of the window sill and head section. Spandrel panels and vision glass are attached within the mullion framework to complete the system. In the unit system, the curtain wall is preassembled into large units complete with spandrel panels and sometimes also with glass panels. The units may be one, two, or sometimes three stories in height. Typical units are designed to snap in place for a sequential interlocking installation.

The unit and mullion system, as the name implies, is a system that attempts to capture the advantages of both the stick and the unit systems. Mullions are installed first, followed by placement of preassembled frame units between them. The panel system is similar in principle to the unit system, the main difference being that the panels do not consist of vertical and horizontal mullions but are integral units formed from sheet metal or laminated aluminum honeycomb panels. Unlike the other systems, the panel system does not have to follow the rigid discipline of vertical and horizontal grid pattern and can be made to represent a wider range of architectural design flexibility.

The column and spandrel cover system can be thought of as a response of the cladding design to the tube system of lateral bracing consisting of closely spaced columns and deep spandrels. Advantage is taken of the close spacing of columns by spanning the glazing units directly between the column covers. The system can be engineered to clearly express the structural skeleton and permits a wide latitude of aesthetic expression for column and spandrels.

12.5.3   Stone Cladding

Stone provides a distinctive alternative to glass and metal curtain walls. Distinctive pinks, reds, grays, and blacks of polished granite and the white, buff, and green of marble along with the black and blue of slate are appearing on the urban skyline. Their popularity has increased in such an extent that major glass and aluminum curtain wall manufacturers have adapted their systems to accept stone inserts. Angles, straps, anchors, and wire ties are common methods of stone erection subject to individual design considerations. New methods have been developed to cut natural stone more precisely and quickly. Because stone is a product of nature with inherent variations in physical properties, proper selection of material and its preparation as a cladding system requires careful evaluation. To this end the C-18 Committee on Natural Building Stones of the American Society of Testing and Materials has developed standards for a variety of building stones.

Stone can be finished in a variety of ways. It can be polished, honed, rubbed, flame or thermal finished, or may be used without any finish to retain the natural sawn appearance. Not all ranges of finishes are applicable to all types of stones. The wide range of finish choices applicable for dense stones such as granite diminishes to a limited choice for soft stones such as limestone and sandstone.

There are four primary types of stone veneer attachment to the building exterior. In the first two, the stone is installed piece-by-piece on site, while in the other two types it is fabricated offsite and installed as an integral panel. In the conventional piece-by-piece system, the stone is installed by using relieving angles, strap anchors, dowels, mortar, or other mechanical systems. In the standard method, which is perhaps the oldest, the stone is attached to a masonry backup using portland cement motor spots. In the mechanical system the stone panel is retained in place by angles fitted into slots cut into the sides of the stone. It is important to note that noncorrosive metals must be protected from electrolytic reaction.

In the prefabricated system, panels are attached off the job site to a concrete back-up system. The stone is lowered, finished-face down, into prepared forms. Stainless steel anchors are placed into precision-drilled holes on the back of the stone. The location and number of anchors used are functions of the spanning capability of the stone. A bond breaker such as a liquid applied membrane, or a sheet of polyethylene is laid on the back of the stone prior to concrete pour. The purpose of bond breaker is to accommodate differential movement between the stone and the concrete backup. After the concrete is cured, the panels are trucked to the job site, hoisted into position, and secured to the structural frame.

Another method of prefabrication is to fasten stone pieces directly onto steel trusses. Yet another is to fasten a fiberglass mesh portland cement backing board to the truss and attaching the stone by using the mortar method.

A system popular in certain parts of North America consists of mullions with precut stone veneer in the non-vision areas.

12.5.4 Brick Veneer Systems

Many problems that occur in a brick façade stem from the expansion of the brick due to moisture and direct sunlight, effects of freezing and thawing, and problems arising from insufficient firing of brick. Brick cladding attached to concrete frame buildings is subjected to serious stress buildups because concrete tends to creep with time while masonry has a tendency to swell. To alleviate cladding problems, it is necessary to provide enough shelf angles for support and sufficient horizontal expansion joints in the wall to allow the two components—concrete and brick—to move independently.

Adequate drainage should be provided to assure that water is not trapped behind the façade of buildings in areas where winters are harsh. Otherwise, water freezes and expands, dislodging the brick. A similar problem occurs if brick is supported on untreated steel structure. When steel rusts it flakes and expands, pushing out the brick. Proper coating of steel to prevent rust is necessary. The cladding joints should be detailed in such a way as not to jostle each other under building movements.

12.5.5 Glass Fiber-Reinforced Concrete Cladding

Glass fiber-reinforced concrete (GFRC) cladding is a composite material consisting of a portland cement base with glass fibers randomly dispersed throughout the material. The fibers add to the tensile and impact strengths, making possible the production of strong yet lightweight architectural panels. The panels can be made much thinner than a conventional precast system thus resulting in a lightweight system. The lightweight GFRC cladding can provide significant savings by minimizing structural framing and foundation costs for multistory construction especially in areas with poor supporting soil.

12.5.6 Curtain Wall Mock-Up Tests

In high-rise construction, it is a normal practice to test a full-scale mock-up of cladding to evaluate the performance of cladding against the various environmental elements. Structural framing, glass, sealants, gaskets, and anchorage devices representing the actual job site conditions. In most cases, the cladding materials for the mock-up are supplied by the individual job supplier under the direction of the curtain wall subcontractor. The height of mock-up may range from one to four stories, typically forming one side of a four-sided roofed test chamber.

Three performance characteristics are commonly investigated in the mock-up tests: resistance to air infiltration, resistance to water penetration, and structural performance. Additionally other tests to measure heat and sound transmission characteristics may also be conducted. Traditionally structural engineers have very little input for the tests except for the one that measures the structural characteristics. Therefore, in this section a brief description of the test for structural performance is given. The reader is referred to other publications, such as *Aluminum Curtain Wall Design Guide Manual*, for a description of tests not covered in this section.

The governing factor in the structural design of framing members and panels is usually the stiffness rather than strength. Analytically it is impossible to account for all the interactions that exist in a curtain wall which is comprised of many interdependent elements such as fastenings, anchors, nonrigid joints, seals, and gaskets. Physical testing is often the most reliable means of verifying the performance. Useful information can be obtained by loading the specimens to failure to get an insight into the ultimate capacity of the system and to identify its weak spots. Structural testing is conducted by the static method using an air chamber, subjecting the test specimen to both positive and negative pressures.

Air is pumped into the chamber to simulate outward-acting, that is, negative wind loads, and is sucked out of the chamber to simulate inward-acting, that is, positive wind loads. A monometer is used to measure the air pressure differential between the outside and inside of the pressure chamber. In the dynamic water test, a propeller is used to create air turbulence to simulate wind storms.

Typically, the structural test consists of positive and negative uniform static design load and a proof load of 150 percent of the design load. Seismic load test is conducted by subjecting the wall to a racking displacement in the plane of the wall equal to the calculated floor-to-floor lateral displacement, and for proof load twice that displacement.

12.6 MECHANICAL DAMPING SYSTEMS

The exact dynamic behavior of tall buildings is impossible to predict with any great certainty because of the complicated nature of wind and the uncertainty in the evaluation of building stiffness. However, this much is well understood: designing a tall building to meet a given drift criterion under equivalent static forces will not automatically preclude creature-comfort problems.

The intrinsic stiffness of buildings can be increased up to a point beyond which it becomes prohibitively costly to do so. Although computer programs are capable of analytically determining the dynamic characteristics of buildings, it should be remembered that the damping characteristics cannot be estimated closer than say, 30 percent until after the building is constructed. From the fundamental dynamic equations, it is well known that the wind-induced building response is inversely proportional to the square root of total damping comprised of aerodynamic plus structural damping. To reduce the response by 50 percent, we have to increase the structural damping by about four times. Because the inherent damping of a building under wind loads is in the range of 0.5 to 1.5 percent, it is impractical to increase this value to achieve a meaningful reduction in response.

Installation of external damping systems offers an effective way of increasing the comfort of occupants. There are two types of externally installed damping systems. One called "passive viscoelastic damper" was first used in the twin towers of the World Trade Center in New York. A more recent application has been in the 76-story building called Columbia Center in Seattle. Viscoelastic dampers used in this building consist of steel plates coated with a polymer compound. The plates are sandwiched between a system of relatively stationary plates. As the building sways under the action of wind loads, the steel plates which are attached to structural members are subjected alternately to compression and tension. The viscoelastic polymer is in turn subjected to shearing deformations absorbing the strain energy created in the structural members. The dissipation of energy into heat reduces the building sway.

In another system called the "tuned mass damper" or TMD, the building oscillations are controlled in high winds by placing near the top of the building a large mass that oscillates in a direction counter to the direction of building deflection. This system is used on two buildings—the City Corp Center in New York and the John Hancock Tower in Boston. The TMD in the Hancock Tower consists of a huge concrete block weighing about 2 percent of the building weight. It rests on a smooth concrete surface with a large spring connected to one side of the concrete block, while a piston similar to a shock absorber is connected to the other side. The building accelerations are continuously monitored and when accelerations exceed 3 milli-g, indicating the possibility of a heavy windstorm, the oil pumps which levitate the block above the concrete surface are automatically activated. The mass of concrete is thus free to move back and forth. The device is said to be tuned because during the installation of the damper the mass of concrete and the stiffness of spring are adjusted to the same frequency as that of the building. As the building sways, the mass begins to move in the opposite direction, creating a force that opposes or dampens the motion of the building.

12.7 FOUNDATIONS

The structural design of a skyscraper foundation is primarily determined by loads transmitted by its many floors to the ground on which the building stands. To keep its balance in high windstorms and earthquakes, its foundation requires special consideration because the lateral loads which must be delivered to the soil are rather large. Where load-bearing rock or stable soils such as compact glacial tills are encountered at reasonable depth, as in Dallas with limestone with a bearing pressure of 50 tons per square foot (47.88×10^2 kPa), Chicago with hard pan at 20 to 40 tons per square foot (19.15×10^2 to 38.3×10^2 kPa), the foundation may be directly carried down to the load-bearing strata. This is accomplished by utilizing deep basements, caissons, or piles to carry the column loads down through poor soils to compact materials. The primary objective of a foundation system is to provide reasonable flexibility and freedom in architectural layout; it should be able to accommodate large variations in column loadings and spacings without adversely affecting the structural system due to differential settlements.

Many principal cities of the world are fortunate to be underlain by incompressible bedrock at shallow depths, but certain others rest on thick deposits of compressible soil. The soils underlying downtown Houston, for example, are primarily clays that are susceptible to significant volume changes due to changes in applied loads. The loads on such compressible soils must be controlled to keep settlements to acceptable limits.

Usually this is done by excavating a weight of soil equal to a significant portion of the gross weight of the structure. The net allowable pressure that the soil can be subjected to is dependent upon the physical characteristics of the soil. Where soil conditions are poor, a weight of the soil equal to the weight of the building may have to be excavated to result in what is commonly known as fully compensated foundation. Construction of deep foundations may create a serious menace to many older neighboring buildings in many ways. If the water table is high, installation of pumps may be required to reduce the water pressure during the construction of basements and may even require a permanent dewatering system. Depending upon the nature of subsoil conditions, the water table under the adjoining facilities can be lowered, creating an adverse effect on neighboring buildings. Another effect to be kept in mind is the settlement of nearby structures from the weight of the new building.

For buildings in seismic zones, in addition to the stiffness and load distribution, it is important to consider the rigidity of the foundation. During earthquakes, the building displacements are increased by the angular rotation of the foundation due to rocking action. The effect is an increase in the natural period of vibration of the building.

Loads resulting at the foundation level due to wind or earthquake must be delivered ultimately to the soil. The vertical component due to overturning effects is resisted by the soil in a manner similar to the effects of gravity loads. The lateral component is resisted by: (i) shear resistance of piles or piers; (ii) axial loads in batter piles; (iii) shear along the base of the structure; (iv) lateral resistance of soil pressure acting against foundation walls, piers, etc. Depending upon the type of foundation, one or more of the above may play a predominant role in resisting the lateral component.

Much engineering judgment is required to reach a sound conclusion on the allowable movements that can be safely tolerated in a tall building. A number of factors need to be taken into account. These are

1. Type of framing employed for the building.
2. Magnitude of total as well as differential movement.
3. Rate at which the predicted movement takes place.
4. Type of movement whether the deformation of the soil causes tilting or vertical displacement of the building.

Every city has its own particular characteristics in regard to the design and construction of foundations for tall buildings that are characterized by the local geology and groundwater conditions. Their choice for a particular project is primarily influenced by economic and soil conditions, and even under identical conditions can vary in different geographical locations. In this section a brief description of two types, namely, the pile and mat foundations, is given, highlighting their practical aspects.

12.7.1   Pile Foundation

Pile foundation using either driven piles or drilled piers (also called caissons) are finding more and more applications in tall building foundation systems. Driven piles usually consist of prestressed precast piles, or steel pipes with pipe, box, or steel H sections. Drilled piers may consist of either straight shafts or may have bells or underreams

at the bottom. The number of different pile and caisson types in use is continually changing with the development of pile-driving and earth-drilling equipment.

Driven piles can be satisfactorily founded in nearly all types of soil conditions. When soils overlying the foundation stratum are soft, normally no problem is encountered in driving the piles. If variations occur in the level of the bearing stratum, it will be necessary to use different lengths of piles over the site. A bearing type of pile or pier receives its principal vertical support from a soil or rock layer at the bottom of the pier, while a friction-type pier receives its vertical support from skin resistance developed along the shaft. A combination pier, as the name implies, provides resistance from a combination of bearing at the bottom and friction along the shaft. The function of a foundation is to transfer axial loads, lateral loads, and bending moments to the soil or rock surrounding and supporting it.

The design of a pier consists of two steps: (i) determination of pier size, based on allowable bearing and skin friction if any, of the foundation material; and (ii) design of the concrete pier itself as a compression member. Piers that cannot be designed in plain concrete with practical dimensions can be designed in reinforced concrete in accordance with the provisions of the applicable codes, such as the ACI code. When tall buildings are constructed with deep basements, the earth pressure on the basement walls may be sufficient to resist the lateral loads from the superstructure. However, the necessary resistance must be provided by the piers when there is no basement, when the depth of basement walls below the surface is too shallow, or when the lateral movements associated with the mobilization of adequate earth pressure are too large to be tolerated. In such cases it is necessary to design the piers for lateral forces at the top, axial forces from gravity loads and overturning, and concentrated moments at the top. One method of evaluating lateral response of piers is to use the theory of beam on elastic foundation by considering the lateral reaction of the soil as an equivalent lateral elastic spring.

The effect of higher concentration of gravity loading over the plan area of a tall building often necessitates use of piles in large groups. In comparison to the stresses in the soil produced by a single pile, the influence of a group of piles extends to a significantly greater distance both laterally and vertically. The resultant effect on both ultimate resistance to failure and overall settlement are significantly different than the summation of individual pile contributions. Because of group action, the ultimate resistance is less while the overall settlement is more.

Oftentimes the engineers and architects are challenged to create a floating effect for the building. This is usually achieved by not bringing the façade right to the ground and by using glass extensively on the ground to create an open feeling in the lobby. A structural system which uses a heavily braced core and a nominal moment frame on the perimeter presents itself as a solution, the core resisting most of the overturning moment and shear while the perimeter frame provides the torsional resistance. Because of the limited width of the core, strong uplift forces are created in the core columns due to lateral loads. A similar situation develops in the corner columns of exterior-braced tube structures. One of the methods of overcoming the uplift forces is to literally anchor the columns into bedrock. A system of post-tensioned, high-strength anchors about 30 ft or so are driven into bedrock, and high-strength grout is injected at the tips to anchor the post-tensioned steel into the rock. A concrete pier constructed below the foundation is secured to the rock by the post-tensioned anchors. Anchor bolts for steel columns cast in the pier transfer the tensile forces from columns to the pier. Another method of securing the columns is to thread the post-tensioned anchors directly through the base plate assemblies of the column.

Figure 12.14 shows the plan and cross-section of a foundation system for a corner column of an X-braced tube building. The spread footing founded on limestone resists the compressive forces while the belled pier under the spread footing is designed to resist uplift forces. To guard against the failure of rock due to horizontal fissures a series of rock bolts are installed around the perimeter of spread footing.

12.7.2 Mat Foundation

General Considerations. The absence of high bearing and side friction capacities of stratum at a reasonable depth beneath the footprint of the building precludes the use of piles or deep underreamed footings. In such circumstances, mat foundations are routinely used under tall buildings, particularly when the soil conditions result in conventional footings or piles occupying most of the footprint of the building. Although it may be possible to construct a multitude of individual or combined footings under each vertical load-bearing element, mat foundations are preferred because of the tendency of the mat to equalize the foundation settlements. Because of continuity, mat foundations have the capacity to bridge across local weak spots in substratums. Mat foundations are predominantly used in two instances: (i) whenever the underlying load-bearing stratum consists of soft,

compressible material with low bearing capacity; and (ii) as a giant pile cap to distribute the building load to a cluster of piles placed under the footprint of the building.

Mat foundations are ideal when the superstructure load is delivered to the foundation through a series of vertical elements resulting in a more or less uniform bearing pressure. It may not be a good solution when high concentrations of loading occur over limited plan area. For example, in a core-supported structure carrying most of the building load, if not the entire load, it is uneconomical to spread the load over the entire footprint of the building because this would involve construction of exceptionally thick and heavily reinforced mat. A more direct solution is to use driven piles or drilled caissons directly under the core.

The plan dimensions of the mat are determined such that the mat contact pressure does not exceed the allowable bearing capacity prescribed by the geotechnical consultant. Typically, three types of allowable pressures are to be recognized: (i) net sustained pressure under sustained gravity loads; (ii) gross pressure under total design gravity loads; and (iii) gross pressure under both gravity and lateral loads.

In arriving at the net sustained pressure, the loads to be considered on the mat area should consist of

1. Gravity load due to the weight of the structural frame.

2. Weight of curtain wall, cooling tower, and other mechanical equipment.

3. An allowance for actual ceiling construction including air conditioning ductwork, lights sprinklers, and fireproofing.

4. Probable weight of partition based on single and multitenant layouts.

5. Probable sustained live load.

6. Loads applied to the mat from backfill, slab, pavings, etc.

7. Weight of mat.

8. Weight of soil removed from grade to the bottom of the mat.

This last item accounts for the reduction in overburden pressure and therefore is subtracted in calculating the net sustained pressure.

In calculating the sustained pressure on mats, typically less than the code-prescribed values are used for items 4 and 5, requiring engineering judgment in their estimation. In the opinion of the author, a total of 20 psf (958 Pa) for these items appears to be adequate. A limit on sustained pressure is basically a limit on the settlement of the mat. In practical cases of mat design, it is not uncommon to have the calculated sustained pressure under isolated regions of

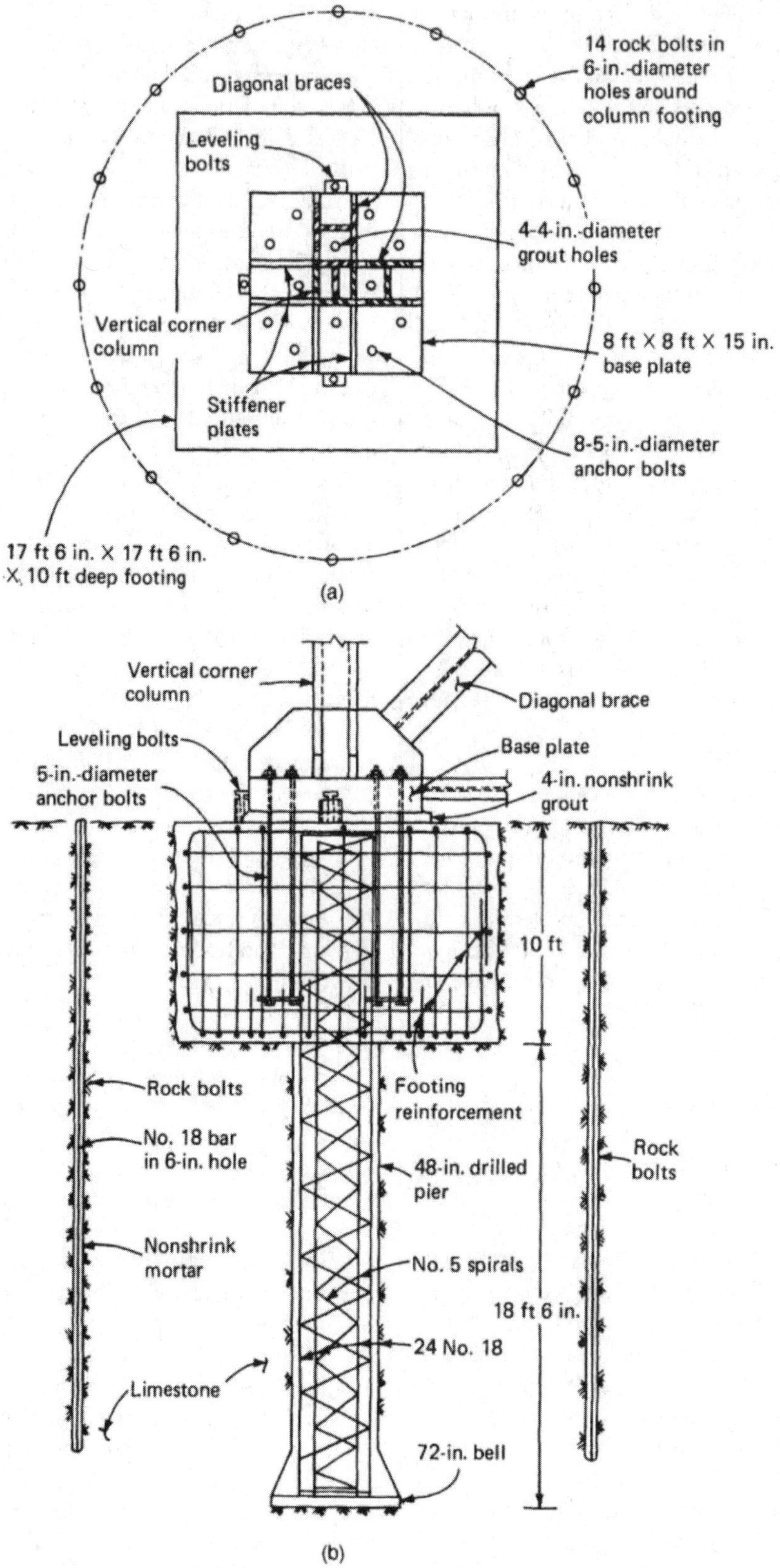

Figure 12.14 Foundation system for a corner column for an *X*-braced tube building: (a) plan; (b) section.

mat somewhat larger than the prescribed limits. This situation should be reviewed with the geotechnical engineer and usually is of no concern as long as the overstress is limited to a small portion of the mat.

The gross pressure on the mat is equivalent to the loads obtained from items 1 through 7. The weight of the soil removed from grade to the bottom of the mat is not subtracted from the total load because the gross pressure is of concern. Also, in calculating the weight of partition and live loads, the code-specified values are used.

The transitory nature of lateral loads is recognized in mat design by allowing a temporary overstress on the soil. This concept is similar to the 33 percent increase in stresses allowed in most building codes for wind and seismic loads. From an academic point of view, the ideal thickness for a mat is the one that is just right from punching shear considerations while at the same time minimum reinforcement of $0.002A_c$ provided for temperature works just right from flexural considerations. However, in practice it is found that it is more economical to construct pedestals or provide shear reinforcement in the mat rather than to increase the basic mat thickness.

In detailing the flexural reinforcement there appear to be two schools of thought. One school maintains that it is more economical to limit the largest bar size to a #11 bar which can be lap-spliced. However, this limitation may force the use of as many as four layers of reinforcement both at the top and the bottom of mat. The other school promotes the use of #14 and #18 bars with mechanical tension splices. This requires fewer bars, resulting in cost savings in the placement of reinforcement. The choice is, of course, a matter of economy as perceived by the contractor.

Analysis. A vast majority of soil-structure interaction takes place under sustained gravity loads. Although the interaction is complicated by the nonlinear and time-dependent behavior of soils, it is convenient for analytical purposes to represent the soil as an equivalent elastic spring. This concept was first proposed by Winkler in 1867 and hence the name Winkler spring. He proposed that the force and vertical displacement relationship of the soil be expressed in terms of a constant K called the modulus of subgrade reaction. It is easy to incorporate the effect of the soil by simply including a spring with a stiffness factor in terms of force per unit length beneath each reaction. However, it should be remembered that the modulus is not a fundamental property of the soil. It depends on many things, including the size of the loaded area and the length of time it is loaded. Consequently, the modulus of subgrade reaction used for calculating the spring constants must be consistent with the type and duration of loading applied to the mat.

Prior to the availability of finite element programs, mat analysis used to be undertaken by using a grid analysis by treating the mat as an assemblage of linear elements. The grid members are assigned equivalent properties of a rectangular mat section tributary to the grid. The magnitude of the Winkler's spring constant at each grid intersection is calculated on the basis of tributary area of the joint.

The preferred method for analyzing mats under tall buildings is to use a finite element computer program. With the availability of computers, analytical solutions for complex mats are no longer cumbersome; engineers can incorporate the following complexities into the solution with a minimum of effort:

1. Varying subgrade modulus.
2. Mats of complex shapes.
3. Mats with nonuniform thickness.
4. Mats subjected to arbitrary loads due to axial loads and moments.
5. Soil-structure interaction in cases where the rigidity of the structure significantly affects the mat behavior.

As in other finite element idealizations, the mathematical model for the mat consists of an assemblage of discretized elements interconnected at the nodes. It is usual practice to use rectangular or square elements instead of triangular elements because of the superiority of the former in solving plate-bending problems. The element normally employed is a plate-bending element with 12 degrees of freedom for three generalized displacements at each node. The reaction of the soil is modeled as a series of independent elastic springs located at each node in the computer model. The behavior of the soil tributary to each node is mathematically represented as a Winkler spring at each node. There is no continuity between the springs other than through the mat. Also, the springs because of their very nature can only resist compression loads although computationally it is not possible to impose this restriction in a linear elastic analysis. Therefore, it is necessary to review the spring reactions for any possible tensile support reactions. Should this have occurred in the analysis, it is necessary to set the spring constant to zero at these nodes and to perform a new analysis. This iterative procedure is carried out until the analysis shows no tensile forces in springs.

In modeling the mat as an assemblage of finite elements, the following key factors should be considered: (i) grid lines that delineate the mat into finite elements should encompass the boundaries of the slab, as well as all openings. They should also occur between elements with changes in thicknesses. Skew boundaries of mat not parallel to the orthogonal grid lines may be approximated by steps that closely resemble the skewed boundary;

(ii) grid lines should intersect preferably at the location of all columns. Minor deviations are permissible without loss of meaningful accuracy; (iii) a finer grid should be used to define regions subjected to severe displacement gradients. This can be achieved by inserting additional grid lines adjacent to major columns and shear walls.

Although it is possible to construct an analytical model consisting of both the mat and superstructure, practical budgetary and time considerations preclude use of such complex analyses in everyday practice. Admittedly the trend, with the availability of computers and general analysis programs, is certainly toward this end. However, the current practice of accounting for superstructure interaction is to simulate the stiffness of superstructure by incorporating artificially stiff elements in mat analysis. Although the procedure is approximate, it has the advantage of being simple and yet capable of capturing the essential stiffness contribution of the superstructure.

The complex soil-structure interaction can be accounted for in the design by the following iterative procedure. Initially, the pressure distribution under the mat is calculated on the assumption of a rigid mat. The geotechnical engineer uses this value to obtain the deformation and hence the modulus of subgrade reaction at various points under the footprint of the mat. Under uniform pressure, the soil generally shows greater deformations at the center than at the edges of the mat. The modulus of subgrade reaction, which is a function of the displacement of the soil, therefore has higher values at the edges than at the center.

The finite element mat analysis is performed using the varying moduli of subgrade reaction at different regions of the mat. A new set of values for contact pressures is obtained and processed by the geotechnical engineer to obtain a new set of values of soil displacements and hence the moduli of subgrade reaction. The process is repeated until the deflections predicted by the mat finite element analysis and the settlement predicted by the soil deflection due to consolidation and recompression of soil stratum converge to a desirable degree.

Two examples are presented following to give the reader a feel for the physical behavior of mats. The first consists of a mat for a 25-story concrete office building and the second example highlights the behavior of an octagonal mat for an 85-story composite building.

Mat for a 25-Story Building. The floor framing for the building consists of a system of haunch girders running between the interior core walls and columns to the exterior. The haunch girders are spaced at 30 ft (9 m) on centers and run parallel to the narrow face of the building. Skip joists spaced at 6 ft (1.81 m) center-to-center span between the haunch girders. A 4 in. (101.6 mm) thick concrete slab spanning between the skip joists completes the floor framing system. Lightweight concrete is used for floor framing members while normal-weight concrete is employed for columns and shear walls.

Shown in Fig. 12.15 is a finite element idealization of the mat. The typical element size of 12 by 10 ft (3.63 by 3.03 m) may appear to be rather coarse, but an analysis

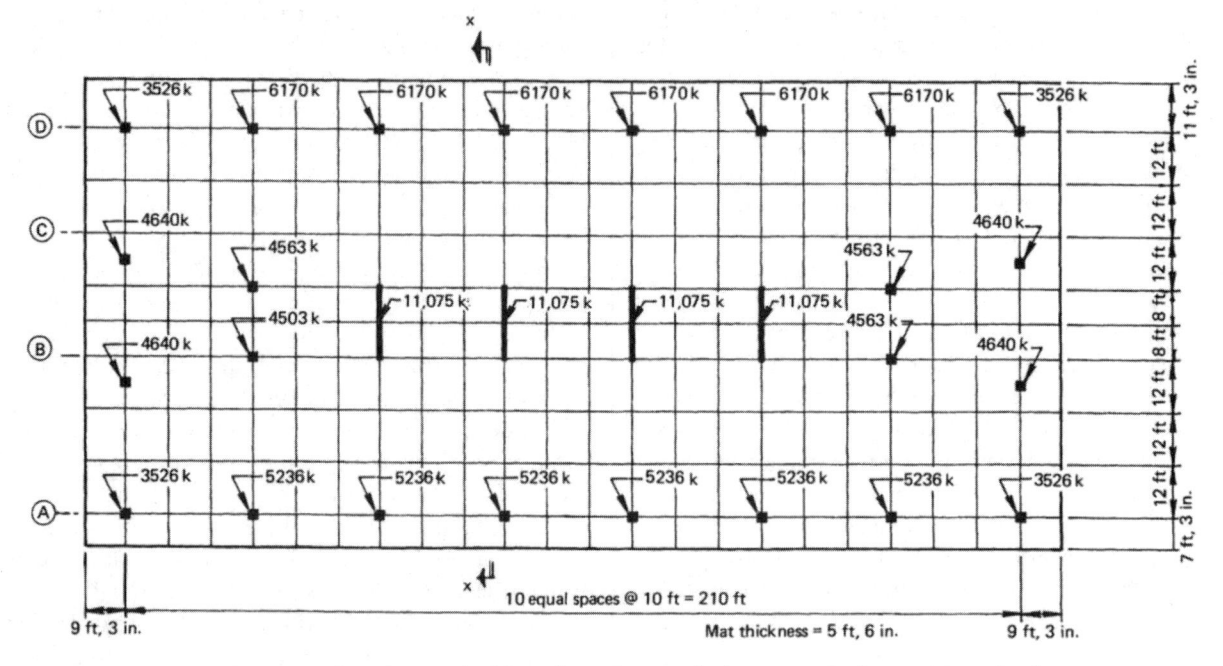

Figure 12.15 Foundation mat for a 25-story building; finite element idealization and column ultimate loads.

which used a finer mesh in which the typical element was 3 by 2.5 ft (0.9 by 0.76 m) showed results identical to that obtained for the coarse mesh. The calculated ultimate loads at the top of the mat are shown in Fig. 12.15. It may be noted that the finite element idealization is chosen in such a manner that the location of almost all columns, with the exception of four exterior columns on the narrow face, coincide with the intersection of the finite element mesh. The loads at these locations are applied directly at the nodes. The loads on the four exterior columns are, however, divided into two equal loads and applied at the two nodes nearest to the column. The resulting discrepancy in the analytical results has very little impact, if any, on the settlement behavior and the selection of reinforcement for the mat.

Assuming a value of 100 lb/in.³ (743 kg/mm³) for the subgrade modules, the spring constant at a typical interior node may be shown to be equal to 1728 kips/in. (196 × 10³ N/m). Figure 12.17 shows the mat deflection comparison for two values of subgrade reaction, namely 100 and 25 lb/in.³ (743 and 185.75 kg/mm³). As can be expected, the mat experiences a larger deflection when supported on relatively softer springs (Fig. 12.16). The variation of curvature, which is a measure of bending moments in the mat, is relatively constant for the two cases. This can be verified further by comparing the bending-moment diagrams shown in Fig. 12.17. Also shown in this figure are the bending moments obtained by assuming the mat as a continuous beam supported by three rows of supports corresponding to exterior columns and interior shear walls and subjected to the reaction of the soil acting vertically upward. The results for the

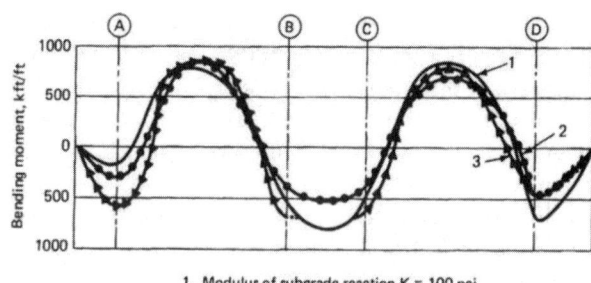

1. Modulus of subgrade reaction K = 100 pci
2. Modulus of subgrade reation K = 25 pci
3. Continuous beam analysis

Figure 12.17   Bending moment variation along section *x-x*.

example mat appear to indicate that mat reinforcement selected on the basis of any of the three analyses will result in adequate design.

Mat for an 85-Story Building. Figure 12.18 shows a finite element idealization of a mat for a proposed 85-story composite building, in Houston, Texas. Note that diagonal boundaries of the mat are approximated in a stepwise pattern using rectangular finite elements. To achieve economy, the thickness of the mat was varied; a thicker mat was proposed under the columns where the loads and thus the bending of the mat were expected

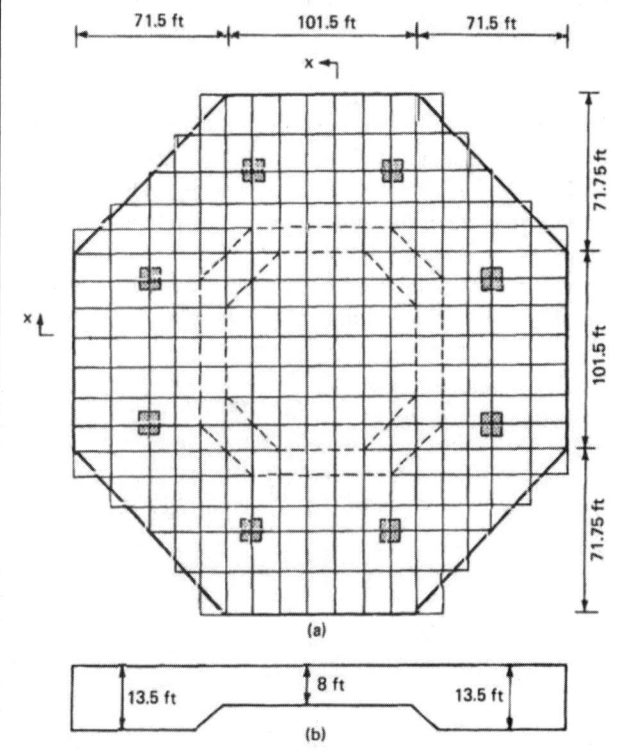

Figure 12.18   Foundation mat for a proposed 85-story office building: (a) finite element idealization; (b) cross-section *x-x* through mat.

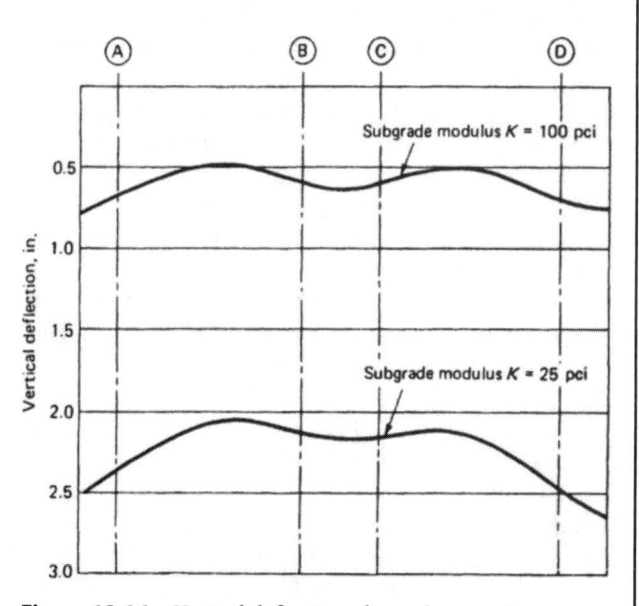

Figure 12.16   Vertical deflection of mat along section *x-x*.

to be severe. A relatively thin mat section was proposed for the center of the mat. The appropriateness of choosing two mat thicknesses can be appreciated by studying the pressure contours in Fig. 12.19a and b. The pressure contours plotted in these figures were obtained from computer analyses for two different loading conditions—gravity loads acting alone, and gravity loads combined with wind loads. No uplift due to wind loads was evident.

12.8 SEISMIC DESIGN OF DIAPHRAGMS

12.8.1 Introduction

Why is it that the design of floor and roof diaphragms is much more significant in seismic design than in wind design? The answer lies in the magnitude of, and the manner in which the loads are originated. Wind is an externally applied load, and although it acts in a dynamic fashion, its basic effect is similar to a statically applied load. Wind buffeting the exterior curtain wall is successively resisted by the glass, the mullions, the floor slab, and finally by the lateral load-resisting elements such as shear walls and frames. To get a feel for the difference in the magnitude of wind and seismic loads consider a building 200×100 ft, with a 14 ft floor-to-floor height, subjected to a uniform wind load of 30 psf. The wind load tributary to the floor in the short direction $= 30 \times 14 \times 200/1000 = 84$ kip. The load per ft of diaphragm $= 420$ plf. Now consider the seismic design of building in Seismic Design Category D per ASCE 7-16 using an Importance factor $I_e = 1.0$. Assume the building is of steel construction, with a floor construction consisting of 3 in. deep metal deck and $3\frac{1}{4}$ in. lightweight concrete topping. The seismic dead load of the floor system including an allowance of 10 psf for ceiling and mechanical, and another 10 psf for partition gives a seismic weight of $46 + 10 + 10 = 66$ psf of diaphragm. The minimum diaphragm design force

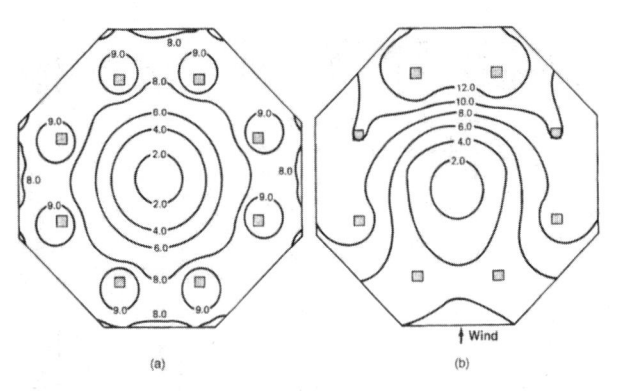

Figure 12.19 Contact pressure contour under the mat (ksf): (a) dead loads plus live loads; (b) dead plus live plus wind loads (note $K = 100$ lb/in.³).

F_{px} as given by the ASCE 7-16 given $S_{DS} = 0.95$ and

$$\left(\sum_{i=x}^{n}F_i \bigg/ \sum_{i=1}^{n}w_i\right) = 0.22$$

$$0.4S_{DS}I_e w_{px} \geq F_{px} = \frac{\displaystyle\sum_{i=x}^{n}F_i}{\displaystyle\sum_{i=1}^{n}w_i}w_{px} \geq 0.2S_{DS}I_e w_{px}$$

$$0.4(0.95)(1.0)(66) = 25.08\,\text{psf} \geq F_{px} = 0.22(66)$$
$$= 14.52\,\text{psf} \geq 0.2(0.95)(1.0)(66)$$
$$= 12.54\,\text{psf}$$

F_{px} per ft $= 14.52 \times 100 = 1452$ lb which is 3.46 times the load caused by wind.

Observe that the comparison just made is between the design values.

12.8.2 Diaphragm Behavior

Buildings are composed of vertical and horizontal structural elements which resist lateral forces. Horizontal forces on a structure produced by seismic ground motion originate at the centroid of the mass of the building elements and are proportional to the masses of these elements. These forces include inertia forces originating from the weight of the diaphragm and the elements attached thereto, as well as forces that are required to be transferred to vertical resisting elements because of offsets or changes of stiffness in vertical resisting elements above and below the diaphragm.

Horizontal forces at any floor or roof level are distributed to the vertical resisting elements by using the strength and rigidity of the floor or roof deck to act as a diaphragm. It is customary to consider a diaphragm analogous to a plate girder laid in a horizontal plane where the floor or roof deck performs the function of the plate girder web, the beams function as web stiffeners, and the peripheral beams or integral reinforcement function as flanges (Fig. 12.20).

The diaphragm chord can take on many forms such as a line of edge beams connected to the floor, or reinforcing in the edge of a slab, or reinforcing in a spandrel (Fig. 12.21). Boundary members at edges of diaphragms must be designed to resist direct tensile or compressive chord stresses, including adequate splices at points of discontinuity. For instance, in a steel frame building the spandrel beams acting as a diaphragm flange component require a splice design at the columns for the tensile and compressive stresses induced by diaphragm action.

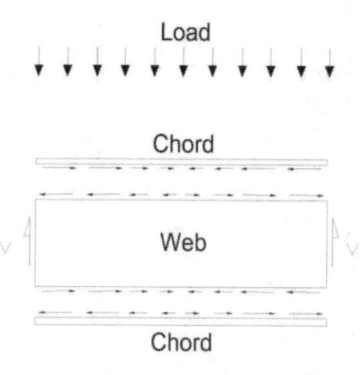

Diaphragm as a Beam

Figure 12.20 Diaphragm as a beam: free-body diagram.

The fundamental requirements for the chord are the continuity of the chord and the connection with the slab.

To continue the similarity with the beam, an opening in the floor for a stair, elevator or a skylight may weaken the floor just as a hole in the web for a mechanical duct weakens the beam. Similarly, a break in the edge of the floor may weaken the diaphragm just as a notch in a flange weakens the beam. In each case, the diaphragm should be detailed such that all stresses around the openings are developed into the diaphragm.

Another beam analogy applicable to diaphragms is the rigidity of the diaphragm compared to the walls or frame that provide lateral support and transmit the lateral forces to the ground. A metal-deck roof is relatively flexible compared to concrete walls while a concrete floor is relatively rigid compared to steel moment frames.

Yet another beam characteristic is continuity over intermediate supports. Consider, for example, a three-bay building with three spans and four supports. If the diaphragm is relatively rigid, the chords may be designed like flanges of a beam continuous over the intermediate supports. On the other hand, if the diaphragm is flexible, it may be designed as a simple beam spanning between walls with no consideration of continuity. However, in the latter case, it should remember that the diaphragm really is continuous, that its continuity is simply being neglected. The consequence of the neglect may well be some damage where adjacent spans meet.

Another essential consideration is that the connection between the beam and its support must be adequate for the transfer of diaphragm shear to the shear wall. As part of the diaphragm design, it may be necessary to add a collector or drag strut to collect the diaphragm shear and drag it into the vertical subsystems. This technique has the desirable effect of precluding a concentration of stress in the diaphragm alongside the vertical subsystem. The collector, however, becomes a critical element: it must be continuous, developing its required capacity across any interrupting elements such as cross beams, and there must be an adequate connection to transfer the collector force into vertical system.

In buildings, the effect of flexible diaphragms on the normal walls must be considered. Figure 12.22 shows the deflected shape of the normal wall of a two-story building. In the building on the left (the more usual case), the floor is stiffer than the roof, and the wall is bent out of its plane, with the maximum movement near the upper floor. In the building on the right (an unusual case), the roof is stiffer, and the bending can be more severe at the floor level.

A common plan irregularity is an odd shape: a single isolated odd shape or one that results from offsetting each

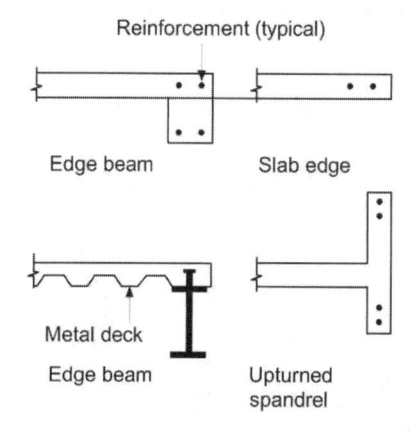

Figure 12.21 Chord sections.

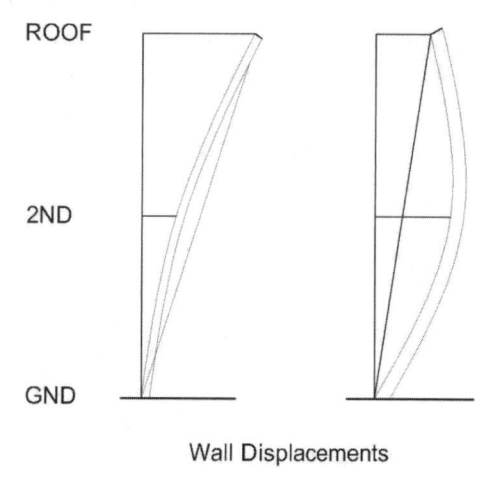

Figure 12.22 Wall displacements.

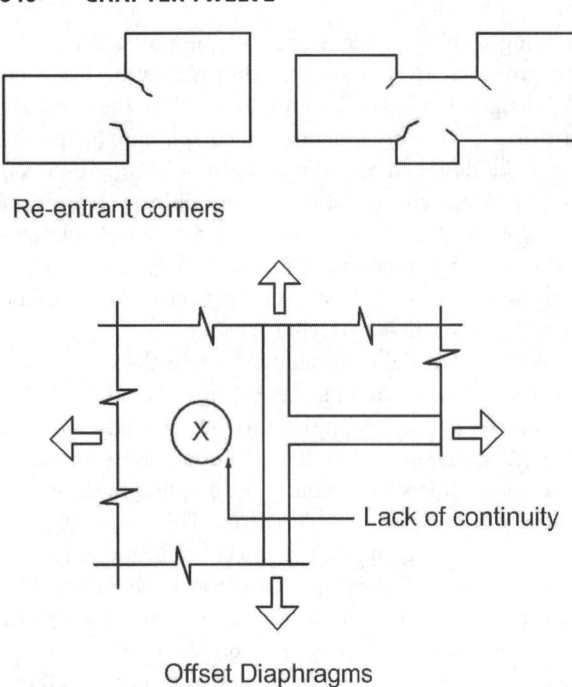

Re-entrant corners

Offset Diaphragms

Figure 12.23   Offset diaphragms.

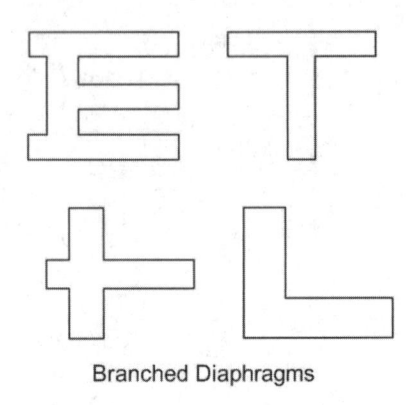

Branched Diaphragms

Figure 12.25   Branched diaphragms.

module in a group of buildings. Shown in Fig. 12.23 are examples of plan shapes with re-entrant corners. If the tensile capacity provided at the re-entrant corners at the points marked "X" is not sufficient, a local concentration of damage may occur, and this could lead to loss of support for beams or slabs.

Another common irregularity is an inset, usually for an entrance or loading bay resulting in reduced chord depth (Fig. 12.24). Using the beam analogy, the inset is like a notch in the beam flange requiring development of chord forces on either side of notch.

Buildings of E-, T-, X-, and L-shaped plans (Fig. 12.25) are also troublesome. The branches of these shapes have modes of vibration that the designer may not have considered in a simple design for north-south and east-west forces. Large tensile forces can develop in the re-entrant corners. Often this condition is aggravated by holes in the diaphragm made for elevators and stairs.

The L- and T-shaped buildings should have the flange, that is, the chord stresses, developed through or into the heel of L or T. This is analogous to a girder with a deep haunch in which the girder flange forces are developed into the haunch.

Small openings have not been a life-safety issue in buildings, but large openings as shown in Fig. 12.26 can have a disastrous effect if not properly accounted for in design.

When an opening in the diaphragm is adjacent to a wall or frame, there are two concerns. The first concern is that if the diaphragm opening is long, it may not be possible to transfer the diaphragm shear to the shear wall. The second concern is that there may not be enough floor to provide lateral support for the wall when the direction of the earthquake loading is perpendicular to the wall.

In Fig. 12.27, the opening reduces the length for shear transfer from L to $A + B$. Stresses are higher than otherwise in regions C and D; moreover, a region such as C

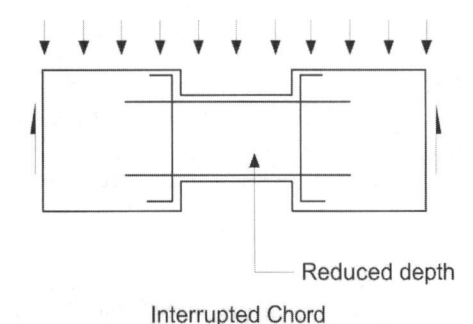

Reduced depth

Interrupted Chord

Figure 12.24   Interrupted chord.

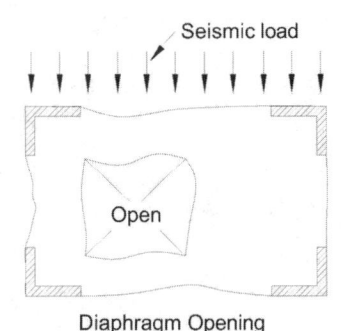

Seismic load

Open

Diaphragm Opening

Figure 12.26   Diaphragm opening.

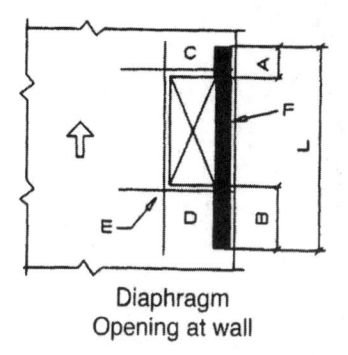

Figure 12.27 Diaphragm opening at wall.

may have so little rigidity compared to *D* that it is relatively ineffective for shear transfer. Reinforcing is necessary at the re-entrant corners such as *E*.

For earthquake loads in the other direction (i.e., perpendicular to the wall), the wall is a normal wall. It should be designed as a beam to span the length of the opening *F*, to collect out-of-plane forces from above and below and deliver them through suitable connections into the diaphragm at the ends of the opening.

12.8.3 Rigid, Semi-Rigid, and Flexible Diaphragms

The total shear, which includes the forces contributed through the diaphragm as well as the forces contributed from the vertical resisting elements above the diaphragm, at any level are distributed to various vertical elements of the lateral force-resisting system such as shear walls or moment-resisting frames, in proportion to their rigidities considering the rigidity of the diaphragm.

The effect of diaphragm stiffness on the distribution of lateral forces is schematically illustrated in Fig. 12.28. For this purpose, diaphragms are classified into three groups of flexibilities relative to the flexibilities of the walls. These are: (i) rigid, (ii) semi-rigid, and (iii) flexible diaphragms. No diaphragm is actually infinitely rigid and no diaphragm capable of carrying a load is infinitely flexible.

1. A rigid diaphragm (Fig. 12.28a) is assumed to distribute horizontal forces to the vertical resisting elements in proportion to their relative rigidities. In other words, under symmetrical loading a rigid diaphragm will cause each vertical element to deflect an equal amount with the result that a vertical element with a high relative rigidity will resist a greater proportion of the lateral force than an element with a lower rigidity.

2. A flexible diaphragm (Fig. 12.28c) is analogous to a shear-deflecting continuous beam or series of beams spanning between supports. The supports are considered non-yielding, as the relative stiffness of the vertical

resisting elements compared to that of the diaphragm is large. Thus, a flexible diaphragm is considered to distribute the lateral forces to the vertical resisting elements on a tributary load basis. A flexible diaphragm is considered incapable of distributing torsional stresses resulting from eccentricity of masses.

3. Semi-rigid diaphragms are those which have significant deflection under load, but which also have sufficient stiffness to distribute a portion of their load to vertical elements in proportion to the rigidities of the vertical resisting elements. The action is analogous to a continuous beam of appreciable stiffness on yielding supports (Fig. 12.28b). The support reactions are dependent on the relative stiffness of both diaphragm and vertical elements. A rigorous analysis is sometimes very time-consuming and frequently unjustified. In such cases, a design based on reasonable limits may be used; however, the calculations must reasonably bracket the likely range of reactions and deflections.

A torsional moment is generated whenever the center of gravity, *CG*, of the lateral forces fails to coincide with the center of rigidity, *CR*, of the vertical resisting elements, providing the diaphragm is sufficiently rigid to transfer torsion. The magnitude of the torsional moment that is required to be distributed to the vertical resisting elements by a diaphragm is determined by the larger of the following: (i) the sum of the moments created by the physical eccentricity of the translational forces at the level of the diaphragm from the center of rigidity of the resisting elements ($M = Fe$, where e = distance between *CG* and *CR*); or (ii) the sum of the moments created by an "accidental" torsion of 5 percent. This is an arbitrary code requirement equivalent to the story shear acting with an eccentricity of not less than 5 percent of the maximum building dimension at that level.

The torsional shears are combined with the direct, translational shears. However, when the torsional shears are opposite in direction to the direct shears, the lateral forces are not decreased.

12.8.4 Metal Deck Diaphragms

Bare metal deck can be used as a diaphragm when the individual panels are properly welded to each other and to the supporting framing. The strength of the diaphragm depends on the profile and gauge of the deck and the layout and size of the welds. Allowable values of metal deck diaphragms are usually obtained from approved data developed by the industry from test and analytical work.

Metal deck used in floors has a concrete fill. In some cases, the strength of the diaphragm is considered to be that of the metal deck acting by itself, with the concrete

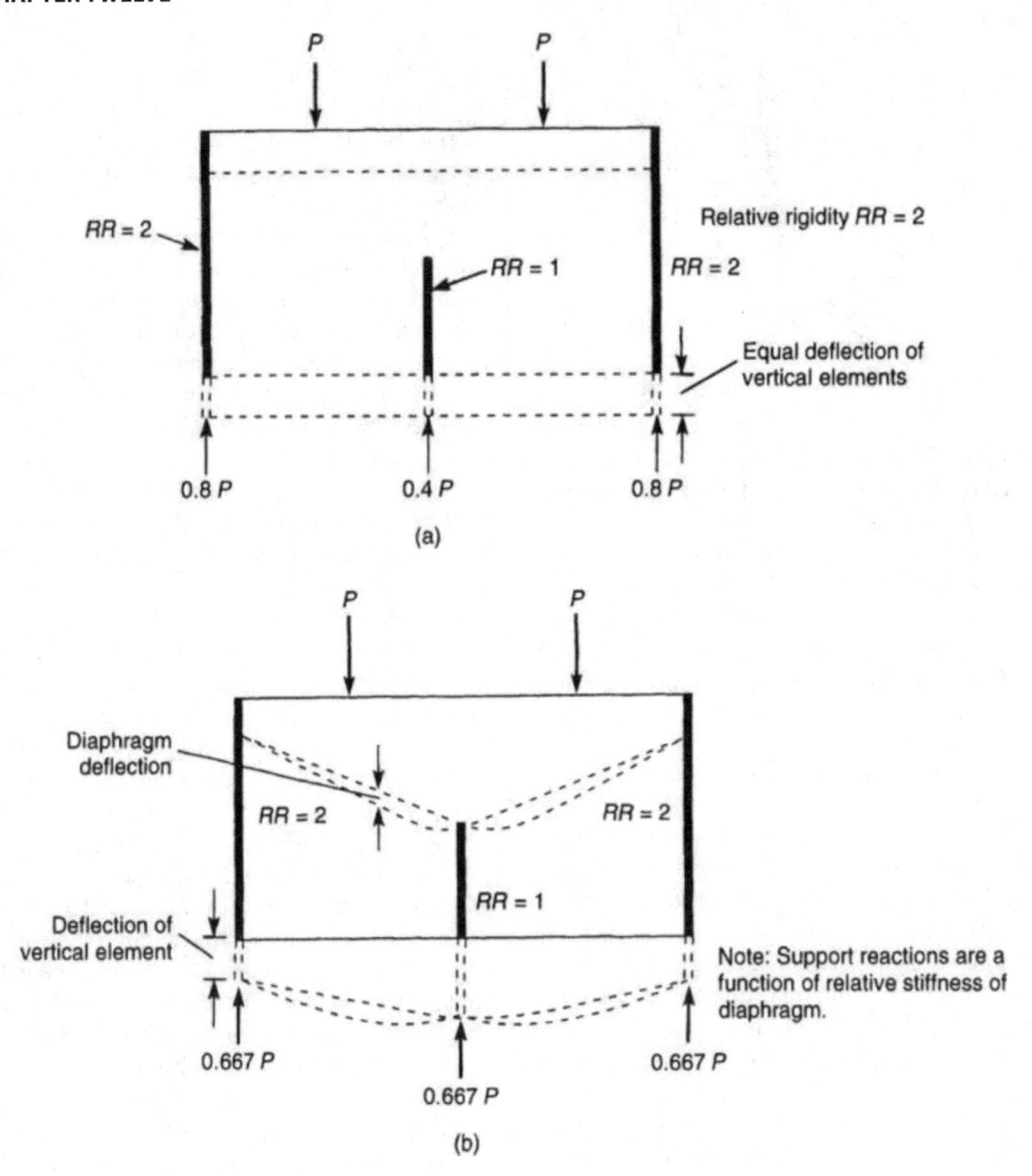

Figure 12.28   Rigid, semi-rigid, and flexible diaphragms: (a) rigid diaphragm; (b) semi-rigid diaphragm; (c) flexible diaphragm.

functioning only as a topping that produces a level floor and covers conduit laid on the deck. In these cases, the concrete can have a stiffening effect that makes the capacity of the system greater than that of the bare deck. In other cases, the metal deck is considered to act only as a form, and the diaphragm is treated as a reinforced concrete diaphragm.

Concrete-filled metal decks generally make excellent diaphragms and usually are not a problem as long as the basic requirements for chords, collectors, and reinforcing around openings are met. However, it is necessary to check for conditions that can weaken the diaphragm: troughs, gutters, electrical raceways, and recesses for architectural purposes which can have the effect of reducing the concrete diaphragm to the bare deck diaphragm.

12.8.5   Design Criteria

The ASCE 7-16 section 12.10 requires that the diaphragm at each level be designed to span horizontally between the lateral load-resisting elements. In addition to the forces F_{px}, the diaphragm should transfer the shears from discontinuous shear walls and frames. The reader is referred to Chap. 3 of this publication for complete design criteria on diaphragms in accordance with ASCE 7-16.

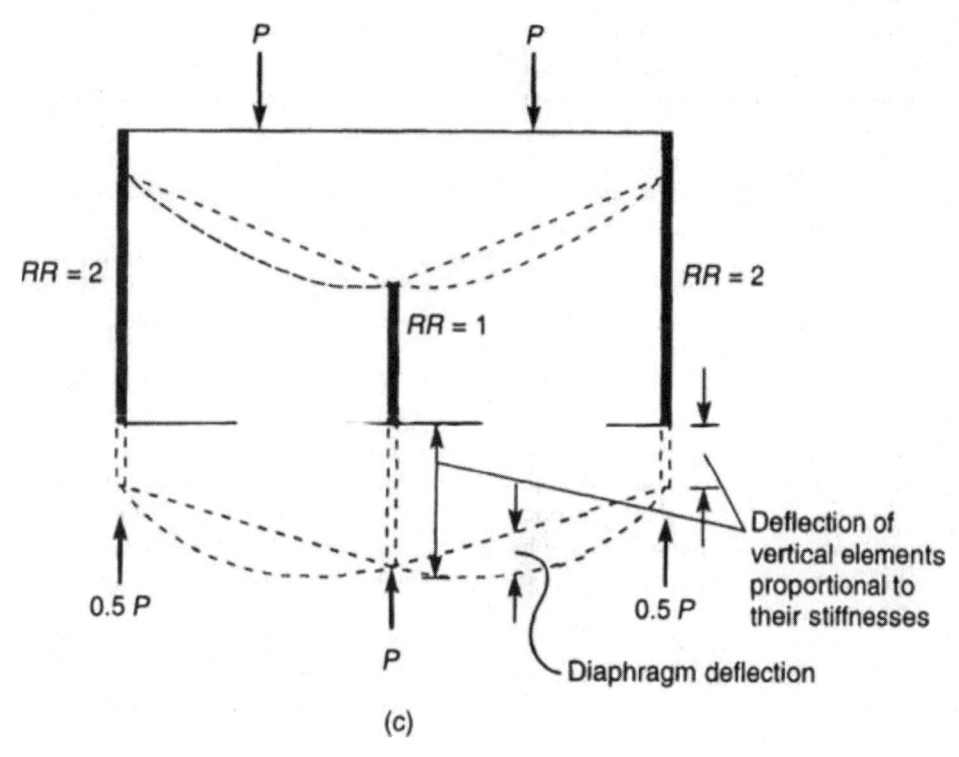

Figure 12.28 *(Continued)*

12.9 EARTHQUAKE HAZARD MITIGATION TECHNOLOGY

Earthquake mitigation technology includes seismic isolation, energy dissipation, ductility, damping systems, and other technologies, which strive to reasonably protect buildings and nonstructural components, building content, and functional capacity from earthquake damage. The potential advantage of this technology lies on its effectiveness in protecting nonstructural components and building contents thereby improving the capability of the facility to function immediately after the earthquake.

Two main concepts used in achieving the above concept are: (i) seismic isolation, and (ii) energy dissipation. The idea of seismic isolation is to shift the fundamental period of vibration of the building away from the predominant period of earthquake-induced ground motion to reduce the forces transmitted into the structure. Isolation provides a means of limiting the earthquake entering the structure by decoupling the building's base from its superstructure. The energy dissipation concept, on the other hand, allows seismic energy into the building. However, by incorporating damping mechanisms into the lateral load-resisting system of structure, earthquake energy is dissipated as the structure sways back and forth due to seismic loading. A brief description of each of these concepts follows.

12.9.1 Seismic Base Isolation

12.9.1.1 INTRODUCTION

The demands of seismic loads far exceed the traditional capacity of members normally associated with building structures. To build a structure to resist seismic loads in the same manner as it would resist gravity or wind loads, results in unacceptably expensive and architecturally unmanageable buildings. The general public may be unaware, but the premise of seismic design is to reduce the likelihood of total collapse by dissipating the seismic energy by allowing the structural members to undergo permanent deformation.

Earthquakes, when they occur, cause high accelerations in stiff buildings and large interstory drifts in flexible structures. These two factors impose difficulties in the design of essential facilities such as hospitals, communication, and emergency centers, etc. which must remain operational immediately after an earthquake when needed most. Although theoretically it may be possible to build-in enough strength in these structures to retain their functionality by resisting the seismic forces elastically by brute force, a relatively new technique termed "seismic isolation" is finding more and more application in the design of essential facility. With this technique, the building is detached or isolated from the ground in such a way that only a very small portion of seismic ground motions is transmitted up through the building

(Fig. 12.29). In other words, although the ground underneath it may vibrate violently, the building itself would remain relatively stable. This results in a significant reduction in floor accelerations and interstory drifts, thereby providing protection to the building contents and components. Therefore, seismic isolation maximizes the possibility that the essential facility will remain operational after a major seismic event.

In simple terms, the seismic isolation consists of mounting the building on seismic isolators which have large flexibility in the horizontal plane. At the same time, dampers may be introduced to reduce the amplitude of motion caused by the earthquake.

Although each earthquake is unique, it can be stated in general that earthquake ground motions result in greater acceleration response in a structure at shorter periods than at longer periods. A seismic isolation system exploits this phenomenon by shifting the fundamental period of the building from the more force-vulnerable shorter periods to the less force-vulnerable longer periods.

The principle of seismic isolation is to introduce flexibility in the basic structure in the horizontal plane, while at the same time adding damping elements to restrict the resulting motions. A practical base isolation system should consist of:

1. A flexible mounting to increase the period of vibration of the building sufficiently to reduce forces in the structure above.

2. A damper or energy dissipater to reduce the relative deflections between the building and the ground to a practical level.

3. A method of providing rigidity to control the behavior under minor earthquakes and wind loads.

Flexibility can be introduced at the base of the building by many possible devices such as elastomeric bearings, rollers, sliding plates, cable suspension, sleeved piles, rocking foundations, etc. Substantial reductions in acceleration with increase in period and consequent reduction in base shear are possible, the degree of reduction depending on the initial fixed-base period and shape of the response curve. However, the decrease in base shear comes at a price; the flexibility introduced at the base will give rise to large relative displacements across the flexible mount. Hence the necessity of providing additional damping at the level of isolators. This can be provided through hysteric energy dissipation of mechanical devices which use the plastic deformation of either lead or mild steel to achieve high damping. New seismic isolation devices such as high-density rubber HDR, lead rubber bearing LRB, friction pendulum system, FPS, and viscous damping devices, VDD are some of the other devices for achieving high damping.

While a flexible mounting is required to isolate the building from seismic loads, its flexibility under frequently occurring wind and minor earth tremors is undesirable because the building motions may be perceived by the building occupants and create discomfort. Therefore, the devices introduced at the base must be stiff enough at these loads, such that the building response is as if it is on a fixed base. Lead rubber bearings and other mechanical energy dissipaters provide the desired rigidity at low loads while providing flexibility at seismic loads.

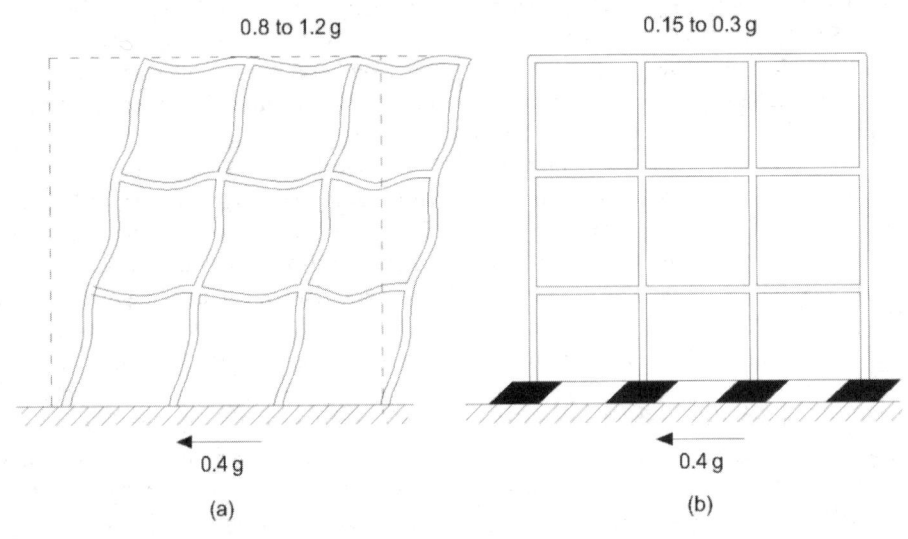

Figure 12.29 Response of a fixed-base and base-isolated building: (a) fixed-base; (b) base-isolated. Base isolation typically reduces the maximum roof acceleration response of low-rise buildings from maximum of 0.8 to 1.2 g to maximum of 0.15 to 0.3 g.

The plane of isolation, meaning the level at which the flexible mountings are introduced, is usually a trade-off between the amount of additional construction, access to the building, utility connections, access to the isolators for inspection, etc. Generally, one isolator per column is used. Occasionally when the column loads are high, of the order of several thousands of kips, more than one isolator may be used at the column location. For shear walls, one or more isolators are used at each end, and if the wall is long, isolators are placed along its length, the spacing depending upon the spanning ability of the wall between the isolators.

12.9.1.2 SALIENT FEATURES

1. Access for inspection and replacement of bearings should be provided at bearing locations.

2. Stub walls or columns to function as backup systems should be provided to support the building in the event of isolator failure.

3. A diaphragm capable of delivering lateral loads uniformly to each bearing is preferable. If the shear distribution is unequal the bearings should be arranged such that larger bearings are under stiffer elements.

4. A moat to allow free-movement for the maximum predicted horizontal displacement must be provided around the building (Figs. 12.30 and 12.31).

5. The isolator must be free to deform horizontally in shear and must be capable of transferring maximum seismic forces between the super substructure and the foundation.

6. The bearings should be tested to ensure that they have lateral stiffness properties that are both predictable and repeatable. The tests should show that over a wide range of shear strains, the effective horizontal stiffness and area of the hysteresis loop are in agreement with those used in the design.

When earthquakes occur, the elastomeric bearings used for base isolation are subjected to large horizontal displacements, as much as 15 in. or greater in a 10-story steel-framed building. They must therefore be designed to carry the vertical loads safely at these displacements.

12.9.1.3 ASCE 7-16 REQUIREMENTS

General requirements for seismic isolated structures can be found in Chapter 17 of ASCE 7-16 Although simple procedures such as an equivalent static procedure and a response spectrum analysis are permissible under the regulations, the restrictions are so many that in all but very simple and regular buildings, it is mandatory to perform response spectrum and time-history analysis with exceptions to few cases where the Equivalent Lateral Force Procedure (ELFP) is permitted.

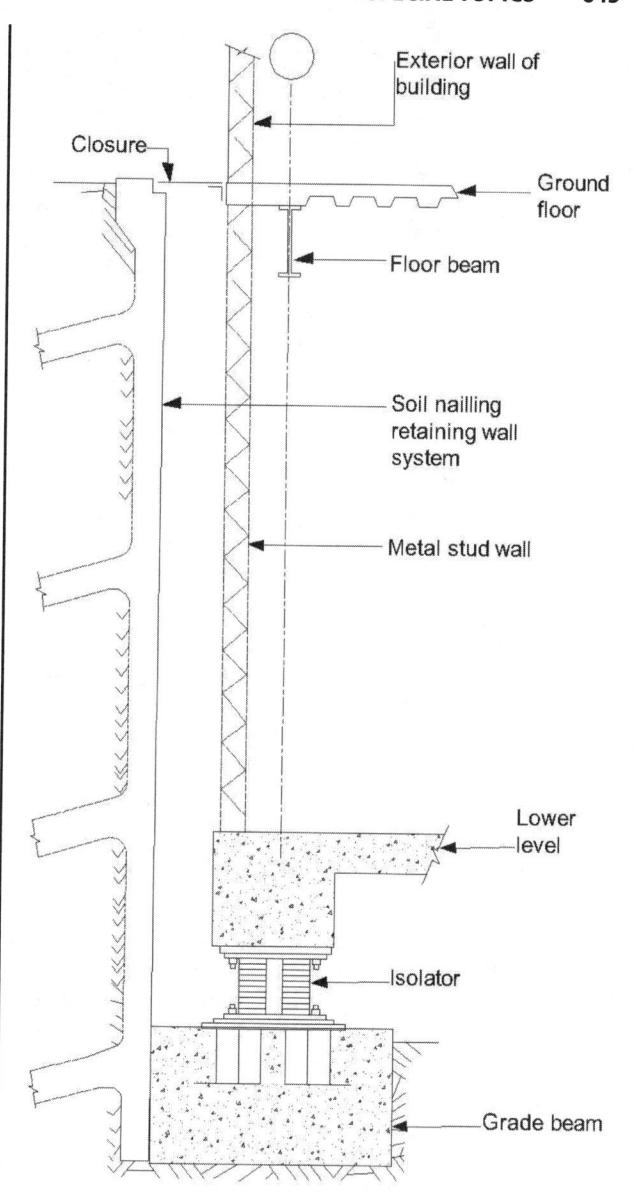

Figure 12.30 Moat around base isolated building.

Chapter 17 of ASCE 7-16 includes detailed provisions for seismically isolated structures, in this section key provisions are described briefly as follows:

• The Importance Factor which is given as $I_e = 1$ per section 17.2.1.

• An isolated structure shall be designated as having a structural irregularity if the structural configuration above the isolation system has a horizontal structural irregularity Type 1b, see Table 12.3-1 of ASCE 7-16, or vertical irregularity Type 1a, 1b, 5a, 5b, see Table 12.3-2 of ASCE 7-16 per section 17.2.2.

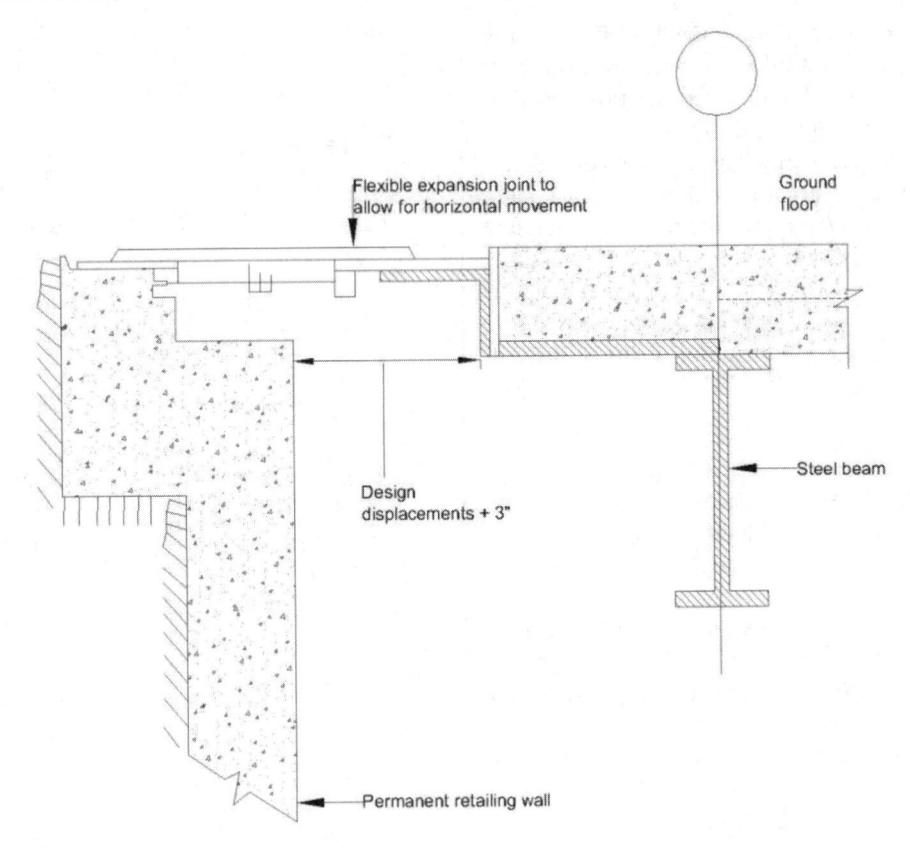

Figure 12.31 Moat detail at ground level.

• A redundancy factor, ρ, shall be assigned to the structure above the isolation system based on the requirements of Section 12.3.4. For isolated structures that do not have a structural irregularity, as defined in Section 17.2.2, the redundancy factor, ρ, is permitted to be equal to 1.0 per section 17.2.3.

• Environmental Conditions such as aging effects, creep, fatigue, operating temperature, and exposure to moisture or damaging substances shall be considered per section 17.2.4.1.

• Wind Forces: Isolated structures shall resist design wind loads at all levels above the isolation interface. A wind-restraint system at the isolation interface shall be provided to limit lateral displacement in the isolation system to a value equal to that required between floors of the structure above the isolation interface according to section 17.5.6 of ASCE 7-16, see section 17.2.4.2.

• Fire Resistance: Fire protection at the same degree required for columns, walls, or other gravity-bearing elements at the same location in the isolation system shall be provided for the isolation systems, see section 17.2.4.3.

• Lateral Restoring Force: The isolation system shall be configured to produce a restoring force such that the lateral force at the corresponding maximum displacement is at least 0.025 W greater than the lateral force at 50 percent of the corresponding maximum displacement per section 17.2.4.4.

• Displacement Restraint: The isolation system shall not be configured to include a displacement restraint that limits lateral displacement caused by risk-targeted maximum considered earthquake (MCE_R) ground motions to less than the total maximum displacement, D_{TM}, unless the seismically isolated structure is designed in accordance with all of the following criteria listed in section 17.2.4.5.

• Stability for each element of the isolation system under design vertical load computed using load combination 2 of section 17.2.7.1 for the maximum vertical load and load combination 3 of section 17.2.7.1 for the minimum where it is subjected to a horizontal displacement equal to the total maximum displacement, see section 17.2.4.6.

• Structural System: Provisions for structural systems used with base isolation are given in section 17.2.5. Note that ordinary concentrically braced frames that have limited height of 35 ft. for SDC D and E and not permitted in SDC F in structures without base isolation can be used with structures with base isolation with height limit up to 160 ft. in SDC D, E, or F provided the requirements of this section are met.

• Seismic ground motion criteria: The seismic design criteria to be used with base-isolated structures is given in Section 17.3. Site-specific seismic hazard is given in section 17.3.1. To determine the MCE_R response spectrum for the site of interest, the MCE_R response spectrum requirements of sections 11.4.5 and 11.4.6 are permitted to be used. The site-specific ground motion procedures given in Chapter 21 are permitted to be used to determine ground motions for any isolated structure. For isolated structures on Site Class F sites, site response analysis shall be performed in accordance with section 21.1.

• According to section 17.3.2, the MCER response spectrum shall be the MCER response spectrum of sections 11.4.6 and 11.4.7. The MCER response spectral acceleration parameters SMS and $SM1$ shall be determined in accordance with section 11.4.4 or 11.4.8. In the case where MCER Ground Motion Records are used, see section 17.2.4 for detailed requirements.

• Analysis procedure: Seismically isolated structures shall be designed using the dynamic procedures given in section 17.6 except those defined in section 17.4.1 where the Equivalent Lateral Force Procedure (ELFP) of section 17.5. Where supplementary viscous dampers are used, the response history analysis procedures of section 17.4.2.2 shall be used.

12.9.2   Energy Dissipation

Unlike base isolation, which involves a period shift of the structure from the predominant period of earthquake motion, passive energy dissipation allows earthquake energy into the building. Through appropriate configuration of the lateral resisting system, earthquake energy is directed toward energy dissipation devices located within the lateral resisting elements to intercept this energy. There are four general types of energy dissipation systems: (i) metallic, (ii) friction, (iii) viscoelastic, and (iv) viscous and semi-viscous systems. The common feature of all these devices is the transformation of earthquake energy into heat, which is dissipated into the structure.

12.9.2.1   METALLIC SYSTEMS

In these systems, energy dissipation is achieved through yielding of metal parts. These can be fabricated from steel, lead or special shape memory alloys. Lead extruders force the material through an orifice within a cylinder for energy dissipation. The shape memory alloys have potential capabilities of dissipating energy without incurring damage, as in the case of steel dampers when they yield. These systems are referred to as amplitude-dependent systems since the amount of energy dissipated, which is hysteretic in nature, is usually proportional to force and displacement. Their force-displacement characteristics

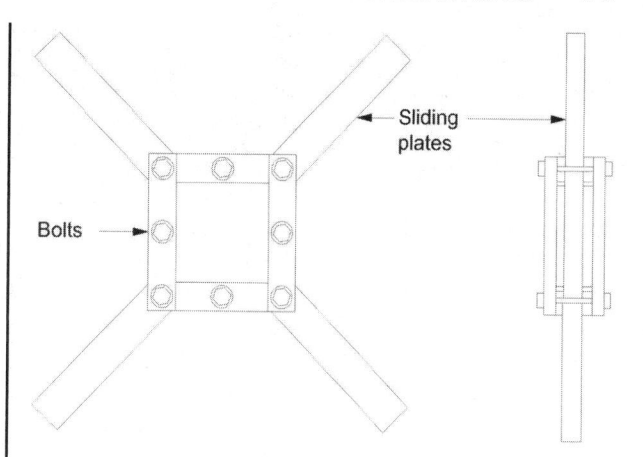

Figure 12.32   Friction damper with sliding plates.

are configured to be highly nonlinear to take full advantage of their energy dissipation potential. The devices are most often located within structural lateral-load-resisting elements such as braced frames.

12.9.2.2   FRICTION SYSTEMS

In this system, the friction surfaces are clamped with prestressing bolts (Fig. 12.32). The characteristic feature of this system is that almost perfect rectangular hysteretic behavior is exhibited. These systems are referred to as displacement dependent systems since the amount of energy dissipated is proportional to displacement. Contact surfaces used are lead bronze against stainless steel, or Teflon against stainless steel.

12.9.2.3   VISCOELASTIC SYSTEMS

These systems use materials similar to elastomer. The materials are usually bonded to steel, and dissipate energy when sheared, similar to elastomeric bearings (Fig. 12.33). The materials also exhibit restoring force capabilities. Stiffness properties of some viscoelastic materials are temperature and frequency dependent. These variations should be taken into consideration in the design of these systems.

12.9.2.4   VISCOUS AND SEMI-VISCOUS FLUID SYSTEMS

These systems dissipate energy by forcing a fluid through an orifice, and in that respect are similar to the shock absorbers of an automobile (Fig. 12.34). The fluids used are usually of high viscosity, such as silicone. The orifices are specially designed for optimum performance. The unique feature of these devices is that their damping characteristics and hence the amount of energy dissipated can be made proportional to the velocity.

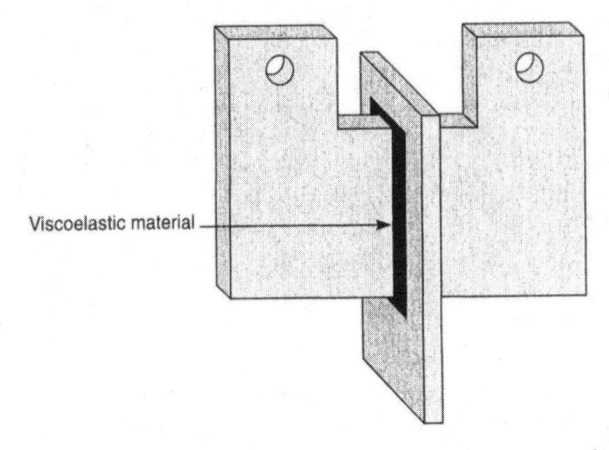

Viscoelastic material

Figure 12.33 Viscoelastic damper: stacked plates separated by inert polymer materials.

Fluid viscous damping reduces stress and deflection because the force from the damping is completely out of phase with stresses due to seismic loadings. This is because the damping force varies with stroking velocity. Other types of damping such as yielding elements, friction devices, plastic hinges, and visco-elastic elastomers do not vary their output with velocity; hence they can increase column stress while reducing deflection. To understand the out of phase response of fluid viscous dampers, consider a building shaking laterally back and forth during a seismic event. The stress in a lateral load-resisting element such as a frame column is at a maximum when the building has deflected a maximum amount from its normal position. This is also the point at which the building reverses direction to move back in the opposite direction. If a fluid viscous damper is added to the building, damping force will drop to zero at this point of maximum deflection. This is because the damper stroking velocity goes to zero as the building reverses direction. As the building moves back in the opposite direction, maximum damper force occurs at maximum velocity, which occurs when the building goes through its normal, upright position. This is also the point where the stresses in the lateral-load-resisting elements are at a minimum. This out of phase response is the most desirable feature of fluid viscous damping.

12.10 STEEL SEISMIC SYSTEMS AND PREQUALIFIED STEEL MOMENT CONNECTIONS

Steel systems are expected to supply high level of ductility when subjected to strong ground motion. This ductility is achieved by proper seismic detailing. Steel seismic systems include moment frames, braced frames, and shear walls. Braced frames can be either concentrically braced, eccentrically braced, or buckling-restrained; these systems have different levels of ductility and are used according to seismic design category (SDC) and the height of the building according to Table 12.2-1 of ASCE 7-16. Moment Frames also have different level of ductility and are used according to seismic design category (SDC) and the height of the building according to Table 12.2-1 of ASCE 7-16. The detailing requirements for each of the seismic steel systems are according to AISC 341-16. The basic seismic detailing requirement for a moment connection in a moment-resisting frame is that moment-resisting frames should sustain large ductility without failure of the beam-to-column connections. The minimum requirements for a beam-to-column connection in special moment frames (SMF) are specified in Chapter K of the AISC 341-16, whereas prequalified moment connections are listed in AISC 358-16 standard, which includes 16 prequalified moment connections (AISC 358). Reduced beam section (RBS) connection is commonly used in Special Moment Frames (SMF) and in intermediate moment frames (IMF) as prequalified connections.

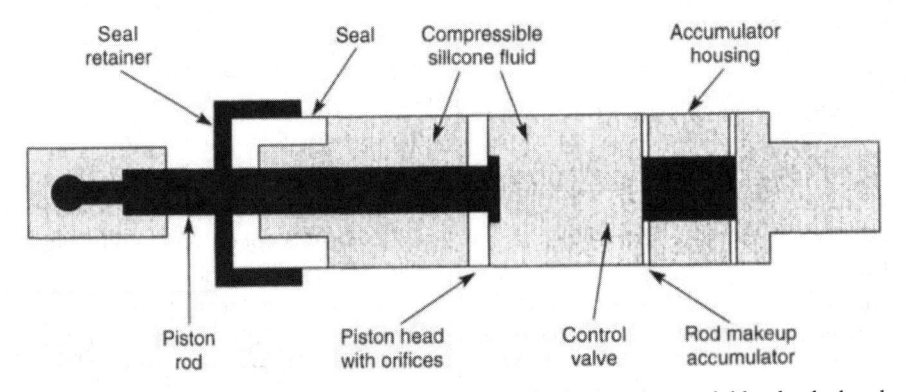

Figure 12.34 Fluid viscous dampers are pistons in metal cylinders that work like shock absorbers.

Selected References

1. Naeim, F. (ed.). *The Seismic Design Handbook*. New York, NY: Van Nostrand Reinhold; 1989.

2. International Code Council. *International Building Code (IBC 2018)*. Falls Church, VA: International Code Council; 2018.

3. American National Standards Institute (ANSI) A58.1; 1982.

4. Structuring Tall Buildings, *Progressive Architecture*, December 1980.

5. Cermak, JE. Applications of Fluid Mechanics to Wind Engineering. *Journal of Fluids Engineering*; March 1975.

6. *Wind Effects on High Rise Buildings, Symposium Proceedings*. Northwestern University, Evanston, IL; March 1970.

7. Tschanz, T. Measurement of Total Dynamic Loads Using Elastic Models With High Natural Frequencies. In: Timothy J. Reinhold (ed.), *Wind Tunnel Modeling for Civil Engineering Applications*. Cambridge: Cambridge University Press; 1982.

8. *Tall Buildings, Conference Proceedings*; 1984, Singapore.

9. Coleman, RA. *Structural System Designs*. Englewood Cliffs, NJ: Prentice-Hall; 1983.

10. *Post-Tensioning Manual*, 6th edition. Phoenix, AZ: Post-Tensioning Institute; 2006.

11. Lin TY, Stotesbury SD. *Structural Concepts and Systems for Architects and Engineers*. New York, NY: John Wiley; 1981.

12. Libby JR. *Modern Pre-Stressed Concrete*. New York, NY: Van Nostrand Reinhold; 1984.

13. Lin TY, Burns AP. *Design of Prestressed Concrete Structures*, 3rd edition. Hoboken, NJ: Wiley; 1991.

14. Darwin D, Dolan CW, Nilson AH, Nilson AH. *Design of Concrete Structures*, 14th edition. New York, NY: McGraw Hill; 2016.

15. Merritt FS, Ricketts JT. *Building Design and Construction Handbook*, 6th edition. New York, NY: McGraw Hill; 2000.

16. Shen J, Akbas B, Seker O, Faytarouni M. *Design of Steel Structures*, 1st edition. New York, NY: McGraw Hill; 2021.

17. Newmark NM, Rosenblueth E. *Fundamentals of Earthquake Engineering*. Englewood Cliffs, NJ: Prentice-Hall; 1971.

18. Coull and Stafford-Smith. *Tall Buildings With Particular Reference to Shear Wall Structures*. New York, NY: Pergamon Press; 1967.

19. Wakabayashi M. *Design of Earthquake Resistant Buildings*. New York, NY: McGraw Hill; 1986.

20. Taranath BS. *Structural Analysis & Design of Tall Buildings*. New York, NY: McGraw Hill; 1988.

21. Leet K, Bernal D. *Reinforced Concrete Design*. New York, NY: McGraw Hill; 1996.

22. Biggs JM. *Introduction to Structural Dynamics*. New York, NY: McGraw Hill; 1964.

23. Mahamid M, Gaylord EH, Gaylord CN. *Structural Engineering Handbook*, 5th edition. New York, NY: McGraw Hill; 2020.

24. Cooper SE, with Chen AC. *Designing Steel Structures—Methods and Cases*. Englewood Cliffs, NJ: Prentice-Hall; 1985.

25. Karlberg J. *Preliminary Design for Post-tensioned Structures, Structural Engineering Practice, Vol. 2*. New York, NY: Marcel Dekker, Inc.; 1983.

26. Schueller W. *High Rise Building Structures*, subsequent edition; 1986.

27. Zbirohowski-Koscia K. *Thin-Walled Beams*. London: Crosby Lockwood & Son Ltd.; 1967.

28. Ketchum MS (ed.), *Structural Engineering Practice*. Vols. 1 & 2, 1982; vol. 4, 1982–83; vol. 1, 2, & 3, 1983; vol. 4, 1983–84. New York, NY: Marcel Dekker.

29. *National Building Code of Canada (NBC), 2020.*

30. Murray NW. *Introduction to the Theory of Thin-Walled Structures*. New York, NY: Oxford University Press; 1984.

31. Reinhold TJ (ed.), *Wind Tunnel Modeling for Civil Engineering Applications*. Cambridge: Cambridge University Press; 1982.

32. Kanchi MB. *Matrix Methods of Structural Analysis*. New York, NY: Wiley Eastern Limited; 1994.

33. Paz M, Kim YH. *Structural Dynamics: Theory and Computation*, 6th edition. Springer; 2019.

34. Choudhury JR, "Analysis of Plain and Spatial Systems of Interconnected Shear Walls." Ph.D. Thesis, University of Southampton; 1968.

35. Vlasov VZ. *Thin-Walled Elastic Beams*. Washington, DC: National Science Foundation; 1961.

36. Taranath BS. "Torsional Behavior of Open-Section Shear Wall Structures." Ph.D. Thesis, University of Southampton; 1968.

37. Qadeer A. "Interaction of Floor Slabs and Shear Walls." Ph.D. Thesis, University of Southampton; 1968.

38. Taranath BS. "A New Look at Composite High-Rise Construction." Our World in Concrete and Structures, Singapore; 1983.

39. Taranath BS, et al. "A Practical Computer Method of Analysis for Complex Shear Wall Structure." *ASCE 8th Conference in Electronic Computation*; 1983.

40. Taranath BS. "Composite Design of First City Tower." *The Structural Engineer*; 1982.

41. Tanarath BS. "Differential Shortening of Columns in High-Rise Buildings." *Journal of Torsteel*; 1981.

42. Tanarath BS. "Analysis of Interconnected Open Section Wall Structures." *ASCE Journal*; 1986.

43. Tanarath BS. "The Effect of Warping on Interconnected Shear Wall Flat Plate Structures." *Proceedings of the Institution of Civil Engineers*; 1976.

44. Tanarath BS. "Torsion Analysis of Braced Multi-Storey Buildings." *The Structural Engineer*; 1975.

45. Tanarath BS. "Optimum Belt Truss Locations for High-Rise Buildings." *AISC*, vol. II, 1974.

46. Lenzen KH. "Vibrations of Steel Joist-Concrete Slab Floors." *AISC Engineering Journal*; July 1966.

47. Wiss JF, Parmalee RH. "Human Perception of Transient Vibrations." *Journal of Structural Division ASCE*, vol. 100, April 1974, pp. 773–783.

48. Murray TM. "Design to Prevent Floor Vibrations." *Engineering Journal–American Institute of Steel Construction*, third quarter, 1975.

49. Halvorson R, Isyumov N. "Comparison of Predicted and Measured Dynamic Behavior of Allied Bank Plaza." In N. Isyumov, T. Tschanz (eds), *Building Motion in Wind*. New York, NY: ASCE; 1982.

50. Smith BS, Coull A. *Tall Building Structures, Analysis and Design*. Hoboken, NJ: John Wiley & Sons, Inc.; 1991.

51. Building Seismic Safety Council (BSSC). 2015. *NEHRP Recommended Seismic Provisions for New Buildings and Other Structures*, FEMA P-1050-1. Federal Emergency Management Agency, Washington, DC.

52. 2015 NEHRP Recommended Seismic Provisions: Design Examples (FEMA P-1051, 2016).

53. *NEHRP Handbook for the Seismic Evaluation of Existing Buildings* (FEMA 178, 1992).

54. *NEHRP Handbook for Seismic Rehabilitation of Existing Buildings* (FEMA 172, 1992).

55. *Vision 2000: A Framework for Performance-Based Engineering of Buildings*, by Ronald O. Hamburger, Anthony B. Court, and Jeffrey R. Sonlager, SEAOC Proceedings 1995 Convention.

56. Coull A, Choudhury JR. "Analysis of Coupled Shear Walls." *ACI Journal*; September, 1967.

57. ETABS, Structural analysis and design software, *Computers and Structures, Inc.*, Berkeley, CA.

58. *Risk Management Series: Designing for Earthquakes—A Manual for Architects* (FEMA 454 / December 2006).

59. *Seismic Design Guidelines for Upgrading Existing Buildings*: Department of Army, Navy, & Air Force; 1985.

60. *Building Code Requirements for Structural Concrete* (ACI 318-19) and Commentary (ACI 318R-19). Farmington Hills, MI: American Concrete Institute.

61. *Manual of Steel Construction*, 15th edition. Chicago, IL: American Institute of Steel Construction; 2016.

62. Seismology Committee, Structural Engineers Association of California, Recommended Lateral Force Requirements and Commentary (Blue Book); 2019.

63. Fanella DA. *Structural Load Determination 2018 IBC and ASCE/SEI 7-16*, International Code Council (ICC); 2018.

64. American Society of Civil Engineers. *Minimum Design Loads and Associated Criteria for Buildings and Other Structures* (ASCE 7-16). Reston, VA: ASCE; 2017.

65. Coull and Stafford-Smith. *Tall Buildings With Particular Reference to Shear Wall Structures*. New York, NY: Pergamon Press; 1967.

66. Lin TY, Stotesbury SD. *Design of Prestressed Concrete Structures*. New York, NY: John Wiley.

67. American Society of Civil Engineers. *Wind Tunnel Studies of Buildings and Structures* (ASCE Manual of Practice Number 67). Reston, VA: ASCE; 1999.

68. American Society of Civil Engineers. *Minimum Design Loads and Associated Criteria for Buildings and Other Structures* (ASCE 7-22). Reston, VA: ASCE; 2021.

69. American Society of Civil Engineers. *Wind Tunnel Testing for Buildings and Other Structures* (ASCE 49-21). Reston, VA: ASCE; 2021.

70. American Society of Civil Engineers. *Prestandard for Performance-Based Wind Design*. Reston, VA: ASCE; 2019.

71. Australasian Wind Engineering Society. *Wind Engineering Studies of Buildings* (AWES-QAM-1-2019); 2019.

72. Davenport AG. *Technical Committee No. 7, Wind Loading and Wind Effects, Theme Report*, American Society of Civil Engineers (ASCE) and International Association for Bridge and Structural Engineering (IABSE) International Conference on Design and Planning of Tall Buildings, Lehigh University, Bethlehem, PA, USA, 21–26 August 1972.

73. Holmes JD, Bekele SA. *Directionality and Wind Induced Response—Calculation by Sector Methods*. 14th International Conference on Wind Engineering, ICWE14, Porto Alegre, Brazil.

74. Lepage MF, Irwin PA. *A Technique for Combining Historical Wind Data With Wind Tunnel Tests to Predict Extreme Wind Loads*, Proceedings of the 5th US National Conference on Wind Engineering, Lubbock, TX, USA, 6–8 November 1985.

75. Irwin P, Denoon R, Scott D. *Wind Tunnel Testing of High-Rise Buildings: An Output of the CTBUH Wind Engineering Working Group*, Council on Tall Buildings and Urban Habitat. Chicago, IL; 2013.

76. ISO. *Bases for Design of Structures – Serviceability of Buildings and Walkways Against Vibrations*, ISO 10137 International Organization for Standardization, Geneva; 2007.

77. Building Seismic Safety Council (BSSC). *Earthquake-Resistant Design Concepts—An Introduction to the NEHRP Recommended Seismic Provisions for New Buildings and Other Structures*, FEMA P-749. Federal Emergency Management Agency, Washington, DC; 2010.

78. International Code Council (ICC). *2018 International Existing Building Code*. Washington, DC: International Code Council; 2017.

79. Federal Emergency Management Agency (FEMA). *Quantification of Building Seismic Performance Factors*. P-695, Applied Technology Council, Washington, DC; 2009.

80. Federal Emergency Management Agency (FEMA). *Quantification of Building Seismic Performance Factors: Component Equivalency Methodology*, P-795, Washington, DC; 2011.

81. Chopra AK. *Dynamics of Structures Theory and Application to Earthquake Engineering*, 2nd edition. Prentice Hall, NJ; 2001.

82. American Society of Civil Engineers (ASCE). *Seismic Analysis of Safety-Related Nuclear Structures*, ASCE 4-8, Reston, VA; 1998.

83. National Council of Structural Engineering Associations (NCSEA). *Guide to the Design of Diaphragms, Chords and Collectors Based on the 2006 IBC and ASCE/SEI 7-05*. International Code Council, Washington, DC; 2009.

84. Mays TW. *Guide to the Design of Out-of-Plane Wall Anchorage Based on the 2006/2009 IBC and ASCE/SEI 7-05*. Washington, DC: International Code Council; 2010.

85. Taranath BS. *Structural Analysis and Design of Tall Buildings*. Boca Raton, FL: CRC Press, Taylor & Francis Group; 2012.

86. Taranath BS. *Reinforced Concrete Design of Tall Buildings*. Boca Raton, FL: CRC Press, Taylor & Francis Group; 2009.

87. Specification for Structural Steel Buildings, AISC 360-16, American Institute of Steel Construction, Chicago, IL; 2016.

88. Seismic Provisions for Structural Steel Buildings, AISC 341-16, American Institute of Steel Construction, Chicago, IL; 2016.

89. Fanella DA. *Design and Detailing of Low-Rise Reinforced Concrete Buildings*. Schaumburg, IL: Concrete Reinforcing Steel Institute; 2017.

90. Next-Generation Performance-Based Seismic Design Guidelines Program Plan for New and Existing Buildings (FEMA-445, 2006).

91. Guidelines for Performance-Based Seismic Design of Tall Buildings, 2017. Pacific Earthquake Engineering Center, Report No. 2017/06.

Index